# THE SEA

## Ideas and Observations on Progress in the Study of the Seas

| | |
|---|---|
| *Volume 1* | HILL/PHYSICAL OCEANOGRAPHY |
| *Volume 2* | HILL/COMPOSITION OF SEA WATER |
| *Volume 3* | HILL/THE EARTH BENEATH THE SEA |
| *Volume 4* | MAXWELL/NEW CONCEPTS OF SEA FLOOR EVOLUTION (IN TWO PARTS) |
| *Volume 5* | GOLDBERG/MARINE CHEMISTRY |
| *Volume 6* | GOLDBERG/MARINE MODELING |
| *Volume 7* | EMILIANI/OCEANIC LITHOSPHERE |
| *Volume 8* | ROWE/DEEP-SEA BIOLOGY |
| *Volume 9* | Le MÉHAUTÉ/OCEAN ENGINEERING SCIENCE |
| *Volume 10* | BRINK AND ROBINSON/THE GLOBAL COASTAL OCEAN: PROCESSES AND METHODS |
| *Volume 11* | ROBINSON AND BRINK/THE GLOBAL COASTAL OCEAN: REGIONAL STUDIES AND SYNTHESES |
| *Volume 12* | ROBINSON, McCARTHY, AND ROTHSCHILD/BIOLOGICAL-PHYSICAL INTERACTIONS IN THE SEA |
| *Volume 13* | ROBINSON AND BRINK/THE GLOBAL COASTAL OCEAN: MULTISCALE INTERDISCIPLINARY PROCESSES |

# THE SEA

## Ideas and Observations on Progress in the Study of the Seas

# THE GLOBAL COASTAL OCEAN

## MULTISCALE INTERDISCIPLINARY PROCESSES

*Edited by*

**ALLAN R. ROBINSON**
*Harvard University*

**and**

**KENNETH H. BRINK**
*Woods Hole Oceanographic Institution*

## THE SEA

**Ideas and Observations on Progress in the Study of the Seas**

**Volume 13**

Harvard University Press
Cambridge, Massachusetts and London

Published with the financial assistance of UNESCO/IOC.



Printed in the United States of America.

***Library of Congress Cataloging-in-Publication Data***

Robinson, Allan R.

The Sea: Ideas and Observations on Progress in the Study of Seas

Volume 13. The Global Coastal Ocean: Multiscale Interdisciplinary Processes

Allan R. Robinson and Kenneth H. Brink

Includes bibliographic information and index.

1. Oceanography 2. Submarine geology

ISBN: 0-674-01526-6

Library of Congress catalog card number: 62018366

# CONTENTS

# THE GLOBAL COASTAL OCEAN MULTISCALE INTERDISCIPLINARY PROCESSES

# FOREWORD

Coastal areas are vital to the life-support system of our planet. They represent 20 percent of the Earth's surface, yet serve as the home for over 50 percent of the entire human population. Coastal populations are expected to account for 75 percent of the total world population by the year 2025. A significant number of the world's megacities, with more than 8 million inhabitants, are located in coastal areas. Coastal ecosystems produce about 25 percent of global biological productivity and yield 90 percent of global fisheries. As a result of these dense human populations, the coasts suffer the effects of intense coastal development that threatens the integrity of coastal ecosystems.

The effective management and protection of coastal ecosystems must be science-based. With this general purpose in mind, the COASTS Programme, sponsored by the Intergovernmental Oceanographic Commission of UNESCO and the Scientific Committee on Oceanic Research, was established *to promote and facilitate research and applications in interdisciplinary coastal and shelf ocean sciences and technology on a global basis to increase scientific understanding of coastal ocean processes.*

The production line of science functions constantly and without pause. It works through a myriad of individual contributions and the publication of scientific "papers," often written in isolation from other work. Each project answers the next question down a logical chain of analysis of a given phenomenon. Each answer reflects and acknowledges the particular perspective of the authors. Apparently, there is no "a priori" plan. But collectively, all these elemental contributions build the edifice of science, and push back the boundary of the unknown.

For this collective knowledge to be used, a major effort of synthesis must be organized on a regular basis. This is the purpose for COASTS for the management of coastal oceans. Through this programme, the IOC is facilitating the synthesis of oceanographic knowledge and communicating those results to a wide audience.

We are fortunate in that the recent advances in interdisciplinary ocean sciences allow us to identify research issues better and to make feasible definitive studies on several topics. These advances need to be understood by all at national, regional and global levels. This need is very effectively served by the monographic volumes of THE SEA, which are a reference point in the advance of oceanography, helping to train future generations of oceanographers around the world.

The results emerging from the present volumes are expected to lead to the sustainable use of the resources of the coastal oceans, producing a knowledge base on the science of the coastal oceans for developed and developing nations alike. The Intergovernmental Oceanographic Commission of UNESCO is proud to be associated with this very meaningful and high quality academic activity.

IOC is grateful to Professor Allan Robinson for his leadership of the COASTS Programme, and to Dr. Kenneth Brink, along with the authors of the introductory chapters and the external reviewers, together with a large international ocean science community, for their contributions towards the realization of Volumes 13 and 14 of THE SEA.

PATRICIO A. BERNAL
ASSISTANT DIRECTOR GENERAL, UNESCO
EXECUTIVE SECRETARY, IOC

*July, 2004*

# PREFACE

## THE INTERDISCIPLINARY GLOBAL COASTAL OCEAN

Companion Volumes 13 and 14 of THE SEA deal respectively with fundamental multiscale interdisciplinary processes that occur generally throughout the coastal seas of our planet, and with the specific combination of processes which govern the dynamics of regions of the coastal seas which together constitute the whole. The term *coastal ocean* here is used for the combined shelf and slope seas and the term *interdisciplinary processes* implies coupled physical-biological-chemical-sedimentological processes and interactions. These volumes complement and supplement Volumes 10 and 11 of THE SEA, which respectively deal generally and regionally with the physical oceanography of the global coastal oceans. The intent of these volumes is to review and overview the status and prospectus of fundamental coastal ocean science in the context of advanced applications and as input to enhanced rational management of coastal seas.

Volumes 13 and 14 are comprised of six parts. Part 1 (13) provides general perspective, and Chapter 1 on multiscale interdisciplinary processes, which has been prepared as a general introduction, refers throughout to all the other chapters of the volume. Part 2 (13) on sediment, biogeochemical and ecosystem dynamics and Part 3 (13) on episodic and long time scale dynamics together constitute the dynamical core of the volume. Part 4 (13) presents scientific issues for applications. The chapters of Part 1 (14) are panregional syntheses for western ocean boundaries (W), eastern ocean boundaries (E), polar ocean boundaries (P), and semi-enclosed seas, islands and Australia (S). These four chapters serve to summarize and introduce the material in the chapters on regional interdisciplinary oceanography, Part 2 (14). The regional segments of the global coastal ocean are indicated on the map following the table of contents of Volume 14 and their region numbers and panregional attributions are indicated following the chapter titles of Part 2 (14) in the table of contents. (Note that the region numbers differ from the chapter numbers.) In order to facilitate coordinated study of material presented in Volumes 11 and 14, the numbering of regional segments is the same. However, here it was necessary to subdivide segment 10 off the coast of eastern Asia into two segments (one coastward of the Oyashio current and the other coastward of the Kuroshio current), and to extend segment 18 northward into the Bay of Biscay, which was not treated in Volume 11. Thus the map in Volume 14 indicates four segments as 10 a, b and 18 a, b.

Note that the coastal segments on the map are numbered geographically, i.e., consecutively moving along a global coastline rather than by panregional groupings. Because Volume 14 comprises 1532 pages of text, it was necessary to publish it in two parts with the geographically numbered coastal segments split between segments 14 and 15. Thus Volume 14, Part A is subtitled "Panregional Syntheses and the Coasts of North and South America and Asia," and Volume 14, Part B is subtitled "The Coasts of Africa, Europe, Middle East, Oceania and Polar

Regions." Both Parts A and B contain all introductory material and a complete index.

In the preface to Volume 1 of THE SEA in 1962, Dr. Maurice N. Hill states, "In collecting the material for these volumes we have had most helpful cooperation from our contributors. In some topics, however, ... we must admit ... to omissions, some of which are conspicuous and important." In Volume 13 we had intended to include a chapter on coral reefs. All thirty-two regional segments of the global coastal ocean are treated in Part 2 (14) except for region 25 (northern North America and West Greenland) because of the withdrawal of the potential author. However, aspects of that region's oceanography are discussed in the chapter on the Northwest Passage (region 26). Additionally, region 17 (the Gulf of Guinea) is treated only by a note, and region 9 (the Northeastern Pacific shelves) only by a bibliography.

The in-depth and comprehensive synthesis of interdisciplinary coastal ocean science presented in these volumes was possible only through the dedicated and scholarly collaboration of the international community of coastal ocean scientists. Chapters were reviewed both by authors of other chapters in the volumes and by external reviewers, whose expertise has contributed essentially to this study. Following the list of contributors is a list of external reviewers in alphabetical order.

This pair of Volumes is a contribution of the COASTS (Coastal Ocean Advanced Science and Technology Studies) program of the Intergovernmental Oceanographic Commission (IOC) of UNESCO and the Scientific Committee on Oceanic Research (SCOR). The scope and structure of the volumes were developed and most chapter topics were presented and discussed at the second COASTS international workshop attended by over sixty scientists, held in Paris in August 2001 under the support of IOC/UNESCO, SCOR, and the U.S. ONR (Office of Naval Research). Subsequent support for the preparation and production of the volumes was provided by IOC and ONR. We sincerely appreciate the support of these agencies which has enabled the substantial international cooperation necessary to accomplish this study.

It is a pleasure to thank Dr. Patricio A. Bernal, Assistant Director General, UNESCO/ Executive Secretary, IOC, for his encouragement and stimulation and for preparing a foreword to these volumes, and Dr. Umit Unluata, Head of the Ocean Science Section of IOC, for his guidance and advice. We are most pleased that Harvard University Press (HUP) is now publisher of *The Sea* series starting with these volumes, and thank Mr. Michael G. Fisher, Executive Editor in Science and Medicine, for his interest and help. The challenging task of preparing for publication this comprehensive study of 60 chapters authored by 170 international scientists was only made possible by the dedicated scientific expository skills of Mr. Wayne G. Leslie and the dedicated administrative editorial expertise of Ms. Gioia L. Sweetland of Harvard University. We thank M. Julian Barbiere and Dr. Maria Hood (IOC), Ms. Sara Davis (HUP), and Mr. Oleg Logoutov and Ms. Margaret S. Zaldivar (Harvard) for their valued contributions to this study and its publication. ARR acknowledges ONR grant N00014-02-1-0989 and KHB acknowledges grant NSF-OCE-0227679 for support throughout this study.

ALLAN R. ROBINSON
KENNETH H. BRINK

*July, 2004*

# CONTRIBUTORS

DANIEL H. ALONGI
Australian Institute of Marine Science
PMB 3, Townsville M.C.
Queensland 4810
Australia

ANDREAS J. ANDERSSON
University of Hawaii at Manoa
School of Ocean and Earth Science and Technology
Department of Oceanography
1000 Pope Road
Honolulu, HI 96822

LARRY P. ATKINSON
Center for Coastal Physical Oceanography
Old Dominion University
Crittenton Hall
768 52nd. St.
Norfolk, VA 23529

ANDREW BAKUN
University of Miami
Center for Sustainable Fisheries
Rosenstiel School of Marine and Atmospheric Science
4600 Rickenbacker Causeway
Miami. FL 33149–1098

JOSE LUIS BLANCO
Center for Coast Physical Oceanography
Old Dominion University
Crittenton Hall
768 52nd Street
Norfolk, VA 23529

HEATHER BOUMAN
Department of Biology
Dalhousie University
Halifax, Nova Scotia
B3H 4JI
Canada

KENNETH H. BRINK
Woods Hole Oceanographic Institution
Department of Physical Oceanography
360 Woods Hole Road-Clark 302A
Woods Hole, MA 02543–1541

FRANCISCO P. CHAVEZ
Monterey Bay Aquarium Research Institute
7700 Sandholt Rd.
Moss Landing, CA 95039–9644

JEREMY S. COLLIE
University of Rhode Island
Oceanography, Bay Campus
South Ferry Road
Narragansett, RI 02882

CLAIRE COPIN-MONTEGUT
Laboratoire d'Oceanographie de Villefranche
Universite Pierre et Marie Curie
CNRS-INSU
B.P. 28, 06230
Villefranche sur mer
France

PHILIPPE CURY
Centre de Recherche Halieutique Méditerranéenne et Tropicale
IRD-IFREMER & Université Montpellier II
Avenue Jean Monnet, BP 171
34203 Sète Cedex
France

HUGH W. DUCKLOW
The College of William and Mary
School of Marine Sciences
Biology Department
Gloucester Point, VA 23062

JOHN FARRINGTON
Woods Hole Oceanographic Institution
360 Woods Hole Road/MS-21
Woods Hole, MA 02543

MICHAEL J. FOGARTY
NOAA - National Marine Fisheries Service
Northeast Fisheries Science Center
166 Water Street
Woods Hole, MA 02543

PETER J.S. FRANKS
Scripps Institution of Oceanography
Integrative Oceanography Division
La Jolla, CA 92093–0218

PIERRE FREON
IRD
CRHMT- Centre de Recherche Halieutique
Méditerranéen et Tropical
Avenue Jean Monnet
BP 171, 34203 Sète Cedex
France

RICHARD GEIDER
Department of Biological Sciences
University of Essex
Wivenhoe Park
Colchester CO4 3SQ
UK

WENDY C. GENTLEMAN
Department of Engineering Mathematics
Dalhousie University
Halifax, Nova Scotia, B3H 2X4
Canada

JONATHAN H. GREGORY
Hadley Centre for Climate Prediction and Research
Met Office
London Road
Bracknell, Berks RG12 2SY
UK

STEPHEN J. HALL
The Worldfish Center
Jalan Batu Maung, Batu Maung
11960 Bayan Lepas
Penang
Malaysa

SUSAN M. HENRICHS
Institute of Marine Science
School of Fisheries and Ocean Sciences
University of Alaska Fairbanks
P.O. Box 757220
137 IRV II
Fairbanks, AK 99775–7220

JOHN M. HUTHNANCE
Proudman Oceanographic Laboratory
Joseph Proudman Laboratory
6 Brownlow Street
Liverpool L3 5DA
UK

RICHARD A. JAHNKE
Skidaway Institute of Oceanography
10 Ocean Science Circle
Savannah, GA 31411

MICHEL J. KAISER
School of Ocean Sciences
University of Wales, Bangor
Menai Bridge, Anglesey LL59 5AB
UK

TONY KNAP
The Bermuda Biological Stations for Research, Inc.
17 Biological Lane
Ferry Reach
St. George's GE 01
Bermuda

ABRAHAM LERMAN
Northwestern University
WCAS Geological Sciences
1847 Sheridan Road – Rm 309
Evanston, IL 60208

FRED T. MACKENZIE
School of Ocean and Earth Sciences and Technology
University of Hawaii
Department of Oceanography
1000 Pope Road, MSB 505
Honolulu, HI 96822

THOMAS C. MALONE
Ocean.US Office for Integrated and Sustained Ocean Observations
2300 Clarendon Blvd. – Suite 1350
Arlington, VA 22201–3667

S. LEIGH MCCALLISTER
The College of William and Mary
School of Marine Sciences
Biology Department
Gloucester Point, VA 23062–1346

BERNARD A. MEGREY
National Marine Fisheries Service
Alaska Fisheries Science Center
7600 Sand Point Way NE
Seattle, WA, 98115

JACK J. MIDDELBURG
Netherlands Institute of Ecology (NIOO)
P.O Box 140 4400 AC Yerseke
The Netherlands

COLEEN L. MOLONEY
Marine Biology Research Institute
Zoology Department
University of Cape Town
Rondebosch, 7701
South Africa

ROBERT J. NICHOLLS
Flood Hazard Research Centre
Middlesex University
Queensway, Enfield
Middlesex EN3 4SF
UK

CHARLES A. NITTROUER
University of Washington
School of Oceanography
Seattle, WA 98195–7940

ANDREA S. OGSTON
University of Washington
School of Oceanography
Box 357940
Seattle, WA 98195

JEFFREY D. PARSONS
School of Oceanography
University of Washington
111A-MSB
Seattle, WA 91895–7940

TREVOR PLATT
Bedford Institute of Oceanography
Ocean Sciences Division
P. O. Box 1006
Dartmouth, Nova Scotia B2Y 4A2
Canada

NANCY N. RABALAIS
Louisiana Universities Marine Consortium
8124 Hwy. 56
Chauvin, LA 70344

ALLAN R. ROBINSON
Harvard University
Division of Engineering and Applied Sciences
29 Oxford Street
Cambridge, MA 02138

KENNETH A. ROSE
Coastal Fisheries Institute
Wetlands Resources Bldg.
Louisiana State University
Baton Rouge, LA, 70803–7503

BRIAN J. ROTHSCHILD
Graduate School of Marine Sciences and Technology
University of Massachusetts Dartmouth
706 South Rodney French Boulevard
New Bedford, MA 02744

JEFFREY A. RUNGE
University of New Hampshire
Ocean Process Analysis Laboratory
Morse Hall, Room 142
Durham, NH 03924

SHUBHA SATHYENDRANATH
Dalhousie University
Department of Oceanography
1355 Oxford Street
Halifax, Nova Scotia B3H 4JI
Canada

ANTOINE SCIANDRA
Laboratoire d'Oceanographie de Villefranche
Universite Pierre et Marie Curie
CNRS-INSU B.P. 28
06234 Villefranche sur mer
France

LYNNE SHANNON
Marine and Coastal Management
Department of Environmental Affairs and Tourism
Private Bag X2
Rogge Bay, 8012
South Africa

YUNNE-JAI SHIN
IRD
CRHMT- Centre de Recherche Halieutique
Méditerranéen et Tropical
Avenue Jean Monnet
BP 171, 34203 Sète Cedex
France

KARLINE SOETAERT
Netherlands Institute of Ecology (NIOO)
P.O Box 140 4400 AC Yerseke
The Netherlands

JOHN H. STEELE
Woods Hole Oceanographic Institution
Marine Policy Center, M.S. 41
Woods Hole, MA 02543–1138

RICHARD W. STERNBERG
University of Washington
School of Oceanography
Box 357940
Seattle, WA 98195–7940

DAVID N. THOMAS
School of Ocean Sciences
University of Wales, Bangor
Menai Bridge, Anglesey LL59 5AB
UK

LEAH MAY VER
University of Hawaii
School of Ocean and Earth Science and Technology
Department of Oceanography
1000 Pope Road
Honolulu, HI 96822

FRANCISCO E. WERNER
University of North Carolina
Marine Sciences, CB# 3300
Chapel Hill, NC 27599–3300

PHILIP L. WOODWORTH
Proudman Oceanographic Laboratory
Bidston Observatory
Birkenhead, Merseyside CH43 7RA
UK

TIM WYATT
Instituto Investigaciones Marinas
CSIC
Eduardo Cabello 6
ES-36208 Vigo
Spain

BRUNO A. ZAKARDJIAN
Institut des Sciences de la Mer de Rimouski
310 Allée des Ursulines
Rimouski, Qc. G5L 3A1
Canada

ADRIANA ZINGONE
Stazione Zoologica 'A. Dohrn'
Villa Comunale
80121 Napoli,
Italy

## EXTERNAL REVIEWERS

ROBERT ALLER
SUNY, Stony Brook
Stony Brook, New York

CAROL ARNOSTI
University of North Carolina at Chapel Hill
Chapel Hill, North Carolina

PETER J. AUSTER
University of Connecticut at Avery Point
Groton, Connecticut

PHILIP BOGDEN
Gulf of Maine Ocean Observing System
Portland, Maine

STEVEN BOGRAD
NOAA/NMFS
Pacific Grove, California

ANNY CAZENAVE
GRGS/Observatoire Midi Pyrenees
Toulouse, France

C. T. ARTHUR CHEN
National Sun Yat-Sen University
Taiwan, R.o.C.

MICHAEL FASHAM
Southampton Oceanography Centre
Southampton, U.K.

CARL T. FRIEDRICHS
The College of William and Mary
Gloucester Point, Virginia

RICHARD W. GARVINE
University of Delaware
Newark, Delaware

PATRICIA M. GLIBERT
Horn Point Laboratory
University of Maryland
Cambridge, Maryland

PETER A. JUMARS
University of Maine
Walpole, Maine

DAVID KARL
University of Hawaii
Hololulu, Hawaii

MICHAEL LANDRY
Scripps Institution of Oceanography
University of California San Diego
La Jolla, California

JAMES J. MCCARTHY
Harvard University
Cambridge, Massachusetts

JAMES G. QUINN
University of Rhode Island
Narragansett, Rhode Island

MICHAEL R. ROMAN
Horn Point Laboratory
University of Maryland
Cambridge, Maryland

VICTOR SMETACEK
Alfred Wegener Institute Foundation for Polar and Marine Research
Bremerhaven, Germany

KENNETH R. TENORE
Chesapeake Biological Laboratory
Solomons, Maryland

PETER TRAYKOVSKI
Woods Hole Oceanographic Institution
Woods Hole, Massachusetts

DAN WALKER
National Research Council
Washington, D.C.

ROLAND WOLLAST
University of Brussels
Brussels, Belgium

# Part 1. PERSPECTIVE

# Chapter 1. INTERDISCIPLINARY MULTISCALE COASTAL DYNAMICAL PROCESSES AND INTERACTIONS

ALLAN R. ROBINSON

*Harvard University*

KENNETH H. BRINK

*Woods Hole Oceanographic Institution*

HUGH W. DUCKLOW

*The College of William and Mary*

RICHARD A. JAHNKE

*Skidaway Institute of Oceanography*

BRIAN J. ROTHSCHILD

*University of Massachusetts Graduate School of Marine Sciences and Technology*

**Contents**



## 1. Introduction

The global coastal ocean is a most interesting interdisciplinary marine system. Ocean science generally is beginning to achieve a new level of realistic interdisciplinary research made feasible by decades of progress in the traditional subdisciplines of physical, biological, chemical and geological oceanography and by recent

*The Sea*, Volume 13, edited by Allan R. Robinson and Kenneth H. Brink
ISBN 0-674-01526-6 

results focused on interdisciplinary processes. The coastal ocean provides a relatively accessible and logistically convenient natural laboratory for the study of fundamental coupled physical-biological-chemical-sedimentological processes. Fundamental processes, similarities and differences among coastal and deep sea processes and the contributions of the global coastal ocean to global ocean dynamics generally, are all research issues of the utmost importance. Through active exchanges and fluxes, coastal ocean waters link together the land, the open sea, the atmosphere and the underlying sediments, and must be expected to affect global processes disproportionately to their relative volume. The coastal zone is undergoing unprecedented changes in response to local and global forcings of natural and anthropogenic forcings and is particularly sensitive to climate change. From a practical viewpoint human populations increasingly utilize and interact with the coastal ocean purposefully, inadvertently and negligently in numerous ways. A deeper scientific understanding of the coastal ocean is essential input to the development of advanced management methods.

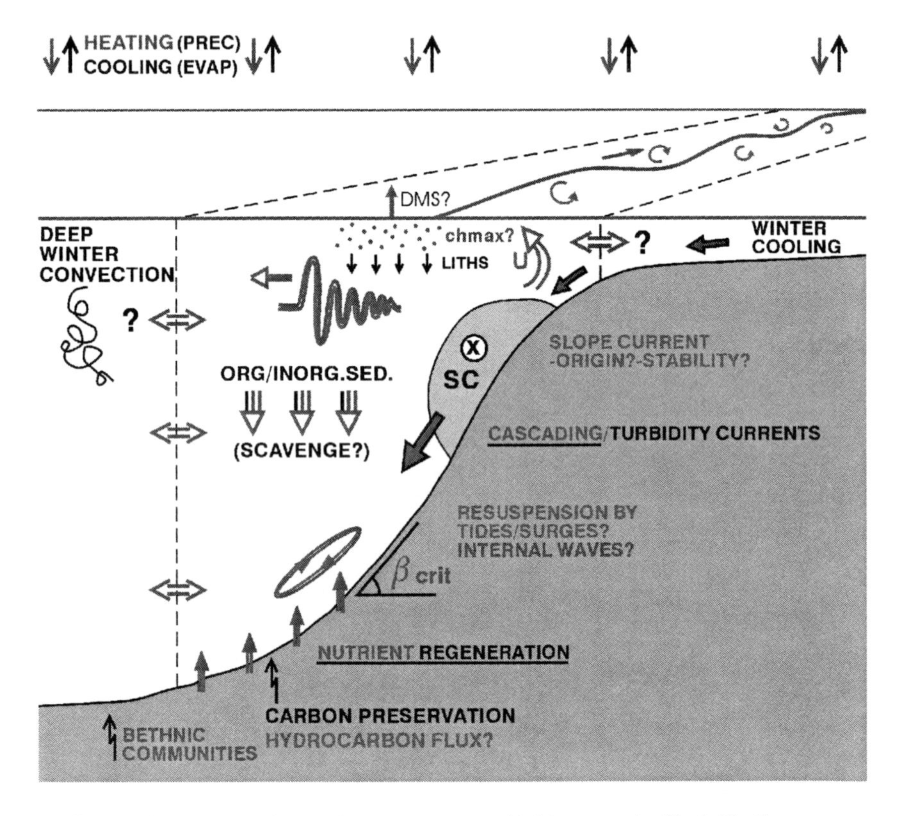


Fig. 1.1 – Schematic depiction of coastal ocean processes. (Atkinson *et al.*– Ch. 8, Fig.1)

A great variety of processes and dynamical interactions occur in the coastal ocean over a myriad of scales in time and space both smoothly and intermittently. Some important processes are schematized in Fig. 1.1. These processes are driven by external forcings and by internal dynamics and instabilities. The narrow width of the coastal sea compared to its alongshore extent assymetrizes the circulation and transports. The shallow depth of the coastal sea amplifies tides, enhances the importance of topographic effects and bottom interactions throughout the water column, and allows the penetration of sunlight to the sea bottom. Water mass and tidally driven fronts occur as well as mesoscale and submesoscale eddies, jets and filaments. The local effects of wind forcing are much more important on the shelf than in the deep sea, and wind induced upwelling from the deep is a major source of nutrients to the shelf.

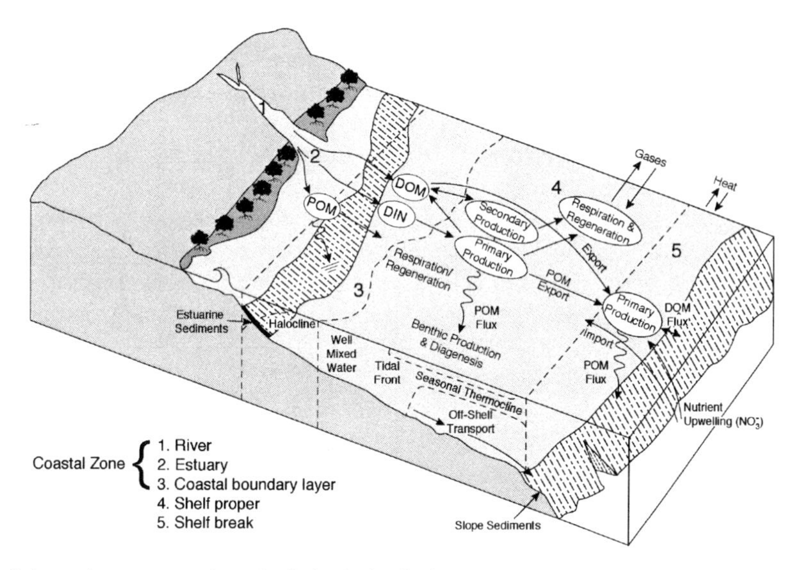


Fig. 1.2 – Schematic representation of relative latitudinal patterns in size spectra and functional characteristics of coastal ocean biota. (Alongi – Ch. 10, Fig.1)

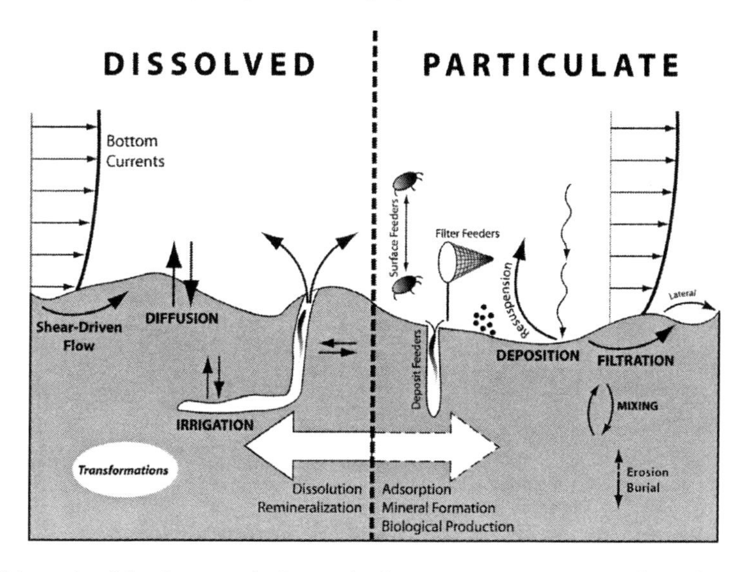


Fig. 1.3 – Schematic of dominant particulate and solute transport processes and reaction pathways in sediments. (Jahnke – Ch. 6, Fig.1)

Upwelling and onwelling events, and as yet poorly understood eddy exchanges and intrusions from the open sea, are the dominant source of nutrients for primary productivity in the coastal ocean. Riverine and estuarine flows and runoffs are also sources as well as atmospheric and bottom sediment fluxes, as depicted in Fig. 1.2. Boundary exchanges and benthic processes include dissolved and particulate inorganic and organic matter, and some important ones are shown in Fig. 1.3. The water column and the sediment interactive processes include: settling, deposition, resuspension, and remineralization of organic matter; dissolution and precipitation of mineral phases; accumulation and burial of materials. In shallow water, benthic primary productivity is now known to contribute significantly to the total productivity. The global coastal ocean is highly fertile. Comprising less than 10% by area of the global ocean, the coastal ocean accounts for almost 30% of the total marine net primary production and at least 90% of the world's fish catch. The global coastal ocean processes are known to contribute importantly to global biogeochemical cycles, and the quantification of the coastal contributions remains a challenging research topic. The role of the coastal ocean in the global carbon cycle is presented in Fig. 1.4. In this chapter we discuss further aspects of the global coastal ocean and its interactive sediment, biogeochemical and ecosystem dynamics and indicate where in the remaining chapters of the volume more details can be found.

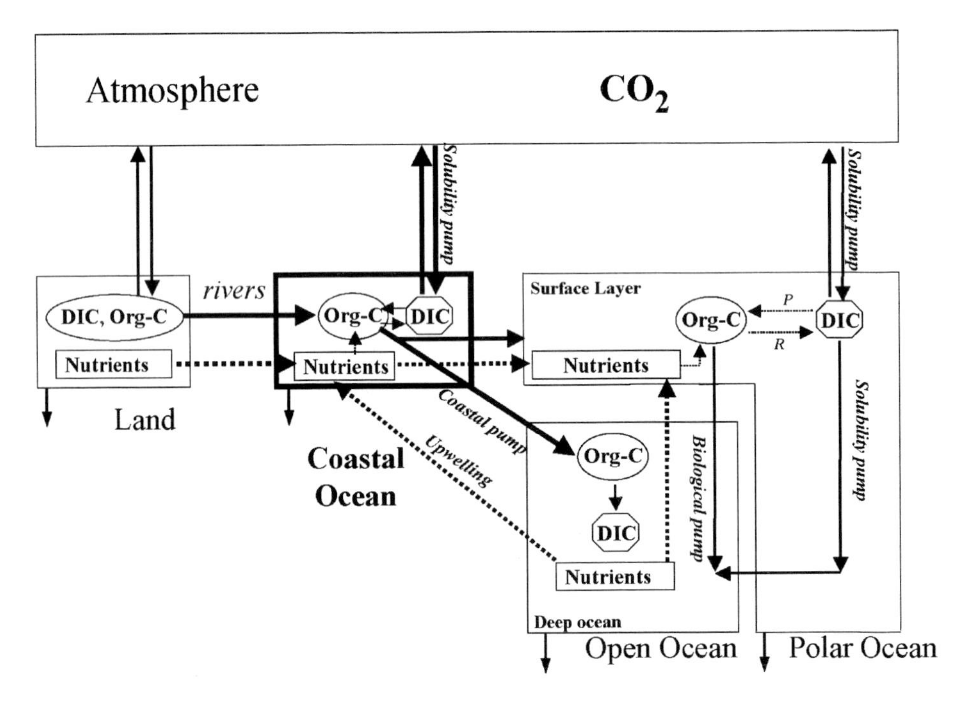


Fig. 1.4 – Carbon and nutrient exchanges between land, the coastal ocean and the open ocean, showing the three carbon (solubility, biological and continental shelf) pumps regulating the allocation of $CO_2$ between the surface and deep ocean. (Ducklow and McCallister – Ch. 9, Fig.1)

## 2. The continental margin

Defining the coastal ocean is not simple. Encompassing all the continents from the tropics to the poles, and amounting to a narrow strip roughly 500,000 km long and just 100 km wide, this energetic and diverse biome defies easy characterization. In this volume it is called different names by different authors: the coastal ocean, coastal zone, coastal margin, continental shelf, continental margin, shelf sea. Estuaries and coastal marshes are generally excluded from consideration. Differences in terminology still confound the discussion of the role of the continent-ocean boundary region in global chemical cycles and ecosystems. Readers should be warned that without common definitions, comparisons of values are often misleading. In general, authors refer to the ocean extending from the coast to the slope waters lying seaward of the shelf break, and subject to the influence of physical oceanographic processes generated at the continental margins such as upwelling, filaments, jets and streamers (Brink, Chapter 2; Atkinson et al., Chapter 8). However, in many other treatments the coastal ocean is taken to be the area of ocean < 200 m deep (e.g., Liu et al., 2000)—a definition that excludes the Antarctic shelf seas 200–700 m deep and that otherwise does not take into account water column processes. The total area of these different domains ranges from 26 (Gattuso et al., 1998) to 37 (Longhurst et al., 1995) x $10^6$ $km^2$.

Here we propose a definition of the coastal ocean as that area, extending offshore from the surf zone and from estuarine mouths, that includes at least the continental shelf and slope, and that also includes waters extending uninterruptedly farther offshore that are (based on temperature or salinity properties only) of shelf or inshore origin. While we readily acknowledge that this definition could be criticized, it is intuitive and specific.

Defining the boundaries of subregions and provinces within the world's coastal oceans is similarly challenging. Many systems exist, depending on the objectives and often, the disciplinary focus involved. At the largest scale, the global coastal ocean can be divided into 4 pan-regions (companion Vol. 14 – The Sea chapters 1–4), based on the dominant geographic or circulation features: Western and Eastern Boundary Current regimes (see also Atkinson et al., Chapter 8), Polar regions, and Islands and Inland Seas. Within this gross regionalization, individual subregions can be divided into as many as 6 depth zones proceeding offshore: the immediate nearshore zone, adjacent regions of freshwater influence, midshelf well-mixed regimes, tidal mixing fronts, offshore regions thermally-stratified in summer and the shelf edge (Atkinson et al., Chapter 8). A more comprehensive basis for defining subregions draws on physical circulation, climate and ecological characteristics (Alongi, Chapter 10), identifying 7 regimes: permanent polar ice, intermittent polar ice, mid-latitude coastal, topographically forced coastal, coastal upwelling, wet and dry tropical coastal. Finally, the coastal ocean may be compartmentalized into regions based on large-scale biogeochemical processes like the air-sea exchange of carbon dioxide ($CO_2$), depending on the source-sink balance of the region and the dominant mechanisms of exchange (physical vs. biological): Subtropical continental shelf $CO_2$ pump, Temperate shelf: biology dominant, Temperate shelf: physics dominant, Upwelling: biology dominant, Upwelling: physics dominant, Coral reefs, Polar ice-rectified $CO_2$ pump (Ducklow and McCallister, Chapter 9). Some unified scheme of characterizing ocean provinces drawing on physical, ecological, biogeochemical and geological criteria would have great utility, for example, in facilitating global integrations for the purposes of elemental budgets or documenting and understanding biodiversity loss (Ducklow, 2003).

In order to help the reader with the wide range of terms used by the various coastal ocean scientific disciplines, we have included, as an Appendix to this chapter, a glossary that is meant to be useful to those readers beginning to explore new areas outside their own discipline. The usage of these terms is, however, not yet uniform across the sub-disciplines.

## 3. Status of knowledge

It is impossible to summarize the content of the various articles in this volume in a few pages, but we do point out some common threads here.

- *The coastal ocean's role in the global carbon cycle* (as well as in other global elemental cycles). This is treated here by Ducklow and McCallister (Chapter 9), Mackenzie et al. (Chapter 7), Alongi (Chapter 10), and Jahnke (Chapter 6). Conclusions in this subject are rather controversial at this time, partly because of varying definitions of exactly what the coastal ocean is, and partly because the balances are so difficult to work out. One part of this problem is the question of whether the coastal ocean is a net source or sink of $CO_2$ to the atmosphere. Deep upwelling source waters are generally oversaturated with $CO_2$, making coastal waters (where upwelling often takes place) a source to the atmosphere. At the same time, coastal waters are highly productive biologically, so that the "biological pump" tends to draw down $CO_2$ concentrations and ultimately leads to sinking or offshore export of particulate carbon in its many forms. Thus, the difficulty of the source/sink problem is created by the difficulty of finding a difference of two large numbers that are very poorly measured on a global basis. In fact, aspects of this problem, like the offshore flux of particulates, are not well known even in the best-studied parts of the world's coastal ocean.
- *High coastal ocean biological production.* The coastal oceans are, on a square kilometer basis, the most biologically productive part of the world's oceans. This has been well known for decades at least. The problem is to explain this productivity (the current state of knowledge is summarized in Atkinson et al. (Chapter 8)). In some locations, the causes of productivity are relatively well known, such as wind-driven upwelling areas or along the Gulf Stream off the eastern United States. In other cases, such as in tidally mixed areas, there are now very sound, quantitatively stated hypotheses to account for nutrient sources, but definitive measurements to support the findings have not yet been made. Finally, there are areas, such as the Gulf of Alaska, that are biologically productive, but where none of the known explanations for productivity appear to match the facts. Our ability to model the coastal ocean, or our ability to do good biogeochemical budgets in the coastal ocean, must start with knowing the processes that drive coastal productivity.
- *Natural ecosystem variability.* We know, based on paleooceanographic data, that substantial shifts (such as the sardine/herring seesaw off western North

America: see Bakun (Chapter 24)) in higher trophic level populations have taken place long before human activity became a factor. The causes for these variations are painfully difficult to discern, but physical causes are sometimes conjectured to be at the root of these swings. Indeed, the ongoing international GLOBEC (Global Ocean Ecosystem Dynamics) program has made it a goal to seek better understanding of these potential couplings. The problem is difficult because of the possible subtle nature of the linkages, and the inherently long (relative to traditional seasonal sampling intervals) time scales often associated with recruitment. Yet, understanding the natural modes of ecosystem variability is critical to better managing living resources in the ocean.

At present, the IMBER (Integrated Marine Biogeochemistry and Ecosystem Research) program is now taking shape (Ducklow and McCallister, Chapter 9) to tackle biological production and its relation with ecosystem structure. This very ambitious international program will effectively address the first three themes mentioned here and their inter-relationship. The sorts of couplings that will be dealt with include how ecosystem structure (i.e. what is living there versus how much is living there) can affect carbon removal from the water column, and the possible effect of climate on ecosystems. Considerations such as these point to the need for broadly defined, integrated projects that can resolve the sorts of feedbacks and couplings inherent to the system.

- *The role of the bottom.* Couplings of bottom processes to those in the water column are not as well characterized as they might be. Water column processes can affect the bottom, for example, through driving sediment transport (e.g., Parsons and Nittrouer, Chapter 17), providing detritus to the bottom or limiting the light that reaches the bottom. Many of these effects are relatively well known, if not always well understood quantitatively. On the other hand, we are still at the learning stage with regard to understanding the effect of the bottom on the water column. The benthic environment is itself biologically, chemically and even physically (through groundwater flow) very active in its own right. Hence materials that reach the bottom can be considerably modified before physical or biological processes allow benthic materials to affect water column processes. It is sometimes estimated (Jahnke, Chapter 6) that benthic processes are in fact major participants in coastal water column carbon or nutrient cycling. The links coupling the bottom to the water column are generally poorly quantified on a global basis, so that it is natural that further efforts should be devoted to this suite of problems. Within the United States, the CoOP (Coastal Ocean Processes) effort is now turning its attention to exactly these issues.
- *Understanding human impacts on the coastal ocean.* Anthropogenic influences on the ocean are becoming more apparent with each year, and can be quite diverse. Nutrients released directly or indirectly by humans are increasingly reaching the coastal ocean and causing effects such as anoxia or even possibly harmful algal blooms (Rabalais, Chapter 21). Fishing practices increasingly change ecosystems, both through direct removal of links in the food web, and through changes in the bottom habitat. The introduc-

tion of exotic species and the process of habitat modification (Kaiser et al., Chapter 23) further affect life in the coastal ocean. Damming rivers changes sediment load, in ways that can have considerable local impacts around river deltas (Kaiser et al., Chapter 23; Mackenzie et al., Chapter 7). The increasing practical importance of these problems has led to a diversity of focused programs (largely on a national basis) to deal with problems and potential remedies.

- *Need for more data.* As the reader studies this volume and its companion, it will become clear that, even in the best-studied areas, some fundamental questions have not yet been resolved through direct observation. It is even more apparent that large parts of the global coastal ocean, through physical isolation or economic strain, have so little *in situ* data available that even some of the simplest questions go unanswered, let alone the quantitative questions that require answers in order to generate meaningful budgets. A basic level of observational coverage, that resolves means as well as seasonal variability, is ultimately required for the whole global coastal ocean. The problems in obtaining such coverage are immense, as they involve both limitations of resources (both financial and personnel) and the understandable sensitivities attached to one nation carrying out research in another nation's waters. It seems possible that the Global Ocean Observing System (GOOS: see Malone et al., Chapter 19) may eventually help to resolve this problem, but we should not expect it to be the complete solution.

As a generality, we point out that our ever-improving ability to observe and model the ocean has certainly led to advances in our understanding, but it has also sometimes led to important deeper or broader problems that require yet more information. Science tends to follow a circuitous path where we may not find answers to our initial questions, but to ones that may not have been originally conceived. As we define current important research areas and assess the efforts needed for their resolution, we should keep in mind that, although research will undoubtedly lead in fruitful directions, they may not be the directions originally envisioned.

## 4. Sediment dynamics

While small in size, the coastal zone is the interface between land and ocean and is also in direct contact with the atmosphere. In addition, many important biogeochemical processes, such as primary production and denitrification, are enhanced and focused in coastal regions. Because it is the exchange across interfaces between the major reservoirs that control large-scale biogeochemical cycles, coastal ecosystems play a critical role in global chemical cycles. Due to anthropogenic influences, dramatic changes have already occurred in the carbon and nutrient balances in near shore coastal zones and it is anticipated that additional changes, especially in developing regions such as Asia, will occur in the future. These changes are disproportionately large compared to those occurring in the open ocean making it difficult to relate present measurements of rates of processes to marine cycles which respond more slowly (Mackenzie et al., Chapter 7).

In general, coastal ecosystems are shallower than their open ocean counterparts. The closer proximity of the seafloor to the surface greatly expands the role of the sediments in the overall ecology and biogeochemical processes that characterize the coastal ecosystem, especially on shorter time scales.

There is a basic paradox in the study of coastal sediments. At locations where sediments are accumulating, characteristics of the sediments reflect to some degree the source of particles and physical processes controlling accumulation, providing a record of the depositional environment. On the other hand, coastal sediments are important sites for the remineralization of organic materials and the regeneration of nutrients. As this is a destructive process, remineralized materials are eliminated from the deposit and materials destined for remineralization often comprise a tiny fraction of the deposit. Many important studies of sediments, therefore, examine the recycling of materials in the sediments, attempting to quantify and understand the processing of the materials that are not retained in the sediment column. Thus, paradoxically, many sediment studies attempt to understand what is not there.

A common theme of recent studies is that increasingly the sea floor is recognized not simply as a bottom boundary condition for the water column ecosystem but rather as an integral part of the dynamics of the coastal system in total. The processes, exchanges and interactions that define this emerging role span a wide range of time and space scales.

- Important space scales range from less than cm, associated with, for example, burrows of infauna and microenvironments within fecal pellets, to 10s or 100s of kilometers reflecting regional variations in shelf geometry, terrestrial inputs, external forces and adjacent coastal currents.
- Important time scales vary from greater than decades and centuries, reflecting long term burial in sediments and frequency of rare tectonic events, to seconds and minutes, representing rate of pressure fluctuations at the sediment surface and in pore fluids due to the passage of surface gravity waves.

*Sources of sediments and/or organic matter.* Rivers are the primary source of lithogenic sediment particles to continental shelves. Often the supply is extremely episodic with individual, large flood events supplying the majority of the materials (Parsons and Nittrouer, Chapter 17). In contrast, the dominant source of organic matter on shelves is marine productivity although there are a few locations, such as directly offshore of the Mississippi/Atchafalaya River system where organic matter of terrestrial origin may dominate (Henrichs, Chapter 5).

Because light reaches the sea floor for much of the world's continental shelf regions, photosynthetic production by benthic microalgae may be a significant source of organic matter and oxygen and sink for nutrients. Because benthic production is often greatest when water column production is lowest, sea floor production may be ecologically important, providing a source of organic material during oligotrophic periods in the water column (Jahnke, Chapter 6).

*Transports.* Transport of fine-grained sediments in river plumes and coastal currents have been well documented in past studies. More recent investigations emphasize the potential role of fluidized mud flows and other lateral transports that decouple the locations of sediment plumes from bottom deposits. Gravity-driven transport of fluid mud may occur on shelves where the benthic layer may reach densities of > 1.3g $l^{-1}$ and may occur more commonly than thought previously. One case study on the Eel shelf (Northern California) documented a fluid mud transport event that, within a three-day period, accounted for 77% of the total material transport for the previous twelve-month period (Ogston et al., Chapter 4).

Within the sediments, particle transport is controlled by commonly recognized transport processes such as accumulation, erosion and biologically-induced particle motions through organism movements, ingestion and defecation. In sediments of appreciable permeability, advective transport of particles, including the filtration of particles from the advecting pore waters may be important (Jahnke, Chapter 6).

The exchange of solutes, such as nutrients, oxidants and inorganic carbon, between the sediments and over lying waters exerts a critical impact on the biogeochemistry and ecology of coastal systems. The minimum rate of solute transport is determined by molecular diffusion. At specific environments this rate may be significantly enhanced by the irrigation of burrows by benthic infauna, by dispersion induced through fluctuations in pore pressure due to the passage of surface gravity waves, by advective transport caused by bottom current - pore water interactions and by groundwater inputs (Jahnke, Chapter 6).

*Deposition/resuspension.* The traditional view of muddy deposits forming on shelves by particles settling vertically out of sediment-laden river plumes and accumulating on the sea floor must be modified to include: 1) influences of density-driven flows (fluid muds) within the benthic boundary layer, 2) other across-shelf transport mechanisms that vary between the inner, mid and outer shelf subregions, and 3) impact of the relative timing of the sediment supply and coastal-ocean energetics that may redistribute the sediment. Deposits and sources are often not in equilibrium with present conditions (Ogston et al., Chapter 4).

Suspended sediment concentration reflects the combination of sediments advected into the area and the difference between local resuspension and deposition which is primarily controlled by the shear stress (Ogston et al., Chapter 4). Advected plumes can be both buoyant (hypopycnal) and negatively buoyant (hyperpycnal) where water mass density within the plume is greater than the surrounding waters due to the addition of sediments.

The balance between resuspension and deposition determines net accumulation. Because larger particle grains require larger energies for movement, wave and current motions commonly sort surface sediments by winnowing the finer particles from many coastal deposits. Quantifying the pore water-particulate framework, as represented by the porosity, tortuosity, and permeability of the sediments, all of which are in part related to grain size, is critical to discussing pore water and sedimentary processes. Because 70% of all continental shelves are classified as geologically relict and most probably sandy with elevated permeabilities, advective transport and elevated rate of dispersion may be common and greatly enhance the exchange of solutes and particles between the water column and sediments. This, in turn, intensifies the coupling between the sedimentary and water column subsystems (Jahnke, Chapter 6).

*Transformations.* Biological processes (microbial, enzymatic, ingestion, transport) combine with abiotic processes such as adsorption, aggregation, and transport, to control the degradation of organic material in sediments (Henrichs, Chapter 5). Utilization of the major oxidants in pore waters follows the well documented order of decreasing free energy release resulting in the sequential utilization of oxygen, nitrate and manganese oxides, iron oxides, and sulfate. As the organic matter input to the sediments increase in coastal sediments, a greater proportion of the organic carbon is oxidized through suboxic and anoxic metabolism and a great proportion of the oxygen flux is used to reoxidize the reduced metabolites, ammonium, sulfide, ferrous iron and manganese (II) (Jahnke, Chapter 6).

*Burial.* Most (about 86%) burial of organic matter occurs in continental margin (deltaic, continental shelf and upper slope) sediments. Dominance of margins as the location of organic carbon burial appear to extend throughout geologic history as the vast majority of organic carbon found in sedimentary rocks appears to have been buried under similar conditions (Henrichs, Chapter 5). The organic matter in most sediments represents a small residual of the organic matter initially deposited. The material has generally been re-worked numerous times and recalcitrant compounds that were a minor component of the initial material may become enriched. Major debates remain as to what combination of factors control the preservation of organic materials in sediments. Important factors include the differences between oxic and anoxic degradation pathways and physical protection through surface adsorption and particle aggregation with mineral grains.

While the sediments may serve as a "tape recorder" of past depositional environments, extreme events dominate sediment signals and features and the record is very biased. This is because rare, large events will obliterate smaller depositional signals. Understanding the magnitude of this bias and the relationship between the sediment record and non-extreme event processes presents a significant challenge. Besides the obvious low probability of directly observing rare, extreme events, instrumentation is often not capable of withstanding the event—although occasionally *in situ* sensors capture specific events (Parsons and Nittrouer, Chapter 17).

Two classes of extreme events dominate the sediment record: climatic events that ultimately obtain their energy from sun (i.e. floods or storms) and tectonic events that rely on geophysically supplied energy (i.e. earthquakes). The study of the deposits associated with these events has greatly been advanced through acoustic and seismic measurements and models. The duration (to a certain extent, thickness) and the recurrence of events can provide clues as to its origin: tectonic events are usually shorter in duration than climatic events and release more energy while climatic events contain less energy than tectonic events and extend over longer periods (Parsons and Nittrouer, Chapter 17).

*Modeling.* There has been significant progress in developing numerical simulations and interactions of benthic and pelagic processes (Middelburg and Soetaert, Chapter 11). Detailed conclusions are dependent on model assumptions. Nevertheless, calculations demonstrate that on continental shelves: benthic processes such as remineralization and denitrification exhibit much less short-term variability than deposition and other water column processes and that there is a lag period in the maximum rates of these processes. This also includes variations in oxygen flux and oxygen penetration depth. The model results demonstrate that sediments can be a very important component in the overall shelf ecosystem.

## 5. Biogeochemical cycles

The most important feature of the global coastal ocean is that its internal processes are dominated by exchanges across its boundaries: the coast, the bottom sediments, the frontal systems bordering the open ocean and the atmosphere-ocean interface. Large amounts of inorganic nutrients, sediment and organic matter enter from the land via rivers, estuaries and groundwater flows (Mackenzie et al., Chapter 7; Atkinson et al., Chapter 8; Ducklow and McCallister, Chapter 9). Currently about 50 megatons (Mton; or 50 teragrams—Tg or $10^{12}$ g) of reactive nitrogen and 8 Mton phosphorus enter the coastal ocean annually from rivers and sewage loading (Mackenzie et al., Chapter 7). This input supports about 350–800 Mton of primary production, or about 2–5% of the total net production in the coastal ocean. The majority of primary production in the coastal ocean is supported by upwelling and onwelling from the open ocean (Ducklow and McCallister, Chapter 9). Inputs from land, including sewage and fertilizer runoff are important especially in the immediate nearshore coastal zone, and will increase by ~50–100% over the next 50 years as a consequence of population growth, development and increased cultural eutrophication (Mackenzie et al., Chapter 7; Rabalais, Chapter 21). But offshore nutrient inputs from the open sea are the dominant source of primary production in the coastal ocean as a whole. This important point is critical for understanding the functioning of the coastal ocean as an integrated system.

The other principal feature that governs coastal ocean processes is shallow depth. The bottom depth ranges from less than 10 m nearshore to 100–200 m over the outer continental shelf in temperate, tropical and north polar regions, to 300–700 m in the Antarctic. This geomorphological feature of the coastal ocean has two consequences. First, most, and often all of the water column as well as the bottom is illuminated, supporting enhanced rates of photosynthesis (Jahnke, Chapter 6; Platt et al., Chapter 12). Production by benthic microalgae can be a significant component of the total primary production, and dominates over water column production at shallow depths. Benthic photosynthesis serves as a source for oxygen and organic matter to the benthos as well as water column and a sink for nutrients, thus modulating the exchanges of these substances (Henrichs, Chapter 5; Jahnke, Chapter 6). The second consequence of shallow depth is that over much of the coastal ocean, bottom sediments can dominate water column processes (Ogston et al., Chapter 4; Middelburg and Soetaert, Chapter 11). For example nutrient fluxes from the bottom sediments support an important fraction of the primary production in coastal ocean waters—a source of nutrition non-existent in the open sea.

Taken together, these features—intense boundary exchanges, nutrient inputs from land, sea and sediment, shallow depth and enhanced light availability support disproportionate rates of primary production, relative to the area of the coastal ocean. Making up just 8–10% of the area of the world ocean, the coastal ocean might account for 800–1200 Tmol C $y^{-1}$, (20–30% of the total marine primary production, or 10–15% of the total terrestrial plus marine production (Ducklow and McCallister, Chapter 9; Platt et al., Chapter 12). Relative to the open sea, a larger fraction of the primary production in the coastal ocean is carried out by large–celled phytoplankton such as diatoms, dinoflagellates and haptophytes, which can form harmful algal blooms (Zingone and Wyatt, Chapter 22). In the open sea most of the primary production is by smaller cells usually less than 5–10 μm in diameter. Most mesozooplankton (copepods and krill) are incapable of grazing these nanoplankton (2–20 μm cells) and picoplankton (0.2–2 μm). Thus in the open sea, plankton foodchains are dominated by a complex assemblage of small phytoplankton and protozoan grazers. Because these organisms at the base of the foodweb are small, several extra trophic levels are necessary to transfer organic production to larger zooplankton and fish. In contrast, coastal ocean food-chains are shorter because they are initiated with larger phytoplankton that are grazed by larger zooplankton (Alongi, Chapter 10). This contrast between coastal vs. oceanic foodchains is greater in temperate and polar systems than in tropical regions, which tend to be more similar to the open sea.

Plants, animals and bacteria respire organic matter back into $CO_2$ and nutrients to generate chemical energy for life processes. The total amount of organic matter formed in photosynthesis by autotrophs (both phytoplankton and benthic algae) prior to autotrophic (plant) respiration ($R_a$) is called the gross primary production or GPP. The amount of plant biomass left after plant respiration is the net primary production or NPP= GPP – $R_a$. Once the NPP is grazed or decomposed, it is re-spired by animals and bacteria (heterotrophs). The sum of the autotrophic and heterotrophic respiration ($R_h$) is the community respiration ($R_a + R_h = CR$) and the organic matter surviving both autotrophic and heterotrophic respiration is the net ecosystem production, or NEP (NEP = GPP – CR) (Mackenzie et al., Chapter 7). Thus the NEP is the net production by the ecosystem functioning as an integrated whole. NEP is the residual between two large quantities, GPP and CR, neither of which are determined easily or accurately. Uncertainties are large. The coastal ocean might contribute ~200–250 Tmol of the annual global ocean NEP of ~500 Tmol. Determining the NEP more precisely is important because it is the NEP that is available for harvest or export to another system or region external to the eco-system in which it was produced. In the open sea, remote from land or other boundaries (e.g., in the oceanic central gyres), much of the NEP is exported verti-cally by gravitational settling or sedimentation into the ocean interior. Vertical export from the illuminated surface layer also occurs in those parts of the coastal ocean that are sufficiently deep to manifest vertical stratification, but the principal export from the coastal ocean is lateral transport across the offshore boundaries to the open sea (Ducklow and McCallister, Chapter 9). Lateral export is driven by the myriad of circulation and sediment processes outlined in Brink (Chapter 2), Ogston et al. (Chapter 4), Henrichs (Chapter 5), Jahnke (Chapter 6), Mackenzie et al., (Chapter 7) and Atkinson et al. (Chapter 8). The formation and fate of coastal filaments, jets and streamers are potentially very important in determining the budgets of organic matter formed in the coastal ocean.

This lateral transfer of organic matter from the coastal ocean is often called shelf export or the continental shelf carbon pump, and quantifying it in different geographic regions has been a major effort in the past two decades (Mackenzie et al., Chapter 7; Ducklow and McCallister, Chapter 9). Coastal ocean NEP not ex-ported laterally to the ocean interior is deposited in shelf sediments and either respired or buried. Most shelf sediments are geologically relict and sandy with low organic content, suggesting that shelf export is efficient. The range of estimates of shelf export is wide, but indicate that about 8–16% of the coastal ocean NPP is exported.

The continental shelf pump is one of the three integrated processes transferring carbon from the surface to the deep ocean where it is stored on the timescale of the oceanic thermohaline circulation. These three processes or carbon pumps are the solubility pump, the biological pump and the continental shelf pump, and together they regulate the level of $CO_2$ in the atmosphere over centennial and longer (e.g., glacial-interglacial) timescales. $CO_2$ exchanges freely across the atmosphere-ocean interface by physical chemical processes that are functions of wind stress and the difference in the partial pressure of $CO_2$ gas ($pCO_2$) in the two adjoining reservoirs. When the $pCO_2$ is lower in seawater than in the air, $CO_2$ passes into the sea, i.e., the sea is a sink for atmospheric $CO_2$. Once dissolved in seawater it can be transported to depth in dense, sinking water (solubility pump) or fixed into organic matter (Platt et al., Chapter 12) and exported laterally (shelf pump) or vertically (biological pump). Initially the coastal ocean was not considered separately in global carbon budgets, primarily because global ocean circulation models did not resolve it. However increased research and more finely resolved surveys show that $pCO_2$ is highly variable in the coastal ocean, ranging from deep undersaturation (<200 μatm or ppm) to great supersaturation (>500 ppm) (Mackenzie et al., Chapter 7; Ducklow and McCallister, Chapter 9).

The source-sink identity of the coastal ocean is a controversial and misunderstood problem. It is often noted that the organic matter cycle in the coastal ocean is a source of $CO_2$ for the atmosphere, because of terrestrial inputs of organic matter that render the region net heterotrophic-respiration exceeds production, that is, NEP is negative and there is a net production of $CO_2$ (Mackenzie et al., Chapter 7). This is believed to have been the case prior to the anthropogenic increase in atmospheric $CO_2$ from its interglacial level of 280 ppm that began about 300 years ago. However this does not mean that the coastal ocean as an integrated physical-biogeochemical system is necessarily a $CO_2$ source region today. Although theoretical modeling studies and some carbon budget approaches consider that it is still a weak net source (Mackenzie et al., Chapter 7), empirical syntheses of coastal ocean $pCO_2$ suggest the region as a whole is a net sink (Ducklow and McCallister, Chapter 9). Integrated globally and over all seasons, the average $pCO_2$ in the coastal ocean is less than the current, anthropogenically-enriched atmospheric level of ~380 ppm. In effect, the net excess of organic matter respiration over autotrophic production is insufficient to overcome the anthropogenically-driven piston force pumping atmospheric $CO_2$ into the coastal ocean. Estimates of the continental ocean $CO_2$ pump range from about 8 to 70 Gt C $y^{-1}$, or 8–45% of the total net ocean uptake. Models suggest the coastal ocean sink will strengthen in the coming decades as increased nutrient inputs from land render the coastal ocean net-autotrophic (Mackenzie et al., Chapter 7).

This brief consideration suggests the importance of considering the coastal ocean as a dynamic and integrated physical-biogeochemical system. Considered in isolation, biological processes (the organic matter cycle) constitute a source of $CO_2$ for the atmosphere. But strictly physical processes actually control the air-sea exchange, and the weight of observations indicates that the net exchange is into the coastal ocean at the present time.

## 6. Ecosystem dynamics

Ecosystems are complex systems of interdependent animals and plants. The interdependence is primarily trophic and relates to the transfers of energy from one organism to another and to the biological and physical variables that influence the transfer (Cury et al., Chapter 14 discusses many aspects of the trophic interaction among organisms). The term ecosystem is generally taken as a bound or boundary that defines a specific group of trophically interacting plants and animals. The bounds or boundaries that define specific ecosystems are characterized in a variety of ways. Ecosystems are generally defined by dominant physical features (e.g., upwelling, inter-tidal, or pelagic ecosystems) or by regions (e.g. northwest Atlantic) or geographic designations (e.g., the Georges Bank or the Barents Sea ecosystem). Terrestrial ecologists often characterize ecosystems by dominant plants and animals such as the "oak-hickory forest".

Classical studies of ecosystems (see Allee et al., 1949) focus upon the demography of populations (see Rothschild – Vol. 12 The Sea) and the plant and animal community or communities that comprise the ecosystem (see Cullen et al. – Vol. 12 The Sea). These features are studied in the context of "community stratification, metabolism, and periodism" (Aspects of classical ecosystem theory are discussed in Steele and Collie (Chapter 20). The issue of community metabolism is particularly important today. The early ecologists took into account the physiological idea of individual metabolism where an individual's metabolism is measured in terms of ingestion, respiration, growth, and excretion. The idea is that whole communities and ecosystems possess the same properties of ingestion, growth, respiration, and excretion. These are potentially convenient measures of ecosystem performance, ecosystem change and variability. These metrics can also be used to compare the function of ecosystems and subsets of ecosystems.

A particularly intriguing aspect of classical ecology is the concept of succession. Open-ocean plankton succession is thought of on more or less geological time scales. However as we move into the coastal ocean, coastline modification and eutrophication become issues of contemporary concern and raise the important scientific challenge of predicting these effects and somehow ameliorating the anthropogenic components.

The definition of ecosystem boundaries and ecosystem performance are important tools in the study of ecosystems. For example, we might think of a generalized ecosystem and study its performance measures in terms of community metabolism. We are interested in how variability in ecosystem performance measures is reduced by considering subsystems. How small should subsystem partitions be? An issue particularly relevant to the coastal ocean is the issue of how open ocean ecosystems change when there volume begins to intersect with the benthic boundary. This approach can be seen in the discussion in Alongi (Chapter 10) where he classifies ecosystems and considers in some case their performance in terms of a characteristic production-respiration ratio. Steele and Collie (Chapter 20) discuss how diversity is modified as ecosystem bounds move from deep water into the shallower coastal seas.

In the simplest terms there are two main components: a primary production component and a secondary production component. The primary production ecosystem component is characterized by populations of plants that utilize both photosynthesis and respiration to produce "plant biomass". In plants the photosynthetic process interacts with the respiratory process to utilize the sun's energy, $CO_2$, and nutrients (e.g. nitrate, silicate, etc.) to produce biomass while at the same time burning oxygen and producing carbon dioxide in the process of respiration. The secondary production ecosystem component is characterized by "animal" populations that utilize only the "respiratory process" to produce animal biomass from ingested or absorbed nutrients and $O_2$. The ingestion process can involve absorption of molecules (by bacteria, for example), grazing by animals on phytoplankton, or other very small organisms or detrital particles, and predation where a larger animal consumes a smaller animal.

The primary and secondary production components are particularly linked via excretion and grazing. Grazing of plants and excretion of ammonium by secondary producers is an important linkage between primary and secondary production. Excretion is a byproduct of both primary and secondary production. Secondary producers contribute to the mortality of primary producers via grazing, while at the same time secondary producers contribute to the nutrition of primary producers by producing the nutrient ammonium (along with some urea).

The organisms that characterize ocean ecosystems are for the most part microscopic plankton (historically, the term plankton was used to describe organisms that were thought to be so small that their mobility was thought to be "at the mercy of the currents"). Corresponding to the generalized ecosystem, the plankton primary producers are phytoplankton or autotrophs. The secondary producers are zooplankton or heterotrophs. In actuality depending on various definitional approaches there are two kinds of secondary producers: the zooplankton and the bacteria. The zooplankton and bacteria can be collectively called heterotrophs. The different types of plankton vary in size by orders of magnitude. For example bacteria might be 1 μm in equivalent spherical diameter (ESD); phytoplankton might be 20–30 μm in ESD; and a large copepod might be of the order of 100 μm ESD. The density (number per unit volume) of this organism is inversely related to their ESD. For example there might be densities of $10^6$ bacteria per cc and $10^{-6}$ copepods per cc. The density of the plankton translate into inter-particle distance. A general idea of the importance of the inter-particle distance and its second order properties (i.e. its variance which is a measure of patchiness) can be considered that for a predator to encounter a prey it needs to reduce the relative distance between itself and the prey to some threshold distance where it can identify, attack, and consume the prey. The energy acquired by the predator should generally be more than the energy expended to reach the prey (attack predators consume more energy than ambush predators). In other words the space scales that govern the inter-particle distances among the plankton particles have consequences in terms of ecosystem energetics.

There is a multiplicity of pathways that involve the flux of energy through the ecosystem. The modality and rates that the plankton ecosystems process energy depend upon the trophic transactions among the interacting living and non-living particles in the ocean ecosystems. There are many different types of trophic transactions that characterize the ocean ecosystem. Just to name a few examples: phytoplankton utilize photons and molecules; very small zooplankton and bacteria consume detritus and molecular algal exudates. Larger zooplankton graze on phytoplankton cells and detritus. Zooplankton captures food particles in a feeding current. Fish larvae prey upon zooplankton nauplii. Herring graze on zooplankton.

One of the most important characteristics of each population and its trophic interactions is the family of non-linear responses that it makes to its own density. This density dependence is exemplified by the fact that populations tend to grow faster, reproduce at a higher rate, and exhibit reduced mortality when they are at a low level of abundance and *vice versa*. The important property stabilizes populations preventing them from exploding in number or becoming extinct. There is a rich array of density dependent stabilizing processes that is difficult to capture in analytic models, particularly those that have been linearized (Runge et al. (Chapter 13) discuss density dependence).

Individual organisms, populations and ecosystems integrate these trophic transactions in various ways on different time and space scales. For example, on the shortest time scale individual organisms encounter and ingest a proportion of the encountered items according to the nature of the functional response. On longer time scales, the ingested food is incorporated into biomass. In yet longer time scales some of the biomass is incorporated into reproductive biomass.

The final important point that relates to the ecosystem concept is the issue of scale. With regard to space scales we can think of several ways to utilize the ecosystem concept. First it is scientifically interesting and practically important to consider the global ocean ecosystem. One of the most important scientific and practical problems relate to scenarios of anthropogenically induced change in climate. A particularly important question that relates to this global ecosystem is how the changes in climate influence the oceans contribution and uptake of carbon dioxide at the sea surface. An important aspect of this question relates to the fact that surface ocean basin scale ecosystems are linked via the wind forcing implies on different time scales by various oscillations such as the NAO and the PDO (Chavez, Chapter 16). A second important scientific and practically important problem involves the basin and sub-basin scale influence of fish on ecosystems and ecosystems on fish. The third problem of particular scientific interest involves viewing ecosystems in terms of its spatial subsystems. For example, we could imagine kilometer volumes, 10 km volumes and so on.

With regard to time scale there are particularly important scientific and practical problems. For example, if we observe a change in a property then we can ask if this change is a random property or part of a definite shift. Without a theoretical basis this is difficult. One of the more interesting time scale problems relates to anthropogenic effect and biodiversity. It is usually thought that various anthropogenic effects limit biodiversity. This is certainly true in many short-term cases. However, the processes need to be studied carefully, reckoning the effects of the influence of anthropogenic effects on mutation and selection.

The conceptual picture of ecosystem structure and functioning enables development of analytic models that can be used to sharpen questions on the structure and functioning of ecosystems. Various models of ecosystems and approaches to ecosystem models are found in Rothschild (Chapter 3), Cury et al. (Chapter 14), Runge et al. (Chapter 13), and Steele and Collie (Chapter 20). A central theme is the NPZ approach, which is ordinarily presented as a set of coupled differential equations representing N (nutrients), P (phytoplankton), and Z (zooplankton). These NPZ equations, which are discussed in detail in Runge et al. (Chapter 13), contain constants or terms that relate to mortality for the living components and rates of uptake or consumption generally modified by a functional response. These differential equations which can be considered as systems of equations and can be imbedded in physical processes (see Gargett (large scale); Flierl and McGillicuddy (mesoscale) and Yamasaki (small scale) chapters in Vol. 12 The Sea) generate various behaviors. By varying parameters and configurations it can be seen how the system is "controlled". In some parameter space nutrients dominate (bottom up control) whereas in other parameter space predators dominate the system (top down control). Likewise, stability of these systems is generally determined by mathematical tests of trajectories to or from stable or unstable nodes (there maybe several stable or unstable nodes) of the linearised system. It is interesting to observe that perturbation of parameters that might be the result of physical forcing might make a stable set of questions but it might also make an intrinsically unstable equation stable.

A difficulty of the NPZ approach is that it is not easy to handle the recruitment problem, which is characteristic of fish populations (Runge et al., Chapter 13) and probably other populations as well. A number of workers have proposed analytic approaches that are promising and represent the flux of biomass or energy through the ecosystem but these have not been widely accepted (Rothschild, Chapter 3). The NPZ approach needs to be extended to handle the small-scale particulate nature of plankton (contrast Rothschild (Chapter 3) and Cury et al. (Chapter 14) giving the stochastic and deterministic version of the functional response respectively).

This conceptual description of the ocean ecosystem has been more or less traditional in the sense that it has described the unforced dynamics of the ocean ecosystem. The forced dynamics, that is the effect of physical forcing is still a major research topic even though there have been many important advances over the past few decades. In fact, there is no question that the most significant questions of the past quarter century relate to the role that physical forcing plays in modulating biological productivity. Physical forcing operates in manifold modalities. For example light operates in at least two fundamental and important ways: it is the fuel for photosynthesis and it affects the competence of visual or light-dependent predators. Temperature sets the pace of metabolism of most organisms in the sea. A change of temperature can move a population from optimal performance to less than optimal performance and *vice versa.* Temperature and light interact strongly in the photosynthetic process. The wind field has a strong effect interacting with the internal dynamic of the ocean to sculpt the thermocline controlling the delivery of remineralized nutrients to plants but also affecting the TKE which is implicated in affecting the interaction between predator (increasing or decreasing feeding rate independent of the functional response and increasing or decreasing mortality (see Vol. 12 The Sea). The wind field via its global teleconnections potential reflects the possibility of global interrelationships given the connections between oscillations in the Pacific and the Atlantic.

A growing interest in physical forcing continues to generate increasing definition in the nexus among temperature, light, and the flow field. Light is particularly important because it is a key factor along with temperature in pacing the connection between nutrients and temperature in the photosynthetic process and P in the NPZ formalism. Platt et al. (Chapter 12) discusses the delivery of light to the sea surface and to phytoplankton cells; Platt's discussion demonstrates that there are important critical issues to be studied in this fundamental process, particularly the behavior of the photosynthetic system in realistic unsteady environments. It is interesting to contrast the synthesis of primary production given by Platt with the synthesis of secondary production given by Runge et al. (Chapter 13).

## 7. Measurement and observations

It is critical to make measurements in the coastal ocean for a range of reasons. These include:

- Establishing a global baseline of coastal information,
- Discovering ocean processes, understanding how they work, and developing parameterizations,
- Testing hypotheses,
- Driving and evaluating numerical models, and
- Observing change, either secular (as in the atmospheric increase of $CO_2$), or low frequency (such as ENSO phenomenology).

Our ability to measure the coastal ocean has, fortunately, improved radically in the last decade. Historically, the best-resolved measurements were those related to physical oceanography: currents, temperature, salinity and sea level. The relative simplicity of these observations allowed the establishment of long time series (especially for sea level) that are well-resolved in time. This capability, in turn, allowed the ability to develop a sophisticated understanding of what represents adequate sampling, given a set objective and the measured inherent space and time scales of the particular quantity.

Advanced sensing technologies have led to capabilities for other ocean disciplines that are rapidly becoming comparable to physical capabilities. For example, optical technologies have advanced from fluorometers and transmissometers (that provide interesting but highly aggregated information), to newer systems that provide useful information on nutrients, plankton growth rates or species. One example, the Video Plankton Recorder, measures zooplankton populations as the cm-scale, continuously, over days or more; the images are analyzed into counts by use of expert systems. Compare this level of zooplankton information to that obtained laboriously by nets that could only sample over vertical or horizontal averages of meters at least! Comparably radical steps forward in sampling are being made in other coastal ocean disciplines as well, through use of biotechnology, reaction chambers, *in situ* wet chemistry or acoustical methods. While not every measurement that we might want to measure automatically is yet becoming possible, the overall trend is extremely encouraging.

It is likely that the entire way that we measure the ocean will change over the coming decades. New developments in *how* we deploy sensors are bound to be important. Of course, we will continue to see a continuing improvement in ship and remote sensing capabilities. In terms of truly new approaches, Ocean Observatories represent one opportunity. These systems, now under development, will provide a permanent, interactive, presence in the ocean with revolutionary capabilities in terms of power and communications. These systems could be connected with a land base either through satellite communications or through submarine fiber-optic cables. In either case, the system will allow the resolution of long-term changes and will make possible the documentation of rare, but important, episodic events such as slides or hurricane passages. A second revolutionary platform will be mobile sensing platforms that can control their positions: either propelled Autonomous Underwater Vehicles (AUVs) or gliders that have much less demanding power requirements. These platforms, which are very capable of intensive spatial coverage, will be an important supplement to the Ocean Observatories, that are necessarily sparsely sampled in space. The data streams from either of these measurement systems will undoubtedly be part of a larger observation/model system.

The opportunity presented by these new tools is enormous. Quantitatively meaningful interdisciplinary observations require the ability to measure all variables with an adequate and identical space and time resolution that captures the smallest important scale of the relevant process. This is a very demanding criterion, which has not yet been made specific in many cases because of our inability so far even to determine what those important scales might be.

Thus, developing the new instruments is really just the beginning. The sorts of measurements to be made overall can be divided broadly into measuring state variables, and those measuring the related fluxes and transformations. Say, for example, that one wants to measure phytoplankton biomass in a volume (the state variable), and the factors that change it (growth, grazing, advection-fluxes). Knowledge of the space and time scales of the quantity is critical. How frequently must one measure in order to resolve changes? Over how many events (which could have periods of seconds to years) must one average to get a reasonable estimate of the quantity and its changes? How finely must one sample in space to be confident that important information is not being missed because of patchiness? To deal with these issues, one needs a sensor that measures the desired quantity with appropriate spatial and temporal resolution as well as duration.

While measuring state variables is difficult enough, the factors of change are likely to be more difficult, since more than one thing needs to be measured compatibly. Take, for example, the advective flux of a nutrient. First, the nutrient and currents need to be measured at the same places over extended times. The spatial resolution (both vertically and horizontally) of the measurements must be adequate to resolve the smallest important scales of the quantity (nutrient or current) that has the shorter inherent scale. The duration of the measurements must be sufficiently long to obtain statistically significant flux estimates. The issue of how long to sample is often complicated by the fact that fluxes can be dominated by large-scale but weak flow patterns that can be largely masked by energetic, but, in context, unimportant variability. For example, strong tidal currents over Georges Bank easily mask the more subtle flow patterns that deliver nutrients to the Bank.

Thus, making good use of the new generation of interdisciplinary sensor systems will require learning a good deal of background information about natural scales, and it will call for deployments that are intensive in both time and space. Further, the demand for the measurements itself is growing dramatically both in terms of things that need to be observed, and in terms of more exacting spatial and temporal needs.

These increasing demands inevitably put a strain on financial and intellectual resources. There are certainly some long-term solutions that will help. One is to develop reliable parameterizations of specific processes that would eventually eliminate the need to measure the details, but rather just some key input variables. A second solution is simply to be more clever about what we measure and how. Technologies related to smart floats, dye tracers or proxy variables all have promise. Further, it might make sense to, say, measure the alongshore fluxes of a nutrient, and calculate its cross-shelf flux as a residual of consumption and alongshore flux divergence, than to measure a very difficult cross-shelf flux directly. Ultimately, though, we expect that sophisticated, well-resolved, data assimilating models will prove to be part of a sensible and cost effective solution for many applications. Getting to this point presents both modeling and observing challenges in its own right.

## 8. Models

Modeling of interdisciplinary marine systems in general and coastal systems in particular involves a number of modeling approaches. Types of models range from attempts at realistic models in four dimensions to highly idealized models of isolated processes, with a host of models of intermediate complexities.

A general and realistic interdisciplinary dynamical model for the global coastal ocean must link biogeochemical and ecosystem dynamical processes and interactions in the water column and in the benthos, and be coupled to a physical circulation model which provides realistic physical forcings over multiscales in time and space. Accurate representations of fluxes and exchanges with the atmosphere, the deep sea and the land are required together with correct high-resolution treatments of the geomorphology of the shelf and slope, the coastline and underlying topography. Efficient and effective representations of the distributions and behaviors of the larger animals are necessary. Such a general model can only be expected to be achieved after decades of future research. However, novel aspects of realistic interdisciplinary modeling have been initiated and serious progress may be anticipated in coming years.

At this time the approach to a realistic interdisciplinary global coastal ocean dynamical model must be to construct it from a set of linked regional interdisciplinary coastal models which focus on dominant regional interdisciplinary processes. Regional studies Chapters 10, 12, 16, 18, 20, 24, 29 and 33 of the companion Vol. 14 on the interdisciplinary global ocean explicitly present regional models. Although a regional process focus simplifies the modeling to some extent, realistic interdisciplinary modeling remains intricate and challenging as evidenced by the linked coupled model concept (Runge et al., Chapter 13) to be discussed below. Interdisciplinary process characterization for coastal ocean regions and the cataloging of regions of the global coastal ocean by processes (Brink, Chapter 2; and Section 1.4) are currently important research problems.

The physical component for a general interdisciplinary coastal dynamical model has a fundamental theoretical basis. Realistic multiscale numerical physical circulation models now exist for some regions and are rapidly being implemented for others. Ultimately a more fundamental theoretical basis for the biological component is desirable (Rothschild, Chapter 3). However, even within the present *ad hoc* and empirical theoretical framework there does not yet exist a general theory that can be ascribed to the functioning of marine ecosystems (Cury et al., Chapter 14). Biological process formulations constitute a major research issue for biogeochemical/ecosystem modeling research.

Marine ecosystems are very complex systems which encompass an extremely large number of species and their associated life and size stages, with a myriad of possible species-species interactions. Interactions include grazing, predation, competition, symbiosis, and parasitism. The intricate complexity of the direct approach to a food web is illustrated in Figure 1.5 for a system consisting of only 75 elements (Cury et al., Chapter 14). There are, however, both natural and heuristic concepts and considerations which facilitate the study of complex ecosystems. The concept of a food web for an ecosystem consisting of a total number $n$ of life stages of component species has potentially $O(n^2)$ interactions or links among them. However, it is usually found that only $O(n)$ linkages are important to define the functionality of the web (Steele and Collie, Chapter 20). Quantitatively the food web represents the flow of nourishment energy among the elements of the web. Groups of species aggregate naturally into trophic (energy) levels, and into functional groups defined by physical factors, nutrient availability and predator-prey relationships. To reduce the number of model state variables aggregates of species rather than individual species are often used. Keystone species, which affect community or ecosystems processes to a larger extent than indicated simply by their abundance, are important to be identified.

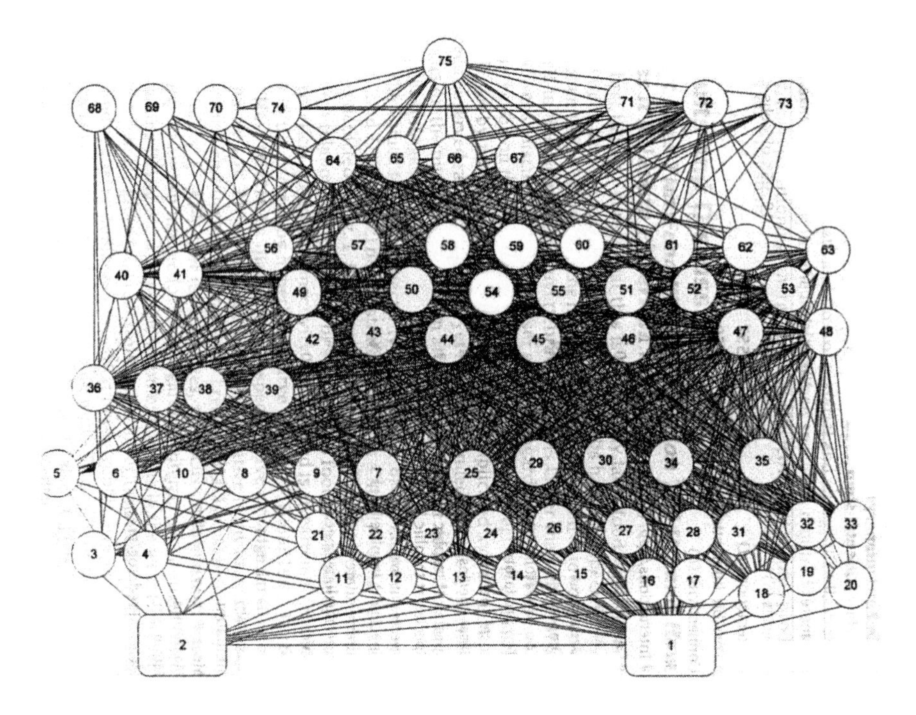


Figure 1.5 – Species and links for a northwest Atlantic food web, assuming that interactions among 75 components have similar strength in time and space. (Cury et al. – Ch. 14, Fig. 4)

The influence of physical forcings on marine ecological processes on multiple scales in time and space and at all trophic levels has long been recognized. The importance of research in this area cannot be overemphasized and such research is accelerating (Robinson et al., 2002 – Vol. 12 The Sea). However, models of aspects of ecosystem and population dynamics in the absence of physical forcing play an important role in elucidating basic processes. Today these time-dependent models are often referred to as zero (spatial) dimensional models. Idealized models of complete ecosystems have been studied in order to study the propagation of assumed change (e.g. nutrient supply, environment variability) at one trophic level throughout the food web. An example is a four-level food chain consisting of: bottom-level phytoplankton; zooplankton; forage fish; top-level predators (Cury et al., Chapter 14). An attempt is made to classify processes as bottom-up, top-down or wasp-waist (mid-level) controls. Such considerations are not generally applicable to realistic ecosystem modeling because another but comparable species can replace the perturbed species and maintain the function of the trophic level in the food web. Also, interactive processes often force real ecosystem dynamics in combination.

Zero-dimensional models of aspects of ecosystem dynamics exist which deal explicitly with a few trophic levels in detail, and which simply parameterize the interaction with lower and/or higher trophic levels. The top-down approach models known as predator-prey nonlinear models originated with the pioneering work of Lotka (1932) and Volterra (1926) and are reviewed by Cury et al. (Chapter 14). The bottom-up approach known as NPZ (nutrient-phytoplankton-zooplankton) non-linear models originated with the pioneering research of Riley (1946, 1947) and is reviewed by Hoffman and Lascara (1998 – Vol. 10 The Sea). Recent research on nonlinear ecosystem dynamics includes studies via the methods of nonlinear dynamical systems and chaos theory (Strogatz, 1994; Murray, 1989). Steele and Collie (Chapter 20) discuss nonlinear models as well as a linear network approach to the modeling of entire food webs which are in near equilibrium. Recent results presented in the chapters cited indicate that the productivity, diversity and stability of marine ecosystems are linked dynamical concepts. Both a holistic system level perspective and the perspective provided by the population dynamics of species and species interactions are relevant and complementary. Complex marine ecosystems have multiple equilibrium states consisting of subsystems of different community structures with some inherent variabilities. The ecosystem spends long periods in such relatively stable structures separated by shorter intervals of rapid non-linear transition to another community structure, or regime shifts. The overall ecosystem is responsive to variabilities at all trophic levels and is therefore adaptable to changes.

Coupled benthic/pelagic modeling has now been initiated in time and one (vertical) spatial dimension, and interesting results are found in Middelburg and Soetaert (Chapter 11). The status and prospectus of four-dimensional realistic modeling are reviewed by Runge et al. (Chapter 13). They advocate an approach via a hierarchy of models for specific complex problems and regions. The concept of linked coupled models is introduced as a method to handle the complexity of real ocean ecosystems. Coupling refers to a physical model forcing of a biogeochemical/ecosystem model. Key processes for a given problem are identified which provide a set of focus points in trophic levels about which appropriate dynamical models can be constructed. This set of models is then linked together. The linked coupled model concept is illustrated in Figure 1.6, which is interesting to compare with Fig. 1.5. The case discussed in detail (Cury et al., Chapter 14) links an NPZ model to a life-stage model of a key zooplankton species, which in turn provides a spatially explicit prey distribution for a bioenergetic model of fish larvae.

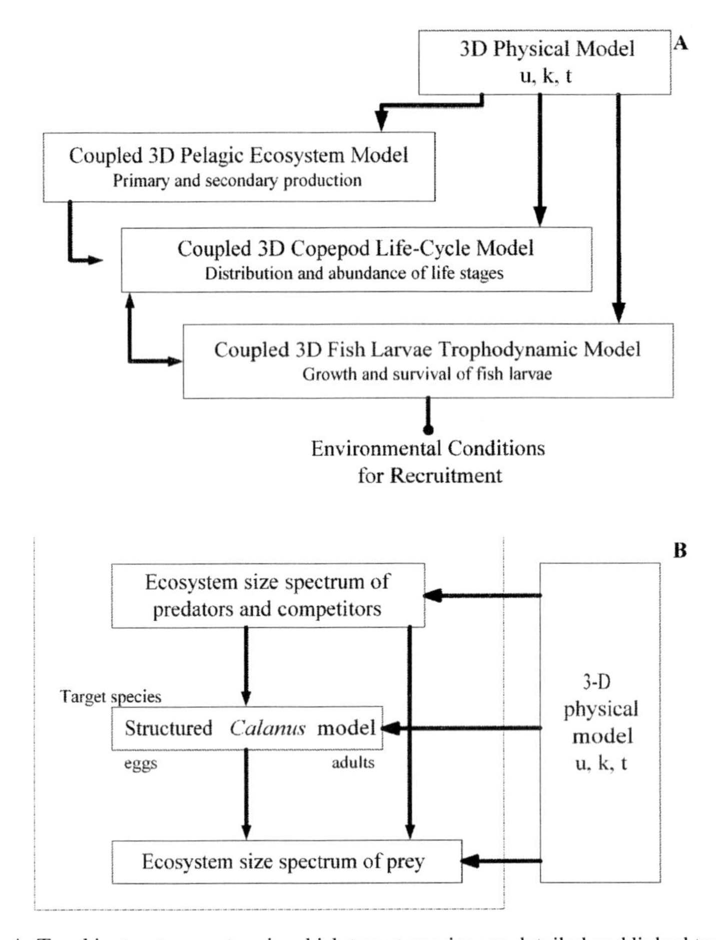


Figure 1.6 – A: Trophic structure system in which target species are detailed and linked to lower resolution models of prey and predator. B: LCM for use in describing environmental conditions. (Runge et al. – Ch. 13, Fig. 16)

There are presently two main approaches to physical coupling. An Eulerian approach treats biological state variables as space-time fields governed by advective-diffusive-reactive (biology) partial differential equations. This approach involves an explicit definition of state variables as time and space averaged concentration distributions of dissolved and particulate matter. A Lagrangian approach follows trajectories of parcels and particles of biological matter in the physical flow field. Both approaches can allow for self-motility and sinking. An individual-based model (IBM) approach is useful for tracking individuals in space, time and age, but the statistics of an ensemble of four-dimensional trajectories can be computationally intensive. The effective modeling of higher trophic level nektons (large fish) with dominant behaviors and strong swimming capabilities is presently particularly challenging.

The major effort required to develop realistic four dimensional coupled physical-biogeochemical ecosystem models suitable for carrying out realistic interdisciplinary simulations and for initiating real-time interdisciplinary forecasting in the coastal ocean is now underway and must be sustained. Together with new and dedicated coastal ocean data sets (Section 1.9, Malone et al., Chapter 19), these models (Hoffman and Friedrichs, 2002; Robinson and Lermusiaux, 2002 – Vol. 12 The Sea) will enable realistic and applicable research otherwise not possible. The data are assimilated into the models for initial and boundary conditions, updating, and parameter estimation. The model components of the nascent first generation of ocean observing and prediction systems are very complex and their successful development and use must involve collective expertise in applied mathematics, computational and computer science, as well as physical, biological, chemical and sedimentological oceanography. Consideration of modelability and measurability must together constrain the choice of state variables and representation of dynamical interactions. Rigorous analyses are necessary to prevent spurious apparent dynamical effects which can arise from: computational errors; lack of compatibility of biological and physical data being simultaneously assimilated; a mismatch in space and time scales across dynamical linkages in ecosystem model components; and other sources. However the potential impact of such models for accelerated scientific progress and a new level of rational management is enormous.

## 9. Summary, conclusions and research directions

Significant progress is now occurring in novel interdisciplinary research on the dynamics of the global coastal ocean. Such interdisciplinary research is challenging from intellectual, scientific, technical and methodological viewpoints. Three important new interdisciplinary concepts are emerging from recent advances in multidisciplinary and multiscale coastal ocean science. The first concept involves the importance of coupled pelagic-benthic dynamical processes interactive between the water column and the sediments. The second involves the significant contribution arising from coastal ocean biogeochemical cycles to global cycles and budgets. The third involves the large variety of dominant structures, trophic interactions and variabilities that occur in the highly productive coastal ocean ecosystems. These three concepts and their linkages provide a framework for further progress in coastal ocean dynamics. Of greatest importance is the formulation of research issues and specific next step problems within this framework.

For the broad range of interdisciplinary research topics treated in this volume it is neither feasible nor particularly useful to provide here detailed and comprehensive lists of such problems and issues. We do hope, however, that readers interested in focused research areas will be able to cull such information from the study of relevant chapters. Here we simply mention a few important research directions in order to convey the flavor of contemporary coastal research. These include: i) the role of fluidized mudflows in the lateral transport that decouples the location of sediment plumes from bottom deposits; ii) the advective transport of particles in permeable sediments and filtering of particles from pore water; iii) the relative importance of oxic and anoxic processes in preserving organic material in sediments; iv) the processes by which physical submesoscale variabilities (filaments, jets, squirts, streamers) both effect nutrient inputs to the coastal sea from the open sea and also contribute to the lateral export offshore of the shelf pump; v) the distribution and variability of the partial pressure of $CO_2$ in the coastal ocean and the role of the coastal ocean as a source or sink of $CO_2$ for the atmosphere; vi) the role of secondary production in biogeochemical dynamics; vii) microbial processes and variabilities in the coastal ocean water column and particularly the benthos; viii) recruitment dynamics linking the environment to trophic interactions in the context of the food web.

Coastal ocean science is initiating a new era of truly interdisciplinary research. The early stages of new interdisciplinary efforts are usually hampered by problems of communication that arise from lack of commonality of terms and from the same word being used in different disciplines with somewhat or completely different meanings. This is certainly the case here as evidenced by the discussion of the glossary in section 1.2 above. Additionally the definition of a few new terms essential for characterizing new interdisciplinary processes is necessary. Moreover the definitions of the extent of the coastal ocean and the subregions of the coastal ocean (vid Section 1.4) vary across the disciplines and need to be worked on from an interdisciplinary viewpoint.

Much work remains to be done in the characterization and categorization of the global coastal ocean. Efficient international cooperation and collaboration in the context of a common scientific vernacular could significantly expedite such research. Classifications via global oceanic geography, geomorphology, and physical and interdisciplinary dynamical processes are all necessary. A substantive starting point for integration, development and generalization is provided by the types of subregions discussed in Section 1.4. These include: the four pan regions (eastern and western boundaries, polar, semi-enclosed seas/islands); the five fundamental physical processes (boundary layers, tides, wind forcing, buoyancy driving, boundary current interactions – Brink, Chapter 2); the six successive offshore regions (near shore, freshwater influence, well mixed, tidal fronts, thermally stratified, shelf-edge – Atkinson et al., Chapter 8); the seven types for biogeochemical processes and exchanges (subtropical shelf pumps, temperate shelf: biology or physics dominant, upwelling: biology or physics dominant, coral reefs, polar ice pump – Ducklow and McCallister, Chapter 9); the seven coastal ecosystem types (permanent and intermittent polar ice, mid-latitude and topographically forced coastal, upwelling, wet and dry tropical – Alongi, Chapter 9).

Methodological advances underway in marine science generally may be expected to accelerate progress in coastal marine science and to enable process and regional research otherwise not feasible. The concept of advanced ocean observing and prediction systems (Malone et al., Chapter 19) generally involves an observational network gathering data that is assimilated into numerical ocean models (Runge et al., Chapter 13). The observational network consists of a mix of platforms and sensors efficient and adequate for special purposes (Section 1.9). The concept is useful and relevant both for a complex set of advanced platforms and novel instruments and for a simple set of traditional ones. Advanced ocean models will consist of a hierarchical system of linked coupled models (Section 1.8) and advanced data assimilation schemes will be fully multivariate and interdisciplinary (Robinson and Lermusiaux, 2002 – Vol. 12 The Sea). The space-time distribution of state variables is estimated by melding new observations with the *a priori* model predictions based on model dynamics and earlier data (sequential estimation or filtering for nowcasting and forecasting). For hindcasting, future data can be also utilized (smoothing). In the context of data assimilation, inverse methods make possible the estimation of dynamical parameters and rates (e.g. advective flux divergence, nutrient uptake, mortality, etc.) that are impossible or difficult and costly to measure directly *in situ*. When the ocean dynamical model is run in real time as a predictive model, powerful feedbacks are possible within the system including adaptive sampling and adaptive modeling. Adaptive sampling sends mobile platforms on sampling tracks to obtain data of greatest impact. Adaptive modeling adjusts model parameters and components of the linked coupled model system to best represent the observations being acquired. The recent initiation of research on the development of an interdisciplinary real time predictive capability for the coastal ocean indicates that substantial progress should be anticipated in the next decade. This should include the formulation of realistic problems for researching the highly nonlinear interdisciplinary limits of predictability for coastal regions.

To advance understanding of the dynamics of the complex global coastal ocean critical processes must be identified, investigated and quantified by critical experiments on dynamical processes (e.g. coastal deep sea exchanges, benthic productivity, $CO_2$ uptake/release, recruitment, etc.). In the design of such experiments, and the concomitant development of advanced theoretical/numerical models, issues of measurability and modelability should be considered as important constraints on the conceptual bases. Research on measurement models that relate sensor readings to state variables of interest is required. Current research of this type includes: i) the interpretation of color observations from satellites in terms of chlorophyll and phytoplankton; ii) the aggregation of several species into model state variables for plankton; and iii) the interpretation of measurements from autonomous underwater vehicles and progressive yo-yoing gliders as information on continuous four dimensional (space-time) fields for the coastal region of interest. There is, of course, a close relationship between the ideas of this and the preceding paragraph. Indirect and novel methods are needed for measuring important quantities either with low signal-to-noise ratios (e.g. cross-shelf fluxes) or for which direct methods involve small differences of large numbers (e.g. biogeochemical budgets). Compelling fundamental issues in modelability are the replacement of ad hoc and empirical biological dynamics with a more fundamental and general dynamics and the definition of novel robust concepts for the representation of diversity and stability of shelf ecosystems.

Some regions of the global coastal ocean are relatively well studied and understood while others are less so. Efficient progress can be made by identifying dynamically analogous regions and sharing and transferring scientific information and effective coastal technologies and methodologies among them. An important step forward will be the achievement of the classification in a common vernacular of the regions of the global coastal ocean via dynamical processes, together with a knowledge of the critical dynamical process experiments and concomitant dynamical numerical models relevant for such regions. This must be, of course, an ongoing and iterative procedure. The result will be useful not only on a regional basis but will also provide essential input for the estimation of global coastal ocean effects for processes and budgets of the global ocean generally and for input into whole earth system analyses.

Human society has many often-overlapping interests in and uses for the coastal ocean. These include fisheries and other living and nonliving resources; nutrient, toxicant and $CO_2$ management; transportation and other marine and maritime operations and recreation. Efficient and holistic management of the global coastal ocean and its interconnected natural and national subregions is essential to ensure the health of the coastal seas and their ecosystems, and the well being of the ever increasing human coastal populations. Such management can only be achieved via effective and well-motivated collaboration and cooperation among the scientific, technical, governmental, commercial, economic and environmental communities.

The material presented in this chapter is indicative both of the great complexity of the requisite scientific input and of the unique opportunity for advanced management now provided by ongoing scientific and methodological progress. The complexity of the coastal marine scientific system is exacerbated by the multiscale variabilities and changes caused by both anthropogenic and natural forcings and by coastal ocean interdisciplinary internal dynamics. These variabilities include pollution, harmful algal blooms, eutrophication, habitat modification, regime shifts, extreme events and climate change.

At every stage in the evolution of our scientific knowledge of the coastal ocean, state-of-the-art numerical dynamical models generalize and summarize our scientific knowledge and serve as vehicles for applications. From a technical and methodological viewpoint we are at the threshold of a new management era. The assimilation into advanced models of dedicated data from novel coastal ocean observing and prediction systems provides a powerful potential for a new level of rational management.

## Acknowledgements

The authors are pleased to acknowledge support for the preparation of this chapter from the following grants: ARR ONR N00014–02–1–0989, KHB NSF OCE-0227679, HWD NSF OPP-0217282 and NSF Grant OCE-0097237, RAJ NSF OCE-9911707.

## Appendix – Glossary

| | |
|---|---|
| Algaenans: | non-hydrolyzable aliphatic biomacromolecules produced by microalgae |
| Aliphatic: | hydrocarbons which do not contain a benzene ring |
| Anaerobic: | metabolism in the absence of oxygen; organisms capable of metabolizing organic matter without the use of oxygen |
| Anoxic: | lacking oxygen |
| Autotrophy: | metabolic process responsible for de novo synthesis of organic matter from inorganic precursors. Photosynthesis is an autotrophic process and phytoplankton are autotrophs. |
| Baroclinic: | a flow which is depth dependent in the absence of turbulent effects. Describing a fluid where density and pressure surfaces are not parallel. |
| Barotropic: | a flow that is not depth dependent. Describing a fluid where density and pressure surfaces are parallel. |

Benthos: usually referring to the organisms which live on or in the sediments of the ocean floor

Biomarker: refractory organic molecules characteristic of specific groups of organisms

Boundary layer: a relatively small portion of a fluid, along an edge, where some particular effect (usually turbulent or inertial) is confined and plays an important role.

Cephalopod: class of the phylum Mollusca, having large head with eyes, horny jaws and many tentacles. Shell may be internal, external or absent.

Chlorophytes: algae containing chlorophyll b as the primary photosynthetic pigment; "green algae"

Ciliate: protozoan group characterized by mobility conferred by cilia. Ciliates are members of the microbial foodweb and ingest bacteria, small phytoplankton and other protozoans.

Clinoforms: sloping depositional surfaces, commonly associated with strata prograding into deep water

Clupeoid: herring-like fishes

Coccolithophore: phytoplankton distinguished by formation of calcareous plates (coccoliths) covering the cell. Important in the fossil record and as micropaleontological indicator species.

Cohort: year class; in fisheries terminology, fish in a stock born in the same year

Copepod: crustacean zooplankton widely distributed throughout the global ocean. Important herbivores in many ecosystems

Crustacean: group within Phylum Arthropoda including shrimps, crabs, copepods, krill

Cyanobacteria: photosynthetic prokaryotes (bacteria) containing characteristic pigments phycocyanin or phycobilin

Debris flow: mudslides, mudflows, lahars, or debris avalanches are common types of fast-moving landslides or as density-driven flows downslope at the sea floor

Demersal: living on or near the bottom

Denitrification: conversion (via reduction) of nitrate to nitrite and thence to elemental nitrogen. The major process removing combined nitrogen from biological availability.

Detritus: dead organic matter

Diagenesis: sedimentary processes that alter materials deposited on the sea floor. This usually includes the breakdown and alteration organic matter and dissolution and precipitation of minerals in bottom sediments

Diapause: a quiescent period (often during winter) during which animals wait for favorable conditions for reproduction and growth

Diatom large and diverse phytoplankton group distinguished by silicon-containing cell walls (frustules). Important in the fossil record and a major group supporting marine food chains.

Dinoflagellates phytoplankton group of often larger cells with distinctive morphology and unique pigments. Most are motile and many also capable of heterotrophic nutrition as grazers.

Epifauna: animals which live at the sediment water interface, either attached or motile

Eulerian: describing a flow field in terms of the spatial fields of flow properties

Eustigmatophytes: generally unicellular algae whose members are characterized by vegetative and motile stages. These algae contain chlorophyll a and violaxanthin, a diagnostic accessory pigment, but do not contain other forms of chlorophyll.

Eutrophication: nutrient enrichment leading to excessive organic matter accumulation and oxygen depletion

*f* ratio: the proportion of total biological primary productivity in the ocean supported by the input of new nutrients to the photic zone. The remaining production is supported by nutrients recycled within the photic zone.

Facies: any areally segregated part of a sediment division which differs significantly from adjacent sediments

Fermentation: metabolic process carried out in the absence of oxygen in which organic compounds serves as electron acceptors in place of oxygen (the electron acceptor during respiration)

Flagellate: a protozoan group characterized by motility conferred by one or more flagellae. Some are autotrophic; many are herbivorous or bacterivorous heterotrophs

Fluid mud: dense fine grained suspensions that form on continental shelves at times of high wave energy during storms

Flysch deposits: thick layers of fine-grained sedimentary deposits characteristic of orogenic geosynclines; principal components are shale or silty shale interbedded with graded dark sandstones

Foreset: inclined layers of sediment deposited on advancing edge of growing deltas or along lee slope of advancing sand dunes

Genotype: the genetic makeup of an organism (see phenotype)

Geodetic: having to do with the shape of the earth

Geostrophic: referring to a balance of forces between the Coriolis force (associated with the earth's rotation) and a pressure gradient. Geostrophic flows follow lines of constant pressure.

Groundwater: water primarily from precipitation which has percolated through the soil into the saturation zone

Heterotrophic: mode of metabolism or nutrition in which preformed organic matter is used as a source of energy for generation of ATP, and synthesis of new organic matter. Grazers, many bacteria and humans are heterotrophs.

Holocene: the last 11,000 years, since the last major glacial epoch

Humic materials: yellow and brown compounds in water derived from alteration (diagenesis) of terrestrial and marine organic matter. Complicated polymeric organic compounds.

Hydrograph: representation of the annual flow and discharge of a waterway

Hydrolyzable: capable of breakdown by hydrolytic enzymes such as carboxylases and proteases

Hyperpycnal: flow produced when the density of the river water entering the basin is greater than the density of the standing water in the ocean basin. This higher density river water will flow below the standing water in the basin because of the difference in density.

Hypopycnal: associated with a lower river water density entering a higher density standing water density in the basin. Under these conditions, the river water will flow out over the standing water gradually depositing the suspended clay portion of the sediment load to the prodelta. The clay particles settle out of suspension through the process of flocculation (the clumping of clay particles together due to a positive-negative charge relationship created by the seawater)

Ichthyoplankton: fish larvae – literally, fish-plankton

Internal waves: waves, with dynamics similar to those of familiar surface waves, that propagate within a fluid along density gradients instead of the air-sea density jump.

Isostatic: the state of gravitational equilibrium between the lithosphere and the asthenosphere of the Earth such that lithospheric plates "float" at a given elevation depending on their thickness. The balance between the elevation of the lithospheric plates and the asthenosphere is achieved by the flowage of the denser asthenosphere. Various hypotheses about isostasy take into account density (Pratt hypothesis), thickness (Airy hypothesis), and pressure variations to explain topographic variations among lithospheric plates. The current model consists of several layers of different density. An equilibrium state where the same pressure is felt everywhere on a horizontal surface.

Kerogen: a solid, waxy, organic substance that forms when pressure and heat act on the remains of plants and animals

Lagrangian: describing a fluid flow field in terms of the motion of individual particles

Lee wave: a stationary wave that forms behind an obstacle in response to an ambient flow

Lignin: biochemically stable, hard to decompose component of wood

Lipid: class of organic matter compounds insoluble in water and soluble in organic solvents such as chloroform. High energy-containing molecules and important components of cell membranes.

Macrofauna: organisms larger than 500 microns in size

Meiofauna: organisms between 45 and 500 microns in size; mostly benthic sediment-inhabiting metazoans

Melanoidin-type condensation: reactions between two simple molecules, for example and amino acid and carbohydrate, to give a more complex compound, potentially leading to highly cross-linked macromolecules. Also called a Maillard reaction, a type of non-enzymatic browning which involves the reaction of simple sugars (carbonyl groups) and amino acids (free amino groups).

Metazoan: multicellular organisms

Microbial loop: a food chain based on bacteria recovery of dissolved organic compounds leaked or released from organisms

Milankovitch periods: variations in the Earth's orbit, including eccentricity, tilt and precession of the equinoxes, which may be responsible for glacial/interglacial cycles

Mineralization: breakdown of organic matte into its inorganic components such as carbon dioxide and nutrient ions

Nepheloid layers: thin, particle-rich microlayers within the water column or within the benthic boundary layer

Nitrification: oxidation of ammonium into nitrite and nitrate

Nitrogen fixation: the biological capture and conversion of elemental nitrogen into biologically useful forms such as nitrate and ammonium by nitrogen-fixing bacteria

Nutrient regeneration: the production of inorganic nutrients from decomposition of organic matter (see mineralization)

Oligotrophy: the opposite of eutrophy, lacking in nutrients with low levels of organic matter and biomass

Ontogenetic Oxic: processes carried out in the presence of oxygen; opposite of anoxic

| | |
|---|---|
| Oxidation: | chemical process in which electrons are transferred from a compound to an electron acceptor leaving the reacting compound in a higher chemical oxidation state. Many oxidation reactions in marine systems are biologically mediated and provide the basis for organism metabolism. Typical important electron acceptors in marine systems are oxygen, nitrate, manganese oxide, iron oxide and sulfate. |
| Pelagic: | living in the water column as opposed to the benthos; having to do with the open ocean as opposed to nearshore (neritic) waters |
| Phenotype: | the outward appearance and functioning of an organism, conferred by the external expression of the genotype in the organism's environment |
| Polyphagous: | capable of ingesting many types of food |
| Population dynamics: | the fluctuation of population numbers and other demographic properties (age structure, mortality curves etc); manifestation of interactions among trophodynamic, genotypic and phenotypic processes |
| Protist: | the simplest eukaryotic organisms including the algae, protozoans, slime molds |
| Protozoan: | mostly single-celled protists, but including some colonial forms and a great diversity of morphological types, autotrophs, mixotrophs and heterotrophs |
| Protozooplankton: | small zooplankton consisting of protozoans that ingest other organisms |
| Prymnesiophytes: | group of phytoplankton characterized by distinctive pigments including the coccolithophorids. A generally unicellular algae that are often externally scaled and exhibit vegetative and motile stages. These algae contain chlorophylls a, c1 and c2, a number of diagnostic carotenoids, and some members may form large blooms. Phaeocystis and Emiliania spp are important prymnesiophytes. |
| Recruitment: | the survival of fish to their final adult life stage |
| Redox: | chemical reactions involving changes in the oxidation – reduction state of a compound |
| Reduction: | the opposite of oxidation, in which compounds accept (gain) electrons and reduce their oxidation state |
| Remineralization: | same as mineralization |
| Secchi depth: | operationally-defined depth of light penetration in the water column defined as the depth at which a white disk on a rope just becomes visible (or invisible) to the naked eye as a result of light attenuation |
| Seismic tomography: | technique for performing tomography on seismic travel-time data which involves parameterising the Earth using an unstructured grid, and numerically solving a linear system |
| Soliton: | a class of internal wave that exists as a single hump or dip in density surfaces, and that, because of momentum advection, can propagate great distances without changing form |
| Steric: | having to do with density changes in water |
| Suboxic: | a vertical layer or stratum in the water column or sediment below the oxic later and above the anoxic layer, containing neither oxygen nor sulfide |
| Taxa: | taxonomically-defined group of organisms. Species, genera, families, orders, phyla are examples of lower and higher taxa |
| Thermal wind: | referring to a geostrophic flow where the pressure gradient is associated with density variations in the fluid |

| | |
|---|---|
| Thermocline: | a portion of the water column characterized by pronounced vertical temperature differences |
| Tortuosity: | the degree of twistedness |
| Trophodynamic: | processes of nutrient uptake and regeneration, organic matter uptake and mineralization, grazing and predation, the mechanisms of energy transfer and element cycling in food chains |
| Turbidite: | graded-bed sedimentary facies characterized by rhythmic alteration of relatively think sandstones with equally thin shale layers |
| Turbidity current: | a slurry of sand-laden muddy water which behaves as a heavy liquid, typically moving shallow-water sediments dislodged by a slump downslope into deep waters |
| Vorticity: | the tendency of a parcel of water to spin |

## Bibliography

Allee, C.C., A.E. Emerson. O. Park and K.P. Schmidt, 1949. Principals of Animal Ecology. Saunders, Philadelphia, 837 pp.

Ducklow, H. W., 2003. Chapter 1. Biogeochemical Provinces: Towards a JGOFS Synthesis. In *Ocean Biogeochemistry: A New Paradigm.* M. J. R. Fasham. ed. Springer-Verlag. New York. pp. 3–18.

Gattuso, J.-P., M. Frankignoulle and R. Wollast, 1998. Carbon and carbonate metabolism in coastal aquatic ecosystems. *Annual Review of Ecology and Systematics,* 29, 405–434.

Hofmann, E.E. and M.A.M. Friedrichs, 2002. Predictive modeling for marine ecosystems. In *The Sea,* Vol. 12, A.R. Robinson and K.H. Brink, eds. Wiley, New York, pp. 537–565.

Liu, K. K., K. Iseki and S. Y. Chao, 2000. Continental margin carbon fluxes. In *The Changing Ocean Carbon Cycle,* R. B. Hanson, H. W. Ducklow and J. G. Field, eds. Cambridge University Press, Cambridge, 187–239.

Longhurst, A., S. Sathyendranath, T. Platt and C. Caverhill, 1995. An estimate of global primary production in the ocean from satellite radiometer data. *Journal of Plankton Research,* **17,** 1245–1271.

Lotka, A.J., 1932. The growth of mixed populations: two species competing for a common food supply. *J. Washington Academy Sciences,* **22,** 461–469.

Murray, J.D., 1989. Mathematical Biology. Springer-Verlag, Berlin, 767 pp.

Riley, G.A., 1946. Factors controlling phytoplankton populations on Georges Bank. *J. Mar. Res.,* **6,** 54–73.

Riley, G.A., 1947. A theoretical analysis of the zooplankton population on Georges Bank. *J. Mar. Res.,* **6,** 104–113.

Strogatz, S.H., 1994. Nonlinear Dynamics and Chaos; with applications to physics, biology, chemistry and engineering. Addison-Wesley, Reading, Mass. 497 pp.

Voltera, V., 1926. Variations and fluctuations of the numbers of individuals in animal species living together. Reedition traduite. In R.N. Chapman (ed.), *Animal Ecology,* pp. 409–448, McGraw Hill, New York.

# Chapter 2. COASTAL PHYSICAL PROCESSES OVERVIEW

K. H. BRINK

*Woods Hole Oceanographic Institution*

**Contents**



## 1. Introduction

Biological, chemical and geological oceanographers often need to understand the physical context of their own studies in the coastal ocean. In some cases, they might simply need to know which way the water is moving or what the temperature at a particular place or time might be. In other cases, including many of the examples in Robinson et al. (2002), there is a need for a more thorough understanding that could even include some form of prediction. In any case, there is some need to understand the state of the physical system.

Physical oceanographers often characterize the state of the ocean in terms of processes. That is, they consider individual aspects of the ocean's behavior that can be studied in an isolated (or nearly so) sense, and that each have recognized names. Examples might include surface tides, mixed layer deepening or coastal-trapped waves. To a non-physical oceanographer, this process orientation may seem at first to be just another level of jargon, or it might appear to be a distraction from concentrating on some overall view of the variability.

The process approach works well, though, because it concentrates on aspects that are simple enough to be dealt with quantitatively, at least in idealized circumstances. The hope is that by understanding a set of dominant processes, we will be able to understand their interplay and their relative importance in a range of settings. The process (as opposed to geographical) approach is an attempt to gain generality in our understanding of the coastal ocean. Further, a process orientation is important in dealing with complexities in both the real world and in our increasingly realistic numerical models. Processes can provide a framework both for planning experiments (numerical and oceanographic) and for classifying, interpreting, and testing the results.

*The Sea*, Volume 13, edited by Allan R. Robinson and Kenneth H. Brink
ISBN 0-674-01526-6 

The following pages are thus meant to provide an overview of coastal ocean physical processes and, along the way, to provide geographic examples where they might be found in particularly pure forms. The reader is also advised to consult Volume 10 of *The Sea* (Brink and Robinson, 1998) for more detailed reviews written by those most familiar with the material.

## 2. Physical Processes in the Coastal Ocean

### *a. Boundary Layer Processes*

The turbulent surface and bottom boundary layers are where the ocean makes contact with the atmosphere and the solid earth, respectively. They are characterized by stress or flux variations in the vertical being of lowest order importance. In these thin (order 20m) layers, the directly stress-driven flow (Ekman transport) takes place, and homogeneous mixed layers tend to form due to the boundary turbulence. One strikingly robust feature of this transport (in water deeper than the boundary layer thickness) is that it always goes at right angles to the applied stress in the limit of steady state dynamics and in the absence of lateral momentum advection (Ekman, 1905). While, in many regards, coastal turbulent boundary layers are similar to those in the open ocean, they tend to have distinctive properties for at least four reasons.

The first distinction is that, since lateral gradients tend to be large in the coastal ocean, advection can play a substantial role in upper ocean dynamics. For example, lateral advection of the cold surface water associated with wind-driven upwelling tends to destabilize the upper ocean and create deep mixed layers (deSzoeke and Richman, 1984). Similarly, numerical modeling results show the same sort of deep mixed layers close to shore in upwelling regions (e.g., Allen et al., 1995). A careful examination of a number of coastal upper ocean data sets (Lentz, 1992) showed, however, that mixed layers were not anomalously deep in these regions, consistent with a tidy compensation of the destabilizing lateral advection by stabilizing surface heat fluxes. As part of the same study, Lentz found that the surface Ekman transport agreed with the simple, one-dimensional formulation so long as the integration was carried out to a depth including the entire shear layer at the base of the mixed layer, i.e. that the offshore transport associated with upwelling takes place over a depth exceeding the mixed layer thickness. That this simple transport prescription works suggests that, at least for these data sets, horizontal momentum advection (in contrast to buoyancy advection) may not be a strong effect.

The second distinction of the shelf and slope is that the bottom is systematically tilted. This means that cross-shelf bottom Ekman transports (associated with alongshore flows) either advect dense water upslope or light water downslope, thus either stabilizing or destabilizing the bottom boundary layer (figure 2.1; Trowbridge and Lentz, 1991). Further, the up- or down-slope advection creates horizontal density gradients so that a thermal wind shear in the bottom boundary layer weakens the bottom flow geostrophically. Through this mechanism, the bottom stress and bottom Ekman transport weaken over an adjustment time scale that depends on the bottom drag, bottom slope, the Earth's rotation and stratification. This time scale is typically of the order of days. One implication of this stress reduction is that steady flows in stratified shelf waters tend to adjust so that they are strictly alongshore (not crossing isobaths), and to be long-lived due to the absence of bottom stress (e.g., Chapman, 2002, and references therein). There are, however, situations in the real ocean where observed bottom stress is perhaps small relative to local forcing but remains finite even at the very low frequencies (e.g., Shearman and Lentz, 2003) where buoyancy shutdown might be expected based on theory to date. Recently, Garvine (2004) has presented strong evidence for the failure of bottom boundary layer arrest off the coast of New Jersey, and has presented a dynamical explanation for the observations. A good deal more thus needs to be done on this topic to account for realistic time dependence (simultaneously over a range of frequencies), forcing, and alongshore and cross-shelf variability. Specifically, we need to understand the natural conditions under which we can expect bottom stresses to shut down, and the conditions under which bottom stress will remain effective.

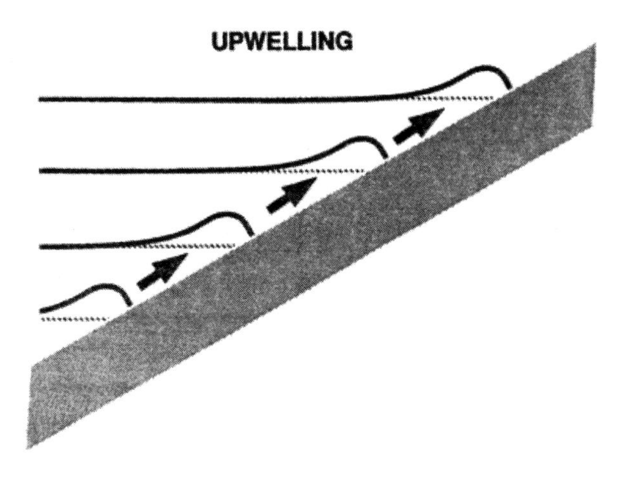


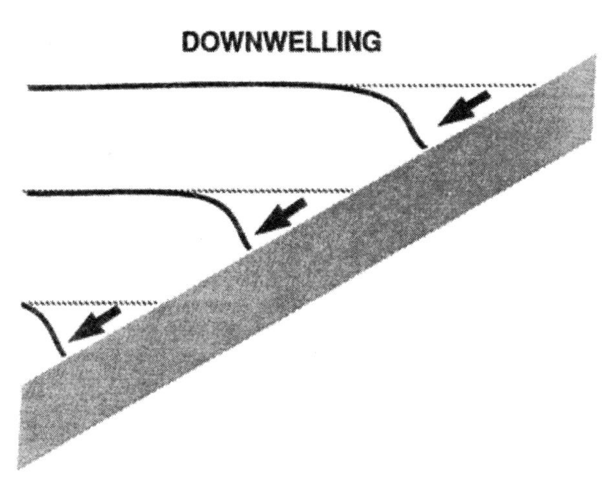


Figure 2.1 Schematic demonstrating the effects of upslope (upper panel) and downslope (lower panel) bottom Ekman transport on density stratification. After Lentz and Trowbridge 1991).

A third distinction is that coastal waters are shallow enough that surface wave orbital velocities can be substantial at the bottom for typical 10–20 second ocean swell. These surface waves produce a thin (centimeters) boundary sublayer immediately above the bottom. The net effect of coupling this boundary layer with the thicker, overlying, one due to lower frequency motions is that the lower frequency flow effectively experiences a rougher bottom, hence an enhanced stress (Grant and Madsen, 1986). The implication of this coupling is that the bottom stress is substantially modulated by the wave climate, so that stress can vary from day to day even with a steady current. Storm events are characterized by both high waves and strong currents, which combine to enhance bottom stress and so assure that sediment resuspension and transport are dominated by only the most powerful storm events (Parsons and Nittrouer, 2004). In a few cases, models have been used to show that wave-enhanced bottom stresses substantially affect lower frequency flows (e.g. Brink et al., 1987).

The fourth thing that distinguishes coastal ocean boundary layer physics is that, as water becomes shallower, the turbulent surface and bottom boundary layers eventually ought to come into contact and merge (e.g., Lentz, 1994). Exactly how far offshore this merged region extends depends strongly on the balance of factors that create turbulence (surface cooling, destabilizing lateral advection, wind stress) and that damp it out (stratification, stabilizing lateral advection, and surface heating). In some cases, however, such as off the east coast of the United States during the summertime, the water appears to remain stratified up to very near the coast (e.g. Muenchow and Chant, 2000), perhaps partly because of stabilizing freshwater outflows. Although a surface-to-bottom boundary layer, when it occurs, implies a sluggish flow (as observed by Lentz, 1994 and modeled by Allen and Newberger, 1996), there is some indirect evidence suggesting that, perhaps in the offshore fringes of the merged boundary layer, water passes through vertically with relatively little resistance but with no net cross-shelf transport (Samelson, 1997). As the innermost part of the shelf is reached, effects associated with the momentum carried onshore by surface waves become important and flows can again be quite energetic (Lentz et al., 1999) as the surf zone is approached. The complex and fascinating processes that typify the surf zone are not discussed here, but the reader may consult Mei and Liu (1993) for an introduction.

### *b. Tides and related processes*

The study of tides is arguably the oldest branch of physical oceanography. Aristotle, for example, was fascinated by them even though he lived where tides are not as dramatic as in other parts of the world. Basic tidal physics was sketched out by the end of the 18$^{th}$ century: Gravitational and centripetal forces act on the ocean to accelerate motions and cause sea level changes. Tides exist at a bewildering range of discrete frequencies that result from sums and differences of the apparent periods of the lunar and solar orbits and of the earth's axis orientation. Of course, solving the equations for a realistic, self-gravitating ocean where solid earth tides play a role is not a trivial undertaking, and the subject has occupied many of the finest oceanographic minds to this day (see Hendershott, 1981 for a fine review of the general subject).

Since the coastal ocean is a fairly narrow, shallow strip, tide-generating forces are often negligible locally and it makes sense to concentrate only on how coastal waters respond to the tides in adjoining offshore waters. This line of reasoning is implicit in the sequence of simple and insightful tidal diagnoses by Clarke and co-workers (e.g., Clarke, 1991). Simply stated, the tides are maximal over the shelf when the geometry (shelf-slope shape and alongshore structure of the open-ocean tide) is such that propagating waves can resonate with the forcing. The most common cases occur at frequencies higher than the inertial ($2\Omega \sin\varphi$, where $\Omega$ is the earth's rotation rate, and $\varphi$ is the latitude), so that resonances with inertia-gravity waves can occur. The result can be extraordinarily high tides such as those found in the Gulf of Maine-Bay of Fundy system (Brown and Moody, 1987). Additionally, there are cases at higher latitudes where diurnal tidal currents can be enhanced due to excitation of subinertial frequency, coastal-trapped waves (e.g. Daifuku and Beardsley, 1983, south of New England; Crawford and Thomson, 1984, off Vancouver Island). In any case, our advanced knowledge of oceanic tides allows us to run practical tidal models that can be very good indeed (e.g., Foreman et al., 1995).

Once the basic tidal variability is established, a number of important secondary effects come into play (see Simpson, 1998 for more detail). A few of these phenomena are discussed.

*Tidal mixing:* Tidal currents passing over the bottom are associated with a bottom stress, and thus generate turbulence. When tides are particularly strong, as they are over Georges Bank or around the British Isles, they generate sufficient turbulence to mix away stratification throughout the water column (Simpson and Hunter, 1974). Stated another way, the turbulent bottom boundary layer thickness, which depends directly on bottom stress, becomes as great as the water depth. If one considers a one dimensional (vertical) energy balance, top-to-bottom mixing will occur when the stabilizing tendency of the surface heat flux is overcome by tidal mixing. The mathematical expression of this imbalance results in the famous Simpson and Hunter (1974) condition for surface to bottom mixing:

$$h/u^3 < a, \tag{1}$$

where $h$ is the water depth, $u$ is the tidal current amplitude, and $a$ depends on the efficiency of mixing and on the surface heat flux. Simply stated, shallow waters, or areas of strong tides, tend to be unstratified. These tidally mixed regions are usually surrounded by a sharp front that separates mixed waters from ambient stratified waters (Simpson, 1998). The front, through the thermal wind relation, is, in turn, associated with a persistent vertically sheared, alongfront jet (Chen et al., 1995). Near-surface drifters tend to accumulate in these fronts on Georges Bank (Brink et al., 2003), consistent with tidal mixing fronts being convergent near the surface.

Within the vertically mixed waters, horizontal mixing is greatly enhanced by shear dispersion (e.g. Simpson, 1998). This process involves a combination of strong vertical mixing and a time-dependent, vertically sheared flow. The vertical mixing is idealized as going on continuously, and the sheared flow means that mixing spreads properties among vertically separated parcels that become horizontally displaced relative to each other with time. One could imagine, for example, a point release of a dye near the surface in a sheared flow that has vertical mixing only. With time, the dye spreads out laterally because of the vertical shear in velocity. If the mixing were strong enough, the dye could still be almost uniformly distributed in the vertical as it spreads horizontally. In a tidally dominated area such as central Georges Bank, the ingredients are all there: strong vertical mixing and sheared (because of turbulent friction) tidal currents

Substantial temporal and spatial variability can be introduced when a river flows into a tidally mixed area. The river-induced buoyancy, under conditions of weaker tides or stronger outflows, overcomes the tidal mixing, and causes locally stratified waters: Regions of Freshwater Influence (ROFIs: see Simpson et al., 1991). Interesting temporal variability in stratification at a point can be induced by changes in the river outflow or by the spring-neap tidal cycle (Simpson, 1998).

One of the most interesting issues about the tidal mixing front is the question of what passes through it. It seems unlikely, for example, that biological productivity could take place over Georges Bank over seasonal time scales if there were not a sustained flux of nutrients into the shallower waters, through the front, from deeper stratified waters (Townsend and Pettigrew, 1997). Simply homogenizing an initially stratified water column will not provide the continuous nutrient source required for sustained production. One promising hypothesis for how the on-bank flux takes place is provided by Franks and Chen (1996). Simply stated, the front is bent back and forth during different (onshore or offshore) phases of the tides. Turbulent vertical mixing through the distorted front then allows the passage of nutrients and heat by a process similar to shear dispersion. The Franks and Chen calculations suggest that this mechanism provides a nutrient flux of the magnitude required to explain Georges Bank's productivity. Observing this flux in nature, however, would be difficult because of its localized and time-dependent character.

*Tidal rectification:* A fluctuating flow across a depth gradient generates a mean along-isobath flow on a rotating planet (e.g., Loder, 1980). The flow tends to be strongest where the fractional change in depth (more accurately: depth gradient divided by depth squared) is largest. Hence, for example, there is a strong eastward tidally rectified flow along the shallow, steeply sloping northern edge of Georges Bank. Over the more gently sloping southern side of the Bank, the mean flow is a good deal less apparent. Near the surface, the sense of the rectified flow is the same as that of the tidal mixing frontal jet, so the two tend to reinforce each other (Chen and Beardsley, 1995). Observations (e.g., Brink et al, 2003) show that there is indeed a sustained jet at the northern side of the Bank, and that the mean flow is seasonally modulated in a manner qualitatively consistent with a combination of rectification and a frontal thermal wind jet. The dynamics of the frontal jet at the edge of a tidally mixed area is demonstrated particularly well through a combination of models and observations in the Irish Sea by Horsburgh et al. (2000), Horsburgh and Hill (2003) and Holt and Proctor (2003).

*Internal tides:* Internal waves are nearly ubiquitous in the ocean (e.g., Garrett and Munk, 1979), propagating with the support of vertical density gradients just as gravity waves propagate on the air-sea density difference. When the ocean bottom is flat, small amplitude internal tides (internal waves at tidal frequencies) can be described in terms of distinct dynamically defined vertical modes that usually do not interact strongly, but a sloping bottom allows the different modes to couple at lowest order. Thus, at the outer edge of the continental shelf, tidal currents (which tend to be depth-independent) interact with the stratification to generate propagating internal waves of tidal frequency (e.g. Holloway et al., 2001). Although the process is essentially linear, density stratification itself and ambient currents vary with time, so internal wave propagation can be rather variable and intermittent in some cases (e.g. Lerczak et al. 2001 at a diurnal frequency). In extreme cases where the tidal amplitudes are large, highly nonlinear short internal waves, having properties similar to bores or solitons, can be generated at the shelf edge. These waves need not have tidal frequencies, and are quite energetic (Colosi et al., 2001). When these nonlinear internal waves break in shallower water, they can contribute to mixing over the shelf at a rate locally comparable to that due to wind mixing (Sandstrom and Elliott, 1984). Another effect of the nonlinear waves can occur when topographic coupling becomes strong, and internal wave energy is converted (under near-resonant circumstances) into energetic seiches that can cause serious perturbations in local sea level (e.g., Geise et al, 1998).

### *c. Coastal-trapped waves and regional wind forcing*

*Coastal-trapped waves:* Coastal-trapped waves, in the broadest sense, propagate only in the alongshore direction. There are two general categories: high frequency (period less than inertial) that are trapped in or near the surf zone, and lower frequency waves that are detectable over most of the shelf and sometimes on the continental slope. As a special case, Kelvin waves exist at all frequencies, and affect the entire shelf and slope. Although a good deal of interesting work has been done on higher frequency trapped waves (e.g., Mei and Liu, 1993), attention here focuses on the lower frequency modes that can be prominent over the mid-shelf. These subinertial coastal-trapped waves usually propagate phase (and usually energy) in one direction: to the right when looking offshore in the northern hemisphere, and the opposite in the southern hemisphere (Huthnance, 1978). There are cases where these subinertial waves can be detected in a relatively pure form, especially in connection with diurnal tides (e.g. Crawford and Thomson, 1984) or with distant forcing (e.g. off Peru: Smith, 1978; Cornejo-Rodriguez and Enfield, 1987). Much of the extensive literature on the subject has been reviewed by Allen (1980) and by Brink (1991).

The most powerful aspect of coastal-trapped wave theory involves regional-scale wind-driven motions. In the limit of low frequency (relative to inertial), and large alongshore scale (relative to the shelf width), the coastal-trapped wave modes represent a mathematically complete orthogonal set that allows the expansion of the wind-forced solution in a remarkably convenient form. The wave modes in this set (and their properties) can be computed with publicly available software, given information on latitude, topography and density stratification. It thus becomes quite straightforward to predict or hindcast regional scale responses to alongshore winds over about a 1000km scale. This approach has been applied in numerous examples, but the study of Chapman (1987) for the California coast is representative. The hindcast does very well for pressure (sea level) and alongshore current variability over the shelf, although it tends to underestimate amplitudes somewhat. The model shows that alongshore current and pressure variations are forced by winds to the south of the observation location, consistent with the free low-frequency waves only propagating northward. The model does very poorly with cross-shelf current and temperature variations. The failure to obtain good hindcasts for cross shelf currents and temperature appears to be related to advective effects and to the tendency for these variables to be dominated by motions having short alongshore scales (e.g. Dever, 1997), hence violating the initial assumptions of the regional wind-forced theory. However, the answer may not be so simple as that, since observations and arguments by Pringle and Riser (2003) demonstrate that bottom temperature changes off southern California are remotely driven on a scale similar to that of regional coastal trapped wave forcing, but the mechanism that links local temperature changes to the larger scale motions is not clear.

Coastal-trapped wave theory has proven to be a remarkably useful tool for understanding regional scale wind driving. Although there remain some interesting problems in this line of theory, particularly involving nonlinear effects, the field has been less active in the last decade than it had been previously.

Coastal-trapped waves are, in a sense, an expression of the natural asymmetry that arises over topography in the coastal ocean (analogous to that imposed by the latitudinal gradient of the earth's effective rotation rate for the open ocean). Specifically, information *tends* to propagate only in one direction: that of long coastal-trapped waves. One result of this is that steady flow over a rough bottom only generates coastal-trapped lee waves, hence form drag, when flow is in the opposite sense to free wave propagation (e.g. Brink, 1986). Flow in the direction of wave propagation experiences no resistance. The implication of this asymmetry is that a zero-mean forcing tends to generate a mean flow in the direction of free coastal-trapped wave propagation (Haidvogel and Brink, 1986) and thus minimizes the form drag. Similar results can be found more generally and more elegantly by means of an entropy formulation commonly referred to as the "Neptune effect" (e.g., Holloway, 2002 and references therein). An interesting consequence of the Neptune effect is the tendency to generate a net near-bottom onshore transport (resulting from the tendency for alongshore changes in depth and pressure to align) that could potentially be significant biologically (Holloway et al., 1989). The observed tendency for mean flows over most continental margins (e.g. Hill et al., 1998) to be in the sense of free wave propagation lends a good deal of credence to this body of ideas.

*Upwelling:* Another aspect of wind forcing involves upwelling and downwelling over the shelf. Specifically, an alongshore wind forces a cross-shelf surface Ekman transport in the upper part- typically 10–20m- of the water column. This transport is balanced (at least when integrated over some large alongshore scale) by an opposite cross-shelf flow below the Ekman layer. In order to complete the flow pattern, there is a vertical transport near the coast. The net effect is that an alongshore wind forces vertical exchanges. In the case of upwelling, cool nutrient-rich waters come to the surface and, where alongshore winds are sustained (such as off Peru, California, northwestern Africa and southwestern Africa), cause enhanced biological primary productivity that propagates up the food chain (e.g. Barber and Smith, 1981). This level of understanding has existed for almost a century now (Ekman, 1905; Thorade, 1909).

There have been a number of obstacles to gaining a more sophisticated view. One is observational. The actual upwelling patterns that exist in the ocean tend to be rather three-dimensional (when viewed locally) and to be substantially less energetic than some of the ambient fluctuations that can mask the relatively weak (order 0.05 m/s or less) flows of interest. Further, measuring the offshore transport in the upper ocean is especially difficult because of the physical problems of measuring flows where surface waves are important (Weller and Davis, 1980). There appears to be only one example of a systematic measurement of upwelling circulation: Lentz (1987) was able to close the volume budget off California consistently over time because the Brobdignagian CODE (Coastal Ocean Dynamics Experiment) moored array involved good alongshore resolution as well as extensive upper ocean measurements. Essentially, he had enough alongshore resolution to average out the "noise" that was later shown to be related to small scale eddies intruding onto the shelf from offshore (Dever, 1997).

The other major hurdle to understanding is that, beyond the basic mechanism, upwelling is a nonlinear process. For example, advecting cold upwelled water offshore eventually means that, in the absence of a stabilizing surface heat flux, denser water will be carried above lighter water. Thus, upper ocean properties, such as the mixed layer thickness, depend critically on lateral advection (deSzoeke and Richman, 1984). Further, the horizontal convergences associated with the flow cause upper ocean fronts to develop (Mooers, Collins and Smith, 1976), and these fronts are themselves unstable, giving rise to meanders and pinched off eddies (Barth, 1989a,b; Barth, 1994). Vorticity conservation suggests that these smaller scale flow features can give rise to secondary, transient vertical circulations.

In some regions, the upwelling front is not usually observed over the shelf, but rather there is a region of cool surface water extending well off the shelf and often forming into cool surface filaments: narrow (10–50km) tongues that can extend far (100s of km) offshore. These filaments have been observed off California, Portugal, Oman and South Africa (Hill et al., 1998), and they are often associated with very strong offshore jets. These features have been particularly well studied off the central coast of California, where they are in fact the offshore flowing segments of a meandering equatorward jet (figure 2.2; Brink and Cowles, 1991 and references therein). It appears that the jet marks the boundary between offshore oligotrophic waters and "coastal" waters, even though this boundary can be far offshore of the shelf itself (Hood et al., 1991). The other upwelling regions where cool filaments are found appear to be consistent with this phenomenology. Further, off the west coast of the United States, it appears that this meandering eastern boundary current jet has its upstream origin over the shelf in the coastal upwelling frontal jet off Oregon (Barth et al., 2000), but it is not clear whether this shelf origin is typical of other regions.

The Leeuwin Current off western Australia (Smith et al., 1991) has rather peculiar dynamics: a study of wind charts would suggest that this ought to be a region of active upwelling with an equatorward current analogous to the California Current. Instead, the predominant current is a poleward shelf-edge flow, and the symptoms of upwelling (high nutrients and cold water near the coast) are absent (Thompson, 1987). Although Thompson (1987) provided an explanation for these phenomena based on the implications of an imposed oceanic onshore flow (associated with a basin scale pressure field), his explanation for the absence of apparent upwelling is now questionable in light of the Trowbridge and Lentz (1991) results. Subsequent numerical models (e.g. Spall, 2003) have, however, reinforced the importance of an oceanic flow for driving the shelf edge alongshore flow. The peculiarity of this case highlights the difficulty of ascribing much commonality to eastern boundary current processes (which can be associated with buoyancy inputs, winds, or basin scale flow patterns), in contrast to western boundary currents which do seem to have dynamical commonality, at least as related to forcing.

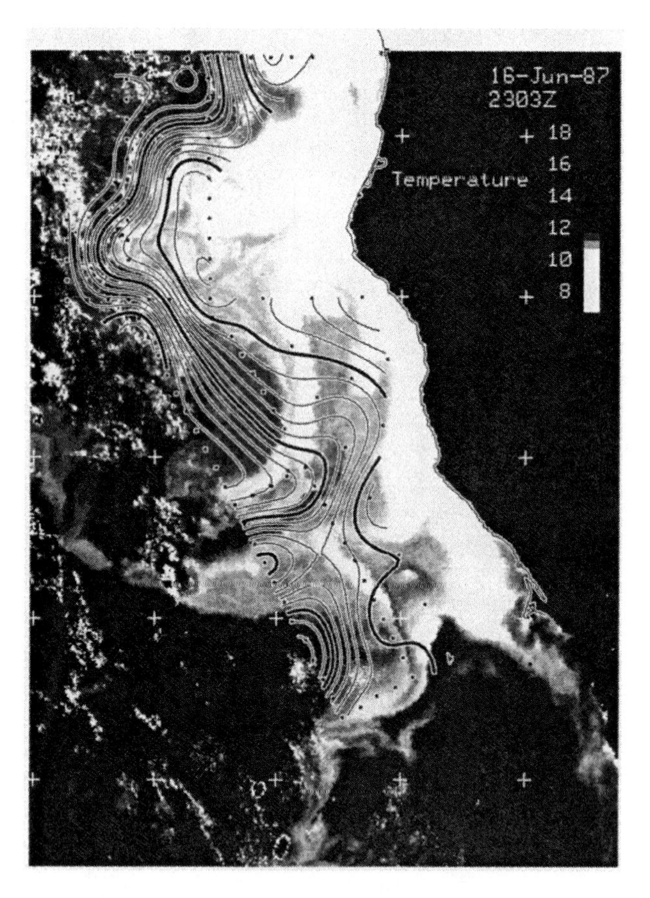


Figure 2.2 Sea surface temperature (lighter shades represent cooler water) and contours of dynamic height (contour interval 0.02 m) off the northern coast of California, June 16, 1987. After Kosro et al. (1991).

Because such fine vertical and horizontal scales occur as part of coastal upwelling, it has only been fairly recently that numerical models have been able to achieve the needed resolution (e.g. Allen et al., 1995; Allen and Newberger, 1996), and then only in two dimensions (vertical, offshore). Although two-dimensional models yield a good deal of information, they are unable to deal with frontal instabilities, topographic irregularities and other potentially important effects. Further, two-dimensional models have a rather unrealistic tendency for cross-shelf flow to be depth-independent far offshore even though observations suggest that upwelling source waters generally come from only the upper couple hundred meters depth (Smith, 1981). Nonetheless, rapid progress can be expected in numerical modeling over the coming years. This progress will be welcome, since the details of the flow field, such as its convergence, appear to play an important role in biological developments.

### d. *Buoyancy driven flows*

*Coastal currents:* The essential physics of buoyancy-driven coastal currents is deceptively straightforward: relatively light, fresh water debouches from rivers or estuaries onto the continental shelf. The resulting density contrast (density increasing offshore) is associated, through geostrophy, with an alongshore flow to the right (looking offshore) in the northern hemisphere. The plume of fresh water and its related current spreads in the alongshore direction through a mechanism that could be idealized as a very nonlinear internal Kelvin wave, but that is more complicated in practice (Lentz et al., 2003). Overall complexity of the problem enters because the flow advects its own driving agency (the density contrast), so that the problem is strongly nonlinear. Hill (1998) provides a good summary and entry into the relevant literature.

Buoyancy flows are strongly affected by a range of factors such as upstream vorticity distributions, instabilities, topography and so forth. Most models, to simplify, have concentrated on cases with vertical walls and layer idealizations. Because of these strong assumptions, and the underlying nonlinearity, simple results that are testable quantitatively against observations have been less common than in the coastal-trapped wave case, for example.

One important direction of increasing realism involves the presence of a sloping, beach-like bottom, as generally found in nature. In this case, how the density current interacts with the topography is quite important, as summarized by Chapman and Lentz (1994) and Lentz and Helfrich (2002). First, if there is bottom friction, the bottom Ekman transport advects the front and current offshore until an equilibrium is achieved when the alongshore density-driven flow is exactly zero at the bottom (figure 2.3), so that there is no bottom stress or Ekman transport (see section 2.a). Nature thus locates the front (when it intersects the bottom) at a location that depends upon the balance of imposed volume transport, the Earth's rotation, the cross-front density contrast, and the depth profile. Once the steady density current locks onto its equilibrium isobath, it then stays on that isobath indefinitely as it moves alongshore.

According to this line of reasoning, the shelfbreak front south of New England is likely just a regional scale response to an upstream outflow (Chapman, 2000), perhaps associated with the Bay of Fundy or the St. Lawrence River. If so, this would explain the front's tendency to stay near a single isobath, especially near the bottom. This front, though, is complicated by sustained instabilities (e.g., Lozier et al., 2002), secondary circulations (Houghton and Visbeck, 1998), interactions with offshore eddies and wind forcing (Houghton et al., 1994).

Buoyancy currents are strongly affected by wind forcing (Chao, 1987; Fong and Geyer, 2001) and outlet strength (Garvine, 2001). A wind-driven upwelling circulation, where winds oppose the direction of the current, moves the front and buoyancy current offshore, and so makes them weaker and more diffuse. A downwelling-favorable alongshore wind reinforces the density current, pushes the front closer to the coast and sharpens the front. Observational evidence for this sort of response is beginning to accumulate (Rennie et al., 1999).

Although our dynamical understanding of buoyancy-driven flows has improved considerably over the last decade, some disturbing aspects remain. For example, most numerical models of outflow-driven currents tend to have unrealistic properties, such as an expanding offshore bulge of freshwater, rather than a turning and formation of the expected alongshore current. Most often, the problem is dealt with by providing a weak ambient flow in the direction of the expected buoyancy driven current. Garvine (2001) pointed out the unsatisfactory nature of this approach, and looked more deeply into the model dynamics to seek explanations. He found that it is extremely important to treat the freshwater source as a realistic estuarine (two-way) flow pattern rather than the more common imposition of a one-way flow out of the coastal wall. Perhaps providing for a realistic estuarine circulation allows the outflow waters to adjust partially before reaching the shelf and so obtain a realistic vorticity distribution that avoids bulge formation. Further, Garvine (2001) found that replacing a coastal wall by a coastal wedge (where the water can become extremely shallow) also tends to alleviate the need for imposing an artificial ambient flow.

*Sill flows and outflows:* When there is a severe topographic constriction, the maximum allowable flow through the gap will be determined locally (e.g., Trowbridge et al., 1998). When the flow is barotropic (density variations are not dynamically significant), the flow tends to be determined by the combined constraints of along-strait pressure gradient (imposed by basin scale effects), time dependence, the Earth's rotation and perhaps bottom friction (Garrett and Toulany, 1982; Pratt, 1991). The dynamics is essentially linear in this case. A much richer phenomenology arises in the stratified case when there is an exchange of water through the strait and the two waters masses have differing densities. The Strait of Gibraltar (Bryden et al., 1994) is a good example of this sort of exchange system.

In the case of Gibraltar, fresher, lighter water from the Atlantic flows into the Mediterranean above the denser outflow. Evaporation within the Mediterranean assures that the outflowing water is saltier. The flow within the Strait is modulated by very strong (of order1 m/sec) tidal currents (Candela et al., 1990): even though tides within the basin are small, what flux there is all has to pass through this Strait. These tides are modified by along-strait topographic variations to give rise to dramatic internal waves (Watson and Robinson, 1990). Further time dependence is contributed by meteorological forcings (Candela et al., 1989). The shear across the inflow/outflow interface, combined with the enhanced internal wave shears, means that there has to be substantial dissipation and interlayer mixing within the Strait (Wesson and Gregg, 1994). Thus, time dependence and scale interactions are important here at lowest order.

Theoretical treatments of cases like Gibraltar have usually been based on steady rotating internal hydraulics applied to two distinct layers that do not mix with each other (e.g., Whitehead et al., 1974; Armi and Farmer, 1988; Farmer and Armi, 1988). These seemingly simple models allow a rich phenomenology, including the establishment of a maximum exchange rate, and of hydraulic jumps when flow is strong (in the sense of an internal Froude number: flow speed relative to internal wave speed). More recent developments have treated the important roles played by time-dependence (Helfrich, 1995; Hogg et al, 2001a) and interfacial mixing (Hogg et al., 2001b), particularly with regard to the determination of net exchange. Allowing for the latter effect, in turn, means that models will likely need to deal eventually with continuous stratification (rather than a layer approximation).

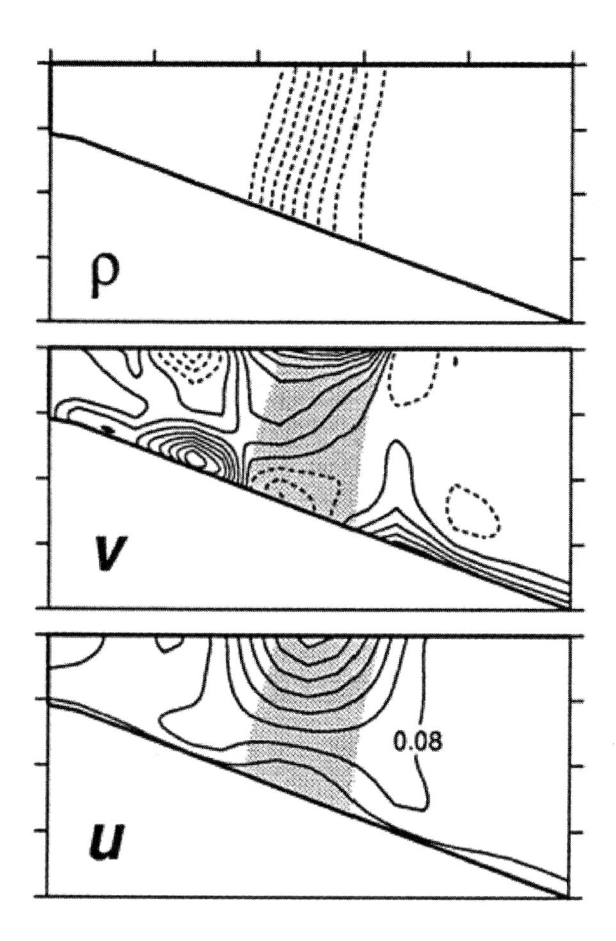


Figure 2.3 Numerical model results showing the inter-relation of structures for an adjusted front over the continental shelf. Upper panel: Density, Middle panel: Cross shelf flow (solid contours for offshore flow), lower panel: alongshore flow. In the middle and lower panels, the frontal location is shaded to help highlight how the front's equilibrium position is that where alongshore flow is near zero at the bottom and that cross-shelf flow converges onto the front in such as way as to allow no net cross-shelf advection. After Chapman and Lentz (1994).

Very often, the dense outflow from the Strait remains coherent as it moves downward and alongshore from the Strait. This confined, energetic current then entrains water across its strongly sheared upper boundary until it reaches an equilibrium depth (Price and O'Neil Barringer, 1994). Downstream observations off Portugal (Bower et al., 1997) make it rather plain that the flow can become unstable and depart from the continental slope in the form of a sequence of coherent subthermocline eddies known as Meddies. Thus the Strait flow is strongly affected by basin conditions, and in turn the waters that pass through the strait affect water properties on a basin scale.

*Densification over the shelf:* Under the influence of wintertime cooling or of evaporative salinity increase, waters over the shelf or in bays can become denser with time. The effect is most pronounced in shallow water because densification from above causes vertical mixing, and shallower, mixed water columns have a greater density increase per unit volume than deeper waters subject to the same buoyancy loss. Densification can take at least two general forms. One is an "inverse estuarine" circulation typified by conditions in the highly evaporative Spencer Gulf of southern Australia (Lennon et al., 1987), where fresher water flows in near the surface and saltier water passes out below it at the mouth of the Gulf (a miniature Mediterranean). On the other hand, models of strong densification in shallower waters over an open shelf demonstrate a rapid growth in baroclinic instabilities that lead to an eddy buoyancy exchange in the cross-shelf direction (Gawarkiewicz and Chapman, 1995; Chapman, 2003; Pringle 2003 and references therein). It is challenging, however, to verify these model results against observations for a number of reasons, not the least of which is carrying out the needed observations under severe wintertime conditions. Perhaps open, highly evaporative shelves will provide a better opportunity for a comprehensive measurement program.

### e. *Western Boundary Currents*

Energetic western boundary currents, such as the Gulf Stream, generally run along-isobath offshore of the shelfbreak. These features affect shelf flow patterns and hydrography either directly or indirectly through a range of mechanisms. One rather obvious one is simply determining the offshore water mass that interacts with the shelf water. Both the Gulf Stream and the Kuroshio tend to have bimodal positions with regard to distance from the shelf break (Boicourt et al., 1998) so that the current's position makes a substantial difference in terms of the temperature, salinity or nutrient content of the water directly influencing the shelf.

A specific mode of coupling is through wind-driven upwelling or downwelling circulations (Boicourt et al., 1998). These are superficially similar to those found in other regions, but their impact is very dependent on the offshore location of the Gulf Stream front, hence the source water properties (Hofmann et al., 1981). The shelf waters inshore of the Gulf Stream are so shallow (<35m) that the whole water column can be in the euphotic zone, so that upwelled water can have a biological impact even if it never reaches the surface mixed layer. During wintertime downwelling, Gulf Stream water can come onto the shelf, forming a distinct front that can vanish due to further advection or to vertical mixing (Oey et al., 1987).

A second boundary current effect is related to hydrodynamic instabilities, known as shingles, on the inshore edge of the Gulf Stream. These are, by now, relatively well-studied features. The shingles appear on satellite temperature images as shoreward-intruding backswept tongues of warm water, that appear first off the coast of Florida, and that propagate northward as their amplitude grows (Miller, 1994). Examination of the subsurface structure reveals an upward doming near the shelfbreak, and that the surface warm tongue is typically only about 20m thick (figure 2.4; Bane et al., 1981). The shingle amplitude seems to be quite dependent on the offshore position of the Gulf Stream axis relative to the shelf edge (Glenn and Ebbesmeyer, 1994). Linearized theory (Luther and Bane, 1985) yields convincing evidence that these features are likely to be rationalized as a simple instability. The regularities of alongshore propagation and growth suggest that these features may be amenable to predictability over time scales of a few days.

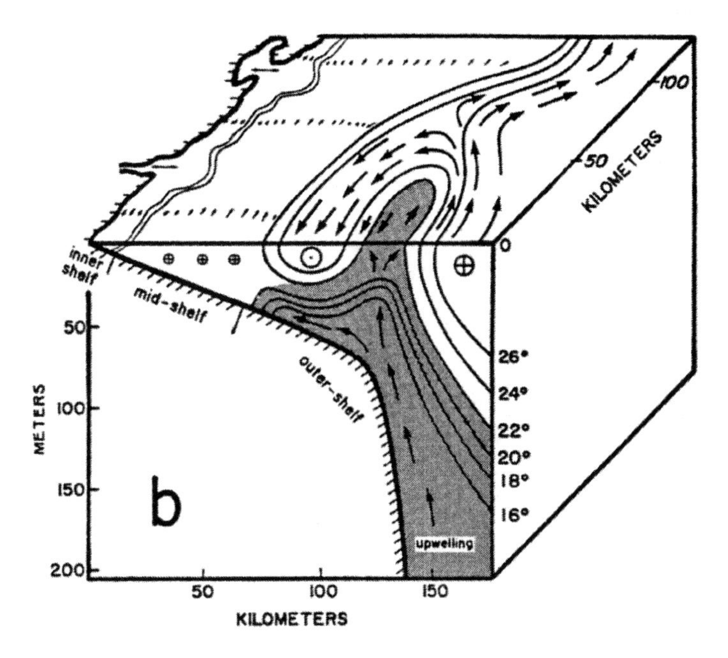


Figure 2.4 Schematic of Gulf Stream, propagating shingle and cold dome structure off the southeastern coast of the United States. After Bane et al. (1981).

The Gulf Stream typically turns eastward, away from the continental slope, near Cape Hatteras (about 35.5° N), but this does not end its impact on shelf waters. First, there is increasing evidence that shelf water is drawn offshore in this separation region and that it tends to be transported rapidly away in a narrow, coherent strip on the poleward side of the separated Gulf Stream (Lillibridge et al., 1990): this mechanism appears to be one of the major venues for cross-shelf exchange off the east coast of North America. The exact physical process involved is not obvious, but the setting suggests rather nonlinear dynamics.

After separation, the Gulf Stream is highly unstable, and it rapidly gives rise to large-amplitude meanders and eddies, called warm-core rings when they appear north of the current axis (e.g., Joyce, 1991). These features usually drift westward after their separation, and so sometimes contact the outer continental shelf, at which time they can draw substantial fluxes of shelf water offshore (Joyce et al., 1992). It appears that normally warm-core rings do not penetrate onto the shelf proper (e.g. Beardsley et al., 1985), but there is some documentation for infrequent Ring penetration onto the continental shelf (e.g., Flagg, 1987). It is not known whether or not these Ring penetrations leave behind any substantial changes in shelf water masses.

It is tempting to regard the well-studied shelf inshore of the Gulf Stream as typical of coastal areas adjoining western boundary currents. Clearly, this is not strictly the case. For example, where the Agulhas flows along the Natal-KwaZulu coast, the shelf is very narrow and the Agulhas influences seem to be felt almost to the coast (Schumann, 1987), in contrast to the Gulf Stream region where shingles never seem to reach all the way to shore. Further, the Kuroshio flows along a topographically complex area studded with islands that the current must negotiate. There are large boundary current meanders onto the shelf (Hsueh et al., 1996) that do not have an analog off the east coast of the United States. While there are common threads in western boundary current regions, such as the tendency to generate warm core ring-like structures, and for fluctuating patterns along the inshore edge of the boundary current, local effects do vary.

## 2. Conclusions

A short summary cannot begin to do justice to the range of interesting and potentially important processes that can take place over continental shelves and slopes. It is hoped, however, that the above survey provides some context, introduction and direction for further study. Some broader questions are suggested, however, and are discussed briefly here because of their importance for interdisciplinary problems.

*What distinguishes the coastal ocean?* A number of aspects make coastal physical oceanography distinctive. For example, wind-driven current fluctuations below the surface mixed layer tend to be far more apparent than in the deep ocean. Coastal-trapped waves propagate energy rapidly though many coastal systems, and so spread the effects of different forcings over large alongshore scales. Tides are generally amplified by coastal topographies, and are almost always more energetic than in deep water. Currents tend to be anisotropic at subinertial frequencies, with more energetic, coherent flow in the alongshore direction than in the cross-shelf direction. Density driven flows, associated with estuarine outflows, are often found in many coastal settings. These and the other distinctive aspects of the coastal ocean can basically be traced back to one feature: the simple fact of a coastal boundary. Flow cannot pass through the shoreline, and so flow patterns have to close over the shelf. As a consequence, wind-driven upwelling and wind-driven currents are pronounced even with winds no stronger than those found far offshore. Further, the coast provides both a source and a support against which density driven flows can propagate.

*Observability:* We often observe the result of a process rather than its actual occurrence. For example, in the case of coastal upwelling (Smith, 1995), it is common to measure surface cooling or increases in biological productivity in association with the wind, but rare to capture the detailed cross-shelf flow field. In the case of Georges Bank, there must be cross-frontal nutrient fluxes, but to date the associated flow field has not been observed. Similarly, there must be exchanges of waters across the shelfbreak front south of New England, but the cross-frontal fluxes have been difficult to measure directly or to relate to a specific process. In all of these cases, as well as others that could be mentioned, short-scale "noise" processes make the desired signal hard to observe.

Is it possible that intruding eddies from offshore (as found off California by Dever, 1997) are the single answer to the "noise" problem? This seems unlikely. For one thing, there are clearly regions were other processes do create large, short-scale cross-shelf currents, for example near unstable shelf edge fronts or, probably, upwelling fronts. Further, on wide shelves, such as the Mid Atlantic Bight, one might expect that any small eddies penetrating from offshore would be attenuated across the shelf, but the observed "noise" does not seem to die out so neatly.

Thus cross-shelf flows associated with some process of interest, such as upwelling, may not be readily observable even though they can be exceedingly important on some integrated, regional scale. This, in turn, may mean that future interdisciplinary programs that need to measure a cross-shelf flux of, say, zooplankton may require considerable resources to meet their objective directly. In this sort of case, it might make a good deal more sense to try to meet objectives indirectly. An indirect approach might call for measuring the difference between two conservative alongshore fluxes (which would balance a cross-shelf flux), or it might mean taking some very different tack, such as using floats or dyes.

*Extreme events:* Since some of the fundamental physical dynamics of the coastal ocean is linear, to at least some extent, physical oceanographers often study the region with linear statistical tools. This approach is less successful to the extent that processes are nonlinear. For example, sediment transport tends to scale with some power of the wind stress, since suspension itself goes as stress to a power, flow is proportional to a power, and the wave field (which also depends on the wind) can also be a factor. The net result is that sediment suspension and transport are dominated by rare, energetic events. In general, if the processes involved are nonlinear, the episodic events will tend to be much more important than lower, but sustained, levels of activity. Since biological and geological processes are often characterized by nonlinear couplings, it stands to reason that practitioners in these fields will tend to be more preoccupied with rare, but energetic, events than will physical oceanographers. Stated another way, physical oceanographers tend to concentrate on what is typical vs. what might dominate some transport process. This difference in viewpoint has sometimes caused poor communications, but it seems likely that the physical community will, with time, come to concentrate on the more extreme events as well. One reason is that this would follow naturally from the growing predominance of interdisciplinary research. A second reason is that, on closer understanding, many physical processes have nonlinear aspects (as discussed above) and so some important mean properties could well be dominated by a few extreme events.

*Global knowledge base:* Physical processes in the coastal ocean are important to understand quantitatively for a number of reasons ranging from scientific to practical (e.g. Malone et al, 2004). However, our knowledge has critical gaps even on the best-studied shelves, and many shelves have hardly been studied at all. How should we proceed?

One approach is as follows. First, we should specify what we want to know. One simple question might be "how do new nutrients reach the coastal ocean from offshore reservoirs?" Surprisingly, this straightforward question has only been answered for a few prototypical shelves, such as those dominated by wind-driven upwelling, or western boundary current instabilities. Next, a catalog of relevant coastal physical processes could be compiled, and a prototypical shelf for each process isolated. At the same time, to the extent possible, it would be useful to catalog all of the world's shelves in terms of what processes are likely to occur in each. For each process, the one prototypical shelf should be observed in detail with enough interdisciplinary measurement systems to provide a thorough, quantitative set of results relevant to the key question. In conjunction with the observations, a numerical modeling program should take place so as to encapsulate the knowledge gained from observation. This approach should be repeated for each prototype shelf. In the end, the model system ought to be capable of functioning almost anywhere in the global coastal ocean, and to be able to deal with a range of superimposed processes. Further, by this point, there should be a clear understanding of the sorts of information required to drive the model accurately. The hope is that, through models, we could gain a global process-oriented understanding that does not require terribly intensive sea-going programs over a wide area.

## Acknowledgements:

This summary has benefited from years of discussion with Dave Chapman and Steve Lentz and from two very helpful reviews. Comments from Rich Garvine were especially helpful. Support from the National Science Foundation through grant OCE-9806445 (GLOBEC) is gratefully acknowledged.

## Bibliography

Allen, J.S., 1980. Models of wind-driven currents on the continental shelf. *Ann. Rev. Fluid Mech.*, **12,** 389–433.

Allen, J.S. and P.A. Newberger, 1996. Downwelling on the Oregon continental shelf. Part I: response to idealized forcing. *J. Phys. Oceanogr.*, **26,** 2011–2035.

Allen, J.S., P.A. Newberger and J. Federiuk, 1995. Upwelling circulation on the Oregon continental shelf. Part I: response to idealized forcing. *J. Phys. Oceanogr.*, **25,** 1843–1866.

Armi, L. and D.M. Farmer, 1988. The flow of Mediterranean water through the strait of Gibraltar. *Prog. Oceanogr.*, **21,** 1–105

Bane, J.M., D.A. Brooks and K.R. Lorenson, 1981. Synoptic observations of the three –dimensional structure and propagation of Gulf Stream meanders along the Carolina continental margin. *J. Geophys. Research,* **86,** 6411–6425.

Barber, R.T., and R.L. Smith, 1981. Coastal upwelling ecosystems. in *Analysis of Marine Ecosystems,* A.R. Longhurst, editor, Academic Press, 31–68.

Barth, J.A., 1989a. Stability of a coastal upwelling front, 1, Model development and a stability theorem. *J. Geophys. Res.*, **94,** 10844–10856.

Barth, J.A., 1989b. Stability of a coastal upwelling front, 2, Model results and a comparison with observations. *J. Geophys. Res.*, **94,** 10857–10883.

Barth, J.A.,1994. Short-wavelength instabilities on coastal jets and fronts. *J. Geophys. Res.*, **99,** 16095–16115.

Barth, J.A., S.D. Pierce and R.L. Smith, 2000. A separating coastal upwelling jet at Cape Blanco, Oregon and its connection to the California Current system. *Deep-Sea Res. II,* **47,** 783–810.

Beardsley, Robert C., David C. Chapman, Kenneth H. Brink, Steven R. Ramp, and Ronald Schlitz, 1985. The Nantucket Shoals Flux Experiment (NSFE79), Part 1: A basic description of the current and temperature variability. *J. Phys. Oceanogr.,* **15**(6), 713–748.

Boicourt, W.C., W.J. Wiseman, A. Valle-Levinson and L.P. Atkinson, 1998. Continental shelf of the southeastern United States and the Gulf of Mexico: in the shadow of the western boundary current. in *The Sea,* volume 11, A.R. Robinson and K.H. Brink, editors, J. Wiley and Sons, Inc., New York, 135–182.

Bower, A.S., L. Armi and I Ambar, 1997. Lagrangian observations of Meddy formation during a Mediterranean undercurrent seeding experiment. *J. Phys. Oceanogr.,* **27,** 2545–2575.

Brink, K. H., 1986. Topographic drag due to barotropic flow over the continental shelf and slope. *J. Phys. Oceanogr.,* **16**(12), 2150–2158.

Brink, K.H., 1991: Coastal-trapped waves and wind-driven currents over the continental shelf. *Ann. Rev. Fluid. Mech.,* **23,** 389–412.

Brink, K. H., D. C. Chapman, and G. R. Halliwell, Jr., 1987. A stochastic model for wind-driven currents over the continental shelf. *J. Geophys. Res.,* **92**(C2), 1783–1797.

Brink, K.H. and T. Cowles, 1991. The Coastal Transition Zone Program. *J. Geophys. Res.,* **96**(C8), 14637–14647.

Brink, K.H., R. Limeburner and R.C. Beardsley, 2003. Properties of flow and pressure over Georges Bank as observed with near-surface drifters. *J. Geophys. Research,* **108**(C11), 8001.

Brink, K.H. and A.R. Robinson, 1998: The Global Coastal Ocean Processes and Methods., in *The Sea,* Volume 10, J. Wiley and Sons, New York, 604pp.

Brown, W.S., and J.A. Moody, 1987. Tides. In *Georges Bank,* R.H. Backus and D.W. Bourne, editors, The MIT Press, Cambridge, Massachusetts, 100–107.

Bryden H.L., J. Candela and T.H. Kinder, 1994. Exchange through the Strait of Gibraltar. *Prog. Oceanogr.,* **33,** 201–244.

Candela, J., C.D. Winant and H.L. Bryden, 1989. Meteorologically forced subinertial flows through the Strait of Gibraltar. *J. Geophys. Res.,* **94,** 12667–12679.

Candela, J., C.D. Winant and A. Ruiz, 1990. Tides in the Strait of Gibraltar. *J. Geophys. Res.,* **95,** 7313–7335.

Chao, S.-Y., 1987. Wind-driven motion near inner shelf fronts. *J. Geophys. Res.,* **92,** 3849–3860.

Chapman, D.C., 1987. Application of wind-forced long coastal-trapped wave theory along the California coast. *J. Geophys. Res.,* **92,** 1798–1816.

Chapman, D.C., 2000. Boundary layer control of buoyant coastal currents and the establishment of a shelfbreak front. *J. Phys. Oceanogr.,* **30,** 2941–2955

Chapman, D.C., 2002. Deceleration of a finite-width, stratified current over a sloping bottom: frictional spindown or buoyancy shutdown?, *J. Phys. Oceanogr.,* **32,** 336–352.

Chapman, D.C., 2003. Comment on "Cross-shelf eddy heat transport in a wind-free coastal ocean undergoing wintertime cooling" by J.M. Pringle, *J. Geophys. Res.,* **108,** 3026.

Chapman, D.C. and S.J. Lentz, 1994. Trapping of a coastal density front by the bottom boundary layer. *J. Phys. Oceanogr.,* **24,** 1464–1479.

Chen, C. and R. Beardsley, 1995. A numerical study of stratified tidal rectification over finite-amplitude banks, part I: symmetric banks. *J. Phys. Oceanogr.,* **25,** 2090–2110.

Chen, C., R.C. Beardsley and R. Limeburner 1995. A numerical study of stratified tidal rectification over finite-amplitude banks. Part II: Georges Bank. *J. Phys. Oceanogr.,* **25,** 2111–2128.

Clarke, A.J, 1991. The dynamics of barotropic tides over the continental shelf and slope (review), in *Tidal Hydrodynamics,* B.B. Parker, editor, J. Wiley & Sons, Inc., 79–108.

Colosi, J., R. Beardsley, J. Lynch, G. Gawarkiewicz, C. Chiu, and A. Scotti, 2001. Observations of Internal Waves on the outer New England Continental shelf during the summer Shelfbreak PRIMER study. *J. Geophys. Res.*, **106,** 9587–9601.

Cornejo-Rodriguez, M. and D.B. Enfield, 1987. Propagation and forcing of high-frequency sea-level variability along the west coast of South America. *J. Geophys. Res.,* **92,** 14,323–14,334.

Crawford, W.R. and R.E. Thomson, 1984. Diurnal period shelf waves along Vancouver Island: a comparison of observations with theoretical models. *J. Phys. Oceanogr.,* **14,** 1629–1646.

Daifuku, P.R and R.C. Beardsley, 1983. The $K_1$ tide on the continental shelf from Nova Scotia to Cape Hatteras. *J. Phys. Oceanogr.,* **13,** 3–17.

DeSzoeke, R.A., and J.G. Richman, 1984. On wind-driven mixed layers with strong horizontal gradients- a theory with application to coastal upwelling. *J. Phys. Oceanogr.,* **14,** 364–377.

Dever, E.P., 1997. Subtidal velocity correlation scales on the northern California shelf. *J. Geophys. Res.,* **102,** 8555–8571.

Ekman, V.W., 1905. On the influence of the earth's rotation on ocean currents. *Ark. Mat. Astron. Fys.,* **12,** 1–52.

Farmer, D.M. and L. Armi, 1988. The flow of Atlantic water through the Strait of Gibraltar. *Progr. Oceanogr.,* **21,** 1–105.

Flagg, C.N., 1987. Hydrographic structure and variability. In *Georges Bank,* R.H. Backus and D.W. Bourne, editors, The MIT Press, Cambridge, Massachusetts, 108–124.

Fong, D.A. and W.R. Geyer, 2001. Response of a river plume during an upwelling favorable wind event. *J. Geophys. Res.,* **106,** 1067–1084.

Foreman, M.G.G., R.A. Walters, R.F. Henry, C.P. Keller, A.G. Dolling, 1995. A tidal model for eastern Juan de Fuca Strait and the southern Strait of Georgia. *J. Geophys. Res.,* **100,** 721–740.

Franks, P.J.S. and C. Chen, 1996. Plankton production in tidal fronts: a model of Georges Bank in summer, *J. Marine Res.,* **54,** 631–651.

Garrett, C.J.R. and W. Munk, 1979. Internal waves in the ocean. *Ann Rev. Fluid Mech.,* **11,** 339–369.

Garrett, C.J.R. and B. Toulany, 1982. Sea level variability due to meteorological forcing in the northeast Gulf of St. Lawrence. *J. Geophys. Res.,* **87,** 1968–1978.

Garvine, R.W., 1999. Penetration of buoyant coastal discharge onto the continental shelf: a numerical model experiment. *J. Phys Oceanogr.,* **29,** 1892–1909.

Garvine, R.W., 2001. The impact of model configuration in studies of buoyant coastal discharge. *J. Marine Res.,* **59,** 193–225.

Garvine, R.W., 2004. The vertical structure and subtidal dynamics of the inner shelf off New Jersey. *J. Marine Res.,* **62,** 337–371.

Gawarkiewicz, G., and D.C. Chapman, 1995. A numerical study of dense water formation and transport on a shallow, sloping continental shelf. *J. Geophys. Res.,* **100,** 4498–4507.

Giese, G.S., D.C. Chapman, M.G. Collins, R. Encarnacion and G. Jacinto, 1998. The coupling between harbor seiches at Palawan Island and Sulu Sea internal solitons, *J. Phys. Oceanogr.,* **28,** 2418–2426.

Glenn, S.M. and C.C. Ebbesmeyer, 1994. Observations of Gulf Stream frontal eddies in the vicinity of Cape Hatteras, *J. Geophys. Res.,* **99,** 5047–5055.

Grant, W.D. and O.S. Madsen, 1986. The continental shelf bottom boundary layer. *Ann. Rev. Fluid Mech.,* **18,** 265–305.

Haidvogel, D. B., and K. H. Brink, 1986. Mean currents driven by topographic drag over the continental shelf and slope. *J. Phys. Oceanogr.,* **16**(12), 2159–2171.

Henderschott, M.C., 1981. Long waves and ocean tides. In *Evolution of Physical Oceanography,* B.A. Warren and C. Wunsch, editors, The MIT Press, Cambridge, Massachusetts, 292–341.

Helfrich, K.R., 1995. Time dependent two-layer hydraulic exchange flows. *J. Phys Oceanogr.,* **25,** 359–373.

Hill, A.E., 1998. Buoyancy effects in coastal and shelf seas. In The Global Coastal Ocean Processes and Methods, *The Sea,* volume 10, K.H. Brink and A.R. Robinson, editors, J. Wiley and Sons, New York, 21–62.

Hill, A.E., B.M. Hickey, F.A. Shillington, P.T. Strub, K.H. Brink, E.D. Barton and A.C. Thomas, 1998. Eastern Ocean Boundaries. In *The Global Coastal Ocean Regional Studies and Syntheses, The Sea,* volume 11, A.R. Robinson and K.H. Brink, editors, J. Wiley and Sons, New York, 29–68.

Hofmann, E.E., L.J. Pietrafesa and L.P. Atkinson, 1981. A bottom water intrusion in Onslow Bay, North Carolina, *Deep-Sea Res.,* **28**(4A), 329–346.

Hogg, A.M., K.B. Winters and G.N. Ivey, 2001a. Linear internal waves and the control of stratified exchange flows, *J. Fluid Mech.,* **447,** 357–375.

Hogg, A.M., G.N. Ivey and K.B. Winters, 2001b. Hydraulics and mixing in controlled exchange flows. *J. Geophys. Res.,* **106,** 959–972.

Holloway, G., 2002. Toward a statistical ocean dynamics. In *Statistical Theories and Computational Approaches to Turbulence,* Y. Kaneda and T. Gotoh, editors, Springer, 277–288.

Holloway, G., K. Brink, and D. Haidvogel, 1989. Topographic stress in coastal circulation dynamics. In: *Poleward Flows Along Eastern Ocean Boundaries,* S. J. Neshyba, C. N. K. Mooers, R. L. Smith, and R. T. Barber, editors, Springer-Verlag, New York, pp. 315–330.

Holloway, P.E., P.G. Chatwin and P. Craig, 2001. Internal tide observations from the Australian north west shelf in summer 1995. *J. Phys. Oceanogr.,* **31,** 1182–1199.

Holt, J.T. and R. Proctor, 2003. The role of advection in determining the temperature structure of the Irish Sea. *J. Phys. Oceanogr.,* **33,** 2288–2306.

Hood, R.R., M.R. Abbott and A. Huyer, 1991. Phytoplankton and photosynthetic light response in the Coastal Transition Zone off northern California in June 1987. *J. Geophys. Res.,* **96,** 14769–14780.

Horsburgh, K.J. and A.E. Hill, 2003. A three-dimensional model of density-driven circulation in the Irish Sea. *J. Phys. Oceanogr.,* **33,** 343–365.

Horsburgh, K.J., A.E. Hill, J. Brown, L. Fernand, R.W. Garvine and M.M.P. Angelico, 2000. Seasonal evolution of the cold pool gyre in the western Irish Sea. *Progr. Oceanog.,* **46,** 1–58.

Houghton, R.W., C.N. Flagg and L.J. Pietrafesa, 1994. Shelf-slope water frontal structure, motion and eddy heat flux in the southern mid Atlantic Bight. *Deep-Sea Res. II,* **41,** 273–306.

Houghton, R.W. and M. Visbeck, 1998. Upwelling and convergence in the middle Atlantic Bight shelfbreak front. *Geophys. Res. Lett.,* **25,** 2765–2768.

Hsueh, Y., H.-J. Lie and H. Ichikawa, 1996. On the branching of the Kuroshio west of Kyushu. *J. Geophys. Res.,* **101,** 3851–3858.

Huthnance, J.M., 1978. On coastal trapped waves: analysis and numerical calculation by inverse iteration. *J. Phys. Oceanogr.,* **8,** 74–92.

Joyce, T.M., 1991. Review of U.S. contributions to warm-core rings. *Rev. Geophys.,* **S29,** 610–616.

Joyce, T.M., J.K.B. Bishop and O.B. Brown, 1992. Observations of offshore shelf water transport induced by a warm-core ring. *Deep-Sea Res.,* **39**(1), S97-S113.

Kosro, P.M., A. Huyer, S.R. Ramp, R.L. Smith, F.P. Chavez, T.J. Cowles, M.R. Abbott, P.T. Strub, R.T. Barber, P. Jesson and L.F. Small, 1991: The structure of the transition zone between coastal waters and the open ocean off northern California, winter and spring 1987. *J. Geophys. Res.,* **96** (C8), 14,707–14,730.

Lennon G.W, D.G. Bowers, R.A. Nunes, B.D. Scott, M. Ali, J. Boyle, C. Wenju, M. Herzfeld, G. Johansson, C. Nield, P. Petrusevics, P. Stephenson, A.A. Suskin and S.E.A Wijffels, 1987. Gravity currents and the release of salt from an inverse estuary, *Nature,* **327,** 695–697.

Lentz, S.J., 1987: A heat budget for the northern California shelf during CODE 2. *J. Geophys. Res.,* **92,** 14491–14509.

Lentz, S.J., 1992. The surface boundary layer in coastal upwelling regions. *J. Phys. Oceanogr.,* **22,** 1517–1539.

Lentz, S.J., 1994. Current dynamics over the northern California inner shelf. *J. Phys. Oceanogr.,* **24,** 2461–2478.

Lentz, S.F., S. Elgar and R.T. Guza, 2003. Observations of the flow near the nose of a buoyant coastal plume. *J. Phys. Oceanogr.,* **33,** 933–943.

Lentz, S.J., R.T. Guza, S. Elgar, F. Fedderson and T.H.C. Herbers, 1999. Momentum balances on the North Carolina inner shelf. *J. Geophys. Res.,* **104,** 18205–18226.

Lentz, S.J. and K.R. Helfrich, 2002. Buoyant gravity currents along a sloping bottom in a rotating fluid, *J. Fluid Mech.,* **464,** 251–278.

Lentz, S.J. and J. H. Trowbridge, 1991. The bottom boundary layer over the northern California shelf. *J. Phys. Oceanogr.,* **21** (8), 1186–1201.

Lerczak, J.A., M.C. Hendershott and C.D. Winant, 2001. Observations and modeling of coastal internal waves driven by a diurnal sea breeze. *J. Geophys. Res.,* **106,** 19715–19729.

Lillibridge, J.L., G. Hitchcock, T. Rossby, E. Lessard, M. Mork and L. Golmen, 1990. Entrainment and mixing of shelf/slope waters in the near-surface Gulf Stream, *J. Geophys. Res.,* **95,** 13065–13087.

Loder, J.W., 1980. Topographic rectification of tidal currents on the sides of Georges Bank., *J. Phys. Oceanogr.,* **10,** 1399–1416.

Lozier M.S., M.S.C. Reed and G. G. Gawarkiewicz, 2002. Instability of a shelfbreak front. *J. Phys. Oceanogr.,* **32,** 924–944.

Luther, M.E. and J.M. Bane, 1985. Mixed instabilities in the Gulf Stream over the continental shelf. *J. Phys. Oceanogr.,* **15,** 3–23.

Malone, T., T. Knap and M. Fogarty, 2004. Overview of Science Requirements. In *The Sea,* Volume 13, A.R. Robinson and K.H. Brink, editors, Harvard University Press, Cambridge, in press.

Mei, C.C. and P. L.-F. Liu, 1993. Surface waves and coastal dynamics. *Ann. Rev. Fluid Mech.,* **25,** 215–240.

Miller, J.L., 1994. Fluctuations of Gulf Stream frontal position between Cape Hatteras and the Straits of Florida. *J. Geophys. Res.,* **99,** 5057–5064.

Mooers, C.N.K., C. A. Collins and R.L. Smith, 1976. The dynamics structure of the frontal zone in the coastal upwelling region off Oregon. *J. Phys. Oceanogr.,* **6,** 3–21.

Muenchow, A. and R.J. Chant, 2000. Kinematics of inner shelf motions during the summer stratified season off New Jersey. *J. Phys. Oceanogr,* **30,** 247–268.

Oey, L.-Y., L.P. Atkinson and J.O. Blanton, 1987. Shoreward intrusion of upper Gulf Stream water onto the U.S. southeastern continental shelf. *J. Phys. Oceanogr.,* **17,** 2318–2333.

Parsons, J.D. and C.A. Nittrouer, 2004. Extreme events moving particulate material on continental margins .. In *The Sea,* Volume 13, A.R. Robinson and K.H. Brink, editors, Harvard University Press, Cambridge, in press.

Pratt, L. J., 1991. Geostrophic vs. critical control in straits. *J. Phys. Oceanogr.,* **21,** 728–732.

Price, J.F. and M. O'Neil Barringer, 1994. Outflows and deep water production by marginal seas. *Prog. Oceanogr.,* **33,** 161–200.

Pringle, J.M., 2003. Reply to comment by D.C. Chapman on "Cross-shelf eddy heat transport in a wind-free coastal ocean undergoing winter time cooling". *J. Geophys. Res.,* **108,** 3027, doi:10.1029/2002JC001398.

Pringle, J.M. and K. Riser, 2003. Remotely forced upwelling in southern California. *J. Geophys. Res.,* **180,** 3131, doi:10.1029/2002JC001447.

Rennie, S., J.L. Largier and S.J. Lentz, 1999. Observations of low-salinity coastal current Pulses downstream of Chesapeake Bay. *J. Geophys. Res.,* **104**(C8), 18,227–18,240.

Robinson, A.R., J.J. McCarthy and B.J. Rothschild, 2002 Biological-Physical Interactions in the Sea. *The Sea,* Vol.12, J. Wiley & Sons, Inc., New York, 634pp.

Samelson, R.M., 1997. Coastal boundary conditions and the baroclinic structure of wind-driven continental shelf currents. *J. Phys. Oceanogr.,* **27,** 2645–2662.

Sandstrom, H., and J.A. Elliott, 1984. Internal tides and solitons on the Scotian shelf: a nutrient pump at work. *J. Geophys. Res.,* **89,** 6415–6426.

Schumann, E.H., 1987. The coastal ocean off the east coast of South Africa. *Trans. Roy. Soc. S. Afr.,* **46**(3), 215–229.

Shearman, R.K. and S.J. Lentz, 2003: Dynamics of mean and subtidal flow on the New England shelf. *J. Geophys. Res.,* **108,** 3281, doi:10.1029/2002JC001417.

Simpson, J.H, 1998. Tidal processes in shelf seas. in *The Sea,* Volume 10, K.H. Brink and A.R. Robinson, editors, J. Wiley and Sons, New York, 113–150.

Simpson, J.H. and J.R. Hunter, 1974: Fronts in the Irish Sea. *Nature,* **250,** 404–406.

Simpson, J.H., J. Sharples and T.P. Rippeth, 1991 A prescriptive model of stratification induced by coastal runoff. *Estuarine and Coastal Shelf Sci.,* **33,** 23–35.

Smith, R.L., 1978. Poleward propagating perturbations in currents and sea level along the Peru coast. *J. Geophys. Res.,* **83,** 6083–6092.

Smith, R.L., 1981. A comparison of the structure and variability of the flow field in three coastal upwelling regions: Oregon, northwest Africa and Peru. In *Coastal Upwelling,* F.A. Richards, editor, American Geophysical Union, Washington D.C., 107–118.

Smith, R.L., 1995. The physical processes of coastal ocean upwelling systems. In *Upwelling in the Ocean Modern processes and Ancient Records,* C.P. Summerhayes, K.-C. Emeis, M.V. Angel, R.L. Smith and B. Zeitzschel, editors, J. Wiley & Sons, Inc., Chichester, 39–64.

Smith, R.L., A. Huyer, J.S. Godfrey and J.A. Church, 1991. The Leeuwin Current off Western Australia, 1986–1987. *J. Phys. Oceanogr.,* **21,** 323–345.

Spall, M. A., 2003. Islands in zonal flow. *J. Phys. Oceanogr.,* **33,** 2689–2701.

Thompson, R.O.R.Y., 1987. Continental-shelf-scale model of the Leeuwin Current. *J. Marine Res.,* **45,** 813–827.

Thorade, H., 1909. Über die Kalifornische Meerströmung. *Ann. D. Hydrogr. U. Mar. Meteor.,* **37,** 17–34.

Townsend, D.W. and N.R. Pettigrew, 1997. Nitrogen limitation of secondary production on Georges Bank, *J. Plankton Research,* **19,** 221–235.

Trowbridge, J.H., D.C. Chapman and J. Candela, 1998. Topographic effects, straits and the bottom boundary layer. in *The Sea,* volume 10, K.H. Brink and A.R. Robinson, editors, J. Wiley and Sons, New York, 63–88.

Trowbridge and Lentz, 1991. Asymmetric behavior of an oceanic boundary layer above a sloping bottom. *J. Phys. Oceanogr.,* **21,** 1171–1185

Watson, G., and I.S. Robinson, 1990. A study of internal wave propagation in the Strait of Gibraltar using shore-based marine radar images. *J. Phys. Oceanogr.,* **20,** 374–395.

Weller, R.A. and R.E. Davis, 1980. A vector measuring current meter. *Deep-Sea Res.,* **27,** 565–582.

Wesson, J.C. and M.C. Gregg, 1994. Mixing at Camarinal Sill in the Strait of Gibraltar. *J. Geophys. Res.* **99,** 9847–9878.

Whitehead, J.A., A. Leetmaa and R.A. Knox, 1974. Rotating hydraulics of strait and sill flows. *Geophys. Fluid Dyn.,* **6,** 101–125.

# Chapter 3. MULTIPLE SCALES IN SPACE AND TIME

BRIAN J. ROTHSCHILD

*University of Massachusetts Graduate School of Marine Sciences and Technology*

## Contents



## 1. Introduction

Explaining the relation between biological productivity and ocean physics is a major challenge. Understanding the biological response to physical forcing is a significant component of the challenge. It is fair to say that much of the understanding linking biological response to physical forcing is based on linear correlations between biological and physical variables. There is general agreement that moving beyond the correlation approach, improving causal understanding, is necessary (Robinson et al., 1999, 2002).

A characterization of the challenge requires consideration of the theoretical-development/observational-validation cycle. Oceanographic science, in the context of the interactions between biology and physics, has made very large strides from a descriptive and observational point of view over the last several decades. Important contributions to theory based upon these descriptions and observations are beginning to advance (see, e.g., Flierl and McGillicuddy, 2002 in Vol. 12, The Sea).

An issue particularly important for theoretical advancement is the status of coherence between biological and physical theory. Material coherence between biological and physical theory is required to substantiate the cause-and-effect relation between physical forcing and biological production. The achievement of coherence involves two main issues. The first relates to the dynamic character of physical and biological theories. The second relates to the physical framework constituting physical and biological theories.

*The Sea*, Volume 13, edited by Allan R. Robinson and Kenneth H. Brink
ISBN 0-674-01526-6 

With regard to the dynamic aspects of physical and biological theories, physical theories of the ocean are based on fundamental dynamics, while biological theories are based on ad hoc dynamics.

For example, the ideas and concepts of ocean-physics dynamics are associated with the Navier-Stokes equations. These dynamic equations are used to deduce the flow field as a function of external forcing constrained by the axiomatic relation between force, mass, and acceleration. In contrast to the universally accepted Navier-Stokes equations, there is no single generally accepted set of biological equations. The biological governing equations (e.g., the NPZ descendents of the Lottka-Volterra equations) that seem to be most popular are considered to be ad hoc because they are not derived from a fundamental physical proposition, such as the balance between force, mass, and acceleration.

The distinction can be subtle in the sense that biological dynamics can be coupled in various ways to fluid-flow models. These models basically distribute the NPZ material in time and space according to the flow, the NPZ entities reacting to the concentrations of their respective components. While, in a sense, these variations in local concentrations result from the forcing of the flow field, the variations are not constrained by thermodynamic or thermodynamic-like properties of biological metabolism that interact with physical forcing. For this reason, the biological equations, even when embedded in a flow field, are still considered as ad hoc.

In fact, in biological oceanography there is no single generally accepted set of biological equations. The biological governing equations (e.g., the NPZ descendants of the Lotka-Volterra equations) that seem to be most popular are not derived from physically-based axioms. This explains, evidently, why there are many versions of biological equations that attempt to explain the same phenomena. No single equation is regarded as being "universal." As a consequence, in contrast to physical theory, scaling arguments derived from biological equations have not had traction.

What is important is not simply the recognition of the difference between biological and physical theories, but rather the way that the difference highlights what needs to be accomplished in biological oceanography with regard to explaining temporal and spatial multiscale variability in ecosystem functioning. Surely biological processes are governed by physical laws. In fact, there are detailed studies of the physiology of *individual* organisms that describe the correspondence. The difficulty is in how all of the individual responses can be integrated to represent an entire ecosystem, accounting for the axiomatically non-linear biological response to physical forcing. Putting it another way, an in-depth appreciation of biological-physical interactions is difficult to attain without a truly fundamental dynamic view of biological processes.

With regard to the physical framework constituting physical and biological theories, physical theory embraces processes that span many orders of magnitude in time and space. The functioning of these processes is difficult to comprehend without partitioning time and space among various scales. Further, some physical processes, such as flow, are materially different depending upon the scale range that they inhabit. In contrast, biological theory represents the interactions of organisms that also function in a time-space framework that covers many orders of magnitude. Biological theory is generally phrased in terms of the concentrations of organisms rather than length and time scales (such as, for example, interorganism separation distances and the inverse of characteristic mortality rates) that are typical in ocean physics. To make biological theory compatible with physical theory, it is necessary to express biological theory in the context of length and time scales.

So to develop compatibility between physics and biology, it is necessary to view the intersection between biology and physics at the same level of dynamics and to have compatibility between the physical frameworks and metrics (time and space) that underlies the interaction of organisms.

The fact that biological and physical processes reside in multiple time and space scales that vary over orders of magnitude necessitates incorporating issues relating to scales into theoretical discussions (Mann and Lazier, 1991). In physical oceanography, the distribution of physical properties in a time-space reference frame is a point of departure for developing governing equations of the flow field. However, this is not generally the case in biological oceanography. To develop a parallel approach in biological oceanography, it is necessary to link biological energetics to time and space. If understanding is to be advanced, then biological and physical phenomena need to be considered in a physically defined reference frame.

An appreciation of the breadth of the physical reference frame can be obtained from the chapters in Robinson et al. (2002, Vol. 12, The Sea). For example, at larger scales Gargett and Marra (2002) describe how the global ocean/atmosphere system sculpts the nutricline. The global details of the cycling of nutrients between the oceans interior and surface layer is described by Gruber and Sarmiento (2002). Flierl and McGillicuddy (2002) give the details of the sculpting effect in terms of mesoscale forcing, which eventually drives processes to smaller dissipation and diffusion scales (Yamazaki et al., 2002). That major components of processes reside at different locations in scale space is exemplified by McCarthy (2002), who discusses how important components of large-scale ocean production are driven by very small phytoplankton that depend upon heterotroph excretion, and thus the very small-scale physics and chemistry of organism-organism nutrient exchange.

A common reference frame displays the issues that link variability in physical, biological, and chemical properties with space and time. Henry Stommel (1963) was among the first to think about multiple scales in space and time. Stommel thought about scales of variability in ocean physics. He pointed out that if one were concerned with variability of the relatively constant temperature of the deep ocean, observations taken only occasionally would give a reasonable assessment of variability. In contrast, if one was concerned with sea level height, a very large number of observations would need to be taken to adequately sample the multiple scales in space and time in which sea-level variability occurred. Stommel presented the first diagram of scale space in which scale-specific variability of a phenomenon (in this case, sea level height) is depicted (Fig. 3.1). He noted that, "No economical plan for mounting an expedition or setting up an observational program can encompass all the scales and periods; each plan must provide a definite significance level within a limited part of the spectrum despite contamination from other parts of the spectrum." He further observed that "...the spectra for all physical variables in the sea are very complex, and that there is a need for more sophisticated and more physically-oriented observational programs...these new programs need to be directed toward revealing the interactions between different portions of the spectra...such new observational material will have a tremendous impact upon theoretical studies..."

Stommel's ideas, which are still relevant to physical oceanography, morphed into biological oceanography with the consideration of spatial patterns and patchiness in plankton communities. An introduction from a mid-1970s perspective can be found in "Spatial Patterns in Plankton Communities" edited by John Steele (1977); Haury et al. (1977) superimposed zooplankton variability on Stommel's scale-space diagram; Okubo (1977) described critical scales for plankton patches; Platt (1977) analyzed the spectral composition of phytoplankton populations; and Fasham (1977) presented spectrums associated with particle point processes.

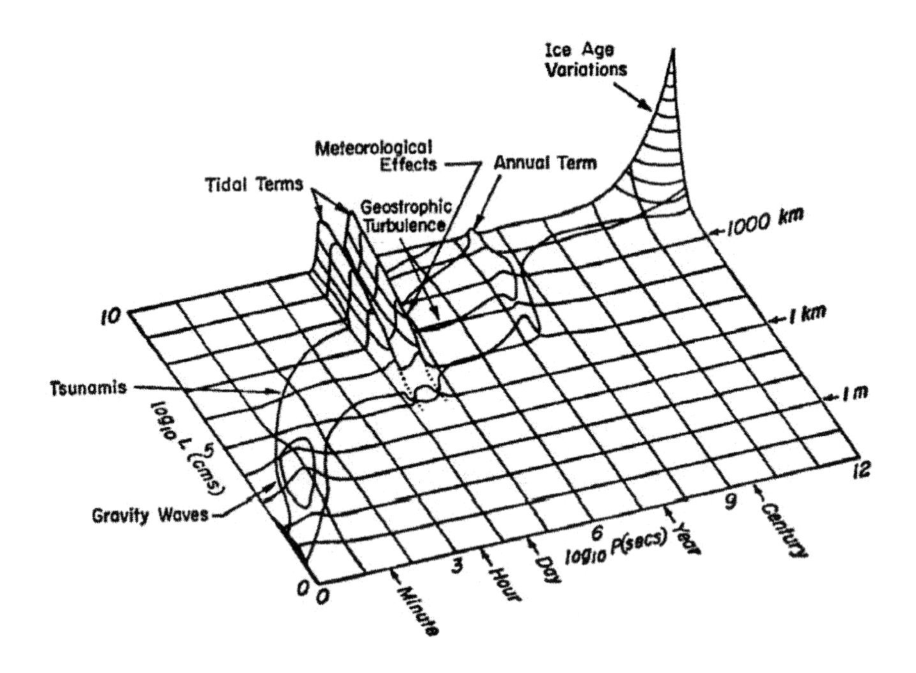


Figure. 3.1. "Scale-space" for sea level height. (Redrawn from Stommel, 1963.)

These studies had a significant impact in their own right, and encouraged, in subsequent decades, classification of phenomena according to location of their statistical power in scale space. However, it appears that in the evolution of knowledge and the present interest in biological-physical interaction that it has been difficult to fully develop Stommel's ideas. This is in many ways not surprising since the investigation of physical-biological variability in a space-time frame is exceedingly difficult from both an observational and theoretical point of view.

Descriptions of where phenomena reside in scale space begin to address the problem. For example, many phenomena are not associated with a small or discrete location in scale space. And, in fact, connections in scale space between different components of the same phenomena might not even be continuous. Understanding complexity of phenomena residing in various interconnected and disjoint sectors of scale space requires an organizing system of axiomatically derived governing equations. Without underpinning axioms, it is difficult to develop universal acceptance of an invariant formalism that can be used as a point of departure to acquire new knowledge. As well, without these axioms, the full power of physical intuition expressed as scaling arguments cannot be applied to understanding the biodynamics of the ocean as a component of the "earth system."

This expository chapter describes issues, ideas, tools, and examples intended to serve as a platform for further inquiry into the problem of linking physical and biological theories and multiple scales in time and space. The discussion begins with stage-setting issues, the definitions of scale, and the continuum hypothesis (Batchelor, 1967). There are several definitions of scale. Some definitions are empirically based, while other definitions are derived from governing equations. From the empirical point of view, identifying and understanding mechanisms for inducing or changing scale-related properties is fundamental. These mechanisms are implicit in the filter configuration of the differential equations that represent the ecological systems. From the theoretical point of view, the primary issue involves the development of governing equations that are axiomatic and expressible in a physical reference frame. Before this can be done, it is necessary to consider applicability of the continuum hypothesis to biological variability. The fact that in a number of instances the continuum hypothesis does not apply in oceanographic biology/chemistry requires considering statistical techniques amenable to studying particles that are many orders of magnitude more distant from one another greater than the fluid molecules in which they are embedded.

Having discussed the stage-setting issues, the chapter goes on to describe specific examples. The first shows how ecosystem-dynamic equations, representing the continuum-hypothesis worldview, yield scale-related non-dimensional numbers. It also shows how the continuum-hypothesis equations can be embedded in modern state-space theory. This facilitates understanding the frequency response or time scales in ecological system representations characterized by a relatively large number of states. The second example is in the non-continuum or particulate setting where particles are separated by particle-separation length scales. The particulate setting enables calculation of the inter-particle length scale. It also enables analysis of patchiness. Availability of particle-separation length scales enables direct coupling between particle-separation length scales and physical length scales, such as the Kolmogorov microscale. Studying the consequence of patchiness or spatial variance enables deeper insight into the relationship between the mean and second order properties of plankton distributions. The study of particle-separation length scales also enables mapping the fundamental idea of the functional response used in all ecosystem models from the continuum setting to the particulate setting. This enables the development of a simple energy balance model that takes into account the energetics of feeding and how it might be coupled to physical forcing. Finally, to resolve the issue of non-axiomatic biological/chemical representations, a recollection of the Silvert-Platt size spectrum model is evoked.

I thank the editors Allan Robinson and Kenneth Brink for challenging me to write this chapter. It was interesting to explore the catenary that hangs between scale-space description and governing-equation derived scaling arguments. Comments from Michael Fasham and Allan Robinson are deeply appreciated.

## 2. Multiple Scales in Space and Time Issues

If a phenomenon is of constant magnitude in space and time, a single observation at any point in the time-space continuum would suffice to completely describe the phenomenon. The idea of scale attempts to define space-time characteristics of phenomena. It would seem that "scale" would be generally thought of in terms of a physical reference frame in time and space. However, the concept of "scale" is used in three ways. In the first, the term "scale" is used in a descriptive context. In the second, "scale" is defined in terms of a statistic (typically resulting from an autocovariance or autocorrelation)—that characterizes the spectral, temporal, or spatial organization of properties or phenomena. In the third, "scale" is defined on the basis of "governing equations." The definitions that relate to scale as a descriptor or as a statistical measure are self-contained. However, the scale definitions that relate to governing equations are not self-contained; the definitions depend upon the governing equations.

These different views of scale are superimposed upon a physical reality; the fundamental nature of phenomena may be characterized by their location in scale space. For example, the physical characteristics of fluid flow in a $cm^3$ box are different than those in a $km^3$ box even though the small-scale characteristics obtain throughout the larger box. As another example, the smallest time-scale window in the kinematics of populations is occupied by feeding or trophic interactions. On a relatively longer scale, growth integrates feeding. On a yet longer time-scale window, reproduction integrates growth. The very nature of scale-dependent variability often requires very different observations, analyses, and approaches as a function of the scale window. Yet, if we are truly interested in understanding the ocean as a system, then we need to understand ways of integrating the phenomena that occur over various scales. In physics the integration is often implicit in the equations of motion where scaling arguments are derived from the equations of motion.

In this section we discuss the characterization of scale. First, we describe the descriptive approach. This is followed by a discussion of the statistical characterization of scale. This involves the idea of time-series models or filters as well as point processes. We call special attention to how scale properties are generated or modified via differential equations, filters, or transfer functions. The third approach, theory-dependent scaling, is discussed in the context of the contrast between physically-realizable and non-physically-realizable approaches. An example of the multiscale variability is given in the context of the relation between the large-scale wind field and the small-scale flow field. After considering the different kinds of scales, we discuss the fundamental issues of the contrast between the continuum hypothesis and the way that it breaks down when considering the role of plankton in physical-biological interactions.

### *2.1 Descriptive Characterization*

The most common use of scale appears to be descriptive. The descriptive approach was evidently first described by Stommel and reported in the context of zooplankton by Haury et al., 1977. Variability in the physics, chemistry, and biology of the sea is observed on a range of scales that vary from planetary to microscopic; and it is often the case that as we view a phenomenon through a moving scale window, the nature of the variability changes. In the descriptive characterization, events are simply classified according to whether they are large or small or whether they are fast or slow. The motivation for considering scale arises from the fact that the phenomena that interest us have length and time scales that vary over many orders of magnitude. For example: 1) a microbe might be 1 μM in length and a fish might be $10^6$ μM in length; 2) the cosmology of the ocean and atmosphere is measured in $10^6$–$10^9$ years, while the life of a plankton might be $10^{-2}$ years; and 3) planetary waves are global in scale, while small-scale turbulent flow is suppressed at a scale smaller than 1 cm. The descriptive approach can be very misleading if it is taken to mean that all scale-space loci are isolated from other scale-space loci.

### 2.2 *Statistical Characterization*

The second scale-related definition involves the notion of spatial or temporal auto-correlation statistics. If a phenomenon is auto-correlated in space or time, then a length or time scale is implied. Spatial and temporal autocorrelations are often studied independently of one another using techniques that tend to differ. Spatial observations are generally made at irregular locations. The spatial continuum is estimated by weighting observations by the distance between adjacent observations and their variance. The relation between variability and distance is called a semivariogram and the procedure is referred to as kriging (e.g., Ripley, 1981; Cressie, 1991).

A completely different spatial technique relates to the number of points or particles in k-dimensional space (e.g., Stoyan et al., 1987). This is called the theory of point processes. It is important to note (and as we shall see subsequently) that the theory of point processes can generally be expressed in dual form: 1) in terms of the number of points in a volume, a density measure or a counting process; or 2) in terms of the intervals between points, an interval process. Autocovariances in the counting and interval process have been used to define spectral properties of distributions (Bartlett, 1963; Fasham, 1978).

It is important to note that the counting property has been traditionally used to consider plankton patchiness. But it is really the interval property, the inter-particle or inter-organism length scale, that is of physical significance. The interval property enables matching inter-particle length scales with physical length scales, such as the Batchelor or the Kolmogorov scale.

If events are autocorrelated in space, then this implies that their autocovariances are positive or negative for specific distances, wave numbers, or lags. The autocovariance property enables computation of a spectrum—a measure of scale—which indicates the distances, wavenumbers, or lags that contain the most variance or power. The distances, wave numbers, or lags are scales that characterize the spatial distribution of the phenomena of interest. The existence of spatial autocorrelation is often referred to as patchiness.

In contrast to spatial autocorrelations, temporal data are usually measured at equally spaced intervals yielding statistics on what might be called temporal patchiness. This is called time-series analysis. Techniques in time series have classically been associated with systems identification in the context of Box and Jenkins (1970). Their methodology has blossomed into systems identification that extends and generalizes the Box-Jenkins methodology to take into account input-output series and explicit modeling of relatively complex noise structures (Ljung, 1987). As physical observations move to smaller and smaller scales, their statistical properties become important, so the autocovariance of velocities are used to develop the theory of small-scale turbulence, for example.

**Generation of Scale Properties**

It is worthwhile to examine in a deeper way how statistical-scale properties can be generated. The autocovariance idea is motivated by the notion that phenomenon become correlated in time or space by an integrating or averaging process. A simple example involves the computation of a moving average from a time series. If we have a time series of random noise, for example, and we compute a moving average of three, then each point will contain information from at least three instants of time. If we use the same time series and generate a "moving average of five," each point will contain information at least five instants of time. The two time series, both depending upon the same identical series of random noise, will be different and contain different autocorrelation information. We can think of a moving-average-of-three population that smoothes it's forcing function by a moving-average-of-three as different from a moving-average-of-five population that smoothes the same identical forcing function by a moving-average-of-five.

To give a specific motivating example, let us consider an autoregressive order two "population" and examine how the autoregressive process converts random noise into a non-zero autocovariance and a spectrum that contains energy at a particular frequency or "scale." Random Gaussian data is filtered with a second order autoregressive filter $A(q)=1-.75q^{-1}+.5q^{-2}$ (where $q$ is the back shift operator). The random input data are shown in Fig. 3.2, and the filtered data are shown in Fig. 3.2. Visually the data look about the same. However, the covariance functions of the two data sets are quite different. The white noise covariance drops to zero, while the AR2 filtered data has a negative covariance at time lag 4 and a positive covariance at time lag 8. After time lag 8, the covariance settles around zero. This demonstrates the powerful notion that filtering data adds time-scale structure, and conversely observing autocovariances in data reflects the nature of the filtering process. The covariances enable computing a spectrum and hence scale-content, which shows not only the specific frequency characteristics of the filter but also the rate of change of variance as a function of frequency. It is clear that the moving average by three and five and the second order autoregressive populations (or processes) are just special cases of a more general scheme of filters or averaging mechanisms that either generate measurable scales from random inputs or otherwise modulate non-random inputs.

It should be intuitively clear that there are many approaches for generating scale-related properties or for averaging or integrating forcing. The number of points "averaged" connotes a time, distance, or length scale depending upon one's point of view. To demonstrate this, note that an averaging period is denoted by a back shift operator or polynomial, viz,

$$A(q)=1+a_1q^{-1}+\cdots+a_{na}q^{-na} \tag{1}$$

where $na$ is the order of the back-shift operator and $a_i$ are weighting or averaging coefficients. For example, to average the past two data points, we use a back shift operative of order two,

$$A(q) = 1 + a_1 q^{-1} + a_2 q^{-2} \tag{2}$$

This yields when combined with observations made at $t$, $t$-$1$, and $t$-$2$,

$$A(q)Y(t) = Y_t + a_1 Y_{t-1} + a_2 Y_{t-2} \tag{3}$$

If we add an error term, then equation 3 becomes the well-known autoregressive model of order two,

$$A(q)Y(t) = e(t) \tag{4}$$

The many approaches (that include the back-shift formalism) are generalized by the general parametric model (Ljung, 1987; see also MATLAB, Systems Identification Toolbox),

$$A(q)Y(t) = \frac{B(q)}{F(q)} u(t - nk) + \frac{C(q)}{D(q)} e(t) \tag{5}$$

where

$$\begin{aligned} A(q) &= 1 + a_1 q^{-1} + \cdots + a_{na} q^{-na} \\ B(q) &= b_1 + b_2 q^{-1} + \cdots + b_{nb} q^{-nb+1} \\ C(q) &= 1 + c_1 q^{-1} + \cdots + c_{nc} q^{-nc} \\ F(q) &= 1 + f_1 q^{-1} + \cdots + f_{nf} q^{-nf} \end{aligned} \tag{6}$$

$B(q)/F(q)$ and $C(q)/D(q)$ are called transfer functions and $u(t$-$nk)$ and $e(t)$ are inputs with delay $nk$ and random noise, respectively. $na$, $nb$, $nc$, and $nf$ are the orders of the polynomial. It is important that this approach partitions process noise and random noise. Setting any of the polynomials to either 0 or 1 results in a particular type of model or filter. For example, setting $B(q)=C(q)=0$ results in the autoregressive model. There is an important literature on choosing the appropriate polynomial for the process, $B(q)/F(q)$ and the noise $C(q)/D(q)$. These alternate model descriptions and procedures for their estimation are discussed by Ljung.

The main point of equation 5 and equation 6 is that they demonstrate the rich possibilities of studying time-scales in both process and associated noise. It is important to understand how a model or an averaging approach generates information in a time or space series and in turn, given the time or space series, how we can extract information on the nature of the model or the averaging process.

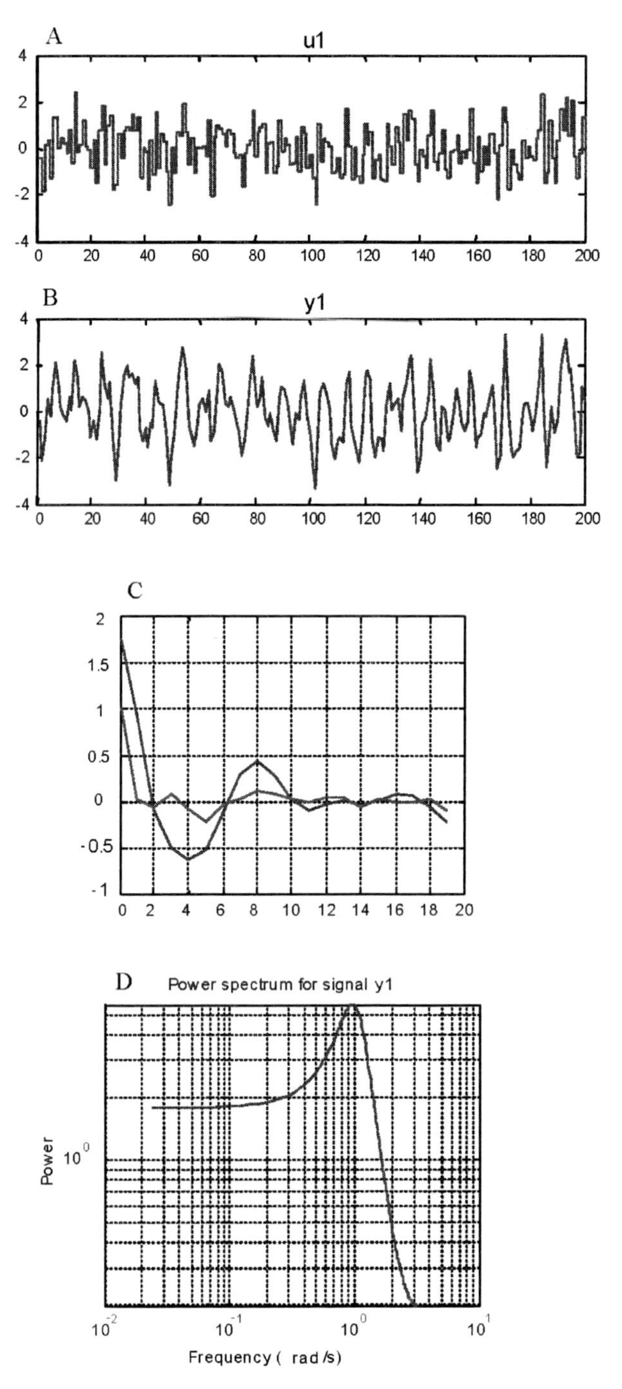


Figure. 3.2. The filter effect is a motivating idea in a discussion of statistical scale. Panel A is a time series of random Gaussian noise. The noise is filtered through an AR2 filter. The resulting time series is shown in Panel B. Panel C gives the covariance to lag 20 of the random input and the filtered output. We see that the covariance in the random series drops to zero at lag 1, while the covariance in the filtered series has a minimum at lag 4 and a maximum at lag 8. This generates a spectral peak at 1 rad/T. Incidentally, this particular model was used in an example by Box and Jenkins (1970).

The information content can be viewed in time-space or in frequency-space (i.e., spectrums). For example, it can be seen that because we are averaging across observations that previously uncorrelated error terms become correlated and that the orders (cf equation 6) all contribute to the lag-specific correlations and covariances. Further, the covariance corresponds directly to the spectrum of the process, which is yet another measure of scale, related, of course, to the slope of the spectrum, which can be taken as a function of frequency or wave number.

The identification of appropriate models and estimating their parameters, yielding information on temporal or spatial multiscale patchiness, can apply to almost any variable of interest. It is important to recognize that certain assumptions need to be taken into account, such as linearity and stationarity. It may be that data do not fit the underlying models, suggesting the use of other approaches, such as neural nets. However, these approaches may not yield scale information in the sense that they do not contain coherent "length" information. The covariance/spectral techniques are of course used in physical oceanography to study *inter alia* the correlation among velocities at particular wave numbers.

Finally, it should be mentioned as well that sampling alone could induce properties in system output as evidenced by the aliasing property of sampling below the Nyquist frequency.

**Duality Between Ecosystem Dynamic Equations and Filters**

An essential theme in this chapter involves the correspondence between differential equations—that might represent population or ecosystem dynamics under the continuum hypothesis—and transfer functions or filters. This correspondence demonstrates in a transparent way the relation between community matrices, such as those used in NPZ formulations, and scale-related frequency response. Take, for example, a linearized, forced ecosystem-dynamic system consisting of the differential equations,

$$\sum_{k=0}^{N} a_k \frac{d^k y(t)}{dt^k} = \sum_{k=0}^{L} b_k \frac{d^k u(t)}{dt^k} \tag{7}$$

If we take the Laplace transform of equation 7, we have,

$$Y(s)\left[\sum_{k=0}^{N} a_2 s^k\right] = U(s)\left[\sum_{k=0}^{L} b_k s^k\right] \tag{8}$$

This leads to the transfer function or filter,

$$H(s) = \frac{Y(s)}{U(s)} = \frac{\sum_{k=0}^{N} a_k s^k}{\sum_{k=0}^{L} b_k s^k} \tag{9}$$

There are several ways of studying equation 8, which is of course equivalent to studying equation 7. For example, a consideration of the poles and zeros of equation 8 indicates the system's relative stability and dynamic properties. By studying the root loci associated with the poles and zeros, it is possible to determine whether the intensity of the input, i.e., the system gain, can destabilize the system. It is worth commenting that many of the phenomena that are of concern are non-linear. However, it is common in engineering practice to consider linear approximations to non-linear processes. The linearization increases to a tremendous degree the scope for analysis and insights to system performance. There is extensive literature on the use of linear models to approximate non-lincar processes.

### *2.3 Physical and Model Dependent Characterization*

A fourth notion of scale relates to well-known physical scales or dimensionless numbers. Kantha and Clayson (2000) list 62 different physical scales or dimensionless numbers (see also, Cushman-Roisin, 1994). The idea of non-dimensionalization is a powerful, simplifying concept used in the analysis of physical variability. Numbers generally derive from physical equations, such as the comparison of advection to Coriolis force, giving the Rossby number, or from intuitive considerations, such as in the development of the Kolmogorov microscale. Statistical oceanography is somewhat related to this approach in the consideration of small-scale velocity correlations and the resulting spectral calculations.

It is interesting to observe that physical scales and non-dimensional numbers are often used in ocean physics but hardly used at all in ocean biology (see, however, Wroblewski et al., 1975; Okubo, 1977; Sverdrup, 1953; Kantha and Clayson, 2000) even though it is possible to derive seemingly powerful biological scaling relationships and non-dimensional numbers.

**The Multiscale**

The use of scale and scale-related arguments becomes richly complicated in the general case, as opposed to the narrowly, focused descriptive case, because, in general, we are more interested in the upward and downward cascades of energy that lie outside the descriptive frameworks. For example, energy in eddies dissipate as length scales become smaller and smaller, but larger eddies can also capture energy from smaller eddies through the process of enstrophy. A particularly interesting example of the multiscale interactions relates to the effects of large-scale to smaller-scale wind forcing on the magnitude of the Kolmogorov and Batchelor scales. Generally, the large-scale wind field affects the turbulent kinetic energy (TKE) dissipation rate (Rothschild, 1989). This affects the Batchelor, Kolmogorov microscale, Ozmidov, and Taylor microscale scales. All of these scales are defined on much smaller lengths than the basin-scale wind field and govern what seem to be extremely important biological-chemical processes, such as the interoganism length scales that are exposed to strictly Fickian diffusion or correlated or uncorrelated velocities in the flow field.

### *2.4 The Continuum Hypothesis as a Fundamental Scaling Argument*

In the consideration of scale, the student of multidisciplinary oceanography is confronted with an assumption that is generally taken for granted: the continuum hypothesis. Under the continuum hypothesis, it is argued in the study of fluid dynamics, that the particle structure of the fluid (i.e., the molecular structure) is irrelevant for the purpose of studying fluid flow, in the sense that any sensor used in measuring properties associated with the flow averages the properties of the many molecules that come in contact with the sensor (Batchelor, 1967). The explicit or implicit use of the continuum hypothesis relates to the situation where the examination of smoothed contour plots of "concentrations," such as salinity or zooplankton, is consonant with the continuum hypothesis.

The use of continuum hypothesis does not always make sense in the study of biological-physical interactions in the sea. This is because the higher-order plankton-separation length scales are critical to population dynamics (the higher order separation scales are the variances and covariances of the separation distance—it is these variances and covariances that measure "patchiness"). Most sampling sensors integrate distributions and therefore suppress the higher order properties of plankton distributions, the patch and inter-patch structures that are thought to be important in ecosystem dynamics. Understanding the variability in the dynamical interactions of the plankton particles requires information on the plankton separation length scales.

So in this particular context, the issue of scale emerges immediately as a key element in our discussion of oceanography. It is interesting that whether or not we are in a continuum or a particulate discussion, the same ideas apply to the analysis of scale. In this context, scale is determined in the context of how phenomena are correlated with one another in time and space. The basic ideas are that if phenomena are spatially or temporally independent, then the location of sampling in time or space is unimportant. One approach to measuring the spatial independence involves covariance in time or space. It is particularly interesting to observe that the process by which the organisms, the physics, or the observer average the temporal or spatial data can create spatial and or temporal covariance.

In the continuum scale, critical information on spatial-temporal structure and, in particular, a capability to respond to intermittency in physical forcing are not considered. For example, in a stochastic sense, if two processes have the same mean, then it will take longer for a high-variance patchy process to reach its statistical equilibrium than a low-variance non-patchy process. If smaller-scale forcing persists, the high-variance process may never equilibrate. We can readily see that even in terms of kinematic description, the omnipresent scale-space diagrams fall short of enabling us to integrate dynamical issues. While these diagrams enable us to describe how big and small various phenomena are, they gloss over the point that most large phenomena are intermittently related to small phenomena and vice versa. For example, the planetary wind field drives local small-scale turbulent flow; large zooplankton and fish have critical stages in their life history that are hundreds of times quicker and smaller than the adult stages.

Before we can study the physical manifestations of scale and multiscale, it is important to understand how scale is measured, assessed, and evaluated independent of specific phenomena. It is important to understand the contrast between observations of the fluid and observations of the biology at the two polar extremes of our knowledge base. For the fluid we have physics and a fundamental dynamical understanding, while for the biology we have an ad hoc dynamical understanding.

## 3. Population and Ecosystem Dynamics in the Continuum Setting—Natural Time Scales

Our comments on the continuum setting are not intended to mean that it should be rejected or bypassed. Rather, that as we move into multidisciplinary oceanography, the particulate setting also needs to be taken into account. In the continuum setting, the population and ecosystem dynamics equations traditionally studied by ecologists are considered. The equations are called coupled ordinary differential equations, and they relate the time derivative of various populations or population groups to population abundance and rates of population growth (called the intrinsic rate of increase) and transfer between populations (the functional response). This class of equations was evidently first developed by Lotka and Volterra (see Murray, 1989). The study of the equations is generally focused upon versions that are considered as homogeneous in zero-dimensional space. The equations are often embedded in continuum flow field equations, and so any expression of concentration must be averaged in space and time. The embedded equivalents of the ecosystem dynamic equations have "internal" temporal scales that need to be studied in the evaluation of numerical results.

To develop the flavor of these ideas, we will show how population-dynamics equations are scaled to yield dimensionless numbers and demonstrate a state-space framework for studying forced responses. The state-space framework yields time-scale related information. A particularly convenient example is provided by Murray (1989). This example provides a framework for demonstrating the natural time scale of population-dynamic or ecosystem-dynamic systems and a non-dimensionalized or scaling argument for such systems.

$$\dot{N} = N\left[r\left(1-\frac{N}{K}\right)-\frac{kP}{N+D}\right] \tag{10}$$

$$\dot{P} = P\left[s\left(1-\frac{hP}{N}\right)\right] \tag{11}$$

Where $P$ is the predator, $N$ is the prey, $K$ is the carrying capacity, $D$ is the half-saturation constant, $r$ is the intrinsic rate of increase of prey, and $s$ is the intrinsic rate of increase of the predator. Murray reduces the number of parameters by non-dimensionalization,

$$u(\tau)=\frac{N(t)}{K}, v(\tau)=\frac{hP(t)}{K}, \tau = rt \tag{12}$$

$$a = \frac{k}{hr}, b = \frac{s}{r}, d = \frac{D}{K} \tag{13}$$

The non-dimensionalization scaling simplifies the interpretations of equation 10 and equation 11. But perhaps more importantly, it lends insight into the underlying dynamics. For example, $u(\tau)$ and $v(\tau)$ are now normalized to the carrying capacity.

The constant $h$ is the proportionality between the predator and prey given the carry capacity, and the constant $b$ characterizes the relative growth capability of the predator and prey. The non-dimensionalization transforms equations 10 and 11 to,

$$\dot{u} = u(1-u) - \frac{auv}{u+d} \tag{14}$$

$$\dot{v} = bv\left(1 - \frac{v}{u}\right) \tag{15}$$

Murray solves for the steady states $(\dot{u} = \dot{v} = 0), u^*, v^*$, to obtain

$$u^* = \frac{(a+d+1) + \sqrt{(a+d+1)^2 + 4d}}{2}, \qquad v^* = u^* \tag{16}$$

The dynamics can now be written in matrix form as,

$$\begin{bmatrix} \dot{u} \\ \dot{v} \end{bmatrix} = \boldsymbol{A} \begin{bmatrix} \mu \\ v \end{bmatrix} \tag{17}$$

where $N$ and $P$ give the state of the system and $\boldsymbol{A}$ is the so-called community matrix. The non-linear system is linearized by taking the Jacobian of $\boldsymbol{A}$ and evaluating it at the steady state, viz,

$$\begin{bmatrix} \frac{\partial f}{\partial u} & \frac{\partial f}{\partial v} \\ \frac{\partial g}{\partial u} & \frac{\partial g}{\partial v} \end{bmatrix} = \begin{bmatrix} u^* \left[ \frac{au^*}{(u^*+d)^2} - 1 \right] & \frac{-au^*}{u^*+d} \\ b & -b \end{bmatrix} \tag{18}$$

As is well known, the dynamics of this linear system depend upon the relative position of the trace and determinant of equation 18. This is described in Fig. 3.3, and it can be seen that changing values in the parameter set changes the steady state output $u^*$, $v^*$, and the dynamics of the system. These dynamics can be exceedingly important to understand not only the behavior of the system, but how the particular system under study relates to other systems.

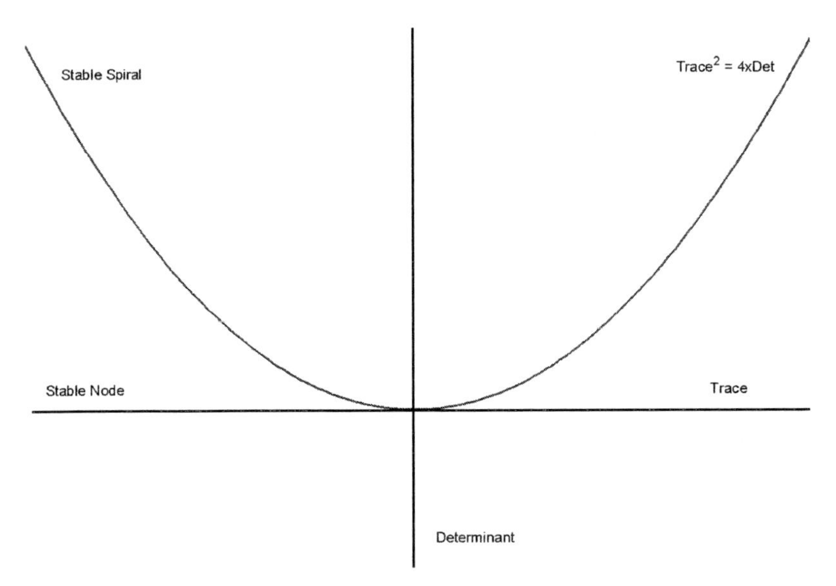


Figure. 3.3. Dynamics associated with trace determinant space. The location of trace-determinant values are indicative of dynamics behavior (e.g., stability versus instability, spirals, saddles, and nodes). An important consideration is that when the trace-determinant point is close to boundary (e.g., the discriminant parabola), the system becomes relatively sensitive to changing its characteristic frequency.

This is particularly important because we can see how varying the parameters *a, b,* and *d* can affect the dynamic behavior of the system. (From a qualitative point of view, how does physical forcing affect *a, b, d?*) Classic ecology generally focuses upon whether a particular parameter set yields stability. However, perturbation of the parameter space also affects system efficiency (e.g., the magnitude of the steady state), and dynamic characteristics, such as whether trajectories oscillate, for example (Fig. 3.4). In addition, the sensitivity of the system's dynamic behavior is related to the location of the trace and determinant on the trace-determinant plane. For example, if the trace is negative but close to zero, we can expect that the system has a relatively high propensity to become unstable.

The analysis of linearized systems in the ecological literature generally stops at pointing out the dynamic properties of systems in terms of stability and, in some cases, in terms of dynamic behavior. This traditional approach tends to concentrate only on the initial condition response of the dynamic equation. But from the perspective of attempting to understand the forced response of the system, the natural frequency of the population-dynamic system needs to be considered (see, however, Nisbet and Gurney, 1982). Put another way, as the parameters of equation 10 and equation 11 change, its frequency response also changes. In this context, we might be interested in how the equation system implied by equations 10 and 11 might vary dynamically with respect to periodic tidal or diurnal forcing or quasi-periodic effects of wind forcing.

This analytic analysis seems particularly important in thinking about embedding ecological models in numerical models of the flow field. This is because physical forcing in either an analytic or a numerical model sense can create or destroy steady states and other dynamics (e.g., oscillations), which are different than those in an unforced system. The difference between the forced and unforced system illustrates the physical-forcing effect. This has important potential usefulness in studying how "global change," for example, might affect the ecosystem.

The extension of the traditional approach from the initial-condition response to forced responses is particularly relevant to utilize methodologies for studying forced responses of systems of coupled differential equations. Considerable insights into forcing, control, stability, complexity, and a myriad of other issues can be developed by embedding the community matrix $\boldsymbol{A}$ in the state-space formalism. Of these properties, the frequency response is particularly relevant to the study of time scales. The state-space formalism is the "modern" approach for studying the effects of forcing and the forced response. There are many books available on this general subject (e.g., Ogata, 1990). The state-space approach facilitates dealing with large community matrices (e.g., 10 x 10), multiple inputs and outputs, process and observation noise, and various forms of feedback. Let us consider the so-called community matrix $\boldsymbol{A}$, which is defined in equation 17. The state-space model is,

$$\dot{x}(t) = \boldsymbol{A}x(t) + \boldsymbol{B}x(t) \tag{19}$$

$$y = \boldsymbol{C}x + \boldsymbol{D}u \tag{20}$$

The elements of $\boldsymbol{B}$ modify the forcing rates, while $\boldsymbol{C}$ and $\boldsymbol{D}$ modify measurements of the system state. There is a large engineering literature on the analysis of state-space systems. It is worth referring to the general solution of equations 19 and 20, particularly inasmuch as it leads to the scale-relevant issue of the frequency response of community matrices, such as $\boldsymbol{A}$. The solution of equations 19 and 20 in the context of forced response is facilitated through the use of the Laplace transforms. Equations 19 and 20 can be solved using Laplace transforms. The transform of equations 19 and 20 is

$$sX(s) - x(o) = \mathbf{A}\mathbf{X}(s) + \boldsymbol{B}U(s) \tag{21}$$

which can be rewritten,

$$x(s) = \Phi(s)x(o) + \Phi(s)\boldsymbol{B}U(s) \tag{22}$$

where

$$\Phi(s) = (s\boldsymbol{I} - \boldsymbol{A})^{-1} \tag{23}$$

Inverting yields,

$$x(t) = \Phi(t)x(o) + \int_o^t \Phi(t-\tau)\boldsymbol{B}u(\tau)dT \tag{24}$$

where the first term on the right is the initial condition response and the second term is the forced response. The transfer function system output to input is given by

$$H(s) = \boldsymbol{C}\Phi(s)\boldsymbol{B} + \boldsymbol{D} \tag{25}$$

A

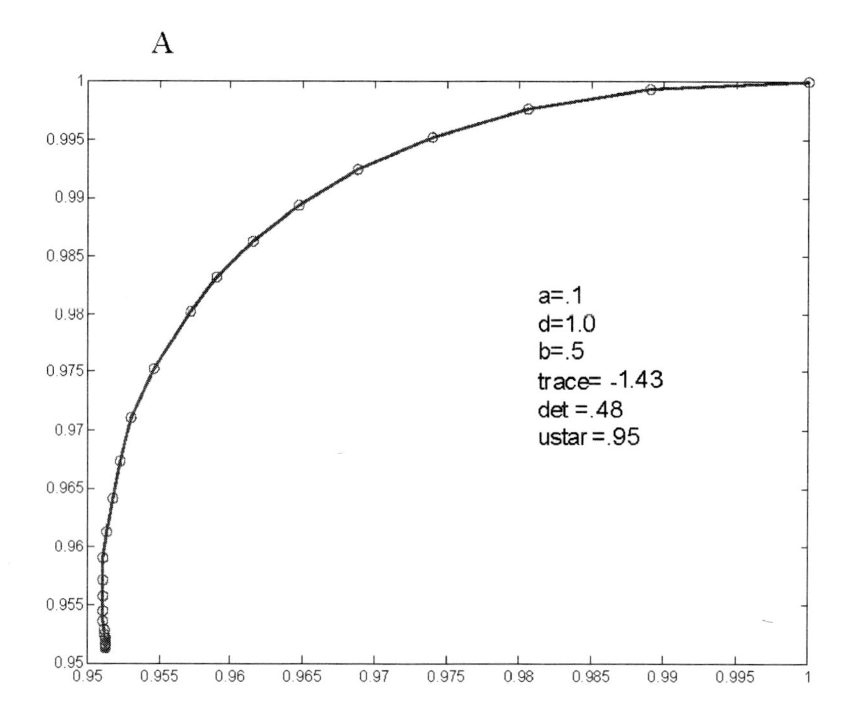


B

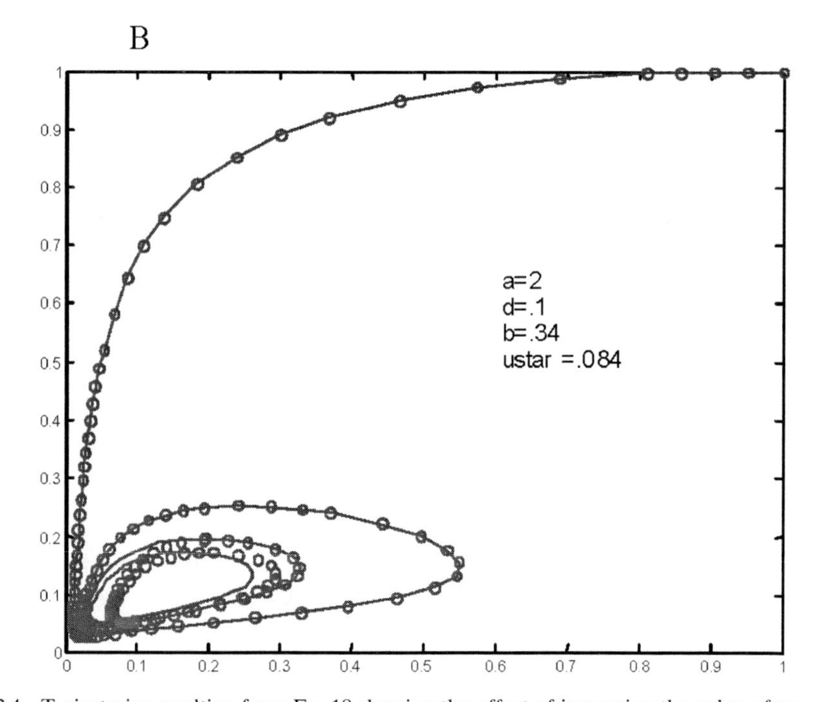


Figure 3.4. Trajectories resulting from Eq. 18 showing the effect of increasing the value of trace. As can be seen from Figure 3.3, increasing the value of trace develops an oscillatory trajectory. Note that in Panel A the trace = 1.43, while in Panel B the trace = ____.

and, most importantly, for the purposes of this chapter, the frequency response or time scale of the system represented by equations 19 and 20 is obtained by replacing $s$ in equation 25 with the frequency $i\omega$.

$$\boldsymbol{H}(i\omega) = \boldsymbol{C\Phi}(i\omega)\boldsymbol{B} + \boldsymbol{D} \tag{26}$$

So now we have shown how all systems of coupled linear differential equations representing ecosystems have an "internal clock" or time scale that responds differently to different forcing frequencies. As an example, Fig. 3.5 shows the frequency response for the parameter set used in Fig. 3.4B. It can be seen that a frequency of .7 rad per unit time is resonant. In this particular case, this means that the system is particularly productive when forced by the resonant frequency. We can ask the inverse question, what rates in $\boldsymbol{A}$ are consistent with diurnal or tidal frequencies, for example? In other words, we can design a pass-band filter that is resonant at .7 rad per unit time and inspect the set of parameters in the ecosystem dynamic equations that yield this particular frequency response.

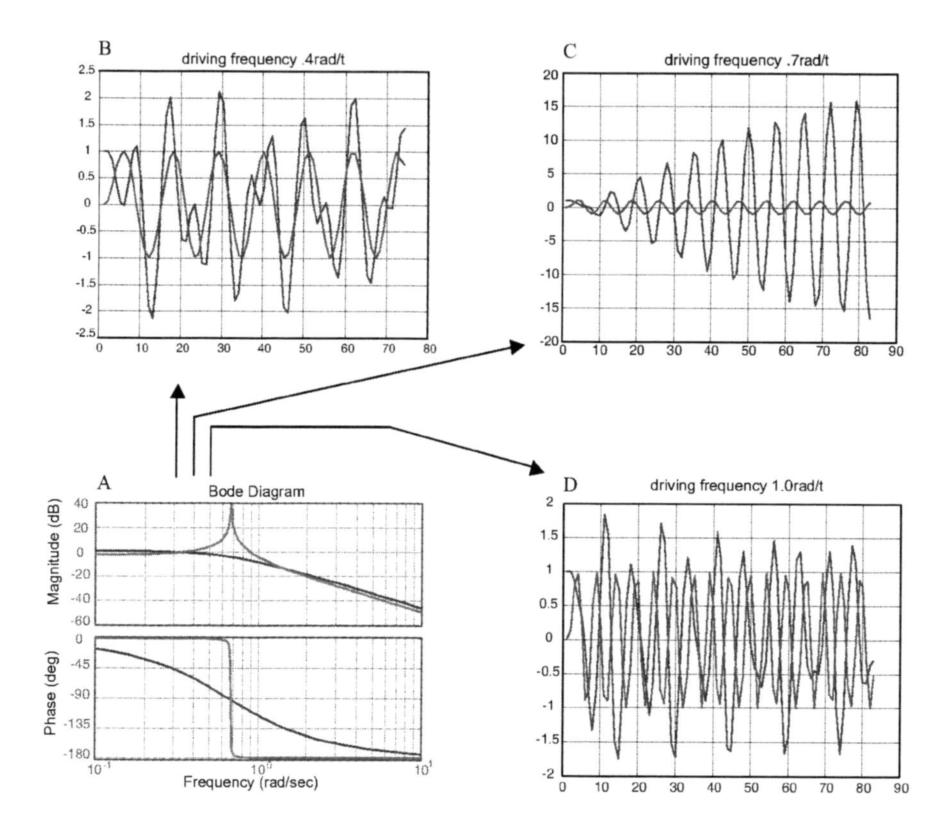


Figure 3.5. The forced response of equations 14 and 15 corresponding to the parameter set in Figure 3.4B. Panel A is a Bode diagram that shows that the parameter set has a resonant response at .7 rad/T. Panel B shows the response to sub-resonant frequency .4 rad/T, Panel D to supra-resonant frequency 1 rad/T, and C to the resonant frequency. Note that the varying vertical scales and that the amplitude of the resonant response is much greater than the non-resonant response.

## 4. Population and Ecosystem Dynamics in the Particulate Setting—Inter-Particle Length Scales

The preceding discussion was phrased in the context of the continuum hypothesis. The continuum hypothesis implicitly assumes that the averaging of concentrations of particulate points that contribute to the continuum does not suppress variability needed to explain population dynamic variability. In particular, we need to take into account the spatial arrangement of organisms on scales critical to understanding population-dynamic or ecosystem-dynamic variability. The importance of these small scales can be judged from the universal use of the functional response (e.g., third term on the right in equation 10) in all continuum ecosystem dynamic equations.

In fact, the continuum class of models further testifies to the need for inter-particle length-scale representations. This is because in continuum representations, nutrient donors and nutrient acceptors are subject to sources of forcing, which are not completely independent. The first is what might be called "scalar forcing" where temperature, for example, affects the rate-of-increase response. This is in contrast with "flow-field forcing" that alters the relative prey-predator velocity to affect the functional response. (An inability to logically separate growth and feeding results in part from the fact that they operate in different time scales.)

To put the idea of inter-particle-separation length scales in context, consider Fig. 3.6. Figure 3.6 gives two examples of encounter times or distances for six particles. In a continuum setting each example would be equivalent, but in a particulate setting the two examples are quite different. From a biological point of view, the functional response would generally be different between the two examples.

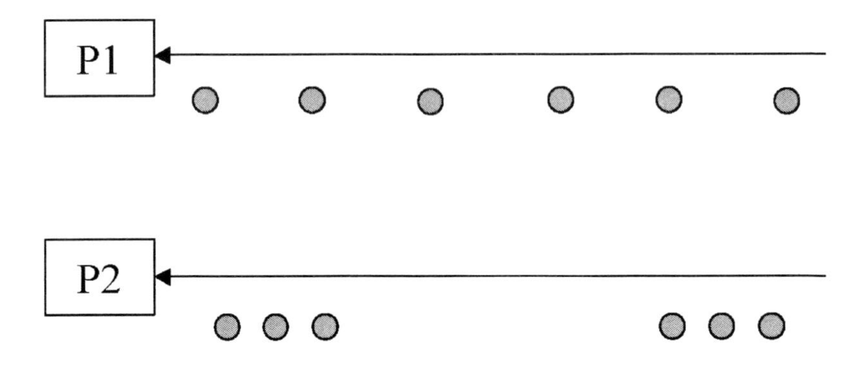


Fig.ure 3.6. Two predators each encountering six prey. The encounter structure for the two predators is fundamentally different. Even though both predators encounter six prey, the variance or second-order properties of the interprey spacing differs.

This observation is of general interest. But, more importantly, it should be recalled (Fig. 3.3) that very small changes in population-parameter values (as might be induced by the groupings in Fig. 3.6) could induce substantial changes in quantitative and qualitative dynamics (e.g., from stable to unstable).

It is worth noting an additional manifestation of scale that we can see from Fig. 3.6. This multiscale manifestation relates to the "observational window" of the observer or the predator. Long observation windows average the inter-particle spacing, while short observation windows emphasize the spacing. Organisms that feed diurnally, for example, have different observation windows than organisms that feed continually. Larger windows in space or time average intermittence and lose resolution. Higher resolution may reveal links between a forcing function and its response that may otherwise be hidden. At the same time, a level of resolution that is too great can shroud a key relationship in noise. Addressing these issues requires an examination of particulate theory.

The distribution of randomly distributed particles or randomly shaped objects in space is analyzed in the realm of stochastic geometry (Stoyan et al., 1987; Rothschild, 1992). In the case of predator-prey interactions, the special case of point processes can be applied. Discussions of the particulate setting generally begin with the completely random process, the Poisson process. The Poisson probability density function is

$$p(k;\lambda t) = e^{-\lambda t}\frac{(\lambda t)^k}{k!} \tag{27}$$

This is the probability density of finding $k$ points in an interval of length $t$ (Feller, 1957). The intensity of the process $\lambda$ can be thought of as the mean number of points per unit length or time.

While the number of points or particles per unit length or time is of considerable biological interest, the distances between points is of more interest, particularly if we are relating the distribution of particles to physical forcing, that is to physical length scales. We move from the counting measure implied by equation 27 to the length-scale measure by noting that the probability that zero points in an interval of length $t$ is

$$p(0;\lambda t) = e^{-\lambda t} \tag{28}$$

This leads us to the fundamental idea used to develop the duality between counts or densities of particles to distances between particles in length or time. If $X$ is a random variable denoting the time or distance from $t_0$ to the first particle,

$$P[X > x] = P(\textit{number of zero points in the interval from } t_0 \textit{ to } t) = e^{-\lambda t} \tag{29}$$

Hence, the probability of at least one point between $0$ and $t$ is $1\text{-}e^{-\lambda t}$. This then gives the distribution and density functions for the intervals, respectively,

$$F_X(x) = 1 - e^{-\lambda t} \tag{30}$$

$$F'_X(x) = f_X(x) = \lambda e^{-\lambda t} \tag{31}$$

So we have derived the exponential distribution and density functions, equations 30 and 31, from the Poisson distribution demonstrating the duality between counts and intervals or "lengths."

While the Poisson distribution is based upon the number of particles in an interval, the exponential distribution gives the probability density of the time or distance between particles. It is precisely this approach that enables us to derive the important length scale, the Poisson distribution nearest neighbor distance (Rothschild, 1992; Pielou, 1969).

As might be appreciated, there are many different point processes that can be derived from the Poisson process. For example, clustering models describe the distribution of "children points" about a "parent point." Doubly stochastic processes possess a parameter $\lambda$ that is a function of some continuous process. For contagious processes, space-time position of the points are interdependent (Cox and Isham, 1980; Daley and Vere-Jones, 1988). These processes are mathematically complex. In addition, an empirical capability to distinguish among the processes may be difficult.

At this point in our understanding of the underlying biological processes, predominantly in the context of scale, it seems particularly worthwhile to build from the simplest idea of the random Poisson point-process distribution of food particles and its natural first generalization, the hyper-exponential distribution. The hyper-exponential distribution of inter-particle distances generally represent patch structures or patchiness. This distribution and its related moments as applied to plankton were studied in some detail in an important paper by Beyer and Nielsen (1996) (see also Rothschild, 1991). Again, to generalize, let us assume that the intervals between the particles are not drawn from a single exponential distribution, rather, they are drawn from a mixture of exponential distributions, with a probability density function,

$$f_T(t) = \theta\rho_1 e^{-\rho_1 t} + (1-\theta)\rho_2 e^{-\rho_2 t}, 0 \geq \theta \geq 1 \tag{32}$$

where $\rho_i$ is the parameter of the $i$-th exponential distribution and $\theta$ is the mixing parameter. The random variable, the inter-particle distance, is denoted by $T$. In our discussion, $T$ is a distance or a length, but it can also be a time, a so-called interannual time for an observer traveling at a fixed velocity through an n-dimensioned field of particles.

Given the density function, the mean and variance are derived in the usual way, yielding

$$E(T) = \frac{\theta}{\rho_1} + \frac{1-\theta}{\rho_2} \tag{33}$$

and

$$VAR(T) = 2\left(\frac{\theta}{\rho_1^2} + \frac{1-\theta}{\rho_2^2}\right) - \left(\frac{\theta}{\rho_1} + \frac{(1-\theta)}{\rho_2}\right)^2 \tag{34}$$

We can see that the mean is now a weighted mean, where the weight is given by the magnitude of $\theta$. Any increase in $\rho_p$, suitably weighted, decreases the mean time or distance between particles (the more particles, the closer they are together). The variance in the inter-particle time or distance, relates to the value of $\theta$. If $\theta$ is small, then the highest variance is contributed by low values of $\rho_2$, but if $\theta$ is large, the highest variance is contributed by low value of $\rho_1$.

Having demonstrated the properties of the hyperexponential length scale, we can now turn our attention to incorporating the hyperexponential length scale in *renewal-process* theory so that questions, such as, given some particular hyperexponential distribution, how long will it take to encounter $r$ particles, or the dual, how many particles will be encountered in an interval of length $T$? Put another way, how many prey are encountered in a particular period of time, or what is the length of time that it will take to encounter a particular number of prey? Put in physically realizable terms, we can consider the total energy acquired for a specific period of time or the time that it takes to acquire a particular level of energy.

Our next step is to solve the dual problem of determining the total time it takes to encounter $r$ prey, $S_r$, or the mean number of prey that will be encountered in a period of time of length $t$ (or a distance equivalent to $t$), $H(t)$, using the flexible hyperexponential distribution formalism. The calculations are based upon renewal theory (Cox, 1962). In fact, $H(t)$ is called a renewal function.

To illustrate the structure of the problem, Fig. 3.6 can be formalized by drawing a diagram of the particle encounter process, Fig. 3.7. In Fig. 3.7 we have three encounters, with a time or distance interval between each encounter of $T_i$ units. The $T_i$ are random variables drawn from the hyperexponential distribution. Symbolically we have the total time or distance to the $r$-th renewal.

$$S_r = T_1 + T_2 + \cdots + T_r \tag{35}$$

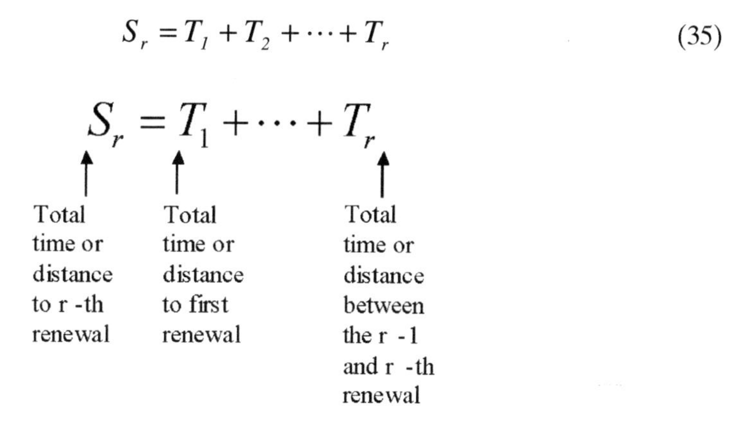


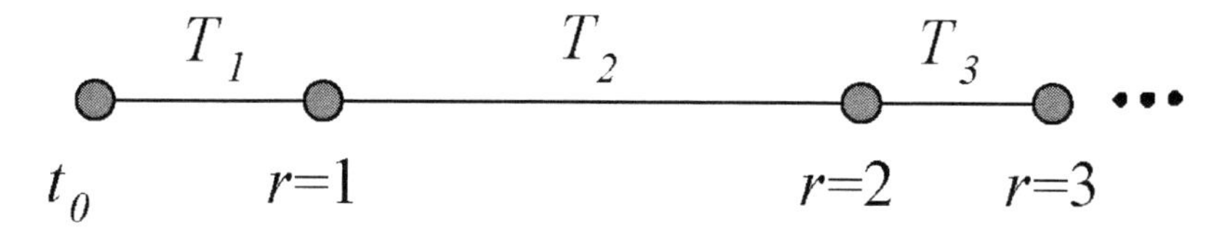


Figure 3.7. Diagram illustrating definitions of a renewal process *Sr* where *Sr* is the sum of *r* random variables. There are two problems: 1) how long in time or distance will it take to encounter *r* particles?; and 2) how many particles will be encountered in a period of time or distance *t*?

It should be mentioned that equation 35 is an *ordinary renewal process* because the clock starts running at a point in time to when an encounter actually occurs. Another type of renewal process is called the *equilibrium renewal process,* where the clock begins to run at a point in time in the distant past. The third type of renewal process is called the *modified renewal process.* It is the same as an ordinary renewal process except for the first interval, which is defined to have a different set of parameters or a different p.d.f. than subsequent intervals.

To specify the dual circumstances of determining the time to the $r$-th encounter, or the number of encounters at time $t$, note that,

$$P[N_t < r] = P[S_r > t] = 1 - K_r(t) \tag{36}$$

This implies that

$$P[N_t = r] = K_r(t) - K_{r+1}(t) \tag{37}$$

where $K_r(t)$ is the distribution function of $S_r$.

To obtain the p.d.f. of the sum of $n$ random variables, we need to obtain the $n$-fold convolution of the p.d.f.s that comprise the sum. But for even small values of $n$, the analytic convolution is computational cumbersome. However, using the fact that the $n$-fold convolution of the R.V.s is the product of the $n$ Laplace transforms of individual p.d.f.s, viz,

$$f^*_{(n)}(s) = \{f^*(s)\}^n \tag{38}$$

facilitates development of computationally feasible renewal functions.

To introduce the idea, it is simpler for expository purposes to work with the ordinary renewal process. This yields the Laplace transform time to the $r$-th renewal as,

$$k^*_r = [f^*(s)]^r \tag{39}$$

To derive the renewal function or the number of renewals at time, $t$, we note that the mean number of encounters or recurrences at time, $t$, is,

$$H(t) = \sum_{r=0}^{\infty} rP[N_t = r] = \sum_{r=1}^{\infty} K_r(t) \tag{40}$$

The Laplace transform of the sum of the p.d.f.s is consequently

$$H^*(s) = \frac{1}{s} \sum_{r=1}^{\infty} k^*_r(s) \tag{41}$$

It follows because $k_r(s)=[f^*(s)]^r$, that the infinite sum in equation 41 becomes

$$H^*(s)=\frac{f^*(s)}{s(1-f^*(s))} \tag{42}$$

Now the flexible (with respect to patchiness) hyperexponential distribution can be used to exemplify the theory.

The Laplace transform of the hyperexponential distribution is

$$f^*(s)=\frac{\theta\rho_1}{s+\rho_1}+\frac{(1-\theta)\rho_2}{s+\rho_2} \tag{43}$$

Substituting equation 43 in equation 42, we have the Laplace transform for the ordinary renewal process where the renewals are governed by the hyperexponential distribution,

$$H^*(s)=\frac{\frac{\theta\rho_1}{s+\rho_1}+\frac{(1-\theta)\rho_2}{s+\rho_2}}{s\left(1-\frac{\theta\rho_1}{s+\rho_1}-\frac{(1-\theta)\rho_2}{s+\rho_2}\right)} \tag{44}$$

Which upon inversion yields,

$$H(t)=\frac{\rho_2\rho_1 t}{-\rho_1+\theta\rho_1-\rho_2\theta}$$

$$+\frac{2\theta(\rho_1-\rho_2)^2(-1+\theta)\sinh\left(\left(-\frac{1}{2}\rho_1+\frac{1}{2}\theta\rho_1-\frac{1}{2}\rho_2\theta\right)t\right)e^{((-1/2\rho_1+1/2\theta\rho_1-1/2\rho_2\theta)t)}}{(-\rho_1+\theta\rho_1-\rho_2\theta)^2} \tag{45}$$

Substituting specific values for $\theta$, $\rho_1$, and $\rho_2$ gives an equation of the form,

$$H(t)=at-b\sinh(ct)e^{-ct} \tag{46}$$

This enables contrasting high-variance and low-variance examples of the renewal process. First, we show individual realizations of the stochastic renewal process and then the theoretical result, which gives the theoretical "average" overall realizations. Figure 3.8 shows two individual realizations from the high-variance and low-variance cases. (Equations 33 and 34 enable choosing a low-variance and high-variance parameter set for demonstration purposes: $\theta=.2$, $\rho_1=2$, $\rho_2=.88$ and $\theta=.2$, $\rho_1=.2$ and $\rho_2=16.8$, both yield means of 1 cm and low and high variances, respectively). Both realizations are constrained to a mean inter-particle distance of 1 time unit (in inter-particle distance terms, this is equivalent to the Kolmogorov microscale where the particles are spaced 1 cm apart). We can see that the high-variance case reflects a series of "patches" in contrast to the low-variance case. The theoretical result is plotted in Fig. 3.9. We can see that the high-variance case given an elevated encounter rate over the low-variance case for relatively short time periods and that the fact that the mean interval is equal to one is evident over a longer time period. This convergence is expected for long periods of time, and the rate of convergence is similar to an integral scale. This points again to the importance of taking into account intermittency in forcing functions in the sense that this intermittency disturbs the equilibrium and restarts the initial condition clock. Thus we have solved an important problem in scale-related point process theory. This problem relates to the difference between renewal functions that have different variances associated with their renewals but the same mean. In other words, how does different degrees of patchiness affect the encounters of the same number of organisms? (This is translated, *inter alia,* into modified encounters for predators and modified survival rates for prey.) This is an interesting length or time-scale problem as well. It addresses the length or time scales during which the patchiness effect is important inasmuch as the mean encounters forget variance or patchiness as time increases, as is reasonable.

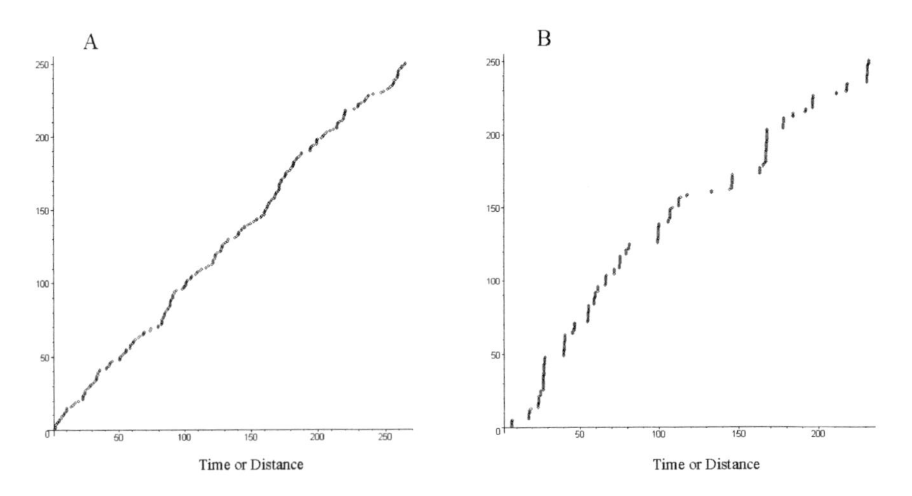


Figure 3.8. Realizations of the hyperexponential renewal process. The number of particles encountered is plotted as a function of time or distance. The distance between particles is drawn from the hyperexponential distribution. The sum of time or distance at any point is the renewed process. The mean inter-particle time or distance was set at unity. Panel A is a low-variance realization. Panel B is a high-variance realization; note its patchy distribution.

An important scale-related point that can now be introduced is the relative length of the predator or observer scale and the inter-particle distance or times of the prey particles. This scale-related issue is exemplified by the forward recurrence time. In other words, in general, if the visual predator (for example) begins its search in the morning with the onset of some light threshold, then the distribution

particle length scales greater than 1 cm allow the particles to be exposed to uncorrelated velocities in the flow field, while inter-particle length scales of less than 1 cm shelter particles from uncorrelated velocities. To examine the biological aspects of this conclusion would be a fascinating problem.

### *4.2 Energetics of Feeding*

Up to this point, we have studied the number of particles that a predator would encounter on a one-dimensional line as it swims at a particular velocity. Encounter is of course only the first part of the problem. Consumption is the second part. According to the idea of the fractional response, not every prey is eaten. Now we can ask, what is the benefit to the predator as it swims along the one-dimensional track? The energy costs of swimming are proportional to the drag of the predator. These costs can be measured in Joules. The benefits accrue from energy gains from ingested particles, also measured in Joules. It is generally agreed that the predator cannot eat all of the particles that it encounters because of the functional response (this is the $NPk/(N+D)$ term on equation 10). The geometry of the encounter theory is sketched in Fig. 3.10.

Now we can derive the particle ingestion versus the particle encounter to define a refractory period when the prey organism cannot eat, an idea aligned with the notion of a functional response. If the refractory period is exponentially distributed, then the refractory period and the inter-particle distance are the convolution of the probability density functions of both. This results in the Laplace transform,

$$H^*(s) = \frac{f_1^*(s)}{s\{1 - f_1^*(s) f_2^*(s)\}} \tag{48}$$

where $f_1^*(s)$ is the Laplace transform of the inter-particle length-scale density function and $f_2^*(s)$ is the density function of the refractory period.

Stochastic Version of Functional Response

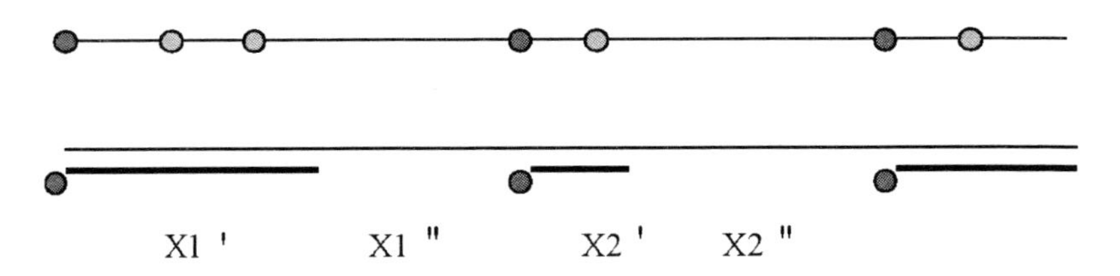


Figure 3.10 Stochastic version of functional response. The PDF of arrivals and PDF of refractory time results in a PDF of "eatens" that is the sum of the two RVs.

If we use the hypergeometric distribution for the inter-particle distances and the exponential distribution for the refractory time or distance, we have

$$H^*(s) = \frac{\dfrac{\theta\rho_1}{s+\rho_1} + \dfrac{(1-\theta)\rho_2}{s+\rho_2}}{s\left(1 - \dfrac{\zeta\left(\dfrac{\theta\rho_1}{s+\rho_1} + \dfrac{(1-\theta)\rho_2}{s+\rho_2}\right)}{s+\zeta}\right)} \tag{49}$$

The inverse transform is a very lengthy algebraic expression (available from the author). When $\zeta$ is taken to be $\zeta=.1$, then the curves in Fig. 3.11 result. $\zeta$ is chosen to give a relatively long refractory time. The figure shows the effect of the functional response and how different levels of variance—the patch structure—affect the amount of food eaten, an interpretation containing more information than in the continuum theory.

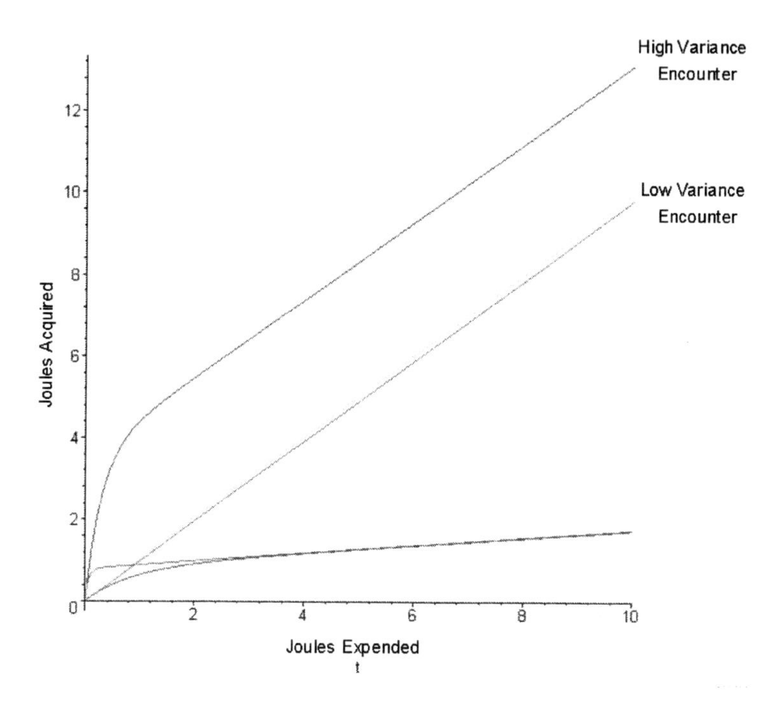


Figure 3.11 The effect of the "functional response" on particle ingestion. The high- and low-variance case in Figure 3.10 are shown in the upper two curves. The consumed-particle process is given by the lower two curves. The expected consumption is slightly higher for the high variance case. The similarity of the high-variance and low-variance consumption is related to the long refractory period implied by setting $\zeta=.1$.

## 5. Alternative Ecosystem Models—Basis for Physically Realizable "Numbers"

As pointed out earlier, issues of scale and non-dimensional numbers are difficult to consider independently from underlying governing equations. In this regard, it is necessary to consider the general ODE approach typified by equations 10 and 11 or numerical approaches that embed equations similar to equations 10 and 11. In this regard, the general ODE approach is not consistent with a physically realizable description of an ecosystem in the sense that the equations are not driven by physical laws (in contrast with the Navier-Stokes equations). While the ODE equations result in kinematic trajectories, they do not account for the metabolic efficiency of the ecosystem. To be fully dynamic, the models must relate to the flux of energy through the biodynamic system and how this flux of energy relates to physical forcing. How efficient is the community metabolism engine? To what degree is the sun's energy captured and portioned among biomass and metabolic byproducts? Answering these questions with respect to entire ecosystems would lend new insights into understanding the dynamics of sinking export, the degree to which iron limitation suppresses ecosystem efficiency, the role of heterotrophs in modifying and shaping primary production, and the fluctuations of fish populations.

Various extensions can be thought of that make the general ODE approach more physically realizable. One approach might be, for example, to simply convert biomass to energy units (i.e., Joules). This transformation can be viewed as simple or complex. The simple approach involves simply multiplying the numbers by constants and the biomass by a measure of energy density. The complex view is that the transformation can involve cubic functions or more accurately approximate-cubic functions. Careful modeling would need to take into account the bookkeeping among numbers of organisms (important at the egg stage, for example), biomass, and energy density. Other improvements can be made in terms of taking into account physical forcing by parameterizing the intrinsic rate of natural increase by temperature (for example) or the functional response by the turbulent kinetic energy dissipation rate (for example). However, this is not so simple, because certainly the scalar and vector properties interact and are confounded with the increase rate and functional response (remembering as well that the particular functional response typically used, the Michaelis-Menten functional response is only one of several conjectures relating to feeding saturation that could be used). Another approach is to use size-specific states; but in this circumstance, the issue of invariant transfer among number size biomass and energy density becomes an issue.

These modifications of the ODEs remain ad hoc, and attempting to resolve this difficulty in the ODE context makes any attempt to address the flux of energy through the ecosystem very complex. A difficulty with the complex approaches is that they constitute a manufactured system that is infinitely simpler than the real world, but infinitely more complex than human capability to understand the manufactured system. The data requirements of complex approaches often extend well beyond data collection resources. Perhaps the best feature of complex systems is that they demonstrate the practical need for a simple approach in terms of both scientific elegance—Occam's razor—and the importance of the multiscale.

Evidently, realizing these points, Platt and Denman (1977, 1978) and Silvert and Platt (1978) studied the energy flux through the ecosystem (Fig. 3.12). Silvert and Platt write "the mass of particles per unit volume of water in the interval of sizes between $w$ and $w + dw$" as

$$b(w,t) = \beta(w,t)dw \tag{50}$$

This implies that for any small interval of particle size (say, approximately 10 to 11 $\mu M$) the energy loss (owing to respiration, growth, excretion, or consumption by another organism) and the rate of flux of energy to the next largest size is a function of the particular size of the particle *w*.

The size-specific reduction in energy is

$$b(w + g(w)dt, t + dt) = b(w,t)[1 - \mu(w)dt] \tag{51}$$

where $g(w)$ and $\mu(w)$ are growth and loss functions, respectively. To complete equation 51, it is necessary to take into account the fact that it takes larger organisms longer to grow an increment in biomass than smaller organisms. Silvert and Platt make this adjustment

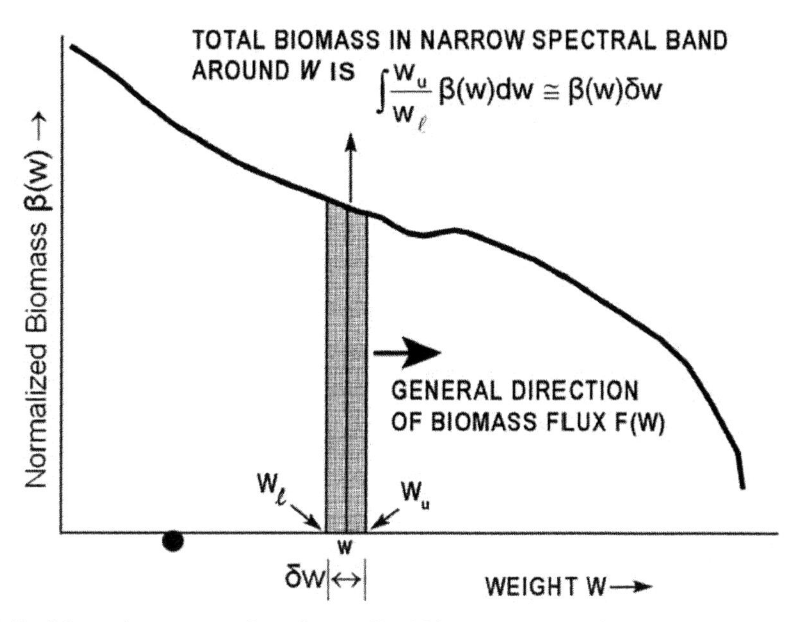


Figure 3.12 Schematic representation of normalized biomass spectrum for pelagic food chain. (Redrawn from Silvert and Platt, 1978.)

$$\begin{aligned} w + dw &\rightarrow w + dw + g(w + dw)dt \\ &\Rightarrow w + g(w)dt + dw + \frac{dg}{dw}dwdt \\ &= (1 - \mu dt)\beta(w,t)dw \end{aligned} \tag{52}$$

which represents the advection of biomass (or energy) along the *w* axis with loss rate $-\mu\beta$. This is written as,

$$g\frac{\partial\beta}{\partial w}+\frac{\partial\beta}{\partial t}+\beta\frac{dg}{dw}=-\mu\beta \tag{53}$$

Silvert and Platt provide both steady-state and general solutions to equation 53. They note that a pulse in biomass (i.e., energy) connecting through the system, as specified by equation 53, would remain intact. They conjecture that this pulse should "smear" across the spectral band as biomass increases.

An attractive and obvious idea is to convert the Silvert-Platt idea to include a diffusion term. This admittedly heuristic approach is attractive from at least three points of view. First, it will generate the "smear" or "blur" among size-spectral bands, accounting for the stochastic properties of energy flow. Second, it allows parameterization of the common situation where the direction of flux includes the flux of energy from large to small organisms as well as from small to large organisms. There are two cases of this reverse flux direction. The first is in the smallest plankton in which acquisition of nutrients is often via molecular diffusion, and the second is reproduction. However, in the smaller organisms the offspring size may be only marginally smaller than the parents—however, this is not the case in some of the large organisms. The third involves the coupling with physical forcing and noise. As an example, consider that turbulent flow smears trophic interactions in a variety of ways.

As an example, consider writing an energy-balance partial differential equation model of energy flux through the heterotrophic portion of the ecosystem. For example, we can sketch such an approach by writing,

$$\begin{gathered}\frac{\partial u}{\partial t}=a\frac{\partial^2 u}{\partial x^2}+b\frac{\partial u}{\partial x}-cu(x,t)\\ u(x,0)=(\textit{exponentially declining function})\\ u(0.5,t)=(\textit{cosinisoid, representing photosynthetic input})\end{gathered} \tag{54}$$

With an additional boundary condition and where for any defined volume of water, $u$ is the ecosystem energy density in Joules; $x$ is the individual biomass energy in Joules; and $a$, $b$, and $c$ are "advection," "diffusion," and respiration constants, respectively. The balance equation for biological energy density, $u$ [Joules] is characterized by the advection of energy through the ecosystem from the smallest organisms to the largest organisms. It is well known that biological energy not only advects from left to right, but because of complex food webs and the fact that energy also flows from larger to smaller organisms, it makes sense to include a term for energy diffusivity. Finally, as energy flows through the ecosystem, it is partitioned among biomass and metabolic by-products, such as respiration and excretion.

The initial and boundary conditions are particularly important. The first constraint sketches the idea that biomass declines with size, the second is that the photosynthetic input is periodic, and the third conjectures that the biomass change at the largest individual organism size is constant.

This sketch illustrates an approach that could be used to structure a physically realizable ecosystem dynamics model. It is important to observe that the photosynthetic process is relatively well known and can be modeled to any degree of detail as an input process.

Important multiscale issues (in this case averaging) can be derived from equation 54. For example, a focus on the fate of primary production and its interaction with heterotrophs would almost certainly consider the hourly temporal and small spatial scales implied by the periodic input. On the other hand, the propagation or suppression of forcing variability would consider larger time and space scales where the periodic input would be averaged away. This is a good example of a multiscale issue. The need to change the constants $a$, $b$, and $c$ into functions is another example, particularly related to physical forcing. For example, it is well known that the length scales separating particles is an increasing function of the size of the organism group (e.g., copepods are generally more distant from one another than bacteria). These separation length scales relate to the Batchelor scale, the Kolmogorov scale, etc. The values of $a$, $b$, and $c$ relate to the separation length scales, illustrating again the multiscale content of equation 54 and its capability to relate to physical processes, such as change in temperature and flow.

The PDE approach that specifies the flux of energy through the ecosystem size spectrum provides insights on addressing the so-called predator-prey trophic transaction. In actuality, there are three types of pair-wise trophic transactions: the predator-prey transaction, where a predator consumes a prey; the grazing transaction, where a grazer consumes single-cell organisms; or molecular transactions, where populations donate molecular of nutrient material to other populations. The types of transactions are linked to the size of the organisms. The predator-prey transactions are linked to the largest organisms, grazing to intermediate size organisms, and molecular exchange to the smallest organisms. But, in general, the dynamics of any population depends on a three-way interaction not a pair-wise interaction. Placing a population(s) of interest in its appropriate place in the size spectrum illuminates the three-way interaction in terms of the relative abundance and flux of a nutrient donor set, a competitor set, and a nutrient acceptor set of organisms.

Insights to at least two important problems might be obtained. The first is the ammonium production and utilization problem. Ammonium is a significant contribution to nutrition of primary producers. The PDE approach lends itself to study factors that relate to the total production of ammonium. It also relates to the delivery of ammonium.

For ammonium-excreting organisms that are distant from one another, ammonium rapidly drops to background levels. However, organisms that are relatively close to one another generate an ammonium concentration field, a nutrient field that is higher than background levels. The PDE model then begins to sketch how the density of ammonium generators contribute to local concentrations of ammonium perceived by nutrient acceptors. A second problem is the recruitment problem. It is fairly well understood that recruitment to fish populations depends upon the nutrition of and predation upon fish eggs and larvae. Nutrition of larvae is relatively easy to understand. However, predation upon fish eggs and larvae is much more complex, particularly owing to the diversity of the predator field. It seems that important insights into the recruitment problem might be obtained by studying the relative abundance and energy flux as it passes from sizes that relate to the prey of fish larvae, to sizes equivalent to competitors of fish larvae, to predators on fish larvae.

Taking these ideas a step further, it is possible to parameterize the concentration or density of organisms into inter-organism distances or length scales. For example, the highest density of organisms are bacteria with $10^{12}$ individuals per $M^3$. At the other end of the plankton spectrum, we might have fish larvae at 1 individual per $M^3$. If the organisms are distributed at random, then the three-dimensional inter-particle distance is $.55\lambda^{-3}$, where $\lambda$ is the density of the organism. While patchiness might increase or decrease the spacing, the random model provides a point of departure. With the ensuing spectrum of inter-particle distances, it is then possible to include the effects of molecular diffusion and correlated and uncorrelated flow velocities to the various types of trophic transactions and the flux of energy through the ecosystem.

## 6. Discussion

The components of this chapter have been assembled from various "disciplines," both in oceanography and applied mathematics; and as a consequence, each has a different provenance. The subject of scale has been considered by physicists and non-physicists from different perspectives. Physicists view scale in terms of categorizing physical behavior as a function of time and length, while biologists view scale in terms of description, which sometimes extends to issues of quantitative patchiness. The main difference is that the physical view results from physical axioms, while the biological view results from empirically based notions. However, there are biological equations, which have deep roots in theoretical ecology. As we have shown, these equations are capable of analyses that yield powerful scaling arguments. However, these arguments have had little traction, evidently because a universally acceptable biological equation has not been agreed upon. The implicit, if not explicit, reason is that the biological equations are not based on physical principles. Nevertheless, the biological equations provide an important architecture for the kinematics of ecosystem components and the application of modern-state space and filter theory mathematics. The state-space methodology also provides a facile methodology for studying the natural time scale of various formulations and parameter sets, including those of relatively high dimensionality (several states). The conventional biological equations also constitute a basis for examining in more detail aspects of the cascades of energy evident in any ecological system. There are basically two components in these cascades. The first relates to dynamical rates that are an intrinsic property of the populations or ecosystem components that represent a "state" in the system. The second relates to rates that govern the interactions between the states. These are the so-called functional responses. (Even though the Michaelis-Menten form is commonly used, there is no agreement upon the appropriate formalism to represent the functional response. In fact, the Michaelis-Menten functional response was based upon "crowding" at a cell wall relative to enzyme kinetics.) The functional response is not usually represented in physical terms, nor does it take into account the second order properties or patchiness of the interactions between the nutrient donor and the nutrient acceptor. Stochastic geometry and point processes enables the discussion of length scales in the interaction between donors and acceptors. In this regard, the interesting work of Fasham can be extended to transit from counts to intervals, which couple directly to length scales. In fact, the development of the idea of length scales from point process theory enables an even deeper coupling with physical reality in the sense that we can begin to think of the balance between the energy consumed and the energy expended in its own right and how this relationship couples with physical forcing. In fact, the analysis reflects an additional feature of the process that requires further study, and this is the intermittency issue. The patch effect is a second order effect, and hence it is only effective over a "short" time. For some phenomena, under some forcing regimes, the effect can be quite important, yet for other phenomena, the effect can "average out."

While the point process analyses reveal important physical properties that underlie ODE kinematics, it still is not completely universal in the sense that we have no particular basis, at the present time, for accepting the hyperexponential distribution of particles over a compound Poisson distribution of particles (for example).

There are of course fundamental feasibility issues that need to be taken into account. First of all, the analytic approach becomes more and more limited as problems become more and more complex. The solutions involve linearization and numerical approximation. However, the capability to quickly compute the small-scale processes that are described here is not generally available. This results in the need for parameterization of the stochastic processes, a task that is challenging because of the second order properties. On the other hand, not taking into account second order properties may lead to misleading conclusions.

The development of physically explicable biological-physical-chemical governing equations is clearly a major challenge. If axiomatically-based biological/chemical equations that couple with the physics are to be developed, then it is important to think about the nature of the axioms, the simplicity of the way that they would be incorporated into governing equations, and the modality of coupling. It seems clear that the axiomatic principle is thermodynamic. The laws of thermodynamics seem too general to apply to the ecosystem. On the other hand, the photosynthetic fixation of "energy" and its flux and dissipation as it moves through the ecosystem appear physically palpable (in the context of the advection of biomass energy; see Silvert and Platt, 1978) and observable.

The obvious next step is to examine formulations that do have a physical basis (time, space, energy). Odum, in fact, concerned himself with ecosystem energy. He extended the views of community metabolism, an important concept in early ecological theory, to involve thermodynamic evolutionary goals. Odum's ideas have been criticized from several points of view but, nevertheless, they lead us almost ineluctably to the conclusion that the physical analysis of the ecological system—the biology and chemistry of the sea, and its physical forcing—can only be addressed in a coherent manner through the energy characteristics of the ecosystem. This will lead to equations and scale relationships that describe the different "phases" of the system and scalings that will result in elegant simplifications facilitating future progress.

## Bibliography

Bartlett, M.S., 1963. The spectral analysis of point processes. *J. Roy. Statist. Soc.,* **25**B, 264–296.

Batchelor, G.K., 1967. *An Introduction to Fluid Dynamics.* Cambridge University Press, Cambridge.

Beyer, J.E. and B.F. Nielsen, 1996. Predator foraging in patchy environments: the interrupted Poisson process (IPP) model unit. Dana, **11**(2), 65–130.

Box, G.E.P. and G.M. Jenkins, 1970. *Time Series Analysis, Forecasting and Control.* Holden-Day, San Francisco, CA.

Cox, D.R., 1962. *Renewal Theory.* Methuen & Co. Sci. Paperbacks (ed. 1970).

Cox, D.R. and V. Isham, 1980. *Point Processes.* (Monographs on applied probability and statistics). Chapman and Hall.

Cressie, N.A.C., 1991. *Statistics for Spatial Data.* Wiley, New York.

Cushman-Roisin, B., 1994. *Introduction to Geophysical Fluid Dynamics.* Prentice Hall, New York.

Daley, D.J. and D.Vere-Jones, 1988. *An Introduction to the Theory of Point Processes.* Springer-Verlag, New York.

Fasham, M.J.R., 1977. The application of some stochastic processes to the study of plankton patchiness. In *Spatial Pattern in Plankton Communities,* Series IV, J.H. Steele, ed. Plenum Press, New York, pp. 131–156.

Fasham, M.J.R., 1978. The statistical and mathematical analysis of plankton patchiness. *Oceanogr. mar. Biol. ann. Rev.,* **16,** 43–79.

Feller, W., 1957. *An Introduction to Probability Theory and its Applications.* Wiley, New York.

Flierl, G. and D.J. McGillicuddy, 2002. Mesoscale and submesoscale physical-biological interactions. In *The Sea,* Vol. 12, A.R. Robinson, J.J. McCarthy, and B.J. Rothschild, eds. Wiley, New York, pp. 113–185.

Gargett, A. and J. Marra, 2002. Effects of upper ocean physical processes (turbulence, advection, and air-sea interaction) on oceanic primary production. In *The Sea,* Vol. 12, A.R. Robinson, J.J. McCarthy, and B.J. Rothschild, eds. Wiley, New York, pp. 19–49.

Gruber, N. and J.L. Sarmiento, 2002. Large-scale biogeochemical-physical interactions in elemental cycles. In *The Sea,* Vol. 12, A.R. Robinson, J.J. McCarthy, and B.J. Rothschild, eds. Wiley, New York, pp. 337–399.

Haury, L.R., J.A. McGowan and P.H. Wiebe, 1977. Patterns and processes in the time-space scales of plankton distribution. In *Spatial Pattern in Plankton Communities,* Series IV, J.H. Steele, ed. Plenum Press, New York, pp. 277–327.

Kantha, L.H. and C.A. Clayson, 2000. *Small Scale Processes in Geophysical Fluid Flows.* Academic Press.

Ljung, L., 1987. *System Identification: Theory for the User.* Prentice-Hall, New Jersey.

Mann, K.H. and J.R.N. Lazier, 1991. *Dynamics of Marine Ecosystems.* Blackwell, Cambridge.

McCarthy, J.J., 2002. Biological responses to nutrients. In *The Sea,* Vol. 12, A.R. Robinson, J.J. McCarthy, and B.J. Rothschild, eds. Wiley, New York, pp. 219–244.

Murray, J.D., 1989. *Mathematical Biology.* Springer-Verlag, Berlin.

Nisbet, R.M. and W.S.C. Gurney, 1982. *Modeling Fluctuating Populations.* Wiley, New York.

Ogata, K., 1990. *Modern Control Engineering.* Prentice-Hall, New Jersey.

Okubo, A., 1977. Horizontal dispersion and critical scales for phytoplankton patches. In *Spatial Pattern in Plankton Communities,* Series IV, J.H. Steele, ed. Plenum Press, New York, pp. 21–42.

Pielou, E.C., 1969. *Mathematical Ecology.* Wiley.

Platt, T., 1977. Spectral analysis of spatial structure in phytoplankton populations. In *Spatial Pattern in Plankton Communities,* Series IV, J.H. Steele, ed. Plenum Press, New York, pp. 73–84.

Platt, T and K.L. Denman, 1977. Organization in the pelagic ecosystem. *Helgol. Wiss. Meeresunters.*, **30,** 575–581.

Platt, T and K.L. Denman, 1978. The structure of pelagic marine ecosystems. *Rapp. P.-V. Reun. Cons. Int. Explor. Mer,* **173,** 60–65.

Ripley, B., 1981. *Spatial Statistics.* Wiley.

Robinson, A.R., J.J. McCarthy and B.J. Rothschild, 1999. Interdisciplinary ocean science is evolving and a systems approach is essential. *Journal of Marine Systems,* **22,** 231–239.

Robinson, A.R., J.J. McCarthy and B.J. Rothschild (eds.), 2002. *The Sea: Biological-Physical Interactions in the Sea.* Wiley, New York.

Rothschild, B.J. and T.R. Osborn, 1988. Small-scale turbulence and plankton contact rates. *J. Plank. Res.,* **10,** 465–474.

Rothschild, B.J., 1991. Food-signal theory: population regulation and the functional response. *J. Plank. Res.,* **13**(5), 1123–1135.

Rothschild, B.J., 1989. On the causes for variability of fish populations in long-term variability of pelagic fish populations and their environment. *Proceedings of the International Symposium,* Sendai, Japan, 14–18 November 1989, T. Kawasaki, S. Tanaka, Y. Toba and A. Taniguchi, eds. Pergamon Press, pp. 367–376.

Rothschild, B., 1992. Application of stochastic geometry to problems in plankton ecology. *Phil. Trans. R. Soc. Lond. B,* **336,** 225–237.

Rothschild, B.J., A.F. Sharov, A.J. Kearsley and A.S. Bondarenko, 1997. Estimating growth and mortality in stage-structured populations. *J. Plank. Res,* **19**(12), 1913–1928.

Silvert, W. and T. Platt, 1978. Energy flux in the pelagic ecosystem: A time-dependent equation. *Limnol. Oceanogr.,* **23**(4), 813–816.

Steele, J.H. (Ed.), 1977. *Spatial Pattern in Plankton Communities,* NATO conference series IV, Marine sciences. Plenum Press, New York.

Stommel, H.M., 1963. Varieties of oceanographic experience. *Science,* **139**(3555), 572–576.

Stoyan, D., W.S. Kendall and J. Mecke, 1987. *Stochastic Geometry and Its Applications.* Wiley, New York.

Sverdrup, H.U., 1953. On the conditions for vernal blooming of phytoplankton. *J. Cons. Int. Explor. Mer.,* **18,** 287–295.

Wroblewski, J.S., J.J. O'Brien and T. Platt, 1975. On the physical and biological scales of phytoplankton patchiness in the ocean. *Memoires Société Royale des Sciences de Liège,* 6th Series, **7,** 43–57.

Yamazaki, H., D.L. Mackas and K.L. Denman, 2002. Coupling small-scale physical processes with biology. In *The Sea,* Vol. 12, A.R. Robinson, J.J. McCarthy, and B.J. Rothschild, eds. Wiley, New York, pp. 51–112.

# Part 2.
# SEDIMENT, BIOGEOCHEMICAL AND ECOSYSTEM DYNAMICS AND INTERACTIONS

# Chapter 4. RECENT ADVANCES IN FINE-GRAINED SEDIMENT-TRANSPORT PROCESSES ON THE CONTINENTAL SHELF

ANDREA S. OGSTON

*School of Oceanography, University of Washington, Box 357940, Seattle WA 98195*

RICHARD W. STERNBERG

*School of Oceanography, University of Washington, Box 357940, Seattle WA 98195*

CHARLES A. NITTROUER

*School of Oceanography, University of Washington, Box 357940, Seattle WA 98195*

**Contents**



## 1. Introduction

A traditional view of river discharge, shelf sediment transport and mud-deposit formation consists of terrestrial sediment settling from a river plume through the water column and depositing at the site on the continental shelf where it is buried. On the shoreward side of this deposit, the mud facies pinches out due to strong oscillatory flows associated with wave shoaling. In this sense, the inner shelf is decoupled from processes forming the shelf mud deposit. On the seaward side of the deposit, the mud pinches out as some function of sediment input, distance from shore, flow conditions, and depth (e.g., Curray, 1965; Nittrouer and Sternberg, 1981; Leithold and Bourgeois, 1989).

Over the past few decades, great progress has been made in evaluating and gaining insights into the basic concepts described above. Instrument development has provided a means to document the physical processes and sediment response in a wide variety of shelf environments and over extended periods of time. Com-

*The Sea*, Volume 13, edited by Allan R. Robinson and Kenneth H. Brink
ISBN 0-674-01526-6 

prehensive observations include in situ particle characteristics (e.g., size, floc nature, settling velocity), benthic boundary-layer flow conditions, suspended-sediment load and flux, sediment accumulation rates, and strata development. Most importantly, knowledge of shelf sedimentary systems has advanced to the degree that large comprehensive numerical models have been developed to predict or infer shelf sedimentation. These models are based on turbulent boundary-layer theory and compute a range of parameters on a shelf-wide basis: boundary-layer shear stress under combined wave and current action, sediment resuspension and flux, sediment sorting, deposition and erosion of the seabed.

Over the past decade new evidence from field studies has started to change our understanding of shelf sedimentary processes that control the fate of muddy sediment (i.e., particle diameters <64 μm). These processes are highlighted in this paper, with a focus both on the spatial processes that control the transfer of sediment across shelf subenvironments (inner shelf to mid shelf to slope) and on the temporal variability associated with transfer processes and burial of sediment. The studies represent a mosaic of results from several continental shelf settings that have been important for expanding our knowledge of shelf sediment dynamics. Ultimately, they may engender additional physical concepts to be incorporated into shelf sedimentation models. The relevant processes include:

- the formation of gravity-driven density flow, or fluid mud, and its role in sediment transport;
- across-shelf transport mechanisms, with particular attention to the linkages between different shelf subenvironments, i.e., the role of the inner shelf in creating the mid-shelf deposit, and the role of the mid-shelf deposit on off-shelf transport by nepheloid layers (zones within the water column with diffuse clouds of increased sediment concentration) and gravity-driven density flows in canyons;
- longer-term (time scales greater than weeks) mechanisms impacting annual and interannual variability of sediment flux, and the sedimentological impact of the relative timing between sediment supply and coastal-ocean energetics.

Sediment resuspension and transport are key aspects of interdisciplinary studies within the global coastal ocean, and the concepts presented here were selected for their importance to understanding the oceanic processes at work. The path an individual particle takes on its way to permanent burial in the seabed will influence the associated chemical and biological constituents. Thus, these new findings have implications beyond just the sedimentological realm.

This chapter includes a brief and relatively simple summary of the fundamentals of clastic sediment resuspension and suspended-sediment transport. For a more comprehensive treatment, the readers are referred to compilations by Dyer (1986) and Wright (1995). The primary focus of this chapter is on recent concepts and discoveries. Shelf examples in which modern mud facies are being formed are emphasized, because these are the ones actively receiving sediment from terrestrial sources. Muddy coastal oceans are of great relevance to interdisciplinary studies involving the transfer of both biological and chemical species. Coarser-grained environments that allow the transport of seabed porewater solutes are discussed in Chapter 6 (Jahnke, this volume).

## 2. Fluid/Sediment Interactions

Fluid/sediment interactions refer to both the characteristics of flow near the sediment-water interface and the seabed response. Flow parameters in the benthic-boundary layer (the region dominated by frictional effects) include current velocity, turbulent structure, and boundary shear stress. Sediment response includes resuspension, upward turbulent diffusion away from the interface, transport and deposition. Transport and deposition of sediment can be evaluated in terms of the sediment flux integral,

$$SedimentFlux = \int u \cdot c\partial z, \tag{1}$$

where $u$ is the instantaneous velocity and $c$ is the instantaneous suspended-sediment concentration. Integration over the vertical dimension, at any point in space and time yields the flux, or transport rate, of sediment past that point at a particular instant. Thus, an estimation of sediment flux requires a description of both the velocity field and the suspended-sediment concentration field.

### *2.1 Currents and Shear Stress*

Currents in the coastal ocean are a combination of multiple external forcing mechanisms that result in site-specific velocity characteristics (see Brink, this volume). These forcing mechanisms encompass a wide range of temporal and spatial scales varying from seconds for surface gravity waves through decadal processes, e.g., El Nino/La Nina processes (Table 1). Sediment forcing mechanisms identified and studied by shelf scientists include waves (surface and internal), tides, wind events, seasonal variations (including geostrophic processes), and most recently, events comprising interannual variability (e.g., Drake and Cacchione, 1986; Sherwood et al., 1994; Wiberg et al., 1994; Ogston and Sternberg, 1999; Guerra, 2004).

TABLE 1
Physical processes considered in the discussion and evaluation of sediment transport (after Mooers, 1976)

| Process | Time Scale | Along shelf Scale (km) | Across-shelf Scale (km) |
|---|---|---|---|
| Long Term (e.g., decadal, El Nino, interannual) | years→decades | $10^3 \rightarrow 10^4$ | margin width |
| Low-frequency flows, e.g.: | | | |
| Wind (including coastal trapped waves) | days→weeks | $10^2 \rightarrow 10^3$ | shelf width |
| Buoyancy-driven flows * | days→months | $10^2 \rightarrow 10^3$ | $10^1 \rightarrow 10^2$ |
| Discharge plumes and fronts | days→months | $\sim 10^2$ | $10^1 \rightarrow 10^2$ |
| Tides * | semidiurnal→ diurnal, fortnightly | $10^3 \rightarrow 10^4$ | margin width |
| Surface-gravity waves * | tens of seconds | $\sim 10^{-1}$ | $10^{-2} \rightarrow 10^{-1}$ |

* see Brink (this volume) for further discussion.

Above the zone of frictional boundary-layer effects, the prediction of free-stream velocity has been the focus of significant study in recent years. For example, community models utilizing the equations of motion at a range of forcing scales (Table 1) can accurately predict geostrophic and buoyancy-driven flows. More specifically for shelf transport studies, modeling efforts have been shown to resolve successfully mesoscale features in response to small-scale bathymetric changes, wind forcing and river-plume interactions (e.g., Pullen and Allen, 2000, 2001). Although presently models may not always accurately predict the location of features unless strongly trapped by bathymetry, the features themselves are reproduced and their variance compares well with measurements.

Within the zone of frictional effects, a bottom boundary layer develops where the free-stream velocity interacts with the seabed (Fig. 4.1a). At the seabed, the "no-slip" condition causes the velocity profile to go to zero at or near the seabed. Very close to the seabed where the viscous forces are large relative to the turbulent forces (i.e., low Reynolds number) due to low velocity, a viscous sublayer (dominated by laminar flow) can form on the order of 1 mm thick. Above this region and ignoring waves, the velocity profile exhibits a logarithmic increase upwards in the zone of constant stress (Dyer, 1980; Grant and Madsen, 1986; Sleath, 1990). The thickness of the logarithmic layer on the continental shelf is typically 1 to 2 m. In the coastal ocean, where waves are also felt near the seabed, the velocity profile includes a wave boundary layer that scales as a function of the wave-generated stress and frequency. The scale is typically on the order of 2—10 cm thick for surface gravity waves (Fig. 4.1b). The wave and current boundary layers are not independent and interact to enhance each other. Velocity profiles with wave-current interaction can be iteratively predicted (e.g., Smith, 1977; Grant and Madsen, 1979; Wiberg and Smith, 1983; Styles and Glenn, 2000).

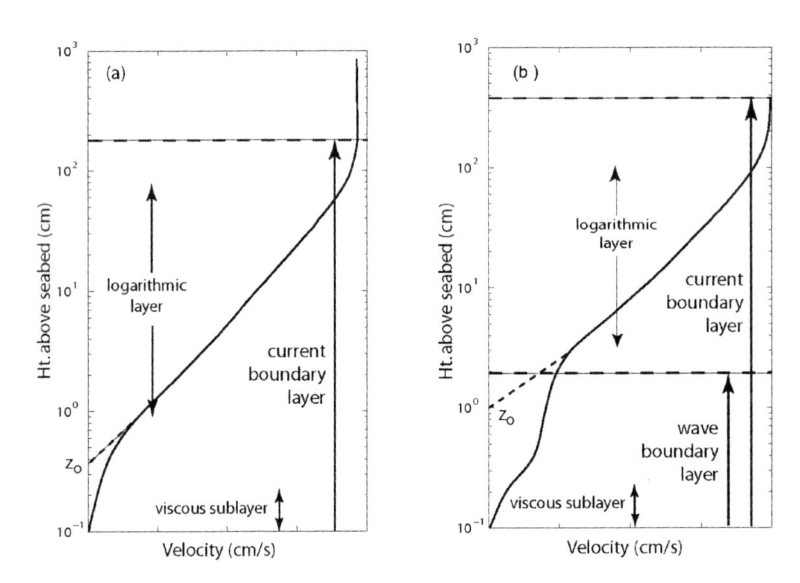


Figure 4.1. Diagramatic velocity structure of the bottom boundary layer (a) without, and (b) with waves. The viscous sublayer, wave and logarithmic layer scales are shown.

Shear stress ($\tau_b$) is generated within the bottom boundary layer due to the velocity gradient. Boundary shear stress acts to initiate sediment motion, and aids in particle diffusion away from the seabed (acting against the gravitational settling of particles). The zone of constant stress occurs where the flow is dominated by bed roughness, yet is relatively unaffected by the earth's rotation. Shear stress can be estimated in at least three ways: 1) quadratic stress laws which utilize a friction, or drag, coefficient (Sternberg, 1968, 1976) and the square of the velocity, 2) from the slope of the velocity profile in the logarithmic region (Sternberg, 1966; Dyer, 1970; Channon and Hamilton, 1971), and 3) the Reynold's stress formulation (Gordon, 1975; Gross and Nowell, 1983; McLean, 1983) where the stress in the fluid is represented by the turbulent stress ($\overline{u'w'}$). Other methods utilizing the turbulent kinetic energy and the turbulence dissipation rate are also becoming available as instrumentation is developed to sample at temporal and spatial scales of turbulence (on the order of 100 Hz, $10^{-3}$ cm) and as our understanding of small-scale turbulence processes expands.

### 2.2 *Suspended-Sediment Concentration (Turbulent Diffusion)*

In the absence of advected sediment input, the suspended-sediment concentration profile responds primarily to the magnitude of the friction velocity ($U_*$, which is related to the bed shear stress by the relationship, $\tau_b = \rho U_*^2$, where $\rho$ is the density of the fluid), and particle settling velocity ($Ws$). The concentrations are a function of the grain-size distribution, particle settling velocity including flocculation effects such as size and density changes, and other related factors affecting the characteristics of the seabed, which will be referred to in the section on seabed roughness and consolidation.

The Rouse equation, in its most simplistic form, describes the vertical profile of a specific suspended-sediment size class,

$$C_z = C_a\left(\frac{z}{a}\right)^{-Ws/KU_*} \tag{2}$$

where the $C_z$ is the concentration at some height, $z$, relative to $C_a$, the concentration at a reference height, $a$. The concept of a reference height was initially formulated for systems where suspended sediment is shed from an underlying bedload layer, the reference height being the distance from the seabed to the top of the bedload layer. Although there may be no bedload layer in muddy sediment, the concept is still used. The ratio of $C_z/C_a$ is a function of the shear velocity ($U_*$) and the settling velocity ($Ws$) for a given size class or, in the case of flocculated muddy sediments, a representative value for the aggregate settling velocity. $K$ is von Karman's constant (~0.40; Nowell, 1983). The concentration profile is also influenced by the effects of density stratification (e.g., Wiberg, et al., 1994; Wright et al., 1999) due to the mass of sediment in the water column. As sediment-induced density stratification in the fluid results from high shear stresses and associated resuspension, turbulence is damped causing sediment to fall out of suspension and thus the resuspension process has been considered to be self-limiting.

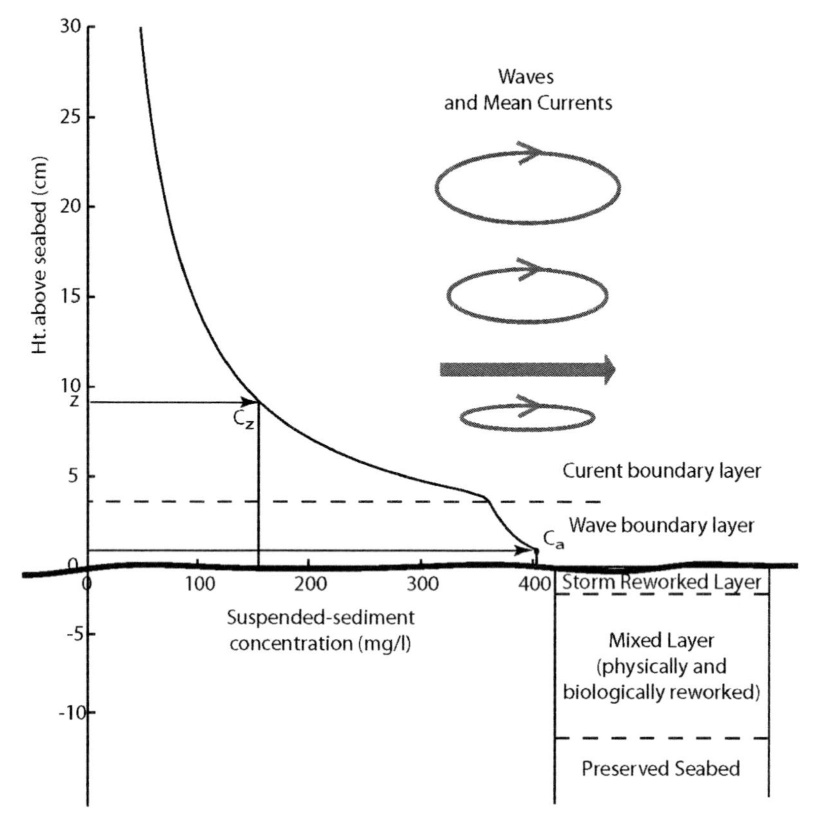


Figure 4.2 Diagramatic suspended-sediment concentration profile evaluated from the Rouse equation (Eq. 2) above a seabed that is reworked by both physical and biological mechanisms (after Wiberg, 2000).

The reference concentration, $C_a$, must be estimated to determine the absolute concentration profile of sediment in suspension and is strongly connected with the state of the seabed (Fig. 4.2). Field studies using measured values of $C_a$ have found the Rouse equation to be a robust descriptor of the suspended-sediment concentration profile except in high concentrations where the volume of water displaced by the particles settling must be considered (e.g., Kineke and Sternberg, 1989). The prediction of $C_a$ is a more difficult task. A common formulation to predict $C_a$ uses a resuspension coefficient, and is referred to as a concentration bottom boundary condition (Smith and McLean, 1977):

$$C_a = \frac{C_o \gamma_o S}{1 + \gamma_o S}, \tag{3}$$

where $\gamma_o$ is the resuspension coefficient, $C_o$ is the concentration of sediment in the seabed, and $S$ is the normalized excess shear stress $(\tau_b - \tau_c)/\tau_c$ where $\tau_c$ is the critical value of shear stress for sediment movement. Estimates of $\gamma_o$ from field and laboratory studies have revealed a large variation ranging from $10^{-2}$ to $10^{-5}$ (e.g. Smith and McLean, 1977; Wiberg and Smith, 1983; Sternberg et al, 1986; Hill et al., 1988;

of time to encounter the first prey is different than if as a special case the prey was always immediately available at the onset of the threshold.

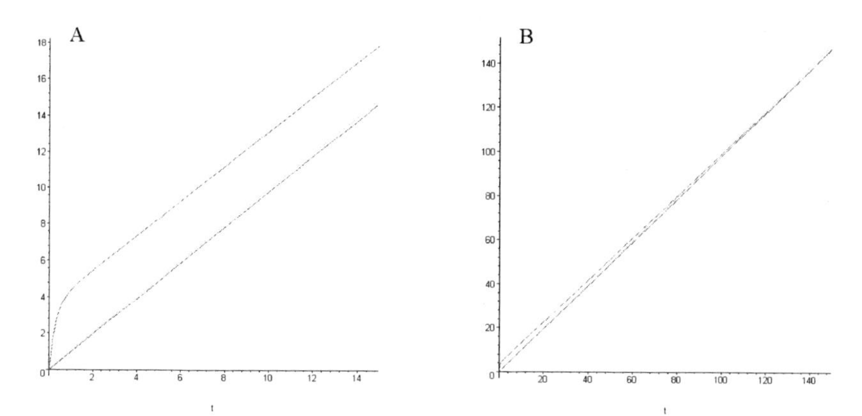


Figure 3.9 Theoretical hyperexponential renewal functions, $H(t)$, for a high-variance and a low-variance case plotted on two scales. In both cases, the mean value of unity is used for either distance or time. Panel A amplifies the transient initial condition and shows that the high-variance patch distribution initially generates higher encounters than for the low-variance case. As the process progresses, both reach the equilibrium level of one encounter per unit distance or time (as expected). This is a good example that shows statistically the circumstances when patchiness is important (short times or distances) and circumstances when it is not important (longer times or distances). What is long or short depends on the underlying process, but it can be seen that patchiness can have an important impact on mortality of prey.

### *4.1 Physical Coupling Between the Hyperexponential Distribution and Small-Scale Turbulent Flow*

Here we can illustrate how casting the plankton density problem in the context of length scales rather than densities (numbers or biomass per unit volume) enables a rudimentary coupling between physical forcing and biological distributions. These calculations enable comparing a variety of cases for various degrees of patchiness (i.e., variance) and particle separation. It is particularly interesting to consider the mean separation as 1 cm because this is approximately the Kolmogorov microscale (see, e.g., Kantha and Clayson, 2000; Rothschild and Osborn, 1988). In other words, particles with separation lengths of less than 1 cm would be sheltered from uncorrelated velocities, where as particles with separation distances greater than 1 cm would be exposed to uncorrelated velocities of the fluid flow. So using the fact that we can solve equation 33 for the mean distance between particles equal to 1 cm as

$$\rho_2 = \frac{(\theta - 1)\rho_1}{\theta - \rho_1} \tag{47}$$

The main point of the example is that if we are interested in particles that have a separation distance of 1 cm, then only subsets of the entire parameter set are consistent with this constraint. Generally values of the parameter set that permit inter-

Drake and Cacchione, 1989). Drake and Cacchione (1989) suggest that variation of $\gamma_o$ appears to be inversely related to excess shear stress ($S$), possibly in response to bed armoring or seabed sediment consolidation. Additionally, modeling studies suggest that the $\gamma_o$ variations can be minimized when both bed armoring and sediment-induced stratification are considered (Wiberg et al., 1994).

Alternately, a flux bottom boundary condition can be employed that does not necessarily make the assumption of an underlying bedload layer, and is more applicable to time-dependent models in which erosion and deposition time scales differ. In this formulation, the vertical flux of sediment in the water column due to settling and turbulent diffusion must balance the flux of sediment into and out of the seabed due to deposition and erosion. The level of complexity to which this condition can be physically described is hindered by our knowledge of processes in this very thin layer between the seabed and water column. A description of these processes and their formulation is described in Hill and McCave (2001).

In both formulations of the bottom boundary condition, it is necessary to know the consolidation state of the seabed as it affects both $C_o$ and $S$. As pore water is released from the seabed, $C_o$ increases and, in muddy sediment, $\tau_c$ increases dramatically (Krone, 1976). Laboratory studies of consolidation in deep-sea marine sediment suggest that removal of pore water occurs over time scales of hours to days, and is dependent on grain-size distribution, mineralogy, and pressure loading (e.g., waves stresses, Lee and Edwards, 1986). Benthic organisms are also known to change the value of $\tau_c$ significantly (Nowell et al., 1981). In view of the wide variability observed in $\gamma_o$, the use of Equation 3 to predict a nearbed reference concentration remains a major problem (Smith 1977; Drake and Cacchione, 1989).

Other properties of the seabed also affect the amount of stress available to resuspend particles. Although theoretically, muddy seabeds cannot sustain ripple structures, only a small amount of sand or coarse silt appears to be necessary for wave-orbital ripples to form and remain on the seabed (Rees, 1966). Bedforms can produce spatially inhomogeneous bed stress in the bottom boundary layer by, for example, locally increasing the roughness, which produces eddies in the lee of bedforms. Bedforms also require consideration when evaluating bed stress available for resuspension of particles from the seabed (skin friction) due to the removal of energy from the flow due to the drag on the bedforms (form drag).

In addition to bedforms created by waves and currents, bed roughness can be created by organisms who produce mounds that influence resuspension much in the same manner as ripples. Burrowing activities also change the consolidation state of the seabed by irrigating sediment to release or introduce porewater. Other possible influences on the resuspension and transport of sediment due to biological action include the formation of bacterial or algal surface binding (Nowell et al., 1981; Noffke et al., 2001), biological controls on flocculation, and mediation of bed armoring through bioturbation.

### 2.3 *Suspended-Sediment Concentration (Advection)*

In addition to resuspension from the seabed, sediment can be advected via plumes and nepheloid layers from a distant part of the seabed or from a river mouth. A three-dimensional approach is necessary to interpret suspended-sediment concentrations at a specific site. Various types of river plumes advect sediment in coastal areas (Fig. 4.3). Positively buoyant freshwater plumes (hypopycnal) are frequently observed (Fig. 4.3a), and more rarely, negatively buoyant plumes (hyperpycnal) have been observed in high-discharge situations for both small and large river systems (Wright et al., 1990; Kineke et al., 2000)(Fig. 4.3b). Particle removal from hypopycnal plumes is enhanced by the processes of flocculation, whereby individual particles form aggregates, which are larger and settle faster. In general, flocculation begins when saline water is mixed with the fresh plume (at ~1–3 ppt salinity, Krone, 1978; Spielman, 1978), although recent studies have observed flocculated particles in fresh water where organic matter and certain ionic species exist (e.g., Fox et al., 2004). Along with the current and wave regime, the supply function determines the plume dynamics and extent of flocculation, all of which contribute to the advection of particles and the subsequent removal of particles to the bottom boundary layer.

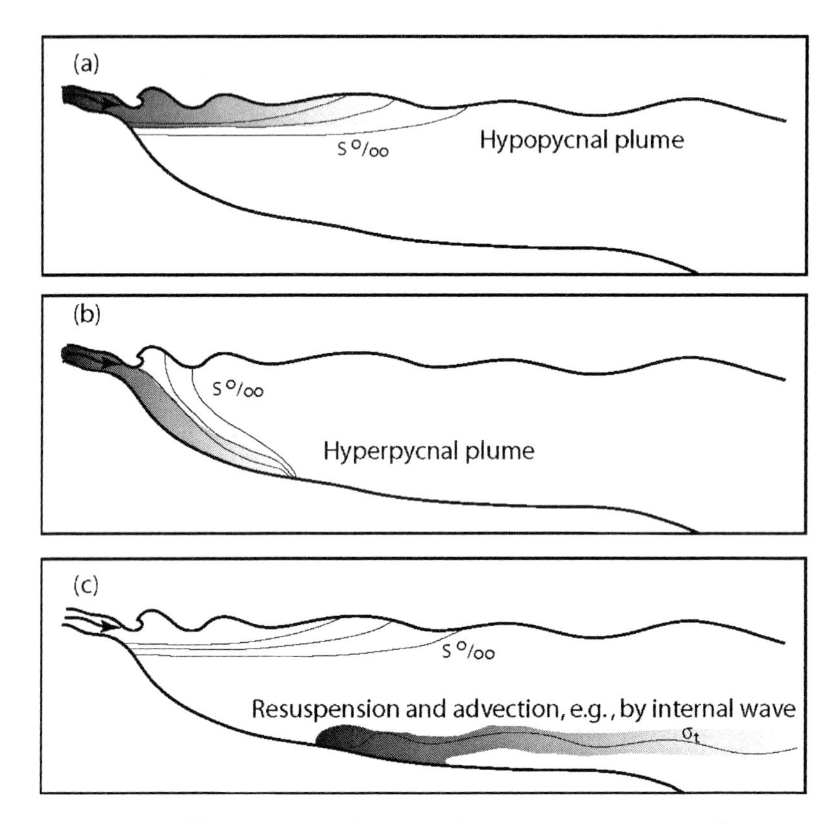


Figure 4.3 Examples of (a) a hypopycnal plume with density less than the surrounding water body, and (b) a hyperpycnal plume with density greater than the surrounding water body. In addition, sediment-laden layers can be formed in coastal ocean, for example, (c) by resuspension and advection due to internal waves.

Hyperpycnal plumes can be defined as freshwater plumes with bulk density greater than that of the receiving water body due to the sediment load (i.e., plume suspended-sediment concentration >40 g $l^{-1}$). High-density suspensions formed in the ocean (i.e., salt water and sediment with concentrations >10 g $l^{-1}$) also have been called hyperpycnal (due to a density higher than the ambient water column) and some confusion has occurred because of this lack of distinction in terminology. Here we refer gravity flows due to high-density river discharge that forms an underflow when discharged into the marine environment as hyperpycnal, and those due to marine nearbed processes that produce concentrations that exceed 10 g $l^{-1}$ as gravity-driven density flow. Whether a river discharges hypo- or hyperpycnally into the coastal ocean depends upon the sediment concentration at the discharge point, which results from the sediment supply function (i.e., small-catchment episodic signal or large-catchment damped signal) and sediment yield within a basin. In an analysis of major rivers, Mulder and Syvitski (1995) identify as many as 150 rivers on a worldwide basis that may become hyperpycnal due to high sediment discharge.

Both hyperpycnal plumes and gravity-driven density flows have significant relevance to across-isobath transport because of the gravity anomaly induced. These dense underflows tend to be associated with an abrupt suspended-sediment interface, the lutocline, which minimizes mixing with overlying water. Thus, large quantities of sediment can be moved across-isobath regardless of the current regime in the overlying water.

In the deeper coastal ocean (e.g., continental slope), advection of particles along density interfaces is frequently a dominant process of transport (Fig. 4.3c). For example, internal waves intersecting the continental shelf or slope can cause resuspension of particles and ejection of those boundary-layer suspensions into the water column along density surfaces, creating layers within the water column of turbid water, known as intermediate nepheloid layers (McPhee-Shaw et al., 2004). Removal of sediment from these layers is thought to be a dominant process in the delivery of sediment to the deep ocean (Drake, 1976).

### *2.4 Deposition and Erosion of the Seabed*

Deposition and erosion of the seabed depends upon the convergence and divergence of the sediment flux integral (Eq. 2). Although a flux of sediment exists past a particular point, unless there is less (more) flux at an adjacent point, there is no deposition (erosion) at that point (Fig. 4.4). Thus, transport gradients are key to linking studies of sediment transport to studies of seabed accumulation and strata formation.

Observations of spatial gradients of sediment flux require multiple instrumentation suites over large areas, and require detection of a small gradient amongst large signals. Few studies have been able to observe spatial gradients and relate them to sediment accumulation on the seabed. Wright et al., (1999) found an across-shelf sediment-flux convergence on the Eel shelf that corresponded to the zone of fine-sediment accumulation on the mid-shelf region (Fig. 4.4a). Additionally, Ogston et al. (2000) reported an along-shelf convergence on the same shelf during major events that was consistent with the development of flood deposits on the mid shelf (Fig. 4.4b).

Spatial variations in bed shear stress have been investigated by two-dimensional modeling of across-shelf processes (Harris and Wiberg, 2001, 2002). Another approach is to couple three-dimensional, high-resolution ocean models to bottom boundary layer models, which together can determine the free-stream velocity and the bed shear stress. Wright et al., (2001) developed an analytical model using the gradient of the shelf and the gradient of wave and current velocity (see section 3.1.2). This model shows that gradients in geomorphology and processes can connect gravity-driven density flow (e.g., fluid mud) with the formation of mud deposits (Scully et al., 2002; Friedrichs and Wright, in press).

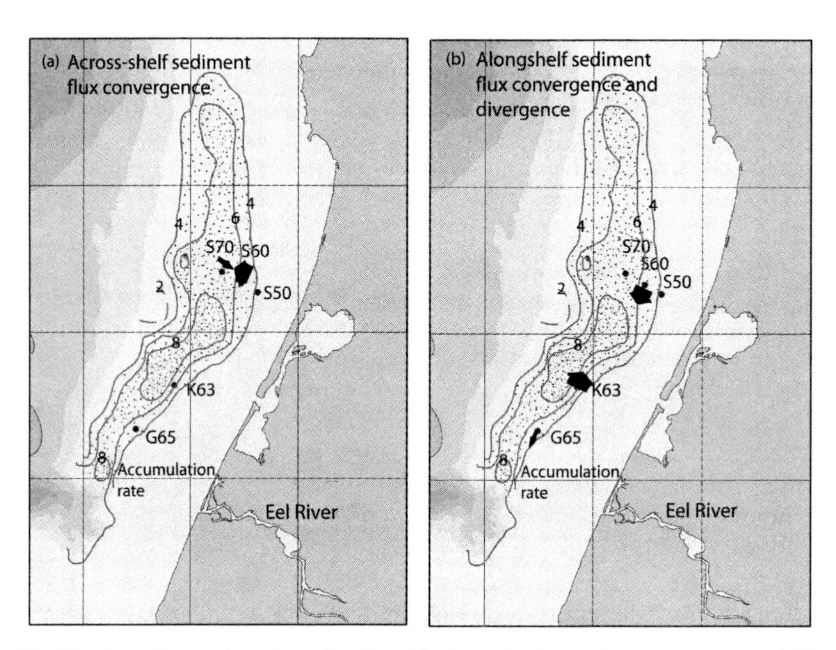


Figure 4.4 The deposition and erosion of sediment is dependent upon the convergence and divergence of the sediment flux integral. In this example from the Eel shelf, (a) a long-term average of the convergence of across-shelf sediment flux is co-located with the axis of the modern mud deposit, and (b) the alongshelf sediment flux convergence also corresponds with the depocenter of the deposit. The contours represent $^{210}$Pb accumulation rates ($10^{-1}$ g cm$^{-2}$ yr$^{-1}$; Sommerfield and Nittrouer, 1999), and the flux arrows are representations of data from Wright et al. (1999)(a) and Ogston et al. (2000)(b).

Typical winter storms on north Pacific continental shelves can erode approximately 0.5 cm into the seabed before coarsening of the active seabed surface (as fine sediment is removed) limits the erosion depth, known as bed armoring (Wiberg, 2000). This assumes that there is a balance between settling and resuspension, and divergent/convergent processes are not in action. In extreme wave conditions where bed armoring is not the limiting factor, but sediment stratification in the bottom boundary layer limits the carrying capacity (Kachel, 1980; Wiberg, 2000), erosion depths may reach 5–6 cm depth. Typically, the resuspension depth (or depth of erosion) is much less, and depends not only on the magnitude of storms creating high shear stresses on the seabed, but also the seabed texture and history leading to bed armoring or consolidation state. The resuspension depth impacts the preservation potential of seabed deposits (Fig. 4.2), and subsequently the chemical diagenesis and remineralization of constituents associated with the sediment particles.

During the present highstand of sea level, much of the sediment supplied by rivers accumulates on continental shelves (Wright and Nittrouer, 1995). Many natural (e.g., carbon) and anthropogenic (e.g., heavy metals) chemical species are transported on the surfaces of sediment particles, and accumulate with them. In fact, the muddy deposits on continental shelves represent the major marine repository for carbon (Hedges and Kiel, 1995). The processes of sediment resuspension and deposition described in this chapter affect the character of chemical constituents that are ultimately buried (e.g., Aller, 1998). In addition, the source function for fluvial sediment discharge to shelves has changed as a result of anthropogenic activities such as agriculture, logging, and river damming/ diversions (Mead, 1996). The result is that both the quantity and the quality of muddy sediment accumulating on continental shelves are sensitive to natural and human influences affecting sediment supply.

## 3. Recent Insights on Shelf Sediment Transport

### *3.1 Fluid Mud—a New Shelf Transport Concept*

Fluid muds are relatively dense, nearbed suspensions of flocculated fine-grained sediment and salt water. Their concentrations are in excess of 10 g $l^{-1}$ (corresponding to a combined fluid/sediment density of 1.03 g $cm^{-3}$), a point at which they begin to exhibit non-Newtonian behavior, and may reach > 330 g $l^{-1}$ (corresponding to a density approaching 1.30 g $cm^{-3}$) (Inglis and Allen, 1957; Wells, 1983; Kineke, 1993). Fluid mud has been observed in many of the world's estuarine environments in association with the turbidity maximum (e.g., Nichols, 1985) (Fig. 4.5a) and is thought to be caused by rapid settling due to flocculation and/or particle trapping in the convergence region of an estuarine salt wedge (Meade, 1972).

In the 1980s, fluid mud was observed off large river systems, where the mixing of the freshwater was pushed out of confined estuarine reaches (e.g., Amazon, Faas, 1986; Kineke and Sternberg, 1995; Huanghe, Wright et al., 1986). The formation of these very high concentrations on the Amazon shelf (Fig. 4.5b) was due to elevated sediment discharge and lateral sediment convergence at inner-shelf frontal zones during energetic conditions (Kineke et al., 1996). Off the Huanghe River during lower discharge periods, internal and surface wave action contributed to remobilization and maintenance of dense underflows (Wright et al., 1990).

Recent observations on the northern California shelf show that fluid mud also occurs on high-energy shelves with only episodic sediment input (Fig 4.5c). These dense suspensions are thought to result from elevated sediment discharge and particle trapping in frontal zones on the inner shelf (Ogston et al., 2000) and also by sediment trapping due to dynamics in the wave-boundary layer (Traykovski et al., 2000). Another process that has recently been observed is discharge of river plumes (Fig 4.5d) that split into a hypopycnal component, which transports sediments according to prevailing surface currents, and a hyperpycnal component moving directly down-slope, transporting sediment to greater depths (e.g., Sepik River, Kineke et al., 2000; Walsh and Nittrouer, 2003).

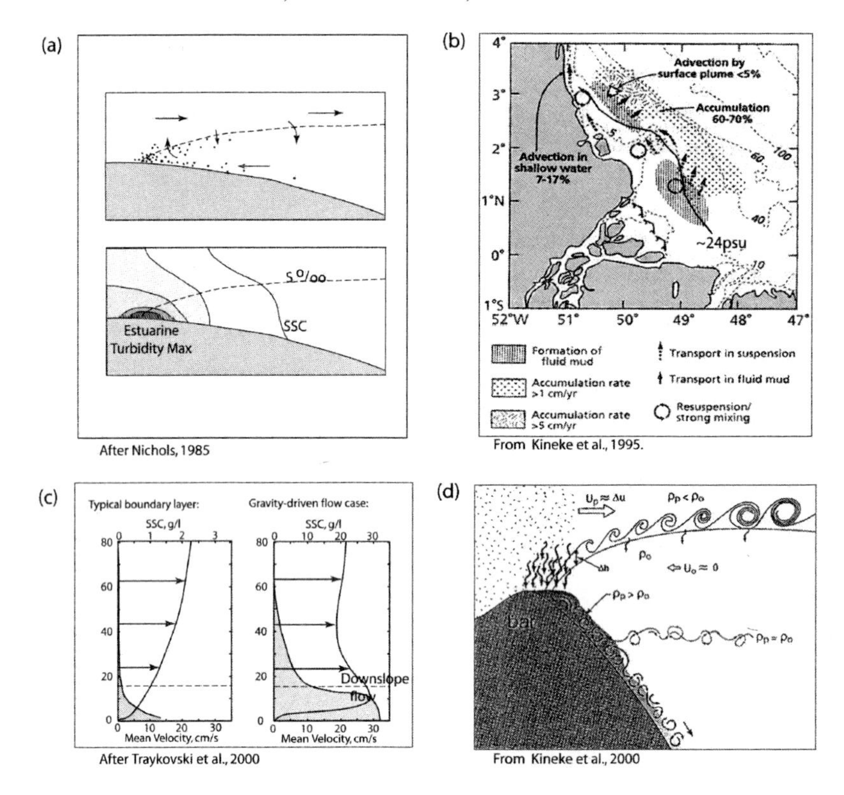


Figure 4.5 Fluid mud has been observed to form under multiple mechanisms: (a) at the estuarine turbidity maximum due to flocculation dynamics and flow convergence, (b) under conditions of lateral flow convergence centered on the 24-ppt isohaline of the Amazon shelf, and (c) within the wave-boundary layer on the Eel shelf. Fluid-mud formation can also be responsible for a divergent plume, where the advection of particles in the surface layer is controlled by shelf circulation, and the particles in the fluid-mud layer flow down slope (d). Parts b and d used with permission of Elsevier.

### 3.1.1. Fluid-Mud Impact on Sedimentary Deposits

Fluid-mud formation and subsequent gravity-driven transport may be more common than previously thought. Because of the high concentrations of suspended sediment in a fluid mud, a single event of fluid-mud transport can account for much or all of the average annual seaward flux due to turbulent boundary-layer processes. For example, a fluid-mud event was documented in the nearbed region (30 cmab; i.e., cm above bed) at 60-m depth on the Eel shelf during a major flood. Sediment flux associated with this single event exceeded other non-gravity-driven events by about two orders of magnitude, and the net seaward flux over the three-day event accounted for 77% of the net transport at that site in the entire previous year (Ogston et al., 2000).

Sedimentary deposits formed by fluid-mud have different sedimentological and geochemical signatures than deposits formed through classic resuspension and advection (Allison et al., 1995; Kuehl et al., 1996). Identification of deposits within the seabed is possible using grain size, radiochemical indicators, and microfabric. Typically, flood deposits can be identified by the high content of clay-size particles and presence of Berylium-7 (Sommerfield et al, 1999). The grain-to-grain fabric can provide clues regarding transport as fluid mud. Typically the sediment fabric created by deposits of fluid mud contains alternating layers of aggregates (flocs) and individual particles sheared by sediment transport (Kuehl et al., 1988).

Fluid-mud processes on continental shelves have far-reaching implications. On the Amazon shelf, fluid mud reaches 7 m in thickness and the shelf-wide suspended-sediment inventory incorporated in fluid mud is approximately equal to the annual sediment discharge of the river itself (Kineke et al., 1996). From a sedimentological view point, fluid-mud processes influence the shallow stratigraphy of river mouth and shelf deposits (Jaeger and Nittrouer, 1995; Kuehl et al., 1996) and are agents of progradation on the subaqueous delta (Nittrouer et al., 1986). Additionally, these nearbed suspensions strongly impact chemical exchange between the seabed and water column (Aller et al., 1996; Moore et al., 1996). They also create reduced bottom drag, which strongly influences the propagation of the tidal wave over the shelf and affects river-mouth mixing processes (Beardsley et al., 1995; Geyer, 1995). Knowledge of fluid-mud processes also may help to explain other unanswered questions, such as the long-observed decoupling of shelf deposits from river plume distributions and the high sediment accumulation rates on the foreset region of prograding clinoforms (Kuehl, et al., 1996; Walsh et al., 2004).

### 3.1.2. Mechanics of Fluid-Mud (Gravity-Driven) Transport

Numerous modeling efforts have been and are being carried out to better understand fluid-mud transport. A model was developed to contrast the role of turbulent mixing and sediment-induced stratification (Trowbridge and Kineke, 1994; Kineke et al., 1996). It used the gradient Richardson number (ratio of sediment-induced stratification to shear) for closure, and has provided important insights into the structure and dynamics of fluid-mud suspensions. Results suggest that at concentrations between 10 and 100 g $l^{-1}$, a condition exists where settling is unimportant and the water column has not been saturated with suspended sediment. At concentrations between 100 and 330 g $l^{-1}$, stratified fluid mud occurs where settling is important, the water column is saturated, and sediment has begun to deposit.

Downslope migration of fluid mud has been modeled using a Chezy relationship in which the downslope component of excess gravity is balanced by the quadratic-stress relationship incorporating underflow velocity and drag coefficients for the seabed and upper interface. Based on this approach, gravity-driven density underflows were estimated off the Huanghe River to be 10–20 cm $s^{-1}$ (Wright et al., 1990); on the Amazon subaqueous delta to be 23–40 cm $s^{-1}$ (Cacchione et al., 1995); and on the Eel inner shelf to be 17 cm $s^{-1}$ (Ogston et al., 2000). The approximations representing the Amazon River and Eel River underflows generally agree with independent evidence from benthic tripod observations.

Numerical models have also been developed that focus on river discharge events and wave-supported fluid-mud layers, and these models constrain along-shelf sediment delivery and downslope sediment flux. An analytical model formulated by Scully et al. (2002) suggests that critical stratification due to fine sediment in the wave-boundary layer dominates the nearbed dynamics when greater amounts of sediment are delivered by floods than can be removed, and demonstrates that gravity-driven density flows can account for the majority of sediment reaching the mid shelf. They also use their model of gravity-driven density flows to explain the equilibrium shape of subaqueous deltas and clinoforms associated with high-yield rivers (Friedrichs and Wright, 2004).

Fluid-mud dynamics on continental shelves represent a recently discovered sedimentary process that has not been fully incorporated into the sedimentologists' tool bag. Great strides have been made in recent years in turbulent boundary-layer modeling of shelf sediment transport (e.g., Harris and Wiberg 2001, 2002; Sherwood et al., 2002) and now a range of shelf environments have been documented where gravity-driven density underflows may at times dominate seaward transport, rather than turbulent boundary-layer processes. As discussed above, modeling efforts of fluid-mud processes are progressing and will help not only to broaden our conceptual framework of shelf sedimentology but ultimately provide improved predictive capabilities of shelf sediment transport and deposition.

### *3.2 Spatial Variation in Mechanisms – from Inner Shelf to Slope*

The physical and sedimentary processes discussed above act on different portions of the continental margin at varying levels of impact. In temperate and tropical regions, shelf environments receiving modern sediment input typically consist of a transgressive sand layer overlain by a recent muddy deposit. The formation of shelf mud deposits are discussed conceptually by McCave (1972), and the locus of deposition on the shelf is explained by the complex interplay between modern sediment supply and energetic physical activity (e.g. wave/current)(Fig. 4.6). Although the picture is not yet complete, in recent years these factors are starting to be evaluated quantitatively.

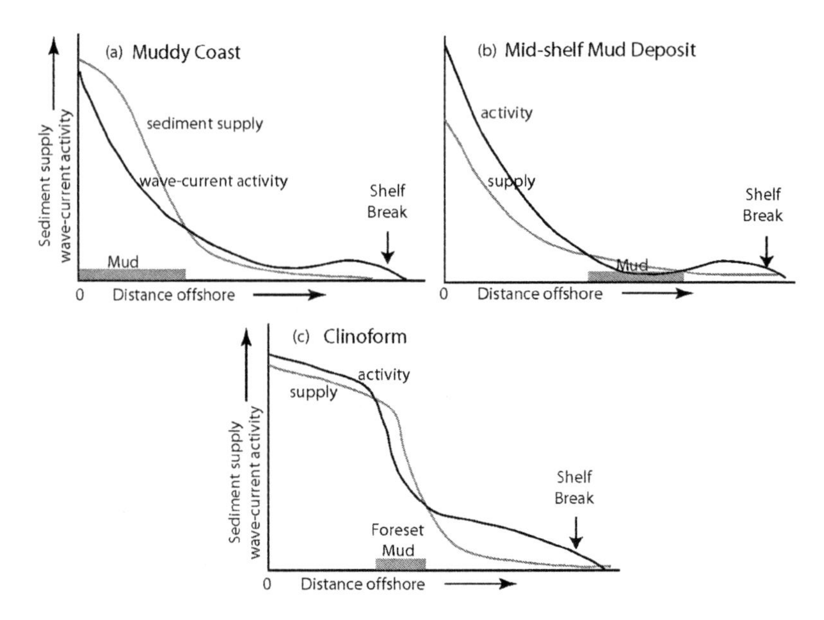


Figure 4.6 Schematic diagram of three cases of shelf mud accumulation where the wave and current activity relative to the sediment supply (or concentration in the terminology of McCave, 1972) controls the type of deposition: (a) muddy coast, (b) mid-shelf mud deposit, and (c) muddy shelf clinoform.

In this paper, the continental shelf is divided into contiguous environments, and new insights are discussed about the processes that dominate each environment and that link the environments. As a simplistic generalization, the four environments of relevance are: the inner shelf, mid shelf, shelf break and continental slope (Fig. 4.7a). In addition, large parts of the global coastal ocean contain morphologic features such as clinoforms and submarine canyons that play a significant role in the transfer of sediment across and off the shelf (Fig. 4.7b, c and d).

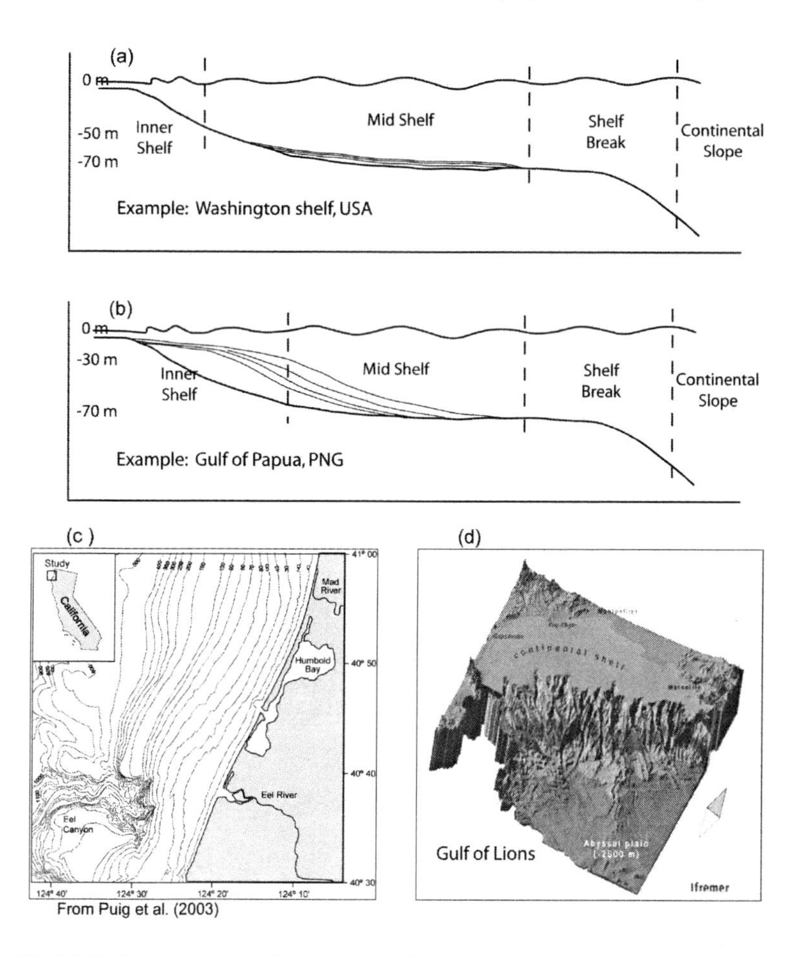


Figure 4.7 (a) Definition sketch of a classic continental shelf from shoreline to the slope showing a mid-shelf mud deposit, and (b) a continental shelf on which a subaqueous delta clinoform is prograding seaward. In plan view, the shelf and slope can be smooth (c) as on the Eel shelf north of the Eel Canyon, or (d) completely incised by canyons as in the Gulf of Lions. Part c used with permission of Elsevier.

### 3.2.1. New Insights on Inner to Mid-Shelf Sediment Transport

The inner shelf is considered to be a zone seaward of the shoreline where waves frequently agitate the seabed (Wright, 1995). In regions of high wave energy relative to sediment supply, little fine-grained sediment can persist (Fig. 4.6b; Fig. 4.7a); in the case of very large sediment supply relative to wave energy, muddy coasts or clinoform topset beds develop (Fig. 4.6a and c; Fig 4.7b)(McCave, 1972; Wright, 1995). The inner shelf has not been studied as extensively as the mid-shelf or surf zones, due to the difficulties in making comprehensive observations in the shallow coastal environment. A recent and comprehensive observation program on a sandy inner shelf occurred during the STRATAFORM program off the Eel River in northern California (Nittrouer and Kravitz, 1996; Nittrouer, 1999). Rapid-response observations were made on the inner shelf to document physical processes and suspended-sediment concentrations during storm events. These were accomplished with helicopter surveys rather than vessels, because of the high sea state (Geyer et al., 2000; Wheatcroft, 2000). On the Eel shelf, as with many shelves, very little of the fine sediment accumulates on the inner shelf (~10%; Crockett and Nittrouer, 2004), but all of the terrestrially derived sediment must pass through this region.

During flood stage, the Eel River plume and its sediment load are confined to the inner shelf (<50 m depth) and thus the plume is decoupled from the mid-shelf mud deposit (Geyer et al., 2000), which is located seaward of the 50-m isobath. The structure and position of the Eel River plume were observed during numerous floods and found to be confined within ~ 7 km of the shore, and to flow northward under wind forcing in the same direction. The flocculated fine sediment settling from the plume was deposited within 10 km of the river mouth, while the largely unflocculated sediment was transported as much as 30 km northward along the coast (Geyer et al., 2000; Hill et al., 2000). Additionally, complex density interactions between the fresh plume and the underlying coastal water may act to bring sediment particles to the seabed (McCool and Parsons, 2004). Plume delivery of sediment is a combination of settling by rapidly flocculated particles and buoyancy-driven circulation dynamics, and all of these processes occur over the inner shelf. Only small amounts of sediment are transported by the plume beyond the inner shelf. The inner shelf is also a zone of fluid-mud formation, due to intense wave shear stresses (Traykovski et al., 2000) and the potential for focussing the supply of sediment settling from plumes (Ogston et al., 2000).

The results observed on the Eel shelf suggest a sediment transport system far different than previously envisioned. The mid-shelf mud deposit is not formed by sediment supplied from an overlying sediment-laden plume. The plume occurs shoreward of the mud deposit and secondary processes (e.g., gravity-driven density flows, seaward sediment flux) move sediment laterally from the inner to the mid shelf. Understanding the mechanisms operating on the inner shelf could increase our understanding of seaward sediment flux and help close short-term sediment budgets.

The mid shelf is commonly a site of net sediment accumulation. Boundary shear stress is decreased at mid-shelf depths, and thus on the mid shelf, the mud deposit that forms is relatively stable because it occurs in a location where the net supply of sediment exceeds the net removal due to physical processes (Fig. 4.6b). Mud that has temporarally resided on the inner shelf or derived from river discharge is transported to the mid shelf and accumulates in zones of flux convergence. Subsequent resuspension and redistribution of sediment occurs during seasonal events (e.g., winter storms in temperate latitudes or intensified trade-wind seasons in tropical latitudes). However, in the long term, located below the depth of continuous resuspension yet close to sediment input, the mid shelf is where deposits tend to be preserved. Sediment and associated chemical components can be buried and removed from the system (Middelburg and Soetaert, this volume), as well as undergo diagenesis. The mid shelf also may be the recipient for downslope gravity-driven density flows that are formed on the inner shelf (Traykovski et al., 2000; Scully et al., 2002), and transported to the mid shelf where the deeper water depths and gentler slopes inhibit continued maintenance of high-concentration flows.

On shelves with rapid sediment accumulation (>1 cm $yr^{-1}$), morphologic features known as clinoforms are created in response to across-shelf variations in sediment supply and shear stresses (Fig. 4.6c). Even in regions of great sediment supply, wave and tide activity in shallow regions (topset beds) inhibits deposition and displaces the sediment farther seaward. At greater water depths, shear stresses decrease and allow rapid sediment accumulation, thus creating foreset beds that prograde seaward (Fig. 4.7b) (Walsh et al., 2004).

### 3.2.2. New Insights on Mid-Shelf to Slope Sediment Transport

The mid-shelf deposit makes sediment available to the offshelf dispersal system, through processes that can resuspend and deliver the sediment to the open slope and submarine canyons. Continental slopes are generally below the depth of wave reworking and are characterized by hemipelagic sedimentation, the slow fall of particles from nepheloid layers. The presence of bottom and intermediate nepheloid layers has been well documented over most continental slopes, providing a mechanism that links mid-shelf deposits to slope deposition. The formation of nepheloid layers has been connected to wave resuspension on the shelf and detachment of bottom turbid layers at the shelf break (Hickey et al., 1986; Nittrouer and Wright, 1994; Walsh and Nittrouer, 1999). This detachment commonly occurs along isopycnal surfaces (Drake et al., 1976; Pierce, 1976). Internal wave dynamics (Brink, this volume), can lead to internal wave breaking at critical angles on the slope and subsequent injection of particles on isopycnal surfaces (McPhee-Shaw et al., 2002). Internal waves operating in this manner have been proposed as a mechanism controlling the gradient of the continental slope. The internal waves would control deposition and erosion, keeping the continental slope at the critical angle (Cacchione et al., 2000).

Submarine canyons act as significant conduits for sediment transferred from the continental shelf to the deep ocean, through the action of turbidity currents. Based on analysis of turbidite stratigraphy, it has been thought that canyons were primarily active in transporting sediment at low stands of sea level when rivers discharged near the shelf break (directly supplying sediment to canyon heads). Turbidity currents in canyons generally have been inactive during the present sea-level high stand (e.g., Griggs and Kulm, 1970; Carson, 1971, 1973).

Studies of flow conditions and sediment transport in several submarine canyons have shown that canyons may be more active in transporting modern sediment seaward than previously thought (e.g., Gardner, 1989; Kineke et al., 2000; Mullenbach and Nittrouer, 2000; Puig et al., 2000). Observations indicate that large quantities of sediment are transported downslope via gravity-driven density flows (e.g., Monterey Canyon, Johnson et al., 2001; Sepik Canyon, Kineke et al., 2000, Walsh and Nittrouer, 2003; Eel Canyon, Puig et al., 2003). These canyons characteristically occur on narrow and steep shelves, with few estuaries, and close to a sediment source. Sediment provided to the heads of canyons in modern environments originates on the continental shelf where reworking of mid-shelf deposits during storms can transport sediment alongshelf over heads of canyons that incise the shelf. Alternately, sediment can be directly supplied to canyon heads through gravity-driven density flows originating at nearby river mouths. Puig et al. (2003) found gravity flows to occur in the Eel canyon contemporaneously with storms felt on the shelf, but not correlated with river discharge events. There are many canyons worldwide that fit this description and the role of modern canyons in cross-margin transport is being re-evaluated (see Parsons and Nittrouer, this volume).

### *3.3 Temporal Variation in Mechanisms – Seconds to Episodic Events*

Processes in the bottom boundary layer vary on multiple scales, causing fluctuations in shear stress, sediment resuspension and subsequent transport. These processes can be examined over time scales that range from turbulent fluctuations to interannual and episodic events. As discussed above, particles are influenced by a range of processes, depending on the forcing and the spatial regime, each of which has its own time scale (Table 1): river-plume trajectory, mixing dynamics of the plume, shear stresses due to wind-driven currents and waves, and mean and low-frequency circulation. Much progress has been made in sediment-transport studies with observational data provided by instrumentation capable of sampling at very high frequencies and resolutions. These short-term studies have been combined with observations on longer time scales, and are beginning to provide new insights about the connections between processes and strata formation.

#### 3.3.1. New Insights on Interannual Fluctuations in Sediment Transport Climatology

It has long been recognized that seasonal patterns such as stormy winters/quiescent summers in temperate regions and tradewind/monsoon seasons in tropical regions determine when sediment is discharged and how it is redistributed. Seasonal patterns of water-column structure also influence plume interactions, flocculation dynamics, and frontal or convergent circulation patterns that affect the transport mechanisms of sediment. Most observational studies have been performed over shorter periods of time (weeks to months), and the connection to longer-term depositional features are then extrapolated from these shorter records. Processes that range from high-frequency waves and tides to seasonal wind-driven events are presently being modeled with successful results.

A five-year record of sediment dynamics on the Eel shelf (as part of the STRATAFORM project) provides a look at longer time scales, which may have impact on our interpretation of shorter-term observations (Guerra, 2004; Ogston et al., 2004). This study showed that low-frequency motions are overlain on seasonal cycles, and play a significant role in determining the fate of shelf sediment. Earlier physical oceanographic studies have shown the importance of large-scale circulation features (mesoscale processes) with low-frequency time scales (Largier et al., 1993; Washburn et al., 1993), as well as interannual to decadal cycles (Table 1). These longer-term cycles cause variability in the strength and direction of storm-driven waves and currents for any specific region as well as the magnitude and timing of sediment delivery to the marine environment, with the result that net sedimentation in a single year (or over a specific experiment period) is potentially different from the long-term sedimentation pattern.

For example, on the Eel shelf, the frequency structure of the five-year record of sediment flux showed significant spectral energy of sediment flux in both the very low frequency band (VLF, with periods from weeks to months) and annual band (with periods of years)(Fig. 4.8a, b). Not only did the strength and frequency of physical events, including storms and river floods, and duration of the active winter season vary interannually (Guerra, 2004), but the net alongshelf direction of sediment flux for annual cycles varied from year to year over a five-year period (Fig. 4.8c) (Ogston et al., 2004). In addition to higher-frequency processes (e.g., tides, winds), the VLF band appears related to mesoscale circulation features, which have been observed by Largier et al. (1993) and modeled by Pullen and Allen (2000), and exert significant control on net sediment flux convergence and divergence. As yet, no observational studies have been performed that cover time scales that span decadal oscillations. However, these extended studies likely will provide the key to relating shelf processes and the preserved sedimentary record.

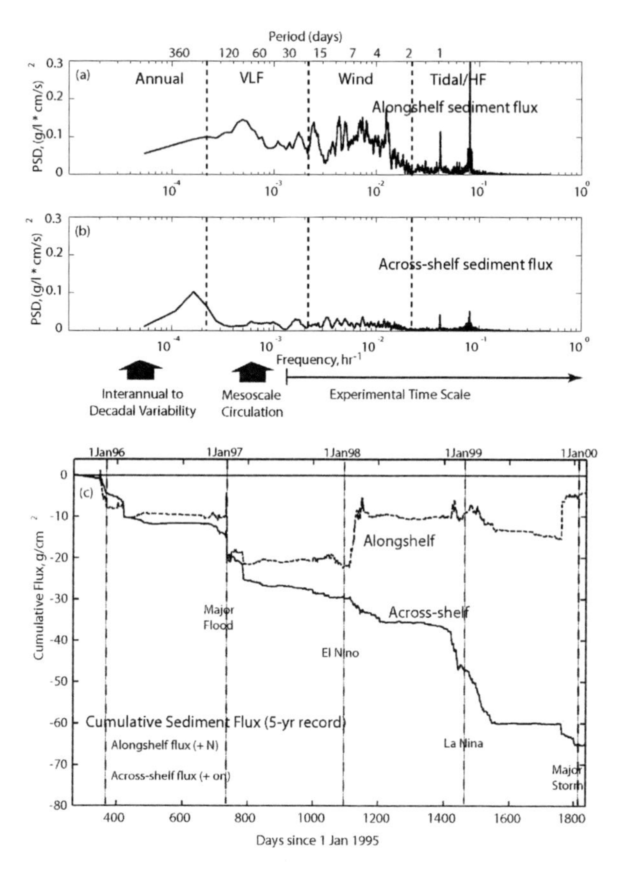


Figure 4.8 Example of the distribution of spectral energy of sediment flux on the Eel shelf (after Ogston et al., 2004) in (a) the along-shelf (b) the across shelf. Note that there is significant energy in both the annual and very low-frequency (VLF) bands representing the effects of temporal scales associated with interannual to decadal variability and mesoscale circulation. The effects in terms of net sediment flux over the longer time scales is seen in (c), where the flux in the alongshelf varies in direction and the quantity of across shelf flux varies in magnitude each yearly period corresponding to varying oceanic conditions (e.g., El Nino, La Nina). Used with permission of Elsevier.

### 3.3.2. New Insights Regarding Temporal Relationships among Sediment Supply, Transport, and Deposition

Not only is the frequency structure of processes important in the transport and deposition of sediment but the relative timing between these processes is important. In terms of the sediment supply, the river hydrograph (including the variability and episodicity of discharge) is chiefly determined by the size of the drainage basin, geographical location, and climate oscillations inducing rainfall and snowmelt. The energetic processes in the ocean basin, as discussed above, have time scales of variability that range from seconds (e.g., waves) to decades (e.g., ENSO oscillations). The interactions of these fluvial and marine processes control the sediment response.

The oceanic conditions at the time of fluvial discharge events are key factors determining the transport mechanism and resulting sedimentary deposit (Fig. 4.9). For example, floods in mountainous rivers with small drainage basins occur almost simultaneously with energetic conditions in the ocean, so that an abundant supply of sediment and highly energetic conditions on the inner shelf are concurrent. Comparison of the Columbia River discharge on the Washington-Oregon coast and the Eel River discharge from northern California illustrates the importance of timing on shelf sedimentology. The Columbia River drains 670,000 $km^2$ and the hydrograph shows floods twice per year (Fig. 4.9a), during the winter rainy season and the spring snowmelt season. During the winter flood, winds force the plume northward along the coast and sediment is discharged to the shelf during both calm and energetic conditions. During the spring flood, northerly winds transport the plume several hundred kilometers offshore and southward (Barnes, et al., 1972). This occurs during relatively calm conditions when sediment can settle from the plume near the discharge location. Thus, over an annual cycle, sediment can be deposited over a large geographic region emanating from the river mouth, but the mid-shelf mud deposit reflects deposition near the river mouth and redistribution northward during winter storms (Fig. 4.9c)(Smith and Hopkins, 1972; Sternberg et al., 1972). Approximately seventy percent of the sediment discharged from the Columbia River can be found in this deposit (Nittrouer, 1978; Sternberg, 1986).

In contrast, the Eel River, 500 km to the south, has a small, low-elevation drainage basin (9500 $km^2$). As a result, the river floods quickly in response to winter storms and intense rainfall and each flood only lasts a few days (Fig. 4.9b). River floods and associated sediment discharge occur contemporaneously with the winter cyclonic storms that sweep the shelf. Southerly winds force the river plume northwards against the coast and the sediment load of the Eel River settles out of the plume on the inner shelf (<50-m water depth) and within 10–30 km of the river mouth. At the same time, energetic conditions keep sediment in motion near the seabed, and creates convergent conditions vital for fluid-mud formation. Seaward transport moves sediment to the mid shelf, where much of the Eel sediment is dispersed beyond the shelf break (either delivered to the Eel Canyon or through nepheloid layers on the slope). This results in a distinct mid-shelf flood deposit disconnected from the river mouth, which contains only approximately twenty percent of the sediment discharged during the flood event (Fig. 4.9d) (Sommerfield and Nittrouer, 1999).

This comparison illustrates that in addition to all of the individual processes that contribute to shelf sediment distribution, resuspension and flux, the relative timing of these processes is likewise important. Ultimately, coastal oceanographers will be able to build on these concepts to construct an understanding of how the infinite variations on coastal morphology develop.

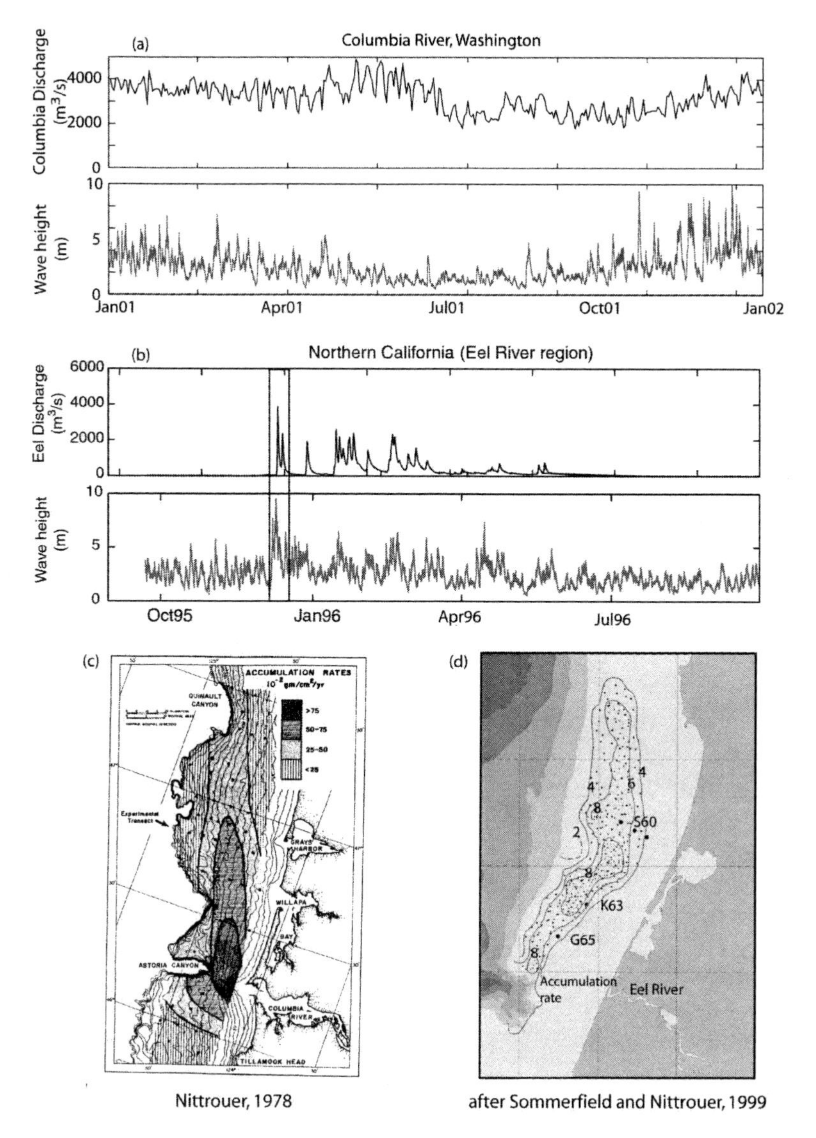


Figure 4.9 The potential impact on depositional processes of timing between the forcing mechanism can be illustrated by comparing (a) the river discharge and offshore wave climate of the Columbia River and (b) the Eel River. On the Columbia River shelf (c), variations in river discharge are at low frequency relative to the storm frequency and the resulting mid-shelf deposit is relatively continuous from river mouth to mid shelf. On the Eel shelf (d), storms and floods are of the same time scale and concurrent (see example in (b), boxed data), resulting in a deposit on the mid shelf disconnected from the river mouth.

## 4. Conclusions

Understanding of sedimentary processes using boundary-layer theory has advanced considerably in recent years, and comprehensive results are being incorporated into modeling efforts that provide insight into the relationship between active sediment-transport processes and the development of shelf sedimentary deposits. Concepts of shelf sedimentation are changing in light of new information garnered from recent studies, with special attention paid to the long-term variability and the shelf-wide coupling of processes, in particular,

- fluid mud (gravity-driven density flows of sediment) can be formed in a range of shelf environments and may transport vast quantities of sediment in the across-shelf direction,
- fluvial sediment must pass across the inner shelf, which represents a region of temporary sediment reprocessing that controls the supply mechanisms to deeper portions of the shelf,
- submarine canyons can be active during high stands of sea level, and might be a locus of gravity-driven density flows in modern times, moving sediment from the shelf environment to the deep sea,
- longer-term processes that have interannual variability and those associated with mesoscale circulation have a major influence on net sediment flux,
- the relative timing of processes has a strong influence on how they combine to cause shelf transport and ultimate deposition.

## Acknowledgments

The authors gratefully acknowledge the Office of Naval Research (N00014–02–1–0082 and N00014–99–1–0028) and the National Science Foundation (OCE-0203351) for their support of the authors while writing this manuscript, as well as for their support over the years of research activities relating to the projects discussed. This manuscript benefited from input by two anonymous reviewers.

## Bibliography

Allison, M.A., Nittrouer, C.A., Kineke, G.C. (1995) Seasonal sediment storage on mudflats adjacent to the Amazon River. *Marine Geology* 125, 303–328.

Aller, R.C., Blair, N.E., Xia, Q., Rude, P.D. (1996) Remineralization rates, recycling, and storage of carbon in Amazon shelf sediments. *Continental Shelf Research* 16, 753–786.

Barnes, C.A., Duxbury, A.C., Morse, B.A. (1972) Circulation and selected properties of Columbia River effluent at sea. In *The Columbia River Estuary and Adjacent Waters. Environmental Studies,* University of Washington Press 41–80.

Beardsley, R.C. Candela, J., Limeburner, R., Geyer, W.R., Lentz, S.J., Castro, B.M., Cacchione, D., Carneiro, N. (1995) The M2 tide on the Amazon shelf. *Journal of Geophysical Research* 100, 2283–2319.

Berne, S., Satra, C. et al. (2002) *Carte morpho-bathymetrique du Golfe du Lion, notice explicative,* Ifremer, Brest.

Brink, K.H. (in press) Coastal physical processes overview. In *The Sea* this volume.

Cacchione, D.A., Pratson, L.F., Ogston, A.S. (2000) The shaping of continental slopes by internal tides. *Science* 296, 724–727.

Cacchione, D.A., Drake, D.E., Kayen, R.W., Sternberg, R.W., Kineke, G.C., Tate, G.B. (1995). Measurements in the bottom boundary layer on the Amazon subaqueous delta. *Marine Geology* 125, 235–257.

Carson, B. (1971) Stratigraphy and depositional history of Quaternary sediments in Northern Cascadia Basin and Juan de Fuca Abyssal Plain, Northeast Pacific Ocean. Ph.D.Dissertation, University of Washington, Seattle, WA 249pp.

Carson, B. (1973) Acoustic stratigraphy, structure, and history of Quaternary deposition in Cascadia Basin. *Deep-Sea Research* 20, 387–396.

Channon, R.D., Hamilton, D. (1971) Sea bottom velocity profiles on the continental shelf southwest of England. *Nature* 231, 383–385.

Crockett. J.S., Nittrouer, C.A. (2004) The sandy inner shelf as a repository for muddy sediment: an example from Northern California. *Continental Shelf Research* 24, 55–73.

Curray, J.R. (1965) Structure of the continental margin off central California. *Transactions of the New York Academy of Sciences* 27, 794–801.

Drake, D.E. (1976) Suspended sediment transport and mud deposition on continental shelves. In *Marine Sediment Transport and Environmental Management,* John Wiley and Sons, New York 127–158.

Drake, D.E., Cacchione, D.A. (1986). Field observations of bed shear stress and sediment resuspension on continental shelves, Alaska and California. *Continental Shelf Research* 6, 415–429.

Drake, D.E., Cacchione, D.A. (1989) Estimates of the suspended sediment reference concentration ($C_\alpha$) and resuspension coefficient ($\gamma_0$) from near-bottom observations on the California shelf. *Continental Shelf Research* 9, 51–64.

Dyer, K.R. (1970) Linear erosional furrows in Southampton water. *Nature* 225, 56–58.

Dyer, K.R. (1980) Velocity profiles over a rippled bed and the threshold of movement of sand. *Estuarine Coastal Marine Sciences* 10, 181–199.

Dyer, K.R. (1986) Coastal and Estuarine Sediment Dynamics. Wiley, New York 342pp.

Faas, R.W. (1986) Mass-physical and geotechnical properties of surficial sediments and dense nearbed sediment suspensions on the Amazon continental shelf. *Continental Shelf Research* 6, 189–208.

Fox, J.M., Hill, P.S., Milligan, T.G., Boldrin, A. (2004) Flocculation and sedimentation on the Po River Delta. *Marine Geology* 203, 95–107.

Friedrichs, C.T., Wright, L.D. (2004) Gravity-driven sediment transport on the continental shelf: implications for equilibrium profiles near river mouths. *Coastal Engineering,* 51, 795–811.

Gardner, W. (1989) Periodic resuspension in Baltimore Canyon by focusing of internal waves. *Journal of Geophysical Research* 94, 18185–18194.

Geyer, W.R. (1995) Tide-induced mixing in the Amazon frontal zone. *Journal of Geophysical Research* 100, 2341–2353.

Geyer, W.R., Hill, P., Milligan, T., Traykovski, P. (2000) The structure of the Eel River plume during floods. *Continental Shelf Research* 20, 2067–2093.

Gordon, C.M. (1975) Sediment entrainment and suspension in a tidal flow. *Marine Geology* 18, M57-M64.

Grant, W.D., Madsen, O.S. (1979). Combined wave and current interaction with a rough bottom. *Journal of Geophysical Research* 84, 1797–1808.

Grant, W.D., Madsen, O.S. (1986) The continental shelf bottom boundary layer. *Annual Review of Fluid Mechanics* 18, 265.

Griggs, G.B., Kulm, L.D. (1970) Sedimentation in Cascadia deep-sea channel. *Geological Society of America Bulletin* 81, 1361–1384.

Gross, T.F., Nowell, A.R.M. (1983) Measurements of the turbulent structure in a tidal boundary layer. *Continental Shelf Research* 2, 109–126.

Guerra, J.V. (2004) Interannual variability of nearbed sediment flux and associated physical proceses on the Eel River Shelf, Northern California, USA. Ph.D. Dissertation, University of Washington, Seattle, WA 114pp.

Harris, C.K., Wiberg, P.L. (2001) A two-dimensional, time-dependent model of suspended sediment transport and bed reworking for continental shelves. *Computers and Geosciences* 27, 675–690.

Harris, C.K., Wiberg, P.L. (2002) Across-shelf sediment transport; interactions between suspended sediment and bed sediment. *Journal of Geophysical Research,* 107, C1, 3008, 10.1029/2000JC000634.

Hedges, J.I., Keil, R.G. (1995) Sedimentary organic-matter preservation-an assessment and speculative synthesis. *Marine Chemistry* 49, 81–115.

Hickey, B., Baker, E., Kachel, N. (1986). Suspended particle movement in and around Quinault submarine canyon. *Marine Geology* 71, 35–83.

Hill, P.S., McCave, I.N. (2001) Suspended particle transport in benthic boundary layers. In *The Benthic Boundary Layer,* Oxford University Press, New York, 78–103.

Hill, P., Milligan, T., Geyer, W.R. (2000) Controls on effective settling velocity of suspended sediment in the Eel River plume. *Continental Shelf Research* 20, 2095–2111.

Hill, P.S., Nowell, A.R.M., Jumars, P.A. (1988) Flume evaluation of the relationship between suspended sediment concentration and excess boundary shear stress. *Journal of Geophysical Research* 93, 12499–12510.

Ingliss, C.C., Allen, F.H. (1957) The regimen of the Thames as affected by currents, salinities, and river flow. *Proceedings of the Institute of Civil Engineers* 7, 827–878.

Jahnke, R.A. (2004) Diagenic and transport processes in coastal sediments. In *The Sea* this volume.

Jaeger, J.M., Nittrouer, C.A. (1995) Tidal controls for the formation of fine-scale sedimentary strata near the Amazon river mouth. *Marine Geology* 125, 259–281.

Johnson, K.S., Paull, C.K., Barry, J.P., Chavez, F.P. (2001) A decadal record of underflows from a coastal river into the deep sea. *Geology* 29, 1019–1022.

Kachel, N.B. (1980) A time-dependent model of sediment transport and strata formation on a continental shelf. Ph.D. Dissertation, University of Washington, Seattle, WA 119pp.

Kineke, G.C. (1993). Fluid muds on the Amazon continental shelf. Ph.D. dissertation. University of Washington, Seattle WA 249 pp.

Kineke, G.C., Sternberg, R.W. (1989) The effect of particle settling velocity on computed suspended sediment concentration profiles. *Marine Geology* 90, 159–174.

Kineke, G.C., Sternberg, R.W. (1995) Distribution of fluid muds on the Amazon continental shelf. *Marine Geology* 125, 193–233.

Kineke, G.C., Sternberg, R.W., Trowbridge, J.H., Geyer, W.R. (1996). Fluid-mud processes on the Amazon continental shelf. *Continental Shelf Research* 16, 667–696.

Kineke, G.C., Woolfe, K.J., Kuehl, S.A., Milliman, J., Dellapenna, T.M., Purdon, R.G., (2000) Sediment export from the Sepik River, Papua New Guinea: evidence for a divergent sediment plume. *Continental Shelf Research 20,* 2239–2266.

Krone, R.B. (1976) Engineering interest in the benthic boundary layer. In *The Benthic Boundary Layer,* Plenum, New York 143–156.

Krone, R.B. (1978) Aggregation of suspended particles in estuaries. In *Estuarine Transport Processes,* University of South Carolina Press 177–190.

Kuehl, S. A., Nittrouer, C.A., Allison, M.A., Ercilio, L., Faria, C. Dukat, D.A., Jaeger, J.M., Pacioni, T.D., Figueiredo, A.G., Underkoffler, E.C. (1996) Sediment deposition, accumulation, and seabed dynamics in an energetic fine-grained coastal environment. *Continental Shelf Research* 16, 787–815.

Kuehl, S. A., Nittrouer, C.A., DeMaster, D.J. (1988) Microfabric study of fine-grained sediments: observations from the Amazon subaqueous delta. *Journal of Sedimentary Petrology* 58, 12–23.

Largier, J.L., Magnell, B.A., Winant, C.D. (1993) Subtidal circulation over the Northern California shelf. *Journal of Geophysical Research* 98, 18147–18180.

Lee, H.J., Edwards, B.D. (1986) Regional method to assess offshore slope stability. *Journal of Geotechnical Engineering, ASCE* 112, 489–509.

Leithold, E.L., Bourgeois, J. (1989) Sedimentation, sea-level change, and tectonics on an early Pleistocene continental shelf, Northern California. *Geological Society of America Bulletin* 101, 1209–1224.

McCave, I.N., 1972. Transport and escape of fine-grained sediment from shelf areas. In: *Shelf Sediment Transport: Process and Pattern,* Dowden, Hutchinson & Ross, Stroudsburg, PA, 225–248.

McCool, W.W., Parsons, J.D. (2004) Sedimentation from buoyant fine-grained suspensions. *Continental Shelf Research* 24, 1129–1142.

McLean, S.R. (1983) Turbulence and sediment transport measurements in a North Sea tidal inlet (the Jade). *North Sea Dynamics* 436–452.

McPhee-Shaw, E.E., Kunze, E. (2002) Boundary layer intrusions from a sloping bottom; a mechanism for generating intermediate nepheloid layers. *Journal of Geophysical Research* 107, C6, 3050, 10.1029/2001JC000801.

McPhee-Shaw, E.E., Sternberg, R.W., Mullenbach, B., Ogston, A.S. (2004) Observations of intermediate nepheloid layers on the Northern California continental margin. *Continental Shelf Research* 24, 693–720.

Meade, R.H. (1972) Transport and deposition of sediments in estuaries. *Geological Society of America, Memoir* 133, 91–120.

Middelburg, J.J., Soetaert, K. (in press) The role of sediments in shelf ecosystem dynamics. In *The Sea* this volume.

Moore, W.S., DeMaster, D.J., Smoak, J.M., McKee, B.A., Swarzenski, P.W. (1996) Radionuclide tracers of sediment-water interactions on the Amazon Shelf. *Continental Shelf Research* 16, 645–665.

Mulder, T., Syvitski, J.P.M. (1995) Turbidity currents generated at river mouths during exceptional discharges to the world oceans. *Journal of Geology* 103, 285–299.

Mullenbach, B.L. and Nittrouer, C.A., (2000) Rapid deposition of fluvial sediment in the Eel Canyon, Northern California. *Continental Shelf Research* 20, 2191–2212.

Nichols, M.M., (1985) Fluid mud accumulation processes in an estuary. *Geo-Marine Letters* 4.

Nittrouer, C.A. (1978) The process of detrital sediment accumulation in a continental shelf environment: an examination of the Washington shelf. Ph.D. Dissertation, University of Washington, Seattle, WA 243pp.

Nittrouer, C.A. (1999). STRATAFORM: overview of its design and synthesis of its results. *Marine Geology,* 154, 3–12.

Nittrouer, C.A., Kravitz, J.H. (1996) STRATAFORM: a program to study the creation and interpretation of sedimentary strata on continental margins. *Oceanography* 9, 146–152.

Nittrouer, C.A., Kuehl, S.A., DeMaster, D.J., Kowsmann, R.O. (1986). The deltaic nature of Amazon shelf sedimentation. *Geological Society American Bulletin* 97, 444–458.

Nittrouer, C.A., Sternberg, R.W. (1981). The formation of sedimentary strata in an allochthonous shelf environment: application to the Washington continental shelf. *Marine Geology* 42, 201–232.

Nittrouer, C.A., Wright, L.D. (1994) Transport of particles across continental shelves. *Reviews of Geophysics* 32, 85–113.

Noffke, N., Gerdes, G., Klenke, T., Krumbein, W.E. (2001) Microbially induced sedimentary structures - a new category within the classification of primary sedimentary structures. *Journal of Sedimentary Research, Section B: Stratigraphy and Global Studies* 71, 649–656.

Nowell, A.R.M. (1983) The benthic boundary layer and sediment transport. *Rev. Geophys Space Phys.* 21, 1181–1192.

Nowell, A.R.M., Jumars, P.A., Eckman, J.E. (1981) Effects of biological activity on the entrainment of marine sediments. *Marine Geology* 42, 133–153.

Ogston, A.S., Cacchione, D.A., Sternberg, R.W., Kineke, G.C., (2000) Observations of storm and river flood-driven sediment transport on the northern California continental shelf. *Continental Shelf Research* 20, 2141–2162.

Ogston, A.S., Guerra, J.V., Sternberg, R.W. (2004). Interannual variability of nearbed sediment flux on the Eel River shelf, northern California. *Continental Shelf Research* 24, 117–136.

Ogston, A.S., Sternberg, R.W. (1999) Sediment transport events on the northern California continental shelf. *Marine Geology* 154, 69–82.

Parsons, J., Nittrouer, C.A. (in press) Extreme events moving particulate material on continental margins. In *The Sea,* this volume.

Pierce, J.W. (1976) Suspended sediment transport at the shelf break and over the outer margin. In *Marine Sediment Transport and Environmental Management,* John Wiley and Sons, New York 437–458.

Puig, P., Ogston, A.S., Mullenbach, B.I., Nittrouer, C.A., Sternberg, R.W. (2003) Shelf-to-canyon sediment transport processes on the Eel continental margin (Northern California). *Marine Geology* 193, 129–149.

Pullen, J.D., Allen, J.S. (2000) Modeling studies of the coastal circulation off Northern California: Shelf response to a major Eel River flood event. *Continental Shelf Research* 20, 2213–2238.

Pullen, J.D., Allen, J.S. (2001) Modeling studies of the coastal circulation off Northern California; statistics and patterns of wintertime flow. *Journal of Geophysical Research* 106, 26959–26984.

Rees, A.I. (1966) Some flume experiments with a fine silt. *Sedimentology* 6, 209–240.

Scully, M.E., Friedrichs, C.T., Wright, L.D. (2002) Application of an analytical model of critically stratified gravity-driven sediment transport and deposition to observations from the Eel River continental shelf, Northern California. *Continental Shelf Research* 22, 1951–1974.

Sherwood, C.R., Drake, D.E., Wiberg, P.L., Wheatcroft, R.A. (2002) Prediction of the fate of *p,p'*-DDE in sediment on the Palos Verdes shelf, California, USA. *Continental Shelf Research* 22, 1025–1058.

Sherwood, C.R., Butman, B., Cacchione, D.A., Drake, D.E., Gross, T.F., Sternberg, R.W., Wiberg, P.L., Williams III, A.J. (1994). Sediment-transport events on the Northern California continental shelf during the 1990–1991 STRESS experiment. *Continental Shelf Research* 14, 1063–1099.

Sleath, J.F.A. (1990) Seabed boundary layers. In *The Sea* 9, 693–727.

Smith, J.D. (1977) Modeling of sediment transport on continental shelves. In *The Sea* 6, 539–577.

Smith, J.D. Hopkins, T.S. (1972) Sediment transport on the continental shelf off to Washington and Oregon in light of recent current measurements: In *Shelf Sediment Transport: Process and Pattern.* Dowden, Hutchinson, & Ross, Stroudsburg, PA, 143–180.

Smith, J.D., McLean, S.R. (1977) Spatially averaged flow over a wavy surface. *Journal of Geophysical Research* 82, 1735–1746.

Sommerfield, C.K., Nittrouer, C.A. (1999) Modern accumulation rates and a sediment budget for the Eel shelf: a flood-dominated depositional environment. *Marine Geology* 154, 227–241.

Sommerfield, C.K., Nittrouer, C.A., Alexander, C.R. (1999) $^{7}$Be as a tracer of flood sedimentation on the Northern California continental margin. *Continental Shelf Research* 19, 335–361.

Spielman, L.A. (1978) Hydrodynamic aspects of flocculation. In *The Scientific Basis of Flocculation,* Sijthoff & Noordhoff, 63–88.

Sternberg, R.W. (1966) Boundary layer observations in a tidal current. *Journal of Geophysical Research* 71, 2175–2178.

Sternberg, R.W. (1968) Friction factors in the tidal channels with differing bed roughness. *Marine Geology* 6, 243–260.

Sternberg, R.W. (1972) Predicting initial motion and bedload transport of sediment particles in the shallow marine environment. In *Shelf Sediment Transport: Process and Pattern.* Dowden, Hutchinson & Ross, Stroudsburg, PA, 61–82.

Sternberg, R.W. (1976) Measurements of boundary-layer flow and boundary roughness over Campeche Bank, Yucatan. *Marine Geology* 20, M25-M31.

Sternberg, R.W. (1986) Transport and accumulation of river-derived sediment on the Washington continental shelf, USA. *Journal of the Geological Society, London* 143, 945–956.

Sternberg, R.W., Cacchione, D.A., Drake, D.E., Kranck, K. (1986) Suspended sediment dynamics in an estuarine tidal channel within San Francisco Bay, California. *Marine Geology* 71, 237–258.

Styles, R., Glenn, S.M. (2000). Modeling stratified wave and current bottom boundary layers on the continental shelf. *Journal of Geophysical Research* 105, 119–124.

Traykovski, P., Geyer, W.R., Irish, J.D., Lynch, J.F., (2000) The role of wave-induced density- driven fluid mud flows for cross-shelf transport on the Eel River continental shelf. *Continental Shelf Research* 20, 2113–2140.

Trowbridge, J.H., Kineke, G.C. (1994) Structure and dynamics of fluid muds on the Amazon continental shelf. *Journal of Geophysical Research* 99, 865–874.

Walsh, J.P., Nittrouer, C.A. (1999) Observations of sediment flux on the Eel Continental slope, Northern California. *Marine Geology* 154, 55–68.

Walsh, J.P., Nittrouer, C.A. (2003) Contrasting styles of off-shelf sediment accumulation in New Guinea. *Marine Geology* 196, 105–125.

Walsh, J.P., Nittrouer, C.A., Palinkas, C.M., Ogston, A.S., Sternberg, R.W., Brunskill, G.J. (2004) Clinoform mechanics in the Gulf of Papua, New Guinea. *Continental Shelf Research* 24, 2487–2510.

Washburn, L., Swenson, M.S., Largier, J.L., Kosro, P.M., Ramp, S.R. (1993) Cross-shelf sediment transport by an anticyclonic eddy off Northern California. *Science* 261, 1560–1564.

Wells, J.T. (1983) Dynamics of coastal fluid muds in low-, moderate-, and high-tide-range environments. *Canadian Journal of Fishery and Aquatic Sciences* 40(suppl. 1), 130–142.

Wheatcroft, R.A. (2000) Oceanic flood sedimentation: a new perspective. *Continental Shelf Research* 20, 2059–2066.

Wiberg, P. (2000) A perfect storm: formation and potential for preservation of storm beds on the continental shelf. *Oceanography* 13, 93–99.

Wiberg, P.L., Drake, D.E., Cacchione, D.A. (1994) Sediment resuspension and bed armoring during high bottom stress events on the Northern California inner continental shelf: measurements and predictions. *Continental Shelf Research* 14, 1191–1219.

Wiberg, P.W., Smith, J.D. (1983) A comparison of field data and theoretical models for wave- current interactions at the bed on the continental shelf. *Continental Shelf Research* 2, 147–162.

Wright, L.D. (1995) Morphodynamics of Inner Continental Shelves. CRC Press, Inc., Boca Raton. 241pp.

Wright, L.D., Friedrichs, C.T., Kim, S.C., Scully, M.E. (2001) Effects of ambient currents and waves on gravity-driven sediment transport on continental shelves. *Marine Geology* 175, 25–45.

Wright, L.D., Kim, S.C. , Friedrichs, C.T. (1999) Across-shelf variations in bed roughness, bed stress and sediment suspension on the Northern California shelf. *Marine Geology* 154, 99–116.

Wright, L.D., Wiseman, W.J., Yang, Z.S., Bornhold, B.D., Keller, G.H., Prior, D.B., Suhayda, J.N. (1990). Processes of marine dispersal and deposition of suspended silts off the modern mouth of the Huange (Yellow River). *Continental Shelf Research* 10, 1–40.

Wright, L.D., Yang, Z.S., Bornhold, B.D., Keller, G.H., Prior, D.B., Wiseman Jr., W.J. (1986) Hyperpycnal plumes and plume fronts over the Huanghe (Yellow River) delta front. *Geo -Marine Letters* 6, 97–105.

Wright, L.D., Nittrouer, C.A., (1995). Dispersal of river sediments in coastal seas: six contrasting cases. *Estuaries* 18, 494–508.

# Chapter 5. ORGANIC MATTER IN COASTAL MARINE SEDIMENTS

SUSAN M. HENRICHS

*University of Alaska Fairbanks*

## Contents



## 1. Introduction

Annually 1.6 X $10^{14}$ g of organic carbon accumulate in marine sediments, and about 86% of that burial occurs in deltaic, continental shelf, and upper continental slope sediments (Hedges and Keil, 1995). The vast majority of organic carbon in sedimentary rocks, 1.6 X $10^{22}$ g C, apparently was laid down in similar environments, under an oxygenated water column at high sediment accumulation rates (Summons, 1993; Hedges and Keil, 1995). Over geologic time, this large reservoir of reduced carbon is a key part of the checks and balances of the global carbon cycle that regulate atmospheric carbon dioxide and oxygen concentrations (Berner and Canfield, 1989; Berner and Caldeira, 1997). However, the organic matter preserved in sedimentary deposits, which typically contain 0.3 to 3% organic carbon, constitutes less than 0.2% of global primary productivity (Hedges and Keil, 1995), and there is great variation in burial efficiency, i.e., the fraction of organic matter deposited at the sediment water interface that is buried long-term (Henrichs and Reeburgh, 1987). Despite decades of research, the processes leading to the sedimentary preservation of a small fraction of primary production remain uncertain. Neither is it understood why the quantity of organic matter buried is reasonably constant through geologic time, although there are large geographic and temporal variations in burial efficiency. However, it is clear that the clues needed to resolve this mystery must lie in the coastal regions where the organic matter is deposited.

*The Sea*, Volume 13, edited by Allan R. Robinson and Kenneth H. Brink
ISBN 0-674-01526-6 

Although the preserved organic matter is important on geologic time scales, the organic matter that decomposes is of at least equal contemporary interest. As pointed out by Jahnke (this volume), coastal sediments are important sites of nutrient regeneration. The decomposition of organic matter consumes oxygen, nitrate, manganese oxides, iron oxides, sulfate, and ultimately carbon dioxide and carbonyl groups of organic molecules as electron acceptors in the oxidation process. Concomitantly, reduced forms of nitrogen, manganese, iron, sulfur, and carbon are produced, leading to striking changes in sediment geochemistry with depth and affecting the cycling of most elements of the periodic table.

Much of the diagenesis of organic matter takes place early, at the sediment-water interface and within the upper 10–20 cm of marine sediments (Berner, 1980). Among indications of this are the steep gradients of oxidizing agents and mineralization products and changes in organic matter composition. Whatever processes ultimately preserve organic matter, crucial initial stages of these processes must be completed within this zone. One of the most striking changes in organic matter composition between the producing organisms and surface sediments is that most sediment organic matter is not identifiable as any of the major biochemical compound groups, proteins, carbohydrates, and lipids, which compose marine and terrestrial plants, with lignin also being a component in the latter case (Hedges and Keil, 1995). The organic matter of sediments cannot be completely characterized by present analytical techniques, although these have given us substantial information about its properties. The uncharacterized fraction of sediment organic matter has been termed humic material, by analogy to similarly difficult to characterize organic matter of terrestrial soils. However, even in coastal sediments the stable isotopic composition of marine sediment organic matter suggests that it is mainly of marine origin, and its chemical characteristics differ from those of organic matter in terrestrial soils (Hedges and Oades, 1997). Consequently, the acronym MUC (molecularly uncharacterized carbon) is preferable to the phrases humic material or humic acids, when referring to marine sediment organic matter (Hedges et al., 2000). Because the identifiable biochemicals in sedimentary organic matter decrease with depth relative to MUC, it is inferred that some of the MUC is refractory to microbial decomposition, but the specific structural characteristics that confer this protection from microbial attack are unclear.

This review will focus on continental shelf and slope sediments, and will not explicitly address estuarine, salt marsh, or freshwater sediments. However, much of the research addressing organic matter decomposition and preservation processes has been done on estuarine sediments, and relatively little in deltas or the open shelf. It is likely that processes in estuarine sediments, with certain reservations, are similar to those of deltaic, shelf, and slope sediments, and so appropriate analogies will be drawn. This review builds upon several earlier papers summarizing organic matter decomposition and preservation processes in marine sediments, particularly Hedges and Keil (1995). That extensive overview remains an excellent reference and is more comprehensive than this chapter. However, there have been some significant recent changes in our view of sedimentary organic matter, particularly in the role that adsorption and other mechanisms of physical protection play in the preservation process (Ransom et al., 1997; Mayer, 1999). There is new information on the importance of highly refractory forms of carbon, such as black carbon and redeposited kerogen, in marine sediments (Blair et al., 2003; Dickens et al., 2004). Further, understanding of the role that oxygen plays in determining the rate and extent of organic matter oxidation is still evolving (Mayer et al., 2002; Hartnett et al., 2003). A variety of modern analytical techniques are offering new insight into sediment organic matter composition and transformations.

## 2. The Starting Material

### *2.1 Marine Phytoplankton*

Influxes of organic matter to the coastal ocean are summarized in Tables 5.1 and 5.2. Stable isotopic and biomarker evidence indicates that much of sedimentary organic matter was produced within the oceans, even in coastal settings. This is probably simply because far more organic matter is produced within coastal seas, $10 \times 10^{15}$ gC $yr^{-1}$ (Liu et al., 2000), than is delivered there by major rivers, $0.4 \times 10^{15}$ gC $yr^{-1}$ (Hedges et al., 1997), even though some of the terrestrial organic matter is refractory and has survived for centuries within soils (Hedges, 1992). Plankton consist predominantly of identifiable, major biochemicals, with 82% of the total being composed of hydrolyzable amino acids, carbohydrates, lipids, and pigments (Wakeham et al., 1997a) (Table 5.3). Wakeham et al. (1997a) did not measure some biochemicals, including nucleic acids, amino sugars, uronic acids, and pigments other than chlorophyll, which therefore constituted part of the uncharacterized 18%. Some phytoplankton contain refractory aliphatic macromolecules, algaenans (Tegelaar et al., 1989; De Leeuw and Largeau, 1993). These are produced by eustigmatophytes and chlorophytes, but apparently not by the diatoms, prymnesiophytes, or most dinoflagellates that dominate marine plankton communities (Gelin et al., 1999). Photochemical processes may also produce complex, refractory compounds from planktonic material in the upper ocean (Kieber et al., 1997).

TABLE 5.1.
Inputs and burial of organic matter in the coastal ocean ($10^{12}$ gC $yr^{-1}$).

| | Primary Productivity | Riverine Input | Burial (% terrigenous) | Export |
|---|---|---|---|---|
| North Sea[a] | 75 | 4.0 | 1.1[d](?) | 0.1 |
| Middle Atlantic Bight[a] | 75 | ? | 0(?) | 4.8 |
| Washington Shelf[a] | 7.5 | 0.6 | 0.4(25%) | 0.2 |
| MacKenzie River Shelf[a] | 3.2 | 2.2 | 1.44[e](94%) | 0.7 (100%) |
| NE Gulf of Alaska Shelf[b] | 10 | 1[f] | 0.5(50%) | ? |
| Mississippi-Atchafalaya Shelf[c] | 2[g] | 3 | 0.4[g](66–79%) | ? |
| Amazon Shelf[a] | 150 | 50 | 4.5 (70%) | ? |

[a]Data for these sites were compiled from a large number of sources by de Haas et al. (2002).
[b]Ding (1998).
[c]Twilley and McKee (1995)
[d]Of this amount, $1 \times 10^{12}$ gC is accumulated in the deep basins, Skagerrak and the Norwegian Channel.
[e]Total for MacKenzie delta and shelf sediments.
[f]Estimated from the sediment delivery (Jaeger et al., 1998), assuming 1% organic C.
[g]For the 5400 $km^2$ area of the sediment plume.

TABLE 5.2

Global total inputs and burial of organic matter in the coastal ocean.

| | $10^{15}$ gC/year | Source |
|---|---|---|
| DOC input, major world rivers | 0.25 | Hedges et al., 1997 |
| POC input, major world rivers | 0.15 | Hedges et al., 1997 |
| Fossil POC input, small mountain rivers | 0.04 | Blair et al., 2003 |
| Total POC input, small mountain rivers | 0.08 | Milliman and Syvitsky, 1992, with the assumption of 10 mg C/g sediment |
| Coastal marine primary productivity | 10 | Liu et al., 2000 |
| Burial in deltaic sediments | 0.07 | Hedges and Keil, 1995 |
| Burial on shelves and upper slopes | 0.07 | Hedges and Keil, 1995 |
| Burial in sediments from mountain rivers | 0.05 | Milliman and Syvitsky, 1992, with the assumption of 6 mg C/g sediment |

TABLE 5.3

Typical organic matter composition in coastal sediments and changes due to decomposition[a]

| Component | Typical marine phytoplankton | Typical surface sediment | Change due to decomposition |
|---|---|---|---|
| C/N (molar) | 5.6 to 6.9 | 8 to 14[b] | increases |
| $C_{org}/P_{org}$ (molar) | 106 | 90 to 400 | increases |
| $\delta^{13}C$ (‰) | -20 to -23 | -20 to -27[b] | usually decreases |
| $\delta^{15}N$ (‰) | 5 to 10 | 0 to 7[b] | usually increases |
| hydrolyzable amino acids or proteins (% of TOC) | 65[c], 67[d] | 8–15 | decreases |
| carbohydrates (% of TOC) | 16[c],12[d] | 3–7[e] | decreases |
| lipids (% of TOC) | 19[c],4[d] | <1 | decreases |
| lignin (% of TOC) | 0 | 0–5[b] | decreases |
| uncharacterized (% of TOC) | ≤18 | ≥70 | increases |

[a]Redfield et al. (1963), Klok et al. (1984); Henrichs and Farrington (1984); Bergamaschi et al. (1997); Wakeham et al. (1997a); Keil et al. (1998); Jennerjahn and Ittekot (1999); Freudenthal et al. (2001); Hedges et al. (2000;2002); Gordon and Goñi (2003); Paytan et al. (2003).

[b]Range reflects the presence of terrigenous organic matter in some samples.

[c]As estimated from $^{13}C$ NMR data (Hedges et al. 2002).

[d]As measured using molecular level analyses (Wakeham et al., 1997a)

[e]Molecular level analysis. Quantities measured using less selective carbohydrate assays can be 2–4X greater.

## 2.2 *Terrestrial Plants*

About 1% of terrestrial primary productivity is delivered to the oceans by major rivers annually, 0.4 X $10^{15}$ g C $yr^{-1}$ (Hedges, 1992; Hedges et al., 1997). This organic matter includes about 40% particulate and 60% dissolved material (Hedges et al., 1997). While fairly intact and identifiable plant remains are present, most of the organic material has undergone preliminary decomposition in soils and has a composition similar to soil organic matter, or in the case of dissolved organic matter, the more soluble fractions thereof. Despite having survived decomposition in soils

(Hedges et al., 1994), most of the organic matter delivered by rivers does not accumulate in marine sediments, since molecular and isotopic evidence indicates that most coastal sedimentary organic matter is of marine origin (Hedges and Mann, 1979; Gough et al., 1993; Hedges and Oades, 1997). Hence, most riverine organic matter is rapidly decomposed in the marine water column or near the sediment water interface (Hedges et al., 1997).

However, for at least one major river system, the Atchafalaya and Mississippi Rivers, terrigenous inputs to the shelf were underestimated using $\delta^{13}C$ data and a two end-member mixing model. Coarse, C3 plant debris depleted in $^{13}C$ were deposited near the river mouth, but other terrigenous organic matter with higher $\delta^{13}C$ and lower C/N, from soils with an important C4 plant-derived component, was found farther offshore. Terrestrial material accounted for 66% of OC in offshore sediments, 85% more than estimated by a two end-member model (Gordon and Goñi, 2003). Highly degraded soil organic matter is an important component of fine-grained POM transported by central U.S. rivers in general, and probably by similar rivers elsewhere. Since this material can be relatively enriched in $^{13}C$ and poor in identifiable lignin residues, it is difficult to distinguish from marine-derived organic matter in shelf sediments (Onstad et al., 2000).

Studies of the northern California shelf indicate that flooding of the Eel River supplies large amounts of terrestrial carbon, mainly coarse plant debris, under conditions of high sediment accumulation that favor its preservation. Such flood conditions may be important to the accumulation of terrestrial carbon on shelves in general (Leithold and Hope, 1999), and the organic matter influx due to small, mountainous rivers is not included in the $0.4 \times 10^{15}$ g C $yr^{-1}$ terrigenous carbon influx estimate, which was based on data from major rivers (Meybeck, 1982; Ittekkot, 1988; Hedges et al., 1997). Terrestrial biomarkers in sediments from the Sea of Okhotsk show increased terrigenous organic matter inputs during glacial periods, but an even greater influx during deglaciation (Ternois et al., 2001). Ice rafting was the inferred mechanism, but deglacial flooding is another potential source of large amounts of terrigenous material.

Terrigenous organic matter inputs have great importance in the Arctic Ocean, with rivers supplying about $20 \times 10^{12}$ g of DOC $yr^{-1}$ and $5 \times 10^{12}$ g of POC (Dittmar and Kattner, 2003). A major reason is that the ratio of freshwater discharge to basin volume is the largest of any ocean; also, ice cover limits marine primary production. A study of the Mackenzie River and Beaufort shelf showed that terrestrial organic carbon dominated, comprising 20–80% (Goñi et al., 2000). On Kara Sea shelves, more than 70% of the carbon preserved is terrestrial (Fernandes and Sicre, 2000). Fluvial organic matter appears to be biogeochemically stable in Arctic estuaries and shelf areas (Dittmar and Kattner, 2003). While most is deposited in margin sediments, some escapes to compose a significant fraction of the organic matter in central Arctic Ocean sediments (Dittmar and Kattner, 2003).

### 2.3 *Black Carbon*

Black carbon is highly refractory reduced carbon that is produced by incomplete combustion of either fossil fuels or biofuels. Black carbon, as operationally defined, can also include graphitic carbon generated during metamorphism of sedimentary rocks, and failing to account for this may lead to overestimates of the importance of combustion sources (Dickens et al., 2004). Because of these varied origins black carbon has diverse properties, which lead to difficulties in its quantitative measurement, particularly because similarly diverse reference standards are not available (Bird and Gröcke, 1997; Masiello, et al., 2002). In a transect off the Washington coast, black carbon concentrations decreased offshore from a maximum of 6.5% of TOC and black carbon was mainly of fossil rather than combustion origin, having a $\delta^{13}C$ characteristic of marine sources and an old radiocarbon age (Dickens et al., 2004). Mitra et al. (2002) found that black carbon comprised >20% of suspended sediment organic carbon in some samples collected from the Mississippi River and adjacent shelf areas. While a fairly small part of TOC accumulation in coastal sediments, black carbon from fossil fuel combustion could account for old radiocarbon ages of some suspended sediment, bottom sediment, or dissolved organic carbon (DOC) samples (Mitra et al., 2002).

### 2.4 *Ancient Organic Carbon from Weathered Sedimentary Rocks*

The carbon in sedimentary rocks is mainly kerogen, a highly insoluble and refractory organic material that is produced from sediment organic matter during the lithification process. Little is known about its chemical structure. Hedges and Keil (1995) described sedimentary rock kerogen as "plastic encased in brick" and marveled at the fact that most kerogen must be destroyed during weathering. Otherwise, there could be nothing but kerogen in marine sediment organic matter, absent a progressive increase in sediment organic matter burial and atmospheric oxygen over geologic time. Exactly how kerogen is destroyed during weathering is not known. Petsch et al. (2001) provided evidence, based on $^{14}C$ content of membrane lipids, for microbial assimilation of ancient macromolecular organic matter during weathering of shales, indicating that bacteria play an important role. Presumably, other aspects of the subaerial weathering process such as extended oxygen exposure, exposure to UV light, microbial activity, and breakdown of the rock into fine particles are also essential, as kerogen survives millions and even billions of years within intact sedimentary rocks.

Under certain circumstances, it is clear that some kerogen survives weathering and is redeposited in sediments (Rowland and Maxwell, 1984; Blair et al., 2003). In particular, Blair et al. (2003) found that about half of the organic matter associated with fine-grained shelf sediments derived from the Eel River, California, originated from ancient sedimentary organic carbon. The Eel River has a small, steep drainage basin and easily eroded bedrock, and hence an unusually short residence time of lithogenic sediments. Further, most of the sediment is delivered in short intervals associated with floods, and is rapidly deposited on the shelf with little reworking (Leithold and Blair, 2001). Major rivers such as the Amazon and Columbia have much longer watershed residence times for the products of rock weathering and extensive reworking of sediments deposited in the coastal ocean (Hedges et al. 1997; Keil et al., 1997; Aller, 1998), which result in the destruction of most kerogen. However, small, mountainous rivers analogous to the Eel are a major source of sediments to the oceans (Milliman and Syvitski, 1992), and so reburial of kerogen could account for accumulation of 40 Tg of organic carbon in coastal marine sediments annually (Blair et al., 2003).

### 2.5 *Redeposited Marine Sediments*

Many coastal sediments include components that have been repeatedly resuspended and redeposited. Such redeposited material is composed of more refractory material than freshly deposited particles. For example, Cape Lookout Bight has high-organic, anoxic sediment that includes a substantial fraction of organic matter that is 500–1500 years old (Martens et al., 1992). Suspended particulate organic carbon (POC) in waters overlying the deep continental slope of the northwest Atlantic continental margin was depleted in both $^{13}C$ and $^{14}C$, indicating the presence of aged terrestrial material (Bauer et al., 2002). Because POC in mid-Atlantic rivers and estuaries is not so depleted in $^{14}C$ (Raymond and Bauer, 2001a,b), the old slope POC probably represents terrigenous material that was deposited, then resuspended later by processes such as slope failure and erosion of submarine canyon walls (Bauer et al., 2002). The $\Delta^{14}C$ of North Carolina continental slope sediments are most consistent with the sediment organic matter having three sources: fresh, planktonic organic matter recently supplied from the ocean surface, redeposited marine organic matter of ca. 100 year age, and refractory marine organic matter with an age of greater than 1000 years (DeMaster et al., 2002).

## 3. Composition of Coastal Marine Sediment Organic Matter

A typical composition of coastal sediment organic matter is summarized in Table 5.3. Major identifiable biochemicals (mainly hydrolyzable amino acids and carbohydrates, with lesser amounts of lipids, lignin, pigments, and other compounds) make up only 20–25% of most coastal ocean surface sediments (Klok et al., 1984; Keil et al., 1998). Based on evidence from laboratory decomposition experiments (Harvey et al., 1995; Harvey and Macko, 1997; Nguyen and Harvey, 1997) and deep sea sediment traps (Wakeham et al., 1997a), the loss of most identifiable biochemicals occurs very rapidly, within weeks to months. On the other hand, evidence from $^{13}C$ NMR (nuclear magnetic resonance spectroscopy) indicates that sinking POC undergoes little structural change, so the loss of identifiable biochemicals is apparently due to fairly subtle changes in the organic matter (Hedges et al., 2001).

The origins and nature of uncharacterized fraction, despite decades of effort, remain unknown (Hedges et al., 2000). It includes refractory aliphatic macromolecules (Tegelaar et al., 1989) produced by some algae, but since nonaliphatic fractions dominate many marine kerogens, other sources of refractory material must be important (Gelin et al., 1999). Melanoidin-type condensation reactions were long ago suggested as a means of rapidly generating uncharacterizable macromolecules in decomposing sediment organic material (Nissenbaum and Kaplan, 1972), and this hypothesis has recently been revitalized (Collins et al., 1995; Zegouagh et al., 1999). However, there is little evidence for major structural modifications to the bulk of organic matter during decomposition (Knicker and Hatcher, 1997; Minor et al., 2003). The increase in MUC relative to identifiable biochemicals during decomposition (Wakeham et al., 1997a; Hedges et al., 2000) indicates that some MUC is less readily decomposed than most of the identifiable compounds, or that some identifiable biochemicals are transformed into MUC. However, most MUC isn't refractory to microbial decomposition in sinking particles or surface sediments, since the particle and sediment concentrations of MUC decrease substantially over time. Bioavailability of some MUC is also evidenced by high rates of microbial metabolism in sediments with very little identifiable organic matter (Arnosti and Holmer, 2003).

Most organic matter in coastal sediments is associated with (and inseparable from) clay minerals (Keil et al. 1998; Mayer et al., 2002). The finest fraction also is the most diagenetically altered. Coarse, low density fractions consist mainly of vascular plant debris (Keil et al., 1998; Leithold and Hope, 1999).

## 4. Processes Affecting Organic Matter Quantity and Composition

### *4.1 Microbial Processes*

The major microbial processes decomposing organic matter are discussed more fully by Jahnke (this volume). Here the effects on organic matter decomposition will be emphasized. As Jahnke describes, there is a vertical pattern in the consumption of oxidizing agents by sediment microbial communities, based upon the energy yield per mole of organic carbon oxidized. Oxygen is used first, then nitrate, manganese oxides, iron oxides, sulfate, and carbon dioxide or carbonyl groups of organic acids (Froelich et al., 1979). Distinct groups of microorganisms carry out the terminal oxidation of organic compounds to carbon dioxide in the different zones of electron acceptor availability (Fenchel et al., 1998). Further, the types of bacteria decomposing organic matter differ among the pelagic environment, particles, and surface sediments (DeLong et al., 1993; Llobet-Brossa et al., 1998) and among different biochemical substrates (Cottrell and Kirchman, 2000).

Decomposition processes are broadly similar, although the organisms carrying out the processes vary. First, organic polymers such as proteins and carbohydrates are hydrolyzed, probably mainly by extracellular enzymes, although such enzymes may be bound to the surfaces of cells. Hydrolytic enzymes are produced by a subset of environmental bacteria. This process has been studied only to a limited extent in sediments, but it is clear that the hydrolysis of small peptides can proceed rapidly (Luo, 1994; Pantoja et al., 1997), while the hydrolysis of larger proteins could require months (Ding, 1998). Because concentrations of dissolved organic substances in most sediments and in experimental decomposition systems are low, hydrolysis is considered a likely rate-limiting step in organic matter decomposition (Kristensen and Holmer, 2001). However, polysaccharide hydrolysis is rapid and fermentation limits the rate of carbohydrate mineralization in some anaerobic sediments (Arnosti et al., 1994; Brüchert and Arnosti, 2003). Matrix-assisted laser desorption/ionization time of flight mass spectrometry (MALDI-TOF-MS) was recently employed in the study of protein hydrolysis in seawater and sediments. A model protein (bovine serum albumen) was hydrolyzed with no functional group selectivity, suggesting that the hydrolysis was accomplished by a mixture of bacterial proteases (Nunn et al., 2003).

In sediments with oxygen or nitrate, bacteria can assimilate and respire the hydrolysis products directly. However, under sulfate reducing or methanogenic conditions, fermentation of monosaccharides and amino acids to low molecular weight (LMW) acids such as acetate and lactate is a necessary preliminary to their oxidation to carbon dioxide (Fenchel et al., 1998). The fermentation is an energy yielding process and is carried out by different groups of bacteria than the terminal oxidation. Fermentation products, mainly acetate but including propionate, butyrate, lactate, and a few other LMW acids, accumulate to quite high (up to millimolar) concentrations in the porewaters of anoxic sediments, while only traces (micromolar) are present in oxidizing sediment porewaters. Porewater DOC in general is greater in anoxic than oxic sediments, a pattern that is not explained by the elevated concentrations of LMW acids alone. Most porewater DOC in estuarine and continental margin sediments is, however, of <3kDa molecular weight; since this LMW organic matter accumulates in porewaters, it appears to be refractory to decomposition (Burdige and Gardner, 2002). Porewater DOC, like seawater DOC, contains a substantial proportion of carbohydrates, and in porewaters the fraction of carbohydrate is often greater than that in the surrounding sediment (Arnosti and Holmer, 1999; Burdige et al., 2000).

Most substances for which there are short-term differences in bacterial decomposition rate under oxic and anoxic conditions are larger, insoluble molecules without hydrolyzable bonds. Examples include lignin (Benner et al., 1984), lipids (Harvey and Macko, 1987)), hydrocarbons (DeLaune et al., 1980), and pigments (Sun et al., 1993 a,b). Decomposition of most such substances can proceed via oxygen-requiring enzymes, such as oxygenases (Sawyer, 1991), or via $H_2O_2$ or other reactive oxygen species produced by aerobic organisms (Sawyer, 1991; Hedges and Keil, 1995). However, most of these substances do eventually degrade even under anoxic conditions.

An issue that is often ignored is that the organic matter remaining in sediments is the net result of both decomposition and synthesis by organisms. Although the faunal and bacterial biomass is typically a small fraction of the organic content of a sediment, much of the organic matter may have been metabolically processed at least once. In laboratory decomposition experiments, 5–25% of the decomposed algal carbon was converted to bacterial carbon, and the bacterial biomass contributed significantly to observed changes in carbohydrates, proteins, and lipids in the residual organic matter (Harvey et al., 1995). Specific biomarkers suggest a significant, but not predominant, bacterial component to sedimentary lipids (e.g., Gong and Hollander, 1997). The major biochemical compound classes of hydrolyzable amino acids and carbohydrates do not have distinctive compositions in many types of bacteria, or include only minor fractions of unusual constituents like D-amino acids. Also, the extent of bacterial origins of the uncharacterized fraction of sediment organic matter is unknown, although destructive analysis (e.g., pyrolysis-gas chromatography-mass spectrometry) generates some fragments that are probably of bacterial origin. As pointed out by Jahnke (this volume) chemosynthesis is a source of new organic matter production in sediments, although in the absence of hydrothermal fluids or methane seeps the quantity produced is limited by the original organic supply to the sediment and the production of reduced substances such as sulfide that resulted from its decomposition.

In both oxic and anoxic sediments, most of the decomposition of organic matter takes place within the upper meter; rates are mainly a function of the lability of available organic matter and generally decrease markedly with depth (Toth and Lerman, 1977; Canuel and Martens, 1996). However, viable bacteria exist hundreds of meters beneath the surface (Fredrickson and Onstott, 1996; Zink et al., 2003), and microbial alteration of some components of sediment organic matter can continue via these deep subsurface communities. Increasing organic matter availability on deep burial can fuel a deep subsurface maximum in bacterial activity (Wellsbury et al., 1997).

### *4.2 Microfaunal, Meiofaunal and Macrofaunal Processes*

A key difference between oxic and anoxic sediments is that the former have much larger populations of animals. Burrowing organisms can penetrate into anoxic layers, but the distance is limited by the requirement for oxygen. This is met either by returning to oxic sediments periodically or by irrigation of the burrow, which can extend to 60–100 cm depth (Thomas et al., 2002). While comparison of metabolic demand of faunal and bacterial populations of sediments indicates that bacteria are responsible for most sediment metabolism (e.g., Aller et al., 2002), animals clearly have several important roles. These include transport of oxidizing agents (mainly dissolved oxygen) into anoxic sediment zones (Aller, 1994; Kristensen, 2000). In sediments with dense macrofaunal populations, all of the surface sediment volume can be processed through invertebrate guts on annual time scales. Benthic fauna are apparently highly selective consumers, as indicated by a 150 per mil difference in $\Delta^{14}C$ between animal tissue and sediment OC on the North Carolina continental slope, which could be due to a combination of selective feeding and more efficient digestion of fresh planktonic remains (DeMaster et al., 2002).

The role of meiofauna and heterotrophic nanoflagellates in sediment organic matter decomposition is less well-defined. Nanoflagellates are thought to be major consumers of bacterial biomass, and as such probably limit its accumulation and contribute to the terminal oxidation of organic matter in sediments (Patterson et al, 1989). Meiofauna may consume bacteria also, but generally require additional food (Montagna et al., 1982; Bott and Borchardt, 1999). Like macrofauna, they are limited to the upper interface of anoxic sediments. In lagoonal sediments, autotrophic inputs of fresh organic matter led to communities with relatively large meiofaunal biomass, and smaller bacterial and nanoflagellate biomass. Lagoons with mainly refractory, detrital carbon inputs had less meiofauna and more bacteria and heterotrophic nanoflagellate biomass (Manini et al., 2003).

### *4.3 Physical and Chemical Processes*

**Adsorption:** Adsorption of organic substances to sediment particles was a leading hypothesis for the mechanism leading to preservation of organic molecules in sediments. Mayer (1994a,b) found a relationship between organic carbon content and mineral surface area in many coastal and estuarine sediments that approximated a monolayer of organic matter coating the mineral particle. This led to the hypothesis that organic matter was preserved by being protected from bacterial attack by mesopores in the clay mineral surface, which are too small for enzymes to penetrate and effect hydrolysis. Keil et al. (1994a) extended the relationship to size fractionated samples of Washington continental shelf and slope sediments, and found a very close correlation (r=0.96) between surface area and organic content. A variety of studies (Sugai and Henrichs, 1992; Luo, 1994; Ding, 1998) showed that adsorption retarded the decomposition of amino acids, peptides, and proteins, albeit not necessarily to the extent required for long term preservation. Henrichs (1995) showed, by means of a simple model, that adsorption can only preserve organic matter indefinitely if very strong (partition coefficients of at least 1000 for typical coastal sediments) and if the adsorbed organic matter is refractory to decomposition.

**Aggregation:** Transmission electron microscopy of sediments showed no evidence for uniform, thin coatings of organic matter (Ransom et al., 1997), as predicted if adsorption was responsible for preservation. Rather, organic matter associated with particle surfaces appeared to be present in thicker, discontinuous aggregations that left much of the surface bare. Using enthalpies of gas adsorption and the Brunauer-Emmett-Teller equation, Mayer (1999) demonstrated that less than 15% of the surface area of coastal sediments with $< 3$ mg C $m^{-2}$ was coated with organic matter. X-ray photoelectron spectroscopy also indicated that organic matter is associated with sediment particle surfaces in discrete locations, and that surface coverage apparently does not increase as organic matter concentration increases (Arnarson and Keil, 2001).

If sediment organic matter is not mainly protected within mesopores, then another mechanism for protecting it from microbial attack is needed. A possibility is that it is encapsulated, either within a matrix of very refractory organic matter (Knicker and Hatcher, 1997; Zang and Hatcher, 2002) or within a mineral matrix. Organic encapsulation is probably of little importance except in highly organic sediments, leaving protection within mineral aggregates as a better explanation for typical coastal sediments (Figure 5.1). Light and electron microscopy reveal that most organic and mineral particles in sediment trap and sediment samples are aggregated (Ransom et al., 1998b). However, to be effective in preserving organic matter, these aggregates must have regions inaccessible to enzymatic attack, and must be strong enough to avoid disaggregation under conditions of deposition and burial, including repeated ingestion by benthic macrofauna and storm wave induced resuspension. Also, to explain the tight correlation between sediment surface area and organic content, they must somehow incorporate and preserve organic matter in a way that is related to mineral grain size.

The aggregations must also be robust on geologic time scales, but compaction and dewatering of sediments probably changes their properties. Salmon et al. (2000) found that kerogen in an ancient black shale was in alternating clay and organic nanolayers, each about 100 nm thick. They also found that extracted kerogen was unstable over two years of storage, showing that the 93 million-year-old material was labile when removed from the protective mineral matrix.

The adhesion that holds aggregates together is likely to be fundamentally similar to the processes that have been studied extensively in the development of industrial adhesives. There are two categories of processes than can bind surfaces together. Chemical adhesive processes include ionic interactions, dipole interactions including hydrogen bonding, and van der Waals forces. Mechanical processes are the result of penetration of the adhesive into pores of the bonded surfaces or dissolution into the cellular structure of woods (Vick, 1999). The chemical forces are identical to those involved in adsorption, and very strong adhesion requires multiple points of attachment, as is the case for strong adsorption (Henrichs, 1995). An effective adhesive must be fluid initially, to conform tightly to the bonded surfaces and cover as much of the interfacial surface as possible. However, a durable bond requires that the adhesive harden. This can involve drying or formation of chemical cross-linkages (Vick, 1999).

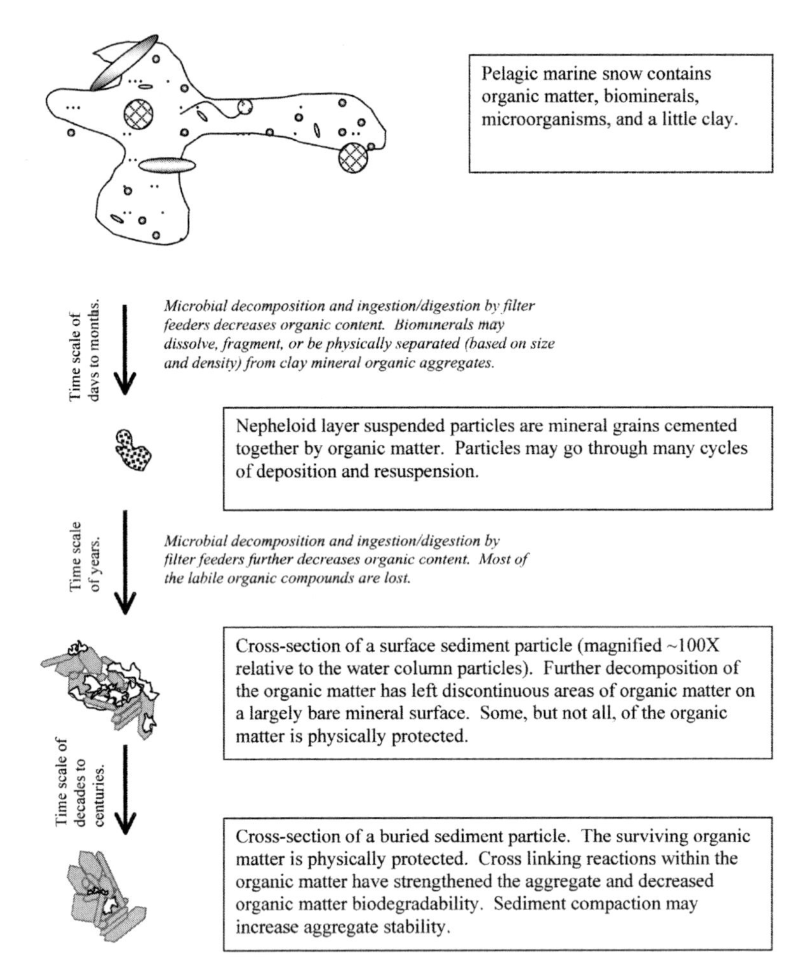


Figure 5.1 Hypothetical formation and diagenesis of a clay-organic matter aggregate, based on Ransom et al. (1997); Ransom et al. (1998b); and Arnarson and Keil (2001).

Marine organisms (plants, animals, and bacteria) produce a wide variety of exceptionally strong and durable adhesives, which often bind these organisms to surfaces for life (Van Schie and Fletcher, 1999; Haemers et al., 2002; Berglin and Gratenholm, 2003). These adhesives naturally bond strongly and harden in water. Marine bioadhesives that have been characterized are often proteins, and the cross-linking during hardening is mainly due to hydrogen bonds. For example, mussel adhesives contain a specific amino acid, dihydroxyphenylalanine (Figure 5.2), which contributes both strong surface adhesion and toughness to the polymer (Yu et al., 1999). There are intriguing reports of phenolic structures in refractory marine-derived organic matter in sediments from the northwest African upwelling system (Zegouagh et al., 1999). While it is tempting to speculate about the role of bioadhesives in forming aggregates, many proteins and carbohydrates have some adhesive properties, and are produced in far greater abundance in the oceans. However, protein and carbohydrate based adhesives are normally easily biodegradable, and if aggregates persist in sediments, the organic "glue" must be inaccessible or resistant to microbial attack.

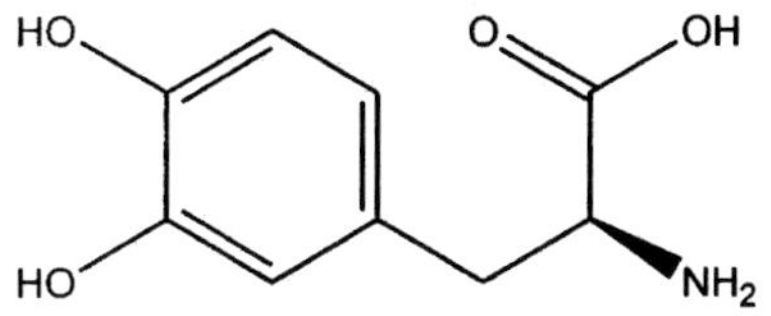

3,4-dihydroxyphenylalanine

Figure 5.2 An amino acid found in mussel adhesive proteins.

**Condensation and cross-linking reactions:** Decades ago it was proposed that some sort of chemical reactions occur within sediments that render the organic matter less susceptible to microbial attack. For example, Nissenbaum and Kaplan (1972) proposed that sugars and amino acids released during hydrolytic decomposition of organic matter condensed to form melanoidin-type compounds, which are quite resistant to bacterial decomposition (Henrichs and Doyle, 1986). However, $^{15}N$-NMR revealed no evidence of such compounds in an organic-rich sediment; instead, sediment organic nitrogen is mainly amide bonded, as is the nitrogen in proteins (Knicker and Hatcher, 1997). However, that does not rule out other types of cross-linking reactions. Zegouagh et al. (1999) suggested that condensation reactions involved mainly protein, rather than carbohydrate, based on structural characteristics of refractory organic matter from northwest African upwelling region sediments. Nguyen and Harvey (2001) extracted proteins by treating organic rich Mangrove Lake, Bermuda, sediments with a reagent that breaks one type of glucose-protein association (Amadori products). Size-exclusion high performance liquid chromatography indicated that hydrophobic and other non-covalent interactions of macromolecules were important (Nguyen and Harvey, 2001). In highly anoxic sediments, reaction with hydrogen sulfide or polysulfides has been proposed as a possible mechanism for producing refractory organosulfides (Kohnen et al., 1989; Ferdelman et al., 1991; Eglinton et al., 1994; Sinninghe Damsté et al., 1998). In sediments cross-linking between or within macromolecules could serve dual purposes, both decreasing biodegradability and increasing the strength of adhesion of mineral grains.

**Biomineralization:** Some organic matter in sediments is also contained within a biomineral matrix, which can preserve it as long as the mineral remains intact (Ingalls et al., 2003). This biomineral bound organic matter can be a significant part of the TOC in biogenic oozes that have low organic content, but it constitutes a minor fraction in coastal lithogenic sediments.

**Diffusion:** As pointed out by Henrichs (1995) any organic molecule that is not extremely insoluble or very strongly adsorbed will tend to diffuse out of sediments, even in the absence of decomposition. Continental margin sediments typically contain DOC concentrations substantially above bottom water concentrations (Alperin et al., 1998; Burdige et al., 1999). This material is predominantly of low molecular weight (Burdige and Gardner, 2002), which may reflect the fact that smaller molecules are, given similar structure and functionality, more soluble and less strongly adsorbed than larger molecules. In California continental margin sediments there is evidence for a linkage between benthic DOC fluxes and sediment carbon preservation that may be mediated by porewater DOC (Burdige et al., 1999). However, in North Carolina continental slope sediments, there was no relationship between porewater DOC concentrations, organic matter sorption, and organic carbon burial efficiency, indicating that total DOC concentrations are not a major factor controlling organic matter preservation (Alperin et al., 1998). In organic-rich sediments, porewater organic matter may affect diffusion and convection processes. In sediments with high organic carbon (>5% by weight) hydrated organic matter can occupy a significant fraction of total sediment porosity (Bennett et al., 1999).

## 5. Changes in Organic Matter Composition During Decomposition

### *5.1 Changes in Elemental and Stable Isotopic Composition*

General trends in organic matter compositional change during decomposition are summarized in Table 5.3. The specific effects of decomposition on organic matter composition are only partly known, and there is a particular lack of information on changes in the uncharacterized fraction of organic matter. Most studies of diagenetic change in organic matter composition have been conducted in estuaries or in unusual, highly productive areas of the coastal ocean, such as the Peru upwelling region. There is no reason to believe that these studies cannot be applied to deltaic and open shelf sediments, but it is prudent to keep in mind several differences in the conditions of diagenesis. In particular, deltaic and open shelf sediments are typically more oxic than estuarine or upwelling region sediments, their average TOC is lower, and porewater sulfide concentrations are lower as well. Sediment accumulation rates of deltaic sediments are markedly greater, while open shelf sediments tend to have lower sediment accumulation rates than estuaries or upwelling regions.

Very early stages in the decomposition of planktonic organic matter, which occur before or shortly after deposition in coastal sediments, are best known from laboratory decomposition experiments and from sediment trap studies in the deep sea. Laboratory decomposition of algae has consistently shown that soluble material is rapidly released and decomposed, regardless of whether conditions are oxic or anoxic, but the more refractory particulate fraction decomposes faster in the presence of oxygen (Kristensen and Holmer, 2001). Within months, major identifiable biochemicals become much less than half of the residual organic matter, for example comprising just 24% of deep-water particles in the central equatorial Pacific (Wakeham et al., 1997a). Pelagic and nepheloid layer marine snow differ significantly. Pelagic marine snow consists of biominerals, organic matter, and microorganisms, while nepheloid layer marine snow consists of clay minerals, clay flocs, and dense clay-organic-rich aggregates in an exocellular organic matrix (Ransom et al., 1998b). This marked change in the physical structure of aggregates is associated with further decreases in identifiable organic constituents, before incorporation in surface sediments.

C/N ratio increases with decomposition in many cases, because proteins and other nitrogen-containing molecules are more easily decomposed than TOC, but this increase is slow, so in rapidly depositing coastal sediments there is often little or no change within the interval sampled by cores (e.g., Henrichs and Farrington, 1987; Klump and Martens, 1987). Often total nitrogen rather than organic nitrogen is measured, and then any increase in $C_{organic}/N_{organic}$ is disguised by increases in adsorbed and fixed inorganic nitrogen. Also, if dissolved inorganic nitrogen is available and the C/N of the starting material is high, as in leaf litter, then decomposition can lead to an initial increase in C/N due to bacterial nitrogen assimilation. Organic P is likewise mineralized more rapidly than organic C (Lückge et al., 1999; Carman et al., 2000; Filippelli, 2001). Because inorganic P tends to associate with iron and manganese oxyhydroxides in oxidizing sediments, much more P is returned to the water column from anoxic sediments and particularly from sediments overlain by anoxic water. This may spur primary production in regions with anoxic bottom waters, and so lead to the accumulation of organic-rich sediments beneath anoxic water columns (Ingall and Jahnke, 1997).

Often, there is little change in $\delta^{13}C$ during early diagenesis, and most changes with depth in stable carbon isotopes are attributed to changes in the isotopic composition of organic matter supplied to sediments. However, changes in isotopic signature due to decomposition can clearly occur. In oxic and anoxic incubation experiments with algae, $\delta^{13}C$ decreased by 1.6‰ over three months, due to slower decomposition of isotopically light organic compounds (Lehmann et al., 2002). Because a greater proportion of total organic N is mineralized compared with C, and because nitrogen can transformed into a variety of organic and inorganic species in sediments, diagenetic effects on $\delta^{15}N$ are more likely. During algal decomposition under anoxic conditions, $\delta^{15}N$ decreased by 3‰, probably due to incorporation of isotopically light nitrogen in bacterial biomass. Under oxic conditions, the $\delta^{15}N$ first increased and then decreased to the initial value (Lehman et al., 2002). During oxic decomposition of suspended particulate matter, $\delta^{13}C$ and $\delta^{15}N$ both increase (Altabet, 1988; Libes and Deuser, 1988). Surface sediments often have greater $\delta^{15}N$ than particles collected by sediment traps at the same site (Saino and Hattori, 1987; Altabet and Francois, 1994; Ostrom et al., 1997). This is apparently because decomposition processes discriminate against the heavier isotope. In sediment analyses where organic and particle-bound inorganic nitrogen are not distinguished, fixation of isotopically light inorganic nitrogen in clay lattices can partly or wholly compensate for enrichment in $^{15}N$ in the organic fraction (Fruedenthal et al., 2001).

### 5.2 *Changes in Major Biochemical Groups*

Of the major biochemical compound classes, total hydrolyzable amino acids (THAA) are the best characterized in terms of diagenetic change. THAA are amino acids that can be extracted from sediments by acid hydrolysis, and include amino acids bound in proteins, adsorbed amino acids, and possibly other forms. The ratios THAA/TOC and THAA/TN (TN = total nitrogen) decrease with decomposition, indicating that THAA are more readily decomposed than the average organic matter. There is a systematic and relatively uniform change in THAA composition with diagenesis. The most easily discerned is an increase in the proportion of non-protein amino acids, in particular, β-alanine and γ-amino butyric acid (Cowie and Hedges, 1994; Keil et al., 2000). In most coastal sediments, however, the proportion of these amino acids is low and their relative enrichment with decomposition is small. Protein amino acid composition also changes, with relative increases in glycine and decreases in other amino acids including glutamic acid, tyrosine, and phenylalanine. (See Figure 5.3a for examples of amino acid structures). Because the compositional changes are similar in a wide range of sedimentary environments, a broad indicator of change in THAA with decomposition, termed the degradation index, was developed based upon Principal Components Analysis of protein amino acid composition (Dauwe and Middelburg, 1998; Dauwe et al., 1999). However, in some coastal sediments the composition of THAA changes very little on 100-year time scales (Henrichs and Farrington, 1987). The subset of amino acids contained in enzyme-hydrolyzed protein (EHAA) has also been examined (Mayer et al., 1995). Although these amino acids are presumably in a labile chemical form, and their concentration decreases relative to TOC with extensive degradation, a fraction of the total resists decomposition. For example, in slope sediments off Cape Hatteras, the concentration stabilized at 0.1 – 0.5 mg EHAA $g^{-1}$ below 1 meter depth in cores, suggesting some physical protection against enzymatic attack (Mayer et al., 2002).

Carbohydrates often do not exhibit selective decomposition in coastal sediments. (See Figure 5.3b for example monosaccharide structures.) For example, total particulate carbohydrates have nearly constant ratios to POC in cores from Dabob Bay (Cowie and Hedges, 1992), Cape Lookout Bight (Martens et al., 1992), Chesapeake Bay, the mid-Atlantic shelf and slope (Burdige et al., 2000), and the continental margin off Japan (Miyajima et al., 2001). In the latter case, this was true of both alkali extractable and residual, acid hydrolyzable polysaccharides (Miyajima et al., 2001). Relatively non-selective assays for carbohydrates, such as the phenol-sulfuric acid method, typically measure significantly greater carbohydrate concentrations than are found by hydrolysis and chromatographic determination of monosaccharides (Klok et al., 1984; Miyajima et al., 2001). This could indicate that sediments contain substantial carbohydrate-like material that is not strictly polysaccharide. As this is not true of plankton, sedimentary carbohydrates could be the result of transformation of the original polymers or selective preservation of a minor component. However, carbohydrate analyses are prone to errors, including loss of charged components during desalting (Miyajima et al., 2001) and incomplete recovery after hydrolysis, which are an alternative explanation for discrepancies between non-specific assays and measurements of specific monosaccharides. Data on specific monosaccharide composition of carbohydrates is limited, and geographic or temporal variations in carbohydrate composition are most often attributed to varying sources of organic matter (e.g., Cowie and Hedges, 1984; Kerhervé et al., 2002). Elevated ribose concentrations have been suggested as indicating a bacterial source and hence extensive decomposition of the original plant material (Jennerjahn and Ittekkot, 1999). Deoxyhexoses, such rhamnose and fucose, that are components of structural polysaccharides appear to increase during decomposition relative to glucose and other components of storage polysaccharides (Hernes et al., 1996; Keil et al., 1998).

Lignin decomposition under aerobic conditions results in well-defined changes in the composition of oxidation products generated during analysis (Figure 5.3b).

Microbial oxidation of the propyl side chains results in increased yield of benzoic acid and decreased yield of methoxylated vannilyl and syringyl phenols relative to *p*-hydroxyl phenols, as well as an increase in the acid/aldehyde ratio (Ertel et al., 1986; Filley et al., 2000). Aerobically, lignin can be largely destroyed within months, but decomposes less rapidly than bulk organic matter, leading to its enrichment in early stages of decomposition (Opsahl and Benner, 1995). However, anaerobic decomposition is slower, accounting for the preservation of some lignin in marine sediments. Anaerobic decomposition probably proceeds via aromatic ring cleavage, and did not result in changes in relative amounts of methoxylated and *p*-hydroxyl phenols in a mangrove sediment (Dittmar and Lara, 2001).

Other evidence concerning the relative degradability of organic material under oxic and anoxic conditions comes from certain turbidite deposits, consisting of organic rich, anoxic continental slope sediments that have been transported in mass flows to the deep sea. Subsequent to their redeposition, oxygen diffuses into the turbidite, causing marked changes in geochemistry (Wilson et al, 1995). Among these are decreases of 75% or more in TOC (Prahl et al., 1989; Thompson et al., 1993), indicating that a large part of the continental margin sediment TOC is degradable under oxic, but not anoxic, conditions. Most of this TOC was uncharacterized. Of the identifiable components in a Madeira Abyssal Plain turbidite, including alkenones (Prahl et al., 1989), sugars, lignin oxidation products, and amino acids (Cowie et al., 1995), nearly all were non-selectively degraded (i.e., made up similar proportions of the TOC in oxidized and unoxidized regions of the turbidite). This suggests that preservation under anoxic conditions is not necessarily a function of the composition of organic matter. An exception was pollen grains, which were extensively degraded in the oxidized layer compared with the unoxidized layer (Keil et al., 1994b).

Density fractionation of sedimentary organic matter has shown marked compositional differences in organic matter associated with clays, sand-sized mineral grains, and floated coarse organic matter, the latter consisting mainly of vascular plant debris (Bergamaschi et al., 1997; Keil et al., 1998). In Washington shelf and slope sediments, the organic matter associated with clays was more altered by decomposition than that in the other two fractions. Non-protein amino acids (especially β–alanine), lignin phenol acid/aldehyde ratios, and the deoxyhexoses fucose and rhamnose were relatively greater in the clay fraction (Keil et al., 1998).

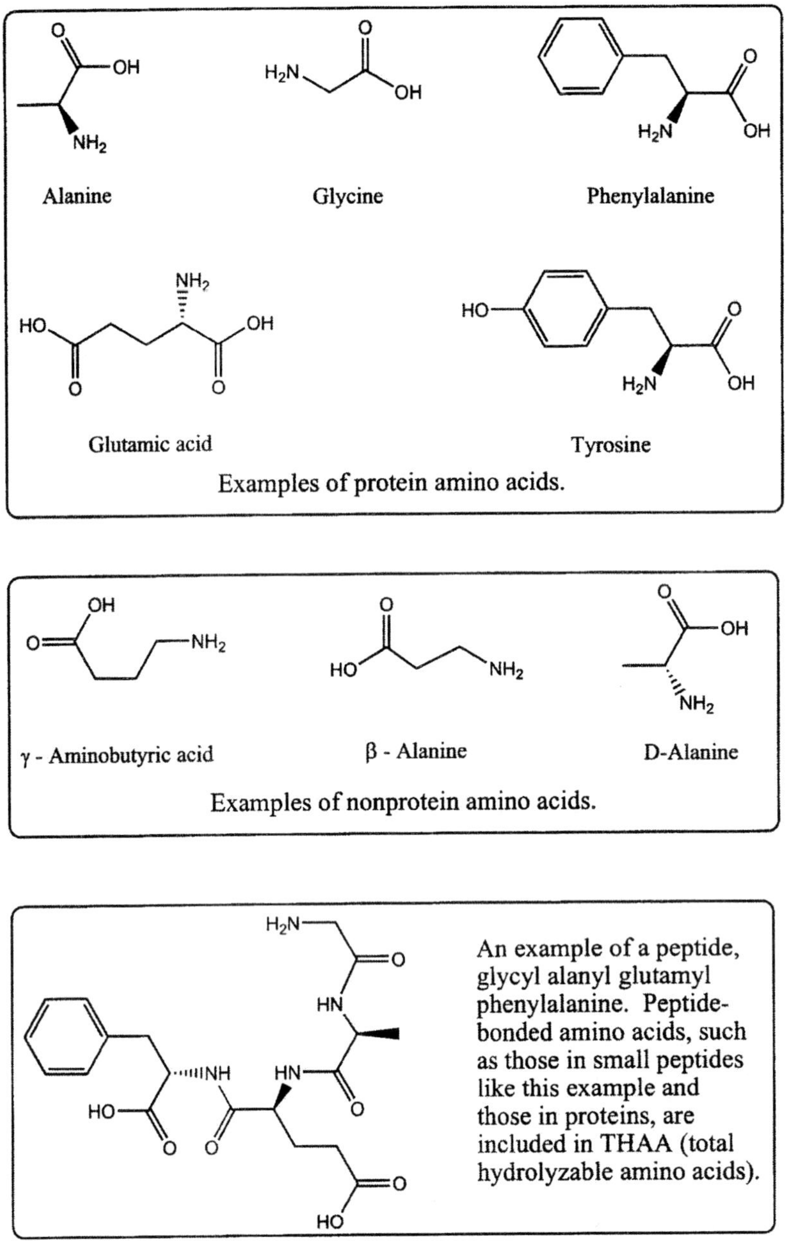

Alanine

Glycine

Phenylalanine

Glutamic acid

Tyrosine

Examples of protein amino acids.

γ - Aminobutyric acid

β - Alanine

D-Alanine

Examples of nonprotein amino acids.

An example of a peptide, glycyl alanyl glutamyl phenylalanine. Peptide-bonded amino acids, such as those in small peptides like this example and those in proteins, are included in THAA (total hydrolyzable amino acids).

Figure 5.3a Examples of protein and non-protein amino acids and a peptide.

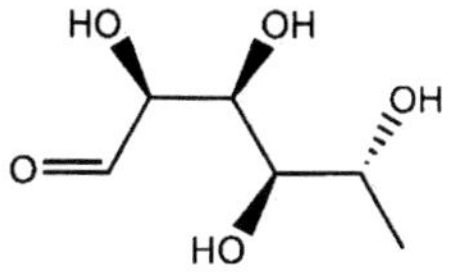

Rhamnose, a deoxyhexose.

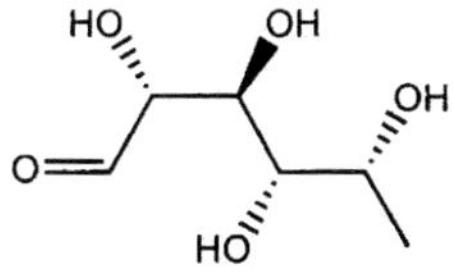

Fucose, a deoxyhexose.

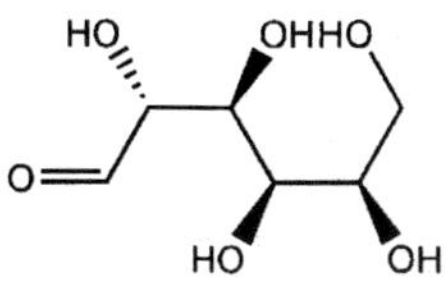

Glucose, a hexose that is a common constituent of carbohydrates, including cellulose, starch, and glycogen.

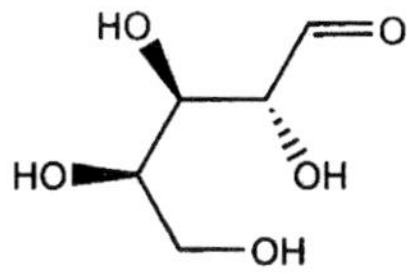

Ribose, a pentose.

Examples of monosaccharides that are constituents of carbohydrates, mucopolysaccharides, RNA and other key biomolecules.

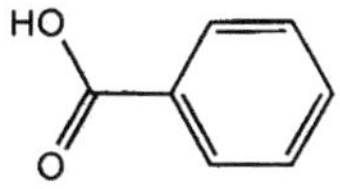

Benzoic acid

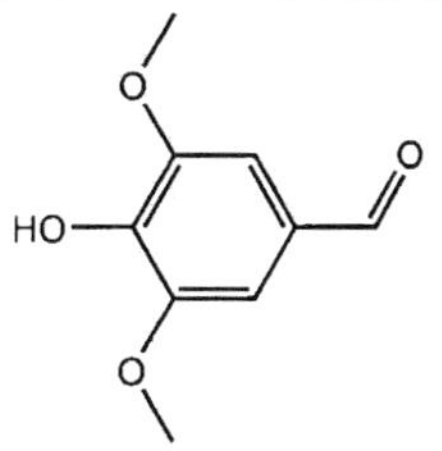

Syringaldehyde

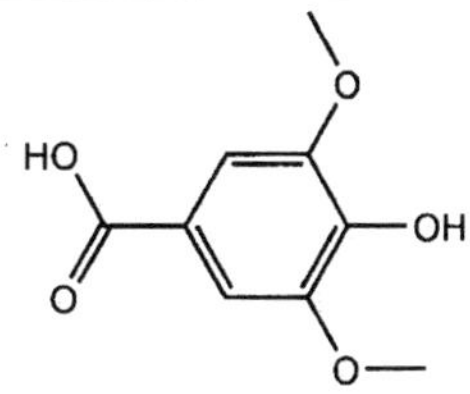

Syringic acid

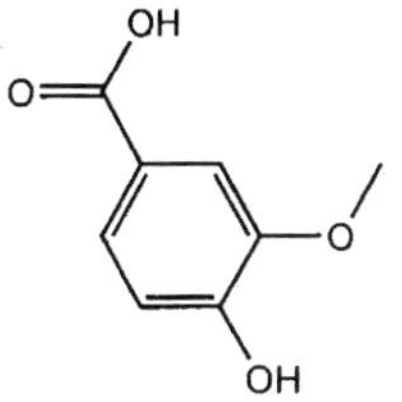

Vanillic acid

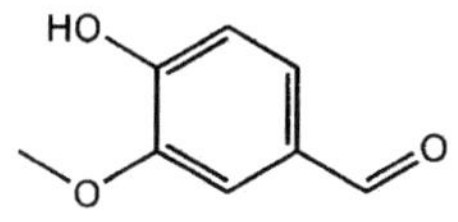

Vanillin, an aldehyde

Examples of oxidation products of lignin or of lignin that has been subject to fungal or bacterial decomposition.

Figure 5.3b Examples of monosaccharides and lignin oxidation products.

Cholesterol

24-Ethylcholest-5-en-3β-ol

HO H H H H HO

Sterols.

O OH O OH

Oleic acid (18:1)

Linolenic acid (18:3)

O HO

Stearic acid (18:0)

Saturated (18:0), monounsaturated (18:1) and polyunsaturated (18:3) fatty acids.

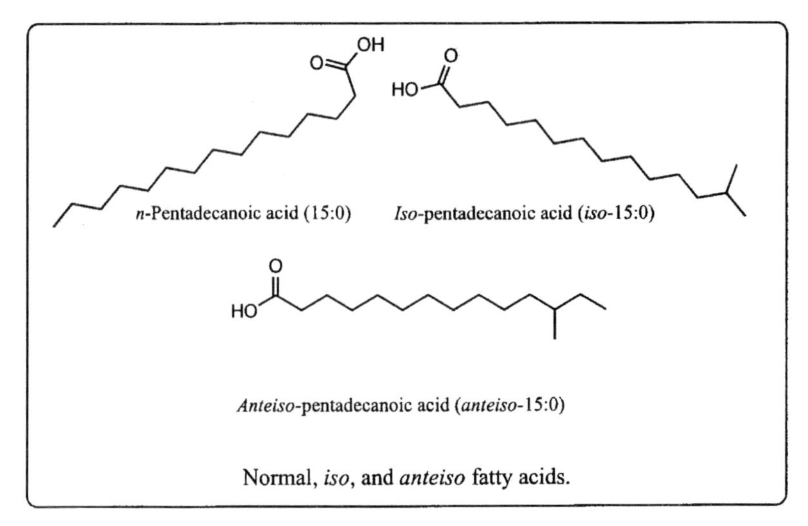


Figure 5.3c Examples of lipids commonly found in coastal sediments.

### 5.3 Changes in Lipids and Other Biomarkers that are Trace Constituents

Lipids and other solvent extractable organic substances are highly diverse among different plants, animals, and microorganisms, which allows them to convey rich information on the sources of organic matter to coastal sediments (e.g., Kaneda, 1991; Albers et al., 1996; Volkman, 1986; Volkman et al., 1998). The varied structures also have differing susceptibilities to biological and chemical modification; nearly all fatty acids, sterols, and pigments decompose faster than the bulk TOC. (See Figure 5.3c for example lipid structures.) Typically (but not universally), decomposition processes in sediments cause relative increases in the proportion of lipids that are derived from terrestrial sources, such as long-chain fatty acids, and from bacteria, such as *iso-* and *anteiso-* fatty acids (e.g., Canuel and Martens, 1996; Budge and Parrish, 1998; Camacho-Ibar et al., 2003). Unsaturated and particularly polyunsaturated compounds are preferentially removed by decomposition (e.g., Budge and Parrish, 1998; Camacho-Ibar et al., 2003). Recently, compound specific $\Delta^{14}C$ analysis has allowed better discrimination of the sources of sediment lipids, which in Santa Monica Basin sediments were predominantly from modern marine phytoplankton, but included terrestrial plants and fossil sources for n-alkanes and chemoautotrophic bacterial sources for isoprenoid lipids (Pearson et al., 2001).

Lipids and chloropigments typically decompose faster under oxic than under anoxic conditions (Harvey and Macko, 1997; Sun et al., 1993b; Sun et al. 2002). Sun et al. (2002) found 10X faster oxic decomposition of algal cell-associated fatty acids and more rapid oxic turnover of a bacterial fatty acid, *iso*-15:0. Degradation of radiolabeled cholesterol, palmitic, and oleic acids was also more rapid under oxic conditions (Sun et al., 1997; Sun and Wakeham, 1998).

### 5.4 Compositional Change and the Mechanism of Preservation

A broader question is whether the residual material after decomposition primarily represents chemical or biological transformation products of the original molecules, selective preservation of the most refractory components, or a non-selective physical protection of a small fraction of the starting material. Examining the patterns of compositional change is one way to address this issue. Sinking particles in deep water columns offer a means to address the initial, rapid stages of decomposition, when >98% of the original material is destroyed (Minor et al., 2003). One-dimensional solid-state $^{13}C$-NMR of the organic matter in sinking particles from the Equatorial Pacific and Arabian Sea indicated that the material consisted of about 40–50% amino acid, 20–40% carbohydrate, and 20% lipid (Hedges et al., 2001), with modest relative decreases in protein and lipid relative to carbohydrate with depth. This composition, and the compositional change with depth, was roughly the same as the composition of the characterized fraction of the organic matter (Wakeham et al., 1997b; Hernes et al., 1996; Lee et al., 2000), although it is important that the molecularly characterized fraction of the total organic matter was as little as 20%. Also, the molecular characterization showed changes in lipid, amino acid, and carbohydrate compositions suggesting selective preservation of more refractory components within a compound class. However, the $^{13}C$-NMR data led Hedges et al. (2001) to conclude that organic matter degradation was not strongly selective for any major compound class, and hence the preservation of organic matter was most likely to be via a physical protective mechanism. Minor et al. (2003) further examined the Equatorial Pacific samples by direct temperature-resolved mass spectrometry (DTMS), which, like $^{13}$C-NMR, partially characterizes the bulk of organic matter in the sample. DTMS showed that the sinking particles were predominantly composed of biochemicals at all depths, indicating that major structural transformations, such as melanoidin-type condensations, had not occurred to a large extent. There were some compositional changes, similar to those found by $^{13}$C-NMR and molecular level analysis, but the initial stages of decomposition were not highly selective by compound class. Loss of ca. 80% of the organic matter from the "analytical window" of molecular-level measurements seems, therefore, to be due to small changes, such as encapsulation in minerals or refractory organic material, selective preservation of the most refractory molecules within a compound class, or small changes in cross-linking or accessible functional groups of the constituent molecules (Minor et al., 2003).

Although not quite as comprehensive, examination of the organic-rich sediments from Mangrove Lake led to similar conclusions, i.e., that preserved organic matter consists of minimally altered biomolecules (Knicker and Hatcher, 1997; Nguyen and Harvey, 2001). A comparably detailed characterization by multiple techniques has yet to be applied to any typical coastal sediment, in part because the mineral matrix makes it impossible unless the organic matter is first extracted. However, it is clear that some minimally altered organic matter, such as enzymatically hydrolyzable protein (Mayer et al., 2002), is preserved in continental margin sediments. Also, the proportions of major biochemical compound classes in the characterized fraction are similar to those in the characterized fraction of sediment trap samples (Table 5.3), and molecular level analysis suggests continuity of the decomposition process between sinking particulate matter and the sediment surface. Therefore, it appears likely that preservation in coastal sediments also involves physical protection, in combination with selection for more refractory forms of biomolecules and small structural changes that disproportionately affect reactivity.

## 6. Global Patterns of Organic Matter Deposition and Accumulation in Coastal Sediments

High sediment accumulation rates were the first identified correlate of high organic matter burial efficiency (Henrichs and Reeburgh, 1987; Betts and Holland, 1991), and this relationship has been found in a number of subsequent data sets (e.g., Epping et al., 2002). However, rates of sediment accumulation *per se* are not necessarily the factor controlling preservation. High deposition rate sediments also are anoxic near the sediment water interface and have large amounts of mineral surface area. Such sediments could tend to preserve organic aggregates better than sediments with lower accumulation rates. The surficial zone of bioturbation and resuspension has a relatively short residence time before burial, for example. Also, although such sediments have high surface porosities, sediments are buried to depths where significant compaction occurs fairly rapidly. Compaction and dewatering could tend to enhance the stability of mineral-organic aggregates.

Primary productivity and flux of organic matter to the sediment-water interface strongly impact organic matter burial patterns in the global ocean (Müller and Suess, 1979; Henrichs, 1992). However, within continental margin areas such relationships are far less obvious, with grain size being a stronger predictor, suggesting that organic matter supply is not the main factor limiting preservation (Mayer, 1994a; Mayer et al., 2002). The association of high burial efficiency (and high TOC) with fine grained sediment or high sediment surface area/unit mass (Keil et al., 1994a; Mayer, 1994a,b) is consistent with the hypothesis that preserved organic matter is protected within clay-organic aggregates. Clays offer large surface areas for attachment of the organic matter that binds aggregates together, and their small grain size allows very small interstitial spacing within the aggregates. In California margin sediments, organic matter appears to preferentially associate with smectite rich sediments rather than those dominated by chlorite (Ransom et al., 1998a).

In many cases anoxia has been found to enhance sediment organic concentration, burial efficiency or OC:SFA (organic carbon to surface area ratios) (Hartnett et al., 1998; Keil and Cowie, 1999; Hartnett and Devol, 2003), but in others no effect was observed (Calvert et al., 1992; Pedersen et al., 1992; Ransom et al., 1998a), or $O_2$ availability was not the overriding control on organic matter distributions (Cowie et al., 1999). In some cases elevated concentrations of OC in anoxic sediments are simply associated with the accumulation of fine-grained sediments under stagnant conditions, and OC:SFA are not always elevated. Northwestern Atlantic Ocean surface sediments have OC:SFA that are comparable to those found beneath anoxic water columns (Mayer et al., 2002), and high burial efficiencies of 3–40% (Thomas et al., 2002), despite well oxygenated bottom waters. Much of the variation in sediment OC concentrations in the northwestern Atlantic continental margin is explained by sediment grain size, as on other margins, but the unusually high OC:SFA apparently results from shallow-water winnowing of organic matter, and transport and redeposition on the upper slope (Mayer et al., 2002). High burial efficiencies on the Iberian margin (up to 48%) can also be explained by the focusing of net sediment deposition in specific areas (Epping et al., 2002). Interestingly, an anoxic sediment off Peru, where dynamic bottom conditions could winnow organic particles, did not have enhanced organic matter preservation (Arthur et al., 1998).

The preponderance of evidence indicates that anoxia slows or prevents the decomposition of certain refractory organic substances (Hulthe et al., 1998), although most major biochemicals (proteins and carbohydrates) are rapidly decomposed anaerobically. Organic matter that is refractory under anoxic conditions probably contributes to the higher burial efficiency in many anoxic sediments, but such sediments also are optimal for preservation of mineral-organic aggregates. They typically have minimal bioturbation and tend to form under quiescent water conditions where sediment resuspension does not occur.

Some sediments with notably high burial efficiencies (e.g., Cape Lookout Bight, Martens et al., 1992; Cape Hatteras margin, Bauer et al., 2002) have substantial old, redeposited sediment organic matter. Such sediments also have other characteristics that enhance preservation, like anoxia and high accumulation rate, and there is no strong evidence of a relationship between sources or composition of organic matter and burial efficiency. However, there are insufficient data to conclude that this relationship does not exist.

Milliman and Syvitski (1992) have pointed out that small rivers draining mountainous coasts are the source of a large fraction of the sediment delivered to the coastal ocean. As such, given the robust association between organic matter preservation and mineral surface area, they are probably also a major driver of organic matter burial on shelves. Ding (1998) has shown this for the Gulf of Alaska shelf, where glacial-marine sedimentation is also a factor. Recent studies of the Eel River shelf (Leithold and Hope, 1999; Blair et al., 2003) confirm substantial shelf sediment preservation of organic matter in conjunction with major flood events, although the character of organic matter preserved (including plant debris and fossil organic matter from sedimentary rock) probably differs from that preserved in major river deltas.

## 7. Present State of Knowledge and Remaining Questions

Coastal marine sediments are the major sink for organic carbon in the global carbon cycle, and as such are a key to understanding the global processes than maintain the Earth system. Most coastal sediment organic matter is of contemporary marine planktonic origin, although it has recently been recognized that fossil carbon is a significant fraction of the carbon content of some coastal sediments. Further, terrigenous contributions to coastal sediment organic carbon may be greater than classical interpretations of stable isotopic or lignin data suggest. Plankton-derived and terrestrial plant-derived organic matter should be readily decomposed, and indeed about 99.8% does decompose before it accumulates in sediments, but about 0.2% of the total supplied to the coastal ocean is preserved. While some of the preserved material is probably refractory biomolecules, much of it requires some chemical or physical protection to survive for millennia. Organic matter composition changes somewhat during decomposition, with selective removal of more labile components, possible modification of reactive functional groups, and probable increases in cross-linking within and between macromolecules. These changes have the important effect of making the bulk of sediment organic matter difficult to analyze, and probably decrease its biodegradability, but it seems likely that much of the original structure survives intact.

The preservation of organic matter is intimately linked to the accumulation of fine-grained sediments. Exactly how clays preserve organic matter is unknown. The leading hypothesis is that organic matter is aggregated with detrital minerals, and that the mineral blocks some of the organic matter to access by bacteria or enzymes. If this is correct, then the aggregates must be very strong. This includes not only strong binding of the organic molecules to the mineral surface, via adsorptive processes, but also mechanical strength. Mechanical strength is essential because most coastal surface sediments are subject to storm-wave or current driven resuspension, bioturbation, and other physical disturbances that would disrupt weak aggregates. Strong adsorption and mechanical strength do not necessarily go together. In addition to forming strong bonds with the mineral surface, the organic matter must form internal cross-linkages to yield physically strong aggregates.

Hence, an appropriate new focus for studies of sediment organic matter is the nature and origins of refractory aggregates. One hypothesis is that proteins are the "glue" that binds mineral aggregates. Proteins have well-known adhesive properties, and are a major biochemical constituent of phytoplankton and bacteria. However, the adhesive binding aggregates need not be derived from specific bioadhesives; any organic material that binds strongly to clay minerals and can form strong internal cross-linkages can fill this role.

## Acknowledgments

I thank the organizers and participants in a recent Friday Harbor Symposium, *New Approaches in Marine Organic Biogeochemistry, A Tribute to the Life and Science of John I. Hedges,* which reinvigorated my thinking on the concepts presented here. Carol Arnosti and an anonymous reviewer provided helpful suggestions for revision. I also thank the National Science Foundation for support under OCE-9530280.

## Bibliography

Albers, C. S., G. Kattner, and W. Hagen. 1996. The composition of wax esters, tracylglycerols and phospholipids in Arctic and Antarctic copepods: evidence of energetic adaptation. *Mar. Chem.,* **55,** 347–358.

Aller, R. C. 1994. Bioturbation and remineralization of sedimentary organic matter: effects of redox oscillation. *Chem. Geol.,* **114,** 331–345.

Aller, R. C. 1998. Mobile deltaic and continental shelf muds as suboxic, fluidized bed reactors. *Mar. Chem.,* **63,** 143–155.

Aller, J. Y., R. C. Aller, and M. A. Green. 2002. Benthic faunal assemblages and carbon supply along the continental shelf/shelf break-slope off Cape Hatteras, North Carolina. *Deep Sea Res. II,* **49,** 4599–4625.

Alperin, M. J., C. S. Martens, D. B. Albert, I. B. Suayah, L. K. Benninger, N. E. Blair, and R. A. Jahnke. 1998. Benthic fluxes and porewater concentration profiles of dissolved organic carbon in sediments from the North Carolina continental slope. *Geochim. Comochim. Acta,* **63,** 427–448.

Altabet, M. A. 1988. Variations in nitrogen isotopic composition between sinking and suspended particles: Implications for nitrogen cycling and particle transformation in the open ocean. *Deep-Sea Res.,* **25,** 535–554.

Altabet, M. A., and R. Francois. 1994. Sedimentary nitrogen isotopic ratio as a recorder for surface ocean nitrate utilization. *Global Biogeochem. Cycles,* **8,** 103–116.

Arnarson, T., and R. G. Keil. 2001. Organic-mineral interactions in marine sediments studied using density fractionation and X-ray photoelectron spectroscopy. *Org. Geochem.,* **32,** 1401–1415.

Arnosti, C., D. J. Repeta, and N. V. Blough. 1994. Rapid bacterial degradation of polysaccharides in anoxic marine systems. *Geochim. Cosmochim. Acta,* **58,** 2639–2652.

Arnosti, C., and M. Holmer. 1999. Carbohydrate dynamics and contributions to the carbon budget of an organic-rich coastal sediment. *Geochim. Cosmochim. Acta,* **63,** 393–403.

Arnosti, C., and M. Holmer. 2003. Carbon cycling in a continental margin sediment: contrasts between organic matter characteristics and remineralization rates and pathways. *Estuar. Coast. Shelf Sci.,* **58,** 197–208.

Arthur, M. A., W. E. Dean and K. Laarkamp. 1998. Organic carbon accumulation and preservation in surface sediment on the Peru margin. *Chem. Geol.,* **152,** 273–286.

Bauer, J. E., E. R. M. Druffel, D. M. Wolgast and S. Griffin. 2002. Temporal and regional variability in sources and cycling of DOC and POC in the northwest Atlantic continental shelf and slope. *Deep-Sea Res. II,* **49,** 4387–4419.

Benner, R., A. E. Maccubbin and R. E. Hodson. 1984. Anaerobic biodegradation of the lignin and polysaccharide components of lignocellulose and synthetic lignin by sediment microflora. *Appl. Environ. Microbiol.,* **47,** 998–1004.

Bennett, R. H., B. Ransom, M. Kastner, R. J. Baerwald, M. H Hulbert, W. B. Sawyer, H. Olsen and M. W. Lambert. 1999. Early diagenesis: impact of organic matter on mass physical properties and processes, California continental margin. *Mar. Geol.,* **159,** 7–34.

Bergamaschi, B. A., E. Tsamakis, R. G. Keil, T. I. Eglinton, D. B. Montluçon, and J. I. Hedges. 1997. The effect of grain size and surface area on organic matter, lignin, and carbohydrate concentration and molecular compositions in Peru Margin sediments. *Geochim. Cosmochim. Acta,* **61,** 1247–1260.

Berglin, M., and P. Gratenholm. 2003. The barnacle adhesive plaque: morphological and chemical differences as a response to substrate properties. *Colloids and Surfaces B-Biointerfaces,* **28,** 107–117.

Berner, R. A. 1980. *Early Diagenesis: A Theoretical Approach.* Princeton University Press, Princeton, 241 pp.

Berner, R. A., and D. Canfield. 1989. A new model for atmospheric oxygen over Phanerozoic time. *Am. J. Sci.,* **289,** 333–361.

Berner, R. A., and K. Caldeira. 1997. The need for mass balance and feedback in the geochemical carbon cycle. *Geology,* **25,** 955–956.

Betts, J. N., and H. D. Holland. 1991. The oxygen content of ocean bottom waters, the burial efficiency of organic carbon and the regulation of atmospheric oxygen. *Paleogeogr. Paleoclimatol. Paleoecol.,* **97,** 5–18.

Bird, M. I., and D. R. Gröcke. 1997. Determination of the abundance and carbon isotope composition of elemental carbon in sediments. *Geochim. Cosmochim. Acta,* **61,** 3413–3423.

Blair, N. E., E. L. Leithold, S. T. Ford. K. A. Peeler, J. C. Holmes, and D. W. Perkey. 2003. The persistence of memory: The fate of ancient sedimentary organic carbon in a modern sedimentary system. *Geochim. Cosmochim. Acta,* **67,** 63–73.

Bott, T. L., and M. A. Borchardt. 1999. Grazing of protozoa, bacteria, and diatoms by meiofauna in lotic epibenthic communities. *J. N. Am. Benthol. Soc.,* **18,** 499–513.

Budge, S. M., and C. C. Parrish. 1998. Lipid biogeochemistry of plankton, settling matter and sediments in Trinity Bay, Newfoundland. II. Fatty acids. *Org. Geochem.,* **29,** 1547–1559.

Brüchert, V., and C. Arnosti. 2003. Anaerobic carbon transformation: experimental studies with flow-through cells. *Mar. Chem.,* **80,** 171–183.

Burdige, D. J., W. M. Berelson, K. H. Coale, J. McManus, and K. S. Johnson. 1999. Fluxes of dissolved organic carbon from California continental margin sediments. *Geochim. Cosmochim. Acta,* **63,** 1507–1515.

Burdige, D. J., A. Skoog, and K. Gardner. 2000. Dissolved and particulate carbohydrates in contrasting marine sediments. *Geochim. Cosmochim. Acta,* **64,** 1029–1041.

Burdige, D. J., and K. G. Gardner. 2002. Molecular weight distribution of dissolved organic carbon in marine sediment pore waters. *Mar. Chem.,* **62,** 45–64.

Calvert, S. E., R. M. Bustin, and T. F. Pedersen. 1992. Lack of evidence for enhanced preservation of sedimentary organic matter in the oxygen minimum of the Gulf of California. *Geology,* **20,** 757–720.

Camacho-Ibar, V. F., L. Aveytua-Alcazár, and J. D. Carriquiry. 2003. Fatty acid reactivities in sediment cores from the northern Gulf of California. *Org. Geochem.,* **34,** 425–439.

Canuel, E. A., and C. S. Martens. 1996. Reactivity of recently deposited organic matter: Degradation of lipid compounds near the sediment-water interface. *Geochim. Cosmochim. Acta,* **60,** 1793–1806.

Carman, R., G. Edlund and C. Damberg. 2000. Distribution of organic and inorganic phosphorus compounds in marine and lacustrine sediments: a $^{31}P$ NMR study. *Chem. Geol.,* **163,** 101–114.

Collins, M. J., A. N. Bishop and P. Farrimond. 1995. Sorption by mineral surfaces: rebirth of the classical condensation pathway for kerogen formation? *Geochim. Cosmochim. Acta,* **59,** 2387–2393.

Cottrell, M. T., and D. L. Kirchman. 2000. Natural assemblages of marine Proteobacteria and members of the *Cytobacter-Flavobacter* cluster consuming low- and high-molecular-weight dissolved organic matter. *Appl. Environ. Microbiol.,* **66,** 1692–1697.

Cowie, G. L., and J. I. Hedges. 1984. Carbohydrate sources in a coastal marine environment. *Geochim. Cosmochim. Acta,* **48,** 2075–2087.

Cowie, G. L., and J. I. Hedges. 1992. The role of anoxia in organic matter preservation in coastal sediments: relative stabilities of the major biochemicals under oxic and anoxic depositional conditions. *Org. Geochem.,* **19,** 229–234.

Cowie, G. L., and J. I. Hedges. 1994. Biochemical indicators of diagenetic alteration in natural organic matter mixtures. *Nature,* **369,** 304–307.

Cowie, G. L., J. I. Hedges, F. G. Prahl, and G. J. De Lange. 1995. Elemental and major biochemical changes across an oxidation front in a relict turbidite: A clear-cut oxygen effect. *Geochim. Cosmochim. Acta,* **59,** 33–46.

Cowie, G. L., S. E. Calvert, T. F. Pedersen, H. Schulz, and U. von Rad. 1999. Organic content and preservational controls in surficial shelf and slope sediments from the Arabian Sea (Pakistan margin). *Mar. Geol.* **161,** 23–38.

Dauwe, B., and J. J. Middelburg. 1998. Amino acids and hexosamines as indicators of organic matter degradation state in North Sea sediments. *Limnol. Oceanogr.,* **43,** 782–798.

Dauwe, B., J. J. Middelburg, P. M. J. Herman, and C. H. R. Heip. 1999. Linking diagenetic alteration of amino aids and bulk organic matter reactivity. *Limnol. Oceanogr.,* **44,** 1809–1814.

de Haas, H., T. C. E. van Weering, and H. de Stigter. 2002. Organic carbon in shelf seas: sinks or sources, processes and products. *Cont. Shelf Res.,* **22,** 691–717.

DeLaune, R. D., G. A. Hambrick III, and W. H. Patrick, Jr. 1980. Degradation of hydrocarbons in oxidized and reduced sediments. *Mar. Poll. Bull.,* **11,** 103–106.

De Leeuw, J. W., and C. Largeau. 1993. A review of macromolecular organic compounds that comprise living organisms and their role in kerogen, coal, and petroleum formation. In *Organic Geochemistry,* M. H. Engle and S. A. Macko, eds. Plenum Press, New York, pp. 23–72.

DeLong, E. F., D. G. Franks, and A. L. Alldredge. 1993. Phylogenetic diversity of aggregate-attached vs. free-living marine bacterial assemblages. *Limnol. Oceanogr.,* **38,** 924–934.

DeMaster, D. J., C. J. Thomas, N. E. Blair, W. L. Fornes, G. Plaia, and L. A. Levin. 2002. Deposition of bomb $^{14}C$ in continental slope sediments of the mid-Atlantic bight: assessing organic matter sources and burial rates. *Deep Sea Res. II,* **49,** 4667–4685.

Dickens, A.F., Y. Gelinas, C.A. Masiello, S. Wakeham, and J.I. Hedges, 2004. Reburial of fossil organic carbon in marine sediments. *Nature,* **427,** 336–339.

Ding, X. 1998. *Organic matter accumulation and preservation in Alaskan continental margin sediments.* Ph.D. thesis, University of Alaska Fairbanks, 298 pp.

Dittmar, T., and R. J. Lara. 2001. Molecular evidence for lignin degradation in sulfate reducing mangrove sediments (Amazonia, Brazil). *Geochim. Cosmochim. Acta,* **65,** 1417–1428.

Dittmar, T., and G. Kattner. 2003. The biogeochemistry of the river and shelf ecosystem of the Arctic Ocean: a review. *Mar. Chem.,* **83,** 103–120.

Eglinton, T. I., J. E. Irvine, A. Vairavamurthy, W. Zhou and B. Manowitz. 1994. Formation and diagenesis of macromolecular organic sulfur in Peru Margin sediment. *Org. Geochem.,* **22,** 781–799.

Epping, E., C. van der Zee, K. Soetaert, and W. Helder. 2002. On the oxidation and burial of organic carbon in sediment of the Iberian margin and Nazare Canyon (NE Atlantic*). Prog. Oceanogr.,* **52,** 399–431.

Ertel, J. R., J. I. Hedges, A. H. Devol, and J. E. Richey. 1986. Dissolved humic substances of the Amazon River system. *Limnol. Oceanogr.,* **31,** 739–754.

Fenchel, T., G. M. King, and T. H. Blackburn. 1998. *Bacterial Biogeochemistry: The Ecophysiology of Mineral Cycling.* Academic Press, New York, 307 pp.

Ferdelman, T. G., T. M. Church, and G. W. Luther. 1991. Sulfur enrichment of humic substances in a Delaware salt marsh sediment core. *Geochim. Cosmochim. Acta,* **55,** 979–988.

Fernandes, M. B., and M. A. Sicre. 2000. The importance of terrestrial organic carbon inputs on Kara Sea shelves as revealed by n-alkanes, OC and $\delta^{13}C$ values. *Org. Geochem.,* **31,** 363–374.

Filippelli, G. 2001. Carbon and phosphorus cycling in anoxic sediments of the Saanich Inlet, British Columbia. *Mar. Geol.,* **174,** 307–321.

Filley, T. R., P. G. Hatcher, W. C. Shortle, and R. T. Praeseuth. 2000. The application of $^{13}C$-labeled tetramethylammonium hydroxide ($^{13}C$-TMAH) thermochemolysis to the study of fungal degradation of wood. *Org. Geochem.,* **31,** 181–198.

Fredrickson, J. K., and T. C. Onstott. 1996. Microbes deep inside the earth. *Scientific American,* **275,** 42–47.

Freudenthal, T., T. Wagner, R. Wenzhöfer, M. Zabel and G. Wefer. 2001. Early diagenesis of organic matter from sediments of the eastern subtropical Atlantic: Evidence from stable nitrogen and carbon isotopes. *Geochim. Cosmochim. Acta,* **65,** 1795–1808.

Froelich, P. N., G. P. Klinkhammer, M. L. Bender, N. A. Luedtke, G. R. Heath, D. Cullen, P. Dauphin, D. Hammond, and B. Hartman. 1979. Early oxidation of organic matter in pelagic sediments of the eastern equatorial Atlantic: suboxic diagenesis. *Geochim. Cosmochim. Acta,* **43,** 1075–1090.

Gelin, F., J. K. Volkman, C. Largeau, S. Derenne, J. S. Sinninghe Damsté, and J. W. De Leeuw. 1999. Distribution of aliphatic, nonhydrolyzable biopolymers in marine algae. *Org. Geochem.,* **30,** 147–159.

Gong, C., and D. J. Hollander. 1997. Differential contribution of bacteria to sedimentary organic matter in oxic and anoxic environments, Santa Monica Basin, California. *Cont. Shelf. Res.,* **15,** 1043–1059.

Goñi, M. A., M. B. Yunker, R. W. Macdonald, and T. I. Eglinton. 2000. Distribution and sources of organic biomarker in arctic sediments from the Mackenzie River and Beaufort shelf. *Mar. Chem.,* **71,** 23–51.

Gordon, E. S., and M. A. Goñi. 2003. Sources and distribution of terrigenous organic matter delivered by the Atchafalaya River to sediments in the northern Gulf of Mexico. *Geochim. Cosmochim. Acta,* **67,** 2359–2375.

Gough, M.A., R. F. C. Mantoura, and M. Preston. 1993. Terrestrial plant biopolymers in marine sediments. *Geochim. Cosmochim. Acta,* **57,** 945–964.

Haemers, S., M. C. van der Leeden, G. J. M. Koper, and G. Frens. 2002. Cross-linking and multilayer adsorption of mussel adhesive proteins. *Langmuir,* **18,** 4903–4907.

Harnett, H.E., R. G. Keil, J. I. Hedges, and A. H. Devol. 1998. Influence of oxygen exposure time on organic carbon preservation in continental margin sediments. *Nature,* **391,** 572–574.

Hartnett, H. E., and A. H. Devol. 2003. Role of a strong oxygen-deficient zone in the preservation and degradation of organic matter: a carbon budget for the continental margins of northwest Mexico and Washington state. *Geochim. Cosmochim. Acta,* **67,** 247–264.

Harvey, H. R., J. H. Tuttle, and J. T. Bell. 1995. Kinetics of phytoplankton decay during simulated sedimentation – Changes in biochemical composition and microbial activity under oxic and anoxic conditions. *Geochim. Cosmochim. Acta,* **59,** 3367–3377.

Harvey, H. R., and S. A. Macko. 1997. Kinetics of phytoplankton decay during simulated sedimentation: changes in lipids under oxic and anoxic conditions. *Org. Geochem.,* **27,** 129–140.

Hedges, J. I. 1992. Global biogeochemical cycles: progress and problems. *Mar. Chem.,* **39,** 67–93.

Hedges, J. I., and D. C. Mann. 1979. The lignin geochemistry of marine sediments from the southern Washington coast. *Geochim. Cosmochim. Acta,* **43,** 1809–1818.

Hedges, J. I., G. L. Cowie, J. E. Richey, P. D. Quay, R. Benner, M. Strom and B. R. Forsberg. 1994. Origins and processing of organic matter in the Amazon River as indicated by carbohydrates and amino acids. *Limnol. Oceanogr.,* **33,** 1116–1136.

Hedges, J. I., and R. G. Keil. 1995. Sedimentary organic matter preservation: an assessment and speculative synthesis. *Mar. Chem.,* **49,** 81–115.

Hedges, J. I., R. G. Keil, and R. Benner. 1997. What happens to terrestrial organic matter in the ocean? *Org. Geochem.,* **27,** 195–212.

Hedges, J. I., and J. M. Oades. 1997. Comparative organic geochemistries of soils and marine sediments. *Org. Geochem.*, **27**, 319–361.

Hedges, J. I., G. Eglinton, P. G. Hatcher, D. L. Kirchman, C. Arnosti, S. Derenne, R. P. Evershed, I. Kögel-Knaber, J. W. de Leeuw, R. Littke, W. Michaelis, and J. Rullkötter. 2000. The molecularly uncharacterized component of nonliving organic matter in natural environments. *Org. Geochem.*, **31**, 945–958.

Hedges, J. I., J. A. Baldock, Y. Gélinas, C. Lee, M. L. Peterson, and S. G. Wakeham. 2001. Evidence for nonselective preservation of organic matter in sinking marine particles. *Nature*, **409**, 801–804.

Hedges, J. I., J. A. Baldock, Y. Gélinas, C. Lee, M. L. Peterson, and S. G. Wakeham. 2002. The biochemical and elemental composition of marine plankton. A NMR perspective. *Mar. Chem.*, **78**, 47–63.

Henrichs, S. M. 1992. Early diagenesis of organic matter in marine sediments: progress and perplexity. *Mar. Chem.*, **39**, 119–149.

Henrichs, S. M. 1995. Sedimentary organic matter preservation: an assessment and speculative synthesis—A comment. *Mar. Chem.*, **49**, 127–136.

Henrichs, S. M., and J. W. Farrington. 1984. Peru upwelling region sediments near 15° S. I. Remineralization and accumulation of organic matter. *Limnol. Oceanogr.*, **29**, 1–19.

Henrichs, S. M., and A. P. Doyle. 1986. Decomposition of $^{14}$C-labeled organic substances in marine sediments. *Limnol. Oceanogr.*, **31**, 765–778.

Henrichs, S. M., and J. W. Farrington. 1987. Early diagenesis of amino acids and organic matter in two coastal marine sediments. *Geochim. Cosmochim. Acta*, **51**, 1–15.

Henrichs, S. M., and W. S. Reeburgh. 1987. Anaerobic mineralization of marine sediment organic matter: Rates and the role of anaerobic processes in the oceanic carbon economy. *Geomicrobiol. J.*, **5**, 191–237.

Hernes, P. J., J. I. Hedges, M. L. Peterson, S. G. Wakeham, and C. Lee. 1996. Neutral carbohydrate geochemistry of particulate material in the central equatorial Pacific. *Deep-Sea Res. II*, **43**, 1181–1204.

Hulthe, G., S. Hulth, and P. O. J. Hall. 1998. Effect of oxygen on degradation rate of refractory and labile organic matter in continental margin sediments. *Geochim. Cosmochim. Acta*, **62**, 1319–1328.

Ingall, E., and R. Jahnke. 1997. Influence of water-column anoxia on the elemental fractionation of carbon and phosphorus during sediment diagenesis. *Mar. Geol.*, **139**, 219–229.

Ingalls, A. E., C. Lee, S. G. Wakeham, and J. I. Hedges. 2003. The role of biominerals in the sinking flux and preservation of amino acids in the Southern Ocean along 170°W. *Deep-Sea Res. II*, **50**, 713–738.

Ittekkot, V. 1988. Global trends in the nature of organic matter in river suspensions. *Nature*, **332**, 436–438.

Jaeger, J. M., C. A. Nittrouer, N. D. Scott, and J. D. Milliman. 1998. Sediment accumulation along a glacially impacted mountainous coastline: Northeast Gulf of Alaska. *Basin Res.*, **10**, 155–173.

Jahnke, R. A. 2004. Chapter 6. Transport processes and organic matter cycling in coastal sediments. In *The Sea*, Volume 13, A. R. Robinson and K. H. Brink, eds. Harvard University Press, Boston.

Jennerjahn, T. C., and V. Ittekkot. 1999. Changes in organic matter from surface waters to continental slope sediments off the São Francisco River, eastern Brazil. *Mar. Geol.*, **161**, 129–140.

Kaneda, T. 1991. Iso- and anteiso-fatty acids in bacteria: biosynthesis, function, and taxonomic significance. *Microbiol. Rev.*, **55**, 288–302.

Keil, R. G., E. Tsamakis, C. B. Fuh, C. Giddings, and J. I. Hedges. 1994a. Mineralogical and textural controls on organic composition of coastal marine sediments: Hydrodynamic separation using SPLITT fractionation. *Geochim. Cosmochim. Acta*, **57**, 879–893.

Keil, R. G., F. S. Hu, E. C. Tsamakis, and J. I. Hedges. 1994b. Pollen degradation in marine sediments as an indicator of oxidation of organic matter. *Nature*, **369**, 639–641.

Keil, R. G., L. M. Mayer, P. D. Quay, J. E. Richey, and J. I. Hedges. 1997. Loss of organic matter from riverine particles in deltas. *Geochim. Cosmochim. Acta,* **61,** 1507–1511.

Keil, R. G., E. Tsamakis, J. C. Giddings, and J. I. Hedges. 1998. Biochemical distributions (amino acids, neutral sugars, and lignin phenols) among size classes of modern marine sediments from the Washington coast. *Geochim. Cosmochim. Acta,* **62,** 1347–1364.

Keil, R. G., and G. L. Cowie. 1999. Organic matter preservation through the oxygen-deficient zone of the NE Arabian Sea as discerned by organic carbon: mineral surface area ratios. *Mar. Geol.,* **161,** 13–22.

Keil, R. G., E. Tsamakis, and J. I. Hedges. 2000. Early diagenesis of particulate amino acids in marine systems. In *Perspectives in Amino Acid and Protein Chemistry,* G. A. Goodfriend, M. J. Collins, M. L. Fogel, S. A. Macko, and J. F. Wehmiller, eds., Oxford University Press, Oxford, pp. 69–82.

Kerhervé, P., R. Buscail, F. Gadel, and L. Serve. 2002. Neutral monosaccharides in surface sediments of the northwestern Mediterranean Sea. *Org. Geochem.,* **33,** 421–435.

Kieber, R. J., L. H. Hydro, and P. J. Seaton. 1997. Photooxidation of triglycerides and fatty acids in seawater: Implication toward the formation of marine humic substances. *Limnol. Oceanogr.,* **42,** 1454–1462.

Klok, J., M. Bass, H. C. Cox, J. W. De Leeuw, W. I. C. Rijpstra, and P. A. Schenck. 1984. Qualitative and quantitative characterization of the total organic matter in a recent marine sediment (Part II). *Org. Geochem.,* **6,** 265–278.

Klump, J. V., and C. S. Martens. 1987. Biogeochemical cycling in an organic-rich marine basin. 5. Sedimentary nitrogen and phosphorus budgets based upon kinetic models, mass balances, and the stoichiometry of nutrient regeneration. *Geochim. Cosmochim. Acta,* **51,** 1161–1173.

Kohnen, M. E. L., J. S. Sinninghe Damsté, H. L. Ten Haven, and J. W. De Leeuw. 1989. Early incorporation of polysulphides in sedimentary organic matter. *Nature,* **341,** 640–641.

Knicker, H., and P. G. Hatcher. 1997. Survival of protein in an organic-rich sediment: possible protection by encapsulation in organic matter. *Naturwissenschaften,* **84,** 231–234.

Kristensen, E. 2000. Organic matter diagenesis at the oxic/anoxic interface in coastal marine sediments, with emphasis on the role of burrowing animals. *Hydrobiologia,* **426,** 1–24.

Kristensen, E., and M. Holmer. 2001. Decomposition of plant materials in marine sediment exposed to different electron acceptors, with emphasis on substrate origin, degradation kinetics, and the role of bioturbation. *Geochim. Cosmochim. Acta,* **65,** 419–433.

Lee, C., S. G. Wakeham, and J. I. Hedges. 2000. Composition and flux of particulate amino acids and chloropigments in equatorial Pacific seawater and sediments. *Deep-Sea Res. I,* **47,** 1535–1568.

Lehman, M. F., S. M. Bernasconi, A. Bargieri, and J. A. McKenzie. 2002. Preservation of organic matter and alteration of its carbon and nitrogen isotope composition during simulated and *in situ* early sedimentary diagenesis. *Geochim. Cosmochim. Acta,* **66,** 3573–3584.

Leithold, E. L., and R. S. Hope. 1999. Deposition and modification of a flood layer on the northern California shelf: lessons from and about the fate of terrestrial particulate organic carbon. *Mar. Geol.,* **154,** 183–195.

Leithold, E., and N. E. Blair. 2001. Watershed control on the carbon loading of marine sedimentary particles. *Geochim. Cosmochim. Acta,* **65,** 2231–2240.

Llobet-Brossa, E., R. Roselló-Mora, and R. Amann. 1998. Microbial community composition of Wadden Sea sediments as revealed by fluorescence in situ hybridization. *Appl. Environ. Microbiol.,* **64,** 2691–2696.

Libes, S. M., and W. G. Deuser. 1988. The isotope geochemistry of particulate nitrogen in the Peru upwelling are and the Gulf of Maine. *Deep-Sea Res. Part A,* **35,** 517–533.

Liu, K.-K., K. Iseki, and S.-Y. Chao. 2000. Continental margin carbon fluxes. In *The Changing Ocean Carbon Cycle,* R. B. Hanson, H. W. Ducklow, and J. G. Field, eds. Cambridge University Press.

Lückge, A., M. Ercegovac, H. Strauss, and R. Littke. 1999. Early diagenetic alteration of organic matter by sulfate reduction in Quaternary sediments from the northeastern Arabian Sea. *Mar. Geol.,* **158,** 1–13.

Luo, H. 1994. *Decomposition and adsorption of peptides in Alaskan coastal marine sediments.* Ph.D. thesis, University of Alaska Fairbanks.

Manini, E., C. Fiordelmondo, C. Gambi, A. Pusceddu, and R. Danovaro. 2003. Benthic microbial loop functioning in coastal lagoons: a comparative approach. *Oceanol. Acta,* **26,** 27–38.

Martens, C. S., R. I. Haddad, and J. P. Chanton. 1992. Organic matter accumulation, remineralization, and burial in an anoxic coastal sediment. In *Organic Matter: Productivity, Accumulation, and Preservation in Recent and Ancient Sediments,* J. K. Whelan and J. W. Farrington, eds. Columbia University Press, New York, pp. 82–98.

Masiello, C. A., E. R. M. Druffel, and L. A. Currie. 2002. Radiocarbon measurements of black carbon in aerosols and ocean sediments. *Geochim. Cosmochim. Acta,* **66,** 1025–1036.

Mayer, L. M. 1994a. Surface area control of organic carbon accumulation in continental shelf sediments. *Geochim. Cosmochim. Acta,* **58,** 1271–1284.

Mayer, L. M. 1994b. Relationships between mineral surfaces and organic carbon concentrations in soils and sediments. *Chem. Geol.,* **114,** 347–363.

Mayer, L. M. 1999. Extent of coverage of mineral surfaces by organic matter in marine sediments. *Geochim. Cosmochim. Acta,* **63,** 207–215.

Mayer, L. M., L. L. Schick, T. Sawyer, C. Plante, P. A. Jumars and R. L. Self. 1995. Bioavailable amino acids in sediments, a biomimetic, kinetics-based approach. *Limnol. Oceanogr.,* **40,** 511–520.

Mayer, L. M., L. Benninger, M. Bock, D. DeMaster, Q. Roberts, and C. Martens. 2002. Mineral associations and nutritional quality of organic matter in shelf and upper slope sediments off Cape Hatteras, USA: a case of unusually high loadings. *Deep Sea Res. II,* **49,** 4587–4597.

Meybeck, M. 1982. Carbon, nitrogen, and phosphorus transport by world rivers. *Am. J. Sci.,* **282,** 401–450.

Milliman, J. D., and J. P. M. Syvitski. 1992. Geomorphic/tectonic control of sediment discharge to the ocean: The importance of small mountainous rivers. *J. Geol.,* **100,** 525–544.

Minor, E. C., S. G. Wakeham, and C. Lee. 2003. Changes in the molecular-level characteristics of sinking marine particles with water column depth. *Geochim. Cosmochim. Acta,* **67,** 4277–4288.

Mitra, S., T. S. Bianchi, B. A. McKee, and M. A. Sutula. 2002. Sources and seasonal delivery of fossil fuel-derived fluvial black carbon from the Mississippi River: Implications for the global carbon cycle. *Environ. Sci. Technol.,* **26,** 2296–2302.

Miyajima, T., H. Ogawa, and I. Koike. 2001. Alkalai extractable polysaccharides in marine sediments: Abundance, molecular size distribution, and monosaccharide composition. *Geochim. Cosmochim. Acta,* **65,** 1455–1466.

Montagna, P. A., B. C. Coull, T. L. Herring, and B. W. Dudley. 1982. The relationship between abundances of meiofauna and their suspected microbial food (diatoms and bacteria). *Est. Coast. Shelf Sci.,* **17,** 381–394.

Müller, P. J., and E. Suess. 1979. Productivity, sedimentation rate, and sedimentary organic matter in the oceans - I. Organic carbon preservation. *Deep Sea Res.,* **26,** 1347–1362.

Nguyen, R. T., and H. R. Harvey. 1997. Protein and amino acid cycling during phytoplankton decomposition in oxic and anoxic waters. *Org. Geochem.,* **27,** 115–128.

Nguyen, R. T., and H. R. Harvey. 2001. Preservation of protein in marine systems: Hydrophobic and other noncovalent associations as major stabilizing forces. *Geochim. Cosmochim. Acta,* **65,** 1467–1480.

Nissenbaum, A., and I. R. Kaplan. 1972. Chemical and isotopic evidence for the *in situ* origin of marine humic substances. *Limnol. Oceanogr.,* **17,** 570–581.

Nunn, B. L., A. Norbeck, and R. G. Keil. 2003. Hydrolysis patterns and the production of peptide intermediates during protein decomposition in marine systems. *Mar. Chem.,* **83,** 59–73.

Onstad, G. D., D. E. Canfield, P. D. Quay, and J. I. Hedges. 2000. Sources of particulate organic matter in rivers from the continental USA: Lignin phenol and stable carbon isotope composition. *Geochim. Cosmochim. Acta,* **64,** 3539–3546.

Opsahl, S., and R. Benner. 1995. Early diagenesis of vascular plant tissues: Lignin and cutin decomposition and biogeochemical implications. *Geochim. Cosmochim. Acta,* **59,** 4889–4904.

Ostrom, N. E., S. A. Macko, D. Deibel, and R. J. Thompson. 1997. Seasonal variation in the stable carbon and nitrogen isotope biogeochemistry of a coastal cold ocean environment. *Geochim. Cosmochim. Acta,* **61,** 2929–2942.

Pantoja, S., C. Lee, and J. F. Marechek. 1997. Hydrolysis of peptides in seawater and sediment. *Mar. Chem.,* **57,** 25–40.

Paytan, A., B. J. Cade-Menun, K. McLaughlin, and K. L. Faul. Selective phosphorus regeneration of sinking marine particles: evidence from $^{31}$P-NMR. *Mar. Chem.,* **82,** 55–70.

Patterson, D. J., J. Larsen, and J. O. Corliss. 1989. The ecology of heterotrophic flagellates and ciliates living in marine sediments. *Prog. Protistol.,* **3,** 185–277.

Pearson, A., A. P. McNichol, B. C. Benitez-Nielson, J. M. Hayes, and T. I. Eglinton. 2001. Origins of lipid biomarkers in Santa Monica Basin surface sediment: A case study using compound-specific $\Delta^{14}$C analysis. *Geochim. Cosmochim. Acta,* **65,** 3123–3137.

Pedersen, T.F., G.B. Shimmield, and N. B. Price. 1992. Lack of enhanced preservation of organic matter in sediments under the oxygen minimum on the Oman margin. *Geochim. Cosmochim. Acta,* **56,** 545–551.

Petsch, S. T., T. I. Eglinton, and K. J. Edwards. 2001. $^{14}$C-dead living biomass: Evidence for microbial assimilation of ancient organic carbon during shale weathering. *Science,* **292,** 1127–1131.

Prahl, F. G. G. J. De Lange, M. Lyle, and M. A. Sparrow. 1989. Post-depositional stability of long-chain alkenones under contrasting redox conditions. *Nature,* **341,** 434–437.

Ransom, B., R. H. Bennett, R. Baerwald, and K. Shea. 1997. TEM study of *in situ* organic matter on continental margins: occurrence and the “monolayer” hypothesis. *Mar. Geol.,* **138,** 1–9.

Ransom, B., D. Kim, M. Kastner, and S. Wainwright. 1998a. Organic matter preservation on continental slopes: Importance of mineralogy and surface area. *Geochim. Cosmochim. Acta,* **62,** 1329–1345.

Ransom, B., K. F. Shea, P. J. Burkett, R. H. Bennett, and R. Baerwald. 1998b. Comparison of pelagic and nepheloid layer marine snow: implications for carbon cycling. *Mar. Geol.,* **150,** 39–50.

Raymond, P. A., and J. E. Bauer. 2001a. DOC cycling in a temperate estuary: a mass balance approach using $^{14}$C and $^{13}$C. *Limnol. Oceanogr.,* **46,** 655–667.

Raymond, P. A., and J. E. Bauer. 2001b. Riverine export of aged organic matter to the North Atlantic Ocean. *Nature,* **409,** 497–500.

Redfield, A. C., B. H. Ketchum, and F. A. Richards. 1963. The influence of organisms on the composition of seawater. In *The Sea,* Volume 2, M. N. Hill, ed. Interscience, New York, pp. 176–192.

Rowland, S. J., and J. R. Maxwell. 1984. Reworked triterpenoid and steroid hydrocarbons in a recent sediment. *Geochim. Cosmochim. Acta,* **48,** 617–624.

Saino, T., and A. Hattori. 1987. Geographical variation of the water column distribution of suspended particulate organic nitrogen and its $^{15}$N natural abundance in the Pacific and its marginal seas. *Deep-Sea Res. Part A,* **34,** 807–827.

Salmon, V., S. Derenne, E. Lallier-Vergès, C. Largeau, and B. Beaudoin. 2000. Protection of organic matter by mineral matrix in a Cenomanian black shale. *Org. Geochem.,* **31,** 463–474.

Sawyer, D. T. 1991. *Oxygen Chemistry.* Oxford, New York, 223 pp.

Sinninghe Damsté, J. S., M. D. Kok. J. Köster, and S. Schouten. 1998. Sulfurized carbohydrates: an important sedimentary sink for organic carbon? *Earth Planet. Sci. Lett.,* **164,** 7–13.

Sugai, S. F., and Henrichs, S. M. 1992. Rates of amino acid decomposition in Resurrection Bay (Alaska) sediments. *Mar. Ecol. Prog. Ser.*, **88**, 129–141.

Summons, R. E. 1993. Biogeochemical cycles: a review of fundamental aspects of organic matter formation, preservation, and composition. In *Organic Geochemistry,* M. H. Engel and S. A. Macko, eds. Plenum Press, New York, pp. 3–21.

Sun, M.-Y., C. Lee, and R. C. Aller. 1993a. Laboratory studies of oxic and anoxic degradation of chlorophyll-a in Long Island Sound sediments. *Geochim. Cosmochim. Acta,* **57,** 147–157.

Sun, M.-Y, C. Lee, and R. C. Aller. 1993b. Anoxic and oxic degradation of $^{14}$C-labeled chloropigments and a $^{14}$C-labeled diatom in Long Island Sound sediments. *Limnol. Oceanogr.* **38,** 1438–1451.

Sun, M.-Y., S. G. Wakeham, and C. Lee. 1997. Rates and mechanisms of fatty acid degradation in oxic and anoxic coastal marine sediments of Long Island Sound, New York, USA. *Geochim. Cosmochim. Acta,* **61,** 341–355.

Sun, M.-Y., and S. G. Wakeham. 1998. A study of oxic/anoxic effects on degradation of sterols at the simulated sediment-water interface of coastal sediments. *Org. Geochem.,* **28,** 773–784.

Sun, M.-Y., R. C. Aller, C. Lee, and S. G. Wakeham. 2002. Effects of oxygen and redox oscillation on degradation of cell associated lipids in surficial marine sediments. *Geochim. Cosmochim. Acta,* **66,** 2003–2012.

Tegelaar, E. W., J. W. De Leeuw, S. Derrenne, and C. Largeau. 1989. A reappraisal of kerogen formation. *Geochim. Cosmochim. Acta,* **53,** 3103–3106.

Ternois, Y., K. Kawamura, L. Keigwin, N. Ohkouchi, and T. Nakatsuka. 2001. A biomarker approach for assessing marine and terrigenous inputs to the sediments of Sea of Okhotsk for the last 27,000 years. *Geochim. Cosmochim. Acta,* **65,** 791–802.

Thomas, C. J., N. E. Blair, M. J. Alperin, D. J. DeMaster, R. A. Jahnke, C. S. Martens, and L. Mayer. 2002. Organic carbon deposition on the North Carolina continental slope off Cape Hatteras (USA). *Deep-Sea Res. II,* **49,** 4687–4709.

Thompson, J., N. C. Higgs, I. W. Croudace, S. Colley, and D. J. Hydes. 1993. Redox zonation of elements at an oxic/post-oxic boundary in deep-sea sediment. *Geochim. Cosmochim. Acta,* **57,** 579–595.

Toth, D. J., and A. Lerman. 1977. Organic matter reactivity and sedimentation rates in the ocean. *Am. J. Sci.,* **277,** 465–485.

Twilley, R. R., and B. McKee. 1995. Ecosystem analysis of the Louisiana Bight and adjacent shelf environments. Vol. II. The fate of organic matter and nutrients in the Louisiana Bight. U.S. Department of the Interior, Minerals Management Service, Gulf of Mexico OCS Regional Office, New Orleans, LA.

Van Schie, P.M., and M. Fletcher. 1999. Adhesion of biodegradative anaerobic bacteria to solid surfaces. *Appl. Environ. Microbiol.,* **65,** 5082–5088.

Vick, C. 1999. Chapter 9. Adhesive bonding of wood materials. In *Wood handbook-Wood as an engineering material.* General technical report FPL-GTR-113, Forest Products Laboratory, U.S. Department of Agriculture, Madison, WI, 463 pp.

Volkman, J. K. 1986. A review of sterol markers for marine and terrigenous organic matter. *Org. Geochem.,* **9,** 83–99.

Volkman, J. K., S. M. Barrett, S. I. Blackburn, M. P. Mansour, E. L. Sikes, and F. Gelin. 1998. Microalgal biomarkers: A review of recent research developments. *Org. Geochem.,* **29,** 1163–1179.

Wakeham, S. G., C. Lee, J. I. Hedges, P. J. Hernes and M. L. Peterson. 1997a. Molecular indicators of diagenetic status in marine organic matter. *Geochim. Cosmochim. Acta,* **61,** 5363–5369.

Wakeham, S. G., J. I. Hedges, C. Lee, M. L. Peterson, and P. J. Hernes. 1997b. Compositions and transport of lipid biomarkers through the water column and surficial sediments of the equatorial Pacific Ocean. *Deep-Sea Res. II,* **44,** 2131–2162.

Wellsbury, P., K. Goodman, T. Barth, B. A. Craig, S. P. Barnes, and R. J. Parkes. 1997. Deep marine biosphere fueled by increasing organic matter availability during burial and heating. *Nature,* **388,** 573–576.

Wilson, T. R. S., J. Thompson, J. Colley, D. J. Hydes, N. C. Higgs, and J. Sorensen. 1985. Early organic diagenesis: The significance of progressive subsurface oxidation fronts in pelagic sediments. *Geochim. Cosmochim. Acta,* **49,** 811–822.

Yu, M., J. Hwang, and T. J. Deming. 1999. Role of L-3,4-dihydroxyphenylalanine in mussel adhesive proteins. *J. Am. Chem. Soc.,* **121,** 5825–5826.

Zang, X., and P. G. Hatcher. 2002. A Py-GC-MS and NMR spectroscopy of organic nitrogen in Mangrove Lake sediments. *Org. Geochem.,* **33,** 201–211.

Zegouagh, Y., S. Derenne, C. Largeau, P. Bertrand, M.-A. Sicre, A. Saliot and B. Rousseau. 1999. Refractory organic matter in sediments from the north west African upwelling system: abundance, chemical structure, and origin. *Org. Geochem.,* **30,** 83–99.

Zink, K.-G., H. Wilkes, U. Disko, M. Elvert, and B. Horsfield. 2003. Intact phospholipids—microbial "life markers" in marine deep subsurface sediments. *Org. Geochem.,* **34,** 755–769.

# Chapter 6. TRANSPORT PROCESSES AND ORGANIC MATTER CYCLING IN COASTAL SEDIMENTS

RICHARD A. JAHNKE

*Skidaway Institute of Oceanography*

**Contents**



## 1. Introduction

The sediments within coastal zones play a significant role in controlling and determining coastal biogeochemistry which in turn exerts a major influence on marine chemical cycles. For example, the removal of biologically available 'fixed nitrogen' from the oceans via microbial denitrification occurs almost exclusively along continental margins and within coastal sediments (Codispoti et al., 2001). The transfer of organic and inorganic carbon from the margins to the open ocean has been identified as an important pathway within the marine carbon cycle (Yool and Fasham, 2001; Jahnke, 1996). The exchanges with surface sediments can also alter the ratios of important biological nutrients such as Fe, N, and Si, influencing biological community composition and the linkage between biological processes and the carbon cycle (Hutchins et al., 1998).

The upper meter of continental shelf sediments comprises the largest pool of organic carbon in the surface ocean. Additionally, shelf sediments are the primary location for the accumulation of anthropogenic contaminants transported to the coastal zone, thereby exerting a substantial influence on the health of near-shore ecosystems. Study of the reactions and processes responsible for transforming and transporting biogenic elements amongst the sedimentary pools and overlying water column is fundamental to furthering our understanding of marine biogeochemistry.

At the simplest level, the sea floor can be viewed as a location where deposited organic matter is remineralized, releasing dissolved inorganic constituents back to

*The Sea*, Volume 13, edited by Allan R. Robinson and Kenneth H. Brink
ISBN 0-674-01526-6 

the overlying waters. This source of nutrients supports biological production within the water column and the cycle is repeated. Because coastal regions are shallower than their deep ocean counterparts, the cycle of deposition, remineralization and efflux is maximized.

Recent studies have revealed, however, that in many coastal regions, the above paradigm must be modified in very significant ways. First, in the majority of coastal areas shallower than 50 m, light reaches the sea floor. Recent measurements have demonstrated that even at relatively low light levels, e.g. 5% of surface irradiance, significant rates of benthic photosynthesis can be supported (Cahoon and Cook, 1992; Jahnke et al., 2000). Thus, in these systems, the traditional paradigm must be modified because the sea floor is also a source of organic matter and oxygen and a sink for nutrients.

Secondly, in most shallow environments (generally less than 100 m) and excluding protected areas and regions of rapid deposition such as directly underneath river plumes, physical processes act to winnow away fine-grained particles, leaving relatively coarse-grained, higher-permeability sediments. Recent studies suggest that bottom current-driven advection and gravity wave-induced dispersion accelerate porewater transport in these systems (Riedl et al., 1972; Huettel et al., 1996; 1998, Shum, 1992; 1993). Such non-diffusive processes can greatly accelerate the exchange of solutes between the sediments and overlying water column. As the bottom waters exchange with the porewaters, small particles (1–10 μm in diameter) may be carried into the sediments and trapped within the pore spaces. Many algae are within this size range, so this process may serve to inject reactive organic matter into the sediments. Thus, in these systems, the influx of particulate materials may not depend solely on gravitational settling of fine particles, but may be significantly augmented by direct advective filtration. Through these series of transport processes, reactive organic particles and oxidants can be rapidly supplied to the sediments and metabolic by-products can be removed. Permeable shallow-water sediments are, therefore, poised to support very rapid metabolic rates. Finally, in addition to the small-scale, localized advective transport introduced above, large-scale groundwater discharge has been observed in many coastal areas (Moore, 1996; 1999). Such inputs can be comprised of recirculated sea water (Jahnke et al., 2003) or primary freshwater discharge from terrestrial aquifers (Moore, 1999) and may represent a significant input pathway of nutrients (Simmons and Lyons, 1994) and terrestrially-derived contaminants (Gallagher et al., 1996) to coastal systems. Thus, in many coastal systems, the traditional paradigm must be modified to include benthic photosynthesis, the direct injection of particulate materials into the sediments and the accelerated rates of porewater-bottom water exchange. Due to these processes the benthic and pelagic systems on continental shelves may be coupled to a much greater extent than previously thought.

In this chapter, the processes that transport porewater solutes and sedimentary particles and control the remineralization of organic materials are reviewed. Emphasis is placed on discussing the processes in general and the overall guiding principles of the remineralization sequence. In the latter sections, the focus is turned to aspects of coastal systems that add complexity to these generalized descriptions. Specifically, the increased importance of the oxidation of secondary metabolites, the potential role of benthic primary production and the non-diffusive transport of porewater solutes and particles are discussed. In the final section, we briefly discuss the methodological and technological challenges presented by these environments.

## 2. Transport Processes

Cycling of bioactive elements in sediments is controlled by the diagenetic and transport processes depicted in Figure 1. It is clear that there are many connections and feedbacks between individual processes and that rates and pathways of specific transport processes should be interpreted within the context of the sedimentary system. Furthermore, the magnitude and details of specific processes depend on the characteristics of the sediments themselves. Therefore, before addressing specific diagenetic processes, a brief review of important sediment characteristics and transport processes is provided.

Transport and reaction processes (Fig. 6.1) are divided into those that transport and transform dissolved constituents (left portion of figure) and particulate constituents (right portion of figure). Particulate constituents are converted into dissolved components through such processes as dissolution of minerals and the remineralization of particulate organic matter; conversely, adsorption, mineral formation via precipitation within the sediments and porewaters and biological production act to remove constituents from the dissolved pools, incorporating them into the particulate phases. Because transport and reaction pathways differ dramatically between particulate and dissolved pools, tracking the partitioning of reactive components between the solid and dissolved phases is fundamental to the description and understanding of sediment processes.

### *2.1 Porosity/Tortuosity/Permeability*

The distribution of solid and void space within a given volume of sediment is represented by the porosity ($\phi$), defined as the volume contained in the voids between sedimentary particles per total volume of sediment. In the absence of gas bubbles, $\phi$ equals $V_{pw}/(V_{pw} + V_{sed})$ where $V_{pw}$ is the volume of porewater and $V_{sed}$ is the volume of sediment particles within a given volume of whole wet sediment. It follows that (1-$\phi$) equals the volume of sediment particles contained in a given volume of whole sediment. In coarse grained sediments (sand-sized and larger), $\phi$ is relatively low (0.3–0.6) and near-surface compaction of the sediments generally causes very minor vertical variations in porosity. In fine grained sediments, however, $\phi$ may vary from values very nearly 1.0 to lower values due to sediment compaction (Boudreau and Bennett, 1999) and the resulting vertical variations in $\phi$ significantly impact the distribution and transport rates of dissolved and particulate phases.

Example fine-scale distributions of porosity, total organic carbon (TOC) and adenosine triphosphate (ATP; a marker for biomass) in fine-grained surface sediments from San Clemente Basin, California Borderlands, are provided in Figure 6.2 (Craven et al., 1986). At this location, $\phi$ near the sediment–water interface is very high (approx. 0.95) and decreases with increasing sediment depth, most likely due to compaction. At these high porosity values, it is important to recall that (1-$\phi$) equals the amount of sediment particles within a volume of sediment. Thus, while the volume of porewater per volume of sediment decreases by about 10% in the upper 3 cm of the sediment column, from 0.95 to 0.85, the volume occupied by sediment particles per volume of whole sediment increases by nearly a factor of three over the same depth horizon, from 0.05 to 0.15. The importance of these variations on the distribution of solid phase constituents is demonstrated by the profiles of TOC and ATP distributions which are displayed both on a per dry weight and per volume basis.

On a weight basis, TOC and ATP are highest at the sediment surface and decrease with sediment depth. That is, the particles near the sediment-water interface are enriched in TOC and ATP relative to the particles below. However, the very high porosities measured at the sediment surface correspond to very low particle abundances. Basically, the sediments at the sediment surface are mostly comprised of water with only a small number of sediment particles. Expressed per volume of whole sediment, therefore, the sediment surface is characterized by decreased levels of TOC and ATP opposite the trend noted above. Thus, it is clear from this example, that variations in $\phi$ can have a dramatic impact on the distribution of sedimentary components.

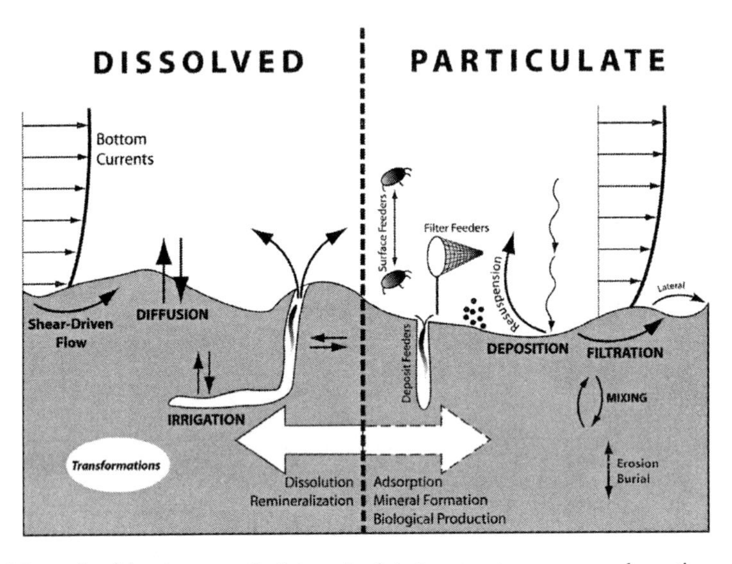


Figure 6.1 Schematic of dominant particulate and solute transport processes and reaction pathways in sediments.

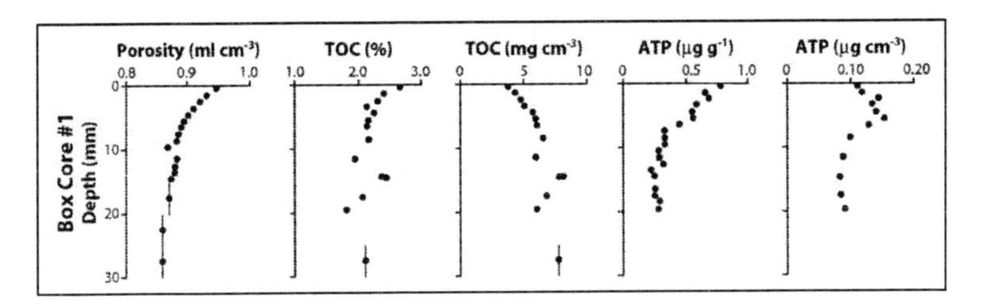


Figure 6.2 Fine-scale porosity, total organic carbon content and ATP content of surface sediments from San Clemente Basin (re-drawn from Craven et al., 1986).

### 2.2 *Deposition/Burial/Sediment Mixing*

Particulate transport in sediments can be approximated by advective and dispersive components. The advection of sediment particles away from the sediment surface is most commonly attributed to net sediment accumulation and burial (Berner, 1980). Although biogenic components including opal, calcium carbonate and particulate organic carbon may contribute to the sediment composition, lithogenic particles generally dominate in coastal environments. Regional exceptions include carbonate banks where biogenic carbonate is the dominant sediment phase. In most coastal areas, sediment accumulation rates are determined by the balance between the deposition and resuspension of lithogenic particles; production or dissolution of biogenic particles, although important, generally exerts a secondary influence on burial rates. At locations where bottom currents are sufficiently energetic, erosion rates may exceed deposition rates and advection is simply reversed, reducing the distance between the sediment surface and a given horizon within the sediments over time.

Over limited vertical horizons, biological processes may also result in advective particle transport. Particular species of benthic organisms, known as conveyor-belt feeders (Rice, 1986), selectively feed at a given horizon within the sediments and defecate ingested sediment particles at the sediment surface. If these organisms are sufficiently abundant and dominate the local community, this activity results in a relatively rapid advective transfer of particles to the depth of feeding and then a wholesale return to the sediment surface and the process repeats. Tracer studies have demonstrated that with a homogeneous population of conveyor-belt deposit feeders, this cycle can repeat several times without significant dispersion of a specific particle layer. Given sufficient organism abundance and feeding rates, bioadvection rates within the feeding zone can greatly exceed rates of burial due to sediment accumulation only. Of course, below the maximum feeding depth, this process does not occur and particle advection rates return to those that can be supported by net particle accumulation.

Sediment accumulation rates in coastal sediments are highly variable. Most continental shelf regions are characterized by sediment classified geologically as relict and, on time scales of hundreds to thousands of years, do not permanently accumulate modern sediments (Emery, 1968). This does not mean that modern, reactive and non-reactive particles are not temporarily deposited and even incorporated within the upper few tens of cm of the sediments. It means that over the long term, deposition is balanced by resuspension, lateral transport, winnowing and remineralization and dissolution (for the reactive components) resulting in no net recent accumulation. Short-lived radioisotopes such as $^{210}Pb$ provide direct evidence that modern sedimentological components do periodically get injected into the non-accumulating surface sediments (Bacon et al., 1994). It is likely that the $^{210}Pb$ is transported into the relict sediments on reactive carrier phases, such as particulate organic matter. Subsequent remineralization or dissolution of the carrier phase releases the $^{210}Pb$ to be re-adsorbed onto mineral surfaces of the bulk relict sediment matrix resulting in a modern $^{210}Pb$ signal within a non-accumulating sediment.

Many of the highest sediment accumulation rate areas also occur in coastal regions (Berner, 1982). In general, these are associated with large sources of sediments such as large turbid rivers whose sediment supply supports the formation of large deltas but also include smaller rivers that are draining steep, easily erodible watersheds such as those found in large islands in the East Indies (Milliman et al., 1999). In general, the accumulation of organic carbon follows that of bulk sediments, with the highest rates of accumulation in river deltas and on the upper continental slope (Hedges and Keil, 1995). High rates of sedimentary organic carbon accumulation can also be found in coastal areas such as the Peru continental shelf (Reimers and Seuss, 1983) where high rates of biological production provide a large source of biogenic debris. However, these locations account for only a small, approx. 5%, portion of the global total (Hedges and Keil, 1995).

The dispersion of particles within surface sediments is generally determined by discrete physical events such as storms or by the burrowing and feeding behaviors of benthic organisms (Boudreau, 1998; Fornes et al., 2001; Muslow et al., 1998). Rates of particle mixing by organisms (bio-mixing) may be influenced by numerous environmental factors. Different organisms affect particle motions in different ways. Some may merely burrow through the sediment and thereby push particles around locally while others may select particles to line their burrows (Fig. 6.3). Most macrobenthic organisms ingest particles. The undigested portion of the particles is then defecated. Through this process the particles are transported the length scale of the organism digestive tract and depending on the orientation of the organism, can be transferred upward, downward or laterally (Wheatcroft et al., 1990). Some organisms feed at depth in the sediments and defecate on the sediment surface; some feed at the surface and defecate at depth; some ingest only specific sizes or types of particles. Although none of these particle motions can be described as random individually, given the myriad of organism types, behaviors and sizes, and when integrated over sufficiently large space and time scales, it is useful to describe sediment dispersion as a diffusion-like process, where the diffusion coefficient is replaced by a 'bioturbation' or 'mixing' coefficient.

Within this relatively simple numerical representation, a variety of correlations between sediment mixing rates and other environmental factors have been observed or calculated. Because bioturbation is dependent on the physical activity and abundance of organisms, one might expect a correlation between the mixing rate and the input of organic carbon that can support the benthic population. In the deep sea, a positive correlation has been observed between the organic carbon deposition rate and the depth of the mixed layer (Smith and Rabouille, 2002) and with the mixing rate based on excess $^{210}$Pb measurements (Shimmield and Jahnke, 1995). These observations are in contrast to the results of a simple feedback model which suggests that the mixed layer depth should be independent of organic carbon input (Boudreau, 1998). In coastal settings, seasonal variations in temperature and other factors may also significantly impact rates of mixing. For example, in intertidal salt marsh sediments, fiddler crabs dominate summertime bioturbation (Botto and Iribarne, 2000) but have little impact on sediment mixing in the winter (McCraith, 1998).

Specific organism types can influence particle mixing in a variety of other complex ways. Organisms have been shown to selectively ingest specific particle types based on such characteristics as size, density, and organic carbon content. Thus, within a given location, different particle types may be mixed at different rates (Miller et al., 2000). Additionally, it has been hypothesized that the rate of mixing decreases with particle age in the mixed layer such that the recently deposited particles that are likely to contain the most labile organic matter are effectively mixed faster than particles that have been within the mixed layer for some period of time (Fornes et al., 2001). This 'age-dependent' mixing hypothesis has been invoked to explain the inverse correlation between estimates of sediment mixing and the half-life of the radiochemical tracer used to estimate mixing rates.

Because macrobenthic organisms require oxygen for respiration, one might expect the availability of oxygen to limit mixing processes. It has been observed that mixed layer depths decrease within an oxygen minimum zone (Smith et al., 2000) although no correlation with the actual mixing intensity was noted. Of course, the complete absence of oxygen in the bottom waters excludes macrobenthic organisms from inhabiting the underlying sediments. At locations with anoxic bottom waters and where the sediments are not periodically mixed by physical events such as storms, sediments retain the chronological fidelity of their deposition. If deposition rates or particle compositions vary periodically, this results in laminated sediments. Because laminated sediments retain an undisturbed history of accumulating particles, these types of sediments play an important role in the study of past environmental conditions, processes and species.

Finally, as will be discussed below, many coastal settings are comprised of permeable sediments. Flow through these types of sediments is capable of transporting particles into and through sediments (Huettel et al., 1996). Thus, the transport of small particles, such as algal cells, within the larger sediment matrix may be influenced by current velocities, sediment permeability and bottom topographic features such as ripple marks and biologically-produced mounds and pits.

## *2.3 Solute Transport*

Solute exchange and transport can be divided into molecular diffusive transport, dispersive transport that exceeds molecular diffusive rates, advective exchange driven by biological activities and advective exchange due to pressure differences within the sediment matrix. A brief description of each is provided below.

### 2.3.1. Molecular Transport

Brownian motions occur continuously, and in the presence of concentration gradients, provide a minimum rate of porewater solute transport. The rates of transport depend on the molecular weight of the diffusing chemical species, rate of Brownian motions (controlled primarily by temperature) and the viscosity of the medium (Li and Gregory, 1974; Himmelblau, 1964; and Burdige et al., 1999). In sediments, the specific rates of diffusion depend not only on the magnitude of the free aqueous diffusion coefficient of the solute but also on the effective mean free diffusion path in the sediments (Boudreau, 1996). That is, to move a specific distance, an ion must not only diffuse that distance but also around any particles that are in its way. The more tortuous the path a solute must travel to go a specific distance, the slower the effective rate of diffusion.

The effective diffusion coefficient in sediments is estimated from the molecular diffusion coefficient and sediment tortuosity using the relationship:

$$D_s = D_m/\theta^2 \tag{1}$$

where $D_s$ is the effective diffusion coefficient in sediments, $D_m$ is the molecular diffusion coefficient in the aqueous medium and $\theta$ is tortuosity (Lerman, 1979; Berner, 1980). The best estimate for $\theta^2$ is obtained by the product of porosity and the formation factor ($\phi$f), where f is the formation factor and represents the ratio of electrical resistance in the bulk wet sediments to that in the aqueous phase alone. In many instances, however, a value for f is not available or has not been measured with the required spatial resolution. Numerous other empirical relationships have been suggested between porosity and toutuosity (Boudreau, 1997). The most commonly employed relationship is Archie's Law (Archie, 1942) where: $\theta^2=\phi^{(1-m)}$. This law has been routinely applied to sands with m=2 and has been shown to apply to finer-grained sediments (porosity > 0.7) with m=2.5–3 (Ullman and Aller, 1982). Recently, Boudreau (1997) demonstrated that the expression $\theta^2=1-2.02(\ln\phi)$ (modified from Weissberg, 1963) does an equally good job of relating porosity and tortuosity.

Utilizing the effective diffusion coefficient, the vertical diffusive flux of dissolved solutes in porewaters can be estimated by Fick's first law:

$$\text{Flux} = -D_s\, \partial C/\partial z \tag{2}$$

where $\partial C/\partial z$ represents the vertical gradient of solute C (z increases with increasing sediment depth). By evaluating the gradients and effective diffusion coefficients in all three directions, this equation can be extended to assess transport in all three dimensions. Note that if estimating the flux at a boundary between layers of different porosity, the porosity differences must be accounted for.

Because Brownian motions are always occurring, estimates of diffusive porewater flux are thought to be minimum estimates of net transport. The only exceptions would be if the effective diffusion coefficient or chemical gradient had been overestimated.

### 2.3.2. Bio-dispersion and Irrigation

Accelerated porewater exchange due to the activities of organisms is common in all coastal regions where macrobenthic organisms exist. Similar to the biological processes that transport particles, transport of porewaters is performed by a variety of organism activities and organism types as depicted in Figure 6.3. Example activities include burrowing and foraging through the sediments, ingestion and elimination of ingested waters at different locations within the sediment column and the active exchange of burrow waters. In the latter process, organisms actively pump bottom water through their burrows to provide bottom water oxygen for their metabolic needs and possibly to remove harmful metabolic byproducts from the burrows. Thus, waters within the burrows are rapidly exchanged with overlying waters. Depending on the rate of irrigation, therefore, the composition of burrow waters reflects some mixture of bulk porewaters and overlying sea water, even at significant distances below the sediment water interface (Aller and Yingst, 1978). Steep chemical gradients across the burrow walls greatly accelerate sea floor-water column solute exchange and support locally-enhanced microbial communities and metabolic rates (Aller and Aller, 1986; Kristensen et al., 1991). Where benthic populations are sufficiently large, irrigation by benthic organisms has been shown to dominate sea floor solute exchange (e.g., Archer and Devol, 1992).

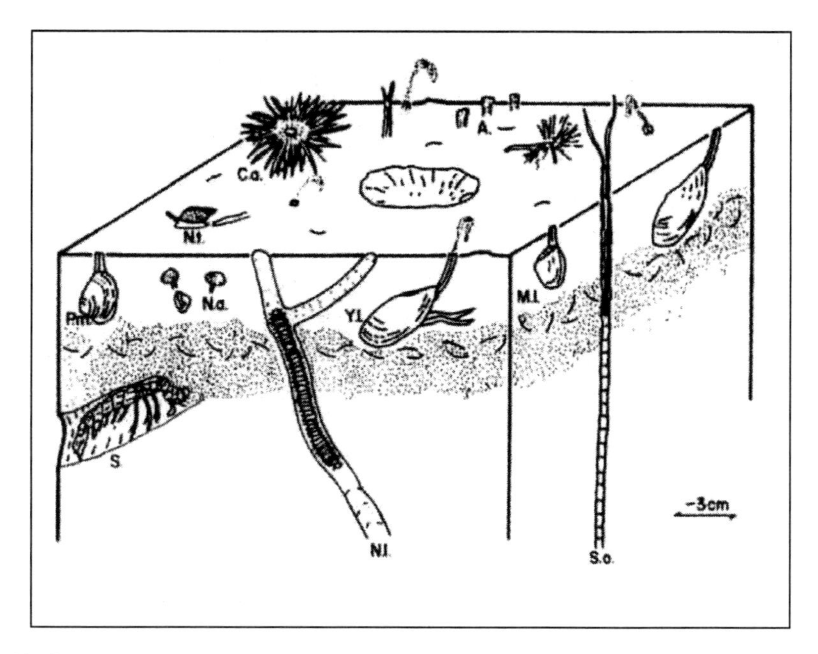


Figure 6.3 Schematic of types of organism—sediment—porewater interactions in marine sediments (re-drawn from Aller, 1977).

Similar to bio-mixing, irrigation rates may be influenced by numerous external factors such as the input of organic materials to the sediments, seasonal variations in bottom water oxygen concentration (Schlüter et al., 2000) and the specific behavior characteristics of the benthic organisms. Burrow geometry is critical to evaluating the effectiveness of exchange between the burrow waters and bulk sediment porewaters. For example, 'U'-shaped burrows exhibit greatest exchange with the surrounding bulk porewaters at their deepest points, where the burrows extend horizontally. Other factors that can influence net exchange rates include the diffusive permeability of the burrow walls (Aller, 1983) and the intensity and frequency of irrigation activities (Marinelli and Boudreau, 1996).

Numerous studies have attempted to model bio-irrigation exchange. Early, simple efforts represented the sum of the complex processes described above with a simple enhanced diffusion coefficient or a non-local exchange term (Emerson et al., 1984). Examples of more mechanistically realistic models of this exchange process include exchange via radially-symmetric vertical burrows (Aller, 1980) and episodically-flushed burrows (Boudreau and Marinelli, 1994). More recent examples of modeling efforts to elucidate irrigation include inverse modeling (Meile et al., 2001) and a stochastic model imploying a set of idealized burrow shapes linked into a burrow network (Koretsky et al., 2002).

### 2.3.3. Physically-forced Porewater Advective Transport

Most studies of biological irrigation have focused on muddy, fine-grained sediments. However, permeable sediments consisting of medium to well-sorted sands are very common in shallow coastal settings. In fact, 70% of the world's continental shelf areas are comprised of sediments that are geologically classified as relict. That is, under present conditions, they are not accumulating modern sediments. In most instances, winnowing of fine-grained materials has left a relatively coarse-grained, permeable sediment. These sediments generally contain very low concentrations of organic carbon and have not been widely studied geochemically. In addition to continental shelves, permeable sediments are common in the main channels of most rivers and estuaries.

Recent studies have demonstrated that significant advective porewater flow can be induced by the interaction between bottom currents and small topographic features on the sea floor such as ripple marks or organism mounds (Fig. 6.4). As water flows over the mound, pressure differences develop between the leading face and the leeward face of the mound. This induces significant advective flow of porewaters through these sediments.

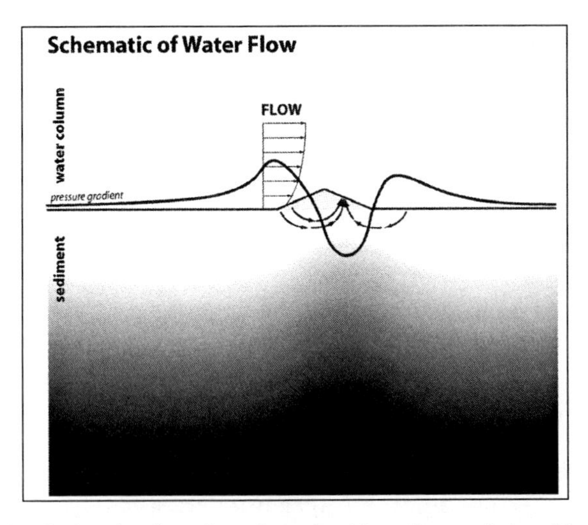


Figure 6.4 Schematic of advective flow through permeable sediments induced by bottom water flow over a sediment mound (after Huettel et al., 1996).

In flume studies, advective exchange to 15 cm was observed using dye to trace porewater flows near topographic bottom features. Advective exchange could be observed associated with structures as small as 700 μm and flow rates as low as 3 cm $s^{-1}$ in sediments with permeability of > 3 Darcy (Huettel and Gust, 1992).

Recent studies provide evidence of enhanced, presumably advective, exchange at numerous field sites. For example, porewater nutrient and oxygen profiles suggest that enhanced exchange occurs in large areas of the North Sea (Gehlen et al., 1995; Lohse et al., 1996), Mediterranean Sea (Ziebis et al., 1996), and South Atlantic Bight (Marinelli et al., 1998; Jahnke et al., 2000). In the Mediterranean Sea, Ziebis et al., (1996) concluded that advective flow was induced by the interactions between bottom currents and mounds constructed by a burrowing shrimp (*Callianassa truncata*), whereas along the South Atlantic Bight, the interaction of currents and flow-generated ripples induced advective porewater transport.

In the field, a variety of other forces may enhance advective transport in surface sediments. These include interactions between bottom ripples and surface gravity waves, tides and storms (Shum, 1992; 1993; Harrison et al., 1983). For example, variations in wave height have been shown to influence vertical distribution of remineralization reactions and porewater redox changes in permeable sediments of a carbonate reef (Falter and Sansone, 2000).

Advective flow through surface sediments is not only important for the exchange of porewater and bottom water solutes but may also play a major role in supplying fresh, reactive particulate organic materials to coastal sediments. In turbulent water columns that are characteristic of many coastal areas, algae that are small (generally in the range of 1–10 μm diameter) and only slightly more dense than sea water are not predicted to routinely sink to the sea floor solely by gravitational settling. However, algal particles may be carried by advective flow into the sediments where they may be caught within the sediment pore spaces. This process might be similar to a sand filter on a swimming pool. Active filtration of particles from the water column could potentially greatly increase the water column to sediment transfer of particulate matter from what might be estimated by gravitational settling alone. This process has been clearly demonstrated in flume studies (Huettel et al., 1996).

Despite the low organic carbon contents, metabolic rates reported for permeable sediments of the South Atlantic Bight (Jahnke et al., 2000) are comparable to those measured in organic-rich, fine-grained muds. It is thought that direct filtration of organic particles may be an important mechanism supplying the organic matter needed to maintain these high metabolic rates.

Finally, advective flow may also play an important role in supplying micronutrients, such as iron, to coastal ecosystems. As will be discussed in a latter section, under diffusive transport conditions, remineralization reactions and ancillary oxidation-reduction reactions occur in a preferred, stratified sequence determined by the order of greatest free energy release. Under these conditions, iron that is reduced and mobilized in the deep, anoxic sediment layers will be reoxidized in the aerobic surface layers and efficiently trapped within the sediments. If sufficiently strong, however, advective transport of reduced porewaters from deeper horizons in the sediments may outcompete the downward diffusion of oxygen and can result in the transport of reduced chemical species past the sediment water interface (Huettel et al., 1998). Under these conditions, mobilized iron can be injected directly into the water column before being reoxidized, where it will be dispersed by turbulence throughout the water column. This can greatly increase the flux of metals such as Fe into the coastal water column where they may have an important influence on coastal biological communities (Hutchins and Bruland, 1998; Hutchins et al., 1998). Because coarse-grained sediments are thought to dominate coastal environments, these observations may force major changes in our conceptual models of benthic-pelagic coupling of biogeochemical cycles in the coastal ocean.

### 2.3.4. Groundwater

In recent years, evidence is increasing that in many coastal regions, there is a substantial input of primary groundwater (Moore, 1996; Moore and Shaw, 1998). In addition to providing a source of fresh water to coastal systems, such input may potentially be important as a source of nutrients and other dissolved chemical constituents. For example, groundwater seepage from shallow, surficial aquifers has been reported to be a significant nutrient source to estuaries and nearshore environments and may also be a conduit for exchange of contaminants such as agricultural pesticides (Gallagher et al., 1996; Simmons and Lyons, 1994; Simmons, 1992). The composition of the groundwaters may be controlled by subsurface mixing and reaction with mineral phases within the permeable layers (Moore, 1999) and by primary inputs.

More work is needed to understand the global input of this source. Groundwater inputs can be difficult to quantify because both slow diffuse seepage and point-source spring discharge may contribute. The principal methods used to quantify submarine groundwater discharge include emplacement of chambers (seepage meters) on the sediment surface and directly measuring the volume of water expelled and the quantification and interpretation of chemical and radiochemical tracers of groundwaters, such as radium isotopes, $^{222}Rn$ and $CH_4$, in the overlying water column (Cable et al., 1996, Moore, 1996). In areas where fresh water discharges into higher salinity estuarine and coastal areas, modeling calculations suggest that electromagnetic detection of discharge locations may be possible (Hoefel and Evans, 2001).

Seepage meters that provide positive relief on the bottom in areas with significant wave and current regimes appear to artificially enhance porewater exchange due to the Bernoulli effect of flow over the device (Shinn et al., 2002). Also, chemical tracers of groundwaters in most cases may be supplied by exchange with sediment mineral surfaces. Shallow, active exchange between groundwaters and overlying waters may then exhibit an enhanced tracer signal relative to the net flow of primary ground water. Despite these concerns, recent comparison studies amongst the different methods have yielded reasonable agreement with the observed differences being attributed to differences in transport pathways and time scales of groundwater-sea water exchange (Burnett et al., 2002).

In many areas, advective groundwater discharge may include or be dominated by a recirculated sea water component. This may be due to the entrainment of sea water in the primary flow through subsurface mixing (Moore, 1999) or through processes such as wave-induced ground water flow across the beach face or tidal pumping in estuarine and salt marsh systems (Li and Barry, 2000; Jahnke et al., 2003). Advective discharge of recirculated sea water from shallow sediments has been shown to be a major conduit for the cycling of nutrients and carbon in estuarine systems (Cai et al., 1999; Jahnke et al., 2003; Whiting and Childers, 1989).

## 3. Remineralization Reactions

### *3.1 General Principles*

Remineralization reactions in sediments represent a complex web of hydrolysis and oxidation reactions whereby the particulate organic materials deposited on the sea bed are degraded, dissolved and eventually oxidized to inorganic constituents.

Some of the solubilized organic materials may be transported out of the sediments before oxidation to inorganic products and may therefore represent a loss of organic matter from the sediments not accounted for in estimates of remineralization and burial. Based on elevated dissolved organic carbon concentrations in porewaters and estimated diffusion coefficients, this loss has been reported to be a significant portion of the organic carbon deposition rate (Martin and McCorkle, 1993; Burdige et al., 1992). Others have suggested that only a small fraction of the deposited organic carbon (nominally <10%) escapes oxidation within the sediments via this pathway (Burdige and Homstead, 1994; Burdige et al., 1999; Alperin et al., 1999; Holcombe et al., 2001). Uncertainties in representing porewater diffusion of this complex mixture of dissolved organic compounds and potential sampling artifacts currently limit the accuracy of these estimates (Burdige and Gardner, 1998; Holcombe et al., 2001).

The parameters that control the exact rate and efficiency with which organic matter is recycled in sediments are not understood. As listed in Hedges et al., (1999), earlier studies have suggested many factors that may influence organic matter preservation in marine sediments such as the source and lability of the deposited organic matter (Hedges et al., 1988, Schubert and Stein, 1996), oxygen content of overlying bottom waters and sedimentation rate (Canfield, 1989). More recent studies have debated the importance of such factors as the association of organic compounds with mineral surfaces (Mayer, 1994a, 1994b; Bennett et al., 1999; Ransom et al., 1997; 1998), redox oscillation of sedimentary environments (Aller, 1994a) and the length of time the organic matter is exposed to oxygen (Hartnett et al., 1998, Hartnett and Devol, 2003; Hedges et al., 1999).

Despite the uncertainties in the controls of organic matter preservation, the sequence of terminal oxidations that characterizes remineralization of organic matter within the sediments is well documented (Table 1). The order in which the reactions proceed follows that initially reported for the water column within permanently anoxic basins (Richards, 1965) and later observed in porewaters (Froelich et al., 1979) and represents the decreasing order of free energy released by each oxidant. That is, the greatest amount of energy is released by the oxidation of idealized organic carbon ($CH_2O$) to $CO_2$ utilizing oxygen. In sediments, therefore, it is envisioned that if $O_2$ is available, the organisms that can utilize $O_2$ will dominate and $O_2$ will be preferentially consumed. Once $O_2$ is exhausted or sufficiently depleted to no longer be easily available, organisms that can utilize $NO_3^-$ and manganese oxides will dominate. Note that manganese oxides can exist in a variety of crystalline forms and that the manganese within a given mineral may be present in a mixture of oxidation states (Burns and Burns, 1979). A range of values may be estimated for free energy release during heterotrophic manganese reduction. Values can range from slightly larger than that of denitrification to somewhat lower. While most investigators conclude that denitrification precedes manganese reduction in most environmental settings, the free energy differences are not very large and overlap as discussed below may be common. Collectively, denitrification and manganese reduction are referred to as suboxic processes. After manganese oxides and nitrate are exhausted, iron oxides and sulfate are the next preferred oxidants. As with the manganese oxides, iron oxides may exist in a variety of forms and compositions and a range of free energies are again possible. Here, however, the difference between sulfate reduction and iron reduction is greater and in general, iron reduction is assumed to precede sulfate reduction. Finally, after iron oxides and sulfate are exhausted, organic matter remineralization continues through methane fermentation.

TABLE 6.1
Major organic carbon oxidation reactions and standard free energies (after Emerson and Hedges, 2003).

| Reaction | $\Delta G_r^o$ (kJ mol$^{-1}$) |
|---|---|
| $O_2 + CH_2O \rightarrow CO_2 + H_2O$ | -518.4 |
| $0.8NO_3^- + CH_2O + 0.8H^+ \rightarrow 0.4N_2 + CO_2 + 1.4H_2O$ | -507.6 |
| $2MnO_2(s) + CH_2O + 4H^+ \rightarrow 2Mn^{2+} + CO_2 + 3H_2O$ | -501.9 |
| $2Fe_2O_3(s) + CH_2O + 8H^+ \rightarrow 4Fe^{2+} + CO_2 + 5H_2O$ | -280.6 |
| $0.5SO_4^{2-} + CH_2O + H^+ \rightarrow 0.5H_2S + CO_2 + H_2O$ | -143.4 |
| $2CH_2O \rightarrow CH_4 + CO_2$ | -34.4 |

In addition to oxidation reactions involving organic carbon, other constituents of organic matter, such as N and P, are released during remineralization. Organic N is ultimately released to the porewaters in the form of $NH_4^+$ where in the presence of $O_2$ (and perhaps Mn-oxides) it is oxidized to $NO_3^-$. The degradation of organic phosphorus, on the other hand, does not involve a change in the oxidation of the phosphorus and will not be addressed here.

In a one-dimensional system where diffusive porewater transport dominates, the above sequence of reactions results in a regular stratified sequence of reactions and porewater profiles (Froelich et al., 1979; Fig. 6.5). This simple conceptual visualization of sediment diagenesis has provided a very powerful tool for the interpretation of porewater profiles. In addition, it must be emphasized that given an average oxidation state for labile sedimentary organic matter, the number of electrons that must be transferred to oxidize the carbon to $CO_2$ is set, thereby fixing the stoichiometric relationship between carbon and oxidant. Thus, this framework also provides a means for quantifying the oxidation of organic carbon in marine sediments. As discussed further in a later section, the re-oxidation of reduced metabolites such as $NH_4^+$ and $S^{2-}$ can alter local stoichiometric ratios but generally overall stoichiometries remain close to theoretical values.

### *3.2 Complexities in High Rain Rate and Coastal Areas*

The above stratified sequence of reactions provides a very useful means of interpreting and discussing sediment diagenesis. However, in coastal sediments a variety of factors conspire to make the situation more complex.

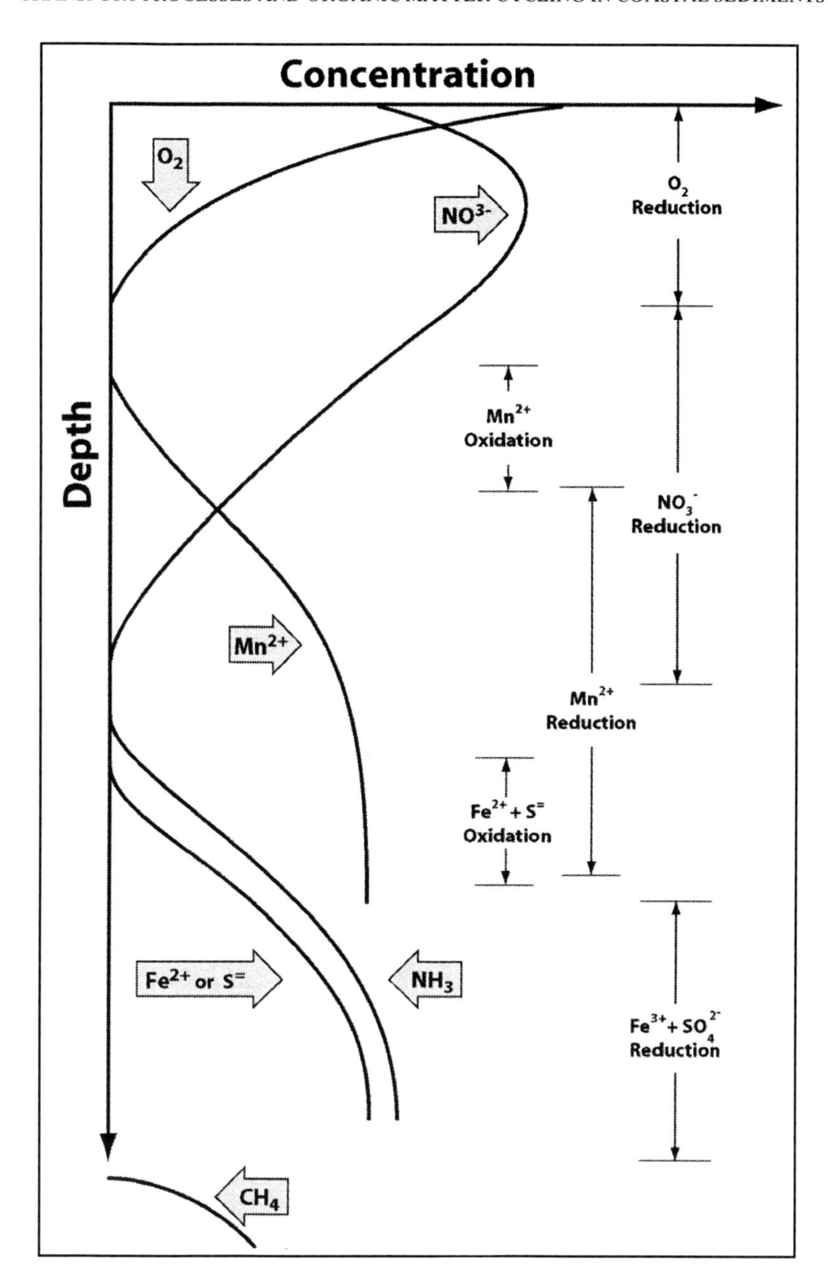


Figure 6.5 Hypothetical porewater profiles for major remineralization oxidants and metabolites under conditions of diffusive porewater transport and low to medium deposition rates of organic carbon.

### 3.2.1. High Organic Input and Respiration Rates

As written in Table 1, metabolic reactions, with the exception of $O_2$ reduction, produce a variety of reduced end-products such as $NH_4^+$, $Mn^{2+}$, $Fe^{2+}$, $HS^-$, and $CH_4$. These reduced species can be re-oxidized by numerous other oxidants. Although most of these reactions will proceed abiotically, in natural sediments these reactions are mediated by a diverse population of chemolithotrophic bacteria. For example, $S^{2-}$ oxidizing bacteria (*Beggiatoa*) are known to produce filamentous mats at the sediment-water interface where they survive utilizing the energy that is released by the oxidation of $S^{2-}$. In general, $O_2$ is the preferred oxidant although it has been reported that these organisms are also capable of storing large quantities of $NO_3^-$, presumably to support metabolic processes during periods when oxygen is absent (Fossing et al., 1995).

An incomplete subset of potentially important re-oxidation reactions are provided in Table 2. Thus, for example, $CH_4$ can be oxidized to $CO_2$ by $SO_4^{2-}$ producing $S^{2-}$; $S^{2-}$ can be re-oxidized by $NO_3^-$ and $O_2$ to $SO_4^{2-}$. While bottom water $NO_3^-$ can directly be taken up into the sediments in specific areas, in most coastal areas the uptake of $O_2$ is ultimately responsible for supporting the re-oxidation reactions. That is, in reducing sediments, these oxidations represent a cascade of electron transfers that ultimately is supported by reaction with $O_2$.

These intense oxidations alter the distribution of the reactions presented in section 3.1 in two important ways. First, the generalized sequence of remineralization is compressed into a very thin reaction zone resulting in very steep gradients in the concentrations of oxidants and metabolic by-products (Anshutz et al., 2000) which have only recently been resolved by direct measurement (Luther et al., 1998). In the absence of sediment inhomogeneity and non-steady state forcings, the order of reaction is not altered. Hypothetically, one might expect the reactions to be compressed at the sediment water interface. The steep gradients that result indicate that oxidized and reduced chemical species are in very close proximity. Mixing of particles or porewaters over even small distances can therefore bring these reactive species in direct contact fueling a complex array of oxidation/reduction reactions. This is discussed briefly below.

TABLE 6.2

Examples of ancillary oxidation/reduction reactions involving metabolically important chemical species in sediments. Note that these are simplified representations. Many of these reactions may occur in discrete steps involving different microorganisms and through chemolithotrophy may result in the fixation of inorganic carbon and slightly altered stoichiometries to those listed below.

| Oxidant | Reaction |
|---|---|
| $O_2$ | $16NH_3 + 32O_2 \rightarrow 16HNO_3 + 16H_2O$ |
| | $Mn^{2+} + 0.5O_2 + 2HCO_3^- \rightarrow MnO_2(s) + 2CO_2 + H_2O$ |
| | $Fe^{2+} + 0.25O_2 + 2HCO_3^- + 0.5H_2O \rightarrow Fe(OH)_3(s) + 2CO_2$ |
| | $HS^- + 2O_2 \rightarrow HSO_4^-$ |
| $NO_3^-/NO_2^-$ | $5NH_3 + 3HNO_3 \rightarrow 4N_2 + 9H_2O$ |
| | $NH_4^+ + NO_2^- \rightarrow N_2 + 2H_2O$ |
| | $2.5Mn^{2+} + NO_3^- + 2H_2O \rightarrow 0.5N_2 + 2.5MnO_2(s) + 4H^+$ |
| | $5Fe^{2+} + NO_3^- + 12H_2O \rightarrow 0.5N_2 + 5Fe(OH)_3(s) + 9H^+$ |
| | $5HS^- + 8NO_3^- + 3H^+ \rightarrow 4N_2 + 5SO_4^{2-} + 4H_2O$ |

| | |
|---|---|
| $MnO_2(s)$ | $NH_4^+ + 1.5MnO_2(s) + 2H^+ \rightarrow 0.5N_2 + 1.5Mn^{2+} + 3H_2O$ |
| | $NH_4^+ + 4MnO_2(s) + 6H^+ \rightarrow NO_3^- + 4Mn^{2+} + 5H_2O$ |
| | $2Fe^{2+} + MnO_2(s) + 2H_2O \rightarrow 2FeOOH + Mn^{2+} + 2H^+$ |
| | $HS^- + MnO_2(s) + 3H^+ \rightarrow S^0 + Mn^{2+} + 2H_2O$ |
| | $HS^- + 4MnO_2(s) + 7H^+ \rightarrow 4Mn^{2+} + SO_4^{2-} + 4H_2O$ |
| FeOOH | $3H_2S + 2FeOOH(s) \rightarrow 2FeS(s) + S^0 + 4H_2O$ |
| $SO_4^{2-}$ | $CH_4 + SO_4^{2-} \rightarrow S^{2-} + CO_2 + H_2O$ |

Second, as organic carbon remineralization rates increase and non-aerobic pathways oxidize a greater proportion of the carbon remineralized, the production of reduced by-products increases. As stated above, whether directly or through intermediates, the majority of these reduced species are re-oxidized by $O_2$ being transported into the sediments. Thus, where organic carbon deposition is large, a greater proportion of organic carbon remineralization is performed through non-aerobic metabolism and a greater proportion of the $O_2$ flux is consumed during the re-oxidation of the reduced end-products. In many coastal locations, even with well oxygenated bottom waters, $SO_4^{2-}$ reduction is estimated to be the primary pathway by which organic carbon is remineralized (Canfield et al., 1993; Thamdrup and Canfield, 1996; Kostka et al., 1999). Note, however, that in this case, the $O_2$ flux still generally corresponds to the total amount of organic carbon oxidized because it ultimately serves as the sink for electrons by continually re-oxidizing the metabolic end-products. The only portion of the organic carbon remineralization rate that would not be accounted for by the $O_2$ flux would be that portion for which the reduced metabolites are not re-oxidized, either by escaping the sediments in reduced form or by being permanently buried in reduced form within the sediments. Note that the production of reduced end-products and re-oxidation rates do not always need to be balanced for the above to be true. In fact, it may be common that during periods of high organic matter input (such as summer) metabolic rates exceed re-oxidation rates and reduced species build up in coastal sediments (Kostka et al., 2002; Sampou and Oviatt, 1991; Graf, 1992). During low organic carbon deposition periods, re-oxidation reactions may deplete the concentrations of reduced species. Thus, it is only when integrated over the appropriate time period (generally annual or greater) that the balance between non-aerobic metabolic rates and re-oxidation reactions is fully evaluated.

### 3.2.2. Particle Mixing and Episodic Irrigation: Redox Oscillation

As noted earlier, in the presence of steep gradients a variety of processes can cause particles to be exposed to altering redox conditions and can bring together reactive redox species that would otherwise be confined to separate specific depth horizons. Particle mixing by benthic organisms routinely exchanges particles between different reaction zones within the sediment. Thus, reduced species such as FeS may be rapidly transported to near surface layers where they will directly contact oxidized species such as $O_2$ or $NO_3^-$. Similarly, oxidized particulate phases such as FeOOH can be transported downward where they may react with reduced organic or inorganic compounds. Through these transport-oxidation-reduction reactions, metals can be used to pump electrons out of the sediments, fueling oxidation of organic matter (Aller, 1994b). Also, through this process individual organic particles can similarly be exposed to a variety of redox states and metabolic pathways.

Alternating redox conditions are thought to promote remineralization and more efficiently break down organic materials than does a simple, linear passage from oxidizing to reducing strata within the sediments (Aller, 1994a).

Solute transport processes may also cause temporal and spatial variations in redox conditions. For example, burrows may extend downward into the sediments traversing a variety of redox environments. During periods of inactivity, consumption of oxidants from the burrow waters will cause more reducing condition to develop. Initiation of irrigation will reverse this trend and the conditions that particles and bacteria along the burrow walls are exposed to will oscillate between more reducing and more oxidizing states (Marinelli and Boudreau, 1996; Boudreau and Marinelli, 1994). In vegetated intertidal areas, analogous situations may arise through transport of $O_2$ by root systems (Kostka et al., 2002). Transport of oxidants in root systems has been shown to influence the distribution of trace elements such as Zn, Pb, Cu and Cd (Cacador et al., 2000). Finally, bubbles rising through sediments can also be an important mechanism for transporting reduced compounds to oxic surface layers (Boudreau et al., 2001a).

Burrowing organisms and resulting redox oscillations not only influence solute and particle transport rates but have also been reported to influence sedimentary reaction rates (Aller and Yingst, 1985). Bacteria that mediate the reduction of iron and manganese oxides have, at least partially, been identified and studied at the organism level (Nealson and Myers, 1992; Kostka et al., 1995). The latter identifies the importance of the extensive interactions between bioturbation of reduced and oxidized particulate forms, transport of oxidants through root systems, and the microbially mediated oxidation/reduction reactions that result in reduced Fe and Mn in porewaters.

Thus diagenesis in coastal sediments is a more complex web of oxidation-reduction reactions than is depicted in the orderly sequence of reactions provided in Table 1. Reaction zones may overlap and reaction 'hot spots' may concentrate diagenetic transformations (Brandes and Devol, 1995). Nevertheless, even in coastal sediments, the generalized reaction sequence still provides a sound framework for interpreting sediment remineralization by providing the basic order of preference for diagenetic reaction.

### *3.3 Denitrification*

The availability of fixed-N (all forms of N except $N_2$) plays a major role in controlling marine biogeochemical processes (Falkowski, 1997). In general, oceanic fixed-N inventories represent the balance between nitrogen fixation and denitrification. Numerous studies have demonstrated high rates of denitrification in coastal marine sediments (Devol, 1991; Devol and Christensen, 1993; Lohse et al., 1993). Summer and winter rates of benthic denitrification have recently been determined in the high-latitude Bering, Chukchi and Beaufort Seas. There was no drastic decrease in rates during winter. This is important since there is a very large area of continental shelf in northern high latitude regions (Devol et al., 1997). Extrapolating the measured rates to the global scale suggests that sedimentary denitrification is the largest marine sink of fixed-N (Codispoti et al., 2001) and therefore of particular importance.

Until recently, denitrification was thought to be an exclusively heterotrophic process in which $NO_3^-$ (or $NO_2^-$) is reduced to $N_2$ by the oxidation of organic carbon. No anaerobic pathways to $N_2$ were known. This is of particular importance in that during the degradation of organic matter, fixed-N is released in the form of $NH_4^+$. Since anaerobic denitrification pathways were not known, it was thought that $NH_4^+$ would first be required to be oxidized to $NO_3^-$ (or $NO_2^-$). Since it was thought that $O_2$ was required for this oxidation (nitrification), all studies of denitrification were focused near the oxic-suboxic boundary, greatly narrowing our view of potential fixed-N transformations.

Within the last ten years, numerous other oxidation-reduction reactions involving fixed-N have been identified that greatly expand the possible number and complexity of fixed-N transformations and denitrification pathways in sediments. Although it was proposed several decades ago (Richards, 1965), recent laboratory studies demonstrate that $NO_3^-$ or $NO_2^-$ can directly react with $NH_4^+$ to form $N_2$ (Mulder et al., 1995; Strous et al., 1999a; see Table 2). This reaction is mediated by autolithotrophic microorganisms that have now been successfully cultured (Strous et al., 1999b). Further studies have identified at least two anaerobic oxidation pathways for $NH_4^+$ using $NO_2^-$ as an oxidant employing hydrazine ($N_2H_4$) and hydroxylamine ($NH_2OH$) as intermediates (Jetten et al., 1999). The importance of these pathways to the evaluation of denitrification in natural sediments is that the overall rate of denitrification can be significantly greater than the rate of nitrification. Recent field studies employing $^{15}N$ tracer techniques have reported that a significant proportion of the total denitrification rate could be attributed to anaerobic ammonium oxidation reactions in sediments from the Skagerrak (Thamdrup and Dalsgaard, 2002). This location is thought to be representative of many continental shelf environments. While similar experiments revealed detectable rates of anaerobic ammonium oxidation in nearby Aarhus Bay sediments, this pathway appeared to account for a very small fraction of the total denitrification in this setting. These differences seem to be related to the relative abundances of other reduced substrates such as Fe and S in these two environments.

Numerous alternative denitrification pathways have also been proposed that utilize the redox transformation of transition metals, predominantly Mn and Fe (Luther et al., 1997; Hulth et al., 1999). Examples of possible reactions are provided in Table 2. Given the earlier discussion of the transport of redox-sensitive species across redox horizons, it is plausible that many of these reactions could be proceeding in coastal sediments.

The emerging complexity of N diagenesis in coastal sediments represents a significant research challenge and will undoubtedly alter long-held assumptions concerning N cycling in sediments. For example, traditionally, nitrate diffusing into the anoxic zone would be thought to support denitrification. Recently, however, $^{15}N\text{-}NO_3^-$ tracer incubations of $S^{2-}$-rich sediments were employed to follow N reaction pathways. Surprisingly, a significant portion of the added tracer was transformed into $^{15}N\text{-}NH_4^+$ suggesting that under these conditions, $NO_3^-$ was reduced to $NH_4^+$ (An and Gardner, 2002).

## 4. Benthic Primary Production

At many locations in coastal zones, photosynthetically active radiation (PAR) reaches the sea floor. Secchi depths from more than 17,000 individual oceanographic stations at water depths less than 200 m indicate that >1% of the surface irradiance reaches the bottom at 2653 of the stations and >0.1% of the surface irradiance reaches the bottom at an additional 2772 stations (Cahoon et al., 1993). If evenly distributed, these results suggest that light sufficient to support significant rates of benthic photosynthesis reaches approximately 30% of the continental shelf sea floor, an area of approximately $3.4 \times 10^8$ km$^2$.

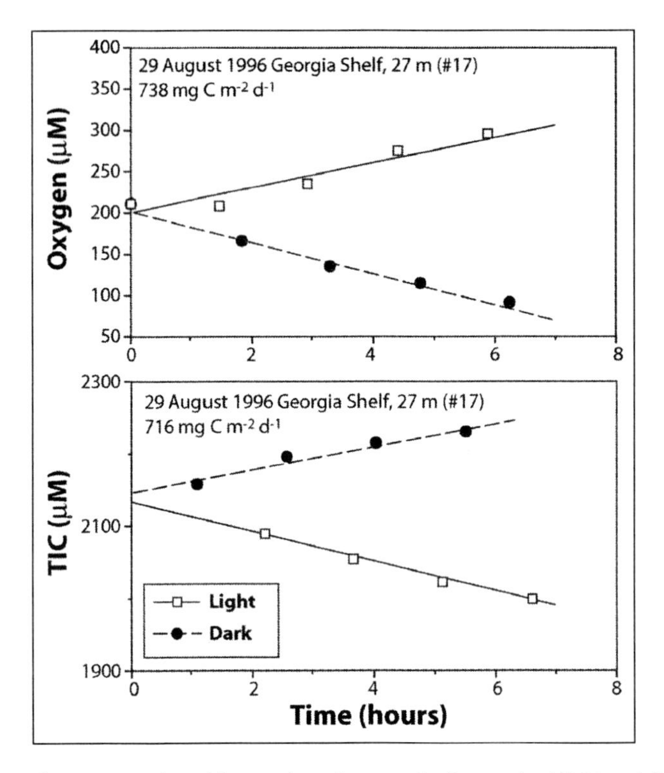


Figure 6.6 Example oxygen and total inorganic carbon results from paired light and dark benthic flux chambers on the southeastern U.S. continental shelf (Jahnke et al., 2000).

Initial studies employed diver operated benthic chambers to estimate in situ the rate of benthic primary production (Hopkinson et al., 1991; Cahoon and Cooke, 1992). More recent estimates have been made utilizing an autonomous paired, light/dark benthic chamber system (Jahnke et al., 2000; Fig. 6.6). In the example results provided in Figure 6, the average PAR flux during the deployment was approximately 60 µE m$^{-2}$ s$^{-1}$. It is obvious that this light flux, which corresponds to 8% of the surface irradiance, supports significant rates of benthic photosynthesis and has a dramatic impact on the benthic exchange of $O_2$ and dissolved total inorganic carbon (TIC). Assuming 12 hours of sun light per day (an underestimate during the summer), the difference between the light chamber and dark chamber

TIC fluxes implies a gross rate of benthic primary production of 716 mg C $m^{-2}$ $d^{-1}$. With a respiratory quotient (ratio of inorganic carbon fixed to oxygen released) of 1.0, the $O_2$ flux results suggest a gross primary production rate of 738 mg C $m^{-2}$ $d^{-1}$. The similarity in the values obtained supports the assumed photosynthetic quotient value of 1.0 and is consistent with ancillary porewater results which suggest that $NH_4^+$ is the dominant form of nitrogen taken up by the benthic microalgae.

While benthic primary production in intertidal settings has received considerable attention (see for example: Austen et al., 1999; Waterman et al., 1999; Goto et al., 1999; Noh et al., 1998; Gould and Gallagher, 1990, and references therein), it has not been well studied in deeper, subtidal coastal environments and on continental shelves. The studies that have been conducted represent diverse locations and suggest that benthic primary production is a wide-spread, common phenomenon. Locations where significant benthic primary production rates have been reported include: the Georgia continental shelf (Hopkinson et al., 1991; Jahnke et al., 2000), the North Carolina continental shelf (Cahoon and Cooke, 1992), the southeastern Kattegat (Sundback et al., 1991) the northern Adriatic Sea shelf (Epping and Helder, 1997) and Stellwagen Bank on the northeastern U.S. continental shelf (Cahoon et al., 1993). Based on results from so few locations, a global estimate of benthic primary production is too uncertain to be useful. Nevertheless, it is important to recognize that significant rates were observed at each of these locations. On the Georgia continental shelf, summertime rates for much of the shelf rivaled integrated water column primary production rates.

Several aspects of benthic primary production should be noted. Significant rates are observed even at low light levels. In the example provided in Figure 6, production rates of >700 mg C $m^{-2}$ $d^{-1}$ were observed even though only 8% of the surface irradiance penetrated to the sea floor. On Stellwagen Bank, Massachusetts Bay, USA, rates equivalent to 250 mg C $m^{-2}$ $d^{-1}$ (assuming 12 hours of sun light per day) were observed even though light levels never exceeded 1% surface irradiance and the absolute photon flux was as low as 4.7 μmol photons $m^{-2}$ $s^{-1}$ (Cahoon et al., 1993). Because of the optical properties of the sediment grains (often quartz sands) and porewaters, the light field around microalgae in sediments differs strongly from the incident light field with respect to intensity and spectral composition (Kühl and Jorgensen, 1994). Nevertheless, a correlation between the light flux and the pennate diatoms that reportedly dominate the benthic microalgae (Cahoon and Laws, 1993) has been observed (Fig. 6.7).

Benthic algae may also provide an important food source for a variety of higher organisms. This aspect may be especially significant when one recogizes that benthic and water column production rates are most likely inversely correlated. That is, when water column production rates are high, a larger portion of the incident radiation is likely to be intercepted before reaching the sea floor, depressing benthic rates. Conversely, at times of water column oligotrophy, more light reaches the sea floor, supporting higher rates of benthic production. Thus, benthic production may supply food at critical periods of water column oligotrophy.

In addition to providing a source of carbon, benthic photosynthesis provides a source of oxygen at the sea floor. As many coastal regions face the deleterious effects of bottom water hypoxia or anoxia, understanding the impacts of water turbidity which limit light penetration may be an important additional step focusing mitigation activities.

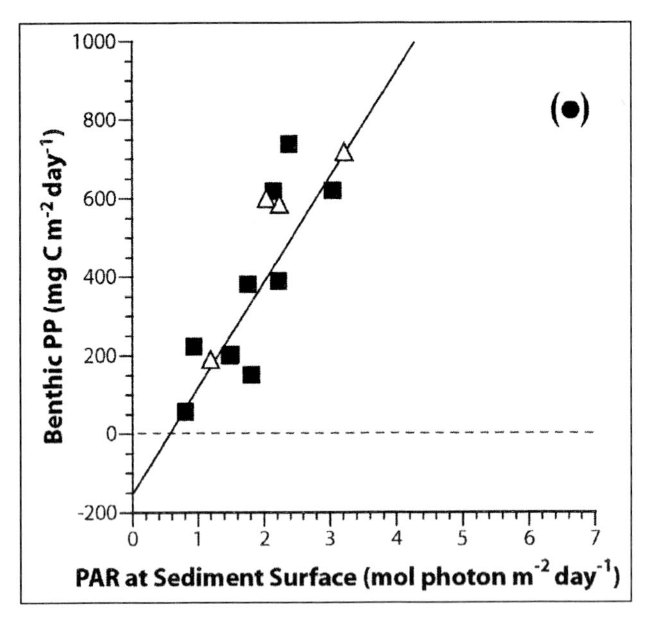


Figure 6.7 Relationship between photosynthetically active radiation levels and benthic primary production on the South Atlantic Bight continental shelf (Jahnke et al., 2000). Open triangles are more recent unpublished results (Jahnke, unpublished).

## 5. Conclusions and Future Challenges

Coastal sediments play a major role in marine biogeochemical cycles. They contain the largest pool of organic matter in the upper 200 m of the marine environment and are the focus for transformations, such as denitrification, that are critical to global biogeochemical cycles. The cycling and exchange of bioelements between coastal sediment pools and other major reservoirs is a critical component of marine biogeochemistry. The physical matrix provided by the sediments has facilitated studies of transport and metabolic processes within the sediments. It is now well established that remineralization reactions tend to proceed in the sequence that provides the greatest free energy release. Furthermore, at sites where molecular diffusion dominates porewater solute transport, quantitative models have represented the basic relationship between reaction and transport accurately.

Despite the success of previous investigations, numerous challenges remain. In fine-grained sediments with steep redox gradients, particle mixing and non-diffusive porewater transport conspire to mix reactive components, producing a mosaic of reactions that vary over very short distance and time scales. It is important for future studies to elucidate the role of this complex web of reactions on the overall biogeochemical functioning of coastal sediments. Permeable sediments represent an especially important challenge (Boudreau et al., 2001b). First, permeable sediments are widespread, comprising the majority of continental shelf environments and many estuarine and river bottoms. Second, most of the existing methodologies for identifying reaction pathways and estimating reaction rates were developed for fine-grained sediments and may not yield accurate results in this setting. For example, benthic flux chambers may interact with bottom currents to either suppress the natural induced porewater flow yielding artificially low results or by generating pressure differences and enhancing porewater exchange. Similarly, shipboard, whole-core incubations may prevent natural advective porewater transport, altering results. Calculations of porewater diffusive fluxes based on vertical concentration gradients only provide minimum estimates because transport may not be dominated by diffusion. Finally, at many coastal locations, light reaches the bottom, supporting significant benthic photosynthetic rates. Even at very low light levels, benthic primary production may alter or even reverse the direction of benthic fluxes and accentuate temporal variability and diel and seasonal time scales. Because photosynthesis is restricted to the upper 1–2 mm of the sediments, it may remain undetected by sampling techniques with less resolution. Thus, in addition to the intellectual challenges discussed above, coastal sediment systems present numerous technological and methodological challenges that must be met if we are to develop an understanding of the biogeochemistry of these systems.

## Acknowledgements

This manuscript benefited from numerous discussions with B. Sundby. Support was provided under NSF Grants OCE-9911707, OCE-9906897 and NASA Grant NAG5–10557.

## Bibliography

Aller, R. C., 1977. The Influence of Macrobenthos on Chemical Diagenesis of Marine Sediments. Ph.D. Thesis, Yale Univ. 600p.

Aller, R. C., 1980. Quantifying solute distributions in the bioturbated zone of marine sediments by defining an average microenvironment. Geochim. Cosmochim. Acta **44,** 1955—1965.

Aller, R. C., 1983. The importance of the diffusive permeability of animal burrow linings in determining marine sediment chemistry, J. Mar. Res., **41,** 299—322.

Aller, J. Y. and Aller, R. C., 1986. Evidence for localized enhancement of biological activity associated with tube and burrow structures in deep-sea sediments at the HEBBLE site, western North Atlantic. Deep-Sea Res., **33,** 755—790.

Aller, R. C. and J. Y. Yingst, 1985. Effects of the marine deposit-feeders Heteromastus filiformis (Polychaeta), Macoma balthica (Bivalvia), and Tellina texana (Bivalvia) on averaged sedimentary solute transport, reaction rates, and microbial distributions. J. Mar. Res. **43,** 615—645.

Aller, R. C., 1994a. Bioturbation and remineralization of sedimentary organic matter: effects of redox oscillation. Chem. Geol. **114,** 331–345.

Aller, R. C., 1994b. The sedimentary Mn cycle in Long Island Sound: Its role as intermediate oxidant and the influence of bioturbation, $O_2$ and $C_{org}$ flux on diagenetic reaction balances. J. Mar. Res. **52,** 259–295.

Aller, R. C. and J. Y. Yingst, 1978. Biogeochemistry of tube-dwellings: A study of the sedentary polychaete Amphitrite ornata (Leidy). J. Mar. Res. **36,** 201—254.

Alperin, M. J., C. S. Martens, D. B. Albert, I. B. Suayah, L. K. Benninger, N. E. Blair and R. Jahnke, 1999. Benthic fluxes and porewater concentration profiles of dissolved organic carbon in sediments from the North Carolina continental slope. Geochim. Cosmochim. Acta, **63,** 427–448.

An, S. and W. S. Gardner, 2002. Dissimilatory nitrate reduction to ammonium (NRNA) as a nitrogen link, versus denitrification as a sink in a shallow estuary (Laguna Madre/Baffin Bay, Texas). Mar. Ecol. Prog. Ser., **237,** 41—50.

Anschultz, P. B. Sundby, L. Lefrancois, G. W. Luther III, and A. Mucci, 2000. Interactions between metal oxides and species of nitrogen and iodine in bioturbated marine sediments. Geochim. Cosmochim. Acta, **64,** 2751—2763.

Archer, D. and A. Devol, 1992. Benthic oxygen fluxes on the Washington shelf and slope: A comparison of in situ microelectrode and chamber flux measurements. Limnol. Oceanogr., **37,** 614—629.

Archie, G.E., 1942. The electrical resistivity log as an aid in determining some reservoir characteristics. Petrol. Tech., **1,** 55—62.

Austen, I., T. J. Andersen and K. Edelvang, 1999. The influence of benthic diatoms and invertebrates on the erodibility of an intertidal mudflat, the Danish Wadden Sea, Estuar. Coast. Shelf Sci., **49,** 99—111.

Bacon, M. P., R. A. Belastock and M. H. Bothner, 1994. $^{210}$Pb balance and implications for particle transport on the continental shelf, U.S. Middle Atlantic Bight. Deep-Sea Res., **41,** 511—535.

Bennett, R. H., B. Ransom, M. Kastner, R. J. Hulbert, W. B. Sowyer, H. Olsen and M. W. Lambert, 1999. Early diagenesis: impact of organic matter on mass physical properties and processes, California continental margin. Mar. Geol., **159,** 7—34.

Berner, R. A., 1980. Early Diagenesis: A Theoretical Approach. Princeton University Press. 241 p.

Berner, R. A., 1982. Burial of organic carbon and pyrite sulfur in the modern oceans: Its geochemical and environmental significance. Am. J. Sci., **282,** 451—475.

Botto, F. and O. Iribarne, 2000. Contrasting effects of two burrowing crabs (Chasmagnathus granulata and Uca uruguayensis) on sediment composition and transport in estuarine environments. Estuar. Coast. Shelf Sci., **51,** 141—151.

Boudreau, B. P., 1998. Mean mixed depth of sediments: The wherefore and the why. Limnol. Oceanogr., **43,** 524–526.

Boudreau, B. P., 1997. Diagenetic Models and their implementation. Springer-Verlag, Berlin, 414 p.

Boudreau, B. P., 1996. The diffusive tortuosity of fine-grained unlithified sediments. Geochim. Cosmochim. Acta, **60,** 3139–3142.

Boudreau, B. P. and R. H. Bennett, 1999. New rheological and porosity equations for steady-state compaction. Am. J. Sci., **299,** 517–528.

Boudreau, B. P., B. S. Gardiner and B. D. Johnson, 2001a. Rate of growth of isolated bubbles in sediments with a giagenetic source of methane, Limnol. Oceanogr., **46,** 616—622.

Boudreau, B. P., M. Huettel, S. Forster, R. A. Jahnke, A. McLachlan, J. J. Middleburg, P. Nielsen, F. Sansone, G. Taghon, W. Van Raaphorst, I. Webster, J. M. Weslawski, P. Wiberg, and B. Sunby, 2001b. Permeable marine sediments: Overturning an old paradigm. EOS, Trans., Am. Geophys. Union, **82,** 133—136.

Boudreau, B. P. and R. L. Marinelli, 1994. A modelling study of discontinuous biological irrigation. J. Mar. Res. **52,** 947—968.

Brandes, J. A. and A. H. Devol, 1995. Simultaneous nitrate and oxygen respiration in coastal sediments: Evidence for discrete diagenesis. J. Mar. Res., **53,** 771—797.

Burdige, D. J., M. J. Alperin, J. Homstead and C. S. Martens, 1992. The role of benthic fluxes of dissolved organic carbon in oceanic and sedimentary carbon cycling. Geophys. Res. Lett., **19,** 1851—1854.

Burdige, D. J. and K. G. Gardner, 1998. Molecular weight distribution of dissolved organic carbon in marine sediment pore waters. Mar. Chem., **62,** 45—64.

Burdige, D. J. and J. Homstead, 1994. Fluxes of dissolved organic carbon from Chesapeake Bay sediments. Geochim. Cosmochim. Acta, **58,** 3407—3424.

Burdige, D. J., W. M. Berelson, K. H. Coale, J. McManus and K. S. Johnson, 1999. Fluxes of dissolved organic carbon from California continental margin sediments. Geochim. Cosmochim. Acta, **63,** 1507–1515.

Burnett, W., J. Chanton, J. Christoff, E. Kontar, S. Krupa, M. Lambert, W. Moore, D. O'Rouke, R. Paulsen, C. Smith, L. Smith and M. Taniguchi, 2002. Assessing methodologies for measuring groundwater discharge to the ocean. EOS, Trans. Am. Geophys. Union, **83,** 117, 122—123.

Burns, R. G. and V. M. Burns, 1979. Manganese Oxides, in Marine Minerals, R. G. Burns, ed. Mineralogical Society of America, pp. 1—46.

Cable, J. E., G. C. Bugna, W. C. Burnett and J. P. Chanton, 1996. Application of $^{222}Rn$ and $CH_4$ for assessment of groundwater discharge to the coastal ocean. Limnol. Oceanogr., **41,** 1347—1353.

Cacador, I., C. Vale and F. Catarino, 2000. Seasonal variation of Zn, Pb, Cu and Cd concentrations in the root-sediment system of Spartina maritima and Halimione portulacoides from Tagus estuary salt marshes, Mar. Environ. Res., **49,** 279—290.

Cahoon, L. B., G. R. Beretich, C. J. Thomas and A. M. McDonald, 1993. Benthic microalgal production at Stellwagen Bank, Massachusetts, USA. Mar. Ecol. Prog. Ser., **102,** 179—185.

Cahoon, L. B. and J. E. Cooke, 1992. Benthic microalgal production in Onslow Bay, North Carolina, USA, Mar. Ecol. Prog. Ser., **84,** 185—196.

Cahoon, L. B. and R. A. Laws, 1993. Benthic diatoms from the North Carolina continental shelf: inner and mid-shelf, J. Phycol., **29,** 257—263.

Cai, W.-J., L. R. Pomeroy, M. A. Moran and Y. Wang, 1999. Oxygen and carbon dioxide mass balance for the estuarine-intertidal marsh complex of five rivers in the southeastern U.S. Limnol. Oceanogr., **44,** 639—649.

Canfield, D. E., 1989. Sulfate reduction and oxic respiration in marine sediments: implications for organic carbon preservation in euxinic environments. Deep-Sea Res. **36,** 121—138.

Canfield, D. E., B. B. Jorgensen, H. Fossing, R. Glud, J. Gundersen, N. B. Ramsing, B. Thamdrup, J. W. Hansen, L. P. Nielsen, and P. O. J. Hall, 1993. Pathways of organic carbon oxidation in three continental margin sediments. Mar. Geol., **113,** 27–40.

Codispoti, L. A., J. A. Brandes, J. P. Christensen, A. H. Devol, S. W. A. Naqvi, H. W. Paerl and T. Yoshinari, 2001. The oceanic fixed nitrogen and nitrous oxide budgets: Moving targets as we enter the anthropocene?, Scientia Marina, **65,** 85–105.

Craven, D. B., R. A. Jahnke and A. F. Carlucci, 1986. Fine-scale verticle distributions of microbial biomass and activity in California Borderland sediments, Deep-Sea Res., **33,** 379–390.

Devol, A. H. 1991. Direct measurement of nitrogen gas fluxes from continental shelf sediments. Nature, **349,** 319—321.

Devol, A. H., L. A. Codispoti and J.P. Christensen, 1997. Summer and winter denitrification rates in western Arctic shelf sediments, Cont. Shelf Res., **17,** 1029—1050.

Devol, A. H. and J. P. Christensen, 1993. Benthic fluxes and nitrogen cycling in sediments of the continental margin of the eastern North Pacific. J. Mar. Res., **51,** 345–372.

Emerson, S. and J. Hedges, in press. Sediment Diagenesis and Benthic Flux in: Geochemical Treatise.

Emerson, S., R. Jahnke and D. Heggie, 1984. Sediment-water exchange in shallow water estuarine sediments. J. Mar. Res. **42,** 709—730.

Emery, K. O., 1968. Relict sediments on continental shelves of world. Bull. Am. Assoc. Petrol. Geol. **52,** 445—464.

Epping, E. G. and W. Helder, 1997. Oxygen budgets calculated from in situ oxygen microprofiles for Northern Adriatic sediments. Cont. Shelf Res., **17,** 1737—1764.

Falkowski, P. G., 1997. Evolution of the nitrogen cycle and its influence on the biological sequestration of $CO_2$ in the ocean. Nature, **387,** 272—275.

Falter, J. L. and F. J. Sansone, in press. Hydraulic control of pore water geochemistry within the oxic-suboxic zone of a permeable sediment. Limnol. Oceanogr.

Fornes, W. L., D. J. DeMaster and C. R. Smith, 2001. A particle introduction experiment in Santa Catalina Basin sediments: Testing the age-dependent mixing hypothesis. J. Mar. Res., **59,** 97—112.

Fossing, H., V. A. Gallardo, B. B. Jorgensen, M. Huettel, L. P. Nielsen, H. Schulz, D. E. Canfield, S. Forster, R. N. lud, J. K. Gundersen, J. Küver, N. B. Ramsing, A. Teske, B. Thamdrup, and O. Ulloa, 1995. Concentration and transport of nitrate by the mat-forming sulfur bacterium Thioploca. Nature **374,** 713—715.

Froelich, P. N., G. P. Klinkhammer, M. L. Bender, N. A. Luedtke, G. R. Heath, D. Cullen, P. Daphin, D. Hammond, B. Hartman and V. Maynard, 1979. Early oxidation of organic matter in pelagic sediments of the eastern equatorial Atlantic: Suboxic diagenesis. Geochim. Cosmochim. Acta, **43,** 1075—1090.

Gallagher, D. L., A. M. Dietrich, W. G. Reay, M. C. Hayes, G. M. Simmons, Jr., 1996. Ground water discharge of agricultural pesticides and nutrients to estuarine surface water, Ground. Water Monit. Remediat., **16,** 118—129.

Gehlen, M., H. Malschaert and W. R. Van Raaphorst, 1995. Spatial and temporal variability of benthic silica fluxes in the southeastern North Sea, Cont. Shelf Res., **15,** 1675—1696.

Goto, N., T. Kawamura, O. Mitamura and H. Terai, 1999. The importance of extracellular organic carbon production in the total primary production by tidal-flat diatoms in comparison to phytoplankton, Mar. Ecol. Prog. Ser., **190,** 289—295.

Gould, D. M. and E. D. Gallagher, 1990. Field measurement of specific growth rate, biomass, and primary production of benthic diatoms of Savin Hill Cove, Boston, Limnol. Oceanogr., **35,** 1757—1770.

Graf, G. 1992. Benthic—pelagic coupling: A benthic view. Oceanogr. Mar. Biol. Annu. Rev. **30,** 149—190.

Harrison, W. D., D. Musgrave and W. S. Reeburgh, 1983. A wave-induced transport process in marine sediments. J. Geophys. Res., **88,** 7617—7622.

Hartnett, H. E., R. G. Keil, J. I. Hedges and A. H. Devol, 1998. Influence of oxygen exposure time on organic carbon preservation in continental margin sediments, Nature, **391,** 572—574.

Hartnett, H. E. and A. H. Devol, 2003. Role of a strong oxygen-deficient zone in the preservation and degradation of organic matter: A carbon budget for the continental margins of northwest Mexico and Washington State. Geochim. Cosmochim. Acta **67,** 247—264.

Hedges, J. I., W. A. Clark and G. L. Cowie, 1988. Fluxes and reactivities of organic matter in a coastal marine bay. Limnol. Oceanogr., **33,** 1116—1136.

Hedges, J. I. and R. G. Keil, 1995. Sedimentary organic matter preservation: an assessment and speculative synthesis. Mar. Chem., **49,** 81–115.

Hedges, J. I., F. S. Hu, A. H. Devol, H. E. Hartnett, E. Tsamakis and R. G. Keil, 1999. Sedimentary organic matter preservation: A test for selective degradation under oxic conditions. Am. J. Sci., **299,** 529—555.

Himmelblau, D. M., 1964. Diffusion of dissolved gases in liquids. Chemical Reviews, **64,** 527—550.

Hoefel, F. G. and R. L. Evans, 2001. Impact of low salinity porewater on seafloor electromagnetic Data: A means of detecting submarine groundwater discharge? Estuar. Coast. Shelf Sci., **52,** 179—189.

Holcombe, B. L., R. G. Keil, and A. H. Devol, 2001. Determination of pore-water dissolved organic carbon fluxes from Mexican margin sediments. Limnol. Oceanogr., **46,** 298—308.

Hopkinson, C. S. Jr., R. D. Fallon, B. O. Jansson, and J. P. Schulauer, 1991. Community metabolism and nutrient cycling at Gray's Reef, a hard bottom babitat in the Georgia Bight. Mar. Ecol. Prog. Ser., **73,** 105—120.

Huettel, M. and G. Gust, 1992. Solute release mechanisms from combined sediment cores in stirred benthic chambers and flume flows. Mar. Ecol. Prog. Ser., **82,** 187—197.

Huettel, M., W. Ziebis, S. Forster, 1996. Flow-induced uptake of particulate matter in permeable sediments, Limnol. Oceanogr., **41,** 309–322.

Huettel, M., W. Ziebis, S. Forster and G. W. Luther, III., 1998. Advective transport affecting metal and nutrient distributions and interfacial fluxes in permeable sediments, Geochim. Cosmochim. Acta, **62,** 613–631.

Hulth, S., R. C. Aller, and F. Gilbert, 1999. Couples anoxic nitrification/manganese reduction in marine sediments, Geochim. Cosmochim. Acta, **63,** 49–66.

Hutchins, D. A. and K. W. Bruland, 1998. Iron-limited diatom growth and Si:N uptake ratios in a coastal upwelling regime, Nature, **393,** 561—564.

Hutchins, D. A., DiTullio, G. R., Y. Zhang and K. W. Bruland, 1998. An iron limitation mosiac in the California upwelling regime, Limnol. Oceanogr., **43,** 1037–1054.

Jahnke, R. A., 1996. The global ocean flux of particulate organic carbon: Areal distribution and magnitude, Global Biogeochem. Cycles, **10,** 71–88.

Jahnke, R. A., C. R. Alexander and J. E. Kostka, in press. Advective pore water input of nutrients to the Satilla River Estuary, Georgia, USA, Estuarine Coastal & Shelf Science.

Jahnke, R. A., J. R. Nelson, R. L. Marinelli and J. E. Eckman, 2000. Benthic flux of biogenic elements on the Southeastern US continental shelf: influence of pore water advective transport and benthic microalgae, Cont. Shelf Res., **20,** 109–127.

Jetten, M. S. M., M. Strous, K. T. van de Pas-Schoonen, J. Schalk, U. G. J. M. van Dongen, A. A. van de Graaf, S. Logemann, G. Muyzer, M. C. M. van Loosdrecht and J. G. Kuenen, 1999. The anaerobic oxidation of ammonium. FEMS Microbiol., **22,** 421–437.

Koretsky, C. M., C. Maile and P. Van Cappellen, 2002. Quantifying bioirrigation using ecological parameters: a stochastic approach, Geochem. Trans., **3,** 17—30.

Kostka, J. E., G. W. Luther III and K. H. Nealson, 1995. Chemical and biological reduction of Mn(III)-pyrophosphate complexes: Potential importance of dissolved Mn(III) as an environmental oxidant, Geochim. Cosmochim. Acta, **59,** 885–894.

Kostka, J. E., B. Thamdrup, R. N. Glud and D. E. Canfield, 1999. Rates and pathways of carbon oxidation in permanently cold Arctic sediments, Mar. Ecol. Prog. Ser., **180,** 7–21.

Kostka, J. E., A. Roychoudhury and P. Van Cappellen, in press. Rates and controls of anaerobic microbial respiration across spatial and temporal gradients in saltmarsh sedimentss, Biogeochemistry,

Kristensen, E., M. H. Jensen and R. C. Aller, 1991. Direct measurement of dissolved inorganic nitrogen exchange and denitrification in individual polychaete (Nereis virens) burrows. J. Mar. Res. **49,** 355—377.

Kühl, M. and B. B. Jorgensen, 1994. The light field of microbenthic communities: radiance distribution and microscale optics of sandy coastal sediments, Limnol. Oceanogr., **39,** 231—257.

Lerman, A., 1979. Geochemical Processes: Water and Sediment Environments. Wiley Interscience.

Li, L. and D. A. Barry, 2000. Wave-induced beach groundwater flow, Adv. Water Res. **23,** 325—337.

Li, Y. and S. Gregory, 1974. Diffusion of ions in sea water and in deep sea sediments, Geochim. Cosmochim. Acta, **38,** 703—714.

Lohse, L., E. H. G. Epping, W. Helder and W. van Raaphorst, 1996. Oxygen pore water profiles in continental shelf sediments of the North Sea: turbulent versus molecular diffusion, Mar. Ecol. Prog. Ser., **145,** 63—75.

Lohse, L., J. F. P. Malschaert, C. P. Slomp, W. Helder and W. van Raaphorst, 1993. Nitrogen cycling in North Sea sediments: interaction of denitrification and nitrification in offshore and coastal areas, Mar. Ecol. Prog. Ser., **101,** 283–296.

Luther, G. W. III, P. J. Brendel, B. L. Lewis, B. Sundby, L. Lefrancois, N. Silverberg, and D. B. Nuzzio, 1998. Simultaneous measurement of $O_2$, Mn, Fe, $I^-$, and S(-II) in marine pore waters with a solid-state voltammetric microelectrode, Limnol. Oceanogr. **43,** 325–333.

Luther, G. W., B. Sundby, B. L. Lewis, P. J. Brendel and N. Silverberg, 1997. Interactions of manganese with the nitrogen cycle: Alternative pathways to dinitrogen, Geochim. Cosmochim. Acta, **61,** 4043–4052.

McCraith, B. J., 1998. The Distribution and Dynamics of Fiddler Crab Burrowing and its Effect on Salt Marsh Sediment Composition and Chemistry in a Southeastern Salt Marsh. Ph.D. Thesis, Univ. South Carolina, 181 p.

Marinelli, R. L. and B. P. Boudreau, 1996. An experimental and modeling study of pH and related solutes in an irrigated anoxic coastal sediment, J. Mar. Res., **54,** 939—966.

Marinelli, R. L., R. A. Jahnke, D. B. Craven, J. R. Nelson and J. E. Eckman, 1998. Sediment nutrient dynamics on the South Atlantic Bight continental shelf. Limnol. Oceanogr., **43,** 1305–1320.

Martin, W. R. and D. C. McCorkle, 1993. Dissolved organic carbon concentrations in marine pore waters determined by high-temperature oxidation, Limnol. Oceanogr., **38,** 1464—1480.

Mayer, L. M., 1994a. Surface area control of organic carbon accumulation in continental shelf sediments, Geochim. Cosmochim. Acta, **58,** 1271—1284.

Mayer, L. M., 1994b. Relationships between mineral surfaces andorganic carbon concentrations in soils and sediments, Chem. Geol., **114,** 347—363.

Meile, C., C.M. Koretsky and P. Van Cappellen, 2001. Quantifying bioirrigation in aquatic sediments: An inverse modeling approach, Limnol. Oceanogr., **46,** 164—177.

Miller, R. J., C. R. Smith, D. J. DeMaster and W. L. Fornes, 2000. Feeding selectivity and rapid particle processing by deep-sea megafaunal deposit feeders: A $^{234}$Th tracer approach, J. Mar. Res., **58,** 653—673.

Milliman, J. D., K. L. Farnsworth and C. S. Albertin, 1999. Flux and fate of fluvial sediments leaving large islands in the East Indies, J. Sea Res., **41,** 97—107.

Moore, W. S., 1996. Large groundwater inputs to coastal waters revealed by $^{226}$Ra enrichments, Nature, **380,** 612—614.

Moore, W. S., 1999. The subterranean estuary: a reaction zone of ground water and sea water, Mar. Chem., **65,** 111–125.

Moore, W. S. and T. J. Shaw, 1998. Chemical signals from submarine fluid advection onto the continental shelf, J. Geophys. Res., **103,** 21,543—21, 552.

Mulder, A., A. van de Graaf, L. A. Robertson and J. G. Kuenen, 1995. Anaerobic ammonium oxidation discovered in a denitrifying fluidized bed reactor. FEMS Microbiol. Ecol. **16,** 177—184.

Muslow, S., B. P. Boudreau and J. N. Smith, 1998. Bioturbation and porosity gradients. Limnol. Oceanogr., **43,** 1–9.

Nealson, K. H. and C. R. Myers, 1992. Microbial reduction of manganese and iron: New approaches to carbon cycling, Appl. Environ. Microbiol., **58,** 439–443.

Noh, J. H. and J. K. Choi, 1998. Ecological role of benthic diatom locomotion in the intertidal mud flat, Ocean Res., **20,** 179—187.

Ransom, B., R. H. Bennett, R. Baerwald and K. Shea, 1997. TEM study of in situ organic matter on continental margins: Occurrence and the "monolayer" hypothesis, Mar. Geol., **138,** 1—9.

Ransom, B., D. Kim, M. Kastner and S. Wainwright, 1998. Organic matter preservation on continental slopes: Importance of minerology and surface area, Geochim. Cosmochim. Acta, **62,** 1329—1345.

Reimers, C. E. and E. Suess, 1983. Spatial and temporal patterns of organic matter accumulation on the Peru continental margin, In Coastal Upwelling, J. Thiede and E. Suess, eds. Plenum Publishing Corp., pp. 311–345.

Rice, D. L., 1986. Early diagenesis in bioadvective sediments: Relationships between the diagenesis of beryllium-7, sediment reworking rates, and the abundance of conveyor-belt deposit-deeders, J. Mar. Res., **44,** 149—184.

Richards, F. A., 1965. Anoxic Basins and Fjords, in Chemical Oceanography, J. P. Riley and G. Skirrow, eds. Academic Press, pp. 611—646.

Riedl, R. J., N. Huang and R. Machan, 1972. The subtidal pump: a mechanism of interstitial water exchange by wave action, Mar. Biol., **13,** 210—221.

Sampou, P. and C. A. Oviatt, 1991. Seasonal patterns of sedimentary carbon and anaerobic respiration along a simulated eutrophication gradient. Mar. Ecol. Prog. Ser. **72,** 271—282.

Schluter, M., E. Sauter, H.-P. Hansen and E. Suess, 2000. Seasonal variations in bioirrigation in coastal sediments: Modelling of field data, Geochim. Cosmochim. Acta, **64,** 821–834.

Schubert, C. J. and R. Stein, 1996. Deposition of organic carbon in Arctic Ocean sediments: terrigenous supply vs marine productivity, Organic Geochem., **24,** 421—436.

Shimmield, G. B. and R. A. Jahnke, 1995. Particle flux and its conversion to the sediment record: Open ocean upwelling systems, in Upwelling in the Ocean: Modern Processes and Ancient Records, C. P. Summerhayes, K.-C. Emeis, M. V. Angel, R. L. Smith, and B. Zeitzschel, eds. John Wiley & Sons, pp. 171—192.

Shinn, E. A., C. D. Reich and T. D. Hickey, 2002. Seepage meters and Bernoulli's revenge, Estuaries, **25,** 126—132.

Shum, K. T., 1992. Wave-induced advective transport below a rippled water-sediment interface, J. Geophys. Res., **97,** 789—808.

Shum, K. T., 1993. The effects of wave-induced pore water circulation on the transport of reactive solutes below a rippled sediment bed, J. Geophys. Res., **98,** 10,289—10,301.

Simmons, G. M. Jr., 1992. Importance of submarine groundwater discharge (SGWD) and seawater cycling to material flux across sediment/water interfaces in marine environments, Mar. Ecol. Prog. Ser., **84,** 173–184.

Simmons, J. A. K. and W. B. Lyons, 1994. The groundwater flux of nitrogen and phosphorus to Bermud'a coastal waters, Water Resour. Bull., **30,** 983 -991.

Smith, C. R., L. A. Levin, D. J. Hoover, G. McMurtry, and J. D. Gage, 2000. Variations in bioturbation across the oxygen minimum zone in the northwest Arabian Sea, Deep-Sea Res., **47,** 227—257.

Smith, C. R. and C. Rabouille, 2002. What controls the mixed-layer depth in deep-sea sediments? The importance of POC flux, Limnol. Oceanogr., **47,** 418—426.

Strous, M., J. G. Kuenen and M. S. M. Jetten, 1999a. Key physiology of anaerobic ammonium oxidation. Appl. Environ. Microbiol., **65,** 3248—3250.

Strous, M., J. A. Fuerst, E. H. M. Kramer, S. Logemann, G. Muyzer, K. T. van de Pas-Schoonen, R. Webb, J. G. Kuenen, and M. S. M. Jetten, 1999b. Missing lithotroph identified as a new planctomycete. Nature, **400,** 446—448.

Sundback, K., V. Enoksson, W. Graneli and K. Pettersson, 1991. Influence of sublittoral microphytobenthos on the oxygen and nutrient flux between sediment and water: a laboratory continuous-flow study, Mar. Ecol. Prog. Ser., **74,** 263—279.

Thamdrup, B. and T. Dalsgaard, 2002. Production of $N_2$ through anaerobic ammonium oxidation coupled to nitrate reduction in marine sediments, Appl. Environ. Microbiol., **68,** 1312 -1318.

Thamdrup, B. and D. E. Canfield, 1996. Pathways of carbon oxidation in continental margin sediments off central Chile, Limnol. Oceanogr., **41,** 1629–1650.

Waterman, F., H. Hillebrand, G. Gerdes, W. E. Krumbein and U. Sommer, 1999. Competition between benthic cyanobacteria and diatoms as influenced by different grain sizes and temperatures, Mar. Ecol. Prog. Ser. **187,** 77—87.

Weissberg, H., 1963. Effective diffusion coefficients in porous media, J. Appl. Phys., **34,** 2636—2639.

Wheatcroft, R. A., P. A. Jumars, C. R. Smith and A. R. M. Nowell, 1990. A mechanistic view of the particulate biodiffusion coefficient: Step lengths, rest periods and transport directions, J. Mar. Res., **48,** 177—207.

Whiting, G. J. and D. L. Childers, 1989. Subtidal advective water flux as a potentially important nutrient input to southeastern USA saltmarsh estuaries, Estuarine, Coast. Shelf Sci., **28,** 417–431.

Yool, A. and M. J. R. Fasham, 2001. An examination of the "continental shelf pump" in an open ocean general circulation model, Global Biogeochem. Cycles, **15,** 831–844.

Ziebis, W., M. Huettel and S. Forster, 1996. Impact of biogenic sediment topography on oxygen fluxes in permeable seabeds, Mar. Ecol. Prog. Ser., **140,** 227—237.

# Chapter 7. BOUNDARY EXCHANGES IN THE GLOBAL COASTAL MARGIN: IMPLICATIONS FOR THE ORGANIC AND INORGANIC CARBON CYCLES

FRED T. MACKENZIE

*University of Hawaii at Manoa*

ANDREAS ANDERSSON

*University of Hawaii at Manoa*

ABRAHAM LERMAN

*Northwestern University*

LEAH MAY VER

*University of Hawaii at Manoa*

## Contents



## 1. Introduction

The area of the global coastal zone, including estuaries, bays, lagoons, banks, and continental and island shelves, is only about 7 to 10% of the surface area of the ocean (25 to $36\times10^6$ $km^2$) and its volume to 100 m mean depth is 2.4 to 3.3% of the volume of the surface ocean layer (300 m deep, $108\times10^6$ $km^3$). Despite its relatively small size, it is an important interface between the land and open ocean, and it is also in direct exchange with the atmosphere. Large river drainage basins connect the vast interiors of continents with the coastal zone through river and groundwa-

*The Sea*, Volume 13, edited by Allan R. Robinson and Kenneth H. Brink
ISBN 0-674-01526-6 

ter discharges. The ocean surface links the coastal ocean to the atmosphere via gas exchange at the air-sea interface, production of sea aerosols, and atmospheric deposition on the sea surface; substances released at the air-sea interface of the coastal zone may be subsequently transported through the atmosphere and deposited on land as wet and dry depositions; conversely, emissions from land to the atmosphere are in part deposited in the coastal zone. Additionally, physical exchange processes at coastal margins, involving for example coastal upwelling (water rising from the deeper ocean) and onwelling (water that moves on and across the shelf), and net advective transport of water, dissolved solids, and particles from the coastal zone offshore connect the coastal zone with the surface and intermediate depths of the open ocean. Some materials biologically and inorganically produced *in situ* or delivered to the coastal zone via river and groundwater discharges, upwelling, and atmospheric deposition may eventually accumulate in coastal marine sediments. The processes of settling, deposition, resuspension, remineralization of organic matter, dissolution and precipitation of mineral phases, and accumulation of materials connect the water column and the sediments of the coastal zone.

In general, interfaces between the larger material reservoirs (i.e., the land, atmosphere, and ocean) are important in the control of the biogeochemical cycling of three of the major bio-essential elements found in organic matter: carbon (C), nitrogen (N), and phosphorus (P) because they act as relatively fast modifiers of transport and perturbation processes at geologically short time scales. Over the past several centuries, activities of humankind have significantly modified the exchange of materials between the land, atmosphere, and ocean on a global scale. Humans have become, along with natural processes, agents of environmental change. For example, because of rapid population growth, increasing population density in the areas of the major river drainage basins and close to oceanic coastlines (about 40% of the world population lives within 100 km of the shoreline; Cohen et al., 1997), socio-economic development, and changes in land-use practices in past centuries, discharges of industrial, agricultural, and municipal wastes into oceanic coastal waters have increased. Land-use activities include the conversion of land for food production (grazing land, agricultural land), for urbanization (building human settlements, roads, and other structures), for energy development and supply (building dams, hydroelectric plants, and mining of fossil fuels), and for resource exploitation (mining of metals, harvest of forest hardwood) (Mackenzie, 2003). These activities on land have contributed to increased soil degradation and erosion, eutrophication of river and coastal ocean waters through addition of chemical fertilizers to agricultural land and sewage discharge, degradation of water quality, and alteration of the coastal marine food web and community structure. It is estimated that only about 20% of the world's drainage basins have pristine water quality at present (Meybeck and Ragu, 1995, 1997). Estuarine and coastal regions showing much human-induced change are located, for example, along the coasts of the North Sea, the Baltic Sea, the Adriatic Sea, the East China Sea, and the east and south coasts of North America (De Jonge et al., 1994; Richardson and Heilmann, 1995).

The coastal zone is not only the oceanic region that is most susceptible to changes in water quality, organic productivity, and biodiversity, but it has also been perturbed disproportionately more than the much larger area of the open ocean (e.g., Galloway et al., 1996; Howarth et al., 1996; Mackenzie et al., 1993, 2002a, 2002b; Rabouille et al., 2001; Ver et al., 1999b; Wollast, 1991; Wollast, 1993; Wollast, 1998). At least 70 to 75% of the mass of terrigenous materials reaching the ocean is deposited in the coastal zone. In addition, 30 to 50% of total carbonate and approximately 80% of total organic carbon accumulation in the ocean occur in the coastal margin, and 10 to 30% of total oceanic biological production takes place in this region (e.g., Berner, 1982; Hu et al., 1998; Mackenzie et al., 1998a; Milliman, 1993; Milliman and Syvitski, 1992; Morse and Mackenzie, 1990; Smith and Hollibaugh, 1993; Vörösmarty et al., 1998; Wollast, 1994, 1998). It is thus understandable why the coastal zone is regarded as both a filter and a trap for natural as well as anthropogenic materials transported from the continents to the open ocean (Mantoura et al., 1991).

In this chapter, we describe how the global coastal zone has responded to changes in the boundary fluxes of carbon, nitrogen, and phosphorus under stresses from major human perturbations on land, the atmosphere, and the open oceans over the past 300 years. In particular, we describe changes in the organic and inorganic carbon balances of the coastal zone from the long-term quasi-steady state condition to the present and future, and discuss implications for the net air-sea exchange of carbon dioxide. The past behavior of this important global environment may provide some clues as to its future course under continuous pressure from those human activities that have become geologically significant factors in the surficial environment of the Earth.

## 2. The present-day organic carbon balance in the global coastal ocean

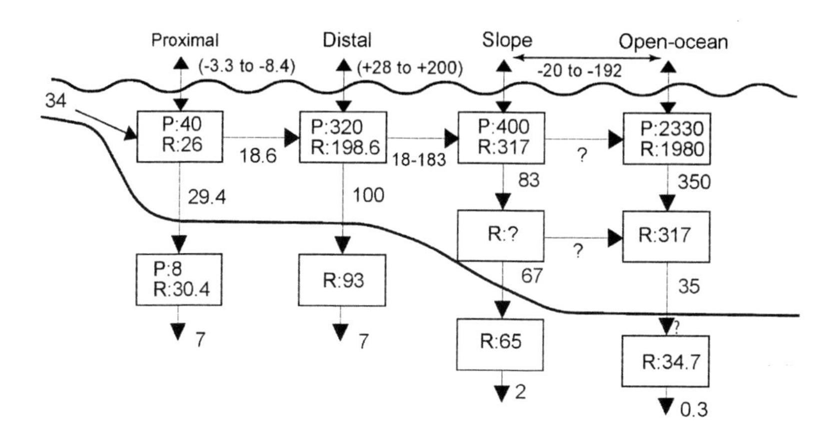


Figure 7.1 Global marine organic carbon cycle (non-steady state). The scheme is modified from Wollast (1998), Rabouille et al. (2001), and Ducklow and McAllister, this volume. *P* denotes production and *R* remineralization ($R_{total}$). Less well known fluxes are denoted by a question mark and/or by a range of values. Negative values indicate a net flux of $CO_2$ from the surface ocean to the atmosphere.

The major driving force of the organic carbon cycle in the global ocean (Figure 7.1) is the production of organic matter by marine primary producers. Estimates of global marine primary production are slightly lower than terrestrial production, and marine primary production constitutes approximately 45% of the organic matter produced globally (Schlesinger, 1997; Mackenzie, 2003). In most oceanic regions, primary production is limited by light intensity and the availability of nutrients, mainly nitrogen and phosphorus, but in some areas by trace elements such as iron. The flux of carbon between the major reservoirs of the global marine organic carbon cycle is controlled by the magnitude of primary production, the extent of respiration and decay, and physical transport processes (Figure 7.1). Coastal environments are regions of higher biological productivity relative to that of average oceanic surface waters, making them biogeochemically important in the global marine cycles of carbon, nitrogen, and phosphorus. Boundary exchange fluxes of C, N, and P between the coastal zone and the land, atmosphere, and open ocean domains are major factors that control the production, storage, and export of organic and inorganic matter in the coastal zone, hence its net organic and inorganic carbon balance and the exchange fluxes of $CO_2$ and volatile N compounds across the coastal ocean-atmosphere interface (the organic carbon balance at steady state is often referred to as net ecosystem production or NEP, to be discussed in more detail in a later section). Observations and global modeling have shown that organic productivity in coastal zone environments, particularly proximal regions, has been heavily impacted by changes in the dissolved and particulate riverine input fluxes of the nutrient elements C, N, and P. Although the latter two elements are considered minor constituents of river water, their fluxes and that of total organic carbon may have nearly doubled over their pristine values on a global scale because of human activities (Wollast and Mackenzie, 1989; Mackenzie et al., 1993; Meybeck, 1982; Figure 7.2). However, within the past century, the detrital sediment fluxes might have been reduced owing to the damming of about half of the world's rivers, although there is evidence that human land-use practices have increased the suspended load of rivers (Hay, 1998; Milliman and Syvitski, 1992; Smith et al., 2001; World Commission on Dams, 2000).

It is clear that in order to understand the importance of the coastal ocean relative to human-induced change in the global carbon cycle and to analyze fully coastal zone system behavior and dynamical response to stress from human perturbations, a comprehensive analysis of the component elements and their interaction in the coupled land-ocean-atmosphere system is necessary. It is important to define not only the pre-anthropogenic (i.e., near steady-state) conditions but also the changes away from the steady state that have occurred over the past 300 years and will continue to occur in the future. One way of studying the biogeochemical components of change is through observations over reasonably long periods of time. Another approach is through modeling and sensitivity analyses that are based on currently available information and values of environmental parameters predictable within some range of uncertainty. Both approaches are necessary and complement one another. Here we discuss how both approaches have been used to analyze the past, present, and future of coastal ocean organic metabolism, the inorganic carbon system, and human-induced change.

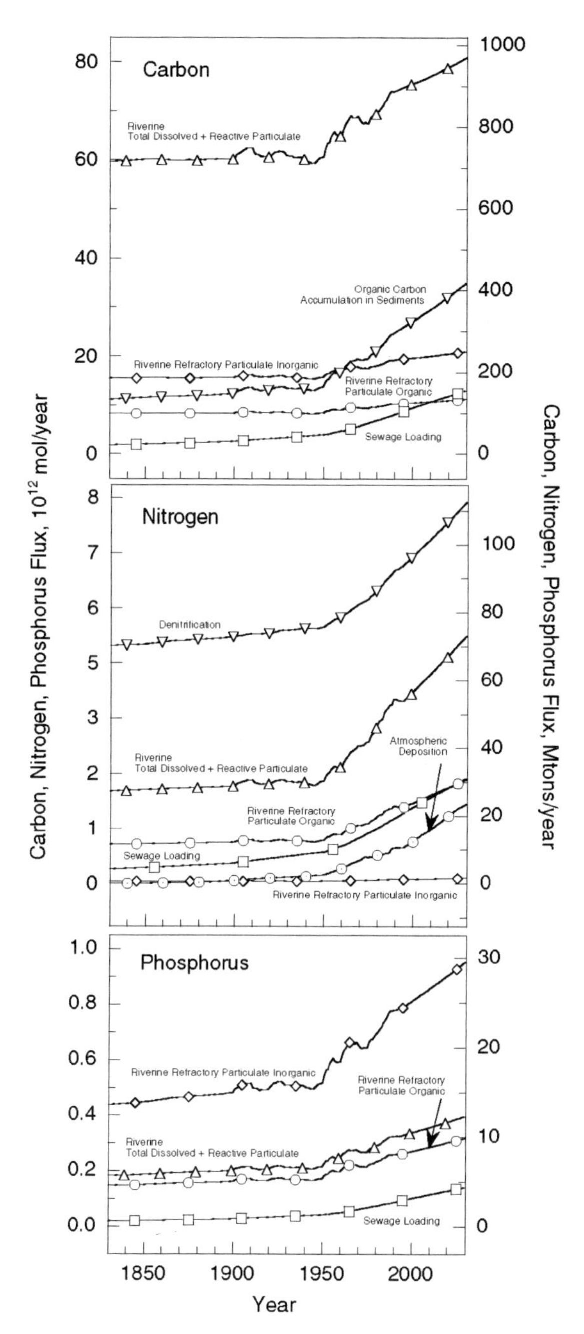


Figure 7.2 Model numerical simulation results for the past, present, and predicted input and output fluxes of carbon, nitrogen, and phosphorus into or out of the global coastal margin in $10^{12}$ mol/yr and Mton/yr. The calculated fluxes were obtained using the model *TOTEM* (e.g., Ver et al., 1999). These fluxes represent the fluxes at steady state in the model for the year 1700 plus the additions owing to anthropogenic activities for the past approximately 150 years with projections to the future (Mackenzie et al., 2002).

The *ecological* concept of net ecosystem production (NEP) is the net change in organic carbon in an ecosystem over a period of time, usually one year. NEP in a system at steady state is the difference between the rates of gross primary production and total respiration; the latter is total ecosystem production of inorganic carbon by autotrophic and heterotrophic respiration, including respiration, decay, decomposition or remineralization (e.g., Smith and Mackenzie, 1987; Smith and Hollibaugh, 1993; Woodwell, 1995; Mackenzie et al., 1998a). In addition, NEP is the difference between the two very large fluxes of gross primary production and total respiration for which data on a global scale collected over a sufficient period of time (time-series data), particularly for coastal zone ecosystems, are still lacking despite the recent advances in the use of incubations employing $O_2$ measurements to obtain such data. Thus, it has been difficult to evaluate global NEP from direct measurements integrated over a long enough time and a large enough geographical scale. Using the notations for the global coastal zone shown in Figure 7.3, NEP can thus be defined for an ecosystem:

$$
\begin{aligned}
\text{NEP} &= F_{o2} - F_{\text{DIC}} \\
&= P - R_{\text{total}}
\end{aligned}
\tag{1}
$$

In (1), NEP can also be defined as a measure of the imbalance between the production *(P)* and total respiration *($R_{total}$)* of organic carbon within the coastal system. Total respiration or remineralizatiion rate, $R_{total}$, is a sum of the respiration rates of autotrophic organic matter photosynthetically produced *in situ* and of allochtonous organic matter imported primarily from land: $R_{autotrophic} + R_{heterotrophic}$. A system at steady state is *net heterotrophic* when the amount of organic carbon respired, decayed, and decomposed is greater than the amount produced by gross photosynthesis: NEP < 0. A system is *net autotrophic* when the amount of carbon fixed by gross photosynthesis exceeds that remineralized by all the respiration processes: NEP > 0. The net result is either a drawdown (net autotrophy) or evolution (net heterotrophy) of $CO_2$ owing only to organic metabolism in the ecosystem.

It is important to clarify at this point that the concept of the NEP discussed here (also variably termed the net trophic status, organic carbon balance, net community production, or organic metabolism; Smith and Hollibaugh, 1993; Williams, 1998; Wollast, 1998) is not equivalent to the term "net primary production" or NPP, dealt with in ecological research: the latter is a balance between primary production and autotrophic respiration. The main difference arises from the consideration of *total* respiration in quantifying NEP, rather than only autotrophic respiration. A closer correlation can be made between the ecological definition of NEP in the photic zone and "new production" (Dugdale and Goering, 1967), i.e., production that results in particulate and dissolved organic matter that is available for export or burial. New production is linked to an external source of nutrients and the fact that new production can be exported or buried is a property of this new production.

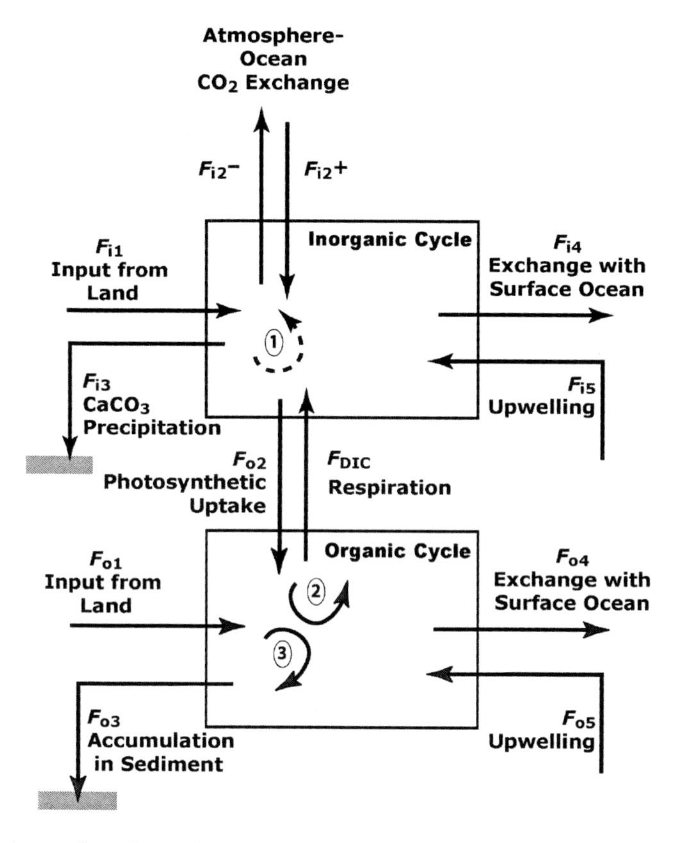


Figure 7.3 The inorganic and organic carbon cycle in the coastal zone (after Mackenzie et al., 1998a). Conceptual model of the main reservoirs and fluxes for organic and inorganic carbon. The inorganic carbon reservoir exchanges $CO_2$ with the atmosphere ($\pm F_{i2}$), and receives inputs of dissolved inorganic carbon from land via rivers, surface runoff and groundwater flow ($F_{i1}$), from the $CO_2$ released by the precipitation of calcium carbonate (curved arrow 1), and by upwelling from the deeper ocean ($F_{i5}$). Outflows of inorganic carbon are via net water outflow from the coastal zone ($F_{i4}$), deposition and accumulation of calcium carbonate, formed in the coastal zone, in coastal sediments ($F_{i3}$), and use of $CO_2$ in photosynthesis ($F_{o2}$). The organic carbon cycle is linked to the inorganic cycle through biologically driven reduction and oxidation processes, corresponding to primary production ($F_{o2}$) utilizing $CO_2$ and respiration and decay ($F_{DIC}$, where DIC stands for total dissolved inorganic carbon). Analogous to the inorganic carbon cycle, there are inputs of organic carbon from land (dissolved and particulate, $F_{o1}$) and by transport of dissolved organic carbon from the deeper ocean, referred to as coastal upwelling ($F_{o5}$). Removal of organic carbon from the coastal zone is through net outflow to the open ocean ($F_{o4}$), accumulation in coastal sediments (*in situ* produced and land-derived, $F_{o3}$), and respiration and decay or remineralization ($F_{DIC}$). Remineralized organic carbon produced *in situ* (curved arrow 2) or imported from land (curved arrow 3) contributes to $F_{DIC}$. Flux ($F$) values are shown in Table 1 for the start of the model analysis in the year 1700 and the present time, computed for the year 2000.

As mentioned above, a survey of estimates of rates of primary production and total respiration shows that $P$ and $R_{total}$ do not differ significantly from one another and the difference $P - R_{total}$ can be practically indistinguishable from zero by evaluation of $P$ and $R_{total}$ individually. Another approach, although not necessarily any more quantitative, involves a budgeting of the organic carbon cycle in the ocean or a portion of the ocean, such as the coastal ocean. Using budgetary reasoning NEP can be derived indirectly from a consideration of all sources and sinks of particulate and dissolved organic carbon, i.e., input from rivers, upwelling from intermediate oceanic waters, export to the open ocean, and burial in sediments. From Figure 7.3, the material balance for the organic carbon reservoir in the coastal margin can be written as (Mackenzie et al., 1998a):

$$\frac{dC_{org}}{dt} = (F_{o1} + F_{o2} + F_{o5}) - (F_{o4} + F_{o3} + F_{DIC}), \tag{2}$$

where $C_{org}$ is the mass of organic carbon in the reservoir and $t$ is time. Gross photosynthesis ($F_{o2}$) and respiration and decay ($F_{DIC}$) are the linkages between the organic and inorganic carbon cycles. Thus, in a system near a steady state ($dC_{org}/dt \approx 0$), NEP can also be defined from equations 1 and 2 as:

$$\text{NEP} = F_{o2} - F_{DIC} = (F_{o3} + F_{o4}) - (F_{o1} + F_{o5}) \tag{3}$$

For a system in transition ($dC_{org}/dt \neq 0$), NEP can be defined as:

$$\text{NEP}^* = F_{o2} - F_{DIC} = \Delta C_{org}\big|_t + (F_{o3} + F_{o4}) - (F_{o1} + F_{o5}), \tag{4}$$

where $\text{NEP}^*$ is NEP for a system in a transient state and $\Delta C_{org}$ is the accumulation or loss of organic carbon in the reservoir in a unit of time. Equations (3) and (4) describe NEP as the net sum of the accumulation or reduction in organic carbon mass and the import and export fluxes of organic carbon in the coastal zone. Equation (2) states that organic carbon accumulates in the coastal zone when, as expected, the inputs are greater than the outputs. Equation (3) gives the balance of organic carbon only at steady state, and if there is more organic carbon imported into the coastal zone than removed by export (i.e., the inputs $F_{o1}+F_{o5}$ are relatively large), then the excess of input would be remineralized and leave the organic carbon pool as DIC in the flux $F_{DIC}$. These equations show that in a system that is net autotrophic ($\text{NEP}^* > 0$), relatively higher rates of primary production ($F_{o2}$) can give rise to a higher storage rate of organic carbon ($F_{o3}$ and/or $\Delta C_{org}$), if export to the open ocean, $F_{o4}$, is smaller than the sedimentation fluxes, $F_{o3}$, which is in agreement with observations that primary production in oceanic surface waters may control to some extent the rate of deposition of organic carbon initially reaching the ocean floor (Lee, 1992; Müller and Suess, 1979). An important conclusion is that the export flux of organic carbon ($F_{o4}$), whose magnitude is related to the rate of water movement and the mass of dissolved and particulate organic carbon available for transport to the open ocean, also determines the trophic state ($\text{NEP}^*$) of the coastal zone.

The prevailing question that must be addressed is how the balance between autotrophy and heterotrophy in the coastal ocean has been modified by human activities. Increased import of inorganic nutrients via loading from land and transport from upwelling zones induces increased primary production, hence potentially a tendency towards net autotrophy ($\text{NEP}^* > 0$). At the same time, increased input from land of detrital and dissolved organic matter, a fraction of which can be highly reactive, and possibly faster recycling of *in situ* produced organic matter can lead to increased remineralization in the coastal zone, hence potentially shifting the coastal system towards increased heterotrophy. Thus these two sets of processes compete one with the other in determining the trophic state of the coastal zone.

### 3. Estimates of long-term (pre-anthropogenic) NEP

Several investigators have made reasonable estimates of NEP for the pre-anthropogenic global ocean calculated from the large biogeochemical fluxes of gross primary production and total ecosystem autotrophic and heterotrophic respiration. Because these earlier estimates shared common sources of data for the biogeochemical flux components used in the calculations, it is not surprising that the estimates agree well with respect to both the magnitude and direction of the net flux of carbon. Garrels and Mackenzie (1972) estimated that the global ocean was a net source of $CO_2$ to the atmosphere in pristine time at the rate of $-27\times10^{12}$ moles C/yr due to organic metabolism. Smith and Mackenzie (1987) concluded that because of aerobic and anaerobic respiration and remineralization of organic carbon exceeding *in situ* gross photosynthesis, the pre-industrial global ocean was a net heterotrophic system, hence a source of $CO_2$ with a calculated flux of $-21\times10^{12}$ moles C/yr. Organic carbon transported from land via rivers and remineralization of a portion of that carbon in the ocean fueled the heterotrophy in the ocean. Calculations by Wollast and Mackenzie (1989) and Smith and Hollibaugh (1993) based on biogeochemical processes involving organic carbon showed a net flux of about $-22\times10^{12}$ moles C/yr from the heterotrophic ocean to the atmosphere.

There have only been a few attempts to estimate either the long-term or present-day NEP of the coastal ocean separately from that of the open ocean. One of the primary reasons is that the transfer fluxes between the coastal margin and the open ocean are poorly constrained (Gattuso et al., 1998). Additionally, it is difficult to model the coastal oceanic region separately from the open ocean because of its large spatial and temporal variability (Mantoura et al., 1991). Despite these difficulties, two of the most notable and frequently cited estimates of coastal ocean NEP (Smith and Hollibaugh, 1993; Wollast and Mackenzie, 1989) independently concluded that the coastal ocean was net heterotrophic in preindustrial times. Wollast and Mackenzie's (1989) estimated heterotrophy in the global ocean included a coastal ocean that was heterotrophic at the rate of $-3.3\times10^{12}$ moles C/yr. Smith and Hollibaugh (1993; Smith, 1995) also calculated the coastal zone to be net heterotrophic at the rate of $-7\times10^{12}$ moles C/yr, equivalent to about 5 times the open ocean net oxidation rate of organic matter. For the late 1990s, coastal zone heterotrophy was estimated at $-8\times10^{12}$ moles C/yr (Mackenzie et al., 1998a). Using the ecosystem-scale approach of balancing inputs and outputs, they estimated NEP from the long-term rates of primary production, respiration, net metabolism, river loading of organic carbon, organic carbon burial, and chemical reactivity of terrestrial organic matter.

## 4. Estimates of present-day coastal zone NEP

Within the past decade, much research has been increasingly devoted to the study of the carbon and nutrient cycles in the coastal zone, primarily to analyze its past, present, and future role in the global balance of carbon (e.g., Andersson et al., 2003; Gattuso et al., 1999; Kempe, 1995; Kleypas et al., 1999; Mackenzie et al., 1998a, 2001, 2002b; Milliman and Droxler, 1995; Rabouille et al., 2001; Smith, 1995; Smith and Wulff, 1999; Ver et al., 1999a, 1999b; Wollast, 1994, 1998; Liu et al., 2000; Feely et al., 2004). Among the growing literature on coastal zone carbon cycles, several attempts have been made to determine present-day NEP by direct or indirect methods and have arrived at widely disparate estimates (e.g., Gattuso et al., 1998; Mackenzie et al., 1998a, 2002a, 2002b; Rabouille et al., 2001; Ver et al., 1999b; Wollast, 1994, 1998). The following summary of these estimates and their derivation highlights the continuing difficulty and uncertainty in evaluating coastal zone NEP even for the present day, indirectly by assessing the inputs and outputs of the organic carbon or directly by obtaining estimates of primary production and total respiration. The resolution of this uncertainty is important to the understanding of the role of the coastal zone relative to the anthropogenically perturbed global C cycle.

In a classic paper presenting a model for the present-day organic carbon cycle, Wollast (1998) estimated global coastal zone NEP to be net autotrophic at the rate of $+200\times10^{12}$ moles C/yr. This would imply an atmospheric $CO_2$ residence time of approximately 250 years with respect to this flux alone, using the preindustrial atmospheric $CO_2$ content of about 50,000 x $10^{12}$ moles C. In Wollast's model, the proximal coastal zone, defined to include the macrophyte-dominated ecosystems, mangroves, marshes, and coral reefs, contributes 25% of the autotrophy ($50\times10^{12}$ moles C/yr ) while the distal zone, the area extending on the continental shelf to a depth of about 130 m, contributes the larger fraction of autotrophy at 75% ($150\times10^{12}$ moles C/yr). The calculated autotrophy apparently is fueled by the high fertility in the distal zone and a low respiration rate.

In Wollast's model (1998), coastal zone primary productivity is estimated at 230 g C $m^{-2}$ $yr^{-1}$ (or $500\times10^{12}$ moles C/yr over a total area of $26\times10^{12}$ $m^2$), and it is supported by significant fluxes of nutrients transferred from the intermediate depths of the open ocean to the shelf by physical exchange processes such as upwelling/onwelling or vertical mixing. This value is at the lower limit in the range of estimates of estuarine and proximal coastal zone primary productivity (230 to 300 g C $m^{-2}$ $yr^{-1}$, Rabouille et al., 2001, and references therein; Smith and Hollibaugh, 1993; Valiela, 1995; Ver et al., 1999a) but larger than the values adopted by some other authors for primary production in distal coastal margins (140 to 160 g C $m^{-2}$ $yr^{-1}$, Smith and Hollibaugh, 1993; Walsh, 1988; Wollast, 1998, and references therein). The primary production estimate of Wollast gives rise to a new production estimate in his model that is equivalent to 40% of the primary production (fraction $f = 0.40$) and is in very close agreement with the value estimated by Liu et al. (2000) of 34% ($f = 0.34$). In addition, Berger et al. (1987) obtained an empirical relationship between new production and primary production from an extensive compilation of literature data indicating that 40% of the primary production of 215 g C $m^{-2}$ $yr^{-1}$ on the shelf is new production ($f = 0.40$). A significant fraction of the production (37%) in Wollast's model is assumed to be exported to the surface open ocean and continental slope where it remineralizes and may upwell back to the coastal zone, ultimately providing the coastal ocean with a source of nutrients. For comparison, in the VERTEX study (Martin et al., 1987; Knauer, 1993), the export of organic carbon amounted to 17% of the primary production and in the East China Sea, Chen and Wang (1999) estimated that organic export amounted to 20% of the primary production. In the case of the SEEP project, the original estimate that only about 6.4% of the primary production is exported is an open question and very likely is an underestimate based on the work of Jahnke (1996) and DeMaster et al. (1994). In the European OMEX project, an average primary production of 200 g C $m^{-2}$ $yr^{-1}$ was obtained, of which 15% was exported to the open ocean by lateral advection, and the system was net autotrophic. Finally, Wollast's (1998, 2002) estimate of the exported organic carbon flux is in reasonable agreement with the analysis of Liu et al. (2000) for the global organic carbon balance. Wollast's calculated export rate is based on eight observational studies of different coastal zone environments.

The calculated net autotrophy in Wollast's (1998) model is based on a mass balance that is essentially driven by adoption of relatively high productivity and low respiration rates ($P >> R_{total}$), and a significant calculated rate of export of organic carbon to the open ocean. Although the estimated respiration at 60% of primary production is reasonable within the possible range, it is lower than the estimates of some other investigators (for example, Rabouille et al., 2001, and references therein; Smith and Hollibaugh, 1993). Finally, in Wollast's (1998) model, the heterotrophic respiration of riverine and atmospheric inputs of organic matter (Table 3 of Wollast, 1998) is not included in the model because these inputs represent only a negligible fraction of the high primary production in the global coastal margin. Wollast nevertheless recognizes that heterotrophic conditions in the coastal zone can be obtained if there is sufficient organic matter of riverine and atmospheric origin that can be intensively respired. Herein lies the one crux of the matter. Much of the riverine particulate organic carbon may be rapidly deposited in coastal marine sediments, although Smith and Hollibaugh argue that about 50% of the flux is remineralized in the coastal zone. In contrast, the riverine dissolved organic carbon appears to be relatively refractory and generally behaves conservatively in estuaries. A significant portion of it could very well be transported to the open ocean before it is respired. Note that the respiration of allocthonous organic matter in the coastal zone ($R_{heterotrophic}$) could contribute up to about $15\times10^{12}$ moles C/yr to the community respiration rate (Hedges et al., 1997; Smith and Hollibaugh, 1993). Thus, it could become a highly significant term in the evaluation of NEP where ($P - R_{autotrophic}$) is within an order of magnitude of this rate, such as seems to be the case in at least some proximal coastal margin ecosystems.

A recent compilation of available data and re-calculation of the global coastal margin NEP using an ecosystem approach (Gattuso et al., 1998) also shows that this region is presently net autotrophic at the rate of $+231\times10^{12}$ moles C/yr. Although this NEP estimate is slightly higher than that of Wollast (1998), it is similarly based on generally high primary production rates and low autotrophic respiration rates.

Rabouille et al. (2001) used a model of the coupled biogeochemical cycles of C-N-$O_2$ to calculate a coastal ocean NEP of $+20\times10^{12}$ moles C/yr, that is not only an order of magnitude lower than estimates of Wollast and Gattuso but is a composite of the proximal zone that is net heterotrophic at $-8\times10^{12}$ moles C/yr and the distal zone that is net autotrophic at $+28\times10^{12}$ moles C/yr. In this model, net autotrophy in the distal zone is attributed primarily to the modestly large particulate organic carbon (POC) export to the open ocean (Bacon et al., 1994; Falkowski et al., 1994), estimated at $40\times10^{12}$ moles C/yr, about 12.5% of the estimated primary production of $320\times10^{12}$ moles C/yr in the distal coastal zone. Finally, Ducklow and McCallister (Chapter 9) estimated recently that the present-day global coastal ocean is net autotrophic based on estimates of inputs and outputs of organic carbon by $+175\times10^{12}$ moles C/yr.

Based on the discussion above, it is probably premature to draw strong conclusions concerning the NEP of the present-day global coastal margin ocean. The most likely situation is that the composite of large bays, the open water part of estuaries, deltas, inland seas, and salt marshes of the proximal coastal margin are slightly net heterotrophic whereas the global open continental shelf is net autotrophic in large part due to export of both dissolved and particulate organic carbon offshore to continental slope and the open ocean. The resolution of the problem to a large extent lies in the acquisition of substantially more and accurate data on the primary productivity and total respiration of coastal margin ecosystems. Finally, it should be pointed out that although most investigators conclude that the present-day ocean as a whole is net heterotrophic, the degree of heterotrophy is not well established. Ducklow and McAllister (Chapter 9) concluded that the present-day open ocean over the entire water column is net heterotrophic by $-200 \times 10^{12}$ mole C/yr. This value is close to their own estimate of the net autotrophy of the global coastal margin, cited above, and to the value of Gattuso et al. (1998). If these values are reasonably correct, then the ocean as a whole is in remarkable balance with relationship to organic production and respiration!

## 5. Model calculations and ecosystem NEP*

The *T*errestrial-*O*cean-a*T*mosphere *E*cosystem *M*odel (*TOTEM;* e.g., Mackenzie et al., 1998a, 2002a, 2002b, 2004; Ver, 1998; Ver et al., 1999a, 1999b) has been used to estimate coastal zone NEP* changes over the time course of the past 300 years with projections into the future. The starting point of the analysis is the year 1700, at which time the coupled biogeochemical cycles of C-N-P in the land-ocean-atmosphere-sediment domains of *TOTEM* are assumed to be in a quasi-steady state condition. Because the organic carbon mass in the coastal zone does not stay constant during the numerical simulations of the model, net ecosystem productivity is defined in terms of equation (4), where NEP* is the net ecosystem metabolism under non-steady state conditions. We recognize the fact that some investigators use NEP for both steady state and non-steady state conditions and may consider the distinction between NEP and NEP* a moot point. However, for modeling purposes, it is important to emphasize that the coastal ocean reservoir masses are changing significantly on decadal to centurial time scales and the use of NEP* enables us to highlight this point. NEP* is essentially equivalent to the term for non-steady state net biome production (NBP) as used by terrestrial ecologists for changes in the size of the terrestrial carbon pool and thus non-zero NBP. Furthermore, it should be emphasized that we employ NEP* in the following model calculations for the coastal zone because there is no *a priori* reason to assume that the coastal zone organic carbon cycle is in steady state when viewed on the time-scale of decades to centuries. Indeed, on a short time scale, direct observations of *P* and *R* may be misleading in terms of interpretation of the trophic status of a system that is evolving with time. In addition, because we know that dissolved organic carbon (DOC) in the ocean has a mean age of about 1000 years (Bauer et al., 1992) and is being stored in certain regions of the ocean (Church et al., 2002), it is unlikely that the organic carbon cycle in the ocean is in a steady state, although whether the situation is a long-term or transient phenomenon is uncertain.

The starting NEP* of the model is taken as $-7\times10^{12}$ moles C/yr, corresponding to net heterotrophy of the coastal zone, following the analysis of Smith and Hollibaugh (1993). It should be emphasized that this is the initial condition for the year 1700 and is not today's situation. Also this model starting condition is not well known; however, regardless of the starting condition, the trends discussed below are probably robust. Numerical calculations using *TOTEM* of the coastal zone system response to 300 years of human-induced perturbations using equation (4) show that the global coastal zone has tended towards increased autotrophy since the late 1990s, with a future projection of NEP* approaching $+18\times10^{12}$ moles C/yr in the year 2100 from an initial value of $-7\times10^{12}$ moles C/yr (Figure 7.4). Notably the most rapid increase toward greater autotrophy occurred in the last 50 years paralleling the exponential increase in human activities adjacent to the coastal margins. Clearly, the magnitude of the model-calculated change in NEP* is small relative to the magnitude and uncertainties in the component fluxes, and one could conclude that global coastal zone NEP* (or NEP for that matter) since the quasi–steady state year 1700 has undergone only modest change despite the large additions to this region of terrestrial organic matter and inorganic N and P from human activities.

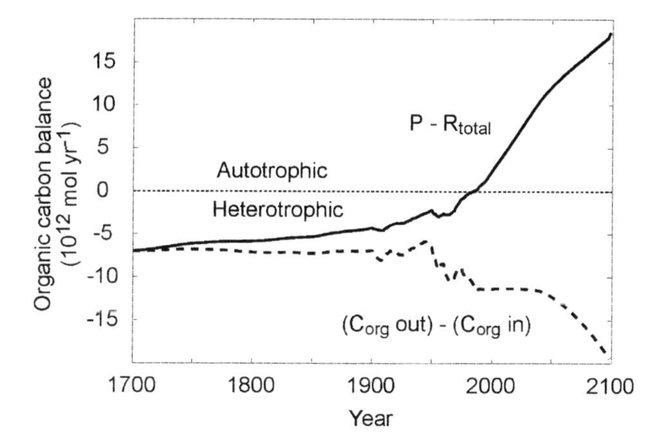


Figure 7.4 Time course of change of the organic carbon balance in the coastal ocean expressed as NEP* and NEP based on a mass balance approach (equations 2–4) owing to sustained perturbations from human activities from year 1700 to 2000 and projected to 2100 under the Business-as-usual scenario of Ver et al. (1999), in units of $10^{12}$ moles C/year. In a general non-steady state case, the algebraic sum of the two curves is the organic carbon rate of change in the coastal zone, $dC_{org}/dt$ or $\Delta C_{org}$ in eqs. (2) and (4). The coastal zone has been heterotrophic for much of the past 300 years and losing $C_{org}$, and future trends indicate decreasing heterotrophy. See also Figure 7.8.

The future projected autotrophy is driven by the significant increase in the riverine input of inorganic nutrients (as documented for example by Caraco, 1995; Meybeck, 2001; Meybeck and Ragu, 1995; Figure 7.2) that stimulate *in situ* production that increasingly exceeds remineralization of organic matter. The flux of organic matter imported from land will also significantly increase, but the increase in organic matter being remineralized would be progressively smaller relative to the increase in new production. However, if the future projections of nutrient input and production were spurious, or if the reservoir of organic carbon were in a steady state (equation 3), model calculations show that the coastal zone would tend toward increasing net heterotrophy as opposed to net autotrophy. In addition, if a steady state were sustained, model sensitivity analyses show that regardless of the trophic status at initial state, the NEP of the modeled coastal margin system under persistent anthropogenic perturbation tends toward increasing heterotrophy (Ver et al., 1999b), consistent with findings by Rabouille et al. (2001).

The importance of the nutrient N and P inputs from land and their role in the coupled cycling of C, N, and P, hence the determination of the trophic status of the coastal margin, can be demonstrated by the following consideration. As human land-use activities increase, the rates of recycling of the soil humus, humic material of an average molar C:N:P ratio of 140:7:1 is mineralized. The remobilized N and P can ideally support growth of land plants of an average molar C:N:P ratio of 510:4:1. With P as the limiting nutrient, about 370 moles (510 less 140) of the required C are biotically sequestered directly from the atmosphere, while about 3 moles of excess N are released to the continental soil-water reservoir. A fraction of this N is lost to the atmosphere through denitrification, and a significant fraction of the remainder is transported to the coastal zone by rivers. Although there are varying estimates of molar C:N:P ratios in land plants and soil organic matter (Lerman et al., 2004), the preceding argument for uptake of $CO_2$ from the atmosphere and release of excess N from humus holds in general when the ratios are $(C{:}P)_{plants} > (C{:}P)_{humus}$ and $(N{:}P)_{plants} < (N{:}P)_{humus}$.

*TOTEM* model results show that the mass of N released through land-use activities dominates other anthropogenic sources of N on land, such as agricultural fertilizer and atmospheric deposition (Figure 7.5 and Table 1) (Mackenzie et al., 2002a). In the year 2000, land use activities accounted for 70% of the anthropogenically mobilized N (about $7.5{\times}10^{12}$ moles N/yr) while fertilizer application and atmospheric deposition each accounted for only about 15% ($1.5{\times}10^{12}$ moles N/yr). N and P exported from land into the coastal zone become available there for new production at an average molar C:N:P ratio of 106:16:1.

The suggestion by Wollast (1998) that the upwelling flux of nutrient N and P from the intermediate ocean to the coastal zone sustains the high primary productivity of the latter region, hence the autotrophy in the distal coastal margin, can also be observed in the numerical simulations using *TOTEM* (Mackenzie et al., 2002a, 2002b; Ver et al., 1999a, 1999b). Model simulations show that employing N fluxes related to physical processes of exchange at coastal margins on the order of Wollast's estimates ($27{\times}10^{12}$ moles N/yr) will lead in the present day to higher coastal zone productivity, enhanced organic matter accumulation in the coastal zone, enhanced coastal zone denitrification flux, primarily from the sediments, greater offshore transport of organic matter, and a tendency towards more autotrophic conditions. Model results also show that the import of nutrient N and P from upwelling could potentially support 30% of total production in the coastal zone. However, it should also be recognized that this exchange flux supplies sufficient DIC to satisfy or exceed the C:N:P Redfield-ratio requirements of the upwelling N and P for coastal primary production, thus conceivably no additional uptake of DIC might arise from this process. Further changes in land-use practices or in ocean circulation patterns in the future are likely to affect the nutrient N and P supply to the coastal zone by fluvial flow or coastal upwelling, thereby affecting the coastal carbon cycle and carbon balance through the C–N–P coupling.

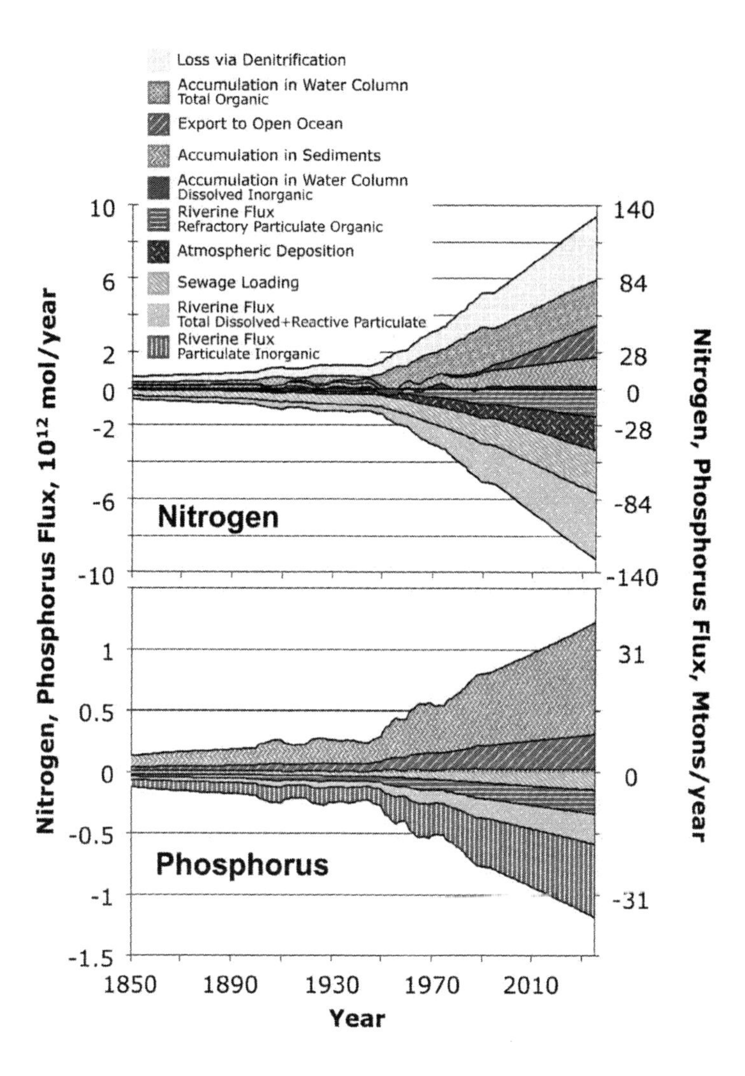


Figure 7.5 Model-calculated partitioning of the human-induced perturbation fluxes in the global coastal margin of (A) nitrogen and (B) phosphorus for the period since 1850 to present (2000) and projected to 2035 under the Business-as-usual scenario, in units of $10^{12}$ moles/year and Mtons/year (1 Mton = $10^6$ ton = $10^{12}$ g). The anthropogenic sources are plotted on the (–) side and the resulting accumulations and enhanced export fluxes are plotted on the (+) side. Model results from Mackenzie et al. (2002).

TABLE 1

Historical and future increment of change in fluxes of carbon, nitrogen, and phosphorus for the global coastal zone (see Figures 7.2 and 7.5 for temporal trends and absolute flux values).‡

| | Initial | ΔCarbon Flux, Mtons/yr | | | |
|---|---|---|---|---|---|
| | 1700 | 1850 | 1950 | 2000 | 2035 |
| INPUT FLUXES: | | | | | |
| Riverine Flux (total dissolved+reactive particulate) | 695 | 24 | 32 | 210 | 287 |
| Riverine Flux (refractory particulate organic) | 96 | 3 | 4 | 29 | 40 |
| Riverine Flux (particulate inorganic) | 180 | 6 | 8 | 54 | 74 |
| Sewage Loading | 0 | 23 | 47 | 113 | 163 |
| Atmospheric Deposition | — | — | — | — | — |
| OUTPUT FLUXES: | | | | | |
| Denitrification | — | — | — | — | — |
| Export to Open Ocean | 5990 | 34 | 101 | 268 | 472 |
| Accumulation in Water Column (organic)* | 0 | 2 | 10 | 24 | 27 |
| Accumulation in Sediments | 108 | 30 | 58 | 216 | 328 |
| Accumulation in Water Column (inorganic) | 0 | 2 | 7 | 50 | 60 |
| | Initial | ΔNitrogen Flux, Mtons/yr | | | |
| | 1700 | 1850 | 1950 | 2000 | 2035 |
| INPUT FLUXES: | | | | | |
| Riverine Flux (total dissolved+reactive particulate) | 25 | 3 | 5 | 32 | 51 |
| Riverine Flux (refractory particulate organic) | 11 | 1 | 2 | 13 | 22 |
| Riverine Flux (particulate inorganic) | 0.6 | 0.1 | 0.1 | 0.7 | 1 |
| Sewage Loading | 0 | 5 | 9 | 22 | 32 |
| Atmospheric Deposition | 0 | 0 | 2.6 | 12 | 25 |
| OUTPUT FLUXES: | | | | | |
| Denitrification | 67 | 4 | 9 | 30 | 49 |
| Export to Open Ocean | 163 | 0.8 | 2 | 6 | 24 |
| Accumulation in Water Column (organic)* | 0 | 2 | 9 | 30 | 35 |
| Accumulation in Sediments | 3 | 2 | 4 | 14 | 22 |
| Accumulation in Water Column (inorganic) | 0 | -0.1 | -4 | 1 | 2 |
| | Initial | ΔPhosphorus Flux, Mtons/yr | | | |
| | 1700 | 1850 | 1950 | 2000 | 2035 |
| INPUT FLUXES: | | | | | |
| Riverine Flux (total dissolved+reactive particulate) | 5 | 0.8 | 2 | 5 | 8 |
| Riverine Flux (refractory particulate organic) | 4 | 0.6 | 1 | 4 | 6 |
| Riverine Flux (particulate inorganic) | 12 | 2 | 4 | 13 | 18 |
| Sewage Loading | 0 | 0.7 | 1 | 3 | 5 |
| Atmospheric Deposition | — | — | — | — | — |

| | | | | | |
|---|---|---|---|---|---|
| OUTPUT FLUXES: | | | | | |
| Denitrification | — | — | — | — | — |
| Export to Open Ocean | 17 | 1 | 2 | 6 | 9 |
| Accumulation in Water Column (organic)* | 0 | 0.05 | 0.3 | 0.6 | 0.7 |
| Accumulation in Sediments | 18 | 3 | 6 | 20 | 28 |
| Accumulation in Water Column (inorganic) | 0 | 0 | -0.3 | -0.8 | -0.9 |

‡ After Mackenzie et al.(2002)

* Net increase in coastal biomass and organic matter in the water column

Consistent with the conclusions from the numerical simulations of *TOTEM* for the global ocean, results from Ianson and Allen's (2002) regional model suggest that the principal role of coastal upwelling systems in the global carbon budget is the ventilation of intermediate-depth oceanic DIC to the coastal zone. The upwelled waters are advected back to the open ocean surface layer bringing with them most of the upwelled DIC, though relatively deplete in dissolved inorganic nitrogen (DIN). The mass of organic carbon exported in this process is small, calculated to be about $2\times10^6$ g C/yr ($1.7\times10^5$ mol C/yr) per 1 m of coastline in the upwelling region. These authors concluded that upwelling regions are much smaller sinks of atmospheric $CO_2$ than the summer season predicts because the strong biological drawdown of $CO_2$ during the summer is roughly balanced by gas evasion during the winter.

If total gross respiration is favored over gross production and the coastal margin is net heterotrophic (that is, $P - R_{total} < 0$), the structure and dynamics of the global cycle suggest that reservoirs downcycle from the coastal margin are likely to be sinks for the C, N, and P produced by heterotrophy. Consider that the net excess dissolved inorganic carbon and nutrients (DIC, DIN, and DIP) produced by a heterotrophic system must be removed in order to prevent their unlimited accumulation in the coastal-zone water column. The mechanisms for their removal involve C, N, and P export to the sediments in inorganic forms, release of gaseous $CO_2$ and N-species to the atmosphere, and inorganic C, N, and P leaking to the open ocean (Figure 7.5 shows the sources and fate of anthropogenic N and P in the coastal zone). However, if gross production is favored over gross respiration and the coastal region is autotrophic, the biogeochemical implications for reservoirs downcycle of the coastal margin are different. An autotrophic coastal margin implies that the organic matter of the system may increase either as biomass or accumulation in the sediments or it may be physically exported to adjacent downcycle oceanic reservoirs. Because the magnitude of both hypothetical scenarios of either heterotrophy or autotrophy is small, the rate of change in the export fluxes from steady-state is slow and would remain below the resolution of most field observations (Rabouille et al., 2001).

In summary, although the present-day trophic status of individual coastal margin environments may vary from net autotrophic to net heterotrophic and the heterotrophic state is likely to be pronounced nearer to the land, in the proximal parts of the coastal margins, the future state of the organic carbon balance in the coastal zone will depend to a considerable extent on the changes in magnitude of input flux of organic matter relative to that of inorganic nutrients to coastal environments. Both of these fluxes are strongly influenced by human activities on land.

## 6. The present-day inorganic C cycle in the global coastal ocean

The shallow-water marine inorganic carbon cycle constitutes a significant part of the global marine inorganic carbon cycle (Fig. 7.6). Approximately 25% of calcium carbonate produced globally is produced within the global coastal ocean and almost 50% of calcium carbonate that accumulates in global marine sediment accumulates within this region (Figure 7.6) (Milliman, 1993; Wollast, 1994, 1998). About half of this accumulation is in regions of coral reefs. According to Milliman (1993), accumulation of calcium carbonate minerals in the coastal ocean is currently remarkably high owing to the significant rise in sea level since the Last Glacial Maximum about 18 ka ago and the subsequent expansion of shallow-water depositional environments. Thus, the global marine inorganic carbon cycle currently could be in a non-steady state, where more calcium carbonate is deposited in marine sediments than is added to the ocean via river input and basalt–seawater interactions (Figure 7.6). The marine inorganic carbon cycle is strongly coupled to the marine organic carbon cycle through primary production and remineralization of organic matter (Figure 7.3). Consequently, direct alterations to either one of these cycles owing to natural or anthropogenic factors are likely to affect the other.

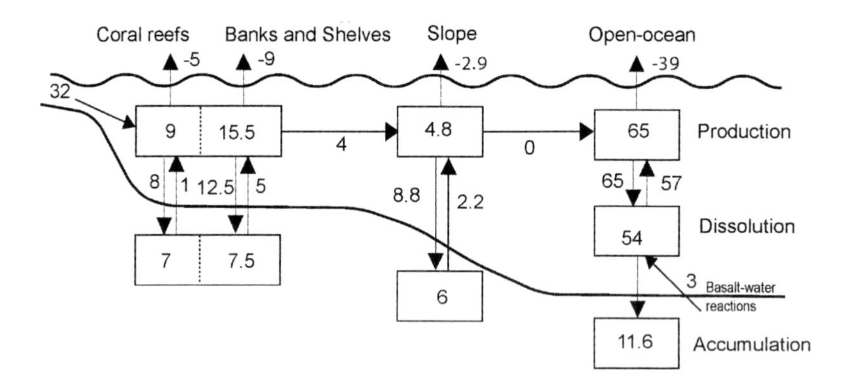


Figure 7.6 Global marine inorganic carbon cycle (non-steady state). Diagram modified from Wollast (1994), based on Morse and Mackenzie (1990) and Milliman (1993). The cycle is not in a steady state because of the assumption that deposition of calcium carbonate currently is unusually high owing to the significant rise in sea level since the Last Glacial Maximum and the subsequent expansion of shallow water depositional environments (Milliman, 1993). Negative values indicate a net flux of $CO_2$ from the surface ocean to the atmosphere.

From experimental evidence and geochemical modeling, it has been suggested that the saturation state of surface ocean waters with respect to carbonate minerals will decline during the twenty-first century owing to increased invasion of atmospheric $CO_2$. Although data are sparse globally, monthly observations from the Hawaii Ocean Time series station (HOTs, located in the North Pacific subtropical gyre) between 1988 and 2000 indicate a consistent annual trend of increasing surface water total DIC and decreasing carbonate saturation state (Hawaii Ocean

Time series, 2003; Winn et al., 1998). This situation has been observed in other regions of the ocean (e.g., Feely et al., 2004). Decreasing surface water carbonate saturation state could negatively affect the ability of calcareous organisms such as corals, coralline algae, coccolithophorids, and other taxa to produce skeletons, shells, and tests out of calcium carbonate. The difficulty posed by lower saturation states to the production of calcium carbonate could result in calcareous organisms and structures being weaker and more vulnerable to environmental stress and physical and biological erosion. Consequently, the role and function of calcareous ecosystems and communities may be altered as a consequence of future environmental conditions (Andersson et al., 2003; Gattuso et al., 1999; Kleypas et al., 1999, 2001; Langdon et al., 2000; Leclercq et al., 2000, 2002; Mackenzie et al., 2000; Riebesell et al., 2000).

At present, calcium carbonate is almost exclusively produced by calcareous organisms and only minor quantities are produced abiotically as cements within sediments or precipitated as whitings from the water column (e.g., Morse and Mackenzie, 1990; Milliman, 1993; Wollast, 1994; Iglesias-Rodriguez et al., 2002). Decreased carbonate saturation state and subsequent decreased production of calcium carbonate could significantly alter the global marine inorganic carbon cycle and consequently also affect the marine organic carbon cycle. Decreased calcification also implies a decreased flux of $CO_2$ to the atmosphere owing to this process (Ware et al., 1992; Frankignoulle et al., 1994) and could act as a minor negative feedback to increasing atmospheric $CO_2$ (Zondervan et al., 2001). However, the magnitude of this feedback is insignificant relative to the total invasion of anthropogenic $CO_2$ into the surface ocean.

Ultimately, decreasing surface and pore water carbonate saturation state may cause increased dissolution of carbonate minerals in the water column and within the pore water–sediment system. Such dissolution, in particular of metastable carbonate minerals such as high magnesian calcite, which is unstable relative to calcite and aragonite, may act as a buffer to neutralize anthropogenic $CO_2$ and prevent any negative effects on calcareous organisms and ecosystems (Garrels and Mackenzie, 1980; Barnes and Cuff, 2000; Halley and Yates, 2000). It is possible that the rate of carbonate dissolution could equal the rate of calcification once atmospheric $CO_2$ concentration reached double pre-industrial levels (Halley and Yates, 2000). However, recent model results indicate that such dissolution will not produce sufficient alkalinity to buffer the surface water from increasing atmospheric $CO_2$ and consequently, anthropogenically-induced changes in coastal ocean saturation state will not be restored by the dissolution of metastable carbonate minerals (Andersson et al., 2003). Thus calcification by calcareous marine organisms and the development of carbonate reefs probably will be negatively affected as a consequence of rising anthropogenic $CO_2$ and lowering of the saturation state of seawater with respect to carbonate minerals (Figure 7.7) (Gattuso et al., 1999; Kleypas et al., 1999; Mackenzie et al., 2000; Langdon et al., 2000; Leclercq et al., 2002).

During early diagenetic modifications on the seafloor, dissolution of carbonate minerals follows a sequence based on mineral thermodynamic stability, progressively leading to removal of the more soluble phases until the stable phases remain (Schmalz and Chave, 1963; Neumann, 1965; Wollast et al., 1980). Consequently, increased dissolution of carbonate minerals owing to anthropogenically-induced changes in carbonate saturation state of the coastal ocean could affect the average $CaCO_3$ composition (its Mg content) and rates of precipitation of carbonate cements in contemporary shallow-water marine sediments (Andersson et al., 2003). It should be pointed out that at present, the extent of carbonate dissolution is mainly controlled by microbial remineralization of organic matter producing $CO_2$ rather than the carbonate reaction kinetics (Morse and Mackenzie, 1990). Increased transport and deposition of organic matter to the sediments of the coastal zone (Mackenzie et al., 1993; Meybeck, 1982) and/or subsequent changes to the $NEP^*$ could therefore have a significant effect on the carbonate content and composition of marine sediments.

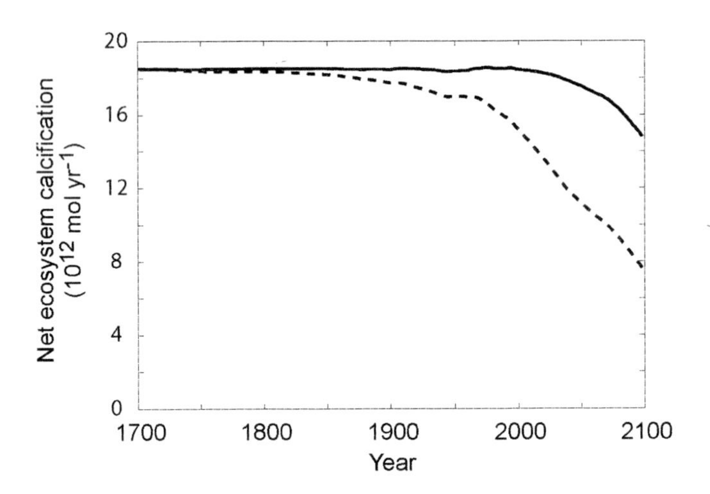


Figure 7.7 Changes in net ecosystem calcification ($CaCO_3$ production – dissolution) in the coastal margin owing to increasing atmospheric $CO_2$ and increased deposition and remineralization of organic matter. Biogenic calcification is related to surface water carbonate saturation state either by a linear (solid line) or a curvilinear relationship (dashed line) saturation dependence (Andersson et al., 2003; Gattuso et al., 1999).

## 7. Carbon flows between the atmosphere and the coastal zone

Net flux of $CO_2$ across the air-seawater interface is a function of oceanic carbonate chemistry, organic metabolism, and atmospheric $CO_2$ concentration (e.g., Morse and Mackenzie, 1990; Andersson and Mackenzie, 2004). Increased autotrophy in the coastal ocean is a sink of DIC and increased heterotrophy is a source of DIC. Because $CO_2$ is a component of DIC, its increase or decrease in coastal ocean waters will affect the pressure gradient across the air-sea interface and the subsequent flux of $CO_2$ between the atmosphere and the surface water. Similarly, production of calcium carbonate minerals produces $CO_2$ in the water column and decreases the DIC concentration, whereas calcium carbonate dissolution consumes $CO_2$ and increases DIC. Consequently, changes in coastal ocean trophic status and/or net production of calcium carbonate minerals will affect the response of the coastal ocean to increasing atmospheric $CO_2$ from human perturbations, modifying the magnitude of the invasion flux of anthropogenic $CO_2$.

Air-sea gas exchange of $CO_2$ between coastal zone waters and the atmosphere on a global scale has been modeled based on the processes described above (Mackenzie et al., 2002a, 2002b; Ver et al., 1999a, 1999b; Andersson and Mackenzie, 2004; see also Figure 7.3) using the following equation:

$$F = -\Psi \cdot \mathrm{NEC} + \mathrm{NEP}^* + F_{\mathrm{DIC,\,out}} - F_{\mathrm{DIC,\,in}} + dC_{\mathrm{DIC}}/dt \quad (5)$$

where $F$ is the flux of carbon in mol C $yr^{-1}$, $\Psi$ ($\Psi > 0$) is the fraction of $CO_2$ released to the atmosphere for each mol of $CaCO_3$ precipitated inorganically or biologically in the coastal zone (Frankignoulle et al., 1994), NEC is the net ecosystem calcification rate (Fig. 7.7), and $NEP^*$ is the net ecosystem metabolism or production rate, as defined in equation (4). At the beginning of the Industrial Age taken as the year 1700, NEC > 0 (Fig. 7.7) and NEP* < 0 (Figs. 7.4, 7.8). The last term in equation (5), $dC_{\mathrm{DIC}}/dt$, represents the flux of carbon (mol/yr) between the surface water and the atmosphere expressed as the time rate of change of total dissolved inorganic carbon content (DIC) of the surface water as a function of changes in the $CO_2$ content of the atmosphere ($C_{\mathrm{atm}}$). This term in the model calculations is a modification of the standard Bacastow-Keeling and Revelle-Munk equations (Bacastow and Keeling, 1973; Revelle and Munk, 1977; see also Zeebe and Wolf-Gladrow, 2001):

$$R = (C_{\mathrm{atm},\,t}/C_{\mathrm{atm},\,0} - 1)/(C_{\mathrm{DIC},\,t}/C_{\mathrm{DIC},\,0} - 1) \quad (6)$$

$$R \approx R_0 + d\times(C_{\mathrm{atm},\,t}/C_{\mathrm{atm},\,0} - 1) \quad (7)$$

where $C_{\mathrm{atm}}$ is $CO_2$ concentration in the atmosphere and $C_{\mathrm{DIC}}$ is the dissolved inorganic carbon concentration in surface ocean water, subscript 0 denotes the initial value and subscript $t$ a value at a later time $t$. Equation (7) with constants $R_0 = 9$ and $d = 4$ is Revelle and Munk's (1977) approximation to the curve of $R$ calculated by Bacastow and Keeling (1973) for an average surface ocean water of total alkalinity $2.435\times10^{-3}$ molequivalent/liter, temperature 19.59°C, chlorinity 19.24 per mil, and initial pH = 8.271. With $R_0 = 9$, as taken in this study, the buffer mechanism of seawater causes a fractional rise of $CO_2$ in coastal zone surface seawater that is one ninth of the increase of $CO_2$ in the atmosphere. The term $dC_{\mathrm{DIC}}/dt$ in (5) is obtained by differentiation of (6) and (7), giving a simpler approximate relationship (8a) and a more complete relationship (8b) with a second-order term:

$$\frac{dC_{\mathrm{DIC}}}{dt} = \frac{C_{\mathrm{DIC},\,0}}{R_0 C_{\mathrm{atm},\,0} + d\times(C_{\mathrm{atm},\,t} - C_{\mathrm{atm},\,0})} \times \frac{dC_{\mathrm{atm}}}{dt} \quad (8a)$$

$$\frac{dC_{\text{DIC}}}{dt} = \left\{ \frac{C_{\text{DIC},0}}{R_0 C_{\text{atm},0} + d \cdot (C_{\text{atm},t} - C_{\text{atm},0})} - \frac{d \times C_{\text{DIC},0} \times (C_{\text{atm},t} - C_{\text{atm},0})}{[R_0 C_{\text{atm},0} + d \times (C_{\text{atm},t} - C_{\text{atm},0})]^2} \right\} \times \frac{dC_{\text{atm}}}{dt} \quad (8b)$$

Model results (Mackenzie et al., 2002a, 2002b; Ver et al., 1999a, 1999b; Andersson and Mackenzie, 2004) indicate that if the organic carbon reservoir of the global coastal ocean is in a steady state (see previous discussion), this region can be a net heterotrophic system due to the organic carbon imbalance between $P$ and $R_{\text{total}}$, in which $R_{\text{total}}$ is greater than $P$ and consequently $CO_2$ potentially can evade coastal zone waters, while at the same time because of the chemical solubility pump, global coastal waters are actually in rough equilibrium or a slight sink for anthropogenic $CO_2$. The net effect of sustained heterotrophy and increased carbonate precipitation is a reduction in the potential sink strength of the coastal ocean for anthropogenic $CO_2$. This conclusion implies that were it not for the accumulation of anthropogenic $CO_2$ in the atmosphere, the global coastal oceans would be sources of $CO_2$ to the atmosphere, reflecting an imbalance favoring gross respiration over gross photosynthesis and calcium carbonate accumulation. However, if the organic carbon reservoir of the coastal ocean were not in a steady state, model results suggest that gross primary production is increasingly favored over remineralization, leading to increasingly autotrophic conditions of this region. Consequently, increased autotrophy (Figure 7.4) and decreased net calcification (Figure 7.7), as well as increasing atmospheric $CO_2$, imply an increased uptake of anthropogenic $CO_2$ in the coastal ocean. Numerical simulations suggest that the coastal ocean has served as a net source of $CO_2$ to the atmosphere throughout most of the past 300 years, but recently has switched or soon will do so to a net sink of $CO_2$ because of rising atmospheric $CO_2$ and increasing inorganic nutrient load from land that stimulates new production (Figure 7.8).

Several global and regional coastal oceanic $P_{CO_2}$ surveys have attempted to resolve the issue of whether the coastal margin is a source or sink for atmospheric $CO_2$ using detailed analysis of $CO_2$ partial pressure in surface waters (Table 2) (e.g., Borges and Frankignoulle, 2002a; Frankignoulle and Borges, 2001; Kempe and Pegler, 1991; Takahashi et al., 2002; Tsunogai et al., 1999). These observational surveys highlight the large spatial and temporal variability of the distribution of $P_{CO_2}$ and the factors that control its variation in coastal surface waters, thus the potential for the global coastal margin to either intensify or weaken the oceanic $CO_2$ sink-strength. Analysis of more than 10 years of direct measurements of $CO_2$ concentrations in the surface waters of the North Atlantic European Shelf (Galician Sea, Gulf of Biscay, Armorican Sea, Celtic Sea, English Channel, and North Sea) show that this coastal region is currently a $CO_2$ sink of +21 to +35 g C $m^{-2}$ $yr^{-1}$ (Frankignoulle and Borges, 2001; flux from the atmosphere into seawater is positive, from water to the atmosphere negative, as shown in Figs. 7.1, 7.3, and 7.6). measurements of $P_{CO_2}$ fluxes in the East China Sea coastal margin show a net influx of $CO_2$ of +14 to +34 g C $m^{-2}$ $yr^{-1}$ (Chen and Wang, 1999; Tsunogai et al., 1999; Wang et al., 2000). Similarly, measurements from certain parts of the U.S. continental shelf such as the Mid-Atlantic Bight (DeGrandpre et al., 2002) and the shelf offshore New Jersey (Boehme et al., 1998) indicate that these regions act as sinks of atmospheric $CO_2$. On the contrary, a recent study from the South Atlantic Bight (Cai et al., 2003) indicates that this region actually acts as a significant source of $CO_2$ at –30 g C $m^{-2}$ $yr^{-1}$ (Table 2).

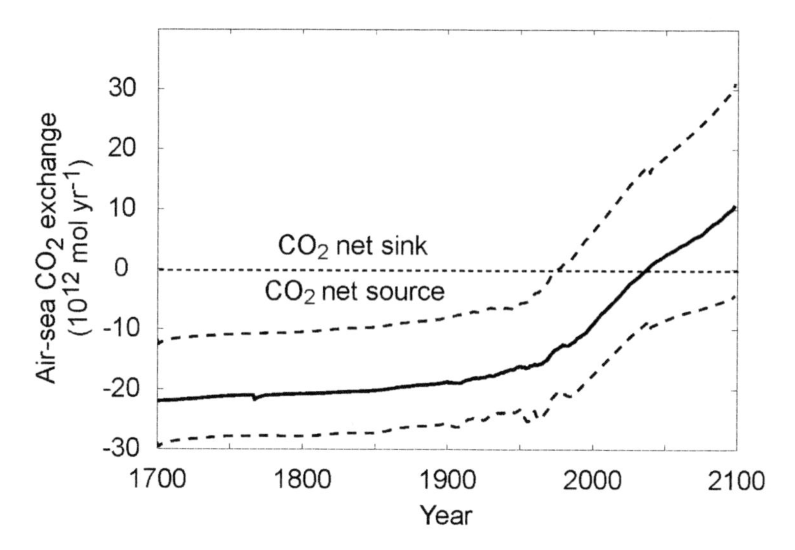


Figure 7.8 Model calculated air-sea $CO_2$ exchange between 1700 and 2100. The results indicate that the coastal ocean has served as a net source of $CO_2$ to the atmosphere for most of the past 300 years, but recently has or soon will change to become a net sink of atmospheric $CO_2$. The dashed lines represent the range of results achieved including a 50% uncertainty with respect to the organic carbon balance (Smith and Hollibaugh, 1993) and the minimum and maximum estimates of net calcification found in the literature (Morse and Mackenzie, 1990; Milliman, 1993).

Coastal upwelling areas are sites of oversaturation of $CO_2$, hence potentially net sources of $CO_2$ to the atmosphere (Goyet et al., 1998; Kortzinger et al., 1997; Simpson and Zirino, 1980). Goyet et al. (1998), for example, reported an annually integrated flux of –11 g C $m^{-2}$ $yr^{-1}$ from the upwelling system off the coast of Oman. However, other studies have found strong spatial variability in the distribution of surface $P_{CO_2}$ off the upwelling coasts of the Iberian Peninsula and Oregon (Borges and Frankignoulle, 2002b; Perez et al., 1999; van Green et al., 2000). Extrapolation of opposing flux estimates to the global upwelling area of 1% of the ocean surface or $3.6 \times 10^6$ $km^2$ (Schlesinger, 1997) suggests either a global sink of $+8.3 \times 10^{12}$ moles C/yr or a global source of $-3.25 \times 10^{12}$ moles C/yr. Although the $CO_2$ source or sink terms owing to upwelling are small relative to the magnitude of the continental shelf pump that removes $CO_2$ from the atmosphere proposed as by Tsunogai et al. ($83 \times 10^{12}$ moles C/yr, 1999), their significance in the carbon budget on the regional scale cannot be overlooked.

TABLE 2.2

Estimates of measured air-sea $CO_2$ exchange from worldwide shallow-water locations (estuaries, reefs, shelves). Negative values denote $CO_2$ flux from ocean to atmosphere; positive values denote flux from atmosphere to ocean.

| Location/Region | Net $CO_2$ flux ( g C $m^{-2}$ $yr^{-1}$) | Reference |
|---|---|---|
| Hog Reef flat, Bermuda | -14.4 | Bates 2001 |
| New Jersey | +5.2 to +10.1 | Boehme *et al.* 1998 |
| Galician Coast | +7.9 to +14 | Borges and Frankignoulle 2001 |
| Scheldt Estuarine plume | -13.2 to -22.8 | Borges and Frankignoulle 2002 |
| South Atlantic Bight (SAB) | -30 | Cai *et al.* 2003 |
| Mid Atlantic Bight (MAB) | +12 | DeGrandpre *et al.* 2002 |
| Gulf of Calvi | -28 | Frankignoulle 1988 |
| Gulf of Biscay | +21 to +34.6 | Frankignoulle and Borges 2001 |
| European estuaries (Elbe, Ems, Rhine, Scheldt, Tamar, Thames, Gironde, Duoro, Sado) | -438 to -3330 | Frankignoulle *et al.* 1998 |
| Moorea, French polynesia | -6.6 | Gattuso *et al.* 1993 |
| Northern Arabian Sea | -5.5 | Goyet *et al.* 1998 |
| Cape Perpetua | +87.6 | Hales *et al.* 2003 |
| North Sea | +16.2 | Kempe and Pegler 1991 |
| Baltic Sea | +10.8 | Thomas and Schneider 1999 |
| East China Sea | +35 | Tsunogai *et al.* 1999 |
| East China Sea | +14.4 to +33.6 | Wang *et al.* 2000 |

Although a substantial fraction of the data currently available from different coastal regions suggest that these regions act as significant sinks of atmospheric $CO_2$, the question whether the global coastal ocean currently serves as a net sink or source is far from certain. Based on numerical modeling, the global coastal ocean was a source of $CO_2$ in pre-industrial and most of industrial time since the year 1700, but the trend is certainly towards a net sink because of the rapid increase in atmospheric $CO_2$ owing to anthropogenic burning of fossil fuels and land-use changes. However, because the magnitude of the initial flux between the atmosphere and coastal ocean waters for pre-industrial time is not well constrained, the timing of the transition from a net source to a net sink is difficult to ascertain.

## 8. Summary and prognosis

The purpose of this chapter is to describe and discuss the response of the global coastal ocean, particularly its trophic status and inorganic carbon system, to changes in boundary exchanges of C, N, and P under stresses from major human perturbations. We acknowledge that because of the paucity of data and limited knowledge of the complex physical and biogeochemical processes at the boundary between the open ocean and the coastal margin, an accurate and comprehensive evaluation of coastal margin NEP* over recent geologic time and into the future is probably not achievable at this stage in our knowledge. In addition, the *TOTEM* calculations presented here assume that the system was net heterotrophic in immediate preindustrial time and because of this feature and the fact that calcification results in emissions of $CO_2$ to the atmosphere, the coastal zone at that time was a net source of $CO_2$ to the atmosphere. However, evaluation of the effects of human-induced perturbations on land, which induce changes in such biogeochemically important processes in the coastal zone as primary production, consumption and respiration of organic matter, and nutrient recycling, relative to the quasi-steady state condition of the year 1700, has allowed us significant insight into the relative importance of the horizontal exchange processes of C, N, and P. The relatively small masses of C, N, P, and the short residence times of water (2–3 years) in the coastal zone with respect to exchange with the open ocean, account for the rapid responses of the coastal zone to the environmental changes in its neighboring reservoirs of land and ocean. Important conclusions of the modeling, supported to some extent by observational evidence, are that: (1) fluxes of riverine nutrients and organic matter will continue to increase during the 21st century and result in changes in the cycling behavior of organic carbon and nutrients in the global coastal zone environment; (2) because of rising atmospheric $CO_2$ and increasing nutrient inputs via rivers and groundwater discharges and atmospheric N inputs over the period of the past 300 years and on into the future, the coastal ocean has progressively become more autotrophic and is now or soon will be a net sink of atmospheric $CO_2$; and (3) the saturation state with respect to carbonate minerals of the coastal ocean and the surface ocean as a whole will continue to decline into this century and probably result in decreasing calcification rates of marine organisms that secret calcareous shells.

To date, fossil fuel $CO_2$ emissions to the atmosphere, changes in land-use practices, and fertilizer and sewage discharges have been the four main factors affecting the carbon cycle in the global coastal zone. Numerical simulations have shown that increases in the inputs of nitrogen and phosphorus from land to the coastal ocean have stimulated new production, but that this increase in gross primary production has been approximately balanced by a similar increase in total ecosystem respiration owing to increased transport and deposition of organic matter of terrestrial origin. Future projections suggest that nutrient input and gross primary production will continue to increase and increasingly exceed organic matter input and remineralization, leading to increased net autotrophy. This increase in $NEP^*$ implies increased storage of organic matter within the coastal ocean and/or export of organic matter to the open ocean, and thereby also the potential role of this region as a sink for atmospheric $CO_2$. However, if the projected increase in inorganic nutrient input is erroneous and the organic carbon reservoir has remained in a steady state with no increase in storage or export of organic matter to the open ocean, total ecosystem respiration is likely to increasingly exceed gross primary production, causing increased net heterotrophy of the coastal ocean and negative values of $NEP^*$ (e.g., Rabouille et al., 2001). Consequently, this would decrease the capacity of this region to absorb anthropogenic $CO_2$. In summation, the direction of future changes in net ecosystem production in the coastal zone strongly depends on changes in the relative magnitudes of organic carbon and nutrient N and P fluxes to the coastal zone via rivers and runoff, provided the upwelling fluxes remain constant.

Projected increases in atmospheric $CO_2$ owing to burning of fossil fuels and land-use changes will affect the inorganic carbon cycle of the coastal ocean. Surface water carbonate saturation state is likely to decrease owing to increased invasion of atmospheric $CO_2$ and could negatively affect the ability of calcareous organisms to calcify. Decreased carbonate saturation state in the surface water or the pore water could also result in increased dissolution of carbonate minerals and subsequently alterations to the average composition and rate of precipitation of carbonate cements in contemporary carbonate sediments.

Based on the projected changes in the organic carbon cycle (i.e., trophic status) and the inorganic carbon cycle (i.e., production and dissolution of calcium carbonate minerals), the global coastal ocean will undergo a transition from its role as a net source to becoming a net sink of $CO_2$ to the atmosphere. Although the exact timing of this transition is uncertain, the trend is robust owing to the significant increase in atmospheric $CO_2$.

Finally what is the prognosis for the future in terms of the water quality and biogeochemical state of coastal zone ecosystems? Certainly continuous increase in the global population and its industrial and agricultural activities will continue to create major perturbations of the biogeochemical cycles of the bio-essential elements C, N, and P, including major alterations in the exchanges of these elements between the land, atmosphere, and sea in any future "business as usual" or "worst case" economic-social scenarios of development (e.g., Mackenzie et al., 1993). During the Industrial Era, the anthropogenic fluxes primarily had their origin in the industrialized and developed countries. Thus some portion of these fluxes mainly affected coastal zone ecosystems of the industrialized world in the Northern Hemisphere, much more so than coastal waters of the developing countries or the open ocean, particularly in terms of water quality, organic productivity, community structure, and biodiversity because of inputs of anthropogenic materials from river and groundwater discharges and from the atmosphere. In addition, industrialized countries' coastal zone waters and sediments were, and continue to be, a major trap or filter for many of these materials as they are transported towards the open ocean. This situation will continue at least into the early decades of the 21st century because emissions and fluxes of C-, N-, and P-bearing materials from human activities into the environment have not slowed dramatically from industrial countries. However, the situation is changing as the industrializing and developing countries move into the 21st century and their C, N, and P emissions increase and global mean atmospheric $CO_2$ concentrations continue to rise at a rate of about 1.5 ppmv/yr, in part due to the increasing industrial, transportation, and agricultural activities of the industrializing countries. One important example of this changing regional situation is the countries of South and Southeast Asia, a region of potentially increasing contributions to the loading of the environment owing to a combination of such factors as its increasing population, increasing industrialization dependent on fossil fuels, concentration of its population along the major river drainage basins and in coastal urban centers, and changing land use practices (Mackenzie et al., 1998b). It is anticipated that a number of countries in this region will experience similar, possibly even greater, loss of storage of C and nutrient N and P on land and increased storage in coastal marine sediments per unit area than experienced by the developed countries during their period of industrialization. With the exponential growth of Asia's population along the oceanic coastal zone, higher inputs of both dissolved and particulate organic nutrients may be expected to enter coastal waters. A similar trend of increasing population concentration in agricultural areas inland, within the drainage basins of the main rivers flowing into the ocean, is also expected to result in increased dissolved and particulate organic nutrient loads that are likely to reach the ocean. However, continuous construction of large dams in these countries may disrupt sediment and water flows to coastal zone ecosystems, as already has been observed for the Yangtze River in China. In addition, emissions of anthropogenic $CO_2$ particularly from fossil fuel burning are rising significantly in several of the industrializing countries, such as China and India. A number of the above changes are also being observed in several countries of the Middle East, South America, and to a much lesser extent in Africa; however, Southeast and South Asian countries seem to be poised for significant development in the 21st century along the lines that the U.S. and European countries followed during the past 150 years (Mackenzie et al., 1998b). Thus the prognosis for any "business as usual" scenario for these countries' coastal zone ecosystems would be for continuous degradation of water quality, increased eutrophication problems, changes in biological community structure and biodiversity, and continuous degradation of coral reef ecosystems found in the coastal waters of some of these countries.

## Acknowledgements

This research was supported by National Science Foundation grants ATM 00–80878 and EAR 02–235090, and additionally by the Arthur L. Howland Fund of the Department of Geological Sciences, Northwestern University. We are especially indebted to the late Prof. *emeritus* Roland Wollast, Université Libre de Bruxelles, for his continuous critical reviews and many thoughtful comments on two initial drafts of this paper. This is University of Hawaii School of Ocean and Earth Science and Technology contribution 6508.

## Bibliography

Andersson, A., F. Mackenzie and L. M. Ver, 2003. Solution of shallow-water carbonates: An insignificant buffer against rising atmospheric $CO_2$. *Geology,* 31, 513–516.

Andersson, A. J. and Mackenzie, F. T., 2004. The shallow-water ocean: a source or sink of atmospheric $CO_2$? *Front. Ecol. Environ,* 7, 348–353.

Bacastow, R. D. and Keeling, C. D., 1973. Atmospheric carbon dioxide and radiocarbon in the natural carbon cycle: II. Changes from A. D. 1700 to 2070 as deduced from a geochemical model. In *Carbon and the Biosphere,* G. M. Woodwell and E. V. Pecan, eds. CONF-720510, National Technical Information Service, Springfield, Va., 86–135, 1973.

Bacon, M. P., R. A. Belastock and M. H. Bothner, 1994. $^{210}Pb$ balance and implications for particle transport on the continental shelf, US Middle Atlantic Bight. *Deep Sea Research,* 41, 511–535.

Barnes, D. J. and C. Cuff, 2000. Solution of reef rock buffers seawater against rising atmospheric $CO_2$. In *Proceedings of the Ninth International Coral Reef Symposium Abstracts,* D. Hopley, M. Hopley, J. Tamelander and T. Done, eds. State Ministry for the Environment, Indonesia, 248 p.

Bates, N. and L. Samuels, 2001. Biogeochemical and physical factors influencing seawater $fCO_2$ and air-sea $CO_2$ exchange on the Bermuda coral reef. *Limnology and Ocenaography,* 46, 833–846.

Bauer, J. E., Williams, P. M. and Druffel, E. R. M.: Super $^{14}C$ activity of dissolved organic carbon fractions in the north-central Pacific and Sargasso Sea. Nature, 357, 667–670, 1992.

Berger, W. H., K. Fisher, C. Lai and G. Wu, 1987. Ocean productivity and organic carbon flux. Part I. Overview and map of primary production and export production. *Scripps Institution of Oceanography,* 87–30.

Berner, R. A., 1982. Burial of organic carbon and pyrite sulfur in the modern ocean: Its geochemical and environmental significance. *American Journal of Science,* 282, 451–473.

Boehme, S. E., C. L. Sabine and C. E. Reimers, 1998. $CO_2$ fluxes from a coastal transect: A time series approach. *Marine Chemistry,* 63, 49–67.

Borges, A. V. and M. Frankignoulle, 2001. Short-term variations of the partial pressure of $CO_2$ in surface waters of the Galician upwelling system. *Progress in Oceanography,* 51, 283–302.

Borges, A. V. and M. Frankignoulle, 2002a. Distribution and air-water exchange of carbon dioxide in the Scheldt plume off the Belgian coast. *Biogeochemistry,* 59, 41–67.

Borges, A. V. and M. Frankignoulle, 2002b. Distribution of surface carbon dioxide and air-sea exchange in the upwelling system off the Galician coast. *Global Biogeochemical Cycles,* 16(2), doi:10.1029/2000GB001385.

Cai W.-J., Z. A. Wang and Y. Wang, 2003. The role of marsh-dominated heterotrophic continental margins in transport of $CO_2$ between the atmosphere, the land-sea interface and the ocean. *Geophysical Research Letters* 30 (16), 1849, doi:10.1029/2003GL017633.

Caraco, N. F., 1995. Influence of human populations on P transfers to aquatic systems: A regional scale study using large rivers. In *Phosphorus in the Global Environment: Transfers, Cycles and Management. SCOPE 54,* H. Tiessen, ed. Wiley, Chichester, UK, 235–244.

Chen, C. T. A. and S. L. Wang, 1999. Carbon, alkalinity, and nutrient budget on the East China Sea continental shelf. *Journal of Geophysical Research,* 104, 20,675–20,686.

Cohen, J. E., C. Small, A. Mellinger, J. Gallup and J. Sachs, 1997. Estimates of coastal populations. *Science,* 278, 1211–1212.

DeGrandpre M. D., G. D. Olbu, C. M. Beatty and T. R. Hammar, 2002. Air-sea $CO_2$ fluxes on the U.S. Middle Atlantic Bight. *Deep-Sea Research Part II,* 49, 4369–4385.

De Jonge, V. N., W. Boynton, C. F. D'Elia, R. Elmgren and B. L. Welsh, 1994. Responses to developments in eutrophication in four different North Atlantic ecosystems. In *Changes in Fluxes in Estuaries,* K. R. Dyer and R. J. Orth, eds. Olsen & Olsen, 179–196.

DeMaster, D., R. H. Pope, L. A. Levin and N. E. Blair, 1994. Biological mixing intensity and rates of organic carbon accumulation in North Carolina slope sediments. *Deep Sea Research Part II,* 41, 735–754.

Dugdale, R. C. and J. J. Goering, 1967. Uptake of new and regenerated nitrogen in primary productivity. *Limnology and Oceanography,* 12, 196–206.

Ducklow, H. W., and McAllister, S. L.: The biogeochemistry of carbon dioxide in the coastal oceans, in The Sea, vol. 13: The Global Coastal Ocean-Multi-scale Interdisciplinary Processes, 5, edited by Robinson, A. R. and Brink, K., Harvard Univ. Press, in press, 2004.

Falkowski, P. G., P. E. Biscaye and C. Sancetta, 1994. The lateral flux of biogenic particles from the eastern North American continental margin to the North Atlantic Ocean. *Deep Sea Research,* 41, 583–601.Frankignoulle, M., 1988. Field measurements of air-sea $CO_2$ exchange. *Limnology and Oceanography,* 33, 313–322.

Feely, R. A., C. L. Sabine, K. Lee, W. Berelson, J. Kleypas, V. J. Fabry, and F. J. Millero, 2004. Impact of anthropogenic $CO_2$ on the $CaCO_3$ system in the oceans. *Science,* 305, 362–366.

Frankignoulle, M., C. Canon and J.-P. Gattuso, 1994. Marine calcification as a source of carbon dioxide: Positive feedback of increasing atmospheric $CO_2$. *Limnology and Oceanography,* 39, 458–462.

Frankignoulle, M. and A. V. Borges, 2001. European continental shelf as a significant sink for atmospheric carbon dioxide. *Global Biogeochemical Cycles,* 15, 569–576.

Frankignoulle, M., G. Abril, A. Borges, I. Bourge, C. Canon, B. DeLille, E. Libert and J.-M. Théate, 1998. Carbon dioxide emission from European estuaries. *Science,* 282, 434–436.

Galloway, J. N., R. W. Howarth, A. F. Michaels, S. W. Nixon, J. M. Prospero and F. J. Dentener, 1996. Nitrogen and phosphorus budgets of the North Atlantic Ocean and its watershed. *Biogeochemistry,* 35, 3–25.

Garrels, R. M. and F. T. Mackenzie, 1972. A quantitative model for the sedimentary rock cycle. *Marine Chemistry,* 1, 27–41.

Garrels, R. M. and F. T. Mackenzie, eds., 1980. Some aspects of the role of the shallow ocean in global carbon dioxide uptake. Workshop report. Carbon dioxide effects research and assessment program. United States Department of Energy.

Gattuso, J.-P., M. Pichon, B. Delesalle and M. Frankignoulle, 1993. Community metabolism and air-sea $CO_2$ fluxes in a coral reef ecosystem (Moorea, French Polynesia). *Marine Ecology Progess Series,* 96, 259–267.

Gattuso, J.-P., P. D. Allemand and M. Frankignoulle, 1999. Interactions between the carbon and carbonate cycles at organism and community levels on coral reefs: A review of processes and control by carbonate chemistry. *American Zoologist,* 39, 160–188.

Gattuso, J.-P., M. Frankignoulle and R. Wollast, 1998. Carbon and carbonate metabolism in coastal aquatic ecosystems. *Annual Review of Ecology and Systematics,* 29, 405–434.

Goyet, C., F. J. Millero, D. W. O'Sullivan, G. Eischeid, S. J. McCue and R. G. J. Bellerby, 1998. Temporal variations of $pCO_2$ in surface seawater of the Arabian Sea in 1995. *Deep Sea Research Part I,* 45, 609– 623.

Hales, B., L. Bandstra, T. Takahashi, P. Covert and J. Jennings, 2003. The Oregon coastal ocean: A sink for Atmospheric $CO_2$? *Newsletter of Coastal Ocean Processes,* 17, 4–5.

Halley, R. B. and K. K. Yates, 2000. Will reef sediments buffer corals from increased global $CO_2$. In *Proceedings of the Ninth International Coral Reef Symposium Abstracts,* D. Hopley, M. Hopley, J. Tamelander, and T. Done, eds. State Ministry for the Environment, Indonesia, 248 p ..

Hawaiian Ocean Time series (HOTs), 2003. *http://hahana.soest.hawaii.edu/hot/hot_jgofs.html*

Hay, W. W., 1998. Detrital sediment fluxes from continents to oceans. *Chemical Geology,* 145, 287–323.

Hedges, J. I., R. G. Keil and R. Benner, 1997. What happens to terrestrial organic matter in the ocean? *Organic Geochemistry,* 27, 195–212.

Howarth, R. W., G. Billen, D. Swaney, A. Townsend, N. Jaworski, K. Lajtha, J. A. Downing, R. Elmgren, N. Caraco, T. Jordan, F. Berendse, J. Freney, V. Kudeyarov, P. Murdoch and Z. Zhao-liang, 1996. Regional nitrogen budgets and riverine N & P fluxes for the drainages to the North Atlantic Ocean: Natural and human influences. *Biogeochemistry,* 35, 75–139.

Hu, D., Y. Saito and S. Kempe, 1998. Sediment and nutrient transport to the coastal zone. In *Asian Change in the Context of Global Change,* J. N. Galloway and J. M. Melillo, eds. Cambridge University Press, 245–270.

Ianson, D. and S. E. Allen, 2002. A two-dimensional nitrogen and carbon flux in a coastal upwelling region. *Global Biogeochemical Cycles,* 16(1), doi:10.1029/2001GB001451.

Iglesias-Rodriguez, M. D., R. Armstrong, R. Feely, R. Hood, J. Kleypas, J.D. Milliman, C. Sabine and J. Sarmiento, 2002. Progress made in study of ocean's calcium carbonate budget. *EOS,* 83, 365, 374–375.

Jahnke, R. A., 1996. The global ocean fluxes of particulate organic carbon: areal distribution and magnitude. *Global Biogeochemical Cycles,* 10, 71–88.

Kempe, S., 1995. Coastal Seas: A Net Source or Sink of Atmospheric Carbon Dioxide? Land-Ocean Interactions in the Coastal Zone Core Project of the IGBP, 27 p.

Kempe, S. and K. Pegler, 1991. Sinks and sources of $CO_2$ in coastal seas: the North Sea. *Tellus,* 43B, 224–235.

Kleypas, J. A., R. W. Buddemeier, D. Archer, J.-P. Gattuso, C. Langdon and B. N. Opdyke, 1999. Geochemical consequences of increased atmospheric carbon dioxide on coral reefs. *Science,* 284, 118–120.

Kleypas, J. A., R. W. Buddemeier and J.-P. Gattuso, 2001. The future of coral reefs in an age of global change. *International Journal of Earth Sciences (Geol Rundschau),* 90, 426–437.

Knauer, G. A., 1993. Productivity and new production of the oceanic system. In *Interactions of C, N, P and S Biogeochemical Cycles and Global Change,* R. Wollast, F. T. Mackenzie and L. Chou, eds. Springer-Verlag, Berlin, 211–231.

Kortzinger, A., J. C. Duinker and L. Mintrop, 1997. Strong $CO_2$ emissions from the Arabian Sea during South-West monsoon. *Geophysical Research Letters,* 24, 1763–1766.

Langdon, C., T. Takahashi, C. Sweeney, D. Chipman, J. Goddard, F. Marubini, H. Aceves, H. Barnett and M. Atkinson, 2000. Effect of calcium carbonate saturation state on the calcification rate of an experimental coral reef. *Global Biogeochemical Cycles,* 14, 639–654.

Leclercq, N., J.-P. Gattuso and J. Jaubert, 2000. $CO_2$ partial pressure controls the calcification rate of a coral community. *Global Change Biology,* 6, 329–334.

Leclercq, N., J.-P. Gattuso and J. Jaubert, 2002. Primary production, respiration, and calcification of a coral reef mesocosm under increased $CO_2$ partial pressure. *Limnology and Oceanography,* 47, 558–564.

Lee, C., 1992. Controls on organic carbon preservation: The use of stratified water bodies to compare intrinsic rates of decomposition in oxic and anoxic systems. *Geochimica et Cosmochimica Acta,* 56, 3323–3335.

Lerman, A., F. T. Mackenzie, and L. M. Ver, 2004. Coupling of the perturbed C-N-P cycles in industrial time. *Aquatic Geochemisrty,* 10, 3–32.

Liu, K. K., K. Iseki and S. Y. Chao, 2000. Continental margin carbon fluxes. In *The Changing Ocean Carbon Cycle,* R. B. Hanson, H. W. Ducklow and J. G. Field, eds. Cambridge University Press, Cambridge, 187–239.

Mackenzie, F. T., 2003. *Our Changing Planet: An Introduction to Earth System Science and Global Environmental Change.* Prentice-Hall, Upper Saddle River, N. J.

Mackenzie, F. T., A. Lerman and L. M. Ver, 1998a. Role of the continental margin in the global carbon balance during the past three centuries. *Geology,* 26, 423–426.

Mackenzie, F. T., L. M. Ver and A. Lerman, 1998b. Coupled biogeochemical cycles of carbon, nitrogen, phosphorus and sulfur in the land-ocean-atmosphere system. In *Asian Change in the Context of Global Climate Change,* J. N. Galloway and J. M. Melillo, eds. Cambridge University Press, 42–100.

Mackenzie, F. T., L. M. Ver and A. Lerman, 2000. Coastal-zone biogeochemical dynamics under global warming. *International Geology Review,* 42, 193–206.Mackenzie, F. T., L. M. Ver and A. Lerman, 2002a. Century-scale nitrogen and phosphorus controls of the carbon cycle. *Chemical Geology,* 190, 13–32.

Mackenzie, F. T., A. Lerman and L. M. B. Ver, 2001. Recent past and future of the global carbon cycle. In *Geological Perspectives of Global Climate Change,* L. C. Gerhard, W. E. Harrison and B. M. Hanson, eds. American Association of Petroleum Geologists Special Publication, 51–82.

Mackenzie, F. T., L. M. Ver and A. Lerman, 2002a. Century-scale nitrogen and phosphorus controls of the carbon cycle. *Chemical Geology,* **190,** 13–32.

Mackenzie, F. T., L. M. Ver and A. Lerman, 2002b. Coastal-zone biogeochemical dynamics under global warming. In *Frontiers in Geochemistry: Organic, Solution, and Ore Deposit Geochemistry. Konrad Krauskopf Volume 2,* W. Ernst, ed. Bellwether Publishing, 27–40.

Mackenzie, F. T., L. M. Ver, C. Sabine, M. Lane and A. Lerman, 1993. C, N, P, S global biogeochemical cycles and modeling of global change. In *Interactions of C, N, P and S Biogeochemical Cycles and Global Change,* R. Wollast, F. T. Mackenzie and L. Chou, eds. Springer-Verlag, 1–62.

Mantoura, R. F. C., J. M. Martin and R. Wollast, 1991. Ocean Margin Processes in Global Change. Wiley-Interscience, New York.

Martin, J., G. Knauer, D. Karl and W. Broenkow, 1987. VERTEX: carbon cycling in the northeast Pacific. *Deep Sea Research,* 34, 267–286.

Meybeck, M., 1982. Carbon, nitrogen, and phosphorus transport by world rivers. *American Journal of Science,* 282, 401–450.

Meybeck, M., 2001. Global alteration of riverine geochemistry under human pressure. In *Understanding the Earth System: Compartments, Processes, and Interactions,* E. Ehlers and T. Krafft, ed. Springer, 97–113.

Meybeck, M. and A. Ragu, 1995. *Water Quality of World River Basins.* UNEP GEMS Collaborating Centre for Fresh Water Quality Monitoring and Assessment, United Nations Environment Programme.

Meybeck, M. and A. Ragu, 1997. Presenting GEMS GLORI: A compendium of world river discharges to the oceans. In *Freshwater Contamination. Proceedings of a symposium held during the Fifth IAHS Scientific Assembly at Rabat, April–May 1997,* B. Webb, ed. International Association of Hydrological Sciences, vol. 243, 3–14.

Milliman, J. D., 1993. Production and accumulation of calcium carbonate in the ocean: Budget of a nonsteady state. *Global Biogeochemical Cycles,* 7, 927–957.

Milliman, J. D. and A. W. Droxler, 1995. Calcium carbonate sedimentation in the global ocean: Linkages between the neritic and pelagic environments. *Oceanography,* 8, 92–94.

Milliman, J. D. and J. P. M. Syvitski, 1992. Geomorphic/tectonic control of sediment discharge to the ocean; the importance of small mountainous rivers. *Journal of Geology,* 100, 525–544.

Morse, J. W. and F. T. Mackenzie, 1990. *Geochemistry of Sedimentary Carbonates.* Elsevier, New York.

Müller, P. J. and E. Suess, 1979. Productivity, sedimentation rate, and sedimentary organic matter in the oceans, I. Organic carbon preservation. *Deep-Sea Research,* 26, 1347–1362.

Neumann, A. C., 1965. Processes of recent carbonate sedimentation in Harrington sound, Bermuda. *Bulletin of Marine Science,* 15, 987–1035.

Perez, F. F., A. F. Rios and G. Roson, 1999. Sea surface carbon dioxide off the Iberian Peninsula (North Eastern Atlantic Ocean). *Journal of Marine Systems,* 19, 27–46.

Rabouille, C., F. T. Mackenzie and L. M. Ver, 2001. Influence of the human perturbation on carbon, nitrogen, and oxygen biogeochemical cycles in the global coastal ocean. *Geochimica et Cosmochimica Acta,* 65, 3615–3639.

Revelle, R. and W. Munk, 1977. The carbon dioxide cycle and the biosphere. In *Energy and Climate,* N.G.S. committee, eds. National Academy Press, 140–158.

Riebesell, U., I. Zondervan, B. Rost, P. D. Tortell, R. E. Zeebe and F. M. M. Morel, 2000. Reduced calcification of marine plankton in response to increased atmospheric $CO_2$. *Nature,* 407, 364–367.

Richardson, K. and J. P. Heilmann, 1995. Primary production in the Kattegat: Past and present. *Ophelia,* 41, 317–328.

Schlesinger, W. H., 1997. *Biogeochemistry: An Analysis of Global Change.* Academic Press, New York.

Schmalz, R. F. and K. E. Chave, 1963. Calcium carbonate: Affecting saturation in ocean waters of Bermuda. *Science,* 139, 1206–1207.

Simpson, J. J. and A. Zirino, 1980. Biological control of pH in the Peruvian coastal upwelling area. *Deep Sea Research,* 27, 234–248.

Smith, S. V., 1995. Net carbon metabolism of oceanic margins and estuaries. In *Biotic Feedbacks in the Global Climatic System: Will the Warming Feed the Warming?* G. M. Woodwell and F. T. Mackenzie, eds. Oxford University Press, 251–262.

Smith, S. V. and J. T. Hollibaugh, 1993. Coastal metabolism and the oceanic organic carbon balance. *Reviews of Geophysics,* 31, 75–89.

Smith, S. V. and F. T. Mackenzie, 1987. The ocean as a net heterotrophic system: Implications from the carbon biogeochemical cycle. *Global Biogeochemical Cycles,* 1, 187–198.

Smith, S. V., W. H. Renwick, R. W. Buddemeier and C. J. Crossland, 2001. Budgets of soil erosion and deposition for sediments and sedimentary organic carbon across the conterminous United States. *Global Biogeochemical Cycles,* 15, 697–708.

Smith, S. V. and F. Wulff, 1999. LOICZ Biogeochemical Modeling Node. In ***http://data.ecology.su.se/MNODE/,*** Vol. 1999. Land-Ocean Interactions in the Coastal Zone.

Takahashi, T., S. C. Sutherland, C. Sweeney, A. Poisson, N. Metzl, B. Tilbrook, N. Bates, R. Wanninkhof, R. A. Feely, C. Sabine, J. Olafsson and Y. Nojiri, 2002. Global sea-air $CO_2$ flux based on climatological surface ocean $pCO_2$, and seasonal biological and temperature effects. *Deep Sea Research II,* 49, 1601–1622.

Thomas, H. and B. Schneider, 1999. The seasonal cycle of carbon dioxide in Baltic Sea surface waters. *Journal of Marine Systems,* 22, 53–67.

Tsunogai, S., S. Watanabe and T. Sato, 1999. Is there a "continental shelf pump" for the absorption of atmospheric $CO_2$? *Tellus,* 51B, 701–712.

Valiela, I., 1995. *Marine Ecological Processes.* Springer-Verlag, New York.

Van Green, A., R. K. Takesue, J. Goddard, T. Takahashi, J. A. Barth and R. L. Smith, 2000. Carbon and nutrient dynamics during coastal upwelling off Cape Blanco, Oregon. *Deep Sea Research Part II,* 47, 975–1002.

Ver, L. M., F. T. Mackenzie and A. Lerman, 1999a. Biogeochemical responses of the carbon cycle to natural and human perturbations: Past, present, and future. *American Journal of Science,* 299, 762–801.

Ver, L. M., F. T. Mackenzie and A. Lerman, 1999b. Carbon cycle in the coastal zone: Effects of global perturbations and change in the past three centuries. *Chemical Geology,* 159, 283–304.

Vörösmarty, C. J., C. Li, J. Sun and Z. Dai, 1998. Drainage basins, river systems, and anthropogenic change: The Chinese example. In *Asian Change in the Context of Global Change,* J. N. Galloway and J. M. Melillo, eds. Cambridge University Press, 210–244.

Walsh, J. J., 1988. *On the Nature of Continental Shelves.* Academic Press.

Wang, S.-L., C.-T. A. Chen, G.-H. Hong and C.-S. Chung, 2000. Carbon dioxide and related parameters in the East China Sea. *Continental Shelf Research,* 20, 525–544.

Ware, J. R., S. V. Smith and M. L. Reaka-Kudla, 1992. Coral reefs: Sources or sinks of atmospheric $CO_2$? *Coral reefs,* 11, 127–130.

Williams, P. J. le B., 1998. The balance of plankton respiration and photosynthesis in the open oceans. *Nature,* 394, 55–57.

Winn, C. D., Y.-H. Li, F. T. Mackenzie and D. M. Karl, 1998. Rising surface ocean dissolved inorganic carbon at the Hawaii Ocean Time-series site, *Marine Chemistry,* 60, 33–47.

Wollast, R., 1991. The coastal organic carbon cycle: Fluxes, sources, and sinks. In *Ocean Margin Processes in Global Change,* R. F. C. Mantoura, J. M. Martin, and R. Wollast, eds. Wiley-Interscience Publishers, 365–381.

Wollast, R., 1993. Interactions of carbon and nitrogen cycles in the coastal zone. In *Interactions of C, N, P and S Biogeochemical Cycles and Global Change,* R. Wollast, F. T. Mackenzie and L. Chou, eds. Springer-Verlag, 195–210.

Wollast, R., 1994. The relative importance of bioremineralization and dissolution of $CaCO_3$ in the global carbon cycle. In *Past and Present Biomineralization Processes: Considerations about the Carbonate Cycle,* F. Doumenge, D. Allemand and A. Toulemont, eds. Musée Océanographique, Monaco, 13–34.

Wollast, R., 1998. Evaluation and comparison of the global carbon cycle in the coastal zone and in the open ocean. In *The Sea: The Global Coastal Ocean,* K. H. Brink and A. R. Robinson, eds. John Wiley & Sons, vol. 10, 213–252.

Wollast, R., 2002. Continental margins-review of geochemical settings. In *Ocean Margin Systems,* G. Wefer, D. Billett, D. Hebbeln, B. B. Jorgensen, M. Schluter and T. Van Weering, eds. Springer-Verlag, Berlin, 15–31.

Wollast, R. and F. T. Mackenzie, 1989. Global biogeochemical cycles and climate. In *Climate and Geo-Sciences,* A. Berger, S. Schneider and J. C. Duplessy, eds. Kluwer Academic Publishers, 453–473.

Wollast, R., R. M. Garrels and F. T. Mackenzie, 1980. Calcite-seawater reactions in ocean surface waters. *American Journal of Science,* 280, 831–848.

Woodwell, G. M., 1995. Biotic feedbacks from the warming of the Earth. In *Biotic Feedbacks in the Global Climatic System: Will the Warming Feed the Warming?* G. M. Woodwell and F. T. Mackenzie, eds. Oxford University Press, 3–21.

World Commission on Dams, 2000. *Dams and Development: A New Framework for Decision Making.* Earthscan Publications Ltd.

Zeebe, R. E. and D. Wolf–Gladrow, 2001. *$CO_2$ in seawater: equilibrium, kinetics, isotopes.* Elseiver.

Zondervan, I., R. E. Zeebe, B. Rost and U. Riebesell, 2001. Decreasing marine biogenic calcification: A negative feedback on rising atmospheric $pCO_2$. *Global Biogeochemical Cycles,* 15, 507–516.

# Chapter 8. CIRCULATION, MIXING AND THE DISTRIBUTION OF REMINERALIZED NUTRIENTS

LARRY P. ATKINSON

*Old Dominion University*

JOHN HUTHNANCE

*Proudman Oceanographic Laboratory*

JOSE L. BLANCO

*Old Dominion University*

**Contents**



## 1. Introduction

The circulation and mixing processes that affect the distribution of re-mineralized nutrients (referred to as 'nutrients' from now on) in the global coastal ocean are the topic of this chapter. The coastal ocean is at the dynamic edge between ocean and land. Thus we are concerned with the two fundamental sources of nutrients: the deep ocean and runoff from land. Processes that control the flow of nutrients from the deep ocean include instabilities in western boundary currents, coastal upwelling, and deep mixing. The inflow of nutrients from land occurs through rivers and estuaries. During this journey nutrients may be chemically altered, deposited and recycled.

*The Sea*, Volume 13, edited by Allan R. Robinson and Kenneth H. Brink
ISBN 0-674-01526-6 

The discussion expands on the chapter by R. Wollast (1998) in Volume 10 of The Sea. In that chapter, Wollast clearly established that about 80% of the new nutrients involved in primary production in coastal waters are from the deep-ocean, the other 20% coming from rivers. This chapter will focus on both processes but emphasize the circulation and mixing processes that affect nutrient flux and place these introduced nutrients in the euphotic zone of coastal waters. We will follow the scheme in Huthnance (1995). His methodology was to examine a variety of processes in regard to circulation, exchange and water mass formation. As he noted, these processes equate on a time-length scale consideration to advection, stirring and mixing. Other authors (e.g., Walsh, 1988) have also focused on processes such as eastern boundary current upwelling. Since most oceanographic studies are related to specific processes, such as western boundary currents or coastal upwelling, we have partitioned this paper in that manner, while using Huthnance's (1995) terminology and scaling.

We must note that we barely understand the processes involved in temperature and salinity exchange, circulation and mixing at the shelf edge; their quantification is especially problematic. Thus, our understanding of how physical processes affect nutrients is even more difficult. Nevertheless, research is progressing and we hope this review will help the cause.

In general terms, the distribution of any constituent nutrient "N" is controlled by a transport equation

$$\partial N/\partial t + \mathbf{u}.\text{grad } N = K_H (\partial_x^2 + \partial_y^2) N + K_V \partial_z^2 N + \text{sources} - \text{sinks} \qquad (1)$$

Thus temporal plus advective change is caused by often unresolved dispersive and mixing processes, sources and sinks. Advection is resolved by flow or circulation, in principle reversible, that is associated with various processes discussed in sections 3 to 5. Dispersion and mixing are irreversible and caused by small-scale processes; vertical mixing is discussed in section 6. For the coastal ocean, important sources of nutrients are the open ocean, remineralization in the water column (from detritus) and in the sea bed, rivers locally, and the atmosphere over extensive shelf seas or estuaries. There may be urban and industrial inputs; groundwater is a general possibility but rarely quantified or even identified. Primary production is an important sink as may be sedimentation and denitrification.

After providing a historical perspective, we give an overview (section 3) of the riverine and open-ocean sources of nutrients in coastal seas. Transport processes in the coastal ocean are discussed in sections 4–6, with examples of their effects on nutrient distributions: along-shelf flows and their adjuncts with an emphasis on western oceanic boundaries in section 4; cross-shelf transports and their adjuncts with an emphasis on eastern oceanic boundaries in section 5; mixing in section 6. Further discussion of successive regions offshore is given in section 7 followed in section 8 by a few coastal ocean "system" cases that illustrate the degree of process understanding and applicability.

## 2. Historical Perspective

The focus on the circulation, mixing and distribution of re-mineralized nutrients results from their critical role in primary productivity. This topic attracted oceanographers nearly a century ago. Redfield (1936) noted:

> "The view has been increasingly accepted, since first put forth by Nathansohn (1906), that the fertility of the sea depends upon the restoration to the surface of plant nutrients such as phosphates and nitrates liberated by the decomposition of organic matter within its depths. In shallow coastal waters the turbulence due to wind and tide, aided in high latitudes by the instability resulting from cooling in winter, suffice to maintain this part of the nutritional cycle. In the deep sea, organic matter generated in the surface as the result of photosynthesis processes appears in large part to sink to great depths before being finally oxidized to its ultimate inorganic products."

Redfield (1936) went on to note the processes that accomplish this in the deep ocean:

> " ... four processes by which these materials are brought again to the surface in the Atlantic Ocean and thus made available for the organic cycle ... [Include] ... (1) upwelling resulting from offshore trade winds off the African coast; (2) upwelling of deep Atlantic water in the Antarctic; (3) upwelling in the boundary between currents as in the Arctic polar front; and (4) winter convection in high latitudes."

The landmark paper by Riley (1967) on nutrients in coastal waters stated clearly stated the importance of the deep-water source of nutrients to coastal waters:

> "Coastal waters generally are more productive than the open sea. Two factors are believed to be responsible, in varying degrees according to local circumstances. The first is shoreward transport, from the edge of the continental shelf, of deep and nutrient rich water, which then becomes available to surface phytoplankton populations in the inshore waters as a result of tidal vertical mixing. The other in enrichment by freshwater drainage."

He further noted that:

> "General conclusions are that the usual pattern of exchange between inshore and offshore waters tends to enrich the coastal zone irrespective of enrichment by freshwater drainage..."

Since these key papers we have learned about the processes, such as eastern and western boundary currents, that cause the enrichment of coastal waters. Additionally, we now better understand the mixing processes that bring those nutrients upward into the photic zone of the coastal ocean.

## 3. Recent work and process overview

The last decade has seen an increase in review papers and books on the coastal ocean. The following are recommended for the topic of this paper: Walsh, 1988; Blanton, 1991; Jickells, 1991; Huthnance, 1995; Brink, 1998; Robinson and Brink,

1998; Wollast, 1998). Recent papers that discuss the global coastal ocean in relation to global carbon and climate issues underscore the importance of understanding these process (Tsunogai et al., 1999; Liu et al., 2000). In addition, several papers are in preparation as part of the IGBP, JGOFS and LOICZ programs.

Of the principal external sources of nutrients in coastal seas, both riverine and oceanic inputs are subject to considerable buffering or constraint.

### *Riverine inputs*

Anthropogenic inputs often enhance riverine concentrations of nitrogen and phosphorus species. However, these inputs are relatively lacking in silicon, which composes a large fraction of river-borne sediments. These inputs are typically buffered by estuarine processes and often have little effect on the wider coastal ocean.

Algal production may severely deplete silica, phosphate and nitrate, as happens in the eutrophic Loire River due to the long residence time O(1 week) (Meybeck et al., 1988). Biomass degradation will generally release dissolved silica, phosphate, and ammonium. This may result in a nutrient maxima if particulate matter is concentrated around a turbidity maximum. Water column nitrification may augment nitrite at very low salinity (e.g. in the Tamar; Knox et al., 1986). For example, nitrification in the upper Forth estuary is aided by ample particulate organic nitrogen (PON) in the turbidity maximum (Balls et al., 1996). This process increases downstream nitrite+nitrate concentrations, especially during low river flows when upper-estuary residence times are longer. In the Humber River, NE England, nitrate, nitrite and silicate have mid-estuarine maxima (>500 μM for nitrate) influenced by physical transport, which is strongly correlated with inflows to the estuary and with the freshwater-saltwater interface location or (for nitrate in summer/autumn) the turbidity maximum (Uncles et al., 1998a; Tappin et al., 2001; a UK contribution to LOICZ). However, the Humber nitrite peak was somewhat upstream of the nitrate, silicate, owing to possible sources from ammonia and bacterial remineralization of organic matter; it decreased through the turbidity maximum and further downstream, possibly through oxidation to nitrate. Around the Humber mouth (Uncles et al., 1998b) bacterial processing of ammonia may be the source of the observed nitrite.

Sediments host denitrification and regeneration of ammonia and nitrite. In Humber sediments, denitrification has been correlated with sediment organic carbon content, macrofauna abundance and nitrate concentrations in the overlying water (Barnes and Owens, 1998). Denitrification rates were significant throughout the estuary with rates up to 10 mmol N $m^{-2}d^{-1}$ in the inner estuary, as low as 1 mmol N $m^{-2}d^{-1}$ elsewhere in the estuary and at most 0.19 mmol N $m^{-2}d^{-1}$ below the plume outside the estuary. Note that 1 mmol N $m^{-2}d^{-1}$ $\equiv$ 0.1 μmol $d^{-1}$ in 10 m water depth. Upper Forth ammonia and nitrite distributions are consistent with PON breakdown processes (e.g. denitrification) in reducing sediments. Rates of their benthic source are 19–44 and 0.39–0.81 mmol $m^{-2}d^{-1}$ respectively with higher concentrations in summer when there is less oxygen in the water and sediments (Balls, 1992). In the Tamar, nitrite and ammonium have maxima indicating sediment production at rates 1–2 mmol $m^{-2}d^{-1}$ (Knox et al., 1986). The intertidal zones of estuaries behave similarly. The Humber overall is a sink for nitrate, a strong source of ammonia, and a small source of nitrite (Mortimer et al., 1998). Large nitrate concentrations favor denitrification and hence this nutrient sink. Some zones were sources of nitrate such as where lower temperatures suppressed denitrification or oxygen levels. Local sinks and sources also prevail on tidal flats in the German Bight, where high ammonium concentration occur (Brockmann et al., 1994).

Transformations of phosphorus in estuaries may be complex, typically beginning with phosphate removal to suspended particulate matter (SPM) in low-salinity reaches. Examples are the Forth and Tay (especially in summer at a large turbidity maximum in the Forth) and the Ouse where colloids may also be a factor (Rendell et al., 1997). In the Chesapeake Bay, biological processes strongly affect phosphorus cycling. Particulate phosphorus is dominant, highest in the upper estuary and suffering most changes of composition in low salinity parts of the estuary. Dissolved organic phosphorus often exceeded dissolved inorganic phosphorus and both had highest concentrations in bottom water during summer anoxia (Conley et al., 1995). Further down-estuary, desorption from SPM often occurs, e.g. around the Humber mouth (Uncles et al., 1998b). In the Forth and Tay, there are mid-estuary inputs from desorption and sediment pore-waters; the summer flux from sediments greatly exceeds the upstream river input (Balls, 1992). In the Delaware River, a half or more of total phosphorus was retained, mainly via geochemical processes (Lebo and Sharp, 1992). Downstream in Delaware Bay, dissolved inorganic phosphorus (DIP) was taken up in phytoplankton and the particulate fraction of flux increased from 35% to 62%. The Bay retained total phosphorus during the winter-spring bloom and acted as a source during fall. Overall, a majority of regeneration occurred in the water column and 84% of the freshwater input of total phosphorus was eventually exported to coastal waters.

Outer-estuary nutrients are typically more conservative. Examples include the outer Loire estuary (Meybeck et al., 1988), the Humber mouth to the North Sea, even though nitrate ranged from 1.4 to 106 μmol (Uncles et al., 1998b), and nine large Chinese estuaries, despite upper-estuary desorption and variable riverine sources (Zhang, 1996).

The overall effect of estuarine processes can be assessed by comparing their rates, e.g.

$$\text{Depletion rate} = \text{(uptake or dissolved-particulate exchange rate) / concentration} \qquad (2)$$

with a flushing rate. If concentrations are moderate and flushing is slow, then even slow uptake may cause a notable reduction in downstream nutrient fluxes, as in the Tweed (Shaw et al., 1998). Fluxes in larger slowly-flushed Scottish estuaries are also affected (Balls, 1994). In the Ouse (Rendell et al., 1997) variations in flow, and hence flushing time (1 to 14 days), control the degree of nutrient conservation or depletion by production. Overall transport numbers for the Humber are shown in Table 1.

Silica and nitrate inputs from the Loire to the ocean are barely modified, but phosphate and ammonium inputs are increased; some large-flow events markedly change the pattern (Meybeck et al., 1988).

TABLE 1
Transport of nitrate and ammonium from the Humber to the North Sea (Tappin et al., 2001). Values in kmol/day.

| | Inputs | To (from) bed | To North Sea | denitrification |
|---|---|---|---|---|
| $NO_3^-$ | 9700 | 1000 | 9900, 10600 | 260 (spring), 600 (summer) |
| $NH_4^+$ | 1500 | (430) | 200 | |

The various processes just mentioned show how the ratios of nutrients eventually entering the coastal sea from the estuary may differ considerably from ratios in their riverine sources. The ratios may also differ between estuaries and with the adjacent coastal sea. The differing ratios of nutrients entering the coastal ocean will cause different limitations to plankton growth and plankton type. For example, phosphorus limitation is found in the Gironde plume (Biscay; Herbland et al., 1998), relative depletion of silicate from the Ouse affecting coastal-sea production species (Rendell et al., 1997), and limitation by nitrogen or phosphorus according to discharge concentrations and location in the Mississippi plume (Lohrenz et al., 1999). Nutrient-salinity plots often help the interpretation of distributions where the estuarine outflow pattern is complex (e.g. the German Bight; Körner and Weichart, 1991).

The flow of nutrients from land to the coastal ocean can vary. The nutrients may flow through a long estuary spending days or weeks such as in the Chesapeake or they may not encounter coastal waters at all but be directly placed in the open ocean as occurs off the Amazon or Mississippi on occasion. The pathway to the coastal ocean will affect the biogeochemical processes the nutrients undergo and, in the end, affect the ratios of the various nutrients when they finally enter the coastal waters. The ratios of the nutrients and their absolute concentration are, of course, critical to plankton growth. The open ocean on the other hand represents a large source of nitrate, phosphate and silicate in rather constant ratios compared to the riverine/estuarine source.

### *Oceanic inputs*

The major source of nutrients to coastal waters is from the deeper oceanic waters offshore over the slope. Currents transporting nutrients tend to flow along isobaths not across them thus other processes than normal along-isobath geostrophic flow must occur. These processes are as follows (Huthnance, 1995):

1. Processes specific to or enhanced at the shelf edge.
2. Relaxation of the geostrophic constraint near the equator (Coriolis term goes to zero).
3. Friction as occurs in Ekman layers.
4. Non-conservative processes. The net flux <uC> may be non-negative along some boundary as C may vary due to phytoplankton uptake of nitrogen for example.
5. Small scale (time O(1 day), length O(2–10 km), speed O(0.2–1 m/s)) local, non-linear and time dependent flows tend towards ageostrophic.

The following summary gives a flavor for the diversity of processes of circulation (advection), exchange (stirring), and water mass formation (mixing) that may enrich or deplete coastal ocean nutrient concentrations. Tables 2 and 3 show the scaling parameters and typical speeds of shelf-edge circulation processes and shelf-edge exchange processes respectively.

Of the various circulation processes (Table 2) those related to coastal currents, western boundary currents and their associated eddies, warm core rings and jets, and tides appear to be the most important for moving large amounts of water, and thus nutrients. Of course the importance depends on the nutrient concentrations.

TABLE 2
Scales and estimated values of process contributions to shelf-edge circulation. From Huthnance (1995) and see that paper for details.

| Process | Scale | e.g. m/s |
|---|---|---|
| Coastal current | ? | 0.1–1 |
| Slope current forced by | | |
| JEBAR | $h_o^2 \lvert \rho^{-1}\, grad\, \rho \rvert g/8k$ | 0.1 |
| Steady wind | $\tau/\rho k$ | 0.1 |
| Unsteady wind | $t\, \tau/\rho h$ | 0.1 |
| Biased form drag | $\tau/(2\pi \rho)\, min(1/k,\, t/h)$ | 0.01 |
| Wave rectification | $u^2 f / L_T \sigma^2$ | 0.01 |
| Eddy momentum | $u\, v\, h/L_T k$ | 0.1 |
| Western boundary current | $(L_X/L_Y)\, \sigma/\rho\, k_I$ | 1 |
| Eddies, warm-core rings, jets | ? | 0.5 |
| Tides | $\zeta\, max\{(g/h)^{1/2},\, \sigma W_S/h\}$ | 0.3 |
| In strait to marginal sea | $\sigma \zeta A / h\, b$ | >1 |

Shelf-edge exchange processes are the most important as they directly affect flow from the deep-ocean to the coastal ocean. The following table (Table 3) shows relative magnitudes of volume flow of the different processes. Note that the flux of nutrients depends on the concentrations of the nutrients times the volume flux. Most of the processes will be discussed in the following text.

Many of the processes shown in the two previous tables bring nutrients into coastal ocean waters but they are often in a bottom layer. Mixing and stirring are required to bring nutrient-rich waters to the upper layers to complete the ascent of nutrients from the deep ocean to the near-surface in the coastal ocean. Table 4 shows the relative importance of different processes that create or destroy stratification. Wind generated surface waves, tidal friction (for strong tidal currents) and especially intensified internal waves can create significant mixing.

TABLE 3
Scales and estimated values of process contributions to shelf-edge exchange. From Huthnance (1995) and see that paper for details. (See also table 4 below for topographic effects).

| Process | Scale | e.g. $m^2/s$ |
|---|---|---|
| Slope current | $kv/f$ | 1 |
| e.g. Atlantic inflow Malin-Lewis | | 0.2Sv/300km |
| Total Scottish slope | | 1 |
| Topographic irregularities | $V \Delta h_I$ | 1 |
| Eddy | © $h_O (h_O / \Delta h) / f$ | (1 Sv × 12 d) |
| Warm-core ring streamer | ? | (1 Sv) |
| Aggregate (Middle Atlantic Bight) | | 0.3 |
| Impulsive wind | $\tau / \rho f$ | 1 |
| Upwelling–wind | $\tau / \rho f$ | 1 |
| – div. W boundary current | $2h_O^2 V_W \partial_x (h_O V_W / \partial_x h_O)\ div / f$ | 20 |
| Jets (narrow-shelf upwelling areas) | ? | (2 Sv) |
| Aggregate | | 2 |
| Cascading (Shapiro et al., 2003) | $0.36\, g\, (\rho^{-1} \Delta\rho)\, h_x / f\, (\nu / f)^{1/2}$ | 1 |
| Front | $\alpha' h\, [g(\rho^{-1}\Delta\rho)\, h']^{1/2}$ | 0.3 |
| e.g. along isopycnals, Middle Atlantic Bight | | 0.2 |
| Tides | $\sigma \zeta W_s$ | 10 |
| Strait to marginal sea | $\sigma \zeta A$ | (> 1 Sv) |
| Shear dispersion (hu = $\sigma\zeta W_S$) | $t_D u h \lvert \boldsymbol{u} \rvert / L_T$ | 0.1 |
| Internal tide solitons | $<\zeta> \lambda / tide$ | 1 |
| Waves' Stokes drift | $0.01 w^2$ | 1 |

TABLE 4
Scales and estimated values of process contributions to energy potentially available for mixing. Mixing energy values ($mW/m^2$) are based on typical values for the quantities in the 'Scale' column given in Huthnance (1955, page 355). From Huthnance (1995); see that paper for details.

| Process | Scale | Typical Mixing Energy, $mW/m^2$ |
|---|---|---|
| Buoyancy flux (heat, cooling, rain) | $\alpha\, g\, h\, H / 2c_p,\ gh\, \Delta\rho(rain\ rate)/2$ | 1 |
| Surface waves | $1.5 \times 10^{-5}\, \rho\, g\, \sigma_W\, a^2$ | 150 |
| | or $5 \times 10^{-7}\, \rho\, w^3$ | 500 |
| Wind | $\tau v$ | 10 |
| Internal tides | $\rho g (\rho^{-1} \Delta\rho) <\zeta^2> \lambda / L_T\ per\ tide$ | 50 |
| Internal waves | $0.1 \times 1\ kW / m / L_T$ | 10 |
| Bottom-reflected internal waves | $Fn\ (h_x, f / N) \times 30\ mW\ m^{-2}\ flux \downarrow$ | 1 |
| Bottom friction | $\rho\, C_D\, v^3$ | 3 |
| Tidal (currents 0.3 or 0.7 m/s) | | 100 or 1000 |
| Canyon-intensified internal waves | $<\rho\, C_D\, u^3>$ | 150 |

Most of the processes mentioned here are shown schematically in Figure 8.1.

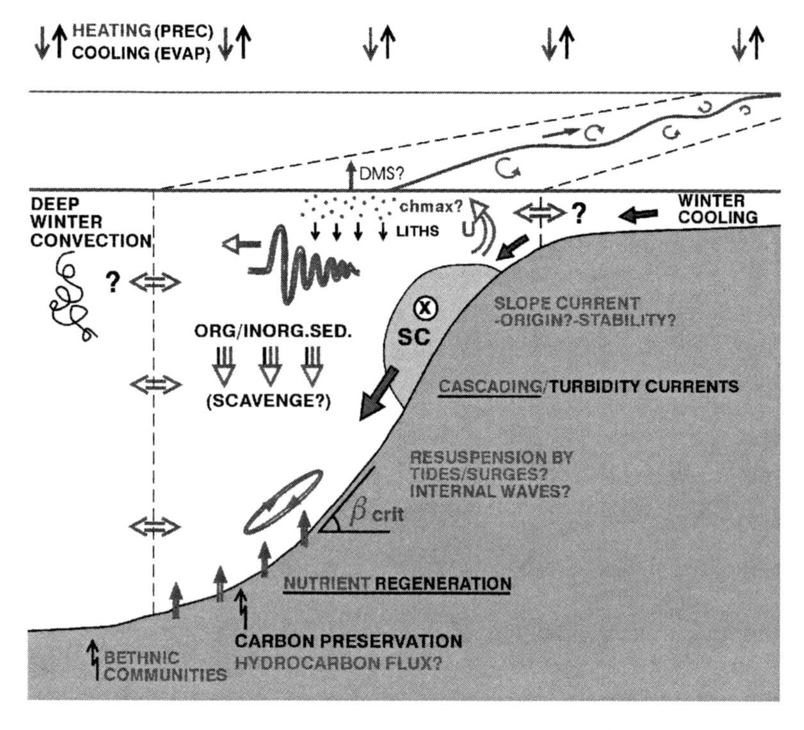


Figure 8.1 Schematic depiction of coastal ocean processes. From an original by John Simpson.

### *Sources, transport and sinks within shelf seas*

The oceanic source of nutrients is present to a greater or lesser extent at all shelf break regions. Typically the deep ocean is the largest source of nutrients however there are exceptions near the outflows of very large rivers. For example ~ 60% of nitrogen input to the northern shelf of the Gulf of Mexico is from the Mississippi River (Ortner and Dagg, 1995). Whether concentrations are comparable with oceanic values depends on the exact location of river outflow to the coastal ocean and on the balance, as in estuaries, between depletion rates and flushing rates. Broad shelves tend to be flushed slowly by oceanic waters, so that the balance determining nutrient concentrations may appear to be relatively local (even though salinity may be near oceanic values and the ocean dominates primary nutrient supply). Riverine sources (via estuaries) are initially localized at estuary mouths. Their domain of influence depends on a balance between the depletion rate and the transport rate. In the Celtic Sea, for example, separate nitrate/salinity mixing lines above and below salinity 35 suggest a long (seasonal) time scale for riverine influence to reach the majority of this shelf with salinity > 35 (Hydes et al., 2001). Rivers flowing into the southern North Sea are a locally important nutrient source limited to the sides of the English Channel (Laane et al., 1993), in the German Bight and west of Denmark (despite relatively small water volume transport). However, the typical transit time to the Skagerrak is O(6 months), allowing time for dissolved inorganic nitrogen (principally) to be taken up by production; these riverine inputs reach the Skagerrak mostly by down-gradient dispersion and after exceptionally large winter discharges of fresh water (Rydberg et al., 1996).

Total atmospheric nitrogen input per unit area, 50 mMol $m^{-2}y^{-1}$ or less (Duce et al, 1991; Cornell et al., 1995), is typically modest compared with water column cycling rates. For example, atmospheric nitrogen input to the North Sea is about 4% of the total, but 38% of that is from river runoff (Chester et al., 1993).

As already implied, production in the water column is a sink for inorganic nutrients. Organic nutrients (not emphasized here) may increase and, in turn, be taken up by some phytoplankton species (e.g. in the English Channel; Butler et al., 1979). Organic particles are subject to water-column remineralization (a direct source of inorganic nutrients), denitrification (a sink) and transport; particulate matter also sinks, at a rate that depends on particle size and hence turbulence (large particles tend to form in stratification with reduced turbulence and break up in strong turbulence). Organic material in the sediment bed is subject to erosion (resuspension), denitrification and remineralization (generally more slowly than in the water column). Resuspension of several centimeters of sediment is needed before there is significant total nitrogen input to the overlying water (of order 1 μM through 1m depth; Blackburn, 1997).

Nitrate exchanges with the bottom depend on sediment C/N ratios. The exchange is into the water column from low C/N sediments with fastest net regeneration and into high C/N (nitrogen-poor) sediments due to denitrification in deposition areas (Skagerrak; Hall et al., 1996). This is consistent with the Humber estuary findings previously mentioned (Barnes and Owens, 1998) except that in the Skagerrak the association is with C/N, not organic carbon content. Ammonium fluxes to the sediment were correlated with nitrate effluxes; take-up by nitrifying bacteria was suggested. Nitrite influxes to the sediments accompanied both high nitrate influxes and high nitrate effluxes. Phosphate fluxes in the Skagerrak appeared to be correlated with clay in the sediments. Nitrogen flux estimates for the Washington shelf (Christensen et al., 1987) are: 28.6 nmol N $m^{-2}s^{-1}$ total regeneration on the basis of carbon oxidized; nitrate influx 8.1 nmol N $m^{-2}s^{-1}$ and ammonium efflux 5.3 nmol N $m^{-2}s^{-1}$ on the basis of pore-water profiles; the "missing" regenerated nitrogen could be accounted for by coupled nitrification and denitrification (11.6 nmol N $m^{-2}s^{-1}$) and higher C:N ratio in the organic matter oxidized. Overall, the annual ammonium efflux was about half of the organic burial rate plus the nitrate influx suggesting a net sink for nitrogen. In 100m water depth, these rates were O(10 nmol N $m^{-2}s^{-1}$) or about O(3 μmol $yr^{-1}$).

In general, strong currents and turbulence favor water-column recycling while more quiescent water flow favors deposition and sea-bed recycling. The character of recycling in either the water column or the sediment medium—regeneration, (de-) nitrification, nutrient fluxes in/out of sediments—depends on redox chemistry. Models exist that represent our present understanding of nutrient behavior in the water column and in the sediment. A benthic module in the European Regional Seas Ecosystem Model (ERSEM) includes N, P, Si nutrient cycles with vertical transport, oxic and anoxic mineralization, silicate dissolution, adsorption, nitrification and denitrification. The model describes the seasonal variation of nutrient fluxes including sediment-water exchanges and (in the North Sea) the influence of organic matter deposition on benthic nitrification and denitrification *via* changes in oxygen availability to the nitrifiers (Ruardij and van Raaphorst, 1995).

In successive regions from the coast offshore, the relative importance of the different sources, transport processes and sinks varies. In turn there may be:

a) near-shore regions: surface-wave currents at the sea bed increase turbulence and mixing;
b) ROFIs: regions of freshwater influence with river/estuary buoyancy and nutrient input, strong offshore salinity and temperature gradients and possible rapidly-varying stratification;
c) regions well-mixed throughout the year by tidal and wind-driven currents;
d) tidal-mixing fronts between (c) and (e);
e) regions that are thermally stratified in summer;
f) the shelf edge adjacent to the open ocean and with particular processes including internal waves that can be an interior source of turbulence and mixing.

Regions a) and f) always exist; the others may be well-expressed on a wide shelf but merge over a narrow shelf. A more specific discussion is given in section 7.

We now discuss the processes that appear important for understanding nutrient distributions.

## 4. Along-slope Currents

High currents in the vicinity of the shelf break is a common feature of most continental shelves. The forcing mechanisms are many (Huthnance, 1992) but an important common aspect is that cross-isobath flow does occur (in particular, in any bottom Ekman layer) as do vertical motions. Thus, along-slope currents are of importance to the distribution of nutrients along and across the coastal ocean. In this section, we discuss the following from Table 8–02: freshwater buoyancy forced flows; western boundary currents; eddies, warm core rings and jets.

### *Buoyancy Forced Coastal Currents*

Freshwater buoyancy fluxes into the coastal ocean cause baroclinic coastal currents. In such regions of freshwater influence (ROFIs), the buoyancy input and strong offshore salinity and temperature gradients may cause rapidly-varying stratification. The baroclinic coastal currents are usually narrow compared with shelf width; thus the influence on nutrient distributions may be localised. On narrow shelves, such as off southeast Alaska or Norway, the coastal current may be over the slope or shelf break. Associated cross-isobath flows are O(0.1 m $s^{-1}$). Classic coastal currents occur off southern Chile, British Columbia and southeast Alaska, the northeast coast of North America, Norway, China, and other locations receiving freshwater input. They are obviously important if the river influx carries a significant nutrient load. In fact, many ROFIs carry high concentrations of riverine nutrients that may support primary production. Some ROFIs carry particulate matter which may block light, inhibiting production. The strong stratification imparted by the buoyancy flux can inhibit vertical mixing and thus affect biogeochemical processes.

In the Adriatic, fresher water from the Po River outflow extends southwards along the Italian coast with strong offshore decreases of nutrient (and phytoplankton) concentrations; phosphorus appears to be the element limiting production (Zoppini et al., 1995). Fresh-water sources of nutrients are illustrated by similar salinity and nutrient contours in the Rhine outflow northwards along the Dutch coast (Simpson et al., 1993).

In the Mississippi plume, closely-linear decreases of nitrate, silicate and phosphate with increasing salinity (Hitchcock et al., 1997) illustrate their riverine source (riverine nitrate and silicate commonly exceed 100 μmol; Ortner and Dagg, 1995). Figure 8.2 shows surface nitrate concentrations in April during high Mississippi River flow and subsequent high nitrate transport. The high concentrations in April are advected westward in the buoyancy- and wind-forced coastal current.

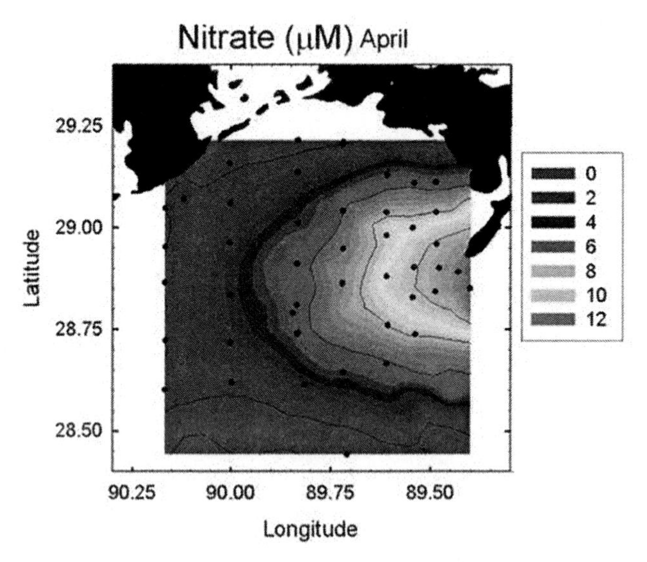


Figure 8.2 Surface nitrate concentrations off the Mississippi River (From Rabalais et al., 2002).

Biological uptake depletes the nutrients in the Mississippi plume, but only by a small fraction in its core; farther from the Mississippi River mouth regenerated nutrients become more significant. Similar maxima of primary production (at salinities 15–30) and biomass (at the convergent plume front) are light-limited in turbid low-salinity water and nutrient-limited outside the plume (Hitchcock et al., 1997; Lohrenz et al., 1999). SPM at these primary production maxima is less than in the river, allowing light penetration; nutrients are still present after near-conservative mixing. Because time scales are short (1–2 days in the plume) the water reaching the plume front has experienced maximal time for phytoplankton growth. Walsh et al. (1989) estimate that less than 25% of the nitrogen effluent from the Mississippi may remain in the sediments, most of the input being rematerialized after several cycles of production and remineralization. To the west on the

Louisiana-Texas shelf, the Mississippi River nutrient source, combined with cross-frontal flows at the plume boundary and production in illuminated upper waters, gives a "dome" of concentrated nutrients near the bottom of the plume front (Chen et al., 1997).

Studies of other large river plumes show similar processes. SPM limitation of light and production in the plume *versus* nutrient limitation outside the plume gives highest primary production bordering the high-nutrient Changjiang plume (Tian et al., 1993) and Amazon plume (DeMaster et al., 1996). These processes tend to sharpen the nutrient front. Around Chinese river plumes, regeneration from organic matter decomposition and nitrification may sustain concentrated nutrients near the bottom (Zhang, 1996), however, phosphate is limiting near the coast. Strong currents in the Amazon outflow inhibit burial of biogenic material, implying that regeneration would occur, there would be little trapping of nutrients, and phytoplankton production would continue by repeated utilization of remineralized nutrients. Sub-surface oceanic water also supplies nutrients in onshore near-bed flows converge (DeMaster and Pope, 1996).

Distributions of reactive phosphorus near the Otago peninsula, New Zealand, provide a contrast. There riverine nutrient concentrations hardly exceed those on the shelf and advection from offshore along the outer shelf dominates the supply. The outcome is nutrient depletion on the shelf relative to further offshore, especially in spring and summer (Hawke and Hunter, 1992). More gradual depletion in one spring was attributed to lower river flow and hence less stable stratification.

Variable stratification also affects nutrient supplies. Stratification in the Rhine River plume may be destroyed by spring tides or wind-mixing. An April example of strong wind-mixing making nutrients available near the illuminated surface (and fuelling a bloom) is described by Joordens et al. (2001). In the Gulf of Gdańsk, where tides are small, winds have a strong effect on movements of its waters and riverine nutrients. High ammonia concentrations in July may have come from regeneration by abundant zooplankton and silicate concentrations in November may have been depleted by a diatom bloom (Pastuszak, 1995).

At higher latitude locations, such as the Gulf of Gdańsk, wind stress is high and variable. Downwelling-favourable winds tend to confine an outflow plume near the coast and mix it strongly downward. Upwelling winds tend to spread it in a surface layer away from the coast. Such varied mixing and vertical motions during repeated upwelling and downwelling events complicate the nutrient distribution regime.

### *Western Boundary Currents*

Western boundary currents affect the distribution of nutrients in three general ways. The first process is the shelf-parallel transport of nutrients in the overall flow of the western boundary current. This has been called the 'nutrient stream'. The second process is inherent in the dynamics of the large-scale boundary current; it combines cross-slope flow along isopycnals (flow along constant density surfaces) and diapycnal mixing (across constant density surfaces). The third, smaller-scale group of processes is related to instabilities in the western boundary current front, interactions with adjacent bathymetric features, Ekman related upwelling and cross-isobath flow.

For context, we look at the distribution of nutrients across a western boundary current such as the Gulf Stream. A typical section shows the uplifting of colder, nutrient-rich water over hundreds of meters in the front (Figure 8.3). The nutrient content of the ascending waters depends on the characteristics of the water masses in the front: the T/S/nutrient relationships. Nutrient-temperature correlations in the Gulf Stream show a strong correlation below the surface layer. Gulf Stream water colder than about 18°C invariably has elevated nutrient concentrations. Similar relationships hold in other western boundary currents.

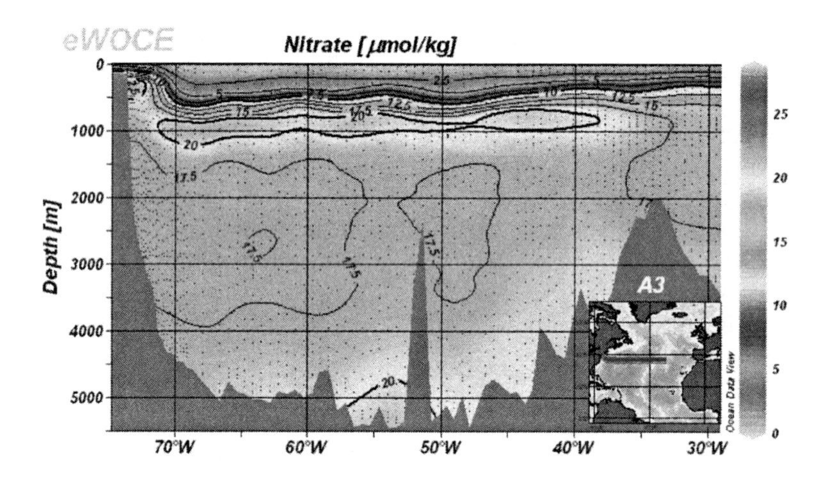


Figure 8.3 Nitrate section across Gulf Stream. Data from WOCE Section A03.

***The Nutrient Stream*** The 'nutrient stream' refers to the nutrient transport inherent in the flow of western boundary currents. It is simply the integrated sum of the nutrient concentrations and along-stream velocities (Figure 8.4). Pelegri and Csanady (1991) made several calculations between 24°N and 35°W in the Gulf Stream. Typical results for the 36°N section were: nitrate 863 kmol $s^{-1}$; phosphate 55 kmol $s^{-1}$; silicate 508 kmol $s^{-1}$. This transport northward into the North Atlantic gyre represents the main source of nutrients to that region (Brewer et al., 1989).

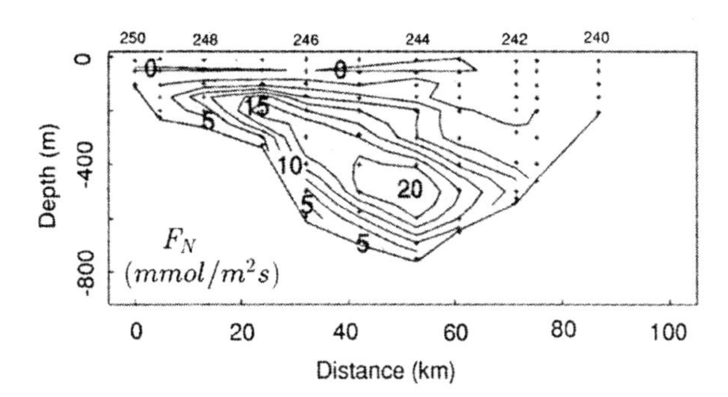


Figure 8.4 Nitrate flux density distribution for section in 24°N across Gulf Stream (From Pelegri and Csanady, 1991).

The shelf-parallel 'nutrient stream' might seem to be unimportant to the distribution of nutrients in the global coastal ocean. However, this process brings nutrient-laden water poleward from the tropics to areas where other processes (to be described next) bring the water laterally into the coastal ocean and upwards into the surface layer.

***Isopycnal Flow and Diapycnal Mixing*** Cross-stream advection along isopycnal layers in the Gulf Stream (Rossby, 1936) effectively transports nutrient-rich deep ocean water (Redfield, 1936) to upper layers and adjacent slope and shelf waters. As noted so succinctly by Yentsch (1974):

> "The combined effects of the earth's rotation and pressure gradients which are associated with ocean current flow produce the effect of drawing water in along the right side of a current ... and discharging into a counter current to the left of the main flow. ... This transfer of waters is along lines of equal density which slope dramatically upward toward the coast. The higher density waters are nutrient rich and stimulate production along the left side of the current. ... This means the biochemical factors of slope and coastal waters are generated from nutrient characteristics of deep open ocean waters."

Studies of the isopycnal flow and diapycnal mixing processes from the thermocline layer to the surface layer show the upward entrainment rate into the surface waters is typically 1.6 $m^3s^{-1}$ per meter along the axis of the Gulf Stream (Pelegri and Csanady, 1991). Over the approximate 80 km width of the Gulf Stream, this amounts to a diapycnal upwelling velocity of $2x10^{-5}$ m $s^{-1}$. Interestingly, the equivalent upwelling velocity including exchange was similar to equatorial upwelling that is about 3.2 $m^2s^{-1}$. The diapycnal velocity causes the advection of high nutrients into the surface waters of the coastal ocean from the deeper waters of the boundary current.

Typical fluxes by this process can be calculated assuming that half the entrained water(0.8 $m^2s^{-1}$) feeds into an onshore flux, rather than directly into production in Gulf-Stream surface waters and a nitrate concentration 20 μmol/l (Pelegri and Csanady, 1991). This would amount to an onshore flux of 16 mmol $s^{-1}$ per meter along the axis of the Gulf Stream. Over a 1000 km long segment of the Gulf Stream this amounts to a nitrate flux 16,000 or 7 Mtons N per year. This nutrient rich water feeds directly into the productive surface waters of the Gulf Stream front and adjacent coastal waters.

Recent studies of advection of nutrients along isopycnals (Schollaert et al., 2003) show that the supply of nutrients into the Slope Sea north of Cape Hatteras is constant regardless of the position of the Gulf Stream (onshore or offshore). However, the concentration of nutrients in the Slope Sea (between the Gulf Stream and the shelf break) varies as the volume of Slope Sea varies: there are higher concentrations when the Gulf Stream is onshore and the volume of the Slope Sea is less; vice-versa for an offshore position of the current.

The magnitude of the flow and exchange in the main part of the western boundary current depends on many factors. The strength of the current causes the slope of the isopycnals to vary. The nutrient content may vary on a given density surface depending on the source of the water (Richards and Redfield, 1955). Surface processes such as heat flux and wind-mixing alter the mixing between the surface layer and the thermocline.

***Meanders*** Large meanders, such as the Charleston Bump off Charleston, South Carolina (Fig. 8–005) or the large gyre off Japan, result in localized upwelling and onshore flow. Upwelling is due to the conservation of potential vorticity; surface waters are carried onshore on the downstream side of the meander (Brooks and Bane 1978). North of Cape Canaveral, onshore flow brings nutrient rich water from depths of several hundred meters in the Gulf Stream into waters 20 m deep on the adjacent shelf (Arthur, 1965; Blanton et al., 1981). As the Gulf Stream passes the Charleston Bump a gyre forms with upwelling in its center (Brooks and Bane, 1978; McClain and Atkinson, 1985). The gyre has elevated nutrient concentrations because of the internal upwelling; the adjacent shelf waters receive these nutrient-rich upwelled waters through advection around the gyre. The shelf waters also receive warm buoyant waters that induce stratification in winter, when these shelf waters are otherwise usually unstratified (Atkinson et al., 1989).

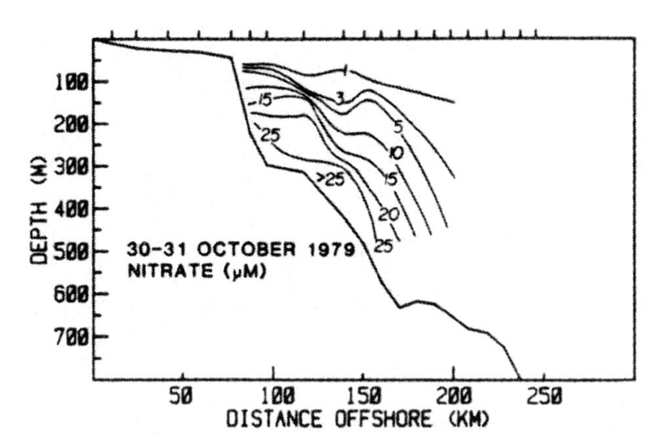


Figure 8.5 Nitrate section across the Charleston Bump off Charleston, South Carolina (from Singer et al., 1983).

***Instability Processes*** Instabilities along the western boundary current front create frontal eddies which migrate poleward in the front. These features have an upwelling core. Upwelled, nutrient rich water can move from the eddy core onto the shelf under special circumstances (Fig. 8.6). Upwelling winds along the shelf break can induce an onshore Ekman flow at depth that brings the nutrient rich water from the eddy core to the shelf.

***Overall nutrient fluxes*** Extensive measurements of nutrient fluxes related to the passage of frontal eddies were summarized by Lee et al. (1991). These observations present the net effect of many processes related to the interaction of western boundary currents with the shelf and slope. Figure 8.7 shows the resultant cross-shore flux.

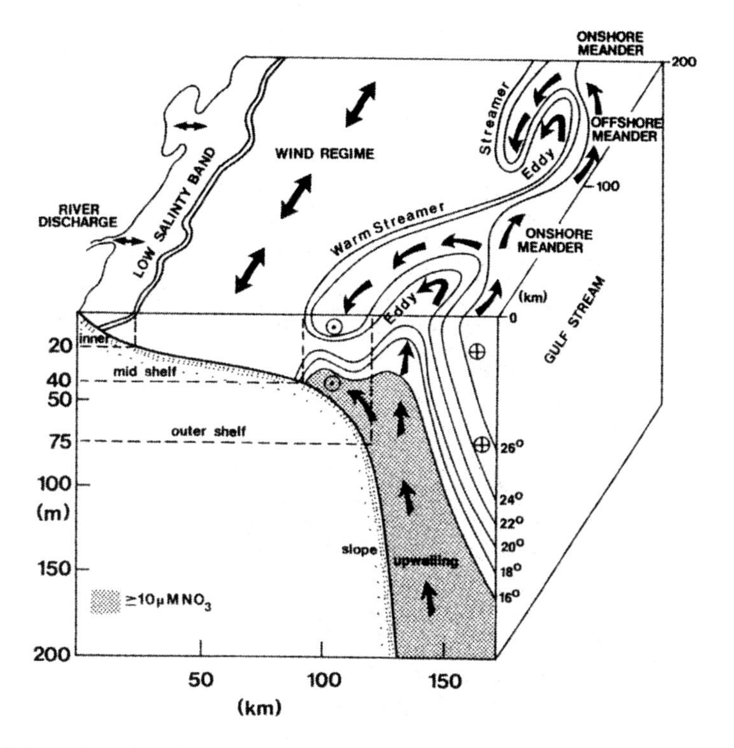


Figure 8.6 Schematic of Gulf Stream frontal eddies and meanders together with shelf flow regimes on the South Atlantic Bight (SAB) (From Lee et al., 1991).

## *Eastern boundary currents*

The ubiquitous combination of oceanic meridional density and pressure gradients with shelf-slope topography provides forcing for poleward along-slope flow. Such flows are most obvious on eastern oceanic boundaries. Poleward currents originate equatorward of upwelling areas and typically reach high latitudes (exchanging water laterally as they go but retaining characteristics of water at lower latitudes). The Peru-Chile Undercurrent reaches at least 48°S (Silva and Neshyba, 1979). Off the west coast of North America, the poleward undercurrent reaches at least 51°N (Pierce et al., 2000). Off western Europe, poleward along-slope flow extends from Portugal to Scotland. Typically the currents are confined over the upper slope, flowing at depths ranging from 200 to 500 m or more (Blanco et al., 2001; Kosro, 2002). Outside upwelling latitudes such poleward currents may reach the surface during certain seasons.

Broader-scale eastern boundary currents typically form part of an ocean gyre (with offshore extent set by the decay distance of baroclinic Rossby waves). In contrast with western boundary currents, eastern boundary currents are typically slow, and the adjacent shelf is often very narrow. The four major eastern boundary current (EBC) systems are the California, Canary, Peru-Chile and Benguela currents.

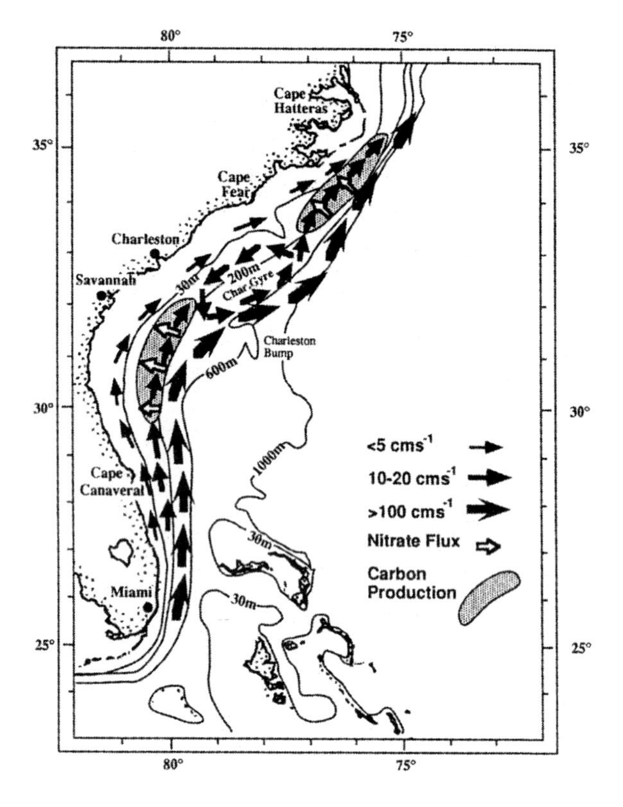


Figure 8.7 Characterization of mean circulation, onshore nitrate flux and sites of potential Gulf Stream-induced new carbon production in the SAB during winter and spring conditions (From Lee et al., 1991).

## 5. Cross-shelf transports

In this section cross-shelf transport processes affecting nutrient distributions are described. A discussion of processes dominant in upwelling systems are followed by discussion of the effects of canyons and capes, embayments, coastal trapped waves, and tidal processes.

### *Upwelling*

All eastern boundary current systems have a strong seasonal upwelling cycle associated with the relative strength of atmospheric pressure systems. The strength and duration of upwelling varies seasonally, latitudinally and interannually. In general, at lowest latitudes the upwelling variability is linked to the ITCZ, where equatorward wind stress predominates all year. In these lower latitudes, rainfall often creates significant freshwater buoyancy flux. At mid-latitudes, upwelling winds vary seasonally and precipitation is scarce (Bakun and Nelson, 1991). It is here that coastal upwelling brings cold and high-nutrient subsurface water into the euphotic zone, causing phytoplankton growth. With increasing latitude, atmospheric pressure fields produce greater seasonality. Fresh-water buoyancy and downwelling often dominate in winter. In higher latitudes, upwelling winds are less frequent and downwelling becomes dominant. Because of the heavy rainfall and runoff, buoyancy-forced coastal flows become more important at higher latitudes (Hill, 1998). On the Washington shelf, for example, summer upwelling causes higher nutrient concentrations in bottom waters, even inshore of the surface Columbia River plume (and supporting high primary productivity); in winter, without upwelling, nutrient concentrations may be higher in the surface river water (Carpenter, 1987); the Columbia River may often be a relatively large source of dissolved silicon, but upwelling contributes most of the nitrate and phosphorus. The Columbia River plume combined with the coastal upwelling presents an interesting and complex nutrient supply situation.

In eastern boundary current systems, the transport of nutrients into the coastal ocean relies on two basic elements: onshore Ekman transport and a deep offshore source of nutrients, often a poleward-flowing undercurrent. The movement onshore and offshore is complicated by bathymetric features, such as capes and canyons, and by cross-shore flowing jets, streamers and intrusions. The following subsections describe these elements of nutrient transport.

***Poleward flowing undercurrent*** In almost all upwelling systems, the source of upwelled water is the nutrient-rich water of the subsurface poleward current (Neshyba et al., 1989; Morales et al., 1996; Blanco at al., 2002). Figure 8.8 shows nutrient and oxygen profiles off central Chile. Note the very low oxygen and high nutrient concentration in the core of the poleward flowing Peru-Chile Undercurrent. Also note the intrusion into the coastal waters as part of the onshore Ekman flow. In near-coastal waters, this transport of new nutrient results in chlorophyll concentration exceeding 3 mg $m^{-3}$ (Thomas et al., 1994) and in annual production rates > 200 g C $m^{-2}$ (Berger et al., 1987). The water also has high salinity, low pH and high $CO_2$. These properties result from the equatorial origin and the cumulative effects of planktonic community respiration as the water moves poleward. High oxidation rates of sinking POM contribute to the nutrient maximum and oxygen minimum. Denitrification may also be an important respiratory mode in these oxygen-deficient waters (Codispoti and Christensen, 1985); organic matter degradation and denitrification were the suggested cause of near-zero nitrate in equatorial water at 50–90 m depth off the Peru shelf (~ 15°S westward of up-tilted isotherms; accompanied by low oxygen and more ammonia; Copin-Montégut and Raimbault, 1994).

***Ekman Process*** The generally alongshore flows of nutrient-laden water over the slope would be inefficient nutrient sources for coastal seas were it not for Ekman transports. Near-surface offshore Ekman transport $\tau/f$ (proportional to alongshore wind stress and inversely proportional to the Coriolis parameter) is compensated by onshore flow at depth. Figure 8.9 shows the deeper compensation flow of water from the undercurrent moving onshore and into shallower waters. The quantity of the nutrient transport depends on the mass flux that is set by the alongshore wind speed and the nutrient content of the undercurrent source waters.

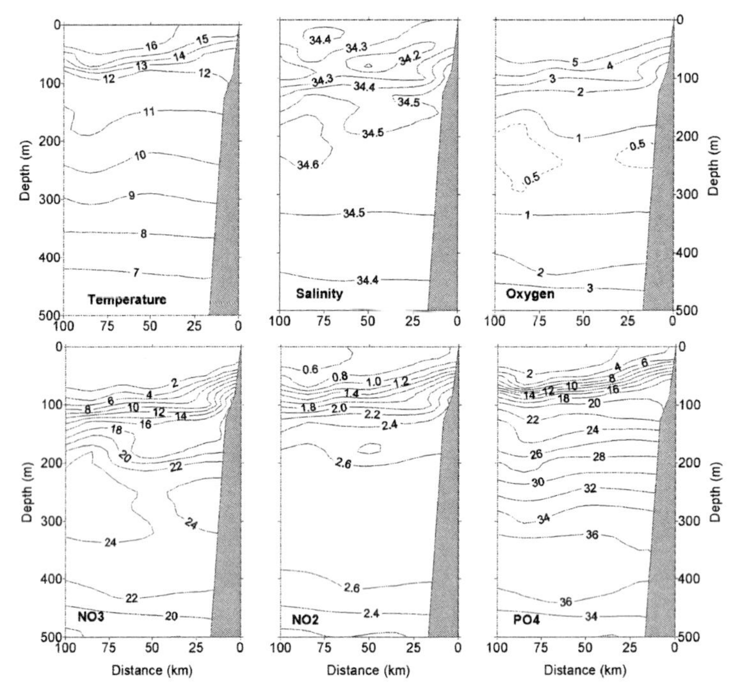


Figure 8.8 Vertical/cross-shelf distribution of temperature (°C), salinity (psu), oxygen (ml/l), nitrate, nitrite and phosphate off Coquimbo (30°S), Chile, May 1992. SCORPIO cruise database.

During upwelling, cold nutrient-rich water moves onshore and shoals, eventually breaking the surface. This is the typical upwelling signature: cold, high-nutrient, high-salinity, and low-oxygen. With further upwelling, the band of upwelled water moves offshore. The front between the upwelled water and the displaced warmer surface water is in approximate geostrophic balance. Flow at the front is geostrophic and in the same direction as the along-shore wind; it may be unstable to on-offshore disturbance. The flow interacts with any capes or canyons and can lead to important cross-isobath flow: jets, squirts and offshore eddies (Fig. 8.10). These cross-isobath transports often exceed the Ekman transport. Off California, the jets may be regarded as meanders in the along-shore geostrophic flow. As a result, e.g. off Point Arena, California, some of the upwelling nutrient may be nitrite resulting from upstream production and regeneration (Kadko, 1993).

### *Filaments*

The interaction between the subsurface upwelling water and the more oceanic surface water has been the object of intensive studies in upwelling areas, especially through the Coastal Transition Zone Program off California (Brink and Cowles, 1991) and the Ocean Margin Exchange (OMEX) project in the Iberian peninsula (Joint and Wassmann, 2001).

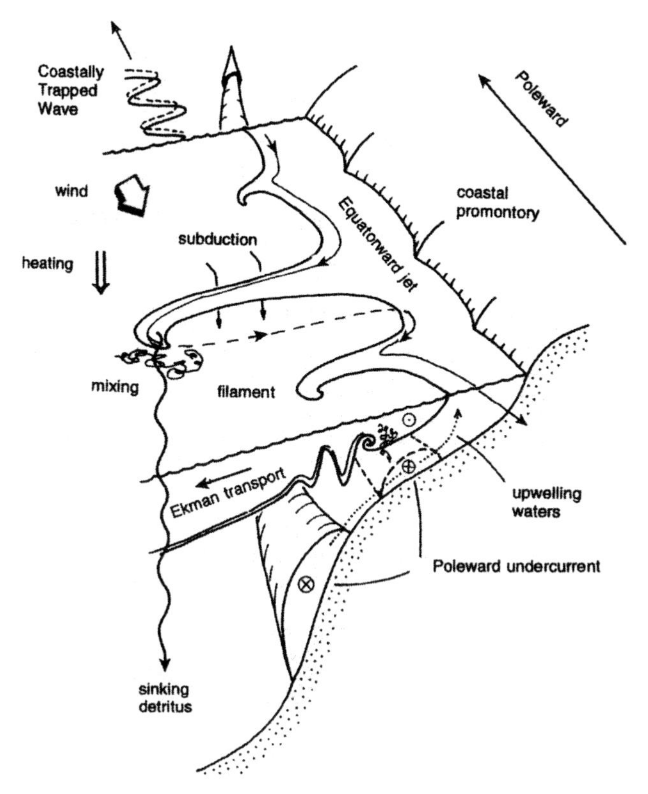


Figure 8.9 Schematic upwelling system for the northern hemisphere (From Hill et al., 1998).

According to Strub et al. (1991), several conceptual models of velocity structures can be associated with cold filaments that have been observed through satellite images in the California Current System and in other eastern boundary currents. The first model corresponds to squirt-like jets, which transport coastal upwelling waters to the open ocean, becoming a pair of vortices that probably rotate in opposite directions. A second conceptual model incorporates a series of mesoscale whirls embedded in a slow equatorward current (Mooers and Robinson, 1984; Rienecker et al., 1987). A third conceptual model consists of a steady equatorward jet, which meanders toward and away from the coast. However, in this model of a meandering jet, the jet is the primary structure and source of energy; this structure also implies that the nutrients (and biomass) of the coastal ocean tend to remain on the inshore side of the jet (Strub et al., 1991).

Filaments are typically cold, less than 100 km wide but several hundreds of kilometers long, extending from the coast towards the open ocean (e.g. off California; Brink and Cowles, 1991; Sobarzo and Figueroa, 2001). These cold filaments are associated with the upwelling season (spring and summer) rather than wintertime; they are mainly associated with capes or promontories. In these areas, meandering currents separate coastal and oceanic waters that differ physically and biologically (Lutjeharms and Stockton, 1987).

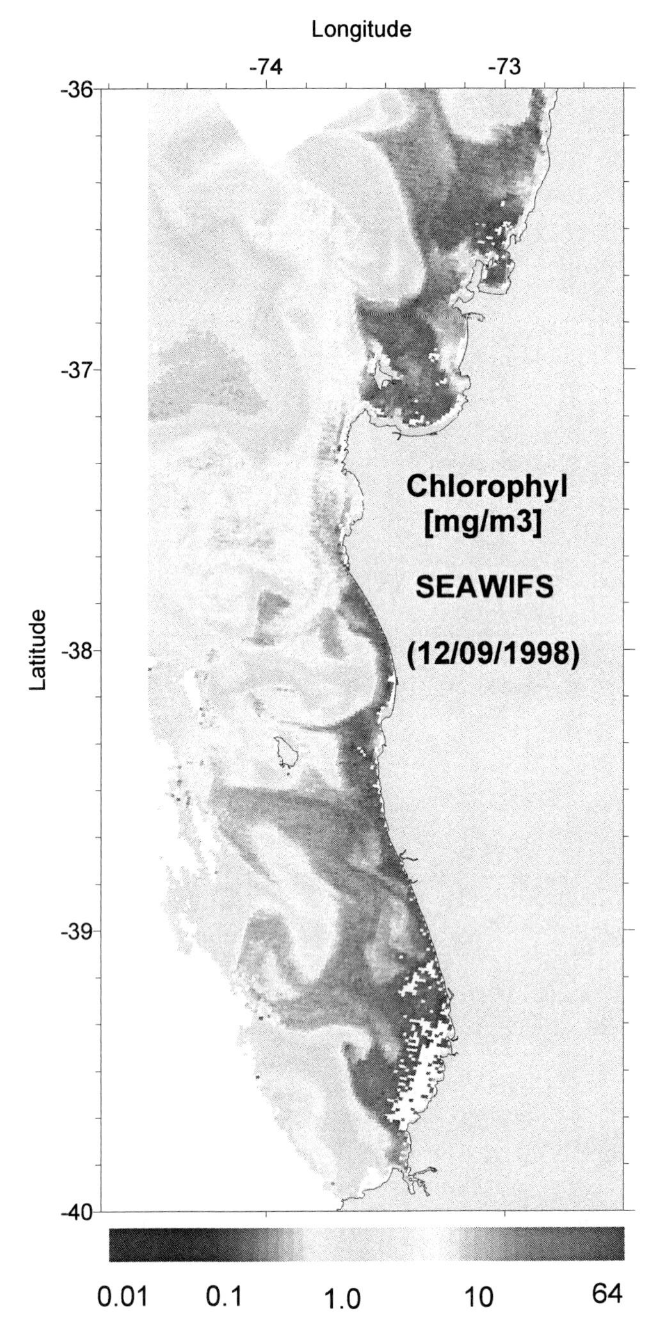


Figure 8.10 Surface chlorophyll SeaWiFS image over the Chilean coast, for 9 December 1998 (From Atkinson et al., 2002).

Exchange flows in filaments after upwelling are variable, but appear to produce more cross-margin exchange than the Ekman transport forced upwelling. Filaments transport water offshore and may also bring water from large distances onto the shelf. Off northwest Iberia, the estimated filament flux is about 1.5 $m^2 s^{-1}$ or half of the overall ocean-shelf exchange, which, in turn, would replace the volume of water on this fairly narrow shelf in about 12 days (Huthnance et al., 2002). Upwelling and, to some extent mixing, is estimated to supply the shelf with the nutrients from the upper 200m of water annually. A patch of upwelled water, followed along the shelf break in early August 1998, showed reduction of near-surface nitrate and nitrite by primary production; depletion deepened to 30 m in 5 days; just below, nitrite then had a marked maximum (Joint et al., 2001a,b). A filament tracked in mid-August had extremely low upper-layer nitrate, nitrite and ammonium but a small nitrite maximum at ~ 50 m depth. In both cases, silicate and phosphate were closely related to nitrate, but some silicate and phosphate remained even where nitrate was exhausted.

### *Coastal Trapped Waves*

Rather than making an independent contribution, coastal trapped waves (CTW) underlie and modify most processes in ocean-margin circulation, exchange and mixing, especially regarding their space-time distribution along the margin (Huthnance, 1995).

Strong, intraseasonal variability (of sea level, coastal currents and sea surface temperature with periods of about 40 to 70 days) has been observed along the eastern boundary (Enfield, 1987; Shaffer et al., 1997; Hormazabal et al., 2001). Much of this variability has been attributed to the presence of free CTW, which are generated when oceanic equatorial Kelvin waves impinge on the eastern boundary and are then deflected to propagate poleward along the coast (Enfield, 1987; Shaffer et al., 1997). Effects of CTW on the circulation, exchange and mixing patterns are more important during summer and particularly during El Niño events (Blanco et al., 2002, Hormazabal et al., 2002).

### *Embayments*

Equatorward facing embayments, such as the Gulf of Arauco and Monterey Bay, cause enhanced cross-shore flux of nutrients and retention within the embayment. In both cases, submarine canyons further complicate the situation. Especially in the Gulf of Arauco (Valle-Levinson et al., 2003), the nutrient transport is apparently very high and other conditions, such as stratification and mixing, facilitate very high production at primary and higher trophic levels.

### *Canyons and Capes*

Capes and submarine canyons have long been recognized as areas of enhanced biological production. This implies that some combination of advection and vertical mixing brings nutrients into the euphotic zone. This process is enabled by small bathymetric scale features such as capes or canyons that may cause along-shore flow to become ageostrophic. Stratification may decouple the flow from the bathymetry (at scales O (10km)), inertia may become important (at scales O(1 km)) or a small cyclonic bend may cause an offshore movement of the current. Locally, small-scale topography can induce tidal rectification and e.g. ridge-associated up- and down-welling (Tee et al., 1993); off Nova Scotia, after mixing to the surface, anomalously high nutrients (and low temperatures) from depth are found in summer.

The cross isobath flow associated with these processes is summarized in the following table:

TABLE 5.
Cross isobath flow processes (from Huthnance 1995).

| Process | Scale | e.g. $m^2/s$ |
|---|---|---|
| Western boundary current and bend | $(h/h_o)^2 (L_X/L_Y)\ \tau/\rho\ \beta$ | (0.5 Sv) |
| Slope current and bend θ | $v\ \theta\ k\ L_T/f$ | (0.01 θ Sv) |
| Cape eddy | $h\ v\ L_T$ | (0.1 Sv) |
| Canyon return flow | $h\ \tau/\rho\ k$ | 10 |
| Ridge-associated upwelling | $(2\Delta h/h_o)\ \tau/\rho\ f$ | 1 |

A schematic of flow in a canyon is shown in Figure 8.11.

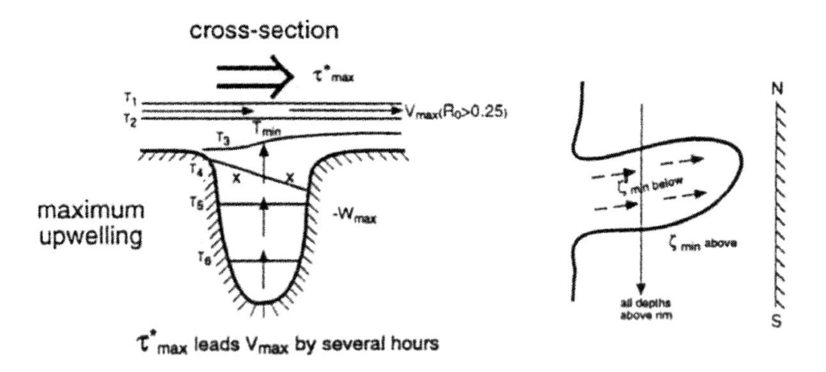


Figure 8.11 Schematic of flow in submarine canyon illustrating data derived characteristic timescales and spatial patterns of alongshelf wind stress, velocity, temperature and vorticity of a maximum upwelling event (From Hickey, 1997).

As with the other processes, the impact on the distribution of nutrients depends on the nutrient content of the water being exchanged.

*Tides, internal tides and shear dispersion*

Tidal currents have along-shore and across-shore components that vary greatly with location. A scaling for the across-shore transport, *hu*, is provided by the water required to raise the surface from low to high tide: $hu \sim \sigma W_S \zeta$ where $\sigma$, $\zeta$ are the tidal frequency and amplitude (in elevation) and $W_S$ is the shelf width (Huthnance, 1995). This may be very large, e.g. 14 $m^2 s^{-1}$ for semi-diurnal $\sigma$, $W_S$ = 100 km, $\zeta$ = 1 m. However, most of this transport returns six hours later (for semi-diurnal tides) and gives little time for any nutrient content to be changed. Hence shear dispersion may be the main process of lateral exchange: nutrients with different depth histories in vertically-sheared tidal currents suffer different net lateral displacements. Comparing modeled to observed tidal-average dispersion over the northwest European shelf, Prandle (1984) found an effective dispersion coefficient $t_D\boldsymbol{u}|\boldsymbol{u}|$ with $t_D = 10^3$s, $\boldsymbol{u}$ being the tidal current above bottom boundary shear, e.g. 20 $m^2s^{-1}$ for the above example and (shelf-edge) depth h = 100 m. This is relatively small: a reduction factor $h/L_T$ gives the equivalent exchange rate, $L_T$ being the topographically related distance for substantial changes in nutrient concentrations.

Internal tides, which are typically induced by tidal flow across the steep continental slope, vertically displace isopycnals, and associated pressure fields and depth-varying currents. Sufficiently large vertical displacements form (non-linear) solitons that transport water and nutrients "bodily" within the wave form. Fluxes per tide are $<\zeta>\lambda$ and may be O(1 $m^2s^{-1}$) for soliton amplitude $\lambda \sim 50$ m and aggregate soliton length $\lambda \sim 1$–2 km. This means of nutrient transport may be significant for production on the Scotian shelf (Sandstrom and Elliott, 1984).

Rectified tidal flow scales as $u^2/(\sigma L)$ and is small except where tidal currents $u$ are strong and vary on a short length scale $L$. This may be the result of topography (cf. the example in *Canyons and Capes* above) or internal tide structure (notably close to the shelf break).

## 6. Mixing Processes

While it is common for exchange processes to bring water from depth into the coastal ocean, those processes often do not bring nutrient-laden water into the euphotic zone itself. To bring the nutrients to the surface layers where phytoplankton can grow, mixing is critical. Processes that are potentially available for mixing are summarized in Table 4. In particular, buoyancy flux can provide the energy to stratify or de-stratify a water column. Heat loss can equal the mixing of winds and waves. Conversely heat gain and rain or freshwater influx enhances stratification inhibiting mixing. Typically, several processes are greater than the wind-mixing rate.

Since this book is meant for a broad audience, it is worthwhile to review the processes that create and destroy stratification. A stratified water column of a given depth has denser water at the bottom than at the top. When the water column is mixed such that the density is increased at the top and decreased at the bottom, the center of gravity of the water column is raised. This potential energy increase requires energy, which comes from buoyancy forcing (heat, freshwater, evaporation) and mixing forces in as shown in Table 4.

Since many parts of the coastal ocean are stratified, it is apparent that during much of the year heating and freshwater influxes overcome the mixing forces of wind, waves and tide. Only in the cooler seasons, with heat loss, stronger winds, and (in some areas) decreased freshwater influx, does the water column mix. Relatively easy calculations can determine what strength and duration of winds or tidal currents will mix a water column. Lund-Hansen et al. (1996) provide equations and examples of the procedure. The following is taken from their paper.

Stratification in coastal waters can be quantified as follows (Simpson and Hunter, 1974):

$$\varphi = \frac{1}{h}\int_{-h}^{0}(\bar{\rho}-\rho)gz\,dz$$

where $\bar{\rho}$ is the vertically averaged density, $g$ is the gravitational acceleration (9.8 m s$^{-2}$), $\rho$ is the water density (kg m$^{-3}$), and $z$ is the sample depth(m). Changes in stratification can be represented by using the following formulation (used generally for coastal waters):

$$\frac{\partial\varphi}{\partial t_{w,h,e,c}} = -\delta K_s \rho_a\left(\frac{W^3}{h}\right) + \frac{\alpha g Q}{2c_p} + \frac{1}{320}\frac{g^2h^4}{N_z\rho}\left(\frac{\partial\rho}{\partial x}\right)^2 - \varepsilon K\rho_w\left(\frac{\bar{u}^3}{h}\right)$$

Time change of φ — Wave Mixing — heating cooling — Estuarine Circulation — Tidal Mixing

where the first term on the right is the energy derived from the wind ($W$), the second is the buoyancy gain(loss) from heating(cooling), the third term is the change related to estuarine circulation, and the fourth is the change related to bottom tidal currents. Of the four terms, the wind is usually an order of magnitude more important than the others. Advection of buoyancy is occasionally very important (Atkinson et al., 1989).

Estimating variations in stratification is relatively straightforward given the energy inputs (heating, cooling, freshwater influx) and advective effects related to estuarine circulation, shelf edge Ekman processes, etc. Such analysis often yields useful insights that assist the understanding of nutrient dynamics in coastal waters.

However, it is also known that primary production and nutrient uptake are very sensitive to the details of mixing through the water column. Apparently similar stratification resulting from slightly different models of vertical mixing (Chen and Annan, 2000) may be accompanied by significant differences in (i) spring-bloom timing (this may vary by weeks, and can occur before seasonal stratification is obvious), (ii) the continuing chlorophyll maximum near the base of the seasonal thermocline, and (iii) total primary production (50% differences). The sensitivity arises because light near the surface is critical (how much of the time plankton are illuminated enough to grow) and nutrient supply is limited by mixing through the thermocline from below. A model has shown that variations in surface wind stresses episodically weaken the thermocline inhibition to allow nutrient input to the photic zone from below (important to forming a mid-water chlorophyll maximum; Sharples and Tett, 1994). Observational support is given by (e.g.) Eppley and Renger (1988): winds of 5–9 ms$^{-1}$ for about 40 hours off Los Angeles caused mixed layer deepening by about 4 m, entraining about 0.5 mmol m$^{-2}$ nitrate from the nitracline (increasing concentrations in the top ~ 20 m from 20 nmol to up to 100 nmol). Ongoing nutrient supply (to the mid-water chlorophyll maximum) is enabled by turbulence from tidal shears (lagging tidal currents) below and in the thermocline. Sharples et al. (2001) estimated an average nitrate flux 2 ± 1.2 mmol m$^{-2}$d$^{-1}$ in the strongly stratified western English Channel; here nitrate decreased from 5.7 mmol m$^{-3}$ below to zero just 2 m above; vertical diffusivity $K_z$ was estimated from turbulent dissipation as averaging $0.8\times10^{-5}$ $m^2s^{-1}$ (only), but varied by a factor of 100 overall and is the main source of uncertainty in the estimate.

## 7. Successive regions offshore

### *Near-shore*

In the near-shore region turbulence, mixing, sediment suspension and light attenuation are increased by surface-wave currents resuspending material at the sea bed. Shallow water emphasizes the role of (current-affected) benthic regeneration and fluxes; nutrients from groundwater can be significant.

Vertical mixing forced by wind and waves is exemplified by coastal waters off Perth, Australia, and in the Humber plume, where tidal currents add to the mixing. Off Perth, lateral variability suggests turnover faster than lateral mixing. Nutrient concentrations are low (little runoff and low concentrations in the adjacent ocean), but a productive benthic community can partly account for lateral variability by reducing nitrate more in shallower water (Johannes et al., 1994).

Light limitation by SPM is the main constraint on primary production in the Humber plume. The Humber River plume is a ROFI in one sense—see next section—but the freshwater influence is through SPM content rather than typical ROFI dynamics. Nitrate and silicate are nearly conservative in winter; primary production and exchanges with the bed affect distributions at other times (Morris et al., 1995). Primary production (hence nutrient depletion) begins when the euphotic-layer depth exceeds 15% of the water-column depth (the diatom bloom is eventually silicate-limited; autotrophic flagellates are eventually limited by grazing because SPM limits nutrient uptake; Allen et al., 1998).

Dissolved inorganic nutrient fluxes from sediments appear to be the major nutrient source (e.g. of order 0.1 μmol $d^{-1}$ nitrogen through the water column) in depths $< 5$m near-shore in the Great Barrier Reef lagoon. For typical adsorbed and pore-water concentrations, resuspension of 0.1 m sediment would add 0.12–0.2 μmol total nitrogen in 10 m water (Ullman and Sandstrom, 1987).

Groundwater input of nitrate in the absence of typical freshwater input is inferred for the Colorado River (an "inverse estuary" with evaporative enhancement of salinity but large tides causing resuspension, turbidity and a near-shore character; Hernández-Ayón et al., 1993). High nitrite, phosphate and silica values are attributed to resuspension of sediments and mixing of pore water into the water column, plus, for nitrite, ammonia oxidation. Leaching of fertilizer has been estimated to exceed natural nutrient flux in groundwater to near-shore waters in several Hawaii locations. Its fate by plankton uptake depends on mixing through the water column (Dollar and Atkinson, 1992).

### *Regions of freshwater influence (ROFIs)*

Buoyancy input from rivers/estuaries, and strong offshore salinity and temperature gradients often create rapidly-varying stratification. ROFIs typically carry high concentrations of riverine nutrients that fuel production and also may carry particulate matter which may block light and inhibit production. Strong offshore gradients combine with strong mixing and rapid exchange out of the Bay of Brest to prevent eutrophication (Le Pape and Menesguen, 1997). Alternatively, upwelling conditions can rapidly advect low-salinity plume waters offshore taking their contents as far as 60 km from the coast of Chesapeake Bay (Reiss and McConaugha, 1999).

Freshwater sources of nutrients are illustrated by several examples: very similar salinity and (nitrate + nitrite) contours in Baie de Seine (France), April-June 1994 (Videau et al., 1998); fluvial and sewage inputs increase winter DIN and DIP in Liverpool Bay (Irish Sea; Gowen et al., 2000); near-surface phosphate and silicate (and chlorophyll-a) decrease rapidly off Georgia as salinity approaches oceanic values (Yoder et al., 1993); near linear decreases of silicate with increasing salinity in the Fly River plume, and a nitrate/phosphate-limited decrease in production further offshore where salinity > 25 (Ayukai and Wolanski, 1997).

The nutrient sources may enhance production, e.g. in spring and summer in Liverpool Bay relative to a location near the Irish coast. However, SPM shading may inhibit production, e.g. in Baie de Seine the spring bloom progresses inshore between April and June, inhibited by Seine-plume turbidity which is most intense near the coast; the Fly River plume has high SPM close inshore in salinity < 10 (Robertson et al., 1998). In this area (Gulf of Papua) rapid detrital decomposition releases benthic nutrients thus satisfying substantial proportions of the phytoplankton requirements for N and P. Inorganic nitrogen is low off Georgia with inputs from riverine and other sources being consumed by primary production near the coast throughout the year but there are high rates of nitrogen recycling (Yoder et al., 1993). Denitrification is inferred from some offshore reduction in Liverpool Bay nitrogen concentrations (Gowen et al., 2002).

The Gulf of Finland has a near-permanent halocline. Estuarine-type exchange with the Baltic Sea supplies phosphorus to the Gulf and takes nitrogen from the Gulf. Nutrient concentrations are dominated by this exchange, discharges (mostly from rivers) and some atmospheric input. A near-balanced budget appears to hold (Perttilä et al., 1995) in which nitrate and phosphate removal play a role and are mediated by oxygen transfer down through the halocline. In the Polish section of the Baltic Sea, nutrient distributions are affected by sources and depleted by primary production in spring-summer; the ratio of nitrogen and phosphorus seasonal cycles is up to 9 in the open sea and 16 in bays; at depth an oxygen deficiency leads to increasing phosphates and silicates but decreasing nitrates and mineralization (Trzosińska, 1990).

The fjord like Clyde Sea receives nutrient-rich water from the river Clyde. Spatial and temporal variations in near-surface production give corresponding variations in nutrient depletion, and from May into the summer, strong vertical gradients develop (Rippeth and Jones, 1997). Nitrate in the deepest basins increased to ~ 14 μM with more marked increases in phosphate and silicate associated with remineralization of organic matter. These mechanisms and exchanges with the external shelf sea result in the Clyde Sea importing nitrate in summer and exporting in the winter and spring.

### *Well-mixed regions*

Regions of the coastal ocean lacking seasonal stratification occur where tidal and wind-driven currents are strong and buoyancy fluxes are weak. River inflow and atmosphere deposition may be important sources of nitrogen (up to 50% of the winter increase in the German Bight). If these shallow areas are suitably located, upwelling can contribute additional nutrients: e.g. by along-shore advection off southern California (Barnett and Jahn, 1987); strong tidal currents and an eastward jet give "centrifugal upwelling" on the northeast flank of Georges Bank (Pastuszak et al., 1982). Shallow depth and mixing will emphasize the role of benthic processes.

Typically, nutrient concentrations increase in winter and continue until there is enough light for production to begin. This occurs, for example, from October to February in the extensive mixed areas of the southern North Sea outside ROFIs (Howarth et al., 1993), and on Georges Bank, with similar distributions in space and time of nitrate, phosphate and silicate. There is a decrease to near zero in summer (Pastuszak et al., 1982). Higher concentrations of nutrients (and chlorophyll), occur in depths < 30 m than further offshore of southern California. The cause is apparently the increased eddy diffusion, including tidal stirring, and from nutrient recycling (Barnett and Jahn, 1987).

For the southern North Sea, a budget suggests that regeneration from the sediment can explain the silicon increase in winter, but that organic detritus in the water column is an important source for the nitrate increase rather than external sources (Hydes et al., 1999). More remineralization in the water column than in sediments is also found by the model of Tett and Walne (1995). The degree of recycling is spatially variable, and largest at fronts (see next section). N:P ratios 2:1 to 6:1 (< Redfield ratio 16:1 and lower than in any of the source waters; Hydes et al., 1999) suggest that nitrogen is the limiting nutrient element, along with silicate (generally < 60% of nitrate concentrations; Tett and Walne, 1995). Hydes et al. (1999) ascribe the nitrogen deficit to denitrification (equivalent loss rate ~ 0.7 mmol N $m^{-2}d^{-1}$).

Benthic efflux of silicon supports the summer diatom production at a station near the Irish coast (Gowen et al., 2000). In the Barents Sea coastal zone, nutrients, except nitrate, are input from the bottom all year except spring, when the benthic community also uses nutrients; in 20 m depth or less, the daily fluxes are comparable with the water-column inventory (of which a majority is organic, especially in summer; Kuznetsov and Volkovskaya, 1994).

### *Tidal-mixing fronts*

Tidal-mixing fronts occur at the boundary between well-mixed regions (previous section) and summer-stratified regions (next section). There is a tendency for water from the mixed side to intrude between the upper and lower layers on the stratified side. With mixing across all interfaces, there is increased chance for some lower-layer nutrients to reach the base of the upper layer and then "ride up" over the intruding mixed layer. Evidence that this process caused enhanced plankton growth was found at central North Sea fronts and in the German Bight (where stratification patterns are complex) by Brockmann et al. (1990) and Howarth et al. (1993). Frontal sites in the North Sea optimally combine light and nutrient supply to be locations of maximum nutrient recycling (Tett and Walne, 1995).

### *Regions that are thermally stratified in summer*

Summer thermal stratification occurs in the deeper more offshore portions of the coastal ocean. Here riverine inputs are reduced and oceanic sources are more important. The lower layer tends to moderate benthic effects. Typically, after winter replenishment, upper-mixed-layer nutrients are depleted by a spring bloom near the time of initial thermal stratification (by increased solar heating of the upper waters). Thereafter production is limited by the amount of continuing nutrient supply to illuminated upper waters.

Replenishment of nutrients to the upper layer may occur because of many process: cross-shelf dispersion from the ocean starting with the bottom layer of the middle and outer shelf; vertical diffusion and mixing by storms; benthic release and possibly nitrification (see Whitledge et al. (1986) for examples from the Bering Sea); and, current-topography interaction (see Macdonald and Wong (1987) for examples from Mackenzie shelf).

Spring blooms and ongoing summer production deplete upper-layer nutrients and are sensitive to physical conditions controlling the supply of nutrients to the euphotic zone. There are many examples. In the southeast Bering Sea (Whitledge et al., 1986) production is variably enhanced by wind-mixing prolonging nitrate uptake and by weather systems driving nutrient-rich water from outer- to mid-shelf (Sambrotto et al., 1986). The Argentinean shelf has spatially variable thermocline timing and advection of lower-salinity coastal waters (Carreto et al., 1995). The southwest Gulf of Maine has relatively later in the season depletion by spring primary production where the (upper) mixed layer is deeper; other variations are introduced by a strong pycnocline in a low-salinity plume—low nutrients with maxima of ammonia and chlorophyll-a below—and mixing over shoals, thus enhancing nutrient and phytoplankton concentrations (Durbin et al., 1995). Remaining (interannually variable) ice cover reduces summer production and nutrient depletion (down to ~ 50 m) on the Mackenzie Shelf (Macdonald and Wong, 1987).

Some regeneration and ammonium are produced in the bottom layer after the spring bloom in the southeast Bering Sea (Whitledge et al., 1986) as well as in the Gulf of Maine. Limited advection in the Gulf of Maine implies that a substantial fraction of the rich-nutrient "reservoir" below ~ 120 m is a result of water-column regeneration. Then higher near-surface concentrations (with cooler temperature) occur repeatedly at a couple of locations near Georges Bank, (Pastuszak et al., 1982). Such regeneration is more widely typical see (section 8).

### *Shelf edge*

The proximity of the shelf edge to the deep ocean lends importance to oceanic sources and the exchange processes discussed in previous sections. Internal waves can cause interior turbulence and mixing.

Off South Carolina, USA, winter wind-driven transports (aside from effects of Gulf Stream eddies, section 4) bring nutrients onto the shelf from the nutrient-rich Gulf Stream (aided by occasional strong wind-mixing throughout the shelf-water depth; Atkinson et al., 1996). Tidal, wind- and wave-forced currents drive cross-slope exchanges O(1 $m^2s^{-1}$) southwest of Britain, and west of Scotland, in a "downwelling" sense on average. Modeled annual nitrate fluxes west of Scotland are on-shelf near the surface: 65 kmol N $m^{-1}y^{-1}$ at the top of the slope (140m depth), 126k mol N $m^{-1}y^{-1}$ over the slope (300m depth). Near the bed, fluxes are 59 and 194 kmol N $m^{-1}y^{-1}$ respectively down the slope (Proctor et al., 2003a; this model lacked several benthic processes). Distributed over the shelf cross-section of O(20 $km^2$), these represent large in/outputs O(5μM $y^{-1}$). Input of sub-thermocline waters and nitrate in the Yucatan Current onto Campeche Bank is attributed to lateral movement along isopycnals that are geostrophically-tilted raising the nutricline above the shelf break (Furnas and Smayda, 1987; this is analogous to advection across the Gulf Stream to the eastern US slope and shelf waters, section 4).

Upwelling is a source of nutrients at more than just eastern boundaries (discussed in section 5). Along the Siberian coast, it appears to be a source of high concentrations of nutrients in the Anadyr Stream (Nihoul et al., 1993). In the Gulf of Maine, upwelling from the Jordan Basin contributes much of the high nutrient concentrations in flow along the Maine coast (Townsend et al., 1987). Upwelling with increased nitrate occurs around the northern Gulf of Mexico in association with the Loop Current (Walsh et al., 1989), and over the Texas-Louisiana shelf edge (in an Ekman layer below north-eastward flow along the outer shelf; Sahl et al., 1993); here, water with high silicate, phosphate and nitrate concentrations (but low nitrite if coming from below the nitrite maximum) reaches the base of the upper mixed layer. Off the Brazil coast at 23°S, upwelling has been observed with nutrient transfer to the shelf. (Regeneration at the bed, perhaps offset by some denitrification, continued to increase nutrients and deplete dissolved oxygen as the water went further onto the shelf; Braga and Müller, 1998). Upwelling from depth in the central Skagerrak (e.g. total nitrogen 7.46 ± 1.43 μmol) is a basis for high production over the whole Skagerrak (Fonselius, 1996).

In many cases, the main influences on nutrient distributions are probably upper-layer production, vertical mixing and entrainment to supply the upper layer from below (showing as an inverse nitrate/surface-temperature correlation). On Campeche bank, sporadic mixing into the surface water creates elevated nitrate concentrations (and phytoplankton patches; Furnas and Smayda, 1987). Southwest of Britain, the top 50 m is depleted by summer production (when nutrients in 100–300 m are greater than in winter; Hydes et al., 2001). Autumn deepening entrains more nutrients to (late-season production in) the surface layer, which is replenished by winter mixing. For mixing and entrainment, Huthnance et al. (2001) estimated O(100m) equivalent entrained water depth near Goban Spur; possibly hundreds of meters depth further east where large internal tides diffuse the thermocline. Wollast and Chou (2001) estimate 18 g N $m^{-2}$ annually entering the euphotic zone from below annually in order to balance the nitrogen budget; in terms of nitrate at (typically) 8 μmol this is a comparable 160 m equivalent entrained depth. Wollast and Chou also show most water column detritus remineralizing to ammonium, much of which nitrifies to nitrate (especially below the euphotic zone).

Benthic fluxes in slope and shelf waters off Washington, USA, were equivalent to 0.4 to 1.7 mol N $m^{-2}y^{-1}$ nitrogen gas to the water column, and exceeded the nitrate flux to the sediments; the difference could be accounted for by oxidation of ammonium to nitrate and subsequent denitrification to $N_2$ (Devol and Christensen, 1993). These rates (aided by macrobenthic irrigation) correspond to ~ 10 μM through 100m water depth (per year), and so can be significant in nitrogen budgets.

## 8. System studies

### *North Sea*

Non-seasonal components of North Sea nutrient distributions indicate specific riverine and broader oceanic sources. Full-depth winter mixing gives similar surface and bottom values (e.g. Brockmann et al., 1990; Howarth et al., 1993). Nutrient minima in the shallower Dogger Bank region of the North Sea indicate some winter production. In spring, concentrations generally decrease, especially in the surface layer above seasonal stratification; production nearly exhausts limiting nutrients there (Tett and Walne, 1995). However, ammonia in the bottom layer can increase as settling biomass decomposes (Brockmann et al., 1990). A seasonal cycle accounts for most of the variance in nutrients and shows summer depletion of nitrate and silicate consistent with their role as limiting nutrients (Prandle et al., 1997). Annual cycles of some nutrients have been modeled quite well (Radach and Lenhart, 1995), but ammonium and nitrate were not depleted enough by the model's primary production in spring (possibly also an effect of excess nutrient dispersion in the model's coarse horizontal scale). As well as production, remineralization is critical to nutrient concentrations and especially to nutrient availability for further production (with recycling several times per year). Nedwell et al. (1993) estimated benthic fluxes of nutrients out of the sediments as fractions of the corresponding element assimilated in net primary production: N, 4–10%; P, 10–26%; Si, 14–38%; for N, this is an excess of nitrate and ammonium out of the sediments over nitrite into the sediments. The North Sea is a large shelf sea; advection plays only a minor role during the season of production; vertical exchanges are all-important. On longer time-scales, however, the North Sea is a net sink for nitrogen (with substantial denitrification, see section 7, ***Well-mixed Regions***) requiring an average oceanic inflow of 0.6 Sv with 7.5 μM nitrate for balance (Hydes et al., 1999).

The ability of models to describe North Sea ecology has been reviewed by Moll and Radach (2003). Models can reasonably represent nutrient sources from land and ocean, mixed and stratified regimes, light attenuation, for at least one nutrient. However, model capabilities should be extended to include multiple nutrient s, algal partitioning and succession, and improved simulation of turbulence and nutrient supply across the thermocline. Present sediment chemistry parameterization is insufficient for modeling long-term trends.

Budgeting is implicit in any correctly functioning model. Thus, Proctor et al. (2003b) modeled advective, cross-thermocline, plankton uptake, benthic, riverine and recycling components for 1995 in sub-areas of the northwest European shelf, and fluxes across selected sections. Overall import from the ocean to offset denitrification is confirmed. The large-scale fluxes are similar to each other, tending to follow large-scale circulation, except that nitrate and phosphate (not silicate) exit the Irish Sea from the south as well as the north to give a net export. In the northern and central North Sea, near-surface depletion causes the majority of nitrogen and phosphate uptake by plankton to be in or below the thermocline. While, below the thermocline and in the mixed southern North Sea, most nitrogen and phosphate supply is benthic or from recycling. Finally, for silicate, the balance is largely between benthic flux and plankton uptake.

*Irish Sea*

The Proctor et al. (2003b) Irish Sea fluxes contradict the following table (in Gmol/yr) from Simpson and Rippeth (1998), which uses the LOICZ budgeting approach.

TABLE 6
Irish Sea Fluxes (Proctor et al, 2003b). Units Gmol/yr.

| | River input | St Georges Channel input | North Channel output | mol $m^{-2}yr^{-1}$ |
|---|---|---|---|---|
| DIP | 0.87 | 0.71 | 1.27 | 0.007 uptake |
| DIN | 9.71 | 10.6–14.1 | 11.4 | 0.3 denitrification |

The LOICZ approach could only export P and N to the south if in- and out-flow concentrations were known to differ, given the net inflow of water. (Sensitivity to P and N inputs and boundary values or gradients render the phosphorus uptake below the level of significance).

*Northern Adriatic*

Northern Adriatic nutrients have been budgeted by Degobbis and Gilmartin (1990). They found that contributions to N, P and Si exceeded losses by 20–50%. The Po river input (see ROFIs) is important with at least 50% of the input. Denitrification in sediments is about 40% of the nitrogen output; for phosphorus and silicon, burial in the sediments is a significant loss. Biological recycling exceeds the inputs and makes the water column the principal location of N and P regeneration; most orthosilicate regeneration is in the sediments.

## 9. Summary

The simple transport of nutrients into and out of the global coastal ocean critically determines a large percentage of primary and higher-level production in the ocean as a whole. Globally, the atmosphere and rivers are minor sources of coastal-ocean nutrients, but riverine inputs are important locally. The deep-ocean source of remineralized nutrients is effectively limitless but supply to the coastal ocean is severely restricted by dynamic processes. Small variations in the relevant forces cause large variations in the nutrient influx. The processes that affect the movement depend on interactions of currents with topography, winds and vertical mixing, all of which can vary. Recycling in the water column and through the seabed is important to nutrient and plankton composition in shallower coastal seas.

Many of the important processes such as upwelling winds, mixing events or boundary current eddies are episodic. Some may occur only a few times during a season or vary strongly with latitude. This variability will no doubt change as climate changes; it behooves us to learn more about these processes. With that knowledge, we may better predict the effect on fisheries and carbon sequestration.

## Acknowledgements

The authors thank their many colleagues who have over the years have provided lively discussion and insightful papers. Funding for this work was provided by the National Science Foundation (NSF-OCE 0234173), the Samuel and Fay Slover Endowment at Old Dominion University and Old Dominion University.

## Bibliography

Allen, J. I., R. J. M. Howland, N. Bloomer and R. J. Uncles (1998). Simulating the spring phytoplankton bloom in the Humber plume, UK. *Marine Pollution Bulletin* **37:** 295–305.

Arthur, R. S. (1965). On the calculation of vertical motion in eastern boundary currents from determinations of horizontal motion. *Journal of Geophysical Research* **70:** 2799–2803.

Atkinson, L. P., E. Oka, W. Yu, T. J. Berger, J. O. Blanton and T. Lee (1989). Hydrographic variability of southeastern United States shelf waters during the GALE Experiment: Winter 1986. *Journal of Geophysical Research* **94:** 10,699–10,713.

Atkinson, L. P., A. Valle-Levinson, D. Figueroa, R. de Pol-Holz, V. A. Gallardo, W. Schneider, J. L. Blanco and M. Schmidt (2002). Oceanographic observations in Chilean coastal waters between Valdivia and Concepción. *Journal of Geophysical Research* **107**(C7): 18.1–18.13 (10.1029/2001JC000991).

Ayukai, T. and E. Wolanski (1997). Importance of biologically mediated removal of fine sediments from the Fly River plume, Papua New Guinea. *Estuarine, Coastal and Shelf Science* **44:** 629–639.

Bakun, A. and C. S. Nelson (1991). The seasonal cycle of wind stress curl in subtropical eastern boundary current regions. *Journal of Physical Oceanography* **21:** 1815–1834.

Balls, P. W. (1992). Nutrient behaviour in two contrasting Scottish estuaries, the Forth and Tay. *Oceanologica Acta* **15:** 261–277.

Balls, P. W. (1994). Nutrient inputs to estuaries from nine Scottish east coast rivers; influence of estuarine processes on inputs to the North Sea. *Estuarine, Coastal and Shelf Science* **39:** 329–352.

Balls, P. W., N. Brockie, J. Dobson and W. Johnston (1996). Dissolved oxygen and nitrification in the upper Forth estuary during summer (1982–1992): patterns and trends. *Estuarine, Coastal and Shelf Science* **42:** 117–134.

Barnes, J. and N. J. P. Owens (1998). Denitrification and nitrous oxide concentration in the Humber estuary, UK, and adjacent coastal zones. *Marine Pollution Bulletin* **37:** 247–260.

Barnett, A. M. and A. E. Jahn (1987). Pattern and persistence of a nearshore planktonic ecosystem off Southern California. *Continental Shelf Research* **7:** 1–25.

Berger, W. , K. Fischer, C. Lai and G. Wu (1987) Ocean productivity and organic carbon flux. Part I: Overview and maps of primary production and export production. Technical Report Reference Series 87–30, SIO, Scripps Institution of Oceanography, University of California.

Blackburn, T. H. (1997). Release of nitrogen compounds following resuspension of sediment: model predictions. *Journal of Marine Systems* **11:** 343–352.

Blanco, J. L., A. Thomas, M. -E. Carr and P. T. Strub (2001). Seasonal climatology of hydrographic conditions in the upwelling region off northern Chile. *Journal of Geophysical Research* **106:** 11451–11467.

Blanco, J. L., M.-E. Carr, A. C. Thomas and P. T. Strub (2002). Oceanographic conditions off northern Chile during the 1996 La Niña and 1997–1998 El Niño: Part 1: Hydrographic conditions. *Journal of Geophysical Research* **107**(C3): 3.1–3.19.

Blanton, J. O. (1991). Circulation Processes along oceanic margins in relation to material fluxes. Ocean margin processes in global change. R. F. C. Mantoura, J. M. Martin and R. Wollast. Chichester, England, Wiley: 145–163.

Blanton, J. O., L. P. Atkinson, L. J. Pietrafesa and T. L. Lee (1981). The intrusion of Gulf Stream water across the continental shelf due to topographically-induced upwelling. *Deep-Sea Research* **28:** 393–405.

Braga, E. S. and T. J. Müller (1998). Observation of regeneration of nitrate, phosphate and silicate during upwelling off Ubatuba, Brazil, 23°S. *Continental Shelf Research* **18:** 915–922.

Brewer, P. G., C. Goyet and D. Dyrssen (1989). Carbon dioxide transport by ocean currents at 25°N latitude in the Atlantic Ocean. *Science* **246:** 477–479.

Brink, K. (1998). Deep-sea forcing and exchange processes. In, The global coastal ocean: processes and methods. K. Brink and A. Robinson, eds. New York, John Wiley & Sons, 617pp. *The Sea* **10:** 151–167

Brink, K. and T. Cowles (1991). The coastal transition zone program. *Journal of Geophysical Research* **96:** 14637–14647.

Brink, K. H. and A. R. Robinson, Eds. (1998). The global coastal ocean: Processes and methods. *The Sea* **10.** John Wiley & Sons, New York, 617pp.

Brockmann, U. H., R. W. P. M. Laane and H. Postma (1990). Cycling of nutrient elements in the North Sea. *Netherlands Journal of Sea Research* **26:** 239–264.

Brockmann, U. H., K. J. Hesse and U. Hentschke (1994). Nutrient gradients in the tidal flats of the German Bight. *Deutsche Hydrographische Zeitschrift* **S1:** 201–224.

Brooks, D. A. and J. M. Bane (1978). Gulf Stream deflection by a bottom feature off Charleston, South Carolina. *Science* **20:** 1225–1226.

Butler, E. I., S. Knox and M. I. Liddicoat (1979). The relationship between inorganic and organic nutrients in sea water. *Journal of the Marine Biological Association of the U.K.* **59:** 239–250.

Carpenter, R. (1987). Has man altered the cycling of nutrients and organic C on the Washington continental shelf and slope?. *Deep-Sea Research* **34A:** 881–896.

Carreto, J. I., V. A. Lutz, M. O. Carignan, A. D. Cucchi Colleoni and A. G. De Marco (1995). Hydrography and chlorophyll *a* in a transect from the coast to the shelf-break in the Argentinean Sea. *Continental Shelf Research* **15:** 315–336.

Chen, C., D. A. Wiesenburg and L. Xie (1997). Influences of river discharge on biological production in the inner shelf: a coupled biological and physical model of the Louisiana-Texas shelf. *Journal of Marine Research* **55:** 293–320.

Chen, F. and J. D. Annan (2000). The influence of different turbulence schemes on modelling primary production in a 1D coupled physical-biological model. *Journal of Marine Systems* **26:** 259–288.

Chester, R., G. F. Bradshaw, C. J. Ottley, R. M. Harrison, J. L. Merrett, M. R. Preston, A. R. Rendell, M. M. Kane and T. D. Jickells (1993). The atmospheric distributions of trace metals, trace organics and nitrogen species over the North Sea. *Philosophical Transactions of the Royal Society of London* **A343:** 543–556.

Christensen, J. P., W. M. Smethie and A. H. Devol (1987). Benthic nutrient regeneration and denitrification on the Washington continental shelf. *Deep-Sea Research* **34A:** 1027–1047.

Codispoti, L. A. and J. P. Christensen (1985). Nitrification, denitrification and nitrous oxide cycling in the eastern tropical South Pacific Ocean. *Marine Chemistry* **16:** 277–300.

Conley, D. J., W. M. Smith, J. C. Cornwell and T. R. Fisher (1995). Transformation of particle-bound phosphorus at the land-sea interface. *Estuarine, Coastal and Shelf Science* **40:** 161–176.

Copin-Montégut, C. and P. Raimbault (1994). The Peruvian upwelling near 15°S in August 1986. Results of continuous measurements of physical and chemical properties between 0 and 200 m depth. *Deep-Sea Research I* **41(3):** 439–467.

Cornell, S., A. Rendell and T. Jickells (1995). Atmospheric inputs of dissolved organic nitrogen to the oceans. *Nature* **376:** 243–246.

Degobbis, D. and Gilmartin, M. (1990). Nitrogen, phosphorus, and biogenic silicon budgets for the northern Adriatic Sea. *Oceanologica Acta* **13:** 31–45.

DeMaster, D. J. and R. H. Pope (1996). Nutrient dynamics in Amazon shelf waters: results from AMASSEDS. *Continental Shelf Research* **16:** 263–289.

DeMaster, D .J., W. O. Smith, D. M. Nelson and J. Y. Aller (1996). Biogeochemical processes in Amazon shelf waters: chemical distributions and uptake rates of silicon, carbon and nitrogen. *Continental Shelf Research* **16:** 617–643.

Devol, A. H. and J. P. Christensen (1993). Benthic fluxes and nitrogen cycling in sediments of the continental margin of the eastern North Pacific. *Journal of Marine Research* **51:** 345–372.

Dollar, S. J. and M. J. Atkinson (1992). Effects of nutrient subsidies from groundwater to nearshore marine ecosystems off the island of Hawaii. *Estuarine, Coastal and Shelf Science* **35:** 409–424.

Duce, R. A., P. S. Liss, J. T. Merrill, E. L. Atlas, P. Buat-Menard, B. B. Hicks, J. M. Miller, J. M. Prospero, R. Arimoto, T. M. Church, W. Ellis, J. N. Galloway, L. Hansen, T. D. Jickells, A. H. Knap, K. H. Reinhardt, B. Schneider, A. Soudine, J. J. Tokos, S. Tsunogai, R. Wollast and M. Zhou (1991). The atmospheric input of trace species to the world ocean. *Global Biogeochemical Cycles* **5:** 193–259.

Durbin, E. G., A. G. Durbin and R. C. Beardsley (1995). Springtime nutrient and chlorophyll *a* concentrations in the southwestern Gulf of Maine. *Continental Shelf Research* **15:** 433–450.

Enfield, D. (1987). The intraseasonal oscillation in eastern Pacific sea levels: How is it forced? *Journal of Physical Oceanography* **17:** 1860–1876.

Eppley, R. W. and E. H. Renger (1988). Nanomolar increase in surface layer nitrate concentration following a small wind event. *Deep-Sea Research* **35A:** 1119–1125.

Fonselius, S. (1996). The upwelling of nutrients in the central Skagerrak. *Deep-Sea Research II* **43:** 57–71.

Furnas, M. J. and T. J. Smayda (1987). Inputs of subthermocline waters and nitrate onto Campeche Bank. *Continental Shelf Research* **7:** 161–175.

Gowen, R. J., D. K. Mills, M. Trimmer and D. B. Nedwell (2000). Production and its fate in two coastal regions of the Irish Sea: the influence of anthropogenic nutrients. *Marine Ecology Progress Series* **208:** 51–64.

Gowen, R. J., D. J. Hydes, D. K. Mills, B. M. Stewart, J. Brown, C. E. Gibson, T. M. Shannon, M. Allen and S. J. Malcolm (2002). Assessing trends in nutrient concentration in coastal shelf seas: a case study in the Irish Sea. *Estuarine, Coastal and Shelf Science* **54:** 927–939.

Hall, P. O. J., S. Hulth, G. Hulthe, A. Landén and A. Tengberg (1996). Benthic nutrient fluxes on a basin-wide scale in the Skagerrak (north-eastern North Sea). *Journal of the Sea Research* **35:** 123–137.

Hawke, D. J. and K. A. Hunter (1992). Reactive P distribution near Otago peninsula, New Zealand: an advection-dominated shelf system containing a headland eddy. *Estuarine, Coastal and Shelf Science* **34:** 141–155.

Herbland, A., D. Delmas, P. Laborde, B. Sautour and F. Artigas (1998). Phytoplankton spring bloom of the Gironde plume waters in the Bay of Biscay: early phosphorus limitation and food-web consequences. *Oceanologica Acta* **21:** 279–291.

Hernández-Ayón, J. M., M. S. Galindo-Bect, B. P. Flores-Báez and S. Alvarez-Borrego (1993). Nutrient concentrations are high in the turbid waters of the Colorado river delta. *Estuarine, Coastal and Shelf Science* **37:** 593–602.

Hickey, B. M. (1997). The response of a steep-sided, narrow canyon to time-variable wind forcing. *Journal of Physical Oceanography* **27**(5): 697–726.

Hill, A. E. (1998). Buoyancy effects in coastal and shelf seas. In, The global coastal ocean—Processes and methods. K. Brink and A. Robinson, Eds. John Wiley & Sons New York. *The Sea* **10:** 21–62.

Hill, A. E., B. M. Hickey, F. A. Shillington, P. T. Strub, K. Brink, E. D. Barton and A. C. Thomas (1998). Eastern Ocean Boundaries (E). In, The global coastal ocean: regional studies and synthesis. A. Robinson and K. Brink, Eds. New York, John Wiley & Sons. *The Sea* **11:** 29–68.

Hitchcock, G. L., W. J. Wiseman, W. C. Boicourt, A. J. Mariano, N. Walker, T. A. Nelsen and E. Ryan (1997). Property fields in an effluent plume of the Mississippi river. *Journal of Marine Systems* **12:** 109–126.

Hormazabal, S., G. Shaffer, J. Letelier and O. Ulloa (2001). Local and remote forcing of sea surface temperature in the coastal upwelling system off Chile. *Journal of Geophysical Research* **106:** 16657–16671.

Hormazabal, S., G. Shaffer and O. Pizarro (2002). Tropical Pacific control of intraseasonal oscillations off Chile by way of oceanic and atmospheric pathways. *Geophysical Research Letters* **29**(6), 10.1029/2001GL013481.

Howarth, M. J., K. R. Dyer, I. R. Joint, D. J. Hydes, D. A. Purdie, H. Edmunds, J. E. Jones, R. K. Lowry, T. J. Moffatt, A. J. Pomroy and R. Proctor (1993). Seasonal cycles and their spatial variability. *Philosophical Transactions of the Royal Society of London* **A343:** 383–403.

Huthnance, J. M. (1992). Extensive slope currents and the ocean-shelf boundary. *Progress in Oceanography* **29:** 161–196.

Huthnance, J. M. (1995). Circulation, exchange and water masses at the ocean margin: the role of physical processes at the shelf edge. *Progress in Oceanography* **35:** 353–431.

Huthnance, J. M., H. Coelho, C. R. Griffiths, P. J. Knight, A. P. Rees, B. Sinha, A. Vangriesheim, M. White and P. G. Chatwin (2001). Physical structures, advection and mixing in the region of Goban Spur. *Deep-Sea Research I* **48:** 2979–3021.

Huthnance, J. M., H. M. van Aken, M. White, E. D. Barton, B. Le Cann, E. Ferreira, E Alvarez, P. Miller and J. Vitorino (2002). Ocean margin exchange water flux estimates. *Journal of Marine Systems* **32:** 107–137.

Hydes, D. J., B. A. Kelly-Gerreyn, A. C. Le Gall and R. Proctor (1999). The balance of supply of nutrients and demands of biological production and denitrification in a temperate latitude shelf sea—a treatment of the southern North Sea as an extended estuary. *Marine Chemistry* **68:** 117–131.

Hydes, D. J., A. C. Le Gall, A. E. J. Miller, U. Brockmann, T. Raabe, S. Holley, X. Alvarez-Salgado, A. Antia, W. Balzer, L. Chou, M. Elskens, W. Helder, I. Joint and M. Orren (2001). Supply and demand of nutrients and dissolved organic matter at and across the NW European shelf break in relation to hydrography and biogeochemical activity. *Deep-Sea Research I* **48:** 3023–3047.

Jickells, T. D. (1991). Group Report: What determines the fate of materials within ocean margins? In, Ocean margin processes in global change. R. F. C. Mantoura, J. M. Martin and R. Wollast, Eds. Chichester, Wiley: 211–234.

Johannes, R. E., A. F. Pearce, W. J. Wiebe, C. J. Crossland, D. W. Rimmer, D. F. Smith and C. Manning (1994). Nutrient characteristics of well-mixed coastal waters off Perth, Western Australia. *Estuarine, Coastal and Shelf Science* **39:** 273–285.

Joint, I. and P. Wassmann (2001). Professor Roland Wollast. *Progress in Oceanography* **51:** 215–216.

Joint, I., M. Inall, R. Torres, F. G. Figueiras, X. A. Álvarez-Salgado, A. P. Rees and E. M. S. Woodward (2001a). Two Lagrangian experiments in the Iberian upwelling system: tracking an upwelling event and an off-shore filament. *Progress in Oceanography* **51:** 221–248.

Joint, I., A. P. Rees and E. M. S. Woodward (2001b). Primary production and nutrient assimilation in the Iberian upwelling in August 1998. *Progress in Oceanography* **51:** 303–320.

Joordens, J. C. A., A. J. Souza and A. W. Visser (2001) The influence of tidal straining and wind on suspended matter and phytoplankton distribution in the Rhine outflow region. *Continental Shelf Research* **21:** 301–325.

Kadko, D. (1993). Excess $^{210}$Po and nutrient recycling within the California coastal transition zone. *Journal of Geophysical Research* **98:** 857–864.

Knox, S., M. Whitfield, D. R. Turner and M. I. Liddicoat (1986). Statistical analysis of estuarine profiles: III. Application to nitrate, nitrite and ammonium in the Tamar Estuary. *Estuarine, Coastal and Shelf Science* **22:** 619–636.

Körner, D. and G. Weichart (1991). Nutrients in the German Bight, concentrations and trends 1978–1990. *Deutsche Hydrographische Zeitschrift Ergänzungsheft* **A17:** 3–41.

Kosro, P. M. (2002). A poleward jet and an equatorial undercurrent observed off Oregon and northern California, during the 1997–1998 El Niño. *Progress in Oceanography* **54:** 343–360.

Kuznetsov, L. L. and L. E. Volkovskaya (1994). Fluxes of nutrients between benthic and planktic communities in the coastal zone of the Barents Sea. *Oceanology* **34:** 510–514.

Laane, R. W. P. M., G. Groeneveld, A. de Vries, J. van Bennekom and S. Sydow (1993). Nutrients (P, N, Si) in the Channel and the Dover Strait: seasonal and year-to-year variation and fluxes to the North Sea. *Oceanologica Acta* **16:** 607–616.

Lebo, M. E. and J. H. Sharp (1992). Modeling phosphorus cycling in a well-mixed coastal plain estuary. *Estuarine, Coastal and Shelf Science* **35:** 235–252.

Lee, T. N., Y. A. Yoder and L. P. Atkinson (1991). Gulf Stream frontal eddy influence on productivity of the southeast U. S. continental shelf. *Journal of Geophysical Research* **96:** 22191–22205.

Le Pape, O. and A. Menesguen (1997). Hydrodynamic prevention of eutrophication in the Bay of Brest (France), a modelling approach. *Journal of Marine Systems* **12:** 171–186.

Liu, K. K., L. Atkinson, C. T. A. Chen, S. Gao, J. Hall, R. W. Macdonald, L. Talaue McManus and R. Quiñones (2000). Exploring continental margin carbon fluxes on a global scale. *EOS* **81:** 641–642, 644.

Lohrenz, S. E., G. L. Fahnenstiel, D. G. Redalje, G. A. Lang, M. J. Dagg, T. E. Whitledge and Q. Dortch (1999). Nutrients, irradiance, and mixing as factors regulating primary production in coastal waters impacted by the Mississippi River plume. *Continental Shelf Research* **19:** 1113–1141.

Lund-Hansen, L. C., P. Skyum and C. Christian (1996). Modes of stratification in a semi-enclosed bay at the North Sea—Baltic Sea transition. *Estuarine, Coastal and Shelf Science* **42:** 45–54.

Lutjeharms, J. R. E. and P. L. Stockton (1987). Kinematics of the upwelling front off southern Africa. South African *Journal of Marine Science* **5:** 35–50.

Macdonald, R. W. and C. S. Wong (1987). The distribution of nutrients in the southeastern Beaufort Sea: implications for water circulation and primary production. *Journal of Geophysical Research* **92:** 2939–2952.

McClain, C. R. and L. P. Atkinson (1985). A note on the Charleston Gyre. *Journal of Geophysical Research* **90:** 11,857–11,861.

Meybeck, M., G. Cauwet, S. Dessery, M. Somville, D. Gouleau and G. Billen (1988). Nutrients (organic C, P, N, Si) in the eutrophic river Loire (France) and its estuary. *Estuarine, Coastal and Shelf Science* **27:** 595–624.

Moll, A. and G. Radach (2003). Review of three-dimensional ecological modelling related to the North Sea shelf system Part 1: models and their results. *Progress in Oceanography* **57:** 175–217.

Mooers, C. N. K. and A. R. Robinson (1984). Turbulent jets and eddies in the California Current and inferred cross-shore transports. *Science* **223:** 51–53.

Morales, C. E., J. L. Blanco, M. Braun, H. Reyes and N. Silva (1996). Chlorophyll-a distribution and associated oceanographic conditions in the upwelling region off northern Chile during the winter and spring 1993. *Deep-Sea Research* **43:** 267–289.

Morris, A. W., J. I. Allen, R. J. M. Howland and R. G. Wood (1995). The estuary plume zone: source or sink for land-derived nutrient discharges? *Estuarine, Coastal and Shelf Science* **40:** 387–402.

Mortimer, R. J. G., M. D. Krom, P. G. Watson, P. E. Frickers, J. T. Davey and R. J. Clifton (1998). Sediment-water exchange of nutrients in the intertidal zone of the Humber estuary, UK. *Marine Pollution Bulletin* **37:** 261–279.

Nathansohn, A. (1906). Abhand. Konigl. sachs Gesel. der Wissensch. Leipsig **39:** 3.

Nedwell, D. B., R. J. Parkes, A. C. Upton and D. J. Assinder (1993). Seasonal fluxes across the sediment-water interface, and processes within sediments. *Philosophical Transactions of the Royal Society of London* **A343:** 519–529.

Neshyba, S. J., C. N. K. Mooers, R. L. Smith, and R.T. Barber Eds. (1989). Poleward Flows along Eastern Ocean Boundaries. New York: Springer-Verlag, 374pp.

Nihoul, J. C. J., P. Adam, P. Brasseur, E. Deleersnijder, S. Djenidi and J. Haus (1993). Three-dimensional general circulation model of the northern Bering Sea's summer ecohydrodynamics. *Continental Shelf Research* **13:** 509–542.

Ortner, P. B. and M. J. Dagg (1995). Nutrient-enhanced coastal ocean productivity explored in the Gulf of Mexico. *Eos, Transactions,* American Geophysical Union **76:** 97, 109.

Pastuszak, M. (1995). The hydrochemical and biological impact of the river Vistula on the pelagic system of the Gulf of Gdansk in 1994. Part 1. Variability in nutrient concentrations. Oceanologia **37:** 181–205.

Pastuszak, M., W. R. Wright and D. Patanjo (1982). One year of nutrient distribution in the Georges Bank region in relation to hydrography, 1975–1976. *Journal of Marine Research* **40S:** 525–542.

Pelegri, J. L. and G. T. Csanady (1991). Nutrient Transport and Mixing in the Gulf Stream. *Journal of Geophysical Research* **96:** 2577–2583.

Perttilä, M., L. Niemistö and K. Mäkelä (1995). Distribution, development and total amounts of nutrients in the Gulf of Finland. *Estuarine, Coastal and Shelf Science* **41:** 345–360.

Pierce, S. D., R. L. Smith, P. M. Kosro, J. Barth and C. D. Wilson (2000). Continuity of the poleward undercurrent along the eastern boundary of the mid-latitude north Pacific. *Deep-Sea Research* **47:** 811–829.

Prandle, D. (1984). A modelling study of the mixing of $^{137}$Cs in the seas of the European continental shelf. *Philosophical Transactions of the Royal Society of London* **A310:** 407–436.

Prandle, D., D. J. Hydes, J. Jarvis and J. McManus (1997). The seasonal cycles of temperature, salinity, nutrients and suspended sediment in the southern North Sea in 1988 and 1989. *Estuarine, Coastal and Shelf Science* **45:** 669–680.

Proctor, R., F. Chen and P. B. Tett (2003a). Carbon and nitrogen fluxes across the Hebridean shelf break, estimated by a 2D coupled physical-microbiological model. *The Science of the Total Environment,* in press.

Proctor, R., J. T. Holt, J. I. Allen and J. Blackford (2003b). Nutrient fluxes and budgets for the north west European Shelf from a 3-dimensional model. The Science of the Total Environment, in press.

Rabalais, N. N., R. E. Turner, Q. Dortch, D. Justic, V. J. Bierman, Jr. and W. J. Wiseman, Jr. (2002). Review: Nutrient-enhanced productivity in the northern Gulf of Mexico. *Hydrobiologia* **475/476:** 39–63.

Radach, G. and H. J. Lenhart (1995). Nutrient dynamics in the North Sea: fluxes and budgets in the water column derived from ERSEM. *Netherlands Journal of Sea Research* **33:** 301–335.

Redfield, A. C. (1936). An ecological aspect of the Gulf Stream. Nature **138:** 1013.

Reiss, C. S. and J. R. McConaugha (1999). Cross-frontal transport and distribution of ichthyoplankton associated with Chesapeake Bay plume dynamics. *Continental Shelf Research* **19:** 151–170.

Rendell, A. R., T. M. Horrobin, T. D. Jickells, H. M. Edmunds, J. Brown and S. J. Malcolm (1997). Nutrient cycling in the Great Ouse estuary and its impact on nutrient fluxes to The Wash, England. *Estuarine, Coastal and Shelf Science* **45:** 653–668

Richards, F. A. and A. C. Redfield (1955). Oxygen-density relationships in the western North Atlantic. *Deep-Sea Research* **2:** 182–199.

Rienecker, M. M., C. N. K. Mooers and A. R. Robinson (1987). Dynamical interpolation and forecast of the evolution of mesoscale features off northern California. *Journal of Physical Oceanography* **17:** 1189–1213.

Riley, G. A. (1967). Mathematical model of nutrient conditions in coastal waters. *Bulletin of the Bingham Oceanographic Foundation* **19:** 72–80.

Rippeth, T. P. and K. J. Jones (1997). The seasonal cycle of nitrate in the Clyde Sea. *Journal of Marine Systems* **12:** 299–310.

Robertson, A. I., P. Dixon and D. M. Alongi (1998). The influence of fluvial discharge on pelagic production in the Gulf of Papua, northern Coral Sea. *Estuarine, Coastal and Shelf Science* **46:** 319–331.

Robinson, A. R. and K. H. Brink, Eds. (1998). The coastal global ocean—regional studies and syntheses. John Wiley & Sons, New York. *The Sea* **11:** 1062pp.

Rossby, C.-G. (1936). Dynamics of steady ocean currents in the light of experimental fluid mechanics. Papers in *Physical Oceanography and Meteorology* **5**(1): 43pp.

Ruardij, P. and W. van Raaphorst (1995). Benthic nutrient regeneration in the ERSEM ecosystem model of the North Sea. *Netherlands Journal of Sea Research* **33:** 453–483.

Rydberg, L., J. Haamer and O. Liungman (1996). Fluxes of water and nutrients within and into the Skagerrak. *Journal of Sea Research* **35:** 23–38.

Sahl, L. E., W. J. Merrell and D. C. Biggs (1993). The influence of advection on the spatial variability of nutrient concentrations on the Texas-Louisiana continental shelf. *Continental Shelf Research* **13:** 233–251.

Sambrotto, R. N., H. J. Niebauer, J. J. Goering and R. L. Iverson (1986). Relationships among vertical mixing, nitrate uptake, and phytoplankton growth during the spring bloom in the southeast Bering Sea middle shelf. *Continental Shelf Research* **5:** 161–198.

Sandstrom, H. and J. A. Elliott (1984). Internal tide and solitons on the Scotian shelf: a nutrient pump at work. *Journal of Geophysical Research* **89:** 6415–6426.

Schollaert, S.E., T. Rossby and J.A. Yoder (2003). Gulf Stream cross-frontal exchange: possible mechanisms to explain interannual variations in phytoplankton chlorophyll in the Slope Sea during the SeaWiFs Years. *Deep-Sea Research.* in Press

Shaffer, G., O. Pizarro, L. Djurfeldt, S. Salinas and J. Rutllant (1997). Circulation and low-frequency variability near the Chilean coast. Remotely forced fluctuations during the 1991–1992 El Niño. *Journal of Physical Oceanography* **27:** 217–235.

Shapiro, G. I., J. M. Huthnance and V. V. Ivanov (2003). Dense water cascading off the continental shelf. *Journal of Geophysical Research,* accepted.

Sharples, J. and P. Tett (1994). Modelling the effect of physical variability on the midwater chlorophyll maximum. *Journal of Marine Research* **52:** 219–238.

Sharples, J., C. M. Moore, T. P. Rippeth, P. M. Holligan, D. J. Hydes, N. R. Fisher and J. H. Simpson (2001). Phytoplankton distribution and survival in the thermocline. *Limnology and Oceanography* **46:** 486–496.

Shaw, P. J., C. Chapron, D. A. Purdie and A. P. Rees (1998). Impacts of phytoplankton activity on dissolved nitrogen fluxes in the tidal reaches and estuary of the Tweed, UK. *Marine Pollution Bulletin* **37:** 280–294.

Silva, N. and S. Neshyba (1979). On the southernmost extension of the Peru-Chile Undercurrent. *Deep-Sea Research* **26A:** 1387–1393.

Simpson, J. H. and J. R. Hunter (1974). Fronts in the Irish Sea. *Nature* **250:** 404–406.

Simpson, J. H. and T. P. Rippeth (1998). Non-conservative nutrient fluxes from budgets for the Irish Sea. *Estuarine, Coastal and Shelf Science* **47:** 707–714.

Simpson, J. H., W. G. Bos, F. Schirmer, A. J. Souza, T. P. Rippeth, S. E. Jones and D. Hydes (1993). Periodic stratification in the Rhine ROFI in the North Sea. *Oceanologica Acta* **16:** 23–32.

Singer, J. J., L. P. Atkinson, J. O. Blanton and Y. A. Yoder (1983). Cape Romain and the Charleston Bump: Historical and recent hydrographic observations. *Journal of Geophysical Research* **88:** 4685–4697.

Sobarzo, M., and D. Figueroa (2001). The physical structure of a cold filament in a Chilean upwelling zone (Península de Mejillones, Chile, 23° S). *Deep-Sea Research* **38:** 2699–2726.

Strub, P., M. Kosro, and A. Huyer (1991). The nature of the cold filaments in the California Current System. *Journal of Geophysical Research* **96:** 14743–14768.

Tappin, A. D., G. E. Millward and J. D. Burton (2001). Estuarine and coastal water chemistry in Land-Ocean Interaction. In Measuring and modelling fluxes from river basins to coastal seas (D. A. Huntley, G. J. L. Leeks and D. E. Walling, eds.). IWA Publishing, London, 241–279.

Tee, K. T., P. C. Smith and D. LeFaivre (1993). Topographic upwelling off southwest Nova Scotia. *Journal of Physical Oceanography* **23:** 1703–1726.

Tett, P. and A. Walne (1995). Observations and simulations of hydrography, nutrients and plankton in the southern North Sea. *Ophelia* **42:** 371–416.

Thomas, A. C., F. Huang, P. T. Strub and C. James (1994). Comparison of the seasonal and interannual variability of phytoplankton pigment concentrations in the Peru and California Current Systems. *Journal of Geophysical Research* **99:** 7355–7370.

Tian, R. C., F. X. Hu and J. M. Martin (1993). Summer nutrient fronts in the Changjiang (Yantze River) estuary. *Estuarine, Coastal and Shelf Science* **37,** 27–41.

Townsend, D. W., J. P. Christensen, D. K. Stevenson, J. J. Graham and S. B. Chenoweth (1987). The importance of a plume of tidally-mixed water to the biological oceanography of the Gulf of Maine. *Journal of Marine Research* **45:** 699–728.

Trzosi•ska, A. (1990). Seasonal fluctuations and long-term trends of nutrient concentrations in the Polish zone of the Baltic Sea. *Oceanologia* **29:** 27–50.

Tsunogai, S., S. Watanabe, and T. Sato (1999). Is there a 'continental shelf pump' for the absorption of atmospheric $CO_2$? *Tellus* **51B:** 701–712.

Ullman, W. J. and M. W. Sandstrom (1987). Dissolved nutrient fluxes from the nearshore sediments of Bowling Green Bay, central Great Barrier Reef lagoon (Australia). *Estuarine, Coastal and Shelf Science* **24,** 289–303.

Uncles, R. J., R. J. M. Howland, A. E. Easton, M. L. Griffiths, C. Harris, R. S. King, A. W. Morris, D. H. Plummer and E. M. S. Woodward (1998a). Seasonal variability of dissolved nutrients in the Humber-Ouse estuary, UK. *Marine Pollution Bulletin* **37:** 234–246.

Uncles, R. J., R. G. Wood, J. A. Stephens and R. J. M. Howland (1998b). Estuarine nutrient fluxes to the Humber coastal zone, UK, during June 1995. *Marine Pollution Bulletin* **37:** 225–233.

Valle-Levinson, A., L. P. Atkinson, D. Figueroa and L. Castro (2003). Flow induced by upwelling winds in an equatorward facing bay: Gulf of Arauco, Chile. *Journal of Geophysical Research* **108**(C2) 36.1–36.14 (10.1029/2001JC001222).

Videau, C., M. Ryckaert and S. L'Helguen (1998). Phytoplancton en baie de Seine. Influence du panache fluvial sur la production primaire. *Oceanologica Acta* **21:** 907–921.

Walsh, J. J. (1988). On the nature of continental shelves. Academic Press, London, 520pp.

Walsh, J. J., D. A. Dieterle, M. B. Meyers and F. E. Müller-Karger (1989). Nitrogen exchange at the continental margin: a numerical study of the Gulf of Mexico. *Progress in Oceanography* **23:** 245–301.

Whitledge, T. E., W. S. Reeburgh and J. J. Walsh (1986). Seasonal inorganic nitrogen distributions and dynamics in the southeastern Bering Sea. *Continental Shelf Research* **5:** 109–132.

Wollast, R. (1998). Evaluation and comparison of the global carbon cycle in the coastal zone and in the open ocean. In, The global and coastal ocean, processes and methods. K. H. Brink and A. R. Robinson, Eds. John Wiley & Sons, New York. *The Sea* **10:** 213–252.

Wollast, R. and L. Chou (2001). The carbon cycle at the ocean margin in the northern Gulf of Biscay. *Deep-Sea Research I* **48:** 3265–3293.

Yentsch, C. S. (1974). The influence of geostrophy on primary production. Tethys **6:** 111–118.

Yoder, J. A., P. G. Verity, S. S. Bishop and F. E. Hoge (1993). Phytoplankton Chl *a,* primary production and nutrient distributions across a coastal frontal zone off Georgia, U.S.A. *Continental Shelf Research* **13:** 131–141.

Zhang, J. (1996). Nutrient elements in large Chinese estuaries. *Continental Shelf Research* **16:** 1023–1045.

Zoppini, A., M. Pettine, C. Totti, A. Puddu, A. Artegiani and R. Pagnotta (1995). Nutrients, standing crop and primary production in western coastal waters of the Adriatic Sea. *Estuarine, Coastal and Shelf Science* **41:** 493–513.

# Chapter 9. THE BIOGEOCHEMISTRY OF CARBON DIOXIDE IN THE COASTAL OCEANS

HUGH W. DUCKLOW
S. LEIGH MCCALLISTER

*The College of William and Mary*

**Contents**



## 1. Introduction

Accumulation of atmospheric $CO_2$ (Keeling and Whorf, 2002) and depletion of atmospheric oxygen (Keeling and Shertz, 1992) from combustion of organic matter (OM) stored over geological time provide conclusive evidence that *H. sapiens* has driven the planet into a metabolic state of net heterotrophy in the past 300 years. That is, the fossil fuel combustion, cement production, anthropogenically-stimulated and natural ecosystem metabolism of the planetary ecosystem are currently consuming more organic matter than is produced in photosynthesis by all the world's terrestrial and marine plants. Rising concentrations of atmospheric $CO_2$ threaten the planet with global warming and associated impacts like rising sea levels (Chapter 18). The ocean and land biospheres also serve to ameliorate these effects by absorbing $CO_2$ from the atmosphere (Sarmiento and Gruber, 2002). In this chapter we provide a comprehensive synthesis of observations of $CO_2$ uptake and release by the coastal oceans, critically evaluate arguments about their metabolic status, and discuss new findings on the sources and character of organic matter inputs that fuel coastal ocean metabolism. We conclude that the global coastal ocean as a whole is autotrophic and potentially a strong sink for atmospheric $CO_2$. Further we suggest that the open sea, over its full depth, is heterotrophic by ~ 200

*The Sea*, Volume 13, edited by Allan R. Robinson and Kenneth H. Brink
ISBN 0-674-01526-6 

Tmol $y^{-1}$, as a consequence of organic matter exported from the coastal ocean. Yet it too is a sink for atmospheric $CO_2$.

Defining the area of the global coastal ocean is a difficult and important issue. Previous debate about the role of these continental margin systems in the global carbon cycle stems in part from inconsistencies in definition and in the areas considered in budgeting exercises. Here we adopt a general encompassing definition articulated recently by an international planning group (Hall et al., 2004). In their definition, the global coastal ocean includes "...the region between the land and the open ocean that is dominated by processes resulting from land—ocean boundary interactions. The exact dimensions vary depending on the research issue or chemical element of interest but draw attention to the unique aspects attributable to the boundary system and generally consist of the continental shelf, slope, rise and adjacent inland seas." (Hall et al. 2004, page 9). This concept of the coastal ocean system also includes regions impacted by boundary currents that may extend significantly offshore, such as jets and filaments. The problem of defining the area of interest is illustrated by comparing Tables 2 and 3, to which we return below. In general, adhering to the definition just articulated, we use the term "coastal ocean" or "global coastal ocean" in deference to the title of this volume, unless we are referring to a particular, more specific marine system.

However defined, the coastal ocean is the junction point in the biosphere where the land, ocean and atmospheric components of the planetary biogeochemical system meet and interact. It is 400,000–500,000 km long but just ~100 km wide (Wollast, 1998; Smith, 2000). Into this region flow annually riverine inputs including 85 Tmol carbon, 2 – 4 Tmol nitrogen and 0.7 Tmol phosphorus. Each element is supplied in both dissolved and particulate, and organic and inorganic forms from natural and anthropogenic terrestrial sources (Table 1; Meybeck 1993; Gattuso et al. 1998). If confined within buoyant river plumes that encompass a surface layer ~25 m deep on the inner third of the global continental shelves, this input would result in an average enhancement of *in situ* N and P concentrations of 10 and 3 μM, respectively. These levels are well above the average half-saturation constants[1] for uptake by phytoplankton and bacteria. Of course, the actual nutrient inputs may be concentrated in smaller areas where the local impact is greater, but this simple calculation gives an idea of the intensity of the biogeochemical exchanges and transformations in coastal ocean. The coastal ocean also receives even larger inputs of nutrients from offshore, oceanic processes like upwelling (see Section 4). The global continental ocean is a powerful reactor mediating biogeochemical exchanges between the land and ocean biospheres.

[1] $K_s$, the half-saturation constant for nutrient uptake by phytoplankton defines the nutrient concentration at which the utilization rate by the cell is exactly half the maximum rate at nutrient saturation. Thus when the ambient nutrient concentration is near $K_s$, any increase in concentration results in an increase in the uptake rate, faster growth and possibly eutrophication.

TABLE 1
Carbon, nitrogen and phosphorus inputs from land to the global coastal ocean (Tmol $yr^{-1}$).
Data taken from summary discussion in Gattuso et al (1998). --, no data.

| Input | Natural | Anthropogenic | Total | Ratio | Natural | Anthropogenic | Total |
|---|---|---|---|---|---|---|---|
| | Dissolved Inorganic fluxes | | | | Dissolved Inorganic ratios | | |
| DIC | 32 | -- | 32 | C:N | 121 | -- | 62 |
| DIN | 0.27 | 0.25[1] | 0.52 | N:P | 22 | 17 | 19 |
| DIP | 0.01 | 0.02[2] | 0.03 | C:P | 2667 | -- | 1185 |
| | Dissolved Organic fluxes | | | | Dissolved Organic ratios | | |
| DOC | 17 | 4[3] | 21 | C:N | 24 | 16 | 22 |
| DON | 0.7 | 0.25[1] | 0.95 | N:P | 58 | 17 | 35 |
| DOP | 0.01 | 0.02[2] | 0.03 | C:P | 1417 | 267 | 778 |
| | Particulate fluxes | | | | Particulate ratios | | |
| PIC | 14 | -- | 14 | | | | |
| POC | 14 | 4[3] | 18 | C:N | 12 | -- | 11 |
| PON | 2.4 | 0.5 | 2.9 | N:P | 4 | -- | 5 |
| PP[4] | 0.6 | -- | 0.6 | C:P | 23 | -- | 30 |
| | Total Organic fluxes | | | | Total Organic ratios | | |
| TOC | 31 | 8 | 39 | C:N | 10 | 11 | 10 |
| TON | 3.1 | 0.75 | 3.85 | N:P | 258 | 50 | 6 |
| TOP[1] | 0.01 | 0.02 | 0.63 | C:P | 2583 | 533 | 62 |
| | Total Input fluxes | | | | Total Input ratios | | |
| C | 77 | 8 | 85 | C:N | 23 | 8 | 19 |
| N | 3.4 | 1.0 | 4.4 | N:P | 5 | 33 | 7 |
| P | 0.62 | 0.03 | 0.65 | C:P | 123 | 267 | 130 |

[1] anthropogenic dissolved nitrogen flux of 0.5 Tmol split evenly between DIN and DON
[2] anthropogenic dissolved phosphorus flux of 0.03 Tmol split evenly between DIP and DOP
[3] anthropogenic POC + DOC flux of 8 Tmol split evenly between fractions
[4] particulate phosphorus assumed to be inorganic

Marine biogeochemistry can trace its origins to the application of chemical methods to the study of plankton ecology by the 'Kiel School' of biological oceanography in the late 19th–early 20th centuries and its later flowering at Plymouth in the 1920's (Mills 1989). The field became global in scope following the advent of satellite-based remote sensing and deployment of a global network of deep-ocean sediment traps in the 1980's (NAS, 1984). Even so, development was concentrated in the open sea, largely as a result of the Joint Global Ocean Flux Study (JGOFS) that conducted basin-scale surveys and large process studies; and established several long-term time series observatories in the 1990's (Hanson et al., 2000). The focus on the open sea derived from the need to understand and quantify the role of the ocean as a sink for anthropogenic carbon dioxide ($CO_2$). $CO_2$ exchange is mediated by gas transfer across the ocean surface (see below), of which 93% (360 x $10^6$ $km^2$) is over the open ocean >200 m deep. John Walsh (Walsh et al., 1981) first suggested that continental shelf systems could sequester significant amounts of atmospheric $CO_2$ in response to anthropogenic eutrophication (see also Walsh et al., 1985; Walsh, 1991). This idea eventually led to a new generation of coordinated research programs over the world's continental shelves. Uncertainties about the oceanic carbon balance, and great intrinsic interest in the coastal ocean stimulated the development of the Land-Ocean Interactions in the Coastal Zone (LOICZ) program (Holligan and Reiners, 1992) and a new focus on coastal ocean biogeochemistry (Liu et al., 2000ab). Of course there are many other strands and themes in marine biogeochemistry which cannot be cited here. It was the combined influence of JGOFS and LOICZ, and interest in the carbon cycle that stirred some of the developments evaluated critically below.

One focus of this research has been the balance between *in situ* net primary production of organic matter (P) and its oxidation by respiratory processes (R). This *metabolic balance* determines whether a system is a net producer or consumer of organic matter. Systems are termed net heterotrophic (R>P) or autotrophic (P>R), depending on the annual metabolic balance. The P:R balance has a venerable tradition in aquatic ecology deriving from the concept of community production (Odum, 1956). Purdy (1917, cited in Odum, 1956) demonstrated net heterotrophy in the Potomac estuary. Smith and Hollibaugh (1993) presented a budgetary approach to the carbon cycle in the global coastal ocean and open sea, arguing that the ocean as a whole is heterotrophic, with a net oxidation (respiration) rate of ~23 Tmol C $yr^{-1}$ (0.3 PgC $yr^{-1}$), 30% of which is concentrated in the coastal ocean (7% of total ocean area). Thus they argue that organic metabolism in the sea is a net source of $CO_2$ release to the atmosphere and not a sink, as is usually understood (see below). 0.3 PgC is a small number when compared to the 48 PgC of annual net primary production in the global ocean (Field et al., 1998), but significant in relation to recent estimates of anthropogenic $CO_2$ uptake of 1 – 2 PgC $yr^{-1}$ by the global ocean (Houghton et al., 2001).

Other recent reviews provide a valuable foundation for our treatment. Walsh (1988) constructed carbon budgets and numerical simulations for a variety of continental shelf systems in different circulation regimes. Wollast (1998) compared the C and N cycles of the coastal and open oceans in a preceding volume of this series, suggesting that the inner margins of the continental shelves were heterotrophic while the outer shelves had positive net community production. Gattuso et al. (1998) critically reviewed primary production and net metabolism in coastal ocean systems, concluding that only estuaries were heterotrophic, whereas the open sea as well as the shelves had positive net production (see below). Atkinson et al. (Chapter 8) assess the physical mechanisms supporting primary production over the global coastal ocean. Liu et al. (2000ab) offer a new interpretation of the role of the coastal ocean as a conduit for the net transport of carbon from the land and atmosphere to the open sea. Finally Chen et al. (2003) provide a comprehensive analysis of nutrient and carbon cycling in the coastal oceans. Together these reviews conclude that the ocean as a whole consumes more organic matter than it produces internally. This problem is the subject of intense debate, and a source of considerable misunderstanding regarding the role of marine biological processes in acting as sources or sinks of atmospheric $CO_2$.

## 2. Overview

The coastal ocean is an open system, continuously exchanging materials with its several interfaces (land, open sea, sediments and atmosphere). In this chapter we are concerned mainly with the exchanges of carbon at three of these interfaces (signified by the thick arrows in Figure 9.1): transport of organic matter from land to the coastal oceans; net transport of organic matter from the coastal oceans into the open ocean ("continental shelf pump"); and exchange of atmospheric $CO_2$

across the air/sea interface. The net transfer of organic matter across these interfaces dictates the metabolic balance of the coastal ocean. If organic matter synthesis by primary production is balanced by *in situ* consumption, the system is in metabolic balance. If organic matter is exported from the coastal ocean, the system leans toward net autotrophy. Ultimately, transports of allochthonous (externally-produced) organic matter from land set the bounds for the metabolic balance of the coastal ocean system (Smith and Hollibaugh 1993). These ideas are also discussed in Chapter 7.

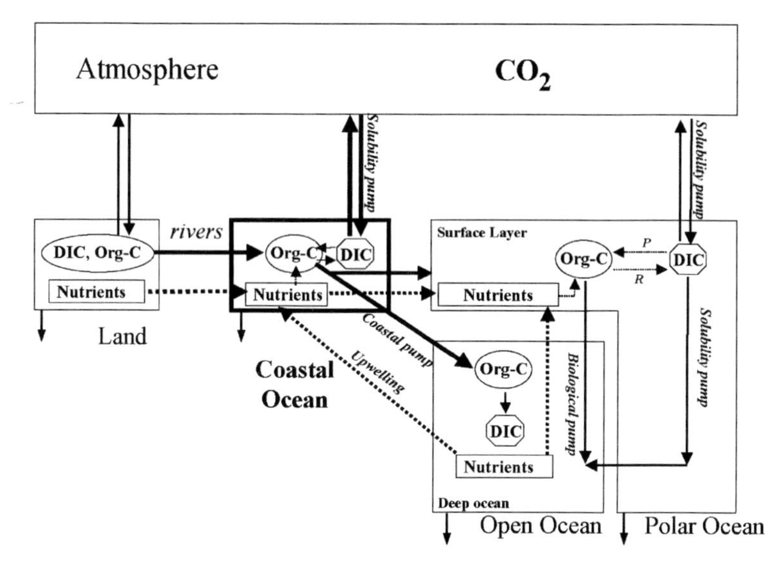


Figure 9.1 Carbon and nutrient exchanges between land, the coastal ocean and the open ocean, showing the three carbon pumps regulating the allocation of CO2 between the surface and deep ocean (solubility, biological and continental shelf pumps). Biologically mediated exchanges (P, photosynthesis; R, respiration) between inorganic and organic carbon pools are indicated by the ovals and hexagons in the ocean boxes. Heavy lines and arrows indicate the primary processes covered in this chapter.

The most important factor regulating the exchange of $CO_2$ between the coastal ocean and atmosphere is the difference in the partial pressure of $CO_2$ ($\Delta pCO_2$) between these two reservoirs. $\Delta pCO_2$ varies in space and time and is influenced on the ocean side by both physical and biological processes. The purely physicochemical process of gas exchange across the air-water interface (the solubility pump in Figure 9.1) proceeds independently of the biological processes governing $\Delta pCO_2$. This separation of physical and biological influences, and the variability of $\Delta pCO_2$ in time and space, mean that the ocean as a whole can simultaneously be net heterotrophic (with biological processes serving as a $CO_2$ source) yet also function as a net sink for atmospheric $CO_2$ (see also Chapter 7), a point often misunderstood by chemists and biologists alike. In the present-day ocean, $\Delta pCO_2$ and the air-sea flux of $CO_2$ are influenced more by the concentration of anthropogenic $CO_2$ in the atmosphere than by heterotrophic $CO_2$ production in the sea.

The source-sink character of the coastal ocean also depends on the magnitude of organic matter export to the open sea. The coastal ocean will only function as an effective long term sink for atmospheric $CO_2$ if the net $CO_2$ taken up is transported into the deep ocean (e.g., by bottom water formation at high latitude) or fixed into organic matter and then exported to the deep ocean by the continental shelf pump (Figure 9.1). Dissolved inorganic carbon (DIC, see below) and carbon fixed into organic matter that are retained in the coastal ocean will likely be returned to the atmosphere due to its shallow depth and relatively short residence time (Tsunogai et al., 1999; Liu et al., 2000b). The key processes affecting $CO_2$ storage (primary production, $\Delta pCO_2$, the metabolic balance and the continental shelf $CO_2$ pump) are addressed in the following sections.

## 3. Inputs from land.

Table 9.1 summarizes the globally averaged fluxes of carbon, nitrogen and phosphorus into the coastal ocean by riverine inputs, a topic reviewed extensively elsewhere (Meybeck 1993; Ludwig et al. 1996; Frankignoulle et al. 1998; Gattuso et al. 1998; Wollast 1998). The carbon input is 40–45% organic, whereas N inputs are 85–90% organic, emphasizing the potential for organic loading in the immediate coastal zone, and for export offshore. Overall, the terrestrial input of dissolved material is enriched in C and N, relative to P, as shown by the elemental ratios presented in Table 9.1. This C:N:P stoichiometry suggests P-limitation of the primary production derived from the terrestrial input (see next section). The particulate fraction is rich in P, due to a large contribution of P adsorbed onto clay minerals but the magnitude of this flux is uncertain (Meybeck, 1993). The fate of the organic and particulate fractions is also poorly understood. Smith and Hollibaugh (1993) estimated that 11 Tmol of the 34 Tmol of organic matter exported to the coastal ocean was buried, leaving 23 Tmol available for oxidation in the water column. The subsidy provided by anthropogenic additions is also uncertain, and known only to within an order of magnitude (Gattuso et al. 1998).

In terms of C, the annual export from rivers is quantitatively significant. Globally, rivers deliver ~20 Tmol of dissolved organic carbon (DOC) per year to the ocean (Meybeck, 1982; Hedges, 1992; Hedges et al., 1997), an amount large enough to sustain the turnover of the entire pool of marine DOC (685 PgC over 5000 yr). Likewise the magnitude of particulate organic carbon (POC) (~13 Tmol) transported by rivers is sufficient to account for the total organic C buried in coastal sediments each year (Hedges and Keil, 1995; Hedges et al., 1997). Despite the magnitude of these inputs and the presumed recalcitrance of riverine organic matter, organic matter dissolved in seawater or buried in sediments bears little isotopic or chemical evidence of a terrigenous origin (Hedges and Keil, 1995; Hedges et al., 1997). Although these findings suggest extensive and efficient remineralization of land-derived OM, the biogeochemical processes responsible for this rapid turnover remain as yet unresolved.

Terrigenous OM is frequently traced through the use of either bulk characteristics (e.g. isotopic values, C:N, etc. ) or molecular (lignin phenols, lipids etc.) tracers. Historically, the resolution of molecular level detail in the dissolved phase (DOM) has not been commensurate with analyses of particulate OM (POM), owing in part to the limitations imposed by the required sample size and interference by salt. However the advent of tangential flow ultrafiltration (Benner et al., 1992; Bianchi et al., 1995; Santschi et al., 1995; Bauer et al., 1996) coupled with lowered detection thresholds for many molecular tracers now allow for greater insights into the composition of and relationships between POM and DOM. Although terrestrial OM in the particulate phase (POM) is primarily deposited within the nearshore coastal zone (Prahl et al., 1994; Shi et al., 2001), adsorptive/desorptive exchange between particulate and dissolved reservoirs (Henrichs and Sugai, 1993; Wang and Lee, 1993; Hedges and Keil, 1999; Komada and Reimers, 2001) simultaneously dispels our once simplistic view and complicates our ability to delineate the seaward flow of terrigenous OM.

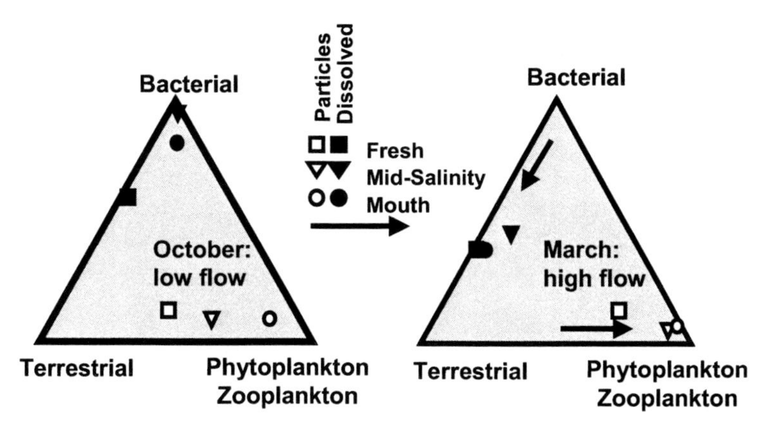


Figure 9.2 Source classification of fatty acid (FA) biomarkers in particulate organic matter (POM) and high molecular weight dissolved organic matter fractions along the York River, Virginia estuary under low and high flow conditions. The vertices of the triangles are the sources of POM and DOM in the estuarine system, characterizing Phytoplankton/ Zooplankton (polyunsaturated FA C18, C20, and C22); Bacteria (iso and anteiso-C13, C15, and C17 FA) and vascular plants (long chain saturated FA: C24-C30). Each particulate and dissolved sample is normalized to the ranges in potential sources (e.g., phytoplankton/zooplankton; bacterial; terrestrial) representative of the dataset as a whole. Figure after McCallister (2002).

Biogeochemical processing in estuaries is the primary control regulating the relocation of terrigenous OM from land to the coastal ocean. The predominantly conservative transport of riverine DOC through estuaries (Mantoura and Woodward, 1983; Ittekkot, 1989; Aminot et al., 1990; Alvarez-Salgado and Miller, 1998) is based on bulk DOC concentrations. Molecular level analysis of DOC composition provides greater sensitivity with which to trace the flow and fate of terrigenous DOM (Hedges et al., 1997; Benner and Opsahl, 2001; Aluwihare et al., 2002). A recent study employing fatty acid biomarkers to trace sources of OM suggest a substantial terrigenous signature persists in the dissolved phase along the estuarine salinity gradient (Figure 9.2). In contrast, the seaward-flowing particulate counterpart is dominated by a mixed phytoplankton/zooplankton source with a nominal terrestrial influence. The physical state of OM (i.e., particulate vs dissolved) inherently predisposes it to select pathways of physico-chemical transformations and sinks (e.g., sedimentation and flocculation, adsorption and desorption etc.) within an estuary. However, the ultimate persistence of OM within each physical state is not predetermined, as both biogeochemical and physico-chemical processes may result in OM exchange between phases (Henrichs and Sugai, 1993; Wang and Lee, 1993; Hedges and Keil, 1999; Komada and Reimers, 2001).

Sharp gradients in physical and chemical properties such as the boundary between the landward intrusion of salt and freshwater (i.e. estuarine turbidity maximum) are zones of active partitioning between the particulate and dissolved OM phases. Analogous gradients in pH, redox potential, and solute concentrations between sediments and porewaters may contribute to the adsorption/desorption of OM (Gu et al., 1995; Gu et al., 1996; Thimsen and Keil, 1998) and disperse formerly sediment-bound OM back to the water column (Keil et al., 1997). Physical resuspension of sediments may likewise produce abrupt physico-chemical gradients and thus initiate extensive exchange between particulate and dissolved OM reservoirs (Komada and Reimers, 2001). The terrigenous (Bianchi et al., 1997; Mannino and Harvey, 2000; Mitra et al., 2000) and/or aged (Guo and Santschi, 2000; Komada et al., 2002) signature of solubilized OM suggests that sediment-derived subsidies of terrigenous OM may preserve an allochthonous signature in the dissolved phase throughout estuarine transit.

The handful of studies that measured lignin phenols in high molecular weight (>1 kDa) DOM (HMW DOM), an unequivocal biomarker for vascular plants and a terrigenous origin (Hedges and Mann, 1979), report terrigenous sources may account for ~5–60% of HMW DOM in coastal ocean systems (Opsahl and Benner, 1997; Kattner et al., 1999; Mannino and Harvey, 2000; Benner and Opsahl, 2001; Aluwihare et al., 2002). In comparison land-derived OM comprised a mere 0.7–2.4% of the open ocean HMW DOM (Opsahl and Benner, 1997). These data suggest a rapid (21–132 yr) turnover (Opsahl and Benner, 1997) of the terrigenous pool when compared to bulk DOC ages of 4000–6000 yr ((Williams and Druffel, 1988; Bauer et al., 1992). In contrast to the POM pool, a major sink for terrestrial OM in the dissolved phase may occur in the coastal ocean. The susceptibility of highly colored, strongly light adsorbing (Morris et al., 1995; Reche et al., 1999) terrestrial humics to photochemical losses (Miller, 1999) may be enhanced in coastal waters as a result of increased sunlight penetration into the water column (Amon and Benner, 1996; Bushaw et al., 1996). Furthermore, the photochemical removal of terrestrial OM components may generate a residual signal of photochemical processing in HMW DOM which has been detected in the waters of the Mississippi River plume and the open Pacific and Atlantic oceans (Benner and Opsahl, 2001). Varying degrees of susceptibility of terrigenous particulate and dissolved OM to physico-chemical processes may ultimately regulate the persistence of terrestrial-derived OM in the hydrosphere and determine the re-location of riverine OM to regions of net heterotrophy (Section 6; Smith and Hollibaugh, 1993; del Giorgio et al., 1997; Duarte et al., 2001; del Giorgio and Duarte, 2002). These processes will also influence the relative amounts of terrestrial and marine OM entering the oceanic carbon inventory, after transport and processing in the coastal ocean.

## 4. Organic matter production within the global coastal ocean.

Primary production in the coastal ocean, as in the open sea, is dominated by unicellular phytoplankton but the composition and physiology of the flora and the relative importance of nutrients, light and grazing as limiting factors on phytoplankton growth differ in important ways from the open sea. The fundamental influence of coastal physical processes on primary production and foodweb structure and dynamics are discussed in detail in Chapter 12 as well as by Cullen et al. (2002), and the effects with regard to coastal ocean systems can be summarized as follows. The relatively shallow depth of most coastal ocean environments and enhanced stratification resulting from terrestrial inputs of freshwater limit the vertical extent of mixing. These processes may ameliorate light limitation, compared to the open sea, although greater turbidity and self-shading due to high plankton biomass and other suspended solids will counteract buoyancy-related effects. Turbulent kinetic energy in the water column is dissipated more rapidly in shallower, more protected inshore areas than in the open sea, so phytoplankton blooms are induced earlier and faster in the coastal ocean, especially in bays and bights (Smetacek and Passow, 1990). Large fluxes of macronutrients transported horizontally across the continental slope and shelf break and into the coastal ocean (Chapter 9) and from terrestrial and anthropogenic sources (Table 9.1) as well as vertical inputs from underlying sediments support high production rates in coastal ocean waters. The flux of nutrients from the oceanic boundary is poorly known and difficult to characterize but certainly dominates as a source of N and P to the coastal ocean (Wollast, 1998). Supply of micronutrients (e.g., iron) is also generally greater due to the proximity of terrestrial dust sources, sediments (Johnson et al., 1999) and anthropogenic inputs. The combined effects of enhanced turbulence and greater nutrient supply interact with the physiological requirements of phytoplankton, leading to predictable dominance (blooms) of larger-celled organisms (Malone, 1971; Margalef, 1978; Cullen and MacIntyre, 1998). Smaller-celled phytoplankton constitute a more uniform background flora (Redalje et al., 2002). In general, coastal ocean systems fall under Cullen et al's (2002) high-turbulence, high-nutrient category and are characterized by high production of large phytoplankton forms (diameter > ~10 μm including diatoms, dinoflagellates and prymnesiophytes – mainly *Phaeocystis*). In spite of this generalization, it is also important to recognize that higher nutrient inputs, though favoring the larger-celled flora, also stimulate growth of the small-celled 'background' plankton, which are dominated by coccoid cyanobacteria and nanoplankton (Malone et al., 1991). Coastal ocean systems have higher stocks and production of both large and small phytoplankton, compared to the open sea, as revealed by ocean color images. Higher biomass and a greater proportion of larger (more rapidly sinking) forms generally means that the vertical transport of surface production to depth will be more efficient. That is, the ratio of sedimentation to primary production will be greater in nearshore and coastal seas than in the ocean. The magnitude of lateral export (i.e., the continental shelf pump in Figure 9.1) will be examined further below, after a synthesis of recent estimates of PP in the global coastal ocean.

Early estimates of net primary production (NPP) over continental shelves were reviewed by Walsh (1988) and Smith and Hollibaugh (1993). These reviews indicated that NPP averaged 160 gC $m^{-2}$ $y^{-1}$ (13 Mol C $m^{-2}$ $y^{-1}$; range for global shelf regions 60 – 600 mgC $m^{-2}$ $y^{-1}$). The magnitude of primary production in continental shelf and other marine systems is summarized in Tables 9.2 and 9.3. Gattuso et al. (1998) compiled recent data from those studies which considered both gross primary production (GPP) and community respiration (R)[2]. Areal rates of gross primary production in the coastal ocean are high and comparable to those measured in estuaries (18 vs 22 Mol C $m^{-2}$ $yr^{-1}$). The coastal ocean, with just 6% of the marine surface area, contributes 10% of the gross primary production but 33% of the global net ecosystem production (NEP = GPP-R), according to Gattuso et al.'s compilation (Table 9.2). The NEP is approximately equivalent to the new production (Dugdale and Goering, 1967) or export production (Ducklow et al., 2001) supported by external inputs of new (allochthonous) nutrients. An independent set of estimates confirms the disproportionate share of net primary production contributed by coastal ocean systems. Longhurst et al. (1995) derived regional estimates of NPP for 57 coastal and oceanic ecological provinces using remotely sensed sea surface color observations, from which global estimates were extrapolated (Table 9.3). In their analysis, the Coastal domain, including Arctic and Antarctic marginal ice zones and covering 11% of the global ocean surface, contributed 29% of the total marine NPP. The discrepancy between the two sets of estimates derives from Longhurst et al.'s large estimate of 32.1 mol C $m^{-2}$ $y^{-1}$ for *net* primary production in the Coastal Domain. This value is about twice as large as Gattuso et al.'s *gross* primary production estimate of 17.4 mol C $m^{-2}$ $y^{-1}$. Longhurst's estimate includes the upwelling provinces with the coastal ocean domain, whereas those areas are in the open ocean category in Table 9.2. NPP and NEP cannot be easily compared, further complicating the comparison, but it seems clear that coastal ocean systems contribute a disproportionate share of the global marine planktonic production. This raises the question of their importance as potential sinks for atmospheric carbon dioxide.

The nutrient demand from external sources supporting NPP in the coastal ocean can be estimated by considering the amount of NPP that sinks or is advected out of the photic zone. This quantity is the export flux and it is equivalent to the new production over sufficiently large time and space scales, i.e. near steady state (Eppley and Peterson, 1979). Leaving aside the question as to whether the global coastal ocean can be so considered, the export or E-ratio (export divided by NPP) gives a rough estimate of the proportion of the net primary production supported by externally supplied nutrients. In Wollast's (1998) summary, the export flux over the continental shelves was given as 13.3 Mol C $m^{-2}$ $y^{-1}$ (70% of his NPP estimate of 19.2 Mol C $m^{-2}$ $y^{-1}$), in contrast to 1.3 Mol C $m^{-2}$ $y^{-1}$ (15% of the NPP) in the open sea. The magnitude of NPP and export quoted by Wollast (1993) are both high, relative to other sources discussed here (see also following section), and an export ratio of 50% is more consistent with most recent analyses (e.g. Walsh 1991). Thus the input of land and oceanic nutrients to coastal ocean could support ~50% of the total NPP, with the remaining 50% sustained by *in situ* nutrient regeneration. The total P input from land (Table 9.1) can be used to estimate an upper limit to the amount of new production supported by terrestrial nutrient sources. Assuming that all the riverine dissolved phosphorus input is regenerated and becomes available for phytoplankton uptake over the shelf, then multiplying by the Redfield C:P

[2] Gross primary production (GPP) is the total carbon fixed in photosynthesis, including that portion subsequently respired during algal cellular metabolism. GPP minus the phytoplankton respiration is the net primary production available to consumers. Community respiration, R, is the sum total of respiratory activity carried out by phytoplankton plus all consumers and microbes. The amount of organic matter remaining after GPP-R is the net ecosystem production.

ratio (106) gives an upper bound of ~7 Tmol C $y^{-1}$ for terrestrially-derived, P-limited NPP. The NPP supported by phosphorus input from land is a small fraction of the total NPP (7/1195, cf. Table 9.3), but is about 22% of the terrestrial input of organic matter (7/31 from Table 9.1). Accordingly, most of the new production in the coastal ocean must be sustained by onshore transport of nutrients from the open sea. The latter input must be large. Possibly 21 Tmol N $y^{-1}$ is denitrified in shelf and slope regions (Codispoti et al., 2001). Just to balance the nitrogen budget for the coastal ocean, and assuming no losses (certainly not true), the N input from the ocean must exceed the river input by a factor of 5.

Regeneration and utilization of the terrestrial C and N (particulate and dissolved) associated with phosphorus uptake during photosynthesis leaves behind ~90% of the C and ~40% of the N supplied in excess of the Redfield ratios. These unutilized nutrients are available for export into the open sea (Hydes et al., 2001). Although it may be relatively small, the terrestrial input plays a large role in setting the sign of the metabolic balance because the P:R of these ecosystems is so finely poised (Table 9.2).

TABLE 9.2.
Areal extent and primary production rates in marine systems (after Gattuso et al., 1998).

| System | Surface area $10^6$ $km^2$ | GPP Mol C $m^{-2}$ $yr^{-1}$ | GPP Tmol C $yr^{-1}$ | NEP Tmol C $yr^{-1}$ | GPP:R[1] |
|---|---|---|---|---|---|
| Estuaries | 1.4 | 22 | 31 | -8 | 0.8±0.05 |
| Saltmarshes[2] | 0.4 | 206 | 83 | 82 | 1.0 |
| Other margins[3] | 3.2 | 119 | 381 | 68 | 1.2±0.06 |
| Shelves | 21.4 | 18 | 377 | 171 | 1.3±0.3 |
| Inner Shelves and bays[2] | 10 | 23 | 230 | -36 | 0.9±0.9 |
| Total Margin[4] | 26 | 31 | 789 | 231 | 1.2±0.05 |
| Open ocean[5] | 334 | 10 | 3396 | 340[6] | 1.11 |

[1]recalculated from data available at www.annualreviews.org/supmat/ using a reduced major axis (Model II) regression. The slope of the GPP vs R regression is the GPP:R ratio. GPP, gross primary production; NEP, net ecosystem production (net community production, analogous to export production).

[2]as summarized in Smith and Hollibaugh (1993) and included for comparison (area taken to be ~0.3 of total for shelves).

[3]mangroves, macroalgae, seagrasses, coral reefs.

[4]Total margin = estuaries + other margins + shelves as reported in Gattuso et al. 1998.

[5]In this compilation high productivity zones over the continental slope and rise, including major upwelling regimes are included in the open ocean category.

[6]derived from Williams' (2000) GPP:R estimates for the eastern North Atlantic and Southern Oceans, Mediterranean and Arabian Seas and North Pacific Subtropical Gyre and Gattuso et al.'s GPP value.

TABLE 9.3.

Recent estimates of primary production over continental shelves and the open sea.

| | Gattuso GPP and NEP[1] | | | | | Longhurst NPP[2] | | | | |
|---|---|---|---|---|---|---|---|---|---|---|
| **Region** | **Area** | **GPP** | **GPP** | **NEP** | **GPP** | **Area** | **NPP** | | | |
| | **$10^6$ km$^2$** | **mol C/m$^2$y$^{-1}$** | **Tmol C yr$^{-1}$** | **Tmol C yr$^{-1}$** | **%[3]** | **$10^6$ km$^2$** | **GtC yr$^{-1}$** | **mol C/m$^2$ y$^{-1}$** | **Tmol C y$^{-1}$** | **%[3]** |
| **Cont Shelf** | 21.4 | 17.6 | 377 | 171 | 10 | 37.4 | 14.4 | 32.1 | 1195 | 29 |
| **Total Margin** | 26 | 30.3 | 789 | 231 | 21 | -- | -- | -- | -- | -- |
| **Open Ocean** | 334 | 10.2 | 3396 | 340 | 90 | 291 | 35.7 | 10.2 | 2963 | 71 |
| **Total Shelf + Open** | 355 | 10.6 | 3773 | 511 | 100 | 328 | 50.1 | 12.7 | 4158 | 100 |

[1] after Gattuso et al (1998). "Total margin" includes estuaries, continental shelves, coral reefs, mangrove, saltmarshes and macrophyte-dominated coastal systems. GPP, gross primary production; NEP, net ecosystem production (net community production, analogous to export production). See also Table 1. Gattuso's shelf domain excludes upwelling zones over slope and rise.

[2] after Longhurst et al. (1995). Cont. shelf is Longhurst's Coastal Domain and includes upwelling zones over slope and rise. Open ocean includes his Polar, Westerlies and Trade Winds domains. NPP, Net primary production.

[3] % of global total GPP in region.

## 5. Export production: what is the fate of organic carbon in the global coastal ocean?

Atmospheric $CO_2$ absorbed by coastal ocean processes will only enter long-term storage if it is transported laterally offshore to the ocean interior. Vertical export from the euphotic zone consists of large, rapidly sinking particles (Alldredge and Silver, 1988) or exportable DOC mixed or advected away by physical processes (Hansell and Carlson, 2001). Greater stocks of larger phytoplankton in the coastal oceans lead in turn to larger consumers and faster particle sedimentation rates, and thus more efficient vertical export by sinking from the surface layer. In the open sea, the export flux is usually considered in a one-dimensional (vertical) context, since lateral transfers between regions have not been addressed. But the issue of lateral export pathways is critical for understanding fluxes between the coastal and open ocean (continental shelf pump in Figure 9.1). Some of the primary production over the continental shelf, slope and rise settles directly to the bottom, where it is oxidized or buried, and some is transported laterally across the coastal ocean boundary into the open sea where it may sink into deep water. Walsh (1991) calculated that the coastal ocean contributed 50% of the total oceanic particle flux at a depth of 2650 m as a consequence of this transport process. Wollast (1998) estimated that about half the total production exported vertically from the surface layer was subsequently transported laterally off the shelf, with the rest being respired or buried in the shelf sediments. Liu (2000a) estimated that just 5% of the export production was buried in shelf sediments, with the remainder (2.7 PgC $yr^{-1}$) entering the ocean interior. Here, we use the term *shelf export* to indicate the lateral export from the global coastal ocean into the ocean interior.

Shelf export is controlled by a complex combination of circulation and transport processes that have seldom been resolved at the appropriate time and space scales to allow precise estimates or budgets. These processes were first studied in the Shelf-Edge Exchange Processes (SEEP) studies, during which sediment traps and moored observing systems were deployed over the NE US continental shelf (Biscaye et al., 1988; Biscaye et al., 1994). The export process is apparently highly episodic (Churchill et al., 1994). Adequate resolution requires arrays of high frequency current meters and transmissometers to monitor particle transport during wind events and the resulting upwelling circulation (Walsh et al., 1988) as well as cruise-based studies of isotopic tracers (Bacon et al., 1988) and biomarkers of specific organic source materials (Venkatesan et al., 1988). Cross-shelf sediment trap arrays deployed normal to the alongshore axis of the shelf revealed that near-bottom particle transport in the Ekman layer resulted in enhancement of settling rates at intermediate depths along the shelf-slope region (Biscaye et al., 1988). In the Middle Atlantic Bight off the NE USA, export is driven by Ekman transport and the repeated settling and resuspension of particles derived from the spring phytoplankton bloom. Particles deposited on the seafloor over the shelf are resuspended and transported seaward, eventually reaching depocenters on the continental slope or rise (Biscaye and Anderson, 1994).

Coastal jets and filaments (Chapter 8) may transport both particulate and dissolved organic matter from shelves into the surface layer of the open ocean. A filament associated with the Gulf Stream transported 0.3 – 2.6 Tmol of riverine

DOC and 'new' DOC produced *in situ* from the northeast US continental shelf/slope into the Gulf Stream (Bates and Hansell, 1999; Vlahos et al., 2002). An upwelling filament from the NW Iberian peninsula exports a total of 0.03 Tmol C $yr^{-1}$ into the NE Atlantic Ocean, equally divided between POC and DOC (Alvarez-Salgado, 2001ab). There is great variability in these estimates, and few filaments have been studied, but their importance as mechanisms of coastal export into the surface layer of the open sea demands further attention. The upper end of this range of export estimates suggests a potentially large role for this mode of carbon export (see below).

The magnitude of offshelf transport and storage was unknown until the past decade. The initial debate on this question (Rowe et al., 1986; Walsh, 1989) directly inspired the SEEP Program off the US east coast, and subsequently influenced the Kuroshio Edge Exchange Processes (KEEP; Wong et al. 2000), Ocean Margin Exchange (OMEX, Wollast and Chou 2001) and Gulf of St. Lawrence, Canada (Roy et al., 2000) programs and other studies. A primary aim of all these studies was to quantify the magnitude of export to the ocean interior of organic matter produced over continental shelves. SEEP set out explicitly to test the hypothesis that export from the shelf to the ocean interior was large in relation to *in situ* NPP. Although the complexities of shelf circulation processes, resuspension and near-bottom transport, seasonal variability and sediment trapping make it very challenging to get a definitive answer, it does not appear that the majority of the coastal ocean regions studied to date are intensive export regions, relative to *in situ* NPP, or in an absolute sense (Biscaye and Anderson, 1994; Antia et al., 2001). In particular, lateral export of phytoplankton was just 5% of PP in the area north of Cape Hatteras in the Middle Atlantic Bight (Falkowski et al., 1994). As Boehme et al. (1998) observed for the Middle Atlantic Bight, USA, the 1 Mt of annual net uptake of atmospheric $CO_2$ represented only a few percent of the annual NPP, suggesting little net export of organic matter from this system.

However not all shelf systems behave similarly. De Haas et al. (2002) reviewed sedimentation and burial budgets for shelf systems worldwide, but only a few have estimates of export to the ocean. More than 90% of the primary production (150 Mt) on the Amazon Shelf (strongly light limited by turbidity in the outflow of the Amazon River) is oxidized in the water column (Aller et al., 1996), but only 50% of the organic matter in the annual river discharge (50 Mt) is buried locally (de Haas et al., 2002). Thus the offshelf flux might be about 40 Mt, equivalent to about 25% of the local PP. 16% of the terrigenous organic carbon entering the Beaufort Shelf in the Canadian Arctic is exported offshelf (Macdonald et al., 1998). 48% of the primary production on the outer shelf in the Bering Sea is exported (Walsh and McRoy, 1986). Organic particle export from the East China Sea off Taiwan may be more important than the other regions studied (Chung and Hung, 2000). There, a cyclonic eddy served as an efficient transport mechanism, moving material offshore into canyons, then to the ocean interior via the Okinawa Trough (Wong et al., 2000). But elsewhere, in areas lacking special circulation and bottom topography, most material produced over the shelves appears to be oxidized and recycled *in situ* (Falkowski et al., 1988; Vezina et al., 2000; Soetaert et al., 2001).

Most of the shelf systems noted above are broad shelves where much of the export production might be expected to be respired; thus the lateral export might be a lower fraction of the total exportable production. Coastal ocean systems at the continental boundary of narrow shelves such as the Peru/Chile upwelling system might be expected to export more material. Walsh's (1981) budget for the Peru system implied that only 8% of the total production was buried or exported before overfishing but almost 60% met that fate after the collapse of the anchovy fishery. As a result of such regional, temporal and anthropogenic effects, estimates of the global shelf export vary by an order of magnitude. Walsh's (1991) estimate of 1 GtC $yr^{-1}$ (83 Tmol C $yr^{-1}$) was derived from a variety of sediment trap, current meter array and geochemical tracer studies. Smith and Hollibaugh (1993) considered only organic matter of terrestrial origin that drives net heterotrophy in the coastal ocean in their budget. They estimated that about half the riverine input, or 18 Tmol C $yr^{-1}$ was exported to the open sea. This estimate neglects organic matter derived from *in situ* new production supported by offshore nutrient inputs (Chapter 9). Wollast (1998), considering the SEEP and OMEX programs in the North Atlantic, derived an estimate of 183 Tmol (2.2 GtC) $yr^{-1}$. Liu et al's. (2000a) recent reanalysis included a pass-through of 29 Tmol C of riverine input in their estimate of 225 Tmol C $yr^{-1}$, so is not much different from the Wollast figure. Thus roughly 100–200 Tmol C $yr^{-1}$ of a total annual NPP of 1200 Tmol (8–16%) is exported into the ocean interior by the continental shelf $CO_2$ pump. This export flux is equivalent to the 158 Tmol of carbon entering the coastal ocean from the net air-sea exchange (75 Tmol; see Table 9.5 below) and the riverine input (83 Tmol). In essence the coastal ocean ecosystem exports its exogenous inputs to the open sea.

## 6. The metabolic status of coastal and open ocean systems: autotrophic or heterotrophic?

In the rather small data set considered by Gattuso et al. (1998; for continental shelves, n = only 10 from 7 sources), the continental shelves have a positive metabolic balance, producing slightly more carbon than they respire (net ecosystem production, NEP = GPP-R = +68 Tmol C $yr^{-1}$ globally). In contrast estuaries receive large external inputs of organic matter (Hopkinson and Vallino, 1995) and respire more carbon than they produce internally (Kemp et al., 1997; Raymond et al., 2000). Estuarine NEP was –8 Tmol C $yr^{-1}$ globally in Gattuso et al's. dataset. Earlier data compiled by Smith and Hollibaugh (1993) showed that inner continental shelves, bays and marginal seas also had a negative metabolic balance (Table 9.2). In Chapter 7, Mackenzie et al. estimate that the coastal ocean is net heterotrophic, with NEP = –20 Tmol C $yr^{-1}$. In Section 8 below, we conclude that the coastal ocean, including offshore zones of coastal upwelling on the slope and rise, is net autotrophic by 175 Tmol C $yr^{-1}$.

The impact of organic matter additions appears to vary with distance from land, as one would expect. Wollast (1998) characterized inner shelves as slightly net heterotrophic (i.e., P<R), but concluded that outer shelves tended toward autotrophy (P>R). Gattuso et al.'s (1998) synthesis is consistent with that scheme, with only estuaries having P<R (Table 9.1). Earlier, Garside et al's (1976) analysis suggested that primary production supported by sewage-derived nutrients dominated over respiration of sewage organic matter and extended well offshore in the New York Bight. A catastrophic anoxic event in the same area was the result of sinking

and decay of a phytoplankton bloom, not oxidation of sewage organic matter (Falkowski et al., 1980). Boehme et al.'s (1998) recent model analysis showed that the effect of organic matter decomposition on $pCO_2$ decreased offshore on the New Jersey shelf, with the largest effect and greatest variability of organic matter influence near the 10-meter isobath.

That nearshore ecosystems would have negative metabolic balance is not at all surprising. Since P:R in most plankton systems is so close to 1 (Table 9.2), organic matter inputs are likely to render these systems net heterotrophic. Kemp et al. (1997) and Hopkinson and Vallino (1995) concluded that the metabolic balance in estuaries and the nearby coastal zone was dependent on the ratio of inputs of dissolved inorganic nitrogen to labile organic matter, a point also demonstrated by Garside et al. (1976). With increased eutrophication, net ecosystem metabolism could shift toward positive or more negative values, depending on the composition of the input (see Chapters 7, 21). Shifts in either direction would have consequences for the efficiency of the coastal oceans as $CO_2$ sinks.

But what is surprising is that some offshore regions, including the oligotrophic ocean surface layer have also been characterized as net heterotrophic. del Giorgio et al. (1997) compiled data on bacterial respiration (BR) and NPP from oceanic, coastal, estuarine and lake systems and showed that bacterial respiration exceeded the NPP when NPP was below about 100 $\mu gC\ l^{-1}\ d^{-1}$. This is a typical value for surface waters in coastal regions and oceanic phytoplankton blooms, comprising about 30% of the global ocean. Subsequently Hoppe et al. (2002), using similar measurements, suggested that bacterial metabolism exceeded NPP over large areas of the tropical Atlantic. Since bacterial metabolism is supported by preformed organic matter, bacterial respiration can only exceed NPP if there are external subsidies of organic matter. The startling conclusion from this analysis is that most oceanic systems, remote from obvious sources of organic matter input, have negative net ecosystem metabolism (P:R < 1). Indeed, Duarte and Agusti (1998) concluded on the basis of a direct comparison of NPP and R in many data sets, that about 80% of the global ocean surface area was heterotrophic. del Giorgio et al's (1997) finding was based on an indirect comparison of independent estimates of NPP and BR, and BR estimates themselves are usually derived from other measurements (del Giorgio and Cole, 1998). Thus this conclusion was criticized and remains controversial (Geider, 1997). NPP and R however, are both based on measurements of oxygen in the same bottles, so the conclusions of Duarte and Agusti (1998) are on firmer technical ground. Even so, their conclusions as well as the del Giorgio et al paper were disputed by Williams and Bowers (1999), referring to (Williams, 1998)'s reanalysis of the NPP:R data, concluding that "...open oceans as a whole are not substantially out of organic carbon balance. There is no evidence of the large regional imbalances observed previously." Williams and Bowers' (1999) statistical analysis was questioned by Duarte et al. (1999) who stated that their conclusions about oligotrophic regions were improperly extrapolated from more productive areas. The debate remains unresolved.

One implication of possible net heterotrophy in open ocean basins is that there must be an import flux of labile organic matter to subsidize the excess respiration. The only two likely possibilities are a vertical flux from deep water or a lateral flux from the coastal ocean. Inputs of labile DOC from atmospheric deposition are poorly known but have been estimated to average less than 58 μmolC $m^{-2}$ $d^{-1}$ (Willey et al. 2002), insufficient to support the necessary level of respiration alone. There is a large reservoir of DOC in the deep ocean that could support heterotrophic metabolism in the surface layer, but it has a mean age of several 1000 years and is generally believed to be highly resistant to biological decomposition (Bauer et al., 1992). However some of this DOC may be rendered more biochemically labile by exposure to ultraviolet radiation when it reenters the surface layer in the thermohaline overturning (Anderson and Williams, 1999). A large supply of labile organic carbon via lateral exchange processes is not a much more likely possibility. The existing data show that the concentration gradients in DOC extending from the ocean margins into the central gyres are not sufficiently steep to support the necessary flux. Williams and Bowers (1999) estimated that the carbon subsidy needed to support the level of net heterotrophy in the open ocean *surface layer* proposed by Duarte and Agusti (1998) amounted to 500 Tmol C $y^{-1}$. This is over twice the amount supplied by the continental shelf pump (Section 5). Thus shelf export seems insufficient to support large oceanic heterotrophy. Moreover, the shelf export is supplied predominantly at a depth of several 100 meters, not at the surface. In Section 8 we conclude that the open ocean *over the entire water* column is heterotrophic by just 200 Tmol C $y^{-1}$, less than half what Duarte and Agusti attribute to the surface layer alone. The question of net heterotrophic metabolism in the oceanic euphotic zone remains tantalizing but unresolved, and presents a challenge demanding a better understanding of carbon exchanges between the coastal and open ocean.

## 7. The carbon balance in the coastal ocean: Is it a sink for atmospheric $CO_2$?

The chemistry of $CO_2$ in seawater has now been studied for over 70 years (Buch et al., 1932; Greenberg et al., 1932). Yet in spite of its great biogeochemical reactivity, relatively little is known about the $CO_2$ balance in the coastal ocean. The state of understanding stands in stark contrast to that of the open ocean, where nearly a million discrete measurements of the partial gas pressure of $CO_2$ exerted in surface water ($pCO_2$) have been made since the International Geophysical Year of 1956–59 (Takahashi et al., 2002).

The global ocean is a sink for 1.9–2.2 Pg atmospheric C annually (Houghton et al., 2001; Sarmiento and Gruber, 2002), as estimated from $CO_2$ (Takahashi et al. 2002) and $O_2$ measurements (Keeling and Shertz, 1992; Bender et al., 1998), global circulation models (Sarmiento et al., 1992; Murnane et al., 1997) and changes in atmospheric and oceanic $\delta^{13}C$ (Quay et al., 1992). The global (open) ocean is a contemporary sink because of the anthropogenic increase in atmospheric $CO_2$ since the 18th century, which increased the difference between sea and air $pCO_2$ ($\Delta pCO_2$) from near equilibrium ($\Delta pCO_2 \approx 0$) to its current global mean value of –4 μatm at a mean atmospheric $pCO_2$ of 360–370 ppm (Takahashi et al. 2002). This anthropogenically-forced gas pressure opposes the potential biogenic flux of $CO_2$ out of the ocean (Section 6). The preindustrial ocean as a whole is believed to have been a slight source of $CO_2$ to the atmosphere (0.2–0.4 PgC $y^{-1}$), partially balancing carbon inputs of 0.8 PgC from the land (Figure 9.1; Smith and Mackenzie 1987; Sarmiento and Sundquist 1992 and Mackenzie et al., Chapter 7 in this volume).

Surprisingly however, the contemporary source/sink identity of the coastal oceans is unresolved (Kempe, 1995). Yet there has been a large amount of new research on $CO_2$ exchange and $pCO_2$ variability in the coastal ocean. Here we provide a comprehensive, critical review of the $CO_2$ balance of the global coastal ocean, concluding it is likely a strong sink for atmospheric $CO_2$. This identity may change in the next century (Chapter 7).

### 7.1 *Controls on $pCO_2$ in the coastal oceans.*

Unlike other dissolved gases (especially oxygen), the ocean is the major global reservoir for $CO_2$ due to the action of the carbonate buffering system in seawater (Figure 9.3). Dissolved $CO_2$ undergoes reaction with seawater to form carbonic acid, $H_2CO_3$, which dissociates into the bicarbonate and carbonate ions, $HCO_3^-$ and $CO_3^{2-}$. At high pH the chemical reactions in Figure 9.3 are forced to the right, so that at the average seawater pH of 8.2, over 95% of the dissolved inorganic carbon (DIC) is in the form of bicarbonate and less than 1% is dissolved $CO_2$ (Butler, 1982). Of every mole of $CO_2$ which enters the ocean, most becomes bicarbonate, explaining the huge capacity of the ocean to take up atmospheric $CO_2$. When the dissolved $CO_2$ is near equilibrium with the atmosphere (currently ~370 μatm or ~ 15 μM), there is about 2000 μM total DIC in the seawater, and 50 times more DIC in the ocean than $CO_2$ in the atmosphere (Feely et al., 2001). However if dissolved $CO_2$ were the only form used by phytoplankton (see below), about 15 μM, rather than 2000 μM of DIC would be directly available for photosynthesis (Riebesell et al., 1993).

$pCO_2$ in the coastal ocean is controlled by a complex array of climatic, physical and biological factors (Boehme et al., 1998). $CO_2$ solubility increases inversely with temperature; thus for a fixed concentration of $CO_2$, the resulting $pCO_2$ will *increase* as the temperature rises. It would increase by a factor of about 4 over the polar to equator temperature gradient (–1.9 to 30 °C), in the absence of other processes. Proximity to land adds a number of factors important for determining coastal ocean (though not necessarily open ocean) $pCO_2$, including river outflows, groundwater and rain inputs of dissolved $CO_2$ gas, and inputs of both dissolved and particulate organic and inorganic carbon from land. Rivers are generally highly supersaturated in dissolved $CO_2$ (Frankignoulle et al., 1998; Raymond et al., 2000), adding $CO_2$ to the inner shelf and extending the range of $pCO_2$ variability compared to the open sea. Several processes transport oceanic properties onto shelves. Coastal upwelling, exchanges mediated by filaments, jets and rings and other frontal exchange processes all directly affect the carbon balance and $pCO_2$, by bringing onto the shelf higher DIC from greater depths (Huthnance, 1995; Wollast, 1998). Many of these physical processes are still poorly understood (Chapter 2). Inputs of upwelled DIC have the direct effect of raising shelf $pCO_2$ (see below). Terrestrial inputs of N and P indirectly affect $pCO_2$ by providing a source of new nutrients for photosynthesis, which draws down the $pCO_2$.

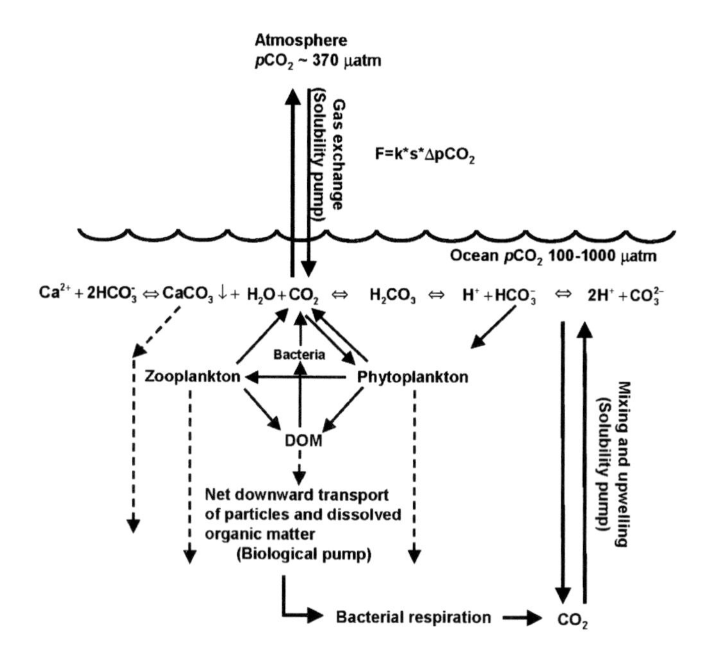


Figure 9.3 Carbon transformations and transport in the ocean water column. See text for details. The biological pump transports organic carbon into the deep sea against the vertical gradient of dissolved inorganic carbon. The solubility pump governs the air-sea exchange and is controlled by physicochemical processes, principally temperature and wind stress. About 95% of the total DIC is seawater is in the form of bicarbonate, explaining the enormous capacity of the ocean to store atmospheric $CO_2$. Biogenic formation of calcium carbonate produces one molecule of $CO_2$ for each molecule of carbonate. The carbonate shells sink, contributing to the biological pump (hard tissue pump).

Exchange of $CO_2$ across the atmosphere-ocean interface is a purely physical process. The air-sea flux, *F,* depends on the $\Delta pCO_2$, the solubility of aqueous $CO_2$ (*s,* a function of temperature and salinity), and the gas transfer coefficient (*k,* a nonlinear function of wind speed):

$$F = k \bullet s \bullet \Delta pCO_2 \tag{1}$$

(Feely et al., 2001). $pCO_2$ varies seasonally and geographically by ±20 μatm in the atmosphere (±5%) but by almost a factor of 10, from ~90 to 900 in the coastal ocean (see below). Therefore, $\Delta pCO_2$ is primarily driven by changes in $pCO_2$ in seawater. The seawater $pCO_2$ is a function of temperature, DIC concentration and alkalinity. The DIC concentration is a function of the balance between photosynthesis and respiration, and alkalinity is primarily controlled by precipitation and dissolution of calcium carbonate. As ocean currents move from lower toward higher latitudes, or as seawater cools seasonally, $CO_2$ solubility increases, the $pCO_2$ declines, and $\Delta$ $pCO_2$ increases, causing atmospheric $CO_2$ to be driven into the water. Conversely, as cold, deep water rises to the surface and warms, $CO_2$ degasses to the atmosphere. Since deep waters contain high concentrations of $CO_2$ as a residue of the action of the **biological pump** (Figure 9.3), coastal upwelling may constitute a powerful source of atmospheric $CO_2$ (Feely et al., 1999). These temperature-driven processes of $CO_2$ exchange form the **solubility pump** for atmospheric $CO_2$ (Figure 9.3). Because most upwelling occurs in low altitudes, these regions tend to be $CO_2$ source areas, whereas high latitudes are sink regions of $CO_2$ uptake (Takahashi et al., 2002).

### *7.2 $CO_2$ uptake by phytoplankton and the biological pump.*

A vertically uniform distribution of DIC in a sterile ocean would yield a vertical profile of $pCO_2$ (ignoring pressure effects) that was slightly higher at the surface and lower at depth as a result of temperature stratification. However DIC and $pCO_2$ both *increase* with depth as a result of biological processes. Phytoplankton take up inorganic carbon during photosynthesis in the light by transporting either dissolved $CO_2$ or bicarbonate ion across the cell membrane prior to fixation with the enzyme ribulose-1,5-biphosphate carboxylase/oxygenase (Rubisco; Falkowski and Raven 1997). There is evidence supporting passage of carbonate ion into cells as well (Raven, 1997), but the primary form taken up is generally believed to be dissolved $CO_2$ (Steeman Nielsen, 1947; Riebesell and Wolf-Gladrow, 2002) and this is the form usually represented in models of the ocean carbon cycle (Antoine and Morel, 1995). The biological processes can be faster than abiotic dehydration of bicarbonate to form $CO_2$ as this latter reaction is biologically catalyzed by the enzyme carbonic anhydrase (Falkowski and Raven 1997), making the large pool of bicarbonate (~1800 μM) potentially available for uptake. The relative importance of $CO_2$ and $HCO_3^-$ in uptake processes varies among phytoplankton taxa and is still not well understood or quantified in nature (Tortell et al., 2000; Riebesell and Wolf-Gladrow, 2002). The competitive advantage of bicarbonate acquisition is obvious, as it affords access to a carbon pool 2 orders of magnitude larger than the dissolved $CO_2$ pool. It is also possible that in coastal regions where $pCO_2$ declines to below 100 ppm (8 μM), cells using only $CO_2$ could become carbon limited. However the net result in either case is lowering of the ambient $pCO_2$ during photosynthesis. $pCO_2$ drawdown occurs because the timescale for equilibration of $CO_2$ in the mixed layer with the atmosphere is about one year, much slower than the biological processes.

Photosynthetic removal of DIC and drawdown of $pCO_2$ forms part of the mechanism of the biological pump for atmospheric $CO_2$ (Ducklow et al., 2001). The other components include biological consumption and respiration processes, and downward transport of organic matter via gravitational sedimentation and advection (Longhurst and Harrison, 1989). Phytoplankton respire organic carbon back into $CO_2$, and so do zooplankton and bacteria following consumption (Figure 9.3). The net balance of production ($CO_2$ uptake) and respiration ($CO_2$ release) processes influence the sign of $pCO_2$ change in the water column. In the lower part of the water column below the euphotic zone, respiration is the dominant biological process affecting $pCO_2$. Biotransformations of calcium carbonate and chemolithotrophic bacterial processes (e.g., nitrification) also affect $pCO_2$. Direct uptake of DIC by chemosynthetic bacteria is minor in significance compared to the large amount of heterotrophic respiration (del Giorgio and Duarte, 2002). Thus the net effect of the biological pump is to remove $CO_2$ from the surface in the form of organic matter, transport it to depth, and respire it back into DIC. The resulting vertical profiles of DIC and $pCO_2$ increase from the surface to the bottom. A third $CO_2$ pump unique to the coastal ocean, the *continental shelf pump,* is discussed below.

### *7.3 Influence of terrestrial inputs.*

$pCO_2$ is also inextricably linked to the inputs and cycling of N, P, S and organic matter through both land and ocean metabolism as described previously (Smith and Hollibaugh, 1993; Ver et al., 1999), and therefore depends on the regional as well as local carbon balance. Primary production is supported by inputs N, P and micronutrients like iron in river runoff. Intense primary production during spring and summer blooms or associated with wind-driven coastal upwelling can counteract the effect of warming, and produce large seasonal decreases in $pCO_2$ (Codispoti et al., 1982; DeGrandpre et al., 2002). N and P inputs from land and sea are accompanied by DIC, and so there will only be a net input of DIC if it is supplied in excess of the Redfield ratio (Table 9.1). The excess DIC may be released into the atmosphere (source) or exported via offshore transport. Conversely, if inorganic inputs are N- or P-rich, the subsequent primary production will draw $CO_2$ from the atmosphere and serve as a sink for atmospheric $CO_2$. This reasoning led Walsh et al. (1981) to postulate that coastal eutrophication could be part of the missing sink in the global carbon budget.

Organic matter loadings from the land also influence the carbon balance in the coastal ocean and thus affects its source/sink characteristics. Dissolved (Raymond and Bauer, 2001) and particulate organic matter exported from rivers (Wollast 1998) result in net additions of $CO_2$ to the coastal ocean if they are decomposed there, raising $pCO_2$ levels and increasing the potential for release to the atmosphere. The extent of organic matter contributions to local carbon balance depends on the biochemical lability of imported compounds (Section 3) and their residence times in coastal ocean waters. Residence times in rivers are relatively short, indicating that they may export fresh (labile) organic constituents into the coastal ocean, but estuaries have extended residence times and serve as filters and traps for organic matter exported from land or produced *in situ* (Hedges et al., 1997). Smith and Hollibaugh (1997) estimated that 34 ± 10 teramoles (0.4 Gt) C $yr^{-1}$ are transported from land by rivers, of which 11 ± 6 is buried in estuaries, marshes and deltas, leaving 23 ± 12 Tmol in the water column of the coastal ocean. They further estimated that 5–15 Tmol of that is labile on short time scales (< 1 yr). It is this fraction that is apparently oxidized and respired in the coastal ocean, compensating to a varying extent for the $pCO_2$ drawdown during photosynthesis and cooling. Boehme et al. (1998) provided a detailed analysis of the factors influencing $pCO_2$ variations in the coastal ocean off New Jersey, USA, using data taken during monthly cross-shelf transects collected over two years. In their analysis, the separate contributions of heating, organic matter production and decomposition, precipitation inputs, carbonate production and dissolution, air-sea exchange and mixing to temporal changes in $pCO_2$ were assessed by calculating $pCO_2$ from DIC and alkalinity using the thermodynamic equations for the carbonate system and varying each parameter value in turn (Sabine and Key, 1998). They found that temperature and organic matter production/decomposition had the largest effects on $pCO_2$, with air-sea exchange and mixing making somewhat smaller contributions. In general effects were larger at inshore stations and smaller offshore, reflecting the decreased variability offshore. The effects of some factors were not symmetrical: changes in photosynthesis and respiration tended to raise $pCO_2$ rather than lower it, pointing out the importance of upwelled remineralized carbon to surface waters (Figures 9.4, 9.5). Further, the effects of some factors were counterbalanced: in some seasons, some factors had the effect of raising $pCO_2$ while others lowered it, leaving no net change (Figure 9.4). Differences between observations and model predictions were ascribed to mixing of water masses with unspecified characteristics. This analysis is the most comprehensive attempt to date at dissecting the interplay of physical and biological factors on coastal $pCO_2$ variability. Boehme et al.'s results show that time series of coastal $pCO_2$ can be accurately reproduced with just a few parameters, indicating an improved fundamental understanding of how this variability is regulated at local to regional scales.

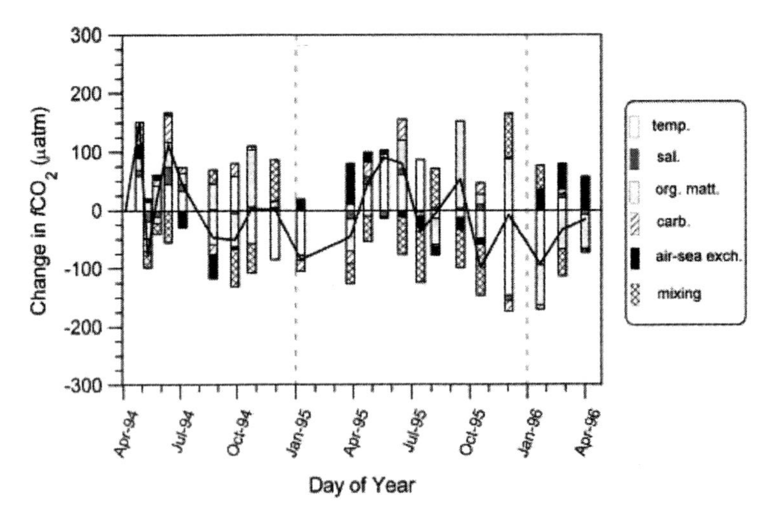


Figure 9.4 Effects of various factors on temporal variations in coastal ocean pCO2 (Figure courtesy Boehme et al. (1998).

### 7.4 *Survey of* $pCO_2$ *in the coastal oceans.*

The relative importance of these factors varies in different regions producing cycles in $pCO_2$ that are just beginning to be characterized. Here we provide a first attempt at such a characterization, following Longhurst's (1995) province-based scheme for defining variations in biological cycles of production and grazing across ocean provinces (Longhurst 1998; Ducklow 2003 and see Chapter 10). Table 9.4 summarizes studies of $pCO_2$ in coastal oceans, including systems influenced by continental margin processes that can extend well beyond the shelf (e.g., upwelling off Peru, Oman, etc.). Several aspects are notable. One is that only a few regions of the global coastal ocean have been surveyed extensively, with good areal and seasonal (annual) coverage: the west European shelves, the East China Sea (ECS), the USA Middle Atlantic Bight (MAB) and the coastal upwelling province of the Arabian Sea off the coast of Oman (Figures 9.5, 9.6). Save for a single study off Peru in the 1970's, and recent work in the Arabian Sea, there are no studies from the tropics other than on coral reefs. The other conspicuous and important feature of these studies of coastal ocean $pCO_2$ is its great variability. The combination of factors influencing $pCO_2$ variability in coastal seas produces both large under- and supersaturation in dissolved $CO_2$, with values ranging from 90–980 ppm (Table 9.4), yielding air-sea $\Delta pCO_2$ values from –270 to +646 μatm. Thus coastal oceans may be from 75% undersaturated to nearly 200% oversaturated, potentially producing large air-sea fluxes in both directions. Closer inspection of the data shows however that only temperate seas have this wide dynamic range of $pCO_2$ variability. Low values in spring and summer are driven by phytoplankton blooms, whereas high values in autumn and winter are forced by net respiration and riverine runoff (Borges and Frankignoulle, 1999). Note that this cycle (high in winter/low in summer) runs counter to that expected by physical effects alone. An annual temperature cycle ranging from 4 to 25°C, imposed on a closed water body with constant DIC and alkalinity would drive a cycle in $pCO_2$ ranging from a wintertime low of ~220 μatm to a summer maximum of ~560 μatm (DeGrandpre et al., 2002). The high degree of variability on temperate shelves is in some contrast to the open sea where the peak seasonal amplitude in $pCO_2$ is generally less than 200 μatm (Takahashi et al. 2002).

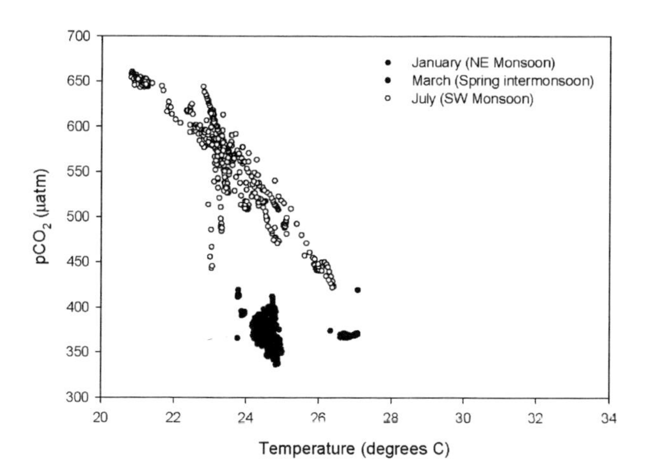


Figure 9.5 pCO2 in the surface coastal ocean of the northwest Arabian Sea, 1995. Data obtained from the US JGOFS database. See also Goyet et al. (1998).

TABLE 9.4
Partial pressure of Carbon dioxide ($pCO_2$) in coastal seas.

| Location | Temp Range (°C)[1] | $pCO_2$ Range (µtm) | Mean $pCO_2$ (µatm)[2] | Air-Sea $\Delta pCO_2$ Range (µatm)[3] | Month Season | Reference |
|---|---|---|---|---|---|---|
| **European Shelves** | | | | | | |
| Portugal Shelf | 14–17 | 360–460 | | 0 to +100 | Summer Upwelling | (Perez et al. 1999) |
| Galicia, Spain | 16–21 | 263–415 | 342 | -46 to +1 | Summer Upwelling | (Borges and Frankignoulle 2002) |
| North Sea | 5–17 | 100–450 | | -250 to +100 | May–June 1986 | (Kempe and Pegler 1991) |
| North Sea (Helgoland)[4] | | 200–900 | 480 | +155 | 1962–1978 | (Kempe 1995) |
| North Sea | 9–13 | 220–340 | | -139 to -19 | June, 1991 | (Schneider et al. 1992) |
| English Channel | 10–12 | 100–300 | | -260 to -10 | April–June, 1994 | (Frankignoulle et al. 1996) |
| English Channel | 13–15 | 230–700 | | -130 to +340 | Sept.–Oct, 1994 | " " |
| Bay of Biscay | 7–22 | 306–367 | 332 | -60 to +2 | Annual, 1993–99 | (Frankignoulle and Borges 2001) |
| Belgian coast | 1–16 | 90–778 | | -270 to +410 | Annual, 1995–96 | (Borges and Frankignoulle 1999) |
| Dutch coast | 2–6 | 150–200 | | -148 to -198 | Mar.–April, 1986 | (Hoppema 1991) |
| Dutch coast | 15–17 | 300–800 | | -25 to -40 | September, 1993 | (Bakker et al. 1996) |
| Celtic Shelf | 12–17 | 320–335 | | | September, 1995 | (Keir et al. 2001) |
| **American Shelves** | | | | | | |
| Bering Sea | -1–8 | 125–440 | | -216 to +100 | March -June, 1980 | (Codispoti et al. 1982) |
| Mid Atlantic Bight (USA) | 2–25 | 211–658 | | -149 to +298 | 1994–1996 | (Boehme et al. 1998) |
| Mid Atlantic Bight (USA) | 4–8 | 190–400 | | -35 to -66[5] | Spring | (DeGrandpre et al. 2002) |
| Mid Atlantic Bight (USA) | 8–22 | 210–630 | | +49to+61[5] | Summer | " " |
| Mid Atlantic Bight (USA) | 17–26 | 660–140 | | +15-+63[5] | Fall | " " |
| Mid Atlantic Bight (USA) | 17–4 | 510–180 | | 0 to -65[5] | Winter | " " |
| Mid Atlantic Bight (USA) | 19–26 | 345–655 | | -13 to +297 | Sept.–Oct. 1996 | (Bates and Hansell 1999) |
| West Florida Shelf (USA) | 22–25 | 301–322 | | -64 to -43 | April, 1996 | (Wanninkhofetal. 1997) |
| Oregon coast (USA, 1995) | 9–18 | 150–690 | | -210 to +330 | Summer upwelling | (van Geen et al. 2000) |
| Monterey Bay CA (USA) | 12–14 | 298–438 | | -80 to +20 | Feb.–Mar. 1993 | (Friederich et al. 1995) |
| Peru upwelling (Cabo Nazca) | 19–21 | 120–980 | | -214 to +646 | May, 1976 | (Simpson and Zirino 1980) |

| | | | | | | |
|---|---|---|---|---|---|---|
| | | | **Asian Shelves** | | | |
| Arabian Sea (Oman coast) | 20–24 | 520–720 | 250 | +160 to +360 | SW Monsoon (7/95) | (Kortzinger et al. 1997) |
| Arabian Sea (Oman coast) | 24–25 | 336–411 | 372 | -24 to +51 | NE Monsoon (1/95) | (Goyet et al. 1998)[6] |
| Arabian Sea (Oman coast) | 23–27 | 365–419 | 374 | -4 to +58 | Intermonsoon (3/95) | " " |
| Arabian Sea (Oman coast) | 20–27 | 392–660 | 565 | +30 to +298 | SW Monsoon (7/95) | " " |
| Bay of Bengal (India coast) | 26–30 | 275–400 | | -75 to +50 | Intermonsoon (3/91) | (Kumar et al. 1996) |
| Bay of Bengal (India coast) | 25–26 | 240–380 | | -110 to +30 | NE Monsoon (11/91) | " " |
| Funka Bay, Japan | 3–20 | 195–345 | 292 | -165 to -20 | Annual, 1995–96 | (Nakayama et al. 2000) |
| East China Sea | 15–24 | 289–392 | | -72 to +21 | Feb.–March, 1993 | (Tsunogai et al. 1997) |
| East China Sea | 23–28 | 320–390 | | -37 to +33 | August, 1994 | (Tsunogai et al. 1999) |
| East China Sea | 22–27 | 307–368 | | -56 to +4 | October, 1993 | (Tsunogai et al. 1997) |
| East China Sea | 27 | 220–370 | | -37 | July, 1992 | (Wang et al. 2000) |
| | | | **Coral Reefs** | | | |
| Bermuda | 18–27 | | | -10 to + 33 | 1994–1996 | (Bates 2002) |
| Enewetak Atoll | | 190–320 | | -138 to -8 | May–June, 1971 | (Smith 1973) |
| Moorea, Tahiti | 27 | 240–400 | | -116 to +44 | July–Aug. 1992 | (Frankignoulle et al. 1996) |
| Great Barrier Reef | 27 | 250–700 | | -107 to +343 | December, 1993 | " " |
| Palau, Caroline Islands | 29 | | 414 | +55 | April 1992 | (Kawahata et al. 1997) |
| Majuro Atoll, Marshall Is. | 30 | | 360 | +5 | September 1994 | " " |
| Ryukyu Islands, Japan | | 200–550 | | -157 to +193 | Oct. 93–July 94 | (Kraines et al. 1997) |
| | | | **Arctic and Antarctic shelves** | | | |
| East Baffin Bay, Canada | -1.8 | | 410 | +30 | April 1998 | (Miller et al. 2002) |
| East Baffin Bay, Canada | -1–+1 | 130–150 | 140 | -230 | June 1998 | " " |
| NW Greenland Sea | -1.8–3 | 167–279 | 218 | -189 to -77 | July–August, 1992 | (Yager et al. 1995) |
| Ross Sea, Antarctica | 0–1 | 166–314 | | -197–49 | January, 1997 | (Sweeney et al. 2000)[7] |
| Ross Sea, Antarctica | -1.5–1 | 150–410 | | -212 to +48 | Oct.–April 96–97 | (Takahashi et al. 2002) |
| Gerlache Strait, Antarctica | -1–0 | 246–363 | | -150 | December, 1995 | (Carillo and Karl 1999) |

[1]obtained from World Ocean Database 2001 (http://www.nodc.noaa.gov/OC5/WOD01/pr_wod01.html) using Ocean Data View (Schlitzer, 2002) if not given in publication.

[2]Calculated from tabulated data where available if not given in publication.

[3]Positive numbers indicate fluxes from sea to air; values as reported or calculated using Mauna Loa monthly atmospheric values (CD Keeling & TP Whorf, http://cdiac.ornl.gov/ftp/ndp001/maunaloa.co2)

[4]Calculated from continuous moored measurements of pH (W. Hidal, Hamburg).

[5]Seasonal means from continuous mooring data at Buzzards Bay, MA and LEO-15, USA.

[6]Data originated by Catherine Goyet and available from http://usjgofs.whoi.edu/jg/dir/jgofs/arabian/

[7]Data from T. Takahashi measured at 4°C and available from http://usjgofs.whoi.edu/jg/dir/jgofs/southern/nbp97_1/

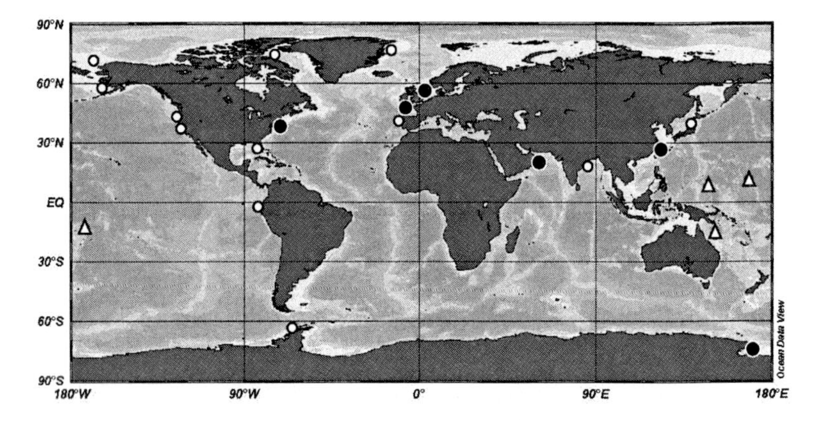


Figure 9.6 Regions in the global coastal ocean where $pCO_2$ has been measured. Figure generated using Ocean Data View (Schlitzer, 2002).

The dramatic effects of biological processes (primary production) in drawing down high $pCO_2$ are exemplified by observations made in upwelling regimes off Peru, the Iberian Peninsula and Oregon (Simpson and Zirino, 1980; Perez et al., 1999; van Geen et al., 2000; Borges and Frankignoulle, 2002). Cold water from depth, supersaturated in $CO_2$ ($pCO_2$ ~400 – 900 μatm) and high in N and P as well, upwells to the surface and then supports intense primary production, which in turn reduces $pCO_2$ to <200 μatm. The upwelling signature can also be seen in the Arabian Sea (Table 9.4, Figure 9.5), but there the upwelling is seldom fully compensated by biology, in spite of high primary production, even during the Southwest Monsoon (Savidge and Gilpin, 1999; Barber et al., 2001; Dickson et al., 2001; Kinkade et al., 2001). Instead, $pCO_2$ remained above the atmospheric equilibrium level or just slightly below ($\Delta pCO_2$ ~ -4 μatm). Goyet et al. (1998) found that the areal average of high-resolution underway $pCO_2$ measurements remained above the atmospheric equilibrium throughout the year. That is, the coastal Arabian Sea was always a net source for $CO_2$ flux to the atmosphere. Goyet et al. (1998) estimated that the coastal ocean off Oman released about 1–2 MtC per year (~10 mmol C $m^{-2}$ $d^{-1}$; Table 9.5), while Kortzinger et al. (1997) suggested that the region outgassed 11–26 MtC (52–119 mmol C $m^{-2}$ $d^{-1}$) in the SW monsoon period (July-September) alone. The latter flux densities seem very high relative to most other estimates (Table 9.5). The discrepancy in flux estimates illustrates our still-poor understanding of the processes governing the $CO_2$ balance, even in a comparatively well-studied area.

TABLE 9.5
Annual mean air-sea $CO_2$ flux in the global coastal ocean.

| Location | Flux density[1] mMol C $m^{-2}$ $d^{-1}$ | Area of shelf $10^6$ $km^2$ | Flux Tmol C $yr^{-1}$ | Reference |
|---|---|---|---|---|
| European Shelves | 7.9 | 5 | 14 | (Frankignoulle and Borges 2001) |
| East China Sea | 7.9 | 0.9 | 2.6 | (Tsunogai et al. 1999) |
| East China Sea | 5.5 | 0.9 | 1.8 | (Wang et al.2000) |
| Baffin Bay, Canada | 20[2] | 0.08 | 0.3 | (Miller et al. 2002) |
| Barents Sea | 3.2[2] | 1.4 | 0.8 | (Fransson et al. 2001) |
| North Sea | 3.8 | 0.51 | 0.7 | (Thomas et al. 2004) |
| Mid Atlantic Bight (USA) | 3.6 | 0.06 | 0.1 | (DeGrandpre et al. 2002) |
| Arabian Sea (Oman coast) | -2.5 | 0.15 | -0.13 | (Goyetetal. 1998) |
| Coral reefs | -5 | 0.6 | -1.1 | (Frankignoulle et al. 1996) |
| Global Shelves (model) | 7.9 | 36 | 49 | (Yool and Fasham 2001) |
| Global coastal ocean | 5.6<br>(4.9) | 36[3] | 73<br>(65) | Area-weighted mean<br>(Arithmetic mean) |

[1]positive fluxes are from air to sea
[2]Assuming 170 day ice-free exchange period (Miller et al., 2002)
[3]the ocean area <200 deep, excluding estuaries, Liu et al., (2000).

Coral reefs are apparently net $CO_2$ source regions, but this generalization is complicated by the diversity of coral reef types and growing anthropogenic influence (Bates, 2002). A large amount of data indicates that although many reef systems have very high rates of gross primary production, their net community production (NEP) is near zero, that is, P:R ratios are not significantly different from 1, and organic metabolism does not contribute to $pCO_2$ disequilibrium (Crossland et al., 1991). However these ecosystems are also areas of intense calcification (Kinsey, 1985). Calcification releases about 0.6 mole of $CO_2$ per mole of $CaCO_3$ precipitated at the current temperature/salinity/alkalinity properties of modern coral reefs (Frankignoulle et al., 1994). When the calcification to net production ratio is 1.67, there is no net $CO_2$ production because the yield from $CaCO_3$ production is balanced by carbon uptake during organic carbon production. But since the NEP is near zero (P:R near 1), the ratio is >1.67 and so reefs are regions of $CO_2$ efflux (Gattuso et al., 1999), with evasion rates of 3 – 7 mmol C $m^{-2}$ $d^{-1}$ (Frankignoulle et al., 1996). The total global surface area of coral reefs is 0.6 x $10^6$ $km^2$, so coral reefs represent a substantial source of $CO_2$ to the atmosphere. But this efflux is dwarfed by $CO_2$ uptake in the larger temperate shelves (see below and Table 9.5). The source/sink balance of coral reefs may change in the future, because anthropogenic $CO_2$ enrichment of the upper ocean is lowering the pH of seawater (Kleypas et al., 1999) and the calcification process is sensitive to pH changes (Langdon et al., 2000).

In contrast to the Arabian Sea and the marked seasonal excursions of the temperate seas, the subtropical East China Sea shows less variability and $pCO_2$ levels are nearly always below the atmospheric equilibrium or just slightly above (Tsunogai et al., 1999; Table 9.4). In the ECS, wintertime cooling drew $pCO_2$ below the atmospheric equilibrium, and high primary production, sustained by upwelling of offshore nutrients (Chen, 1996), counteracted the effects of summer warming, maintaining low $pCO_2$ throughout the year. Based on their observations of an annual mean fugacity deficit ($\approx \Delta pCO_2$) of 55 μatm in the region, Tsunogai et al. (1999) hypothesized a "continental shelf pump" for absorption of atmospheric $CO_2$. In their model, surface cooling on the shelf produces denser water and this process in conjunction with primary production enhances absorption of $CO_2$ from the atmosphere. The absorbed $CO_2$ is transported below the pycnocline (as organic matter or dissolved gas) and then exported from the shelf region into the open sea. They further suggested that if the global continental shelves all behaved like the ECS, they would account for ~ 1 PgC (1 GtC) of uptake annually. This amount could be in addition to the 2.2 PgC estimated by observations and models to be taken up in the open sea (Liu et al., 2000a and see below). Clearly the ECS is not typical of all the world's shelf seas. But whether or not the ECS is a suitable model for the global coastal ocean, it is an important site of $CO_2$ uptake on its own, being the third-largest of the world's shelf seas, comprising 3% of the total shelf area shallower than 200 m, and accounting for 20–30 Mt (1.8 – 2.6 Tmol) of $CO_2$ uptake annually (Table 9.5). One weakness of this study is that site-specific empirical relationships were used to extrapolate from a sparse set of observations to produce regional scale monthly maps of surface $pCO_2$ and derive the annual fugacity deficit and pumping rate. Comprehensive mass balance studies in the ECS indicate lower fluxes and do not support a high $CO_2$ sink (Chen and Wang, 1999).

Is there a continental shelf pump? In some areas there clearly is. The west European shelves are huge, 5 million km$^2$ in area. Although they exhibit large variations in $pCO_2$, many of the higher values are concentrated near river mouths and coastal plumes (Bakker et al., 1996; Borges and Frankignoulle, 1999), whereas the low values are found throughout the observed shelf area. Even in the Bay of Biscay, where the seasonal amplitude is low, the mean annual $\Delta pCO_2$ is ~ –30 μatm (Frankignoulle and Borges, 2001). The West European shelves appear to function like the East China Sea, with winter $pCO_2$ drawdown maintained in summer by inputs of riverine nutrients which support high primary production. Thus these shelves, among the best studied on the planet, also appear to function as $CO_2$ pumps, taking up a startling 170 MtC (14 Tmol) annually (Table 9.5). However a recent study in the North Sea, including monthlong surveys in consecutive seasons, found a lower areal flux rate more comparable to other temperate regions (Thomas et al., 2004) (Table 9.5).

DeGrandpre et al. (2002) questioned the universality of the shelf-pump hypothesis, basing their argument on analysis of high frequency, cross-shelf measurements of $pCO_2$ made by moored sensors and underway sampling systems in the Middle Atlantic Bight, USA (see also Boehme et al. 1998). High-resolution observations in time and space addressed the problem of extrapolation noted in the East China Sea. DeGrandpre et al. (2002) observed large amplitude, short-term changes in $pCO_2$ >100 μatm (Figure 9.7), forced by local upwelling and mesoscale circulation. In this region, unlike the ECS and European margin, the thermally driven rise in $pCO_2$ in summer was not compensated by primary production, and waters became supersaturated. Apparently, nutrient inputs are not sufficient to maintain high PP, and the region becomes depleted of N and P following the spring bloom (Malone et al., 1983). The $pCO_2$ cycle appears to be driven primarily by the annual heating and cooling cycle, and the asymmetry between the thermal cycle and the winds renders it a weak sink region. Because winds are stronger in winter when $pCO_2$ is low, uptake predominates slightly, and the area takes up about 0.1 Tmol C per year (DeGrandpre et al., 2002; Table 9.5), acting as a weak $CO_2$ sink. High-resolution measurements of $pCO_2$, DIC and alkalinity extending from the Georgia salt-marsh estuaries to the Gulf Stream, across the South Atlantic Bight continental shelf revealed a similar pattern with an important addition. Export of DIC from the salt marsh system is sufficiently large that the SAB shelf exports DIC to the open ocean even during the summer when (due to heating) the flux of $CO_2$ is to the atmosphere (i.e. $\Delta pCO_2$ is positive). Thus, in this type of shelf system, studies of $\Delta pCO_2$ are not sufficient to quantify the sign or magnitude of the flux of DIC from the shelf to the open ocean (Wang 2002 and R. Jahnke, personal communication). Further generalization about whether other continental shelves act more like the ECS or MAB must await exploration and surveying of more shelf systems.

The continental shelf pump (CSP) hypothesis was examined by Yool and Fasham (2001) using a general circulation model. They parameterized the CSP by adding atmospheric $CO_2$ to model cells intersecting continental shelf areas (depths < 200 m) at the rate estimated by Tsunogai et al. (1999), then used the circulation model to trace the transport and fate of the $CO_2$ injected into the coastal ocean. Like other GCM's, their model did not explicitly include continental shelf areas, due to relatively coarse spatial resolution. Defining CSP export efficiency as the fraction of $CO_2$ taken up in a given shelf area which was exported and outgassed elsewhere, they found that the global coastal ocean was 53% efficient at exporting absorbed $CO_2$ into the open sea. Export efficiencies ranged up to 90% for small, narrow shelves (e.g., Hawaiian shelf), with larger temperate shelves ranging from 28% (North Sea) to 74% (Patagonia). Interestingly, they found that the East China Sea had an export efficiency of 52%, nearly identical to the global mean, and lending support to Tsunogai et al's CSP hypothesis. The results also suggested an global, annual uptake rate of 49 Tmol C, somewhat lower than the amount estimated by Tsunogai et al. (1999). A different modeling approach, emphasizing the coupled cycles of C-N-P-S suggests the coastal ocean is a very weak sink (–5 Tmol C $y^{-1}$), with strong net heterotrophy lessening the net $CO_2$ uptake C (Chapter 7).

The modeling approach of Yool and Fasham has the advantage of providing a consistent, defined way of following atmospheric $CO_2$ from its entry into the coastal ocean to its ultimate fate as outgassing or long-term storage in the ocean interior. But it is a very simplified approach, with no biological transformation of the absorbed $CO_2$; rather, $CO_2$ remains in its inorganic form, and acts as a passive tracer of transport in the ocean circulation. Yool and Fasham (2001) addressed this point by performing separate simulation experiments in which the $CO_2$ was transformed into a DOC-like tracer that did not exchange with the atmosphere and was respired back to $CO_2$ over seasonal (0.033 $d^{-1}$) to annual (0.003 $d^{-1}$) time scales. When the CSP was parameterized with this delay term, it allowed the absorbed $CO_2$ to survive for a longer period before outgassing; thus the export efficiency of the shelf regions was increased, and 8% more carbon was outgassed in the open sea. It is unclear if a model with more realistic biology, for example including a CSP dominated by sinking particles, would be still more efficient, owing to the uncertainty of representing near-bottom offshelf particle transport (Chapters 6, 17).

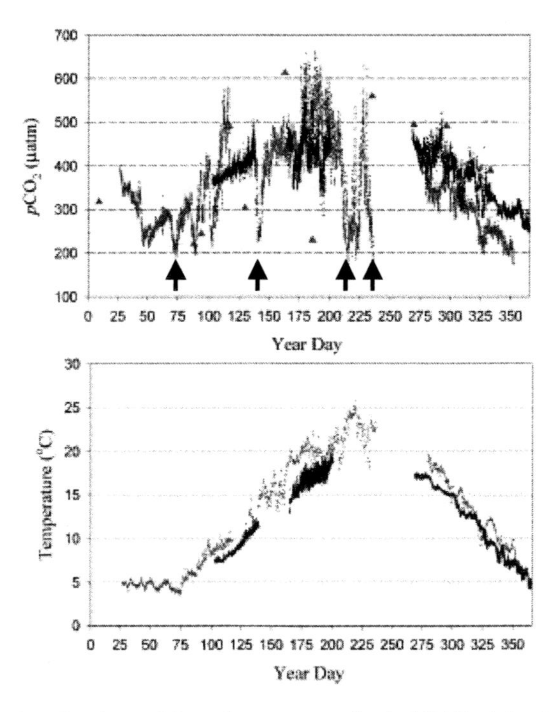


Figure 9.7 Annual cycles of surface $pCO_2$ and temperature in the Middle Atlantic Bight, USA. Arrows denote large short-term changes due to local upwelling. Figure courtesy deGrandpre et al. (2002).

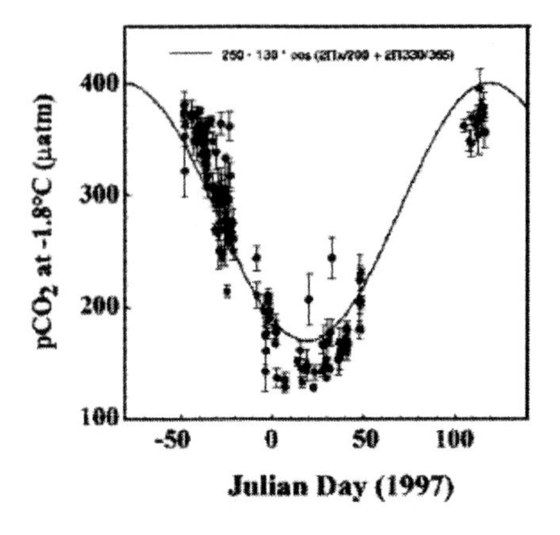


Figure 9.8 Figure 8. Seasonal changes in surface $pCO_2$ along longitude 76.5° in the Ross Sea, Antarctica. The solid trend line is approximated by a simple sine function. Figure courtesy Takahashi et al. 2002).

Shelves covered by seasonal sea ice also act as $CO_2$ pumps, but for another reason. The Ross Sea is a deep shelf sea with bottom depths ranging from 300–800 m due to isostatic adjustment from the weight of the ice mass on the Antarctic continent. The region was sampled throughout the ice-free season during the US JGOFS AESOPS Program, revealing a seasonal $pCO_2$ range of 150–410 ppm (Figure 9.8; Takahashi et al. 2002). Sea surface temperature is almost constant near Antarctica and the $pCO_2$ excursion is almost entirely due to biological drawdown and respiration. Most other ice-margin seas have been sampled only in summer, with similar levels of undersaturation (Table 9.4). Yager et al. (1995) found that the Northeast Water polynya on the Greenland Shelf was strongly undersaturated in the summer, ice-free season. They put forward the "seasonal rectification hypothesis," stating that in marginal ice zones, the ice-free season coincides with the main period of low $pCO_2$, when the regions act as atmospheric sinks. At other times of the year, when $pCO_2$ could be well above saturation, the water is covered by sea ice and gas exchange is prevented. In spring, primary production appears to consume excess dissolved inorganic carbon before the ice cover recedes. The extent to which high $CO_2$ escapes through leads, cracks and persistent polynyas is not known, but cannot be great, since only small areas are exposed. There is also uncertainty in estimates of gas transfer through the sea ice itself (Miller et al., 2002). Yager et al (1995) and Miller et al (2002) assumed complete capping of the air-sea interface by sea ice. Thus in effect they use estimates of $CO_2$ exchange in the ice-free season as an annual average, leading to very high estimates of the air to sea flux (Table 9.5). Miller et al. (2002) pointed out that if net respiration in autumn (September-October) resulted in increases in $pCO_2$ prior to restoration of ice cover, the strength of these polynyas as $CO_2$ sinks would be reduced. They also noted that horizontal advection can move low $pCO_2$ waters under surrounding ice. Complexity in circulation, timing of biological cycles and interannual variability combine to regulate the efficiency of seasonal polynyas as $CO_2$ sinks. Fransson et al. (2001) used a budget approach to derive a flux of 9.2 MtC $y^{-1}$ for the Barents Sea (Table 9.5). Yool and Fasham (2001) noted in their model that large concentrations of $CO_2$ absorbed by the continental shelves in the coastal Arctic were trapped under the permanent sea ice in the central Arctic Ocean, further enhancing the role of these ice-covered seas as regions of atmospheric $CO_2$ storage. Both the Arctic and Antarctic polar regions are a complex mosaic of sea ice cover in time and space, and few areas have been sampled well even in summer. The potentially large fluxes driven by intense ice-edge blooms bear further study of these remote regimes.

### *7.5 Synthesis: air-sea $CO_2$ exchange in the coastal oceans.*

It appears from the observations of air-sea $pCO_2$ made to date, that some shelves do act as $CO_2$ pumps for a variety of reasons; while others are only weak sinks, or else sources of atmospheric $CO_2$. Observations of $pCO_2$ seasonality and its causes are still comparatively scarce, and in many areas insufficient to resolve the sign of the annual flux balance. Some of the largest shelf seas on the planet have been only poorly surveyed (Bering Sea, Arctic Ocean) or not studied at all (Patagonian Shelf, Indonesia/northern Australian marginal seas). Based on the available data

(Table 9.4) we present in Table 9.6 a preliminary typology of regional $pCO_2$ variability in the global coastal ocean, another step toward segmentation of the ocean into true biogeochemical provinces (Platt and Sathyendranath, 1988; Longhurst, 1998; Ducklow, 2003). Seven distinct province types can be recognized. Takahashi et al. (2002) differentiated the open ocean into areas where the seasonal amplitude of $pCO_2$ excursions was influenced primarily by biological processes or temperature. Their analysis showed that $pCO_2$ variability in most high latitude regions (>40° N or S lat.) and the equatorial zone was controlled primarily by biological processes, whereas temperature effects predominated in the subtropical gyres. Coastal ocean $pCO_2$ variability is roughly consistent with this general scheme. Biology dominates the air-sea exchange in high latitudes (marginal ice zones) and in the tropics (coral reefs, Peru). In the Arabian Sea, biology modulates the $CO_2$ efflux, which would be greater but for intense primary production. The balance between physical and biological processes is more complicated in the subtropics and temperate zones. The two sets of effects enhance each other in the East China Sea and European shelves, producing strong sink regions, but are compensating in the US Mid-Atlantic Bight, rendering this region a 50% weaker sink than other temperate areas. If we could classify the rest of the global coastal ocean according to this scheme, it would be possible to extrapolate its contribution to the total $CO_2$ exchange. But a lack of seasonal and areal coverage still prevents a true objective mapping.

TABLE 9.6

Characteristic patterns of $pCO_2$ variability in the global coastal ocean.

| Pattern | Characteristics | Sink / Source | Examples |
|---|---|---|---|
| Subtropical continental shelf $CO_2$ pump | Temperature and biology effects both favor persistent low/$pCO_2$; strong sink regions. | Sink | East China Sea |
| Temperate shelf: biology dominant | Low/$pCO_2$ in spring-summer, high in winter and near river mouths, strong net sink regions. | Sink | West European shelves |
| Temperate shelf: physics dominant | Low/$pCO_2$ in winter-spring, high in summer and near river mouths, weak net sink regions due to wind- temperature asymmetry (net annual flux unresolved). | ??? | USA Middle Atlantic Bight |
| Upwelling: biology dominant | Intense primary production in upwelled water consumes upwelled $CO_2$. | ??? | Iberia, Peru, Oregon, USA |
| Upwelling: physics dominant | Upwelled high/$pCO_2$ not fully compensated by biology; weak source regions. | Source ? | Arabian Sea |
| Coral reefs | Strong diurnal $pCO_2$ cycle; calcification compensates organic production, source regions. | Source | Great Barrier Reef |
| Polar, ice- rectified $CO_2$ pump | Sea ice cover in winter prevents air- sea exchange, low $pCO_2$ in summer in open water; weak sink regions | Sink | Greenland, Antarctic |

TABLE 9.7

Carbon budgets[a] for the coastal and open oceans.

| Region | Budget components | | Tmol $y^{-1}$ | Reference[b] |
|---|---|---|---|---|
| Coastal Ocean | External Sources | Rivers | 83 | This |
| | | | 34 | S&H |
| | | | 67 | Liu |
| | | Open ocean | 167 | This |
| | | | -- | S&H |
| | | | 167 | Liu |
| | | Atmosphere (net) | 73 | This |
| | | | 8 | S&H |
| | | | 8 | Liu |
| | Internal Cycling | Photosynthesis | 1200 | This |
| | | | 400 | S&H |
| | | | 833 | Liu |
| | | Respiration | -1025 | This |
| | | | -507 | S&H |
| | | | -667 | Liu |
| | Internal Sink | Coastal sediments | -13 | This |
| | | | -9 | S&H |
| | | | -13 | Liu |
| | | Accumulation | -61 | This |
| | | | -15[c] | S&H |
| | | | -4 | Liu |
| | External Sinks | Open oceans | -249 | This |
| | | | -18 | S&H |
| | | | -225 | Liu |
| Open Ocean | External Sources | Coastal ocean | 249 | This |
| | | | 18 | S&H |
| | | | 225 | Liu |
| | | Atmosphere (net) | 83 | This |
| | | | 169 | S&H |
| | | | 108 | Liu |
| | Internal Cycling | Photosynthesis | 3000 | This |
| | | | 3600 | S&H |
| | | | 3333 | Liu |
| | | Respiration | -3196 | This |
| | | | -3616 | S&H |
| | | | -2667 | Liu |
| | Internal Sink | Oceanic sediments | -4 | This |
| | | | -2 | S&H |
| | | | -4 | Liu |
| | | Accumulation | -161 | This |
| | | | -185[c] | S&H |
| | | | -163 | Liu |
| | External sink | Coastal ocean | -167 | This |
| | | | -- | S&H |
| | | | -167 | Liu |

[a]Budgets are balanced by subtracting sinks (including accumulation) from sources, neglecting the internal cycling terms (photosynthesis and respiration).

[b]This study; S&H: Smith and Hollibaugh 1993 (budget for river input of organic matter only); Liu et al. 2000.

[c]Not stated explicitly; calculated as input minus output

A rough global extrapolation that is very much influenced by the European and ECS shelf fluxes, suggests that $CO_2$ exchange is 1–10 mmol C $m^{-2}$ $d^{-1}$, into or out of the sea (Table 9.5). Air-sea exchange in Table 9.5 is reported as calculated in the original references. The original authors used a variety of assumptions about the relationship between exchange and wind velocity and the resulting estimates are not consistent. The global coastal ocean is potentially a net sink of 0.5 – 1 PgC annually. The fluxes in better-studied systems suggest an area-weighted mean annual carbon sink of 73 Tmol (0.9 Pg) in the coastal ocean (Table 9.5). This analysis notwithstanding, a common misunderstanding must be kept in mind. A net influx by itself does not render the global coastal ocean a sink for atmospheric $CO_2$. The subsequent fate of the absorbed $CO_2$ determines whether, or to what extent these coastal ocean systems are truly atmospheric $CO_2$ sinks. The absorbed $CO_2$ must be buried in coastal ocean sediments or else transported to the subsurface waters of the open ocean if it is to be stored over longer time scales (Sarmiento and Siegenthaler, 1992; Liu et al., 2000a). Otherwise the $CO_2$ stored in organic matter will be respired and released to the atmosphere during organic matter decomposition and denitrification, while any excess dissolved $CO_2$ could be lost through gas exchange. In the final section we present a new carbon budget for the coastal and open sea, showing how the continental shelf pump exports atmospheric $CO_2$ (as organic matter) into the open sea where it supports a moderate level of oceanic net heterotrophy. Both the coastal and open oceans are sinks for atmospheric $CO_2$ in our synthesis.

## 8. Synthesis: coastal and open ocean carbon budgets.

Smith and Hollibaugh (1993) presented a thorough analysis of organic metabolism in the coastal ocean, concluding that 18 Tmol of terrigenous organic matter was exported to the open sea. As a result of terrestrial inputs, both systems had negative metabolic balance in their scheme. The 18 Tmol of river-derived organic carbon constituted the entire shelf pump in their analysis. Subsequently Liu et al. (2000a) constructed a new budget, based on a diagnosis of the global carbon cycle by Siegenthaler and Sarmiento (1993). The principal differences in these budgets are given in Table 9.7. In some respects the budgets are not directly comparable because Smith and Hollibaugh focus exclusively on organic carbon, whereas Liu et al present a full carbon budget (organic plus inorganic). In Liu et al's analysis, the coastal ocean was autotrophic, and the continental shelf pump amounted to about 200 Tmol C annually. Both Smith and Hollibaugh (1993) and Liu et al. (2000a) postulated that the coastal ocean was a weak net sink for atmospheric $CO_2$, taking up 8 Tmol C annually. However the levels they assigned for NPP in the coastal ocean differed significantly: 400 vs 833 Tmol C $yr^{-1}$, with Smith and Hollibaugh providing the more conservative figure.

Our new analysis (Figure 9.9, Table 9.7), based on this review, differs in several ways from the earlier budgets presented by Smith and Hollibaugh (1993) and Liu et al (2000a). The annual river input is increased to 83 Tmol C (45% organic), reflecting data on the anthropogenic subsidy reviewed above. Most of the increase appears to be DIC; thus Smith and Hollibaugh's estimate of 38 Tmol of riverine organic carbon entering the coastal ocean is retained in our budget. Although the global total NPP is similar in all three budgets, we adopt Longhurst et al's (1995) estimate of 1200 Tmol C for the coastal ocean NPP. We further assume an annual mean f- (or export) ratio of 0.5 for the coastal ocean (recall this is for vertical export from the surface layer; Section 4) and 0.2 for the open sea (Eppley and Peterson 1979). From these assumptions about the total NPP and export ratios it follows that the vertical exports from the euphotic zone of the coastal ocean and the open sea are equivalent at 600 Tmol C $yr^{-1}$. The global net $CO_2$ uptake is increased to 1.9 GtC (158 Tmol) annually (Sarmiento and Gruber 2002), of which 0.9 GtC (73 Tmol) is absorbed in coastal ocean waters (Table 9.5). The remaining 83 Tmol is absorbed in the open sea.

Our estimate for global oceanic $CO_2$ uptake, including coastal ocean uptake of 73 Tmol C $yr^{-1}$ is a conservative assumption; the alternative would be that the full 158 Tmol of C uptake is allocated to the open sea, with an *additional* 73 Tmol C taken up by the coastal oceans. The justification for such an allocation would be that previous global estimates tended to ignore the coastal ocean, usually because it is not resolved in global GCM's (Yool and Fasham 2001). The resulting total (233 Tmol or 2.8 PgC) is slightly greater than the high estimate provided by Takahashi et al (2002) of 2.6 PgC. Broecker (2001) noted that if a new cubic relationship between wind velocity and air-sea exchange (Wanninkhof and McGillis, 1999) were valid, the total ocean uptake might be as high as 3.7 PgC $yr^{-1}$. This global total would easily accommodate an open ocean uptake of 1.9–2.2 PgC $yr^{-1}$ and an additional coastal ocean uptake of 0.9 GtC, although the latter estimate would need to be revised upward as well. As Broecker cautioned, these higher values for the global uptake need to be reconciled with existing tracer ($^{13}C$, $^{14}C$) budgets. So we adopt the most conservative assumption here. Although conservative, it represents a significant increase in coastal ocean $CO_2$ uptake over previous estimates (Table 9.7).

Finally, we assumed the continental shelf pump delivers 18 Tmol of riverine organic C to the open ocean surface layer and 182 Tmol C to the oceanic interior via downslope transport processes (Section 5). The small coastal and oceanic burial terms are left unchanged, as there is little evidence they have altered in response to the anthropogenic $CO_2$ transient (but see Chapter 7 for another view; and Chapters 5, 6, 11 on burial processes generally). The 167 Tmol C of net DIC flux back into the coastal ocean was derived from Walsh's (1991) estimate of a net ocean to shelf onwelling of ~0.6 Pg $NO_3$-N $yr^{-1}$ (40 Tmol N x 6.7 = 268 Tmol C, adjusted to 167 to balance budget).

Respiration is derived by mass balance. In the lower layer, respiration is not divided between water column and benthos but benthic respiration is implicit. The resulting budget shows that the coastal ocean is autotrophic by 175 Tmol C $y^{-1}$ (Table 9.7, NPP = 1200, R=1025 and cf. Table 9.2). This excess production is a consequence of exporting 200 Tmol C, which would otherwise be respired *in situ,* increasing R to 1225 and rendering the system very slightly heterotrophic. The excess NEP of 175 Tmol is about 3 times the estimate of 68 derived earlier by Gattuso et al. (1998). The export by the continental shelf pump is 15% of the coastal ocean NPP (not including 18 Tmol of riverine DOM), consistent with the observations (Section 5). This export renders the open sea heterotrophic if respiration in the whole water column is considered (Figure 9.9, NPP=3000, R=3214). The oceanic surface layer is autotrophic (Figure 9.9; P:R=1.24), consistent with some observations (Table 9.2 and Karl et al., 2003), but in contrast to the hypothesis that it is heterotrophic (Duarte and Agusti, 1998) by up to 500 Tmol C $yr^{-1}$, as calculated by Williams and Bowers (1999). As noted in section 6, this latter estimate of net heterotrophic metabolism in the oceanic surface layer would require a labile C input of 1082 Tmol C $yr^{-1}$, over five times the amount estimated here, via unknown transport processes. The observed net $CO_2$ uptake in the open sea is a strong constraint on the $CO_2$ produced by oceanic respiration; here it contributes to an annual storage rate of 161 Tmol C $yr^{-1}$, as in Liu et al. (2000a). The storage would be greater except for 167 Tmol C of net return flow into the coastal ocean (see above). Carbon is also being stored in central gyres as DOC (Church et al., 2002), but it remains to be determined if this is a long-term or transient, decade-scale phenomenon. Finally, the global ocean as a whole (coastal ocean plus open sea) is slightly heterotrophic by 38 Tmol, as a result of respiration of the terrestrial input (Smith and Hollibaugh, 1993), yet it remains a net sink for atmospheric $CO_2$. The identity of both the coastal and open oceans as $CO_2$ sinks is a consequence of anthropogenic $CO_2$ addition to the atmosphere, which has had the effect of decoupling ocean metabolism and $CO_2$ exchange. It is necessary to acknowledge this decoupling to understand the carbon balance in the present-day ocean system.

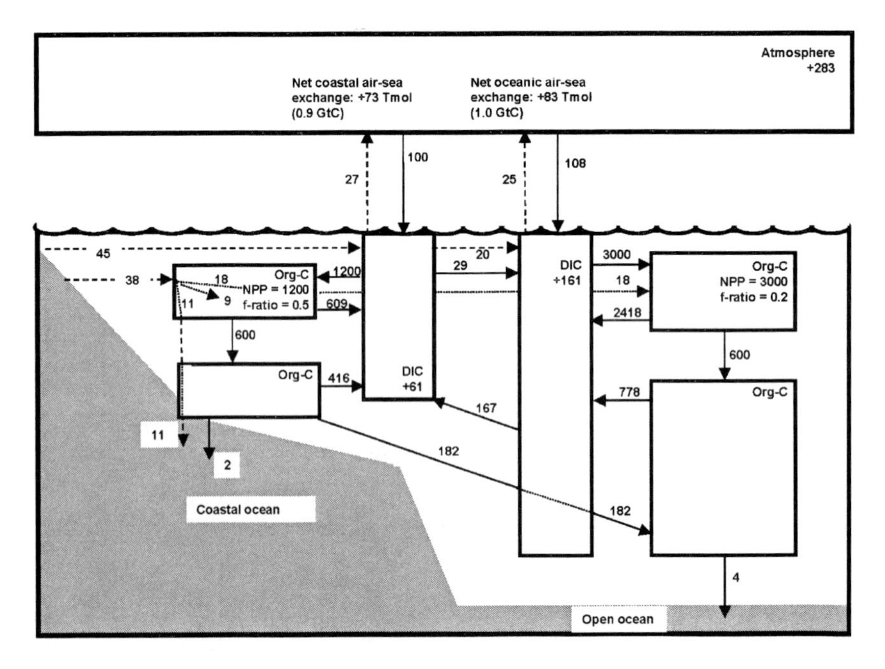


Figure 9.9 A new carbon budget for the global coastal ocean, incorporating fluxes and transports discussed in this chapter. The total (coastal + open ocean) $CO_2$ air-sea exchange is referenced to Sarmiento and Gruber (2002). The principal differences from previous budgets are annual net primary production (NPP) of 1200 Tmol and net air-sea exchange of 73 Tmol in the coastal ocean. As a consequence of the continental shelf pump (182 + 18 Tmol), the open ocean is net heterotrophic by about 200 Tmol per year, although this imbalance is expressed in the midwater column where the continental shelf pump is believed to enter the open sea. Dashed lines trace terrestrial inputs of DIC and organic carbon through the ocean system. Fluxes only terminate at arrowheads. Dotted lines signify pass-throughs of transport from a remote source (e.g., the 18 Tmol of riverine input to the open sea).

## Acknowledgements

We thank Mike DeGrandpre, Lisa Miller and Tish Yager for contributing unpublished data and for very helpful comments. Two reviewers' comments were instrumental in improving our chapter. Dennis Hansell, David Kirchman and Peter LeB. Williams provided separate comments on the manuscript. Support was provided by NSF Grants OCE-0097237 and OPP-0217282 to HWD and DEB-0073243 to SLM.

## Bibliography

Alldredge, A. L. and M. W. Silver. 1988. Characteristics, dynamics and significance of marine snow. *Progress in Oceanography*, **20**, 41-82.

Aller, R. C., N. E. Blair, Q. Xia and P. D. Rude. 1996. Remineralization rates, recycling, and storage of carbon in Amazon shelf sediments. *Continental Shelf Research*, **16**, 753-786.

Aluwihare, L. I., D. J. Repeta and R. F. Chen. 2002. Chemical composition and cycling of dissolved organic matter in the Mid-Atlantic Bight. *Deep Sea Research II*, **49**, 4421-4437.

Alvarez-Salgado, X. A., M. D. Doval, A. V. Borges, I. Joint, M. Frankignoulle, E. M. S. Woodward and F. G. Figueiras. 2001a. Off-shelf fluxes of labile materials by an upwelling filament in the NW Iberian Upwelling System. *Progress in Oceanography*, **51**, 321-337.

Alvarez-Salgado, X. A., J. Gago, B. M. Miguez and F. F. Perez. 2001b. Net ecosystem production of dissolved organic carbon in a coastal upwelling system: the Ria de Vigo, Iberian margin of the North Atlantic. *Limnology and Oceanography*, **46**, 135-147.

Alvarez-Salgado, X. A. and A. E. J. Miller. 1998. Dissolved Organic Carbon in a Large Macrotidal Estuary (the Humber, UK): Behaviour During Estuarine Mixing. *Marine Pollution Bulleti*, **37**, 3-7.

Aminot, A., M. A. El-Sayed and R. Kerouel. 1990. Fate of natural and anthropogenic dissolved organic carbon in the macrotidal Elorn Estuary (France). *Marine Chemistry*, **29**, 255-275.

Amon, R. M. W. and R. Benner. 1996. Photochemical and microbial consumption of dissolved organic carbon and dissolved oxygen in the Amazon River System. *Geochim. et Cosmochim. Acta*, **60**, 1783-1792.

Anderson, T. R. and P. J. l. Williams. 1999. A one-dimensional model of dissolved organic carbon cycling in the water column incorporating combined biological-photochemical decomposition. *Global Biogeochemical Cycles*, **13**, 337-349.

Antia, A. N., J. Maassen, P. Herman, M. Voss, J. Scholten, S. Groom and P. Miller. 2001. Spatial and temporal variability of particle flux at the N.W. European continental margin. *Deep Sea Research II*, **48**, 3083-3106.

Antoine, D. and A. Morel. 1995. Modelling the seasonal course of the upper ocean $pCO_2$ (1). Development of a one-dimensional model. *Tellus. Series B: Chemical and Physical Meteorology*, **47B**, 103-121.

Bacon, M. P., R. A. Belastock, M. Tecotzky, K. K. Turekian and D. W. Spencer. 1988. Lead-210 and Polonium-210 in ocean water profiles of the continental shelf and slope south of New England. *Continental Shelf Research*, **8**, 841-854.

Bakker, D. C. E., H. J. W. De Baar and H. P. J. De Wilde. 1996. Dissolved carbon dioxide in Dutch coastal waters. *Marine Chemistry*, **55**, 247-263.

Barber, R. T., J. Marra, R. C. Bidigare, L. A. Codispoti, D. Halpern, Z. Johnson, M. Latasa, R. Goericke and S. L. Smith. 2001. Primary productivity and its regulation in the Arabian Sea during 1995. *Deep Sea Research II*, **48**, 1127-1172.

Bates, N. R. 2002. Seasonal variability of the effect of coral reefs on seawater $CO_2$ and air-sea $CO_2$ exchange. *Limnology and Oceanography*, **47**, 43-52.

Bates, N. R. and D. A. Hansell. 1999. A high resolution study of surface layer hydrographic and biogeochemical properties between Chesapeake Bay and Bermuda. *Marine Chemistry*, **67**, 1-16.

Bauer, J. E., K. C. Ruttenberg, D. M. Wolgast, E. Monaghan and M. K. Schrope. 1996. Cross-flow filtration of dissolved and colloidal nitrogen and phosphorus in seawater: Results from an intercomparison study. *Marine Chemistry*, **55**, 33-52.

Bauer, J. E., P. M. Williams and E. R. M. Druffel. 1992. super(14)C activity of dissolved organic carbon fractions in the north-central Pacific and Sargasso Sea. *Nature*, **357**, 667-670.

Bender, M. L., M. Battle and R. F. Keeling. 1998. The $O_2$ balance of the atmosphere: A tool for studying the fate of fossil-fuel $CO_2$. *Annual Review of Energy and the Environment*, **23**, 207-223.

Benner, R., J.D. Pakulski, M. McCarthy, J.I. Hedges, and P.G. Hatcher. 1992. Bulk chemical characteristics of dissolved organic matter in the ocean. *Science*, **255**, 1561-1564.

Benner, R. and S. Opsahl. 2001. Molecular indicators of the sources and transformations of dissolved organic matter in the Mississippi river plume. *Organic Geochemistry*, **32**, 597-611.

Bianchi, T. S., C. Lambert, P. H. Santschi, M. Baskaran and L. Guo. 1995. Plant pigments as biomarkers of high molecular weight dissolved organic carbon. *Limnology and Oceanography*, **40**, 422-428.

Bianchi, T. S., C. D. Lambert, P. H. Santschi and L. Guo. 1997. Sources and transport of land-derived particulate and dissolved organic matter in the Gulf of Mexico (Texas shelf/slope): The use of lignin-phenols and loliolides as biomarkers. *Organic Geochemistry*, **27**, 1-2.

Biscaye, P. E. and R. F. Anderson. 1994. Fluxes of particulate matter on the slope of the southern Middle Atlantic Bight: SEEP-II. *Deep-Sea Research II*, **41**, 459-509.

Biscaye, P. E., R. F. Anderson and B. L. Deck. 1988. Fluxes of particles and constituents to the eastern United States continental slope and rise: SEEP-I. *Continental Shelf Research*, **8**, 855-904.

Biscaye, P. E., C. N. Flagg and P. G. Falkowski. 1994. The shelf edge exchange processes experiment, SEEP-II: An introduction to hypotheses, results and conclusions. *Deep-Sea Research II*, **41**, 231-252.

Boehme, S. E., C. L. Sabine and C. E. Reimers. 1998. $CO_2$ fluxes from a coastal transect: a time-series approach. *Marine Chemistry*, **63**, 49-67.

Borges, A. V. and M. Frankignoulle. 1999. Daily and seasonal variations of the partial pressure of $CO_2$ in surface seawater along Belgian and southern Dutch coastal areas. *Journal of Marine Systems*, **19**, 251-266.

Borges, A. V. and M. Frankignoulle. 2002. Aspects of dissolved inorganic carbon dynamics in the upwelling system off the Galician coast. *Journal of Marine Systems*, **32**, 181-198.

Broecker, W. S. 2001. A Ewing Symposium on the contemporary carbon cycle. *Global Biogeochemical Cycles*, **15**, 1031-1034.

Buch, K., Harvey, H. W., H. Wattenberg and S. Gripenberg. 1932. Uber das Kohlensauresystem im Meerwasser. *Conseil Permanent International pour l'Exploration de la Mer, Rapports et Proces-Verbaux*, **79**, 1-70.

Bushaw, K. L., R. G. Zepp, M. A. Tarr, D. Schulz-Jander, R. A. Bourbonniere, R. E. Hodson, W. L. Miller, D. A. Bronk and M. A. Moran. 1996. Photochemical release of biologically available nitrogen from aquatic dissolved organic matter. *Nature*, **381**, 404-407.

Butler, J. N. 1982. *Carbon dioxide equilibria and their applications*. Reading, MA, Addison-Wesley.

Carillo, C. J. and D. M. Karl. 1999. Dissolved inorganic carbon pool dynamics in northern Gerlache Strait, Antarctica. *Journal of Geophysical Research*, **104C**, 15873-15884.

Chen, A. C.-T. 1996. The Kuroshio Intermediate Water is the major source of nutrients on the East China Sea continental shelf. *Oceanologica Acta*, **1**, 523-527.

Chen, A. C.-T., K. K. Liu and R. Macdonald. 2003. Continental Margin Exchanges. In *Ocean Biogeochemistry: The Role of the Ocean Carbon Cycle in Global Change*. M. J. R. Fasham. ed. Springer-Verlag. Berlin. pp. 53-98.

Chen, C.-T. A. and S.-L. Wang. 1999. Carbon, aklinity and nutrient budgets on the East China Sea continental shelf. *Journal of Geophysical Research*, **104C**, 20675 - 20686.

Chung, Y.-C. and G.-W. Hung. 2000. Particulate fluxes and transports on the slope between the southern East China Sea and the South Okinawa Trough. *Continental Shelf Research*, **20**, 571-597.

Church, M., H. W. Ducklow and D. M. Karl. 2002. Temporal Variability in Dissolved Organic Matter Stocks in the Central North Pacific Gyre. *Limnology and Oceanography*, **47**, 1-10.

Churchill, J. H., C. D. Wirick, C. N. Flagg and L. J. Pietrafesa. 1994. Sediment resuspension over the continental shelf east of the Delmarva Peninsula. *Deep-Sea Research II*, **41**, 341-363.

Codispoti, L. A., J. A. Brandes, J. P. Christensen, A. H. Devol, S. W. A. Naqvi, H. W. Paerl and T. Yoshinari. 2001. The oceanic fixed nitrogen and nitrous oxide budgets: Moving targets as we enter the anthropocene? *Scientia Marina*, **65**, 85-105.

Codispoti, L. A., G. E. Friedrich, R. L. Iverson and D. W. Hood. 1982. Temporal changes in the inorganic carbon system of the southeastern Bering Sea during spring, 1980. *Nature*, **296**, 242-245.

Crossland, C. J., B. G. Hatcher and S. V. Smith. 1991. Role of coral reefs in global ocean production. *Coral Reefs*, **10**, 55-64.

Cullen, J. J., P. J. S. Franks and D. M. Karl. 2002. Physical influences on marine ecosystem dynamics. In *The Sea, Volume 12. Biological-Physical Interactions in the Sea.* A. R. Robinson, J. J. McCarthy and B. J. Rothschild. eds. Wiley. New York. pp. 000-000.

Cullen, J. J. and J. G. MacIntyre. 1998. Behavior, physiology and the niche of depth-regulating phytoplankton. In *Physiological Ecology of Harmful Algal Blooms.* D. M. Anderson, A. D. Cembella and G. M. Hallegraeff. eds. Springer-Verlag. Heidelburg. pp. 559-580.

de Haas, H., T. C. E. van Weering and H. de Stigter. 2002. Organic carbon in shelf seas: sinks or sources, processes and products. *Continental Shelf Research*, **22**, 691-717.

DeGrandpre, M. D., G. J. Olbu, C. M. Beatty and T. R. Hammar. 2002. Air-sea $CO_2$ fluxes on the US Middle Atlantic Bight. *Deep Sea Research II*, **49**, 4355-4367.

del Giorgio, P. A. and J. J. Cole. 1998. Bacterial Growth Efficiency in Natural Aquatic Systems. *Annual Review of Ecology and Systematics*, **29**, 503-541.

del Giorgio, P. A., J. J. Cole and A. Cimberlis. 1997. Respiration rates in bacteria exceed phytoplankton production in unproductive aquatic systems. *Nature*, **385**, 148-151.

del Giorgio, P. A. and C. M. Duarte. 2002. Respiration in the open ocean. *Nature*, **420**, 379 - 384.

Dickson, M.-L., J. Orchardo, R. T. Barber, J. Marra, J. J. McCarthy and R. N. Sambrotto. 2001. Production and respiration rates in the Arabian Sea during the 1995 Northeast and Southwest Monsoons. *Deep Sea Research II*, **48**, 1199-1230.

Duarte, C. M. and S. Agusti. 1998. The $CO_2$ balance of unproductive aquatic ecosystems. *Science*, **281**, 234-236.

Duarte, C. M., S. Agusti, J. Aristegui, N. Gonzalez and R. Anadon. 2001. Evidence for a heterotrophic subtropical Northeast Atlantic. *Limnology and Oceanography*, **46 (Suppl.)**, 425-428.

Duarte, C. M., S. Agusti, P. A. del Giorgio and J. J. Cole. 1999. Regional carbon imbalance in the oceans. *Science*, **284**, 1735b.

Ducklow, H. W. 2003. Chapter 1. Biogeochemical Provinces: Towards a JGOFS Synthesis. In *Ocean Biogeochemistry: A New Paradigm.* M. J. R. Fasham. ed. Springer-Verlag. New York. pp. 3-18.

Ducklow, H. W., D. K. Steinberg and K. O. Buesseler. 2001. Upper Ocean Carbon Export and the Biological Pump. *Oceanography*, **14**, 50-58.

Dugdale, R. C. and J. J. Goering. 1967. Uptake of new and regenerated forms of nitrogen in primary production. *Limnology and Oceanography*, **12**, 196-206.

Eppley, R. W. and B. J. Peterson. 1979. Particulate organic matter flux and planktonic new production in the deep ocean. *Nature*, **282**, 677-680.

Falkowski, P. G., P. E. Biscaye and C. Sancetta. 1994. The lateral flux of biogenic particles from the eastern North American continental margin to the North Atlantic Ocean. *Deep-Sea Research II,* **41**, 583-601.

Falkowski, P. G., C. N. Flagg, G. T. Rowe, S. L. Smith, T. E. Whitledge and C. D. Wirick. 1988. The fate of a spring phytoplankton bloom: Export or oxidation? *Continental Shelf Research*, **8**, 457-484.

Falkowski, P. G., T. S. Hopkins and J. J. Walsh. 1980. An analysis of factors affecting oxygen depletion in the New York Bight. *Journal of Marine Research*, **38**, 479-506.

Falkowski, P. G. and J. A. Raven. 1997. *Aquatic Photosynthesis*. Malden, MA, Blackwell Scientific.

Feely, R. A., C. L. Sabine, T. Takahashi and R. Wanninkhof, 2001. Uptake and storage of carbon dioxide in the ocean. *Oceanography*. **14:** 18-32.

Feely, R. A., R. Wanninkhof, T. Takahashi and P. Tans. 1999. Influence of El Nino on the equatorial Pacific contribution to atmospheric $CO_2$ accumulation. *Nature*, **398**, 597-601.

Field, C. B., M. J. Behrenfeld, J. T. Randerson and P. Falkowski. 1998. Primary production of the biosphere: Integrating terrestrial and oceanic components. *Science*, **281**, 237-240.

Frankignoulle, M., G. Abril, A. Borges, I. Bourge, C. Canon, B. Delille, E. Libert and J. M. Theate. 1998. Carbon dioxide emission from European estuaries. *Science (Washington)*, **282**, 434-436.

Frankignoulle, M. and A. Borges. 2001. European continental shelf as a significant sink for atmospheric carbon dioxide. *Global Biogeochemical Cycles*, **15**, 569-576.

Frankignoulle, M., C. Canon and J. P. Gattuso. 1994. Marine calcification as a source of carbon dioxide: Positive feedback of increasing atmospheric $CO_2$. *Limnology and Oceanography*, **39**, 458-462.

Frankignoulle, M., J. P. Gattuso, R. Biondo, I. Bourge, G. Copin-Montegut and M. Pichon. 1996. Carbon fluxes in coral reefs. 2. Eulerian study of inorganic carbon dynamics and measurement of air-sea $CO_2$ exchanges. *Marine Ecology Progress Series*, **145**, 123-132.

Fransson, A., M. Chierici, L. G. Anderson, I. Bussmann, G. Kattner, E. Peter Jones and J. H. Swift. 2001. The importance of shelf processes for the modification of chemical constituents in the waters of the Eurasian Arctic Ocean: implication for carbon fluxes. *Continental Shelf Research*, **21**, 225-242.

Friederich, G. E., P. G. Brewer, R. Herlien and F. P. Chavez. 1995. Measurement of sea surface partial pressure of $CO_2$ from a moored buoy. *Deep Sea Research I,* **42**, 1175-1186.

Garside, C., T. C. Malone, O. A. Roels and B. A. Sharfstein. 1976. An evaluation of sewage-derived nutrients and their influence on the Hudson Estuary and New York Bight. *Estuarine Coastal. Marine Science*, **4**, 281-289.

Gattuso, J., M. Frankignoulle and S. V. Smith. 1999. Measurement of community metabolism and significance in the coral reef $CO_2$ source-sink debate. *Proceedings of the National Academy of Sciences, USA*, **96**, 13017-13022.

Gattuso, J. P., M. Frankignoulle and R. Wollast. 1998. Carbon and carbonate metabolism in coastal aquatic ecosystems. *Annual Review of Ecology and Systematics*, **29**, 405-434.

Geider, R. J. 1997. Photosynthesis or planktonic respiration? *Nature*, **388**, 132.

Goyet, C., F. J. Millero, D. W. O'Sullivan, G. Eischeid, S. J. McCue and R. G. J. Bellerby. 1998. Temporal variations of $pCO_2$ in surface seawater of the Arabian Sea in 1995. *Deep Sea Research II*, **45**, 609-623.

Greenberg, D. M., E. G. Moberg and E. G. Allen. 1932. Determination of carbon dioxide and titratable base in sea water. *Industrial and Engineering Chemistry Analytical Edition*, **4**, 309-313.

Group, I. P., 2003. Integrated Marine Biogeochemistry and Ecosystem Research Program: Science Plan. Stockholm, International Geosphere- Biosphere Program**:** 1-53.

Gu, B., T. L. Mehlhorn, L. Liang and J. F. McCarthy. 1996. Competitive adsorption, displacement, and transport of organic matter on iron oxide: 1. Competitive adsorption. *Geochimica et Cosmochimica Acta*, **60**, 1943-1950.

Gu, B., J. Schmitt, Z. Chen, L. Liang and J. F. McCarthy. 1995. Adsorption and desorption of different organic matter fractions on iron oxide. *Geochimica et Cosmochimica Acta*, **59**, 219-229.

Guo, L. and P. H. Santschi. 2000. Sedimentary sources of old high molecular weight dissolved organic carbon from the ocean margin benthic nepheloid layer. *Geochimica et Cosmochimica Acta*, **64**, 651-660.

Hall, J. A., P. Monfray, A. Bucklin, D. A. Hansell, C. Heip, R. A. Jahnke, P. Kumar, A. Körtzinger, W. Miller, S. W. A. Naqvi, R. Murtugudde, H. Saito, S. Sundby and Y. E. F., 2004. Integrated Marine Biogeochemistry and Ecosystem Research (IMBER) Science and Implementation Plan. Stockholm, International Geosphere-Biosphere Program (IGBP): 1-96.

Hansell, D. A. and C. A. Carlson. 2001. Marine dissolved organic matter and the carbon cycle. *Oceanography*, **14**, 41-49.

Hanson, R., H. W. Ducklow and J. G. Field. 2000. *The Changing Carbon Cycle in the Oceans*. Cambridge, U. K., Cambridge University Press.

Hedges, J. I. 1992. Global biogeochemical cycles: progress and problems. *Marine Chemistry*, **39**, 67-93.

Hedges, J. I. and R. G. Keil. 1995. Sedimentary organic matterpreservation: An assessment and speculative synthesis. *Marine Chemistry*, **49**, 81-115.

Hedges, J. I. and R. G. Keil. 1999. Organic geochemical perspectives on estuarine processes: sorption reactions and consequences. *Marine Chemistry*, **65**, 55-65.

Hedges, J. I., R. G. Keil and R. Benner. 1997. What happens to terrestrial organic matter in the ocean? *Organic Geochemistry*, **27**, 195-212.

Hedges, J. I. and D. C. Mann. 1979. The lignin geochemistry of marine sediments from the southern Washington coast. *Geochimica et Cosmochimica Acta*, **43**, 1809-1818.

Henrichs, S. M. and S. F. Sugai. 1993. Adsorption of amino acids and glucose by sediments of Resurrection Bay, Alaska, USA: Functional group effects. *Geochimica et Cosmochimica Acta*, **57**, 823-835.

Holligan, P. M. and W. A. Reiners. 1992. Predicting the responses of the coastal zone to global change. *Advances in Ecological Research*, **22**, 211-255.

Hopkinson, C. S., Jr. and J. J. Vallino. 1995. The relationships among man's activities in watersheds and estuaries: A model of runoff effects on patterns of estuarine community metabolism. *Estuaries*, **4**, 598-621.

Hoppe, H.-G., K. Gocke, R. Koppe and C. Begler. 2002. Bacterial growth and primary production along a north–south transect of the Atlantic Ocean. *Nature*, **416**, 168-171.

Hoppema, J. M. J. 1991. The seasonal behavior of carbon dioxide and oxygen in the coastal North Sea along The Netherlands. *Netherlands Journal of Sea Research*, **28**, 167-179.

Houghton, J. T., Y. Ding, D. J. Griggs, M. Noguer, P. J. v. d. Linden and D. Xiaosu, Eds. (2001). *Climate Change 2001: The Scientific Basis Contribution of Working Group I to the Third Assessment Report of the Intergovernmental Panel on Climate Change (IPCC)*. Cambridge, UK, Cambridge University Press.

Huthnance, J. M. 1995. Circulation, exchange, and water masses at the ocean margin: the role of physical processes at the shelf edge. *Progress in Oceanography*, **35**, 353-431.

Hydes, D. J., A. C. Le Gall, A. E. J. Miller, U. Brockmann, T. Raabe, S. Holley, X. Alvarez-Salgado, A. Antia, W. Balzer and L. Chou. 2001. Supply and demand of nutrients and dissolved organic matter at and across the NW European shelf break in relation to hydrography and biogeochemical activity. *Deep Sea Research II*, **48**, 3023-3047.

Ittekkot, V. 1989. Global trends in the nature of organic matter in river suspension. *Nature*, **332**, 436-438.

Johnson, K. S., F. P. Chavez and G. E. Friederich. 1999. Continental-shelf sediment as a primary source of iron for coastal phytoplankton. *Nature*, **398**, 697 - 700.

Karl, D. M., E. A. Laws, P. Morris, P. J. l. Williams and S. Emerson. 2003. Global carbon cycle (communication arising): Metabolic balance of the open sea. *Nature*, **426**, 32.

Kattner, G., J. M. Lobbes, H. P. Fitznar, R. Engbrodt, E. M. Nothig and R. J. Lara. 1999. Tracing dissolved organic substances and nutrients from the Lena River through Laptev Sea (Arctic). *Marine Chemistry,* **65**, 1-2.

Kawahata, H., A. Suzuki and K. Goto. 1997. Coral reef ecosystems as a source of atmospheric $CO_2$: evidence from $PCO_2$ measurements of surface waters. *Coral Reefs*, **16**, 261-266.

Keeling, C. D. and T. P. Whorf. 2002. Atmospheric $CO_2$ Concentrations--Mauna Loa observatory, Hawaii, 1958-2001 (revised June 2002). In *Trends Online: A Compendium of Data on Global Change.* C. D. I. A. Center. ed. Oak Ridge National Laboratory, U.S. Department of Energy. Oak Ridge, TN. pp.

Keeling, R. F. and S. R. Shertz. 1992. Seasonal and interannual variations in atmospheric oxygen and implications for the global carbon cycle. *Nature*, **358**, 723-727.

Keil, R. G., L. M. Mayer, P. D. Quay, J. E. Richey and J. I. Hedges. 1997. Loss of organic matter from riverine particles in deltas. *Geochimica et Cosmochimica Acta*, **61**, 1507-1511.

Keir, R. S., G. Rehder and M. Frankignoulle. 2001. Partial pressure and air-sea flux of $CO_2$ in the Northeast Atlantic during September 1995. *Deep Sea Research II*, **48**, 3179-3189.

Kemp, W. M., E. M. Smith, M. M. DiPasquale and W. R. Boynton. 1997. Organic carbon balance and net ecosystem metabolism in Chesapeake Bay. *Marine Ecology Progress Series*, **150**, 229-248.

Kempe, S. 1995. Coastal seas: A Net Source or sink of atmospheric Carbon dioxide? *LOICZ Reports and Studies*, **95-1**, 1-27.

Kempe, S. and K. Pegler. 1991. Sinks and sources of $CO_2$ in coastal seas: The North Sea. *Tellus. Series B: Chemical and Physical Meteorology*, **43B**, 224-235.

Kinkade, C. S., J. Marra, T. D. Dickey and R. Weller. 2001. An annual cycle of phytoplankton biomass in the Arabian Sea, 1994-1995, as determined by moored optical sensors. *Deep Sea Research II*, **48**, 1285-1301.

Kinsey, D. W. (1985). *Metabolism, calcification and carbon production I. System Level studies*. 5th International Symposium on Corals and Coral Reefs, Tahiti.

Kleypas, J. A., R. W. Buddemeier, D. Archer, J. P. Gattuso, C. Langdon and B. N. Opdyke. 1999. Geochemical consequences of increased atmospheric carbon dioxide on coral reefs. *Science (Washington)*, **284**, 118-120.

Komada, T. and C. E. Reimers. 2001. Resuspension-induced partitioning of organic carbon between solid and solution phases from a river-ocean transition. *Marine Chemistry*, **76**, 155-174.

Komada, T., O. M. E. Schofield and C. E. Reimers. 2002. Fluorescence characteristics of organic matter released from coastal sediments during resuspension. *Marine Chemistry*, **79**, 81-97.

Kortzinger, A., J. C. Duinker and L. Mintrop. 1997. Strong $CO_2$ emissions from the Arabian Sea during the south-west monsoon. *Geophysical Research Letters*, **24**, 1763-1766.

Kraines, S., Y. Suzuki, T. Omori, K. Shitashima, S. Kanahara and H. Komiyama. 1997. Carbonate dynamics of the coral reef system at Bora Bay, Miyako Island. *Marine Ecology Progress Series*, **156**, 1-16.

Kumar, M. D., S. W. A. Naqvi, M. D. George and D. A. Jayakumar. 1996. A sink for atmospheric carbon dioxide in the northeast Indian Ocean. *Journal of Geophysical Research*, **C101**, 18121-18125.

Langdon, C., T. Takahashi, C. Sweeney, D. Chipman, J. Goddard, F. Marubini, H. Aceves, H. Barnett and M. J. Atkinson. 2000. Effect of calcium carbonate saturation state on the calcification rate of an experimental coral reef. *Global Biogeochemical Cycles*, **14**, 639-654.

Liu, K. K., L. Atkinson, C. T. A. Chen, S. Gao, J. Hall, R. W. MacDonald, L. T. McManus and R. Quinones. 2000a. Exploring continental margin carbon fluxes on a global scale. *EOS: Transactions American Geophysical Union*, **81**, 641-642, 644.

Liu, K. K., K. Iseki and S.-Y. Chao. 2000b. Continental margin carbon fluxes. In *The Changing Ocean Carbon Cycle.* R. Hanson, H. W. Ducklow and J. G. Field. eds. Cambridge University Press. Cambridge, UK. pp. 187-239.

Longhurst, A. 1995. Seasonal cycles of pelagic production and consumption. *Progress in Oceanography*, **36**, 77-167.

Longhurst, A. 1998. *Ecological Geography of the Sea.* San Diego, CA, Academic Press.

Longhurst, A., S. Sathyendranath, T. Platt and C. Caverhill. 1995. An estimate of global primary production in the ocean from satellite radiometer data. *Journal of Plankton Research*, **17**, 1245-1271.

Longhurst, A. R. and W. G. Harrison. 1989. The biological pump: Profiles of plankton production and consumption in the upper ocean. *Progress in Oceanography*, **22**, 47-123.

Ludwig, W., P. Amiotte-Suchet and J. L. Probst. 1996. River discharges of carbon to the World Ocean: Determining local inputs of alkalinity and of dissolved and particulate organic carbon. *Comptes rendus de l'Academe des Sciences Series II*, **323**, 1007-1014.

Macdonald, R. W., S. M. Solomon, R. E. Cranston, H. E. Welch, M. B. Yunker and C. Gobeil. 1998. A sediment and organic carbon budget for the Canadian Beaufort Shelf. *Marine Geology*, **144**, 255-273.

Malone, T. C. 1971. The relative importance of nannoplankton and netplankton as primary producers in tropical oceanic and neritic phytoplankton communities. *Limnology and Oceanography*, **16**, 633-639.

Malone, T. C., H. W. Ducklow, E. R. Peele and S. E. Pike. 1991. Picoplankton carbon fluxes in Chesapeake Bay. *Marine Ecology Progress Series*, **78**, 11-22.

Malone, T. C., T. S. Hopkins, P. G. Falkowski and T. E. Whitledge. 1983. Production and transport of phytoplankton biomass over the continental shelf of the New York Bight. *Continental shelf research. Oxford, New York NY*, **1**, 305-337.

Mannino, A. and H. R. Harvey. 2000. Terrigenous dissolved organic matter along an estuarine gradient and its flux to the coastal ocean. *Organic Geochemistry*, **31**, 1611-1625.

Mantoura, R. F. C. and E. M. S. Woodward. 1983. Conservative behavior of riverine DOC in the Severn Estuary: chemical and geochemical implications. *Geochimica et Cosmochimica Acta*, **7**, 1293-1309.

Margalef, R. 1978. Life-forms of phytoplankton as survival alternatives in an unstable environment. *Oceanol. Acta*, **1**, 493-509.

McCallister, S. L., 2002. Organic matter cycling in the York River estuary, Virginia: An analysis of potential sources and sinks. *School of Marine Science*. Gloucester Point, VA, The College of William and Mary: 217 pp.

Meybeck, M. 1982. Carbon, nitrogen, and phosphorus transport by world rivers. *Am. J. Sci.*, **282**, 401-450.

Meybeck, M. 1993. C, N. P and S in rivers: From sources to global inputs. In *Interactions of C,N,P and S biogeochemical cycles and global change.* R. Wollast, F. T. Mackenzie and L. Chou. eds. Springer-Verlag. Berlin. pp. 163-193.

Miller, L. A., P. L. Yager, K. A. Erickson, D. Amiel, J. Bacle, J. Kirk Cochran, M.-E. Garneau, M. Gosselin, D. J. Hirschberg and B. Klein. 2002. Carbon distributions and fluxes in the North Water, 1998 and 1999. *Deep Sea Research II*, **49**, 5151-5170.

Miller, W. L. 1999. An overview of aquatic photochemistry as it relates to microbial production. In *Microbial Biosystems: New Frontiers. Proceedings of the 8th International Symposium on Microbial Ecology.* C. R. Bell, M. Brylinsky and P. Johnson-Green. eds. Canadian Society for Microbiology. Halifax, NS Canada. pp.

Mills, E. L. 1989. *Biological Oceanography, An Early History, 1870-1960.* Ithaca, NY, Cornell Univ. Press.

Mitra, S., T. S. Bianchi, L. Guo2 and P. H. Santschi. 2000. Terrestrially derived dissolved organic matter in the chesapeake bay and the middle atlantic bight. *Geochimica et Cosmochimica Acta*, **64**, 3547-3557.

Morris, D. P., H. Zagarese, C. E. Williamson, E. G. Balseiro, B. R. Hargreaves, B. Modenutti, R. Moeller and C. Queimalinos. 1995. The attenuation of solar UV radiation in lakes and the role of dissolved organic carbon. *Limnology and Oceanography*, **40**, 1381-1391.

Murnane, R., J. L. Sarmiento and C. Le Quere. 1997. The spatial distribution of the air-sea $CO_2$ fluxes and interhemispheric transport of carbon by the oceans. *Global Biogeochemical Cycles*, **13**, 287-305.

Nakayama, N., S. Watanabe and S. Tsunogai. 2000. Difference on O2 and $CO_2$ gas transfer velocities in Funka Bay. *Marine Chemistry*, **72**, 115-129.

NAS, 1984. Global Ocean Flux Study: Proceedings of a Workshop. Washington, DC, National Academy of Sciences: 1-360.

Odum, H. T. 1956. Primary production in flowing waters. *Limnology and Oceanography*, **1**, 102-117.

Opsahl, S. and R. Benner. 1997. Distribution and cycling of terrigenous dissolved organic matter in the ocean. *Nature*, **386**, 480-482.

Perez, F. F., A. F. Rios and G. Roson. 1999. Sea surface carbon dioxide off the Iberian Peninsula (North Eastern Atlantic Ocean). *Journal of Marine Systems*, **19**, 27-46.

Platt, T. and S. Sathyendranath. 1988. Oceanic primary production: Estimation by remote sensing at local and regional scales. *Science*, **241**, 1613-1622.

Prahl, F. G., J. R. Ertel, M. A. Goni, M. A. Sparrow and B. Eversmeyer. 1994. Terrestrial organic carbon contributions to sediments on the Washington margin. *Geochimica et Cosmochimica Acta*, **58**, 3035-3048.

Purdy, W. C. 1917. Results of algal activity, some familiar, some obscure. *Journal of the American Waterworks Association*, **27**, 1120-1133.

Quay, P. D., B. Tilbrook and C. S. Wong. 1992. Oceanic uptake of fossil fuel $CO_2$: Carbon-13 evidence. *Science (Washington)*, **256**, 74-79.

Raven, J. A. 1997. Inorganic carbon acquisition by marine autotrophs. In *Advances in Botanical Research*. J. A. Callow. ed. Academic Press. San Diego, CA. pp. 86-210.

Raymond, P. A. and J. E. Bauer. 2001. Riverine export of aged terrestrial organic matter to the North Atlantic Ocean. *Nature*, **409**, 497-500.

Raymond, P. A., J. E. Bauer and J. J. Cole. 2000. Atmospheric $CO_2$ evasion, dissolved inorganic carbon production, and net heterotrophy in the York River estuary. *Limnology and Oceanography*, **45**, 1707-1717.

Reche, I., M. L. Pace and J. J. Cole. 1999. Relationship of trophic and chemical conditions to photobleaching of dissolved organic matter in lake ecosystems. *Biogeochemistry*, **44**, 259-280.

Redalje, D. G., S. E. Lohrenz, P. G. Verity and C. N. Flagg. 2002. Phytoplankton dynamics within a discrete water mass off Cape Hatteras, North Carolina: the Lagrangian experiment. *Deep Sea Research II*, **49**, 4511-4531.

Riebesell, U. and D. A. Wolf-Gladrow. 2002. Chapter 5. Supply amd uptake of inorganic nutrients. In *Phytoplankton productivity. Carbon assimilation in marine and freshwater environments.* P. J. B. Williams, D. N. Thomas and C. S. Reynolds. eds. Blackwell Scientific. Oxford, UK. pp. 386.

Riebesell, U., D. A. Wolf-Gladrow and V. Smetacek. 1993. Carbon dioxide limitation of marine phytoplankton growth rates. *Nature*, **361**, 249-251.

Rowe, G. T., S. Smith, P. Falkowski, T. Whitledge, R. Theroux, W. Phoel and H. W. Ducklow. 1986. Do continental shelves export organic matter? *Nature*, **324**, 559-561.

Roy, S., B. Sundby, A. F. Vezina and L. Legendre. 2000. A Canadian JGOFS study in the Gulf of St. Lawrence: An introduction. *Deep-Sea Research II*, **47**, 377-384.

Sabine, C. L. and R. M. Key. 1998. Controls on $fCO_2$ in the South Pacific. *Marine Chemistry*, **60**, 95-110.

Santschi, P. H., L. Guo, M. Baskaran, S. Trumbore, J. Southon, T. S. Bianchi, B. Honeyman and L. Cifuentes. 1995. Isotopic evidence for the contemporary origin of high-molecular weight organic matter in oceanic environments. *Geochimica et Cosmochimica Acta*, **59**, 625-631.

Sarmiento, J. L. and N. Gruber. 2002. Sinks for anthropogenic carbon. *Physics Today*, **55**, 30-36.

Sarmiento, J. L., J. C. Orr and U. Siegenthaler. 1992. A perturbation simulation of $CO_2$ uptake in an ocean general circulation model. *Journal of Geophysical Research. C. Oceans*, **97**, 3621-3645.

Sarmiento, J. L. and U. Siegenthaler. 1992. New Production and the Global Carbon Cycle. In *Primary Productivity and Biogeochemical Cycles in the Sea*. P. G. Falkowski and W. A. D. eds. Plenum Press. New York. pp. 317-332.

Sarmiento, J. L. and E. T. Sundquist. 1992. Revised budget for the oceanic uptake of anthropogenic carbon dioxide. *Nature*, **356**, 589-593.

Savidge, G. and L. Gilpin. 1999. Seasonal influences on size-fractionated chlorophyll a concentrations and primary production in the north-west Indian Ocean. *Deep Sea Research II*, **46**, 701-723.

Schlitzer, R., 2002. Ocean Data View. Bremerhaven, Germany, Alfred-Wegener Institute for Polar Research.

Schneider, B., K. Kremling and J. C. Duinker. 1992. $CO_2$ partial pressure in Northeast Atlantic and adjacent shelf waters: Processes and seasonal variability. *Journal of Marine Systems*, **3**, 453-463.

Shi, W., M.-Y. Sun, M. Molina and R. E. Hodson. 2001. Variability in the distribution of lipid biomarkers and their molecular isotopic composition in Altamaha estuarine sediments: implications for the relative contribution of organic matter from various sources. *Organic Geochemistry*, **32**, 453-467.

Siegenthaler, U. and J. L. Sarmiento. 1993. Atmospheric carbon dioxide and the ocean. *Nature*, **365**, 119-125.

Simpson, J. J. and A. Zirino. 1980. Biological control of pH in the Peruvian coastal upwelling area. *Deep-Sea Research*, **27**, 733-743.

Smetacek, V. and U. Passow. 1990. Spring bloom initiation and Sverdrup's critical-depth model. *Limnology and Oceanography*, **35**, 228-234.

Smith, S. V. 2000. C, N, P fluxes in the coastal zone: The LOICZ approach to budgeting and global extrapolation. *http://data.ecology.su.se/MNODE/Methods/powerpoint/BGCProcedures&Examples3.ppt*, 1-46.

Smith, S. V. and J. T. Hollibaugh. 1993. Coastal metabolism and the oceanic organic carbon balance. *Reviews of Geophysics*, **31**, 75-89.

Smith, S. V. and F. T. Mackenzie. 1987. The ocean as a net heterotrophic system: implications from the carbon biogeochemical cycle. *Global Biogeochemical Cycles*, **1**, 187-198.

Soetaert, K., P. M. J. Herman, J. J. Middelburg, C. Heip, C. L. Smith, P. Tett and K. Wild-Allen. 2001. Numerical modelling of the shelf break ecosystem: reproducing benthic and pelagic measurements. *Deep Sea Research II*, **48**, 3141-3177.

Steeman Nielsen, E. 1947. Photosynthesis of aquatic plants with special reference to the carbon-sources. *Dansk Botanisk Archiv*, **12**, 1-71.

Takahashi, T., S. C. Sutherland, C. Sweeney, A. Poisson, N. Metzl, B. Tilbrook, N. Bates, R. Wanninkhof, R. A. Feely, C. Sabine, J. Olafsson and Y. Nojiri. 2002. Global sea-air $CO_2$ flux based on climatological surface ocean $pCO_2$, and seasonal biological and temperature effects. *Deep-Sea Research II*, **49**, 1601-1622.

Thimsen, C. A. and R. G. Keil. 1998. Potential interactions between sedimentary dissolved organic matter and mineral surfaces. *Marine Chemistry*, **62**, 65-76.

Thomas, H., Y. Bozec, K. Elkalay and H. J. W. de Baar. 2004. Enhanced Open Ocean Storage of $CO_2$ from Shelf Sea Pumping. *Science*, **304**, 1005-1008.

Tortell, P. D., G. H. Rau and F. M. M. Morel. 2000. Inorganic carbon acquisition in coastal Pacific phytoplankton communities. *Limnology and Oceanography*, **45**, 1485-1500.

Tsunogai, S., S. Watanabe, J. Nakamura, T. Ono and T. Sato. 1997. A preliminary study of carbon system in the east China Sea. *Journal of Oceanography*, **53**, 9-17.

Tsunogai, S., S. Watanabe and T. Sato. 1999. Is there a "continental shelf pump" for the absorption of atmospheric $CO_2$? *Tellus. Series B: Chemical and Physical Meteorology*, **51B**, 701-712.

van Geen, A., R. K. Takesue, J. Goddard, T. Takahashi, J. A. Barth and R. L. Smith. 2000. Carbon and nutrient dynamics during coastal upwelling off Cape Blanco, Oregon. *Deep Sea Research II*, **47**, 975-1002.

Venkatesan, M. I., S. Steinberg and I. R. Kaplan. 1988. Organic geochemical characterization of sediments from the continental shelf south of New England as an indicator of shelf edge exchange. *Continental Shelf Research*, **8**, 905-924.

Ver, L. M. B., F. T. Mackenzie and A. Lerman. 1999. Carbon cycle in the coastal zone: effects of global perturbations and change in the past three centuries. *Chemical Geology*, **159**, 283-304.

Vezina, A. F., C. Savenkoff, S. Roy, B. Klein, R. Rivkin, J. C. Therriault and L. Legendre. 2000. Export of biogenic carbon and structure and dynamics of the pelagic food web in the Gulf of St. Lawrence Part 2. Inverse analysis. *Deep-Sea Research II*, **47**, 609-635.

Vlahos, P., R. F. Chen and D. J. Repeta. 2002. Dissolved organic carbon in the Mid-Atlantic Bight. *Deep Sea Research II*, **49**, 4369-4385.

Walsh, J. and C. P. McRoy. 1986. Ecosystem analyses in the southern Bering Sea. *Continental Shelf Research*, **5**, 259-288.

Walsh, J. E. 1981. A carbon budget for overfishing off Peru. *Nature*, **290**, 300-304.

Walsh, J. J. 1988. *On the nature of continental shelves*. San Diego, Academic.

Walsh, J. J. 1989. How much shelf production reaches the deep sea? In *Productivity of the ocean: present and past.* W. H. Berger, V. S. Smetacek and G. Wefer. eds. Wiley-Interscience. Chichester, UK. pp. 175-191.

Walsh, J. J. 1991. Importance of continental margins in the marine biogeochemical cycling of carbon and nitrogen. *Nature*, **350**, 53-55.

Walsh, J. J., E. T. Premuzic, J. S. Gaffney, G. T. Rowe, G. Harbottle, R. W. Stoenner, W. L. Balsam, P. R. Betzer and S. A. Macko. 1985. Organic storage of $CO_2$ on the continental slope off the Mid-Atlantic bight, the southeastern Bering Sea, and the Peru coast. *Deep-Sea Research*, **32**, 853-883.

Walsh, J. J., G. T. Rowe, R. L. Iverson and C. P. McRoy. 1981. Biological export of shelf carbon is a neglected sink of the global $CO_2$ cycle. *Nature*, **291**, 196-201.

Walsh, J. J., C. D. Wirick, L. J. Pietrafesa, T. E. Whitledge, F. E. Hoge and R. N. Swift. 1988. *High-frequency sampling of the 1984 spring bloom within the Mid-Atlantic Bight: Synoptic shipboard, aircraft, and in situ perspectives of the SEEP-I Experiment.*

Wang, S.-L., C.-T. A. Chen, G.-H. Hong and C.-S. Chung. 2000. Carbon dioxide and related parameters in the East China Sea. *Continental Shelf Research*, **20**, 525-544.

Wang, X.-C. and C. Lee. 1993. Adsorption and desorption of aliphatic amines, amino acids and acetate by clay minerals and marine sediments. *Marine Chemistry*, **44**, 1-23.

Wang, Z., 2002. Biogeochemical Changes of Chemical Signals in the Georgia Land-to-Ocean Continuum. *Ph. D Thesis, Department of Marine Science.* Athens, University of Georgia.

Wanninkhof, R., G. Hitchcock, W. J. Wiseman, G. Vargo, P. B. Ortner, W. Asher, D. T. Ho, P. Schlosser, M. L. Dickson, R. Masserini, K. Fanning and J. Z. Zhang. 1997. Gas exchange, dispersion, and biological productivity on the west Florida shelf: Results from a Lagrangian tracer study. *Geophysical Research Letters*, **24**, 1767-1770.

Wanninkhof, R. and W. R. McGillis. 1999. A cubic relationship between air-sea $CO_2$ exchange and wind speed. *Geophysical Research Letters*, **26**, 1889-1892.

Williams, P. J. l. 1998. The balance of plankton respiration and photosynthesis in the open oceans. *Nature*, **394**, 55-57.

Williams, P. J. l. and D. G. Bowers. 1999. Regional carbon imbalances in the oceans. *Science*, **284**, 1735b.

Williams, P. M. and E. R. M. Druffel. 1988. Dissolved organic matter in the ocean: Comments on a controversy. *Oceanography*, **1**, 14-17.

Wollast, R. 1998. Evaluation and comparison of the global carbon cycle in the coastal zone and in the open ocean. In *The Sea, Volume 10. The global coastal ocean.* K. H. Brink and A. R. Robinson. eds. Wiley. New York. pp. 213-252.

Wollast, R. and L. Chou. 2001. Ocean Margin Exchange in the Northern Gulf of Biscay: OMEX I. An introduction. *Deep-Sea Research II*, **48**, 2971-2978.

Wong, G. T. F., S.-Y. Chao, Y.-H. Li and F.-K. Shiah. 2000. The Kuroshio edge exchange processes (KEEP) study - an introduction to hypotheses and highlights. *Continental Shelf Research*, **20**, 4-5.

Yager, P. L., D. W. R. Wallace, K. M. Johnson, W. O. Smith, Jr., P. J. Minnett and J. W. Deming. 1995. The Northeast Water Polynya as an atmospheric $CO_2$ sink: A seasonal rectification hypothesis. *Journal of Geophysical Research. C. Oceans*, **100**, 4389-4398.

Yool, A. and M. J. R. Fasham. 2001. An examination of the "continental shelf pump" in an open ocean general circulation model. *Global Biogeochemical Cycles*, **15**, 831-844.

# Chapter 10. ECOSYSTEM TYPES AND PROCESSES

DANIEL M. ALONGI

*Australian Institute of Marine Science*

**Contents**



## 1. Introduction

The coastal ocean represents a dynamic equilibrium between land and the deep ocean, encompassing only about 8% of the global ocean yet accounting for nearly 20% of total net primary production and at least 90% of the world's fish catch. The high fertility of coastal ocean ecosystems is driven by nutrient inputs from rivers and groundwater, upwelling, exchanges at the shelf edge and inputs from the atmosphere. The complexity of the coastal ocean is illustrated in Figure 10.1. Boundaries between zones and the energetics of coastal organisms are not divorced from the surrounding environment or each other, but are subtly and intricately woven into the machinery of a dynamic ocean (Alongi, 1998). The tight physical coupling at the land-sea boundary is crucial for the ecology of shelf seas. It is on the shelf proper that estuarine and oceanic boundaries, and food webs, frequently intermingle. At the shelf edge, tongues of cold, nutrient-rich water regularly or irregularly intrude onto the outer shelf margins promoting conditions favorable for higher fertility on the shelf than in the open ocean.

The structure and function of coastal ocean ecosystems differs greatly among continental shelves, being driven largely by differences in net primary production that are ultimately determined by the interplay of many factors such as boundary currents, shelf geomorphology, river runoff, upwelling (if present) and water and sediment chemistry, that are unique to each shelf margin. Early work of coastal ecosystems necessarily focused on descriptive studies of net plankton, epibenthos and fish, being the most obvious organisms in the sea (Mills, 1989). This early phase of oceanography provided abundant evidence of the complexity of pelagic

*The Sea*, Volume 13, edited by Allan R. Robinson and Kenneth H. Brink
ISBN 0-674-01526-6 

and benthic food webs, but it was the development of sophisticated methods to measure pelagic production and consumption cycles that became a catalyst for the growth of ideas and concepts of factors regulating marine food chains and their role in the ecology of the sea (Longhurst, 1998).

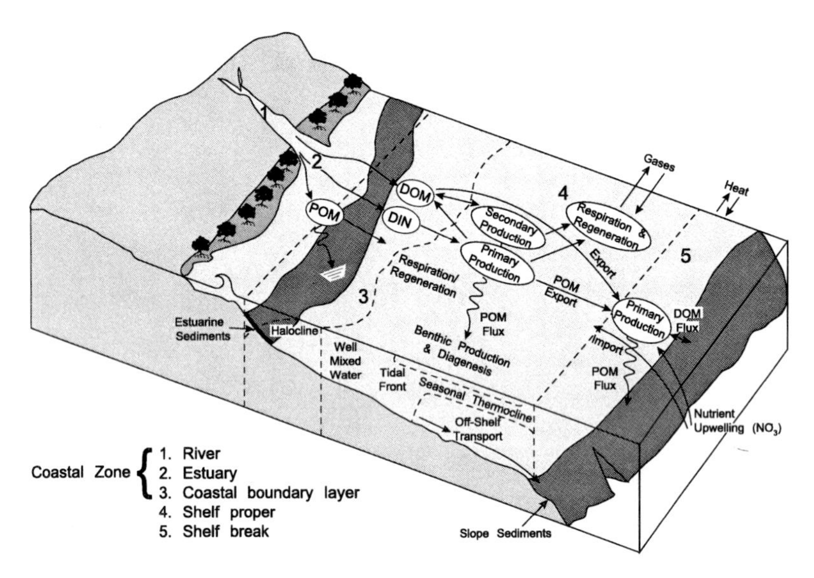


Figure 10.1 Idealized scheme of the coastal ocean with key processes linking land, sea and atmosphere (modified from Alongi, 1998). Processes are not to scale.

Hardy's (1924) classic description of the trophic relations of the herring, *Clupea harengus,* in the North Sea emphasized early that marine food chains are not simple or linear. The traditional concept of pelagic food webs, where phytoplankton is consumed by herbivorous zooplankton which are in turn eaten by carnivores, was repeatedly advocated and confirmed for more than half a century (Steele, 1974). Benthic food chains were quickly perceived as less simple, based on plant-derived detritus and associated microbes (Petersen, 1918). Complex trophic relationships between microbes, meiofauna and macrofauna were deduced early last century (Mare, 1942) but the role of benthos was overshadowed, until recently, by the dominance of the role of pelagic food chains in supporting the production of commercially harvestable species (Cushing, 1988). Benthos was often presented in food web models only as "tertiary consumers". There is no doubt that the main difficulty lay in reliably extracting and separating benthic organisms from mud and sand.

Our understanding of ecosystem types and processes in the global coastal ocean has undergone a paradigm shift with the rise of microbial ecology, acknowledgement of the role of benthos, and the growth of technology and multi-disciplinary studies of physical-biological interactions in the sea. The development of techniques to measure bacterial abundance and productivity in the plankton rapidly allowed for a reassessment of their role in the sea. Evidence soon accumulated indicating that pelagic bacteria and protozoa were more numerous and productive than thought earlier. Moreover, it was deduced that the abundance of bacteria was controlled by ubiquitous nanoplanktonic flagellates that were in turn consumed by microzooplankton, suggesting that some energy was returned via this "microbial loop" (Azam et. al., 1983). This development had important consequences for our understanding of the structure and function of pelagic food webs, particularly with feeding relations and energy flow.

Pomeroy (1979) was among the first to recognize that earlier conceptual and compartmental models of energy flow through shelf ecosystems would have to incorporate microbial pathways of production and consumption, higher ecological efficiencies, and greatly modify the concept of trophic levels. The subsequent growth of ecosystem-scale studies conducted jointly by physicists, chemists, and biologists further crystallized a new paradigm as to the complexity of ecosystem structure and function in the coastal ocean. Simply put, the new integrated view is that marine organisms, their food webs and trophic networks interact with and are greatly influenced by external forces that affect the movement of water and sediment (Mann and Lazier, 1996). These forces and ecosystems, varying greatly in space and time, and their categorization, are the subject of this chapter.

## 2. External forcing of coastal ocean ecosystems

In the sea as on land a small number of physical factors regulate primary producers (e.g., phytoplankton, macrophytes) and, consequently, the distribution of associated organisms from microbes to whales. It is important to distinguish this external forcing from the secondary effects of biological interactions that assist in sculpting an ecosystem. In the open ocean the physical forcing of ecosystem function is easier to generalize than on continental shelves where basin-scale circulation interacts with coastal winds and shelf morphology. This interaction determines both regional oceanography and the nature of the ecosystem of each continental shelf. It is important to emphasize that the boundaries between ecosystems of continental shelves and the open ocean are leaky. Biota encountered on continental shelves, even routinely and abundantly, may be vagrants from oceanic ecosystems. Likewise the flow of energy and materials through food webs differs greatly among continental shelves, driven largely by differences in rates of carbon fixation. Primary production is ultimately driven by the interplay of local-and regional-scale patterns of water circulation, chemistry and shelf geomorphology.

To understand the physical factors that force ecosystem processes on continental shelves, as in the open ocean, there is no better starting point than Sverdrup (1953). He directs our attention to those factors that control the depth of the wind-mixed layer and the penetration of irradiance, principally surface dilution, wind stress and irradiance. The diminution of the Coriolis parameter (f) towards the equator and the consequent reduction of the slope of the sea surface required for motion must also be considered. Wind stress at high latitudes generally induces local deepening of the mixed layer while at low latitudes motion is preferentially induced. In low latitudes an excess of precipitation over evaporation reinforces the resistance of mixed layers to wind deepening, while in high latitudes ice cover may also induce a near-surface low-salinity layer, again constraining seasonal changes in mixed layer depth.

But in the coastal seas with which we are concerned these processes are strongly modified by the characteristic features of each region—the effects of coastal winds, tides and river runoff as constrained by water depth. The energy required to circulate and mix coastal waters is derived from solar heating or gravity, barometric pressure and water density. Shallow seas are characteristically occupied by two water masses: (a) offshore, an intrusion of variously modified oceanic water (and carrying oceanic biota) and (b) inshore, a low-salinity, high-turbidity water mass significantly modified by terrestrial run-off. In principle, river plumes will turn in an anticyclonic fashion on encountering the sea, and be maintained by this mechanism within the inshore water mass. Breakdown of such fronts requires major physical forcing.

No clear categorization of shelf-sea fronts exists, but there are many different types — tidal fronts, estuarine fronts or plumes, shelf-break fronts, island wakes and fronts, and upwelling fronts. Upwelling is a complex process driven mainly by winds causing surface waters to diverge from the coast and to be replaced by upwelling of cold, nutrient-rich water from the adjacent deep ocean. Buoyancy-driven upwelling (e.g. off the Amazon shelf), tidally-driven upwelling (e.g., Celtic Sea, Georges Bank) and intrusions driven by instability of western boundary currents (e.g., South Atlantic Bight) are other forms of upwelling. At least one comprehensive review of upwelling systems exists and should be consulted for further information (Summerhayes et. al., 1995).

Depending on shelf morphology and climate, various types of fronts can exist simultaneously on the same shelf. On the Great Barrier Reef shelf, for example, mixing and trapping of estuarine waters occurs and is maintained as a coastal boundary layer during the summer wet season, while topographically controlled fronts are sustained around islands and individual reefs further seaward as well as by upwellling from the Coral Sea (Wolanski and Hamner, 1988; Brinkman et. al., 2001). These fronts greatly influence the ecology of coastal plankton and benthic communities such as the species composition and abundance of ichthyoplankton (Thorrold and McKinnon, 1995) and help to maintain benthic assemblages at the pioneering stage (Alongi and Robertson, 1995).

The composition and distribution of bottom sediments are arguably as influenced by physical forces as are overlying water masses. Shallow sea sediments are derived from continental rock, from erosion of the coastline, from organic material deposited by river water and from biological sources, both organic and inorganic. Calcium carbonate derived from skeletal fragments is the dominant sediment source in some low-latitude regions with minor river flow or coral reefs or both (Reading, 1996). Tropical and subtropical regions are greatly influenced by high rates of evaporation and rainfall. It is estimated that ≈70% of freshwater and ≈75% of sediment discharging onto the world's shelves occurs within the tropics (Milliman and Syvitski, 1992). Consequently most mud is deposited on tropical shelves. The latitudinal distribution of sediment types on the world's inner continental shelves was first analyzed by Hayes (1967) who noted the tropical dominance of mud and carbonate derived from reefs, which are proportionally displaced by gravel and rock with higher latitude. The global distribution of quartz sand and relict shell are not related to climate. Hayes (1967) indicated that the major climatic factors responsible for the global patterns were weathering, presence or absence of major rivers, glaciation and ice rafting. There are enormous differences in distribution patterns of sediment types with latitude and often on the same shelf, but Hayes' (1967) analysis still defines a global trend reflecting differences in climate.

All sediments are subject to physical and biological reworking and to transportation by bottom currents and gravity. Soft, organic sediments tend to dominate in deeper basins and in shoal water where quiescent conditions facilitate deposition. The offshore extent of fine sediment often defines the physical limit of terrigenous input. The mosaic of sediment types and associated inputs of organic matter is the primary determinant of the distribution and abundance of shelf benthic communities. Complex mixtures of dead and decaying phytoplankton, epiphytes, microalgae, macrophytes, and metazoans constitute the richest sources of proteins, carbohydrates, lipids, fatty acids, and other organic compounds that sustain benthic life.

An interactive hierarchy of mechanisms regulates benthic food webs. These mechanisms include the quality and quantity of food resources, physiological constraints and tolerances, and behavioral and life history strategies. All of these mechanisms operate within the confines of an exceedingly complex environment of chemical, geological and biological gradients. Sediment type, temperature, salinity, intensity of bioturbation, rate of organic input and mass sediment accumulation rate are just some of the factors that regulate the rate of return of nutrients back to the overlying water column.

## 3. Characteristics of shelf biota

Although phytoplankton comprise the familiar three guilds of coccoid cyanobacteria, nano-flagellates and larger algal cells, a far wider variety of biota are involved in primary production in shallow seas than in the open ocean. Benthic production by macroalgal kelp and marsh grasses in mid to high latitudes, and by coralline algae, seagrasses and mangroves in lower latitudes, are very significant carbon fixers. To these must be also added the contribution of symbiotic algal cells in the tissues of many species of benthic invertebrates (e.g., hard corals) on some low latitude shelves. More recently, widely distributed and diverse assemblages of phototrophic bacteria have been found in the plankton and may be productive carbon fixers (Kolber et. al., 2001).

The above-mentioned patterns of primary producers are a presage of latitudinal changes in the structure and function of shelf sea biota (Figure. 10.2). It has been observed that bacterial biomass increases in proportion to other sediment infauna towards the tropics (Alongi and Robertson, 1995; Aller and Stupakoff, 1996) because of the greater prevalence of muds and types of physical disturbance on low latitude shelves that are, on average, shallower than ocean margins poleward (Alongi, 1998). The development of opportunistic tropical infaunal communities of variable diversity and low biomass is fostered by episodes of physical disturbance, erratic food supply, and dilution of riverine organic matter with highly weathered inorganic material. Bacteria and other microbes dominate tropical infauna because with short generation times and high growth rates they are better adapted to respond to and recover more quickly from these events than metazoans.

For the plankton, it is the various responses of shelf organisms to regional oceanography that determines the floristics and seasonal production cycles of each region (Longhurst, 1998). As in the open ocean, intermittent relief from nutrient limitation favors large cells in temperate coastal waters, hence their dominance in spring blooms or seasonal outbursts in river plumes. Under nutrient limitation the smaller cell fractions dominate phytoplankton biomass, being more efficient at recycling nitrogen. In tropical coastal waters, the small cells usually dominate year-round, except when large pulses of nutrients derived from heavy monsoonal rains, lead to rapid blooms of net phytoplankton.

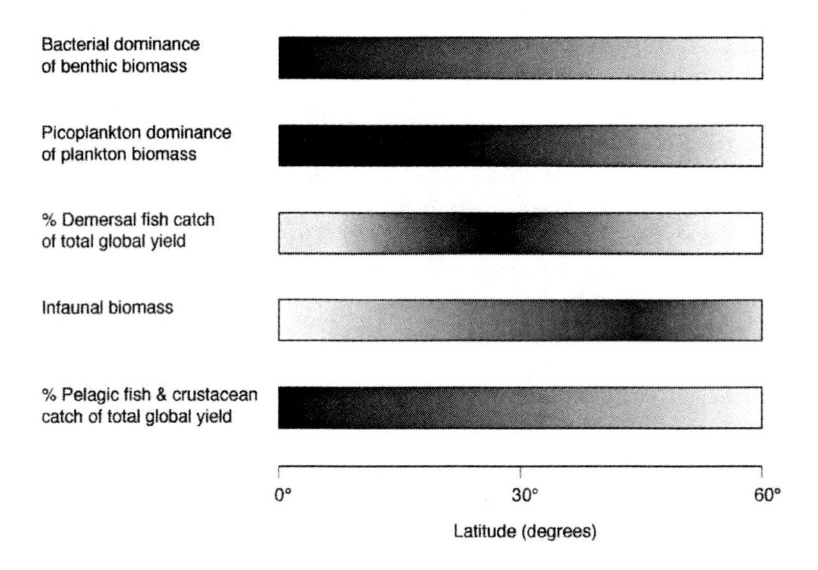


Figure 10.2 Schematic representation of relative latitudinal patterns in size spectra and functional characteristics of coastal ocean biota. Based on data compiled from various sources in Alongi (1998).

Benthic macroalgae, constrained by illumination, occur only in shoal water. The largest accumulation of their biomass occurs in mid-latitude, coastal upwelling regions where kelp may be anchored by holdfasts in relatively deep water (30–40m), reaching up into the illuminated zone on elongated stipes, buoyed by gas-containing bladders which reach the surface. In tropical water of the greatest clarity, encrusting benthic algae may photosynthesize down to mid-shelf depths. Activity of symbiotic algal cells in benthic invertebrates is restricted by sufficient light and by the opacity of their host's tissues.

Loss of standing biomass of primary producers to herbivores is also complex and difficult to assess because we must quantify not only the role of protozooplankton and zooplankton, but also that of the many benthic invertebrates and fish which directly consume plant biomass. Echinoderms graze kelp fronds, scarid fish consume coral biomass containing algal cells, while shoaling clupeid fish and clams in muddy deposits filter-feed on diatom blooms. As in other ecosystems, rates of herbivory and carnivory vary greatly in time and space, constraining generalizations on levels of energy flow on continental shelves.

As noted in Figure 10.2, pelagic fish and crustaceans appear to make up a greater fraction to total fisheries production than demersal species from the high to low latitudes (Longhurst and Pauly, 1987; Sharp, 1988). On the basis of area

(Pauly and Christensen, 1995), tropical shelves yield greater numbers of total fish catch (2.15 t $km^{-2}$) than non-tropical shelves (1.64 t $km^{-2}$). Such generalizations are fraught with problems due to the fact that that there are always notable exceptions (e.g. upwelling regions) plus there are constraints because of the lack of tropical data and the lack of understanding of life history strategies of many commercially important species.

What we do know is that some pelagic species maintain populations both over continental shelves and sometimes in the open ocean, as in the Atlantic herring *Clupea harengus,* which migrates seasonally between the two habitats. Most pelagic organisms, plants and animals, occurring in continental shelf seas have one of two possible origins: (a) they may be members of populations centered in shallow-water habitats of the shelf (*in sensu* Sinclair's (1988) 'member-vagrant' hypothesis) or (b) they may be vagrants from populations characteristic of and having traits appropriate for deep-water oceanic habitats. Taking copepods as an example, some deep-water species have traits (e.g. deep interzonal diel migration, as in *Pleuromamma*) that prevent their survival over continental shelves. Other deep-water species, such as *Calanus finmarchicus,* may maintain populations seasonally, or even – where deep basins exist for deep overwintering – permanently over shelves (Hind et. al., 2000). Those who are familiar with *Calanus* species (and other oceanic vagrants) on continental shelves must, therefore, keep in mind that they have a limited shelf life!

Despite what appear to be emerging global patterns (Figure. 10.2), both taxonomic and functional diversity of shallow-sea biota must be generalized and characterized for each type of ecosystem. Because Atlantic biota was significantly attenuated by post-Tertiary climatic events, there are only 35 Atlantic species of reef-building corals in 26 genera, but 700 Indo-Pacific species in 80 genera. Not only that, several important coral-associated guilds of benthic organisms are absent in the Atlantic, so fluxes must be modeled differently in the two oceans. Otherwise, diversity responds significantly to latitude in plankton and fish of shallow seas. There are few fish species of low functional diversity in polar seas, but very many in the low latitudes. Paradoxically this attenuation does not occur in benthic macroinfauna and meiofauna. On sedimentary gradients of sand and mud mixtures, the benthic infauna is characterized by identical genera of molluscs and echinoderms appropriate to each admixture — essentially, the lower the sediment organic content, the higher the percentage of filter-feeding benthos therein – and vice versa. This basic pattern, although clarified more recently (e.g., Snelgrove and Butman, 1994), has been known since a description was first attempted by Thorson (1957). These "parallel" communities have similar taxonomic diversity and k-dominance from the tropics to polar seas. The benthic epifauna of more mobile invertebrates, on the other hand, is most diverse both taxonomically and functionally in low latitudes. In the tropics, variation in diversity within a region is often as great as or greater than variations between regions.

Many constructs of shelf-sea ecosystems lack an adequate representation of the complexity of the heterotrophic bacteria that respire most of the organic material in the sea (Pomeroy, 1979). Bacteria communities in water and sediment respond very rapidly to change. Within days of the onset of a phytoplankton bloom, and the sinking of cells, there is an increase in bacterial biomass actively engaged in the remineralization of cell debris. In coastal seas, bacteria, protozoa and microalgae form the basis of pelagic and benthic food webs and are the chief biological regulators of nutrient recycling. Bacteria are usually represented as a single heterotrophic entity, but their activities (and species) are as diverse as the substrates (e.g., chitin, lignin, amino acids, proteins and carbohydrate) they degrade. The numbers of bacterial cells are relatively constant in oceanic waters, but bacteria are very large and numbers are more variable in organic-rich sediments and suspensions. Nearly all bacterial types are deeply involved in recycling carbon, nitrogen, sulphur, and phosphorus. A varied protist fauna exists in both water column and sediments, largely fuelled by consumption of heterotrophic bacteria, though the smaller fractions of phytoplankton are also eaten. These microorganisms interact with each other in complex networks (the "microbial loop") and with larger organisms in trophic cascades in which regulation works either from the "bottom-up" or the "top-down" (Fenchel, 1988). Trophic interactions among pelagic and benthic biota oscillate in duration and intensity in coastal waters particularly in response to changes in physical forces such as light, temperature or tidal flow.

## 4. Linking land-coastal ocean-atmosphere biogeochemical cycles and ecosystem function

Microbes play the major role in the processing of carbon and nutrients in shelf sea ecosystems, but the magnitude of the flux of energy and materials between the shallow seas and the deep ocean remains unclear. This is not surprising given the complexity of the processes involved. River-borne terrestrial organic material passes across the shelf and is partially respired by heterotrophic processes during transit, while autotrophic production on the shelf is fuelled partly by inorganic nutrients from river water, but mostly from the offshore nutrient reservoir (Atkinson et. al., Chapter 8). In short, the intermingling of riverine, coastal and open-ocean waters results in a complex interplay of energetic processes and material flow. Critically, our ecological knowledge is well short of our ability to reliably model the rates and pathways of carbon flow on tropical shelves.

The actual mechanism of transport from rivers to the shelf and to the deep sea is not easily generalized, because each individual shelf region is somewhat unique and must be treated as a special case. Flux is forced not only by special regional oceanography but also takes routes determined by the geomorphology of each individual shelf. It is not enough to study a single type of shelf at mid-latitude for extrapolation to the global scale as has been done so uncritically in the past. The boundaries between coastal and oceanic zones are often blurred by the energetics of a very dynamic sea. With this caveat in mind, three generalizations about the energetics of shelf ecosystems can be offered:

- Outwelling of materials from the continents is usually restricted to the coastal zone by complex physical, chemical and geological processes, but this material greatly influences inshore food chains and nutrient cycles;
- Impingement of nutrient-rich, open-ocean water enhances primary and secondary productivity on the outer shelf and shelf edge, including major fisheries, particularly along boundary currents; and
- Little organic matter is exported from the continental margin.

The continental shelves are areas of high primary production and large standing stocks of particulate organic material compared with the open ocean, so it is not surprising that great effort has been made to determine the fate of this material. As noted earlier, the classical food webs were cited to account for the fate of this matter as well as in supporting relatively high fisheries production on continental shelves. By the mid-1970's it was suggested that zooplankton grazing and benthic respiration were too low to account for the fate of the remaining fixed carbon. Walsh et. al. (1981) suggested that this discrepancy was accounted for if the majority of carbon fixed by phytoplankton during the spring bloom was exported from the shelf to the adjacent slope or central ocean basins. They suggested that about 50% of shelf carbon production is shunted to the slope, with nutrients regenerated from anthropogenic sources used to cyclically sequester carbon. Deposition of carbon on the slope could, they reasoned, represent the missing billion tons of carbon in global carbon budgets.

Since the seminal work of Walsh et. al. (1981), a number of large-scale interdisciplinary projects (e.g., SEEP, JGOFS) has taken place to deduce the fate of organic carbon on continental shelves. The results of most of these studies concluded that the "missing carbon" hypothesis failed to incorporate microbial heterotrophy adequately— based on observations from a very limited range of ecotypes.

TABLE 10.1
Global Mass Balance Model of Organic Carbon Flow on the Continental Shelf and Slope

| | Flux ($10^{15}$ g C $yr^{-1}$) |
|---|---|
| *Continental Shelf* | |
| Primary production | 6.9 |
| | ⇓ |
| Pelagic respiration | 3.7 |
| Flux to sediments | 0.2 ⇒ 2.0 respired + 0.2 buried |
| Available for expor | 1.0 |
| | ⇓ |
| *Continental Slope* | |
| Import from shelf | 1.0 |
| Pelagic respiration | 0.8 |
| Flux to sediments | 0.2⇒ 0.18 respired + 0.02 buried |

*Source:* Modified from Alongi (1998) based on original data in Wollast (1991, 1993)

A few workers have attempted mass balances of carbon in the global coastal ocean. Wollast (1991, 1993) constructed a preliminary global mass balance of organic carbon on the continental shelf and slope (Table 10.1). His model implies that more than 82% of carbon fixed by shelf phytoplankton is respired by shelf benthic and pelagic consumers, with the remainder exported to the slope where most of the carbon is respired by slope heterotrophs below the thermocline. Only 4% of the original fixed carbon is buried in shelf and slope sediments. This model implies low carbon burial in coastal sediments. In a speculative review, Hedges and Keil (1995) assessed the available evidence for carbon transport and deposition on the shelf, and concluded that most terrestrially-derived organic matter is buried in marine deltaic and inshore sediments. They estimated a global burial rate of 0.16 ×

$10^{15}$ g C $yr^{-1}$ with <0.5% of global ocean productivity and <10% of organic matter depositing to sediments being preserved.

A more detailed carbon mass balance of the global ocean constructed by Smith and Hollibaugh (1993) reveals some significant differences with Wollast's (1991, 1993) estimates. First, the Smith and Hollibaugh (SH) budget indicates that respiration exceeds primary production by 1.4%, whereas Wollast (W) calculated that coastal ecosystems have a production to respiration ratio of 1.03, that is, are slightly heterotrophic. In the SH model, the net heterotrophy of the coastal ocean is compensated by terrigenous loading of DOC and POC. Second, less carbon deposits to sediments in the SH model (≈ 15% of NPP) compared to the W model (≈32% of NPP). Both models agree that comparatively little carbon is preserved in shelf sediments. The sediment carbon cycle in the SH model is balanced by the return mostly of DOC to the water column. Third, the SH model indicates that only 4% of carbon equivalent to NPP is exported off the shelf, nearly four times less than estimated by Wollast (1991, 1993). More recently, Wollast (1998) modified his budget to indicate that proportionally more carbon (37% of NPP compared with 16% in previous W budget) is available for export to the slope and the open ocean, assuming that proportionally less carbon (30% of NPP) is respired than estimated previously (54% of NPP).

As implied from the differences in budgets, the computations of the flux of coastal ocean carbon remain tentative, though it is now thought that the balance between production (plus import) and consumption (plus burial) is much closer than previously calculated (Liu et. al., 2000). Recent computations have tended to converge on much lower values than before. If one incorporates the inherent variability among shelves, primary production in the coastal ocean represents 10–29% (mean = 19%) of global ocean production and new production or shelf export represents 3–64% of new global carbon production (Table 10.2).

TABLE 10.2
Estimates of Coastal Ocean Primary Production ($CO_{PP}$), New Production ($OC_{NP}$) or Organic Carbon Export ($OC_E$) and Their Contribution to Global Ocean Primary Production ($GO_{PP}$)

| $CO_{PP}$ | $CO_{PP}$/ $GO_{PP}$ | $OC_{NP}$ | $OC_E$ | $OC_{NP}$ / $CO_{PP}$ | $OC_E$/ $CO_{PP}$ | $OC_{NP}$/ $GO_{PP}$ | $OC_E$/ $GO_{PP}$ |
|---|---|---|---|---|---|---|---|
| 4.1 | 13.2% | 1.5 | 1.25 | 31% | 14-20% | 40-44% | 3.2% |
| 4.8 | 20.2% | 1.5 | 1.00 | 17% | 14% | 20% | 18% |
| 7.4 | 24.9% | 3.0 | 0.35 | 25% | 4.5% | 50% | |
| 9.0 | 17.6% | 4.7 | 0.21 | 34% | 4.9% | 64% | |
| 7.7 | 28.5% | 0.9 | 0.44 | 11% | 5.6% | 40% | |
| 8.3 | 18.4% | | | | | | |
| 4.8 | 10.0% | | | | | | |
| 7.8 | 19.5% | | | | | | |

*Source:* Modified from Liu et. al. (2000). Units: Gt C $yr^{-1}$.

The coastal ocean may serve as a 'continental shelf pump' (Ducklow and McCallister, Chapter 9), that is, as a transfer mechanism of $CO_2$ from the atmosphere to the open ocean. As pointed out by Liu et. al. (2000), new production as defined by uptake of nitrate not derived from recycled sources, is only a good measurement of the pump on shelves with narrow margins or where advective processes are dominant. On wide shelves or on margins where longshore transport predominates, new production is overestimated. On these latter shelves, the shelf pump is better represented by cross-shelf export to subsurface water below the thermocline. Liu et. al. (2000) estimated the capacity of the 'continental shelf pump' to be $1 \times 10^{15}$ g C $yr^{-1}$, or 20% of the estimated global pump. Although the contribution from tropical rivers was included in the Smith and Hollibaugh (1993) model, it is likely that more recent estimates of carbon flux from tropical river-dominated shelves would result in higher global average rates of carbon input and more absolute amounts of carbon being remineralized or preserved in shelf, slope and continental rise sediments.

The extent to which shelf carbon is available for export to the adjacent deep ocean remains controversial. A recent analysis of data largely from temperate shelves and off the Amazon suggests that >95% of fixed carbon on the shelf is mineralized in the water column or in sediments (de Haas et. al., 2002). This estimate implies that the role of the coastal ocean as a site for carbon storage has been overestimated. Large areas of many continental shelves do not show any substantial accumulation of organic matter. de Haas et. al. (2002) suggests that organic carbon accumulates and is buried locally under favorable conditions. On a larger area basis and over long time scales, the role of continental shelves as sinks for organic carbon is limited. They suggest that continental slopes, deep-sea fans and canyons are the main sinks.

A recent analysis of carbon mass balance on the central Great Barrier Reef shelf (Brunskill et. al. 2002) supports the idea of de Haas et. al. (2002) that many shelves may not be significant storage sites of organic carbon. On the central Great Barrier Reef shelf roughly 1% of riverine and marine organic carbon production is buried in sediments as nearly all organic carbon supplied to the shelf either from rivers or from fixed production *in situ* is balanced by pelagic and benthic respiration. The storage of organic carbon is small and limited spatially to protected embayments inshore of 20 m water depth. Burial rates of carbonate carbon were high but restricted to deep reef lagoons at mid- and outer- shelf that represent only 13% of total shelf area.

Tropical shelves are thought to be the main producers and sinks of carbonate carbon, but the global coastal ocean estimates of carbonate carbon flow are poorer than the estimates of organic carbon flux (Milliman, 1993; Wollast, 1998). Milliman (1993) calculated that current production of $CaCO_3$ in the world ocean is about 5 billion tons $yr^{-1}$, of which about 60% accumulates in sediments and the remainder dissolves in the sea. Nearly 50% of the carbonate sediment accumulates on reefs, banks and on tropical shelves. Most carbonate is in the form of metastable aragonite and magnesium calcite. The problem with this production estimate is that it equates to twice as much calcium brought in by rivers and hydrothermal activity, suggesting either that outputs have been overestimated or inputs underestimated or a combination of both, or that the ocean is not in steady-state with respect to calcium. To reconcile the discrepancy, Wollast (1998) constructed a global estimate of carbonate flux in the coastal ocean assuming $CaCO_3$ dissolution rates equivalent to the observed vertical profiles of total dissolved carbon and alkalinity. His model shows that one-half of carbonate accumulation in the world ocean is occurring in shallow tropical seas, with nearly 20% of shelf production exported to the slope. The rate of carbonate accumulation is still very high supporting Milliman's (1993) idea that the cycle is not presently in steady-state. During periods of low sea-level stand the rates of accumulation were probably slower (Milliman and Droxler, 1996).

The carbonate system is closely linked to atmospheric $CO_2$ but up until very recently there has been considerable disagreement over whether or not the ocean is a net sink or source of $CO_2$ (Ducklow and McCallister, Chapter 9). The Smith and Hollibaugh (1993) and Wollast (1993) models estimated net fluxes of $CO_2$ of 8.5 gC $m^{-2}$ $yr^{-1}$ from the shelf, 0.9 gC $m^{-2}$ $yr^{-1}$ from the slope and 0.5 gC $m^{-2}$ $yr^{-1}$ from the open ocean. These models provide estimates of net export of biologically mediated carbon gases from the ocean to the atmosphere on the order of $\approx$ 20–25 $\times$ $10^{12}$ mol C $yr^{-1}$ which is much less than present atmospheric inputs from the burning of fossil fuels, estimated at about 200 $\times$ $10^{12}$ mol C $yr^{-1}$ (Takahashi et. al. 2002, Watson and Orr, 2003).

More recent analyses (Walsh and Dieterle, 1994, Ducklow and McCallister, Chapter 9) indicate that the global coastal ocean is a sink for atmospheric $CO_2$ due to fossil fuel burning and increased deforestation. Moreover, significant differences in atmosphere-ocean exchange of $CO_2$ occur with latitude because of natural changes in partial pressure, upwelling and the warming of shelf waters resulting in net release of $CO_2$. Walsh and Dieterle (1994) calculated that temperate and polar shelves sequester from 8.3 to 9.7 $\times$ $10^{13}$ mol C $yr^{-1}$, but tropical shelves release 3.1$\times$ $10^{13}$ mol $yr^{-1}$ of $CO_2$ to the atmosphere, for a net input on the order of 5.2 to 6.6 $\times$ $10^{13}$ mol $yr^{-1}$. The reliability of these estimates depends on assumptions that POC is buried, that DOC is advected below the photic zone and that total dissolved inorganic carbon crosses the lysocline of adjacent slope and basin waters.

Increased atmospheric $CO_2$ inputs to the ocean may translate into decreased rates of carbon fixation by autotrophs due to a decrease in seawater pH with attendant $CO_2$ invasion (Ducklow and McCallister, Chapter 9). The extent to which this scenario eventuates will be complicated by increases in the severity of eutrophication and of river runoff as a result of global climate change.

There is little doubt that phytoplankton and macrophytes, and bacteria and protozoa, are respectively the main producers and consumers of fixed carbon in the coastal ocean. What is less understood is whether or not a net balance exists between production and consumption of carbon in coastal ecosystems. Is the coastal ocean in steady- state? Gattuso et. al. (1998) maintain that the data is too limited to answer this question. In addition to the problem of limited data, the question also depends on the scale of the ecosystem under consideration. For instance, mangroves are net autotrophic if one considers their high rates of net primary production, but scaling up to an entire estuary, the balance shifts to net heterotrophy. Increases in carbon gas concentrations in the atmosphere as a result of industrialization have now likely tipped the balance in favor of the coastal ocean being a sink for $CO_2$ (Alongi, 1998; Rabouille et. al., 2001). Predictions of future trends are uncertain as are many functional aspects of the coastal ocean, including ocean-atmosphere exchange, so it is important to understand whether or not a balance exists between carbon production and consumption.

A plot (Figure. 10.3) of mean rates of gross primary production versus total system respiration for various habitats that comprise the coastal ocean (rivers, estuaries, lagoons, bays, shelf zones) suggests that most ecosystems are either slightly autotrophic (seagrass beds, kelp and mangrove forests, coral reefs, salt marshes) or slightly heterotrophic (beaches and intertidal flats, estuaries, bays, lagoons, inner shelf). Given the inherent variability within ecosystems and the very sparse data, a reasonable, conservative summation of available evidence indicates that the coastal ocean is currently in overall carbon balance. The coastal zone (including estuaries) is slightly net heterotrophic and a source of $CO_2$. This flow is not balanced by the slight net autotrophy of habitats such as coral reefs, mangrove forests and seagrass and kelp beds, given the greater overall area of open water habitats on the shelf.

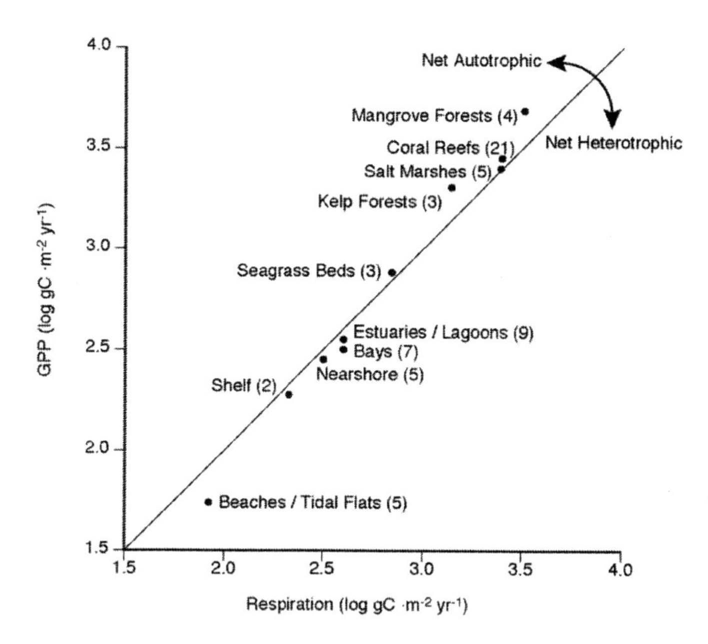


Figure 10.3 A summary of the balance between gross primary production and total respiration in various coastal ecosystems (modified from Alongi, 1998). Each point is the arithmetic mean of data obtained from several systems (numbers in parentheses). Data are $\log_{10}$-transformed.

In a review of the current literature and a modeling exercise of human disturbance of coastal biogeochemical cycles, Rabouille et. al. (2001) found that before anthropogenic impacts, the global coastal ocean was a net autotrophic system, with a net export to sediments and the open ocean of 20 Tmol organic C $yr^{-1}$. Again, such conclusions depend on the scale and grouping of habitats. In their model the coastal ocean was divided into a proximal zone (large bays, open waters of estuaries, deltas, inland seas and salt marshes) that was slightly heterotrophic at 8.4 Tmol C $yr^{-1}$, and a distal zone (open shelves to 200 m water depth) that was net autotrophic at 28.4 Tmol C $yr^{-1}$. Using this differentiation of zones, Rabouille et. al. (2002) simulated human impacts over the past half century and found that their model suggested a doubling of primary production and an accumulation of biomass. The coastal ocean became more heterotrophic in response to increased terrestrial flow via rivers. Future projections suggest that the coastal ocean may become increasingly heterotrophic. Seitzinger et. al. (2002) indicate that nitrogen fluxes are projected to increase over the next 50 years from 21 Tg N $yr^{-1}$ in 1990 to 47 TgN $yr^{-1}$ in 2050. Most of this increase will be felt in Asia.

These models and projections come with great uncertainty in time and space. Like the other models, the Rabouille et. al. (2002) model is subject to considerable constraints. For instance, it is likely that marine angiosperm production and its role in the coastal ocean is underestimated. Duarte and Cebrian (1996) have calculated that marine macrophytes, whose production represents approximately 4% of marine primary production, export and store a considerable fraction of their net fixed carbon in the coastal zone.

Despite the uncertainty, one fact is clear — high productivity of coastal seas equates to nearly 20% of total net oceanic primary production and to at least 90% of the global fish catch. Duarte and Cebrian (1996) point to why this latter fact may be so. In a global assessment of the fate of marine primary production in the coastal ocean they calculated that, on average, greater than 50% of net primary production in coastal seas is consumed initially by herbivores for transfer to higher trophic networks, with 36% shunted directly to decomposers. Only ≈ 10% of this material is stored in the short term as most will be eventually consumed and decomposed. Only 0.25% of this material is buried, most derived from angiosperms. Thus, the vast bulk of carbon fixed in the coastal ocean is dissipated through food webs to produce the levels of commercially valuable finfish and shellfish production currently observed on continental shelves. As speculated by Nixon (1988) and Mann (1993), the more intensive yields of harvestable organisms in the sea than in freshwater ecosystems may be indirectly caused by greater temporal and spatial variations in physical forces, leading to such phenomena as upwelling, downwelling, vertical mixing, tidal exchanges, tidal fronts, and stratification-destratification of water masses. These great physical events may enhance transfer of energy and materials up the food chain.

## 5. Typology of Coastal Ocean Ecosystems

It is appropriate to use the principles discussed in the previous sections to define the types of ecosystems in the global coastal ocean. There are undoubtedly a number of ways coastal ecosystems can be typed, but the regional biogeographical approach of Longhurst (1995, 1998) coupled with my own analyses based on functional responses to latitudinal differences in climate (Alongi, 1998) seem to be a good starting point for an overview (see summary, Table 10.3).

### *5.1 Type 1 - Permanent Polar Ice*

Fast-ice at the coast is permanent on parts of the E. Siberian and Laptev Sea coasts of Siberia, NE and N coasts of Greenland, N coasts of Canadian archipelago to W. Beaufort Sea, and along essentially the whole coast of Antarctica (Table 10.3). Here, productivity is light limited, its seasonal cycle being symmetrical about a local irradiance maximum which may correspond not with solar maximum, but rather with minimal snow cover. Primary production is low also because of limiting concentrations of trace metals such as iron. Most (60–90 %) algal biomass is contributed by large diatoms and coccolithophorids; cyanobacteria and prochlorophytes are limited by the low temperatures. The contribution of the picoplankton size fraction to phytoplankton production is highly variable. In the central Artic Ocean, primary production in areas of high ice coverage can range from 5–60 mgC $m^{-2}$ $d^{-1}$, but ice algae can contribute roughly 50–60% of the entire primary production (Gosselin et. al., 1997). Total primary productivity at both poles has recently been revised upwards to 10–15 gC $m^{-2}$ $yr^{-1}$ (Gosselin et. al., 1997; Ishii et. al., 2002).

TABLE 10.3
Types of Coastal Ocean Ecosystems

| Ecosystem Type | Examples |
|---|---|
| Type 1- Permanent Polar Ice | Laptev Sea; E. Siberia; N. Greenland; W. Beaufort Sea; Antarctica |
| Type 2-Intermittent Polar Ice | S. Greenland; northern N. America (Newfoundland to Aleutians); northern Asia (Finland to Bering Sea); Sea of Okhotsk; E. Ross Sea; Dumont d'Urville Sea |
| Type 3-Mid-latitude Coastal | Europe (Norway to Iberia); N. America (Newfoundland to Florida); southern Australia |
| Type 4-Topography-forced Coastal | southern New Zealand; Falklands; Gulf of Alaska; southern North Sea |
| Type 5-Coastal Upwelling | Canary Current, NW Africa; Benguela Current, SW Africa; Peru Current, Chile; Somali Current, N. Africa and Arabia; California Current |
| Type 6-Wet Tropical Coastal | Papua New Guinea; Indonesia; Amazon; Niger; Indus;Ganges; Mekong |
| Type 7-Dry Tropical Coastal | Caribbean Sea; Red Sea; NW Australia, southern Great Barrier Reef; south Indonesian archipelago |

Below ice, phytoplankton biomass and zooplankton biomass are usually low, but abundant sympagic flora (often pennate diatoms) may occur on ice surfaces in the infiltration zone of the ice-seawater matrix. Undersides of ice <2m thick, especially if landfast, can support dense epontic growth of diatoms associated with large populations of polychaetes, copepods, nematodes, and amphipods (Gradinger et. al. 1999).

The ecology of zooplankton in coastal waters with permanent ice-cover is dominated by their response to the seasonal pulse of phytoplankton biomass. Most species often are restricted in the sense that there may not be sufficient time within a single season for a complete generation to occur. Growth rates are very slow and individuals are large compared with their congeners of lower latitude. Vertical distribution is dominated by seasonal ontogenetic migration, mostly by copepods.

The food network model of Nemoto and Harrison (1981) illustrates how the Antarctic has comparatively few keystone species (Figure. 10.4), but it also shows that the pack ice food chains are linked to the Type 2 ecosystem (below) where ice-cover is not permanent. In the pack ice regions the most important zooplankter is *Euphausia crystallophorias,* a herbivore that feeds on diatoms and epontic algae. This zooplankter is a key link to higher trophic groups serving as food for pelagic fish (e.g., *Pleurogramma*), crabeater seals, penguins and minke whale. Leopard, Ross and Weddell seals share the apex of the trophic network with killer whales.

In the seabed where most of the primary production and ice-algae deposits, extremely rich epibenthic communities of suspension feeders such as sponges and echinoderms proliferate, supporting abundant, but species-poor, fish and squid assemblages, and certain crustaceans, which are in turn taken up by penguins and seals. As pointed out by Hempel (1985), the brief and irregular pulses of primary production are incapable of sustaining a rich pelagic community but do support a suite of benthic assemblages that thrive on the rain of detritus derived mostly from sinking ice-algae.

### *5.2 Type 2 – Intermittent Polar Ice*

This type of coastal ecosystem occurs on the coasts of Greenland, off N. America from Newfoundland to the Aleutians, and off northern Asia from Finland to the Bering Sea and the Sea of Okhotsk, and on short sectors of the Antarctic coast in midsummer in eastern Ross Sea, to east of the Ronne ice shelf and in the Dumont d'Urville Sea (Table 10.3; Longhurst, 1998). In such places a shallow polar halocline induces stability very early in open-water season and productivity is light limited, its seasonal cycle being symmetrical about the local irradiance maximum. Where pack-ice remains, conditions may resemble a Type 1 ecosystem.

After ice-melt, phytoplankton communities in open water accumulate during the period when productivity increases. Phytoplankton is dominated by diatoms and both production rate and standing stocks are relatively high. A subsurface chlorophyll maximum is often observed. In shoal water, benthic macroalgae develop significant biomass especially at southern Greenland and along the Labrador coastline. In marginal ice zones characterized by weakly stratified waters, rates of carbon fixation in midsummer can approach values of up to 2.9 gC $m^{-2}$ $d^{-1}$ (Saggiomo et. al., 2000).

Planktonic herbivores are represented by abundant large copepods, euphausiids and salps. Some crustacean species form swarms and, in so doing, support major stocks of baleen whales and seals (cf Figure 10.4). Both salps and krill are recognized as playing an important role in pelagic-benthic coupling in polar marginal ice regions repackaging small particles into large, rapidly-sinking feces that channel carbon to the seafloor. Despite the trophic importance of krill in the Southern Ocean, an estimate of their total standing stocks is uncertain owing to their extremely patchy distribution. The most recent estimate suggests that the average biomass of *E. superba* ranges from 400–37,800 mgC $m^{-2}$ in regions of dense aggregation and from 1–1200 mg C $m^{-2}$ in waters of low concentration (Voronina, 1998). The pelagic tunicate *Salpa thompsoni* can be an important filter feeder in the Southern Ocean during the warmer months when it forms extensive blooms, but its trophic importance remains uncertain (Pakhomov et. al., 2002).

The fish fauna lacks diversity, especially in the Antarctic, where small Notothenids dominate. The wider Arctic shelves support a greater diversity of Gadidae, Sebastidae, wolf-fish and halibut. Some fish, like most benthic invertebrates, avoid a planktonic larval stage to stabilize recruitment. Excess of production over consumption in the water column supports rich and diverse macrobenthos, especially in boreal regions, where shelf areas uncovered by ice are much more extensive than around Antarctica. Grey whale, walrus and bearded seal are boreal benthic feeders having no austral equivalents. The decline of these mammals and some birds can be traced to the abrupt decline of krill populations (de la Mare, 1997). The decline in krill may be related to regional warming and decreased frequency of winters with extensive sea-ice development (Loeb et. al, 1997). More extensive analysis of the region reveals that phytoplankton, krill, whales and seabirds are concentrated where sea-ice is maximal, salps are located where sea-ice was minimal, but enhanced biological activity occurs south of the southern boundary of the Antarctic Circumpolar Current. In essence, ocean circulation drives biological productivity off east Antarctica (Nicol et. al., 2000).

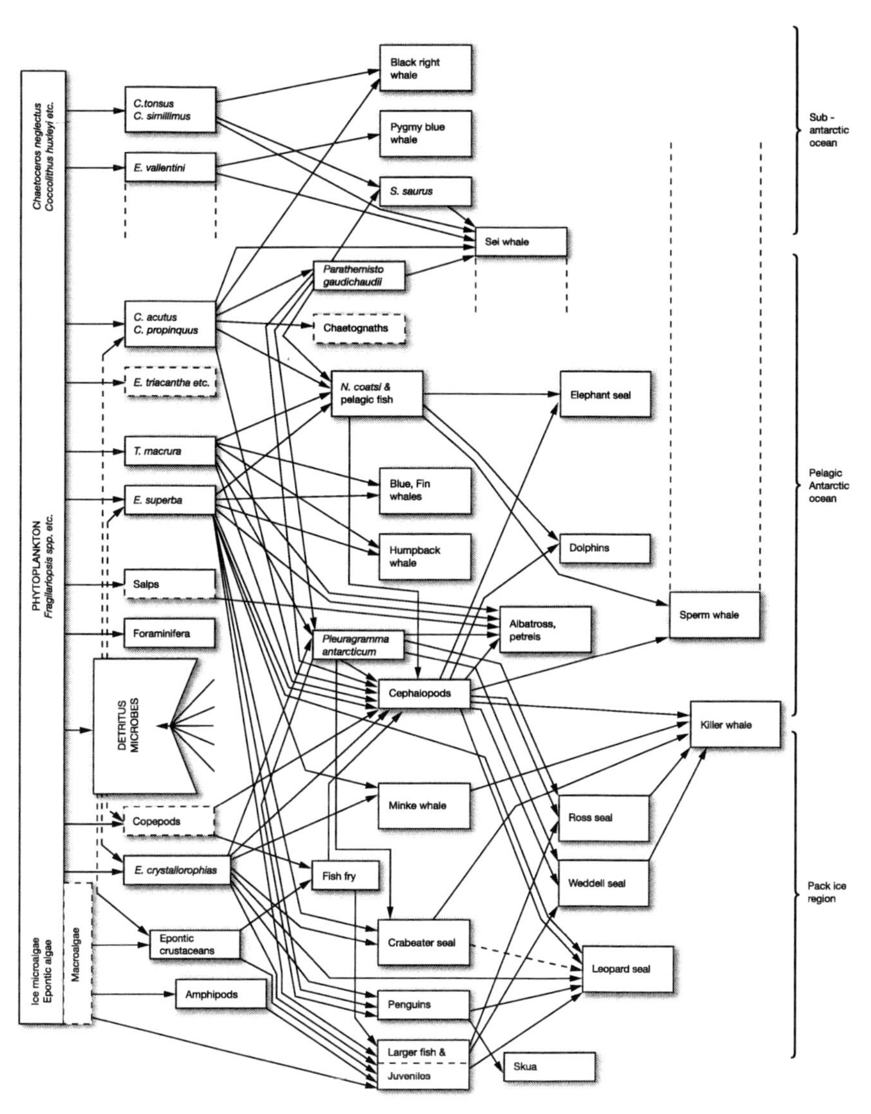


Figure 10.4 Conceptual model of the keystone players of Antarctic food webs. Modified from Nemoto and Harrison (1981).

In the Type 2 ecosystems, pack ice also plays an important role as much biological activity is associated with sea-ice systems, its internal matrix and brine channels, the sea ice surface and underside and waters in the vicinity that are modified by freezing and thawing of pack ice. Brierley and Thomas (2002) argue that pack ice functions as islands, foci for abundant and productive microbial and microalgal assemblages that attract a wide variety of protozoa and macrozooplanktonic consumers. Enhanced secondary production associated with sea ice is exploited by aggregations of many higher predators such as seals, seabirds and whales, and explains why the marginal ice zone was an intense region of pelagic whaling.

Arguably the most productive if not the most spectacular Type 2 ecosystem is found on the broad interlocking shelves of the northern Bering and Chukchi Seas, which are among the world's largest trawling grounds for pelagic and demersal fish and a rich nursery ground for whales, walrus, seals and seabirds (see results of the ISHTAR program, Coachman and Hansell, 1993). Essentially, the high biological productivity is sustained by complex spatial and temporal interactions of the Alaskan coastal (ACW), Bering Sea (BSW), and Anadyr (ADW) water masses that entrain upon this broad shelf. The ecosystem of the western shelf is fueled by the ADW. Production is high with rates of primary production exceeding 10 gC $m^{-2}$ $d^{-1}$ and an oceanic fauna is upwelled onto the shelf resulting in a east-west dichotomy of food webs. The passage and convergence of the ADW and ACW water masses at the Bering Strait induce a succession of productivity peaks in the Chukchi Sea. A cyclical pattern of ice conditions is superimposed on these physical forces. Consequently, secondary consumers such as zooplankters and macrobenthos are abundant and productive, fueled by the high rates of pelagic carbon fixation that result in high rates of sedimentation of labile carbon on the inner and middle shelf regions. At the outer shelf edge, the so-called "Green Belt", the food webs are predominantly pelagic in character (esp. *Neocalanus*), sustaining the pelagic fishery of the Alaskan pollock (*Theragra chalcogramma*) now the world's largest single species fishery.

More recent analyses of time series data from the southeastern Bering Sea suggests that energy flow within the walleye pollock fishery grounds is greatly influenced by changes in climate. Hunt and Stabeno (2002) reviewed changes in the variability in the persistence of sea ice and the timing of its retreat since the mid-1970's and found that control of pollock abundance switches from bottom-up limitation during periods of late ice retreat to top-down control in warmer periods when the ice retreats in mid-March. They found some evidence that the abundance of adult Pollock negatively affects the availability of forage fishes to top predators. This is due to the sensitivity of the growth and production of zooplankton and larval and juvenile fish to changes in water temperature.

### *5.3 Type 3 - Mid-latitude Coastal*

On some mid-latitude continental shelves— for example, from Norway to Iberia, from Newfoundland to Florida, and along the southern coast of Australia (Table 10.3) — after winter mixing, a vernal pulse of productivity and chlorophyll is induced when water-column stability is established. In summer when these shelf waters are stratified and nutrients are depleted, primary productivity is relatively low. Progressive mixing in autumn may induce renewed productivity, fuelled by nutrients accumulated below the summer pycnocline. Other processes significantly modify this sequence: intermittent wind-induced coastal convergence and divergence, and persistent mixing in regions where tidal velocities exceed the critical value. Effects of estuarine turbidity plumes may mask the canonical sequence, which is also generally weaker towards the equator. Temperate shelves constitute the best-studied marine ecosystems, so only a précis is offered here. The reader should consult the books of Cushing (1988), Walsh (1988), Mann and Lazier (1996), Alongi (1998) and Longhurst (1998) for more detailed treatments.

In temperate shelf waters the balance between pico-autotrophs and larger algal cells is more equitable than in very high latitudes, the latter dominating vernal blooms. In shoal water especially at higher latitudes (e.g., off Norway, Iceland, Newfoundland, Tasmania), there is significant autotrophic production in kelp beds. Small copepods dominate the inshore herbivorous plankton with a seasonal cycle of abundance following that of phytoplankton. Larger species occur near the shelf edge and often overwinter in deep water. In the North Sea and adjacent waters as elsewhere off heavy industrialized nations, there has been a trophic shift in pelagic food webs as a result of human-induced rises in nutrient concentrations (Reid et al., 1990; Radach, 1992). From 1962 to 1984, there has been a strong increase in phytoplankton biomass and a dominance shift from diatoms to flagellates. There has also been a change in the annual cycle. Phytoplankton growth normally centers around the spring diatom bloom which is followed by a flagellate bloom often persisting throughout the summer. A second diatom bloom occurs in some years in fall. More recently the onset of the spring bloom has occurred earlier (March instead of April) and flagellate biomass remains high year-round, with dinoflagellate blooms persisting from May to September. Phytoplankton production has increased incrementally since the 1950's.

Dramatic changes have also occurred in the rest of the pelagic community with marked declines in zooplankton, fish and seabird numbers and biomass since the first measurements early last century (Jackson et. al., 2001). These changes may not necessarily be due solely to eutrophication and other human disturbance such as overfishing (Boesch et. al., 2001). A series of analyses by Aebischer et al. (1990) has shown remarkable synchrony of long-term changes in abundance of phytoplankton, zooplankton, herring (*Clupea harengus*) and an open-ocean seabird (the kittiwake *Rissa tridactyla*) with the frequency of dominant westerly weather cells in the North Sea. Although the mechanism by which climatic changes might be coupled to marine populations is unclear, it is apparent that each trophic group is directly affected by the weather. There may be a simple regulatory effect up the food chain. It is equally likely that the reasons are more complicated. Climatic forcing of herring populations could be indirect, linked to changes in larval transport by shifts in wind-driven circulation or through climate-induced changes in vertical mixing or stratification. Regardless, the climate signal is strong for marine food webs of the northeastern Atlantic.

Much of the autotrophic production in temperate waters passes directly or indirectly to the benthos whose organisms are generally more abundant, larger and diverse than on tropical shelves. Macrofaunal activities and the sedimentation of organic matter stimulate both bacterial and protozoan abundance (Rhoads 1974). High densities of benthic nanoflagellates imply that protozoans exert some control over bacterial abundance and production in surface sediments. In deeper sediments where anoxic conditions limit metazoan life, benthic food webs are simple, composed of highly diverse suites of obligate and facultative anaerobic microbes that busily decompose organic matter and participate closely in biogeochemical cycles of elements (Canfield, 1993; Middelburg et. al., 1993).

Benthic food chains on temperate shelves contribute greatly to the sustainability of demersal fish stocks, such as plaice (*Pleuronectes platessa*) and other flatfish. Evidence for a trophic link comes directly from analysis of fish diets and also from the similarity in the large-scale patterns of abundance of macrobenthos and demersal fish. Fish fauna diversity is greater than in polar ecosystems, characteristically of around 200 species of >50 families. In boreal regions major stocks of shoaling Clupeidae and Scombridae occur together with demersal Gadidae, Percidae and Pleuronectidae. In much more restricted austral shelf regions, clupeids occur as in the north, together with a more-difficult-to-specify demersal fauna. Energy flow from pelagic invertebrates is mainly to clupeids and scombrids, and from benthos mainly to demersal fish fauna.

The trophic transfer of energy through food webs on shelves of this type is best illustrated from data obtained from the northeast shelf of North America, arguably the best-studied continental shelf ecosystem. A recent assessment of production estimates from this region (Sherman et. al., 1996) indicates clear differences between different sections of the shelf (Figure 10.5). In the Gulf of Maine and on the Nova Scotian shelf, primary production is lower than on Georges Bank and the Mid-Atlantic Bight. Secondary production is similar in all sections except the Mid-Atlantic Bight where macrobenthic production is high and macrozooplankton production is low. These results imply a lower trophic transfer efficiency of primary to secondary producers on the Georges Bank and Mid-Atlantic Bight. This does not however necessarily translate to, or explain, high maximum sustainable yield of fish and squid on Georges Bank. Some other factors may come into play, including changes in fisheries stocks in relation to prey, or differences in community structure and species composition or advective losses of pelagic prey by circulation patterns and shelf-slope exchange. As on other non-upwelling shelves, no one mechanism is crucial in supporting fish yields.

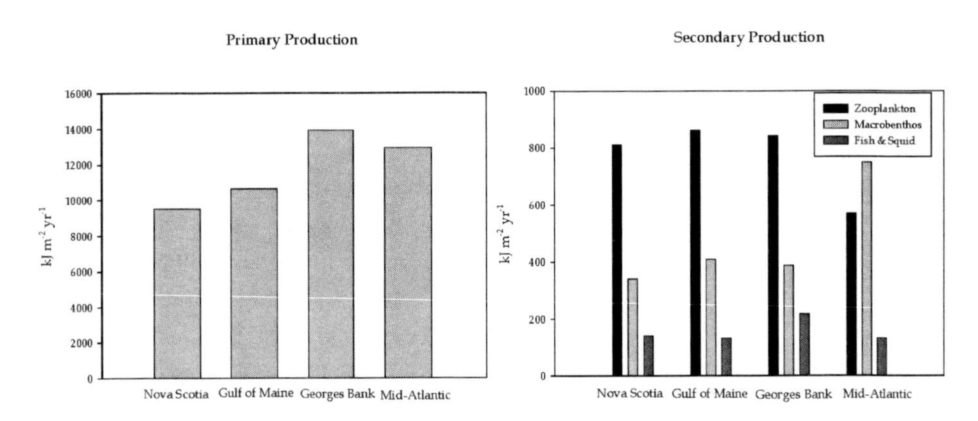


Figure 10.5 Differences in primary and secondary production along different sections of the Northeast Shelf of North America. Data taken from Sherman et. al. (1996).

Food web structure and the main pathways of carbon flux can differ greatly between bloom and non-bloom periods without significantly altering the total flow of carbon and energy. In a detailed study of pelagic food webs in the Gulf of St. Lawrence, Rivkin et. al. (1996) found that during the spring bloom, herbivorous zooplankton dominate, whereas in the post-bloom period the microbial loop dominates trophic structure. The pelagic food web thus shifts from net autotrophy to net heterotrophy, but despite this shift the composition but not the amount of material exported from the euphotic zone to the benthos is altered. During the bloom, bacteria are fueled via DOC derived from exudates of the large phytoplankton community and by sloppy feeding of protozooplankton and the herbivorous zooplankton. Most phytoplankton remains ungrazed and settles as phytodetritus to the seabed. In non-bloom periods, the mesozoooplankton community is omnivorous, feeding mostly on protozooplankton because autotrophs are scarce. Dissolved nutrients originating from within the microbial loop continue to drive bacterioplankton growth at a slower rate, but fecal pellets settle out of the eutrophic zone at a rate equivalent to the rain of phytodetritus. Thus, vertical flux is maintained but derived from very different sources. Structure and function of food webs on continental shelves may therefore not necessarily be directly related in any obvious way.

### *5.4 Type 4 - Topography-forced Coastal*

This type of shelf ecosystem constitutes a ‘special case’ of Type 3 where tidal mixing consistently dominates the stability of the shelf water mass. The principal areas are at relatively high latitude on the Falklands and southern New Zealand shelves, the shelf of the Gulf of Alaska, and to a lesser extent, in the southern North Sea (Table 10.3). It may also be appropriate where the oceanic permanent halocline of the temperate North Pacific passes inshore across the shelf, so constraining winter mixing to water above the pycnocline.

In these shelf regions the spring increase in phytoplankton production is coincident with increased light and the shoaling of the mixed layer. Moreover, primary production starts increasing about one month after the winter solstice and peaks in mid-summer through the continued entrainment of nutrients by the interaction between topography and circulation induced by high wind stress (Longhurst, 1995). In the southern hemisphere wind mixing is deep in winter, but constrained on the Alaskan shelf by the shallow halocline of the Alaska gyre. In each region, seasonal changes in phytoplankton production and consumption are approximately balanced year-round, being more closely coupled than in open ocean ecosystems at the same latitudes. River discharge onto the shelf at La Plata and Bahia Blanca in winter usually results in a phytoplankton biomass and production peak in June-July. The autotrophic and heterotrophic biota, and their energy flows, is broadly similar to those appropriate to Type 3 ecosystems, although comparatively little such data exists for the Falklands shelf (Longhurst, 1998).

In the southern North Sea, tidal fronts play a strong role in regulating benthic food webs and mineralization rates (Cramer, 1990; Upton et. al., 1993). The Frisian Front, a remarkable transition zone between sandy sediments of the Southern Bight and mud between the German Bight and Dogger Bank, is characterized by a very large benthic biomass. This enriched transition zone is thought to be the direct result of enhanced sediment due to a combination of a sharp decline in tidal current speeds and a steep bottom profile. Increased heat input in summer results in stratification of the water column in the northern section, and residual tidal currents produce a well-mixed water mass to the south, producing a tidal front. Benthic respiration is highest underneath the front immediately after the spring bloom supporting the idea that the deposition of fresh phytodetrtius fuels benthic production and consumption.

Tidal fronts and downwelling along the Gulf of Alaska shelf enhance biological activity, as in the other 'special case' regions. Water circulation in this province is complex, where numerous shelf edge and tidal fronts impinge on the permanent halocline of the Gulf of Alaska. In the Strait of Georgia along the western coast of Canada, the spring phytoplankton community is dominated by the diatoms *Thalassiosira* and *Skeletonema,* followed in summer by *Chaetoceros, Ditylum, Nitzschia* and *Leptocylindricus* and by *Cosinodiscus* in autumn (Harrison et. al., 1983). Zooplankton in spring is dominated by a large influx of larvae and early copepodite stages of *Neocalanus plumchrus* as well as by the ocean immigrants, *Calanus pacificus, C. marshallae* and *M. pacifica.* In summer when primary production usually peaks, predators abound: *Sagitta* and numerous species of amphipods. Further up the food chain it has been speculated that enhanced zooplankton production and biomass at the convergent front between the Alaska Coastal Current and inshore water is coincident with the migration run of juvenile salmon bound for offshore feeding ground to the north (Cooney, 1986; Longhurst, 1998).

The physics and biology of the waters off New Zealand are equally complex (Vincent et. al., 1991). In this region where mass island effects are evident, warm, saline subtropical waters from the north meet cooler, less salty waters from the south. There are many fronts, wakes and eddy streams around the north and south islands, but the principal feature is the Subtropical Convergence Zone, a permanent circumglobal front located to the east coast of New Zealand. Within this front, phytoplankton assemblages are dominated by diatoms especially *Lauderia annulata, Nitzschia* and *Pseudonitzschia* spp. Further to the north in subtropical waters, dinoflagellates dominate in winter and diatoms in spring and to the south, nanoflagellates dominate year-round in subantarctic waters (Hoe Chang and Gall, 1998). New Zealand waters are productive and fuel commercially important fisheries especially of the New Zealand hake or hoki (*Macruronus novaezelandiae).* It is known that hoki eggs are spawned during winter on two major spawning grounds, one off the west coast of the South Island and the other centered in Cook Strait between North and South Islands. The young hake feed initially on phytoplankton, tintinnids and copepod larvae, but as adults feed on copepod adults. It has been hypothesized that deep mixing within the various fronts along the New Zealand coast promotes the growth of zooplankton species important to the diet of hoki larvae.

The strength of hoki recruitment in both western and eastern waters has been linked to variations in the Southern Oscillation. Favorable recruitment in the western stock is correlated with a negative SOI, coincident with lower SSTs, early onset of deep mixing and greater southwest flow. Eastern stock recruitment is correlated with periods of westerly flow (Livingston, 2000).

In energy budget for the west coast of New Zealand indicates that most of the energy for benthic production is derived from phytodetritus, as rivers are a minor carbon source (Probert, 1986). For the west coast the relationship between pelagic and benthic production resembles that of upwelling ecosystems where very high phytoplankton production does not translate into high benthic production. This discrepancy may reflect a highly efficient pelagic food web that supports a substantial pelagic fishery.

### *5.5 Type 5 - Coastal Upwelling*

Five great coastal upwelling regions bordering eastern boundary currents, some at relatively low latitudes (Table 9.3), are known:

- The Canary Current off Northwest Africa
- The Benguela Current off Southwest Africa
- The Peru Current off Peru and Chile
- The Somali Current off Northeast Africa and Arabia
- The California Current off Oregon and California.

All of these systems intrude onto shelves that are relatively very narrow and where effects of river discharge are very minor. Coastal upwelling is a complex process driven mainly by trade winds inducing strong and persistent offshore Ekman drift away from the coast and replaced by the upwelling of cold, nutrient-rich deep water. Upwelling is usually strongest during summer as in the Indian Ocean when the southwest monsoon forces offshore drift, principally off Somalia. There are other forms of upwelling such as buoyancy-driven and tidally driven upwelling and intrusions driven by Western boundary current instability, but our discussion here is restricted to wind-driven upwelling. A large literature exists of upwelling ecosystems and so the reader is urged to consult two works on the topic (Summerhayes et. al. 1992, 1995).

Upwelling results (after a significant time-lag) in a rapid increase in primary production of phytoplankton, principally diatoms, and chlorophyll accumulation coincides with the duration of upwelling periods. Biota are characteristically of low diversity and high biomass. Pelagic grazers are overwhelmed by the increase in diatom biomass, leading to excess production sedimenting to the sea floor. Smaller algal cells have their appropriate protistan and invertebrate consumers. Specialized invertebrate herbivores are usually large calanoid copepods (typically *Calanus* or *Calanoides*), euphausiids and filter-feeding anomuran crabs, each having life history tactics that take them to deep water or to the shallow sea-floor during non-upwelling periods. Vertebrate herbivores are often anchovies (*Engraulis, Cetengraulis*) and in the Indian Ocean also oil sardines (*Sardinella longiceps*). Appropriate populations of predators – mostly pelagic clupeids (*Sardina, Sardinops*), mackerel (*Scomber*), hake (*Merluccius, Micromesistius*), sea lions and birds, are characteristic of these regions.

A new view has emerged of coastal upwelling systems as sites in which tongues or filaments of upwelled and downwelled water masses intermingle and extend out to hundreds of kilometers from the shelf edge (Chavez et. al., 1991). In the coastal transition zone of northern California, highest concentrations of plankton occur as narrow filaments in response to jet-induced upwelling. Such processes rather than the classical view of nutrient-laden cold water being advected along the shelf edge may be responsible for the high productivity. Food web processes within these ecosystems are still poorly understood. For instance, it is not clear whether the elevated algal biomass commonly observed is the result of enhanced production or the net effect of entrainment and aggregation of phytoplankton induced by converging water masses. Temporal and spatial dynamics are also poorly known.

Despite these uncertainties, most coastal upwelling ecosystems exhibit some common characteristics:

- Closed circulation cells
- Strong lateral and benthic advection
- High carbon deposition and burial
- Low rates of carbon recycling
- Large stocks of apex predators
- Low biodiversity

Energy flow analysis of the Peruvian and Benguela upwelling systems suggest that they are several times more efficient in producing fish compared with non-upwelling coastal ecosystems. Baird et. al. (1991) ran models of energy flow for these two eastern boundary systems and found that the higher yields of planktivorous (in the Benguela system) and carnivorous (in the Peruvian system) fish may be due to higher production-to-biomass ratios and shorter food chains, as lower transfer of energy and materials would result in more carbon available to sustain high fish production. Fish yields are very variable in these ecosystems due to variations in climate and other oceanographic factors.

It is still not clear how upwelled water impinges on shallow, inshore shelf areas. In at least one system, cycles of primary and secondary producers appear to be uncoupled. In the inner shelf off central Chile, there is both temporal and spatial separation of pelagic primary producers and secondary consumers, controlled by distinct upwelling phases, each eliciting a different response (Peterson et. al. 1988). Newly upwelled water is advected into the coastal zone resulting in seaward transport of plankton to be replaced by different communities transported laterally. As upwelling wanes, shearing of currents separates this new assemblage from offshore assemblages advected inshore from deeper water.

Off the Gulf of Guinea, upwelling events are climatically forced by oscillations of the inter-tropical convergence zone (ITCZ), the doldrums area separating northern and southern tradewinds at the equator. This upwelling sustains a pelagic fishery, the most abundant species being *Sardinella aurita, S. maderensis, Engraulis encrasicolus, Brachydeuterus auritus* and rarely *Scomber japonicus.* The collapse and rapid recovery of this fishery in the late 1970s and early 1980s was probably not due to overfishing but to changes in equatorial climate (Binet and Marchal, 1993).

Upwelling can also be triggered by the onset of monsoons as in the northern Arabian Sea where equatorial, open-ocean and coastal upwelling events co-occur. During the southwest summer monsoon, both coastal and open ocean upwellings driven by a combination of atmospheric, oceanic, and continental winds, intrude cold, nutrient-laden water onto and along the shelf bordering the northern Arabian Sea. This event promotes an intense oxygen minimum zone between 200–1500m between Arabia and India that is formed by microbial decay of high phytoplankton production induced by the upwelling, and facilitated by deep Indian Ocean water of low oxygen content. These monsoon-driven events have a great impact of pelagic food webs in coastal waters bordering the Arabian Sea. High rates of primary production decline eastwards until discharge from the Indus and smaller rivers along the west coast of India facilitate an increase in pelagic and benthic production (Qasim and Wafar, 1990).

Cold upwelled water penetrating into rivers, bays and estuaries of very narrow coasts can enhance biological productivity and fish and shellfish production. The best-studied example of this type of coastal intrusion is the Finisterre coastal upwelling system off Galicia, northwest Spain, where intrusions of North Atlantic Central water into the interior of the rias stimulates high rates of primary production which supports intensive raft cultivation of the mussel *Mytilus edulis* (Tenore et. al., 1982). Consequently, detritus derived from the mussel rafts outwells to the adjacent shelf, contributing to the enrichment of the benthic regime. An important demersal fishery for hake (*Merluccinus merluccius*), blue whiting (*Micomesistius poutassou*) and Norwegian lobster (*Nephrops norvegicus*) are sustained by these enriched benthic communities (Lopez-Jamar et. al., 1992).

### *5.6 Type 6 – Wet Tropical Coastal*

These are the wet tropical coasts (Table 10.3), dominated by inputs of a few major rivers (e.g., Amazon, Niger, Indus, Ganges, Mekong) or cumulative inputs of many smaller rivers (e.g. Papua New Guinea, Indonesia). On these river-dominated shelves, trade wind regimes force only weak seasonality in the mixed layer depth, the minor changes that are observed representing geostrophic response of the pycnocline to seasonality in trade wind stress, rather than mixing. Coriolis forces are small. River discharges into the low salinity surface layer have strong seasonality, reflecting regimes of wet and dry seasons. Consequently, the seasonal schedule of primary production is principally forced by nutrient input from land and perhaps by reduced irradiance due to very heavy cloud cover during the wet season and heavy suspended loads year-round. Extensive "estuarization" or extensive buoyant plumes on the entire inner continental shelf occurs in many regions (West Africa, NE Indian Ocean, SE Asia) as a consequence of run-off from land.

There are two general patterns of phytoplankton production in relation to salinity on these wet tropical shelves (Figure. 10.6): algal blooms occur either within plumes as off the Amazon and Fly Rivers, or further offshore as in the Changjiang and Zaire systems. The difference in the location of the algal bloom is best related to shelf topography. A shoal area at or near the mouths of the Amazon and Fly Rivers results in rapid sedimentation and lower turbidity permitting rapid phytoplankton growth inshore. Off the Changjiang and Zaire Rivers, sediments remain in suspension at much higher salinities owing to a deeper water column on their respective shelves. Despite these differences, the relative magnitude of primary production is similar among rivers (Figure. 10.6).

Rates of heterotrophic respiration and bacterioplankton production in tropical estuarine waters are not well-known but presumably are high, considering the relatively high nutrient concentrations and rates of fixed carbon production. Off the Fly River in the Gulf of Papua, rates of pelagic respiration (0.8–36 gC $m^{-2}$ $d^{-1}$) far exceed rates of primary production. Zooplankton biomass is very variable although peak densities coincide with peak phytoplankton biomass. Bacterial biomass is low implying intense grazing within the microbial loop. Pelagic food chains are fueled primarily by detritus outwelled from the extensive freshwater wetlands and mangroves (Robertson et. al., 1998). Similarly close associations between bacteria and phytoplankton have been found off the Changjiang, but it remains to be seen if pelagic food chains dominated by microbes and small phytoplankters as off the Fly River are characteristic of other tropical river-dominated ecosystems.

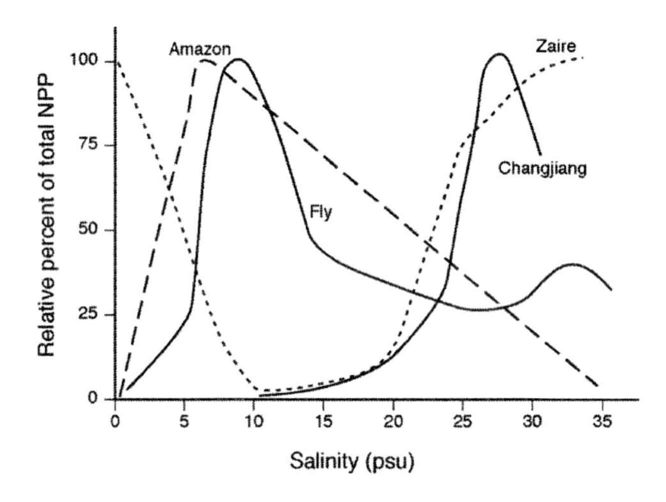


Figure 10.6 Patterns of phytoplankton production in relation to salinity in the plumes of some of the world's major tropical rivers (Adapted from Alongi, 1998).

Episodes of physical disturbance, erratic food supply and dilution of river-derived, particulate organic matter, foster the development of opportunistic benthic communities of variable diversity and low biomass, dominated by bacteria (Table 10.4). Densities of meiofauna and macroinfauna are very low. Some delta areas are devoid of metazoan benthos (Alongi and Robertson, 1995; Aller and Stupakoff, 1996). Off the Fly delta, for example, bacteria account for 75% of total biomass and at least 80% of benthic production (Table 10.4). Faunal abundance and productivity are highest under zones of high primary production.

Within these deltas, benthic food chains are short. The pioneering seres are the main food for penaeid shrimp which dominate the demersal trawl fisheries in these tropical river deltas. Alongi and Robertson (1995) hypothesized that the main trophic pathway off these deltas is

$$\text{Detritus} + \text{bacteria} \rightarrow \text{shrimp}.$$

Predation is not a major factor regulating tropical delta benthos because of (1) the absence or low abundance of large, bottom-feeding fish or other obvious major predators, and (2) most epibenthos is small composed mostly of sponges, crinoids, hydroids and bryozoans. Off the Fly delta, demersal nekton abundance is low, dominated by small fish such as the leatherjacket *Paramonacanthus filicauda,* the pony fish, *Leiognathus splendens* and the grunter, *Pomodasys argyreus* (Alongi et. al., 1992). Fisheries information for the Amazon, Fly and Changjiang shelves indicates a coincidence of major shrimp and finfish (off the Changjiang) catch over the most bioturbated sedimentary facies, where benthic production and biomass are maximal.

TABLE 10.4
Estimates of mean biomass and secondary production of benthic size classes in the Gulf of Papua, northern Coral Sea.
(Adapted from Alongi and Robertson, 1995).

| | Biomass (gC $m^{-2}$) | Production (gC $m^{-2}$ $yr^{-1}$) |
|---|---|---|
| Bacteria | 6.5 | 40.0 |
| Meiofauna | 0.3 | 4.5 |
| Macroinfauna | 0.7 | 2.1 |
| Epifauna | 0.9 | 1.8 |

It is clear that the major regulatory factor of benthos is massive physical disturbance of the delta seabed. These disturbances are periodic, probably related to the wax and wane of tides, waves and rainfall. Shear stresses associated with these physical factors resuspend sediment frequently and cause the greatest bedload transport while during quiescent phases, sediment is trapped inshore until the next disturbance cycle (Kuehl et. al., 1996). Analysis of benthic communities off the Amazon has revealed further evidence of periodic physical disturbance. Aller (1995) found evidence of cycles of surface exposure and burial of invertebrate death assemblages, mostly mollusks, out to a water depth of 40 m. She also found signs of large-scale, onshore transport of shells and fragments of relict coral and bryozoans from the outer shelf, followed by deposition and downward mixing in inshore deposits.

Such disturbance events allow the biogeochemical development of extensive vertical and lateral regions of iron and manganese cycling (Aller et. al., 1996). This is in contrast to most temperate shelf sediments where sulfate reduction dominates early diagenesis. Despite massive dilution by terrigenous debris, the sediment carbon cycle in these tropical deltas is driven by carbon fixed by marine plankton, and rates of benthic carbon flow are comparable to other eutrophic shelf environments. Benthic mineralization is most rapid during periods of low or falling river discharge. Roughly 90% of the carbon remineralized in Amazon shelf sediments escapes to the overlying water. The remainder is buried, mostly in the form of carbonate minerals. Data from the Amazon, Changjiang and Papuan shelves indicate that the spatial patterns of rapid decomposition mirror plankton blooms, suggesting close pelagic-benthic coupling. Tropical river-dominated shelves of this type appear to violate the current paradigm of estuarine systems being highly efficient traps for inorganic and organic matter with little, if any, material available for export to the adjacent coastal or open ocean.

### 5.7 Type 7 - Dry Tropical Coastal

Type 7 shelves occur along the coasts of the dry tropics where river runoff is insignificant. It is for these areas that arguably the least ecological information exists. These areas include the Caribbean, parts of the Arabian Sea, the Red Sea, northwest and parts of northeast Australia, including the southern Great Barrier Reef, and southern coasts of the Indonesian archipelago (Table 10.3). Many isolated islands and archipelagos in tropical seas are surrounded by Type 7 ecosystems, of which the dominant characteristic is the development of coral reefs (where topography permits). Elsewhere there are unconsolidated sediments, dominated by pure and impure carbonate sand. The reader is urged to consult the two exhaustive volumes on dry coastal ecosystems edited by van der Maarel (1993a, 1993b) for more detailed information.

There is weak seasonality in mixed layer depth in these dry tropical ecosystems, and the nutricline is usually shoaler than the photic depth except during exceptional events such as hurricanes and cyclones. Autotrophy is exceptionally partitioned among benthic and pelagic organisms, in many instances, such in reefal areas, most primary production occurs on hard or soft bottom where water clarity permits. Macroalgae (*Sargassum*), encrusting coralline green (*Halimeda*), red algae, cyanobacterial mats and seagrass meadows dominate community production. The activity of symbiotic dinoflagellates within the tissues of many invertebrates such as scleractinian corals, giant clams (*Tridacna*), coelenterates (alcyonians, anthozoans and scyphozoans), large ascidians and encrusting sponges, are prominent.

The phytoplankton community is dominated by the pico- and nano-fractions as in oligotrophic ocean ecosystems. These are consumed by protists and small zooplankters themselves the prey of many filter- and tentacular-feeding polyps of corals and other coelenterates. In the Caribbean, oligotrophic profiles of chlorophyll, phytoplankton and zooplankton biomass are prominent. The vertical distribution of the > 150-μm zooplankton is dominated by warm-water copepods (*Clausocalanus, Euchaeta, Scolecithrix,* and *Nannocalanus*) and diel migrant genera such as *Pleuromamma* and *Sergestes.* These latter genera migrate between 300–400m and the photic zone (Hopkins, 1982).

Nutrient sources and fluxes are many and highly variable: advection, upwelling, vertical flux in fractured basement rocks, some terrestrial runoff, etc. In the Gulf of Mexico and on the Great Barrier Reef shelf, complex topography and regional-scale divergences and countercurrents induce upwelling at the continental margin (Walsh et. al., 1989; Brinkman et. al., 2002). Most dry coastal shelf areas are topographically complex mainly due to reefs and relict carbonate platforms. These features are often the site of enhanced biological productivity in part due to the formation of complex wakes and eddies, that break down strong and permanent gradients in water temperature and density. For example, off Australia's arid northwest shelf, pelagic microbial production is rapid with mean phytoplankton and bacterioplankton production rates of 0.75 and 0.31 g C $m^{-2}$ $d^{-1}$, respectively (Furnas and Mitchell, 1999). The sources of nutrients fueling this rapid productivity appears to be tidal resuspension of predominantly carbonate sediments and *in situ* formation of particles. Further inshore, flooding events during cyclones probably play an important role, with bays and gulfs functioning in the dry season as ecosystems where new nitrate is more important then recycled nutrients for biological productivity (McKinnon and Ayukai, 1996). In other areas such as in the Red Sea, the strong pycnocline is a permanent barrier to subsurface mixing of nutrients, resulting in low primary productivity. Even offshore Red Sea water has extremely low ( $\approx$ 100 mgC $m^{-2}$ $d^{-1}$) phytoplankton production (Gradinger et. al., 1992) with the 0.2–2.0 μm size fraction accounting for 75% of the productivity.

A significant source of nitrogen in Type 7 ecosystems is the nitrogen-fixing bacteria that occur in the tissues of some corals, and the cyanobacteria that fix nitrogen within algal mats and on the water surface as blooms. Internal exchanges of nutrients within and between organisms are highly complex, especially on coral reefs and hard bottom areas. In the Red Sea, blooms of *Oscillatoria thraeum* are frequent in the open areas although the significance of their nitrogen-fixing abilities is not well known. This is true for other dry tropical shelves where *Oscillatoria* blooms occur often, especially in calm, hot conditions (Longhurst, 1998).

Export from the benthic ecosystem, except in the form of carbonate eroded to sand and gravel factions, is a small but complex flux. The macrobenthos associated with coral reef formations is exceptionally diverse at all taxonomic levels. On open sandy sediments unencumbered with reefs, very high densities of filter-feeding crabs (*Pinnixa, Xenophthalmus*) are typical, together with filter-feeding bivalves. In general it appears that the benthic ecosystem on dry shelves is dominated by epifauna, especially on rocky or consolidated bottom, as little food can be derived from unconsolidated sediments of low organic content as compared with particles in the water column (Alongi, 1990).

Fish fauna is also diverse, both taxonomically and functionally. Parrotfish (Scaridae) are among the most important herbivores, directly consuming coralline and other algal mats, and these may form a large fraction of total fish biomass in coral-dominated waters. An intense and complex network of trophic links between fish and epibenthic invertebrates is characteristic of this ecosystem. This trophic complex supports a wide variety of large demersal and pelagic predators, especially tunas, billfish, sharks, skates and rays (Longhurst and Pauly, 1987).

## 6. Conclusions

Greater certainty of the role of coastal ocean ecosystems in global mass balances can be achieved only after two principal problems have been solved. First, better measurements are needed of the carbon inputs and losses onto tropical shelves, especially the flux of river-borne terrestrial organic material and *in situ* carbon fixation and pelagic respiration. These fluxes, often intermittent and large during the wet season, are very difficult to quantify even with modern techniques. Export routes to the adjacent deep ocean are poorly understood and are not simple to map, though some of the major fluxes are known to pass down the canyons that are often associated with the mouths of major rivers, to form the deep-sea cones or fans of muddy sediments; the Ganges and Niger fans are the examplars. To what extent transport parallel to the coast from other regions contributes to these deposits is also unquantified.

The final problem is the lack of observational data on regional carbon fluxes in many important regions. Most modern investigations have been done in temperate, upwelling and boreal regions. For most of the world's shelf ecosystems we must currently rely on what may be derived from models of 'typical' ecosystem types such as those presented here. As an example of the extent of current uncertainty, south Asian rivers alone carry more organic material to the ocean than all the rivers of the Americas and Africa combined, yet we still known exceedingly little about them (Milliman and Syvitski, 1992).

Some final comments, nevertheless, are warranted:

1. The upwelling coasts in eastern boundary currents (Type 5 ecosystems) are relatively more significant exporters of organic material than other shelves at similar latitude (e.g., Type 3 or 4 ecosystems). It is here that persistent offshore Ekman drift will, more effectively than elsewhere, transport suspended organic material out beyond the shelf edge and, for another, here there tends to be little cross-shelf transport of terrestrial material;
2. That microbes process the bulk of organic material in the coastal ocean is clear, but understanding how they vary structurally and functionally to regional changes in climate and water motion, and relate to higher trophic groups, especially in the benthos, is still in its infancy;
3. How much organic carbon and nitrogen gets buried, consumed and exported from continental shelves is known for some regions but the factors regulating the degree of consumption and burial require further attention; and
4. The poorly studied Type 6 ecosystems of the wet tropics make the greatest contribution to carbon flux of the global coastal ocean. How these immense fluxes balance against, for instance, the production of carbonates in Type 7 ecosystems remains a challenge for the future, as does the global impact of climate change on the ecosystems of the world's coastal ocean.

## Acknowledgments

I thank Alan Longhurst for his help in earlier stages of production of this chapter and for being an inspiration to me and many other marine ecologists for many years. Much of the discussion on ecosystem types is based on Alan's latter books and journal publications, although any errors or omissions are mine.

## Bibliography

Aebischer, N.J., J.C. Coulson and J.M. Colebrook, 1990. Parallel long-term trends across four marine trophic levels and weather. *Nature* **347,** 753–755.

Aller, J.Y., 1995. Molluscan death assemblages on the Amazon shelf: implications for physical and biological controls on benthic populations. *Palaeogeogr. Palaeoclim. Palaeoecol.* **118,** 181–212.

Aller, J.Y. and I. Stupakoff, 1996. The distribution and seasonal characteristics of benthic communities on the Amazon shelf as indicators of physical processes. *Cont. Shelf Res.* **16:** 717–751.

Aller, R.C., N.E. Blair, Q. Xia and P.D. Rude, 1996. Remineralization rates, recycling, and storage of carbon in Amazon shelf sediments. *Cont. Shelf Res.* **16,** 753–786.

Alongi, D.M., 1990. The ecology of tropical soft-bottom benthic ecosystems. *Oceanogr. Mar. Biol. Annu. Rev.* **28,** 381–496.

Alongi, D.M., 1998. *Coastal Ecosystem Processes.* CRC Press, Boca Raton, Fl., 419 pp.

Alongi, D.M. and A.I. Robertson, 1995. Factors regulating benthic food chains in tropical river deltas and adjacent shelf areas. *Geo-Mar. Lett.* **15**:145–152.

Alongi, D.M., P. Christoffersen, F. Tirendi and A.I. Robertson, 1992. The influence of freshwater and material export on sedimentary facies and benthic processes within the Fly delta and adjacent Gulf of Papua. *Continental Shelf Res.* **12,** 287–326.

Azam, F., T. Fenchel, J.G. Field, J. S. Gray, L.A. Meyer-Reil and F. Thingstad. 1983. The ecological role of water-column microbes in the sea. *Mar. Ecol. Prog. Ser.* **10:** 257–263.

Baird, D., J.M. McGlade and R.E. Ulanowicz, 1991. The comparative ecology of six marine ecosystems. *Phil. Trans. Roy. Soc. Lond B* **333,** 15–29.

Binet, D. and E. Marchal, 1993. The large marine ecosystem of shelf areas in the Gulf of Guinea: long-term variability induced by climatic changes. In *Large Marine Ecosystems: Stress, Mitigation and Sustainability,* K. Sherman, L.M. Alexander and B.D. Gold, eds. Am Soc. Adv. Sci. Press,Washington D.C., pp. 104–118.

Boesch, D.F., E. Burreson, W. Dennison, E. Houde, M. Kemp, V. Kennedy, R. Nedweel, K. Paynter, R. Orth and R. Ulanowicz, 2001. Factors in the decline of coastal ecosystems. *Science* **293,** 1589–1590.

Brierley, A.S. and D.N. Thomas, 2002. Ecology of Southern Ocean pack ice. Adv. Mar. Biol. **43,** 171–276.

Brinkman, R., E. Wolanski, E. Deleersnijder, F. McAllister and W. Skirving, 2002. Oceanic inflow from the Coral Sea into the Great Barrier Reef. *Estuar. Cstl. Shelf Sci.* **54:** 655–668.

Brunskill, G.J., I. Zagorskis and J. Pfitzner, 2002. Carbon burial rates in sediments and a carbon mass balance for the Herbert River region of the Great Barrier Reef continental shelf, North Queensland, Australia. *Estuar. Coast. Shelf Sci.* **54,** 677–700.

Canfield, D.E., 1993. Organic matter oxidation in marine sediments. In *Interactions of C, N, P and S Biogeochemical Cycles,* R. Wollast, F. Mackenzie and L. Chou, eds. Springer, Berlin, pp. 333–363.

Chavez, F.P., R.T. Barber, P.M. Kosro, A. Huyer, S.R. Ramp, T. P. Stanton and R. de Mendiola, 1991. Horizontal transport and the distribution of nutrients in the coastal transition zone off Northern California: effects on primary production, phytoplankton biomass and species composition. *J. Geophys. Res.* **96,** 14,833–14,848.

Coachman, L.K. and D.A. Hansell, 1993. ISHTAR-Inner shelf transfer and recycling in the Bering and Chukchi Seas. *Cont. Shelf Res.,* **13,** 473–704.

Cooney, R.T., 1986. The seasonal occurrence of *Neocalanus cristatus, N. plumchrus* and *Eucalanus bungii* over the shelf of the Gulf of Alaska. *Cont. Shelf Res.,* **5,** 541–553.

Cramer, A., 1990. Seasonal variation in benthic metabolic activity in a frontal system in the North Sea. In *Trophic Relationships in the Marine Environment,* M. Barnes and R.N. Gibson, eds. Aberdeen University Press, Aberdeen, pp. 54–76.

Cushing, D.H., 1988. The flow of energy in marine ecosystems, with special reference to the continental shelf. In *Ecosystems of the World, Vol. 27, Continental Shelves,* H. Postma and J.J. Zijlstra, eds. Elsevier, Amsterdam, pp. 203–230.

de Haas, H., T.C.E. van Weering and H. de Stigter, 2002. Organic carbon in shelf seas: sinks or sources, processes and products. *Contin. Shelf Res.* **22,** 691–717.

de la Mare, W.K., 1997. Abrupt mid-twentieth century decline in Antarctic sea ice extent from whaling records. *Nature* **389,** 57–60.

Duarte, C.M. and J. Cebrian, 1996. The fate of marine autotrophic production. *Limnol. Oceanogr.* **41,** 1758–1766.

Fenchel, T., 1988. Marine plankton food chains. *Ann. Rev. Ecol. Syst.* **19:** 19–38.

Furnas, M.J. and A.W. Mitchell, 1999. Wintertime carbon and nitrogen fluxes on Australia's northwest shelf. *Estuar. Coastal Shelf Sci.,* **49,** *165–175.*

Gattuso, J.P., M. Frankignoulle and R. Wollast, 1998. Carbon and carbonate metabolism in coastal aquatic ecosystems. *Annu. Rev. Ecol. Syst.* **29,** 405–434.

Gosselin, M., M. Levasseur, P.A. Wheeler, R.A. Horner and B. C. Booth, 1997. New measurements of phytoplankton and ice algal production in the Artic Ocean. *Deep-Sea Res. II* **44,** 1623–1644.

Gradinger, R., C. Friedrich and M. Spindler, 1999. Abundance, biomass and composition of the sea ice biota of the Greenland Sea pack ice. *Deep-Sea Res.II,* **46,** 1457–1472.

Gradinger, R., T. Weiss and T. Pillen, 1992. The significance of picoplankton in the Red Sea and Gulf of Aden. *Bot. Mar.* **35,** 245–250.

Hardy, A.C., 1924. The herring in relation to its animate environment. I. The food and feeding habits of the herring with special reference to the east coast of England. *Fish. Invest., Lond.,* ser. 2, vol. **7(3),** 1–53.

Harrison, P.J., J.D. Fulton, F.J.R. Taylor and T.R. Parsons, 1983. Review of the biological oceanography of the Straits of Georgia. *Can. J. Fish. Mar. Sci.* **40,** 1064–1094.

Hayes, M.O., 1967. Relationships between coastal climate and bottom sediment type of the inner continental shelf. *Mar. Geol.,* **5,** 111–132.

Hedges, J.I. and R.G. Keil, 1995. Sedimentary organic preservation: an assessment and speculative synthesis. *Mar. Chem.* **49,** 81–115.

Hempel, G., 1985. Antarctic marine food webs. In *Antarctic Nutrient Cycles and Food Webs,* W.R. Siegfried, P.R. Condy and R.M. Laws, eds. Springer-Verlag, Berlin, pp. 266–270.

Hind, A., S.C. Gurney, M. Heath and A.D. Bryant, 2000. Overwintering strategies in *Calanus finmarchicus. Mar. Ecol. Prog. Ser.,* **193,** 95–207.

Hoe Chang, F. and M. Gall, 1998. Phytoplankton assemblages and photosynthetic pigments during winter and spring in the Subtropical Convergence region near New Zealand. *N.Z. J. Mar. FreshwaterRes.* **32,** 515–530.

Hopkins, T.L., 1982. The vertical distribution of zooplankton in the Gulf of Mexico. *Deep-Sea Res.,* **29,** 1069–1083.

Hunt, G.L. and P.J. Stabeno, 2002. Climate change and the control of energy flow in the southeastern Bering Sea. *Prog. Oceanogr.* **55,** 5–22.

Ishii, M., H.Y. Inoue and H. Matsueda, 2002. Net community production in the marginal ice zone and its importance for the variability of the oceanic $pCO_2$ in the Southern Ocean south of Australia. *Deep-Sea Res. II* **49,** 1691–1706.

Jackson, J.B.C., M.X. Kirby, W.H. Berger, K.A. Bjorndal, L.W. Botsford, B.J. Bourque, R.H. Bradbury, R.Cooke, J. Erlandson, J.A. Estes, T.P. Hughes, S. Kidwell, C.B. Lange, H.S. Lenihan, J.M. Pandolfi, C.H. Peterson, R.S. Steneck, M.J. Tegner and and R.R. Warner, 2001. Historical overfishing and the recent collapse of coastal ecosystems. *Science,* **293,** 629–638.

Kolber, Z.S., F.G. Plumley, A.S. Lang, J.T. Beatty, R.E. Blankenship, C.L. Van Dover, C. Vetriani, M. Koblizek, C. Rathgeber and P.G. Falkowski, 2001. Contribution of aerobic photoheterotrophic bacteria to the carbon cycle in the ocean. *Science* **292,** 2492–2495.

Kuehl, S. A., C.A. Nittrouer, M.A. Allison, L.E.C. Faria, D.A. Dukat, J.M. Jaeger, T.D. Pacioni, A.G. Figueiredo and E.C. Underkoffler, 1996. Sediment deposition, accumulation, and seabed dynamics in an energetic fine-grained coastal environment. *Cont. Shelf Res.* **16,** 787–816.

Liu, K.K., K. Iseki and S.Y. Chao, 2000. Continental margin carbon fluxes. In *The Changing Ocean Carbon Cycle: A Midterm Synthesis of the Joint Global Ocean Flux Study,* R.B. Hanson, H.W. Ducklow and J.G. Field. Cambridge University Press, Cambridge, pp. 187–239.

Livingston, M. 2000. Links between climate variation and the year class strength of New Zealand hoki (*Macruronus novaezelandiae*) Hector. *N.Z. J. Mar. Freshwater Res.* **34,** 55–69.

Loeb, V., V. Siegel, O. Holm-Hansen, R. Hewitt, W. Fraser, W. Trivelpiece and S. Trivelpiece, 1997. Effects of sea-ice extent and krill or salp dominance on the Antarctic food web. *Nature* **387,** 897–900. Longhurst, A.R., 1995. Seasonal cycles of pelagic production and consumption. *Prog. Oceanog.* **36,** 77–167.

Longhurst, A. R., 1998. *Ecological Geography of the Sea.* Academic Press, San Diego, Ca., 398 pp.

Longhurst, A.R. and D. Pauly, 1987. *Ecology of Tropical Oceans.* Academic Press, San Diego, Ca., 407 pp.

Lopez-Jumar, E., R.M. Cal, G. Gonzalez, R.B. Hanson, J. Rey, G. Santiago and K.R. Tenore, 1992.

Upwelling and outwelling effects on the benthic regime of the continental shelf off Galicia, northwest Spain. *J. Mar. Res.* **50,** 465–488.

Mann, K.H., 1993. Physical oceanography, food chains, and fish stocks: a review. *ICES J. Mar. Sci.* **50,** 105–119.

Mann, K.H. and J.R. N. Lazier, 1996. *Dynamics of Marine Ecosystems-Biological-Physical Interactions in the Oceans.* Blackwell Scientific Publications, Boston, 495 pp.

Mare, M.F., 1942. A study of a marine benthic community with special reference to the micro-organisms. *J. Mar. Biol. Assoc. U.K,* **25,** 517–554.

McKinnon, A.D. and T. Ayukai, 1996. Copepod egg production and food resources in Exmouth Gulf, Western Australia. *Mar. Freshw. Res.* **47,** 595–603.

Middelburg, J.J., T. Vlug and F.J. W. A. van der Nat, 1993. Organic matter mineralization in marine systems. *Global Planetary Change* **8,** 47–58.

Milliman, J.D., 1993. Production and accumulation of calcium carbonate in the ocean: budget of non-steady-state. *Global Biogeochem. Cycles,* **7** (4), 927–957.

Milliman, J. D. and A.W. Droxler, 1996. Neritic and pelagic carbonate sedimentation in the marine environment: ignorance is not bliss. *Geol. Rundsch.* **85,** 495–503.

Milliman, J.D. and J.P.M. Syvitski, 1992. Geomorphic/tectonic control of sediment discharge to the ocean: The importance of small mountainous rivers. *J. Geol.* **100,** 325–344.

Mills, E.L., 1989. *Biological Oceanography-An Early History, 1870–1960.* Cornell University Press, Ithaca, NY, 378 pp.

Nemoto, T. and G. Harrison, 1981. High latitude ecosystems. In *Analysis of Marine Ecosystems,* A.R.

Longhurst, ed. Academic Press, New York, pp. 95–126.

Nicol, S., T. Pauly, N.L. Bindoff, S. Wright, D. Thiele, G.W. Hosie, P.G. Strutton and E. Woehler, 2000. Ocean circulation off east Antarctica affects ecosystem structure and sea-ice extent. *Nature* **406,** 504–507.

Nixon, S.W., 1988. Physical energy inputs and the comparative ecology of lake and marine ecosystems. *Limnol. Oceanogr.* **33,** 1005–1025.

Pakhomov, E.A., P.W. Froneman and R. Perissinotto, 2002. Salp/krill interactions in the Southern Ocean: spatial segregation and implications for carbon flux. *Deep-Sea Res. II* **49,** 1881–1907.

Pauly, D. and V. Christensen, 1995. Primary production required to sustain global fisheries. *Nature* **374,** 255–257.

Petersen, C.G.J., 1918. The sea bottom and its production of fish food. *Rep. Danish Biol. Stat.,* **53:** 1–62.

Peterson, W.T., D.F. Arcos, G.B. McManus, H. Dam, D. Bellantoni, T. Johnson and P. Tiselius, 1988. The nearshore zone during coastal upwelling: daily variability and coupling between primary and secondary production off central Chile. *Prog. Oceanogr.* **20,** 1–40.

Pomeroy, L.R., 1979. Secondary production mechanisms of continental shelf communities. In *Ecological Processes in Coastal and Marine Systems,* R. J. Livingston, ed. Plenum Press, New York, pp. 163–186.

Probert, P.K., 1986. Energy transfer through the shelf benthos off the west coast of South Island, New Zealand. *N.Z. J. Mar. Freshwater Res.* **20,** 407–417.

Qasim, S.Z. and M.V.M. Wafar, 1990. Marine resources in the tropics. In *Tropical Resources: Ecology and Development,* J.I. Furtado, W.B. Morgan, J.R. Pfafflin and K. Ruddle, eds. Harwood Academic Publ., Chur, pp. 141–169.

Rabouille, C., F.T. Mackenzie and L.M. Ver, 2001. Influence of the human perturbation on carbon, nitrogen and oxygen biogeochemical cycles in the global coastal ocean. *Geochim. Cosmochim. Acta* **65,** 3615–3641.

Radach, G., 1992. Ecosystem functioning in the German Bight under continental nutrient inputs by rivers. *Estuaries* **15,** 477–490.

Reading, H.G., 1996. *Sedimentary Environments: Processes, Facies, and Stratigraphy, 3rd Edition.* Blackwell Publishers, Oxford, 688 pp.

Reid, P.C., C. Lancelot, W.W.C. Gieskes, E. Hagmeier and G. Weichart, 1990. Phytoplankton in the North Sea and its dynamics. *Neth J. Sea Res.* **26,** 295–331.

Rhoads, D.C., 1974. Organism-sediment relations on the muddy sea floor. *Oceanogr. Mar. Biol. Annu. Rev.* **12,** 263–300.

Rivkin, R.B., L. Legendre, D. Deibel, J.E. Tremblay, B. Klein, K. Crocker, S. Roy, N. Silverberg, C. Lovejoy, F. Mespie, N. Romero, M.R. Anderson, P. Matthews, C. Savenkoff, A. Vezina, J.C. Therriault, J. Wesson, C. Berube and R.G. Ingram, 1996. Vertical flux of carbon in the ocean: is there food web control? *Science* **272,** 1163–1166.

Robertson, A.I., P. Dixon and D.M. Alongi, 1998. The influence of fluvial discharge on pelagic production in the Gulf of Papua, Northern Coral Sea. *Estuar. Coastal Shelf Sci.* **46,** 319–331.

Saggiomo, V., G.C. Carrada, O. Mangoni, D. Marino and M. Ribera d-Alcala, 2000. Physiological and ecological aspects of primary production in the Ross Sea. In *Ross Sea Ecology,* F.M. Faranda, L.

Guglielmo and A. Ianora, eds. Springer-Verlag, Berlin, pp. 247–258.

Seitzinger, S.P., C. Kroeze, A.F. Bouwman, N. Caraco, F. Dentener and R.V. Styles, 2002. Global patterns of dissolved inorganic and particulate nitrogen inputs to coastal systems: recent conditions and future projections. *Estuaries* **25,** 640–655.

Sharp, G.D., 1988. Fish populations and fisheries-their perturbations, natural and man-induced. In *Ecosystems of the World, Vol. 27, Continental Shelves,* H. Postma and J.J. Zijlstra, eds. Elsevier, Amsterdam, pp. 155–202.

Sherman, K., M. Grosslein, D. Mountain, D. Busch, J. O'Reilly and R. Theroux, 1996. The Northeast Shelf ecosystem: an initial perspective. In *The Northeast Shelf Ecosystem: Assessment, Sustainability and Management,* K . Sherman, N.A. Jaworski and T. Smayda, eds. Blackwell Sci., Cambridge, MA, pp. 103–126.

Sinclair, M., 1988. *Marine Populations.* University of Washington Press, Seattle, 252 pp.

Smith, S.V. and J.T. Hollibaugh, 1993. Coastal metabolism and oceanic organic carbon balance. *Rev. Geophys.,* **31**(1), 75–89.

Snelgrove, P.V. R. and C. A. Butman, 1994. Animal-sediment relationships revisited: cause versus effect. *Oceanogr. Mar. Biol. Annu. Rev.* **32:** 111–177.

Steele, J.H., 1974. *The Structure of Marine Ecosystems.* Harvard University Press, Cambridge, Mass., 128 pp.

Summerhayes, C.P., K.C. Emeis, M.V. Angel, R.L. Smith, and B. Zeitzschel, 1995. *Upwelling in the Ocean: Modern Processes and Ancient Records.* J. Wiley & Sons, Chicester, 422 pp.

Summerhayes, C.P., W.L. Prell and K.C. Emeis, 1992. *Upwelling Systems: Evolution Since the Early Miocene.* Geol. Soc. Sp. Publ. No. 64, London, 459 pp.

Sverdrup H.U., 1953. On the conditions for vernal blooming of the phytoplankton. *J. Cons. Perm.Int. Explo. Mer.* **18:** 287–295.

Takahashi, T., S.C. Sutherland, C. Sweeney, A. Poisson, N. Metzl, B. Tilbrook, N. Bates, R. Wanninkhof, R.A. Feely, C. Sabine, J. Olafsson and Y. Nojiri. 2002. Global sea-air $CO_2$ flux based on climatological surface ocean $pCO_2$, and seasonal biological and temperature effects. *Deep-Sea Research II,* **49,** 1601–1622.

Tenore, K.R., L.F. Boyer, R.M. Cal, J. Corral, C. Garcia-Fernandez, N. Gonzalez-Gurriaran, R.B. Hanson, J. Iglesias, M. Krom, E. Lopez-Jumar, J. McClain, M.M. Pamatmat, A. Perez, D.C. Rhoads,

G. deSantiago, J.H. Tietjen, J. Westrich and H.L. Windom, 1982. Coastal upwelling in the Rias Bajas, northwest Spain: contrasting the benthic regimes of the Ria de Arosa and de Muros. *J. Mar. Res.* **40,** 701–772.

Thorrold, S.R. and A. D. McKinnon, 1995. Response of larval fish assemblages to a riverine plume in coastal waters of the central Great Barrier Reef lagoon. *Limnol. Oceanogr.* **40:** 177–181.

Thorson, G., 1957. Bottom communities (sublittoral and shallow shelf). In *Treatise on Marine Ecology and Paleoecology,* J. Hedgpeth, ed. Geol. Soc. Am. Mem., **67:** 461–534.

Upton, A.C., D.B. Nedwell, R.J. Parkes and S.M. Harvey, 1993. Seasonal benthic microbial activity in the southern North Sea: oxygen uptake and sulfate reduction. *Mar. Ecol. Prog.Ser.* **101:** 273–281.

van der Maarel, E., 1993a. *Ecosystems of the World 2A. Dry Coastal Ecosystems. Polar Regions and Europe.* Elsevier, Amsterdam, 600 pp.

van der Maarel, E., 1993b. *Ecosystems of the World 2B. Dry Coastal Ecosystems. Africa, America, Asia and Oceania.* Elsevier, Amsterdam, 616 pp.

Vincent, W.F., Howard-Williams, C., Tildsley, P. and Butler, E., 1991. Distribution and biological properties of ocean water masses around the South Island. *N.Z. J. Mar. Freshwater Res.* **25,** 21–42.

Voronina, N.M., 1998. Comparative abundance and distribution of major filter-feeders in the Antarctic pelagic zone. *J. Mar. Syst.* **17,** 375–390.

Walsh, J.J., 1988. *On the Nature of Continental Shelves.* Academic Press, San Diego, 520 pp.

Walsh, J.J. and D.A. Dieterle, 1994. $CO_2$ cycling in the coastal ocean. I. A numerical analysis of the southeastern Bering Sea with applications to the Chukchi Sea and the northern Gulf of Mexico. *Prog. Oceanogr.* **34** 335–392.

Walsh, J.J., D.A. Dieterle, M.B. Meyers and F.E. Muller-Karger, 1989. Nitrogen exchange at the continental margin: a numerical study of the Gulf of Mexico. *Prog. Oceanogr.* **23,** 245–301.

Walsh, J.J., G.T. Rowe, R.L. Iverson, and C.P. McRoy, 1981. Biological export of shelf carbon is a neglected sink of the global $CO_2$ cycle. *Nature* **291,** 196–201.

Watson, A.J. and J.C. Orr, 2003. Carbon dioxide fluxes in the global ocean. In *Ocean Biogeochemistry. The Role of the Ocean Carbon Cycle in Global Change,* M.J.R. Fasham, eds. Springer, Berlin, pp. 123–143.

Wollast, R., 1991. The coastal organic carbon cycle: fluxes, sources, and sinks. In *Ocean Margin Processes in Global Change,* R.F.C. Mantoura, J.M. Martin and R. Wollast, eds. Dahlem Workshop Report. Wiley, Chicester, England, pp. 365–381.

Wollast, R., 1993. Interactions of carbon and nitrogen cycles in the coastal zone. In *Interactions of C, N, P and Sbiogeochemical Cycles and Global Change,* R. Wollast, F.T. Mackenzie and L. Chou, eds. NATOASI Series, Vol. 14. Springer-Verlag, Berlin, pp. 195–210.

Wollast, R. 1998. Evaluation and comparison of the global carbon cycle in the coastal zone and in the open ocean. In *The Global Coastal Ocean, The Sea, Vol. 10,* K.H. Brink and A.R. Robinson, eds. Wiley, New York, pp. 213–252.

Wolanski, E. and W.M. Hamner, 1988. Topographically controlled fronts in the ocean and their biological influence. *Science* **241:** 177–181.

# Chapter 11. THE ROLE OF SEDIMENTS IN SHELF ECOSYSTEM DYNAMICS

JACK J. MIDDELBURG
KARLINE SOETAERT

*The Netherlands Institute of Ecology, Yerseke*

**Contents**



## 1. Introduction

Continental shelves are characterized by an active benthic compartment in close contact and interaction with the pelagic compartment. These interactions are two-way: there is an effect of the pelagic system on the ecology and biogeochemical functioning of the benthic system and vice versa, organisms living at or in the sediments affect the pelagic system. The benthic-pelagic interactions can be direct or indirect and operate at tidal to interannual timescales. For instance, benthic suspension feeders can affect phytoplankton concentration directly by their filtering activities and indirectly by enhancing recycling of nutrients, thus supporting pelagic primary production.

Traditionally, the benthic compartment of shelf ecosystems is considered to receive detritus settling out of the water column and to return most of these materials (with some delay) in the form of dissolved inorganic carbon and nutrients and to consume significant quantities of oxygen (Soetaert et al., 2000). A small but significant fraction of the incoming organic matter is buried (Middelburg et al., 1993; Hedges and Keil, 1995). Shelf sediments are also known to transform fixed N into $N_2$ by denitrification either directly by nitrate uptake from the water column or indirectly by coupled nitrification-denitrification following mineralization of

*The Sea*, Volume 13, edited by Allan R. Robinson and Kenneth H. Brink
ISBN 0-674-01526-6 

organic nitrogen (Rysgaard et al., 1994). This $N_2$ escapes from the shelf ecosystem and this loss of fixed N by denitrification in shelf sediments accounts for ~7 Tmol N $yr^{-1}$ or about $^{1}/_{3}$ of total nitrogen loss from the ocean (Middelburg et al., 1996; Brandes and Devol, 2002).

Shelf ecosystems are biologically active. Estimates of primary production range from 375 to 1195 Tmol C $yr^{-1}$ (Gattuso et al., 1998; Wollast, 1998; Alongi, Chapter 10; Longhurst et al., 1995; Ducklow and McCallister, Chapter 9) and estimates of shelf ecosystem respiration vary from ~200 to ~300 Tmol C $yr^{-1}$ (Gattuso et al., 1998; Wollast, 1998; Alongi, Chapter 10). The difference between production and respiration is either buried in shelf sediments (~13 Tmol C $yr^{-1}$; Hedges and Keil, 1995) or transferred to the open ocean (by difference: 77 to 895 Tmol C $yr^{-1}$).

A large proportion of organic matter processing in shelf ecosystems occurs in the benthic compartment because these systems are shallow, and the proportion of benthic to total respiration depends primarily on water depth (Heip et al., 1995). The relationship reported by Heip et al. (1995) predicts the benthic contribution to respiration to vary from about 40 % at shallow depths to a few % at a 100 m. Middelburg et al. (2004) reported a global respiration rate of shelf sediment of about 166 Tmol $yr^{-1}$, consistent with the 160 Tmol $yr^{-1}$ of Jørgensen (1983). Wollast (1998) reported global shelf production and respiration rates of 500 and 300 Tmol $yr^{-1}$ with sediments accounting for 150 Tmol $yr^{-1}$ or about 1/2 of shelf respiration (range 65 to 333 Tmol $yr^{-1}$). Similarly, Rabouille et al. (2001) reported that sediments accounted for about 40 % of total respiration in the distal zone of the coastal ocean (mean depth of 130 m). Alongi (Chapter 10) reported a total shelf respiration of 307 Tmol $yr^{-1}$ with 54 % occurring in the sediments.

Notwithstanding the prominent position of sediments in the nutrient and carbon cycling of shelf ecosystems, many studies only consider the pelagic compartment. On the basis of this incomplete, biased picture, inferences are made about the balance between autotrophic and heterotrophic processes and the export of organic matter to the open ocean. This transport of organic matter from the shelf to the open sea amounts to about (77 to 895 Tmol C $yr^{-1}$) and fuels the inferred open-ocean heterotrophic status (del Giorgio and Duarte, 2002).

In this chapter we will use numerical, coupled benthic-pelagic biogeochemical models to study the flow of organic matter and nutrients through shelf ecosystems. We have chosen nitrogen as our model currency and will give special attention to nitrogen dynamics of the shelf ecosystem as a function of water depth, and of sediment type. In section 11.2 we will first summarize the basic ecological and biogeochemical characteristics of shelf sediments; Jahnke (Chapter 6) provides a more detailed treatment. In section 11.3 we derive some simple parameterization of shelf-specific biogeochemical processes. In the next section we provide a short description of our modeling platform (Soetaert et al. 2001, 2000), before discussing simulations on shelf nitrogen cycling in full detail (section 11.5). In section 11.6 we address the potential importance of benthic primary producers, benthic suspension feeders and advective flow through permeable, sandy sediments for shelf biogeochemistry. In the final section, we summarize our conclusions and identify a few outstanding, pressing questions.

## 2. Characteristics of shelf sediments

During the last glacial advance, sea level dropped more than 150 m, and the majority of shelves were exposed and subject to erosion. At present, sediments cover most continental shelf systems. The type of sediment on the continental shelf is governed by the tectonic setting, the availability and river input of sediment and transport by waves and currents. Seasonal, storm-driven events are also important on many shelves. The inner continental shelf (< 65 m) are covered mainly with muds (37 %) and sands (47 %), the remainder is covered by corals (6%), shell debris (4%) and rocks/gravel (6%) (Hall, 2002).

These differences in sediment characteristics have a direct influence on the density, biomass, distribution and diversity of benthic communities (Heip and Craeymeersch, 1995) and on sediment biogeochemistry (Dauwe and Middelburg, 1998; Jahnke, Chapter 6). There exists duality in the way biogeochemists and ecologists perceive and investigate benthic-pelagic coupling (Soetaert et al., 2002). While benthic ecologists have studied both the muddy and sandy substrates, most biogeochemists and microbial ecologists studying carbon and elemental cycling have focused on muddy and silty sediments. Traditionally, ecologists are mainly interested in the transfer of organic matter (food) from the water column to the benthic organisms and its cascading within the benthic food web. Both the role of bacteria and remineralization of nutrients are given only modest attention. The impact of biota on benthic-pelagic coupling is studied from the point of view of the higher organisms (Thomsen and Flach, 1997).

In contrast, biogeochemists traditionally view the seafloor as passively receiving detritus from the overlying water. Heterotrophic bacteria then mineralize this deposited organic matter, and the dissolved inorganic substances diffuse back to the water column (with some delay). These regenerated nutrients support primary production in the water column, and the nutrient cycle is closed. There are several reasons why this traditional, biogeochemical view of the role of sediment in shelf ecosystem functioning requires revision (Boudreau et al., 2001; Jahnke, 2004).

First of all, in the more shallow parts of shelf systems with clear waters, light may reach the bottom, and populations of benthic microalgae may proliferate. Cahoon (1999) and Jahnke (Chapter 6) have summarized the few data available, indicating that benthic microalgal production may approach that of phytoplankton in oligotrophic, clear water systems such as the Georgia continental shelf (Jahnke et al., 2000). The global rate of benthic microalgal production (28 Tmol $yr^{-1}$) reported by Charpy-Roubaud and Sournia (1990) is almost an order of magnitude lower than estimates of shelf phytoplankton production (~375 Tmol $yr^{-1}$; Gattuso et al., 1998; 575 Tmol $yr^{-1}$, Alongi, Chapter 10). However, the data set at hand is too sparse too allow derivation of a robust estimate of global shelf benthic primary production. Benthic microalgae have important consequences for the functioning of the shelf ecosystem because they (1) affect the exchange of nutrients and oxygen across the sediment-water interface, (2) constitute a major food resource for heterotrophic organisms and moderate benthic carbon flows and (3) may affect sediment stability (Paterson and Black, 1999; Middelburg et al., 2000).

Secondly, benthic suspension feeders have relative high biomass on the shelf (Heip and Craeymeers, 1995; Bryant et al., 1995) and these animals actively filter algae from the lower layer of the water column, the benthic boundary layer. This clearance of algae from the water can be a major term in benthic-pelagic exchange, by its effects on both algal loss and nutrient recycling. Benthic suspension feeders often dominate benthic-pelagic exchange and algal consumption in shallow ecosystems. This has been well documented for tidal inlets and other more enclosed basins (Herman et al. 1999; Heip et al., 1995).

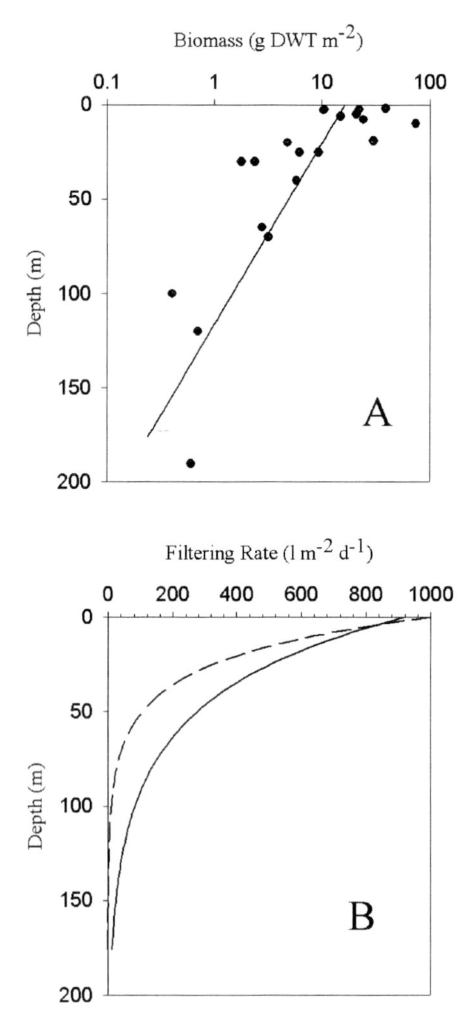


Figure 11.1 A. The depth dependence of benthic suspension-feeder biomass (data from Heip et al. 1995 and Bryant et al., 1995)
B. The depth dependence of filtering rates by benthic suspension feeders (solid line) and permeable beds (dashed line)

Thirdly, sandy sediments dominate continental shelves (de Haas et al., 2002). These sandy sediments have been subject to repeated cycles of deposition and erosion, resulting in well-sorted, permeable deposits. Their high permeability in combination with high current velocities and wave activities (another characteristic of coastal systems) results in advection through these permeable beds and enhanced dispersion (Huettel and Webster, 2001; Jahnke, Chapter 6). Bedform topography, whether resulting from physical or biological processes, causes additional exchange of water between sediments and water column (Huettel et al., 1996; Precht and Huettel, 2003). These non-diffusive processes enhance exchange of solutes between the sediments and overlying water, relieving or eliminating diffusive limitations on oxygen and nitrate fluxes and mineralization product effluxes. This permeability-related exchange of water parcels between the water column and the sediments also results in the transfer of small particles from the water column to the sediment where they may be trapped depending on the relative sizes of pore spaces and the particles, and the energy of transport processes (Huettel and Rusch, 2001; Fries and Trowbridge, 2003). The overall effect is enhanced input of particles from the benthic boundary layer to the sediments and increased influxes of oxidants.

## 3. Quantifying enhanced exchange rates and primary production in shelf sea sediments.

The potential for filtration and primary production, and additional characteristics of shelf ecosystems contribute to the complexities of benthic-pelagic coupling in these regions. We are not yet able to include most of these processes as mechanistic equations in a shelf ecosystem model. However, we may describe their effect on ecosystem behavior using approximate dependencies and assuming steady-state conditions (i.e. change in biomass of neither benthic primary producers nor filter feeders).

The enhanced exchanges between water and sediment induced by filter-feeding animals or by physical advection through sands can be quantified in terms of the number of liters of water, with its dissolved and particulate contents, that is exchanged per unit area and time. These enhanced exchange rates can also be compared with filtration rates of ~10 l $m^{-2}$ $d^{-1}$ for diffusive exchange (Precht and Huettel, 2003). Heip et al. (1995) reported filter-feeder biomasses ranging between 9 and 74 g ash-free dry weight (ADWT, about 50% is C) for water depths shallower than 25 m. Bryant et al. (1995) tabulated values of 0.2 and 3 g C $m^{-2}$ in the North Sea, at depths between 20-180 m. These data can be fitted with an exponential decay as a function of water depth (Figure. 11.1A). A filter feeder of 1 g AFDWT can filter on average some 57 l $d^{-1}$ of water (Heip et al., 1995). Using this clearance rate and the above-derived biomass-depth relationship, we obtain a potential filtering of about 950 l $m^{-2}$ $d^{-1}$ at 0 m water depth, rapidly declining to 285 l $m^{-2}$ $d^{-1}$ at 50 m depth (Figure 11.1B). The best-fit function, relating filtration rate (l $m^{-2}$ $d^{-1}$) to water depth (z, m) is:

$$FilterFeederRate = 945.e^{-0.024 \cdot Z} \quad (1)$$

Precht and Huettel (2003) calculated that filtering rate in beaches was about 1000 l $m^{2}$ $d^{-1}$, whereas at 50 m depth on the continental shelf near the Eel River the filter rate was 103 l $m^{-2}$ $d^{-1}$. Fitting these points, we obtain the following regression, expressing the filtration rate through sands (l $m^{-2}$ $d^{-1}$) as a function of water depth (z, m):

$$SandFilterRate = 1000.e^{-0.045 \cdot Z} \quad (2)$$

Although based on only 2 (!) points, the depth-decay coefficient for filtering by sands (0.045 $m^{-1}$) falls well within decay coefficients of 0.15 $m^{-1}$ to 0.03 $m^{-1}$ as estimated from exchange rates reported by Riedl et al. (1972) for wave periodicities varying from < 6 to > 13 s.

Based on this first-order estimate, it is clear that the potential filtering activity of biota and sands is of comparable importance, although the latter attenuates faster with water depth (Figure. 11.1B). However, there are also more subtle differences. Benthic suspension feeders have a preference for algae and often utilize only a small amount of detritus as food (Heip et al., 1995). Such particle selection is probably less pronounced in permeable beds, but data on this subject are scarce (Fries and Trowbridge, 2003). In addition, there is a feedback between algal clearance and the rate of benthic suspension feeding via biomass changes, which may not be the case for permeable bed filtration. However, clogging of permeable sediments may occur until the next sediment transport event.

Benthic primary production can be described using data from the Georgia continental shelf (Jahnke et al., 2000). We fitted a semi-mechanistic formulation that relates benthic primary production (BPROD, mmol C $m^{-2}$ $d^{-1}$) to photosynthetically active irradiation at the sediment surface (I, μmol $m^{-2}$ $s^{-1}$; Figure. 11.2):

$$BPROD = P\max \cdot (1 - e^{-\alpha \cdot I} - r) \tag{3}$$

where Pmax, the maximal photosynthetic rate equals 100 mmol C $m^{-2}$ $d^{-1}$; $\alpha$, the photosynthetic efficiency = 0.018 (μmol $m^{-2}$ $s^{-1})^{-1}$ and r, the activity respiration fraction = 0.27. Both the values of $\alpha$ and activity respiration are within ranges observed for pelagic phytoplankton.

In what follows we will use a set of coupled physical-biogeochemical, mathematical models to exemplify the role of benthic-pelagic exchanges in shelf nitrogen dynamics. We will first deal with the nature of benthic-pelagic coupling in a setting patterned on a turbid, temperate shelf at 50 m water depth (section 11.5). We then use the model as a tool to isolate the effect of benthic primary production, benthic suspension feeders and permeable sediments on shelf ecosystem dynamics along a depth transect from 2 to 150 m (section 11.6). Although equations 1-3 only provide only a rough estimate of the potential for filtration and benthic primary production, it suffices for our exploratory modeling. We will not address the role of deposition-erosion cycles in shelf ecosystem dynamics. These intense, repeated cycles of sedimentation-resuspension interact with residual currents and result in strong lateral gradients of organic matter delivery, mineralization and burial (Dauwe and Middelburg, 1998; de Haas et al., 2002). Some parts of the shelf may act as depocenters, while other parts are net exporters. This lateral transfer causes a decoupling between local pelagic production and benthic response. We will also do not cover the regional or local groundwater input to shelf ecosystems and their ecological consequences (Moore, 1999). Moreover, we will focus our modeling efforts on biogeochemical dynamics and will use simple parameterizations for the dynamics of zooplankton.

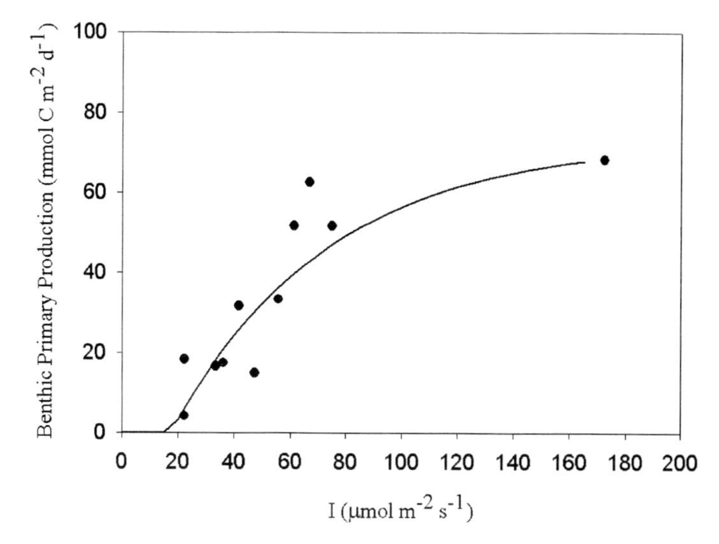


Figure 11.2 Benthic primary production versus irradiation. Data are from Jahnke et al. (2000). The solid line corresponds to eq. 3.

## 4. Model description

Models, and in particular biogeochemical models, are a simplified representation of the complexity found in nature. With respect to benthic-pelagic coupling, the main task of the pelagic model is to provide upper boundary conditions to the sediment model, i.e. the solute concentrations (nutrients and oxygen) of the water overlying the sediment and the deposition of particles on the sediment surface. These properties are then used to update the corresponding sediment state variables (if appropriate). In return, the sediment model calculates the exchange of various solutes across the sediment-water interface, and these fluxes are used to update the respective pelagic state variables.

Soetaert et al.(2000) have proposed a hierarchical classification of the formulation of benthic-pelagic exchanges in biogeochemical models ranging from the simplest models where the bottom is plainly ignored to the most complex ones in which a depth-resolving pelagic model is coupled on-line with a depth-resolving diagenetic model (Boudreau, 1997). At intermediate levels of complexity one finds mass-conserving models that either consider a depth-integrated benthic model, or assume immediate equilibrium between particle deposition from the water column and the return flux of dissolved constituents from the sediment. Other, very simple descriptions where the return flux of solutes is imposed fail to conserve mass and are not suitable to represent benthic-pelagic coupling. It was argued that the computationally efficient zero-dimensional benthic models capture much of the dynamics inherent in benthic-pelagic coupling (Soetaert et al., 2000), so we will use this type of model when we calculate the effect of active filtering or benthic primary production in a water depth gradient. However, it is also instructive to examine how the variability in pelagic properties propagates towards and is smoothed by the sediments and this examination requires a fully resolved diagenetic model. Inclusion of such a generic diagenetic model also allows assessment of denitrification and partitioning of nitrogen regeneration and nitrification between the sediment and the water column. Moreover, simulations with this complete model can be used as a benchmark solution for the exploratory modeling.

The basic modeling framework was extensively described in Soetaert et al. (2000, 2001). In its most complex form it couples a Nutrient-Phytoplankton-Zooplankton-Detritus (NPZD) model for water-column biogeochemistry, a turbulence-closure and aggregate formulation for the physical part and a fully resolved diagenetic model in a 1-D setting (Figure 11.3 model 1, Table I). In short, the pelagic model is based on the algal model and detritus dynamics as propagated by Tett and Droop (1988) and Smith and Tett (2000), but including an explicit description of zooplankton dynamics. The algal model is a so-called unbalanced growth model, which explicitly describes the uncoupling of C and N assimilation through the build-up and consumption of an intracellular nitrogen pool (Droop, 1973). Turbulence intensity and temperature, the physical variables of the water column, were generated using a turbulence-closure formulation as described by Gaspar et al. (1990). In contrast to the previous modeling, sinking of detrital matter was made somewhat more realistic by taking into account aggregate formation (Kriest and Evans, 1999). The model is patterned on turbid shelf regions. Initial average nitrogen concentrations in the water column were 10.5 mmol N $m^{-3}$. The background extinction coefficient ($m^{-1}$) was made a function of total water depth and decreased from 0.1 at 5 m to 0.05 at the 150 m. While the pelagic compartment formulations are identical in all simulations, different sedimentary biogeochemical processes were included. In section 11.5 the full diagenetic model of Soetaert et al., (1996) is coupled, which essentially describes the C, N, and O cycle in the sediment (Figure 11.3 model 1). The effect of reduced species other than ammonium is lumped by describing the amount of oxygen that is needed to reoxidise them (so-called 'oxygen demand units' sensu Soetaert et al., 1996).

The exploratory model simulations in section 11.6 are based on simplified representations of the benthic compartment (Figure 11.3 model 2, Tables I, II). Briefly, it involves a vertically integrated dynamic sediment model in which the temporal evolution of one organic carbon and nitrogen pool is resolved and where sediment-water exchanges of constituents are parameterized (see Soetaert et al., 2000, 2002 for details). The effect of active filtration was included via the simple empirical relations derived in section 11.3: i.e. equation 1 and 2 for suspension-feeder and permeable bed filtration respectively. Sand filtering rates add to the sinking (advection) of particulate detritus and algae to the sediment surface, while suspension-feeding rates add only to algal removal from the water column. The pelagic model includes a light conservation equation (Soetaert et al., 2001), and calculates the light reaching the bottom. This fuels the net benthic primary production, modeled using eq. 3. To prevent benthic primary production from continuing to rise in the absence of nutrients, a nutrient-limitation term is added. Primary production adds organic matter to the sediment detritus pools and removes nitrate and ammonium from the overlying water column. All simulations presented include at least two full annual cycles, but our discussion focuses on only the annually integrated or averaged rates.

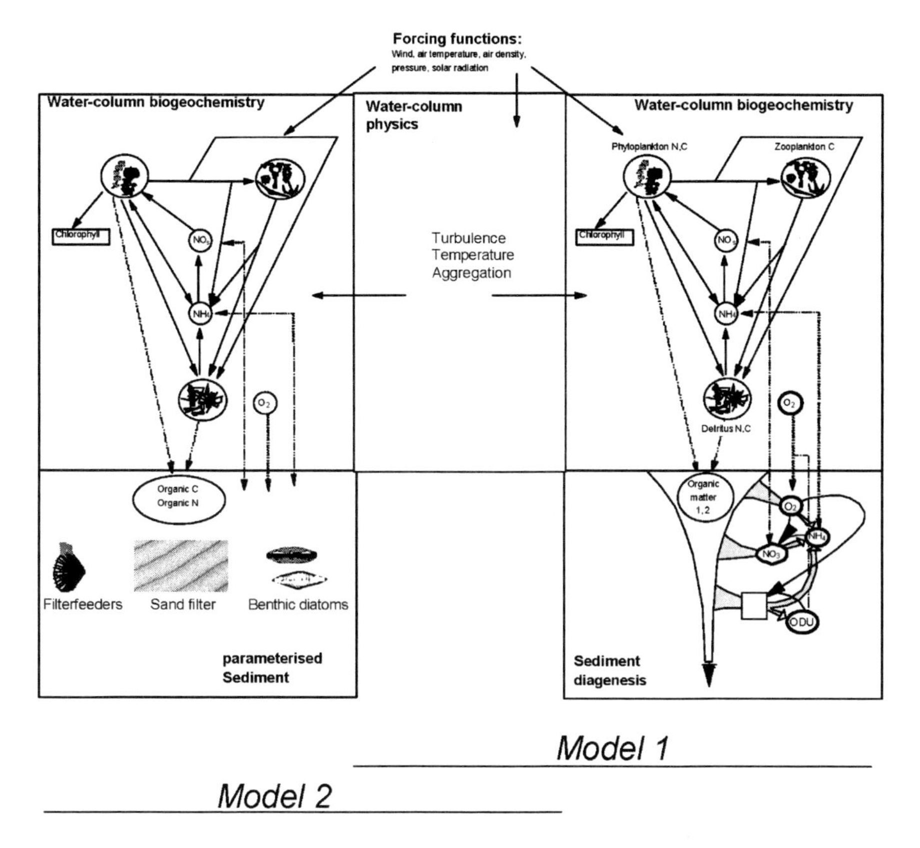


Figure 11.3 The structure of the two models, both encompassing the same physical and pelagic biogeochemical submodels, but differing in their benthic models. Model 1 (section 5) includes a vertically resolved diagenetic model, whereas model 2 (section 6) employs some simple parameterisation for benthic filter activities and benthic primary production

## 5. Nitrogen dynamics of continental shelf ecosystems

Nitrogen is an essential nutrient for autotrophic and heterotrophic organisms and is involved in a number of microbiological transformations including assimilation, regeneration, nitrification and denitrification. To assess its cycling in shelf ecosystems we have modeled in detail the major nitrogen flows for a 50-m deep station in a temperate shelf with depositional, silty/muddy sediments and relatively high total pelagic nitrogen content (> 7 mmol N $m^{-3}$). The parameter settings from the 50-m deep station were taken in accordance to the relationships derived by Soetaert et al. (1996), with a net sedimentation rate of 1 cm $yr^{-1}$, a bioturbation rate of 15 $cm^2$ $yr^{-1}$ and diffusion enhancement by irrigation by a factor of 3 (Figures. 11.4-11.6). To prevent nitrogen loss due to sedimentary denitrification from draining the water column of nitrogen, we assumed that the $N_2$ flux was directly compensated by a nitrate flux towards the deepest water box. Previous tests have shown that the actual mechanism by or depth range at which this nitrogen is returned to the water column is not important (Soetaert et al., 2000).

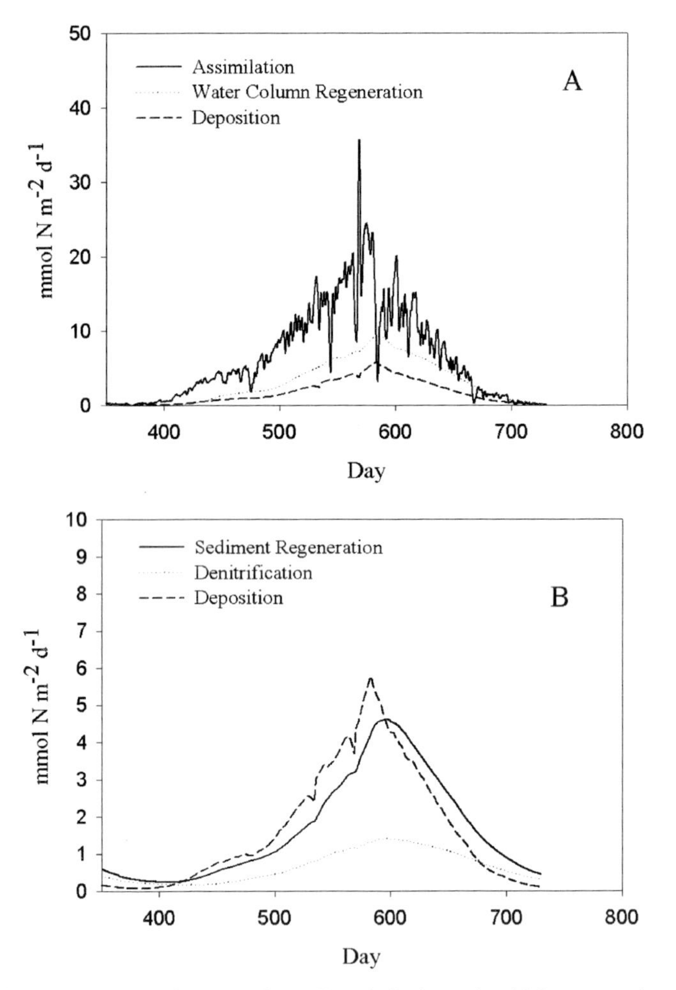


Figure 11.4 A.) Simulated rate of water-column N-assimilation, microbial regeneration and N deposition. B.) Simulated rate of N-deposition, sedimentary N-mineralisation and denitrification.

The physical model is forced with weather and as a consequence primary production and nitrogen assimilation by phytoplankton show not only seasonal variability, but also short-term variability due to wind events (Figure 11.4A). Simulated microbial nitrogen regeneration from the detrital pool in the water column and delivery of particulate nitrogen (algae and detritus) to the sediments show much less short-term variability, and there is some delay in maximum activity (Figure 11.4A). This attenuation and delay is due to redistribution among and transfer to various members of the pelagic food web (phytoplankton, zooplankton and detritus with associated bacteria). The simulated sediment properties display far less short-term variability than the pelagic system (Figure 11.4B). Pulsed delivery of organic N to the sediments (as algae and detritus) induces a sedimentary response that is attenuated and delayed (Figure 11.5, Plate). All labile organic N delivered to the sediments is eventually regenerated, and annual integrated rates of N deposition and sediment mineralization are equal, but the seasonality in sedimentary N mineralization is smoother and has a lower amplitude. Over the year oxygen fluxes and penetration depths vary from 2.5 to 23 mmol $m^{-2}$ $d^{-1}$ and from 0.5 to 3.5 cm, respectively (Figure 11.5, Plate). Similarly, nitrate and ammonium effluxes vary from 0.16 to 1.6 and 0.02 to 1.4 mmol $m^{-2}$ $d^{-1}$, respectively. Sedimentary denitrification varies from 0.17 to 1.4 mmol $m^{-2}$ $d^{-1}$ (Figure 11.4B) and contributes between 6 and 15 % to mineralization.

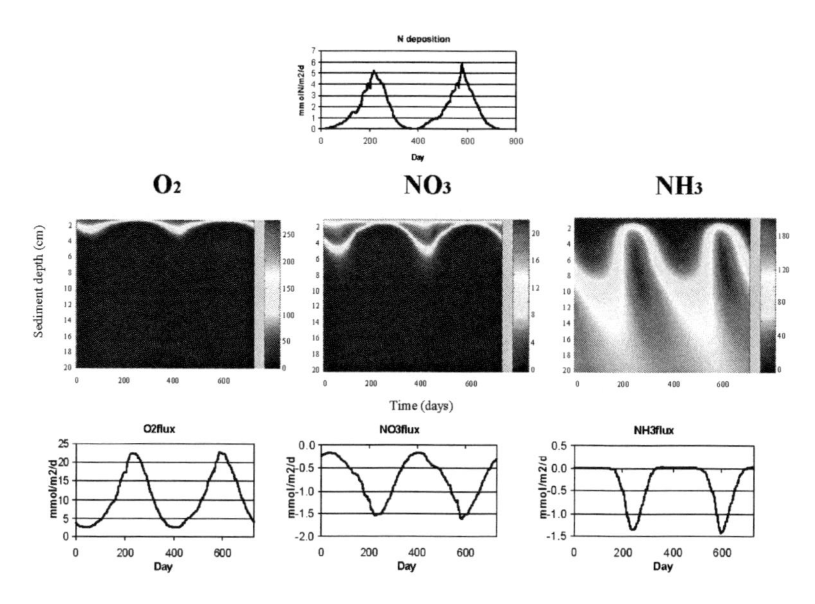


Figure 11.5 Spatio-temporal plots of oxygen, nitrate and ammonium in the sediment and time-series of nitrogen delivery to sediments and sediment-water exchange fluxes of oxygen, nitrate and ammonium.

The model predicts (Figure. 11.6) that each year phytoplankton assimilates ~ 2404 mmol N $m^{-2}$, 33% (791 mmol N $m^{-2}$) in the form of nitrate and 67% (1613 mmol N $m^{-2}$) as ammonium. The average phytoplankton stock (47.7 mmol N $m^{-2}$, corresponding to an average *Chl* of 1.83 mg $m^{-3}$) is turned over about 50 times per year. This rapid recycling is due to phytoplankton losses by settling (~7%, 165 mmol N $m^{-2}$), mortality (~ 15%, 351 mmol N $m^{-2}$) and grazing by metazoans (78%, 1889 mmol N $m^{-2}$). In the model, zooplankton (inventory of 8 mmol N $m^{-2}$) graze on phytoplankton only, 33% (619 mmol N $m^{-2}$) of the food is not assimilated and enters the detrital pool, another part of the food is respired (25%, 472 mmol N $m^{-2}$) and the remaining 42% (798 mmol N $m^{-2}$) is assimilated.

The detritus pool (46.5 mmol N $m^{-2}$) comprises dead organic nitrogen as well as microbial nitrogen and each year 1620 mmol N $m^{-2}$ flows through this pool: this pool is turned over 35 times per year. Detritus is produced from algal mortality (22%, 351 mmol N $m^{-2}$), zooplankton mortality (40%, 650 mmol N $m^{-2}$) and by sloppy zooplankton grazing (38%, 619 mmol N $m^{-2}$). Part of the detritus (29%, 464 mmol N $m^{-2}$) settles to the sediment, another part is mineralized to ammonium (71%, 1157 mmol N $m^{-2}$). Ammonium is rapidly recycled (average turnover time of about 4 d) because its stock (22 mmol N $m^{-2}$ or 0.44 mmol $l^{-1}$) is rather low compared to its total regeneration of 1886 mmol N $m^{-2}$: 61% (1157 mmol N $m^{-2}$) from detritus, 33% (622 mmol N $m^{-2}$) from zooplankton and 6% (107 mmol N $m^{-2}$) from the sediments. The majority (85%, 1613 mmol N $m^{-2}$) of ammonium is taken up by phytoplankton; the remaining 15% (273 mmol N $m^{-2}$) is nitrified. In the model universe, each year algae assimilate about 791 mmol N $m^{-2}$ of nitrate and the nitrate stock (277 mmol N m-2 or 5.5 mmol $l^{-1}$) is refueled by the sediments (34%, 270 mmol N $m^{-2}$), by nitrification (35%, 273 mmol N $m^{-2}$) and by lateral advection (31%, 245 mmol N $m^{-2}$). The phytoplankton (165 mmol N $m^{-2}$, 26%) and detritus (464 mmol N $m^{-2}$, 74%) delivered to the sediment surface are added to the pool of sedimentary organic matter and are mineralized to ammonium (620 mmol N $m^{-2}$), the majority (83 %, 514 mmol N $m^{-2}$) of which is nitrified, the rest is buried (9 mmol N $m^{-2}$) or fluxes out of the sediment (107 mmol N $m^{-2}$). The nitrate produced either fluxes out of the sediment (53 %, 270 mmol N $m^{-2}$) or is denitrified (47 %, 245 mmol N $m^{-2}$).

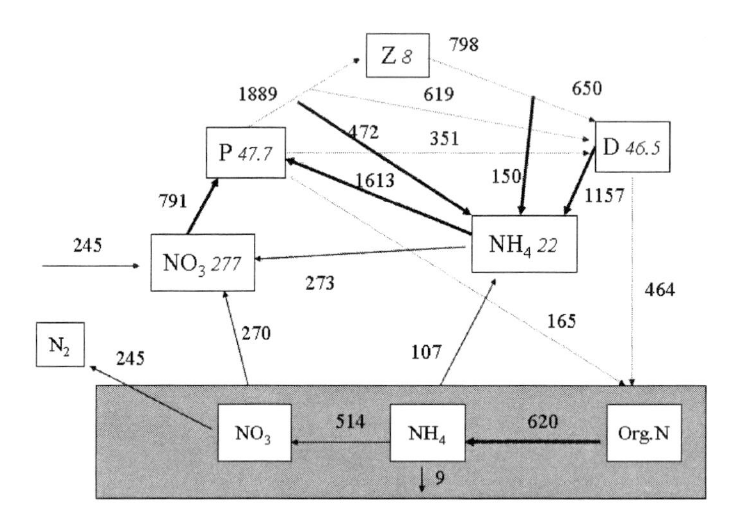


Figure 11.6 Annual nitrogen budget for a 50-m deep shelf system. Italic numbers in boxes refer to inventories (mmol N $m^{-2}$). Fluxes are given in mmol N $m^{-2}$ $yr^{-1}$; organic to inorganic transformation are in bold; organic-to-organic transformations are dashed and inorganic-to-organic transformations are solid arrows.

Oceanographers are increasingly studying the continental shelf because of its high primary production and important role in the oceanic biological pump (Ducklow and McCallister, Chapter 9). Net and recycled production on the continental shelf are often analyzed using approaches that have shown to be useful in the open ocean. However, the prominent role of sediment in shelf ecosystem dynamics complicates the application of some of these approaches. Primary production is often partitioned into net and regenerated production based on nitrogen substrate use and $^{15}N$ uptake measurement can be used to quantify this in terms of an f-ratio (Dugdale and Goering, 1967). Nitrogen isotope-based f-ratios for continental shelf systems are typically ~ 0.4 (Chen, 2003), consistent with our simulation (~ 0.33, Figure 11.6). However, a significant fraction of the nitrate is recycled within the water column (nitrification) or comes from the sediments and therefore cannot be considered as an external input. The nitrate that is denitrified in shelf sediments is truly removed from the system and provides a better estimate of new nitrate required for steady state. The rationale is as follows: denitrification in shelf sediments is mainly fuelled by nitrification (Middelburg et al., 1996), hence eventually by organic nitrogen input to the sediment. The organic nitrogen input to sediments is largely returned in a fixed form (Figure 11.6), but a small fraction is lost by denitrification and must be resupplied externally to keep the system running. Our simulations indicate that the ratio of external nitrate (245 mmol N $m^{-2}$) to total nitrogen assimilation (2404 mmol N $m^{-2}$) corresponds to an export ratio of ~ 0.1 (Figure 11.6), more in line with carbon budget estimates (Chen, 2003).

The importance of sediments to shelf ecosystem functioning is clear from the nitrogen budget. Denitrification is restricted to shelf sediments and about 65% of nitrification and 26% of nitrogen regeneration occur in the benthic compartment (Figure. 11.7). Although there is pronounced seasonality in shelf sediment nitrogen cycling and sediment-water exchange fluxes, the short-term variability found in water-column properties is attenuated in the sediment. Thus the sediment provides some memory to shelf ecosystems. One of the main factors governing the relative importance of water-column and sedimentary processes is the water depth, and we will therefore address this question in the next section.

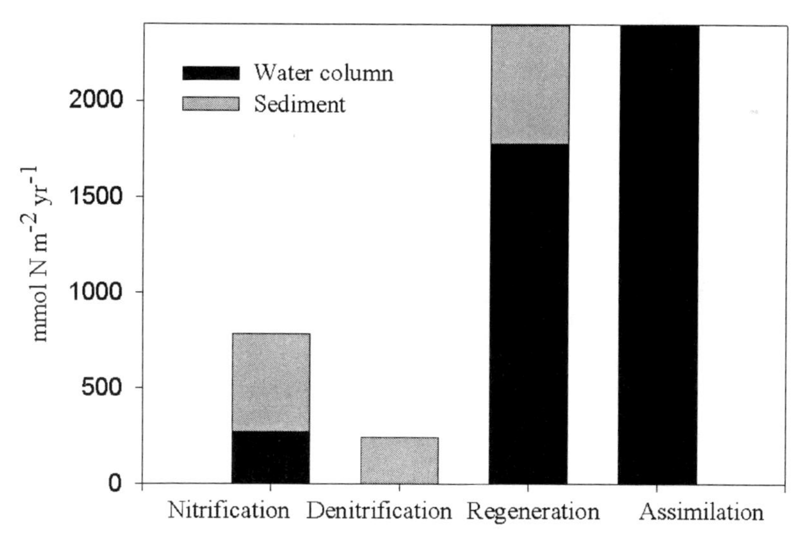


Figure 11.7 Partitioning of nitrification, denitrification, N-regeneration and assimilation between water column and sediment for a 50-m deep shelf.

## 6. Dynamics of a coupled benthic-pelagic ecosystem with an active benthic compartment

In order to assess how ecosystem behavior is modulated by the action of filter feeders, physical filtering through sands and benthic primary production, the same coupled model, excluding and including the effect of these processes but with simplified benthic-pelagic exchange (Table I, II) was applied along a shelf depth gradient (2-150 m; Figures. 11.8-11.10). In the absence of benthic filtering activity (Figure 11.8 A, B), pelagic integrated photosynthesis peaks at 20 m (67.1 mmol C $m^{-2}$ $d^{-1}$), while phytoplankton N-assimilation is maximal at 30 m water depth (8.6 mmol N $m^{-2}$ $d^{-1}$). Both decrease with increasing station depth to about 31.5 mmol C $m^{-2}$ $d^{-1}$ and 6 mmol N $m^{-2}$ $d^{-1}$ at 100 m depth and also decrease toward the 2-m deep station (12.9 mmol C $m^{-2}$ $d^{-1}$, 1 mmol N $m^{-2}$ $d^{-1}$).

Integrated water-column production is a function of the total N-content of the water, the standing stock of algae and the light regime experienced by the algae, which depends on the depth of the euphotic zone and the mixed layer. With decreasing water depth, the average light regime improves, hence their capacity for biosynthesis and photosynthesis, but the total nitrogen load decreases, which lowers the total amount of nitrogen that can be converted into their biomass. Conversely, with increasing water depth, the euphotic zone becomes smaller than the mixed layer and the integrated amount of light received by algae decreases while the total nitrogen load increases. As a consequence of these opposing trends in light and N availability there is an optimum for phytoplankton assimilation around 20 to 40 m. The optimum for C fixation is shallower than that of N assimilation (Figure 11.8A,B), because the former more closely tracks the light regime than the latter.This spatial separation of optimum C and N assimilation can be compared to the uncoupling of C fixation and N assimilation observed in open-ocean, mixed-layer vertical profiles showing maximum N assimilation at greater depths than C fixation.

Simulated benthic primary production was limited to depths < 5 m and does not play a role of significance except for the shallowest station at 2 m, where it accounts for about 7% of all primary produced matter (Figure 11.8C). This limited role of benthic primary production in the absence of benthic filtration is consistent with observations in turbid, temperate estuaries and is a direct consequence of our model parameters, i.e. a relatively high background attenuation coefficient and high N concentrations in the water column: > 7 mmol $m^{-3}$. This large N-content allows the development of substantial phytoplankton biomass with accompanying shading.

With increasing water depth, a decreasing fraction of the net assimilated carbon and nitrogen is mineralized in the sediment, 80% and 88% at 2 m, 15% and 13% at 100 m depth for carbon and nitrogen, respectively. Total benthic mineralization rates (Figure 11.9) peak at 15 m for C (29.2 mmol C $m^{-2}$ $d^{-1}$) and at 25 m for N (3.4 mmol N $m^{-2}$ $d^{-1}$). This difference in depth of maximum sedimentary mineralization of C and N is due to difference in the depth-integrated assimilation of C and N (Figure 11.8). The depth of maximum sediment mineralization is shallower than the depth of maximum C and N assimilation because a larger proportion of assimilated C and N is mineralized in the water column with increasing depth.

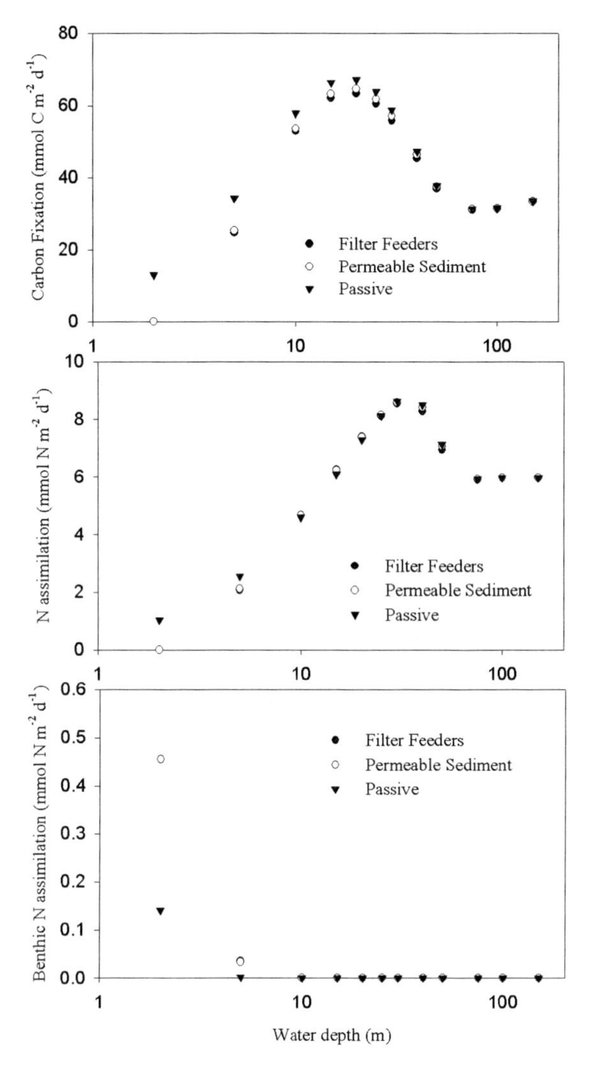


Figure 11.8 A) Simulated phytoplankton carbon fixation as a function of water depth for impermeable, passive sediments, permeable sands and sediments with benthic suspension feeders. B) Simulated phytoplankton nitrogen assimilation as a function of water depth for impermeable, passive sediments, permeable sands and sediments with benthic suspension feeders. C)Simulated nitrogen assimilation by benthic algae a function of water depth for impermeable, passive sediments, permeable sands and sediments with benthic suspension feeders.

The simulated depth dependence of sedimentary C and N mineralization (at depths more than 25 m) can be described by exponential relationships (Figure 11.9 insets) with depth attenuation coefficients of -0.017 and -0.016 $m^{-1}$ for C and N, respectively. These attenuation coefficients based on our modeling exercises are very similar to those based on compilations of experimental results for estuarine systems (-0.015 $m^{-1}$: Heip et al., 1995) and shelf ecosystems (-0.018±0.006 $m^{-1}$: Middelburg et al., 2004). This similarity is not inherited from a common database because the settling velocity of particulate matter in our simulations is modeled with a particle aggregation module.

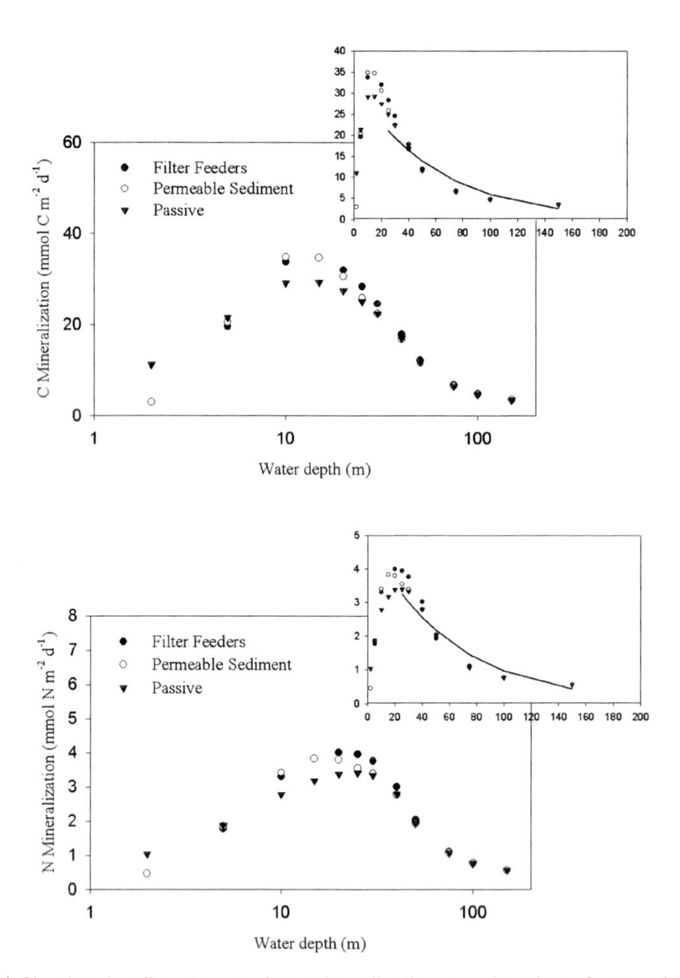


Figure 11.9 A) Simulated sedimentary carbon mineralization as a function of water depth for impermeable, passive sediments, permeable sands and sediments with benthic suspension feeders. The inset shows an exponential fit to sediment mineralization for depths ≥ 25 m. B) Simulated sedimentary nitrogen mineralisation as a function of water depth for non-permeable, passive sediments, permeable sands and sediments with benthic suspension feeders. The inset shows an exponential fit to sediment mineralisation for depths ≥ 25 m.

The effect of benthic filtering activity, whether biotic or abiotic is discernible only for those stations where the water column is shallower than 50 m (Figure. 11.8, 11.9). The effect of filter feeders or filtering through sands is largely similar, so we will discuss them together. While the direct effect of detrital fall and algal filtering is an increase of material transfer from the pelagic to benthic system, there are indirect effects via their impacts on the functioning of the pelagic systems that may result in lower benthic mineralization rates. Simulated benthic filtration reduces phytoplankton concentrations and as a result integrated carbon assimilation rates in the water column are depressed to 73% at 5 m depth, 92% at 10 m depth and then slowly attain 100% at 100 m. For the 2-m deep water column, filtering rates are sufficient to annihilate pelagic production, whilst increasing benthic production from 0.91 mmol C $m^{-2}$ $d^{-1}$ and 0.14 mmol N $m^{-2}$ $d^{-1}$ in the absence of filtration to 2.9 mmol C $m^{-2}$ $d^{-1}$ and 0.46 mmol N $m^{-2}$ $d^{-1}$ in the presence of filtration (Figure 11.8, 11.9). The effect on phytoplankton nitrogen assimilation rates is different, as here filtering not only reduces algal biomass, but also relaxes the nitrogen limitation via increased benthic mineralization rates. Consequently, phytoplankton nitrogen assimilation is depressed only at 2 and 5 m depth, to 0% and 80% of its value in the absence of filtering activity.

At water depths > 50 m, filtering of pelagic systems by benthic animals and/or sandy, permeable beds has minor effect on the functioning of the ecosystem (Figure 11.8–11.10). At these larger depths, most of the phytoplankton production is grazed by metazoans (~ 80%) or ends up in the microbial food chain after mortality (~ 15-19%) with little settling of algal material following aggregation (Figure 11.10A). Shoaling of the water column results in an increase in the proportion of algal settling to the sediment (~ 54% at 5 m) at the expense of zooplankton grazing (~ 35% at 5 m) and algal mortality (~ 11% at 5 m). However, the fate of phytoplankton production depends not only on water depth, but also on the nature of the benthic compartment. Filtering of the water column augments these differences between shallow and deep shelf ecosystems and at 5 m ~72% of the algal reach the bottom, ~ 22% is grazed and only 6% of phytoplankton losses can be attributed to mortality (Figure 11.10B).

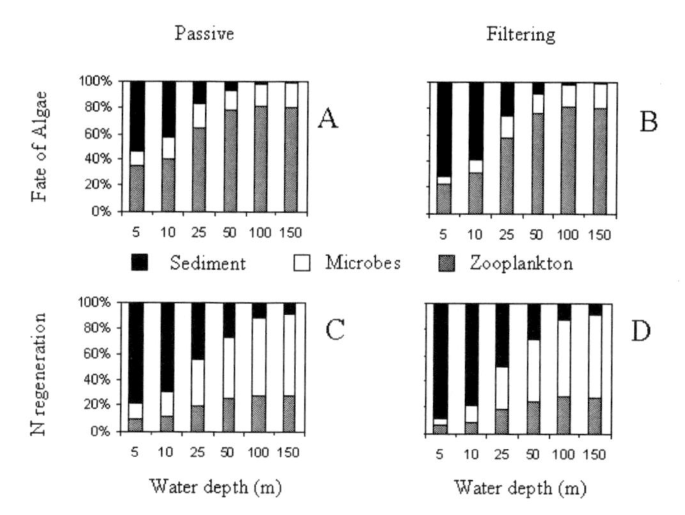


Figure 11.10 A) The fate of phytoplankton production as a function of water depth in the absence of benthic filtering activity: grazing, settling to the sediment and mortality. B) The fate of phytoplankton production as a function of water depth in the presence of benthic filtering activity: grazing, filtering and settling to the sediment and mortality. C) The partitioning of the nitrogen regeneration among sediments, zooplankton and heterotrophic microorganisms for passive sediments. D) The partitioning of the nitrogen regeneration among sediments, zooplankton and heterotrophic microorganisms with benthic filtering activity.

Similar conclusions regarding depth dependence and the role of filtering are reached when nitrogen regeneration is analyzed (Figure 11.10 C, D). At 150 m water depth, ~ 28% of the regeneration is due to zooplankton, 63 % is due to pelagic bacteria and only 9 % of the regeneration can be attributed to sediments. At 5 m depth and in the absence of filtering activities, zooplankton and pelagic bacteria each account for about 10% of the nitrogen regeneration with sediment mineralization contributing about 80% (Figure 11.10C). Active filtering of the water column by benthic animals or permeable beds further increases the share of sediment to 88% (Figure 11.10D).

## 7. Conclusions

The presence of a benthic compartment in direct contact with the pelagic compartment has major consequences for shelf ecosystem functioning. We are the first to admit that our computational approach has severe limitations in reproducing actual rates because of the many simplifications underlying the model. Moreover, model parameterization is based on a rather turbid, nutrient-rich shelf in the temperate zone. As a consequence we might have underestimated the role of benthic microalgae for clear-water, nutrient-poor shelf systems (e.g. Blackford, 2002; Jahnke et al., 2000). Nevertheless it will be clear that the benthos is an essential part of shelf ecosystem functioning. Respiration and N regeneration in sediments account for > 80 % of total system activity in shallow sediments to about 10 % at the shelf-break and can not be ignored in studies and budgets of the continental shelf. Moreover, some processes are limited to the sediment because they depend on suboxic or anoxic conditions, e.g. denitrification. Regeneration of material in the sediment also results in an attenuation and delay of the response of shelf ecosystems to short-term pulses of primary production (Figures 11.4, 11.5).

Biogeochemists traditionally focus on clays and silts that passively receive detritus from the overlying water (Boudreau et al., 2001). However, sandy sediments are dominant in continental shelves and act as filter beds that remove particles from the water column (Jahnke, Chapter 6). Our simulations were based on a very simple parameterization of benthic filtering activity (because of a lack of basic data) and should be considered as exploratory exercises rather than as quantitative reproductions. These simulations suggest that the effect of the benthic filter is insignificant if the water depth is > 50 m. However, at shallower depths, both the benthic and pelagic compartments are affected by filtering.

Pelagic ecosystem dynamics is influenced directly by removal of algal biomass by benthic filtering and may result in a lower phytoplankton biomass and production. At shallow depths (< 10 m), this effect is more important than the increased efficiency of input to the sediment and leads to lower supply of material to the sediments. At intermediate depths (10-50 m), the presence of a benthic filter has a smaller effect on algal mortality but significantly increases the net flux of organic matter to the sediment. We hope that we have conveyed that, in order to further understanding, a continental-shelf ecosystem study should explicitly consider the benthos. Since the median shelf depth is about 60-70 m, it is also clear that benthic filtering due to suspension feeders or permeable beds requires further study because of its consequences for ecosystem functioning and stability.

TABLE I.

Sediment-water exchange formulations.

Water-column equations are the same for all models and are of the form:

$$\frac{\partial C_w}{\partial t} = \frac{\partial}{\partial z} Kz \frac{\partial C_w}{\partial z} - \frac{\partial w_s \cdot C_w}{\partial z} + sources - \sin ks$$

$$\frac{\partial NO3_w}{\partial t} = \frac{\partial}{\partial z} Kz \frac{\partial NO3_w}{\partial z} + sources - \sin ks$$

Here sources and sinks are as in Soetaert et al. (2001).

Equations used in the sediment and the boundary conditions at the sediment-water interface

| Dynamic equations | Upper Boundary conditions |
|---|---|
| $\frac{dC_S}{dt} = Cdeposition + C_S \,\mathrm{Prod} - \lambda \cdot f(T) \cdot C_S$ | $Cdeposition = (w_s + SandFilterRate) \cdot DetritalC_w\|_0 +$ $+(w_A + SandFilterRate + ClearanceRate) \cdot AlgalC_w\|_0$ |
| $\frac{dN_S}{dt} = Ndeposition + N_S \,\mathrm{Prod} - \lambda \cdot f(T) \cdot N_S$ | $Ndeposition = (w_s + SandFilterRate) \cdot DetritalN_w\|_0$ $+(w_A + SandFilterRate + ClearanceRate) \cdot AlgalN_w\|_0$ |
| $C_S \,\mathrm{Prod} = \mu_{\max} \cdot MIN(LightLim, Nut\lim)$ | |
| $N_S \,\mathrm{Prod} = C_S \,\mathrm{Prod} \cdot \gamma_C^N$ | $NO3efflux = p_{NIT} \cdot \lambda \cdot f(T) \cdot N_S - p_{NO3} \cdot N_S \,\mathrm{Prod} = -Kz \left.\frac{\partial NO3_w}{\partial z}\right\|_0$ |
| $LightLim = 1 - e^{I_0 \alpha} - r$ | |
| $NutLim = \frac{(NO3_w + NH3_w)\|_0}{(NO3_w + NH3_w)\|_0 + ksDIN}$ | $NH3efflux = (1 - p_{NIT}) \cdot \lambda \cdot f(T) \cdot N_S - (1 - p_{NO3}) \cdot N_S \,\mathrm{Prod} = -Kz \left.\frac{\partial NH3_w}{\partial z}\right\|_0$ |
| $f(T) = Q10^{\frac{(T-20)}{10}}$ | $O2efflux = \gamma_{O2}^C \cdot (C_S \,\mathrm{Prod} - \lambda \cdot f(T) \cdot C_S) - 2 \cdot p_{NIT} \cdot \lambda \cdot f(T) \cdot N_S = -Kz \left.\frac{\partial O2_w}{\partial z}\right\|_0$ |
| | where: |
| | $SandFilterRate = Filter\|_{D=0} \cdot e^{cF \cdot D}$ |
| | $ClearanceRate = SpecificClearance \cdot Biomass\|_{D=0} \cdot e^{cC \cdot D}$ |
| | $p_{NO3} = \frac{NO3_w\|_0}{NO3_w\|_0 + NH3_w\|_0}$ |

TABLE II
Symbols used.

| Variable | Units | Description |
|---|---|---|
| D | m | Water depth of the station |
| z | m | Vertical dimension, axis origin at the sediment-water interface |
| Kz | $m^2 d^{-1}$ | Vertical diffusion coefficient, generated by a turbulence-closure model (Gaspar et al., 1990). |
| $w_s$ | $m d^{-1}$ | Sinking velocity of detritus, calculated by the aggregation formulation of Kriest and Evans (1999). |
| Cw | mmol $m^{-3}$ | Water-column concentration of organic matter, vertically resolved |
| DetritalCw, AlgalCw | mmol $m^{-3}$ | Water-column concentration of detritus or algal organic carbon, vertically resolved. |
| T | flC | Temperature |
| f(T) | — | Temperature scaling function (Q10) |
| $O2_w$ | mmol $m^{-3}$ | Water-column concentration of oxygen, vertically resolved |
| $NH3_w$ | mmol $m^{-3}$ | Water-column concentration of ammonium, vertically resolved |
| $NO3_w$ | mmol $m^{-3}$ | Water-column concentration of nitrate, vertically resolved |
| $C_S$, $N_S$ | mmol $m^{-2}$ | Vertically integrated sediment organic C and organic N |
| $C_S$Prod, $N_S$prod | mmol $m^{-2} d^{-1}$ | Benthic primary production |
| LightLim | — | Light limitation factor for benthic primary production |
| NutLim | — | Nutrient limitation factor for benthic primary production |
| $I_0$ | μEinst $m^{-2} s^{-1}$ | Photosynthetically active radiation at the sediment-water interface |
| SandFilterRate | $d^{-1}$ | Rate of filtering due to advection through sandy sediment |
| ClearanceRate | $d^{-1}$ | Rate of filtering due to filterfeeders |
| O2efflux | mmol $O_2$ $m^{-2} d^{-1}$ | Calculated flux of oxygen out of the sediment |
| NO3efflux | mmol N $m^{-2} d^{-1}$ | Calculated flux of nitrate out of the sediment |
| NH3efflux | mmol N $m^{-2} d^{-1}$ | Calculated flux of ammonium out of the sediment |
| $_P$NO3 | — | Part nitrate in overlying water DIN |

Parameter values

| Parameter | Value | Units | Description |
|---|---|---|---|
| $\lambda$ | 0.1 | $d^{-1}$ | First-order decay rate of sediment organic matter at 20 ° |
| Q10 | 2 | — | Factor of increase of decay rate for every increase in temperature of 10 °C |
| $\mu_{max}$ | 100 | mmol C $m^2 d^{-1}$ | Maximal sedimentary photosynthesis rate |
| $\alpha$ | –0.018 | (μEinst $m^{-2} s^{-1})^{-1}$ | Light dependency parameter for benthic photosynthesis |
| r | 0.27 | — | Dark respiration fraction of benthic photosynthesis |
| $Ks_{DIN}$ | 0.5 | mmol N $m^{-3}$ | Half-saturation nitrogen concentration for benthic primary production |
| $\gamma_C^N$ | 0.156 | mol N (mol C)$^{-1}$ | N:C ratio of benthic production |
| $p_{NIT}$ | 0.9 | — | Part of sedimentary ammonium production that is nitrified |
| $\gamma_C^{O2}$ | 1 | mol $O_2$ (mol C)$^{-1}$ | O:C ratio in organic matter |
| $Filter_{D=0}$ | 1 | $m^3 m^{-2} d^{-1}$ | Rate of filtering due to advection through sandy sediment at 0 m water depth (beaches) |
| cF | –0.045 | $m^{-1}$ | Decline of sand filtering rate with increasing water depth |
| $w_A$ | 0.5 | m $d^{-1}$ | Sinking rate of algae |
| SpecificClearance | 0.057 | $m^3$ gADWT$^{-1}$ $d^{-1}$ | Rate of filtering per gram of filterfeeder ash free dry weight (ADWT) |
| $Biomass_{D=0}$ | 16.59 | gADWT $m^{-2}$ | Filter feeder ash-free dry weight per $m^2$ at 0 m water depth |
| cC | –0.024 | $m^{-1}$ | Decline of filtering feeder biomass with increasing water depth |

## Acknowledgements

We thank Adri Knuyt and Celia Berlizot for graphic and manuscript preparation support, our NIOO colleagues for discussion and Peter Jumars for a constructive review. Support was provided by the European Union (ORFOIS EVK2-CT2001-00100; COSA EVK3-2002-00076) and the Netherlands Organisation for Scientific Research (PIONIER 833.02.002). This is NIOO-KNAW publication 3259.

## Bibliography

Boudreau, B. P., 1997. *Diagenetic models and their implementation.* Springer Verlag, Berlin, 441p.

Boudreau B.P., M. Huettel, S. Forster, R.A. Jahnke, A. McLachlan, J.J. Middelburg, P. Nielsen, F. Sansone, G, Taghon, W. van Raaphorst, I. Webster, J.M. Weslawski, P. Wiberg, B. Sundby, 2001. Permeable marine sediments: overturning an Old Paradigm. *EOS* **82 (11)**: 133-136.

Blackford J.C., 2002. The influence of microphytobenthos on the Northern Adriatic Ecosystem: A Modelling study. *Estuarine Coastal Shelf Science* **55**, 109-123.

Brandes, J. A., and A. H. Devol, 2002. A global marine-fixed nitrogen isotopic budget: Implications for Holocene nitrogen cycling. *Global Biogeochemical Cycles*, **16**(4), art. no.-1120.

Bryant, A.D., M. R. Heath, N. Broekhuizen, J.G. Ollason, W.S.C. Gurney, and S.P.R. Greenstreet, 1995. Modelling the predation, growth and population dynamics of fish within a spatially-resolved shelf-sea ecosystem model. *Netherlands Journal of Sea Research*, **33**(3/4), 407-421.

Cahoon, L. B., 1999. The role of benthic microalgae in neritic ecosystems. *Oceanography and Marine Biology*, **37**, 47-86.

Charpy-Roubaud, C., and A. Sournia, 1990. The comparative estimation of phytoplanktonic, microphytobenthic and macrophytobenthic primary production in the oceans. *Marine Microbial Food Webs*. **4**, 31-57.

Chen, C-T. A., 2003. New vs. export production on the continental shelf. *Deep-Sea Res.* II **50**, 1327-1333.

Dauwe, B., and J. J. Middelburg, 1998. Amino acids and hexosamines as indicators of organic matter degradation state in North Sea sediments. *Limnology and Oceanography*, **43**(5), 782-798.

de Haas, H., T. C. E. van Weering, and H. de Stigter, 2002. Organic carbon in shelf seas: sinks or sources, processes and products. *Continental Shelf Research*, **22**(5), 691-717.

del Giorgio, P. A., and C. M. Duarte, 2002. Respiration in the open ocean. *Nature*, **420**(6914), 379-384.

Droop, M.R., 1973. Some thoughts on nutrient limitation in algae. *J. Phycol.* **9**, 264-272.

Dugdale, R. C., and J. J. Goering, 1967. Uptake of new and regenerated forms of nitrogen in primary productivity. *Limnology and Oceanography*, **12**, 196-206.

Fries, J. S., and J. H. Trowbridge, 2003. Flume observations of enhanced fine-particle deposition to permeable sediment beds. *Limnology and Oceanography*, **48**(2), 802-812.

Gaspar, P., Grégoris, Y., Lefevre, J-M., 1990, A simple eddy-kinetic energy model for simulations of the oceanic vertical mixing : tests at Station Papa a long-term upper ocean study site. *Journal of Geophysical Research* **95** (C9), 16179-16193.

Gattuso, J.-P., M. Frankignoulle, and R. Wollast, 1998. Carbon and carbonate metabolism in coastal aquatic ecosystems. *Annual Review of Ecology and Systematics*, **29**, 405-434.

Hall, S. J., 2002. The continental shelf benthic ecosystem: current status, agents for change and future projects. *Environmental Conservation*, **29**(3), 350-374.

Hedges, J. I., and R. G. Keil, 1995. Sedimentary organic-matter preservation - an assessment and speculative synthesis. *Marine Chemistry*, **49**(2-3), 81-115.

Heip, C. H. R., N. K. Goosen, P. M. J. Herman, J. Kromkamp, J. J. Middelburg, and K. Soetaert, 1995. Production and consumption of biological particles in temperate tidal estuaries. *Oceanography and Marine Biology - an Annual Review*, **33**,1-149.

Heip, C. H. R. and J. Craeymeersch, 1995. Benthic community structures in the North Sea. *Helgoländer Meeresuntersuchungen* **49**, 313-328.

Herman, P. M. J., J. J. Middelburg, J. Van de Koppel, and C. H. R. Heip, 1999. Ecology of estuarine macrobenthos. *Advances in Ecological Research*, **29**, 195-240.

Huettel, M., W. Ziebis, and S. Forster, 1996. Flow-induced uptake of particulate matter in permeable sediments. *Limnology and Oceanography*, **41**(2), 309-322.

Huettel, M., and A. Rusch, 2000. Transport and degradation of phytoplankton in permeable sediment. *Limnology and Oceanography*, **45**(3), 534-549.

Huettel, M., and I. T. Webster, 2001. Porewater flow in permeable sediments. In *The Benthic boundary layer*, B.P. Boudreau and B.B. Jorgensen, eds. Oxford University Press, pp. 144-179.

Jahnke, R. A., J. R. Nelson, R. L. Marinelli, and J. E. Eckman, 2000. Benthic flux of biogenic elements on the Southeastern US continental shelf: influence of pore water advective transport and benthic microalgae. *Continental Shelf Research*. **20**, 109-127.

Kriest, I. and G.T. Evans, 1999. Representing phytoplankton aggregates in biogeochemical models. *Deep-Sea Research* I **46**, 1841-1859.

Jørgensen, B.B., 1983. Processes at the sediment-water interface, In The major biogeochemical cycles and their interactions, B. Bolin and R.B. Cook [eds.], SCOPE. p. 477-515

Longhurst, A., S. Sathyendranath, T. Platt, and C. Caverhill, 1995. An estimate of global primary production in the ocean from satellite radiometer data. *Journal of Plankton Research*, **17**(6), 1245-1271.

Middelburg, J. J., T. Vlug, and F.J.W.A Van der Nat, 1993. Organic-matter mineralization in marine systems. *Global and Planetary Change*, **8**(1-2), 47-58.

Middelburg, J. J., K. Soetaert, P. M. J. Herman, and C. H. R. Heip, 1996. Denitrification in marine sediments: A model study. *Global Biogeochemical Cycles*, **10**(4), 661-673.

Middelburg, J. J., C. Barranguet, H. T. S. Boschker, P. M. J. Herman, T. Moens, and C. H. R. Heip, 2000. The fate of intertidal microphytobenthos carbon: An in situ C-13-labeling study. *Limnology and Oceanography*, **45**(6), 1224-1234.

Middelburg J. J., C.M. Duarte, and J.P. Gattuso, 2004. Respiration in coastal benthic communities. In *Respiration in aquatic ecosystems*, P. A. del Giorgio and P.J. leB. Williams, eds., Oxford University Press, p. 206–224

Moore, W.S., 1999. The subterrenean estuary: a reaction zone of ground water and sea water. *Marine Chemistry* **65**, 111-125.

Paterson, D. M., and K. S. Black, 1999. Water flow, sediment dynamics and benthic biology. *Advances in Ecological Research*, **29**, 155-193.

Precht, E., and M. Huettel, 2003. Advective pore water exchange driven by surface gravity waves and its ecological implications. *Limnology and Oceanography* **48**(4), 1674-1684.

Rabouille, C., F. T. Mackenzie, and L. M. Ver, 2001. Influence of the human perturbation on carbon, nitrogen, and oxygen biogeochemical cycles in the global coastal ocean. *Geochimica Et Cosmochimica Acta*, **65**(21), 3615-3641.

Riedl, R., N. Huang, and R. Machan, 1972. The subtidal pump: a mechanism of intertidal water exchange by wave action. *Mar. Biol*, **13**(3), 210-221.

Rysgaard, S., N. Risgaard-Petersen, N.P. Sloth, K. Jensen and L.P. Nielsen, 1994. Oxygen regulation of nitrification and denitrification in sediments. *Limnology Oceanography* **39**, 1643-1652.

Smith, C.L. and P. Tett, 2000. A depth-resolving numerical model of physically forced microbiology at the European shelf edge. *Journal of Marine Systems*, **26**, 1-36

Soetaert, K., P. M. J. Herman, and J. J. Middelburg, 1996. A model of early diagenetic processes from the shelf to abyssal depths. *Geochimica Et Cosmochimica Acta*, **60**(6), 1019-1040.

Soetaert, K., J. J. Middelburg, P. M. J. Herman, and K. Buis, 2000. On the coupling of benthic and pelagic biogeochemical models. *Earth-Science Reviews*, **51**(1-4), 173-201.

Soetaert, K., P.M.J. Herman, J.J. Middelburg, C.H.R. Heip, C.L. Smith, P. Tett, and K. Wild-Allen, 2001. Numerical modelling the shelf break ecosystem: reproducing benthic and pelagic measurements. *Deep-Sea Res. II* **48**, 3141-3177.

Soetaert, K., J. J. Middelburg, J. Wijsman, P. Herman, and C. Heip, 2002. Ocean margin early diagenetic processes and models. In: *Ocean Margin Systems* G. Wefer, D. Billet, D. Hebbeln, B.B. Jørgensen, M. Schlüter, and T. van Weering, Eds, Springer-Verlag, Heidelberg, pp. 157-177.

Tett, P., and M.R. Droop, 1988. Cell quota models and planktonic primary production. In *CRC Handbook of Laboratory Model Ecosystems for Microbial Ecosystems*, J.W.T. Wimpenny, Ed., CRC Press, Boca Raton, FL, pp. 177-233.

Thomsen, L, and E. Flach 1997. Mesocosm observations of fluxes of particulate matter within the benthic boundary layer. *Journal of Sea Research* **37 (1-2)**: 67-79

Wollast, R., 1998. Evaluation and comparison of the global carbon cycle in the coastal zone and in the open ocean. In *The Sea, The Global coastal ocean, processes and methods*, K.H. Brink, and A.R. Robinson, eds. Wiley and Sons, New York, pp. 213-252.

# Chapter 12. DYNAMICS AND INTERACTIONS OF AUTOTROPHS, LIGHT, NUTRIENTS AND CARBON DIOXIDE

TREVOR PLATT

*Bedford Institute of Oceanography*

RICHARD GEIDER

*University of Essex*

ANTOINE SCIANDRA
CLAIRE COPIN-MONTÉGUT

*Laboratoire d' Océanographie de Villefranche*

HEATHER A. BOUMAN
SHUBHA SATHYENDRANATH

*Dalhousie University*

## Contents



## 1. Introduction

In the partition of the ocean for biogeochemical analysis by Longhurst (1998), the coastal zone is designated as one of four major domains or biomes. In this biome, the characteristic forcing mechanism, in so far as phytoplankton[1] are concerned, is the modification of the effect of wind and currents by topography. The continental margin, although representing a modest fraction of the surface area of the global

[1] We exclude here autotrophic production by attached macroalgae, a minor contribution to the primary production of the entire continental margin.

*The Sea*, Volume 13, edited by Allan R. Robinson and Kenneth H. Brink
ISBN 0-674-01526-6 

ocean, is responsible for an appreciable proportion of the marine primary production. Hence, on a unit-area basis the continental margin is highly productive. The principal reasons include a typically-shallow water column; ready access to nutrients through upwelling, vertical mixing caused by tides or wind or through runoff from land; and outside of the tropics, the creation of a stably-stratified surface layer in the warm season that permits rapid phytoplankton growth rates and complete utilisation of available nutrients.

Among the factors that control the growth of autotrophic taxa in the sea, the dominant one is light: after all, their mode of nutrition is photosynthesis. In the coastal zone, the submarine light field is particularly difficult to diagnose. Whereas, in the open ocean, the attenuation of light is dominated by the presence of phytoplankton itself, in the coastal zone the contribution of other factors has to be taken into account. These include inorganic sediments, detrital organic material carried in from the land and coloured, dissolved organic material (yellow substance) arising as a product of catabolism. None of the contributing factors are spectrally-neutral: a complete description of the submarine light field must take wavelength into account.

The submarine light field, acting on the photosynthetic pigments in the autotrophic biomass, forces primary production. The relationship between irradiance and photosynthesis (the photosynthetic response of unit pigment biomass to light) is called the light–saturation curve. It is a rectangular hyperbola that requires two parameters in the simple case of no photoinhibition. These are the slope of the curve at low light (the initial slope) and the amplitude of the plateau at high light (the assimilation number). Their magnitudes at a particular time and place reflect the influence of secondary determinants of primary production, such as temperature, nutrients, community composition, light history and so on.

Beyond these secondary determinants of photosynthesis, the formation of a stably-stratified surface layer is also of profound importance for coastal primary production. In the surface mixed layer, production is strongly dependent on the quantity $KZ_m$, where $Z_m$ is the depth of the mixed layer, and $K$ is a representative optical attenuation coefficient (Platt and Sathyendranath, 1991). Shallow mixed layers favour high rates of production per unit volume in the layer. At the beginning of the spring heating season, there is a tendency for a shallow pycnocline to be developed. Typically, the water in this layer is replete with nutrients following winter mixing. These conditions are ideal for the initiation of a bloom, and biomass will increase rapidly provided that calm conditions prevail. Turbulence due to wind tends to erode incipient stratification and redistributes the biomass over a broader vertical range. In any layer, biomass increases through production and decreases through respiration. The balance between the two sets the sign of the net change in biomass. The layer depth for which the net change in one day would be zero is called the critical depth (Sverdrup, 1953). A mixed layer shallower than the critical depth is a necessary condition for development of a bloom.

In coastal waters, it is particularly important to understand the processes controlling the initiation of blooms (Zingone and Wyatt, Chapter 22). The incidence of harmful blooms is of increasing concern (Rabalais, Chapter 21). Also, the timing, duration and intensity of the spring phytoplankton blooms, and their variation between years (Platt et al., 2003), are of fundamental interest to fishery managers, given that many commercially-important species spawn in this season.

It is generally accepted that phytoplankton dynamics are, ultimately, under control of physical forcing. This is particularly true in coastal regimes, where the forcing is varied and often intense. An especially important phenomenon is the formation of fronts, topographically or tidally induced. These are usually the sites of enhanced biological production. Less generally recognised is the influence of biological dynamics on the seasonal evolution of mixed-layer physics and frontal dynamics. Models to diagnose the temperature and depth of the mixed layer usually strike a balance between the tendency for solar heating to stratify the water column and the opposing tendency for wind to erode vertical structure.

In specifying the heating side, it is required to assign the diffuse, vertical attenuation coefficient $K$, a quantity dependent on the pigment biomass, which itself changes rapidly in the spring season. Increased biomass localises the heating by the penetrative component of the sun nearer the sea surface, and so reduces the depth interval over which convection must take place to supply the surface heat loss to the atmosphere. Thus, during the season when the surface mixed layer is forming, pigment biomass and mixed layer depth evolve in a coupled manner (Platt et al., 1994b). These, in turn, can impact the formation and evolution of fronts in coastal waters.

In the context of the ocean carbon cycle, we distinguish between new production (using nitrogen supplies external to the local photic layer) and regenerated production (using reduced products of metabolism as a nitrogen source): the sum of these two components is the total primary production. Regenerated production is the component that meets the demands of community metabolism. Only new production is relevant to the issue of carbon export. In the coastal zone, externally-supplied nutrients come from river drainage and especially from upwelling. The upwelled water may be rich in carbon dioxide as well as nitrogen: the carbon supplied to the surface by this mechanism may exceed the consumption of carbon in photosynthesis. An open question concerns the role of iron, which may be transported from sediments to the surface layer, as a limiting resource for primary production in coastal areas.

In the sections that follow, we shall explore in more detail some of the issues presented above, attempting to show how the various determinants of photosynthesis reinforce or oppose each other, leading to the rates of primary production and the partial pressures of carbon dioxide that we measure in the natural coastal environment.

## 2. Primary Production and Light

There are two principal issues to consider in discussion of the effect of light on marine photosynthesis. One is the instantaneous, photosynthetic response at a particular depth to the light available at that depth. The other is the photosynthetic response of the entire water column to the surface irradiance, either instantaneously or integrated over the light day.

To calculate primary production at a discrete depth, we employ the light-saturation curve, for which there are several mathematical options (Platt et al., 1977). A representative one is:

$$P^B(I) = P_m^B(1 - \exp(-\alpha^B I / P_m^B)). \quad (1)$$

This equation describes a rectangular hyperbola. It relates the instantaneous rate of primary production $P$, normalised to the pigment biomass $B$ to the irradiance $I$. The curve is linear at the origin with slope $\alpha^B$ and will rise to a plateau of amplitude $P_m^B$, called the assimilation number. In this representation, we have ignored the possible influence of photoinhibition (see Platt et al., 1990).

More importantly, we have ignored all complications associated with the spectral structure of the light field and with the spectral dependence of photosynthetic response. Such spectral effects are real and significant. They will be particulary important in coastal regions, given that the optical properties of coastal zones are variable and complex compared with those of the open ocean.

### 2.1 *Spectral Models of Photosynthesis*

Non-spectral representations of the light-saturation curve can be converted to spectral ones without difficulty (Sathyendranth and Platt, 1989). The key is that, no matter what the representation used for the light -saturation curve, the irradiance $I$, wherever it occurs, will always be multiplied by the initial slope $\alpha^B$. Let us define this product as $\alpha^B I = \Pi^B$. To find the corresponding representation of $\Pi^B$ for a spectrally-resolved model, we have to add the contributions to $\Pi^B$ from all spectral elements in the photosynthetically-active range (PAR). Let us call the result $\Pi_\lambda^B$. Then, we have

$$\Pi_\lambda^B = \int_{PAR} \alpha^B(\lambda) I(\lambda)\, d\lambda. \quad (2)$$

This equation amounts to a spectrally-weighted integral of photosynthetic response to different spectral components of the submarine light field. The spectral equivalent of, for example, equation (1) can then be found by replacing the product $\alpha^B I$ with $\Pi_\lambda^B$, as follows

$$P^B(I) = P_m^B(1 - \exp(-\Pi_\lambda^B / P_m^B)). \quad (3)$$

### 2.2 *The Submarine Light Field*

Computation of production at any depth $z$ thus requires knowledge of the forcing variable, $I$, at that depth. Given the irradiance at the surface, light at any given depth $z$ can be computed if we know $K$, the diffuse attenuation coefficient for downwelling irradiance, applicable to the layer from the surface to depth $z$. This attenuation coefficient is a spectral property, and the most rigourous computations account for the spectral composition of light at the sea surface, and for spectral variations in $K$. The coefficient $K$ determines the rate of decrease of downwelling light with depth, and for a wavelength $\lambda$, it is defined as

$$K(z,\lambda) = \frac{-1}{I(z,\lambda)} \frac{\mathrm{d}I(z,\lambda)}{\mathrm{d}z}. \tag{4}$$

Generally, especially for open-ocean waters, we may parameterise $K$ using the approximation of Sathyendranath and Platt (1988) as follows

$$K(z,\lambda) = \frac{a(z,\lambda) + b_b(z,\lambda)}{\langle \mu_d(z,\lambda) \rangle}, \tag{5}$$

where $a$ is the absorption coefficient, and $b_b$ is the backscattering coefficient. The quantity $\langle \mu_d \rangle$ is the mean cosine of the downwelling light field at depth $z$, that is to say the average of the cosine of downwelling irradiance weighted by the angular distribution of the downwelling stream. It accounts for the increased path length of photons per unit vertical excursion for photons incident away from the normal to the surface.

Equation (5) works reasonably well in waters where absorption coefficient is much larger than the back-scattering coefficient. In coastal waters, however, where highly-scattering conditions may prevail, it might be preferred to use a more complete formulation for $K$, such as presented by Priesendorfer and Mobley (1984; see also Sathyendranath and Platt, 1991)

$$K(z,\lambda) = \frac{a(z,\lambda) + b_u(z,\lambda)}{\langle \mu_d(z,\lambda) \rangle} - \frac{b_d(z,\lambda)}{\langle \mu_u(z,\lambda) \rangle} R(z,\lambda), \tag{6}$$

where $\langle \mu_u \rangle$ is the mean cosine for the upwelling irradiance, $b_u$ is the upward scattering coefficient for downwelling light, and $b_d$ is the downward scattering coefficient for upwelling light. The quantity $R$ is the reflectance, defined as the ratio of the upwelling irradiance to the downwelling irradiance. Note also that, in turbid coastal waters, $\langle \mu_u \rangle$ and $\langle \mu_d \rangle$ may vary more than in open-ocean waters, due to multiple scattering.

Generally, the second term in equation (6) may be neglected because the reflectance is small compared to unity. Moreover, in typical open-ocean waters, $b_d \ll a$. However, the second term may not be negligible in highly turbid coastal waters. Note also that the first terms of equations (5) and (6) differ in that the term $b_b$ of the one is replaced by $b_u$ in the other. Particle scattering in sea water is highly asymmetrical, with 99% being in the forward direction, such that, for light not incident in the normal direction, upward scatter exceeds backscatter. Therefore, equation (5) will underestimate $K$ in the highly-scattering waters commonly found in coastal environments. It is recommended to use equation (6), even if the second term on the right-hand side is dropped.

When the photosynthesis model used is a nonspectral one, we can improve its performance by care in assigning the value of $K$ that would be applicable to the entire range of photosynthetically-active radiation (PAR). The $K$ so defined is often referred to as $K_{PAR}$. Then the $K_{PAR}$ we seek is the one that satisfies the equation

$$\int_{PAR} I(z,\lambda)d(\lambda) = \exp(-zK_{PAR})\int_{PAR} I(0,\lambda)d\lambda, \tag{7}$$

where the integrals are taken over the entire range of PAR. In other words, the $K_{PAR}$ we want is

$$K_{PAR} = \frac{1}{z}\times\log_e\left(\int_{PAR}(I(0,\lambda)d\lambda / \int_{PAR}(I(z,\lambda)d\lambda)\right). \tag{8}$$

Here, we take advantage of our need for a $K$ that is representative only of a layer as a whole (from surface to depth $z$) without regard to the variation of $K_{PAR}$ within the layer (Platt et al., 1994a). What is often not recognised, however, is that this $K_{PAR}$ is not a property of the entire water column, even in the case of a uniform water body. It can be defined only for a particular depth interval, in this case the interval $(0, z)$.

These equations emphasise the premium placed on characterisation of the submarine light field, here represented by $K$. Another matter in computation of light attenuation concerns the representation of the absorption coefficient $a$. Here again, in coastal waters, more care is needed than in open-ocean waters. The contributions to $a$ arise from the water itself; the suspended, inert matter (detritus); the living, suspended matter (phytoplankton); and the coloured, dissolved organic matter (yellow substances). In the open ocean, the detrital component is usually not significant, or at least, may be assumed to co-vary with phytoplankton as a first approximation. Thus, in the open-ocean waters, it is often possible to estimate the total absorption coefficient as a function of a single independent variable, the concentration of chlorophyll-a.

In coastal waters, however, the contribution of sediments to the detrital component can be substantial and variable. The same is true of the yellow substances: because of the supply of decaying organic matter to the coastal zone from river drainage, it may not be secure to model their contribution as a quantity dependent on phytoplankton pigment concentration. Furthermore, given that phytoplankton biomass and production can reach high values at certain seasons in the coastal zone, the absolute values of absorption due to phytoplankton and yellow substances may, at times, be far higher than those encountered in the open ocean. To illustrate the relative importance of yellow substances in coastal waters compared with waters removed from terrestrial influences, we show the total absorption spectrum of phytoplankton, yellow substances and detritus for two stations; one located in a coastal inlet (Bedford Basin, Nova Scotia, Canada) and the other on the Nova Scotian Shelf (Figure 12.1). For the coastal station, clearly, yellow substances are the dominant contributors to total absorption in the blue wavebands, whereas for the shelf station, phytoplankton is the principal contributor to the absorption of blue light.

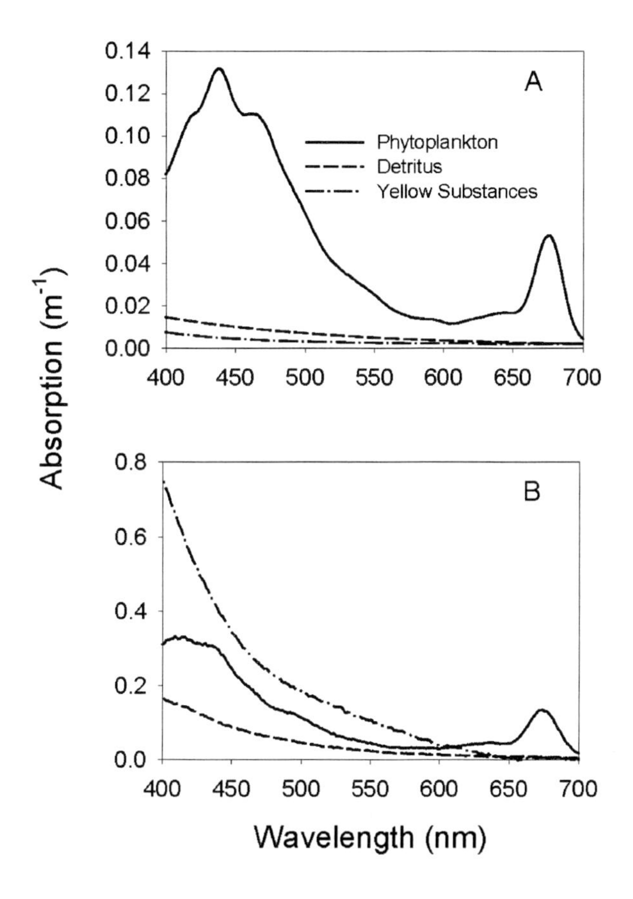


Figure 12.1 Total absorption spectra of phytoplankton, yellow substances and detritus for A) a station located on the Nova Scotian Shelf (Sathyendranath and Devred, unpublished data) and B) a station located in a coastal inlet (Bedford Basin, Nova Scotia). (Bouman, unpublished data).

The absorptive properties of phytoplankton are the most studied of the optical components. Variability in the spectral shape of phytoplankton absorption is known to be influenced by pigment composition, cell size and the intracellular concentration of pigment. Phytoplankton pigments, which absorb within discrete windows of the visible spectrum and have characteristic spectral shapes, contribute jointly to the overall shape of the phytoplankton absorption spectrum. Differences in the shape of the absorption spectrum of phytoplankton due to differences in pigment composition can be clearly seen in Figure 12.2*a*, which shows two absorption spectra normalised to 440 nm. The spectrum with strong absorption in the green region was obtained from a sample collected during a red tide of the non-toxic photosynthetic ciliate *Mesodinium rubrum* (= *Myrionecta rubra*). The broad absorption peak centred at 545 nm denotes the presence of the phycobiliprotein phycoerythrin. The absorption spectrum with the relatively low absorption at the 545 nm waveband was collected just one day before the red tide event. Pigment data indicated that the phytoplankton composition was a mixture of prymnesiophytes and diatoms.

Phytoplankton cell size and intracellular pigment concentration also influence the absorptive efficiency of phytoplankton by the package or flattening effect. Since pigments are packaged within chloroplasts, which in turn are contained within individual cells, the efficiency with which they absorb light is lower than it would be if they were distributed uniformly in solution. The efficiency factor for absorption ($Q_a$) at a given wavelength can be written as a function of the intracellular chlorophyll-a concentration, $C_i$ (mg chlorophyll-a $m^{-3}$), the chlorophyll-specific absorption coefficient of the cellular material $a_{cm}^{B}$ ($m^2$ [mg chl-a]$^{-1}$), and the equivalent spherical diameter of the cell, $d$ (m) (Morel and Bricaud, 1981):

$$Q_a = (2/3) a_{cm}^{B} C_i d. \tag{9}$$

To illustrate the effect of cell size on phytoplankton absorption, we show two absorption spectra collected in Bedford Basin normalised to chlorophyll-a concentration (Figure 12.2*b*). One spectrum was measured from a sample taken during a spring diatom bloom, whereas the other spectrum was collected in the summer, when small flagellate cells were dominant. The flattening effect is apparent, the diatom-dominated sample having a much flatter shape and lower absorption coefficients than the flagellate-dominated sample.

Compared with phytoplankton, variability in the spectral shapes of the non-algal components of light attenuation (detritus and yellow substances) is less well known. Like phytoplankton, detritus and CDOM are strong absorbers of blue light and weak absorbers of green light. The shapes of the absorption spectra of CDOM and detritus can both be described by the following exponential function:

$$a_x(\lambda) = a_x(\lambda_0)\exp[-S(\lambda - \lambda_0)], \tag{10}$$

where $a_x(\lambda)$ is the absorption coefficient of either detritus ($x$=$d$) or yellow substances ($x$=$y$) at wavelength $\lambda$, $a_x(\lambda_0)$ is the absorption coefficient at reference wavelength $\lambda_0$ and $S$ is the slope of the exponential curve. Højerslev (1998) has shown that the slope parameter for yellow substances can vary significantly in coastal systems, with values of $S_y$ ranging from 0.008 to 0.042 $nm^{-1}$. Kopelovich et al. (1989), who analysed over 400 absorption spectra from a number of studies conducted in oceanic, coastal and freshwater systems, reported much less variability in $S_y$, with a mean value of 0.017 $nm^{-1}$ and a standard deviation of 0.001 $nm^{-1}$. In a study of the coastal waters around the San Juan Islands, Roesler et al. (1989) reported the same mean value of $S_y$ of 0.017 $nm^{-1}$ (SD = 0.003 $nm^{-1}$). A recent study by Babin et al. (2003) conducted in a wide range of European coastal regions obtained a similar mean $S_y$ value of 0.0176 $nm^{-1}$ (SD = 0.0020 $nm^{-1}$). Slope values of detritus spectral absorption have been shown to be lower than that of yellow substances. Studies by Roesler et al. (1989) and Babin et al. (2003) reported mean values 0.011 $nm^{-1}$ (SD = 0.002 $nm^{-1}$) and 0.012 $nm^{-1}$ (SD = 0.001 $nm^{-1}$), respectively.

Figure 12.3 shows a histogram of the slope parameters for both yellow substances and detritus for an optical dataset collected from the Bedford Basin. The figure emphasises the relatively low variance in slope values for the two optical components. Both the mean and standard deviation of the slope values for yellow substances ($S_y$ = 0.017 nm$^{-1}$, S.D. = 0.002 nm$^{-1}$) and detritus ($S_d$ = 0.010 nm$^{-1}$, S.D. = 0.002 nm$^{-1}$) are consistent with those reported in the studies of Kopelovich et al. (1989), Roesler et al. (1989) and Babin et al. (2003).

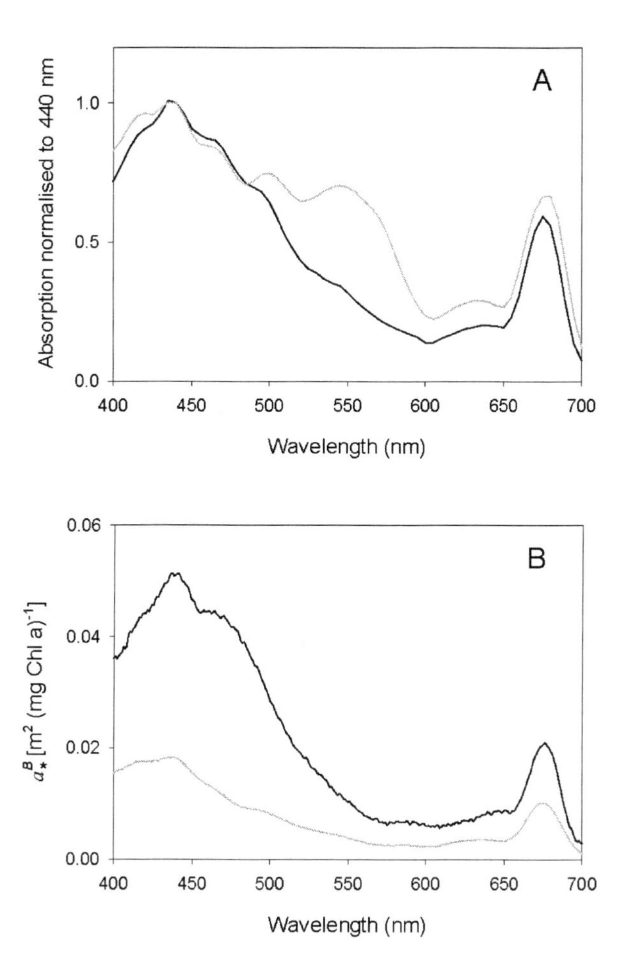


Figure 12.2 A) Phytoplankton absorption spectra normalised to 440 nm. The dashed line represents an absorption spectrum measured during a bloom of the red-tide, photosynthetic ciliate Mesodinium rubrum (=Myrionecta rubra). The solid line represents an absorption spectrum measured prior to the red-tide event, when diatoms and flagellate cells were the dominant contributors to the total algal biomass. B) Chlorophyll-specific absorption spectra of a diatom bloom (dashed) and a flagellate-dominated phytoplankton community (solid). (Bouman, unpublished data).

A final point about the absorption $a$ is related to the distinction between two general approaches to modelling the effect of light on photosynthesis. The models that we have presented are available-light models. They treat the photosynthetic response to irradiance available at the particular depth concerned. The relevant parameter is the initial slope $\alpha^B$. Another class of models is called the absorbed-light models. They deal with the response to the light *already absorbed* by the cells. In this case, the appropriate parameter is the (dimensionless) maximum quantum yield of photosynthesis $\phi_m$. It is possible to show (Platt and Sathyendranath, 1988) that these two approaches are entirely equivalent. The advantage of available-light models is that the parameter $\alpha^B$ is measurable in the field, whereas $\phi_m$ is not.

Because the light absorbed by the cells is determined only by the pigments contained therein, we can formulate a useful relation between these two parameters (Platt and Jassby, 1976). If $B$ is the pigment biomass, the absorption by pigments is $a_p = a_*^B B$, where $a_*^B$ is the pigment-specific absorption coefficient. Then

$$\alpha^B = a_*^B \phi_m. \tag{11}$$

This fundamental result allows us to convert any available-light model into an absorbed-light model (or the reverse). Further, it can be used to determine, in principle and in particular circumstances, whether variations in $\alpha^B$ are due mainly to variations in $\phi_m$ or to changes in $a_*^B$.

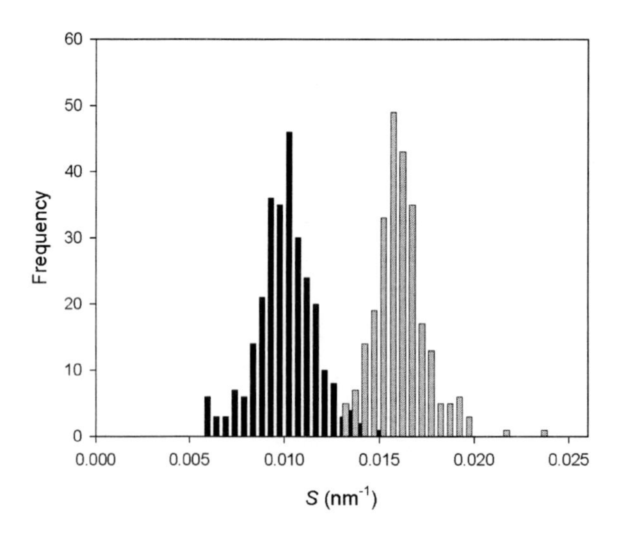


Figure 12.3 Frequency histograms of the exponential slope parameter, S, used to describe the wavelength-dependence of yellow substance (grey) and detrital absorption (black) (see equation (10)). (Bouman, unpublished data).

A potential difficulty in applying the formalism of photosynthesis-irradiance curves in real situations is assignment of the parameters $\alpha^B$ and $P_m^B$. They are not universal constants, but vary according to ambient conditions, including the ambient nutrient concentrations. This matter will be dealt with at length below. A second difficulty is more technical. Whereas use of non-spectral models often leads to mathematically-analytic results, application of spectral models almost always forces numerical treatments.

### 2.3 *Daily Primary Production for the Water Column*

Often, we need to model the daily, watercolumn, autotrophic production, or the daily production of the mixed layer. Under the assumption that the pigment biomass is uniformly distributed with depth (often a good assumption for the entire water column in coastal waters and certainly a good assumption for the mixed layer), we can write the daily production of the water column, $P_{Z,T}$ in the (non-spectral) form

$$P_{Z,T} = \frac{BP_m^B D}{K_{\mathrm{PAR}}} f(I_*^m), \tag{12}$$

where $B$ is the pigment biomass, $D$ is the daylength, $P_m^B$ is the assimilation number, $K_{\mathrm{PAR}}$ is the diffuse attenuation coefficient for downwelling PAR and $f(I_*^m)$ is a function of the dimensionless irradiance whose form is determined once the mathematical representation of the light — saturation curve has been selected (Platt and Sathyendranath, 1993). The quantity $I_*^m$ is given by $I_*^m = I_0^m / I_k$ where $I_0^m$ is the local irradiance at noon, and $I_k$ is the photoadaptation parameter (the ratio of the two photosynthesis parameters, $I_k = P_m^B / \alpha^B$ where $\alpha^B$ is the initial slope: the superscripts $B$ indicate normalisation to biomass). Equation (12) is so basic to the field that we refer to it (Platt and Sathyendranath, 1993) as the canonical form for the daily, watercolumn production. It is robust no matter what the choice of the light-saturation curve. It is often written in the compact form

$$P_{Z,T} = Af(I_*^m), \tag{13}$$

where the scale factor $A = BP_m^B D / K$.

For convenience of application, we usually replace $f(I_*^m)$, which may be a slowly-converging series, by a low-order polynomial (Platt et al., 1990; Platt and Sathyendranath, 1993). Thus

$$f(I_*^m) = \sum_{x=1}^{N} \Omega_x (I_*^m)^x, \tag{14}$$

where the weights $\Omega_x$ are determined by least-squares fitting to the numerical solution of equation (13). The magnitudes of the weights will depend on the number of terms $N$ in the polynomial series and on the range of $I_*^m$ for which they are computed. Usually, it is sufficient to take $N$=5 and compute for a range $0< I_*^m <20$. The weights need be determined only once.

### 2.4 *Primary Production in the Mixed Layer*

If we are interested in the daily production of only the surface mixed layer, it can be developed easily (Platt and Sathyendranath, 1993) from the compact canonical form stated in equation (13). Thus,

$$P_{Z_m,T} = Af(I_*^m) - Af(I_*^m e^{-K_{\mathrm{PAR}} Z_m}), \tag{15}$$

where $Z_m$ is the mixed-layer depth. The interpretation of equation (15) is that the mixed-layer production can be found as the difference between the production for two infinite water columns, one forced with the surface irradiance and one forced with the irradiance prevailing at depth $Z_m$. In other words, the second column is forced with the surface irradiance reduced by a factor $\exp(-K_{\mathrm{PAR}} Z_m)$. Then the argument of the second term on the right of equation (15) is also reduced by the same factor. The function $f(.)$ itself remains unchanged.

The polynomial representation also applies readily to the case of the finite mixed layer, with no change of weights. By analogy with equations (14) and (15), we can write

$$P_{Z_m,T} = A\sum_{x=1}^{N} \Omega_x (I_*^m)^x - A\sum_{x=1}^{N} \Omega_x (I_*^m e^{-K_{\mathrm{PAR}} Z_m})^x. \tag{16}$$

In this equation, the scale factor $A$ remains unchanged and the weights $\Omega_x$ are the same as before.

## 3. Nutrient Limitation in Coastal Waters

Phytoplankton growth requires the macronutrients N and P for biosynthesis of proteins, nucleic acids, phospholipids (Geider and LaRoche, 2002) and micronutrients such as Fe, Zn, Mn and Cu which serve as co-factors in enzymes. In addition to these nutrients, diatoms require Si for their cell walls (frustules). The particulate Si:N ratio of diatoms ranges from about 0.5 to 2 (Kudo 2003) depending on species under consideration and growth conditions. Silicate limitation has often been invoked to account for the termination of the Spring diatom bloom in coastal waters (Rocha et al., 2002, Wu and Chou, 2003). Inorganic carbon is not normally considered to be a limiting factor for phytoplankton because of the high total inorganic carbon concentration in sea water. Although inorganic carbon is unlikely to limit the yield of phytoplankton biomass, there is evidence that changes in the speciation of inorganic carbon can affect photosynthesis and growth rate. Hein and Sand-Jensen (1997) found that altering the $CO_2$ concentration in sea water from 10 μM under natural conditions to 3 μM by adding NaOH, or 36 μM by adding Cl, changed $^{14}C$ assimilation by up to 100% in oceanic phytoplankton from the North Atlantic. However, most investigators report significant reductions in growth rate only at $CO_2$ levels <5 μM (Riebesell et al., 1993; Burkhardt et al., 1999; Tortell et al., 2000). This concentration is well below current air-equilibrium levels, but may be achieved in nature under bloom conditions.

### *3.1 Limitation of yield versus limitation of rate*

Nutrient availability may limit the biomass and/or the growth rate of phytoplankton. Limitation of the biomass (or yield) that a system can support may be considered as a modification of Liebig's Law of the Minimum, whereas limitation of growth rate may be considered as an extension of Blackman's treatment of optimum and limiting factors. Nutrient limitation is most often inferred from bioassay experiments. In such experiments, the increase of biomass that accompanies addition of suspected limiting nutrients to small volumes (typically <20 L) is measured over periods of one to several days. Care should be taken in interpreting bioassay experiments owing to possible artifacts that may arise from isolating phytoplankton in small volumes, the most significant of which may be uncoupling phytoplankton growth from losses to grazing. Thus, an increase in biomass in incubation bottles during a bioassay experiment may not necessarily mean that growth rate was limiting *in situ,* given possible complications arising from grazing and recycling of nutrients.

In nature, the distinction between yield and rate limitation may also be complicated. For example, limitation of the rate of photosynthesis by inorganic carbon or available Fe may increase the time required for phytoplankton biomass to reach the yield limitation imposed by other factors, such as inorganic nitrogen. This is because the forms of carbon or iron that can be assimilated by phytoplankton may comprise only a small proportion of the total pool of these elements in sea water. Slow physical/chemical exchange between unavailable and available pools of these elements may limit the rate at which they can be accessed by the phytoplankton.

Although nitrogen is generally considered to be the nutrient most likely to limit phytoplankton productivity and biomass accumulation in coastal waters (Paerl et al., 1999), recent data suggest that iron may limit growth rates in some high-nutrient coastal regions (Hutchins et al., 1998, 2002).

### *3.2 Cultural Eutrophication*

The nutrient load to the coastal ocean has increased dramatically over the past 300 years as a result of intensification of agricultural practices and population growth near the coast (Mackenzie et al., 2002; see also the treatment by Rabalais (Chapter 21)). Not only has the load increased, but the ratio of N:P:Si has changed as well. Turner et al. (2003) concluded that increased nitrogen loading associated with human development of the landscape leads to higher N:P and N:Si ratios in rivers. At the same time that N loading has been increasing, the dissolved silicate load to coastal waters has decreased due to damming and eutrophication inland (Humborg et al. 2000). Changes in nutrient loading and nutrient supply ratios may in turn lead to shifts in phytoplankton community composition. Specifically, diatoms, which require Si, become silicate limited at high N:Si ratios (Wu & Chou 2003). Increased nutrient loads may lead to increased organic matter production and bottom water anoxia and hypoxia (Justic et al. 2002). However, with increased nutrient loading, production in turbid near shore systems may become light-limited. For example, Sanders et al. (2002) recently suggested that low silicate concentrations limit diatom growth in the Thames estuary, whereas phytoplankton taxa in this otherwise nutrient replete estuary are limited by light.

Although changes in anthropogenic nutrient loads may lead to increased phytoplankton growth in estuaries and near the coast, the nutrient budget of the continental shelf is typically dominated by exchange with the open ocean (Mackenzie et al., 2002). Changes in nutrient concentrations that have been observed at some long term coastal sampling sites may reflect anthropogenic loading (Foch 2003), but the effect of climate-related changes in salinity distributions cannot always be eliminated. For example, changes in the position of the freshwater-seawater mixing zone have been invoked to account for changes in nutrient concentration observed in the Irish Sea (Evans et al., 2003).

## 4. Models of Nutrient Uptake and Growth

### *4.1 Nutrient uptake*

Nutrient uptake by phytoplankton is described typically by a hyperbolic equation relating nutrient transport $\rho$ to the external nutrient concentration $N$:

$$\rho = \frac{\rho_{\max} N}{N + K_\rho}. \tag{17}$$

Application of this equation, the Michaelis-Menten equation, requires specification of the half-saturation constant for nutrient transport $K_\rho$ and the maximum transport rate $\rho_{\max}$. For any particular transport system, $K_\rho$ is constant, but $\rho_{\max}$ varies in proportion to the number of transporters per cell. Acclimation to nutrient limitation often involves an increase in the number of transporters and hence an increase in $\rho_{\max}$, but with unchanged $K_\rho$. As such, equation (17) cannot be used to model phytoplankton growth rate because $\rho_{\max}$ is a variable that is related to growth rate. The upper and lower limits to $\rho_{\max}$ have been designated as $\rho_{\max}^{\text{high}}$ and $\rho_{\max}^{\text{low}}$ by Morel (1987).

### *4.2 Nutrient-limited Growth*

Under steady nutrient limitation, phytoplankton growth can be described in terms of the external concentration of the limiting nutrient by the Monod equation, which has the same form as the Michaelis-Menten equation:

$$\mu = \frac{\mu_{\max} N}{N + K_\mu}, \tag{18}$$

where μ is the growth rate, $\mu_{max}$ is the maximum growth rate, and $K_\mu$ is the half saturation constant for growth. In contrast to the Michaelis-Menten equation in which $\rho_{max}$ is variable, $\mu_{max}$ is a constant in the Monod equation. Typically, $K_\mu << K_\rho$ and $\mu_{max} << \rho_{max}$ (Morel 1987).

Alternatively, growth rate can be described as a function of the intracellular concentration of the limiting nutrient using the Droop equation:

$$\mu = \mu'_{max} \frac{Q - Q_{min}}{Q}, \tag{19}$$

where $Q$ is the intracellular nutrient content which is also referred to as the cell quota, $Q_{min}$ is the minimum cell quota, $\mu'_{max}$ is a constant chosen such that the maximum growth rate is obtained when $Q$ reaches its maximum value, $Q_{max}$. Alternatives to this form of the cell quota model have been proposed, most recently by Flynn (2002) who advocates that quotas should be expressed relative to cell-carbon content (*e.g.*, as N:C or P:C ratios for N-, or P-limited, conditions) rather than in terms of cell contents (with units of pg $cell^{-1}$).

Under conditions of balanced growth, equations (17), (18) and (19) can be reconciled provided that the parameters are chosen such that the following equalities hold.

$$K_\mu = K_\rho \frac{\rho_{max}^{low}}{\rho_{max}^{high}} \frac{Q_{min}}{Q_{max}}; \tag{20}$$

$$\mu_{max} = \frac{\rho_{max}^{low}}{Q_{max}}; \tag{21}$$

$$\mu'_{max} = \mu_{max} \frac{Q_{max}}{Q_{max} - Q_{min}}. \tag{22}$$

Equations (17) through (22) summarise the expected relationship between external and intracellular nutrient concentrations, nutrient transport and growth under steady conditions of nutrient limitation.

### 4.3 *Non-steady Conditions and Surge Nutrient Uptake*

The capacity for cells to store nutrients internally and to regulate transport by both biochemical and genetic mechanisms (Flynn et al., 1997) means that nutrient transport is a highly dynamic process. Physiological investigations have shown that nutrient transport can respond rapidly to changing nutrient concentrations, often leading to surge uptake when an excess of nutrients is presented to the cell. It has been argued that such processes may be significant in patchy environments (Flynn,

2002). The significance of fluctuations in nutrient concentration on assimilation of nutrients will depend on the frequency and magnitude of the fluctuations that occur in nature (Seuront et al., 2002) relative to the response times of enzyme activation/deactivation, *de novo* enzyme synthesis and coupling of energy supply to nutrient availability.

In phytoplankton, growth and photosynthesis are coupled on time scales of hours to days associated with cell-division cycle (*i.e.,* the generation time). Short-term uncoupling of photosynthesis from growth can occur in response to variations of irradiance associated with vertical mixing and diel cycle of solar radiation. For example, in the diatom *Skeletonema costatum,* photosynthate that accumlates during the day fuels nitrogen assimilation and biosynthesis at night under nutrient replete conditions (Anning et al., 2000).

### *4.4 Interaction of Limiting Factors*

When the rates of supply of two nutrients are varied independently under controlled conditions in the laboratory, there is typically a sharp transition point (or threshold) between limitation by one or the other nutrient (Droop, 1983). The evidence for the threshold interaction comes from the examination of the relationship between growth rate and intracellular nutrient content. Given the threshold nature of N-, and P-limitation (Droop, 1983), the transition between P-limited and N-limited growth occurs at the point where the cell contents of both N and P limit growth rate simultaneously (Terry et al., 1985). This is the only point where growth rate is co-limited by the two nutrients and is called the critical ratio (Terry et al., 1985). The critical N:P ranges between 20 and 40 in *Pavlova lutheri* and *Phaeodactylum tricornutum* (Terry et al., 1985) to more than 100 in *Aureoumbra lagunensis* (Liu et al., 2001). The critical N:P is not a biological constant, but is expected to depend upon the biochemical composition of phytoplankton cells (Geider and La Roche, 2002), which may in turn be regulated by environmental conditions. In particular, the critical N:P is expected to be influenced by irradiance, because light-harvesting pigment-protein complexes and photosynthetic electron-transfer-chain components account for a large, but variable, fraction of cell mass (Geider et al., 1996).

Consideration of the possible light and temperature dependencies of the critical N:P described above raises a more general issue related to the choice of parameter values for the Droop and Monod models. The values of the parameters of these models have been found to depend upon the prevailing irradiance and temperature. This brings us to the consideration of models of algal growth based on energy balance, described below.

### *4.5 Energy-balance Models of Phytoplankton Growth*

In 1983, Kiefer and Mitchell presented a model that could account for both nutrient-replete and nutrient-limited phytoplankton growth in terms of the light absorption cross-section and quantum efficiency of photosynthesis. In this model of balanced growth, the quantum efficiency of photosynthetic carbon assimilation (at steady state) was assumed to depend only on irradiance, independently of the degree of nutrient limitation. Limitation of growth rate by nutrients was assumed to induce a decrease in the absorption cross-section, whereas limitation by light was assumed to induce an increase in the absorption cross-section. These metabolic adjustments in response to light- and/or nutrient-limitation operated in such a way that the energy required for growth and respiration always balanced the energy supplied by photosynthesis:

$$\mu + r = \phi_m \left\{ \frac{K_\phi}{K_\phi + I} \right\} \frac{C'}{C} a_*^B I, \tag{23}$$

where $\mu$ is the growth rate, $r$ is the respiration rate, $\phi_m$ is the maximum quantum efficiency of photosynthesis, $K_\phi$ is a parameter that is used in the description of the light dependence of quantum efficiency, $C'$ is the cell chlorophyll content, $C$ is the cell carbon content, $a_*^B$ is the chlorophyll-specific light absorption cross-section and $I$ is the photon flux density. The model was latter modified by Sakshaug et al. (1989) to allow the empirical dependence of quantum efficiency on light, contained within the braces in equation (23), to be replaced by a formulation based on target theory:

$$\mu + r = \phi_m \left\{ \frac{1 - \exp(-\sigma\tau I)}{\sigma\tau I} \right\} \frac{C'}{C} a_*^B I. \tag{24}$$

This model, which is parameterized in terms of the absorption cross-section $\sigma$, quantum efficiency $\phi_m$ and turnover time $\tau$ for light-saturated photosynthesis, was recast in terms of the photosynthesis light response curve by Geider (1990) as follows:

$$\mu + r = P_m^B \frac{C'}{C} \left\{ 1 - \exp\left( \frac{-\alpha^B I}{P_m^B} \right) \right\}, \tag{25}$$

where $P_m^B$ is the chlorophyll-a-specific light-saturated photosyntheic rate, and $\alpha^B$ is the chlorophyll-a-specific initial slope of the light-saturation curve and the parameters used in equations (24) and (25) are related by the identity:

$$\frac{\alpha^B}{P_m^B} = \sigma\tau; \tag{26}$$

and equation (11).

A model that combined the energy-balance approach developed by Kiefer and coworkers (equations (23) and (24)) with the Monod model for nitrogen-limited growth (equation (18)) was developed by Geider et al. (1997). This model assumed that nitrogen-limitation could be treated as a multiplicative factor on the carbon-specific, light-saturated photosynthetic rate. However, it also included further feedback between nitrogen limitation and photosynthetic rates through regulation of the chlorophyll-a-to-carbon ratio. This model was later extended to account for variability in the cellular nitrogen-to-carbon ratio for N-limited phytoplankton (Figure 12.4). This dynamic-balance model (Geider et al., 1998) can account for the interaction of photon flux density with growth rate and cellular N:C. The model does not provide a simple solution for phytoplankton growth rate, which arises as a consequence of the interaction of nutrient assimilation, carbon fixation and regulation of chlorophyll synthesis (Geider et al., 1998).

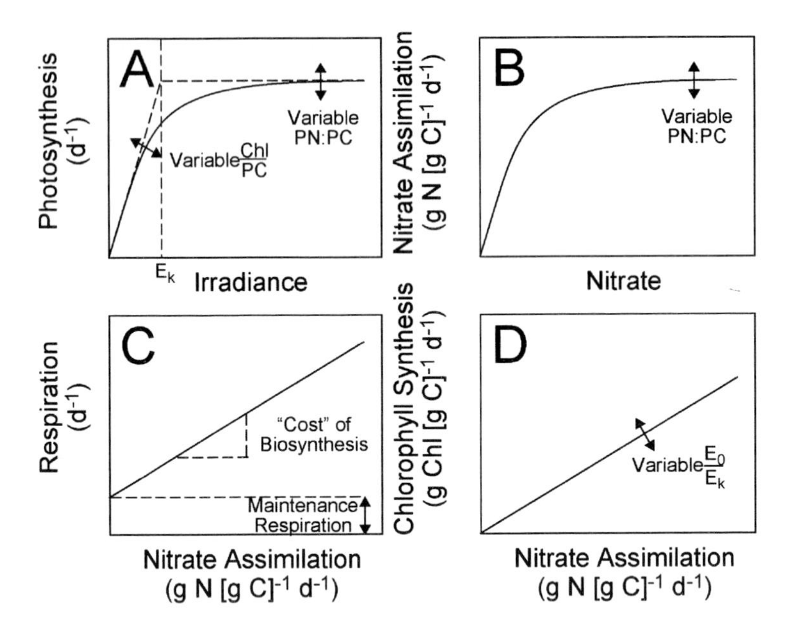


Figure 12.4 Graphical summary of the dynamic acclimation model of Geider et al. (1998) illustrating the dependencies of photosynthesis, nitrate assimilation, chlorophyll a synthesis and respiration on environmental and physiological variables. A) Photosynthesis is a saturating function of irradiance where the initial slope increases with increasing Chl:C and the light-saturated rate increases with increasing nitrogen-to-carbon ratio (N:C). The light saturation parameter (Ik) is given by the irradiance at which the initial slope intercepts the light-saturated rate. B) The carbon-specific nitrate assimilation rate is a saturating function of nitrate concentration where the maximum uptake rate is down-regulated at high values of N:C. C) The rate of chlorophyll-a synthesis is obligately coupled to protein synthesis and thus to nitrate assimilation. However, the magnitude of the coupling depends on the ratio of irradiance to the light saturation parameter (I/Ik). At a given rate of nitrate assimilation the carbon-specific rate of chlorophyll-a synthesis declines as I/Ik increases. D) The carbon-specific respiration rate is a linear function of the rate of nitrate assimilation. The model assumes that there is no lag between nitrate assimilation and protein synthesis. Major respiratory costs are associated with reduction of nitrate to ammonium, incorporation of ammonium into amino acids and polymerization of amino acids into proteins and other respiratory costs are assumed to scale with the rate of protein synthesis.

The dynamic-balance model (Geider et al., 1998) makes some simplifying assumptions regarding the photosynthesis parameters. These include the assumptions that $\alpha^B$ is a species-specific constant, independent of the degree of light-, or nitrogen, -limitation (Figure 12.5*a*), and that $P_m^B$ varies largely as a consequence of variability of the chlorophyll-to-carbon ratio in nutrient-replete cells. Under nutrient-limiting conditions, the model considers $P_m^B$ to vary as a result of the interaction of limitation between carbon-specific light-saturated photosynthesis by N:C (Figure 12.5*b*) and regulation of chl:C. Observations for *Pavlova lutheri,* one of the few phytoplankton species for which appropriate measurements are available at more than one photon flux density, are broadly consistent with the assumptions (Figure 12.4) and predictions (Figure 12.6) of this dynamic model.

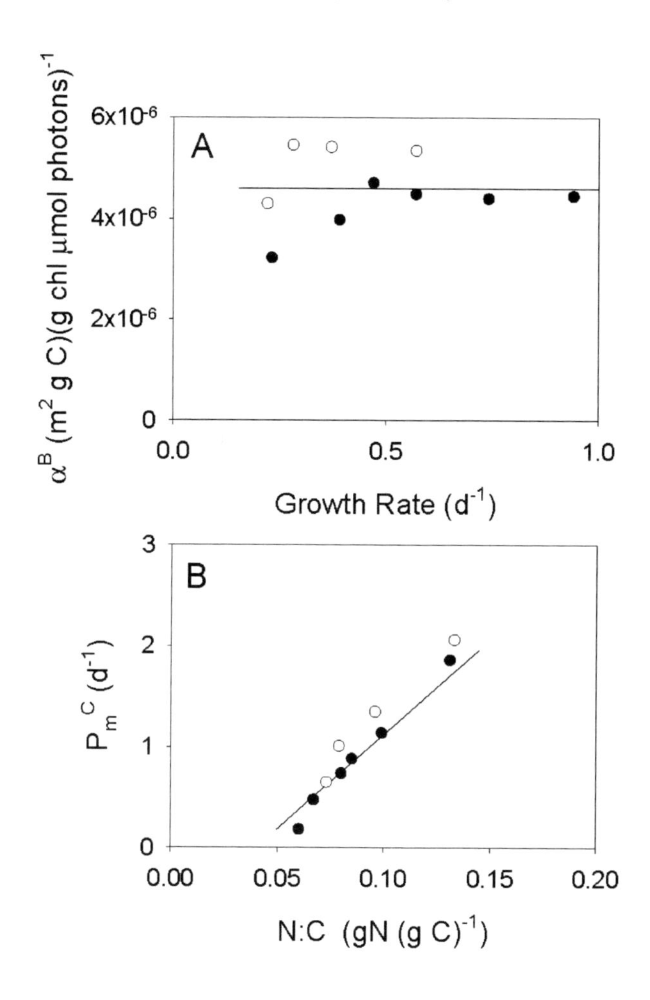


Figure 12.5 A) Comparison of observations for the light-limited initial slope of the PI curve ($\alpha^B$) for *Pavlova lutheri* (data from Chalup and Laws, 1990) with the assumption of constant $\alpha^B$ used in the dynamic acclimation model of Geider et al. (1998). Low-light cultures grown at a photon flux density of 63 μmol photons m$^{-2}$ s$^{-1}$ (open circles). High-light cultures grown at a photon flux density of 189 μmol photons m$^{-2}$ s$^{-1}$ (filled circles). B) Comparison of observations of the dependence of the carbon-specific light-saturated photosynthesis rate on the nitrogen-to-carbon ratio in N-limited *Pavlova lutheri* (data from Chalup and Laws, 1990) with the dependence used in the acclimation model of Geider et al. (1998). Symbols as in Figure 12.5A.

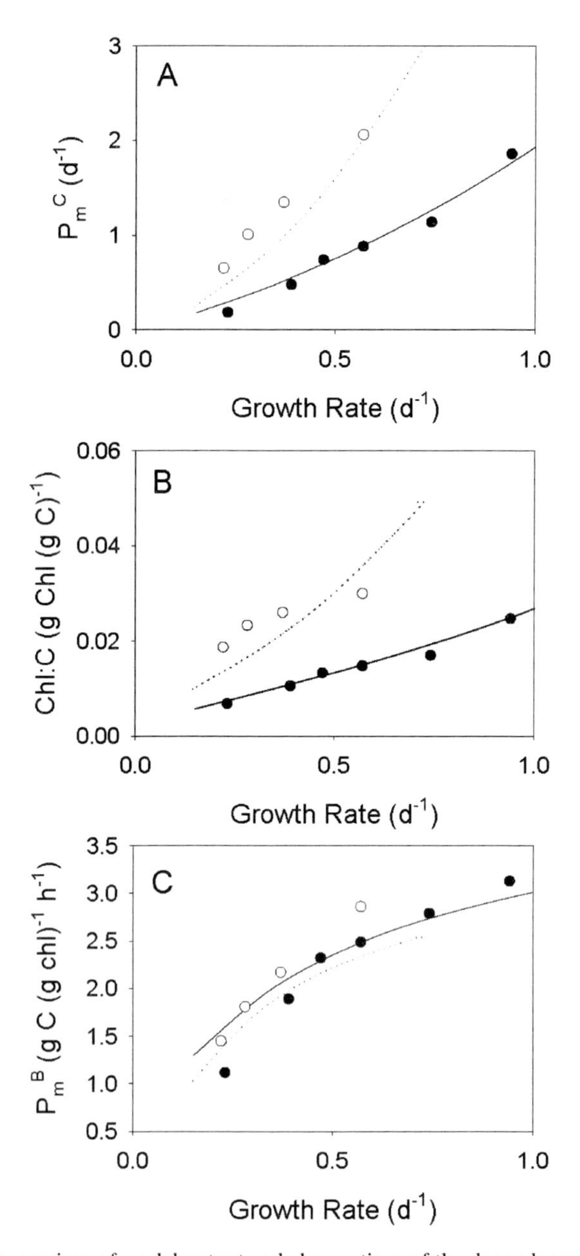


Figure 12.6 A)Comparison of model output and observations of the dependence of the carbon-specific light-saturated photosynthesis rate on growth rate in low-light and high-light cultures of *Pavlova lutheri* (data from Chalup and Laws, 1990). Symbols as in Figure 12.5. B) Comparison of model output and observations of the dependence of the chlorophyll-a-to-carbon ratio on growth rate in low-light and high-light cultures of *Pavlova lutheri* (data from Chalup and Laws, 1990). Symbols as in Figure 12.5. C) Comparison of model output and observations of the dependence of the chlorophyll-a-specific light-saturated photosynthesis rate on growth rate in low-light and high-light cultures of *Pavlova lutheri* (data from Chalup and Laws, 1990). Symbols as in Figure 12.5A.

## 5. Modelling Interactions between Growth Factors in Unsteady Environments

Conventionally, the saturating relationships between one particular factor, such as light, temperature or nutrient, and the growth rate, are demonstrated experimentally under two general conditions: first, the other growth factors are maintained constant and optimal; and second, the data that define the relationship are acquired for different, but steady, values of the limiting factor. For example, the parameters $K_s$ and $\rho_m$ of the Michaelis-Menten curve are representative only for those conditions where growth is limited by only one unchanging nutrient.

Evidently, the real pattern of the growth conditions in the sea is much more complex. First, owing to sporadic physical forcing, steady state is rarely observed, and cells are subjected to transient and variable nutrient, light and temperature limitations. Second, the different growth factors, of which some or all may be at sub-optimal levels at the same time, act concomitantly rather than independently. Typically, the maximum of production in a stratified water column occurs at the sub-surface where both light and nutrient allow cell division, but at suboptimal rates. In the context of modelling the effects of nutrient and light on growth, one must then question to what extent the classical nutrient- or light-growth relationships, determined in steady, single-factor experiments, can be applied to calculate phytoplankton growth inside physical-biological models where light and nutrients are unsteady and concomitantly limiting.

### *5.1 Concomitancy of Growth Factor Effects*

In biochemical models, two empirical devices are adopted to solve the problem of concomitant limiting factors: the minimum and multiplicative laws. The Law of Minimum considers that, even if several factors are at sub-optimal levels at the same time, the growth rate will depend only on the factor for which the corresponding calculated growth rate is the lowest:

$$\mu = \mu_m \times \min\{(f(F_1), g(F_2), h(F_3), \ldots)\}, \tag{27}$$

where $f$, $g$ and $h$ are saturating functions of the factors $F_1$, $F_2$ and $F_3$, varying in the interval between 0 and 1. On the contrary, the Multiplicative Law considers that the different factors act simultaneously through the product of their corresponding functions:

$$\mu = \mu_m \times f(F_1) \times g(F_2) \times h(F_3). \tag{28}$$

Both these formalisations suppose implicitly that the different factors act on growth independently of each other, since the functions $f$, $g$ and $h$ are generally taken as invariant, whatever the levels of the other factors.

Experimental evidence suggests that such a simplification is acceptable only in specific cases, and generally does not correspond to reality (Sciandra et al., 1997). Owing to internal metabolic coupling, the response curves for each factor are not independent of those for the other ones. It has been known for a long time that the $P_m^B$ parameter of the light-saturation curve is temperature-dependent because of the temperature sensitivity of the dark reactions driven by the RuBISCO enzyme. Interactions between the C and N metabolisms are also multiple (Turpin, 1991). Nutrient uptake requires energy, as does the subsequent metabolism of that nutrient to form macromolecules. Carbohydrates and lipids may be thought of as forms of energy storage in that ATP and reducing power may be released by the catabolism of these molecules. For example, assimilation of $NO_3$ requires reducing power and cofactors at the level of the nitrite-to-ammonium reduction, that are provided ultimately through photosynthetic activity. Owing to the large quantity of apoproteins incorporated into active pigment complexes, photoadaptation efficiency, adaptation of pigment composition, and energy transfer from accessory pigments to chlorophyll-a are nitrogen-dependent (Sciandra et al., 2000). At the level of the dark reactions of photosynthesis, nitrogen deficiency may also deregulate the light induction of RuBISCO activity (Osborne and Geider, 1986). Nitrogen, temperature and light conditions have been shown to affect differently the chlorophyll-a-specific absorption of diatoms, notably through the packaging effect (Stramski et al., 2002), and consequently the initial slope of the light-saturation curve. These interacting effects make phytoplankton adaptation very difficult to formalise mathematically using the simplistic relationships stated above, even at steady state. It is remarkable that no consistent body of experimental support exists in favour of such formalisations, currently used in modelling. On the contrary, co-limitation experiments have sometimes demonstrated their inconsistency (Rhee and Gotham, 1981). More complex models allow us to approach reality by representing the response of variables, such as chlorophyll, that integrate the interacting effects of external factors and acclimation capacity (Geider et al., 1998).

### 5.2 *Modelling the Effects of Variability in Growth Conditions*

Owing to the spatio-temporal variability of the oceanic environment (diel cycle and vertical mixing), it is probable that the growth of phytoplankton is rarely balanced. Moreover, the cost of energy involved in acclimation is difficult to evaluate. The somewhat-slower rates of growth measured in perturbed physical conditions means that the disturbances in the environment can reach a point at which it is difficult for the organisms to respond to the variability and that the populations are limited by the variance in resources (Harris, 1984). It takes at least one generation time for phytoplankton populations to adapt to changed conditions. Time lags or delays are characteristics of acclimation to variable inputs, and induce a mismatch between the actual production and the production calculated by static-curve responses. Under non-steady state conditions, the uptake and storage of the major nutrients becomes a highly temporal sequence of events controlled by the integrative nature of cellular physiology and the priorities placed on the synthesis of cellular constituents. In such conditions, the parameters of the response curve are no longer constant, but are variables depending upon internal and external properties. For example, surge uptake of nutrient by phytoplankton at higher rates than the steady saturation rate is commonly enhanced by low internal cell quota, high enzymatic absorption capacity and sufficient external availability of the limiting nutrient. Since low internal quota supports low growth rate, an important uncoupling may occur between absorption and growth in pulsed-nutrient environments, when $K_s$ and $\rho_{max}$ may vary significantly (Sciandra, 1991). It is not clear how these deviations from stability affect the growth efficiency, but it is sure that excessively-simple models such as those of Monod or Droop overestimate this efficiency, since they suppose that organisms are always adapted to external conditions. Surely, one of the most exciting fields of investigation for the future will be the development of sophisticated experimental devices able to mimic the real environment, with a view to providing data proper to model validation.

## 6. Carbon Dioxide and Phytoplankton

### *6.1 Introduction*

The oceans play a major role in the planetary carbon cycle in that their carbon reservoir is not only the largest one in the biosphere, but it can exchange freely with the atmospheric reservoir of carbon, the agent of exchange being the gas carbon dioxide. The transfers of carbon dioxide into and out of the ocean are forced by chemical imbalances resulting from physicochemical and biological processes such as the photosynthetic fixation of inorganic carbon by phytoplankton (see also the treatment by Ducklow and McCallister (Chapter 9)). Pelagic primary production is the macroscopic demonstration of an organic biosynthesis conducted by the unicellular microflora of the ocean. The substrates of this biosynthesis are inorganic molecules dissolved in seawater. For more than one hundred years, the macro-nutrients such as nitrogen N, phosphorous P (and silicate Si for diatoms) were the principal targets for study, given their capacities to limit phytoplankton production. More recently, the processes of limitation by trace metals such Fe and Zn, and by $CO_2$ in particular cases, have been intensively researched. Dissolved inorganic carbon (DIC), necessary for photosynthesis, is always abundant in seawater and can be subdivided into several chemical species, collectively constituting the carbonate system, whose relative proportions depend upon the physicochemical characteristics of sea water. For the strict autotrophs, photons constitute the only source of energy available to support absorption and assimilation of nutrients into the organic macromolecules that ensure the growth of these microphytes.

The biogenic production of organic matter in the oceanic surface layers, and its export to the depths, constitute what is referred to as the biological carbon pump, because it generates a positive gradient of DIC from the surface towards the ocean floor. The major part of photosynthetically-fixed C remains only temporarily in the surface waters. On a global scale, it is considered that 95% of phytoplanktonic cells die, or are consumed by the zooplankton, above the seasonal thermocline. The processes of mortality and excretion, which intervene at all the levels of the microbial loop, produce *in fine* a dissolved and particulate organic material accessible to bacteria, which themselves continue degradation until remineralisation is complete. The major part of the particulate organic carbon (POC) produced is thus recycled within the trophic network above the thermocline. The remaining 5% is exported to depth where it is isolated from contact with the atmosphere for long periods (approximately 1000 years). Only a small fraction, less than 0.5%, forms a deposit that is sequestered on the ocean floor for geological time scales. Ultimately, it is this fraction that regulates the content of $CO_2$ in the surface layers of the oceans which are, in the long term, in equilibrium with the atmosphere (Schlesinger, 1991; Butcher et al., 1992).

According to their effects on the $CO_2$ exchange between the ocean and the atmosphere, two distinct biological pathways can thus be considered. First, the suite of processes that ensure the export of POC towards the ocean bottom, referred to conventionally as the organic carbon pump, constitute a sink of carbon on geological scales (Falkowski, 1997):

$$6CO_2 + 12H_2O \rightarrow C_6H_{12}O_6 + 6O_2 + 6H_2O.$$

Second, the formation of calcareous plates by certain phytoplanktonic species ensures the export of particulate inorganic carbon (PIC) towards the ocean bottom:

$$Ca^{2+} + 2HCO_3^- \rightarrow CaCO_3 + CO_2 + H_2O.$$

The carbon that is precipitated in $CaCO_3$ leaves the DIC pool, so that calcification, even if it constitutes a $CO_2$ source for the environment (0.63 mole $CO_2$ per mole of fixed DIC), is nevertheless a sink for the DIC reservoir.

### 6.2 *Variability of Dissolved Inorganic Carbon*

The $pCO_2$ in seawater, as measured by (often widely-spaced) point sampling from ships, reveals significant spatial heterogeneity. There are several possible explanations. Typically, spring blooms of phytoplankton can occur as soon as the conditions of light and temperature and vertical stability permit nutrient assimilation by phytoplankton. Under such conditions, where the rate of the C uptake may be higher than the rates of diffusion and mixing, significant spatial structure in the $pCO_2$ fields can appear within a few days (Figure 12.7*a,b*). Diel changes of $pCO_2$ can also result from a day-night alternation of photosynthesis and respiration. Many expeditions have been programmed to consider local variability of the $pCO_2$ field, and considerable effort is being made to understand its origins and consequences (Bates et al., 1998a; Bates et al., 1998b; Bates et al., 1998c; Murphy et al., 1998; Murphy et al., 2001). Nevertheless, increasing the number of $pCO_2$ data has only partially solved the problem of spatial resolution. By automating the measuring device, it has been possible to acquire a semi-continuous signal from seawater pumped on board oceanographic (Copin-Montégut and Avril, 1995; Metzl et al., 1995) and commercial vessels (Zeng et al., 2002). Automated, fixed and drifting buoys have also been developed and are providing high-frequency information on Lagrangian variability. For example, the CARIOCA buoys measure, simultaneously, $pCO_2$, temperature, chlorophyll, and wind velocity, the real-time data being sent via satellites to the processing centres (Bates et al., 2000; Hood et al., 2001).

The DIC content in seawater is relatively constant (1.9 to 2.4 mmol C per kg). On the other hand, the pCO2 in surface waters can vary widely, particularly in the coastal zones. Values of $pCO_2$ close to 100 µatm were recorded in the English Channel (Frankignoulle et al., 1996) and values exceeding 1000 µatm were measured along the Peruvian coast (Copin-Montégut and Raimbault, 1994). Other measurements of $pCO_2$ between 300 and 800 ppm were reported off the Dutch coast by Bakker et al. (1996). Offshore, the surface waters are generally supersaturated and under-saturated with $CO_2$ at low and high latitudes respectively (Takahashi et al., 2002). In the temperate zones one can observe significant seasonal cycles in $pCO_2$ (Bates et al., 1998c; Bégovic and Copin-Montégut, 2002), but the deviations from saturation seldom exceed 100 μatm. These differences in $pCO_2$ induce significant variations in C fluxes, not only at the air-to-sea interface, but also at the ocean bottom, as attested by the relative values of biomass collected in sediment traps, which can vary by a factor of ten (Walsh 1991). Certain coastal zones, in particular the western edges of the continents, are favourable areas for the development of upwellings. The cold waters from the depths, enriched with nutrients and DIC, can support a significant production at the surface. The $pCO_2$ of these coastal surface waters, can exceed significantly that of the atmosphere (360 ppm), and are often regarded as sources of C for the atmosphere. The phytoplankton can influence this flux when optimal growth conditions occur at the sea-surface, and if the POC production is exported rapidly from the surface layer before being recycled. The assessment of C exchanges at the air-to-sea interface is thus a complex function of the residence time of water masses at the sea surface, the $pCO_2$ gradient at the interface, and the net C fixation achieved by phytoplankton. The differences in relative contributions of each of these terms are responsible for the great variability of the $pCO_2$ measured in coastal zones (Gattuso et al., 1998).

### *6.3 Acquisition of DIC by Phytoplankton*

Recent models indicate that in the absence of biological activity, the DIC content in the superficial ocean would be approximately 20% higher, and that this would double the $CO_2$ concentration in the atmosphere (Sarmiento and Orr, 1991). Such calculations, even if uncertain (Gurney et al., 2002), illustrate that in the context of the total heating of the planet, biological processes in the oceans, in particular primary production, have a controlling effect on the carbon cycle and on biogeochemical equilibria in general (Sarmiento, 1991). Sarmiento and Le Quere (1996) found that, according to the assumptions used to model primary production, the calculated production could vary widely, and that in the extreme, the ocean could appear to be a significant source of $CO_2$ to the atmosphere. Currently, biogeochemical models are the only tools available for long term prediction. However, the representation of physical as well as biological phenomena in these models remains too uncertain, and thus intensive experimental studies on phytoplankton are required, independent of these models, to increase their reliability.

Determining the impact of phytoplankton on DIC in sea water requires elucidating the mechanisms by which phytoplankton use carbon for photosynthesis and calcification. Conversely, these mechanisms condition the adaptation of autotrophic communities to variation in DIC, and thus to the potential limitation of primary production by reduced availability of this substrate. Although phytoplankton, and in particular large diatoms, play a significant rôle in the sequestration of oceanic $CO_2$ *via* significant organic matter sedimentation (Sarmiento and Le Quere, 1996), the consequences of natural or anthropogenic $CO_2$ variations in efficiency of

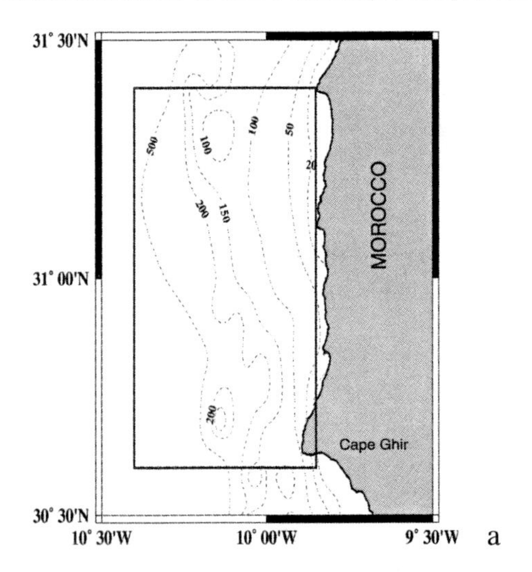


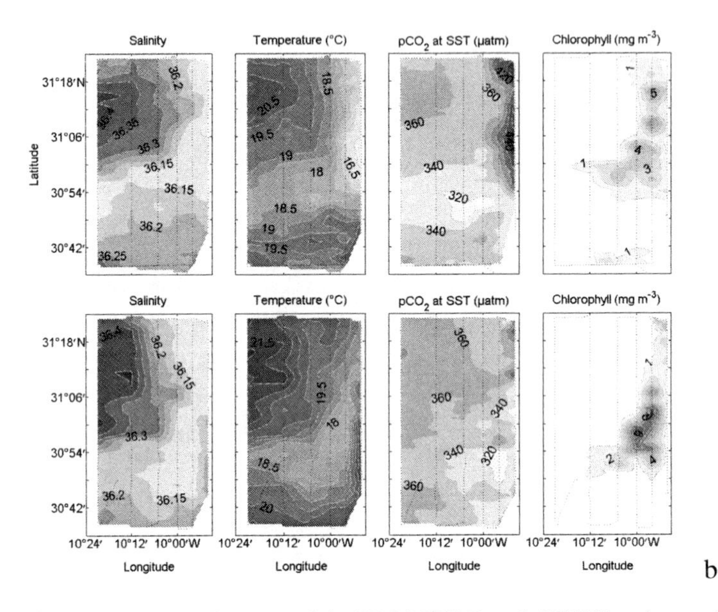


Figure 12.7 A) Map of the sampling area of the PROSOPE French JGOFS cruise conducted in September 1999 in an upwelling region off Morocco. B) Contour plots of salinity, temperature, $pCO_2$ and chlorophyll-a along two tracks observed on 8 September 1999 (upper four plots) and on 12 September 1999 (lower four plots) in the Morocco upwelling region. The upwelled waters are characterised by salinities lower than 36.15 ppt and low temperatures. During the first series of measurements, the highest $pCO_2$ values (up to 460 µatm) were observed along the most eastern transect at about 3 miles off the coast. Despite the rise in temperature of the upwelled water during its offshore progression, $pCO_2$ decreased below the saturation level, which may be attributed to biological production. The highest values of chlorophyll-a (about 6 mg $m^{-3}$) were not found alongshore, but between 6 and 10 miles from the coast. During the second series of measurements, the temperature of upwelled water was higher, indicating a decline in the upwelling intensity. The $pCO_2$ values were low and chlorophyll-a was very high (up to 11 mg $m^{-3}$) in the coldest part of the plume. These results confirm the observations of MacIsaac et al. (1985), showing that a period of adaptation to high irradiance is necessary for the development of the phytoplanktonic cells.

the biological pump remain obscure. This is partly due to the fact that dissolved carbon used for primary production exists in different forms that are exchanged following complex chemical equilibria (Frankignoulle, 1994; Zeebe and Wolf-Gladrow, 2001), and the biochemical uptake mechanisms are still unclear.

### *6.4 Carbon Dioxide and Primary Production*

Oceanic production occurs typically within an environment where DIC, composed of dissolved CO2, carbonate and bicarbonate, is plentiful. The bicarbonates are the most abundant species (approximately 2 mmole $kg^{-1}$) and also the least variable of the carbonates, whereas dissolved $CO_2$, much more variable, represents only a small proportion (approximately 5 to 25 mol $kg^{-1}$). During the last two decades, a strong experimental effort was made to dissect the processes of DIC acquisition by phytoplankton. Given its relative abundance compared with that of other nutrients, dissolved C was regarded for a long time as not limiting to oceanic photosynthesis (Clark and Flynn, 2000). This idea was reassessed following publication of various experimental and theoretical results. First, it appeared that the *in vivo* affinity of many algal species for $CO_2$ was higher than the affinity of the enzyme RuBISCO (see below) isolated *in vitro*. The difference between the *in vivo* and *in vitro* photosynthetic kinetics was interpreted as resulting from the existence of a concentrated intracellular pool of C (Raven and Johnston, 1991). Second, theoretical calculations suggested that $CO_2$, although exchanged with other carbonate-system species, could be temporarily growth-limiting because of its low diffusion rate in seawater, in particular in the vicinity of large cells such as diatoms (Riebesell et al., 1993). In addition, other work showed that bicarbonate could also constitute a photosynthetic source of carbon, on condition that trace elements such as Fe and Zn were sufficiently available to promote the synthesis of bicarbonate-assimilating metallo-enzymes (Morel et al., 1994). These works, as well as several later studies, (Raven, 1997; Beardall et al., 1998; Kaplan and Reinhold, 1999) confirmed previous results (Badger et al., 1978; Beardall and Raven, 1981) according to which phytoplanktonic cells could concentrate their internal pool of inorganic C (cell concentration mechanism, CCM), by taking up external $CO_2$, or bicarbonates, or both (Raven, 1997).

During the dark reaction of photosynthesis, $CO_2$ is fixed by RuBISCO (Ribulose Bisphosphate Carboxylase Oxygenase), an enzyme that probably constitutes the most abundant protein of the vegetable kingdom. Owing to the large size of the molecule (550 kDa), and its high N/C ratio, RuBISCO activity has a high cellular cost (Raven and Johnston, 1991). Its mutually antagonistic activities to carboxylation and photorespiration are controlled, respectively, by the concentrations of $CO_2$ and $O_2$ in its vicinity. An increase of the $CO_2/O_2$ ratio favours the biosynthesis of carbon chains to the detriment of photorespiration, which constitutes a cellular loss of energy and carbon (Canvin, 1990). Given that the concentration of $CO_2$ in sea water in gaseous equilibrium with the atmosphere is close to 10 μM, and that this concentration is very much lower than the half-saturation constant of RuBISCO, which can vary from 20 to 200 μM (Badger et al., 1998), the CCM appears, for the majority of species studied until now, as an essential function maintaining optimal growth under average concentrations of $pCO_2$ in seawater. By modulating their CCM activity according to external conditions, phytoplanktonic cells are able to preserve an optimal carbon fixation in an environment where the $pCO_2$ may vary over a broad range.

It is now accepted that both $CO_2$ and $HCO_3^-$ are used by microalgae (Colman and Rotatore, 1995; Rotatore et al., 1995; Korb et al., 1997; Tortell et al., 1997; Elzenga et al., 2000; Burkhardt et al., 2001). $CO_2$ can enter the cell by passive diffusion or active transport through the plasma membrane (Miller et al., 1991; Li and Canvin, 1998). Although the charged $HCO_3^-$ ion constitutes 95% of the DIC in sea water, its diffusion into the cell is limited by its low solubility in the lipid membrane, and by the electric potential difference across the membrane (Raven, 1980, 1997). Consequently, bicarbonate influx implies either active transport (Nimer et al., 1997; Tortell et al., 1997), or its prior catalytic conversion into $CO_2$ outside the cell (Sueltemeyer, 1998). Therefore, the CCM requires internal and/or external carbonic anhydrase activity (CA) to catalyse the interconversion $CO_2 + H_2 \leftrightarrow HCO_3^- + H^+$. Activity of CA was not detected in early experiments because the phytoplankton cultures used to prepare cellular extracts were often bubbled with growth-stimulating $CO_2$-enriched air. It was only when experiments were performed with cultures in equilibrium with air that a CA activity could be clearly demonstrated. The majority of phytoplanktonic species possess one or more CA enzymes. The various physiological functions of CA have been discussed in reviews (Badger and Price, 1992, 1994; Raven, 1995) which probe the relations between CA activity and other resources such as nitrogen, light, or trace metals (Cullen et al., 1999). There are two distinct types of CA enzymes in microalgae. One is intracellular, soluble, found in chloroplasts and increases the concentration of $CO_2$ in the vicinity of RuBISCO. Intra-mithocondrial activities have also been suggested as contributing to maintenance of high intracellular concentration of C (Raven, 2001). The other CA activity is found in the plasma membrane (external, with a soluble part in the external medium, and a soluble part in the bilipidic layer of the plasmic membrane) (Aizawa and Miyachi, 1986; Nimer et al., 1997). This latter CA is generally synthesized when the external $CO_2$ concentration becomes limiting for growth.

To date, the majority of studies on C uptake by phytoplankton were conducted with monospecific cultures maintained under optimal and steady conditions of macro- and micro-nutrients *in vitro*. In only few studies has the attempt been made to determine the precise modes of acquisition of carbon *in situ* and its actual capacity to limit primary production, despite various reports indicating that $CO_2$ concentrations as low as 5 $\mu$M (in the Bering Sea) can be measured in highly productive coastal waters (Codispoti et al., 1982). The difficulty in estimating a potential DIC limitation of photosynthesis *in situ* is inherent in the type of protocol generally employed. By measuring the kinetics of $^{14}C$ incorporation by phytoplankton communities over a range of DIC concentrations and during a relatively short time interval, it is obviously impossible to take into account the processes of physiological adaptation that allow cells gradually to modulate their CCM, and to increase their affinity for DIC. Nor is it possible to estimate the effects resulting from change in the taxonomic composition of natural communities. In other words, studies undertaken over the long term and over the short term cannot address the same questions and should not be expected to lead to same conclusions. Despite this, *in situ* studies support the view that, globally, DIC does not limit primary production. For example, results obtained by Tortell et al. (2000) in coastal waters of the Pacific show that, in spite of substantial production of phytoplankton dominated by diatoms, $CO_2$ did not appear to limit the production, precisely because of a CCM activation (see also Tortell et al., 1997).

These findings do not mean, however, that the large variations of $CO_2$ that can occur in coastal zones are without effect on phytoplankton. In fact, the activation of the CCM reflects an intrinsic response to metabolic stress. Since cells must mobilize significant resources (light, macro-, and micro-nutrients) to activate the CCM and to take up inorganic C, all variations of ambient $CO_2$ that affect the mechanisms of this acquisition also modify the efficiency with which the cells exploit these potentially limiting resources. For example, the question of the coupling between metabolic pathways and the problem of colimitation were addressed for Zn by Morel et al. (1994), who showed that, in case of deficiency, this element necessary for synthesis of the CA enzyme can limit the growth of the diatom *Thallassiosira weissflogii* in an environment where $CO_2$ is low. In the same way, laboratory (Badger and Andrews, 1982) and field experiments showed that a reduction of $pCO_2$ could increase significantly the cells' photosynthetic quantum requirement. Work from Tortell et al. (2000) and Burkhardt et al. (1999) also suggested that the availability of $CO_2$ conditions the cellular demand for nitrogen.

It follows that in environments with low nutrients or low light levels, an additional limitation can strongly restrict the CCM, and render the adaptation of phytoplankton more dependent on variablity in $CO_2$. Recent experimental studies to elucidate the influence of the $CO_2$ concentration on the mechanisms of DIC acquisition revealed, in addition, that marked differences could exist not just between the species of the same taxonomic family, as in the diatoms, but also between the strains within the same species, such as *Phaeodactylum tricornutum*, which makes it extremely difficult to formulate any generalizations (Burkhardt et al., 2001). The various phytoplanktonic species have significant differences in their ability to develop a CCM (Badger et al., 1998; Tortell, 2000), and one can suppose that the variability of coastal $CO_2$ concentrations contributes to diversity of community structure, and *vice versa.* The elucidation of such a relationship is not simple, because it would require long term *in situ* observations.

### *6.5 Carbon dioxide and Calcification*

In the context of the relation between autotrophic species and DIC, special mention must be made of the coccolithophorid group represented in particular by the species *Emiliania huxleyi* (Lohmann) and *Gephyrocapsa oceanica. E. huxleyi* is probably one of the most-studied marine species in the world, in the laboratory as well as *in situ,* because of its capacity to convert DIC into solid calcium carbonate. This species forms extensive oceanic and coastal blooms (Holligan et al., 1983) and is found everywhere, except in the Arctic and Antarctic oceans (Brand, 1994; Winter et al., 1994). Those ecotypes adapted to coastal conditions have been studied especially along the Norwegian coast (Young, 1994), and in the North Sea (Van Bleijswijk et al., 1991; Holligan et al., 1993b). *G. oceanica* is associated with the neritic environments of tropical seas (Brand, 1994; Winter et al., 1994). The functional relationships between calcification and photosynthesis have been the subject of many studies (see Paasche, 2002 for a compilation). The reason is that calcification, which uses bicarbonate as a substrate, constitutes a potential source of $CO_2$ for the environment (Holligan et al., 1993a; Frankignoulle et al., 1994; Sikes and Fabry, 1994), and furthermore not only conditions the carbon cycle on diurnal and seasonal time scales, but also on a geological time scale (Westbroek et al., 1993). In this context, the influence of coccolithophorids on the carbonate system differs completely from that exerted by the other autotrophic groups. This influence is difficult to quantify because the calcification-to-photosynthesis ratio (C/P) varies for multiple interfering reasons (Berry et al., 2002). According to the schema suggested by Nimer and Merrett (1995), the process of calcification can produce $CO_2$ if calcification exceeds photosynthesis (Paasche, 1964). The importance of this production is thus a function of the C/P ratio, which can vary between 2 (equation (29)) and 1 (equation (30)) according to the following equilibria:

$$4\,HCO_3^- + 2Ca^{++} \rightarrow 2CaCO_3 + "CHO" + CO_2 + O_2 + H_2O, \qquad (29)$$

$$2\,HCO_3^- + Ca^{++} \rightarrow CaCO_3 + "CHOH" + O_2. \qquad (30)$$

Probably these schematic equations remain hypothetical since, as shown in numerous experimental studies, the C/P ratio can vary between 0 and more than 2 (Paasche, 2002).

According to whether the phytoplankter is a non-calcifying, slightly or strongly calcifying species, the net variation of $pCO_2$ induced by blooms will be, respectively, negative, null or positive. Although such patterns have been verified occasionally by *in situ* investigations (Holligan et al., 1993a; Sikes and Fabry, 1994), this simple schema cannot be easily generalized, in particular in the long term, and it would be inaccurate to affirm that coccolithophorid blooms are accompanied systematically by a net production of $CO_2$. For example, investigations of Buitenhuis et al. (1996) concluded that coccolithophorid blooms observed in the North Sea act as a sink rather than as a source of C. Variation in the net C production associated with the development of *E. huxleyi* has to be explained in relation to the causes that affect the C/P ratio, such as the mean light intensity received by the cells (Zondervan et al., 2002) which depends on their position in the water column, and on the degree of nutrient limitation (Paasche, 1998; Riegman et al., 2000). The sign of the net production is difficult to determine because it depends not only on processes occurring at the cellular level, such as respiration (Crawford and Purdie, 1997), but also at the level of the community, such as the heating of the water column by scattering of light by free coccoliths (Holligan et al., 1993a; Gordon and Du, 2001). Losses associated with cell sedimentation out of the photic zone (Van Der Wal et al., 1995), or coccolith dissolution (Milliman et al., 1999) are also to be considered.

Since they are at the origin of more than 50% of the planktonic carbonate precipitation (but only about 20% of the $CaCO_3$ produced in the surface layers of the ocean is actually sequestered in the sediments), coccolithophorids are also studied in relation to the long-term evolution of atmospheric $pCO_2$, which has increased continuously since the beginning of the industrial era (Houghton et al., 1995). Although still uncertain, contemporary models predict that, by the end of this century, the accumulation of atmospheric $CO_2$ resulting from anthropogenic activity will result in average $CO_2$ concentrations at the ocean surface approximately three times higher than those of the pre-industrial era. This should involve reductions in carbonate concentration and pH of approximately 50% and 0.35 units, respectively, in surface waters (Wolf-Gladrow et al., 1999). These long-term changes will affect both carbonate chemistry and the processes of calcification. An increase in $pCO_2$, by changing the carbonate-system speciation, reduces seawater pH and thus its buffering capacity, and results in an increase of $CO_2$ production through calcification (Frankignoulle et al., 1994). It is thus expected that, at constant C/P, the increase in $pCO_2$ predicted for the next centuries will exert a positive feedback on the $CO_2$ produced by the calcifying communities. However, recent studies suggest that calcification (C/P) could respond negatively to a rise in seawater $pCO_2$. Such conditions, simulated through experiments in which the $pCO_2$ of water was increased artificially by the addition of strong acids (Riebesell et al., 2000; Zondervan et al., 2002), caused a significant decrease in the calcification rate and the C/P ratio in *G. oceanica,* and, to a lesser extent, in *E. huxleyi.* Ultimately, it is possible that this reduction in the $CO_2$-producing process of calcification compensates for the $CO_2$ increase induced by the feedback mentioned above (Zondervan et al., 2001). These results are still too preliminary for us to infer the precise changes expected in the long term. Moreover, these experiments are far from reproducing the real conditions that control primary production in the natural environment. In particular, perturbing the $pCO_2$ with other factors such as nutrients, which *in fine* control the productivity, most probably modifies the responses observed under optimal experimental conditions (Sciandra et al., 2003).

## References

Aizawa, K., and S. Miyachi. 1986. Carbonic anhydrase and CO2 concentrating mechanisms in microalgae and cyanobacteria. *FEMS Microbiol. Rev.,* **39,** 215–233.

Anning, T., S. M. Pratt, P. J. Sammes, H. L. MacIntyre, S. Gibb, and R. J. Geider, 2000. Photoacclimation in the marine diatom *Skeletonema costatum. Limnol. Oceanogr.,* **45,** 1807–1817.

Babin, M., D. Stramski, G.M. Ferrari, H. Claustre, A. Bricaud, G. Obolensky and N. Hoepffner, 2003. Variations in the light absorption coefficients of phytoplankton, non-algal particles, and dissolved organic matter in coastal waters around Europe. *J. Geophys. Res.,* 108.

Badger, M. R., A. Kaplan and J. A. Berry, 1978. A mechanism for concentrating $CO_2$ in *Chlamydomonas reinhardtii* and *Anabaena varaibilis* and its role in photosynthetic $CO_2$ fixation. *Carnegie Inst. Wash. Year Book,* **77,** 251–261.

Badger, M. R. and T. J. Andrews, 1982. Photosynthesis and inorganic carbon usage by the marine cyanobacterium *Synechococcus* sp. *Plant Physiol.,* **70,** 517–523.

Badger, M. R. and G. D. Price, 1992. The $CO_2$ concentrating mechanism in cyanobacteria and microalgae. *Physiol. Plant.,* **84,** 606–615.

Badger, M. R. and G. D. Price, 1994. The role of carbonic anhydrase in photosynthesis. *Annual Reviews of Plant Physiology and Plant Molecular Biology,* **45,** 369–392.

Badger, M. R., T. J. Andrews, S. M. Whitney, M. Ludwig, D. C. Yellowlees, W. Leggat, and G. D. Price, 1998. The diversity and coevolution of Rubisco, plastids, pyrenoids, and chloroplast-based $CO_2$-concentrating mechanisms in algae. *Can. J. Bot./Rev. Can. Bot.,* **76,** 1052–1071.

Bakker, D. C. E., H. J. W. De Baar and H. P. J. De Wilde, 1996. Dissolved carbon dioxide in Dutch coastal waters. *Mar. Chem.,* **55,** 247–263.

Bates, N. R., D. A. Hansell, C. A. Carlson and L. I. Gordon, 1998a. Distribution of CO2 species, estimates of net community production, and air-sea $CO_2$ exchange in the Ross Sea polynya. *J. Geophys. Res.,* **103,** 2883–2896.

Bates, N. R., A. H. Knap and A. F. Michaels, 1998b. Contribution of hurricanes to local and global estimates of air-sea exchange of $CO_2$. *Nature,* **95,** 58–61.

Bates, N. R., T. Takahashi, D. W. Chipman and A. H. Knap, 1998c. Variability of $pCO_2$ on diel to seasonal timescales in the Sargasso Sea near Bermuda. *J. Geophys. Res.,* **103,** 567–515.

Bates, N. R., L. Merlivat, L. Beaumont and A. C. Pequignet, 2000. Intercomparison of shipboard and moored CARIOCA buoy seawater $fCO_2$ measurements in the Sargasso Sea. *Mar. Chem.,* **72,** 2–4.

Beardall, J. and J. A. Raven, 1981. Transport of inorganic carbon and the $CO_2$ concentrating mechanism in *Chlorella emersonii* (Chlorophyceae). *J. Phycol.,* **17,** 134–141.

Beardall, J., A. Johnston, and J. Raven, 1998. Environmental regulation of $CO_2$-concentrating mechanisms in microalgae. *Can. J. Bot./Rev. Can. Bot.,* **76,** 1010–1017.

Bégovic, M. and C. Copin-Montégut, 2002. Processes controlling annual variations in the partial pressure of CO2 in surface waters of the central northwestern Mediterranean Sea (Dyfamed site). *Deep-Sea Res.,* **49,** 2031–2047.

Berry, L., A. R. Taylor, U. Lucken, K.P. Ryan and C. Brownlee, 2002. Calcification and inorganic carbon acquisition in coccolithophores. *Funct. Plant Biol.,* **29,** 1–11.

Brand, L.E., 1994. Physiological ecology of marine coccolithophores. In *Coccolithophores,* A. Winter and W. G. Siesser eds. Cambridge University Press, Cambridge, pp. 39–49.

Buitenhuis, E., J. Bleijswijk, D. Bakker, and M. Veldhuis, 1996. Trends in inorganic and organic carbon in a bloom of *Emiliania huxleyi* in the North Sea. *Mar. Ecol. Prog. Ser.,* **143,** 271–282.

Burkhardt, S., I. Zondervan and U. Riebesell, 1999. Effect of $CO_2$ concentration on C:N:P ratio in marine phytoplankton: a species comparison. *Limnol. Oceanogr.,* **44,** 683–690.

Burkhardt, S., G. Amoroso, U. Riebesell and D. Sueltemeyer, 2001. $CO_2$ and $HCO3^-$ uptake in marine diatoms acclimated to different $CO_2$ concentrations. *Limnol. Oceanogr.,* **46,** 1378–1391.

Butcher, S. S., R. J. Charlson, G. H. Orians, and G. V. Wolfe, 1992. *Global Biogeochemical Cycles,* Academic Press, New York.

Canvin, D. T., 1990. Photorespiration and CO2 concentrating mechanisms. In *Plant physiology, biochemistry and molecular biology,* D. T. Dennis and D. H. Turpin eds. John Wiley and Sons, New York, 253–273.

Chalup, M. S. and E. A. Laws, 1990. A test of the assumptions and predictions of recent microalgal growth models with the marine phytoplankter *Pavlova lutheri. Limnol. Oceanogr.,* **35,** 583–596.

Clark, D. R. and K. J. Flynn, 2000. The relationship between the dissolved inorganic carbon concentration and growth rate in marine phytoplankton. *Proc. R. Soc. Lond. B,* **267,** 953–959.

Codispoti, L. A., G. E. Friederich, R. L. Iverson and D. W. Hood, 1982. Temporal changes in the inorganic carbon system of the Southeastern Bering Sea during Spring 1980. *Nature,* **296,** 242–245.

Colman, B. and C. Rotatore, 1995. Photosynthetic inorganic carbon uptake and accumulation in two marine diatoms. *Plant Cell Environ.,* **18,** 919–924.

Conley, D. J., 2002. Terrestrial ecosystems and the global biogeochemical silica cycle. *Global Biogeochem. Cycles* **16,** 1121.

Copin-Montégut, C and B. Avril, 1995. Continuous $pCO_2$ measurements in surface water of the northeastern Tropical Atlantic. *Tellus Ser. B: Chem. Phys. Meteorol.,* **47B,** 86–92.

Copin-Montégut C. and M. Bégovic, 2002. Distributions of carbonate properties and oxygen along the water column (0–2000m) in the central part of the NW Mediterranean Sea (Dyfamed site): influence of winter vertical mixing on air-sea $CO_2$ and $O_2$ exchanges. *Deep-Sea Res.,* **49,** 2049–2066.

Copin-Montégut, C. and P. Raimbault, 1994. The Peruvian upwelling near 15 degree S in August 1986. Results of continuous measurements of physical and chemical properties between 0 and 200 m depth. *Deep-Sea Res.*, **41,** 439–467.

Crawford, D. W. and D. A. Purdie, 1997. Increase of $pCO_2$ during blooms of *Emiliania huxleyi:* theoretical considerations on the asymmetry between acquisition of $HCO_3$ and respiration of free $CO_2$. *Limnol. Oceanogr.*, **42,** 365–372.

Cullen, J. T., T. W. Lane, F. M. M. Morel and R. M. Sherrell, 1999. Modulation of cadmium uptake in phytoplankton by seawater $CO_2$ concentration. *Nature,* **402,** 165–167.

Droop. M. R., 1983. 25 Years of algal growth kinetics. A personal view. *Bot. Mar.*, **26,** 99–112.

Elzenga, J. T. M., H. B. A. Prins and J. Stefels, 2000. The role of extracellular carbonic anhydrase activity in inorganic carbon utilization of *Phaeocystis globosa* (Prymnesiophyceae): a comparison with other marine algae using the isotopic disequilibrium technique. *Limnol. Oceanogr.*, **45,** 372–380.

Evans, G. L., P. J. L., Williams and E. G. Mitchelson-Jacob, 2003. Physical and anthropogenic effects on observed long-term nutrient changes in the Irish Sea. *Estuar. Coast. Shelf Sci.*, **57,** 1159–1168

Falkowski, P. G., 1997. Evolution of the nitrogen cycle and its influence on the biological sequestration of $CO_2$ in the ocean. *Nature,* **387,** 272–275.

Flynn, K. J., M. J. R. Fasham and C. R. Hipkin, 1997. Modelling the interactions between ammonium and nitrate uptake in marine phytoplankton. *Philo. Trans. Roy. Soc.*, **352,** 1–22.

Flynn, K. J., 2002. How critical is the critical N:P ratio? *J. Phycol.*, **38,** 961–970.

Fock, H. O., 2003. Changes in the seasonal cycles of inorganic nutrients in the coastal zone of the southeastern North Sea from 1960 to 1997: effects of eutrophication and sensitivity to meteoclimatic factors. *Marine Pollution Bulletin,* **46,** 1434–1449.

Frankignoulle, M. (1994) A complete set of buffer factors for acid/base $CO_2$ system in seawater. *J. Mar. Sys.*, **5,** 111–118.

Frankignoulle, M., C. Canon and J.-P. Gattuso, 1994. Marine calcification as a source of carbon dioxide - Positive feedback of increasing atmospheric $CO_2$. *Limnol. Oceanogr.*, **39,** 458–462.

Frankignoulle, M., I. Bourge, C. Canon and P. Dauby, 1996. Distribution of surface seawater partial $CO_2$ pressure in the English Channel and in the Southern Bight of the North Sea. *Cont. Shelf Res.*, **16,** 381–395.

Gattuso, J. P., M. Frankignoulle and R. Wollast, 1998. Carbon and carbonate metabolism in coastal aquatic ecosystems. *Annu. Rev. Ecol. Syst.*, **29,** 405–434.

Geider, R. J., 1990. The relationship between steady state phytoplankton growth and photosynthesis. *Limnol. Oceanogr.*, **35,** 971–972.

Geider, R. J. and J. La Roche, 2002. Redfield revisited: variability in the N:P ratio of phytoplankton and its biochemical basis. *Eur. J. Phycol.*, **37,** 1–17.

Geider, R. J., H. L. MacIntyre and T. M. Kana, 1996. A dynamic model of photoadaptation in phytoplankton. *Limnol. Oceanogr.*, **41,** 1–15.

Geider, R. J., H. L. MacIntyre and T. M. Kana, 1997. A dynamic model of phytoplankton growth and acclimation: responses of the balanced growth rate and chlorophyll a: carbon ratio to light, nutrient-limitation and temperature. *Mar. Ecol. Prog. Ser.*, **148,** 187–200.

Geider, R. J., H. L. MacIntyre and T. M. Kana, 1998. A dynamic regulatory model of phytoplankton acclimation to light, nutrients and temperature. *Limnol. Oceanogr.*, **43,** 679–694.

Gordon, H. R., and T. Du, 2001. Light scattering by nonspherical particles: application to coccoliths detached from *Emiliania huxleyi. Limnol. Oceanogr.*, **46,** 1438–1454.

Gurney, K. R., R. M. Law, A. S. Denning, P. J. Rayner, D. Baker, P. Bousquet, L. Bruhwiler, Y. H. Chen, P. Ciais, S. Fan, I. Y. Fung, M. Gloor, M. Heimann, K. Higuchi, J. John, T. Maki, S. Maksyutov, K. Masarie, P. Peylin, M. Prather, B. C. Pak, J. Randerson, J. Sarmiento, S. Taguchi, T. Takahashi and C. W. Yuen, 2002. Towards robust regional estimates of $CO_2$ sources and sinks using atmospheric transport models. *Nature,* **415,** 626–630.

Harris, G. P., 1984. Phytoplankton productivity and growth measurements: past, present and future. *J. Plankton Res.,* **6,** 219–237.

Hein, M. and K. Sand-Jensen, 1997. $CO_2$ increases oceanic primary production. *Nature,* **388,** 526–527.

Højerslev, N. K., 1998. Spectral light absorption by gelbstoff in coastal waters displaying highly different concentrations. In *Proceedings, Ocean Optics XIV,* S.G. Ackelson and J. Campbell, eds. Office of Naval Research, Washington, DC.

Holligan, P. M., M. Viollier, D. S. Harbour, P. Camus and M. Champagne-Philippe, 1983. Satellite and ship studies of coccolithophore production along a continental shelf edge. *Nature,* **304,** 339–342.

Holligan, P. M., E. Fernandez, J. Aiken, W. M. Balch, P. Boyd, P. H. Burkill, M. Finch, S. B. Groom, G. Malin, K. Muller, D. A. Purdie, C. Robinson, C. C. Trees, S. M. Turner and P. Van der Wal, 1993a. A biogeochemical study of the coccolithophore *Emiliania huxleyi* in the North Atlantic. *Global Biogeochem. Cycles,* **7,** 879–900.

Holligan, P. M., S. B. Groom and D. S. Harbour, 1993b. What controls the distribution of the coccolithophore, *Emiliania huxleyi,* in the North Sea? *Fisheries Oceanography,* **2,** 175–183.

Hood, E.M., R. Wanninkhof and L. Merlivat, 2001. Short timescale variations of $f_{CO2}$ in a North Atlantic warm-core eddy: Results from the Gas-Ex 98 carbon interface ocean atmosphere (CARIOCA) buoy data. *J. Geophys. Res.,* **106,** 2561–2572.

Houghton, J. T., L. G. Meira Filho, J. Bruce, H. Lee, B. A. Callander, E. Haites, N. Harris and K. Maskell, 1995. *Climate Change, 1994: Radiative Forcing of Climate and an Evaluation of the IPCC IS92 Emission Scenarios.* Cambridge University Press. p. 339.

Humborg, C., D. J. Conley, L. Rahm, F. Wulff, A. Cociasu, and V. Ittekkot, 2000. Silicon retention in river basins: Far-reaching effects on biogeochemistry and aquatic food webs in coastal marine environments. *AMBIO,* **29,** 45–50.

Hutchins, D. A., G. R. DiTullio, Y. Zhang and K. W. Bruland, 1998. An iron limitation mosaic in the California upwelling regime. *Limnol. Oceanogr.,* **43,** 1037–1054.

Hutchins, D. A., C. E. Hare, R. S. Weaver, Y. Zhang, G. F. Firme, G. R. DiTullio, M. B. Alm, S. F. Riseman, J. M. Maucher, M. E. Geesey, C. G. Trick, G. J. Smith, E. L. Rue, J. Conn and K. W. Bruland, 2002. Phytoplankton iron limitation in the Humboldt Current and Peru Upwelling. *Limnol. Oceanogr.,* **47,** 997–1011.

Justic, D., N. N. Rabalais, and R. E. Turner, 2002. Modeling the impacts of decadal changes in riverine nutrient fluxes on coastal eutrophication near the Mississippi River Delta. *Ecological Modelling* **152,** 33–46.

Kaplan, A. and L. Reinhold, 1999. CO2 concentrating mechanisms in photosynthetic organisms. *Annual Reviews of Plant Physiology and Plant Molecular Biology,* **50,** 359–570.

Kiefer, D. A. and B. G. Mitchell, 1983. A simple, steady state description of phytoplankton growth based on absorption cross section and quantum efficiency. *Limnol. Oceanogr.,* **28,** 770–776.

Kopelovich, O. V., S. V. Lutsarev and V. V. Rodionov, 1989. Light spectral absorption by yellow substance of ocean water. *Okeanologiya,* **29,** 409–414.

Korb, R. E., P. J. Saville, A. M. Johnston and J. A. Raven, 1997. Sources of inorganic carbon for photosynthesis by three species of marine diatom. *J. Phycol.,* **33,** 433–440.

Kudo, I., 2003. Change in the uptake and cellular Si : N ratio in diatoms responding to the ambient Si:N ratio and growth phase. *Marine Biology,* **143,** 39–46.

Li, Q. and D.T. Canvin, 1998. Energy Sources for $HCO_3^-$ and $CO_2$ Transport in Air-Grown Cells of *Synechococcus* UTEX 625. *Plant Physiol.* **116,** 1125–1132.

Liu, H. B., E. A. Laws, T. A. Villareal and E. J. Buskey, 2001. Nutrient-limited growth of *Aureoumbra lagunensis* (Pelagophyceae), with implications for its capability to outgrow other phytoplankton species in phosphate-limited environments. *J. Phycol.,* **37,** 500–508.

Longhurst, A., 1998. *Ecological Geography of the Sea,* Academic Press, San Diego, 398 p.

MacIsaac, JJ., R.C. Dugdale, R.T. Barber, D. Blasco, and T.T. Packard, 1985. Primary production in an upwelling center. *Deep-Sea Res.,* **32,** 503–529.

Mackenzie, F. T., L. M. Vera, and A. Lerman, 2002. Century-scale nitrogen and phosphorus controls of the carbon cycle. *Chemical Geology,* **190,** 13–32.

Metzl, N., A. Poisson, F. Louanchi, C. Brunet, B. Schauer and B. Bres, 1995. Spatio-temporal distributions of air-sea fluxes of $CO_2$ in the Indian and Antarctic oceans. A first step. *Tellus Ser. B: Chem. Phys. Meteorol.,* **47B,** 56–69.

Miller, A. G., G. S. Espie and D. T. Canvin, 1991. Active $CO_2$ transport in cyanobacteria. *Can. J. Bot./J. Can. Bot.,* **69,** 925–935.

Milliman, J. D., P. J. Troy, W. M. Balch, A. K. Adams, Y. H. Li and F. T. Mackenzie, 1999. Biologically mediated dissolution of calcium carbonate above the chemical lysocline? *Deep-Sea Res.,* **46,** 1653–1669.

Morel, A. and A. Bricaud, 1981. Theoretical results concerning light absorption in a discrete medium, and application to the specific absorption of phytoplankton. *Deep-Sea Res.,* **28,** 1375–1393.

Morel, F. M. M., 1987. Kinetics of nutrient uptake and growth in phytoplankton. *J. Phycol.,* **23,** 137–150.

Morel, F. M. M., J. R. Reinfelder, S. B. Roberts, C. P. Chamberlain, J. G. Lee and D. Yee, 1994. Zinc and carbon co-limitation of marine phytoplankton. *Nature,* **369,** 740–742.

Murphy, P. P., D. E. Harrison, R. A. Feely, T. Takahashi, R. F. Weiss and R. H. Gammon, 1998. Variability of Delta $pCO_2$ in the subarctic North Pacific. A comparison of results from four expeditions. *Tellus Ser. B: Chem. Phys. Meteorol.,* **50B,** 185–204.

Murphy, P. P., Y. Nojiri, D. E. Harrison and N. K. Larkin, 2001. Scales of spatial variability for surface ocean $pCO_2$ in the Gulf of Alaska and Bering Sea: toward a sampling strategy. *Geophys. Res. Lett.,* **28,** 1047–1050.

Nimer, N. A. and M. J. Merrett, 1995. Calcification rate in relation to carbon dioxide release, photosynthetic carbon fixation and oxygen evolution in *Emiliania huxleyi. Bull Inst. Oceanogr. Monaco,* **14,** 37–42.

Nimer, N. A., M. D. Iglesias-Rodriguez and M. J. Merrett, 1997. Bicarbonate utilization by marine phytoplankton species. *J. Phycol.,* **33,** 625–631.

Osborne, B. A. and R. J. Geider, 1986. Effect of nitrate-nitrogen limitation on photosynthesis of the diatom *Phaeodactylum tricornutum* Bohlin (Bacillariophyceae). *Plant Cell Environ.,* **9,** 617–625.

Paasche, E., 1964. A tracer study of the inorganic carbon uptake during coccolith formation and photosynthesis in the coccolithophorid *Coccolithus huxleyi. Physiol. Plant.,* **3,** 1–82.

Paasche, E., 1998. Roles of nitrogen and phosphorus in coccolith formation in *Emiliania huxleyi* (Prymnesiophyceae). *Eur. J. Phycol.,* **33,** 33–42.

Paasche, E., 2002. A review of the coccolithophorid *Emiliania huxleyi* (Prymnesiophyceae), with particular reference to growth, coccolith formation, and calcification-photosynthesis interactions. *Phycologia,* **40,** 503–529.

Paerl, H. W., J. D. Willey, M. Go, B. L. Peierls, J. L. Pinckney and M. L. Fogel, 1999. Rainfall stimulation of primary production in western Atlantic Ocean waters: roles of different nitrogen sources and co-limiting nutrients. *Mar. Ecol. Prog. Ser.,* **176,** 205–214.

Platt, T., K. L. Denman and A. D. Jassby, 1977. Modeling the productivity of phytoplankton. In: *The Sea: Ideas and Observations on Progress in the Study of the Seas. Vol. VI,* E.D. Goldberg, ed. John Wiley, New York, 807–856.

Platt, T., C. Fuentes-Yaco, and K. Frank, 2003. Spring algal bloom and larval fish survival. *Nature,* **423,** 398–399.

Platt, T. and A. D. Jassby, 1976. The relationship between photosynthesis and light for natural assemblages of coastal marine phytoplankton. *J. Phycol.,* **12,** 421–430.

Platt, T., and S. Sathyendranath, 1988. Oceanic primary production: estimation by remote sensing at local and regional scales. *Science,* **241,** 1613–1620.

Platt, T., and S. Sathyendranath, 1993. Estimators of primary production for interpretation of remotely sensed data on ocean color. *J. Geophys. Res.,* **98,** 14,561–14,576.

Platt, T., and S. Sathyendranath, 1991. Biological production models as elements of coupled, atmosphere-ocean models for climate research. *J. Geophys. Res.,* **96,** 2585–2592.

Platt, T., S. Sathyendranath and P. Ravindran, 1990. Primary production by phytoplankton: analytic solutions for daily rates per unit area of water surface. *Proc. R. Soc. Lond. Ser. B,* **241,** 101–111.

Platt, T., S. Sathyendranath, G. N. White III, and P. Ravindran, 1994a. Attenuation of visible light by phytoplankton in a vertically-structured ocean: solutions and applications. *J. Plankton Res.,* **16,** 1461–1487.

Platt, T., J. D. Woods, S. Sathyendranath and W. Barkmann ,1994b. Net primary production and stratification in the ocean. *Geophysical Monographs,* **85,** 247–254.

Raven, J. A., 1980. Nutrient transport in microalgae. *Adv. Microbiol. Physiol.,* **21,** 47–226.

Raven, J. A. and A. M. Johnston, 1991. Mechanisms of inorganic-carbon acquisition in marine phytoplankton and their implications for the use of other resources. *Limnol. Oceanogr.,* **36,** 1701–1414.

Raven, J. A., 1995. Photosynthetic and non-photosynthetic roles of carbonic anhydrase in algae and cyanobacteria. *Phycologia,* **34,** 93–101.

Raven, J. A., 1997. Inorganic carbon acquisition by marine autotrophs. *Advances in Botanical Research,* **27,** 85–209.

Raven, J. A., 2001. A role for mitochondrial carbonic anhydrase in limiting $CO_2$ leakage from low $CO_2$-grown cells of *Chlamydomonas reinhardtii. Plant, Cell Environ.,* **24,** 261–265.

Rhee, G. Y. and I. J. Gotham, 1981. The effect of environmental factors on phytoplankton growth: light and the interactions of light with nitrate limitation. *Limnol. Oceanogr.,* **26,** 649–659.

Riebesell, U., D. A. Wolf-Gladrow and V. Smetacek, 1993. Carbon dioxide limitation of marine phytoplankton growth rates. *Nature,* **361,** 249–251.

Riebesell, U., I. Zondervan, B. Rost, P. D. Tortell, R. E. Zeebe and F. M. M. Morel, 2000. Reduced calcification of marine plankton in response to increased atmospheric $CO_2$. *Nature,* **407,** 364–367.

Riegman, R., W. Stolte, A. A. M. Noordeloos and D. Slezak, 2000. Nutrient uptake and alkaline phosphatase (EC 3:1:3:1) activity of *Emiliania huxleyi* (Prymnesiophyceae) during growth under N and P limitation in continuous cultures. *J. Phycol.,* **36,** 87–96.

Rocha, C., H. Galvao, and A. Barbosa, 2002. Role of transient silicon limitation in the development of cyanobacteria blooms in the Guadiana Estuary, south-western Iberia. *Mar. Ecol. Prog. Ser.,* **228,** 35–45

Roesler, C. S., M. J. Perry and K. L. Carder, 1989. Modeling *in situ* phytoplankton absorption from total absorption spectra in productive inland marine waters. *Limnol. Oceanogr.,* **34,** 1510–1523.

Rotatore, C., B. Colman and M. Kuzma, 1995. The active uptake of carbon dioxide by the marine diatoms *Phaeodactylum ticornutum* and *Cyclotella* sp. *Plant Cell Environ.,* **18,** 913–918.

Sakshaug, E., K. Andresen and D. A. Kiefer, 1989. A steady state description of growth and light absorption in the marine planktonic diatom *Skeletonema costatum. Limnol. Oceanogr.,* **34,** 198–205.

Sanders, R., T. Jickells, and D. Mills, 2002. Nutrients and chlorophyll at two sites in the Thames plume and southern North Sea. *J. Sea Res.* **46,** 13–28.

Sarmiento, J. L., 1991. Oceanic uptake of anthropogenic $CO_2$: The major uncertainties. *Global Biogeochem. Cycles,* **5,** 309–313.

Sarmiento, J. L. and J. C. Orr, 1991. Three-dimensional simulations of the impact of Southern Ocean nutrient depletion on atmospheric $CO_2$ and ocean chemistry. *Limnol. Oceanogr.,* **36,** 1928–1950.

Sarmiento, J. L. and C. Le Quere, 1996. Oceanic carbon dioxide uptake in a model of century-scale global warming. *Science,* **274,** 1346–1350.

Sathyendranath, S and T. Platt, 1988. The spectral irradiance field at the surface and in the interior of the ocean: A model for applications in oceanography and remote sensing. *J. Geophys. Res.*, **93,** 9270–9280.

Sathyendranath, S. and T. Platt, 1989. Computation of aquatic primary production: extended formalism to include effect of angular and spectral distribution of light. *Limnol. Oceanogr.*, **34,** 188–198.

Schlesinger, W. H. 1991. Biogeochemistry: an analysis of global change. Academic Press, New York.

Sciandra, A., 1991. Coupling and uncoupling between nitrate uptake and growth rate in *Prorocentrum minimum* (Dinophyceae) under different frequencies of pulsed nitrate supply. *Mar. Ecol. Prog. Ser.*, **72,** 261–269.

Sciandra, A., J. Gostan, Y. Collos, C. Descolas-Gros, C. Leboulanger, V. Martin-Jézéquel, M. Denis, D. Lefèvre, C. Copin-Montégut and B. Avril, 1997. Growth compensating phenomena in continuous cultures of *Dunaliella tertiolecta* limited simultaneously by light and nitrate. *Limnol. Oceanogr.* **42,** 1325–1339.

Sciandra, A., L. Lazzara, H. Claustre and M. Babin, 2000. Responses of the growth rate, pigment composition and optical properties of *Cryptomonas* sp. to light and nitrogen stresses. *Mar. Ecol. Prog. Ser.*, **201,** 107–120.

Sciandra, A., J. Harlay, D. Lefèvre, R. Lemée, P. Rimelin, M. Denis and J. P. Gattuso, 2003. Response of the coccolithophore *Emiliania huxleyi* to elevated partical pressure of $CO_2$ under nitrate limitation. *Mar. Ecol. Prog. Ser.* **261,** 111–122.

Seuront, L., V. Gentilhomme and Y. Lagadeuc, 2002. Small-scale nutrient patches in tidally mixed coastal waters. *Mar. Ecol. Prog. Ser.*, **232,** 29–44.

Sikes, C. S. and V. J. Fabry, 1994. Photosynthesis, $CaCO_3$ deposition, coccolithophorids, and the global carbon cycle. In *Photosynthetic carbon metabolism and regulation of atmospheric* $CO_2$ *and* $O_2$, N. E. Tolbert and J. Preiss, eds. Oxford University Press, New York, pp. 217–233.

Stramski, D., A. Sciandra and H. Claustre, 2002. Effects of temperature, nitrogen, and light limitation on the optical properties of the marine diatom *Thalassiosira pseudonana. Limnol. Oceanogr.*, **47,** 392–403.

Sueltemeyer, D., 1998. Carbonic anhydrase in eukaryotic algae: characterization, regulation, and possible function during photosynthesis. *Can. J. Bot./Rev. Can. Bot.*, **76,** 962–972.

Sverdrup, H. U., 1953. On conditions for the vernal blooming of phytoplankton. *Journal du Conseil,* **18,** 287–295.

Takahashi, T., S. C. Sutherland, C. Sweeney, A. Poisson, N. Metzl, B. Tilbrook, N. Bates, R. Wanninkhof, R. A. Feely, C. Sabine, J. Olafsson and Y. Nojiri, 2002. Global sea-air $CO_2$ flux based on climatological surface ocean $pCO_2$, and seasonal biological and temperature effects. *Deep-Sea Res.*, **49,** 9–10.

Terry, K. L., E. A. Laws and D. J. Burns, 1985. Growth rate variation in the N:P requirement ratio of phytoplankton. *J. Phycol.*, **21,** 323–329.

Tortell, P. D., J. R. Reinfelder and F. M. M. Morel, 1997. Active uptake of bicarbonate by diatoms. *Nature,* **390,** 243–244.

Tortell, P. D., 2000. Evolutionary and ecological perspectives on carbon acquisition in phytoplankton. *Limnol. Oceanogr.*, **45,** 744–750.

Tortell, P. D., G. H. Rau, and F. M. M. Morel, 2000. Inorganic carbon acquisition in coastal Pacific phytoplankton communities. *Limnol. Oceanogr.*, **45,** 1485–1500.

Turner, R. E., N. N. Rabalais, D. Justic, and Q. Dortch, 2003. Global patterns of dissolved N, P and Si in large rivers. *Biogeochem.* **64,** 297–317.

Turpin, D. H., 1991. Effects of inorganic N availability on algal photosynthesis and carbon metabolism. *J. Phycol.*, **27,** 14–20.

Van Bleijswijk, J., P. Van der Wal, R. Kempers, M. Veldhuis, J. R. Young, G. Muyzer, E. de Vrind-de-Jong and P. Westbroek, 1991. Distribution of two types of *Emiliania huxleyi* (Prymnesiophyceae) in the northeast Atlantic region as determined by immunofluorescence and coccolith morphology. *J. Phycol.,* **27,** 566–570.

Van der Wal, P., R. S. Kempers and M. J. W. Veldhuis, 1995. Production and downward flux of organic matter and calcite in a North Sea bloom of the coccolithophore *Emiliania huxleyi. Mar. Ecol. Prog. Ser.,* **126,** 247–265.

Walsh, J. J., 1991. Importance of continental margins in the marine biogeochemical cycling of carbon and nitrogen. *Nature,* **350,** 53–55.

Westbroek, P., C. W. Brown, J. Van Bleijswijk, C. Brownlee, G. J. Brummer, M. Conte, J. Egge, E. Fernandez, R. Jordan, M. Knappertsbusch, J. Stefels, M. Veldhuis, P. Van der Wal and J. Young, 1993. A model system approach to biological climate forcing. The example of *Emiliania huxleyi. Global and Planetary Change,* **8,** 27–46.

Winter, A., R. W. Jordan and P. H. Roth, 1994. Biogeography of living coccolithophores in ocean waters. In *Coccolithophores,* A. Winter and W. G. Siesser, eds. Cambridge University Press, Cambridge, pp 161–177.

Wolf-Gladrow, D. A., U. Riebesell, S. Burkhardt and J. Bijma, 1999. Direct effects of $CO_2$ concentration on growth and isotopic composition of marine plankton. *Tellus,* **51,** 461–476.

Wu, H. T., and T. L. Chou, 2003. Silicate as the limiting nutrient for phytoplankton in a subtropical eutrophic estuary of Taiwan. *Estuar. Coast. Shelf Sci.* **58,** 155–162.

Young, J. R., 1994. Functions of coccoliths. In *Coccolithophores,* A. Winter and W. G. Siesser, eds. Cambridge University Press, Cambridge, pp. 63–82.

Zeebe, R. E. and D. A. Wolf-Gladrow, 2001. *$CO_2$ in seawater: equilibrium, kinetics, isotopes.* Elsevier, Amsterdam.

Zeng, J., Y. Nojiri, P. P. Murphy, C. S. Wong and Y. Fujinuma, 2002. A comparison of Delta $pCO_2$ distributions in the northern North Pacific using results from a commercial vessel in 1995–1999. *Deep-Sea Res. II,* **49,** 5303–5315.

Zondervan, I., R. E. Zeebe, B. Rost and U. Riebesell, 2001. Decreasing marine biogenic calcification: a negative feedback on rising atmospheric $pCO_2$. *Global Biogeochem. Cycles,* **15,** 507–516.

Zondervan, I., B. Rost, U. Riebesell, 2002. Effect of $CO_2$ concentration on the PIC/POC ratio in the coccolithophore *Emiliania huxleyi* grown under light-limiting conditions and different daylengths. *J. Exp. Mar. Biol. Ecol.,* **272,** 55–70.

# Chapter 13. DIAGNOSIS AND PREDICTION OF VARIABILITY IN SECONDARY PRODUCTION AND FISH RECRUITMENT PROCESSES: DEVELOPMENTS IN PHYSICAL-BIOLOGICAL MODELING

JEFFREY A. RUNGE

*University of New Hampshire*

PETER J. S. FRANKS

*University of California, San Diego*

WENDY C. GENTLEMAN

*Dalhousie University*

BERNARD A. MEGREY

*Alaska Fisheries Science Center*

KENNETH A. ROSE

*Louisiana State University*

FRANCISCO E. WERNER

*University of North Carolina, Chapel Hill*

BRUNO ZAKARDJIAN

*Université du Québec*

## Contents



*The Sea*, Volume 13, edited by Allan R. Robinson and Kenneth H. Brink
ISBN 0-674-01526-6 

## 1. Introduction

The use of coupled 3-D physical-biological models has been envisaged as an essential component, combined with observations of ecological processes at sea, for the development of quantitative diagnostic and predictive understanding of recruitment in fish populations (e.g., Cushing, 1996). Such models could provide integrated, "holistic" simulations of impacts on ecosystems and recruitment processes of interannual and decadal scale variability in climate forcing.

The convergence of scientific advancement in several domains of coastal ocean research has now made application of 3-D, coupled physical-biological models more feasible. Increased emphasis in observing long-term changes within ocean systems has fostered the development of coastal ocean observing programs around the world. Impressive advances have been made in the application of operational oceanography to provide synoptic visualization of characteristics of the ocean state. (e.g. Schofield et al., 2002). At the same time, advances in mathematical models for coastal marine ecosystems (e.g. Hofmann and Lascara, 1998), data assimilation methodology (e.g. Robinson et al., 1998; Robinson and Lermusiaux, 2002) and developments in computer technology are leading to the interpretative, diagnostic and forecast model systems for coastal climate-ocean processes. The challenge is to integrate models of different types across trophic levels, embed them into representations of the physics of the coastal ocean, and devise approaches to incorporating uncertainty into simulation projections (de Young et al., 2004).

Much of the emphasis in data acquisition systems and interpretative/predictive physical-biological modeling has been on the characterization of the physical ocean state and biological conditions at the lower trophic level, i.e. primary production (Hofmann and Friedrichs, 2002). Many pressing, real-world questions, including the conservation and management of the coastal fisheries, also require understanding of the physical-biological interactions controlling dynamics of the heterotrophic plankton, i.e. secondary production and recruitment into the exploited populations (e.g., Harrison and Parsons, 2000). Increasing knowledge of the dynamics and interactions determining the distribution and abundance of zooplankton and ichthyoplankton, combined with the developments in physical circulation modeling and computational capacity, is also leading to improved understanding of variation at these higher trophic levels. Understanding in this context implies the ability to diagnose sources and mechanisms of observed variation and the capability to make predictions of change in the future.

In this chapter, we review progress in the development of coupled physical-biological models and in their contribution to the understanding of secondary production and recruitment processes. In the context here, the term *coupled* denotes the embedding of a biological model describing the dynamics of components of pelagic marine ecosystems in time-varying flow and temperature fields provided by a physical circulation model of the region and time period of interest. The two models need not be dynamically coupled, i.e. run at the same time; the currents and hydrographic output from the physical simulation can be stored and used to force the biological model. We describe the recent advances; identify gaps and weaknesses in the approach and outline a vision as to where the coupled models are going and how they might be applied to fisheries and other management issues in coastal and shelf seas.

In Sections 2–4, we examine approaches to the coupled modeling of key processes controlling the dynamics and interactions of heterotrophic populations in the coastal and shelf oceans. No single model or modeling approach can include all relevant processes; a hierarchy of models needs to be developed and applied according to the region or regions and particular questions under consideration (deYoung et al., 2004). The nature of real-world problems is such that consideration of individual species becomes more important at higher trophic levels. This introduces a complexity that can be dealt with by taking a "middle out" approach, in which the dynamics of species or species groups of interest are described in detail and the linked state variables at trophic levels above and below the targeted species are more coarsely resolved or are supplied from other models (e.g. Fig. 13.1).

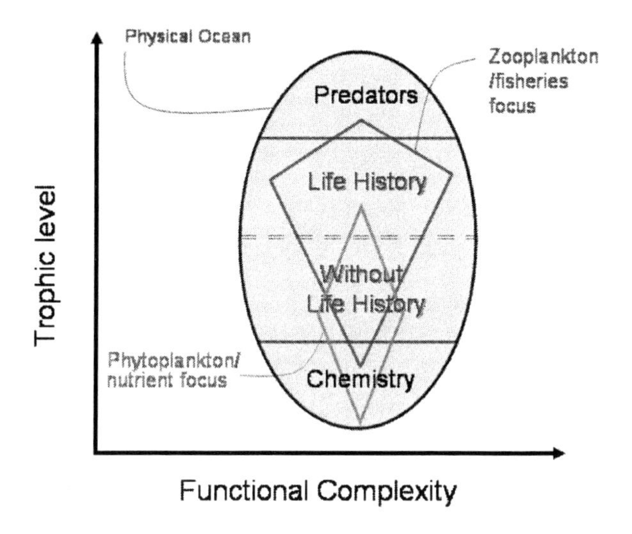


Figure 13.1 A diagramatic representation of the *middle-out* approach, in which the dynamics of individual species at higher trophic levels are modelled at high resolution and forced by physical processes as well as higher and lower level biological processes modelled at coarser resolution (adapted from de Young et al. 2004).

In section 2, we discuss coupled physical nutrient-phytoplankton-zooplankton (NPZ) models, in the context of describing primary production cycles for linking with models of higher trophic levels. In many NPZ oceanographic models, zooplankton is treated as a closure term without much attention paid to individual species or to the dynamics of higher trophic levels (i.e., fish) as sources of mortality (Edwards and Yool, 2000). The Z compartment in NPZ models represents an ensemble of species each with potentially very different physiological and ecological functional relationships. Therefore, while NPZ models are useful for analysis of spatial and temporal patterns of primary production, they are not sufficient for resolving the population dynamics in the detail necessary to understand variability in secondary production and recruitment. In Section 3, we discuss more realistic secondary production models in the form of coupled physical-zooplankton life history models. These models focus on the mesozooplankton (approx. 0.2–20mm in size: see Lenz, 2000), particularly planktonic copepods, which serve as a major link between hydrodynamic effects of climatic variability on primary producers and higher trophic levels (Cushing, 1982, 1996; Runge, 1988; Hansen *et al.*, 1994; Skreslet, 1997). Coupled physical-zooplankton life history models characterize the essential dynamics of the target species with decreasing resolution of predator and prey state variables. In section 4, we examine spatially-explicit models that focus on the growth, survival and spatial movement of the planktonic, early life stages of fish and harvested invertebrates such as shrimp, crab and lobster. Because the survival of these early life stages is a crucial (although not exclusive) factor determining recruitment (e.g. Rothschild, 1986), considerable effort has been made to accurately describe the feeding and movement of these early life stages in simulated flow and temperature fields.

Section 5 reviews examples of the present status of development of coupled biological-physical models in three coastal areas: the southern Benguela upwelling system, the western Gulf of Alaska and the northwest Atlantic. These case studies illustrate programmatic approaches to the multidisciplinary collaboration needed to develop the coupled models and acquire knowledge and data for their implementation in a given coastal region. In the concluding Section 6, we discuss the possibility that coupled model simulations of interactions with physical processes and dynamics at different levels of the pelagic ecosystem may be linked together to provide a comprehensive tool for analysis of coastal observing system data, habitat quality for early life stages of exploited species, recruitment and other coastal zone environmental issues.

## 2. NPZ Models

### *2.1 Overview*

Nutrient-Phytoplankton-Zooplankton (NPZ) models have been widely used in biological oceanography for over 4 decades. While the basic three-compartment NPZ model has changed little over this time (Franks, 2002), modifications to this simple structure by increasing the number of functional groups (e.g., NNPPZZD) have made the ecosystem model a central tool in investigations of processes as diverse as global biogeochemical cycles and red tides in lagoons.

NPZ models are usually formulated using nitrogen as the models' currency. Nitrogen was chosen because (at that time) it was considered to be the main nutrient limiting oceanic primary production. Most field sampling programs do not measure biological properties in terms of nitrogen content, but rather in units such as chlorophyll *a* concentration, cell numbers and zooplankton displacement volume. This necessitates conversion from these measured properties to nitrogen for the model, potentially adding an additional level of inaccuracy and uncertainty to the calculations and model-data comparisons.

Nevertheless, the close coupling of nitrate and phytoplankton to physical forcings makes it possible for (simple) NPZ models to reproduce seasonal or annual cycles of chlorophyll and nitrate, and the spatial details of their distributions in coastal areas (Figure 13.2). Assimilation of data into relatively simple ecosystem models in open ocean conditions has shown that they can also capture the seasonal dynamics of total chlorophyll at oceanic stations quite well (Spitz et al., 2001, Hurtt and Armstrong, 1999; Figure 13.3).

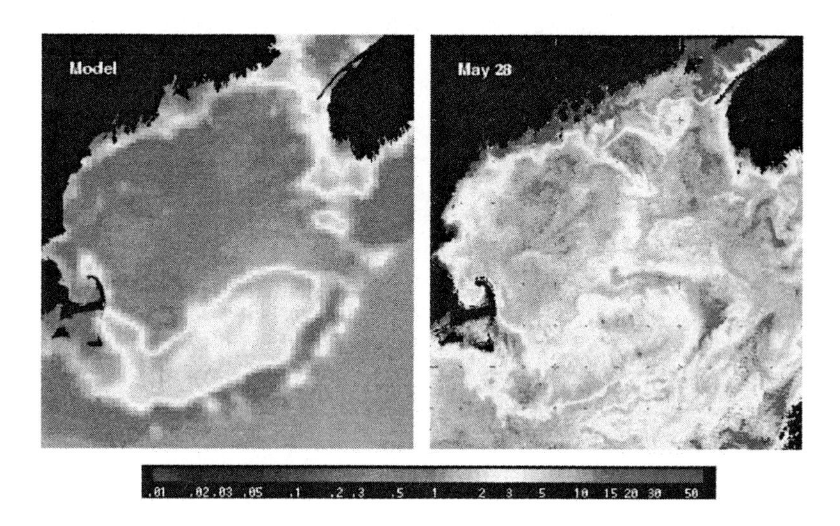


Figure 13.2 Surface chlorophyll *a* in the Gulf of Maine and on Georges Bank from a physical-NPZ model (Franks and Chen, 2001), and from satellite remote sensing. Note that this particular model did not include the nearshore dynamics, such as rivers and nutrient inputs, that lead to high phytoplankton biomass in these areas.

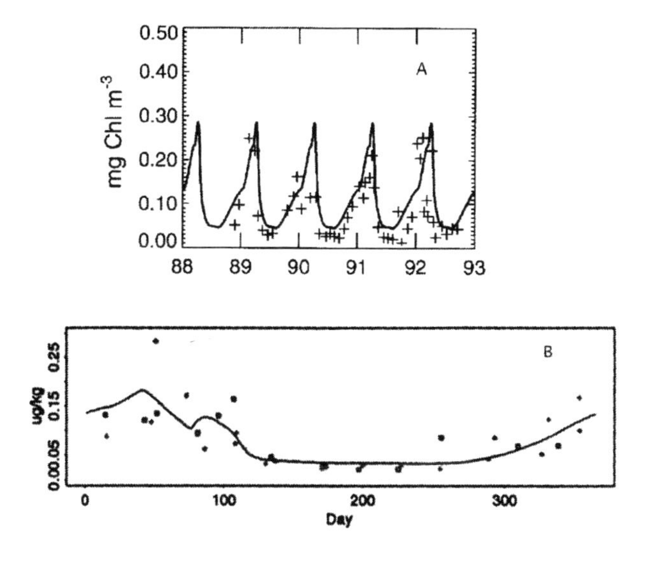


Figure 13.3 Simulation (lines) and data (points) of mixed-layer chlorophyll *a* at the BATS site using data assimilation. A) From Spitz et al. (2001). B) From Hurtt and Armstrong (1999).

The diversity of zooplankton assemblages and their size structure renders more difficult the task of reproducing zooplankton densities in space and time. Models are often parameterized for crustacean zooplankton. However, the dominant grazers in the ocean are the heterotrophic protists – single celled organisms that have substantially different growth and grazing rates than the crustaceans. More highly resolved models may include separate compartments for protist and crustacean zooplankton, allowing comparison of the model results to these two (or greater) broad categories (Doney, 1999; Denman, 2003).

When modeling the growth and population processes of zooplankton or larval fish, the basic role of an NPZ model is to supply a sufficiently accurate prey field: when is the food available, where is it, and how much of it is there? This information can be passed to more complex models of zooplankton and prescribed as its feeding component. Food is often represented in terms of the phytoplankton concentration, or chlorophyll *a*, which NPZ-type models can successfully reproduce because, as stated above, the spatial and temporal patterns of the phytoplankton are strongly coupled to the ambient physical forcings: an accurate physical model is essential to re-creating phytoplankton patchiness and structure particularly in regions of strong physical forcing such as the coastal waters.

*2.2 Approaches to coupling NPZ models with physical models and higher trophic levels*

An NPZ model is usually coupled to a physical model as though it were a set of tracers being moved by the advection-diffusion equation. Each state variable of the NPZ model will have a separate equation describing its motion in space and time, of the form

$$\frac{\partial C}{\partial t} + u\frac{\partial C}{\partial x} + v\frac{\partial C}{\partial y} + (w + w_s)\frac{\partial C}{dz} = \kappa_h\left(\frac{\partial^2 C}{\partial x^2} + \frac{\partial^2 C}{\partial y^2}\right) + \kappa_v\frac{\partial^2 C}{\partial z^2} + \text{biological dynamics} \tag{13.1}$$

where $C$ is the concentration of the state variable ($N$, $P$, or $Z$), $u$, $v$ and $w$ are the horizontal and vertical water velocities determined by the physical model, $w_s$ is the vertical swimming or sinking speed of the state variable, and $\kappa_h$ and $\kappa_v$ are the horizontal and vertical eddy diffusivities. The *biological dynamics* of equation (13.1) are the equations of the NPZ model. Typically $u$, $v$ and $w$ are obtained from a physical model run simultaneously with the biological dynamics. These models range from simple one-dimensional (1D) models with biological dynamics averaged horizontally over the mixed layer, to full 3D models with high-order turbulence-closure submodels (Franks, 2002; Denman, 2003).

Models of the dynamics of the higher trophic levels (large zooplankton and fish) are generally run separately from the coupled physical-NPZ model. The coupling between the physical-NPZ model and the higher trophic level model is through advection and diffusion of the higher trophic levels, the prey available to them, and the temperature field that affects the vital rates (e.g., ingestion, assimilation) of the higher trophic levels. In this case the "Z" term comprises total zooplankton, and the spatially-explicit abundances or biomass of species in the higher trophic levels are depicted in a separate coupled dynamical model.

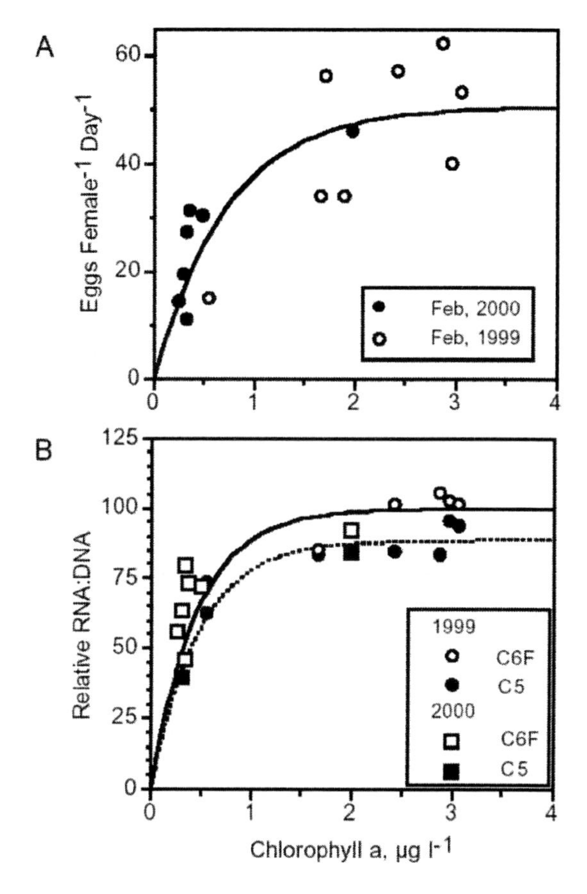


Figure 13.4 Observed egg production rate (A: eggs female$^{-1}$ d$^{-1}$) and relative RNA:DNA ratios (B: adult females and stage C5) of *C. finmarchicus* in the Gulf of Maine as a function of mean chlorophyll *a* concentration (adapted from Durbin et al., 2003).

In the field, bulk measures of food availability (such as chlorophyll *a*) may be sufficient to adequately represent food conditions in higher trophic level models in certain situations. For example, for copepods with a life cycle tied to the spring phytoplankton bloom (see Heinrich, 1962; Plourde and Runge, 1993; Melle and Skjoldal, 1998; Head et al., 2000), an NPZ model could be applied to generate the timing of the bloom. In turn, the phytoplankton bloom would drive the initiation of spawning in the copepod population whose life history and distribution may be described with a spatially explicit population dynamics model (Section 3). Moreover, linear or asymptotic relationships of egg production rate with chlorophyll *a* concentration, have been observed in the sea for a number of copepod species and coastal seas (e.g. Checkley, 1980; Runge, 1985; Peterson and Bellantoni, 1987; Armstrong et al. 1991; Dam et al., 1994; Harris et al., 2000; Calbet et al., 2002;

Durbin et al., 2003). In the Gulf of Maine and on Georges Bank, where high levels of phytoplankton production may occur even in winter, the relationship of weight-specific egg production rate to integrated chlorophyll *a* concentrations is well represented by an Ivlev function, even if chlorophyll *a* may only be a proxy for feeding on phytoplankton and microheterotrophs (e.g. Fig. 13.4). Observations from Georges Bank, the Gulf of Maine, the Scotain Shelf and Labrador Sea indicate that chlorophyll *a* concentrations > 80 mg $m^{-2}$ are generally sufficient to sustain maximum, or near-maximum egg production rates of *C. finmarchicus* (Campbell and Head, 2000; Head et al., 2000; Durbin et al., 2003).

Empirical determination of functional relationships (as illustrated in Fig. 13.4) that include critical food concentrations at which growth and reproductive rates are maximal are useful in linking species-specific zooplankton life-history models to NPZ models. Integrated chlorophyll fields generated by NPZ models could be used to show where and when food is limiting the vital rates of key species. Simple linear or non-linear models relating growth or reproduction to food below the critical concentration may be sufficient for many applications. Given the growing evidence that the critical concentration itself is function of grazer body size (the smaller the grazer, the lower the critical concentration, e.g., Richardson and Verheye, 1999; McKinnon and Duggan, 2001 and references therein), NPZ models may aid in predicting or diagnosing crustacean zooplankton size structure based on food availability.

### *2.3 Present challenges to linking coupled NPZ models with higher trophic level models*

It is tempting to conclude that NPZ models are sufficiently advanced to the point of providing spatial and temporal distributions of food supply for application in higher trophic-level models. However, while relationships of growth and reproduction to bulk estimates of food concentration seem to work for some species in some regions, there are also many examples where such relationships do not hold. There is need for more work determining the relationships between phytoplankton and zooplankton growth and reproduction in key zooplankton taxa. For species and regions where relatively simple functional relationships do not work or are insufficient to capture variability in growth and reproduction, investigation of the relationship between food and zooplankton vital rates is a continuing need. In some cases phytoplankton is not a proxy for food for key zooplankton species—for example, the carnivorous euphausid, *Meganyctiphanes norvegica*—in which case a relationship is not expected. In other cases, food type, chemical composition (food quality) and size distribution, and the spatial/temporal distribution of food are important sources of variability in functional relationships with food concentration.

**Prey composition and size structure**

While it has been known for some time that prey type and quality are important to grazers, it has not yet been fully incorporated in modeling studies. Simple NPZ models cannot possibly capture the variety of prey available to higher trophic levels. With increasing understanding of the dynamics of the microbial loop, and recognition that many if not most copepods are omnivorous (e.g. Paffenhofer and Knowles, 1980; Harris, 1996), it has become evident that in certain situations it may be necessary to include the dynamics of heterotrophic protists and microzooplankton in models of zooplankton prey. A heterotrophic protist may already be several trophic levels removed from the primary producers, which would have important implications for modeling of the transfer of primary production to higher trophic levels. Several models exist that incorporate microheterotrophs (e.g. Frost, 1987; Moloney and Field, 1991); unfortunately the complex versions can have undesirable numerical properties (oscillations). These oscillations are explored in detail in Armstrong (1999), and arise through the opposing effects of phytoplankton nutrient uptake (stabilizing), predator limitation (e.g., Steele and Henderson, 1981, 1992; stabilizing), and predator functional response (stabilizing or destabilizing, depending on the form of the functional response). Simplified versions of these size-structured models have been included in physical models (e.g. Carr, 1998), but such models still require significant development before being widely applicable. Furthermore, gathering field data on rates and biomasses of microheterotrophs is difficult, and is often not done in field programs. Thus there are few good data sets available to formulate and test such models.

Because of the implications for global biogeochemical cycles, variation in the chemical composition of phytoplankton is a core modeling activity in several research programs (Doney, 1999; Doney et al. 2002). Recent attempts to simulate the annual cycles at the Bermuda Atlantic Time Series (BATS) station using a range of NPZ-type models has shown that inclusion of a variable C:chlorophyll ratio of the phytoplankton is necessary to accurately reproduce the cycles of chlorophyll and nitrate (Hurtt and Armstrong, 1999; Spitz et al. 2001). Similarly, interest in iron limitation of phytoplankton growth has led to several models of phytoplankton that include the dynamics of the nitrogen, carbon, iron and silica content of the cells (Moore et al., 2002; Fennel et al., 2002). These models can sometimes reproduce the increased silicification of diatoms under iron stress (Hutchins and Bruland, 1998), and the varying N:C ratios of phytoplankton in different stages of bloom. Such variations are an important beginning in reproducing the prey fields actually encountered by foraging zooplankton. Amino and fatty acids and protein concentration are also known to influence growth and reproduction (e.g. Kleppel et al., 1998; Jonasdottir et al., 2002; Hazzard and Kleppel, 2003), though these are seldom included explicitly in ecosystem models. We still require a great deal of work in elucidating the chemical factors determining prey choice among zooplankton, and how to include these factors in models.

The size structure of the prey field is also an important determinant of zooplankton feeding (e.g. Steele and Frost, 1977; Berggreen et al., 1988). Discovery of the fundamental importance of photosynthetic cyanobacteria to primary production in the ocean and elucidation of the dynamics of the microbial loop have led to a new understanding of the dynamics structuring the planktonic ecosystem. One important empirical finding has been that, as the concentration of phytoplankton (measured as chlorophyll) increases in the ocean, phytoplankton are added in increasingly larger size classes—the abundance of phytoplankton in the smaller size classes first increases and then saturates. This appears to be a consequence of competition for nutrients (small phytoplankton are less limited by diffusion of nutrients to their cell surface), and grazing (grazers of small phytoplankton are small, fast growing, and efficient grazers). Modeling efforts have attempted to reproduce such size spectra in a variety of ways. Some models explicitly include many state variables, one for each size of phytoplankton (e.g., Moloney and Field, 1991). However, this approach can greatly prolong the numerical calculation and, as mentioned above, usually results in unrealistic oscillations of the model's state variables. Armstrong (1999) found that a multi-phytoplankton model could be stable if the grazing pressure was modeled as a single grazer eating all size classes, rather than separate grazers for each size class. Another approach to including multiple size classes of phytoplankton is semi-empirical. An empirical relationship relating total phytoplankton biomass to the size spectrum of the autotrophs is used (e.g., Hurtt and Armstrong, 1999; Denman and Peña, 2002) to give an approximate spectrum while modeling only a single phytoplankton variable (total biomass). This temporally and spatially varying size spectrum can then be used to generate a locally varying average autotrophic growth rate, and a size-based grazing rate. This method is quick and simple to implement numerically, tends to be stable (there is only one phytoplankton state variable), and can give a more realistic representation of the food available to zooplankton.

**Spatial and temporal scaling issues**

All models of plankton dynamics produce results that are averaged both spatially and temporally. The degree of averaging usually depends on the questions being asked and the computer power available. Typically, models average temporally over a day, and/or spatially over the mixed layer. Vertical spatial averaging reduces out-of-phase oscillations in vertically adjacent layers that could generate unrealistically strong (and evanescent) gradients of properties in the euphotic zone. However, such averaging removes any ability of the model to reproduce vertical variations of food for higher trophic levels, including the deep chlorophyll maximum. Given the growing recognition of the stable microscale (< 1 m) vertical structuring of the euphotic ecosystem (Rines et al., 2002), and the plasticity of behavioral responses of zooplankton and ichthyoplankton to their food and physical environment (Tiselius, 1992, 1998; Davis, 1996, 2001; Franks, 2001; Incze et al., 2001), there is a growing need to assess the effects of this vertical structure on the growth and reproduction of zooplankton.

The addition of vertical resolution to simple models will add little except integration time. However, models with advanced turbulence-closure schemes and multi-compartment biological models have the potential to benefit from higher vertical resolution simulations. The trade off is always between the insights gained vs. the extra computational costs. While copepods may migrate hundreds of meters vertically each day, the scales over which they actually feed are still poorly known. Laboratory experiments have shown that zooplankton respond to microscale gradients of phytoplankton concentration (Price, 1989; Paffenhofer and Lewis, 1990) and it is possible that the important trophic interactions take place at this scale. However, as discussed earlier, in many cases vertically integrated chlorophyll standing stock is a workable proxy of food available for growth and reproduction. Whether a spatially averaged prey field is an appropriate measure for food available to zooplankton must be carefully considered, particularly if IBM's (Section 3) of zooplankton and ichthyoplankton are being used.

Temporal averaging (usually over a diel cycle) may help stabilize a model, but also removes any ability of the model to reproduce short (<1 d) timescale fluctuations in prey fields or grazing pressure. While it is clear that phytoplankton only photosynthesize during the day, and many zooplankton—particularly the crustaceans—graze at night to avoid predation, the importance of these dynamics has been largely unexplored in ecosystem models (but see McAllister, 1971). A recent exception is a study by Flynn and Fasham (2003), which examines the sensitivity of a simple plankton model to the inclusion of a diel light cycle. They found that models with only a single phytoplankton group could be simulated well without a light cycle. When the model included two phytoplankton groups with different abilities to assimilate nitrogen in the dark, the inclusion of a light-dark cycle led to increased variability of N:C ratios, changes in nutrient regeneration and $f$ ratios, and changes in the relative abundances of the different phytoplankton groups. This could have immediate effects on the quality of food available to the higher trophic levels.

**Predictions using NPZ models**

Models of plankton dynamics are increasingly being used to attempt to predict the effects of short- and long-term environmental perturbations (Hofmann and Friedrichs, 2002). Underlying these predictions is the assumption that the models have captured the appropriate dynamics to respond to such changes. Obtaining good agreement of models with past or current data, particularly when we must use empirical relationships in place of detailed mechanistic modeling, does not guarantee that the model will accurately reproduce the ecosystem response to future changes. The model is inherently structurally limited by the variables, transfer functions, and parameters that are used to build it. An NPZ model could not, for example, be used to predict a change from a crustacean-dominated ecosystem to a gelatinous zooplankton-dominated ecosystem under environmental stress.

One aid to prediction is data assimilation (Robinson et al. 1998). A promising data source is satellite-derived products such as surface chlorophyll concentrations. A few models have successfully assimilated remotely sensed pigment data to enhance their simulations of observations (Ishizaka, 1990; Armstrong et al., 1995; Friedrichs, 2002). Data assimilation can help improve the accuracy of short-term predictions (in the ocean this is probably up to 1–2 months for mesoscale circulations), but may not help in predicting the ecosystem response to decadal changes. Probably the best "predictive" models will continue to rely on the conceptual understanding gained through dynamic analysis of existing long-term data sets, and the ability to mechanistically reproduce observed long-time scale changes such as decadal oscillations and regime shifts in the plankton.

## 3. Coupled Zooplankton Life History Models

### *3.1 Overview*

In contrast to phytoplankton, bacteria and protozooplankton, which are unicellular organisms with short life-cycles (one to ten days), planktonic metazoans such as copepods or gelatinous zooplankton have longer (weeks to years), more complex life cycles and diverse, species-specific physiological rates (e.g. ingestion, respiration and excretion) and behaviors (e.g. ontogenetic and vertical migration, feeding). Fish larvae do not feed equally on all developmental stages or species, but rather select their prey according to size, among other variables, typically preferring copepod naupliar stages. On the other hand, planktivorous juvenile and adult fish, marine mammals, such as right whales, and sea birds feed preferentially on later developmental stages. A nitrogen-equivalent mesozooplankton biomass calculated by an NPZ type model usually cannot adequately represent very different food conditions for different predators. Thus, a first-order task in quantitative description of metazoan zooplankton population dynamics is the description of stage or size structure at the species level, allowing more accurate portrayal of food availability to higher trophic levels.

Modeling the life-history for all zooplankton species of a given ecosystem, and coupling those models with climatically driven circulation models, is unrealistic. One approach to the study of mesozooplankton population dynamics involves as a first step, the identification of key species (e.g. Fogarty and Powell, 2002). In temperate, subarctic and polar pelagic environments, the major zooplankton contributors to the pelagic ecosystem and recruitment dynamics can be narrowed down to a relatively small list. A feature of a relatively small number of key species is that it is possible, through field and laboratory study, to acquire quantitative knowledge of the physiology and behavior of the targeted species. This understanding can be applied to the formulation of a biological model that is, in a sense, custom fit for that species in a given geographic region. The key species targeted in these systems belong mainly to the Copepoda and Euphausiacea, which are prominent in the zooplankton, both by their presence in plankton net tows and their ecological roles (e.g. Banse, 1994; Eckman, 1994). Gelatinous zooplankton have also received a greater attention recently (e.g., Haskell et al., 1999). To illustrate the approach, particular reference will be made to copepod modeling, in particular to the planktonic copepod, *Calanus finmarchicus,* which dominates zooplankton assemblages in the subarctic North Atlantic.

After hatching, the development of a typical planktonic copepod such as *C. finmarchicus* passes through six naupliar stages and six copepodid stages (Fig. 13.5). The physiological rates and feeding behaviors vary non-linearly with stage and with environmental conditions (e.g., Campbell et al., 2001a). For example, the first naupliar stages (NI-II) develop using internal reserves and generally do not feed. In later stages, growth and development may be partially uncoupled as growth is mainly food dependent while development is generally temperature dependent (e.g., Vidal 1980). Furthermore, copepods exhibit other stage-dependent behaviors such as ontogenetic vertical migration and diapause, which are typical in many species, particularly at high and moderate latitude (e.g., Conover, 1988). Mortality, which generally results from predation or starvation, also depends on environmental conditions and may vary with developmental stage (e.g., Ohman et al., 2002), and is likely correlated with changes to the zooplankton predator assemblage.

The challenge for modelling copepod life cycles is therefore to quantitatively describe the change of population size and structure while accounting for environmental variability and transport. The following sections discuss the different approaches to coupling copepod population dynamics to circulation models. In Part 2 we describe existing structures of coupled copepod models and how they address demography, physiological rates, vertical migration and diapause, respectively. In Part 3 we outline directions for future research.

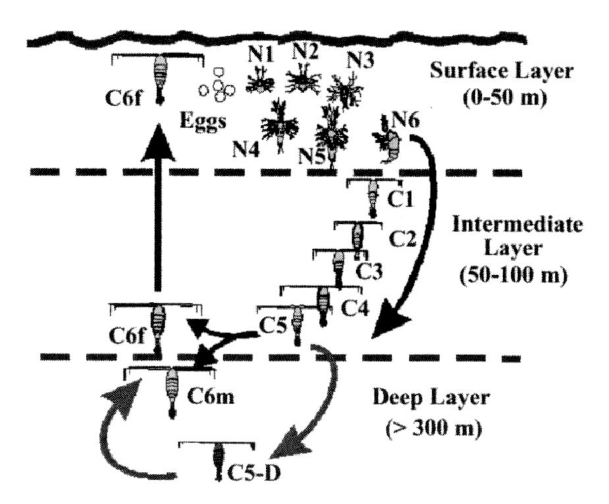


Figure 13.5 Schematic representation of *Calanus finmarchicus* life history. N1-N6: naupliar stages. C1-C6: copepodid stage 1 to copepodid stage 6 (adult, male or female). Stage C5-D: typical diapausing stage. Non-diapausing stage C1-C6, shown at approximate daytime depth, may undergo vertical migration to the surface layer at night.

### 3.2 *Approaches to coupling copepod life history models to physical models*

**The representation of population demographics**

Copepod life history models have traditionally considered the demographic structure of the population using variables to describe the biomass in different weight categories (i.e. weight-based or WBMs) or the abundance in different life-stages (i.e. stage-based models, or SBM; see review in Carlotti et al., 2000). In WBMs, the transfer between the different weight classes depends on the growth rates, which are usually modeled as bioenergetic formulations of ingestion, respiration and egestion that can be complicated (Steele and Frost, 1977; Hofmann and Ambler, 1988, Slagstad and Tande, 1990, Bryant et al., 1997). Weights are usually related to stages through a prescribed classification based on observations. The variables in SBMs quantify stage abundances directly. The rate of change of the population size within a stage is typically based on empirical relations for development, using the reciprocal of stage durations for given temperature and food conditions. Such models can result in developmental diffusion (Lewis et al., 1994); to minimize this problem, some authors refined the demographic resolution to include age-within-stage (e.g. Davis, 1984, Lynch et al., 1998). Other models have also sub-divided the stages to describe weights within stage (e.g. Carlotti and Sciandra, 1989).

In contrast to modeling the population in terms of its abundance in a particular stage or class, there is an increasing trend towards modeling individual copepods as discrete entities, using a large number of individuals to represent the population. This approach is known as i-state configuration (e.g. Caswell and John, 1992) or more commonly in marine literature as individual-based modelling or IBM. IBMs keep track of a number of variables related to each individual (e.g. age, weight, stage, lipid content). Changes to these variables arise from the physiological responses to the environment (e.g. growth, development, reproduction), often described by formulations that are analogous with the WBMs and SBMs (e.g., Voronina et al, 1979, Batchelder and Miller, 1989; Batchelder and Williams, 1995; Miller et al., 1998; Pedersen et al., 2001). However, since IBMs can keep track of an individual's history, these formulations can be adapted to account for individual variation in vital rates, such as that due to the length of time it has been starving (e.g. Crain and Miller, 2001, Batchelder et al., 2002). While IBMs provide for representation of individual variability, they can be computationally taxing and are not necessarily the approach of choice for representing linkage to higher trophic levels.

**IBM vs Eulerian models**

There are two basic approaches to coupling copepod life history models with circulation models. The first uses an Eulerian framework, akin to spatially-explicit NPZ models, which consider copepod population densities as concentrations. The biology and circulation are thus coupled using an advection-diffusion-reaction equation, which describes the local rate of change of population abundance affected by biological rates, currents and turbulence (e.g., Davis, 1984; Lynch et al., 1998; Zakardjian et al., 2003). Vital rates may be dependent on local food conditions and temperature derived from the circulation models, and behavioral movement can be included through a behavioral advection and/or diffusion term (Zakardian et al., 2003)

While the coupling with circulation models is straightforward, the advection-diffusion-reaction equation solves for copepod abundance is subject to the same kinds of numerical issues as can arise in circulation models (e.g. negative concentrations in advection-dominated flows). The transport/mixing routine in these models is presently costly computationally, so that the number of stage, age or weight classes may be a strong limitation. For example, while circulation models typically have 5–10 variables being transported, the number of variables in the higher resolution population models can be very high (e.g. 87 variables in the Lynch et al, 1998 study). Spatially structured population models (e.g., Gurney et al., 2001) are attempts to use an age-structured matrix model of population dynamics in an Eulerian framework. This approach permits the use of a complex weight or age structured matrix for the biological part of the model, which have been shown to clearly resolve the copepod demography and dynamics (see the review of Carlotti et al, 2000) but requires specific preprocessing of the physics, i.e., defining the mixing and transport matrix operator. In Gurney et al. (2001) the mixing or transport matrix operator is defined from multiple particle trackings with surface velocity fields from a 3D circulation model. A model by Bryant et al. (1997) simply used daily horizontal exchange coefficients between grid cells calculated from a 3D circulation model. Both models are however limited to 2D (horizontal) grids.

A second approach for coupling the copepod life history with the circulation uses a Lagrangian framework (i.e. particle tracking) wherein particular particles are followed as they move to different locations. Each particle can represent a cohort with the same history (e.g. ensemble approach as in Carlotti and Wolf, 1998; population sub-sampling as in Miller et al., 1998), or it can represent an individual copepod (Batchelder et al., 2002). Trajectories for a number of particles are simulated by considering the displacement due to local currents and turbulence, the physiological history of the individual or cohort is computed as the particle moves through the environment. With an IBM, the particle can also be given behavior specific to its unique state.

The simulation of numerous trajectories, usually $10^3$–$10^5$ particles, needed to gain insight into the population dynamics is computationally demanding. Hence, while the Lagrangian approach may perform well for qualitative study, its applicability to quantitative estimation of transport, exchange between sub-population and simulations over long time periods may be limited by the cost of analysis, although this situation continues to improve with advances in technology. Lagrangian water column models use an approach that follows the development of a cohort in a water mass that is assumed to be isolated (e.g., Heath et al., 1997). In these models, a time series of physical conditions (turbulent mixing, temperature) prevailing in the water column are extracted from a drift scenario and are used to drive the copepod life history model. This approach allows the use of complex physiological models of copepods life history (as an age or weight structured model) and trophic coupling with primary producers (e.g., Heath et al., 1997). However, the assumption of an isolated water column transported as a whole is only valid perhaps for the upper layer of the ocean (0–100 m). Vertically sheared circulation that prevails for example in the North Atlantic or on coastal shelves, in conjunction with the range of possible diel and ontogenetic vertical distributions of copepods like *C. finmarchicus,* restricts their applicability.

The above-mentioned models can incorporate demographic resolution with varying degrees of computer efficiency. The question is how realistic must the population demography be in order to achieve the research objectives. It is clear from Table 13.1 that few studies have ventured into extension of coupling the population models with circulation models in three dimensions. Increasing the spatial dimensions of Eulerian models has limited the use of population models to the simplest formulations, i.e., SBM (Lewis et al., 1994; Zakardjian et al., 2003). Models with higher demographic resolution (e.g., age-within-stage) have to-date been applied only in 1D or 2D (e.g., Davis, 1984; Carlotti and Radach, 1996; Lynch et al., 1998). A detailed application of a coupled IBM is the work of Batchelder et al. (2002) in a 2D physical framework.

Since each approach has its own advantages and disadvantages, it is better to consider them as complementary. Local process studies may need good resolution of individual growth and population demography, such as source regions of copepods to Georges Bank (e.g., Miller et al., 1998; Hannah et al., 1998), for which the IBM approach has been applied. For larger scale studies, seasonal evolution of stage abundance, described with a simpler stage-based model coupled with realistic circulation patterns, can capture within the right order of magnitude the effect of advection on regional population dynamics of a target species (e.g., Zakardjian et al., 2003).

### *3.3 Present challenges and issues facing coupled copepod life-history models*

**Food limitation of growth and reproduction**

An important frontier in the development of coupled copepod life history models is the inclusion of the effects of food limitation on vital rates (growth, reproduction, mortality) i.e., how to link to the output of the NPZ models. The effect of food supply occurs more on growth and reproductive rates (e.g. Runge, 1985; Campbell

TABLE 13.1.
Characteristics of coupled physical-copepod life history models (list not all-inclusive)

| Authors | Species | Type of life-cycle model | Type of coupling | Dimensions | Lower trophic level coupling | Egg production | Develop-ment times or growth rate | Mortality | Diapause | Vertical migra-tion behavior | Length of simulations |
|---|---|---|---|---|---|---|---|---|---|---|---|
| Voronina et al., 1979 | *Calancides acutus* (Antarctic) | IBM | Lagrangian probabilistic | x,z,t | Seasonaly varying phyto-plankton bio-mass | f(food) | f(age,depth) | f(food,age, depth) | Coupled with the migration behaviors | Seasonaly and latitudinaly varying stage-specific migration behavior | 1 year |
| Miller et al., 1998 | *Calanus fin-marchicus* (Georges Bank) | IBM | Lagrangian | x,y,t | No | f(temp) | f(temp) | Imposed stage-specific mortality rates, increased for 2nd generation | Imposed probability of ascent and descent, in-creased for 2nd generations | No | 250 days |
| Batchelder et al., 2002 | *Metridia spp* (Ore-gon Coast) | IBM | Lagrangian | x,z,t | Driven by a NPZ model | f(food, temp) | f(food) | Imposed stage-specific mortality rates | No | Yes | 120 days |
| Heath et al., 1997 | *Calanus finmarchicus* (Northern Scotland) | ABM | Lagrangian water column drift | x(t), y(t), t | Coupled with a NPZD model | f(food) | f(food) | Imposed mortality rates on eggs and CVI used as fitting parameters | No | No | 50 days |
| Pedersen et al., 2001 | *Calanus finmarchicus* (Norwegian coast) | ABM | Lagrangian | x,y,z,t, | No | Limited to a imposed time-window | f(temp) | No | Imposed time of ascent and decsent | yes | 180–210 days |
| Wroblewski, 1982 | *Calanus marshallae* (Oregon coast) | SBM | Eulerian | x,z,t | No | No | Constant and equivalent to development times at 10°C | Imposed stage-specific mortality rates | No | yes | 1 month |
| Davis, 1984 | *Pseudocalanus spp* (Georges Bank) | AwSB M | Eulerian | Θ*,t (angular coordinates) | No | f(temp) | f(temp) | tuned stage specific mortality rates | No | No | 60 days |

| | | | | | | | | | | | |
|---|---|---|---|---|---|---|---|---|---|---|---|
| Lewis et al., 1994 | *Pseudocalanus spp* (Georges Bank) | SBM | Eulerian | x,y,z,t | No | Imposed and constant | Imposed stage-specific molting rates | Imposed stage-specific mortality rates | No | No | 20 days (wind events) |
| Gupta et al., 1994 | *Coullana canadensis* (Saco Estuary) | SBM | Eulerian | x,t | No | Imposed with sesonal variations | f(temp) | Imposed stag-specific mortality rates | No | No | 1 year |
| Carlotti and Radach, 1996 | *Calanus finmarchicus* | IBM and WwSBM | Eulerian | z,t | Coupled with a NPZD model | f(weight, temp, food/ | f(food, temp) | Imposed stage-specific mortality rates | Imposed diapause at stage CV for imposed temperature and food conditions | No | 1 year |
| Carlotti and Wolf, 1998 | *Calanus finmarchicus* | IBM and WwSBM | Lagrangian | z,t | Coupled with a NPZD model | f(weight, temp, food/ | f(food, temp) | Imposed stage-specific mortality rates | Descent of CIV and CV depending on stored lipids | DVM = f(light,food) | 240 days (spring to autumn) |
| Lynch et al. 1998 | *Calanus finmarchicus* (Gulf of Maine) | AwSBM | Eulerian | x,y,t in 3 layers | Monthly phytoplankton biomass fields from observations (MARMAP) | f(temp, food) | f(temp, food) | Imposed tage-specific mortality rates | Imposed rate of emergence from diapause | C5s move between layers | 90 days (growing season) |
| Zakardjian et al. 2003 | *Calanus finmarchicus* (Eastern Canadian waters) | SBM | Eulerian | x,y,z,t | No | Imposed seasonaly varying eggs production rates | Temp-dependant | Stage-specific, seasonally-varying and temperature-dependent mortality rates | Imposed seasonaly and latitudinaly varying probability of ascent and descent | Stage-specific vertical swimming behaviors | 1 year |
| Bryant et al., 1997 | *Calanus finmarchicus* (North Sea) | WwSBM | Box-transport model | x,y,t | Seasonaly varying phytoplankton biomass fields from CPR data | f(food) | f(food, temp) | f(foodtemp) | No | No | 8 months (Apr to Dec) |
| Gurney et al., 2001 | *Calanus finmarchicus* (NE Atlantic) | ASBM | Spatially structured model | x,y,t | | No | Temp-dependant | No | No | No | Some weeks |

et al., 2001b; Jonasdottir et al., 2002) rather than development times, on which the first order influence is temperature (e.g. McLaren, 1978; McLaren and Corkett, 1981; McLaren et al., 1989; Vidal, 1980). Nevertheless, experimental and field data show that starvation can also affect development (e.g Campbell et al., 2001b) and starvation has been hypothesized to be a source of mortality, especially in naupliar stages. In some regions (e.g. the Gulf of St. Lawrence) there is empirical justification for use of a food-independent reproductive rate (Runge and Plourde, 1996) as a first order approximation (e.g., Zakardjian et al, 2003). Food limitation is nevertheless a fundamental issue for studies of inter-annual or inter-decadal variability in the context of coastal ecosystem change (e.g. Lynch et al. 1998).

Of the sixteen coupled models in Table 13.1, eight do not address food limitation. Three of the remaining models (Voronina et al., 1979; Bryant et al, 1997; Lynch et al., 1998) use as forcing functions averaged phytoplankton fields (as chlorophyll) derived from field observations in the region of study. Five models include detailed parameterization of food effects, with phytoplankton as the prey field in terms of carbon or nitrogen concentrations derived from coupled NPZ models run jointly with the coupled copepod life history model. These latter models have been applied to a one-dimensional water column (Carlotti and Radach, 1996; Carlotti and Wolf, 1998, Heath et al., 1997), and to a two-dimensional coastal upwelling area by Batchelder et al. (2002).

As discussed in Section 2, relationships between chlorophyll *a* (or estimates of microplankton nitrogen or carbon concentrations) and growth and reproduction of planktonic copepods (e.g. Fig. 13.4) are one way to address food limitation in population dynamics models. These relationships are obtained empirically by measurement at sea of growth and reproductive rates of the target species and ambient food concentration in the region of study. As such they need to be determined for each different key species and coastal region. As the empirically determined rates represent means for the population food concentrations typically presented as integrated water column standing stock (or a mean water column concentration), they would be particularly suitable in Eulerian approach. In cases where bulk measures of food do not yield significant relationships with growth or reproduction, or to address growth and reproduction as functions of size and chemical composition of prey as well as issues related to fine-scale spatial and temporal scaling, then the details of feeding behavior could be modeled, and growth and reproduction expressed as the difference between assimilated ingestion rates and metabolic rates, as reviewed by Carlotti et al. (2000). These approaches would be particularly suitable for the Lagrangian approach to coupled models, in which individual growth rates and responses to fine scale variability in prey concentration and composition could be tracked. Comparison at the same location and time period of an Eulerian approach using empirically determined relationships between food and growth and reproductive rates with a coupled IBM model that deduces growth and reproduction from functions expressing feeding, assimilation and metabolic rates would provide insight into the sensitivity of simulated abundance and spatial distribution to the way in which food limitation is addressed in the model.

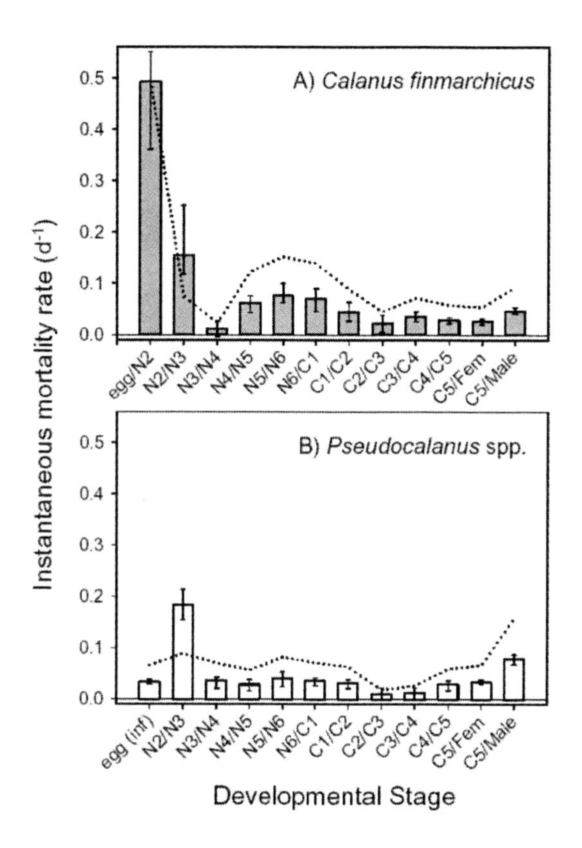


Figure 13.6 Calculated mortality rates for life stage pairs of *C. finmarchicus* and *Pseudocalanus* species based on over 2000 measurements of egg production rate and stage abundance at U.S. GLOBEC Georges Bank broadscale survey stations, 1995–1999. Dotted lines show median rates; histograms show calculated extremes (from Ohman et al., 2002).

**Mortality, ecological closure of copepod life history models**

Accurate depiction of mortality schedules is one of the greatest challenges in the modeling of marine population dynamics. Good demographic studies of mortality in the sea are rare and in many models, mortality rates are crudely imposed as values from limited observations. Tuning (i.e., an *a posteriori* adjustment) of these mortality rates is often needed in order to scale the overall development of the population (e.g., Davis, 1984; Heath et al., 1997; Miller et al., 1998). In copepod populations, mortality is observed to vary with developmental stage (Fig. 13.6), season and across regions (*e.g.,* Ohman and Hirche, 2001; McLaren et al., 2001 Ohman et al., 2002). Predation is certainly a major cause of mortality; other sources include starvation, temperature (which may be especially important at the geographic limits of a species' range), viruses and parasites. A recent study of the *C. finmarchicus* population in the Norwegian Sea showed that high mortality rates of *C. finmarchicus* eggs and early naupliar stages were positively correlated with the depth-integrated abundance of *Calanus* stage C5 and C6f, suggesting cannibalism as an important factor in the density dependent control of zooplankton populations (Ohman and Hirche, 2001; Ohman et al., 2004). Various invertebrate and vertebrate predators of copepods such as krill, fishes, birds and whales may locate and aggregate in dense patches of their prey (Wishner et al., 1988, Lavoie et al., 2000), from which we can infer that mortality rates may be locally higher than the mean schedules that are generally used in the models listed Table 13.1.

Demographic analyses of copepod mortality can be realized. Methods to estimate mortality rates, including Wood's Population Surface method and the Vertical Life Table method (Wood, 1994; Ohman and Wood, 1996; Aksnes and Ohman, 1996; Ohman et al., 2002; Ohman et al., 2004) and have been applied to study mortality of *C. finmarchicus* across its range. In addition to quantitative estimates of abundance of life stages, accurate knowledge of temperature dependent development rates and, for mortality of planktonic eggs of broadcast spawners, determination of egg production rate is also required. Alternate methods of estimating mortality of zooplankton size fractions from automated particle counters such as the Optical Plankton Counter (Edvardsen et al., 2002; Zhou, 1998) may also prove useful. Until we are capable of dynamically modeling mortality rates from predator abundance and their feeding behavior and from other sources, we will have to rely on empirical determination of seasonal and spatial variation in mortality based on field studies of the populations of the key species of interest.

**Vertical distribution and migration: copepods are not just drifters**

In response to vertical gradients in time and space, copepods may vertically migrate. Copepod diel vertical migration allows escape from vertebrate or invertebrate predators (e.g., Ohman et al., 1983; Frost, 1988; Bollens and Frost, 1989; Dodson, 1988) or unfavorable warm temperature (e.g., Wishner and Allison, 1986). Seasonal vertical movements play an important role in the life cycles of many species. Typically, older pre-adult stages inhabit deep water in winter then migrate to surface waters to spawn in spring (Conover, 1988). Both ontogenetic and diurnal migrations may be stage-specific and can range from several meters to several hundred meters (e.g., Wishner and Allison, 1986; Heath et al., 2000). This implies the capacity to swim at several body lengths per second. *C. finmarchicus,* for example, may swim at velocities of more than 50–100 m $h^{-1}$ (e.g., Heywood, 1996). Such vertical movements are one order of magnitude higher than typical vertical velocity fluctuations in the ocean but one order of magnitude less than typical horizontal velocities. Hence, copepods may be free of vertical advection or turbulent mixing but are always subject to transport by horizontal currents prevailing at their resident depth.

The ability to vertically migrate has great implications for population dynamics in areas of strongly sheared circulation, such as coastal seas, where its interaction with the swimming ability of marine planktonic copepods can yield stage-specific patterns of transport (e.g., Gupta et al., 1994; Hannah et al., 1998; Zakardjian et al., 1999). Results from 2D and 3D coupled models that include vertical migration behavior show first order effects of vertical migration on local copepod population dynamics (e.g., Wroblewski, 1982; Hill, 1991, Lynch et al., 1998; Batchelder et al., 2002; Zakardjian et al., 2003), as previously suggested by studies of Lagrangian trajectories of passive particle both in the coastal sea (Hannah et al., 1998) and the open ocean (Bryant et al., 1998; Harms et al., 2000). In special cases, such as shallow vertically homogeneous water of the littoral zone, the interactions between swimming behavior and circulation can be neglected (e.g., Lewis et al., 1994).

The challenge is to build both the observational data base and the fundamental biological understanding of control of stage-specific diel vertical migration and seasonal migrations of target species in each region of study in order to adequately represent vertical distribution in spatially explicit models. There is considerable knowledge of copepod vertical migration behavior (e.g. Frost, 1988; Frost and Bollens, 1992; Bollens et al. 1993, 1994; Kaartvedt, 1996; Falkenhaug et al., 1997). However, vertical migration behavior and consequently vertical distribution may be both stage- and site specific, depending for example on the type of predator (e.g., Tande, 1988), suggesting that a site-independant deterministic formulation may not be possible at this time. We must rely on acquiring extensive observations of vertical distribution at sea, even for a well known species like *C. finmarchicus*. Coupled 3D modeling can be used to identify areas in which the interaction between circulation and vertical distribution are especially critical, for example Cabot Strait and the continental slope of Nova Scotia to *C. finmarchicus* population dynamics in the coastal NW Atlantic (Zakardjian et al., 2003), which can contribute to the design of field studies.

**Diapause: a biologically induced seasonality**

In temperate and polar regions, the life-cycle of planktonic copepods often includes a diapause phase: a quiescent period, generally in winter, during which the animals wait (at depth) for favorable conditions for reproduction and growth, which usually follow the development of the spring phytoplankton bloom. *C. finmarchicus* undergoes diapause mainly in stage C5 (Fig. 13.5), but diapause may occur in stages C3-C4 (Conover,1988) in other species of the genus and even in the egg stage other calanoid copepod species, such as *Labidocera aestiva* (Marcus, 1982). The interruption in individual development can even lead to multi-year life cycles, as for *Calanus hyperboreus* (Conover, 1998). The match between the timing of arousal from diapause and the seasonal variation of environmental conditions, mainly temperature and food abundance, may crucially influence interannual variation in population dynamics (e.g., Hirche, 1996; Head et al., 2000). The coupling of the diapause response with seasonal variations in circulation can be another important factor controlling interannual variability of population abundance over large areas (de Young et al. 2004). For example, *C. finmarchicus* overwinters in the deep North Atlantic Ocean at depths between 500 to 1000 m (Miller et al., 1991). Colonization of the Norwegian, Faroe and North Sea coastal shelves depends on the timing and speed of ascent in relation interannual variations in the northeast Atlantic circulation (Aksnes and Blindheim, 1996; Slagstad and Tande, 1996, Bryant et al., 1998; Heath et al., 1999; Gaard, 1999, 2000; Pedersen et al., 2001).

While the physiological responses to diapause are reasonably well known (e.g., Hirche, 1996), there is much less understanding of the environmental or physiological cues that induce entry into diapause or the processes that control emergence from diapause. In present models (Table 13.1) diapause is generally ignored or parameterized simply. Diapause can be neglected in seasonal models that address processes during the active period (e.g. Davis, 1984; Lewis et al., 1994; Batchelder et al., 2002). In other cases, diapause is generally prescribed through imposed probabilities of ascent and descent based on empirical observations of life cycle timing (e.g., Miller et al., 1998; Lynch et al., 1998; Zakardjian et al., 2003). Carlotti and Radach (1996) used prescribed temperature and food conditions and Carlotti and Wolf (1998) used lipid allocation criteria but without any experimental rationale. Recently, more sophisticated, empirically based models or genetic algorithms for timing of entry into and exit from diapause are promising (e.g. Hind et al., 2000; Fiksen, 2000) for application in the coupled models. Nevertheless, both field observations of the timing of entry and emergence from diapause and new experimental studies are warranted in order to deterministically formulate diapause in models of population dynamics.

## 4. Trophodynamic models of the early life history of fish

### *4.1 Overview*

Recognition of the critical role played by the fine scale dynamics of the physics and lower trophic levels to understanding larval fish growth and survival (e.g. Heath, 1992) has led to increasing effort devoted to developing coupled physical-biological models (see reviews in Heath and Gallego, 1997; Werner et al. 1997; Lynch, 1999; Werner and Quinlan, 2002). Larval fish dynamics are highly dependent on the physical and biological conditions experienced by the individual (e.g., Rice et al., 1987; Cowan et al., 1997). These conditions in turn vary in time and space (Heath, 1992). Slight differences in the trajectories of individual larvae can mean the difference between success and death (e.g., Bartsch et al., 1989; Hare et al., 1999). Subtle differences in the encounter rates of larvae with their zooplankton prey or with their predators can lead to small changes in growth and mortality rates, which in turn can have order of magnitude effects on larval fish cohort survival (Houde, 1987).

Because the spatial distributions of temperature, zooplankton, and other factors important to larval fish growth and survival are greatly influenced by the physics of water movement (i.e., hydrodynamic transport), coupling individual-based fish larvae to spatially-explicit NPZ and zooplankton life history models is a logical approach for simulating larval fish growth and survival. Fisheries scientists have often developed models of larval fish growth and survival that go into great detail on larval feeding behavior, while the zooplankton prey are treated as a forcing variable (e.g., Vlymen, 1977; Beyer and Laurence, 1980) or by side-stepping the issues of fine-scale spatial variability by using implicit approaches such as statistical distributions that mimic encounter rates in patchy environments (e.g., Rose et al., 1999; Winemiller and Rose, 1993).

While not a necessity, the larval fish component of many coupled physical-biological models uses a Lagrangian approach, which tracks individual larvae through space and time. Individual-based modeling has gained enormous popularity in the past ten years (Judson 1994; DeAngelis et al., 1994; Grimm, 1999; Werner et al., 2001a). A Lagrangian individual-based approach is useful in coupled bio-physical models because few larvae survive (Winemiller and Rose, 1992; Sale, 1990), and it is the history of experiences of the larvae and local interactions (between larvae and its prey and larvae and its predators) that can be important to determining the rare survivors. Accurately simulating individual experiences that vary among individuals and representing the effects of local interactions is difficult with other, more aggregated modeling approaches (e.g., matrix projection modeling–Caswell, 2001) but is, at least conceptually, relatively straightforward to implement with an individual-based approach.

Here we first review the various ways fish larvae are coupled to hydrodynamic and lower trophic level models. These range from using the hydrodynamics only to transport fish larvae to individual fish larvae being imbedded into the physical models and fully participating in the ecological dynamics (i.e., zooplankton and fish affecting each other). We then discuss three general issues that arise when fish larvae are coupled to physical and lower trophic level models: reconciling temporal and spatial scales between fish larvae and the physical models, how to include the effects of the fish on the lower trophic levels (i.e., feedbacks), and how to represent behavior-related movement of the larvae in a spatially explicit setting.

### *4.2 Approaches to coupling fish larvae to physical models and lower trophic levels*

We begin with Lagrangian (particle tracking) approaches, and then present progressively more complex models that include larval growth via either abiotic effects such as temperature or via bioenergetic formulations of larval feeding and growth.

**"Simple" Lagrangian models.**

In its simplest form, the coupling is achieved through the advection of passive and/or behaviorally active larvae to determine retention, transport pathways to nursery grounds, etc., through the use of spatially-explicit IBMs (Individual Based Models). Taking advantage of the advent of sophisticated and robust circulation models that capture realism on relevant spatial and temporal scales (e.g., Blumberg and Mellor, 1987; Backhaus, 1989; Lynch et al., 1996; Haidvogel and Beckmann, 1998), perhaps the best established use of IBMs focuses on determining Lagrangian trajectories of planktonic stages of marine organisms in realistic flow fields. The simplest of these studies ignore biotic factors such as feeding and predation; but include imposed swimming behaviors, spawning locations, etc. Among the topics successfully investigated by these studies are the space-time trajectories of larval fish from spawning grounds to nursery areas (Bartsch et al., 1989; Ådlandsvik and Sundby, 1994; Quinlan et al., 1999; Epifanio and Garvine, 2001), retention on submarine banks (Foreman et al., 1992; Werner et al., 1993; Page et al., 1999), effects of interannual variability of physical forcing on dispersal of larval fish populations (Lough et al., 1994; Hermann et al., 1996; Rice et al., 1999), identification of spawning locations (Hare et al., 1999; Stegmann et al., 1999) and the implied long-term dispersal by tidal currents (Hill, 1994). Several of these studies included an imposed behavior on the particles (e.g., Werner et al., 1993; Hill, 1994; Hare et al., 1999) and/or considered the effect of turbulent *kicks* on the larvae's location in the water column (e.g., Werner et al., 2001b) and found that the vertical position of the fish larvae in the water column can affect the results obtained through simple passive advection.

Although lacking in key biological variables, the use of these simple Lagrangian models has been clearly established as a necessary first step in describing the environment sensed by larval fish, their dispersal/retention, the identification of source or spawning regions (Quinlan et al., 1999; Stegmann et al., 1999) and potentially their recruitment at a population level (Page et al. 1999). The Georges Bank study of Page et al. (1999) compares empirical (field) observations on season and location of cod and haddock spawning with (model-derived) seasonal and geographic patterns of residence times. They conclude that fish populations may select areas and times of the year for spawning that enhance the probability of retention on Georges Bank, thus finding support for the member-vagrant hypothesis (Sinclair, 1988) of the regulation of geographic pattern in populations for marine species. While this particular example was simplified by not including aspects of the organisms' feeding environment and growth characteristics, it is clear that we are on the verge of using Lagrangian models to answer biologically complicated population dynamics questions.

**Environmental effects on growth.**
Approaches that consider feeding environment implicitly through its relation to temperature include those of Hinckley et al., (1996) and Heath and Gallego (1998). Hinckley et al. (1996) showed the sensitivity of the population's size distribution as a function of trajectories through variable temperature fields (where growth was based on a $Q_{10}$ relationship), as well as the differences that arise in horizontal dispersal due to differences in rates of growth and vertical behavior. In Heath and Gallego (1998), temperature (resulting from a circulation model) was used as a proxy for feeding environment: prescription of the 3-D temperature field was used to determine individual growth rates of larval haddock. It was found that the *model-derived* spawning locations resulting in the highest larval growth rates (as the larvae are advected in the model domain) coincided with the *observed* preferred spawning locations.

**Bioenergetic models: including prey fields explicitly.**
The next level of complexity commonly introduced into spatially-explicit physical-biological models of larval fish is an imposed spatially-dependent (but temporally fixed) prey distribution based on field observations.

The typical bioenergetics model is based on an equation that relates the new weight ($W$) of a larva to its previous weight plus gains and minus losses (e.g., Beyer and Laurence, 1980; Rudstam, 1988):

$$\frac{dW}{dt} = [C - (R + S + F + E)] \cdot W \qquad 13.2$$

where $C$ is consumption, $E$ is excretion or losses of nitrogenous excretory wastes, $F$ is egestion or losses due to feces, $R$ is respiration or losses through metabolism, $S$ is specific dynamic action (or losses due to energy costs of digesting food), and $t$ is time. The units of C, E, F, and S are weight prey$\cdot$ weight fish$^{-1}\cdot$ time$^{-1}$. Consumption and respiration are weight- and temperature-dependent. Water temperature experienced by a larva affects maximum consumption and respiration. The number of prey encountered depends on prey type, the swimming speeds of the prey and the fish larva and turbulence.

These studies are a first step to introducing realistic representations of the spatial distribution of key variables such as temperature, oxygen, light levels, prey availability, etc. In Brandt et al. (1992) a bioenergetic model was embedded in a spatially heterogeneous representation of its physical (estuarine) habitat, as determined by field measurements, to obtain the spatial distribution of growth rates of the target fish. From the spatial distribution of fish growth potential, Brandt et al. (1992) were able to define the portion of the habitat volume that will support various levels of fish growth. Furthermore the resultant *growth volumes* can provide a mechanism for assessing the suitability of a particular habitat to support a species introduction and can aid in the definition and monitoring of ecosystem *health*.

Similar approaches, based on model-derived spatial structure of prey and habitat (circulation, turbulence and temperature) are discussed in Fiksen et al. (1998) and Lynch et al. (2001) who produced Eulerian maps of potential larval fish growth rates. Fiksen et al. (1998) examined the interactions between vertical profiles of wind-induced turbulence and light to define regions in the water column where highest ingestion rates can occur for certain fish larvae. Lynch et al. (2001) found that, during early spring on Georges Bank, the distribution of certain prey (*Calanus finmarchicus*) is better matched spatially with the spawning location and subsequent drift of cod and haddock larvae than other potential prey (*Pseudocalanus* spp.). Additionally, it was found that spawning in regions of high turbulence is detrimental to young larvae, suggesting that for survival of the earliest larval life stages spawning should occur away from these regions.

A natural extension to these diagnostic approaches is the computation of individual Lagrangian trajectories within the prey field. In this manner, trajectories that are considered favorable for retention or appropriate for transport into nursery areas are more narrowly identified to include only those trajectories where the individuals encounter favorable feeding environments. Studies of this type include Hermann et al. (2001) and Werner et al. (1996 and 2001b). These studies have also been used to explore other spatially-dependent interactions between predators and their prey. For example, the perceived prey field by fish larvae can be effectively increased or reduced as a consequence of local variation in turbulence levels, which alter volume searched (MacKenzie et al., 1994; Dower et al., 1997; Werner et al., 2001b). This requires models to capture not just the spatial distribution of biotic components, but also their modulation by certain abiotic environmental factors. An example of the intersection of large and small scale physics affecting recruitment is given in Werner et al. (1996) in which the effect of the feeding environment, modified by turbulence at the smallest scales, on larval growth and survival was examined. They found that regions of larval survival (with growth rates comparable to field values) coincided with the hydrodynamically retentive subsurface regions of Georges Bank. However, these retentive regions were a subset of those defined by Werner et al. (1993) and Lough et al. (1994). The increase in larval survival in these smaller areas was due to an enhancement of contact rates and effective prey concentrations by turbulence within the tidal bottom boundary layer.

### *4.3 Present challenges and issues facing coupled models of fish early life history*

Coupling fish larvae to physical models raises important issues about scaling of physical and life history processes, trophic feedbacks, and movement and behavior.

**Scaling**

The time step and spatial resolution of the physical model are dictated (or limited) by numerical considerations, which may not coincide with the ideal time step and spatial resolution for simulating growth and survival of fish larvae. Simulating purely advective and dispersive transport of passive particles is in a relatively advanced state of development. Scaling issues arise when the particles have behaviors, and especially when simulated growth and mortality are dependent on dynamic prey and predators. The generally fine-scale of the physical models would involve simulating fish larval behavior on the scale of minutes at the spatial resolution of the hydrodynamics. Larval fish ecologists are often more comfortable simulating larval dynamics on scale of hours to daily, and therefore also at a coarser spatial resolution that matches the longer time step. Output from the hydrodynamics and lower trophic models are often aggregated to some extent to permit coarser simulation of the larval dynamics. How to aggregate the output without losing dynamically-relevant variability is an important consideration. An illustration of the complexity of this scaling is the walleye pollock coupled biophysical model (Hermann et al., 2001), which meshed together a 257 x 97 x 9 nonuniform grid, (with approximately 4 km resolution in the horizontal, variable resolution in the vertical) and 135 sec time step for hydrodynamics), a 20 x 20 x 100 grid (with 20 km resolution in the horizontal, 1 m resolution in the vertical, and a 0.1 day time step) for the NPZ component and a continuous spatial domain for the larval fish IBM (with a combination 24hour/1hour time step) for the larval fish dynamics.

**Trophic feedbacks**

Including the effects of the larval fish on the lower trophic levels (consumption of zooplankton and addition of nutrients via excretion and egestion) can additionally complicate the modeling. Most coupled biophysical models use the hydrodynamics and lower trophic level predictions as input to the larval fish component. This enables the lower trophic level models to be solved independently of the larval fish dynamics. However, separate solution of the lower trophic and larval fish models prevents any density-dependent effects from operating. Perhaps under most average conditions, the effects of larval fish, on their prey and on nutrients, is small enough to be ignored. But it may very well be that the rare set of conditions when such feedback effects are important is of most use to those interested in fish recruitment. Density-dependent effects may be elusive as they may operate only under certain conditions (e.g., years of high egg production and low food production), but such effects are not possible to predict (or dismiss) if the coupled biophysical model do not include the capability to include these feedbacks. Yet, including these feedbacks would require that the lower trophic level component and the larval fish component models be solved simultaneously. For some situations, this can create computational limitations on the analyses and heartache for the programmer. Linking larval fish models where full hydrodynamics and fully evolving population dynamics co-occur will likely be attempted in the next five years (see Ault et al. 1999 for an example of such a linkage using McKendrick-von Forester equations and a 2D hydrodynamic model). We are now beginning to see the first attempts at fully-evolving models (e.g., Hermann et al., 2001; Hinckley et al., 2001).

**Movement and swimming behavior**

Like other zooplankton, fish larvae exhibit active movement not related to advection. Active movement can be critical because such movement can greatly affect the transport, environmental conditions and prey experienced by the individual larvae (Tyler and Rose, 1994). Modeling active movement by fish remains a difficult area. We do not really know why larval fish move, especially on the scale of minutes to hours and over relatively short distances (e.g., meters). Externally imposed (and/or passive) behaviors (e.g., Werner et al., 1993; Hare et al., 1999) may not make sense as the coupled biophysical model move more and more towards simulating the growth and survival of the larvae. Such static approaches to movement will be likely replaced by model-derived behaviors that include components maximizing some biological characteristic, such as reproductive value (Giske et al., 1994; Fiksen and Giske, 1995; Fiksen et al., 1995), survival to maturity (Railsback and Harvey, 2002), or short-term tradeoffs between growth and mortality (Tyler and Rose, 1997). Dynamic programming methods allow organisms to "find" optimal habitats by balancing risks of predation, growth, and advective loss. The issue of how to represent active movement on fine scales is important but remains unresolved at this time. The realism of predicted growth and mortality from coupled models may very well rely on how well we can model fish movement.

## 5. Case Studies: The application of coupled physical-biological models in coastal environments

Within the past decade, considerable progress has been made in the application of coupled physical-biological models of secondary production and recruitment processes to a number of coastal regions around the world, including the Baltic Sea (e.g. Hinrichson et al., 2002), the northeast and northwest Atlantic Ocean (e.g. Bryant et al., 1997; Heath and Gallego, 1998; references below), the northeast and northwest Pacific Ocean (e.g. Batchelder et al., 2002; Kishi et al., 2001), the Gulf of Alaska (references below), and the Benguela upwelling system (references below). Common features to all the regional applications are developments of region-specific 3-D circulation models and the involvement of large, multidisciplinary programs to facilitate cross-disciplinary collaboration and data collection. Differences in approach and emphasis and focus of the research questions are also evident, depending on the particularities of the regional system and the objectives and structure of the multidisciplinary programs, including how the large programs are funded. Here we provide brief summaries of coupled physical-biological modeling in three regions, the southern Benguela upwelling system, the coastal northwest Atlantic Ocean, and the western Gulf of Alaska, to illustrate the present status of application of models to coastal systems.

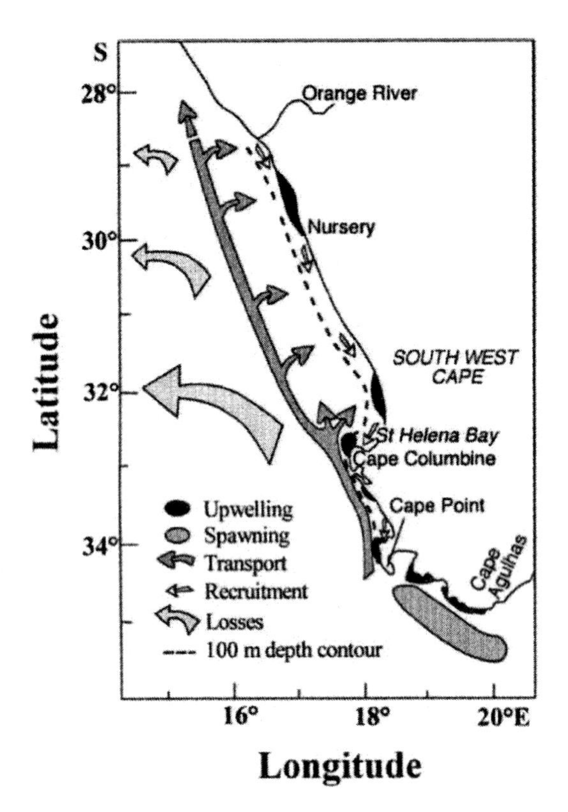


Figure 13.7 Conceptual model of the advective influence of the Benguela Current System on the life history of anchovy (from Pillar et al., 1998 and Parada et al., 2003). Anchovy eggs and larvae spawned on Agulhas Bank are transported north, some to the inshore nursery area.

### *5.1 The southern Benguela upwelling system*

The Benguela Current and its associated upwelling system is one of the most productive ocean environments in the world (Shannon and O'Toole, 2003; Field and Shillington, in press). Conceptual understanding and data acquisition within the region advanced considerably during the late 1980's and 1990's during the Benguela Ecology Program (e.g. Pillar et al., 1998). The dominant, harvested fish in the Benguela upwelling system are species of anchovy (*Engraulis encrasicolus)* and sardine (*Sardinops sagax*), both of which are adapted to the dynamics of upwelling environments. The emergent paradigm of the anchovy life cycle in the southern Benguela is that adults spawn on the Agulhas Bank during austral spring and summer. The eggs and larvae are transported by a coastal jet to a nursery area located approximately 500 km to the north along the west coast of S. Africa and Namibia (Fig. 13.7). Successful recruits are believed to be larvae that find themselves in the inshore nursery area, where they grow into juveniles and mature, then move southward back to Agulhas Bank to spawn at age 1, completing the life cycle. Variability in transport of eggs and larvae from spawning to nursery grounds is considered to be a first order determinant of anchovy recruitment success (Huggett et al., 2003; Parada et al., 2003 and references therein). The coupled physical-biological models reported here were developed as part of the multi-disciplinary IDYLE program, a bilateral research program funded jointly by agencies and institutions within the Republic of South Africa and France.

The regional circulation model to which biological processes are coupled is a high-resolution, 3D, eddy resolving hydrodynamic model (Penven et al., 2001) based on the Regional Ocean Modeling System (ROMS: Haidvogel et al., 2000). The model uses stretched, terrain-following coordinates in the vertical and orthogonal, curvilinear coordinates in the horizontal, yielding a resolution of 9 km at the coast and 16 km offshore, with 20 variable levels in the vertical. The model is forced with averaged winds, heat and salinity fluxes determined from the COADS ocean surface monthly climatology. It is forced at the offshore boundaries by seasonal, time-averaged outputs of a larger, basin-scale model. In studies published to date, the hydrodynamic model was run for 10 years and forced by a repeated climatology. While there was no interannual variability in the winds and other forcing processes, there were differences in simulation outputs between years, attributed to intrinsic mesoscale activity resulting from oceanic instability processes within the model structure (Penven et al., 2001).

The biological applications of the hydrodynamic model focus on the interaction between advection and characteristics of spawning and distribution of eggs and larvae of anchovy in the southern Benguela Current System (e.g., Mullon et al., 2002; Huggett et al., 2003; Parada et al., 2003). The approach and results are summarized in Mullon et al. (2003). Individual-based models of eggs and larvae of anchovy were coupled to the output from individual years of the hydrodynamic model. In a series of trial runs, parameters were varied sequentially in order to assess the relative importance of factors that may influence successful transport to the inshore nursery area. Variables included timing, location and frequency of spawning activity on Agulhas Bank (Huggett et al., 2003), release depth and buoyancy of eggs (and early larval stages, assumed to be the same as egg buoyancy) (Parada et al., 2003), growth and mortality as a function of temperature (Mullon et al., 2003) and active vertical migration behavior of larvae (Mullon et al., 2003). Success was measured by the number of larvae older than 14 d (the age at which larvae are considered capable of maintaining their location) arriving in the inner nursery area within the duration of the trial (typically 90 d). A typical model run was initiated with 5000–10,000 individual eggs; in all, over 20,000 trials were conducted. Success percentages from the trials were analyzed statistically using multifactor analyses of variance. Among the results of these simulation experiments are that eggs spawned on western Agulhas Bank and having a density of 1.025 g $cm^{-3}$ are most successfully transported to the nursery area (Fig. 13.8) and that transport success is greatest among actively swimming larvae keeping a target depth of 40–60 m (Mullon et al., 2003). In general, the impacts of variability in spawning location, egg buoyancy, the direction and intensity of the jet current, the effects of temperature on growth and mortality, and vertical migration behavior on transport success were quantified, although it is not possible to quantify precisely the relative importance of each factor.

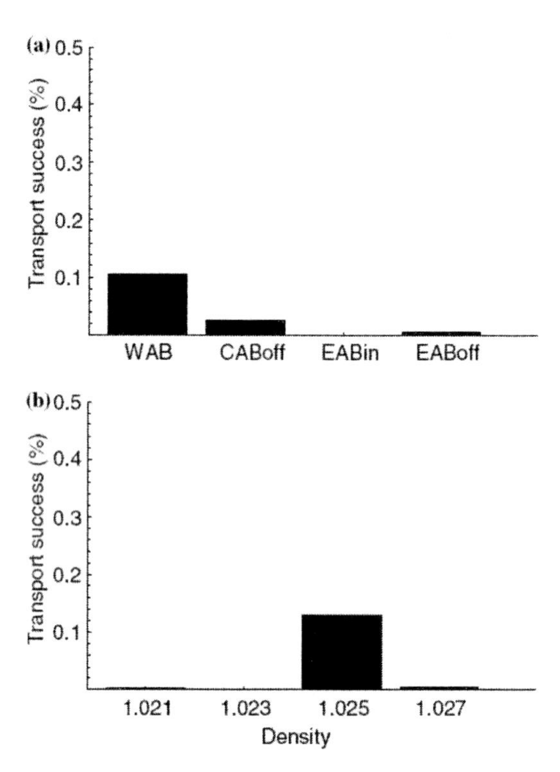


Figure 13.8 Transport success (%) of anchovy larvae in the southern Benguela Current System in relation to (a) location of spawning area (WAB=Western Agulhas Bank, CAB = Central Agulhas Bank, EAG = Eastern Agulhas Bank and (b) changes in specific density (g $cm^{-3}$) of anchovy eggs. From Mullon et al. (2003).

While the focus has been on impacts of advection on retention of larvae in nursery areas, research in the Benguela Current System has laid the groundwork for development of coupled physical-biological modeling simulating dynamics of phytoplankton and zooplankton production. For example, Plaganyi et al. (1999) describe a population dynamics model of a dominant planktonic copepod, *Calanoides carinatus,* that uses satellite derived estimates and chlorophyll a concentration and sea surface temperature as primary inputs. Plaganyi et al. (2000) develop a model of growth rates of juvenile anchovy as a function of abundance of *C. carinatus,* a major component of their diet.

Improvements to the hydrodynamic model of the Benguela Current System, including incorporation of finer scale variation in winds (e.g., Blanke et al., 2002), will allow greater power for addressing interannual and interdecadal variation in factors identified as having first-order impact on larval survival. An analysis of measurements of extreme oceanographic events in the southern Benguela during 1999–2000 (Roy et al., 2001) indicates that record high recruitment of anchovy in 2000 resulted from a synergistic temporal sequence of upwelling events leading to favorable transport and production of phytoplankton and zooplankton prey. There is the potential for future development of the Benguela Current System coupled models to address impacts of the climate-forced variation in upwelling events and transport processes on secondary production and anchovy recruitment.

### 5.2 *The northwest Atlantic coastal ocean*

The northwest Atlantic (NWA) coastal ecosystem extends southward from the Labrador Sea, along the Scotian Shelf, including outflow from the Gulf of St Lawrence, and through the Gulf of Maine and around Georges Bank (Townsend et al., in press). It is a mid-latitude, subarctic region with marked north-south variation in the seasonal cycle of ocean temperature. Important freshwater river inputs influence seasonal stratification and mixing and drive an extended estuarine circulation that advectively couples the region's shelf systems. The banks and inshore areas of the northwest coastal shelves serve as retention and nursery areas for early life stages of fish. Because it is a flow-through system, interannual and decadal scale variations in climate influencing the Labrador Sea and Gulf of St. Lawrence have impacts for the Scotian Shelf and Gulf of Maine (e.g. Loder et al., 1988; Greene and Pershing, 2003).

Here we focus on developments in coupled physical-biological modeling of Georges Bank and the Gulf of Maine. Annual fish production on Georges Bank has been among the highest of the world's shelf ecosystems (Cohen and Grosslein, 1987). The Georges Bank GLOBEC program (e.g. Wiebe et al., 2002), initiated in the mid-1990's, targeted study of physical-biological processes controlling population dynamics of several dominant species: the planktonic copepods *Calanus finmarchicus* and *Pseudocalanus* spp. and the early life stages of Atlantic cod (*Gadus morhua*) and haddock (*Melanogrammus aeglefinus*). The conceptual model (Colton and Temple, 1961) for the early life history of cod and haddock on Georges Bank is that spawning takes place in late winter-early spring (February-April) in the northeast peak of the Bank. A semi-closed gyre transports the eggs and larvae in a clockwise direction around the Bank, particularly along the southern flank (Fig. 13.9). At some point, the larvae are either retained or presumably lost as vagrants (as defined in Sinclair, 1988) in a drift exiting the Bank at its southwest corner (e.g., Werner et al., 1993; Page et al., 1999, and references therein). The planktonic copepods, *Pseudocalanus* spp., *Oithona* spp and *Calanus finmarchicus* are observed to be dominant prey during the cod and haddock larval phase (e.g. Kane, 1984; Lough and Mountain, 1996).

Numerical modeling of the regional circulation began well before GLOBEC (e.g. Greenberg, 1983; Isaji and Spaulding, 1984), but quantitative description improved with larger and more readily available hydrographic datasets and increasing computational power. A number of 3D models have been developed for the area, including QUODDY (Lynch and Naimie, 1993; Naimie et al., 1994, Naimie, 1996; Lynch et al., 1996; Loder et al., 1997, Loder et al., 2001), ECOM and FVCOM (Chen et al., 2001; Lewis et al., 2001) and CANDIE (Sheng et al., 1998). These models all solve primitive equations for heat, fluid, mass and momentum transport, incorporating advanced turbulence closure schemes (e.g. Mellor-Yamada level 2.5), and solve the equation of state relating temperature and salinity to density. Their main differences lie in the numerical methods employed to integrate the governing equations (i.e. finite difference, finite element, finite volume), their spatial and temporal resolution, and their domain of application. By prescribing realistic topography and forcing with historical data for tides, temperature, salinity, and winds, the various models generated representations of the regional currents, turbulence and temperature fields.

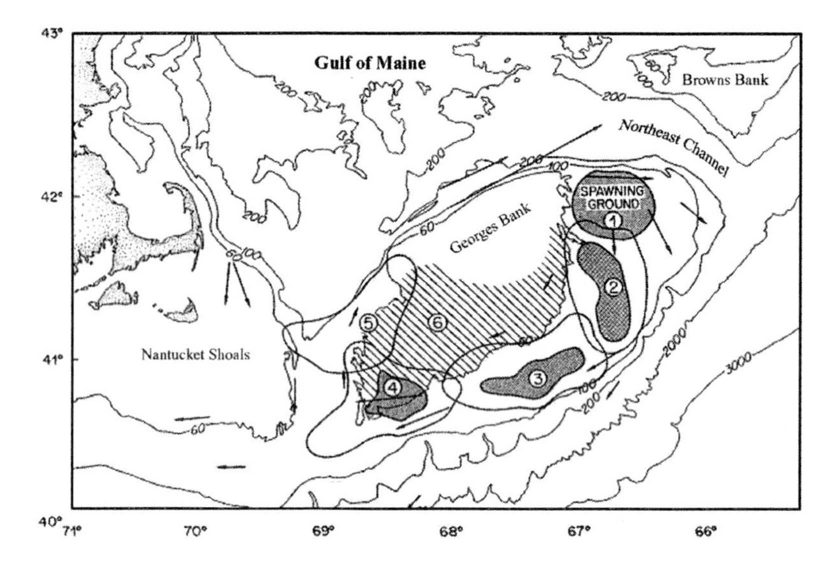


Figure 13.9 Conceptual model of advective influences on the early life history of cod on Georges Bank (from Colton and Temple, 1961). Cod eggs and larvae spawned on the northeast peak are transported to the southwest along the southern flank and are either retained on the crest or advected off the Bank. For main circulation features in the coastal northwest Atlantic Ocean, refer to Townsend et al. (in press).

Coupled physical-biological modeling in the northwest Atlantic includes NPZ, copepod life history and larval fish trophodynamic approaches. Circulation and mixing effects on planktonic production have been studied by coupling NPZ-type ecosystem models with the regional circulation fields. Winds typical of winter storms were shown to result in large exchanges of bank waters and large losses of plankton production on the bank (Lewis et al., 1994). Simulations for early summer yielded phytoplankton distributions consistent with satellite chlorophyll images. Predicted phytoplankton fields were relatively insensitive to removal of modeled advection, indicating the primary importance of vertical mixing in structuring the plankton around the Bank (Franks and Chen, 2001).

A number of models have investigated the transport of copepod populations onto and around Georges Bank. Early simulations used idealized flows (e.g. Davis, 1984; Lewis et al., 1994), whereas later more realistic circulation fields were used. The advective supply of *Calanus finmarchicus* to the Bank and the relative importance of transport vs. *in situ* biological processes was explored with Lagrangian particle tracking (Hannah et al., 1998) and with Eulerian-based models (Lynch et al., 1998; McGillicuddy et al., 1998). Biological components of these models quantified processes such as reproduction, development and mortality at a particular demographic resolution, including stage-based (Lewis et al., 1994), age-within-stage (Lynch et al., 1998), and individual-based (Miller et al., 1998). Because chlorophyll levels on the Bank were generally high, the initial hypothesis was that copepods were food-satiated. Hence, most models assumed vital rates were maximal, and based them on temperature-dependent empirical relationships developed in the lab (e.g. Miller et al., 1998; Lynch et al., 1998). These studies indicate that seasonal variation in the flow and processes influencing timing and emergence from diapause and vertical distribution of the copepods strongly affects rates of supply from the Gulf of Maine and retention on the Bank. Simulations showed that spatial and temporal patterns of *Calanus* recruitment of the first generation were consistent with data only when the model considered food-limitation of populations in the low-chlorophyll Gulf of Maine (Lynch et al., 1998), a prediction that could be tested with the appropriate field study. Evidence for food limitation of *Calanus* has since been found on Georges Bank (Campbell et al., 2001b).

Zakardjian et al. (2003) developed a 3D physical-biological model depicting the population dynamics of *Calanus finmarchicus* in the Gulf of St. Lawrence and Scotian Shelf and Gulf of Maine (Fig. 13.10). One of the objectives of the study was to examine how the local *Calanus* populations are maintained in this interconnected, advective regime (see Townsend et al., in press). A stage-based life history model consisting of the 13 distinct life stages as well as diapausing CV and immature female stages was coupled to the nonlinear z-level ocean circulation model (CANDIE: Sheng et al., 1998). CANDIE was run in diagnostic mode over a full one-year cycle and the stored output was called up at each time step for the calculation of fluxes through grid boundaries, abundance, total spawning production and mortality. The model results indicate that the GSL population is self-sustaining, but that the fluxes of *Calanus,* notably an export to the Gulf of Maine and potentially important input from offshore slope waters, play an important role in controlling abundance on the shelf.

One way of testing the accuracy of the coupled models is to initialize the model with one set of observations, say the abundance of a copepod species on a spatial grid, run the model forward in time to the date of the next set of observations, and then compare the simulated results with the observed data. Small differences between the predictions and observations would impart confidence that the coupled model is capturing the essential biological and physical processes controlling the population dynamics. McGillicuddy et al. (1998) introduced an alternative technique, called adjoint data assimilation, for cases when the discrepancies are large, or when limitations in the model or data do not allow proper investigation of the forward problem. The adjoint data assimilation technique essentially iterates backward in time from the second set of observations in order to find the biological reaction term, R (in two dimensions, R (x,y)), that minimizes the misfit between the observations and predictions. Assuming that the advective fields are correctly characterized, the sign of R(x,y) indicates, for that point in the grid, whether population growth or mortality was dominant. In combination with ancillary data on, for example egg production rates and predator abundance, this method provides insight into the population control processes across the spatial grid.

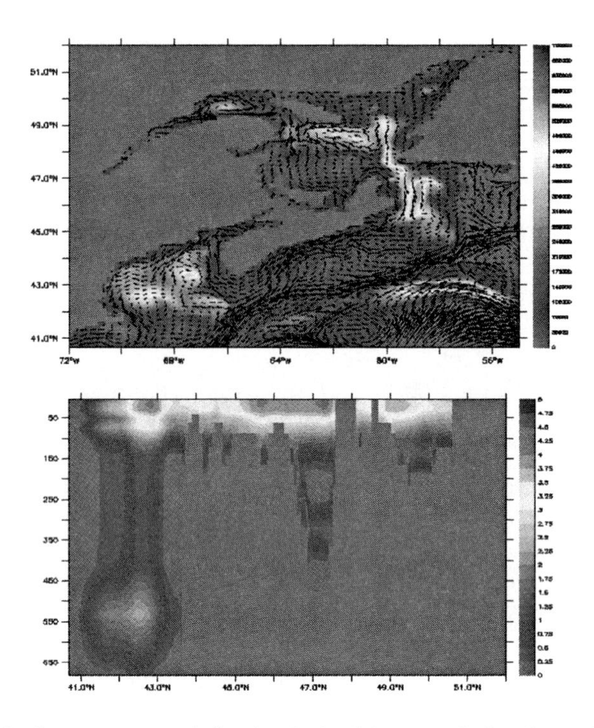


Figure 13.10 A) Surface currents and simulated, depth-integrated abundance (no. m$^{-2}$) of *C. finmarchicus* (all stages) in early June in the Gulf of St. Lawrence, Scotian Shelf and Gulf of Maine system based on a *Calanus* life history model coupled to the 3-D mean climatological circulation calculated from the CANDIE finite difference model (redrawn from Zakardjian et al., 2003). B) Vertical section at 59°30 W of *Calanus* total abundance (log[no. m$^{-3}$]) from (A) showing the stage-specific vertical distribution in the simulated model results. The deepest abundance mode (300 m in the Gulf and 550 in the slope water) represents diapausing CV and males.

The newly developed circulation models have been used to investigate the role of advection for the retention, dispersal and growth rates of larval fish. The fate of larvae spawned on the northeast peak of the Bank was explored through the use of Lagrangian particle tracking (Werner et al., 1993), wherein particles represented early life stages of fish and exhibited various swimming behaviors (passive, fixed-depth, age-dependent migrations, and lateral swimming). Simulations demonstrated large advective losses, such that most larvae were carried out of the region. Particle tracking was also used to estimate the seasonal variation in residence times for water parcels (Page et al., 1999). Particles, initially distributed Bank-wide, were advected with seasonal mean flows for 60 days, revealing large geographic and seasonal differences in residence times. The relative role of advection vs. food limitation was explored by coupling a particle-tracking mode with a trophodynamic model (Werner et al., 1996; Werner et al., 2001b). Larval growth rates are related to encounter, ingestion and assimilation of prey, and incorporated effects of turbulence on contact rates (Fig. 13.11). Prey density was estimated from data, and turbulence levels were derived from the circulation model. Advective losses were found to be on the order of 20% of the larvae spawned, and the region of highest retention coincided with the region of highest growth. Near-bed turbulence played a big role in increasing contact rates, but high values close to the bottom decerased the larvae's capture success and thus the smaller number of survivors close to the bottom.

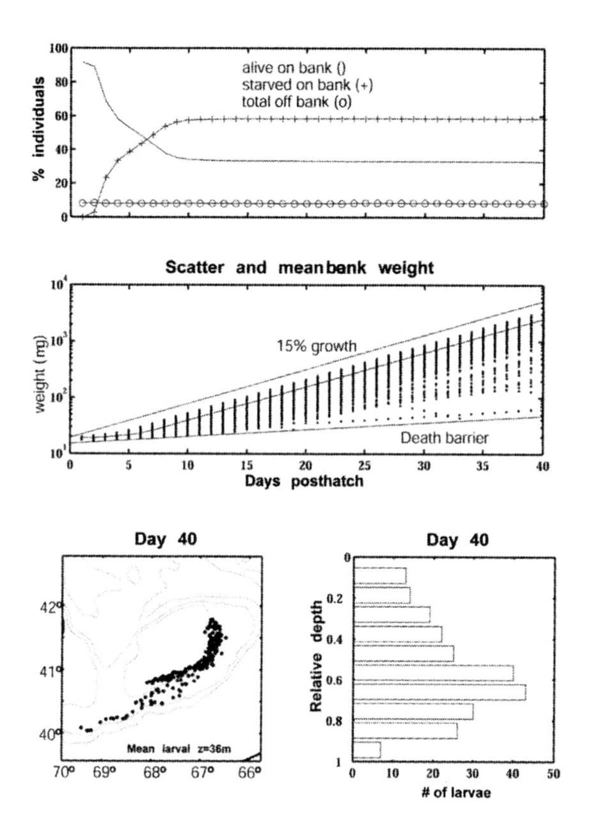


Figure 13.11 Post-hatch history of modeled cod larvae on Georges Bank (from Werner et al. 2001b) including advection, feeding, and effects of turbulence on encounter rates and capture success. Top panel, the percentage of larvae alive (solid line), starved on-Bank (solid line with crosses), and advected off the Bank (solid line with open circles); second panel, the daily size distribution (micrograms) for the live larvae on the Bank, the 15% per day growth curve, the death barrier and the mean daily weight of those live larvae still on the Bank. Bottom left and right panels show, respectively, the horizontal distribution of the live larvae 40 days post-hatch and their vertical distribution relative to the local bottom-depth.

Historical data indicated a large degree of spatial variation in the seasonal prey fields (Meise and O'Reilly, 1996; McGillicuddy et al., 1998). Seasonal and geographic variation in food limitation of larval fish growth rates was investigated from an Eulerian perspective (Lynch et al., 2001). Prey fields were estimated using spatially-explicit zooplankton data and a zooplankton population dynamics model to generate maps of egg and naupliar abundance. Turbulence fields were derived from the circulation model. These spatially-varying fields were then used as input for a simplified model of fish growth to produce maps of the instantaneous fish growth rates as derived from different prey types. Model simulations indicate that small cod larvae could not survive on a diet of *Calanus finmarchicus* alone, whereas large larvae could. It showed that *Pseudocalanus* abundance was sufficient in certain areas to sustain high growth rates, but these could be mismatched with areas of where cod spawning and advection occurred.

Regions of the NWA have a good foundation of physical models. Considerable progress has been made toward the development of region-specific, coupled physical-biological NPZ, copepod life history and larval fish trophodynamic models. Extensive physical and biological data collected over the multi-year field studies in the NWA coastal regions is available for investigation of interannual variability. Field and process studies of vital rates, including reproduction and mortality, reveal empirical patterns relationships and patterns that offer simplifications of complex ecological interactions. The development of alternate modeling approaches such as adjoint data assimilation offer the prospect of new insights into processes controlling population dynamics. The challenge is to advance and link models of the three trophic levels into a comprehensive depiction of ecosystem dynamics and influences of climate variablity.

### *5.3 The western Gulf of Alaska*

Walleye pollock, *Theragra chlacogramma,* a widely distributed and dominant groundfish species in ecosystems across the North Pacific Ocean, supports one of the world's largest fisheries (Bailey et al., 1999). Spawning in this species is constrained to a relatively small number of locations during certain, predictable times of year. In the Gulf of Alaska, most of the walleye pollock constitute one stock that migrates to a limited area (40 km by 80 km) in Shelikof Strait to spawn during the first week in April. Larvae hatching from the free-floating eggs released at depths of 150–250 m drift downstream in the Alaska Coastal Current, a very strong (25–100 cm $s^{-1}$) current that transports larvae either to an inshore juvenile nursery or into the offshore Alaskan Stream (Fig. 13.12), where they are presumably lost to the stock (Kendall et al., 1996a). Larvae are frequently found in large patches associated with eddies. There is considerable potential for interannual and interdecadal forcing functions (e.g. storms, freshwater input, wind-generated turbulence) to influence the successful survival of the walleye pollock from egg to juvenile stages.

Understanding of walleye pollock dynamics in the western Gulf of Alaska has advanced considerably since establishment in 1984 of The Fisheries Oceanography Coordinated Investigations (FOCI) program by the U.S. National Oceanic and Atmospheric Administration. At the outset, knowledge on the life history of pollock and the dynamics of the physical environment in this region were very limited. Initial field and process studies focused on understanding the life history of pollock, identifying important biological processes, and examining the nature of the regional circulation (see, for example, references in Kendall et al., 1996b). As these aspects became better understood, field operations switched to maintenance of time series of selected biological and physical characteristics, studies of biophysical processes, and development of methods for analysis of factors influencing interannual changes in rates of survival. Since 1992, FOCI has analyzed biological and physical time series relative to a working conceptual model of the recruitment process (Megrey et al., 1995) and used these data and models to forecast future recruitment on a qualitative basis (i.e. weak, average, strong). This prediction significantly simplifies the stock projection analysis and facilitates interpretation by fisheries managers by limiting the number of viable recruitment scenarios. While present forecasts of year-class strength (Megrey et al., 1996) do not use coupled physical-biological models directly, they are used to identify gaps in knowledge, synthesize information, organize conceptual ideas, expand the spatial and temporal characteristics of field observations, generate hypotheses of linkages and interactions among sets of biological and physical factors, and provide information into the fisheries management stream.

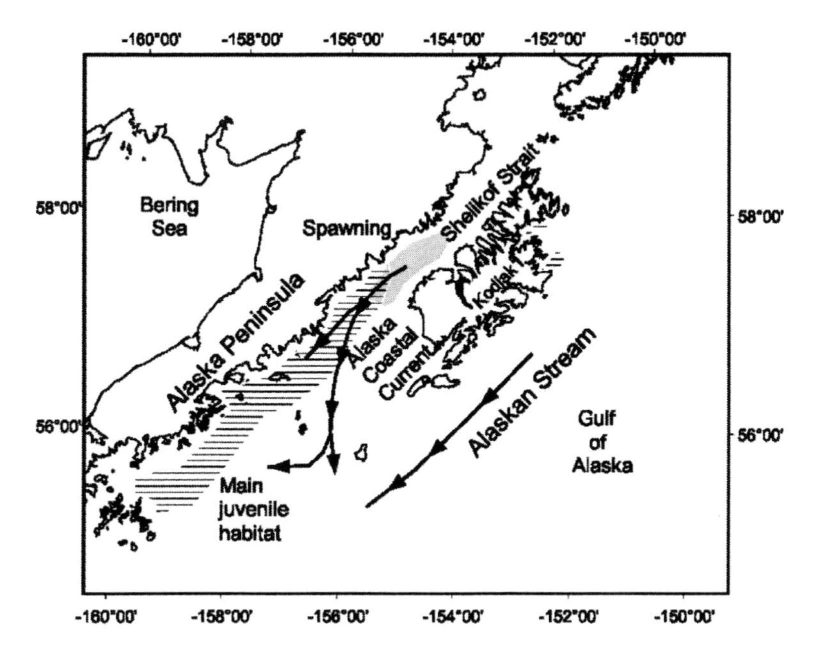


Figure 13.12 Conceptual model of the early life history and advective influences of walleye pollock in Shelikoff Strait, Gulf of Alaska. Eggs and larvae spawned in Shelikof Strait are transported to the southwest to either the inshore nursery area or the offshore Alaska Stream. (from Kendall et al., 1996b)

The circulation model used in the western Gulf of Alaska system is a primitive equation rigid-lid hydrodynamic model based on the Semispectral Primitive Equation Model (SPEM) of Haidvogel et al. (1991) and modified for this region (Stabeno et al., 1995a; Hermann and Stabeno, 1996). The spatial domain of the model comprises the northern Gulf of Alaska from east of Shelikof Strait to west of the juvenile habitat area (vicinity of Shumigan Islands) and contains a total of ~250,000 grid points with approx. 4 km spacing and nine vertical levels in the finely resolved area between Kodiak and the Shumigan Islands. The model is forced by twice-daily winds and monthly fresh-water runoff. Physical factors that pose challenges to proper representation of the area included complex bathymetry with many islands, mesoscale (~20 km) meanders and eddies, strong vertical shear (estuarine-like flow), and strong forcing by winds and freshwater runoff. SPEM has reproduced the observed general spatial features of circulation (Stabeno et al.,

1995b). The physical model has been validated with Eulerian (moored current meters) and Lagrangian (drogued drifters) data from field experiments (Hermann and Stabeno, 1996). The model is capable of generating eddies similar to those observed in terms of spatial scales and frequency of occurrence.

The biological model consists of a stochastic individual-based model (IBM) of egg and larval development (Hinckley et al., 1996). The IBM follows the unique life history of each fish from spawning to September of their 0-age year and therefore provides specific information about survivors. Velocity fields from the hydrodynamic model provide the physical spatially-explicit context to move individuals through space. The model also employs a spatial tracking algorithm for each individual that includes vertical migration according to life stage. Horizontal transport, egg development time, growth, and behavior are governed by velocity, salinity and temperature fields generated by SPEM. Low-pass filtered velocity and scalar fields from SPEM are stored once per model day, then used as input for multiple runs of the biological model. More recent advancements include the addition of an NPZ model to provide food for feeding pollock larvae (Hermann et al., 2001: Fig. 13.13). The zooplankton term in this model comprises the large planktonic copepods, *Neocalanus spp.,* the mesozooplankton species that graze much of the primary production and the smaller *Pseudocalanus* spp., the dominant prey for larval pollack in this region. The submodel for *Pseudocalanus* is stage structured, with egg, nauplius, copepodid and adult compartments.

The hydrodynamic, NPZ and IBM larval fish models have been linked into a coupled physical-biological model system. For each simulated year, the SPEM model is run to obtain velocity, temperature and salinity fields that are subsequently stored. The NPZ model is run through the spring period covering the plankton phase of the walleye pollock life history, using the SPEM-generated physical forcing and storing the 3-D output of stage-specific abundances of *Pseudocalanus* prey. The IBM larval fish model is then run using the SPEM and NPZ stored output to force advection and growth of individuals. Hence, in this coupled system, there is no dynamic linkage between the larval fish predators and their copepod prey.

This coupled model system has been used to explore the relationship between physical oceanographic processes and their effect on biological variability, hindcast the early life history of walleye pollock, and to assess the possible physical causes of interannual variability in recruitment. Modeled spatial distributions qualitatively compare favorably with observed distributions of larvae and juveniles (Hermann et al., 1996: Fig. 13.14). Interannual differences in wind and freshwater runoff lead to differences in the modeled spatial paths of individuals, and in the distributions of population attributes such as growth rates and size distributions. Results from SPEM show that during 1978 (the strongest year class) larvae were more likely transported into coastal waters along the Alaska Peninsula, while in 1990 (a below average year class) they remained in the sea valley where currents then result in transport offshore (Stabeno et al., 1995b). This latter scenario supports FOCI's original transport hypothesis as it implies a loss of recruits.

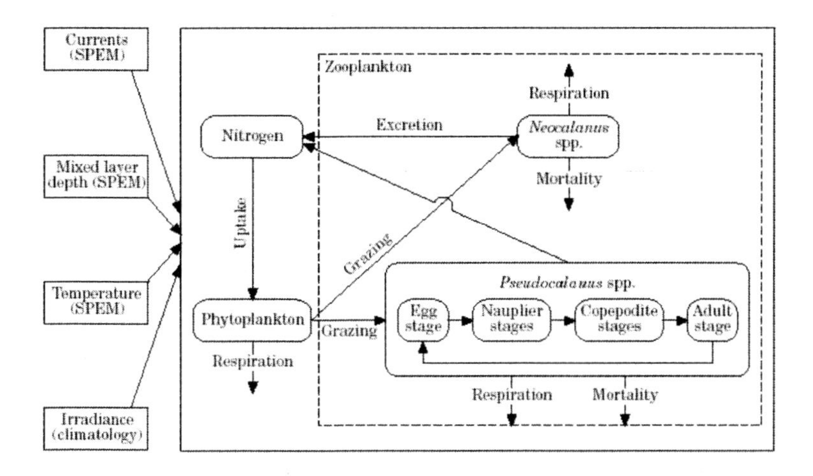


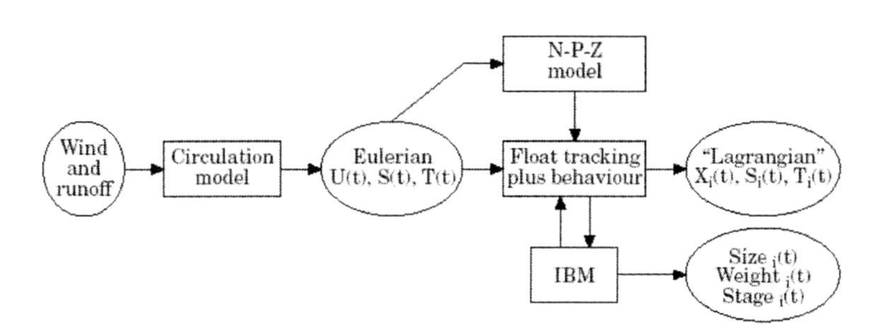


Figure 13.13 Summary of the coupling between the hydrodynamic model, the individual-based model and the NPA model (lower panel) and the flowchart of the nutrient-phytoplankton-zooplankton (NPZ) model (upper panel) (from Hermann et al., 2001).

The coupled model has also been effective in conducting "model experiments" to test the viability of competing hypotheses. For example, Hinckley et al. (2001: Fig. 13.15) used the coupled model system with winds and runoff forcing representative of two years of good recruitment, 1978 and 1994 to examine hypotheses about the timing and location of pollock spawning. In an approach similar to the Benguela upwelling system studies, five regions (1–5) and four spawning times (Early, Middle, Late, Very Late) were considered, where "1-Middle" represents typical observed spawning. Results show that fish spawned to the south of Kodiak Island (3-Middle) or much earlier or later than the observed spawning period (e.g. 1-Very Late) do not reach the Shumagin Island nursery area as juveniles by early September. However, the region and time of spawning which did allow successful transport to the nursery area (e.g. 4-Late) was much broader than the observed region and time. Hence simulation experiment results indicate that factors other than physical transport alone must be considered to explain the spawning location and timing of this stock.

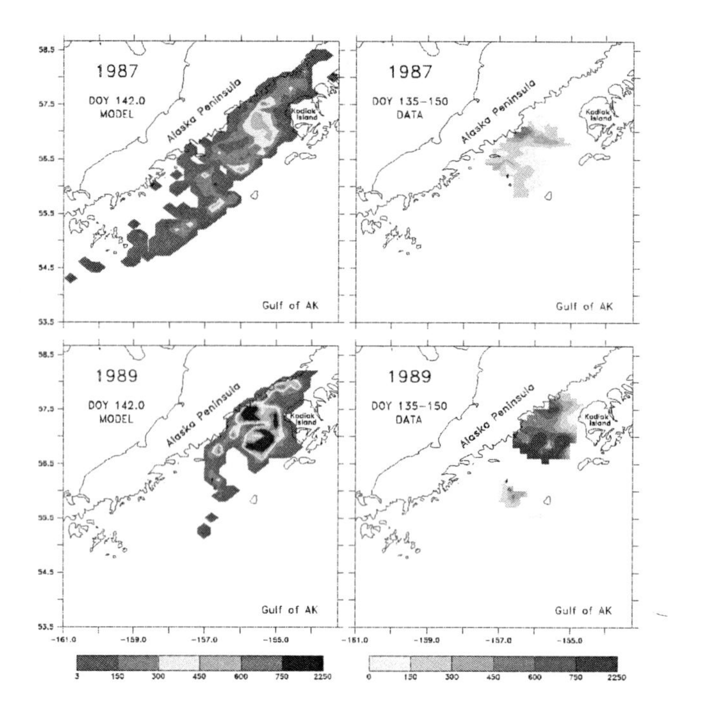


Figure 13.14 Comparison of model output to field observations, Shelikov Strait. Plotted is depth-integrated abundance of larvae (#/10m$^2$) in mid-May of 1987 and 1989 (from Hermann et al., 1996)

In a newer initiative, a suite of nested physical and biological models is under development as part of the West Coast U.S. GLOBEC program. Components of this suite include: 1) multiply nested circulation models spanning basin to regional to local scales; 2) a lower trophic level (NPZ) model including salmon prey items, driven by those circulation fields; 3) an individual-based salmon model which receives circulation and NPZ model output. A preliminary implementation of this approach (Hermann et al., 2002) consisted of two physical models. The Spectral Element Ocean Model (SEOM) (Haidvogel and Beckmann, 1999) was implemented for a global domain, and provided both tidal and subtidal boundary conditions for the S-Coordinate Rutgers University Model (SCRUM) (Song and Haidvogel, 1994), used as the regional model of the Coastal Gulf of Alaska. Recently, the nested GLOBEC circulation models have been based on the Regional Ocean Modeling System (ROMS) (Haidvogel et al., 2000). Output from the regional circulation models are used to drive the NPZ model (presently for salmon prey) for the near-coastal area encompassing the shelf and shelf break. Resulting circulation and prey fields are used as input to a spatially-explicit IBM fish model (of juvenile salmon). The goal behind this new work is to create a tool to investigate interannual and decadal changes in the physical environment of the central Gulf of Alaska, while exploring linkages between physical forcing and biological production.

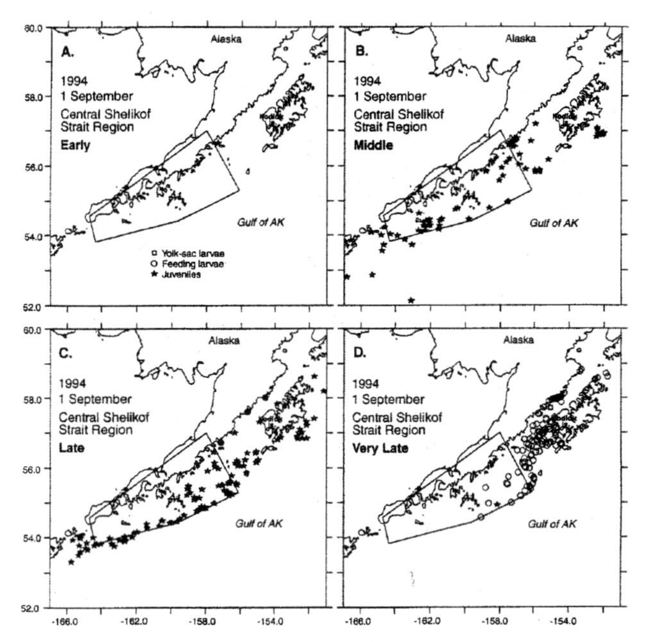


Figure 13.15 Results of model experiments looking at the effect of spawning location, time of spawning, depth of spawning, on the success of spawning products reaching the nursery area. (from Hinckley et al., 2001).

## 6. Applications and the future of coupled physical-biological modeling in coastal marine ecosystems

Sustainable use of marine resources and the management of coastal marine ecosystems are contemporary, real-world problems placing demands on the research community for greater understanding of marine ecosystem functioning (e.g. Barange, 2003). The complexity of the issues, already daunting, is exacerbated by predictions of global change, manifestations of which may include biologically significant increases in ocean temperature and changes in circulation at the regional and basin scale. The scientific contribution to marine resource management will continue to require advances in understanding of the coastal ocean that cut across disciplines, notably but not exclusively physics, biology and geochemistry. Understanding must be distilled into quantitative language for incorporation into models that link variation and change in the physical environment with the dynamics of species and species groups from the level of individuals to communities.

In Sections 2–4 we reviewed the approaches and status of three classes of physical-biological models. NPZ models, in the context of the problem here, address the simulation of primary production as a food resource for higher trophic levels. The Z in these models represents total zooplankton biomass and serves as a closure term for the model. Needs for finer spatial and temporal scale predictions of prey for exploited fish and invertebrate resources place challenging demands on the NPZ models to address the biochemical composition and size structure of the P term and to resolve the heterotrophic and microzooplankton. The second class of models, which we have called coupled zooplankton life history models, resolves the spatially explicit population dynamics of key zooplankton (typically copepod) species or species groups. Two general approaches, Eulerian (advective-diffusive models) or Lagrangian (Individual Based Models) may be used; issues facing development of these models include characterization of food limitation, vertical migration behavior and diapause. The models of fish larvae focus on the processes controlling growth and survival of larvae in the simulated temperature and flow fields. Individual Based Models are generally used in this context, and understanding of the complexities of feeding behavior of fish larvae and their vertical movements are two of the issues facing their development. Common challenges to the development of models in all three classes include temporal/spatial scaling and characterization of mortality. The former is relevant to the linkage of hydrodynamics link to phytoplankton/microplankton (as food), food availability to zooplankton and zooplankton to fish. Improper scaling can distort models results by inaccurately depicting interactions and creating artificial model responses. The mortality issue is critical for realistic simulation of zooplankton and larval fish dynamics. For the time being, we will have to rely heavily on empirical relationships derived for the region, time period and species of interest, not only for mortality but also for functional linkage relationships such as the dependency of zooplankton growth and reproduction on indices of food availability.

In Section 5, we examined the application of coupled models in coastal waters of southwest Africa, the northwest Atlantic and the Gulf of Alaska. Notable is the need for multidisciplinary effort and considerable resources (both manpower and time) in order to develop sophisticated coupled models for site-specific application. Questions were addressed with a diverse tool box of models capable of evolving with increasing data and knowledge of the system of interest. In the northwest Atlantic, 3-D coupled models depicting secondary production and growth and survival of fish larvae are in an active research phase but are not yet applied in a management context. In the Benguela Current System and Gulf of Alaska, coupled models have been developed and applied to synthesize, guide and interpret research and understanding of recruitment processes, and to supply information in a limited sense to fisheries management. Coupled models of the type described here have also been developed for application in other coastal ecosystems, but in general have not yet entered operationally into the activity of marine resource management.

There is, nevertheless, an emerging synthesis of data and knowledge that will continue to move coupled modeling toward the mainstream of marine resource management. Systems of independent models in the three broad classes discussed above, each of which is coupled to a hydrodynamic model describing current and temperature fields for a particular coastal ocean or shelf sea, can be developed. Each of these coupled biophysical models represents a quantitative integration of processes and interactions in one part of the pelagic ecosystem. Examples of how the various coupled models could be linked in a hierarchical structure, which could be called linked coupled models (LCMs), are shown in Fig. 13.16. In Fig. 13.16a, a coupled physical-NPZ model simulates spatially explicit food resources for key zooplankton species that produce prey for target species of fish larvae; the distribution of both fish and zooplankton are also by the flow and temperature generated by the regional circulation model. Fig. 13.16b illustrates linked coupled models for simulation of regional trophic structure with greater resolution of predator sources of mortality on target species (in this case *Calanus* species).

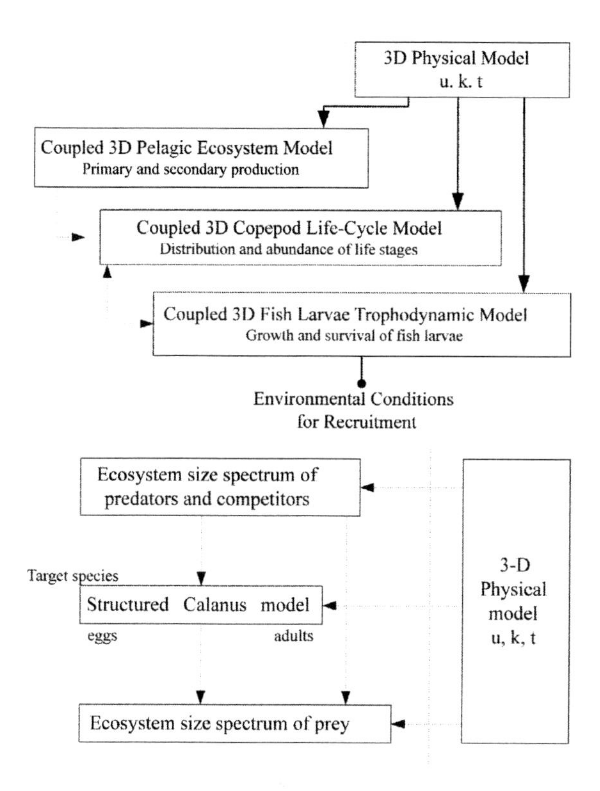


Figure 13.16 Examples of possible interrelationships among various components of linked coupled models (LCMs) for coastal shelves. A): Trophic structure system in which population dynamics of target species are parameterised in detail, linked to lower resolution models of describing prey and predator abundance. B: LCM for use in describing environmental conditions for survival of fish or invertebrate larvae in a coastal system

Given these possibilities, the following subsections examine the potential, data needs and limitations of coupled simulation model systems in the application to resource management issues.

**Coastal Ocean Observing Systems and model-data interactions**

There is world-wide movement toward establishment of a global ocean observing system (e.g. Holland and Nowlin, 2001). In the United States, for example, development of an Integrated Ocean Observing System (IOOS) for U.S. coastal regions has been initiated, with plans to become operational within the next decades. Among the seven major goals of IOOS are to improve predictions of climate change and its effects on coastal populations, protect and restore healthy coastal marine ecosystems and sustain living marine resources.

A vital role for LCMs is in the organization, analysis and interpretation of coastal ocean observing data. Functions of the coupled biophysical models include contributions to identification of measurement variables, design of sampling frequency and location, and interpretation of data with respect to the goals of observing systems, such as the evaluation of habitat quality and environmental conditions for recruitment success (as discussed below).

The interaction between coastal ocean observing systems and LCMs also address data needs of the coupled biophysical models. The model-data interactions can be categorized as follows:

1. Data are needed to define boundary conditions and external forcing, to specify initial conditions and to meld (Robinson et al. 1998) with dynamic components of the LCMs in order to run site and time specific simulations. For example, wind and boundary hydrographic data for a given year would be used to run simulations of surface circulation to track, say, the distribution of fish and invertebrate larvae in the Gulf of Maine for that year. Section 2 mentions satellite-derived surface chlorophyll data that can be assimilated into NPZ models for diagnostic and predictive simulation of prey fields.
2. Data are needed for verification of the model simulations. In addition to high frequency hydrographic and current data (independent of observations used for data assimilation) for validation of circulation models, coastal observing system measured variables may include chlorophyll and high frequency zooplankton abundance data from moored sensors, target copepod life history data from fixed stations for comparison with simulated life cycles and surveys of larval fish abundance for validation of simulated distribution in, say, marine protected areas.
3. Data are needed for adjustment of empirical relationships supporting the biological dynamics. As discussed in sections, many biological relationships must be determined empirically, following the early modeling approaches described by Riley (1963 and references therein). For example, the relationship between egg production rate and chlorophyll a standing stock in the Gulf of Maine (Fig. 13.4) was determined from measurements at sea during process cruises. The mean vertical distributions of *Calanus* life stages used in the coupled copepod life cycle-physical model of Zakardjian et al. (2003) were derived from data collected over a decade starting in the early 1990's. Given the potential for genotypic and consequently phenotypic change of target species in response to global climate change, these relationships may need to be periodically verified, as there is presently no suitable theory to allow dynamic adjustments of physiological or behavioral depictions in the biological models. The need to update these relationships for accurate simulation by LCMs of observational data should be considered in the development of coastal ocean observing systems.

**The role of secondary production in biogeochemical cycles in the coastal ocean**

Continued and planned (e.g., Doney et al., 2004) investigation of biogeochemical cycling and elemental fluxes in the coastal ocean requires greater resolution of the role of secondary production. The *middle-out* coupled physical-biological modeling approach for key zooplankton species or species groups can contribute information on life cycle and spatial distribution needed to resolve temporal and spatial variability in grazing and elemental vertical fluxes in the coastal ocean, for example the flux of carbon to the sediment in areas of coastal upwelling. Additionally, development of coupled models for prediction of shifts in composition and distribution of key zooplankton species or species groups is a direction identified for further development of the coupled biological-physical zooplankton models.

**Habitat and environmental conditions for recruitment success**

Better understanding of how spatial and temporal variation in environmental factors affects larval fish growth and survival is not only of general research interest but also has application to ecosystem-based resource management. Coupled physical-biological models provide a formal method to quantify how environmental conditions affect larval growth, retention and mortality. For example, coupled simulation systems outlined in Fig. 13.16a can be used to diagnose whether environmental conditions in a given year were poor, average or exceptional for growth and survival of larvae. As mechanistic understanding and prediction of climate-induced physical and secondary production variation improves, LCMs can also be used to predict whether environmental conditions for larvae will improve or deteriorate in a multi-year time frame. This information can be fed into the decision-making process for determining appropriate quotas for sustainable fisheries. Coupled models are ideal for identifying the environmental conditions that result in the rare individuals surviving the larval period (e.g., Werner et al., 1996), and may be used in a sensitivity analysis mode to examine the effects of different processes on larval growth (e.g., turbulence: Megrey and Hinckley, 2001; Werner et al., 2001b).

Coupled physical- biological models can make an important contribution to understanding fish habitat quality. Quantifying habitat quality is of vital importance to effective long-term fisheries sustainability, and consideration of essential fish habitat is required by law (Sustainable Fisheries Act, Public Law 104–297) to be included in fisheries management plans. Most previous analyses of habitat quality have generally relied on a correlation approach; if organisms are found in locations then the environmental conditions in those locations must be good habitat (e.g., Minello, 1999). Whereas this approach is clearly unsatisfactory for determining habitat quality of fish larvae, linked physical-biological models, as discussed above, provide a tool for both quantifying and predicting future retention and growth characteristics of a given location. A limitation may be that habitat quality must be quantified on a grand scale (many species in many locations), whereas the development of coupled biophysical models proceeds on a site-specific basis, and each application involves a significant effort. Other issues related to habitat, such as determining the effectiveness of Marine Protected Areas, may also be addressed by coupled model analyses.

**Recruitment prediction**

For most of the past century, a branch of fisheries science has focused on the general problem of predicting the number of young at birth that will survive to some size or age, termed recruitment (Bradford, 1992; Cushing, 1996; Needle, 2002). The relationship between spawners and the subsequent survival of their progeny to recruitment is fundamental to fisheries management (Rose and Cowan, 2003). Understanding the causes of variation in growth and survival of larval fish, as discussed earlier, is an important component of recruitment prediction because much of the eventual interannual variation observed in recruitment in marine fish can be attributed to variation in growth and mortality during early life stages (Shepard et al., 1984; Houde, 1987; Rose and Summers, 1992).

By enabling better understanding of the causes of variation in larval growth and mortality, coupled physical-biological models can play a major role in helping to understand the relationship between spawners and recruitment. One of the major problems with spawner-recruit data is that the high degree of variability can confound interpretation of the relationship between the two (Walters and Ludwig, 1981). Very noisy data can be incorrectly interpreted as no relationship between spawners and recruits, which would imply almost infinite density-dependence (constant recruitment regardless of the number of spawners) and therefore lead to over harvesting (Hilborn and Walters, 1992). Coupled physical-biological models can play a role in the recruitment question by providing a tool for deciphering the variation in the larval stage contribution to recruitment variability.

However, because of the complexity of recruitment prediction, it remains to be seen whether coupled physical-biological models can be of general value beyond the assessment of environmental conditions and habitat quality for larval stages. Major progress on the recruitment question will require that the models include potential density-dependent effects on larval growth and mortality. There are very few examples of coupled physical-biological models that allow for density-dependent mortality via numerical or functional responses of the predators of the larvae. Predation mortality of larvae is critical to the recruitment question (Bailey and Houde, 1989; Leggett and DeBlois, 1994). Furthermore, the larval life stage is but one component of process recruitment, and only one part of the life cycle. Fish exhibit complex life histories that involve life stages that use different habitats (Rose, 2000).

Sissenwine (1984) and others consider that the juvenile stage is critical to understanding density-dependence and recruitment variability. Cowan et al. (2000) concluded that density-dependent growth in marine fish was most likely to occur in the late larval and juvenile life stages. Recent analyses document density-dependent mortality in the juvenile stage (Myers and Cadigan, 1993), and suggest density-dependent adult growth may be more widespread than previously thought (Lorenzen and Enberg, 2002). Population dynamics and population regulation are likely the result of the complex mix of different processes occurring in different life stages. Improving our knowledge of larval stage growth and mortality (even with density-dependence) may not help in predicting overall recruitment if significant density-dependent effects occur after the larval stage.

Extending the coupled physical-biological approach to post-larval life stages is likely not appropriate for many situations, as the rationale for coupling larvae to physical models diminishes for post-larval life stages. Movement of juvenile and adult fish depends more on biological factors and individual decisions and less related to the physics (Tyler and Rose, 1994). In addition, juvenile and adult fish generally spend less time feeding than larvae, and tend to eat larger prey that are themselves less controlled by physical transport. Indeed, many harvested fish species eventually become piscivorous, where the prey becomes larval and juvenile fish. Thus, the important role played by the physical-biological models for larval fish of generating transport, temperature, and prey fields will be much more difficult to achieve for simulating juvenile fish and likely impossible for simulating adult fish.

In the long run, then, recruitment prediction may require models (necessarily not one model) that together encompass the full life cycle. These models should be capable of predicting, at least in qualitative terms, trends in growth and mortality rates and possible density-dependent effects. Coupled physical-biological models can likely be extended to include invertebrate predators of larvae (i.e., mortality) and density-dependent growth, but not much beyond the early larval life stage. We know enough about the ctenophores and medusae (e.g., Brietburg et al., 1999) to at least attempt to include them in coupled models. We envision that coupled physical-biological models including density-dependence and invertebrate predators will be used for the larval life stage. A series of linked models, each temporally and spatially scaled to their particular life stages, would enable full life simulations over multiple generations; the physical-biological modeling being one component in this chain. Individual-based models (albeit spatially simple) of early life stages of fish have been coupled to age-structured matrix projection models for adults (Rose et al., 1996; Rose et al., 2003).

**Concluding remarks**

The rapid increases in our ability to make detailed measurements, coincident with advances in numerical modeling, data assimilation approaches, etc., portend great advances in coupled physical-biological modeling over the next decades. Coupled physical-biological models are ready now to provide information on habitat quality for larval fish. In situations when recruitment is set in or just after the portion of the larval stage, then insights into recruitment variability and forecasts of potential recruitment may be possible. Coupled physical-biological models should also play a role in understanding the dynamics of many fish species in which recruitment is not fully determined until after the larval stage, although non-trivial improvements in the current state of the models will be required. Allowing for density-dependent feedbacks and including predators to permit prediction of mortality are significant challenges for the next generation of coupled models. Effort is needed on creative ways to mesh models that operate on different biological, spatial, and temporal scales.

Continued developments of coupled physical-biological models that include fish, requires close collaboration between oceanographers (bottom-up view of the food chain) and fisheries researchers (top-down view of the food chain). The marriages of spatially-explicit physical, plankton production models, the biologically-complex but spatially simple models of larval fish growth and survival and perhaps eventually including full life cycle models of resource species require a multidisciplinary effort. The computer becomes the medium through which the quantitative synthesis of distilled understanding of different disciplinary perspectives is expressed. Simulation with validated dynamics present hypotheses and predictions that can be tested against data, and therefore become an element of scientific methodology (Robinson et al., 1998; Robinson and Lermusiaux, 2002). Computer-generated graphics and animations become the means of communication to both experts and non-experts of the complex, integrated synthesis of system knowledge.

## Acknowledgements

We thank the National Science Foundation, the National Oceanic and Atmospheric Administration's Coastal Ocean Program and Coastal Observation Technology System, and the Natural Sciences and Engineering Research Council of Canada for support from several research grants. This work represents contributions from the U.S. GLOBEC (number 488) and GLOBEC Canada programs and is contribution number 0530-ROAO from NOAA's Fisheries-Oceanography Coordinated Investigations. We gratefully acknowledge J. D. DuPrie and K. Pehrson-Edwards for assistance in preparation of the chapter.

## Bibliography

Ådlandsvik, B. and S. Sundby, 1994. Modeling the transport of cod larvae from the Lofoten Area. *ICES mar. Sci. Symp.*, **198,** 379–392.

Aksnes, D. L., and J. Blindheim, 1996. Circulation patterns in the North Atlantic and possible impact on population dynamics of *Calanus finmarchicus. Ophelia,* **44,** 7–28.

Aksnes, D.L. and M.D. Ohman, 1996. A vertical life table approach to zooplankton mortality estimation. *Limnol. Oceanogr.,* **41,** 1461–1469.

Armstrong, D.A., H.M. Verheye, and A.D. Kemp, 1991. Short-term variability during an anchor station study in the southern Benguela Upwelling system: Fecundity estimates of the dominant copepod, *Calanoides carinatus. Prog. Oceanogr.* **28,** 167–188.

Armstrong, R.A., 1999. Stable model structures for representing biogeochemical diversity and size spectra in plankton communities. *J. Plankton Res.* **21,** 445–464.

Armstrong, R.A., J.L. Sarmiento and R.D. Slater, 1995. Monitoring ocean productivity by assimilating satellite chlorophyll into ecosystem models. In *Ecological Time Series,* J. Steele and T. Powell, eds. Chapman and Hall. pp. 371–390.

Ault J.S., J. Luo, S.G. Smith, J.E. Serafy, J.D.Wang, R. Humston and G.A.Diaz, 1999. A spatial dynamic multistock production model. *Can. J. Fish. Aquatic Sci.* **56**(Suppl. 1), 4–25.

Backhaus, J.O., 1989. On the atmospherically induced variability of the circulation of the Northwest European shelf sea and related phenomena. In *Modeling Marine Systems,* Vol I., C Davies, A.M., ed. RC Press, Inc, Boca Raton, Florida, 93–134.

Banse, K., 1994. Grazing and zooplankton production as key controls of phytoplankton production in the open ocean. *Oceanogr.,* **7,** 13–20.

Bailey, K.M., and E.D. Houde. 1989. Predation on eggs and larvae of marine fishes and the recruitment problem. Advances in Marine Biology, **25,** 1–83.

Bailey, K.M., N.A. Bond and P.J. Stabeno, 1999. Anomalous transport of walleye pollock larvae linked to ocean and atmospheric patterns in May 1996. *Fisheries Oceanography.* **8,** 264–273.

Barange, M., 2003. Ecosystem science and the sustainable management of marine resources: from Rio to Johannesburg. *Front. Ecol. Environ.* **1,** 190–196.

Bartsch, J., K. Brander, M. Heath, P. Munk, K. Richardson and E. Svendsen, 1989. Modeling the advection of herring larvae in the North Sea. *Nature,* **340,** 632–636.

Batchelder, H.P. and C.B. Miller, 1989. Life history and population dynamics of *Metridia pacifica:* results from simulation modelling. *Ecol. Mod.,* **48,** 119–136.

Batchelder, H.P. and R. Williams, 1995. Individual-based modeling of the population dynamics of *Metridia lucens* in the North Atlantic. *ICES J. Mar. Sci.,* **52,** 469–482

Batchelder, H.P., C.A. Edwards and T.M. Powell, 2002. Individual-based models of copepod populations in coastal upwelling regions: implications of physiologically and environmentally influenced diel vertical migration on demographic success and nearshore retention. *Prog. Oceanogr.,* **53,** 307–333.

Berggreen, U., B. Hansen, and T. Kiøboe, 1988. Food size spectra, ingestion and growth of the copepod *Acartia tonsa* during development: implications for determination of copepod production. *Marine Biology,* **99,** 341–352.

Beyer, J.E. and G.C. Laurence, 1980. A stochastic model of larval growth. *Ecological Modeling,* **8,**109–132.

Blanke, B., C. Roy, P. Penven, S. Speich, J. McWilliams and G. Nelson, 2002. Assessing wind contribution to the southern Benguela interannual dynamics. *GLOBEC International Newsletter,* **8,** 15–18.

Blumberg, A.F. and G.L. Mellor, 1987. A description of a three-dimensional coastal ocean circulation model. In *Three-Dimensional Coastal Ocean Models.* N.S. Heaps, ed., American Geophysical Union. Washington, D.C., pp. 1–16.

Bollens, SM and B.W. Frost, 1989. Zooplanktivorous fish and variable diel vertical migration in the marine planktonic copepod *Calanus pacificus. Limnol. Oceanogr.,* **34,** 1072–1083.

Bollens S.M., K. Osgood, B.W. Frost and S.D.Watts, 1993. Vertical distributions and susceptibilities to vertebrate predation of the marine copepods *Metridia lucens* and *Calanus pacificus. Limol. Oceamogr.,* **38,** 1827–1837.

Bollens S.M., B.W. Frost and J.R. Cordell, 1994. Chemical, mechanical and visual cues in the vertical migration behavior of the marine planktonic copepod *Acartia hudsonica. J. Plankton Res.,* **16,** 555–564.

Bradford, M.J., 1992. Precision of recruitment predictions from early life stages of marine fishes. *Fishery Bulletin,* **90,** 439–453.

Brandt, S.B., D.M. Mason and E.V. Patrick, 1992. Spatially-explicit models of fish growth rate. *Fisheries,* **17,**23–33.

Breitburg, D., K. Rose, and J. Cowan, 1999. Linking water quality to larval survival: predation mortality of fish larvae in an oxygen-stratified water column. *Marine Ecology Progress Series,* **178,** 39–54.

Bryant, A.D., M. Heath, W.S.G. Gurney, D.J. Beare and W. Robertson, 1997. The seasonal dynamics of *Calanus finmarchicus:* Development of a three-dimensional structured population model and application to the northern North Sea. *J. Sea Res.,* **38,** 361–379.

Bryant, A. D., D. Hainbucher, and M. Heath, 1998. Basin-scale advection and population persistence of *Calanus finmarchicus. Fish. Oceanogr.,* **7,** 235–244.

Calbet, A., E. Saiz and M. Alcaraz, 2002. Copepod egg production in the NW Mediterranean: effects of winter environmental conditions. *Mar. Ecol. Prog. Ser.* **237,** 173–184.

Campbell R.G., M.M. Wagner, G.J. Teegarden, C.A. Boudreau and E. G. Durbin, 2001a. Growth and development rates of the copepod *Calanus finmarchicus* reared in the laboratory. *Mar. Ecol. Progr. Ser.,* **221,**161–183.

Campbell, R. G., J.A. Runge, and E. G. Durbin, 2001b. Evidence of food limitation of *Calanus finmarchicus* production rates on the southern flank of Georges Bank during April 1997, *Deep-Sea Res. II,* **48,** 531–549.

Campbell, R.W. and E.J.H. Head, 2000. Egg production rates of *Calanus finmarchicus* in the western North Atlantic: effect of gonad maturity, female size, chlorophyll concentration and temperature. *Can. J. Fish. Aquat. Sci.,* **57,** 518–529.

Carlotti, F. and A. Sciandra, 1989. Population dynamics model of *Euterpina acutifrons* (Copepoda: Harpacticoida) coupling individual growth and larval development. *Mar. Ecol. Prog. Ser.* **56,** 225–242.

Carlotti, F. and G. Radach, 1996. Seasonal dynamics of phytoplankton and *Calanus finmarchicus* in the North Sea as revealed by a coupled one dimensional model. *Limnol. Oceanogr.,* **41,** 522–539.

Carlotti, F. and K.U. Wolf, 1998. A Lagrangian ensemble model of Calanus finmarchicus coupled with a 1-D ecosystem model. *Fish. Oceanogr.,* **7,** 191–204.

Carlotti, F., J. Giske and F. Werner, 2000. Modeling zooplankton dynamics. In *ICES Zooplankton Methodology Manual,* R.P. Harris, P. Wiebe, J. Lenz, M. Huntley, H.R. Skjoldal, eds. Academic Press, pp. 571–667.

Carr, M.E. 1998. A numerical study of the effect of periodic nutrient supply on pathways of carbon in a coastal upwelling regime. *J. Plankton Res.,* **20,** 491–516.

Caswell, H. and A.M. John, 1992. From the individual to the population in demographic models. In *Individual-based models and approaches in ecology: populations, communities and ecosystems.* D.L. DeAngelis and L.J. Gross, eds, Chapman and Hall, pp. 36–61.

Caswell H. 2001. *Matrix Population Models: Construction, Analysis, and Interpretation,* Second Editon, Sunderland, Mass., Sinauer Associates. 722 pp.

Checkley, D.M. Jr., 1980. The egg production of a marine planktonic copepod in relation to its food supply: laboratory studies. *Limnol. Oceanogr.,* **25,** 430–446.

Chen, C., R. Beardsley and P. J. S. Franks, 2001. A 3-D progostic numerical model study of the Georges Bank ecosystem. Part I: physical model. *Deep Sea Research.* **48,** 419–456.

Cohen, E. and M. D. Grosslein, 1987. Production on Georges Bank compared with other shelf ecosystems. In *Georges Bank.* R.H. Backus and D.W. Bourne, eds. MIT Press. Cambridge, pp. 383–391.

Colton, J.B., Jr. and R.F. Temple, 1961. The enigma of Georges Bank spawning. *Limnol. and Oceanogr.,* **6,** 280–291.

Conover, R.J., 1988. Comparative histories in the genera *Calanus* and *Neocalanus* in high latitudes of the northern hemisphere. *Hydrobiolgia,* **167/168,** 127–142.

Cowan, J.H., K.A. Rose, and E.D. Houde, 1997. Size-based foraging success and vulnerability to predation: selection of survivors in individual-based models of larval fish populations. In *Early Life History and Recruitment in Fish Populations,* R Chambers and E. Trippel, eds., Chapman and Hall, pp. 357–386.

Cowan, J.H., Jr., K.A. Rose, D.R. DeVries, 2000. Is density-dependent growth in young-of the-year fishes a function of critical weight? Reviews in Fisheries and Fish Biology **10,** 61–89.

Crain, J.A. and C.B. Miller, 2001. Effects of starvation on intermolt development in *Calanus finmarchicus* copepodites: a comparison between theoretical models and field studies. *Deep-Sea Res. II,* **48,** 551–566.

Cushing, D.H., 1982. *Climate and Fisheries.* Academic Press, London, 373 pp.

Cushing, D.H., 1996. Towards a Science of Recruitment in Fish Populations. In *O. Kinne (ed) Excellence in Ecology.* Ecology Institute, Oldendorf/Luhe.

Dam, H.G., W.T. Peterson and D.C. Bellantoni, 1994. Seasonal feeding and fecundity of the calanoid copepod Acartia tonsa in Long Island Sound: Is omnivory important to egg production? *Hydrobiologia,* **292/293,** 191–199.

Davis, C.S., 1984. Interaction of a copepod population with the mean circulation on Georges Bank. *J. Mar. Res.,* **42,** 573–590.

Davis, M.W., 1996. behavioral determinants of distribution and survival in early stages of walleye pollock, *Theragra chalcogramma:* a synthesis of experimental studies. *Fish. Oceanogr.,* **5,**167–178.

Davis, M.W., 2001. Behavioral responses of walleye pollock, *Theragra chalcogramma,* larvae to experimental gradients of se water flow: implications for vertical distribution. *Environ. Biol. Fishes,* **00:**1–8.

DeAngelis, D.L., K.A. Rose and M.A. Huston. 1994. Individual-oriented approaches to modeling populations and communities." In *Frontiers in Mathematical Biology,* S.A. Levin, editor. Lecture Notes in Biomathematics, Vol. 100, New York: Springer-Verlag, pp. 390–410.

Denman, K.L. and M.A. Peña, 2002. The response of two coupled one-dimensional mixed layer/planktonic ecosystem models to climate change in the NE subarctic Pacific Ocean. *Deep-Sea Res. II,* **49,** 5739–5757.

Denman, K.L., 2003. Modelling planktonic ecosystems: parameterizing complexity. Progress in Oceanography, **57,** 429–452.

deYoung, B., M. Heath, F. Werner, F. Chai, B. Megrey and P. Monfray, 2004. Challenges of Modeling Decadal Variability in Ocean Basin Ecosystems. *Science,* **304,** 1463–1466.

Dodson, J.J., 1988. The nature and role of learning in the orientation and migratory behavior of fishes. *Environ. Biol. Fish.,* **23,** 161–182.

Doney, S.C., 1999. Major challenges confronting marine biogeochemical modeling. *Global Biogeochemical Cycles,* **13,** 705–714.

Doney, S.C., J.A. Kleypas, J.L. Sarmiento and P.G. Falkowski, 2002. The US JGOFS Synthesis and Modeling Project – An introduction. *Deep Sea Research II,* **49,** 1–20.

Doney, S.C., R. Anderson, J. Bishop, K. Caldeira, C. Carlson, M.-E. Carr, R. Feely, M. Hood, C. Hopkinson, R. Jahnke, D. Karl, J. Kleypas, C. Lee, R. Letelier, C. McClain, C. Sabine, J. Sarmiento, B. Stephens, and R. Weller, 2004. *Ocean Carbon and Climate Change (OCCC): An Implementation Strategy for U. S. Ocean Carbon Cycle Science,* UCAR, Boulder, CO, 108pp.

Dower, J.F., T.J. Miller and W.C. Leggett, 1997. The role of microscale turbulence in the feeding ecology of larval fish. *Adv. Mar. Biol.,* **31,**169–220.

Durbin, E.G., R.G. Campbell, M. C. Casas, M. D. Ohman, B. Niehoff, J. Runge and M. Wagner, 2003. Interannual variation in phytoplankton blooms and zooplankton productivity and abundance in the Gulf of Maine during winter. *Mar. Ecol. Prog. Ser.,* **254,** 81–100.

Eckman, J.E., 1994. Modeling physical-biological coupling in the ocean: the U.S. GLOBEC Program. *Deep-Sea Res.,* **41,** 1–5.

Edvardsen, A., M. Zhou, K.S. Tande and Y. Zhu, 2002. Zooplankton population dynamics: measuring in situ growth and mortality rates using an Optical Plankton Counter. *Mar. Ecol. Prog. Ser.* **227,** 205–219.

Edwards, A.M., and A. Yool, 2000. The role of higher predation in plankton population models. *J. Plankton Res.* **22,** 1085–1112.

Epifanio, C.E. and R.W. Garvine, 2001. Larval transport on the Atlantic continental shelf of North America: a review. *Est. Coastal Shelf Sci.* **52,** 51–77.

Falkenhaug, T., K.S. Tande and T. Semenova, 1997. Diel, seasonal and ontogenetic variations in the vertical distributions of four marine copepods. *Mar. Ecol. Progr. Ser.,* **149,** 105–119.

Fennel, K., Y.H. Spitz, R.M. Letelier, et al. 2002. A deterministic model for N-2 fixation at stn. ALOHA in the subtropical North Pacific Ocean. *Deep-Sea Res. II.* **49,** 149–174.

Field, J.G. and F.A. Shillington, in press. Variability of the Benguela Current System. In *The Sea. Vol. 14,* A.R. Robinson and K.H. Brink, eds. Harvard University Press, Cambridge, MA.

Flynn, K.J. and M.J.R. Fasham, 2003. Operation of lightdark cycles within simple ecosystem models of primary production and the consequences of using phytoplankton models with different abilities to assimilate N in darkness. *J. Plankton Res.* **25,** 83–92.

Fiksen Ø., 2000. The adaptive timing of diapause—a search for evolutionarily robust strategies in *Calanus finmarchicus. ICES J. Mar. Sci.,* **57,** 1825–1833.

Fiksen, Ø. and J. Giske, 1995. Vertical distribution and population dynamics of copepods by dynamic optimization. *ICES J. mar. Sci.,* **52,** 483–503.

Fiksen, Ø., J. Giske and D. Slagstad, 1995. A spatially explicit fitness-based model of capelin migrations in the Barents Sea. *Fisheries Oceanography,* **4,** 193–208.

Fiksen, Ø., A.C.W. Utne, D.L. Asknes, K. Eiane, J.V. Helvik and S. Sundby, 1998. Modeling the influence of light, turbulence and ontogeny on ingestion rates in larval cod and herring. *Fisheries Oceanography,* **7,** 355–363.

Foreman, M.G.G., A.M. Baptista and R.A. Walters, 1992. Tidal model of particle trajectories around a shallow coastal bank. *Atmosphere-Ocean,* **30,** 43–69.

Fogarty, M.J. and T. M. Powell, 2002. An Overview of the U.S. GLOBEC Program. *Oceanography.* **15,** 4–12.

Franks, P.J.S., 2002. NPZ models of plankton dynamics: Their construction, coupling to physics, and application. *J. Oceanogr.* **58,** 379–387.

Franks, P.J.S., 2001. Turbulence avoidance: An alternate explanation of turbulence-enhanced ingestion rates in the field. *Limnol. Oceanogr.* **46,** 959–963.

Franks, P.J.S. and C.S. Chen, 1996. Plankton production in tidal fronts: A model of Georges Bank in summer. *J. Mar. Res.* **54,** 631–651

Franks, P.J.S. and C.S. Chen, 2001. A 3-D prognostic numerical model study of the Georges bank ecosystem. Part II: biological-physical model. *Deep-Sea Res. II* **48,** 457–482.

Friedrichs, M.A.M., 2002. Assimilation of JGOFS EqPac and SeaWiFS data into a marine ecosystem model of the central equatorial Pacific Ocean. *Deep-Sea Res. II,* **49,** 289–319.

Frost, B.W., 1987. Grazing control of phytoplankton stock in the open subarctic Pacific Ocean: a model assessing the role of zooplankton, particularly large calanoid copepods *Neocalanus* spp. *Mar. Ecol. Prog. Ser.* **39,** 49–68.

Frost, B.W., 1988. Variability and possible adaptive significance of diel vertical migration in *Calanus pacificus,* a planktonic marine copepod. *Bull. Mar. Sci.,* **43,** 675–694.

Frost B.W. and S. M. Bollens, 1992. Variability of diel vertical migration in the marine planktonic copepod Pseudocalanus newmani in relation to its predators. *Can. J. Fish. Aquat. Sci.,* **49,** 1137–1141.

Gaard, E., 1999. The zooplankton community structure in relation to its biological and physical environment on the Faroe shelf, 1989–1997. *J. Plankton Res.,* **21,** 1133–1152.

Gaard, E., 2000. Seasonal abundance and development of *Calanus finmarchicus* in relation to phytoplankton and hydrography on the Faroe shelf. *ICES J. Mar. Sci.,* **57,** 1605–1611.

Giske, J., D.L. Aksnes and Ø. Fiksen, 1994. Visual predators, environmental variables and zooplankton mortality risk. *Vie Milieu,* **44,** 1–9.

Greenberg, D. A., 1983. Modeling the mean barotropic circulation in the Bay of Fundy and the Gulf of Maine. *J. Phys.Oceanogr.* **13,** 886–904.

Greene, C.H. and A. J. Pershing, 2003. The flip-side of the North Atlantic Oscillation and modal shifts in slope-water circulation patterns. *Limnol. Oceanogr.,* **48,** 319–322.

Grimm, V. 1999. Ten years of individual-based modeling in ecology: what have we learned and what could we learn in the future? *Ecol. Modell.,* **115,** 129–148.

Gupta, S., D.J. Lonsdale and Dong-Ping, 1994. The recruitment patterns of an estuarine copepod: A biological-physical model. *J. Mar. Res.,* **52,** 687–710.

Gurney, W.S.C., D.C. Speir, S.N. Wood, E.D. Clarke and M.R. Heath, 2001. Simulating spatially and physiologically structured populations. *J. Anim. Ecol.,* **70,** 881–894.

Haidvogel, D. B., J. L. Wilkin and R. Young, 1991. A semi-spectral primitive equation ocean circulation model using vertical sigma and orthogonal curvilinear horizontal coordinates, *J. Comput. Phys.,* **94,** 151–185.

Haidvogel, D.B. and A. Beckmann, 1998. Numerical models of the coastal ocean. In: Brink, K.H. and A.R. Robinson, eds., *The Sea,* **10,** 457–482.

Haidvogel, D.B., Beckmann, A., 1999. *Numerical Ocean Circulation Modeling.* Imperial College Press, 318 pp.

Haidvogel, D.B., H.G. Arango, K. Hedstrom, A. Beckmann, P. Malanotte-Rizzoli and A.F. Shchepetkin, 2000. Model Evaluation Experiments in the North Atlantic Basin: Simulations in Nonlinear Terrain-Following Coordinates, *Dynamics of Atmospheres and Oceans,* **32,** 239–281

Hannah, C.G., C.E. Naimie, J.W. Loder, and F.E. Werner, 1998. Upper-ocean transport mechanisms from the Gulf of Maine to Georges Bank, with implications for *Calanus* supply. *Continental Shelf Res.* **17,** 1887–1911.

Hansen, B., P. Verity, T. Falkenhaug, K.S. Tande, and F. Norrbin, 1994. On the trophic fate of Phaeocystis pouchetti (Harriot). 5. Trophic relationships between Phaeocystis and zooplankton: An assessment of methods and size dependence. *J. Plankton Res.,* **16,** 487–511.

Hare, J.A., J.A. Quinlan, F.E. Werner, B.O. Blanton, J.J. Govoni, R.B. Forward, L.R. Settle and D.E. Hoss, 1999. Influence of vertical distribution on the outcome of larval transport during winter in the Carolina Capes Region: results of a three-dimensional hydrodynamic model. *Fisheries Oceanography,* **8,** 57–76.

Harms, I.H., M.R. Heath, A.D. Bryant, J.O. Backhaus and D.A. Hainbucher, 2000. Modeling the Northeast Atlantic circulation: implications for the spring invasion of shelf regions by *Calanus finmarchicus. ICES J. Mar. Sci.,* **57,** 1694–1707.

Harris, R.P., 1996. Feeding ecology of *Calanus. Ophelia,* **44,** 1–3, 85–109.

Harris, R.P., X. Irogoien, R.N. Head, C. Rey, B.H. Hygum, B.W. Hansen, B.Niehoff, B. Meyer-Harms and F. Carlotti, 2000. Feeding, growth, and reproduction in the genus *Calanus. ICES J. Mar. Sci.,* **57,** 1798–1726.

Harrison, P.J. and T.R. Parsons, eds., 2000. Fisheries Oceanography: An integrative approach to fisheries ecology and management. Blackwell Science. 347 pp.

Haskell, A.G.E., E.E. Hofmann, G.A. Paffenhoefer and P.G. Verity, 1999. Modeling the effects of doliolids on the plankton community structure of the southeastern US continental shelf. *J. Plankton Res.,* **9,** 1725–1752.

Hazzard, S.E. and G.S. Kleppel, 2003. Egg production of the copepod Acartia tonsa in Florida Bay: role of fatty acids in the nutritional composition of the food environment. *Mar.Ecol.Prog.Ser.,* **252,** 199–206.

Head, E.J., L.R. Harris and R.W. Campbell, 2000. Investigations on the ecology of *Calanus spp.* In the Labrador Sea. I. Relationship between the phytoplankton bloom and reproduction and development of *Calanus finmarchicus* in spring. *Mar. Ecol. Prog. Ser.,* **193,** 53–73.

Heath, M. R., 1992. Field investigations of the early life stages of marine fish. *Advances in Marine Biolog,* **28,** 1–174.

Heath, M.R. and A. Gallego, 1997. From the biology of the individual to the dynamics of the population: bridging the gap in fish early life histories. *Journal of Fish Biology,* **51,** 1–29.

Heath, M., W. Robertson, J. Mardaljevic and W.S.G. Gurney, 1997. Modeling the population dynamics of *Calanus* in the Fair Isle Current off northern Scotland. *J. Sea Res.,* **38,** 381–412.

Heath, M.R. and A. Gallego, 1998. Biophysical modeling of the early life stages of haddock (*Melanogrammus aelgefinus*) in the North Sea. *Fisheries Oceanography,* **7,**110–125.

Heath, M.R., J.O. Bakhaus, K. Richardson, E. McKenzie, D. Slagstad, D. Beare, J. Dunn, J.G. Fraser, A. Gallego, D. Hainbucher, S. Hay, S. Jonasdottir, H. Madden, J. Mardaljevic and A. Schacht, 1999. Climate fluctuations and the spring invasion of the North Sea by *Calanus finmarchicus. Fish. Oceanogr.,* **8** (Suppl.1), 163–176.

Heath M. R., J.G. Fraser, A. Gislason, S.J. Hay, S.H. Jonasdottir and K. Richardson, 2000. Winter distribution of *Calanus finmarchicus* in the Northeast Atlantic. *ICES J. Mar. Sci.,* **57,** 1628–1635.

Heinrich, A.K., 1962. The life histories of planktonic animals and seasonal cycles of plankton communities in the oceans. *J. Cons. perm. int. Explor. Mer.,* **27,** 15–24.

Hermann, A.J. and P.J. Stabeno, 1996. An eddy resolving model of circulation on the western Gulf of Alaska shelf. I. Model development and sensitivity analysis. *J. Geophys. Res.* **101,** 1129–1149.

Hermann, A.J., S. Hinckley, B.A. Megrey, P.J. Stabeno, 1996. Interannual variability of the early life history of walleye pollock near Shelikof Strait as inferred from a spatially-explicit, individual-based model. *Fisheries Oceanography,* **5,** 39–57.

Hermann, A.J., S. Hinckley, B. Megrey and J.M. Napp, 2001. Applied and theoretical considerations for constructing spatially-explicit individual-based models of marine fish early life history which includes multiple trophic levels. *ICES J. Mar. Sci.,* **58,** 1030–1041.

Hermann, A.J., D.B. Haidvogel, E.L. Dobbins, P.J. Stabeno and P.S. Rand, 2002. Coupling Global and Regional Circulation Models in the Coastal Gulf of Alaska . *Prog. Oceanog.,* **53,** 335–367.

Heywood, K.J., 1996. Diel vertical migration of zooplankton in the Northeast Atlantic. *J. Plank. Res.,* **18,** 163–184.

Hilborn, R., and C.J. Walters, 1992. *Quantitative fisheries stock assessment: choice, dynamics, and uncertainty.* Chapman and Hall, New York.

Hill, A.E., 1991. A mechanism for horizontal zooplankton transport by vertical migration in tidal currents. *Mar. Biol.* **111,** 485–492.

Hill, A.E., 1994. Horizontal zooplankton dispersal by diel vertical migration in $S_2$ tidal currents on the northwest European continental shelf. *Continental Shelf Research,* **14,** 491–506.

Hinckley, S., A.J.Hermann and B.A. Megrey. 1996. Development of a spatially-explicit, individual-based model of marine fish early life history. *Mar. Ecol. Progr. Ser.,* **139,** 47–68.

Hinckley, S., A.J. Hermann, K.L. Meir, and B. A. Megrey, 2001. The importance of spawning location and timing to successful transport to nursery areas: a simulation modeling study of Gulf of Alaska walleye pollock. *ICES J. Mar. Sci.,* **58,** 1042–1052.

Hind, A., W.S.C. Gurney, M. Heath, and A. D. Bryant, 2000. Overwintering strategies in *Calanus finmarchicus. Mar. Ecol. Progr. Ser.,* **193,** 95–107.

Hinrichsen, H.-H., C. Mollmann, M. Voss, F.W. Koster and G. Kornilovs, 2002. Biophysical modeling of larval Baltic cod (*Gadus morhua*) growth and survival. *Canadian Journal of Fisheries and Aquatic Sciences,* **59,** 12, 1858–1873.

Hirche, H. J., 1996. Diapause in the marine copepod, *Calanus finmarchicus* – a review, *Ophelia,* **44,** 129–143.

Hofmann, E.E. and J.W. Ambler, 1988. Plankton dynamics on the outher southeastern U.S. continental shelf. Part II: a time-dependent biological model. *J. Mar. Res.* **46,** 883–917.

Hofmann, E.E. and C.M. Lascara, 1998. Overview of interdisciplinary modeling for marine ecosystems. In *The Sea. Vol. 10,* A.R. Robinson and K.H. Brink, eds. Wiley, New York, pp. 507–540.

Hofmann, E.E. and M.A.M. Friedrichs, 2002. Predictive modeling for marine ecosystems. In *The Sea,* Vol. 12, A.R. Robinson and K.H. Brink, eds.Wiley, New York, pp. 537–565.

Holland, G. and W. Nowlin, 2001. Principles of the Global Ocean Observing System (GOOS) capacity building. *GOOS Report,* **69,** 1–10.

Houde, E.D., 1987. Fish early life dynamics and recruitment variability. *American Fisheries Society Symposium,* **2,** 17–29.

Huggett, J., P. Fréon, C. Mullon and P. Penven, 2003. Modeling the transport success of anchovy (*Engraulis encrasicolus*) eggs and larvae in the southern Benguela: the effect of spatio-temporal spawning patterns. *Mar. Ecol. Progr. Ser.* **250,** 247–262.

Hurtt G.C. and R.A. Armstrong, 1999. A pelagic ecosystem model calibrated with BATS and OWSI data. *Deep-Sea Res., I,* **46,** 27–61.

Hutchins, D.A. and K.W. Bruland, 1998. Iron-limited diatom growth and Si : N uptake ratios in a coastal upwelling regime. *Nature,* **393,** 561–564.

Incze, S. L. and C. E. Naimie, 2000. Modeling the transport of lobster (Homarus americanus) larvae and postlarvae in the Gulf of Maine. *Fish.Oceanogr.,* **9,** 99–113.

Incze, L.S., D. Hebert, N. Wolff, et al., 2001. Changes in copepod distributions associated with increased turbulence from wind stress. *Mar Ecol-Prog Ser.,* **213,** 229–240.

Isaji, T. and M. L. Spaulding, 1984. A model of the tidally induced residual circulation in the Gulf of Maine and Georges Bank. *J. Phys. Oceanogr.,* **14,** 1119–1126.

Ishizaka, J., 1990. Coupling of coastal zone color scanner data to a physical-biological model of the Southeastern U.S. Continental Shelf ecosystem 3. Nutrient and phytoplankton fluxes and CZCS data assimilation. *J. Geophys. Res.,* **95,** 20201–20212.

Jonasdottir, S.H., H.G. Gudfinnsson, A. Gislason and O.S. Astthorsson, 2002. Diet composition and quality for Calanus finmarchicus egg production and hatching success off south-west Iceland. *Mar. Biol.,* **140,** 1195–1206.

Judson, O.P., 1994. The rise of the individual-based model in ecology. *Trends in Ecology and Evolution,* **9,** 9–14.

Kaartvedt S., 1996. Habitat preference during overwintering and timing of seasonal vertical migration of *Calanus finmarchicus. Ophelia,* **44,** 145–156.

Kane, J., 1984. The feeding habits of co-occurring cod and haddock larvae from Georges Bank. *Marine Ecology Progress Series,* **16,** 9–20.

Kendall, A.W. jr., J.D. Schumacher, and K. Suam, 1996a. Walleye pollock recruitment in Shelikof strait: applied fisheries oceanography. *Fish. Oceanogr.,* **5,** 4–18.

Kendall, A.W. Jr, R.I. Perry and S. Kim (eds)., 1996b. Fisheries oceanography of walleye pollock in Shelikof Strait, Alaska. *Fish. Oceanography,* **5,** 1–203.

Kishi, M.J., H. Motonu, M. Kashiwai and A. Tsuda, 2001. An ecological-physical coupled model with ontogenetic vertical migration of zooplankton in the Northwestern Pacific. *J. Oceanogr.,* **57,** 499–507.

Kleppel, G.S., C.A. Burkart, and L. Houchin, 1998. Nutrition and the regulation of egg production in the calanoid copepod *Acartia tonsa. Limnol. Oceanogr.,* **43,** 1000–1007.

Lavoie, D., Y. Simard, and F.J. Saucier, 2000. Aggregation and dispersion of krill at channel heads and shelf edges: the dynamics in the Saguenay – St. Lawrence Mrine Park. *Can. J. Fish. Aquat. Sci.,* **57,** 1853–1869.

Leggett, W.C. and E. DeBlois. 1994. Recruitment in marine fishes: Is it regulated by starvation and predation in the egg and larval stages? *Neth. J. Sea Res.,* **32,** 119–134.

Lenz, J., 2000. Introduction. In *ICES Zooplankton Methodology Manual,* R.P. Harris, P. Wiebe, J. Lenz, M. Huntley, H.R. Skjoldal, eds., Academic Press, London, pp. 1–32.

Lewis, C.V.W., C. S. Davis and G. Gawarkiewicz, 1994. Wind forced biological-physical interactions on an isolated offshore bank. *Deep Sea Research II,* **41,** 51–73.

Lewis, C.V.W., C. Chen and C. S. Davis, 2001. Effect of winter wind variability on plankton transport over Georges Bank. *Deep Sea Research,* **48,** 137–158.

Loder, J.W., G. Han, C.G. Hannah, D.A. Greenberg and P.C. Smith, 1997. Hydrography and baroclinic circulation in the Scotian Shelf region: winter vs. summer. *Can. J. Fish. Aquat. Sci.,* **54,** Suppl.1, 40–56.

Loder, J.W., B. Petrie and G. Gawarkiewicz, 1998. The coastal ocean off northeastern North America: A large scale view. In: *The Sea,* 11, Robinson, A,R, and K.H. Brink, eds., Wiley and Sons, NY, pp. 105–133.

Loder, J.W., J.A. Shore, C.G. Hannah and B.D. Petrie, 2001. Decadal-scale hydrographic and circulation variability in the Scotia-Maine region. *Deep-Sea Res. II,* **48,** 3–35.

Lorenzen, K., and K. Enberg, 2002. Density-dependent growth as a key mechanism in the regulation of fish populations: evidence from among-population comparisons. *Proc. R. Soc. Lond. B,* **269,** 49–54.

Lough, R.G., W.G. Smith, F.E. Werner, J.W. Loder, F.E. Page, C.G. Hannah, C.E. Naimie, R.I. Perry, M. Sinclair and D.R. Lynch, 1994. Influence of wind-driven advection on interannual variability in cod egg and larval distributions on Georges Bank: 1982 vs 1985. *ICES mar. Sci. Symp.,* **198,** 356–378.

Lough, R.G. and D.G. Mountain, 1996. Effect of small-scale turbulence on feeding rates of larval cod and haddock in stratified water on Georges Bank. *Deep-Sea Res. II,* **43,** 1745–1772.

Lynch, D.R. 1999. Coupled Physical/Biological Models for the Coastal Ocean, In *Naval Research Reviews.*

Lynch, D. R. and C.E. Naimie, 1993. The M2 tide and its residual on the outer banks of the Gulf of Maine. *J. Phys. Oceanogr.,* ***23,*** 2222–2253.

Lynch, D.R., J.T.C. Ip, C.E. Naimie and F.E. Werner. 1996. Comprehensive Coastal Circulation Model with Application to the Gulf of Maine. *Cont. Shelf Res.,* **16,** 875–906.

Lynch, D.R., W.C. Gentleman, D. J. McGillicuddy Jr. and C. S. Davis, 1998. Biological/physical simulations of *Calanus finmarchicus* population dynamics in the Gulf of Maine. *Mar. Ecol. Prog. Ser.,* **169,** 189–210.

Lynch, D. R, C.V.W. Lewis and F E. Werner, 2001. Can Georges Bank larval cod survive on a calanoid diet? *Deep Sea Research,* **48,** 609–630.

MacKenzie, B.R., T.J. Miller, S. Cyr and W.C. Leggett, 1994. Evidence for a dome-shaped relationship between turbulence and larval fish ingestion rates. *Limnol.Oceanogr.,* **39,** 1790–1799.

Marcus N.H., 1982. Photoperiodic and Temperature Regulation of Diapause in Labidocera aestiva (Copepoda: Calanoida). *Biol. Bull. Mar. Biol. Lab., Woods Hole.,* **162,** 45–52.

McAllister, C.D., 1971. Some aspects of nocturnal and continuous grazing by planktonic herbivores in relation to production studies. *Fish. Res. Bd. Can. Tech. Rep.,* **248,** 1–281.

McGillilcuddy, D.J. Jr, D.R. Lynch, A.M. Moore, W.C. Gentleman, C.S. Davis and C.J. Meise, 1998. An adjoint data assimilation approach to diagnosis of physical and biological controls on *Pseudocalanus* spp. in the Gulf of Maine-Georges Bank region. *Fish. Oceanogr.,* **7,** 205–218.

McKinnon, A.D. and S. Duggan, 2001. Summer egg production rates of paracalanoid copepods in subtropical waters adjacent to Australia's NorthWest Cape. *Hydrobiologia,* **453/454,** 121–132.

McLaren, I.A., 1978. Generation lengths of some temperate marine copepods: estimation, prediction, and implications. *J. Fish. Res. Board Can.,* **35**(10), 1330–1342.

McLaren, I.A. and C.J. Corkett, 1981. Temperature-dependent growth and production by a marine copepod. *Can. J. Fish. Aquat. Sci.,* **38**(1), 77–83.

McLaren, I.A., Tremblay, M.J., Corkett, C.J. and J.C. Roff, 1989. Copepod production on the Scotian Shelf based on life-history analyses and laboratory rearings. *Can. J. Fish. Aquat. Sci.,* **46**(4), 560–583.

McLaren, I.A., E. Head and D.D. Sameoto, 2001. Life cycles and seasonal distributions of *Calanus finmarchicus* on the central Scotian Shelf. *Can. J. Fish. Aquat. Sci.,* **58,** 4, 659–670.

Megrey, B.A., 1990. Population dynamics and management of walleye pollock stocks in the Gulf of Alaska, 1976–1986. *Fish. Res.,* **11,** 321–354.

Megrey, B.A., S.J. Bograd, W.C. Rugen, A.B. Hollowed, P.J. Stabeno, S.A. Macklin, J.D. Schumacher and W.J. Ingraham Jr., 1995. An exploratory analysis of associations between biotic and abiotic factors and year-class strength of Gulf of Alaska walleye pollock. In *Climate Change and Northern Fish Populations.* R.J. Beamish,ed., *Can. Spec. Publ. Fish. and Aquat. Sci.,* **121,** 227–243.

Megrey, B.A., A.B. Hollowed, S.R. Hare, S. Allen Macklin and P.J. Stabeno, 1996. Contributions of FOCI research to forecasts of year-class strength of walleye pollock in Shelikof Strait, Alaska. *Fish.Oceanogr.* **5 (Suppl.1),** 189–203.

Megrey, B.A., and S. Hinckley, 2001. Effect of turbulence on feeding of larval fishes: a sensitivity analysis using an individual-based model. *ICES J. Mar. Sci.,* **58,** 1015–1029.

Meise, C.J. and J. E. O'Reilly, 1996. Spatial and seasonal patterns in abundance and age composition of Calanus finmarchicus in the Gulf of Maine and on Georges Bank: 1977–1987. *Deep Sea Res. II,* **43,** 1473–1501.

Melle, W. and H.R. Skjoldal, 1998. Spawning and development of *Calanus* spp. In the Barents Sea. *Marine Ecology Progress Series,* **169,** 211–228.

Miller, C. B., Cowles, T. J., Wiebe, P. H., Copley, N. J. and H. Grigg, 1991. Phenology in *Calanus finmarchicus:* hypotheses about control mechanisms. *Mar. Ecol. Prog. Ser.,* **72,** 79–91.

Miller, C.B., D.R. Lynch, F. Carlotti, W. Gentlemen and C.V.W. Lewis, 1998. Coupling of an individual-based population dynamic model of *Calanus finmarchicus* to a circulation model for the Georges Bank region, *Fish. Oceanogr.,* **7,** 219–234

Minello, T. J., 1999. Nekton densities in shallow estuarine habitats of Texas and Louisiana and the identification of essential fish habitat. *American Fisheries Society Symposium,* **22,** 43–75.

Moloney, C.L. and J.G. Field. 1991. Thes size-based dynamics of plankton food webs. 1. A simulation model of carbon and nitrogen flows. *J. Plankton Res.,* **13,** 1003–1038.

Moore, J.K., S.C. Doney, J.A. Kleypas, et al., 2002. An intermediate complexity marine ecosystem model for the global domain. *Deep-Sea Res II,* **49,** 403–462.

Mullon, C., P. Cury and P. Penven, 2002. Evolutionary individual-based model for the the recruitment of anchovy (*Engraulis capensis*) in the southern Benguela. *Can. Jour. Fish. Aquatic Sci.,* **59,** 910–922.

Mullon, C., P. Freon, C. Parada, C. van der Lingen, and J Hugget, 2003. From particles to individuals: modelling the early stages of anchovy (*Engraulis capensis/encrasicolus)* in the southern Benguela. *Fish. Oceanogr.,* **12,** 396–406.

Myers, R.A., and N.G. Cadigan, 1993. Density-dependent juvenile mortality in marine demersal fish. *Can. J. Fish. Aquat. Sci.,* **50,** 1576–1590.

Naimie, C. E., 1996. Georges Bank residual circulation during weak and strong stratification periods: prognostic numerical model results. *J. Geophys. Res.,* **101,** 6469–6486.

Naimie, C. E., J. W. Loder and D. R. Lynch, 1994. Seasonal variation of the three-dimensional circulation on Georges Bank. *J. Geophys. Res.,* **99,** 15,967–15,989.

Needle, C. L., 2002. Recruitment models: diagnosis and prognosis. *Rev. Fish Biol. Fish.,* **11,** 95–111.

Ohman, M.D., G.C. Anderson and E.Ozturgut, 1983. A multivariate analysis of planktonic interactions in the eastern tropical North Pacific. *Deep-Sea Res.,* **29,** 1451–1469.

Ohman, M.D. and S.N. Wood, 1996. Mortality estimation for planktonic copepods: *Pseudocalanus newmani* in a temperate fjord. *Limnol. Oceanogr.,* **41,** 126–135.

Ohman, M.D. and H.J. Hirche, 2001. Density-dependent mortality in an oceanic copepod population. *Nature,* **412,** 638–641.

Ohman, M.D., J.A. Runge, E.G. Durbin, D.B. Field and B. Niehoff, 2002. On birth and death in the sea. *Hydrobiologia,* **480,** 55–68.

Ohman, M.D., K., Eiane, E.G. Durbin, J.A. Runge and H.J. Hirche, 2004. A comparative study of *Calanus finmarchicus* mortality patterns in five localities in the North Atlantic. ICES J. Mar. Sci. **61,** 687–697.

Paffenhofer, G-A., and S. Knowles, 1980. Omnivorousness in marine planktonic copepods. *J. Plankton Res.* **2,** 355–365.

Paffenhofer, G.A. and K.D. Lewis, 1990. Perceptive performance and feeding behavior of calanoid copepods. *J.Plankt.Res.,* **12,** 933–946.

Page, F.H., M. Sinclair, C.E. Naimie, J.W. Loder, R.J. Losier, P. Berrien and R.G. Lough, 1999. Cod and haddock spawning on Georges Bank in relation to water residence times. *Fish. Oceanogr.,* **8,** 212–226.

Parada, C., D.D. van der Lingen, C. Mullon and P. Penven, 2003. Modeling the effect of egg buoyancy on the transport of anchovy (*Engraulis capensis*) eggs from spawning to nursery grounds in the southern Benguela: an IBM approach. *Fish. Oceanogr.,* 12, 170–184.

Pedersen, O.P., K.S. Tande, D. Slagstad, 2001. A model study of demography and spatial distribution of *Calanus finmarchicus* at the Norwegian coast. *Deep-Sea Res. II. Top. Stud. Oceanogr.,* **48,** 567–587

Penven, P., C. Roy, G. Brundrit et al., 2001. A regional hydrodynamic model of upwelling in the Southern Benguela. *S. Afr. J. Sci.,* **97,** 472–475.

Peterson, W.T. and D.C. Bellantoni, 1987. Relationships between water-column stratification, phytoplankton cell size and copepod fecundity in Long Island Sound and off central Chile. In *The Benguela and Comparable Ecosystems,* A.I.L. Payne, J.A. Gulland and K.H. Brink, eds., *S.Afr.J.mar.Sci.,* **5,** 411–421.

Pillar, S.C., C.L. Moloney, A.I.L. Payne and F.A.Shillington (eds), 1998. Benguela Dynamics. Impacts of variability on shelf-sea environments and their living resources. *S. Afr. J. mar. Sci.* **19, 1–225.**

Plaganyi, E.E., L. Hutchings, and J. G. Field, 2000. Anchovy foraging: simulating spatial and temporal match/mismatches with zooplankton .. *Can. J .. Fish. Aquatic Sci.,* **57,** 2044–2053.

Plaganyi,E.E., L. Hutchings, J.G. Field and H. M. Verheye, 1999. A model of copepod population dynamics in the southern Benguela upwelling region. *J.Plankton Res.,***21,** 1691–1724.

Plourde, S. and J.A. Runge, 1993. Reproduction of the planktonic copepod *Calanus finmarchicus* in the lower St. Lawrence Estuary: Relation to the cycle of phytoplankton production and evidence for a *Calanus* pump. *Mar. Ecol. Prog. Ser.,* **102,** 217–227.

Porter, S.M, 2001. Effects of size and light on respiration and activity of walleye pollock (*Theragra chalcogramma*) larvae. *J. Exp.Mar.Biol.Ecol.,* **256,** 253–265.

Price, H.J., 1989. Swimming behavior of krill in response to algal patches: A mesocosm study. *Limnol. Oceanogr.,* **34,** 649–659.

Quinlan, J.A., B.O. Blanton, T.J. Miller and F.E. Werner, 1999. From spawning grounds to the estuary: using linked individual-based and hydrodynamic models to interpret patters and processes in the oceanic phase of Atlantic menhaden *Brevoortia tyrannus* life history. *Fish.Oceanogr.,* **8,** 224–246.

Railsback, S.F. and B.C. Harvey, 2002. Analysis of habitat-selection rules using an individual-based model. *Ecology,* **83,** 1817–1830.

Rice, J.A., L.B. Crowder and M.E. Holey, 1987. Exploration of mechanisms regulating larval survival in Lake Michigan bloater: a recruitment analysis based on characteristics of individual larvae. *Transactions of the American Fisheries Society,* **116,** 703–718.

Rice, J.A., J.A. Quinlan, S.W. Nixon, W.F. Hettler, S.M. Warlen and P.M. Stegmann, 1999. Spawning and transport dynamics of Atlantic menhaden: inferences from characteristics of immigrating larvae and predictions of a hydrodynamic model. *Fisheries Oceanography,* **8,** 93–110.

Richardson, A.J. and H.M. Verheye, 1999. Growth rates of copepods in the southern Benguela upwelling system: The interplay between body size and food. *Limnol.Oceanogr.,* **44,** 382–392.

Riley, G.A., 1963. Theory of food-chain relations in the ocean. In *The Sea,* Vol. 2, M.N. Hill, ed., John Wiley and Sons, New York, pp. 438–463.

Rines, J.E.B., P.L. Donaghay, M.M. Dekshenieks, J.M. Sullivan and M.S. Twardowski, 2002. Thin Layers and Camouflage: Hidden Pseudo-nitzschia populations in a fjord in the San Juan Islands, Washington, USA. *Marine Ecology Progress Series,* **225,** 123–137

Robinson, A.R., P.F.J. Lermusiaux and N. Quincy Sloan III. 1998. Data assimilation. In Vol. 10 of *The Sea.* K.H. Brink and A.R. Robinson, eds.Wiley, New York, pp. 541–565.

Robinson, A.R. and P.F.J. Lermusiaux, 2002. Data assimilation for modeling and predicting coupled physical-biological interactions in the sea. In Vol. 12 of *The Sea.* Robinson A.R., J.R. McCarthy and B.J. Rothschild, eds.Wiley, New York, pp. 475–536.

Rose, K.A. and J.K. Summers, 1992. Relationships among long-term fish abundances, hydrographic variables, and gross pollution indicators in northeastern US estuaries. *Fish Oceanogr.,* **1,** 281–293.

Rose, K.A. and J.H. Cowan, 2003. Data, models, and decisions in US marine fisheries management: lessons for ecologists. *Ann. Rev. Ecol. Evol. Syst.,* **34,** 127–151.

Rose, K.A., 2000. Why are quantitative relationships between environmental quality and fish populations so elusive? *Ecological Applications,* **10,** 367–385.

Rose, K.A., J.A. Tyler, R.C. Chambers, G. MacPhee and D.J. Danila, 1996. Simulating winter flounder population dynamics using coupled individual-based young-of-the-year and age-structured adult models. *Can. J. Fish.Aquatic Sci.,* **53,** 1071–1091.

Rose, K.A., J.H. Cowan, M.E. Clark, E.D. Houde and S-B Wang, 1999. Individual-based modeling of bay anchovy population dynamics in the mesohaline region of Chesapeake Bay. *Mar. Ecol. Prog. Ser.,* **185,** 113–132.

Rose, K.A., C.A. Murphy, S.L. Diamond, L.A. Fuiman and P. Thomas, 2003. Using nested models and laboratory data for predicting population effects of contaminants on fish: a step towards a bottom-up approach for establishing causality in field studies. *Human and Ecological Risk Assessment,* **9,** 231–247.

Rothschild. B., 1986. *Dynamics of Marine Fish Populations.* Harvard Univ. Press, Cambridge.

Roy, C., S. Weeks, M. Rouault, G. Nelson, R. Barlow and C. van der Lingen, 2001. Extreme oceanographic events recorded in the Southern Benguela during the 1999–2000 summer season. *S. African J. Sci.,* **97,** 465–471.

Rudstam, L.G., 1988. Exploring the dynamics of herring consumption in the Baltic: applications of an energetic model of fish growth. *Kieler Meeresforschung Sonderheft,* **6,** 312–322.

Runge, J.A., 1985. Relationship of egg production of Calanus pacificus to seasonal changes in phytoplankton availability in Puget Sound, Washington. *Limnol. Oceanogr.,* **30,** 382–396.

Runge, J.A., 1988. Should we expect a relationship between primary production and fisheries? The role of copepod dynamics as a filter of trophic variability. *Hydrobiologia,* **167/168,** 67–71.

Runge, J. A. and S. Plourde, 1996. Fecundity characteristics of *Calanus finmarchicus* in coastal waters of Eastern Canada. *Ophelia,* **44,** 171–187

Rugen, W.C., 1990. Spatial and temporal distribution of larval fish in the western Gulf of Alaska, with emphasis on the period of peak abundance of walleye pollock (*Theragra chalcogramma*) larvae. *NWAFC Processed Rep. 90–01,* Alaska Fish. Sci. Cent., 162 pp.

Sale, P.F., 1990. Recruitment of marine species: is the bandwagon rolling in the right direction? *TREE,* **5,** 25–27.

Schofield, O., T. Bergmann, W.P. Bissett, F. Grassle, D. Haidvogel, J. Kohut, M. Moline and S. Glenn, 2002. The Long-Term Ecosystem Observatory: An Integrated Coastal Observatory. *Journal of Oceanic Engineering,* **27,** 146–154.

Shannon, L.V. and M.J. O'Toole, 2003. Sustainability of the Benguela: ex Africa semper aliquid novi. In: K. Sherman and G. Hempel, *Large Marine Ecosystems of the World-Trends in Exploitation, Protection and Research.*

Sheng, J., D. Wright, R. Greatbatch and D. Dietrich, 1998. CANDIE: A new version of the DieCAST Ocean Circulation Model,. *J. Atm. and Oceanic Tech.,* **15,** 1414–1432.

Sheng, J., K.R. Thompson and M. Dowd, 2002. Circulation in the eastern Canadian shelf seas with special emphasis on the Gulf of St. Lawrence and Scotian Shelf. *Journal of Geophysical Research,* (Under revision).

Shepard, J. G., J. G. Pope, and R. D. Cousens, 1984. Variations in fish stocks and hypotheses concerning their links with climate. *Rapports et Proces-Verbaux des Reunions, Conseil International pour l'Exploration de la Mer,* **185,** 255–267.

Sinclair, M., 1988. *Marine Populations: An Essay on Population Regulation and Speciation.* Seattle, WA., Washington Sea Grant Program, 252 pp.

Sissenwine, M.P., 1984. Why do fish populations vary? In *Exploitation of Marine Communities,* R.M. May, ed., Springer-Verlag, New York, pp. 59–94.

Skreslet, S., 1997. A conceptual model of the trophodynamical response to river discharge in a large marine ecosystem. *Journal of Marine Systems,* **12,** 187–198.

Slagstad, D. and K.S. Tande, 1990. Growth and production dynamics of *Calanus glacialis* in an arctic pelagic food web. *Mar. Ecol. Prog. Ser.,* **63,** 189–199.

Slagstad, D. and K.S. Tande, 1996. The importance of seasonal vertical migration in across shelf transport of *Calanus finmarchicus. Ophelia,* **44,** 189–205.

Spitz, Y.H., J.R. Moisan and M.R. Abbott, 2001. Configuring an ecosystem model using data from the Bermuda Atlantic Time Series (BATS). *Deep-Sea Rest II,* **48,** 1733–1768.

Stabeno, P.J., R.K. Reed and J.D. Schumacher, 1995a. The Alaska Coastal Current: Continuity of transport and forcing. *J. Geophys. Res.,* **100,** 2477–2485.

Stabeno, P.J., J.D. Schumacher, K.M. Bailey, R.D. Brodeur and E.D. Cocklet. 1995b. Observed patches of walleye pollock eggs and larvae in Shelikof Strait, Alaska: Their characteristics, formation and persistence. *Fisheries Oceanography,* **5,** 81–91.

Steele, J.H. and B.W. Frost, 1977. The structure of plankton communities. *Phil.Trans. Royal Soc. London, B,* **280,** 485–534.

Steele, J.H. and E. Henderson, 1981. A simple plankton model. *Am. Nat.*, **117,** 676–691.

Steele, J.H. and E. Henderson, 1992. The role of predation in plankton models. *J. Plankton Res.*, **14,** 157–172.

Stegmann, P.M., J.A. Quinlan, F.E. Werner and B.O. Blanton, 1999. Projected transport pathways of Atlantic menhaden larvae as determined from satellite imagery and model simulations in the South Atlantic Bight. *Fisheries Oceanography,* **8,** 111–123.

Song, Y. and D. Haidvogel, 1994. A semi-implicit ocean circulation model using a generalized topography-following coordinate system. *J. Com. Physics,* **115,** 228–244.

Tande, K. S., 1988. An evaluation of factors affecting vertical distribution among recruits of *Calanus finmarchicus* in three adjacent high-latitude localities, *Hydrobiologia,* **167–168,** 115–126.

Tande, K.S. and C.B. Miller, 2001. Population Dynamics of Calanus in the North Atlantic: Results from the Trans-Atlantic Study of Calanus finmarchicus. *ICES J. Mar. Sci.,* **57,** 1527–1527.

Tiselius, P., 1998. Short term feeding responses to starvation in three species of small calanoid copepods. *Mar. Ecol. Prog.Ser.,* **168,** 119–126.

Tiselius, P., 1992. Behavior of *Acartia tonsa* in parchy food environments. *Limnology and Oceanography,* **37,** 1640–1651.

Townsend, D.W., A. C. Thomas, L. M. Mayer, M.A. Thomas and J.A. Quinlan, In press. Oceanography of the Northwest Atlantic Continental Shelf. In *The Sea. Vol. 14,* A.R. Robinson and K.H. Brink, eds. Wiley, New York.

Tyler, J.A., and K.A. Rose, 1994. Individual variability and spatial heterogeneity in fish population models. *Reviews in Fish Biology and Fisheries,* **4,** 91–123.

Tyler, J.A., and K.A. Rose, 1997. Effects of individual habitat selection in a heterogenous environment on fish cohort survivorship: a modeling analysis. *Journal of Animal Ecology,* **66,** 122–136.

Vidal, J., 1980. Physiolecology of zooplankton. I. Effects of phytoplankton concentration, temperature and body size on the growth rate of *Calanus pacificus* and *Pseudocalanus* sp. *Mar. Biol.,* **56,** 111–134.

Vlymen, W., 1977. A mathematical model of the relationship between larval anchovy (*Engraulis mordax*) growth, prey microdistribution, and larval behavior. *Env. Biol. Fish,* **2,** 211–233.

Voronina, N.M., V.V. Menshutkin and V.B. Tseytlin, 1979. Mathematical Simulation of the Space-Time Distribution and Age Structure of an Antarctic Copepod Population. *Oceanol.,* **19,** 76–81.

Walters, C.J. and D. Ludwig, 1981. Effects of measurement errors on the assessment of stock-recruitment relationships. *Can. J.Fish. Aquatic Sci.,* **38,** 704–710.

Werner, F.E., F.H. Page, D.R. Lynch, J.W. Loder, R.G. Lough, R.I. Perry, D.A. Greenberg and M.M. Sinclair, 1993. Influence of mean 3-D advection and simple behavior on the distribution of cod and haddock early life stages on Georges Bank. *Fish. Oceanogr.,* **2,** 43–64.

Werner, F.E., R.I. Perry, R.G. Lough and C.E. Naimie, 1996. Trophodynamic and Advective Influences on Georges Bank Larval Cod and Haddock. *Deep Sea Research II,* **43,** 1793–1822.

Werner, F.E. and J.A. Quinlan, 2002. Fluctuations in marine fish populations: physical processes and numerical modeling. *ICES Marine Science Symposia,* **215,** 264–278.

Werner, F.E., J.A. Quinlan, B.O. Blanton and R.A. Luettich, Jr., 1997. The Role of Hydrodynamics in Explaining Variability in Fish Populations. *Journal of Sea Research,* **37,** 195–212.

Werner, F.E., J.A. Quinlan, R.G. Lough and D.R. Lynch, 2001a. Spatially-explicit individual based modeling of marine populations: a review of the advances in the 1990s. *Sarsia,* **86,** 411–421.

Werner, F.E., B.R. MacKenzie, R.I. Perry, R.G. Lough, C.E. Naimie, B.O. Blanton and J.A. Quinlan. 2001b. Larval trophodynamics, turbulence, and drift on Georges Bank: a sensitivity analysis of cod and haddock. *Scientia Marina,* **65,** 99–115.

Wiebe, P., R. Beardsley, D. Mountain and A. Bucklin, 2002. U.S. GLOBEC Northwest Atlantic/Georges Bank Program. *Oceanography,* **15,** 13–29.

Winemiller, K.O. and K.A. Rose, 1992. Patterns of life-history diversification in North American fishes: implications for population regulation. *Can. J.Fish. Aquatic Sci.,* **49,** 2196–2218.

Winemiller, K.O., and K.A. Rose, 1993. Why do most fish produce many tiny offspring? *Am. Nat.,* **142,** 585–603.

Wishner, K., E. Durbin, A. Durbin, M. Macaulay, H. Winn and R. Kenney, 1988. Copepod patches and right whales in the Great South Channel off New England. *Bull. Mar. Sci.,* **43,** 825–844.

Wishner, K. F., and S. K. Allison, 1986. The distribution and abundance of copepods in relation to the physical structure of the Gulf Stream. *Deep-Sea Res.,* **33,** 705–731.

Wood, S.N., 1994. Obtaining birth and mortality patterns from structured population trajectories. *Ecol. Monogr.,* **64,** 23–44.

Wroblewski, J.S.,1982. Interaction of currents and vertical migration in maintaining *Calanus marshallae* in the Oregon upwelling zone – a simulation, *Deep-Sea Res.,* **29,** 665–686.

Yamazaki, H. and K.D. Squires, 1996. Comparison of oceanic turbulence and copepod swimming. *Mar. Ecol. Prog. Ser.,* **144,** 299–301.

Zakardjian, B. A., J.A. Runge, S. Plourde and Y. Gratton, 1999. A biophysical model of the interaction between vertical migration of crustacean zooplankton and circulation in the Lower St. Lawrence Estuary. *Can. J. Fish. Aquat. Sci.,* **56,** 2420–2432.

Zakardjian, B. A., J. Sheng, J.A. Runge, K.R. Thompson, Y. Gratton, I.A. McLaren and S. Plourde. 2003. Effects of temperature and circulation on the population dynamics of *Calanus finmarchicus* in the Gulf of St. Lawrence and Scotian Shelf: Study with a coupled, three-dimensional hydrodynamic, stage-based life-history model. J. Geophys. Res., **108,**1–22.

Zhou, M., 1998. An objective interpolation method for spatio-temporal distribution of marine plankton. *Mar. Ecol. Prog. Ser.,* **174,** 197–206.

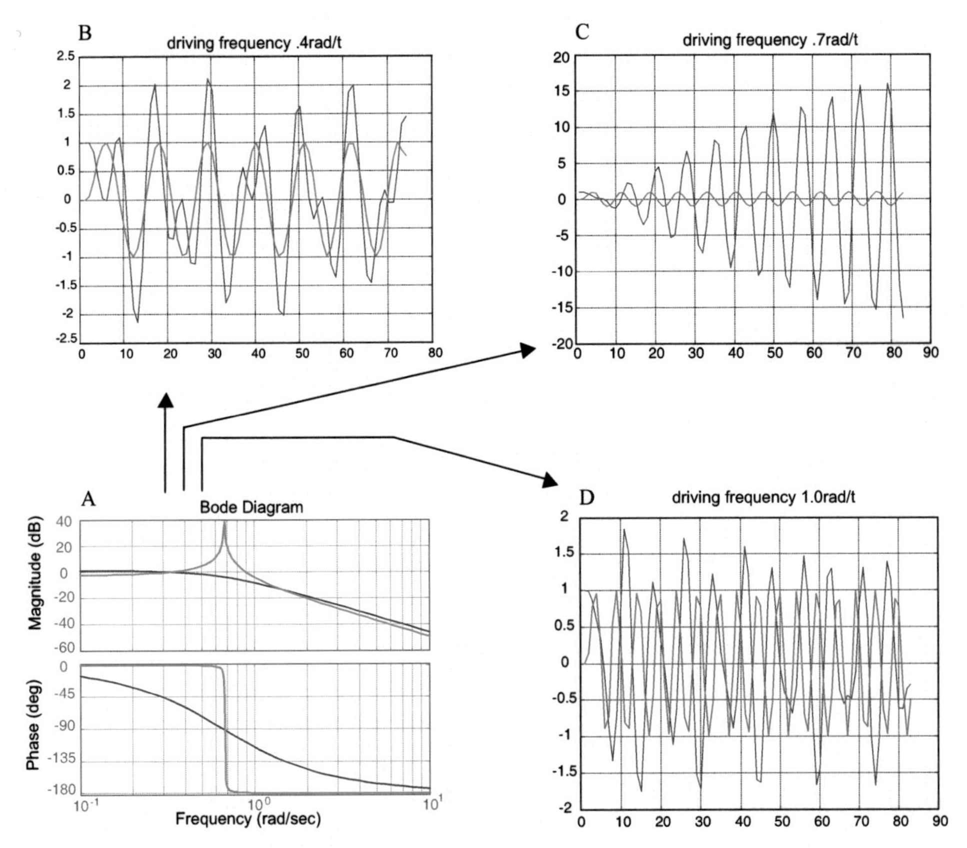


Figure 3.5. The forced response of equations 14 and 15 corresponding to the parameter set in Figure 3.4B. Panel A is a Bode diagram that shows that the parameter set has a resonant response at .7 rad/T. Panel B shows the response to sub-resonant frequency .4 rad/T, Panel D to supra-resonant frequency 1 rad/T, and C to the resonant frequency. Note that the varying vertical scales and that the amplitude of the resonant response is much greater than the non-resonant response.

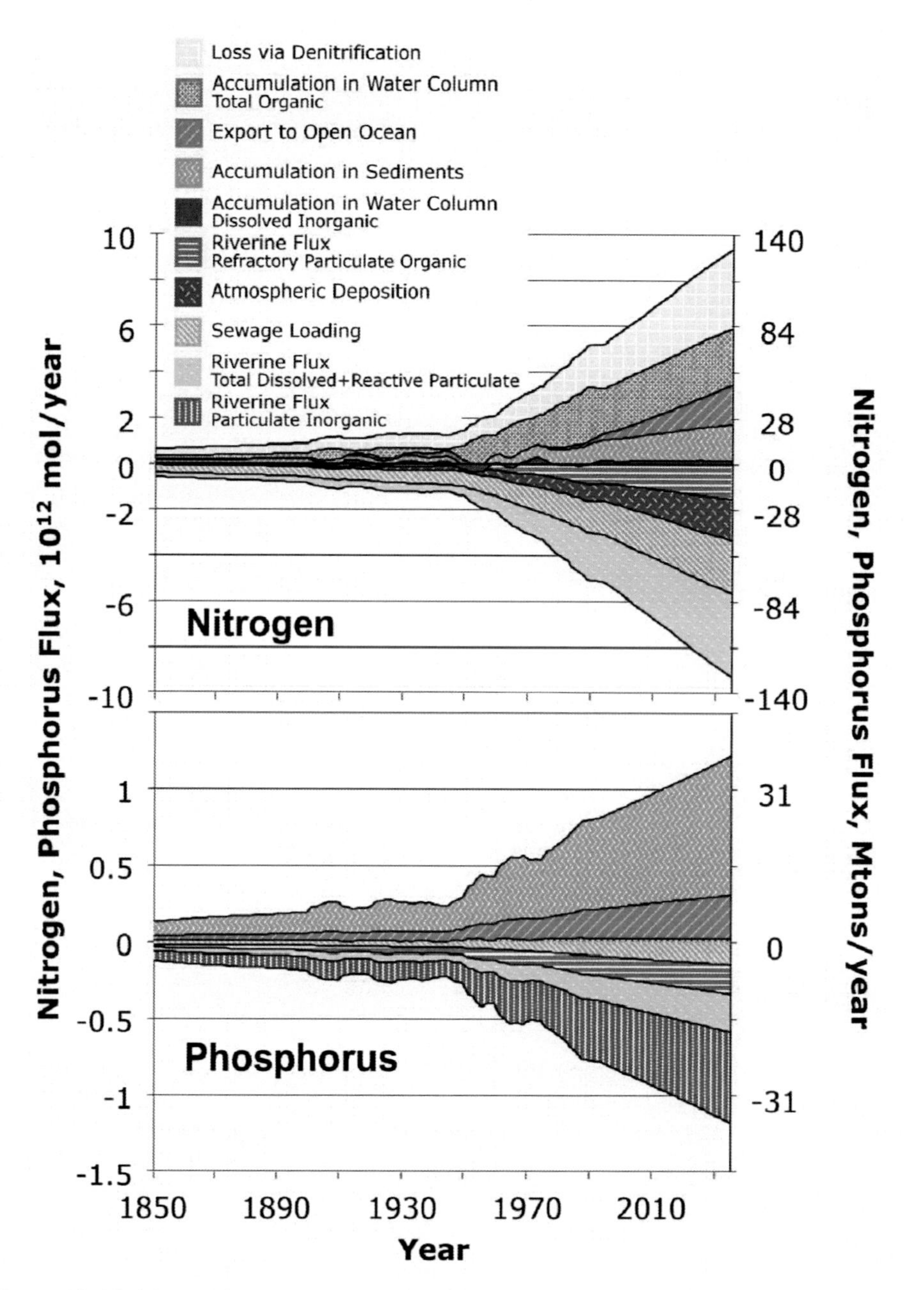


Figure 7.5. Model-calculated partitioning of the human-induced perturbation fluxes in the global coastal margin of (A) nitrogen and (B) phosphorus for the period since 1850 to present (2000) and projected to 2035 under the Business-as-usual scenario, in units of 1012 moles/year and Mtons/year (1 Mton = $10^6$ ton = $10^{12}$ g). The anthropogenic sources are plotted on the (–) side and the resulting accumulations and enhanced export fluxes are plotted on the (+) side. Model results from Mackenzie et al. (2002).

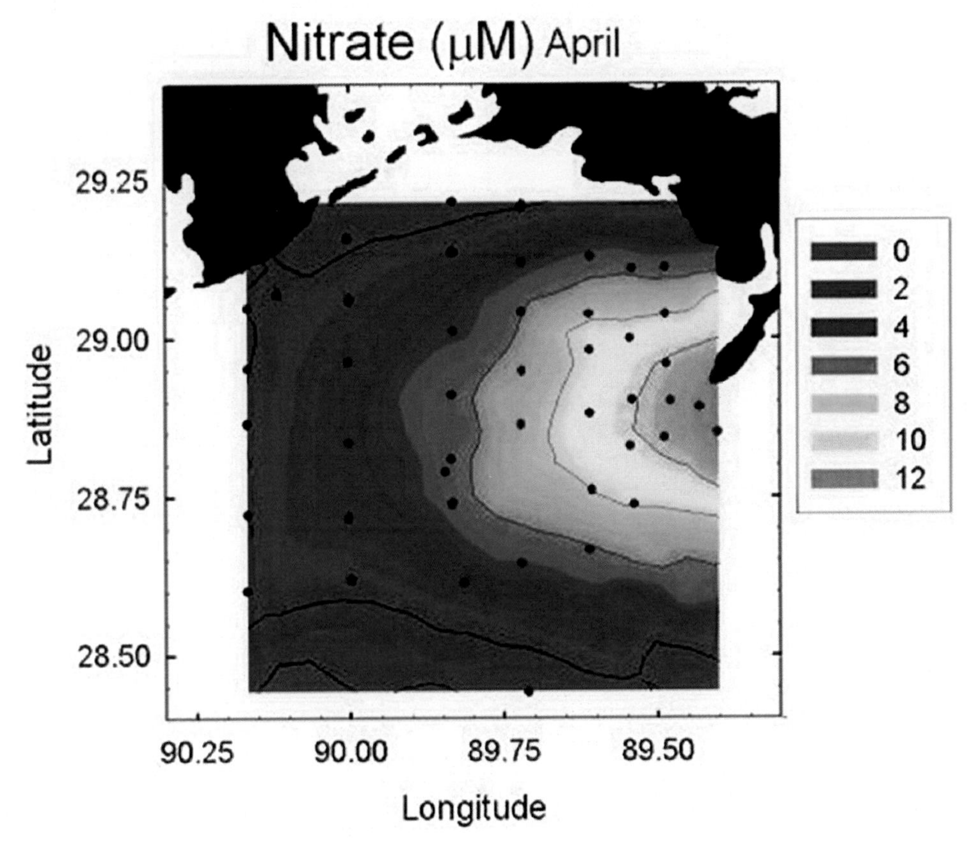


Figure 8.2. Surface nitrate concentrations off the Mississippi River (From Rabalais et al., 2002).

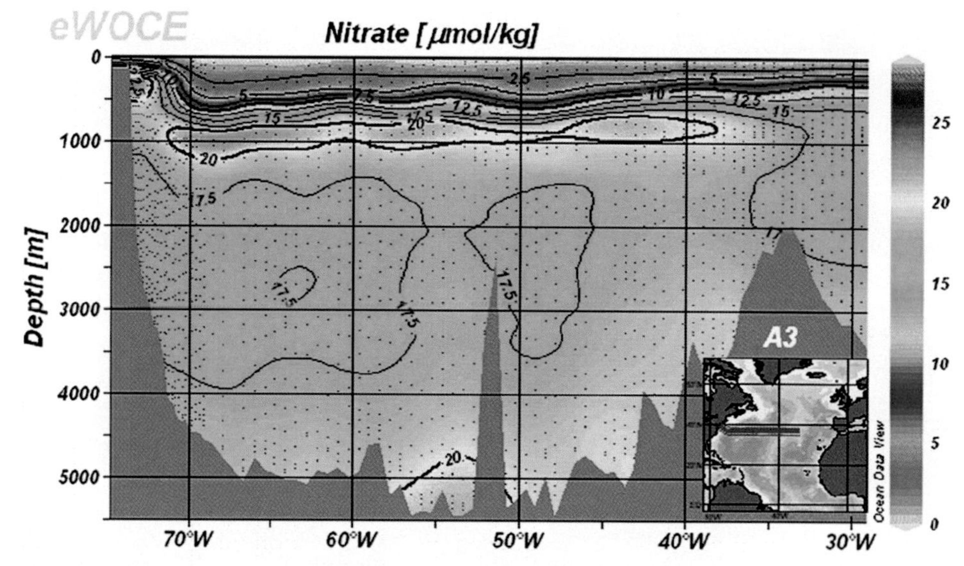


Figure 8.3. Nitrate section across Gulf Stream. Data from WOCE Section A03.

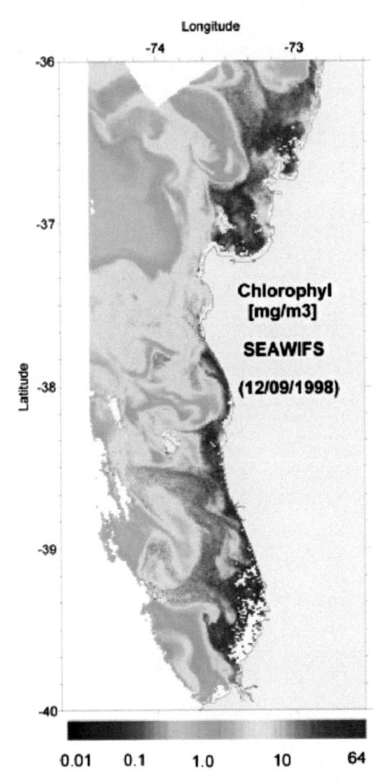


Figure 8.10. Surface chlorophyll SeaWiFS image over the Chilean coast, for 9 December 1998 (From Atkinson et al., 2002).

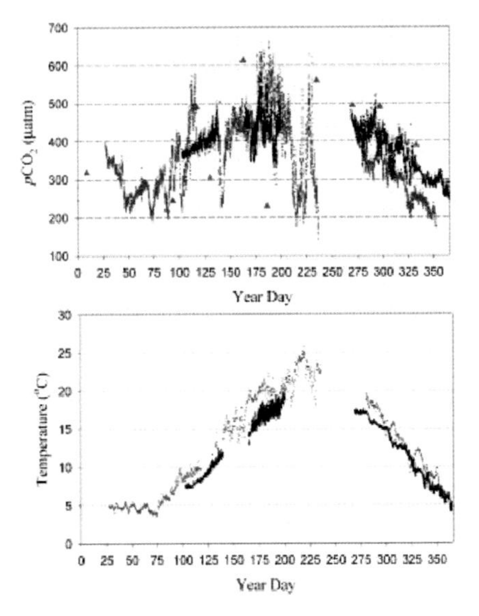


Figure 9.7. Annual cycles of surface $pCO_2$ and temperature in the Middle Atlantic Bight, USA. Arrows denote large short-term changes due to local upwelling. Figure courtesy deGrandpre et al. (2002).

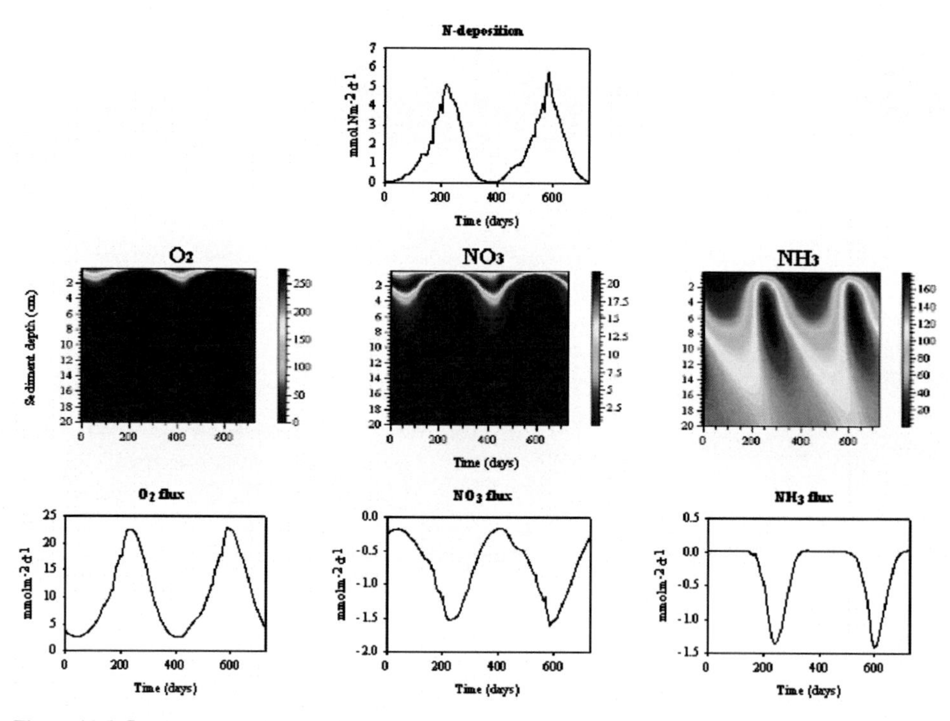


Figure 11.5. Spatio-temporal plots of oxygen, nitrate and ammonium in the sediment and time-series of nitrogen delivery to sediments and sediment-water exchange fluxes of oxygen, nitrate and ammonium.

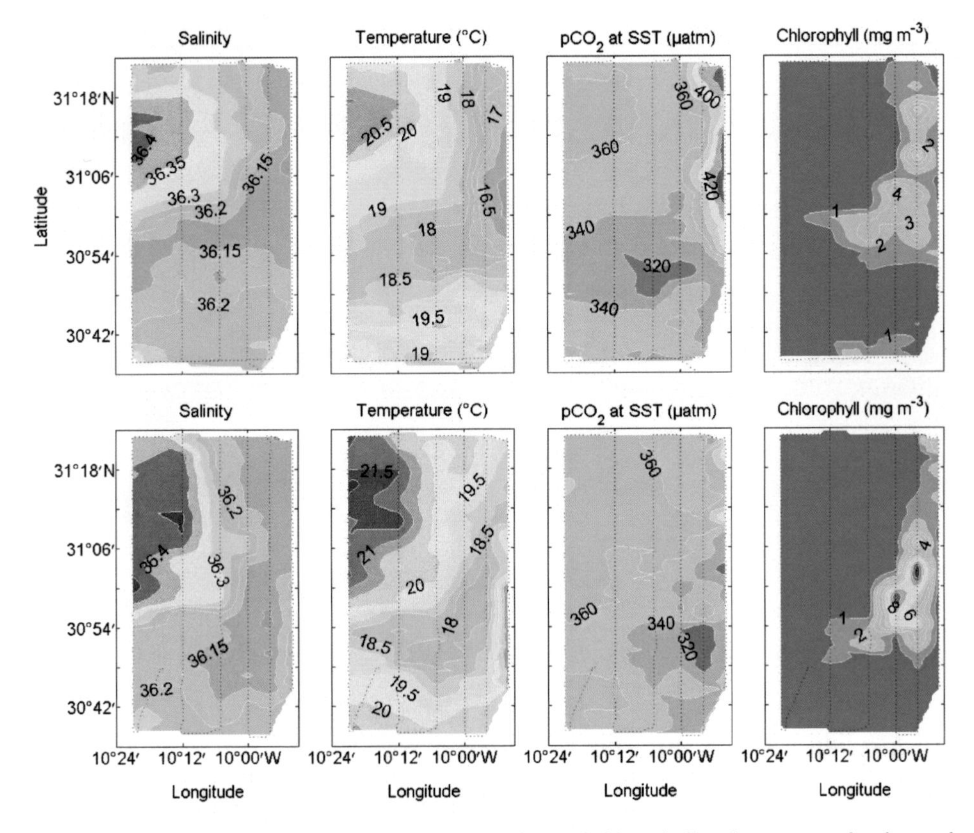


Figure 12.7b. Contour plots of salinity, temperature, $pCO_2$ and chlorophyll-a along two tracks observed on 8 September 1999 (upper four plots) and on 12 September 1999 (lower four plots) in the Morocco upwelling region. The upwelled waters are characterised by salinities lower than 36.15 ppt and low temperatures. During the first series of measurements, the highest $pCO_2$ values (up to 460 matm) were observed along the most eastern transect at about 3 miles of the coast. Despite the rise in temperature of the upwelled water during its offshore progression, $pCO_2$ decreased below the saturation level, which may be attributed to biological production. The highest values of chlorophyll-a (about 6 mg $m^{-3}$) were not found alongshore, but between 6 and 10 miles from the coast. During the second series of measurements, the temperature of upwelled water was higher, indicating a decline in the upwelling intensity. The $pCO_2$ values were low and chlorophyll-a was very high (up to 11 mg $m^{-3}$) in the coldest part of the plume. These results confirm the observations of MacIsaac et al. (1985), showing that a period of adaptation to high irradiance is necessary for the development of the phytoplanktonic cells.

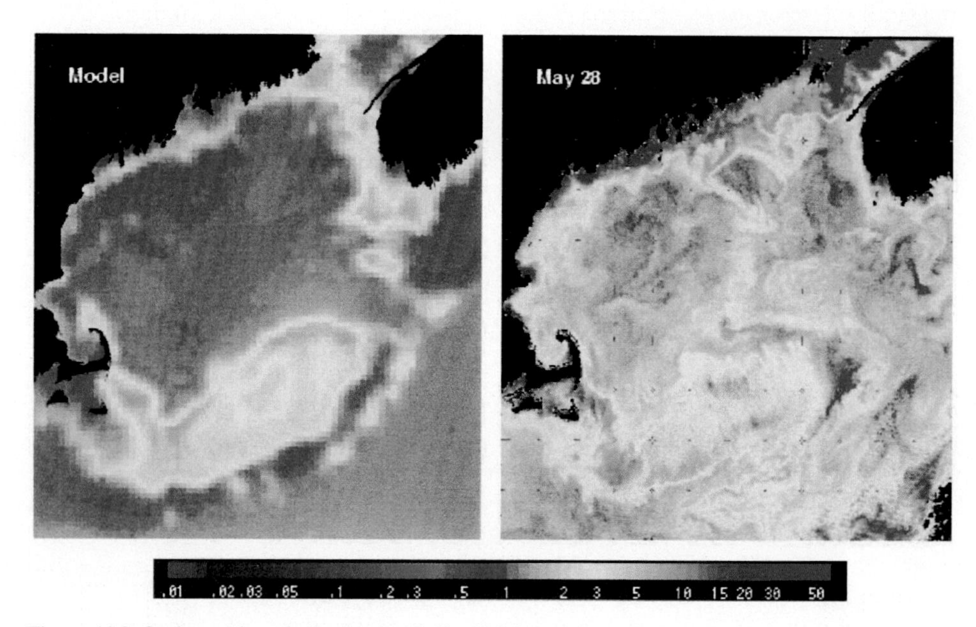


Figure 13.2. Surface chlorophyll a in the Gulf of Maine and on Georges Bank from a physical-NPZ model (Franks and Chen, 2001), and from satellite remote sensing. Note that this particular model did not include the nearshore dynamics, such as rivers and nutrient inputs, that lead to high phytoplankton biomass in these areas.

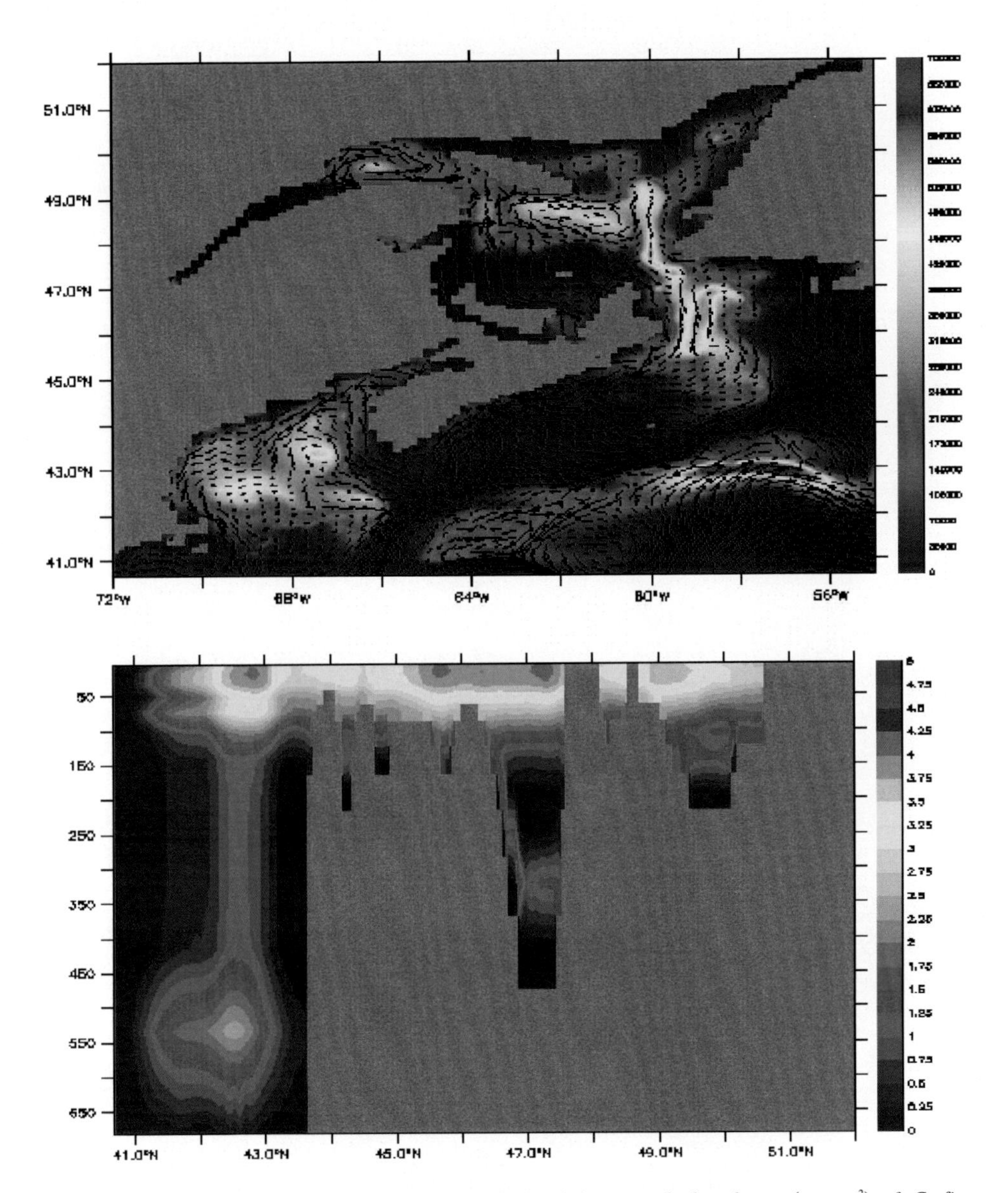


Figure 13.10. A) Surface currents and simulated, depth-integrated abundance (no. $m^{-2}$) of *C. finmarchicus* (all stages) in early June in the Gulf of St. Lawrence, Scotian Shelf and Gulf of Maine system based on a *Calanus* life history model coupled to the 3-D mean climatological circulation calculated from the CANDIE finite difference model (redrawn from Zakardjian et al., 2003). B) Vertical section at 59°30 W of *Calanus* total abundance (log[no. m-3]) from (A) showing the stage-specific vertical distribution in the simulated model results. The deepest abundance mode (300 m in the Gulf and 550 in the slope water) represents diapausing CV and males.

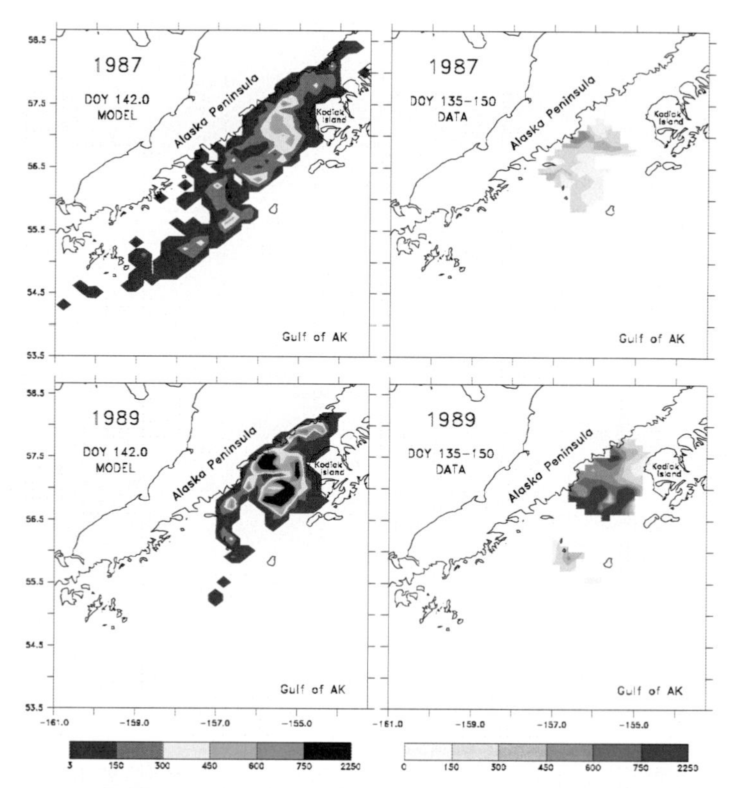


Figure 13.14. Comparison of model output to field observations, Shelikov Strait. Plotted is depth-integrated abundance of larvae (#/m$^2$) in mid-May of 1987 and 1989 (from Hermann et al., 1996)

Figure 17.9. Ancient storm deposit (tempestite) from Coaledo Formation in southwestern Oregon. Darker material is finer-grained sediment, while lighter material is coarser sediment. The combination of depositional and erosional sedimentary structures is typical of storm deposits, when much sediment is in motion under conditions of high shear stresses. Photo appears courtesy of J. Crockett.

Figure 17.11. A trench from a lagoonal environment in Kamchatka. The sediment record in this location is dominated by tsunamis (light brown, sand) and volcanic eruptions (white to gray fine-grained ash). Photo and interpretation appears courtesy of J. Bourgeois.

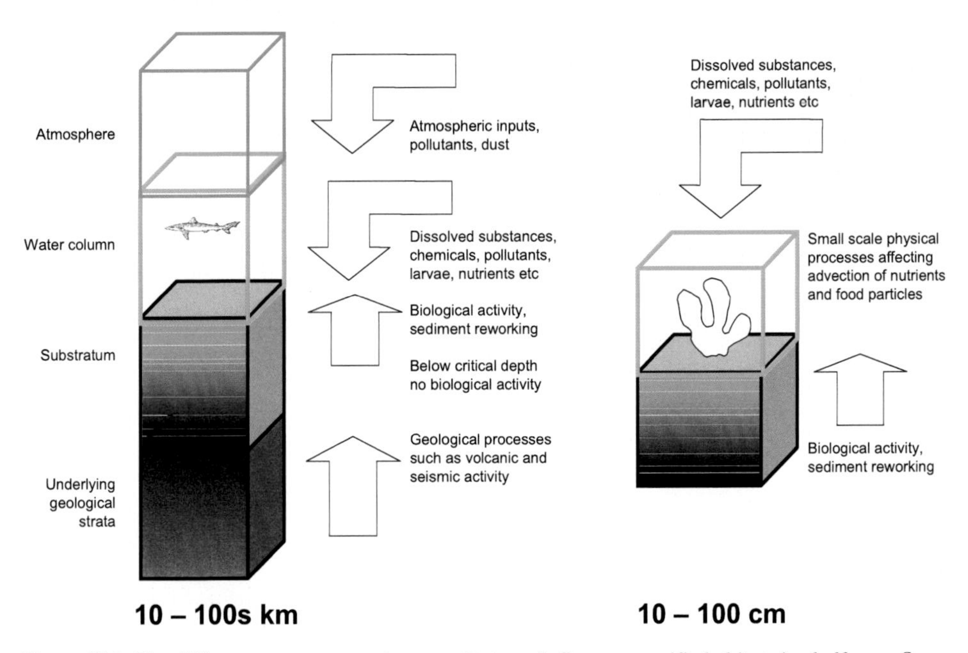


Figure 23.1. The different components that constitute or influence specific habitats in shelf seas. Some influences are very localised (e.g. seismic activity) while others are spread across a much wider area (e.g. oceanic influences).

Figure 23.5. Examples of four habitats that differ in their stability and their exposure to natural forms of disturbance which influences their vulnerability to human activities. Surf beaches (A) are exposed to high levels of chronic natural disturbance by wave action and have an unconsolidated coarse sediment structure and low species abundance and diversity. Gravel lag (B) sediments can provide a stable substratum but can occur in areas of high tidal scour resulting in an impoverished fauna as in this example. Granite bedrock at a depth of 140 m provides a stable substratum protected from wave action, upon which slow growing sessile biota can become established such as the sponges in this illustration. Reef building fauna develop into a stable substratum and enhance local productivity through the production of faeces as seen in this rich horse mussel habitat. Photograph A copyright MJ Kaiser, photograph B,C,D copyright E.I.S. Rees with permission.

# Chapter 14. PROCESSES AND PATTERNS OF INTERACTIONS IN MARINE FISH POPULATIONS: AN ECOSYSTEM PERSPECTIVE

PHILIPPE CURY

*Institut de Recherche pour le Développement (IRD) and University of Cape Town*

PIERRE FRÉON

*IRD and Marine & Coastal Management*

COLEEN L. MOLONEY

*University of Cape Town*

LYNNE J. SHANNON

*Marine & Coastal Management*

YUNNE-JAI SHIN

*IRD Centre de Recherche Halieutique Méditerranéenne et Tropicale* *Contents*



## 1. Introduction

A marine ecosystem has no apparent boundaries and lacks the clear objective or purpose that can be ascribed to other, more tractable, biological or ecological entities (e.g. cells, individuals or populations). It contains water, nutrients, detritus, and numerous kinds of organisms of different sizes and with different life history traits, ranging from bacteria, phytoplankton, zooplankton, and fish to mammals and birds. All these living and non-living components are connected in complex food webs through changing and evolving interactions, which make ecological systems extraordinarily complex (Polis, 1999). In addition, humans have been exploiting marine organisms for many centuries. Within the last half-century in

*The Sea*, Volume 13, edited by Allan R. Robinson and Kenneth H. Brink
ISBN 0-674-01526-6 

particular, fishing impacts have considerably altered natural ecosystems (ICES, 2000).

Scientists have long studied ecosystems, particularly in the terrestrial environment, resulting in numerous mathematical scale-and time-dependent models and theories. Reductionism and holism are two basic approaches that are used to understand processes and patterns (defined here as regularities in what we observe in nature (Lawton, 1999)). Reductionist approaches focus on the dynamics of a limited number of species (limited for practical and technical reasons). In general, these studies are based on single-species dynamics, but might be enriched by adding and formalizing species interactions. Holistic approaches focus on aggregated entities and processes, without necessarily considering the dynamics and interactions of individual species. These approaches are inclusive in terms of the number of species components, but are not detailed at the species level.

Because simple patterns can emerge from complex interactions and simple interactions can produce complex patterns, there is a continuing ecological debate between those who argue that all systems are different and unique, i.e. contingency and chaos reign, and others who argue that patterns are observable and reproducible, i.e. there is order and generalities occur. In the marine environment, fishing adds complication to the dynamics of the ecosystems, and has the potential to perturb existing patterns and generate new ones. In the present review we describe different attempts and several recent advances that have been made in ecology to formulate processes and patterns of interactions in fish populations, and that can help to understand marine ecosystem dynamics. Although this chapter is devoted to marine fish, some of the reviewed studies belong to fresh-water ecosystems, which are sometimes better documented than marine ecosystems. We critically review these different approaches to describe and understand the changing nature of marine ecosystems by analysing processes and patterns. Finally we advocate integrative studies that assemble all accumulated knowledge in a multidisciplinary way and that could provide an efficient framework for studying marine ecosystems.

## 2. Species interactions: from processes to patterns

Species interactions include competition, predation, disease- and parasite- transmission, parasitism and mutualism. Fishing is a special case that involves interactions between fish and humans. In marine fish populations, these interactions usually require individuals to be found at short distances from one another. This matching in time and space of different species is related to their own migration patterns and habitat selection, which in turn depend on a number of factors. These factors include the searching for (prey) or avoidance of (predators, competitors) contact with individuals of other species. Therefore, individuals interact because of their particular biology, and their biology is a result of species interactions—a typical "chicken-and-egg" situation, which makes it difficult to interpret many observations. In the context of fishing impacts, it is also sometimes difficult to disentangle "natural" processes and patterns from those that have been caused by fishing.

The processes involved in species interactions can be simplified in models, which can then be used to identify patterns that might emerge at the community or ecosystem level. However, we also recognise that fish species' interactions can be complex, because of the behaviour of fish, their ability to learn, the fact that they live in an environment that is largely unpredictable, and their evolutionary history. This means that emergent patterns might be the result of a number of different causal processes, and a variety of process models should be used in order to fully understand community dynamics. In this section, we describe some of the models that are used to describe species interactions (including those involving fishers), the underlying behavioural processes affecting fish, and some of the attributes of fish that might modify the patterns generated by these models. We focus on feeding interactions (predation and competition relationships) because almost all multi-species models that are applied in fisheries science refer to these types of interactions; other potential interactions (parasitism and mutualism) are implicitly considered not to be important in determining marine ecosystem dynamics.

### *2.1 Predator-prey systems*

Two-species predator-prey systems have been the subject of investigation since the 1920s, with the pioneer studies of Volterra (1926) and Lotka (1932), both of whom independently proposed the first predation model. This classical model of predation is based on some simplifying assumptions (Appendix), including that the birth and death processes respectively in the prey and predator populations are exponential. Analysis of the Lotka-Volterra model system shows that, apart from the point where both predator and prey populations are extinct, there exists another positive equilibrium point, and the dynamics of the system are such that the predator and prey populations exhibit cycles of abundance. This model system has didactic interest but has not been satisfactorily applied to real situations in the marine environment.

Since the development of the Lotka-Volterra model, numerous studies have proposed different formulations for predator-prey interactions, resulting in systems that exhibit contrasted dynamics. In a general formulation of a predator-prey system, three processes must be specified:

1. the intrinsic growth rate of the prey population in the absence of predators,
2. the functional response of the predator, which describes the factors affecting the number of prey consumed by a predator per unit of space and time, and
3. the numerical response of the predator, which describes the rate of conversion of prey into predators.

For these three processes, the choice of the mathematical functions, which contain important biological information, is crucial to the dynamics of the studied systems (Yodzis, 1994). We will examine each process in more detail below.

**Intrinsic growth rates of prey populations**

The Lotka-Volterra model assumes that prey populations exhibit a Malthusian (exponential) growth rate, which is not realistic. Subsequent predator-prey models assumed logistic, density-dependent growth (Verhulst, 1838), where a "carrying capacity" of the habitat is specified for the prey populations.

In fish populations, the concept of having a constant carrying capacity is seldom realistic, but population growth rates of many species are found to vary inversely with population density because of competition for food. In upwelling ecosystems, consumption by pelagic fish is usually exceeded by plankton production (review in Bakun, 1989), and competition is not expected to be a limiting factor. Nonetheless, density-dependent growth is often observed. A possible explanation concerns the effect of schooling. Most pelagic fish species occur in dense schools of several ten thousand individuals, at least during the daytime, and usually maintain a high level of aggregation during the night. Many demersal species do the same during their early stages of development. Sometimes two or more species of pelagic fish school together, forming a "mixed-school" of individuals of similar size and body shape. Average school size generally increases in response to an increase in population size (Fréon and Misund, 1999), although recent unpublished data from the South African purse seine fishery suggest that this relationship might not be linear. Bakun (1989) proposed that any increase in mean school size in response to an increase in population density would increase local competition for food and oxygen inside the school, thereby decreasing population growth rates.

Growth rates of prey populations are not only limited by the carrying capacity of their environments, but can also be reduced when their densities decrease. The "school-trap hypothesis" (Bakun and Cury, 1999) states that a fish species that occurs in a mixed-school with a more abundant species must effectively subordinate its specific needs and preferences to the "corporate volition" of the school. This hypothesis received recent support from field observations on estuarine clupeoid fish by Maes and Ollevier (2002). School-traps could promote large amplitude, out-of-phase population oscillations of small pelagic fish species. The school-trap hypothesis implies that adaptive changes in population dynamics ("school-mix feedback"; Bakun, 2001) could occur much faster than those related to genetic evolutionary processes.

Growth rates can also be affected by predators which, apart from having a direct, consumptive effect on their prey, can also cause them to move to safer but less productive habitat. As a result, the prey will experience both reduced individual growth and decreased survival (e.g. Tonn et al., 1992). Similarly, Power (1987) gave observational and experimental evidence that large armoured catfish (Loricariidae), although severely resource-limited in the deeper part of pools in a Panamanian stream, avoided shallow, rich areas because of the greater risk of predation by birds.

**Functional responses of predators to their prey**

There are a number of different ways of modelling the functional responses of predators to their prey (see Appendix), depending mainly on what one assumes about how predators interact with one another. When predators do not interfere with one another in their feeding activities (*"laissez-faire"*, Caughley and Lawton (1976)), the functional response of the predators depends only on their prey densities, not on predator density. The Lotka-Volterra model assumes that the number of prey consumed per unit of time is a linear function that is not limited by prey numbers (Czaran, 1998). According to Holling (1959), feeding consists of two types of activity: searching for prey and handling them. He assumed (i) that the total time dedicated to feeding is the sum of search time and handling time, and (ii) that the handling time is a constant. The Holling type II response assumes that the attack rate is proportional to the number of prey in the environment (Begon et al., 1996). The Holling type III response assumes that predators are inefficient at handling prey when prey are not abundant (Yodzis, 1994).

Some of these assumptions are nullified if the prey populations are schooling species. Encounter rates between predator and prey will remain constant with an increase in prey biomass unless there is a substantial increase in the number of schools. There is always more than enough prey in a school to satiate a predator or a group of them, and predators are not observed to follow a given school for many hours. As a result, an increase in schooling fish prey does not necessarily result in a proportional increase in prey accessibility and therefore consumption. Density-dependent mortality in prey can also result from changes in schooling behaviour of predators. Anderson (2001) showed experimentally that predatory kelp bass (*Paralabrax clathratus*) responded to an increase in their prey density (kelp perch *Brachyistius frenatus*) by a strong increase in aggregation, and an increased predation rate. This experimental result was consistent with observed patterns of density-dependent mortality of prey in field studies.

In contrast to Holling's (1959) models, another family of models assumes the existence of predator interference through trophic or reproductive competition, disease transmission, cannibalism, or density-dependent emigration (Yodzis, 1994). In this category, a common functional response is the one of Hassel and Varley (1969), which assumes that the predation rate decreases when predator abundance increases and, for a given density of prey, the greater the abundance of predators, the slower the rate of consumption by a predator. This general formulation can be applied to a large number of predator-prey systems, but their behaviour becomes unrealistic in the particular case when the predator population tends to zero, because the predation rate tends towards infinity.

This problem also occurs for ratio-dependent formulations of the predator functional response. Initially proposed by Arditi and Ginzburg (1989), ratio-dependent formulations assume that consumption rate decreases proportionally as predator abundance increases, because the same resource must be shared by a greater number of consumers. The choice between a prey-dependent versus a ratio-dependent functional response is a controversial and topical subject. Supporters of ratio-dependence endeavour to show that this formulation better accounts for the behaviour of natural ecosystems. For example, Ginzburg and Akçakaya (1992) showed marked differences between prey- and ratio-dependent models in the response of trophic food chains following an increase in primary production in the system. In the ratio-dependent model, all the trophic levels responded proportionally to the increase, whereas in the prey-dependent model, the responses differed according to the trophic level considered and the number of trophic levels in the system. According to the same authors, the first response appears to be common in the food chains of lakes.

The different models of functional responses of predators to their prey aim at representing simple, specific foraging strategies. However, some variations on the general foraging strategy can occur which produce different patterns of interactions. Foraging strategy defines species-specific behaviour that has evolved through natural selection to maximize individual fitness (Hart, 1997). Foraging tactics (or modes of foraging) are behavioural variations within the strategy that allow animals to vary their behaviour in response to local conditions or to variations in prey type. Tactics can be seen as behavioural variations used by a predator to overcome the detection and handling problems posed by the anti-predator defences of the prey. Therefore, strategies and tactics are two levels in a hierarchy (Wootton, 1984). The interaction between predator and prey is dynamic through evolutionary time, with each actor evolving behaviour to outwit the other. According to Endler (1991), the interaction is not true coevolution, but rather an "arms race". Foraging tactics might change during the development of individuals. Ontogenic shifts in morphology are often associated with ontogenic niche shifts and, in some species, there is interplay between behaviour and morphology (Meyer, 1987).

Hart (1997) established that tactical variability in fish foraging behaviour exists and variations have functional significance. Some species are more flexible than others in their ability to vary their foraging tactics, mainly because of morphological constraints. Some species are highly specialized in their hunting behaviour, e.g. stationary search (sit-and-wait) and prey capture by ambush. Other species can alternate ambush and chase in open water, and solitary or cooperative chasing. Different hunting activities are related to significant differences in growth rates of the species, especially when they are forced to use the same tactic for a long time (e.g. Eklöv, 1992; Eklöv and Diehl, 1994). There are limits set by morphological structures such as the jaws, but these structures are not totally rigid, and significant behavioural flexibility can be accommodated (Galis et al., 1994; Hart, 1997). The same applies for tactics of prey capture, manipulation and handling by predators (e.g. Vinyard, 1982; Helfman, 1990). This flexibility can generate greater than expected variability in fish diet, as shown in the next sub-section. Furthermore, the level of hunger modifies feeding behaviour, which becomes less efficient and exposes the predator itself to a greater risk of predation (Miyazaki et al., 2000).

From a review of different search paths used by different species of fish, Hart (1997) suggested that, rather than a set of discrete tactics, there is a continuum between the extremes of movement speeds, with one extreme being continuous search and the other sit-and-wait search. An interesting and apparently common tactic is the saltatory search in which the forager stops and searches the entire volume of water in front of it and then moves on until a new unsearched volume is available (O'Brien et al., 1990). The saltatory search allows the encounter of several prey simultaneously and hence prey selection.

Learning aspects in fish behaviour are poorly known but are likely to play a significant role (reviews in Fréon and Misund, 1999; Brown and Laland, 2001). Learning can allow foragers to adapt to spatial and temporal variation in prey properties (Hart, 1997; Ehlinger, 1989). Hart (1997) hypothesises that fish can use a suite of foraging tactics involving searching for and handling prey to cope with the variability of prey or patch types encountered and their spatial distribution. Furthermore, the effects of learning on prey handling and predator hunger state combine additively and interactively (hungrier fish learn more efficiently than less hungry ones).

**Numerical responses of predators**

The most widely used predator numerical response function corresponds to a "laissez-faire" situation (see Appendix), representing a balance between gains to the predator population (proportional to the quantity of prey consumed by a predator), and its losses. Interference can also be taken into account, affecting the predator consumption rate (the functional response) or the predator growth rate.

The predator numerical response function of Leslie (1948) is based on the assumption that the predator population grows logistically, with a carrying capacity that is proportional to the abundance of prey. A problem with this formulation is that the carrying capacity of the predators is zero when prey abundance is zero. Because predators incur maintenance costs, their carrying capacity should be zero at some positive threshold of prey abundance (Yodzis, 1994).

**Patterns that result from predator-prey interactions**

Different theoretical patterns can emerge from different combinations of processes affecting predator-prey interactions. For example, a predator-prey system that has density-dependent growth of the prey and a type II Holling functional response can exhibit very different dynamics, depending on the predators' handling efficiency (Figure 14.1). When handling efficiency is small, the predator goes extinct (Figure 14.1a), when it is intermediate both populations coexist in a stable equilibrium (Figure 14.1b), and when it is large the predator and prey populations exhibit cyclic oscillations (Figure 14.1c).

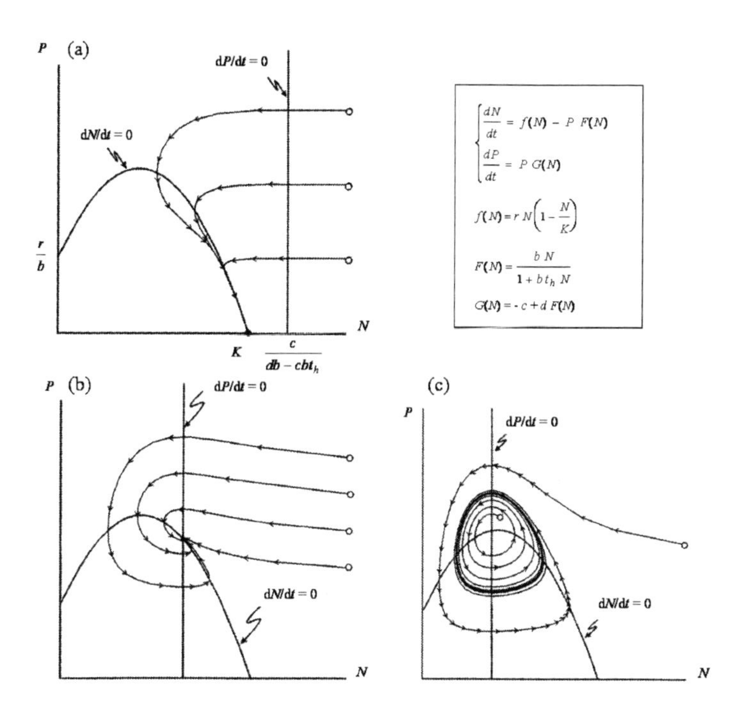


Figure 14.1 Stability properties of a predator-prey model that has density-dependent growth of the prey population (formulae presented in the upper-right panel), a type II Holling functional response and a "laissez-faire" numerical response for the predator (adapted from Czaran, 1998). The effect of the interactions on the dynamics of the system depends on the efficiency of prey handling by their predators. (a) Stable equilibrium point with predator extinction, (b) stable coexistence of both species, and (c) stable limit cycle. $N$ = prey density; $P$ = predators density; $f(N)$ = intrinsic growth rate of prey with carrying capacity $K$; $F(N)$ = functional response of predators to prey; $G(N)$ = numerical response of predators to prey; $t_h$ = handling time (the inverse of handling efficiency); $r$, $b$, $c$ and $d$ are constants.

Michalski and Arditi (1995a; b) conducted a theoretical study of complex food webs, in which they proposed a generalisation of predator-prey models that included a Holling type II functional response and a numerical response with interference in the consumption rate. Using arbitrary values of the parameters, the authors applied the model to a theoretical system consisting of 11 species (Michalski and Arditi, 1995a). At equilibrium, only a few links persisted (Figure 14.2a). By imposing a variation of 60% in the values of parameters that describe prey selectivity and competition among predators, the effective structure of the trophic web changed radically (Figure 14.2b-f). When the system at equilibrium was disturbed (e.g. by changing the abundance of a species), the same authors (Michalski and Arditi, 1995b) showed that their multi-species model leads to systems that are rich in interspecies links when they are far from equilibrium, and poor in links when they approach it. A consequence of these results is that the structure of food webs can vary with seasonal variations in species abundance as well as variations in the parameters for competition efficiency and food preferences. The difficulty is to determine the temporal variation of these parameters in applying this kind of model to real systems.

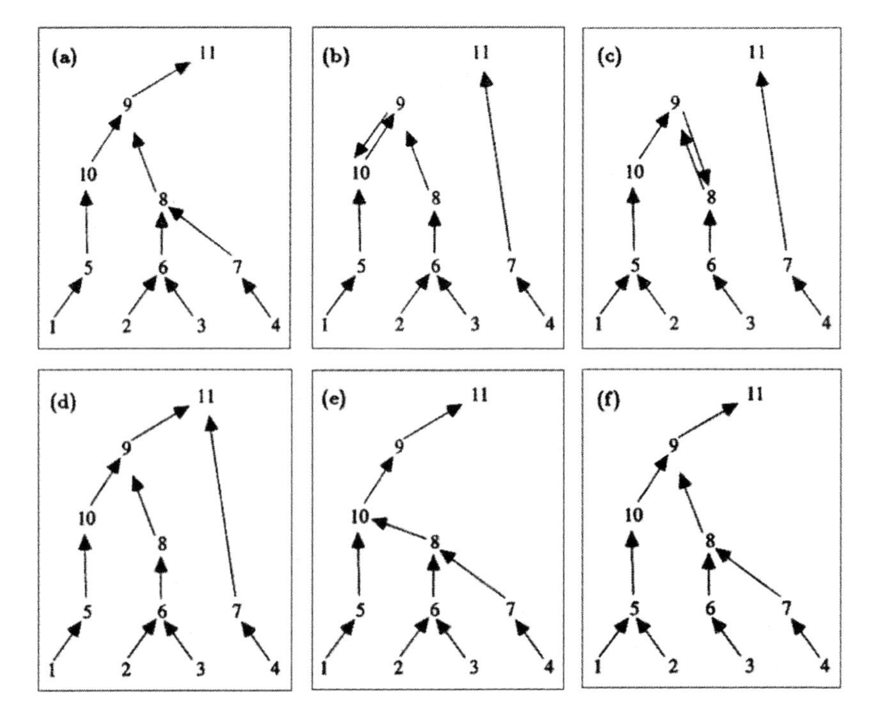


Figure 14.2 Representations of equilibrium states for a trophic web of 11 species, modelled using a system of coupled equations that assume logistic growth for the "forage" species (species 1 to 4), and Holling type II functional responses for predators, and numerical responses with interference by predators for consumption rates. The different trophic structures are obtained from different values for food preferences and competition efficiencies. The arrows indicate the presence of effective interspecies links (from Michalski and Arditi, 1995a).

These dynamic predator-prey models are not easily applied to real predator-prey systems, but they are important tools for exploring complexity in the behaviour of interacting populations. Different studies using such models show their sensitivity to parameter values, to initial conditions, and to the formulation of functional and numerical responses of the predators. Furthermore, they exhibit a wide variety of behaviours, from periodical dynamic to stable coexistence of two species. The results produced by these models have to be cautiously interpreted, because the same equations can lead to very different dynamics, even qualitatively. Predator-prey models have been used to explore the stability of multispecies systems, which is a topical field of investigation in ecology and fisheries science. The use of different concepts or definitions of the stability of ecological systems is at the origin of most of the controversy existing on the subject. It is therefore essential to specify that, here, the stability of mathematical systems of differential equations corresponds to the existence of a stable steady state for all variables of the system (alternatively, stability can be defined as the opposite of temporal variability, and in this case, it is measured by the coefficient of variation of the state variables of the system—Tilman, 1999). Gardner and Ashby (1970) and May (1972) are among the first to have examined the mathematical stability of complex multispecies systems. They tested whether systems that consist of a large number of interacting populations are stable, i.e. all trajectories of the variables converge towards an equilibrium. The authors showed that stability decreases when the size of the system (species richness) and its connectance increase. Using different formulations of predator-prey systems, DeAngelis (1975), Gilpin (1975) and Pimm (1979) reached similar conclusions. More recently, Michalski and Arditi (1999) showed that the risk of population explosions increases with the connectance and species richness of a system.

## Effects of fishing in predator-prey systems

### *The Schaefer model and its derivatives*

The Schaefer (1954) model is a simple stock assessment model, still used in many instances, e.g. tropical and tuna fisheries (Hilborn and Walters, 1992). It is a modified logistic (Verhulst) model, and its dynamics are such that a fished population will grow rapidly when it is reduced far below carrying capacity, but the population growth rate will be zero at carrying capacity. Any removal by fishing will always maintain the population below carrying capacity, and the maximum sustainable yield (MSY) occurs when the population is at half its carrying capacity.

The Schaefer model can be extended to a multi-species case by coupling single-species Schaefer equations through species interaction terms, which are simple multiplicative terms like those found in the Lotka-Volterra model. The main results of the single-species Schaefer model can only be extended to the multi-species case under restricted circumstamces. Kirkwood (1982) showed that the total MSY will correspond to exploitation at half the total pristine biomass only in the restrictive case of a multi-species assemblage where the species are independent and/or competitive and/or mutualistic, and predation interactions are close to zero.

Ströbele and Wacker (1991) considered a two-species system using the Schaefer model. They explored the theoretical consequences, in terms of catch, when different types of interactions (predation, competition, mutualism) and different fishing scenarios are modelled. They showed that, in the case of mutualistic species, the theoretical fishing yield would be greater than the one obtained with independent species. If the species are competitive, the yield is smaller (Figure 14.3). In the case of predation interactions, if the prey are selectively fished the effect is similar to the competition case. This result would be expected, because a decrease in the biomass of prey will cause a decrease in the biomass of its predators. If predators are selectively fished, the same effect is obtained as for the mutualism case. When both predators and prey are fished, the system is more complex, and the results depend on the relative magnitudes of the parameters of predation, catchability and intraspecific competition (Appendix).

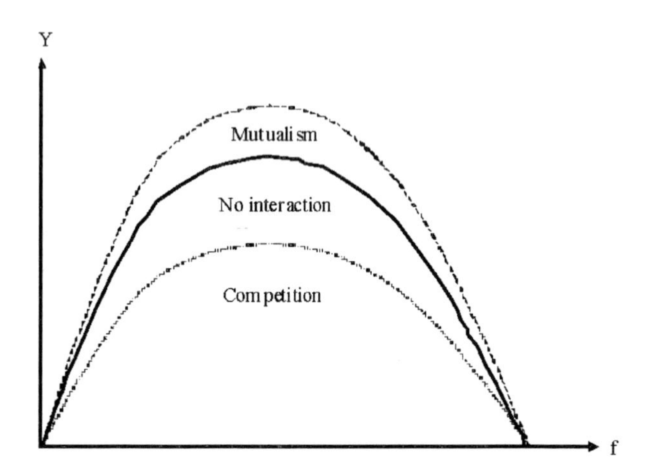


Figure 14.3 Total catch ($Y$) as a function of fishing effort ($f$) for a two-species system (adapted from Ströbele and Wacker, 1991).

*A krill-whale system under exploitation*

There are few applications of predator-prey system models in fisheries science. Among the fisheries applications is May et al.'s (1979) study on multi-species assemblages in the Antarctic. They (and thereafter Beddington and May (1980) and Flaaten (1988)) used the predator-prey model of Leslie and Gower (1960), which consists of a combination of a Lotka-Volterra-type linear functional response of predators to prey, and a Leslie numerical response of the predators. They first investigated the behaviour of a classical predator-prey (whale-krill) model under fishing.

Both populations were exploited at rates that were multipliers of the intrinsic growth rates of the respective populations. Simulations showed that when fishing mortality rates exceeded population growth rates, the whale population collapsed or the whole system collapsed. In contrast, when fishing mortality rates were inferior, there existed a unique equilibrium where the presence of whales had the trivial effect of decreasing the fishing yield and the abundance of krill. When the exploitation of whales increased, the abundance of krill increased, which caused an increase in the growth rate of the remaining whales. Thus, when half the pristine biomass of krill was removed by predation, the maximum sustainable yield (MSY) for whales occurred when fishing effort was 19% greater than in the single-species case. The MSY for krill increased with fishing on whales, the maximum being reached when the whales went extinct.

May et al.'s (1979) analysis was made more complex by adding to the model another krill predator and competitor of the whales (seals), or by adding another intermediate trophic level between whales and krill (cephalopods). One of the conclusions of this study was that, in different multi-species configurations, the only possible generalisation is that the concept of MSY can be applied to the top trophic levels, and that their MSY occurs at effort values that are greater than those corresponding to half the pristine biomass.

Yodzis (1994) criticized the Leslie model used in May et al.'s (1979) study because the combination of a Leslie numerical response and a *laissez-faire* functional response (type II) is biologically paradoxical, and invalidates the structure of the model. Indeed, in this case, the interference between predators is strongly linked to available prey (numerical response), but the predators have no interference among themselves for consumption rates (functional response). Yodzis (1994) showed that the results obtained by May et al. (1979), and subsequently by Flaaten (1988), depend fully on the choice of the Leslie model. This is important to highlight, because Flaaten (1988) advocated intense reductions in the stock of marine mammals in order to increase fishing yields of fish of commercial interest. In addition, we showed how erratic the behaviour of the Leslie model (and also the Hassel-Varley and Arditi-Ginzburg models) becomes when prey densities are low.

### *2.2 Diet selection and food webs*

Foraging behaviour can involve «decisions» (sensu Dill, 1987) such as where to feed, when to change feeding grounds, and which food items to select. Because successful feeding is obviously one of the major components of "fitness" (others being predator avoidance and successful reproduction), behavioural ecology applied to feeding behaviour has been based on the assumption that decisions made while foraging maximise the net rate of energy gain per unit time. This is the basis of optimal foraging theory, proposed by MacArthur and Pianka (1966) and Emlen (1966). It is unclear how fish manage to assess the profitability of their prey, which depends on a lot of factors related to the energetic return from the prey (prey mass, energetic content and digestibility) and handling time (which varies with a number of factors including size relative to forager, body shape, presence or absence of appendages or defensive mechanisms, swimming speed and agility, etc.). Hughes (1997) suggests that this assessment is possible because, through experience or genetic programming, foragers come to respond to a small set of key stimuli indicating the probable profitability of their prey. Empirical *in situ* studies suggest that the effects of morphological constraints, such as predator and prey sizes, on encounter rates and handling times are key factors determining predators' diets (reviews in Persson and Diehl, 1990; Mittelbach and Osenberg, 1994; Hart, 1997; Hughes, 1997).

There are a number of existing models of trophic interactions in aquatic communities or ecosystems. Attempts at modelling trophic interactions (see Appendix for details of the models) make different assumptions or simplifications (usually acknowledged by the authors of those models) about diet compositions. We describe below four ways in which diet composition can be represented.

**Fixed diets**

In a fixed diet, the composition of prey items making up the diet is fixed according to an average derived from stomach-content data. For example, fixed diets are used in Ecopath models (Christensen and Pauly, 1992), which represent foodwebs at a theoretical equilibrium state. However, a fixed diet is an oversimplification of fish behaviour. Feeding can change as a result of changes in foraging tactics, changes in prey abundance and predator experience. Hughes et al. (1992) indicate that performance in all phases of handling prey, from initial recognition to final ingestion, can improve as fish gain experience of specific prey. By comparing hatchery-reared naive fish to wild-caught fish, Reiriz et al. (1998) found that learning had a visible effect on the pattern of selection for three live prey types by juvenile Atlantic salmon (*Salmo salar*). Finally, feeding behaviour might be altered in several ways by the risk of predation while foraging. These alterations include timing and location of feeding, searching mode, feeding rate, prey-handling tactics, vigilance and social behaviour (Lima and Dill, 1990; Connell, 2002). As a result, risk of predation can influence diet selection, causing a forager to broaden its diet or to prefer safer but less profitable prey (Godin, 1990; Sih, 1993). It is likely that the optimal diet is a trade-off between maximising energy gain and minimising mortality (Milinski, 1993; Hughes, 1997).

**Proportional diets**

Proportional diets result from indiscriminate feeding, so that the diets represent the availability of prey items in the environment; the diet is proportional to prey abundance and availability. Proportional feeding is a simple, attractive assumption, but unfortunately it is frequently not true. Many species of predators feed disproportionately on any acceptable prey whose relative abundance is high (Hughes and Croy, 1993). This can be interpreted by a number of untested hypotheses, of which the most common is that there is a benefit from learning and from the development of a "search image" (*sensu* Tinbergen, 1960) favouring detection and capture of prey. There are also many mechanisms that allow prey to avoid, deter or evade predators (reviews in Smith, 1997 and Godin, 1997), thereby violating the assumption of equal vulnerability or accessibility.

Predator avoidance can be achieved by habitat selection; crypsis (body shape and colour resembling the background or mimicking another object like a plant); behavioural avoidance of detection such as "freezing" behaviour or immobility, diel timing of movements, predator recognition (which is related to genetic differences and learning), predator labelling and predator inspection. Predator deterrence mechanisms can be subdivided into morphological deterrents such as body armour, size and arrangement of spines, production of strong electrical discharges; behavioural deterrents like aposematic (=warning) signals in poisonous fish, mimicry of a predator or a poisonous species, pursuit deterrence and pursuit invitation displays, alarm signalling, distress signals, mobbing of predators, shoaling and schooling (but see Connell (2000) for possible exceptions). Evading predators can be achieved by defences mitigating predator capture of prey such as fleeing, hiding, deflecting the attack to certain parts of the prey's body (for instance by displaying eyespots (*ocelli*) on the tail) and shoaling, which has numerous advantages like dilution of risk, predator confusion, possibility of synchronised shoal evasive manoeuvres, transmission of information among shoal members (reviews in Pitcher and Parrish, 1993; Godin, 1997; Fréon and Misund, 1999).

In order to relax the assumption of permanent free access to prey by predators, Walters and Juanes (1993) proposed the concept of a "foraging arena", which assumes that predation takes place largely in spatial patches. The foraging arena concept is now implemented in the ECOSIM II modelling software (Walters et al., 1997), which attempts to model vulnerability distributions by treating the prey as being in one of two behavioural states: "invulnerable" and "vulnerable". Exchange between these states could represent both behavioural and physical processes. The model ignores predator handling time/satiation, following the observation that predators with full stomachs are not a common field observation. This approach represents a great improvement on previous approaches, although the assumption of absence of satiation is probably not equally applicable for different species and habitats.

**Size-based diets**

In some instances, diets are based on size constraints and are proportional to potential prey, as is assumed for the OSMOSE model (Shin and Cury, 2001). The assumption, made in most size-structured models, that predation by piscivores is size-dependent with an increase in prey size range with the size of the forager, could be impaired if handling efficiency for smaller items decreases in large fish. Available evidence on this topic, although limited (Persson and Greenberg, 1990; Juanes, 1994), does not support such a decrease in efficiency. More worrying can be the concern that prey size at maximum profitability is expected to increase as fish grow from larvae to adult and, as a result, large fish are expected to prefer large prey (Galis, 1990). Such size preferences can be specified in most size-structured models.

**Optimal diets**

If diets are assumed to be optimal, they are necessarily based on the most profitable prey type, thereby maximising the net energy gained per unit time. The basic prey model (BPM) of diet choice (Stephens and Krebs, 1986) assumes that foragers encounter prey types sequentially. Two major sources that violate assumptions of the BPM have been identified by Hughes (1997): hunger state and learning. Hungrier individuals are less selective than partly satiated ones (Ivlev, 1961; Kislalioglu and Gibson, 1976) and this may reflect the priority given to the restoration of a positive energy budget as quickly as possible, even at the cost of reduced foraging efficiency. An alternative explanation is that packing constraints become more critical as the stomach of the predator fills (Hart and Gill, 1992). Furthermore, profitability in BPM and related models is usually estimated without taking into account digestion time, because many foragers continue to feed while digestion is in progress. Nonetheless, because digestibility varies according to prey type, profitability calculated by taking into account physiological processes gives opposite ranking of prey compared to ranking based only on behaviour (Kaiser et al., 1992).

**Competition among predators**

As far as we know, none of the current tropho-dynamic models takes into account the effect of competition among predators on diet selection. They also currently ignore the influence of predators on the foraging behaviour of their prey. Competition among predators might not only affect the foraging rate on a given species, as in the functional response model of Hassel and Varley (1969), but also diet selection according to two processes: acceleration of the depletion of prey and alteration of foragers' behaviour. The acceleration of depletion will obviously reduce the encounter rates and thereby discourage selective feeding. Depletion could be anticipated by predators, having noticed the presence of competitors. As a result, predators could adopt an opportunistic strategy of 'first-come-first-served', resulting in a broadening of their diets independently of encounter rates with prey (Dill and Frazer, 1984; James and Poulin, 1998). The alteration of foragers' behaviour can result from contest competition and could cause differential changes in foraging behaviour according to their competitive rank or dominance (Milinski, 1982).

**Patterns that result from trophic interactions**

The mechanisms controlling foraging and predation at the individual level can be complex and variable. As a result, it is easy to criticize any trophic model, because all aspects of fish behaviour cannot be taken into account. Fish behaviour is a result of interplays among adaptive evolutionary processes between predators and prey in the long-term, habitat selection and ontogenic changes in the medium-term, and flexibility in behaviour using learning and memory in the short-term. Even highly complex and sophisticated trophic models that make use of several thousand parameters (e.g. Fulton, 2001) represent gross oversimplifications of reality, because they are based on many hypotheses and assumptions, not all of which can be tested or analysed by sensitivity analyses.

More difficult than criticising is finding a constructive approach that can identify key processes that control trophic flows at the level of populations. Key parameters at the level of individuals or small groups and at high time- and space- resolutions do not always result in key parameters at the level of populations and at low spatio-temporal resolutions, because averaging effects can smooth most of the variability. According to Persson et al. (1997) "size-structured competitive and predator-prey interactions, in combination with resource-dependent individual growth, have the potential to enhance the diversity of behaviours, both with respect to individual and population processes".

In complex system modelling exercises, one has to distinguish between processes that appear as output or emerging properties of the model, and those that are incorporated as input. Typically, factors that control migration and habitat selection at large scales would be treated better as forcing factors in spatially-resolved trophodynamic models, in order to reflect the frequent mismatch in space and time between predator and prey. In contrast, factors that control micro-habitat selection would be considered best as output variables from rules or equations describing predator-prey interactions.

From theoretical considerations, Fryxell and Lundberg (1994) analysed the effect of adaptive diet selection by predators on population stability, showing that this selection enhances the stability properties of a predator-prey system only under a small range of parameter values. Otherwise, adaptive diet choice was found not to be an important stabilizing factor. In contrast, Gleeson and Wilson (1986) showed theoretically that a predator foraging on two competing prey species according to the optimal diet model (Stephens and Krebs, 1986) could prevent the weaker competitor from becoming extinct if the dominant competitor was the more profitable prey in terms of energy maximisation. However, this result does not differ substantially from predictions of models based on frequency-dependent prey selection.

Pioneer ecological studies considered that diet selection played an essential role in community stability. MacArthur (1955) argued that population densities should be more stable when they pertain to complex trophic webs, i.e. those consisting of many species and/or many interactions. Making an analogy with the diversity index of Shannon and Weaver, he quantified the stability of trophic webs by the amount of information circulating along their different trophic pathways. He found that species assemblages are stable in two cases (here, the notion of stability refers to resistance of a system to perturbation): either when the assemblage consists of a large number of species for which the diets are not diversified, or when the assemblage consists of few species which are largely polyphagous. Thus, according to MacArthur (1955), when the trophic energy can flow through many different pathways, the consequences of the disappearance of an interaction or of a species component should be less detrimental for the community than when there are few pathways. In marine ecosystems, diet selection depends largely on the appropriate size ratio between a predator and its prey (Scharf et al., 2000). This leads to complex trophic webs with most predator species having multiple prey species and most prey species having multiple predator species (Cury et al., 2003). Some modelling studies suggest that size-based diet selection can explain the relative stability (as opposed to temporal variability) of the biomass and size spectrum of fish communities compared to the dynamics of individual species (Shin and Cury, 2001; 2004).

One of the difficulties in modelling predator-prey relationships lies in the fact that most fishes are simultaneously predators and prey (Werner and Gilliam, 1984; Connell, 2002), whereas most models focus on the effects of habitat selection or other parameters in just one species. These models do not always reflect the individual's trade-off between their foraging activities and their avoidance of predation. Predators have been considered only as a source of risk to which prey respond (fixed risk assumption of constant attack rates over time or patch-specific risks of predation), whereas in nature predators respond to prey behaviour, especially by moving in search of patchy prey areas or changing their timing. Reviewing the few existing models that incorporate both prey and predator mobility in a predator-prey model based on game theory, Lima (2002) showed that those models can reveal new and unexpected classes of behavioural phenomena that occur at large spatial scales. Under some conditions, predators might appear to ignore prey distributions and distribute themselves according to the distribution of the resources of the prey. Such complex and sometimes counter-intuitive interactions in population dynamics, with "negative switching", were previously described by

Abrams (1992) and Abrams and Matsuda (1993) who incorporated in their models equations for the dynamics of the resources and flexible behaviour in more than one species.

The stability of predator-prey structure in individual-based models incorporating spatial configurations is determined by the relative mobility of predators and prey, and prey mobility in particular has strong effects on stability (McCauley et al., 1993). Other factors of stability in models are provided by aggregative behaviour in predators, high variance in prey abundance and some degree of segregation between prey species when alternative prey species are incorporated (Holt, 1984; Comins and Hassell, 1987). The influence of refuge on model stability is more variable, with some kinds of refuge providing stabilisation whereas others are destabilising (McNair, 1986).

**Effects of fishing in trophic models**

*Multi-species virtual population analysis (multi-species cohort analysis)*

Single-species virtual population analysis (VPA) is currently the method that is most often used in fish stock assessments when historical data of catch by age are available (Hilborn and Walters, 1992). It allows the estimation, for a given stock, of matrices of fish numbers and fishing mortality rates by age for each past year from matrices of natural mortality rates and past catch by age and time.

One of the hypotheses on which VPA is based is that natural mortality is known and generally constant over year and age. However, natural mortality varies with fish age; in particular, young fish are more subject to predation than older and larger ones (Stokes, 1992). The basis for extending cohort analysis to a multispecies case is to better estimate natural mortality by taking into account fish diets. MSVPA was first established and applied by Andersen and Ursin (1977), Pope (1979) and Helgason and Gislason (1979) in the North Sea ecosystem. It consists of dividing the natural mortality rate of a cohort into two components: the mortality rate due to predation by other species included in the model and the residual mortality rate due to other natural causes. To estimate the predation mortality rate, the food of each age group is partitioned among the different potential prey, with the available food for a predator only a fraction of the potential food biomass. A coefficient is introduced, which lies between 0 and 1, and represents the suitability of a prey class as food for a predator class. Suitability is determined by prey size, the overlap of the predator and prey in time and space, and the probability of encounter linked to the respective behaviour of the predator and its potential prey. The coefficients represent a substantial synthesis of biological knowledge concerning the different stages of species' life cycles, their spatial distributions, their behaviour, and their feeding habits (Ursin, 1982).

Since its formation in 1984, the "Multi-species Assessment Working Group" of ICES (International Council for the Exploration of the Sea) has been charged with evaluating, each year, North Sea fish stocks using MSVPA, thereby complementing traditional analyses using VPA (Pope, 1989). MSVPA is also applied in the Baltic Sea (Sparholt, 1991). Their analyses have shown that predation mortalities are high (Pope, 1989), and taking them into account substantially modifies the biological reference points that are defined using single-species analyses (Gislason, 1999).

Despite its usefulness in stock assessment, there are some limitations to MSVPA. Mortality rates are assumed to be constant, but projections of the models simulate management measures that could affect the biomass of predators and, consequently, the predation mortality of their prey. A limitation in the model structure is that the choices of species included in the analyses are essentially determined by data constraints. The species that are studied are those that are exploited, i.e. those for which catch data are available, but other species could affect the dynamics of the exploited species. In addition, estimation of the suitability coefficients requires comprehensive ecological knowledge. Since 1981, this has necessitated large annual sampling surveys in the North Sea for stomach content analyses of the main exploited species (Stokes, 1992).

*Ecopath and Ecosim models*

Polovina (1984) and Christensen and Pauly (1992; 1995) developed the Ecopath model to estimate trophic fluxes within an ecosystem. Knowing the magnitudes of the fluxes should allow estimates to be made of exploitable biomass and, reciprocally, the fraction of the system production that is consumed by predators and by fishing activities. In this compartmental approach, some species are aggregated into functional groups that are linked by biomass fluxes. The modelled system is assumed to be stationary, which implies that the gains in biomass equal the losses from each species group. Ecopath models of marine ecosystems provide estimates of biomass, catches, production, consumption, diets, and ecotrophic efficiency for each trophic group at equilibrium. In 1998, Ecopath was used in about sixty applications in different marine ecosystems (Pauly et al., 1998). The different studies synthesize an important amount of local knowledge. Comparisons of the models have yielded some interesting results about the state and functioning of marine exploited ecosystems, with the calculation of some useful trophic indicators like fractional trophic levels or mixed trophic impacts.

Ecopath provides a static representation of the trophic structure of an ecosystem, but does not allow one to explore the consequences of management measures or variations in trophic fluxes. Ecosim was developed by Walters et al., (1997), and is the dynamic version of Ecopath. It re-expresses the linear equilibrium equations of Ecopath as differential equations. In their modelling work, Walters et al. (1997) succeeded in establishing links between the parameters estimated by Ecopath and those included in the differential equations of Ecosim. The use of existing data without requiring additional measurements and experiments is one of the advantages of Ecopath and Ecosim models. However, this is also one of the weaknesses of Ecosim, which is evident at two different levels. First, Ecosim is not applicable over a large range of biomass values because the parameters are estimated from the Ecopath model, which is at equilibrium. Second, the link with Ecopath constrains the choice of functions for growth of the primary producers, and functional responses of the predators. These functions affect the dynamics of predator-prey models, as was shown in a previous section.

*Size-structured models*

With the use of partial differential equations, some authors modelled the biomass flux through ecosystems, from the smallest organisms to the largest (e.g. Silvert and Platt, 1978). These equations generally do not consider the actual species, but only the dynamics of the biomass per size group (length or weight).

Generally, these models were not used when investigating the effects of fishing on fish communities. Recently, Benoît and Rochet (2004) proposed an improved time- and size-dependent continuous model of biomass flux, and they investigated the effects of fishing on size spectra. They found that fishing should affect the curvature and the regularity of the size spectrum. Using a MSVPA-type model that was size-structured, Gislason and Rice (1998) found a linear relationship between the slope of the size spectrum and fishing mortality. This is consistent with empirical results which suggest that the slope and the intercept of size spectra vary quasi-linearly with fishing mortality when the size spectrum is obtained by plotting the log(fish number) against log(fish length) (e.g. Rice and Gislason, 1996; Bianchi et al., 2000). Consequently, linear models can be used to understand and predict the effects of fishing at the level of communities. An individual-based model (Shin and Cury, 2004) that assumed that predation is an opportunistic and size-based process, also resulted in a linear relationship between the slopes of size spectra and fishing mortality. However, when small fish were included in the model, the simulated marine size spectrum appeared to be curved towards the small sizes of fish, suggesting that the smallest fish undergo the greatest predation mortality. The attributes of curved size spectra should provide information about the level of overexploitation in a given ecosystem.

## 3. System-Level Perspectives: from patterns to processes

### *3.1 Ecosystems, food webs and food chains*

Ecosystems are viewed in many ways ranging from complex and changing adaptive systems (Allen, 1988; Mullon et al., 2002) to simplified ecological components or assemblages that interact through known processes (such as predation) acting on the structure of food webs or food chains. There is also a long and continuing controversy in community ecology among those who focus on idiosyncratic aspects of natural systems and their uniqueness, and those who perceive the general principles that structure natural assemblages (Hairston and Hairston, 1997). This controversy is particularly intense when trophic levels are used to represent food web dynamics and food chains.

Food web theory remains controversial largely because of the intricate complexity resulting from numerous interactions and the resulting lack of prediction about population dynamics in natural systems (Figure 14.4). Polis and Strong (1996) believe that communities are too complex to show general patterns and that no natural groupings of organisms into trophic levels are possible; trophic levels are non-operational concepts with no useful correspondence to reality (Polis and Winemiller, 1996).

The quantitative study of food webs has only recently come into its own, stimulated by theoretical work on factors that might constrain patterns of trophic connection within food webs (Morin and Lawler, 1995). According to Hairston and Hairston (1993), food webs provide the theoretical possibility of obtaining detailed information about the flow of energy through a community and can be used to partition the influences of species that feed at more than one trophic level. However, Hairston and Hairston (1993) found some serious difficulties in the use of food webs. Among these is the fact that quantitative assessments of the strength of most of the links is rarely practical and that competition is of necessity either ignored or assumed to exist.

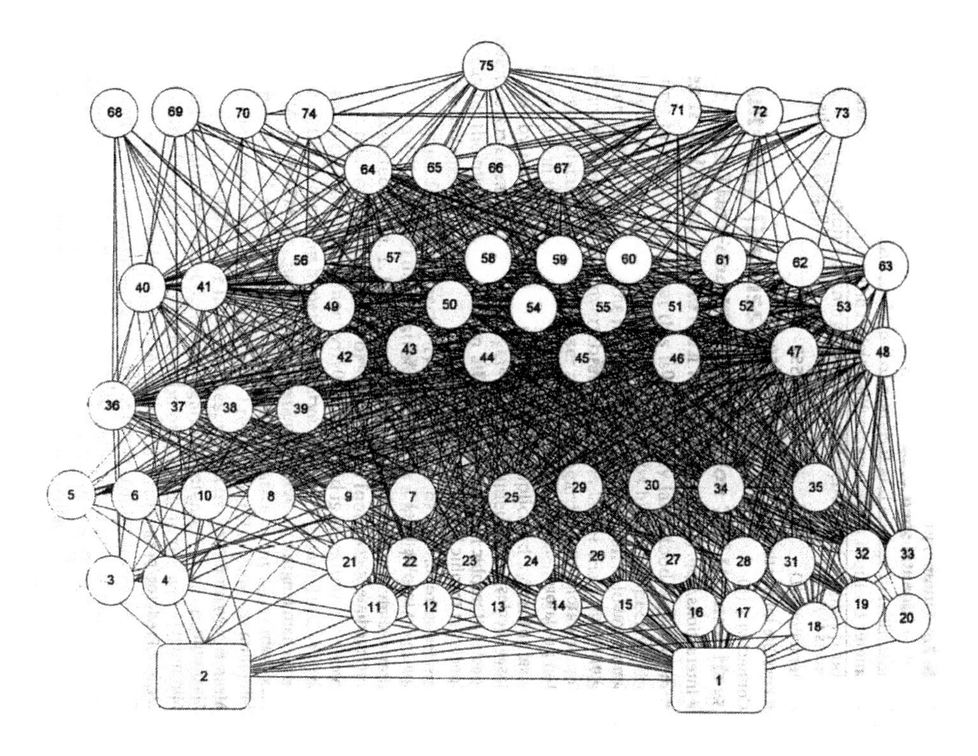


Figure 14.4 Species and links for a northwest Atlantic food web, assuming that interactions among 75 components have similar strength in time and space (from Link, 1999).

Despite the complexity of natural ecosystems, which is real, observable patterns emerge at different levels of organisation and these can be helpful for ecosystem-based management (ICES, 2000). Polis (1999) insisted that food web dynamics appear to be much more important in water than on land, because aquatic systems appear less reticulate and less diverse than terrestrial systems (Strong, 1992). In their famous paper, which has inspired many ecologists, Hairston et al. (1960) interpreted the greenness of the world using a simple holistic argument combined with a reductionist approach. They predicted cascading trophic effects across the food chain and that whether or not organisms are predator- or resource-limited depends on their position in the food chain.

In our quest to understand and generalise ecosystem functioning we need to identify causal factors, and when and where each factor assumes importance. For this purpose we critically review different theories and ways of extracting and analysing patterns in aquatic ecosystems. In the following paragraphs we present different types of patterns, and the associated processes that control them.

## *3.2 Bottom-up control or the control by primary production*

**Ecosystem responses to drastic environmental changes**

Using an analogy with agriculture where crop yields can be predicted from the control of the input, Hensen (1887, in Smetacek, 1999) made the assumption that food supply regulates adult fish stocks, and quantitative studies of phytoplankton and zooplankton production would permit predictions of fish yields (Verity, 1998). From this deduction was born the concept that ecosystems were 'bottom-up' controlled, i.e. the regulation of food-web components is made by either primary producers or the input of limited nutrients (Pace et al., 1999) (Figure 14.5).

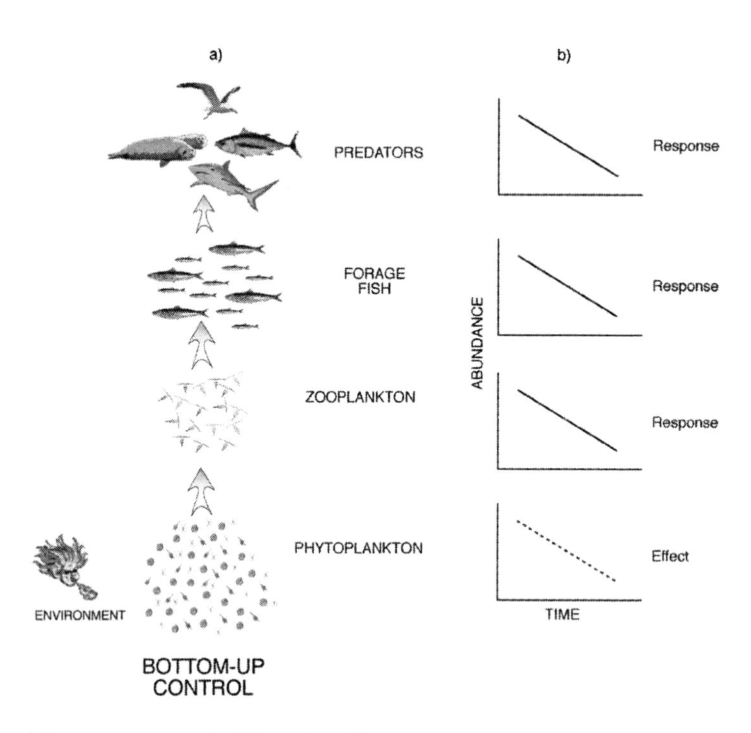


Figure 14.5 a) Bottom-up control within a simplified 4-level food chain in a marine ecosystem; b) The physical environment (sensu Cushing, 1996) being less favourable controls the decrease in abundance of the phytoplankton, which in turn has a negative impact on the abundance of the zooplankton. The decline in zooplankton abundance controls the decrease in abundance of the prey fish, which itself leads to a decrease in the abundance of the predators (the control factor is the dashed line and the responses are the solid lines) (from Cury et al. 2003).

Micheli (1999) analysed twenty natural marine systems, and found that nutrients generally enhance phytoplankton biomass. Plants dominate terrestrial ecosystems but the ocean contains less than one per cent of global (terrestrial and aquatic) plant biomass (Smetacek, 1999). Consequently, nutrient limitation is generally thought to be much more severe in water than on land (Polis, 1999). Large parts of the ocean are still considered to be 'blue deserts', despite the fact that this view is being challenged by modern primary production estimates. For example in the previously-designated "oligotrophic" subtropical Pacific, primary production estimates are 170–220 gC $m^{-2}$ $y^{-1}$, and are potentially as large as 350 gC $m^{-2}$ $y^{-1}$ if one includes dissolved organic carbon production (Karl et al., 1998). In the most productive areas, such as the upwelling areas of the four eastern boundary current regions, primary production averages 672g $C.m^{-2}.y^{-1}$ (Carr, 2002).

Despite certain limitations in productivity for several systems and despite large interannual environmental variability, it appears that the effects of changes in primary productivity rarely cascade upwards to affect biomass of marine pelagic fish consumers (Micheli, 1999). This weak link between primary producers and herbivores can partly explain why parallel long-term trends are only rarely observed across several marine trophic levels, as would be expected under a theoretical bottom-up control scenario (Figure 14.5).

Records of sardine- and anchovy- scale-deposition from anaerobic sediments show that large-amplitude population fluctuations occurred even in the absence of any fishery (Baumgartner et al., 1992), and they are most probably related to environmental changes. However, this has not been formally demonstrated and other hypotheses, involving competition, can account for such patterns (Ferrière and Cazelles, 1999). The effect of the environment on population dynamics has been emphasized in recruitment studies. Food availability (Cushing, 1996) and physical processes (Bakun, 1996) exert a significant pressure on larval fish survival, which determines subsequent fish abundance (Cury and Roy, 1989). In many cases, fish recruitment in the marine environment is viewed as being controlled by bottom-up forces (e.g. Menge, 2000), implying that the availability to the fish of primary production often determines recruitment (Cushing, 1996). Recruitment of fishes that feed on different trophic levels at the adult stage is thus controlled by the environment experienced by their larvae, which usually feed on similar trophic levels at the bottom of the food chain (i.e. fish larvae mostly feed on small zooplankton).

The structure and function of marine ecosystems can respond drastically to interannual changes and interdecadal climatic variations. This has been documented for the California Current, the Gulf of Alaska (McGowan et al., 1998), the North Atlantic (Aebisher et al., 1990) and off Chile (Hayward, 1997). Parallel long-term trends across four marine trophic levels, ranging from phytoplankton, zooplankton and herring to marine birds, have been related to environmental changes in the North Sea (Aebisher et al., 1990). Hollowed et al. (2001) analysed the effect of ENSO (El Niño Southern Oscillation) and PDO (Pacific Decadal Oscillation) on northeast Pacific marine fish production and found that these climatic events, occurring on two principal time scales, play an important role in governing year-class strength of several fish stocks. The North Pacific climatic regime shifts in the mid 1970s and late 1980s (Hare and Mantua, 2000) affected the dynamics of the Korean marine ecosystem (Zhang et al., 2000). Primary production in Korean waters increased after 1988, and was followed by a significant increase in zooplankton biomass after 1991. The 1976 regime shift off Korea manifested itself as reduced biomass and production of saury, but biomass and production of sardine and filefish increased. After 1988, Korean sardine collapsed and were replaced by mackerel. Trends in abundance of zooplankton and salmon in the North Pacific also correspond to changes in the intensity of the Aleutian Low Pressure System (Polovina et al., 1994). It has been suggested that water column stability, determined by strength of the Aleutian Low Pressure System, influences phytoplankton production, which in turn affects species at higher trophic levels.

Interannual environmental fluctuations such as El-Niño events affected the structure of the plankton community, the spatial distribution of fish and invertebrates, the recruitment success of pelagic fish and the mortality of birds and mammals in the northern Pacific (McGowan et al., 1998). In the mid-latitudes of the western Pacific, correlations between SOI (Southern Oscillation index), SST, chlorophyll-a, zooplankton and catch of pelagic fish suggest a bottom-up effect that is more pronounced during certain months of the year, and that modifies the dynamics of the ecosystem in the South Sea of Korea (Kim and Kang, 2000). Large-scale perturbations have taken place during the past twenty years in the Pacific, where a dramatic shift of the atmospheric forcing occurred in the mid-1970s (Hayward, 1997). Interdecadal regime shifts, such as the one experienced in the entire North Pacific Basin and the California Current in the late 1970s to the early 1980s, appear to have altered the productivity of marine ecosystems at various trophic levels (Polovina et al., 1994; Francis et al., 1998). There has been a general increased frequency of southern species moving north, a substantial lowering of secondary productivity and fish landings, a major decline in seabirds, and changes in species composition in most sectors of these ecosystems (McGowan et al., 1998). However, the biological response to the interdecadal regime shift in the Gulf of Alaska is thought to have been in the opposite direction to that of the California Current.

There are large-scale biological responses in the ocean to low-frequency climatic variations. However, the mechanisms by which the climate exerts its influence vary as components of the ecosystem are constrained by different limiting environmental factors. Thus similar species at the same trophic level may respond quite differently to climate change (Hayward, 1997) according to the structure of the ecosystem. For example, in the northern Benguela ecosystem, sardines have been found to respond more quickly to environmental change than anchovy. According to McFarlane et al. (2000), one method of measuring climate change and regime shift is to observe the dynamics of species that potentially could be affected. Obvious environmentally-induced ecological changes are expected in ecosystems, although findings in one system cannot necessarily be extrapolated to others, and predicting the effects of global-scale environmental change on ecosystems does not appear to be a straightforward exercise.

Bottom-up control is the conventional trophic flow control that seems to dominate most ecosystems (Cury et al., 2003). Environmental changes are pervasive, but in most cases difficult to detect at different trophic levels. Literature documenting the relationship between abundance of pelagic fish and environmental variability is plentiful. However the role of the environment on food abundance versus the direct effect of the environment on early life stages are entangled concepts. Several strong patterns, linked to the environment, appear in the form of regime shifts and synchrony between remote fish populations.

**Regime shifts and synchronised large scale fluctuations**

Daan (1980) was the first to review the concept of replacement of a depleted stock by other species. He used a very tight definition of replacement and concluded that the only true replacement of species occurred in the North Sea. Many analyses of alternating trends in abundance of fish stocks have since been undertaken, with species replacement being more loosely defined, and many hypotheses being formulated to explain these "regime shifts". An early indicator of a change in species dominance is an increase in the abundance of the sub-dominant species; a decrease in the dominant species is often only observed at a later stage (Lluch-Belda et al., 1992). Catch statistics (e.g. Kawasaki, 1991), fish scale deposits in sediments (e.g. Baumgartner et al., 1992), biomass research surveys (e.g. Hampton, 1992) and records of seabird guano harvests (Crawford and Jahncke, 1999) have revealed changes in the abundance of anchovy or sardine stocks in many regions of the world. Initially, there were heated debates as to whether collapses of pelagic species were caused by overfishing, thereby allowing competing species to dominate (Francis and Hare, 1994). However, fishing was soon shown to be secondary to other causes when evidence of these fluctuations was found in scale deposits for periods prior to commercial fishing off California (Soutar and Isaacs, 1974; Baumgartner et al., 1992), Peru and Chile and off southern Africa (Shackleton, 1987). Coherent patterns of abundance in the northeastern Pacific Ocean over the past 2,200 years were observed for salmon, sardine and anchovy (Finney et al., 2002). These long-term changes across large regions demonstrate the strong role of large-scale climate forcing on lower trophic levels, which subsequently affects fish and ecosystem changes.

Off western South Africa, guano records suggest that anchovy *Engraulis encrasicolus* was the dominant pelagic fish in the 1920s (Crawford and Jahncke 1999). Horse mackerel *Trachurus trachurus capensis* was abundant in the 1940s and early 1950s, sardine *sardinops sagax* in the late 1950s and early 1960s, chub mackerel *Scomber japonicus* in the late 1960s, *E. encrasicolus* in the 1970s and 1980s and *S. sagax* in the mid 1990s (Crawford, 1999). The sequence of pelagic species succession differed off Namibia; *S. sagax* was dominant in the 1960s, *Trachurus capensis,* pelagic goby *Sufflogobius bibarbatus* and to a lesser extent, also *E. encrasicolus,* were abundant in the late 1970s and early 1980s, whereas *S. japonicus* was dominant in the late 1970s and early 1980s (Crawford et al., 1985; Crawford et al., 1987).

Despite plentiful data showing changes in abundance phases of sardine and anchovy populations in the productive regions of the world's oceans, the mechanisms responsible for initiating, sustaining and terminating sudden increases in population sizes on a decadal time-scale still remain much of a mystery (Lluch-Belda et al., 1992). It is likely that many of the factors accounting for variability in fish stocks play important roles in effecting regime shifts. The mechanisms involved must act on large spatial scales because there is coherence in stock fluctuations in these regions (Crawford et al., 1991; Schwartzlose et al., 1999). Matsuda et al. (1992) modelled environmental effects on pelagic species replacements, listing five possible mechanisms explaining pelagic species dominance in ecosystems: environmental change impacting different species directly, density-dependence in changes of the intrinsic reproductive rate, phase polymorphism of species, competition between species and fluctuations in a one-predator-two-prey-species situation.

There are five regions in the world where anchovy (*Engraulis* spp.) and sardine (*Sardinops sagax*) stocks co-exist and are intensively fished. These are the Japanese system (the western boundary of the North Pacific), the California Current system (the eastern boundary of the North Pacific), the Humboldt Current system (the eastern boundary of the South Pacific), the Canary Current system (the eastern boundary of the North Atlantic) and the Benguela system (the eastern boundary of the South Atlantic). It is generally accepted that sardines in all these regions except the Canary system are the same species, namely *S. sagax* (Parrish et al., 1981). The sardinellas *Sardinella aurita* and *S. eba* occur in the Canary Current system. Anchovies belong to different species of the genus *Engraulis*. In addition to co-occurring in these five regions, both anchovy and sardine (*S. sagax*) are also found off Australia. However, harvesting of these species is limited in this region, therefore data are scarce and stocks in this region have been largely omitted from comparative studies in the literature. The anchovy *E. encrasicolus* and sardine *Sardina pilchardus* co-occur in the Mediterranean Sea. However, catch per unit effort data of the two species in the fishery off the northeast coast of Spain have been less variable than those from oceanic regions where the main pelagic fisheries of the Mediterranean operate (Morales-Nin and Pertierra, 1990). The authors suggest that the environment fluctuates less in this region than in productive upwelling regions, such as off California.

*Linkages between regions*

Shifts in dominance of sardine and anchovy have been discussed in depth by Skud (1982) and Lluch-Belda et al. (1992) (see also Bakun, Chapter 24). Fluctuations in the size of the populations of sardine in the Japan, California and Humboldt Current regions are well matched and are influenced by global scale environmental variation (Kawasaki et al. 1991) (Figure 14.6). In contrast, the species groups of the Benguela and Canary Current systems are out of phase with these three Pacific regions (Schwartzlose et al., 1999). Recently, Alheit and Hagen (1997) showed that alternating periods dominated by herring *Clupea harengus* and sardine *Sardina pilchardus* in the eastern Skagerrak, English Channel and Bay of Biscay are governed by the same climate variations.

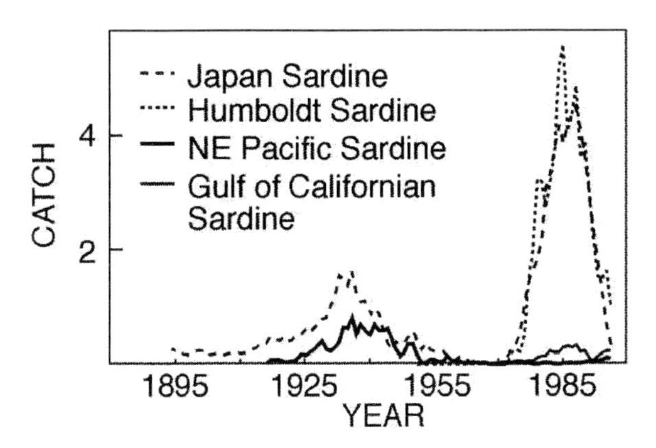


Figure 14.6 Synchronized long-term catches (t) of sardine in the North-Western Pacific by Japan, in the Humboldt off Peru and Chile combined, and in the Gulf of California from 1895–1996. (Data from Schwartzlose et al., 1999).

In the Pacific Ocean (Japan, California and Humboldt systems), fluctuations of sardine populations in the three regions are more closely linked than fluctuations of anchovy or chub mackerel stocks (Crawford et al., 1991). It is possible that the climate influences biological aspects of sardine, which differ from those of the other two species (Crawford et al., 1991). Sardine was found to be more abundant during periods of increased global air and sea temperatures, and anchovy stocks declined during such periods (Lluch-Belda et al., 1992). However, manifestations of other events associated with changes in temperatures, rather than temperature changes themselves, may cause fluctuations in temperature and stocks to coincide (Lluch-Belda et al., 1992). Matsuda et al. (1992) put forward a cyclic model of dominance of pelagic fish species off Japan, and MacCall (1996) showed a similar sequence of pelagic fish dominance off California; a peak in abundance of planktivorous *Sardinops sagax* is followed by a peak in *Engraulis* sp., then the more predatory *Scomber japonicus* dominates and the cycle repeats itself.

There is a significant negative relationship between temperature in the northern Benguela and the Canary Current systems (Crawford et al., 1991). The abundance of sardine in these two regions is negatively related. The two systems seem to be linked largely through the influence of Benguela Niños, which result from the southward intrusion of tropical water into the northern Benguela system off Namibia, associated with cooling in the equatorial Atlantic (Shannon and Pillar, 1986). Anchovy population sizes in the northern Benguela system are related to those in the Canary system one year later, and anchovy population sizes in the southern Benguela are related to those in the Mediterranean the same year (Crawford et al., 1991).

Crawford et al. (1991) investigated the trans-oceanic linkages between the Atlantic and the Pacific through global climate. They postulated that three factors influence trends in abundance of fish species in the two oceans, namely solar radiation, sea surface temperature and ecosystem changes. Solar radiation influences sea surface temperature in the North Pacific in the same year, but influences sea surface temperature in the North Atlantic and air temperature in the Northern Hemisphere two years later. This is reflected in a two-year lag between sardine catches off Japan and England. Sardine in the California, Humboldt, Canary and Benguela current regions have been found to extend into cool waters during warm periods, and also sometimes into warm waters when cooling occurs. Japanese sardine, although always the dominant species, tend to become more abundant as the Kuroshio Current cools. Hence it is possible that warming of cool areas or vice versa can allow sardine to extend its range into new areas (Crawford et al., 1991). Climatic impacts on epipelagic prey species are likely to influence predators and competitors too.

*Changes in spatial distribution*

Associated with changes in the relative abundance of anchovy and sardine is spatial variation of the two species. Both species expand and contract the area across which they occur as stocks increase or decrease in size; sardine spawning distributions usually change in the alongshore direction whereas the spawning ranges of anchovy seem to expand or contract about a geographic centre. MacCall's (1990) "basin hypothesis" states that spawners are expected to contract to the most favourable habitats at low levels of abundance, when effects of density dependence are low. Modelling of anchovy and sardine in the southern Benguela region showed that anchovy and sardine spawned in areas less favourable to survival when they were at low abundances (Shannon, 1998), suggesting that other environmental factors may have restricted suitable areas available for spawning.

**Alternating steady states and pelagic fish assemblages**

Strong environmental effects on fish populations result in large fluctuations in species composition. It also appears that alternating steady states are observed at the level of fish assemblages on decadal scales. For example, upwelling systems tend to be dominated by one species of sardine and one species of anchovy, but most often only one of the two is dominant at any particular time. Alternating patterns between small pelagic fish species have been observed in most upwelling ecosystems over past decades (Figure 14.7). The mechanisms that are generally invoked involve the environmental effects that will favour one or the other species. Analyzing changes in abundance of pelagic species in response to environmental changes, Skud (1982) concluded that dominant species respond to environmental factors, and subordinate species respond to the abundance of the dominant ones. Thus, from an ecosystem perspective, climatic factors are thought to affect fluctuations in abundance of a species whereas its absolute density is rather controlled by intraspecific competition (Skud, 1982; Serra et al., 1998). Recently the competition between species was shown to be magnified by schooling behaviour within mixed-species schools (Bakun and Cury, 1999). Thus the 'school trap' hypothesis constitutes a potential mechanism of competition that could drive one species to ever-lower abundance, rationalizing observed patterns of alternation. These multi-year patterns of alternation are important for long-term management, as exploitation reduces the biomass of the dominant species, which is usually the target species at the time, and sometimes precipitates its collapse. Within a pelagic community, the removal of the dominant species should favour the subordinate species, provided that the latter is only lightly exploited.

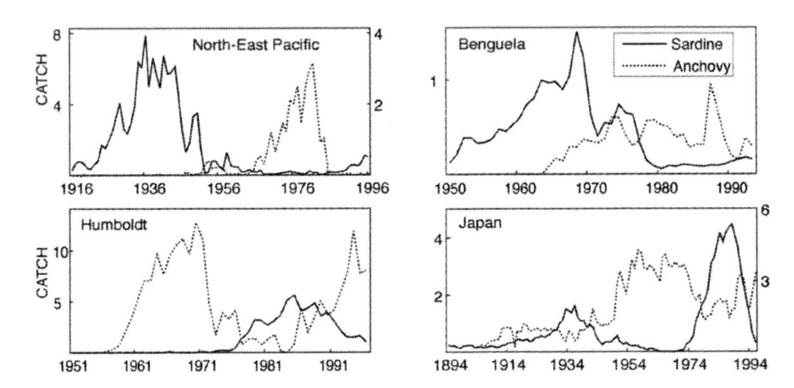


Figure 14.7 Alternation between sardine (solid line) and anchovy (dashed line) as illustrated by the changes of species abundance in catch time series for the north-east Pacific, Japan, the Humboldt and the Benguela currents (Schwartzlose et al., 1999).

### *3.3 Top-Down control or the control by predation*

Trophic interactions between fish species play an important role in ecosystem structure. Predation mortality is believed to be the major source of mortality for marine exploited species (Bax, 1991). By analysing six marine ecosystems (Benguela Current, Georges Bank, Balsfjord, East Bering Sea, North Sea, Barents Sea), Bax (1991) estimated that predation amounted to between two and thirty-five times fishing mortality. This does not imply that effects of fishing are negligible, but rather that fishing has the potential to affect the whole ecosystem because species are tightly linked through trophic interactions. Therefore, it is thought that regulation of ecosystem components at low trophic levels by species at higher trophic levels (termed top-down control) may be critical to marine ecosystem functioning (Figure 14.8).

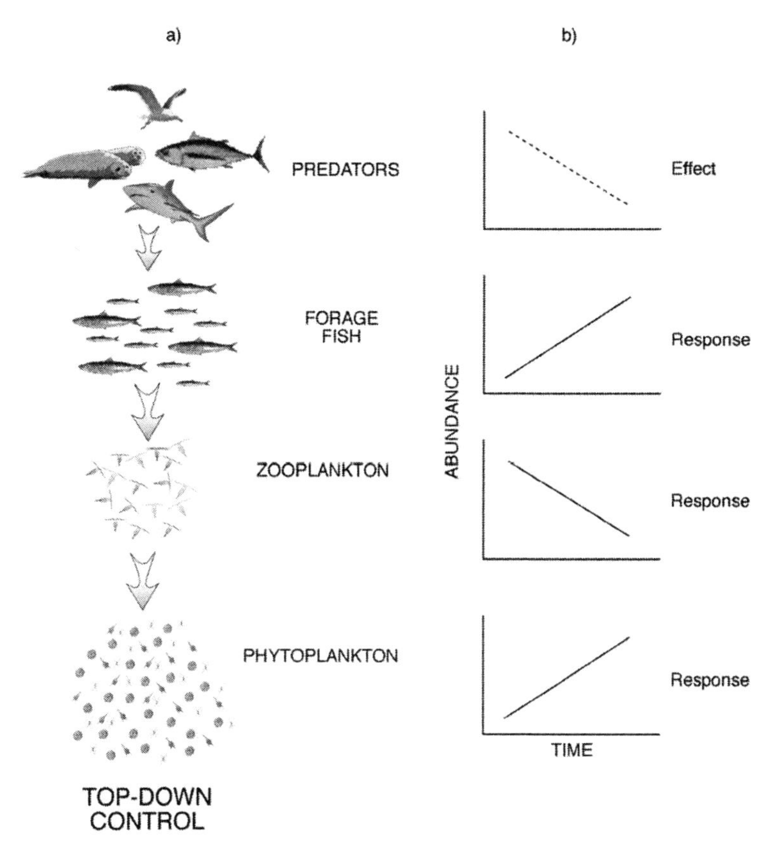


Figure 14.8 a) Top-down control within a simplified 4-level food chain in a marine ecosystem b) The decreasing size of the top predator populations leads to reduced predation on the prey, which in turn leads to an increase in abundance of the prey fish. Increased predation of fish prey on zooplankton leads to a decrease in the zooplankton population size. The smaller zooplankton abundance reduces the grazing pressure on the phytoplankton, which consequently becomes more abundant (the control factor is in dashed line and the responses are in solid lines) (from Cury et al. 2003).

## Size based predation process: an empirical approach

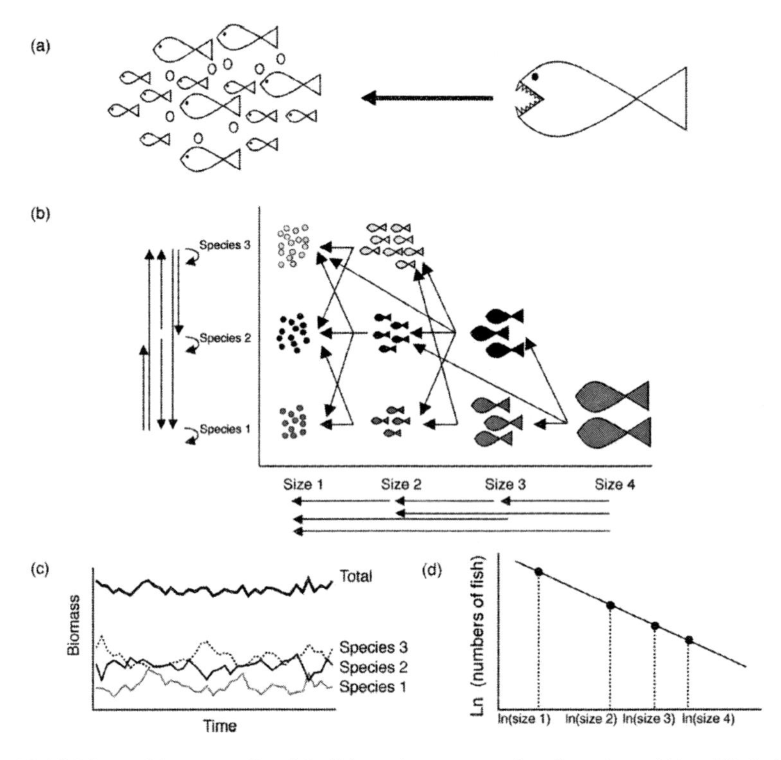


Figure 14.9 (a) Bigger fish eat smaller fish: fish prefer prey smaller than about 1/4 to 1/3 their own size as predators are constrained by the size of their jaw. (b) Who is eating whom? This simple opportunistic feeding behaviour generates complex trophic webs, wherein fish have multiple predators, multiple prey and multiple competitors. A fish can feed on different trophic levels (omnivory), on its own progeny (cannibalism), and on early-life stages of its predators (e.g. eggs and larvae). Three species are represented on the vertical axis, and four size classes on the horizontal axis. Along the axes, the thin arrows correspond to the potential predation interactions between species and size classes. Cannibalism is represented by loops along the vertical axis. Within this framework, the arrows relating fish correspond to a theoretical example of a trophic web. (c) Complexity-stability: a recurrent pattern is the relative stability of the total fish biomass compared to that of individual species. Size-based predation implies multiple and weak trophic interactions between species, which have been theoretically proved to favour stability. (d) Size-based predation provides an explanation for observed size spectra in marine ecosystems. A remarkably linear relationship is obtained when the logarithm of the numbers of fish in a size class is plotted versus the logarithm of the median size of the size class (from Cury et al. 2003).

Ursin (1973) wrote that "[fish] stomach contents are a simple function of local prey availability and suitability, this latter often simply being a function of size". This is the case in marine food webs, where feeding may be considered opportunistic and less dependent on prey taxonomy than on prey size. In aquatic species, the ability of a predator to catch its prey is mainly constrained by body size of the predator compared to that of the prey (Lundvall et al., 1999). Because water is eight hundred times denser than air, a streamlined morphology enables fish to move efficiently through this medium. Appendages for handling and capturing large prey are uncommon among fish. Therefore, size-based predation dominates in aquatic communities (Sheldon et al., 1977). Unlike a lion in the terrestrial world, a predatory fish can only prey on items that are small enough to be swallowed whole. Therefore, because jaw size is related to fish size, it is generally accepted that the size ratio between predator and prey determines whether predation occurs (Figure 14.9a).

Strong patterns emerge from size-based predation (Fig 14.9b). Fish tend to prey on a diversity of species and have a diversity of predators. As larvae, fish feed at the base of the foodweb, but as adults, they have higher trophic levels, feeding on organisms at one or several trophic levels below their own (Rice, 1995). During ontogeny as a fish grows, it moves from one trophic level to another and the relationship between trophic level and the logarithm of the body length is linear, with a steeper slope for top predator species (Pauly et al., 2001). Teleost eggs and larvae are found at the base of the piscivorous food chain. In addition, teleost eggs are mostly of uniform size, about one millimetre in diameter (Cury and Pauly, 2000). Thus "community predation" (Sissenwine, 1984) may occur because each fish species potentially competes with every other fish species. Gulland (1982) aptly compares predation in aquatic versus terrestrial systems by stating that "fish have no direct terrestrial counterparts—a fox or lion does not start competing with mice."

Cannibalism often occurs in aquatic systems and may cause large mortalities of pre-recruits. For example, in the Eastern Baltic cod stock (*Gadus morhua*), cannibalism may be the cause of death of between 31% and 44% of individuals in the first 2 years of life (Neuenfeldt and Köster, 2000). In contrast to the terrestrial situation, two aquatic species can be simultaneously a predator or a prey of each other, depending on their size. For example, North Sea cod preys on herring but adult herring are also able to feed on cod larvae (Stokes, 1992). This suggests that, on a species basis, there are two top-down control mechanisms acting in competition with one another, whereas on the basis of size, there is one top-down control mechanism operating (Figure 14.9b).

Dynamics of marine food webs are complex and evolving, because of the large number of potential interactions between species, size groups or age groups, and across different trophic levels. Despite these complex trophic interactions, strong patterns have emerged at the ecosystem level. These patterns were already recognized by May (1974), who stated that "if we concentrate on any one particular species, our impression will be one of flux and hazard, but if we concentrate on total community properties [ . . . ] our impression will be of pattern and steadiness". One commonly observed pattern in this respect is the relative stability of total fish biomass in spite of the large variations in biomasses of individual marine species (Figure 14.9c). An example is the North Sea ecosystem during the 1970s, when there were large changes in the species composition of catches (herring and mackerel catches sharply declined whereas gadoid catches increased), yet total catch was maintained at a relatively stable level (May et al., 1979). May et al. (1979) believe that year-class strength has been controlled from the top down, i.e. that herring and mackerel biomasses were reduced by fishing and that, as a result of reduced predation by these species on gadoid larvae, gadoids have become more abundant. It has been shown theoretically that stability is favoured by top-down control operating by means of multiple and weak trophic interactions between species (McCann, 2000; Shin and Cury, 2001). Weak interactors may stabilize ecosystems by dampening oscillations caused by strongly interacting species (Polunin and Pinnegar, 2002). This is explained by size-based predation in the marine ecosystem; fish consume a wide variety of prey, thereby exerting stabilizing forces at the population level (Bax, 1998).

Stability extends beyond the population level to size spectra at the community level. Fish abundance and biomass decrease with fish size; depending on the underlying parameters assumed, the relationships are described by linear or dome-shaped functions (e.g. Bianchi et al., 2000). This suggests that energy transfer through marine ecosystems occurs by means of interactions that are size-based rather than species-dependent. Accordingly, bottom-up control by primary producers can affect the scale of an ecosystem's productivity, but top-down control can be the stabilizing force. By targeting the large size classes of an ecosystem, fishing may be considered analogous to apex predation. Variations in the slopes and intercepts of the size spectra of an ecosystem may reflect this top-down control. For example, Pope and Knights (1982) found that heavier exploitation in the North Sea than in the Faroe Bank ecosystem was reflected in the steeper slope of the observed size spectrum for the North Sea ecosystem.

**Keystone species and trophic cascades**

The impact of a species on other species depends on its environment, its abundance and how it interacts with other species in the same ecosystem (Lawton, 1999). Because all interactions are not equally strong, it is not necessary to measure or understand each and every interaction, but rather to determine which interactions are most significant and to focus attention on these. Once this approach was adopted, the important role of certain key species in structuring ecosystems was recognized. Further, it was recognized that representing ecosystems as a network of complex interactions may be misleading (e.g., Figure 14.4).

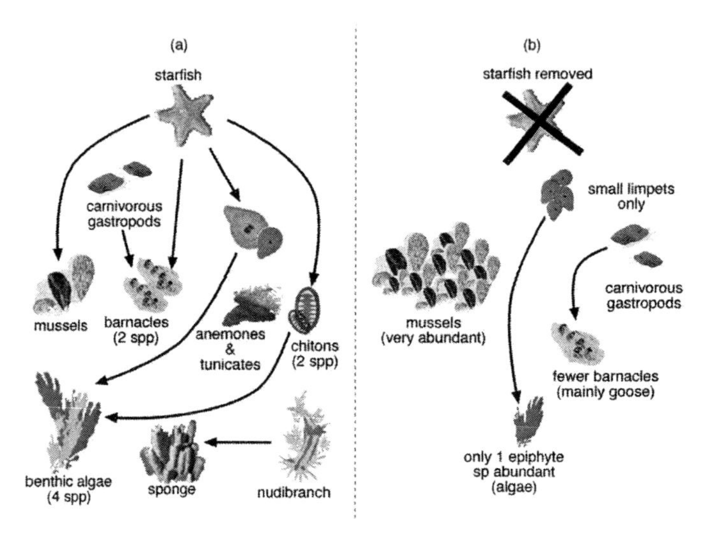


Figure 14.10 Schematic representation of the keystone role of predatory starfish *Pisaster ochraceus* in an intertidal ecosystem in Washington (based on Paine, 1966). (a) *Pisaster sp.* predation maintains a diverse community. (b) removal of *Pisaster ochraceus* allows mussels to dominate, and reduces species diversity.

*Keystone species*

A variety of definitions have been used to describe the term "keystone species", but the most widely accepted is that a keystone is a species "whose impact on its community or ecosystem is large, and disproportionately large relative to its abundance" (Power et al., 1996). In other words, a keystone species affects community or ecosystem level processes to a larger extent than would be assumed likely if one considered only its relative abundance (Bond, 1993); by definition, keystone species are not abundant species. Keystone species interact strongly with other species through predation and competition and also by ecosystem engineering (physical modification of habitat). Although keystone species are often found high in the food chain, they are not exclusively found at the highest trophic levels (Power et al., 1996). Invasive alien species may be considered to be keystone species as their impact in a newly invaded ecosystem may be disproportionately large relative to their initial biomass. Subsequently, such a species may proliferate and become dominant in the absence of predators and diseases (Power et al., 1996). It has been suggested that in the future, conservation management would benefit most from identification and maintenance of keystone species as opposed to attempts to manage all species considered to be important or vulnerable in a given ecosystem (Power et al., 1996).

Most examples of marine keystone species are from the intertidal zone. Paine (1966) was the first to suggest that some species are keystones by showing that predatory starfish are able to prevent other species from monopolizing a limited resource. When the starfish *Pisaster ochraceus* was present in the intertidal region in Mukkaw Bay, Washington, algae, mussels, barnacles, chitons, limpets, sponges and nudibranchs coexisted (Figure 14.10a). However, removal of the starfish allowed its most important prey species, the mussel *Mytilus californianus,* to become abundant. This in turn reduced species diversity and the ecosystem was effectively converted to a mussel monoculture (Figure 14.10b). Although a species can be a keystone in one place over a given period, it is unlikely that it always remains a keystone species (Mills et al., 1993; Menge et al., 1994, Power et al., 1996, Lawton 1999). For example, Sanford (1999) concluded that the environment can have large effects on keystone interactions; during cool upwelling periods, the interaction between *Pisaster and Mytilus* is weakened.

An example of a marine keystone species from beyond the intertidal region is the jellyfish *Aurelia aurita,* which was shown to exert top-down control on zooplankton, thereby determining zooplankton community structure in a shallow cove in Denmark (Oleson, 1995).

Keystone species are rarely positively identified in marine ecosystems. However, in some instances they may operate in conjunction with dominant species, causing major changes in ecosystem structure and functioning to cascade down marine foodwebs.

*Trophic cascades*

Trophic cascades occur when the abundance, biomass or productivity of a population or trophic level is altered across more than one link in a food web, as a result of reciprocal predator-prey effects (Pace et al., 1999) (Figure 14.8). Strong (1992) discussed evidence that trophic cascades were mostly observed in aquatic systems containing few species, representing an exceptional case of food web mechanics, indicating that structure can only be perceived for a subset of the whole ecosystem. Thus the trophic architecture of highly diversified terrestrial ecosystems is more likely a complex web than a trophic ladder. For several authors these assertions provoke questions about why, despite omnivory and the complex linkages of real food webs, manipulations of top predators in communities sometimes trigger chain-like trophic cascades (Power, 1992; Hairston and Hairston, 1997).

A keystone species is usually involved when a true trophic cascade takes place (Paine, 1980), because removal of a species with strong top-down effects causes major perturbations to propagate through a food web, resulting in inverse patterns of abundance or biomass across trophic links. Trophic cascades in lakes (see Carpenter and Kitchell, 1993 for a review) and intertidal zones (Paine, 1980; Estes and Duggins, 1995) were the first to be documented. Initially, it was thought that trophic cascades involved fairly abnormal food web mechanisms and that they were a manifestation of biological instability (Strong, 1992) restricted to certain kinds of marine ecosystems (Hall, 1999). More recently, occurrences of trophic cascades have been reported from a variety of ecosystems, including the open ocean (Pace et al., 1999).

Ecosystems can be strongly impacted by trophic cascades, being stabilized in alternate states through these types of trophic interactions. Predation can result in cascading effects by limiting grazer abundance, thereby enhancing biomass of producers (Posey et al., 1995). The suite of interactions between sea otters, urchins and kelp in Alaska (Estes and Duggins, 1995) is a good example of how the appearance and properties of an ecosystem can be affected by trophic cascades (Pace et al., 1999). Sea otters are considered to be keystone species in the Alaskan ecosystem and, when abundant, their predation pressure on urchins reduces urchin grazing on kelp, stabilizing the kelp forest system. When sea otter abundance is low, urchins proliferate, heavily grazing the kelp and reducing its productivity. Another example of a trophic cascade has been described in the sub-Arctic North Pacific, where pink salmon (*Oncorhynchus gorbuscha*) feed on macrozooplankton and phytoplankton, thereby controlling the biomass of these plankton groups in summer. Shiomoto et al. (1997) found inverse relationships between biomass of the planktivorous pink salmon and biomass of zooplankton, and between zooplankton and phytoplankton biomass. A third example involves trophic cascades in marine pelagic ecosystems in which changes in the abundance of consumers can have cascading effects down the food chain, sometimes all the way down to phytoplankton (Micheli, 1999). However, weak plant-herbivore interactions make true trophic cascades difficult to detect in these systems; zooplanktivores negatively affect mesozooplankton biomass, but mesozooplankton seldom affects phytoplankton (Micheli, 1999). Polunin and Pinnegar (2002) suggest that omnivory, which is common in marine systems, may mask trophic cascades.

Coastal fisheries and their management appear to be having profound effects on ecosystems by altering trophic cascades (Pace et al., 1999). The reduction in abundance of pinnipeds in western Alaska may be related to overfishing of their prey, as well as to climatic change (Estes et al., 1998). This could have forced killer whales (*Orincus orca*), which prefer to prey on marine mammals, to consume more sea otters, with severe consequences for urchins and kelp. In addition to the potential direct effects of fishing on fish species, there is also competition between fisheries and top predators for valuable prey resources. For example, off South Africa the Cape fur seal preys on many commercially exploited fish species. A suggested solution to reduce this competition with fisheries has been to curb seal population growth by culling. However, there is controversy about the success of such a proposal; seal culling has not been shown likely to cause cascades through the ecosystem, nor to have any beneficial effects on stock sizes of fish species consumed by these seals (Yodzis, 2001). In fact, direct competition between marine mammals and fisheries generally seems to be limited (Trites et al., 1997). Despite this, there may be indirect competition for primary production to sustain both marine mammal stocks and fisheries catches. This may result in what has been termed "food-web competition", particularly as fisheries continue to expand at a fast rate (Trites et al., 1997).

We have seen that there are difficulties in identifying keystone species, particularly because certain species may only be keystones for certain states of an ecosystem or over certain periods. Similarly, we have seen that trophic cascades are transitory and dynamic. Their effects are not necessarily felt all the way to the lower trophic levels of an ecosystem (Pace et al., 1999). Decades of intense fishing may destroy cascades that previously occurred; it is possible that reduction in populations of top predators as a result of fishing may allow prey populations to increase and in that way enhance their effects in the marine community (Steneck, 1998). Managing fisheries such as these solely on the basis of the keystone characteristics of a species and whether trophic cascades occur would be impractical, unless there is strong evidence to support these mechanisms (Hall, 1999). Despite these complications, a top-down approach can be helpful in understanding many of the ecological patterns observed, and in estimating the possible consequences, at an ecosystem level, of removing top predators.

To improve fisheries and an ecosystem approach to management, Mills et al. (1993) propose that scientists focus their attention on determining the strength of interactions between species, rather than attempting to identify which species are keystones in each ecosystem, as has often been the focus in the past. The former makes more sense given that exploited species are not usually keystones and therefore changes in their abundance often have little impact on their prey or competitors (Jennings and Kaiser, 1998). For example, removal of substantial proportions of abundant forage fish may have large impacts on their prey, competitors and predators, in a similar manner to removal of a keystone species (Shannon and Cury, 2004).

**Top-down effects of fishing and 'fishing down the marine food web'**

Elton (1927) first noted that food chains are short and their length is variable among natural systems within a rather low range. Although five- to six—link food chains are possible, energy transfer between trophic levels is inefficient, and most communities fall in the range of one– to four-link food chains (Morin and Lawler, 1995; Menge, 2000). Mean chain length is shorter in terrestrial than aquatic ecosystems (Hairston and Hairston, 1993). Thus, Pimm (1991) reports that the number of trophic levels in terrestrial systems is typically three whereas aquatic pelagic systems are typically characterized by four trophic levels. Shelf and coastal systems have longer food chains than do upwelling and oceanic ones (Christensen and

Pauly, 1993). However, there are exceptions. For example, in models of the southern Benguela upwelling system, which spans both upwelling and shelf areas, the maximum trophic level was found to be 5.2 compared to 4.7 in the northern Benguela (Shannon, 2001). Thus adding or removing one trophic level can strongly affect the control operating at a particular trophic level in an ecosystem.

By altering productivity levels in a food-web in a northern California river, Wootton and Power (1993) observed alternating control by trophic level, and between food chains of three and four links. Menge and Olson (1990) predicted that, rather than alternating, predation increases in importance and competition decreases in importance from high to low trophic levels, and that, in addition, food chain length decreases with increases in environmental stress. Fretwell (1987) hypothesized that food chains varied in length as a consequence of variable environmental gradients of nutrients and productivity, emphasizing the role of productivity gradients as the prime determinant of food-chain dynamics in terrestrial systems. Exploring food-chain length in lakes, Post et al. (2000) found that ecosystem size, and not resource availability, determines food-chain length in natural ecosystems.

Food chain length is a topic that has received a lot of attention in terrestrial ecology but consensus has not been reached regarding its role in ecosystem functioning. The debate now shifts to a search for when and where a suite of interacting constraints operates to determine variation in food-chain length (Post, 2002). Because of the intricate nature of predation in the marine environment and the difficulty in assigning a trophic level to a particular fish species or group of fish species, the literature on the length of marine food-chain is scarce. However examination of the global fisheries suggested that the mean trophic level of the catch has been decreasing as a result of 'fishing down marine food webs' (Pauly et al., 1998; 2000). By removing top predators first, i.e. shortening food chain length, predatory pressure was released on small forage fish that constituted the food of top predators. This could potentially lead to a subsequent increase in small forage fish biomass, and hence of harvests. When analysing global fish catches, it appears that pelagic fish species reached a plateau in the mid-1980s, i.e. ten years after the stabilization of the demersal fish catch, suggesting a possible increase in forage fish species during the heavy exploitation of their predators in the 1970s and 1980s (Cury et al., 2000).

As noted by Caddy and Garibaldi (2000), such a decline in trophic level could in some cases be a 'bottom-up' effect due to an increase in nutrients in naturally nutrient-limited marine production systems, even if all levels of the food web are being exploited at a constant rate. In fact, an increase in forage fish abundance, in turn, might even lead to subsequent increases in fish predator biomass and harvest. Analysing the FAO database, Caddy and Garibaldi (2000) found contrasting results suggesting that some ecosystems are controlled by a change in marine productivity, i.e. they are bottom-up controlled, whereas others seem to be controlled by the abundance of the predators, i.e. a top-down control. Since the decline of piscivores in the Northwest Atlantic in the early 1970s, increased biomass of benthivores has been observed. In the North Sea the 'outburst' of gadoids, which was favoured by increasing zooplankton (*Calanus*) and pelagic fish abundance, was followed by a decline in the biomass of planktivores (herring and mackerel), related to high gadoid abundance (the so called 'gadoid dome'). Intensive fishing competed with sea bird breeding colonies for sand eels. In the Mediterranean Sea, the supply of nutrients from land run-off has been proposed as a major factor affecting zooplanktivores and resulting in peak landings of piscivores (e.g. swordfish and hake). Sharp increases in planktivores in the Eastern Central Atlantic and in the Southeast Pacific do not appear to be primarily related to depletion of predators and in the North Atlantic, removing top predators may be the likely cause of the increase in landings of shelf planktivores. By contrast, in the Mediterranean, the increase in planktivores seems to be a bottom-up response to increased marine productivity. These analyses are based on data that were not collected specifically to study ecosystems, and therefore that basically do not have the resolution to do so. Overall they suggest contrasting patterns, as opposed to gradual and continuous changes. Caddy and Garibaldi (2000) propose that 'punctuated equilibria' or 'regime shifts', involving actual changes in ecosystems, are important phenomena that reflect ecological changes but also changing exploitation strategies. In addition, different strong and changing patterns seem to exist. Caddy and Garibaldi (2000) concluded from their analysis that the release of predatory pressure on zooplanktivores, leading to expanded biomasses, is ambiguous at the high scale of aggregation of the FAO database.

In the Black Sea, several authors attributed the numerous ecological changes that have occurred during the last decades to anthropogenic eutrophication (Zaitev and Mamaev, 1997). Using both a statistical and a mass-balance dynamic modelling approach, Daskalov (2002) explored the relative contribution of bottom-up versus top-down control. Time series of piscivorous and planktivorous fish, zooplankton, phytoplankton and phosphate content in surface waters of the Black Sea were considered between the 1950s and the 1990s. Alternating trends across these consecutive trophic levels were found. The author attributed this cascading effect to the overexploitation of large predators, such as dolphins, mackerel, bonito and bluefish. As the fish predators were depleted, the planktivorous fish biomass increased considerably in the early 1970s, and consumption of the zooplankton increased. Jellyfish biomass also increased considerably during the 1980s, affecting zooplankton abundance. The subsequent increase in phytoplankton biomass apparently resulted in the depletion of nutrients in the surface layer since 1975. Comparing different scenarios using Ecosim, Daskalov (2002) concluded that top-down control was the more likely determinant of the structure of the Black Sea ecosystem, given that both overfishing and anthropogenic eutrophication are responsible for these observed changes.

Ecosystem changes in the Northwest Atlantic have been drastic during the last three decades. The cod biomass decreased from 2.5 to 0.05 million tons due to overexploitation (Bundy, 2001). A trophic cascade operated in the 1990s, during which time capelin abundance, released from strong predation pressure from finfish, increased; zooplankton decreased in abundance, whereas primary producers increased in abundance. Meanwhile the abundance of harp seal increased from 1.9 to 5.0 million animals, which, it is believed, might prevent cod from recovery (Bundy, 2001).

The effect of removing top-predators by overexploitation and the resulting top-down effect or "fishing down the food web" is sometimes well illustrated, like in the Black-Sea or in the Northwest Atlantic. However, fishing out a top predator can also cause complex changes in community dynamics. Such an effect of the overexploitation of large apex fish species was recently illustrated. Ecologists have long speculated how predatory fish species are able to achieve large body size, given that their juveniles must grow through a predation-competition phase involving the very species that will be their prey later in life. According to Walters and Kitchell (2001), large, dominant fish species that are the basis of many fisheries may be naturally successful due partly to "cultivation" effects, where adults crop down forage species that are potential competitors/predators of their own juveniles. This "predator-prey role reversal" (Barkai and McQuaid, 1988), or cultivation effect, is apparently common in freshwater communities, and may also explain low recruitment success due to depensatory effects of some major marine stocks following severe declines (such as Newfoundland cod). Implications for fisheries are twofold: by depressing the abundance of large fish predators, fisheries release the top-down control on small fish, and in doing so they increase the risk of depensatory effects due to competition/predation by small fish, preventing the rebuilding of stocks of large fish species (Figure 14.9b). In the Central Baltic clupeid predation on cod eggs (Koster and Schnack, 1994) has resulted in the system being either cod-dominated or clupeid-dominated.

Shortening the food chain by removing top predators can consequently result in a lack of resilience, which is observed for populations of most large fish species (Hutchings, 2000; 2001). However top-down control mechanisms are sometimes dampened by redundant species and complex interactions that are unfortunately rarely documented. For example, Link and Garrison (2002) noticed a shift in the abundance and size composition of fish predators during the last four decades in the Georges Bank fish community. These changes were attributed to fishing pressure. A remarkable shift from cod to spiny dogfish was observed but there were no apparent changes in total fish consumption by the six major predators, despite changes in predator size, structure and abundance. There was a shift in the dominant piscivore in the Georges Bank ecosystem, revealing that exploitation and competition between cod and spiny dogfish were certainly high and that no cascading effects were found in this particular ecosystem. A cross-ecosystem comparison suggests that trophic cascades are stronger in the aquatic environment than in the terrestrial one, despite high variability among systems (Shurin et al., 2002).

### *3.4 Wasp-waist control or the control by dominant species*

Recently the role of dominant pelagic fish has been emphasized as they might exert a major control on energy flows, and this has been termed wasp–waist control (Figure 14.11). In upwelling systems few pelagic fish species occupy the intermediate trophic level, feeding mostly on phytoplankton and/or zooplankton. These species can exhibit large biomasses, which vary radically in size according to environmentally driven recruitment strength. These characteristics are thought to inflict constraints on lower and higher trophic levels. Hairston and Hairston (1993) mentioned that planktivorous fish could reduce herbivorous zooplankton, which would lead to an increase in phytoplankton density. This idea was further developed using data on lakes (e.g. Carpenter and Kitchell, 1993) and the topic received considerable attention from both terrestrial (Schoener, 1989) and aquatic ecologists (Persson et al., 1991). Consumption efficiencies of herbivores on primary producers are much higher in freshwater pelagic communities (32%) than in terrestrial communities (3%) (Hairston and Hairston, 1993). By comparison, the topic only recently received attention in the context of marine ecosystems. Pauly and Christensen (1995) estimated that for the period 1988–1991, 8% of global aquatic primary production was required to support the sum of the mean reported world fish catches (94.3 million tons) and discards (27 million tons). This value is higher if one considers upwelling systems alone; Pauly and Christensen (1995) calculated that a mean of 25.1% (confidence interval 17.8–47.9%) of primary production in upwelling systems was required to sustain catches and discards, suggesting strong links between trophic levels. Micheli (1999) found that interannual fluctuations in mesozooplankton biomass were negatively correlated with those of zooplanktivorous fish, indicating that fish predation can potentially control mesozooplankton biomass. By means of meta-analysis, top-down control of zooplankton by sardine, sardinellas, herring or anchovy was also detected off South Africa, Ghana, Japan, in the Black Sea (Cury et al., 2000), as well as in the northern Baltic (Arrhenius, 1997). In the Central Baltic Sea, Kornilovs et al. (2001) also showed an influence of sprat biomass on the production of cladocerans in summer.

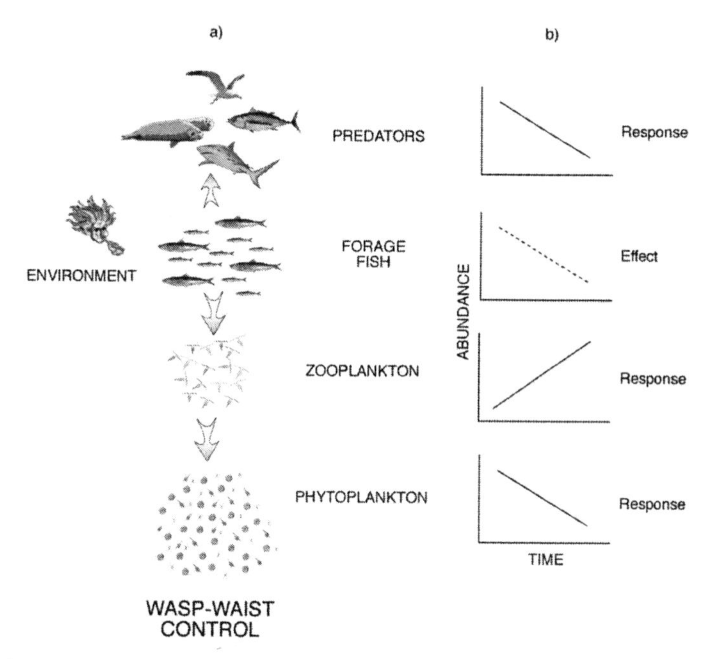


Figure 14.11 a) Wasp-waist control within a simplified 4-level food chain in a marine ecosystem b) The abundance of the prey fish (small pelagic fishes), which is dependent on the environment, controls both the abundance of predators and primary producers. A decrease in abundance of prey fish negatively affects abundance of the predators. The same decrease in abundance of prey fish reduces predation on zooplankton, which becomes more abundant. A larger zooplankton population increases grazing pressure and diminishes phytoplankton abundance (the control factor is a dashed line and the responses are a solid line). The environment (sensu Sinclair, 1988) is considered to have a direct physical effect on pelagic fish recruitment, but no effect on the whole food-chain (from Cury et al. 2003).

Conversely, bottom-up control of fish predators by small pelagic fish has been noticed, as several (but not all) predatory fishes suffer when their prey stocks collapse in the Benguela, the Guinea, and the Humboldt currents (Cury et al., 2000). When pelagic fish stocks recover, the depleted predators may recover quickly, or with delays of a few years to several decades, highlighting the complex response of the ecosystem to change. Despite great plasticity in life-history characteristics, many bird populations are unable to counter the effects of longer-term fluctuations in prey resources (Crawford, 1999). An example is birds off Namibia in the 1970s; horse mackerel and pelagic goby replaced sardine in the diets of seabirds after the sardine stock collapse. These fish were unavailable to penguins and gannets at colonies situated south of Lüderitz as they were either distributed too far north or they occurred too deep in the water column. The result was massive declines in seabird populations in this region (Crawford et al., 1985). A comparative analysis of trends in seabird abundance in Peru and southern Africa emphasizes the dependence of seabirds on anchovy and sardine for reproductive success and growth (Crawford and Jahncke, 1999).

These few examples illustrate a wasp-waist control, under which abundant small pelagic fish constitute mid-trophic-level populations that exert both bottom-up control on top predators and, more surprisingly, top-down control on zooplankton. These effects can eventually cascade down to phytoplankton in freshwater ecosystems, but this is rarely the case in the marine environment. Reproduction of marine animals is mediated by spatially and temporally varying oceanographic processes, affecting most of their life cycles. Marine animals with a pelagic larval stage are likely to offer ample opportunities for differential reproductive success. Under wasp-waist control, the environment plays a direct role (sensu Sinclair, 1988 and Bakun, 1996) in determining the strength of pelagic fish recruitment, i.e. it has an effect on any particular trophic level of the food chain (Figure 14.11).

### *3.5 Mixed and/or changing controls*

Scientists who agree that trophic interactions are important debate whether primary control is by resources or predators. The issue of the relative contributions of top-down versus bottom-up control is not new; it has been debated extensively in terrestrial ecology but consensus has not yet been reached (Power, 1992; Matson and Hunter, 1992). There is general agreement that top-down and bottom-up forces act on populations and communities simultaneously, and that understanding their relative contributions is an important step for future ecosystem approaches to management (Cury et al., 2003). Hunter and Price (1992) offer a compelling argument for the primacy of bottom-up forces in food webs: "... the removal of higher trophic levels leaves lower levels intact (although perhaps greatly modified), whereas the removal of primary producers leaves no system at all". But one must recognize that this assertion leaves almost unanswered the challenging question of what are the factors that can change consumer efficiency. The discussion is no longer about which type of control occurs, but rather about what controls the strength and relative importance of the various forces under varying conditions (Matson and Hunter, 1992).

Link and Brodziak (2002) have compiled a comprehensive report on a suite of metrics that are available for analysing changes and functioning of the Northeast

U.S. continental shelf ecosystem over the last forty years. They categorized these metrics into abiotic and biotic. Abiotic metrics include physical metrics such as the North Atlantic Oscillation, a measure of the air pressure difference between two sites, and temperature anomalies. A possible shift is suggested from a cool to a warm temperature phase. Chemical metrics were not readily available as they require synoptic coverage. Link and Brodziak (2002) propose that key chemical indicators be identified from the array of important nutrients, metals, and toxins that can be monitored. Biotic metrics were plentiful, and from these the authors concluded that the observed changes in the Northeast U.S. continental shelf ecosystem are attributable to top-down forcing (fishing is the dominant top-down effect), as well as strong bottom-up environmental forcing of the zooplankton community. The environment is listed as a second key forcing factor which can determine recovery. There has been a shift in relative biomasses of demersal and pelagic fish communities: demersal fish abundance has decreased whereas pelagic fish have increased. Although species composition has changed over time, the standing biomass of the whole ecosystem and its main sub-components (e.g. phytoplankton, zooplankton, different feeding guilds) have remained relatively constant. Available biotic metrics included a) biomass trends (e.g. zooplankton, separate species), b) species diversity indices (species dominance measures, species evenness etc.), c) food web indices (linkage, density, number of species eating or being eaten by a given species), total consumption, diet compositions, ratio of consumption of a fish species and the landings of that species) and d) system-level indices (energy, exergy, free energy, information content, system-level consumption, flux rates, resilience, stability, persistence, resistance). The third category of metrics considered is human metrics. Human metrics are those derived from effort, landings and profits, and include landings, trawl revenues, fishing effort, CPUE, fishing activity and landed values. Link and Brodziak (2002) write "changes in fishing practices and fishing communities, such as diversification to target non-groundfish resources, have probably contributed to sustaining the fishing fleet while target species abundances have declined and regulations have increased".

Flexible views of the varying roles of top-down and bottom-up forces need further examination to reach any possible generalization (Power, 1992). As discussed previously, in fisheries ecology there has been a long tradition of relating climate and fisheries (Cushing, 1982), assuming that recruitment strength is limited by food availability. With overexploitation, the importance to ecosystem functioning of removing large apex predators has also been emphasized. In rocky intertidal habitats the role of predation has been noted for a long time by marine ecologists. These studies, among others, have been developed by scientists who were interested in evaluating the impact of a particular type of control in ecosystems. For marine ecosystems, the context and complexity of the discussion is expanding, with new case studies being presented and new ways emerging of considering the relative importance of top-down versus bottom-up control. In the following section we discuss ecological factors that can affect controls, the evidence of mixed and changing controls for fish communities, and how better spatial and temporal resolutions can help to understand the contributions of the different controls.

**How ecosystem structure affects trophic controls**

A particular structure or configuration of an ecosystem can reveal the type of trophic control operating. This can be observed, for example, in contrasted environmental contexts. Most studies in marine rocky intertidal habitats have demonstrated the existence of strong top-down control by consumers. Menge (2000) reviews twenty examples of top-down control in different intertidal habitats documented during the 1980s, many of them involving mussel and barnacle interactions. In those examples, strong top-down effects of predators were capable of controlling prey communities on temperate and tropical rocky coasts. However, in many comparative studies, key physical environmental gradients can mitigate the impact of top-down forces, such as hydrodynamic forces and thermal/desiccation stress. Assuming that recruitment of a species at the basal level of food webs is a bottom-up effect when it increases the abundance of prey organisms, Menge (2000) documented five case studies in South Africa, New England, Oregon and New Zealand where both top-down and bottom-up controls were acting simultaneously on rocky intertidal community structure (no fish species were involved). Menge (2000) concluded that bottom-up effects are in fact tightly linked to top-down processes and can be important determinants of community structure in rocky intertidal habitats, although present knowledge is limited and the field needs to be expanded.

Changing control in fish communities has been documented in several other cases. In the northern Baltic the increase in herring abundance in the late 1980s appeared to be the result of weakened top-down regulation as a result of a sharp decline of its main predator, the cod. Meanwhile the herring was affected by the availability of suitable-sized plankters, which was environmentally controlled, i.e. by salinity. A possible top-down effect of cod on herring and a bottom-up process mediated via changes in mesozooplankton species composition seem both to have contributed to the dynamics of the herring in the Baltic (Flinkman et al., 1998). A second example of changing control is the overexploitation of the sparid and serranid fish communities off North-West Africa in the 1960s releasing predation pressure on octopuses, which in turn became abundant successively in Morocco and Mauritania at the beginning of the 1960s, and in Senegal in the mid-1980s (Gulland and Garcia, 1984). Weakened top-down trophic control led to strong relationships between upwelling intensity and octopus recruitment in the three upwelling systems (Faure et al., 2000).

Just as trophic cascades discussed previously are transient and dynamic, it appears that bottom-up and/or top-down control may operate according to the structure of the ecosystem. In some instances described, both top-down and bottom-up control appear to be acting together on the same species (e.g. both control types have been reported to affect the northern Baltic herring), whereas in other instances, one type of control has replaced another when the "dominant" control mechanism regulating a species group has diminished (e.g. overfishing has disturbed the usual top-down control of octopus by sparid fish, so that environmental factors became more important regulators of octopus). Trophic controls can be changed by the structure of the ecosystem, but also by spatial or seasonal constraints.

**Spatial and temporal dependence of trophic controls**

The interactions that occur within an ecosystem produce structure and determine the way the system functions; the temporal and spatial scales of the processes occurring in marine ecosystems are inextricably linked (Murphy, 1995). The marine environment is a dispersive and heterogeneous one and natural variability in ocean circulation and mixing plays a major role in generating fluctuations in marine productivity, as well as in the distribution of populations (Sinclair, 1988). Food availability but also physical constraints such as retention, concentration or enrichment processes that are associated with currents and turbulence are now considered as important factors that affect larval survival, fish recruitment and ultimately stock abundance (Cury and Roy, 1989; Bakun, 1996). Fish populations have geographical closure of their life cycles and climatic factors can affect the spatial context of marine populations in many ways by modifying the dynamics of the spawning or feeding areas, consequently changing recruitment success or migration patterns (Sinclair, 1988). Fish have large-scale ontogenetic habitat shifts in marine ecosystems and predation is one of several ecological constraints that shape their distribution. Since interactions can be spatially patchy (highly localized), most processes are scale–dependent. For example the population dynamics of predators and prey within a specific region can be uncoupled (Murphy, 1995) and rather than addressing the whole ecosystem, we can instead concentrate on a particular region at a particular time period.

Seasonal or year-to-year environmental fluctuations as well as spatial heterogeneities have consequences for ecosystem dynamics. A preliminary approach would be to visualize the distribution of different biomasses per trophic level for different species (or group of species having the same food requirements) as potential patterns of interactions within an ecosystem. This representation can provide a baseline for exploring local trophic control between two or more different species (Drapeau et al., 2004). For example, on an island where marine seabirds breed, the foraging range and distribution of seabird prey (e.g. anchovy) in time and space can provide valuable indicators of potential interactions.

Time lags are obviously more difficult to appreciate. In the "match-mismatch" hypothesis, Cushing (1969) developed the idea that the production of fish larvae matches or mismatches the production of their food, resulting in successful or unsuccessful recruitment. Consequently the control of the abundance of fish recruits depends on the abundance but also the availability of their prey, the zooplankton. In this particular case the temporal variability of the physical environment may or may not be translated into effects at higher trophic levels in the food chain. Several recent papers tentatively explore the relationship between recruitment and dynamics of the food chain in order to explain fluctuations in fish stocks. The importance of seasonality on ecosystem dynamics was illustrated in British Columbia by exploring the links between time series from oceanographic data, zooplankton data and seabird breeding data (Bertram et al., 2001). When spring is early and warm, the duration of the overlap of seabird breeding and copepod availability in surface waters is reduced, creating a mismatch of prey and predator populations. Cooney *et al.* (2001) identify a critical time-space linkage between the juvenile stages of pink salmon and herring in shallow-water nursery areas and seasonally-varying oceanic states, the availability of appropriate zooplankton forage, and the kinds and numbers of predators. These ecosystem-level mechanisms influenced the mortality of the fish, which were shown to be habitat-dependent and to exhibit strong food-type preferences.

Using different dynamic models of interactions in a spatially structured ecosystem, Murphy (1995) explored predator-prey linkages in Southern Ocean food webs. Krill is strongly influenced by large-scale abiotic factors, i.e. over the ocean basin scale. The introduction of spatial structuring changes the relative prey availability for the different predator groups and increases the complexity of the responses. This analysis reveals that there are probable shifts between the influence of food concentration and through-flow systems within and between seasons. It also reveals that analysing interactions only in the temporal dimension may result in misleading inferences being made about key processes. This reinforces the need for considering both spatial and temporal dimensions in an ecosystem approach.

A satellite-based estimate of potential primary production in the four eastern boundary currents (i.e., California, Humboldt, Canary, and Benguela Currents) produced values 4–150 times larger than the ones derived from the observed current fish catch, using an idealized food chain of 2.6 links and an average trophic efficiency of 10% (Carr, 2002). This discrepancy between potential and observed productivity was attributed in the Benguela to the temporal mismatch between feeding and high primary production (Shannon and Field, 1985). For the different eastern boundary currents the explanation apparently lies in the different trophic structures and spatial accessibility of the food in the ecosystems. Carr (2002) estimated that the yield for each upwelling system was an upper bound that would be decreased to 10–20% by environmental accessibility. Thus, despite large primary production found in upwelling systems, food availability affected by spatial and temporal constraints could lead to the unexpected result that small pelagic fish populations are likely to be food-limited (Micheli, 1999; Cury et al., 2000).

Hunter and Price (1992) offered a synthetic framework regarding the controversy around top-down versus bottom-up control and suggested that instead of asking "Do resources or predators regulate this particular population?" ecologists should rather ask "what factors modulate resource limitation and predation in this system, determining when and where predators or resources will dominate in regulating populations?" The relatively few studies on trophic controls in different parts of the ocean, compared to lakes or terrestrial ecosystems, do not allow generalizations to be made. However, a number of recent studies on trophic interactions should help to understand the respective role of top-down versus bottom-up controls in marine ecosystems. The examples presented above are recent, tentative ways of seeking mechanistic rather than correlative understandings of complex natural systems. They are surely the first examples of a long, future list that will recognize the importance of interactions in time and space as one of the key elements in understanding dynamical trophic controls in marine ecosystems.

## 4. Discussion

Fish interact strongly, mainly through trophic interactions. Predation is pervasive and plays an important role in shaping marine communities. However everything is not strongly connected to everything else, and not all production is suitable for the next trophic level. Spatial and temporal constraints, particularly those associated with habitat selection, partly determine the strength of the interactions. Components of the ecosystems are linked in a variable and changing manner, suggesting that there is no need to measure or understand everything, but rather to determine the significant interactions, i.e., when, where and how they can potentially structure the dynamics of the ecosystems. In the speciose marine environment, predation interactions among fish appear to be common, analogous to freshwater systems (Pace et al., 1999), rocky intertidal habitats (Menge, 2000), or benthic marine ecosystems (Pinnegar et al., 2000) where trophic cascades have been described. Patterns of interactions that involve fish in coastal or open marine systems, most of which were presented above, are summarized in Tables 14.1 to 14.4. All the described patterns concern interactions between only a few selected species that comprise the ecosystem, and that are usually of economic interest for fisheries. Despite the quality of the data and of the evidence, which varies greatly between case studies, controls can be found to be bottom-up as well as top-down in marine open systems, with a tendency of recent studies to favour an interplay between the two types of control, as in the case of the wasp-waist control. Alterations of resource availability and consumers often result in general patterns of community change, but not always. The species composition of an ecosystem may change, although the structure, in terms of the composition of the different trophic levels, may remain the same. It also appears difficult to draw generalities from case studies that sometimes contradict previous studies. For example, and surprisingly, top-down processes do not always control the upper trophic levels, as predicted in most theoretical studies, and lower trophic levels appear to be bottom-up controlled in most cases, but not all.

Ecosystems can be resilient to exploitation and do not always exhibit drastic changes in their composition or structure, a situation exemplified by the North Sea (the proportion of large fish has diminished, average fish sizes have decreased . . .). In these cases, species redundancy at intermediate trophic levels (planktivorous fish) or at high trophic levels (large fish predators) plays an important role in delaying any drastic changes in the functioning of ecosystems. However, in many cases, regime shifts and alternative steady-states can substantially modify the structure and dynamics of marine systems. The effect of predator removal offers a compelling example of induced changes at lower trophic levels. The shift from a cod-dominated towards a pelagic-dominated ecosystem in the Atlantic is a documented example that has received many theoretical interpretations.

For a given species within the ecosystem, the nature of the control can change through time according to the configuration of the ecosystem. The dynamics of the octopus populations off West Africa are illustrative of a top-down control released following the overexploitation of predatory fishes, and which later became controlled by bottom-up forces (i.e., the strength of the upwelling). Overexploitation of top predators and 'fishing down the food web' may not only lead to an increase in the biomass of planktivorous fish, but also to cascading effects throughout the different trophic levels, as has been observed in the Black Sea.

Fish recruitment is mainly considered to be bottom-up controlled, but in many cases the formal link between recruitment strength and the dynamics of primary production is not sufficiently documented, casting doubt on the way the environment is controlling recruitment, i.e., directly through dynamic environmental structures (*sensu* Sinclair) or indirectly through the food chain (*sensu* Cushing).

Recruitment studies should focus on linking the environment to trophic interactions in the context of marine food-webs.

Process-oriented models often predict a rich variety of dynamic behaviours that depend on the complexities of the interactions involved. At the other end of the spectrum, pattern-oriented studies predict that only a portion of the habitat or a particular assemblage within the ecosystem can exhibit a strong dynamic structure. Both approaches reveal the transient nature of ecosystems. Ecologists have been analysing ecological interactions in two different and most often exclusive ways using reductionist or holistic approaches respectively; but as stated by Elton (1927), a combination of the two methods would be the best procedure. However seventy-five years after this ecological wish formulated by Elton, it is still more a direction for future research on ecosystem dynamics than actual research objectives.

No general theory can be ascribed to the functioning of marine ecosystems, which results in poor predictive power for fisheries management. Past observational experiences can provide guesses or conjectures of the potential dynamics of the system. Recently, tentative and partial generalizations have been proposed (Cury et al., 2003). For example, trophic cascades are mostly found in lakes or in hard substrata marine ecosystems and mainly for less complex food webs, whereas wasp-waist control is most probable in upwelling systems. This constrains the field of possibilities and introduces opportunities for stimulating comparisons and generalizations.

Stating that "ecological patterns and the laws, rules and mechanisms that underpin them are contingent on the organisms involved, and their environment," Lawton (1999) recognized that this contingency is manageable at a relatively simple level of ecological organisation or for large sets of species, over large spatial scales, or over long time periods, but that it becomes overwhelming for intermediate scales, such as at the level of the community. Currently, neither the empirical database, nor insight gained from interaction models, appear sufficient to permit a synthesis relating cause and effect at the community level (Lawton, 1989).

Patterns of interaction do exist and should be regarded as the focal entry point to ecosystem approaches to management. Nevertheless, ecological understanding and models of ecosystem functioning are provisional and subject to change (Christensen et al., 1996). As stressed by Hairston and Hairston (1997), any attempt to understand broad ecological patterns will be challenged by the complexity of nature.

This scepticism reflects concerns about the realism of basic models used to analyse the properties of food webs. Even though it is possible to identify interaction processes that are involved and several patterns that emerge, only very few generalisations across systems have been made and predictions are mostly out of reach when considering the dynamics of ecological systems. Moreover, and as recently stated by Mace (2001), in terms of making realistic predictions about the future, ecosystem models have not yet proven themselves as management tools. Hall (1999) stated that we must admit our ignorance of the true importance of the effects of fisheries acting through species interactions in marine systems. However, since then many studies have provided new syntheses and field examples showing that fishing effects on ecosystems are paramount and can help fisheries management (Jackson et al., 2001; Walters and Kitchell, 2001; Daskalov, 2002). Several years to decades will be necessary for marine ecologists to refine concepts and to find the appropriate data, which are mostly lacking at present, to strengthen their hypotheses on the functioning of marine ecosystems.

Ecosystems ecology is an emerging discipline that still needs to find its way by formalizing principles, and by building theories for the marine environment. Exploring complexity of ecosystems by linking processes and patterns together is a necessary step to integrate our fragmented and disciplinary knowledge on ecosystems into a framework. This integration could provide the "ecoscope" to study marine ecosystems (a term proposed by Ulanowicz, 1993), i.e. a powerful multidisciplinary tool that will explore the complexity of their dynamics. It is challenging, as it requires that we:

- Develop macroecological studies of the oceans to characterize patterns of ecosystem components, based on large amounts of data (Parsons, 2003).
- develop process- and pattern-oriented studies coupled through integrated field, experimental and modelling studies;
- provide a broad view of ecosystem studies where collection of data at different scales is explicitly prescribed in sampling design of field studies and where top-down versus bottom-up control are simultaneously assessed;
- provide accepted, clear and testable definitions of the terms that are used for characterizing ecological processes;
- develop new observation systems by recognizing that ecological and biological data that are collected for single-species fisheries management are necessary but insufficient for understanding ecosystem dynamics. Detailed predictions require detailed knowledge and an understanding of interaction processes;
- promote comparative and retrospective studies among marine ecosystems to evaluate regime shifts and types of controls, and develop models to evaluate ecological changes within ecosystems and assess anthropogenic changes;
- evaluate states and changes in marine ecosystems by defining new ecosystem-based indicators for fisheries management, assess the usefulness of these indicators for management purposes and apply them to various fisheries.

Ecosystems ecology should become a multidisciplinary field of research of the marine environment and a central focus for fisheries management. This represents a new framework that would challenge the difficulties of understanding the dynamics of complex systems at appropriate scales by enabling repeatable patterns to be tracked by indicators, and by incorporating existing scientific knowledge on processes into models and ultimately into fisheries management.

TABLE 14.1
Documented bottom-up control that involves fish species in marine ecosystems.

| Location | System | Species interactions involved | Reference | Comments |
|---|---|---|---|---|
| Mediterranean Sea | Coastal and open systems | Nutrients (land runoff)-planktivores (sardine-anchovy-sprat)-piscivorous (horse mackerel, whiting)-large piscivorous (swordfish, hake) | Caddy and Garibaldi (2000) | FAO data base |
| Mediterranean Sea | Northern coastal systems | Land run-off-zooplanktivores-piscivorous fish (hake, swordfish) | Skud (1983) | |
| Western Gulf of Alaska | Coastal system | Shrimp and capelin- harbor seal and sea lion | Hansen (1997) | |
| North Sea | Coastal system | Sandeels-sea birds | Furness and Tasker (1997 ) | |
| North Atlantic Georges Bank | Coastal system | Environment (NAO)-zooplankton-cod and Haddock | Hofmann and Powell (1998) | Indirect environmental effect sensu Cushing. Large scale environmental effects |
| North Atlantic Georges Bank | Gulf stream rings | Environment (ring dynamics)-fish recruitment (cod, haddock,...) | Hofmann and Powell (1998) | Direct environmental effect sensu Sinclair.: Small scale environmental effects |
| Guinea and Humboldt currents | Upwelling systems | Pelagic fish (anchovy and sardinella)-piscivorous fish (*Scomber* and *Sarda*) | Cury et al. (2000) | Dampening effects for most predator species |
| Northern and southern Benguela | Upwelling system | Pelagic fish (sardine and anchovy)-marine sea birds (penguins) | Crawford et al. (2001) | |
| North-western Mediterranean Sea | Coastal systems | Environment (global: NAO and local: river runoff and wind mixing)-recruitment of 13 commercial fish and invertebrate species | Lloret et al. (2000) | Indirect environmental effect sensu Cushing proposed but not demonstrated |
| North Atlantic | Coastal system | Environment (westerly weather)–phytoplankton-zooplankton-herring-marine birds | Aebisher et al. (1990) | Observed parallel long-term trends across four marine trophic levels |

| | | | | |
|---|---|---|---|---|
| North-east Pacific | Alaskan coastal systems | Environment (decadal scale change, regime shift)-zooplankton-salmon | Francis and Hare (1994) | |
| Northeast Pacific | California and Alaska | Environment (decadal scale change, low frequency variation)-zooplankton-benthic and pelagic communities | McGowan et al. (1998) | Complex responses of the different components of the ecosystems<br>Change in spatial distribution of many species due to environmental changes |
| Northeast Pacific | Coastal systems including Gulf of Alaska and Bering Sea | Environment (ENSO-PDO)-recruitment of groundfish and pelagic fishes | Hollowed et al. (2001)<br>McFarlane et al. (2000) | Indirect environmental effect sensu Cushing proposed<br>Large scale environmental effects |
| Central North Pacific | North-western Hawaiian Islands | Environment (regime shift mid-1970s)-phytoplankton-reef fishes & spiny lobsters- marine birds and monk seals | Polovina et al. (1994) | |
| Central Baltic Sea | Pelagic Ecosystem | Cod-pelagic fish (Sprat) | Mollman and Koster (1999) | No top down control of pelagic fish on mesozooplankton |
| Pacific Ocean | Pelagic ecosystems: California, Kuroshio-Oyashio, Peru currents, Central North Pacific and Subartic Pacific | Environment (regime shift in the mid-1970s) – macrozooplankton – higher trophic levels | Hayward (1997) | Complex response of high trophic levels to environmental changes |
| Northern Pacific | Coastal systems | Large scale climatic change (ocean atmosphere linkage)-benthic taxa-sardine-anchovy versus Alaskan salmon | Finney et al. (2002)<br>Baumgartner et al. (1992) | Paleoecological reconstruction over 2,200 years<br>Anchovy and sardine vary out of phase with salmon over low frequency |
| Pacific Ocean | Coastal systems | El Niño disturbance (ENSO)-primary production-benthic production-pelagic fish- marine birds and pinnipeds | Glynn (1988) | Complex responses due to confounding effect of numerous interactions acting upon local conditions |

*continued next page*

| | | | | |
|---|---|---|---|---|
| World Ocean | Coastal and open systems | Phytoplankton-higher trophic levels | Ryther (1969) | Comparative study between observed and predicted fish yields from estimates of photosynthetic organic production rates |
| North Sea | Thames Estuary (nursery areas) | North Atlantic Oscillation (NAO)-Juvenile fish (flatfish, clupeids, eel, invertebrates) | Attrill and Power (2002) | Climatic variability affects the structure of the fish assemblage, recruitment and life history traits |
| South Sea of Korea | Coastal system | SOI-SST-Chl-a-zooplankton-pelagic fish (anchovy, sardine, mackerel) | Kim and Kang (2000) | Exploratory analysis using correlations during certain months |
| Barents Sea, Norwegian Sea | Coastal systems | Salinity-zooplankton-herring/capelin-cod-seals/seabirds | Blindheim and Skjoldal. (1993) | |

TABLE 14.2
Documented top-down control that involves fish species in marine ecosystems.

| Location | System | Species interactions involved | Reference | Comments |
|---|---|---|---|---|
| Northwest Atlantic (area 21) | Coastal system | Piscivores (notably cod)- benthivores (invertebrates such as blue crab, lobster, prawns...) | Caddy and Garibaldi (2000) | FAO database |
| Northeast Atlantic | Coastal system | Fish predators (gadoids)-pelagic fish (herring, saithe, whiting, capelin) | Caddy and Garibaldi (2000) | FAO database; (possible reverse trophic flow or cultivation effect between pelagic fishes and cod) |
| Central Baltic Sea | Pelagic ecosystem | Cod-pelagic fish (sprat and herring)-zooplankton | Kornilovs et al. (2001) | Top –down effect between sprat and Cladocera is strong in autumn. Food competition between sprat and herring |
| Yellow Sea | Pelagic system | Demersal fish species and large pelagic fish-small pelagic fish | Jin and Tang (1996) | |
| South Africa, Ghana, Japan, and Black Sea | Coastal system | Pelagic fish-Zooplankton | Cury et al. (2000) | Meta-analysis |
| Sub-Arctic North Pacific | Open ocean | Pink salmon- macrozooplankton-phytoplankton | Shiomoto et al. (1997) | Seasonal control in summer |
| Black Sea | Coastal pelagic domain | Fishers-pelagic predators- Planktivorous fish- zooplankton--phytoplankton- Phosphates | Daskalov (2002) | Strong top-down control coupled with weak bottom-up effect (eutrophication) |
| Northern Baltic Sea | Northern Baltic proper (ICES areas 28-29) | 0-group herring-zooplankton | Arrhenius (1996; 1997) | |
| Baltic Sea | Coastal system | Fishery- cod- planktivorous fish (clupeids); Planktivorous fish-cod recruitment (clupeid-dominated system) | Rudstam et al. (1994) | Existence of two competing top-down control processes (cultivation effect) |

| | | | | |
|---|---|---|---|---|
| World Ocean | Coastal and open systems | Fishers-different trophic levels-primary production | Pauly and Christensen (1995) | Primary production estimated from trophic model |
| Northwest Atlantic | Newfoundland-Labrador area | Fishers/(Harp seal) -Finfish (cod/salmon/plaice)-capelin-zooplankton-phytoplankton | Carscadden et al. (2001); Bundy (2001) | Overexploitation of cod<br>No overexploitation of capelin<br>Changes in distribution of capelin due to the environment |
| North Atlantic Ocean | Barents Sea, Iceland, Skagerrak, Labrador, Northern Newfoundland, Northern Gulf of St. Lawrence, Flemish Cap, Eastern Scotian Shelf, Gulf of Maine | (Fishers)-Cod-shrimp | Worm and Myers (2003) | Strength of the interaction modulates with temperature<br>Meta-analysis |

TABLE 14.3

Documented Mixed or Wasp-waist Controls that involves fish species in marine ecosystems.

| Location | System | Species interactions involved | Reference | Comments |
|---|---|---|---|---|
| Eastern Boundary Currents | Upwelling systems | Environment (upwelling intensity)-primary production (bottom-up control)<br>Small pelagic fish-primary production (deducted top-down control) | Carr (2002) | Compare primary production and fish production from estimated yield and the observed fish catch |
| Experimental and natural systems | Diverse | Planktivorous fish- mesozooplankton (top-down control)<br>Nutrients (N availability)- phytoplankton (bottom-up control) | Micheli (1999) | Comparative empirical analysis of 47 marine mesocosm experiments and from 20 natural marine systems |
| Northern Baltic | Coastal ecosystem | Cod- herring (top-down control)<br>Salinity-mesozooplankton-herring (bottom-up control) | Flinkman et al. (1998) | |
| North Sea | Coastal system | Zooplankton – Herring- Gadoid (bottom-up control)<br>Gadoid- pelagic fish (Herring/Atlantic mackerel) (top-down control) | Cushing (1982)<br>Caddy and Garibaldi (2000) | |
| Eastern Boudary Currents | Pelagic systems | Pelagic fish (sardine, anchovy)-pelagic predators (sarda, mackerel)-marine sea birds (bottom-up control)<br>Pelagic fish (sardine, anchovy)-zooplankton (top-down control) | Cury et al. (2000) | Comparative empirical analysis |
| Northern Baltic Archipelago Sea | Archipelago | Environment (NAO) -small zooplankton (bottom-up control)<br>Herring-large zooplankton (top-down control) | Dippner et al. (2001) | |

| | | | | |
|---|---|---|---|---|
| North-east Pacific | Alaska and California currents | Environment (decadal scale change, regime shift)-phytoplankton-zooplankton (bottom-up)<br>Forage fishes-birds and mammals (bottom-up)<br>Predatory fishes-forage fishes (top-down) | Francis et al. (1998) | Complex response at higher trophic levels |
| North Atlantic Georges Bank | Northeast U.S. continental shelf ecosystem | Environmental forcing-zooplankton community<br>Groundfish-squid and American lobsters<br>Atlantic mackerel and herring – predators and fishing | Link and Brodziak (2002) | Predatory release on zooplankton not apparent when planktivores severely reduced by fishing implies primarily environmental control of zooplankton.<br>Large decline in groundfish abundance under intense fishing benefited groundfish prey.<br>Low predator abundance and fishing pressure benefited pelagic fish.<br>Fishing is the dominant forcing factor. |
| Bohai Sea (Yellow Sea) | China Continental shelf ecosystem | Primary production/phytoplankton-zooplankton-fish productivity | Tang et al. (2003) | Top-down, bottom-up and wasp-waist controls act during different periods |

TABLE 14.4

Documented spatial and temporal changing controls that involve fish species in marine ecosystems.

| Location | System | Species interactions involved | Reference | Comments |
|---|---|---|---|---|
| All ecosystems | All systems | Zooplankton-fish recruitment (bottom-up control) | Cushing (1990) | Match-mismatch hypothesis between food and larvae |
| British Columbia | Coastal ecosystem | Temperature-zooplankton-(sand lance)-marine seabirds (bottom-up control) | Bertram et al. (2001) | Match-mismatch between prey and predators |
| Alaska Prince William Sound | Shallow-water nursery areas | Environment-zooplankton- juvenile herring and juvenile pink salmon (Bottom-up control)<br>Marine bird and predator fish (Pollock and adult herring) -juvenile pink salmon (role of alternative prey) (top-down control) | Cooney et al. (2001) | Time and spatial dependent trophic control:<br>Seasonal availability of prey to predators |
| Southern Ocean South Georgia area | Open systems | (Abiotic factors)-krill-(fish)-whales-seals-penguins (mixed controls) | Murphy (1995) | Large spatial scale effect of abiotic factors<br>Small spatial scale for biotic interactions<br>Ecosystem models considering spatial structure |
| Eastern Central Atlantic | Upwelling ecosystem | Fishers-groundfish (sparids, serranids)-Octopus (Top-down control)<br>Environment (upwelling strength)-Octopus (bottom-up control) | Faure et al. (2000)<br>Caverivière and Demarcq (2001)<br>Gulland and Garcia (1984)<br>Caddy and Garibaldi (2000) | Changing control through time and ecosystem structure |

## Appendix: Models that are referred to in the text

### *Lotka-Volterra model*

With $N$ and $P$ designating the respective abundance of prey and predator, the so-called Lotka-Volterra model consists of the following system of two differential equations:

$$\frac{dN}{dt} = N(r - aP) \qquad r, a, c, d > 0,$$
$$\frac{dP}{dt} = P(acN - d)$$

where $r$ is the intrinsic growth rate of the prey in the absence of the predator, $a$ is the attack rate of the predator (or the number of prey consumed by a predator per unit of time), $c$ is the conversion efficiency of the prey consumed into numbers of predators, and $d$ is the mortality rate of the predators in the absence of prey.

This classical model is based on some symplifying assumptions (Czaran, 1998): the environment is homogeneous, the population is represented by a state variable that is not structured according to age or size, the predator eats only one type of prey, the birth and death processes respectively in the prey and predator populations are exponential, and the functional response of the predator (term $aN$) is a linear function that is not limited by prey numbers (Figure 14.A1a).

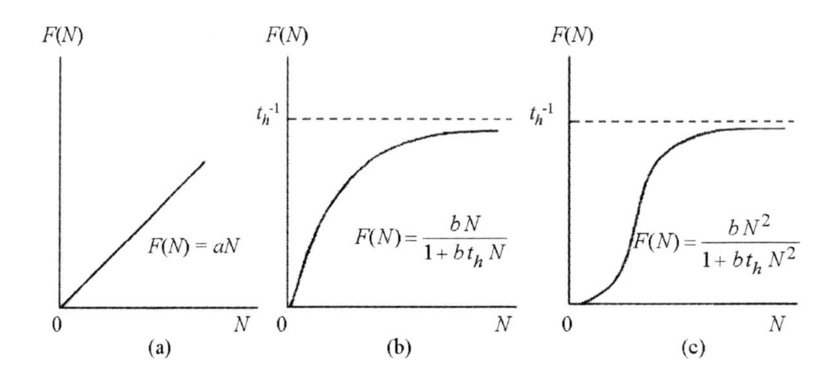


Figure 14.A1 Examples of functional responses of predators in the "laissez-faire" situation (Caughley and Lawton, 1976): no interference between predators. (a) Lotka-Volterra type, (b) type II Holling, (c) type III Holling (adapted from Murray, 1989).

Analysis of the model system shows that, apart from the point ($N$=0, $P$=0), there exists another positive equilibrium point, and the dynamics of the system are such that the predator and prey populations exhibit cycles of abundance.

*General predator-prey models*

Since the development of the Lotka-Volterra model, numerous studies have proposed different formulations for population growth rates and interactions. Consider the general formulation of a predator-prey system:

$$\frac{dN}{dt} = f(N) - P\,F(N,P)$$

$$\frac{dP}{dt} = P\,G(N,P)$$

Where $N$ and $P$ represent respectively the densities of prey and predator.

In such a generalized system, three functions must be specified:

- $f(N)$ : the intrinsic growth rate of the prey population in the absence of predators,
- $F(N,P)$ : the functional response of the predator, *i.e.* the number of prey consumed by a predator per unit of space and time,
- $G(N,P)$ : the numerical response of the predator, describing the production of predators, *i.e.* the rate of conversion of a prey into predators.

The Lotka-Volterra model assumes that the prey population exhibits a Malthusian growth rate, which is not realistic. Subsequent predator-prey models instead assumed logistic, density-dependent growth (Verhulst, 1838):

$$f(N) = r\,N\left(1 - \frac{N}{K}\right)$$

where $K>0$ represents the carrying capacity of the habitat with respect to the prey (note that the Schaefer model is based on this logistic function).

There are a number of different ways of modelling predator functional responses (Table 14.A1), depending on what one assumes about how predators interact with one another. The situation called *laissez-faire* by Caughley and Lawton (1976) corresponds to predators that do not interfere with each other in their feeding activities. In such a system, neither the functional response nor the numerical response of the predators depends on their density, they depend only on prey density. The functional reponses proposed by Holling (1959) belong to this category.

According to Holling (1959), feeding is composed of two types of activity: searching and handling prey. He assumed (i) that the total time dedicated to feeding is the sum of search time and handling time, and (ii) that the handling time $t_h$ is a constant.

If the attack rate or the number of prey consumed by a predator per unit of search time is $a$, and the prey search time is $t_r$, then

$$F = \frac{\text{Numbers of prey consumed by a predator}}{\text{total time of feeding}} = \frac{at_r}{t_r + at_r t_h},$$

TABLE 14.A1:
Different formulations of functional responses of predators. Functional response is a function of the attack rate $a$ on prey by predators: $F = a/(a+t_h)$. The constant $t_h$ corresponds to prey handling time. $b$, $Q$, $P_0$ are positive parameters. $N$ and $P$ are the prey density and the predator density respectively.

| Hypothesis | Predation rate | Functional response | References |
|---|---|---|---|
| No interaction between predators (laissez-faire) | $a = bN$ | $F = \frac{bN}{1+bt_hN}$ | Type II, Holling (1959) |
| | $a = bN^2$ | $F = \frac{bN^2}{1+bt_hN^2}$ | Type III, Holling (1959) |
| Interference between predators | $a = \frac{QN^n}{p^m}$ | $F = \frac{t_h^{-1}N^n}{(Qt_h)^{-1}p^m + N^n}$ | Hassel and Varley (1969) |
| | $a = Q\left(\frac{N}{P_0 + P}\right)$ | $F = \frac{t_h^{-1}N}{(Qt_h)^{-1}(P_0 + P) + N}$ | De Angelis et al. (1975) |
| | $a = Q\left(\frac{N}{N + P}\right)$ | $F = \frac{Q(1+t_h)^{-1}N}{(1+t_h)^{-1}P + N}$ | Getz (1984) |
| Interference and ratio-dependence | $a = Q\left(\frac{N}{P}\right)$ | $F = \frac{QN}{P + Qt_hN}$ | Arditi and Ginzburg (1989) |

and the functional response of the predator can be expressed as follows (Yodzis, 1994; Begon, et al., 1996):

$$F = \frac{a}{1 + a\,t_h}$$

The simplest hypothesis for the attack rate corresponds to a type II Holling response, with the attack rate proportional to the number of prey in the environment (Begon et al., 1996; Figure 14.A1b). The type III Holling response assumes that predators are inefficient at handling prey when preys are not abundant (Yodzis, 1994; Figure 14.A1c).

In contrast to Holling's (1959) models, another family of models assumes the existence of predator interference through trophic or reproductive competition, disease transmission, cannibalism, or density-dependent emigration (Yodzis, 1994). In this category, a common functional response is the one of Hassel and Varley (1969), which assumes that the attack rate decreases when predator abundance increases and, for a given density of prey $N$, the larger the abundance of predators, the slower the rate of consumption by a predator (Table 14.A1). This general formulation can be applied to a large number of predator-prey systems, but their behaviour becomes unrealistic in the particular case when P◊0, when the attack rate $a$ tends towards $\infty$. This problem also occurs for the ratio-dependent formulation of the functional response (Table 14.A1). Initially proposed by Arditi and Ginzburg (1989), this formulation assumes that the consumption rate decreases proportionally with predator abundance, because the same resource must be shared by a greater number of consumers.

The most widely used predator numerical response function corresponds to a "laissez-faire" situation (Table 14.A2). It represents a balance between gains to the predator population (term $dF(N)$, proportional to the quantity of prey consumed by a predator, and its losses (term $-c$). An interference effect can also be taken into account, affecting the predator consumption rate (the functional response) or the predator growth rate (Table 14.A2).

Structurally different, the predator numerical response function of Leslie (1948) is based on the assumption that the predator population grows logistically, with a carrying capacity $hN$ that is proportional to the abundance of prey. A problem with this formulation is that the carrying capacity of the predators is zero when prey abundance is zero. Because predators incur maintenance costs, their carrying capacity should be zero at some positive threshold of prey abundance (Yodzis, 1994). The model of Hassel and Varley (1969) has this same problem—the predators are very resistant to a lack of food.

TABLE 14.A2:
Different formulations of the numerical response $G$ of predators $P$ to prey $N$ (Yodzis, 1994). Parameters $c$, $d$, $e$, $g$, $h$ and $S$ are positive.

| Hypothesis | Numerical response |
|---|---|
| *laissez-faire* | $G = -c + dF(N)$ |
| Interference between predators affecting consumption rate | $G = -c + dF(N,P)$ |
| Interference between predators affecting growth rate | $G = -c + dF(N) - eP^S$ |
| Numerical response of Leslie (1948) | $G = g(1 - \frac{P}{hN})$ |

### *May et al's (1979) krill-whale fishery model*

May et al. (1979) used the predator-prey model of Leslie and Gower (1960), which consists of a combination of a Lotka-Volterra-type linear functional response of predators to prey, and a Leslie numerical response of the predators (see also Tables 14.A1 and 14.A2). If $N_1$ is the abundance of the krill population (prey) and $N_2$ the abundance of the whale population (predator), the system dynamics are described by:

$$\frac{dN_1}{dt} = r N_1 (1 - \frac{N_1}{K}) - a N_1 N_2$$

$$\frac{dN_2}{dt} = g N_2 (1 - \frac{N_2}{hN_1})$$

The krill population is exploited at a rate $r_1F_1$, such that $F_1$ represents a fishing effort multiplier equal to 1 when fishing mortality rate is equivalent to the intrinsic growth rate of the prey population $r_1$. Then, the corresponding fishing yield is $Y_1 =$

$r_1 F_1 N_1$. By applying the same arguments to the predator population and transforming the equations, the following system is obtained:

$$\frac{dX_1}{dt} = r_1 X_1 (1 - F_1 - X_1 - \nu X_2)$$

$$\frac{dX_2}{dt} = r_2 X_2 (1 - F_2 - \frac{X_2}{X_1})$$

where $X_1 = N_1/K$, $X_2 = N_2/(h\,N_1)$ and $\nu = a\,h\,K/r_1$.

If $F_2$>1, the whale population collapses and if $F_1$>1, the whole system collapses. If $F_1$ and $F_2$ are less than 1, there exists a unique equilibrium where the presence of whales has the trivial effect of decreasing the fishing yield and the abundance of krill. The production and the abundance of the whale population decrease linearly with fishing effort on krill.When the exploitation of whales increases, the abundance of krill increases, which causes an increase in the growth rate of the remaining whales. Thus, for $\nu$=1 (predation of half the pristine biomass of krill), MSY occurs when $F_2$=0.59, whereas in the single-species case ($\nu$=0) the corresponding fishing effort equals 0.5. For the krill population, MSY is obtained at $F_1$=0.5. The MSY for krill increases with $F_2$, the maximum being reached when $F_2$=1, which corresponds to extinction of the whales.

### *Complex food-web models*

Michalski and Arditi (1995a; b) proposed a generalisation of predator-prey models that included a Holling type II functional response and a numerical response with interference in the consumption rate (Tables 14.A1 and 14.A2):

$$\frac{dX_i}{dt} = f_i(X) + \frac{A_i \sum_{k\in R(i)} e_{ki} X_k^{r(i)}}{X_i + \sum_{k\in R(i)} B_{ki} X_k^{r(i)}} X_i - \sum_{j\in C(i)} \frac{A_j \sum_{k\in R(j)} e_{kj} X_k^{r(j)}}{X_j + \sum_{k\in R(j)} B_{kj} X_k^{r(j)}} - \mu_i X_i$$

where $A_i$, $B_{ki}$, $e_{ki}$, $\mu_i$ are constant positive parameters, $R(i)$ is the set of possible resources for species $i$, $C(i)$ is the set of consumers of species $i$ and $f_i(X)$ is the growth rate of species $X_i$ in the absence of predators. For the species that are not at the base of the trophic web, $f_i(X)$ is null.

Two auxiliary variables are used:

$$X_i^{r(j)} = \frac{\beta_{ij} X_j^{c(i)}}{\sum_{k\in C(i)} \beta_{ik} X_k^{c(i)}} X_i$$

$$X_j^{c(i)} = \frac{h_{ij} X_i^{r(j)}}{\sum_{k\in R(j)} h_{kj} X_k^{r(j)}} X_j$$

where $h_{ij}$ is the relative preference of consumer $X_j$ for the resource $X_i$ (among other prey species) and $\beta_{ij}$ is the efficiency of the relative competition of consumer $X_j$ (among other consumers) for the resource $X_i$.

Using arbitrary values of the parameters, the authors applied the model to a theoretical system consisting of 11 species (Michalski and Arditi, 1995a). At equilibrium, only a few links persisted. By imposing a variation of 60% in the values of $h_{ij}$ and $\beta_{ij}$, the effective structure of the trophic web changed radically. By disturbing the system at equilibrium (for instance by changing the abundance of a species), the same authors (Michalski and Arditi, 1995b) show that their multi-species model leads to systems that are rich in interspecies links when they are far from equilibrium and poor in links when they approach it. A consequence of these results is that the structure of food webs can vary with seasonal variations in species abundance as well as variations in the parameters for competition efficiency and food preferences.

### *Schaefer model*

The Schaefer (1954) model is a simple stock assessment model derived from the logistic model, still used in many instances, e.g. tropical and tuna fisheries (Hilborn and Walters, 1992):

$$\frac{1}{B}\frac{dB}{dt} = k - aB - F$$

where $B$ is the biomass of a given species, $F$ its rate of fishing mortality, and $a$ and $k$ are two positive constants. If $F$=0, $B_v = k/a$ is the pristine biomass of the species at equilibrium. When the species is exploited, the equilibrium biomass is $B_e = (k - F)/a$.

If $F$ is expressed as the product of $q$, the catchability of the species, and $f$, the fishing effort, the fishing yield $Y_e$ at equilibrium is:

$$Y_e = FB_e \iff Y_e = \frac{q}{a}(kf - qf^2)$$

$Y_e$ is a parabolic function of $f$ and the maximum sustainable yield (MSY) corresponds to $B_{MSY} = \frac{1}{2}B_v$.

### Multispecies Schaefer model

The extension of the Schaefer model to the multispecies case consists of coupling single-species Schaefer equations by including species interaction terms. For $m$ interacting species (Pope, 1976; Kirkwood, 1982) the model system consists of $m$ equations of the form:

$$\frac{1}{B_i}\frac{dB_i}{dt} = k_i - \sum_{j=1}^{m} a_{ij} B_j - q_i f \qquad \text{with } i = 1 \ldots m$$

If an equilibrium exists at $B_{1e}, B_{2e} \ldots B_{me}$ $(B_{ie} \geq 0)$, then:

$$k_i - \sum_{j=1}^{m} a_{ij} B_{j_e} - q_i f = 0 \qquad i = 1 \ldots m$$

This set of equations can be written in the following matrix form:

$$K - A\ B_e - Q\,f = 0$$

where $K$, $B_e$, and $Q$ are $(m,1)$ matrices and $A$ is an $(m,m)$ matrix.

At equilibrium, the total yield of the system $Y_e$ is given by:

$$Y_e = f\ Q^t\ B_e = f\ Q^t A^{-1} K - f^2\ Q^t A^{-1} Q$$

*Virtual population analysis (VPA)*

Matrices of fish numbers $N$ and fishing mortality rates $F$ by age $a$ for each past year $t$ are estimated from matrices of natural mortality rates $M$ and past catch $C$ by age and time. The dynamics of the age-structured population are described by:

$$\frac{dN_{a,t}}{dt} = -Z_{a,t}\ N_{a,t}$$

where $Z_{a,t} = F_{a,t} + M_{a,t}$ is the total mortality rate.

This leads to the following set of difference equations, where $M_{a,t}$ and $F_{a,t}$ are assumed constant over the time interval considered (usually one year):

$$N_{a+1,t+1} = N_{a,t}\ e^{-(F_{a,t}+M_{a,t})}$$

$$C_{a,t} = \frac{N_{a,t}\ F_{a,t}}{M_{a,t}+F_{a,t}}\left(1-e^{-(F_{a,t}+M_{a,t})}\right)$$

The history of the cohort is reconstituted by initializing the system with a recruitment value or the fishing mortality in the terminal age class.

**Multi-species virtual population analysis MSVPA (multi-species cohort analysis)**
One of the hypotheses on which VPA is based is that $M$ is known and generally constant over year and age. However, $M$ varies with fish age; in particular, young fish are more subject to predation than older and larger ones (Stokes, 1992). The basis for extending cohort analysis to a multispecies case is to better estimate $M$ by taking into account predation relationships among species. MSVPA was first es-

tablished and applied by Andersen and Ursin (1977), Pope (1979) and Helgason and Gislason (1979) in the North Sea ecosystem. It consists of dividing the natural mortality rate $M_{i,a}$ of the cohort of age $a$ of species $i$ into two components: $P_{i,a}$ the mortality rate due to predation by the other species included in the model and $D_{i,a}$ the residual mortality rate due to other natural causes. To estimate the predation mortality rate, the food of each age group is partitioned among the different potential prey. If $\phi^{ia}$ is the biomass of class *jb* (species *j*, age class *b*) available for class *ia*, the biomass of class *jb* consumed over a year by *ia* can be expressed as follows:

$$\frac{\phi_{jb}^{ia}}{\phi^{ia}} R_{ia} N_{ia}$$

where $N_{ia}$ is the abundance of *ia* and $R_{ia}$ its rate of consumption. $R_{ia}$ is supposed to depend linearly on a power function of the mean individual weight of class *ia*. The total biomass of fish of class *jb* consumed during a year is then:

$$\sum_{i=1}^{s} \sum_{a=0}^{A_i} \frac{\phi_{jb}^{ia}}{\phi^{ia}} R_{ia} N_{ia}$$

where $s$ is the number of species considered in the model, and $A_i$ is the maximum age reached by species $i$. Hence the annual predation mortality rate of class *jb* becomes:

$$P_{jb} = \frac{1}{B_{jb}} \sum_{i=1}^{s} \sum_{a=0}^{A_i} \frac{\phi_{jb}^{ia}}{\phi^{ia}} R_{ia} N_{ia}$$

where $B_{jb}$ represents the total biomass of class *jb*. The available food for a predator $\phi_{ia}$ is only a fraction of the potential food biomass. A coefficient $G_{jb}^{ia}$ is introduced, which lies between 0 and 1 and represents the suitability of class *jb* as prey for class *ia:*

$$\phi_{jb}^{ia} = G_{jb}^{ia} B_{jb}$$

$G_{jb}^{ia}$ is the product of three coefficients that have values between 0 and 1: the size suitability of the prey, the overlap of the classes in time and space and the probability of encounter linked to the respective behaviour of the predator and its potential prey. The coefficients $G$ represent a substantial synthesis of biological knowledge concerning the different stages of species' life cycles, their spatial distributions, their behaviour, and their feeding habits (Ursin, 1982).

*Ecopath model*

Polovina (1984) and Christensen and Pauly (1992, 1995) developed the Ecopath model to estimate trophic fluxes within an ecosystem. The modelled system is assumed to be stationary, which implies that the gains in biomass equal the losses for each species group. Two mass conservation equations form the basis of Ecopath calculations:

Biomass consumption of group $i$ is described by:

*Consumption (i) = Production (i)+Non-assimilated food (i)+Respiration (i)*

For each group $i$, biomass production is modelled as:

Production (i) = Losses by predation on (i) + Catches of (i)

+ Export of (i) to adjacent systems + Losses by mortality of (i).

The terms of this equation can be expressed as follows:

*Production (i)* = $B_i\,(P/B)_i$

*Losses by predation on (i)* = $\sum_j (B_j\,(Q/B)_j\,DC_{j,i}$

*Catches of (i)* = $Y_i$

*Export of (i)* = $EX_i$

*Losses by mortality of (i)* = $(1\text{-}EE_i)\,B_i\,(P/B)_i$

where:

$B_i$ is the biomass of $i$

$(P/B)_i$ is the production of $i$ per unit of biomass

$(Q/B)_i$ is the consumption of $i$ per unit of biomass

$j$ designates a predator group of group $i$

$DC_{j,i}$ is the mean fraction of $i$ in the diet of $j$, in biomass terms

$EE_i$ is the ecotrophic efficiency of $i$ (or the fraction of the total production consumed by predation or exported from the system)

$(1\text{-}EE_i)$ then represents the fraction of the production of $i$ that goes to detritus.

The linear equation that describes the biomass flux for each species group $i$ is:

$$B_i(P/B)_i EE_i - \sum_{j=1}^{n}(B_j(Q/B)_j DC_{j,i}) - Y_i - EX_i = 0 \qquad (1)$$

The resulting ecosystem model consists of $n$ groups of species and is represented by a system of $n$ linear equations, which allows the estimation of at least one unknown parameter per species group. Under conditions of stationarity and for a closed system (no immigration or emigration), the ratio $(P/B)_i$ equals the mortality rate $Z_i$, which is often a more accessible parameter (Pauly and Moreau, 1997).

Ecopath models represent the trophic structure of an ecosystem at equilibrium, as shown in Figure A2, which results from application of Ecopath to the southern Benguela ecosystem (Jarre-Teichmann et al., 1998).

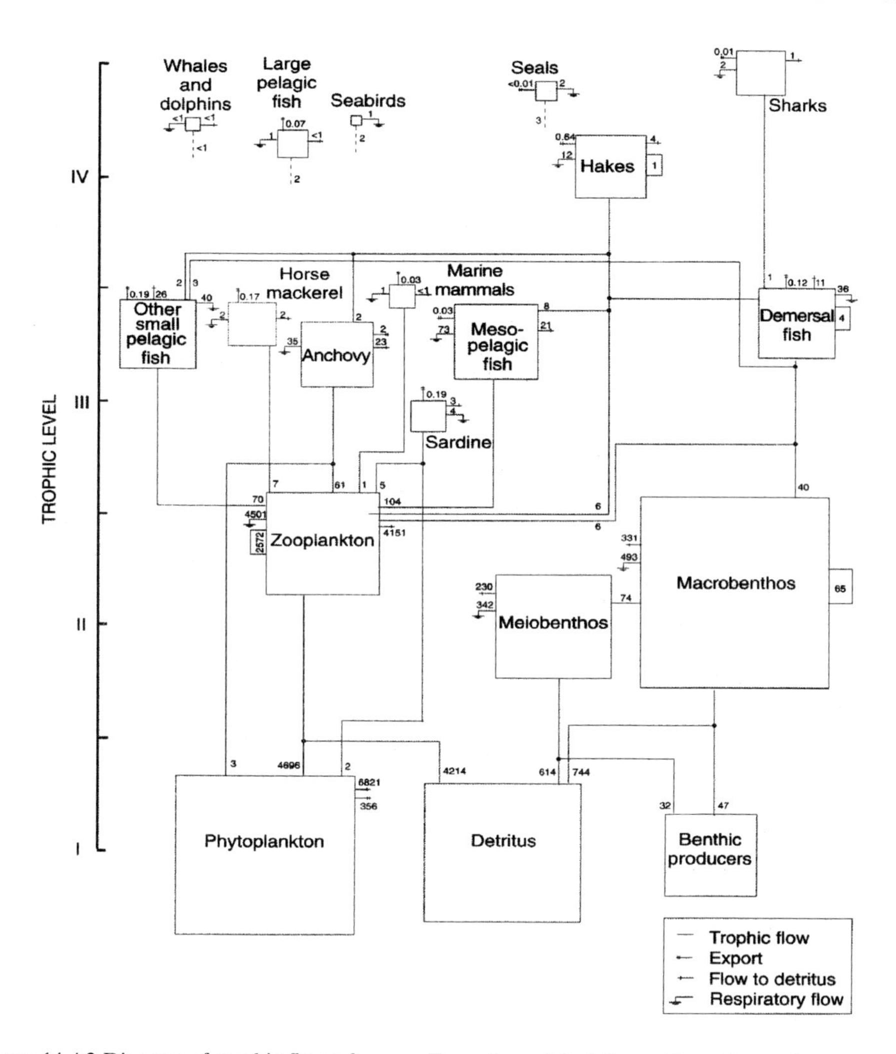


Figure 14.A2 Diagram of trophic fluxes from an Ecopath model of the southern Benguela ecosystem, 1980–1989. The surface areas of the boxes are proportional to the logarithm of their biomass. Compartments are arranged along the vertical axis in terms of their trophic levels. Fluxes are expressed in tons of wet matter per km2 and per year. Trophic fluxes smaller than 0.1% of the total consumption in the system are not represented. Output fluxes are located on the upper side of the compartments and input fluxes on the lower side (from Jarre-Teichmann et al., 1998)

Trophic levels calculated in Ecopath are fractional. Detritus and primary producers (phytoplankton and benthic producers) have, by definition, a trophic level equal to 1. For all other groups, the trophic level *TL* is a function of their prey and the proportions of these in their diets. For a group *i:*

$$TL_i = 1 + \sum_{j=1}^{n} DC_{i,j} TL_j$$

where $DC_{i,j}$ is the proportion of prey *j* in the diet of *i*, *n* is the number of groups in the system.

Other indices, calculated in the Ecopath software, characterize the trophic organisation of the ecosystem. Aggregation of fractional trophic levels into discrete trophic levels (*sensu* Lindeman, 1942) allows representation of the system as a trophic pyramid, and trophic efficiencies represent the proportion of material transferred from one discrete trophic level to the next.

Another index quantifies the mixed trophic impacts, i.e. the direct and indirect interactions between trophic groups. Inspired by Ulanowicz and Puccia (1990), the method successively calculates reciprocal impacts of a predator and its prey. A positive impact ($k_{ij}$) that a prey *j* exerts on its predator *i* is expressed as the proportion contributed by the prey to the diet of the predator:

$$k_{ij} = DC_{ij}$$

A negative impact ($l_{ij}$) that a predator *i* exerts on its prey *j* is expressed as the fraction that the predator *i* contributes to the total predation on the prey:

$$l_{ij} = \frac{B_i (Q/B)_i DC_{i,j}}{\sum_{k=1}^{n} B_k (Q/B)_k DC_{k,j}}$$

The net impact ($q_{ij}$) of prey *j* on predator *i* is then defined as the difference between the positive and the negative impacts:

$$q_{ij} = k_{ij} - l_{ij}$$

*Ecosim model*

Ecosim was developed by Walters et al. (1997), and is the dynamic version of Ecopath. It re-expresses the linear equilibrium equations of Ecopath as differential equations. The same basic equation is used, but export of biomass out of the system is assumed null:

Production (*i*) = Catches on (*i*) + Losses by mortality of (*i*) + Losses by predation on (*i*)

By including fishing and natural mortality rates (respectively $F_i$ and $M_0$), equation (1) becomes:

$$0 = B_i (P/B)_i - F_i B_i - M_0 B_i - \sum_{j=1}^{n} Q_{ij} \tag{2}$$

where $B_i$ is the biomass of *i*, $B_i(P/B)_i$ is the production in biomass of *i*, and $Q_{ij}$ is the biomass of *i* consumed by *j*.

Equation (2) is transformed into a differential equation:

$$\frac{dB_i(t)}{dt} = f(B_i(t)) - F_i(t)B_i(t) - M_0B_i(t) - \sum_{j=1}^{n} c_{ij}(B_i(t), B_j(t)) \tag{3}$$

where two new terms are introduced: $f(B_i(t))$, which is a function representing biomass production in the absence of predation and $c_{ij}(B_i(t),B_j(t))$, which predicts the biomass of prey *i* consumed by predator group *j* at time *t*.

If *i* is not a primary producer, $f(B_i(t))$ is assumed to be proportional to the quantity of ingested food:

$$f(B_i(t)) = g_i \sum_{j=1}^{n} c_{ji}(B_i(t), B_j(t)) \tag{4}$$

where $g_i$ represents the efficiency of conversion of the ingested food. The multiplicative parameter $g_i$ is assumed to account for the losses by respiration and non-assimilation that were introduced in the equilibrium equation (1) of Ecopath.

If *i* is a primary producer, biomass production is calculated using the following saturation function:

$$f(B_i(t)) = \frac{r_i B_i(t)}{1 + h_i B_i(t)}$$

where $r_i/h_i$ is the maximum primary production of *i* when $B_i$ is high, and by analogy with Ecopath, $r_i$ is a maximum $(P/B)_i = Z_i$ (or $f(B_i)/B_i$) when $B_i$ approaches 0. Parameter $h_i$ can also be approximated using values of Ecopath parameters by considering that the system is close to equilibrium. Thus, if $f(B_i) \approx B_i(P/B)_i$ then $h_i \approx [r_i/(P/B)_i - 1]/B_i$.

For the interaction functions $c_{ij}(B_i(t),B_j(t))$, Walters *et al.* (1997) use the expression proposed by Lotka and Volterra, namely:

$$_{ij}(B_i(t),B_j(t)) = a_{ij}\, B_i(t)B_j(t)$$

Parameter $a_{ij}$ is estimated from Ecopath, assuming that the system is close to equilibrium:

$$a_{ij} \approx \frac{B_j(Q/B)_j DC_{j,i}}{B_i B_j}$$

This expression is modified to account for the possible existence of prey biomass that is not accessible to the predator. If each prey group *i* has an unavailable portion $V_{ij}$ of $B_i$ for each predator *j*, the dynamic of the available biomass is described by the following equation:

$$\frac{dV_{ij}(t)}{dt} = \underbrace{v_{ij}(B_i(t) - V_{ij}(t))}_{\text{Input from the unavailable compartment}} - \underbrace{v_{ij}V_{ij}(t)}_{\text{Loss to unavailable compartment}} - \underbrace{a_{ij}V_{ij}(t)B_j(t)}_{\text{Loss by predation}}$$

where $v_i$ is the rate of biomass exchange between the accessible and inaccessible parts of $i$ to predator $j$.

Assuming that the exchange dynamic between $V$ and $B$ operates at a small scale compared to changes between $B_i$ and $B_j$, the following expression is obtained by making the previous derivative equal zero:

$$V_{ij}(t) = \frac{v_{ij}B_i(t)}{2v_{ij} + a_{ij}B_j(t)}$$

and

$$c_{ij}(B_i(t), B_j(t)) = \frac{a_{ij}v_{ij}B_i(t)B_j(t)}{2v_{ij} + a_{ij}B_j(t)}$$

When predator biomass $B_j$ is small, $V_{ij} \rightarrow B_i/2$ and $c_{ij}(B_i, B_j) \rightarrow (a_{ij}/2)B_iB_j$, *i.e.* towards the initial expression of Lotka-Volterra. When $B_j$ is large ($a_{ij}B_j >> 2\,v_{ij}$) and $c_{ij}(B_i, B_j) \rightarrow v_{ij}B_i$. Parameter $v_{ij}$ represents the maximum mortality rate exerted by $j$ on $i$, and can be estimated by Ecopath.

In their modelling work, Walters et al. (1997) succeeded in establishing links between the parameters estimated by Ecopath and those included in the differential equations of Ecosim. The use of existing data without requiring additional measurements and experiments is one of the advantages of Ecopath and Ecosim models. However, this is also one of the weaknesses of Ecosim, which is evident at two different levels. First, equation (3) is not applicable over a large range of biomass since the parameters are estimated from the Ecopath model, which at equilibrium. Second, the fixed link with Ecopath is likely to have constrained the choice of functions for growth of the primary producers, and functional responses of the predators. To understand in which category of models Ecosim belongs, an analogy with the predator-prey models described above is instructive. This comparison is necessary because the choice of functional responses is crucial and strongly determines the output dynamics. In the case of an ecosystem without exploitation, the main equation of Ecosim, which describes the biomass dynamics of group $i$, is:

$$\frac{dB_i(t)}{dt} = f(B_i(t)) - M_0B_i(t) - \sum_{j=1}^{n} B_j(t)\frac{a_{ij}v_{ij}\;B_i(t)}{2v_{ij} + a_{ij}B_j(t)} \tag{5}$$

Comparison with the general formulation of a predator-prey system (which was presented in a previous section of the Appendix) shows that the predator functional response $F$ used in Ecosim corresponds to the last term of equation (5), which is of the form:

$$F(B_i(t), B_j(t)) = \frac{aB_i(t)}{b + cB_j(t)} \tag{12}$$

where $a$, $b$, $c$ are three positive constants, $B_i$ the biomass of the prey and $B_j$ the biomass of the predator.

This formulation does not correspond to any of the functional responses that are commonly used in general ecology (see Table A1). It takes into account interference between predators ($B_j$ in denominator) and the existence of unavailable biomass, this last point being justified by the authors. What is maybe less justified is the choice of a Lotka-Volterra type response (when predator biomass $B_j$ is low, the functional response tends towards the Lotka-Volterra expression), which is currently considered to be unrealistic (Murray, 1989). Other kinds of functional response expressions (e.g. "laissez-faire", ratio-dependent or interference between predators) might give different results from Ecosim.

Still by analogy, the term $f(B_j(t)) - M_0B_j(t)$ of equation (5) corresponds, in the case where $j$ is a predator, to the product of a predator group $j$'s biomass and its numerical response, and in the case of primary producers, the term corresponds to their intrinsic growth rate.

According to equation (4), the numerical response of predator $j$ corresponds to $(g_jF(B_i,B_j)\text{-}M_0)$, which takes into account interference between predators in determining the growth rate. For primary producers, the function describing the population growth rate in the absence of predators corresponds to the following expression:

$$\frac{r_iB_i}{1+h_iB_i} - M_0B_i = \frac{(r_i - M_0)B_i - M_0hB_i^2}{1+h_iB_i}$$

Growth rate in biomass is definitely considered to be density –dependent because, for a given group $i$, it reaches a maximum for $B_i^*=(r_i\text{-}M_0)/M_0\,h_i$, and decreases beyond $B_i^*$. But the choice and the behaviour induced by such a saturation function should be discussed in relation to the Verhulst logistic growth function, which is the most common formulation.

**Acknowledgements**

This paper constitutes an overview of numerous works and comprehensive syntheses. We thank Cathy Boucher who drew the figures and Penny Krohn for the references. This study was financed by the IRD as part of the IDYLE Research Unit dedicated to the study and modelling of marine ecosystems. This is a SCOR/IOC WG-119 contribution.

## Bibliography

Abrams, P.A. 1992. Predators that benefit prey and prey that harm predators: unusual effects of interacting foraging adaptations. *Am. Nat.* **140:** 573–600.

Abrams, P.A. and H. Matsuda. 1993. Effects of adaptive predatory and anti-predator behaviour in a two-prey-one-predator system. *Evol. Ecol.* **7:** 312–326.

Aebischer, N.J., Coulson, J.C. and J.M. Colebrook. 1990. Parallel long-term trends across four marine trophic levels and weather. *Nature* **347:** 753–755.

Alheit, J. and E. Hagen. 1997. Long-term climate forcing of European herring and sardine populations. *Fish. Oceanogr.* **6:** 130–139.

Allen, P.M. 1988. Evolution: why the whole is greater than the sum of the parts? In W. S. Drepper [P.K.1](ed.), *Ecodynamics. Proceedings of an international workshop held at the nuclear research Centre Jülich, GDR 19–20 October Contribution to theoretical ecology,* pp. 213–237. Berlin[P.K.2]: Springer Verlag.

Andersen, K.P. and E. Ursin. 1977. A multispecies extension to the Beverton and Holt theory of fishing, with accounts of phosphorus circulation and primary production. *Meddelelser fra Danmarks Fiskeriog Havundersogelser* **7:** 319–435.

Anderson, T.W. 2001. Predator responses, prey refuges, and density-dependent mortality of a marine fish. *Ecology* **82:** 245–257.

Arditi, R. and L. Ginzburg. 1989. Coupling in predator-prey dynamics: ratio-dependence. *J. Theor. Biol.* **139:** 311–326.

Arrhenius, F. 1996. Diet composition and food selectivity of 0-group herring (clupea harengus L.) and sprat (Sprattus sprattus L.) in the northern Baltic Sea. *ICES Journal of Marine Science* **53:** 701–712.

Arrhenius, F. 1997. Top-down controls by young-of-the-year herring (*Clupea harengus*) in the Northern Baltic proper. In *Forage fishes in marine ecosystems. Proceedings of the International Symposium on the Role of Forage Fishes in Marine Ecosystems,* pp. 77–86. Alaska Sea Grant College Program Publication AK-SG-97–01.

Attrill, M.J. and M. Power. 2002. Climatic influence on a marine fish assemblage. *Nature* **417:** 275–278.

Bakun, A. 1989. Mechanisms for density-dependent growth in Peruvian anchoveta: alternatives to impact on the regional-scale food supply. In D. Pauly, P. Muck, J. Mendo and I. Tsukayama (eds.), *The Peruvian upwelling ecosystem: dynamics and interactions,* pp. 235–243[P.K.3]: ICLARM, Philippines.

Bakun, A. 1996. *Patterns in the Ocean: Ocean Processes and Marine Population Dynamics.* San Diego: University of California Sea Grant in cooperation with Centro de Investigaciones Biologicas de Noroeste, La Paz, Baja California Sur, Mexico, 323 pp.

Bakun, A. 2001. "School-mix feedback": A different way to think about low frequency variability in large mobile fish populations. *Progress in Oceanography* **49:** 485–511.

Bakun, A. and P. Cury. 1999. The "school trap": a mechanism promoting large-amplitude out-of-phase population oscillations of small pelagic fish species. *Ecology Letters* **2:** 349–351.

Barkai, A. and C. McQuaid. 1988. Predator-prey role reversal in a marine benthic ecosystem. *Science* **242:** 62–64.

Bax, N.J. 1991. A comparison of the fish biomass flow to fish, fisheries and mammals in six marine ecosystems. *ICES Mar. Sci. Symp.* **193:** 217–224.

Baumgartner, T. R., A. Soutar and V. Ferreira-Bartrina. 1992. Reconstruction of the history of Pacific sardine and northern anchovy populations over the past two millennia from sediments of the Santa Barbara basin, California. *Californian Oceanic Fishery Investment Report* **33:** 24–40.

Bax, N.J. 1998. The significance and prediction of predation in marine fisheries. *ICES J. Mar. Sci.* **55:** 997–1030.

Beddington, J.R. and R.M. May. 1980. Maximum sustainable yields in systems subject to harvesting at more than one trophic level. *Mathematical BioSciences* **51:** 261–281.

Begon, M., J.L. Harper and C.R. Townsend. 1996. *Ecology. Individuals, Populations and Communities.* Blackwell Science, Oxford.

Benoît, E. and M.-J. Rochet. 2004. A continuous model of biomass size spectra governed by predation and the effects of fishing on them. *J. Theor. Biol.* **226:** 9–21.

Bertram, D.F., D.L. Mackas and S.M. McKinnell. 2001. The seasonal cycle revisited: Interannual variation and ecosystem consequences. *Progress in Oceanography* **49:** 283–307.

Bianchi, G., H. Gislason, K. Graham, L. Hill, K. Koranteng, S. Manickchand-Heileman, I. Paya, K. Sainsbury, F. Sanchez, X. Jin and K. Zwanenburg. 2000. Impact of fishing on demersal fish assemblages. *ICES J. Mar. Sci.* **57**(3): 558–571.

Blindheim, J. and H.R. Skjoldal. 1993. Effects of climatic changes on the biomass yield of the Barent's Sea, Norwegian Sea and West Greenland large marine ecosystem. In L.M. Alexander and B.D. Golds (eds.), *Large marine ecosystem: stress, mitigation and sustainability,* pp. 185–189[P.K.4]: AAAS.

Bond, W.J. 1993. Keystone species. In E.-D. Schultz and H.A. Mooney (eds.), *Biodiversity and ecosystem function,* pp. 237–253. Berlin: Springer Verlag.

Brown, C. and K. Laland. 2001. Social learning and life skills training for hatchery reared fish. *Journal of Fish Biology* **59**(3): 471–493.

Bundy, A. 2001. Fishing on ecosystems: The interplay of fishing and predation in Newfoundland–Labrador. *Can. J. Fish. Aquat. Sci.* **58:** 1153–1167.

Caddy, J.F. and L. Garibaldi. 2000. Apparent changes in the trophic composition of world marine harvests: The perspective from the FAO capture database. *Ocean & Coastal Management* **43:** 615–655.

Carpenter, S.R. and J.F. Kitchell. 1993. *The Trophic Cascade in Lake Ecosystems.* Cambridge University Press, Cambridge.

Carr, M.-E. 2002. Estimation of potential productivity in Eastern Boudary Currents using remote sensing. *Deep-Sea Research II* **49:** 59–80.

Carscadden, J.E., K.T. Frank and W.C. Leggett. 2001. Ecosystem changes and the effects on capelin (*Mallotus villosus*), a major forage species. *Can. J. Fish. Aquat. Sci.* **58:** 73–85.

Caughley, G. and J.H. Lawton. 1976. Plant-herbivore systems. In R.M. May (ed.), *Theoretical Ecology,* pp. 132–166. Oxford, Angleterre: Blackwell Scientific.

Caverivière, A. et H. Demarcq. 2001. Indices d'abondance du poulpe commun et intensité de L'upwelling côtier au Sénégal. In A. Caverivière, D. Jouffre et M. Thiam (eds.), *Le poulpe (Octopus vulgaris). Sénégal et côtes nord-ouest africaines,* pp. 143–156. Paris : IRD, Colloques et Séminaires.

Christensen, V. and D. Pauly. 1992. ECOPATH II - a software for balancing steady-state ecosystem models and calculating network characteristics. *Ecol. Model.* **61:** 169–185.

Christensen, V. and D. Pauly, (eds.) 1993. Trophic models of aquatic ecosystems. *ICLARM Conf. Proc.* **26** (390pp).

Christensen, V. and D. Pauly. 1995. Fish production, catches and the carrying capacity of the world oceans. *Naga, The ICLARM Quaterly* **18**(3): 34–40.

Christensen, N.L. et al. 1996. The report of the Ecological Society of America Committee on the Scientific Basis for Ecosystem Management. *Ecological Applications* 6(3): 665–691.

Comins, H.N. and M.P. Hassell. 1987. The dynamics of predation and competition in patchy environments. *Theor. Popul. Biol.* **31:** 393–421.

Connell, S.D. 2000. Is there safety-in-numbers for prey? *Oikos* **88:** 527–532.

Connell, S.D. 2002. Effects of a predator and prey on a foraging reef fish: implications for understanding density-dependent growth. *Journal of Fish Biology* **60**(6): 1551–1561.

Cooney, R.T., J.R. Allen, M.A. Bishop, D.L. Eslinger, T. Kline, B.L. Norcross, C.P. Mcroy, J. Milton, J. Olsen, V. Patrick, A.J. Paul, D. Salmon, D. Scheel, G.L. Thomas, S.L. Vaughan and T.M. Willette. 2001. Ecosystem controls of juvenile pink salmon (*Onchorynchus gorbuscha*) and Pacific herring (*Clupea pallasi*) populations in Prince William Sound, Alaska. *Fish. Oceanogr.* **10**(Suppl. 1): 13.

Crawford, R.J.M. 1999. Seabird responses to long-term changes of prey resources off southern Africa. In N.J. Adams and R.H. Slotow (eds.), pp. 688–705. *Proc. 22nd Intl. Ornithol. Congr., Durban.* Johannesburg: BirdLife South Africa.

Crawford, R.J.M., R.A. Cruikshank, P.A. Shelton and I. Kruger. 1985. Partitioning of a goby resource amongst four avian predators and evidence for altered trophic flow in the pelagic community of an intense, perennial upwelling system. *South African Journal of Marine Science* **3:** 215–228.

Crawford, R.J.M., L.V. Shannon and D.E. Pollock. 1987. The Benguela Ecosystem. 4. The major fish and invertebrate resources. *Oceanography and Marine Biology: An Annual Review* **25:** 353–505.

Crawford, R.J.M., L.G. Underhill, L.V. Shannon, D. Lluch-Belda, W.R. Siegfried and C.A. Villacastin-Herrero. 1991. An empirical investigation of trans-oceanic linkages between areas of high abundance of sardine. In T. Kawasaki, S. Tanaka, Y. Toba and A. Taniguchi (eds.), *Long-term variability of pelagic fish populations and their environment,* pp. 319–332. Oxford: Pergamon Press.

Crawford, R.J.M. and J. Jahncke. 1999. Comparison of trends in abundance of guano-producing seabirds in Peru and Southern Africa. *S. Afr. J. mar. Sci.* **21:** 145–156.

Crawford, R.J.M. J.H.M. David, L.J. Shannon, J. Kemper, N.T.W. Klages, J.P. Roux, L.G. Underhill, V.L. Ward, A.J. Williams and A.C. Wolfaardt. 2001. African penguins as predators and prey – coping (or not) with change. In A.I.L. Payne, S.C. Pillar and R.J.M. Crawford (eds.), *A decade of Namibian Fisheries Science. S. Afr. J. mar. Sci.* **23:** 435–447.

Cury, P., A. Bakun, R.J.M. Crawford, A. Jarre-Teichmann, R. Quinones, L.J. Shannon and H.M. Verheye. 2000. Small pelagics in upwelling systems: patterns of interaction and structural changes in "wasp-waist" ecosystems. *ICES J. mar. Sci.* **57:** 603–618.

Cury, P. and D. Pauly. 2000. Patterns and propensities in reproduction and growth of marine fishes. *Ecological Research* **15**(1): 101–106.

Cury, P. and C. Roy. 1989. Optimal environmental window and pelagic fish recruitment success in upwelling areas. *Canadian Journal of Fisheries and Aquatic Sciences* **46:** 670–680.

Cury, P., L. Shannon and Y.-J. Shin. 2003. The functioning of marine ecosystems: a fisheries perspective. In M. Sinclair and G. Valdimarsson (eds.), *Responsible Fisheries in the Marine Ecosystem,* pp. 103–123. Wallingford: CAB International.

Cushing, D. H. 1969. The regularity of the spawning season of some fishes. *J. Cons. int. Explor. Mer.* **33:** 81–92.

Cushing, D.H. 1990. Plankton production and year-class strength in fish populations: an update of the match-mismatch hypothesis. *Advances in Marine Biology* **26:** 249–293.

Cushing, D. H. 1982. Climate and Fisheries. London, Academic Press, 373 pp. Cushing, D.H. 1996. Towards a science of recruitment in fish populations. *Excellence in Ecology. Vol. 7.* Germany, Ecology Institute, 175 pp.

Czaran, T., 1998. *Spatiotemporal models of population and community dynamics.* Chapman and Hall, London.

Daan, N. 1980. A review of replacement of depleted stocks by other species and the mechanisms underlying such replacement. In A. Saville (ed.), *The assessment and management of pelagic fish stocks. Rapp. P.-v. Réun. Cons. int. Explor.* **177:** 405–421.

Daskalov, G. 2002. Overfishing drives a trophic cascade in the Black Sea. *Marine Ecology Progress Series* **225:** 53–63.

DeAngelis, D.L. 1975. Stability and connectance in food web models. *Ecology* **56:** 238–243.

DeAngelis, D.L., R.A. Goldstein and R.V. O'Neill. 1975. A model for trophic interaction. *Ecology* 56: 881–892.

Dill, L.M and A.H.G. Frazer. 1984. Risk of predation and the feeding behaviour of juvenile coho salmon (*Oncorhynchus kisutch*). *Behaviour Ecology Sociobiology* **16:** 65–71

Dill, L.M. 1987. Animal decision making and its ecological consequences: the future of aquatic ecology and behaviour. *Canadian Journal of Zoology* **65:** 803–811

Dippner, J.W., J. Hanninen, H. Kuosa and I. Vuorinen. 2001. The influence of climatic variability on zooplankton abundance in the Northern Baltic Archipelago Sea (SW Finland). *ICES Journal of Marine Science* **58:** 569–578.

Drapeau, L., L. Pecquerie, P. Fréon and L. Shannon. 2004. Quantification and representation of potential spatial interactions in the Southern Benguela ecosystem. *Afr. J. mar. Sci.* (In press).

Ehlinger, T.J. 1989. Learning and individual variation in bluegill foraging: habitat-specific techniques. *Anim. Behav.* **38:** 643–658

Eklöv, P. 1992. Group foraging versus solitary foraging efficiency in piscivorous preditors: the perch, *Perca fluviatilis,* and pike, *Esox lucius,* patterns. *Animal Behaviour* **44:** 313–326.

Eklöv, P. and S. Diehl. 1994. Piscivore efficiency and refuging prey: the importance of predator search mode. *Oecologia* **98:** 344–353.

Elton, C. 1927. *Animal ecology.* Sidgwick & Jackson, London.

Emlen, J.M. 1966. The role of time and energy in food preference. *Am. Nat.* **100:** 611–617.

Endler, J.A. 1991. Interactions between predators and prey. In J.R. Krebs and N.B. Davies (eds.), *Behavioural ecology: an evolutionary approach,* pp. 169–196. Oxford: Blackwell Scientific Publishers.

Estes, J.A. and D.O. Duggins. 1995. Sea otters and kelp forest in Alaska: generality and variation in a community ecological paradigm. *Ecological Monographs* **65**(1): 75–100.

Estes, J.A., M.T. Tinker, T.M. Williams and D.F. Doak. 1998. Killer whale predation on sea otters linking oceanic and nearshore ecosystems. *Science* **282:** 473–476.

Faure, V., C.A.O. Inejih, H. Demarcq and P. Cury. 2000. The importance of retention processes in upwelling areas for recruitment of *Octopus vulgaris:* the example of the Arguin Bank (Mauritania). *Fisheries Oceanography* **9**(4): 343–355.

Ferrière, R. and B. Cazelles. 1999. Universal power laws governing intermittent rarity in communities of interacting species. *Ecology* **80**(5): 1505–1521.

Finney, B.P., I. Gregory-Eaves, M.S.V. Douglas and J.P. Smol. 2002. Fisheries productivity in the northeastern Pacific Ocean over the past 2,200 years. *Nature* **416:** 729–733.

Flaaten, O. 1988. The economics of multispecies harvesting. Springer-Verlag, Berlin.

Flinkman, J., E. Aro, I .. Vuorinen and M. Viitasalo. 1998. Changes in northern Baltic zooplankton and herring nutrition from the 1980s to 1990s: top-down and bottom-up processes at work. *Mar. Ecol. Prog. Ser.* **165:** 127–136.

Francis, R.C. and S.R. Hare. 1994. Decadal-scale regime shifts in the large marine ecosystems of the North-East Pacific: a case for historical science. *Fish. Oceanog.* **3**(4): 279–291.

Francis, R.C., S.R. Hare, A.B. Hollowed and W.S. Wooster. 1998. Effects of interdecadal climate variability on the oceanic ecosystems of the NE Pacific. *Fish. Oceanog.* **7**(1): 1–21.

Fréon, P. and O.A. Misund. 1999. Dynamics of pelagic fish distribution and behaviour: effects on fisheries and stock assessment. Blackwell Science, Oxford. Fretwell, S.D. 1987. Food-chain dynamics: the central theory of ecology? *Oikos* **50:** 291–301.

Fryxell, J.M. and P. Lundberg. 1994. Diet choice of predator-prey dynamics. *Evol. Ecol.* **8:** 407–421.

Fulton, E.A. 2001. *The effects of model structure and complexity on the behaviour and performance of marine ecosystem models.* PhD Thesis, University of Tasmania, Hobart, Australia.

Furness, R.W. and M.L. Tasker. 1997. Sea bird consumption in sand lance MSVPA models of the North Sea, and the impact of industrial fishing on seabird population dynamics. *American Fisheries Society* **14:** 147–149.

Galis, F. 1990. Ecological and morphological aspects of changes in food uptake through the ontogeny of *Haplochromis piceatus.* In R.H. Hughes (ed.), *Behavioural mechanisms of food selection,* pp. 281–302. Berlin: Springer-Verlag.

Galis, F., A. Terlouw and J.W.M. Osse. 1994. The relationship between morphology and behaviour during ontogenetic and evolutionary changes. *Journal of Fish Biology* **45**(suppl. A): 13–26.

Gardner, M.R. and W.R. Ashby. 1970. Connectance of large dynamic (cybernetic) systems: critical values for stability. *Nature* **228:** 784.

Getz, W.M. 1984. Population dynamics: a unified approach. *J. Theor. Biol.* **108:** 623–643.

Gilpin, M.E. 1975. Stability of feasible predator-prey systems. *Nature* **254:** 137–138.

Ginzburg, L.R. and H.R. Akçakaya. 1992. Consequences of ratio-dependent predation for steady-state properties of ecosystems. *Ecology* **73**(5): 1536–1543.

Gislason, H. 1999. Single and multispecies reference points for Baltic fish stocks. *ICES J. Mar. Sci.* **56:** 571–583.

Gislason, H. and J. Rice. 1998. Modelling the response of size and diversity spectra of fish assemblages to changes in exploitation. *ICES J. Mar. Sci.* **55:** 362–370.

Gleeson, S.K. and D.S. Wilson. 1986. Equilibrium diets: optimal foraging and prey coexistence. *Oikos* **46:** 139–144.

Glynn, P.W. 1988. El Nino-Southern oscillation 1982–1983: nearshore population, community, and ecosystem responses. *Ann. Rev. Ecol. Syst.* **19:** 309–345.

Godin, J.-G.J. 1990. Diet selection under the risk of predation. In R.N. Hughes (ed.), *Behavioural mechanisms of food selection,* pp. 739–770. NATO ASI Series Vol. G20. Berlin: Springer-Verlag.

Godin, J.-G.J. 1997. Evading predators. In J.-G.J. Godin (ed.), *Behavioural Ecology of Teleost Fishes,* pp. 191–239. Oxford: Oxford University Press.

Gulland, J.A. 1982. Why do fish numbers vary? *Journal of Theoretical Biology* **97:** 69–75.

Gulland, J.A. and S. Garcia. 1984. Observed patterns in multispecies Fisheries. In R.M. May (ed.), *Exploitation of Marine Communities,* pp. 155–190. Berlin: Springer.

Hairston, N.G.J. and N.G.S. Hairston. 1993. Cause-effect relationships in energy flow, trophic structure, and interspecific interactions. *American Naturalist* **142:** 379–411.

Hairston, N.G.J. and N.G.S. Hairston. 1997. Does food web complexity eliminate trophic-level dynamics? *American Naturalist* **149:** 1001–1007.

Hairston, N.G.J., F.E. Smith and L.B. Slobodkin. 1960. Community structure, population control and competition. *American Naturalist* **154**(879): 421–425.

Hall, S.J. 1999. *The Effects of Fishing on Marine Ecosystems and Communities.* Fish Biology and Aquatic Resources Series, No.1. Blackwell Science, London.

Hampton, I. 1992. The role of acoustic surveys in the assessment of pelagic fish resources on the South African continental shelf. In A.I.L. Payne, K.H. Brink, K.H. Mann and R. Hilborn (eds), *Benguela Trophic Functioning. South African Journal of Marine Science* **12:** 1031–1050.

Hansen, D.J. 1997. Shrimp fishery and capelin decline may influence decline of harbor sea (*Phoca vitulina*) and Northern Sea lion (*Eumetopias jubatus*) in western Gulf of alaska. In *Forage Fishes in Marine Ecosystems. American Fisheries Society* **14:** 197–208.

Hare, S.R. and N.J. Mantua. 2000. Empirical evidence for North Pacific regime shifts in 1977 and 1989. *Progress in Oceanography* **47:** 103–145.

Hart, P.J.B and A.B. Gill. 1992. Constraints on prey size selection by the three-spined stickleback: energy requirements and the capacity and fullness of the gut. *Journal of Fish Biology* **40:** 141–314

Hart, P.J.B. 1997. Foraging tactics. In J.-G.J. Godin (ed.), *Behavioural Ecology of Teleost Fishes,* pp. 104–133. Oxford: Oxford University Press.

Hassel, M.P. and G.C. Varley. 1969. New inductive population model for insect parasites and its bearing on biological control. *Nature* **223:** 1133–1136.

Hayward, T.L. 1997. Pacific Ocean climate change: atmospheric forcing, ocean circulation and ecosystem response. *Trends in Ecology & Evolution* **12:** 150–154.

Helfman, G.S. 1990. Mode selection and mode switching in foraging animals. *Adv. Study. Behav.* **19:** 249–298

Helgason, T. and H. Gislason. 1979. VPA analysis with species interaction due to predation. *CIEM C.M. 1979/G,* 52 p.

Hilborn, R. and C.J. Walters. 1992. *Quantitative fisheries stock assessment: choice, dynamics and uncertainty.* Chapman & Hall, New York.

Hofmann, E.E. and T.M. Powell. 1998. Environmental variability effects on marine fisheries: four case histories. *Ecological Application* **8**(Suppl.): S23-S32.

Holling, C.S. 1959. The components of predation as revealed by a study of small mammal predation of the European pine sawfly. *Canadian Entomologist* **91:** 293–320.

Hollowed, A.B., S.R. Hare and W.S. Wooster. 2001. Pacific basin climate variability and patterns of Northeast pacific marine fish production. In S.M. McKinnell, R.D. Brodeur, K. Hanawa, A.B. Hollowed, J.J. Polovina and C.I. Zhang (eds.), *Pacific climate variability and marine ecosystem impacts from the tropics to the Arctic. Progress in Oceanography* **49:** 257–282.

Holt, R.D. 1984. Spatial heterogeneity, indirect interaction and the coexistence of prey species. *American Naturalist* **124:** 377–406.

Hughes, R.N. 1997. Diet selection. In J.-G.J. Godin (ed.), *Behavioural ecology of teleost fishes,* pp. 134–162. Oxford University Press, Oxford.

Hughes, R.N. and M.I. Croy. 1993. An experimental analysis of frequency-dependent predation (switching) in the 15-spined stickleback, *Spinachia spinachia. Journal of Animal Ecology* **62:** 341–352.

Hughes, R.N., M.J. Kaiser, P.A. Mackney and K. Warburton. 1992. Optimising foraging behaviour through learning. *Journal of Fish Biology* **41**(suppl. B): 77–91.

Hunter, M.D. and P.W. Price. 1992. Playing chutes and ladders: heterogeneity and the relative role of bottom-up and top-down forces in natural communities. *Ecology* **73:** 724–732.

Hutchings, J.A. 2000. Collapse and recovery of marine fishes. *Nature* **406:** 882- 885.

Hutchings, J.A. 2001. Conservation biology of marine fishes: perceptions and caveats regarding assignment of extinction risk. *Canadian Journal of Fisheries and Aquatic Sciences* **58:** 108–121.

ICES. 2000. Ecosystem effects of fishing. *ICES Journal of marine Science* **57:** 465–791.

Ivlev, V.S. 1961. *Experimental ecology of the feeding of fishes.* Yale University Press, New Haven.

Jackson, J.B.C., et al. 2001. Historical overfishing and the recent collapse of coastal ecosystems. *Science* **293:** 629–638.

James, K.E.S. and R. Poulin. 1998. The effects of perceived competition and parasitism on the foraging behaviour of the upland bully (Eleotridae*). Journal of Fish Biology* **53**(4): 827–834.

Jarre-Teichmann, A., L.J. Shannon, C.L. Moloney and P.A. Wickens. 1998. Comparing trophic flows in the Southern Benguela to those in other upwelling ecosystems. In S.C. Pillar, C.L. Moloney, A.I.L. Payne and F.A Shillington (eds.), *Benguela Dynamics. S. Afr. J. mar. Sci.* **19:** 391–414.

Jennings, S. and M.J. Kaiser. 1998. The effects of fishing on marine ecosystems. *Advances in Marine Biology* **34:** 201–352.

Jin, X. and Q. Tang 1996. Changes in fish species diversity and dominant species composition in the Yellow Sea. *Fisheries Research* **26:** 337–352.

Juanes, F. 1994. What determines prey size selectivity in piscivorous fishes? In D.J. Stouder, K.L. Fresh and R.J. Feller (eds.), *Theory and application in fish feeding ecology,* pp. 79–100. University of South Carolina Press, Columbia, South Carolina.

Kaiser, M.J., A.P. Westhead, R.N. Hughes and R.N. Gibson. 1992. Are digestive characteristics important contributors to the profitability of prey? *Oecologia* **90:** 61–69.

Karl, D.M., D.V. Hebel, K. Bjorkman and R.M. Letelier. 1998. The role of dissolved organic matter release in the productivity of the oligotrophic North Pacific Ocean. *Limnology and Oceanography* **43**(6): 1270–1286.

Kawasaki, T., S. Tanaka, Y. Toba and A. Taniguchi. 1991. *Long term variability of pelagic fish populations and their environment.* Pergamon Press, Oxford, UK.

Kim, S. and S. Kang. 2000. Ecological variations and El Niño effects off the southern coast of the Korean Peninsula during the last three decades. *Fisheries Oceanography* **3:** 239–247.

Kirkwood, G.P. 1982. Simple models for multispecies fisheries. In D. Pauly and G.I. Murphy (eds.), *Theory and Management of Tropical Fisheries. ICLARM Conf. Proc.* **9:** 83–98.

Kislalioglu, M. and R.N. Gibson. 1976. Some factors governing prey selection by the 15-spined stickleback, *Spinachia spinachia. J. Exp. Mar. Biol. Ecol.* **25:** 159–169.

Kornilovs, G., L. Sidrevics and J.W. Dippner. 2001. Fish and zooplankton interaction in the Central Baltic Sea. *ICES Journal of Marine Science* **58:** 579–588.

Koster, F.W. and D. Schnack. 1994. The role of predation on early life stages of cod in the Baltic. *Dana* **10:** 179–201.

Lawton, J.H. 1989. Food webs. In J.M. Cherett (ed.), *Ecological concepts: the contribution of ecology to an understanding of the natural world,* pp. 43–78. Blackwell Scientific, Oxford.

Lawton, J.H. 1999. Are there general laws in ecology? *Oikos* **84:** 177–192.

Leslie, P.H. 1948. Some further notes on the use of matrices in population mathematics. *Biometrika* **35:** 213–345.

Leslie, P.H. and J.C. Gower. 1960. The properties of a stochastic model for two competing species. *Biometrika* **45:** 316–330.

Lima, S.L. 2002. Putting predators back into behavioural predator-prey interactions. *Trends in Ecology and Evolution* **17**(2): 70–75.

Lima, S.L. and L.M. Dill. 1990. Behavioural decisions made under the risk of predation: a review and prospectus. *Canadian Journal of Zoology* **68:** 619–640.

Lindeman, R.L. 1942. The trophic-dynamic aspect of ecology. *Ecology* **23:** 399–418.

Link, J.S. and L.P. Garrison. 2002. Changes in piscivory associated with fishing induced changes to the finfish community on Georges Bank. *Fisheries Research.* **55:** 71–86.

Link, J. S. and J.K.T Brodziak (eds.) 2002. Status of the Northeast U.S. Continental Shelf Ecosystem. Northwest Fisheries Science Centre Reference Document 02–11.

Link, J. S. 1999. (Re)constructing food webs and managing fisheries. In *Ecosystem Approaches for Fisheries Management.* Proceedings of the 16th Lowell Wakefield Fisheries Symposium. AK-SG-99-01, pp. 571–588. University of Alaska Sea Grant. Fairbanks, Alaska, USA.

Lloret J., J. Lleonart, I. Solé. 2000. Time series modelling of landings in Northwest Mediterranean Sea. ICES Journal of Marine Science, 57:171–184.

Lluch-Belda, D., R.A. Schwartzlose, R. Serra, R.H. Parrish, T. Kawasaki, D. Hedgecock and R.J.M. Crawford. 1992. Sardine and anchovy regime fluctuations of abundance in four regions of the world oceans: a workshop report. *Fish. Oceanogr.* **1**(4): 339–347.

Lotka, A.J. 1932. The growth of mixed populations: two species competing for a common food supply. *Journal of the Washington Academy of Sciences* **22:** 461–469.

Lundvall, D., R. Svanbäck, L. Persson and P. Byström. 1999. Size-dependent predation in piscivores: interactions between predator foraging and prey avoidance abilities. *Canadian Journal of Fishery and Aquatic Sciences* **56:** 1285–1292.

MacArthur, R. 1955. Fluctuations of animal populations and a measure of community stability. *Ecology* **36:** 533–536.

MacArthur, R.H. and E.R. Pianka. 1966. On optimal use of a patchy environment. *Am. Nat.* **100:** 603–609.

MacCall, A. 1990. *Dynamic geography of marine fish populations.* (Books in Recruitment Fishery Oceanography). Washington Sea Grant Program, University of Washington Press, Washington.

MacCall, A.D. 1996. Patterns of low-frequency variability in fish populations of the California Current. *CalCOFI Report* **37:** 100–110.

Mace, P.M. 2001. A new role for MSY in single-species and ecosystem approaches to fisheries stock assessment and management. *Fish and Fisheries* **2:** 2–32.

Maes, J. and F. Ollevier. 2002. Size structure and feeding dynamics in estuarine clupeoid fish schools: field evidence for the school trap hypothesis. *Aquat. Liv. Res.* **15:** 211–216.

Matson, P.A. and M.D. Hunter. 1992. Special Feature: the relative contributions to top-down and bottom-up forces in population and community ecology. *Ecology* **73:** 723.

Matsuda, H., T. Wada, Y. Takeuchi and Y. Matsumiya. 1992. Model analysis of the effect of environmental fluctuation on the species replacement pattern of pelagic fishes under interspecific competition. *Res. Popul. Ecol.* **34**(2): 309–319.

May, R.M. 1972. Will a large complex system be stable? *Nature* **238:** 413–414.

May, R.M. 1974. *Stability and complexity in model ecosystems.* Princetown University Press, Princetown NJ.

May, R.M., J.R. Beddington, C.W. Clark, S.J. Holt and R.M. Laws. 1979. Management of multispecies fisheries. *Science* **205:** 267–277.

McCann, K. 2000. The diversity-stability debate. *Nature* **405**(11): 228–233.

McCauley, E., W.G. Wilson and A.M. de Roos. 1993. Dynamics of age-structured and spatially structured predator-prey interaction: individual-based models and population-level formulations. *Am. Nat.* **142:** 412–442.

McFarlane, G.A., J.R. King and R.J. Beamish. 2000. Have there been recent changes in climate? Ask the fish. *Progress in Oceanography* **47:** 147–169.

McGowan, J.A., D.R. Cayan and L.M. Dorman. 1998. Climate-Ocean variability and ecosystem response in the Northeast Pacific. *Science* **281:** 210- 217.

McNair, J.N. 1986. The effects of refuges on predator-prey interactions: a reconsideration. *Theor. Popul. Biol.* **29:** 38–63.

Menge, B.A. 2000. Top-down and bottom-up community regulation in marine rocky intertidal habitats. *J. Exp. Mar. Biol. Ecol.* **250:** 257–289.

Menge, B.A. and A.M. Olson. 1990. Role of scale and environmental factors in regulation of community structure. *Trends Ecol. Evol.* **5:** 52–57.

Menge, B.A., E.L. Berlow, C.A. Blanchette, S.A. Navarrete and S.B. Yamada. 1994. The keystone species concept: variation in interaction strength in a rocky intertidal habitat. *Ecological Monographs* **64:** 249–286.

Meyer, A. 1987. Phenotype plasticity and heterochrony in *Cichlasoma managuense* (Pisces, Cichlidae) and their implications for speciation in cichlid fishes. *Evolution* **41:** 1357–1369.

Michalski, J. and R. Arditi. 1995a. Food webs with predator interference. *Journal of Biological Systems* **3**(2): 323–330.

Michalski, J. and R. Arditi. 1995b. Food web structure at equilibrium and far from it : is it the same? *Proc. R. Soc. Lond. B* **259:** 217–222.

Michalski, J. and R. Arditi. 1999. The complexity-stability problem in food web theory. What can we learn from exploratory models? In A. Weill (ed.), *Advances in environmental and ecological modelling,* pp. 1–20. Elsevier, Paris.

Micheli, F. 1999. Eutrophication, fisheries, and consumer-resource dynamics in marine pelagic ecosystems. *Science* **285:** 1396–1398.

Milinski, M. 1982. Optimal foraging: the influence of intraspecific competition on diet selection. Behav. Ecol. Sociobiol. 11: 109–115.

Milinski, M. 1993. Predation risk and feeding behaviour. In T.J. Pitcher (ed.), *Behaviour of teleost fishes,* pp. 285–305. Chapman & Hall, London

Mills, L.S., M.E. Soule and D.F. Doak. 1993. The keystone-species concept in ecology and conservation. *BioScience* **43**(4): 219–224.

Mittelbach, G.G. and C.W. Osenberg. 1994. Using foraging theory to study trophic interaction. In D.J. Stouder, K.L. Fresh and R.J. Feller (eds.), *Theory and application in fish feeding ecology,* pp. 45–59. Columbia: University of South Carolina Press.

Miyazaki, T., R. Masuda, S. Furuta and K. Tsukamoto. 2000. Feeding behaviour of hatchery-reared juveniles of the Japanese flounder following a period of starvation. *Aquaculture* **190**(1–2): 129–138.

Mollmann, C. and F.W. Koster. 1999. Food consumption by clupeids in the Central Baltic: evidence for top-down control? *ICES Journal of Marine Science* **56** (Suppl.): 100–113.

Morales-Nin, B. and J.P. Pertierra. 1990. Growth rates of anchovy (*Engraulis encrasicholus)* and sardine (*Sardina pilchardus*) in the Northwestern Mediterranean Sea. *Marine Biology* **107:** 349–356.

Morin, P.J. and S.P. Lawler. 1995. Food web architecture and population dynamics: theory and empirical evidence. *Annu. Rev. Ecol. Syst.* **26:** 505–529.

Mullon, C., P. Cury and P. Penven. 2002. Evolutionary Individual-Based Model for the recruitment of the anchovy in the southern Benguela. *Can. J. Fish. Aquat. Sci.:* **59:** 910–922.

Murphy, E.J. 1995. Spatial structure of the Southern Ocean ecosystem: predator-prey linkages in Southern Ocean food webs. *Journal of Animal Ecology* **64:** 333–347.

Murray, J.D. 1989. *Mathematical biology. Biomathematics texts.* Springer-Verlag, Berlin-Heidelberg.

Neunfeldt, S. and F.W. Köster. 2000. Trophodynamic control on recruitment success in Baltic cod: the influence of cannibalism. *ICES J. Mar. Sci.* **57**(2): 300–309.

O'Brien, W.J., H.I. Browman and B.I. Evans. 1990. Search strategies in foraging animals. *Amer. Sci.* **78:** 152–160.

Oleson, N.J. 1995. Clearance potential of jellyfish *Aurelia aurita,* and predation impact on zooplankton in a shallow cove. *Mar. Ecol. Prog. Ser.* **124**(1–3): 63–72.

O'Neill, R.V., D.L. DeAngelis, J.B. Waide and T.F.H. Allen. 1986. *A hierarchical concept of ecosystems.* Princeton University Press, Princeton, New jersey.

Pace, M.L., J.J. Cole, S.R. Carpenter and J.F. Kitchell. 1999. Trophic cascades revealed in diverse eocsystems. *Trends in Ecology & Evolution* **14:** 483–488.

Paine, R.T. 1966. Food web complexity and species diversity. *American Naturalist* **1000:** 65–75.

Paine, R.T. 1980. Food webs: linkage, interaction strength and community infrastructure. *Journal of Animal Ecology* **49:** 667–685.

Parrish, R.H., C.S. Nelson and A. Bakun. 1981. Transport mechanisms and reproductive success of fishes in the California Current. *Biol. Oceanogr.* **1:** 175–203.

Parsons T.R. 2003. Macroecological studies of the Oceans. *Oceanography in Japan,* **12** (4): 370–374.

Pauly, D., V. Christensen, R. Froese and M.L. Palomares. 2000. Fishing down aquatic food webs. *American Scientist* **88:** 46–51.

Pauly, D., M. Lourdes Palomares, R. Froese, S. Pascualita, M. Vakily, D. Preikshot and S. Wallace. 2001. Fishing down Canadian aquatic food-webs. *Can. J. Fish. Aquat. Sci.* **58:** 1–12.

Pauly, D. and V. Christensen. 1995. Primary production required to sustain global fisheries. *Nature* **374:** 255–257.

Pauly, D., V. Christensen, J. Dalsgaard, R. Froese and F. Torres Jr. 1998. Fishing down marine food webs. *Science* **279:** 860–863.

Pauly, D. and J. Moreau. 1997. Méthodes pour l'évaluation des ressources halieutiques.

Toulouse: Collection Polytech de l'I.N.P. de Toulouse, Cépaduès Editions,Toulouse.

Persson, L. and S. Diehl. 1990. Mechanistic, individual-based approaches in the population/community ecology of fish. *Ann. Zool. Fennici.* **27:** 165–182.

Persson, L. and L.A. Greenberg. 1990. Interspecific and intraspecific size class competition affecting resource use and growth of perch, *Perca fluviatilis. Oikos* **59:** 97–106.

Persson, L., S. Diehl, L. Johansson, G. Andersson and S.F. Hamrin. 1991. Shifts in fish communities along the productivity gradient in temperate lakes - patterns and the importance of size-structured interactions. *Journal of Fish Biology* **38:** 281–293.

Persson, L., S. Diehl, P. Eklöv and B. Christensen. 1997. Flexibility in fish behaviour: consequences at the population and community levels. In J.-G.J. Godin (ed.), *Behavioural ecology of teleost fishes,* pp. 316–343. Oxford University Press, Oxford.

Pimm, S.L. 1979. Complexity and stability: another look at MacArthur's original hypothesis. *Oikos* **33:** 351–357.

Pimm, S.L. 1991. The balance of nature? University of Chicago Press, Chicago.

Pinnegar, J.K., N.V.C. Polunin, P. Francour, F. Badalamenti, R. Chemello, M.-L. Harmelin-Vivien, B. Hereu, M. Milazzo, M. Zabala, G. D'Anna and C. Pipitone. 2000. Trophic cascades in benthic marine ecosystems: lessons for fisheries and protected-area. *Environmental Conservation* **27:** 179–200.

Pitcher, T.J. and J.K. Parrish. 1993. Functions of shoaling behaviour in teleosts. In T.J. Pitcher (ed.), *Behaviour of teleost fishes,* pp. 363–439. Chapman and Hall, London.

Polis, G.A. 1999. Why are parts of the world green? Multiple factors control productivity and the distribution of biomass. *Oikos* **86:** 3–15.

Polis, G.A. and D. Strong. 1996. Food web complexity and community dynamics. *American Naturalist* **147:** 813–846.

Polis, G.A. and K.O. Winemiller, (eds.) 1996. *Food webs integration of patterns & dynamics.* Kluwer, Boston.

Polovina, J.F., G.T. Mitchum, N.E. Graham, M.P. Craig, E.E. Demartini and E.N. Flint. 1994. Physical and biological consequences of a climate event in the Central North Pacific. *Fish. Oceanogr.* **3:** 15–21.

Polovina, J.J. 1984. Model of a coral reef ecosystem I. The ECOPATH model and its application to French Frigate Shoals. *Coral Reefs* **3:** 1–11.

Polunin, N.V.C. and J.K. Pinnegar. 2002. Trophic ecology and the structure of marine food webs. In P.J.B. Hart and J.C. Reynolds (eds.), *Handbook of Fish and Fisheries, Volume I,* pp. 310–320. Blackwell, Oxford:.

Pope, J.G. 1976. The effect of biological interaction on the theory of mixed fisheries. *Int. Comm. Northw. Atlant. Fish. Sel. Pap. 1*: 157–162.

Pope, J.G. 1979. A modified cohort analysis in which constant natural mortality is replaced by estimates of predation levels. *ICES C.M.* 1979/H, 16 p.

Pope, J.G. 1989. Multispecies extensions to age-structured assessment models. *Am. Fish. Soc. Symp. 6*: 102–111.

Pope, J.G. and B.J. Knights. 1982. Simple models of predation in multi-age multispecies fisheries for considering the estimation of fishing mortality and its effects. In M.C. Mercer (ed.), *: Multispecies approaches to fisheries management advice. Can. Spec. Publ. Fish. Aquat. Sci.* **59:** 64–69.

Posey, M., C. Powell, L. Cahoon and D. Lindquist. 1995. Top down vs. bottom up control of benthic community composition on an intertidal tideflat. *Journal of Experimental Marine Biology and Ecology* **185:** 19–31.

Post, D.M. 2002. The long and short of food-chain length. *Trends in Ecology & Evolution* **17**(6): 269–277.

Post, D.M., M.L. Pace and N.G. Hairston Jr. 2000. Ecosystem size determines food-chain length in lakes. *Nature* **405:** 1047- 1049.

Power, M.E. 1992. Top-down and bottom-up forces in food webs: do plants have primacy. *Ecology* **73:** 733–746.

Power, M.E., D. Tilman, J.A. Estes, B.A. Menge, W.A. Bond, L.S. Mills, G. Daily, J.C. Castilla, J. Lubchenco and R.T. Paine. 1996. Challenges in the quest for keystones. *Bioscience* **46**(8): 609–620.

Power, M.E.1987. Predator avoidance by grazing fishes in temperate and tropical streams: importance of stream depth and prey size. In W.C. Kerfoot and A. Sih (eds.), *Predation: direct and indirect impacts in aquatic communities*, pp. 333–351. University Press of New England, Hanover.

Reiriz, L., A.G. Nicieza and F. Braña. 1998. Prey selection by experienced and naive juvenile Atlantic salmon, *Journal of Fish Biology* **53**(1): 100–114.

Rice, J. 1995. Food web theory, marine food webs, and what climate change may do to northern marine fish populations. In : R.J. Beamish (ed.), *Climate change and northern fish populations. Canadian Special Publication of Fisheries and Aquatic Sciences* **121:** 561–568.

Rice, J. and H. Gislason, 1996. Patterns of change in the size spectra of numbers and diversity of the North Sea fish assemblage, as reflected in surveys and models. *ICES J. mar. Sci.* **53:** 1214–1225.

Rudstam, L.G, G. Aneer and M. Hilden. 1994. Top-down control in the pelagic Baltic ecosystem. *Dana* **10:** 105–129.

Ryther, J.H. 1969. Photosynthesis and fish production in the sea. *Science* **166:** 72–76.

Sanford, E. 1999. Regulation of keystone predation by small changes in ocean temperature. *Science* **283:** 2095–2097.

Schaefer, M.B. 1954. Some aspects of the dynamics of populations important to the management of commercial marine fisheries. *Bull. Inter-Am. Trop. Tuna Comm.* **1:** 27–56.

Scharf, F.S., F. Juanes and R.A. Ruantree. 2000. Predator size-prey size relationships of marine fish predators: interspecific variation and effects of ontogeny and body size on trophic-niche breadth. *Mar. Ecol. Prog. Ser.* **208:** 229–248.

Schoener, T.W. 1989. Food webs from the small to the large. *Ecology* **70:** 1559–1589.

Schwartzlose, R.A., et al. 1999. Worldwide large-scale fluctuations of sardine and anchovy populations. *South African Journal of Marine Science* **21:** 289–347.

Serra, R., P. Cury and C. Roy. 1998. The recruitment of the Chilean sardine Sardinops sagax and the "optimal environmental window." In M.H. Durand, P. Cury, R. Mendelssohn, C. Roy, A. Bakun and D. Pauly (eds.), *From local to global changes in upwelling systems,* pp. 267–274. ORSTOM Editions, Paris.

Shackleton, L.Y. 1987. A comparative study of fossil fish scales from three upwelling regions. In A.I.L. Payne, J.A. Gulland and K.H. Brink (eds.), *The Benguela and Comparable Ecosystems. S. Afr. J. mar. Sci.* **5:** 79–84.

Shannon, L.J. 1998. Modelling environmental effects on the early life history of the South African anchovy and sardine: a comparative approach. In S.C. Pillar, C.L. Moloney, A.I.L Payne and F.A. Shillington (eds.), *Benguela Dynamics: impacts of variability on shelf-sea environments and their living resources. S. Afr. J. mar. Sci.* **19:** 291–304.

Shannon, L.J. and P. Cury. 2004. The functional role of small pelagic fish in an exploited upwelling area: indices for keystone species and functional similarity in the southern Benguela ecosystem. *Ecological Indicators.* (In press).

Shannon, L.V. and J.G. Field. 1985. Are fish stocks food-limited in the southern Benguela pelagic ecosystem? *Mar. Ecol. Prog. Ser.* **22:** 7–19.

Shannon, L.J. 2001. *Trophic models of the Benguela upwelling system: towards an ecosystem approach to fisheries management.* PhD thesis, University of Cape Town, Cape Town.

Shannon, L.V. and S.C. Pillar. 1986. The Benguela ecosystem. Part III: Plankton. *Oceanogr. Mar. Biol. Ann. Rev.* **24:** 65–170.

Sheldon, R.W., W.H. Sutcliffe and M.A. Paranjape. 1977. Structure of pelagic food chains and the relationship between plankton and fish production. *Journal of the Fisheries Research Board of Canada* **34:** 2344–2353.

Shin, Y.-J., P. Cury. 2001. Exploring fish community dynamics through size-dependent trophic interactions using a spatialized individual-based model. *Aquat. Liv. Res.* **14**(2): 65–80.

Shin, Y.-J. and P. Cury. 2004. Using an individual-based model of fish assemblages to study the response of size spectra to changes in fishing. *Canadian Journal of Fisheries and Aquatic Sciences* **61**(02): 414–431.

Shiomoto, A., K. Tadokoro, K. Nagasawa and Y. Ishida. 1997. Trophic relations in the subartic North pacific ecosystem: possible feeding effect from pink salmon. *Marine Ecology Progress Series* **150:** 75–85.

Shurin, J.B., E.T. Borer, E.W. Seabloom, K. Anderson, C.A. Blanchette, B. Broitman, S.D. Cooper and B.S. Halpern. 2002. A cross-ecosystem comparison of the strength of trophic cascades. *Ecology Letters* **5:** 785–791.

Sih, A. 1993. Effects of ecological interaction on forager diets: competition, predation risk, parasitism and prey behaviour. In R.N. Hughes (ed.), *Diet selection: an interdisciplinary approach to foraging behaviour,* pp. 182–211. Blackwell Scientific Publishers, Oxford.

Silvert, W. and T. Platt. 1978. Energy flux in the pelagic ecosystem: a time-dependent equation. *Limnol. Oceanogr.* **23:** 813–816.

Sinclair, M. 1988. *Marine Populations: An Essay on Population Regulation and Speciation.* Washington Sea Grant Program. Univ. Wash. Press, Washington.

Sissenwine, M.P. 1984. Why do fish populations vary? In R.M. May (ed.), *Exploitation of marine communities,* pp. 59–94. Springer Verlag, Berlin.

Skud, B.E. 1982. Dominance in fishes: the relation between environment and abundance. *Science* **216:** 144–149.

Skud, B.E. 1983. Interactions of pelagic fishes and the relation between environmental factors and abundance. In G.D. Sharp and J. Csirke (eds.), *Proceedings of the Expert Consultation to Examine Changes in Abundance and Species Composition of Neritic Fish Resources. FAO Fish. Rep.* **291**(2,3): 1133–1140.

Smetacek, V. 1999. Revolution in the ocean. *Nature* **401:** 647.

Smith, R.J.F. 1997. Avoiding and deterring predators. In J.-G.J. Godin (ed.), *Behavioural ecology of teleost fishes,* pp. 163–190. Oxford University Press, Oxford.

Soutar, A. and J.D. Isaacs. 1974. Abundance of pelagic fish during the 19th and 20th centuries as recorded in anaerobic sediment off California. *Fishery Bulletin* **72:** 257–274.

Sparholt, H. 1991. Multispecies assessment of Baltic fish stocks. *ICES mar. Sci. Symp.* **193:** 64–79.

Steneck, R.S. 1998. Human influences on coastal ecosystems: does overfishing create trophic cascades? *Trends in Ecology & Evolution* **13:** 429–430.

Stephens, D.W. and J.R. Krebs. 1986. *Foraging theory.* Princeton University Press, Princeton.

Stokes, T.K. 1992. An overview of the North Sea multispecies work in ICES. In A.I.L. Payne, K.H. Brink, K.H. Mann and R. Hilborn (eds.), *Benguela Trophic Functioning. S. Afr. J. mar. Sci.* **12:** 1051–1060.

Ströbele, W.J. and H. Wacker, 1991. The concept of sustainable yield in multi-species fisheries. *Ecol. Model.* **53:** 61–74.

Strong, D.R. 1992. Are trophic cascades all wet? Differentiation and donor-control in speciose ecosystems. *Ecology* **73:** 747–754.

Tang, Q., X. Jin, J. Wang, Z. Zhuang, Y. Cui and T. Meng. 2003. Decadal-scale variations of ecosystem productivity and control mechanisms in the Bohai Sea *Fish. Oceanogr.* **12**(4/5): 223–233.

Tilman, D. 1999. The ecological consequences of changes in biodiversity: a search for general principles. *Ecology* **80**(5): 1455–1474.

Tinbergen, N. 1960. The natural control of insects in pinewoods. I. Factors influencing the intensity of predation by songbirds. *Arch. Neerland. Zool.* **13:** 265–343.

Tonn, W.M., C.A. Paszkowski and I.J. Holopainen. 1992. Piscivory and recruitment: mechanisms structuring prey populations in small lakes. *Ecology* **73:** 951–958.

Trites, A., Christensen, V. and D. Pauly. 1997. Competition between fisheries and marine mammals for prey and primary production in the Pacific Ocean. *Journal North West Atlantic Fisheries Science* **22:** 173–187.

Ulanowicz, R.E. 1993. Inventing the ecoscope. In V. Christensen and D. Pauly (eds.), *Trophic models of aquatic ecosystems. ICLARM Conf. Proc.* **26:** ix-x.

Ulanowicz, R.E. and C.J. Puccia. 1990. Mixed trophic impacts in ecosystems. *Coenoses* **5:** 7–16.

Ursin, E. 1973. On the prey size preferences of cod and dab. *Meddr. Danm. Fisk.- og Havunders* **7:** 85–98.

Ursin, E. 1982. Multispecies Fish Stock and Yield Assessment in ICES. In M.C. Mercer (ed.), *Multispecies approaches to fisheries management advice. Can. Spec. Publ. Fish. Aquat. Sci.* **59:** 39–47.

Verhulst, P.F. 1838. Notice sur la loi que la population suit dans son accroissement. *Correspondances Mathématiques et Physiques* **10:** 113–121.

Verity, P.G. 1998. Why is relating plankton community structure to pelagic production so problematic? In S.C. Pillar, C.L. Moloney, A.I.L. Payne and F.A. Shillington (eds.). *Benguela Dynamics. South African Journal of Marine Science* **19:** 333–338.

Vinyard, G.L. 1982. Variable kinematics of Sacramento perch (*Archoplites interruptus*) capturing evasive and nonevasive prey. *Canadian Journal of Fish Aquatic Science* **39:** 208–211.

Volterra, V. 1926. Variations and fluctuations of the numbers of individuals in animal species living together. Réédition traduite. In R.N. Chapman (eds.), *Animal ecology,* pp. 409–448. McGraw Hill, New York.

Walters, C. and J.F. Kitchell. 2001. Cultivation/depensation effects on juvenile survival and recruitment: implications for the theory of fishing. *Canadian Journal Fisheries and Aquatic Sciences* **58:** 39–50.

Walters, C., V. Christensen and D. Pauly. 1997. Structuring dynamic models of exploited ecosystems from trophic mass-balance assessments. *Rev. Fish Biol. Fish.* **7:** 139–172.

Walters, C.J. and F. Juanes. 1993. Recruitment limitation as a consequence of natural selection for use of restricted feeding habitats and predation risk taking by juvenile fishes. *Can. J. Fish. Aquat. Sci.* **50:** 2058–2070.

Werner, E.E. and J.F. Gilliam. 1984. The ontogenetic niche and species interaction in size-structured populations. *A. Rev. Ecol. Syst.* **15:** 393–425.

Wootton, J.T. and M.E. Power. 1993. Productivity, consumers, and the structure of a river food chain. *Proc. Natl. Acad. Sci. USA* **90:** 1384–1487.

Wootton, R.J. 1984. Introduction: tactics and strategies in fish reproduction. In G.W. Potts and R.J. Wootton (eds.), *Fish reproduction: strategies and tactics,* pp. 1–12. Academic Press, London.

Worm, B. and R.A. Myers. 2003. Meta-analysis of cod-shrimp interactions reveals top-down control in oceanic food webs. *Ecology* **84:** 84: 162–173.

Yodzis, P. 1994. Predator-prey theory and management of multispecies fisheries. *Ecol. Appl.* **4**(1): 51–58.

Yodzis, P. 2001. Must top predators be culled for the sake of fisheries? *Trends in Ecology & Evolution* **16:** 78–88.

Zaitsev, Y. and V. Mamaev. 1997. *Marine biological diversity in the Black Sea: a study of change and decline.* UN Publications, New York.

Zhang, C.I., J.B. Lee, S. Kim and J.-H. Oh. 2000. Climatic regime shifts and their impacts on marine ecosystem and fisheries resources in Korean waters. *Progress in Oceanography* **47:** 171–190.

# Chapter 15. THE BIOGEOCHEMISTRY OF ORGANIC CHEMICALS OF ENVIRONMENTAL CONCERN IN THE COASTAL OCEANS.

JOHN W. FARRINGTON

*Woods Hole Oceanographic Institution*

**Contents**



## 1. Introduction

### *1.1 Brief History*

Modern civilization has benefited enormously from the use of human mobilized naturally occurring chemicals, e.g., the substances that make up petroleum and coal, iron, mercury, lead, copper and tin. Synthetic chemicals have also been produced by modern civilization for a variety of uses. Chlorinated pesticides, especially DDT – (*d*ichloro*d*iphenyl*t*hrichloroethene –formal chemical name 1,1'-(2,2,2-trichlorethylidene)-bis(4-chlorobeneze), have been legitimately credited with reducing malarial disease worldwide and with combating crop pests to increase agricultural production. PCBs (*p*oly*c*hlorinated *b*iphenyls) were a boon to the effective delivery of electricity to industries, farms and homes, because they had ideal properties to serve as insulation in capacitors and transformers.

The mobilization, production and use of these chemicals had a downside. Human mobilization of natural chemicals alters natural biogeochemical cycles in the environment to varying degrees, often resulting in elevated concentrations of these chemicals in components of ecosystems relative to pre-industrial revolution concentrations. Production and use of several synthetic chemicals resulted in deliberate or unintentional releases of these chemicals to the environment. During the late 1950s and into the 1960s, specific events and observations began to raise

*The Sea*, Volume 13, edited by Allan R. Robinson and Kenneth H. Brink
ISBN 0-674-01526-6 ©2004 by the President and Fellows of Harvard College

awareness that human civilization was contaminating the environment with some chemicals in ways that might prove to be detrimental. The dispersion of radioactive fallout in the northern hemisphere and then into the southern hemisphere as a result of atmospheric nuclear weapons testing in the 1950s and 1960s raised alarms about adverse effects of the widely dispersed radioactive chemicals such as strontium 90, cesium 137, and the plutonium isotopes 239/240/241. (SCEP, 1970; NAS, 1971). "The artificially produced radio-active species, arising from nuclear detonations or as waste products from nuclear plants, were the first group of substances recognized by the marine scientists as a potential challenge to the resources of the sea on a global basis" (Goldberg, 1976). The poisoning of people by methyl mercury contaminated fish from Minimata Bay resulting from mercury in wastes discharged to the coastal ocean from a chloralkali plant in the 1950s and 1960s, and discovered in the 1960s was a warning about using the oceans indiscriminately for waste disposal. (Goldberg, 1976).

During the same time interval, Rachel Carson (1962) in her book, *Silent Spring,* summarized her own observations, and those of several pioneering scientists doing research during the late 1940s and the 1950s (see Hall, 1987), tracking effects of DDT (and other chlorinated pesticides). Her much heralded book raised awareness of the toxic effects on the non-target organisms—particularly birds. In the mid 1960s, analyses for DDT family compounds uncovered a series of interfering chemicals that were identified as polychlorinated biphenyls (PCBs) by Jensen (Anon, 1966), Risebrough et al, (1968), and Koeman et al, (1969). An informative account of the early discoveries and concerns is provided by Jensen (1972). Surveys of a few selected samples worldwide soon revealed the widespread distribution of some persistent, not easily degraded, synthetic organic chemicals such as DDT and PCBs in a variety of ecosystems, including marine ecosystems. The combination of these survey observations, coupled with the relatively rapid nature of the atmosphere-land-oceans interactions in global biogeochemical cycles traced by using radionuclides produced during nuclear weapons testing, gave rise to serious concerns about the extent and severity of contamination of all ecosystems. Oceanic ecosystems were included in these concerns, especially since time scales for the coastal ocean are months to decades and the open ocean are hundreds to millions of years for natural processes (Goldberg, 1976). In addition, there were several pathways from oceanic ecosystems through commercial and recreational fisheries back to humans.

The increased awareness by scientists of the real and potential problems with chemicals of environmental concern led to several scientific and scientific policy conferences and workshops to assess state of knowledge and to develop plans for programs of more detailed assessment of the inputs, fates and effects of these chemicals, including coastal and open-deep ocean programs (e.g. SCEP, 1970, NAS, 1972). The Joint Group of experts on the Scientific Aspects of Marine Environmental Protection (GESAMP), sponsored by eight United Nations bodies, was formed in 1969 (Wells et al, 2002) to provide scientific advice in the international arena about marine environmental quality problems. In the early years, the focus of GESAMP was on chemicals of environmental concern (Wells et al, 2002).

Stimulated by concerns of scientists and their reports to governmental and intergovernmental organizations, and by visually spectacular pollution events—for example the Santa Barbara Oil Well Blowout in 1968 off the coast of California—

there was an increase in public awareness of the environmental problems that led to the first Earth Day in 1970.

### *1.2 Organic Chemicals of Environmental Concern in the Ocean: Persistent Organic Pollutants (POPs).*

Chemicals of environmental concern generally are in three categories: metals such as lead, mercury, tin, copper; radionuclides such as plutonium- 238,239, 240; cesium-137, and strontium-90 introduced to the environment by nuclear weapons production and testing, and leaking from the nuclear fuel cycle; and organic chemicals. It is beyond the scope of this chapter to address all three, although all three are important. The principles in understanding the inputs, fates and effects of chemicals in the coastal ocean are similar for all three groups of chemicals. By addressing one group we gain some insight into processes important for the other two groups.

The term organic contaminant is used in some instances in the literature rather than the term organic pollutant. The word pollutant has been interpreted by several authors and organizations to indicate a proved adverse biological effect at the concentrations encountered in the environment. In the following sections, a wide range of concentrations in air, water, sediments, and organisms will be reported, without reference to adverse biological effects. Many of the compounds to be discussed could have an adverse effect at elevated concentrations and therefore for simplicity they are designated as pollutants. The ability to detect compounds at concentrations well below threshold levels for adverse biological effects is a distinct advantage, in that it provides early warning that a chemical of potential concern is present and provides a means to predict whether or not concentrations will increase in the near future and causes adverse effects.

The organic chemicals discussed in this chapter do not include the more volatile compounds, such as low molecular weight hydrocarbons, freon and other volatile organohalogen compounds. Obviously, freon is of environmental concern in the atmosphere with respect to reactivity with ozone (Rowland and Molina, 1975) and volatile organohalogens such as tertrachloroethylene are a concern in groundwater used for drinking water (Schwarzenbach et al, 2003). Freon is used as a tracer for water mass circulation in the oceans (e.g. Doney et al, 1997). Tetrachloroethelyene, and other similar compounds, will readily exchange between the oceanic and coastal water and the atmosphere (e.g. Wakeham et al, 1983). Certainly a gasoline spill in a coastal area would be of immediate concern. However, experience has shown that the gasoline is short lived because of volatility and the portion that does dissolve in seawater transfers rapidly to the atmosphere (NRC, 2003). In general, these volatile compounds are not of long-term concern in the coastal oceans from the perspective of persistence, biological uptake and long-term effects, including transfer through the food web back to humans

Estimates in the late 1970s and early 1980s suggested that 1000 to 1500 new substances were entering daily use each year and added to the approximately 60,000 already in daily use (Stumm et al, 1983). Not all of these substances are of environmental concern from the perspective of adverse impacts on biota or for toxicity to humans. The number of chemicals of concern, manufactured in quantities requiring assessment of environmental fates and effects, was estimated in 1978 to be about 11,000 (Butler, 1978). A recent assessment of the state of knowledge about impacts of pollution on coastal and marine ecosystems listed 96 pesticides and agrochemicals in worldwide use that are of major environmental concern (Islam and Tanaka, 2004), and those are only two categories of chemicals of environmental concern. Added to this number are chemicals of environmental concern generated by treatment of industrial and domestic wastes by chlorination.

The preceding are only estimates, but the numbers are large enough to provide a clear sense that the assessment of each individual chemical's input, fate and effect in myriad oceanic ecosystems, including coastal ecosystems, would be a formidable task unless there are some underlying principles that can be applied. Thus, the need for better underlying knowledge of the aquatic chemistry of organic chemical contaminants was and is apparent (e.g. Stumm et al, 1983, Farrington, 1991). Fortunately, considerable research has been undertaken for this purpose. Much of this research yielding fundamental underlying principles is admirably summarized in a landmark textbook "Environmental Organic Chemistry" by Schwarzenbach et al (2003). That said, testing how well underlying principles explain the actual fates and bioavailability of chemicals in the environment, by pursuing carefully designed and implemented field and monitoring programs, is also critical for effective assessment, communication and management of risks associated with the presence of these organic contaminants (NRC, 2001).

The past three decades have seen an explosive growth of literature on chemicals of environmental concern in marine ecosystems. The journals *Marine Pollution Bulletin,* and *Marine Environmental Research* were established in 1970 and 1978 respectively, and other scientific journals established over the past several decades publish papers about all aspects of the environment, including the oceans. Hundreds if not thousands of technical reports about field programs have been published in many countries. Some of these reports are from "compliance monitoring" or measurements of a suite of chemicals in designated samples at stations around a specific activity such as an effluent or an offshore oil and gas platform. The goal of monitoring in this instance is to ensure that concentrations of chemicals of concern that are, or might be, deliberate or accidentally discharged from the activity do not exceed guidelines or regulatory limits.

A complete review of the literature, including peer reviewed papers and technical reports, is beyond the scope of this chapter. Rather, a summary of the current state of knowledge about the biogeochemical cycles of organic chemicals of environmental concern in the coastal ocean is presented. Only a limited number of examples will be used to illustrate the general state of knowledge. I take as a given that the favorite chemical or favorite ecosystem example of several readers will be missing. I can only apologize and state that the lack of space and time prevented inclusion of many excellent studies in the examples and references cited.

The focus in this chapter will be Persistent Organic Pollutants (POPs), and petroleum hydrocarbons and polycyclic aromatic hydrocarbons (PAHs) as examples of organic chemicals of environmental concern and their biogeochemical cycles in the coastal ocean (Figures 15.1 and 15.2).

POPs are a focus because their release to the environment is now subject to an international accord and there is more knowledge about their environmental distributions than for any other group of synthetic organic chemicals. Petroleum hydrocarbons and PAHs have been chosen because these fossil fuel compounds are distributed globally and their presence and entry in the environment is connected to the use of fossil fuels as a source of energy. In addition, there is a reasonably extensive set of data for environmental distributions compared to other chemicals of environmental concern. However, the fates and effects of oil spills will not be discussed except as the general knowledge of biogeochemical cycles of hydrocarbons pertain. Oil spills are a special case with details that vary according to type of oil, the environment in which the oil is spilled, season, and weather. The subject has been assessed and reviewed several times over the years (NRC, 1975, 1985, 2003; Royal Commission, 1981).

Early survey assessments in the late 1960s and early 1970s of the concentrations and composition of POPs in the oceans sampled relatively few locations in comparison with the larger data sets available three decades later. Chlorinated pesticides (e.g. the DDT family of compounds, DDT, DDE and others), PCBs, and fossil fuel hydrocarbons were found in open ocean waters, organisms and sediments at lower concentrations than those found in coastal organisms, waters and sediments. Generally, the highest concentrations were present in waters, organisms, and sediments of near shore and estuarine environments near urban areas, river deltas or river mouths draining agricultural lands, sewage and dredge spoil dumpsites, or large urban sewer outfalls. (e.g. IDOE, 1972; Ketchum, 1972; Duce et al, 1974; Farrington and Meyer, 1975; Rhead, 1975). Given the need to further understand the situation in the areas with highest concentrations, less effort was thereafter devoted to continental shelf and continental slope areas compared to estuarine ecosystems. Fortunately, much of what was learned in the estuarine and near shore studies is transferable in terms of fundamental principles to continental shelf and slope areas. In addition, field studies in some areas of the world's continental shelf and slope areas have been pursued to the extent that important aspects of the biogeochemical cycles of POPs in these areas are now understood, at least in a qualitative or semi-quantitative manner.

A risk assessment involving connections between level of exposure and tissue concentrations, and the various types of effects on the organisms in question needs to be established. Knowledge and understanding of the biogeochemical processes governing exposures, including the inputs, cycling in coastal ecosystems and ultimate fates, is an essential part of this risk assessment. Of course, the biological effects on organisms are an important component of the overall risk assessment. However, this chapter focuses on the biogeochemical cycles, with most emphasis on the physical chemistry aspects and environmental distributions in the coastal oceans. Only a brief summary of microbial degradation processes and metabolism in marine organisms will be presented. I recognize the great importance of both of these processes in the fate and biogeochemical cycle of organic chemicals of environmental concern, but the chapter had to be limited in scope because of space and time constraints.

## 2. Organic chemicals of environmental concern

International accords recognize that there are at present 12 Persistent Organic Pollutants (POPs) of special concern and these are listed in Table 15.1 (UNEP, 2004). There are sufficient concerns with the adverse effects of these chemicals on

*ORGANOCHLORINE STRUCTURES*

o,p'-DDT o,p'-DDD o,p-DDE

p,p'-DDT p,p'-DDD p,p'-DDE

PCB Toxaphene

* Cl may be substituted at various positions; 209 possible molecular configurations

* Cl may be substituted at various positions; Toxaphene is a poorly characterized mixture of Chlorinated Camphenes

(a)

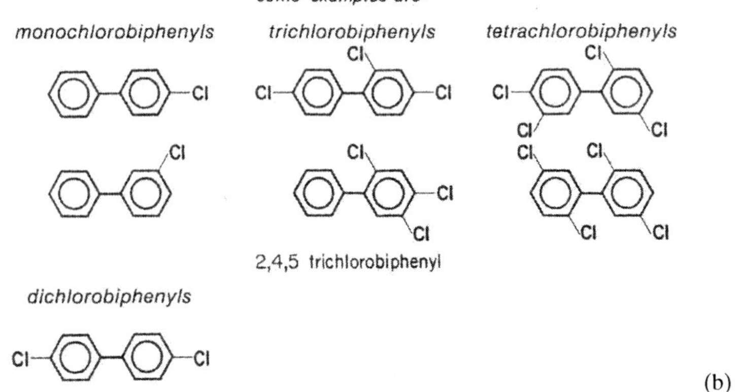


(b)

Figure 15.1 Molecular structures of chlorinated pesticides and polychlorinated biphenyls.
a DDT family, general structure PCB, general structure of toxaphene.
b Examples of specific molecular structures of individual mono-, di-, tri-, and tetra-chlorobiphenyls.
c Examples of detailed structures of penta- and hexa-chlorobiphenyls.
d Examples of molecular structures of chlorinated pesticides.
e Examples of molecular structures of the chlorinated pesticides: hexachlorocyclohexane structures, α-, β-, γ- HCH; and α-, and γ-chlordane.

(*figure continued next page*)

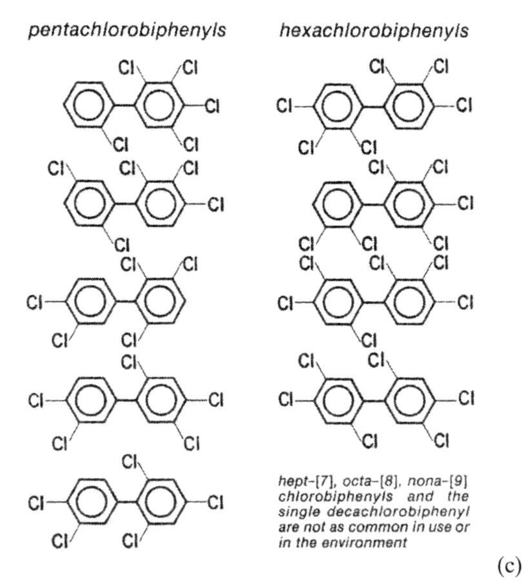


(c)

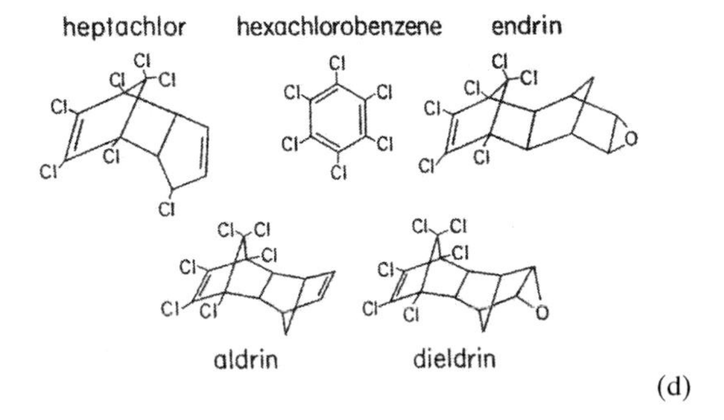


(d)

CYCLOHEXANE

alpha beta gamma

alpha gamma

CHLORDANE

(e)

humans and wildlife that diplomats from 122 countries finalized a treaty that will require governments to minimize and eliminate these chemicals from production and use, i.e., an immediate ban. An exception has been made for DDT because of the continuing need to use this pesticide to control malarial mosquitoes in several countries. Once an environmentally compatible and cost-effective alternative is found, DDT will be banned in those countries. Governments may also maintain PCBs in leak proof existing PCB containing equipment until 2025 to allow time to deploy PCB-free replacements.

TABLE 15.1
Persistent Organic Pollutants according to the United Nations (UNEP, 2004)

| |
|---|
| Aldrin |
| Cholordane |
| Dieldrin |
| Endrin |
| DDT |
| Heptachlor |
| Hexchlorobenzene |
| Mirex |
| PCBs |
| Toxaphene |
| Polychlorinated Dibenzodioxins (PCDDs) – known as Dioxins in general public press. |
| Polychlorinated Dibenzofurans (PCDFs)– known as Furans in general public press. |

Various national lists of chemicals of environmental concern have been promulgated over the years. For example the U. S. Environmental Protection Agency's "Priority Pollutants List" contained 129 chemicals, of which 114 were organic chemicals. Of these chemicals, 81 are in the volatility and molecular weight range that are of interest for this chapter, including the examples listed in Table 15.2. The twelve POPs specified by the recent international treaty, and those chemicals in national lists such as US EPA's Priority Pollutant List, are a few of the many organic chemicals of environmental concern. Molecular structures of several of these chemicals are depicted in Figure 15.1. Note that DDT has often been produced as a technical mixture that contains mainly DDT but also some DDE and DDD. While most of the isomers in the mixture are the *p,p'* isomer, the mixtures also contain lesser, but variable amounts of the *o,p'* isomer.

The chlorinated dibenzodioxins and chlorinated dibenzofurans that are in Table 15.1 as POPs will not be discussed in this paper because there are too few measurements of these compounds in coastal ocean samples to warrant a discussion at this time. They are very important compounds from a toxicity perspective, especially with respect to transfer through the food web to humans. However, we will discuss what has been learned about the biogeochemical cycles of PCBs and PAH in the coastal ocean. This provides a reasonable starting point for the chlorinated dibenzodioxins and chlorinated bibenzofurans with respect to expectations for their biogeochemical cycles.

Advances in our understanding of the properties that cause chemicals to persist in the environment and of their biological effects, coupled with analytical methodology advances, have led to the discovery of more chemicals of environmental concern. for example, knowledge gained about the environmental persistence and biogeochemical pathways for chlorinated pesticides and PCBs, has been useful to understanding the environmental biogeochemistry of the fire retardants, polybrominated diphenyl ethers that are of more recent concern. These compounds have been found widely dispersed in the environment, including marine ecosystems. (e.g. WHO/IPCS, 1994; Boon et al, 2002; Lacorte et al, 2003: Ter Schure et al, 2004; Hites, 2004).

TABLE 15.2
Polychlorinated compounds and aromatic hydrocarbons in the United States EPA List of Priority Pollutants excluding the volatile organics. (excerpted from Keith and Telliard, 1979).

| | |
|---|---|
| **Pesticides and PCBs** | |
| Aldrin | Hexchlorocyclopentadiene |
| α BHC | Hexachlorbutadiene |
| β BHC | Hexachlorobenzene |
| Δ BHC | PCB 1221[a] |
| γ BHC | PCB 1232 |
| Chlordane | PCB 1242 |
| Dieldrin | PCB 1248 |
| Endosulfan sulfate | PCB 1254 |
| Endosulfan, β | PCB 1260 |
| Endosulfan, α | PCB 1016 |
| Endrin | Toxaphene |
| Endrin aldehyde | 4,4' (or *p,p'*) DDD |
| Heptachlor epoxide | 4,4' (or *p,p'*) DDE |
| Heptachlor | 4,4' (or *p,p'*)DDT |
| **Polycyclic Aromatic Hydrocarbons** | |
| Acenaphthalene | Indeno(1,2,3-cd) pyrene |
| Acenaphthene | Naphthalene |
| Anthracene | Flourene |
| Benz(e)acephenanthrylene | Phenanthrene |
| Benzo(k)fluoranthene | Pyrene |
| Benzo(a)pyrene | Benzo(ghi)perylene (1,12 benzoperylene) |
| Chrysene | Benzo(ghi)anthracene (1,2 benzathracene) |
| Fluoranthene | 1,2,5,6-Dibenzanthracene |

[a] numbers are industrial code designations. (See NTIS, 1976; Erickson, 1997)

The exact structural information of individual compounds is important because the ultimate issue of biological effects is tied to the exact structures. In addition, physical chemical properties such as aqueous solubility, partitioning between particulate and dissolved state, and uptake, retention, release and metabolism by organisms are significantly influenced by chemical structure (Schwarzenbach et al, 2003; Farrington, 1991).

POPs are often complex mixtures and this complicates questions about the fates and effects of individual compounds. For example, in theory, industrial mixtures of PCBs could contain as many as 209 congeners. In practice, the industrial mixtures contain somewhat fewer congeners, but they still are complex mixtures. Congeners are a group of like chemicals with the same backbone structure and varying numbers of substituents on this structure. In the case of PCBs, they are chemicals composed of the biphenyl rings and varying numbers of chlorines substituted for hydrogen at any of the carbon atoms that constitute the biphenyl ring (Figure 15.1b and c). PCB industrial mixtures have various compositions and have different trade names in different countries (WHO, 1976; NTIS, 1976; Erickson, 1997)

Polychlorinated naphthalenes (structures not shown) are also ubiquitous environmental contaminants that were produced in technical mixtures such as Halowaxes, Nibren waxes, Sekay waxes, Clonacire waxes and Cerifal material for industrial applications. Production has ceased in many countries, but the PCNs are still present in electrical equipment in use and are formed in incineration and chloraklali production (Lundgren et al, 2002). There are 75 possible congeners of PCNs

Toxaphene (Figure 15.1a) is a chlorinated pesticide produced by chlorination of camphene, yielding mainly chlorobornanes with several hundred components present in technical mixtures (Vetter et al, 2001). PCBs and Toxaphene, and also PCNs can occur along with other chlorinated pesticides listed in Table 15.1 (and shown in Figure 15.1) in samples of the coastal ocean. Thus, there is a significant analytical chemistry challenge associated with accurate measurement of these compounds or mixtures of compounds in environmental samples.

The analyses and risk assessments are further complicated by the presence of other organic contaminants such as fossil fuel compounds. Petroleum is a complex mixture of as many as several thousands of chemicals, including polycyclic aromatic hydrocarbons. Crude oils from the same oil reservoir can vary depending on the location of the well, and compositions vary significantly between different oil reservoirs (Hunt, 1995). A few examples of the molecular structures of chemicals present in petroleum are presented in Figure 15.2. We will include discussions of fossil fuel compounds, especially polycyclic aromatic hydrocarbons (PAH), because of their wide ranging distributions and the toxic, mutagenic, and carcinogenic properties of several PAH.

## 3. Sampling and analyses

### *3.1 Sampling*

Research vessels, especially those of the 1960s and 1970s, are often floating clouds or local sources of the very contaminants to be measured, especially when attempting to measure low concentrations of contaminants (Grice et al, 1972; Farrington, 1974). The need to avoid sampling artifacts was recognized early in the first survey assessments of organic contaminants (Grice et al, 1972, Goldberg, 1972). In cases where precautions were taken such as securing bilge water and sewage pumping on research vessels while sampling on station and moving down current a significant distance when bilges and sewage had to be pumped, samples of biota and sediments were obtained that had minimal or no sampling contamination detected. This seems obvious now, decades later, but avoiding artifacts during sampling requires the constant attention and cooperation of the scientific group on vessels, officers and crew.

## n-ALKANES

| $CH_4$ | $CH_3-CH_3$ | $CH_3-(CH_2)_n-CH_3$ (n = 1-58) |
|---|---|---|
| METHANE | ETHANE | |

## ISOALKANES

$CH_3-CH(CH_3)-CH_2-CH_3$

ISOBUTANE

$CH_3-C(CH_3)_2-CH_2-CH(CH_3)-CH_3$

ISOOCTANE

$CH_3-C(CH_3)-(CH_2)_n-C(CH_3)-(CH_2)_n-C(CH_3)-(CH_2)_n-C(CH_3)-CH_3$ (n = 3)

PRISTANE
(an isoprenoid hydrocarbon)

## CYCLOALKANES

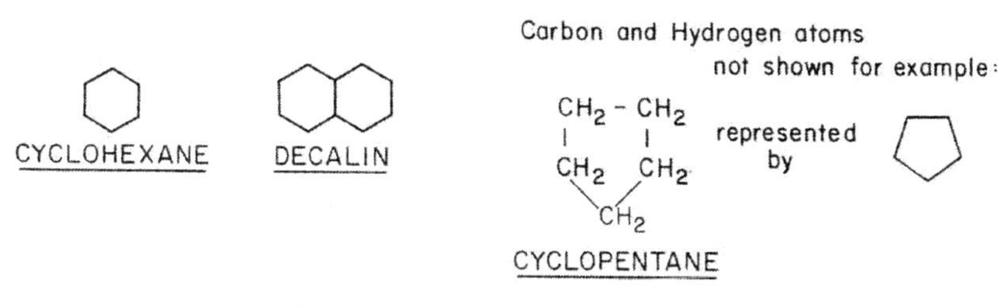


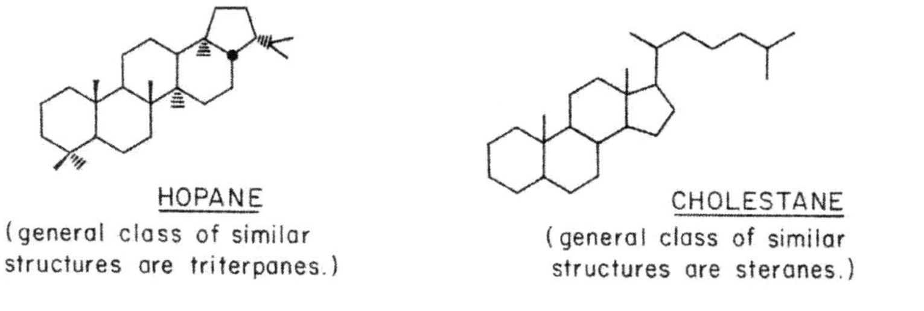


(a)

Figure 15.2 Molecular structures of representative hydrocarbons from petroleum.
a Alkanes and cycloalkanes.
b Examples of one to three rings aromatic hydrocarbons. Two and three ring compounds are included in the designation of polycyclic aromatic hydrocarbons (PAH).
c Examples of four and five ring aromatic hydrocarbons- polycyclic aromatic hydrocarbons (PAH).
(*continued next page*)

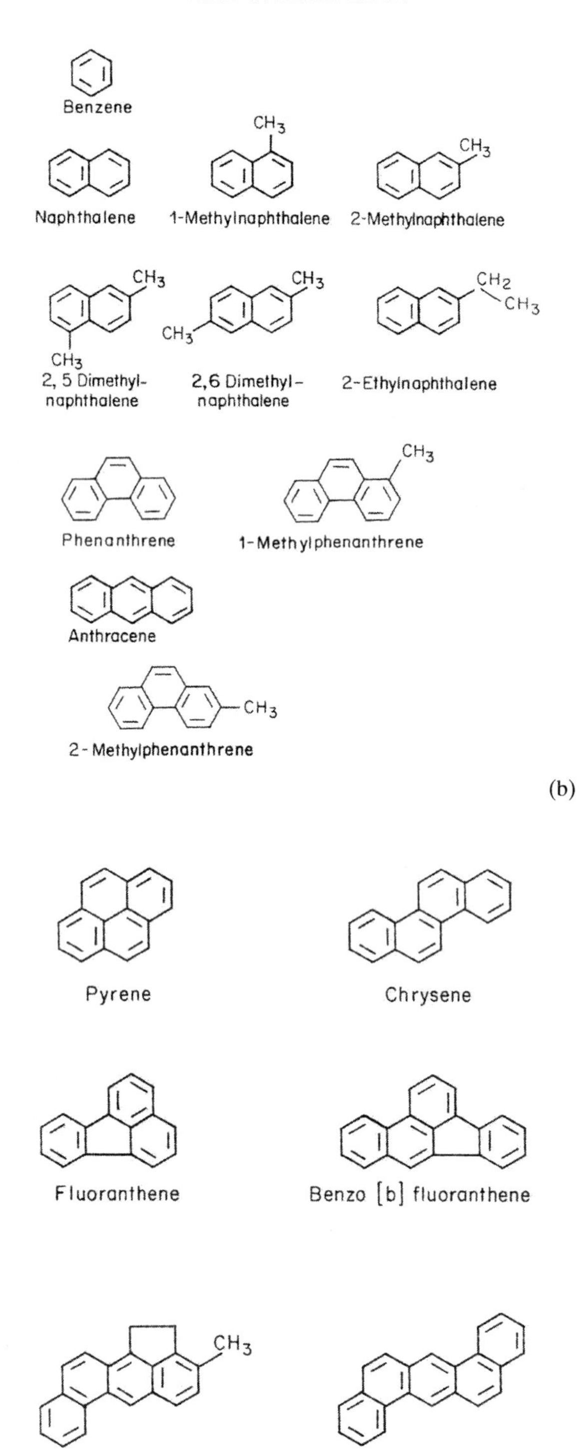
Benzene
Naphthalene
1-Methylnaphthalene
2-Methylnaphthalene
2, 5 Dimethyl-naphthalene
2,6 Dimethyl-naphthalene
2-Ethylnaphthalene
Phenanthrene
1-Methylphenanthrene
Anthracene
2-Methylphenanthrene
(b)
Pyrene
Chrysene
Fluoranthene
Benzo [b] fluoranthene
3-Methylcholanthrene
Dibenz [a, h] anthracene
(c)

Generally, over the years the more difficult analyses have been associated with sampling and analysis of ocean waters and of atmospheric samples. In open shelf, open ocean and deeper waters, sample sizes of between 10s of liters to as much as 1,000 to 2,000 liters are needed to obtain sufficient analyte amounts above blanks and detection limits of the instrumental methods available for quantitative measurement of fossil fuel hydrocarbons and PCBs.

There has been considerable controversy about some of the early 1970s data for concentrations of PCBs in open ocean surface sea water as noted by Fowler (1990). Petrick et al (1996) have provided an excellent review of the challenges and solutions to sampling continental shelf, continental margins and open ocean sea water for PCB, chlorinated pesticide and PAH analyses tried during the 1970s to the early 1990s. A major problem with water samples has been the fact that the more traditional sampling bottles, such as PVC Niskin bottles and even stainless steel and aluminum 90 liter Bodman bottles (Bodman, 1961) were open when they were deployed on a wire from a ship and entered the ocean through the air-sea interface where it was possible for organic compounds concentrated in the sea surface microlayer (e.g. Seba and Corcoran, 1969; Duce et al, 1972) to contaminate the sampling device. Development of large water sampling devices that were deployed closed through the sea surface and opened approximately one half meter below the sea surface partially addressed these problems (e.g. De Lappe et al, 1983).

It was recognized very early that measuring fossil fuel hydrocarbons and chlorinated hydrocarbons in sea water via the solvent extraction of the water—a traditional method for extracting hydrophobic or lipid compounds from sea water—was a monumental task if 10s of liters of sea water were to be extracted, and completely impractical, for larger volumes. Although I have personally witnessed the heroic efforts of several colleagues shaking 20 liter separatory funnels for extractions of sea water with hexane or dichloremethane (and participated a few times myself), I am very happy this is no longer done. Solvent extraction methods were cumbersome, costly, potentially hazardous, required specialized portable laboratories at sea with fume hoods and had no benefits other than a few good data points and excellent aerobic exercise. Quickly it became apparent that sorption of the analytes of interest from the sea water sample to a solid sorbent and subsequent elution of the concentrated analytes for further separation and measurement procedures was the approach of choice (Harvey and Giam, 1976). Isolating the same classes of organic contaminants from air samples was also a formidable task and relied on high volume cascade impactors, filters, and sorbents such as XAD resins or polyurethane foams (e.g. Biddleman and Olney, 1974; Harvey and Steinhauer, 1974; Duce et al, 1976; Atlas and Giam,1984;Peltzer et al, 1984).

De Lappe, et al (1983) developed a system for intake of surface seawater through the bow of a ship using an underway system that involved particle filtration and then sorption on polyurethane foam plugs. This system was based on experience of other investigators in earlier efforts, as cited in this review, and provided high quality reliable data for low surface water concentrations of chlorinated hydrocarbons such as the DDT family compounds, PCBs, and petroleum hydrocarbons. Petrick et al (1996) developed the Kiel in situ Pump (KISP) for filtration and extraction of ultra trace amounts of organic chemicals such as chlorobiphenyls and polycyclic aromatic hydrocarbons in deep waters to 6,000 m. They determined that the system is capable of isolating and detecting chlorobiphenyls at concentrations of 0.01 pg $dm^{-1}$ in sample volumes of 1000 $dm^3$. The system has been tested and deployed on several cruises (Schulz-Bull et al, 1988, 1991, 1995). Thus, we now have a reliable sampling and isolation system for ultra trace amounts of hydrophobic organic compounds, but the number of samples obtained and analyzed is limited at this time.

### *3.2 Analyses*

The scope of this section is to provide a brief overview of the essential steps involved in the analyses and any peculiarities specific to marine samples. Numerous books, chapters and papers have been written about the principles, practices, and challenges of methods for analysis of organic chemicals in environmental matrices, including samples from the marine environment, e.g. Albaiges, et al (1983); Erickson (1997). The Intergovernmental Oceanographic Commission of UNESCO, and the United Nations Environment Program have issued several manuals, guides and related documents about such analyses, e.g. UNEP/IOC/IAEA (1992); IOC (1993). Figure 15.3 is a general flow chart outline of one example of an analysis method for polychlorinated biphenyls (PCBs, linear alkyl benzenes (LABs), polycyclic aromatic hydrocarbons (PAHs) and DDT, DDE, and DDD.

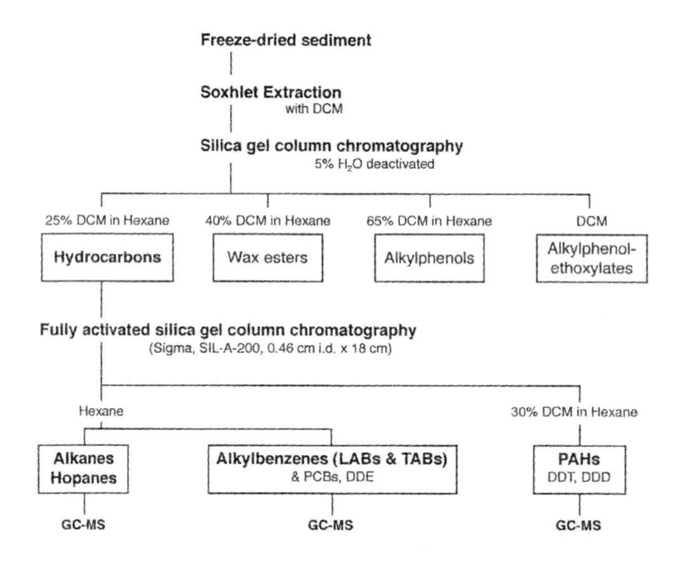


Figure 15.3 General analysis flow diagram for measurement of POPs and PAH in samples from air, water, biota and sediment.

The samples of water or sorbents through which sea water has been passed, particulate matter obtained by filtering sea water, biota or tissue samples from biota, and sediment samples are subjected to solvent extraction (or solvent elution in the case of the sorbents). The non-polar solvents hexane and dichloromethane or mixtures of solvents that have both non-polar and polar components (e.g. toluene/methanol, or hexane/acetone) are normally used. The resulting extract contains the analytes of interest (e.g. PCBs or PAHs) as well as a complex mixture of other solvent extractable, naturally occurring chemicals that are mainly lipid class compounds. Further separation from the naturally occurring lipids is needed to isolate the analytes of interest prior to final separations and identification/quantitation by instrumental techniques such as gas chromatography (GC). That step may be preceded by chemical treatment to reduce the amount of lipids present, e.g. saponification (base hydrolysis) followed by further solvent extraction. Chromatography using silica columns or thin layer chromatography, or now more often high performance liquid chromatography (HPLC) (e.g. Petrick et al, 1988) are used for this separation.

The routine use of gas chromatographs equipped with high resolution glass or fused silica capillary columns and electron capture and/or flame ionization detectors has greatly improved the resolution of complex organic mixtures, compared with the larger diameter packed column chromatography used in the 1960s and 1970s. Computerized data systems make it much easier to collect and process the signals from hundreds of compounds in a single GC analytical run. Even with the high resolution fused silica gas chromatography columns, there were still chlorobiphenyls that co-eluted, i.e., could not be separated, on one fused silica column. Gas chromatography techniques with switching between two fused silica columns, sometimes designated as GC-GC or multidimensional GC, have been developed for complete separation and quantification of chlorobiphenyl congeners (Schulz-Bull et al, 1989). The richness of data available during the past two decades, because of advances in analytical methodology, has played a key role in advancing our understanding of how various biogeochemical processes act on POPs and PAH (and other fossil fuel compounds and organic chemicals) in the coastal ocean.

The complex mixtures of hydrocarbons present in petroleum confounded complete separation and analyses for decades. Indeed, the very fact that there was an UCM (*u*nresolved—by gas chromatography—*c*omplex *m*ixture) of hydrocarbons in the gas chromatographic analyses was one of the prime indicators of the presence of petroleum hydrocarbons in a sample (Farrington and Quinn, 1973; Frysinger et al, 2003). Recently, Frysinger et al, (2003) reported the use of a novel gas chromatographic-gas chromatographic separation and quantification that resolved the "unresolved complex mixture" and provided an amazingly rich set of data for hydrocarbons in environmental samples.

Gas chromatography-mass spectrometers interfaced with computerized data systems became generally available for use in research laboratories in the early 1970s and revolutionized the analyses for POPs and for organic chemicals of environmental concern in general. Verification of PCBs and chlorinated pesticides in certain matrices became easier. The use of mass spectrometers as quantification tools for PAH in marine environmental samples was pioneered by Giger and Blumer (1976) and Youngblood and Blumer (1977) nearly simultaneously with the use of GC-MS-Computer systems for identification and quantification of polycyclic aromatic hydrocarbons (e.g. Farrington et al, 1977; Hites et al, 1977; Teal et al, 1977, LaFlamme and Hites, 1978). These techniques were also used during these same years (1977–1979) for marine organism tissue samples (Farrington et al, 1982a, b).

A truly remarkable recent development is the capability to obtain stable isotopic $^{13}C/^{12}C$ and radioisotope $^{14}C$ data for individual PAH compounds and the same data plus stable chlorine isotope rations for chlorinated organic compounds such as chlorobiphenyls and DDE (O'Malley et al, 1994; Eglinton et al, 1996; Currie et al, 1997 Lichtfouse et al, 1997; Okuda et al, 2002; Reddy et al, 2002a, b). Limited data are now available from these types of studies. I believe they will lead to significant advances in our knowledge of the biogeochemical cycles of POPs and PAH as they are applied to key samples in a variety of coastal ocean ecosystems.

### *3.3 Quality Control/Quality Assurance*

The need for care in analyses of organic contaminants in marine samples through quality control/quality assurance was recognized early in survey and research programs. There were explicit recommendations for the preparation of standard reference materials for analyses of organic contaminants (IDOE, 1972). In the interim, the practice of interlaboratory, intercomparison or intercalibration exercises was begun (e.g. Farrington et al, 1976) in order to have some basis for comparison and integration of data from various laboratories. Guidelines for trace analyses were generally recognized and available by the mid 1970s (LaFleur, 1976) and reaffirmed in the 1980s (ACS, 1980; Taylor, 1985).

Often a major challenge was simply having sufficient standard reference materials of the POPs or PAH in organic solvent solutions that could be used to calibrate instrument detectors. Although it is beyond the scope of this chapter to go into further details, it is important to note that significant effort went into synthesis of standard compounds, preparation of dilutions for standard reference materials, collection of sufficient amounts of matrices such as sediment and shellfish tissue, homogenization to prepare replicate sub-samples, and then analyses of these sub-samples to document replication and the "reference value". It took several years before the first standard reference materials were made available by internationally recognized agencies such as the U. S. National Bureau of Standards—now the National Institute of Standards and Technology. In the interim working reference materials were collected and distributed by interested organizations such as the International Council for the Exploration of the Seas (Law and Portmann, 1980) or individual academic and government research laboratories collaborating with each other (e.g. Wise et al, 1980; Galloway et al, 1983). As one whose laboratory was involved in some of the interim steps, involving collection, shucking, and homogenizing approximately 4,000 blue mussels (*Mytilus edulis*), sub-sample homogeneity verification, and then distribution of sub-samples and collection, collation and interpretation of the data (Farrington et al, 1988), I can attest to the significant effort involved. This latter worldwide exercise illustrates the need for such intercomparisons among laboratories for trace organic contaminant analyses (petroleum hydrocarbons in this specific case). Quoting the abstract of the paper (Farrington et al, 1983):

> "Between 22 and 30 laboratories reported data for n-alkanes, pristane, and phytane determined by GC analyses giving between +/- 69% and +/- 297% R.S.D. depending on the compound. Eliminating the outliers reduced the range of R.S.D.s to +/-67 to +/- 104% R.S.D. A similar experience was noted

for individual polynuclear aromatic hydrocarbons when combining data from HPLC, GC and GCMS analyses. There is a need for much improvement for within and between laboratory precision."

The situation is much improved now because of the availability of Standard Reference Materials for analyses of trace organic contaminants in sediments and marine organism tissues (e.g. the U. S. National Institute or Standards and Technology world-wide-web site https://srmors.nist.gov/detail.cfm). Nevertheless, when combining data from various sources to develop an understanding of biogeochemical cycles of organic contaminants in a specific area or region or to assess the status of geographic or time trends of contamination, great care must be exercised to understand the limitations imposed by uneven quality control/quality assurance procedures in various laboratories that publish the data—even in peer reviewed publications. As one example, Buchholtz ten Brink et al (2002) experienced this problem in assembling a data-base for sediment contaminant analyses for the Gulf of Maine, U.S.A.

I have attempted in the following sections not to cite data that in my opinion might be compromised by significant concerns about quality control and quality assurance. This does not mean that if a particular study is not cited that there are concerns about QC/QA. As noted above, there are too many papers and reports about organic contaminants in the coastal ocean to be comprehensibly reviewed in this chapter.

## 4. Key components of the biogeochemical cycles

The general form of the biogeochemical cycle of organic contaminants in the coastal ocean is the same as that for the larger scale biogeochemical cycles in the oceans for carbon, nitrogen and similar bioreactive elements as depicted in Figure 9.10 of Gruber and Sarmiento (2002). The specific important processes for organic contaminants in the coastal ocean are depicted in the schematic cartoon in Figure 15.4. Unraveling and understanding the details of these processes has progressed since the late 1960s to provide first qualitative understanding and now quantitative understanding of many of the processes depicted in Figure 15.4. Progress has been achieved by pursuit of five simultaneous approaches:

i. Assembling and assessing data for production and use of chemicals in individual countries and on a worldwide basis. This might initially seem like an easy task, but it has proved to be a challenge that was recognized early in the 1970s (e.g. NRC, 1975b).
ii. Laboratory or bench scale experiments—e.g to determine solubility or partitioning behavior between soluble, particulate, and colloid phases; photochemical reactivity, microbial degradation of compounds in cultures.
iii. Mescosm or microcosm experiments—systems with multiple components from real world ecosystems (mesocosms), or microcosms with microbial scale multiple components e.g. water, sediments, microbes, and polychaetes.
iv. Field observations—transects of coastal and continental margin areas with reference to known or suspected chronic input sources or oil spills.

v. Simulation models of "what if" scenarios sometimes interactive with field observations and data from laboratory and mescosm studies to constrain initial parameterization of the models.

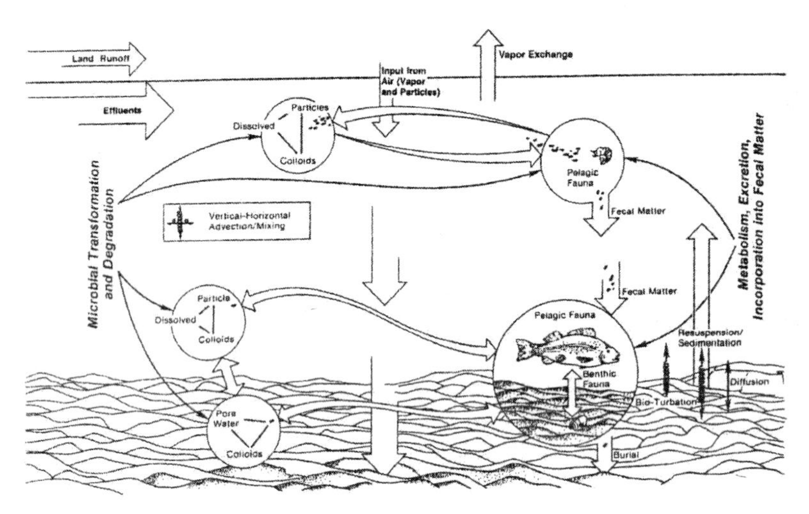


Figure 15.4 Biogeochemical cycle of POPs and PAH in coastal ocean.

### *4.1 Inputs*

Inputs of pollutants into the coastal waters are of several types. Direct industrial or indirect effluent discharges, municipal discharges, and urban runoff to coastal ecosystems can be a significant source of organic pollutants. Two extensively studied examples of this in the United States are the sewage effluent discharges in the Southern California coastal waters and Narragansett Bay (studies by Quinn and coworkers e.g., Hoffman et al, 1983). Riverine inputs carry the accumulations of organic contaminants from air borne deposition on land and run off to the streams and rivers as well as the municipal and industrial effluents from interior areas. An example of this type of input would be the discharge of the Rhine River to the North Sea or the Mississippi River to the Gulf of Mexico. Atmospheric inputs from dry deposition and from precipitation also are significant inputs, especially away from the immediate coastal areas. Dredging and disposing of sediments contaminated with POPs (and other chemicals) from ports and harbors has been, and continues to be, a very controversial issue (NRC, 1989, 1997, 2001). Ships discharging ballast waters or bilge waters during normal operations result in organic contaminant inputs, mainly petroleum hydrocarbons, and smaller water craft are also sources, as we will see below for petroleum hydrocarbons/oil inputs. In several areas over the past few decades, but much less frequently now, barges or ship loads of sewage sludge and industrial wastes were dumped into coastal and continental shelf areas. The New York Bight area has been studied extensively with respect to sewage sludge and dredge spoil disposal as well as disposal of acid-iron wastes.

Production, use and environmental emissions of synthetic organic chemicals have changed over time. New production and uses of PCBs and DDT have been banned in many countries. Therefore new inputs from production have declined dramatically from the late 1960s to the present. However, the residual chemicals remaining in the environment from earlier uses and discharges, e.g. in soils or in river sediments, remain sources that can be remobilized and make their way to the coastal ocean by riverine and atmospheric transport. Actual quantitative estimates of changing inputs over time, on a global and regional basis for the coastal ocean, are not available.

**Chlorinated Compounds**

A logical starting point for assessing inputs would be to assess the total production of a given chemical. However, it has proved difficult to assess the global manufacture and use of several of the POPs because rough estimates for only a limited number of countries are available (Simonich and Hites, 1995). Three examples from the past five years of attempts to get estimates for inputs of chlorinated organic chemicals to the environment are instructive in terms of the current state of the art in providing such assessments. Li (1999) report on global uses of HCH and estimate 10 million tons between 1948 and 1997 with the majority of the use occurring in China and India. Li (1999) also reported that these two countries plus Japan (another significant user of HCH) are also the countries, among those for which HCH environmental data are available, that are the most heavily contaminated with HCH as of 1990.

Breivik et al (2002a) conducted a significant study of available information for production of PCBs in various countries. They used compositional information to arrive at total global production for ten homolog groupings of chlorobiphenyls and from these data proceeded to estimate total global production for 22 congeners from 1930 to 1993 at which point production effectively ceased. They also estimated where the PCBs were consumed by assessing international trade records. From these data, Breivik et al (2002b) used a dynamic mass balance model that took into account amounts in use (open and closed systems), disposal (open burning, waste incineration, destruction), landfills, and accidental releases, to estimate historical emissions to the environment as a result of widespread uses of PCBs for 70 years. They note that there are still major uncertainties in the temporal and spatial scales involved and that their estimates are at best order of magnitude estimates. Nevertheless, their study provides valuable data for total global historical usage and total emissions of the sum of the 22 chlorobiphenyl congeners, both distributed by latitude. How much of the emissions are input to the oceans is unknown, although as we shall discuss later, the amount of PCBs present in continental shelf sediment has been estimated and compared with the emissions (Jönsson et al, 2003).

Li et al (2003) provide an interesting example of a process to estimate emissions in a global gridded manner for β-HCH with available data (and recognizing the uncertainties noted by authors cited in the preceding). They state that β- HCH emissions have shown a southern tilt over time towards the equatorial region, as more northern countries have banned the use of technical HCH.

**Petroleum or Fossil Fuel Hydrocarbon Inputs**

Petroleum inputs to the world's oceans have been assessed from time to time by national and international organizations (e.g. NRC, 1973, 1985, 2003; Royal Commission, 1981; GESAMP, 1993). The latest estimate is provided by the U.S. National Research Council (NRC, 2003) in a comprehensive assessment of use of petroleum on a global basis and various sources to the oceans as well as an overview of fates and effects of the various inputs. The estimates on a worldwide basis are assembled in Table 15.3 taken from that report. The relative proportions of the inputs from the various categories can vary in a given year and for specific locations over several years. Natural oil seeps are significant inputs in areas such as the Gulf of Mexico and off the southern California coast in the U.S. In coastal areas where there has been one large oil spill (e.g. the *Exxon Valdez* oil spill in Prince William Sound, Alaska in 1989 or the *Prestige* oil spill off the coast of Spain in 2003), accidental spills will dominate the inputs if averaged over a decade or sometimes longer.

TABLE 15. 3
Worldwide Average Annual Inputs (1990-1999) of Petroleum to the Marine Environment. (adapted from table 3-2 in NRC, 2003).

| | (in thousands of tonnes) | | |
|---|---|---|---|
| **Categories of Input** | **Best estimate** | **Minimum** | **Maximum** |
| Natural Seeps | 600 | 200 | 2000 |
| Extraction of Petroleum | 38 | 20 | 62 |
| Platforms | 0.86 | 0.29 | 1.4 |
| Atmospheric deposition | 1.3 | 0.38 | 2.6 |
| Produced waters | 36 | 19 | 58 |
| Transportation of Petroleum | 150 | 120 | 260 |
| Pipeline spills | 12 | 6.1 | 37 |
| Tank vessel spills | 100 | 93 | 130 |
| Operational discharges (cargo washings) | 36 | 18 | 72 |
| Coastal facility spills | 4.9 | 2.4 | 15 |
| Atmospheric deposition | 0.4 | 0.2 | 1 |
| Consumption of Petroleum | 480 | 130 | 6000 |
| Land-based (river and runoff) | 140 | 6.8 | 5000 |
| Recreational marine vessels | nd [a] | nd | nd |
| Spills (non-tank vessels) | 7.1 | 6.5 | 8.8 |
| Operational discharges (vessels 100GT) | 270 | 90 | 810 |
| Atmospheric deposition | 52 | 23 | 200 |
| Jettisoned aircraft fuel | 7.5 | 5.0 | 22 |
| **TOTAL** | **1300** | **470** | **8300** |

[a] World-wide population of recreational vessels not available. For the United States the best estimate for this input on an annual average basis is 5.6 thousands of tonnes.

Chronic oil inputs from routine operations such as transportation by ship, or routine consumption such as hydrocarbons from automobile exhaust/road runoff, are significant. While some chronic inputs are point sources, others are non-point sources (Table 15.3). As with the chlorinated organic compounds such as the PCBs and chlorinated pesticides, atmospheric inputs to the oceans are also significant (Table 15.3).

Although data for some contaminant categories were less reliable three decades ago than at present, it is instructive to examine estimates of oil inputs to the world's oceans over time (Table 15.4). A comparison of the estimates of inputs in Table 15.4 with those in Table 15.3 indicates that there appears to be less input per barrel of oil consumed. Inputs have not increased despite the increase in the worldwide petroleum consumption over the three decades of about 1.6 percent per year, e.g. about 45 million barrels of oil per day in 1970 to about 72 million barrels of oil per day in 1999 (NRC, 2003). This is consistent with more stringent regulation of operational discharges from tankers, and greater attention to minimizing spills and to minimizing chronic losses of oil, e.g. better recycling programs for used motor oil (NRC, 2003).

TABLE 15.4
Worldwide Estimates of total petroleum input to the sea and comparison of some assessments over time. (in thousands of tonnes). (Adapted from NRC, 2003 which reported these as compiled from GESAMP, 1993).

| | Reference Cited | | |
|---|---|---|---|
| | **NRC, 1975** | **Kornberg, 1981** | **NRC, 1985** |
| **Source of Input** | | | |
| Natural seeps | 600 | 600 | 200 (20-2,000) |
| Extraction of petroleum | 80 | 60 | 50 (40-60) |
| Transportation of petroleum | 1,580 | 1,100 | 1,250 |
| *Spills (tank vessels)* | *300* | *300* | *400* |
| *Operational discharges(cargo washings)* | *1,080* | *600* | *700 (400-1500)* |
| *Coastal facility spills*[a] | *200* | *60* | *not reported separately* |
| *Other coastal effluents* | *150* | *50(30-80)* | *not reported separately* |
| Consumption of petroleum | 3,850 | 2.900 | 1,700 |
| *Urban runoff and discharges* | *2,500* | *2,100* | *1,080(500-2,500)* |
| *Losses from non-tanker shipping* | *750* | *200* | *320(200-6,000)* |
| *Atmospheric deposition* | *600* | *600* | *300(50-500)* |
| **TOTAL** | **6,110** | **4,760** | **3,200** |
| Percentage of Totals | | | |
| Natural seeps | 10% | 13% | 6% |
| Extraction of Petroleum | 1% | 1% | 2% |
| Transportation of Petroleum | 26% | 24% | 39% |
| Consumption of Petroleum | 63% | 62% | 53% |

[a] Pipeline spills were not reported separately and seem to have been included in coastal facilities inputs.

An examination of the sources of oil to the marine environment indicates that much of the input has been, and continues to be, in the coastal ocean, both near shore and in the continental margin areas. The recent NRC (2003) report provides a detailed breakout of estimates for various coastal and continental shelf areas of the United States in a series of color-coded figures that are too extensive to reproduce here. It suffices to summarize that for the Northeast U.S. coastal areas, consumption related inputs dominate. In the Gulf of Mexico region, natural seep inputs are a significant source. for Southern California it depends significantly on whether the area of focus is the coast where consumption inputs dominate or offshore where natural seep inputs dominate. It is reasonable to assume that similar patterns of inputs occur in other nations worldwide.

### *4.2 Accumulation of POPs in Sediments.*

Early in the studies of POPs, it became clear that their apolar nature and thus association with particulate matter in the water column resulted in accumulation of POPs in coastal sediments. Early surveys of sediments near sources of inputs noted that surface sediment concentrations were elevated significantly compared to surface sediments sampled in transects away from the input sites or in more remote pristine locations ( IDOE, 1972; Farrington and Quinn, 1973; Tissier and Oudin, 1975; Gearing et al, 1976; Farrington and Tripp, 1977).

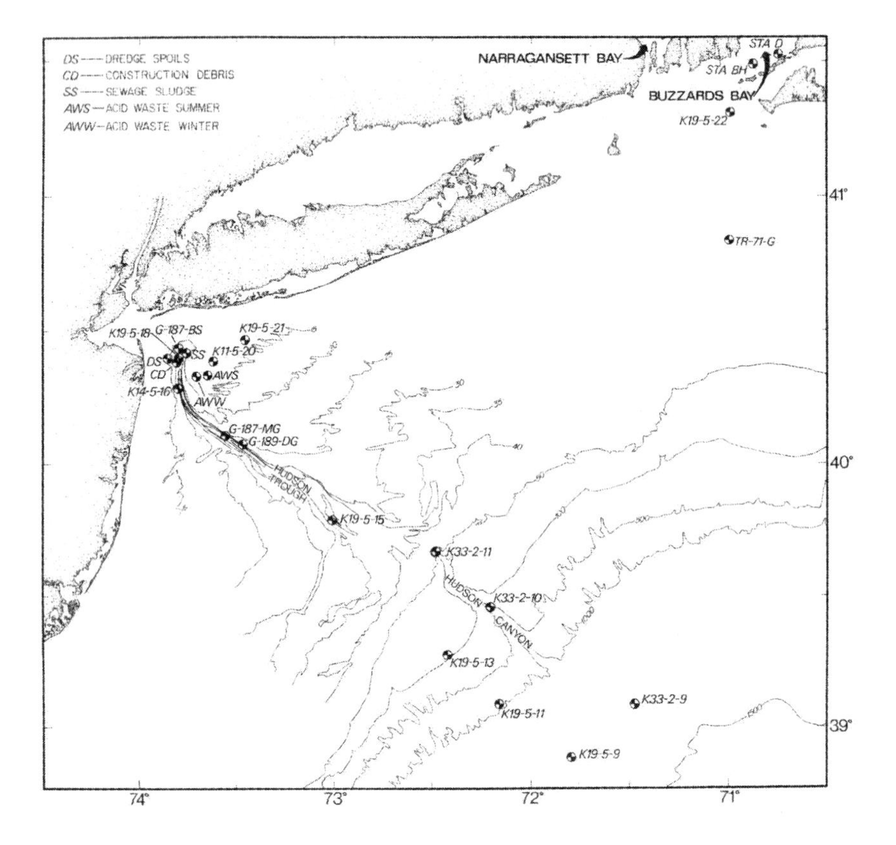


Figure 15.5 Western North Atlantic Coastal, Continental Margin and Open Ocean Surface Sediment Sampling Stations for Hydrocarbon Analyses. (Farrington and Tripp, 1977)

Hydrocarbon concentrations in the upper few cm of surface sediments collected along an offshore transect of stations in the NE Atlantic, sampled from 1972 to 1975, illustrate this point (Farrington and Tripp, 1977). The transect was from the New York Bight across the shelf and down the continental slope, inside and outside the Hudson Channel and Hudson Canyon and to the abyssal plain of the western North Atlantic (Figure 15.5). Hydrocarbon concentrations are given in

Table 15.5 for stations on these transects. Elevated concentrations of parts per thousand of hydrocarbons on a dry weight basis (Stations K19–5–18, K47–1–6 and G.B.S.) were present in the surface sediments of both the sewage sludge dump site and the harbor dredge spoil dump site areas. Elevated concentrations were noted in the Hudson Channel at station K19–5–16, and deeper depressions in the Hudson Channel at G187 MG and DG. Concentrations were also high in surface sediments at station K33–2–10 in the Hudson Canyon. Gas chromatographic analyses (not shown) of these surface sediments indicated degraded petroleum as a significant component of the hydrocarbons in the New York Bight samples and the other samples with elevated concentrations. Progressively with lower concentrations, the fraction one ($f_1$) alkanes were dominated by n-alkanes and a few other hydrocarbons of biogenic origin. Sufficient hydrocarbons were present in the alkane and cylcoalkane fraction of the dump site sediments to obtain a carbon -14 activity. The hydrocarbons had very little C-14 activity, confirming the hydrocarbons were fossil in origin, consistent with their being from partially degraded petroleum. Farrington and Tripp (1977) estimated that about 4, 000 tons per year of petroleum or fossil fuel hydrocarbons were being delivered to the New York Bight by the combination of dredge spoils and sewage sludge dumping.

The hypothesis advanced to explain these observations of elevated fossil fuel or petroleum hydrocarbons in surface sediments of the Hudson Channel areas involved resuspension and transport of the New York Bight polluted sediments to these locations (Farrington and Tripp, 1977). Several sections of a short core of sediments in a fine grained sediment area at K33–2–10 in the Hudson Canyon were analyzed. Comparison of the surface sections with deeper sections indicated petroleum hydrocarbons or fossil fuel hydrocarbon contamination of the upper section, but in deeper sections biogenic hydrocarbons predominated (Farrington et al, 1977). The movement of resuspended sediment from the contaminated sediment areas in the New York Bight and Hudson Channel areas to the Hudson Canyon and deposition at this fine grained sediment accumulation site was a reasonable explanation, in accord with observations and data available at that time (Farrington et al, 1977).

A series of estuarine mescosm experiments with water accommodated No.2 fuel oil and separate experiments with $^{14}C$ –labeled benzanthracene and $^{14}C$ –labeled 7, 12 dimethly benzanthracene demonstrated unequivocally that petroleum or fossil fuel hydrocarbons introduced to the water column would reach the sediment water interface within a few days or less in the 5 m water column in a 1.8 m diameter tank (e.g. Gearing et al, 1979; Gearing et al, 1980; Farrington et al, 1982b; Hinga and Pilson,1987; Hinga et al,1980, 1986). In fact those experiments demonstrated an involvement of almost all of the processes illustrated in Figure 15.4.

Fowler (1990) has reviewed and compiled data for PCBs and DDT in coastal and near shore surface sediments (Table 15.6). These data are from various studies and range over the 1970s and 1980s. They provide a sense of the range of the concentrations reported and the fact that such sediment contamination is substantial in several areas of the world and not just a few localities. Fowler (1990) also reviewed literature to date and provided a summation of concentrations of PCBs and DDT in the seawater and in organisms. That paper is recommended as an excellent compilation.

TABLE 15.5

Hydrocarbons Concentrations in μg/g dry wt. in Surface Sediments from the NewYork Bight, Hudson Channel and Hudson Canyon and nearby Continental Shelf and Slope and Abyssal Plain of the western North Atlantic for stations noted in Figure 15.5.(Farrington and Tripp, 1977).

| | | | (μg/g dry wt) | | | |
|---|---|---|---|---|---|---|
| **Station**[a] | **Sampling Date** | **Water Depth (m)** | **$f_1$**[b] | **$f_2$** | **$f_3$** | **Sum** |
| **Abyssal Plain** | | | | | | |
| K19-4-9 | 3/29/71 | 5250 | 1.2 | nd[c] | nd | -- |
| K33-2-6 | 10/2/73 | 5465 | 2.0 | nd | 4.0 | -- |
| K19-5-3 | 4/8/71 | 4950 | 1.0 | 0.2 | 0.1 | 1.3 |
| K19-5-4 | 4/11/71 | 4900 | 5.0 | 1.9 | 0.6 | 7.5 |
| K19-5-5 | 4/11/71 | 3950 | 4.7 | nd | 5.1 | -- |
| K19-5-6 | 4/12/71 | 3923 | 2.4 | 2.1 | nd | -- |
| **Continental Slope** | | | | | | |
| K19-5-7 | 4/13/71 | 2950 | 5.9 | 4.6 | 4.5 | 15 |
| K19-5-9 | 4/14/71 | 1830 | 3.6 | 1.7 | 5.2 | 11 |
| K19-5-13 | 4/14/71 | 190 | 5.6 | 2.4 | 8.1 | 16 |
| **Hudson Canyon** | | | | | | |
| K33-2-8 | 10/5/73 | 3785 | 5.2 | 4.3 | 9.7 | 19 |
| K33-2-9 | 10/5/73 | 2626 | 4.7 | 1.7 | 7.1 | 14 |
| K33-2-10 | 10/6/73 | 986 | 55 | 6 | 7 | 68 |
| K33-2-11 | 10/6/73 | 137 | 4.4 | 1.9 | 9.0 | 15 |
| **Hudson Channel** | | | | | | |
| K19-5-15 | 4/16/71 | 79 | 8.6 | 4.1 | 6.0 | 19 |
| G187DG | 4/17/72 | 78 | 58 | 15 | 40 | 113 |
| G187MG | 4/18/72 | 67 | 93 | 32 | 66 | 191 |
| K47-1-11 | 2/5/75 | 38 | 3 | 10 | 6 | 19 |
| K47-1-12 | 2/5/75 | 72 | 25 | 7 | 28 | 60 |
| K47-1-8 | 2/5/75 | 60 | 48 | 11 | 31 | 90 |
| K19-5-16 | 4/18/71 | 54 | 399 | 66 | 95 | 560 |
| **New York Bight** | | | | | | |
| K19-5-18 | 4/18/71 | 39 | 1800 | 620 | 480 | 2900 |
| K47-1-6[d] | 2/5/75 | 28 | 440 | 132 | 132 | 704 |
| G.B.S. | 4/17/72 | 27 | 1200 | 250 | 293 | 1743 |
| K19-5-20 | 4/18/71 | 23 | 25 | 4.5 | 5.8 | 35 |
| **Continental Shelf** | | | | | | |
| K19-5-21 | 4/18/71 | 23 | 6.6 | 2.3 | 7.0 | 16 |
| K19-5-22 | 4/19/71 | 37 | 3.5 | 1.2 | 0.29 | 5.0 |

[a] K19-4-9 is R/V *Knorr*, cruise 19, leg 4, station 9. G187is R/V *Gosnold* cruise 187, MG is mid gully, DG is Deep Gully, and B.S. is Buoy Station in sewage sludge dump site.

[b] Column chromatography fractions – f1 is fraction 1 – alkanes and cycloalkanes, f2 is 2-4 ring aromatic hydrocarbons, f3 is 4-6 ring aromatics plus traces of polar lipid such as long chain ketones. (See Farrington and Tripp, 1977 for details).

[c] nd. Is non-detected above blank.

[d] About 30% of the sample was lost prior to weighing.

TABLE 15.6
Chlorinated Hydrocarbon Concentrations (ng/g dry wt.) in Surface Sediments from Coastal and Nearshore Areas. (See Fowler, 1990 for references for data sources. This is a synopsis and not a complete listing from Fowler, 1990). Reprinted with permission from Elsevier.

| General Location | Site | ΣPCB | ΣDDT |
|---|---|---|---|
| Northwest Atlantic | Gulf of Maine | 40-340 | n.r.[a] |
| | New Bedford Harbor, MA | 8 400 | n.r. |
| | Escambia Bay, Florida | <30-480 000 | n.r. |
| | New York Bight | 0.5-2 200 | n.r. |
| | Gulf of mexico, USA | 0.2-34 | <0.3 |
| | Gulf of Mexico, Carribean (Mexico) | n.r. | 0.3-2.27 |
| | Chesapeake Bay | 4-400 | n.r. |
| | Hudson-Raritan Estaury | 286-1950 | 116-739 |
| Southwest Atlantic | Rio de Janeiro, Brazil | 10-38 | 6-22 |
| North Sea | Dunkerque | 134+/-134 | <0.1 |
| | Norway | 14-28 | 0.16-0.36 |
| Baltic Sea | Sweden | 40-160 | n.r. |
| | Finland | <10-20 | |
| | Eckernforde Bight | 134-212 | 28-46 |
| Northeast Atlantic | Brittany | <0.5 | n.r. |
| | Brest, France (subtidal) | 0.4-185 | <0.1-17.6 |
| | Irish Sea, UK | <2-2 890 | n.r. |
| | Ivory Coast | 2-213 | 2-997 |
| Mediterranean | Adriatic | 1-17 | <1-8 |
| | Northwestern | 0.3-1 200 | 0.7-44 |
| | Tyrrhenian | 0.6-3 200 | 4 |
| | Ionian | 0.8-457 | n.r. |
| | Aegean. | 1.3-775 | 7.1-1 893 |
| | Marseille Bay | 157+/-12 | 10-50 |
| Northeast Pacific | San Pedro Basin, California | 1-13 | 5-30 |
| | Santa Monica Basin, California | 0-9 | 30-160 |
| | Palos Verdes, California | 80-7420 | 1 600-100 000 |
| | Puget Sound, Washington | 80-640 | n.r. |
| | San Francisco Bay | 30-50 | n.r. |
| Northwest Pacific | Gulf of Thailand | n.d.[b] | 22-56 |
| | Osaka Bay, Japan | 40-2 000 | n.r. |
| | Harimanada Bay | 50-4 00 | n.r. |
| Southwest Pacific | Port Philip Bay, Australia | <10-390 | n.r. |
| | Bass Strait, Australia | <10 | |
| | Manukua Harbour, New Zealand | 0.5-14.2 | 1.2-2.3 |
| Indian Ocean | East India (Bay of Bengal) | n.r. | n.d.-980 |
| | West India (Arabian Sea) | n.r. | 14-358 |

[a] n.r. means not reported.
[b] n.d. means not detected. (no detection limit was given).

The accumulation of POPs in sediments has left a legacy of pollutants in the environment, even after the significant sources have been reduced and/or eliminated (NRC, 1989, 1997, 2001; Li and Wiberg, 2002). As the inputs from effluents, land runoff and atmospheric deposition decline in some of the areas due to regulatory actions, coastal ecosystems with high concentrations of POPs in sediments are now receiving significant inputs to other segments of the ecosystem from the sediments. For example, there is a flux from the sediments to the overlying water and food web transfer from sediments to benthic animals and throughout the food web

(Figure 15.4). In simple terms, the coastal ecosystems in many areas have converted from mainly "top down" sources, from the atmosphere, effluents, and spills, to a "bottom up" source from the sediments.

Risk assessment and risk management of POPs in the oceanic environment and coastal environments require further detailed understanding of the processes, including the underlying chemistry, biology and physics, between inputs and the accumulation of some POPs in surface sediments and the long term fate of the POPs accumulating in sediments. The next sections provide an overview of these processes.

### *4.3 Important Chemical Properties of POPs and Partitioning in Coastal Ecosystems Assuming Equilibrium.*

A brief discussion of the properties of POPs and PAHs such as solubility, vapor pressures, and partitioning between various phases is in order. However, the reader is encouraged to consult Schwarzenbach et al (2003) for the detailed underlying chemistry and physics that pertain. Important points are excerpted and summarized here from that detailed and informative text. In addition, there are now on line web resources available to provide various useful parameters such as solubility in water and partition coefficients for chemicals in air, water and particle systems. (SRC, 2004).

**Solubility, and Salinity and Temperature Influences on Solubility.**
We know that differences in molecular structure have an influence on solubility and activity coefficients of organic compounds. The simplest comparisons from everyday life provide the lesson. Ordinary household sugar—sucrose—has appreciable solubility in aqueous systems while the hydrocarbons in petroleum do not. Hence the popular saying "oil and water do not mix". Contrary to that saying, hydrocarbons (and other compounds) in petroleum and other relatively non-polar or partially polar organic compounds do have measurable, albeit low solubility in aqueous systems. Solubility is a key characteristic of chemicals in aquatic systems. During the past three decades solubility for POPs and similar, lower solubility compounds, have been determined. This required new or modified methodologies for determining solubility (e.g. Mackay and Shiu, 1977; May et al, 1978; May, 1980; Whitehouse, 1984). Methodology for calculating solubility from molecular structure considerations has evolved (e.g. Banerjee, 1985, Lyman et al, 1982; Schwrazenbach et al, 2003) and has proved useful.

Increasing temperature usually increases the solubility and especially affect solubility for the larger, apolar compounds such as the POPs of interest in this chapter; PAHs, PCBs, and chlorinated pesticides (Schwarzenbach et al, 2003).

As a general rule, increased salinity reduces solubility, commonly referred to by aqueous chemists as the "salting out" effect. The magnitude of the effect varies from compound to compound. An empirical formula for converting the solubility of an organic compound in pure water to its solubility in a saline aqueous solution was developed by Setschenow (1889).

$$C_{\mathrm{iw,salt}}^{sat} = C_{\mathrm{iw}}^{sat} \bullet 10^{-K_i^s [\mathrm{salt}]_{\mathrm{tot}}} \tag{1}$$

$C^{sat}_{\mathrm{iw,salt}}$ is compound $i$ solubility in saline aqueous solution.

$C^{sat}_{\mathrm{iw}}$ is compound $i$ solubility in pure water.

$[\mathrm{salt}]_{\mathrm{tot}}$ is total molar salt concentration.

$K^s_i$ is the Setschenow constant.

Some example values for Setschenow constants are in Table 15.7.

TABLE 15.7
Salting Constants for Persistent Organic Pollutants and Related Compounds. [Excerpted from table 5.7 of Schwarzenbach et al, (2003). Original references to sources of data in that table.] Reprinted with permission of John Wiley and Sons.

| Compounds | $K^s_i$ (L $mol^{-1}$) |
|---|---|
| Various PCBs (dichloro to hexachloro) | 0.3-0.4 |
| Naphthalene | 0.28(+/-0.04) |
| Fluorene (NaCl)[a] | 0.27 |
| Phenanthrene | 0.30(+/- 0.03) |
| Anthracene | 0.30(+/- 0.02) |
| Fluoranthene (NaCl) | 0.34 |
| Pyrene | 0.30(+/-0.02) |
| Chrysene (NaCl) | 0.34 |
| Benz(a)pyrene | 0.34 |
| Benz(a)anthracene | 0.35 |

[a] Determined in NaCl aqueous solution of similar molar salt concentration as seawater.

Xie et al (1997) reviewed the effects of salt concentrations on solubility and noted that Setschenow constants for representative POPs are similar for NaCl solutions at 3.0 -3.5% (0.5M),, artificial sea water and sea water at salinity of 30–35°/oo. They further noted, "a simple correlation is suggested for estimating the Setschenow constants for a variety of organic solutes in seawater which typically yields a reduction in solubility by a factor of 1.36. The hydrophobicity of organic solutes is therefore increased by this factor, as is the air-water partition coefficient, implying an increased partitioning from aqueous solution into air, organic carbon, and lipid phases". We will discuss such partitioning below.

Some of the results from a laboratory experiment measuring the PAH solubility in seawater as a function of salinity and temperature, over a range found in the oceans and coastal waters, are shown in Figure 15.6 (Whitehouse, 1984), one of the few illustrations available of combined temperature and salinity effects. Note that the solubility of the two isomers—phenanthrene and anthracene—differ markedly, although the structures of these two compounds differ only in the relative positioning of the three aromatic rings. Also, the addition of a methyl group to the three rings also changes the solubility. This supports the statement above that exact molecular structure does matter in the biogeochemical cycle of POPs and PAH, and other organic chemicals.

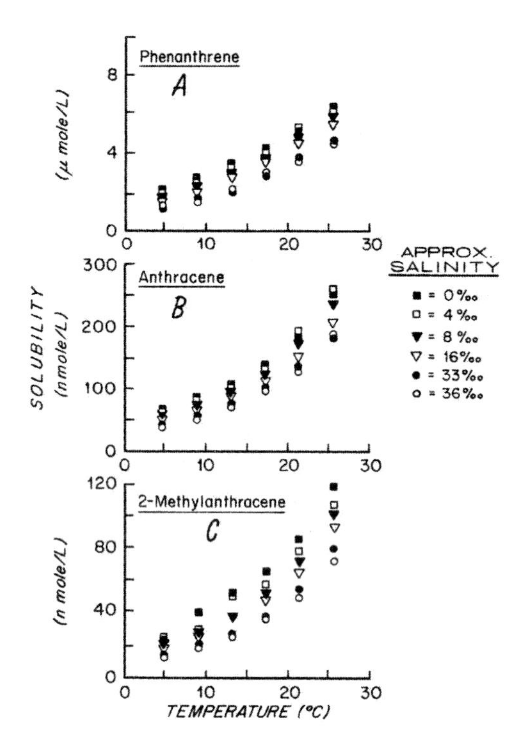


Figure 15.6 Effects of Temperature and Salinity on the Solubility of Some Polycyclic Aromatic Hydrocarbons. (adapted from Whitehouse, 1984). Reprinted with permission from Elsevier.

*Air-Seawater Partitioning.*

Air –seawater partitioning is also important in the biogeochemical cycle of POPs. The partitioning can be estimated reasonably using Henry's Law by knowing Henry's Law Constant and the concentration in either the water or air. There are a number of reasonable assumptions involved as explained by Schwarzenbach et al (2003). Two equations are useful in a practical sense and usually do not deviate more than a factor of 2 from experimentally determined $K_{iaw}$ (air water partition for compound *i*) (Schwarzenbach et al, 2003). $K_{iaw}$ is also called $K_{iH}$ or Henry's Law Constant. Eq. 2 is the equation for liquid aqueous solubility and Eq. 3 is for the solid compounds.

$$K_{iH} \approx \gamma_{iw}^{sat} \bullet p_{iL^*} \bullet \overline{V}_w \approx \frac{p_{iL*}}{C_{iw}^{sat}(L)} \quad (2)$$

$$K_{iH} \approx \frac{P_{is^*}}{C_{iw}^{sat}(s)} \quad (3)$$

$K_{iH}$ is the Henry's law constant for compound *i*.

$p_{iL^*}$ is vapor pressure of compound $i$ liquid.
$P_{is^*}$ is vapor pressure of compound $i$ solid.
$C_{iw^{sat}}$ is concentration of compound $i$ in water at saturation.
$\gamma_{iw^{sat}}$ is the activity coefficient of compound $i$ at saturation in water.
$\overline{V}_w$ is the partial molar volume.

The effect of salt on $K_{iaw}$ in air-water partitioning is taken into account by:

$$K_{iaw,\, saltwater} = K_{iaw} \bullet 10^{+K_i^s\,[salt]_{tot}} \tag{4}$$

$K_{iaw}$ is the dimensionless Henry constant and is related to $K_{iH}$ by:

$$K_{iaw} = \frac{K_{iH}}{\mathrm{RT}} \tag{5}$$

The simplified equation for describing flux across the air-sea interface may be written:

$$F_{ia/w} = \mathrm{v}_{ia/w}\,(C_{iw} - C^{eq}_{iw})\ [\mathrm{ML}^{-2}\,\mathrm{T}^{-1}] \tag{6}$$

$\mathrm{v}_{ia/w}$ is the transfer velocity air-sea.

$$C^{eq}_{iw} = \frac{C_{ia}}{K_{iaw}} \tag{7}$$

and $\mathrm{F}_{ia/w}$ is positive for a net flux from water to the atmosphere.

**Partitioning Between Particulate Matter and Solution.**

A key process in the biogeochemical cycle of POPs in the coastal ocean is the sorption and desorption of POPs between particulate matter and solution. The terms sorption and desorption are used because it is often not clear in most cases whether absorption or adsorption is involved alone or in combination. It is clear that there is an association of the POP in some manner with particulate matter, either in the water column or in sediments (Figure 15.4). The partitioning can be described in a simple equation as,

$$K_{\mathrm{id}} = \mathrm{C}_{\mathrm{ip}}/\mathrm{C}_{\mathrm{iw}} \tag{8}$$

This can be further modified because we know that the organic carbon portion of the particulate matter in either the water column or sediment is the most important phase of particulate matter (i.e. in comparison to mineral surfaces) for sorption of POPs and many other organic compounds. We can therefore write the equation:

$$K_{\mathrm{ioc}} = K_{\mathrm{id}}/\mathrm{f}_{\mathrm{oc}} = \mathrm{C}_{\mathrm{ioc}}/\mathrm{C}_{\mathrm{iw}} \tag{9}$$

However, organic matter in sediments and particles is made of various different types of naturally occurring organic matter. Johnson (1974) showed that there were at least eighteen different types of organic matter-mineral particles in surface sediments of Buzzards Bay, MA. Farrington et al (1977) noted that we should think of these particles as each having their own biogeochemical reactions. In this sense, there is a strong connectivity between the biogeochemistry of POPs and the biogeochemistry of the naturally occurring organic matter (Kujawinski et al., 2001). For example, what happens in diagenesis of naturally occurring organic matter in sediments (Henrichs, Chapter 5) i.e. the change in the quality of the organic matter, changes the sorbents to which POPs sorb. As Schwarzenbach et al (2003) note, the actual sorption of POPs can be written as a series of terms:

$$K_{id} = [C_{ioca} + C_{iocb} + C_{iocc} .....]/C_{iw} \tag{10}$$

where $C_{ioca}$ is concentration of compound $i$ in organic matter type a, $C_{iocb}$ is concentration of compound $i$ in organic matter type b, and $C_{iocc}$ is concentration of compound $i$ in organic matter type c, and so forth for all organic matter types.

In essence each type of organic matter would have its own $K_{oc}$ value for a given compound. In most cases, as a first approximation, we can use Eq. 19.

It has proved useful to estimate the $K_{oc}$ from a surrogate measurement or parameter $K_{ow}$ or octanol-water partition coefficient. For various reasons, explained in detail in Schwarzenbach et al (2003), octanol proved to be a reasonable surrogate for lipids and for naturally occurring organic matter when estimating $K_{oc}$ for various compounds. We can write an equation relating $K_{oc}$ to $K_{ow}$:

$$\log K_{ioc} = a \cdot \log K_{iow} + b \tag{11}$$

One approach used early in the process of determining $K_{ow}$ values was experimentation in the laboratory using sealed flasks with octanol and water as the two liquid media and compound $i$ added to this system and allowed to equilibrate between the two liquids. Ratios of the measured concentration of compound $i$ in the octanol and in the water provided an experimentally determined $K_{ow}$. How the two phase system was mixed and allowed to come to equilibrium was an important factor. There are also ways to estimate $K_{ow}$ by using HPLC or by taking into account additive molecular and/or atomic properties (Schwarzenbach et al 2003; Pontolillo and Eganhouse,2001; and references therein).

A common practice now is to use one parameter LFERs (linear free energy relationships) for compounds within a given class of organic compounds. That is, the compounds have the same functionality, e.g. alkanes, polycyclic aromatic hydrocarbons, chlorobiphenyls, but a part of he structure affecting polarity varies e.g. individual n-alkanes of ten to 30 carbon atoms; naphthalene, phenanthrene, anthracene, pyrene; or trichlorbiphenyls, tetrachlorobiphenyls, pentachlorobiphenyls). The equation is:

$$\log K_{iow} = a \cdot \log \gamma_{iw} + b \tag{12}$$

In the case of low solubility compounds such as the POPs for which $\gamma_{iw}$ is > *ca.*50), i.e. $\gamma_{iw}$ is approximately equal to $\gamma_{iw}^{sat}$, and $\gamma_{iw} = (V_w \cdot C_{iw}^{sat}(L))^{-1} + b'$. Then we can write the equation:

$$\log K_{iow} = -a \cdot \log C_{iw}^{sat}(L) + b' \quad (13)$$

where $b' = b - a \cdot \log \bar{V}_w = b + 1.74\,a$ at 25° C. (Schwarzenbach et al, 2003)

The range of $K_{ow}$ and aqueous activity coefficients or liquid solubility for some compounds of interest for this chapter are given Table 15.8.

TABLE 15.8
Relationship Between Octanol-Water Partition Constants and Aqueous Activity Coefficients or Liquid Aqueous Solubilities at 25° C for Various Sets of Compounds: Slopes and Intercepts of Eqs. 12 and 13. From Schwarzenbach, 2003. Reprinted with permission of John Wiley and Sons.

| Set of Compounds | $a$ [a,b] | $b$ [a] | $b'$ [b] | log $K_{iow}$ range [c] | $R^2$ | $n$ [d] |
|---|---|---|---|---|---|---|
| Alkanes | 0.85 | -0.87 | 0.62 | 3.0 to 6.3 | 0.98 | 112 |
| Polycyclic aromatic hydrocarbons | 0.75 | -1.04 | 0.60 | 3.3 to 6.3 | 0.98 | 11 |
| Chlorobenzenes | 0.90 | -0.95 | 0.62 | 2.9 to 5.8 | 0.99 | 10 |
| Polychlorinated biphenyls | 0.85 | -0.70 | 0.78 | 4.0 to 8.0 | 0.92 | 14 |
| Polychlorinated dibenzodioxins | 0.84 | -0.79 | 0.67 | 4.3 to 8.0 | 0.98 | 13 |

[a] Eq. 12, b. Eq.13, c. Range of experimental values for which relationship has been established. d. Number of compounds used for establishing the relationship.

*Source:* Excerpted from Schwarzenbach et al (2003), table 7.3.

The relationship between $K_{ioc}$ and $K_{iow}$ is given by:

$$\log K_{ioc} = a \cdot \log K_{iow} + b \quad (14)$$

Examples of the constants $a$ and $b$ for ranges of log $K_{iow}$ for four classes of compounds are given in Table 15. 9.

TABLE 15.9
Relationship between particulate organic matter-water partition coefficients and octanol-water partition coefficients at 20-25° C. for some sets of neutral (non-ionized) organic compounds.

The constants a, b, and $K_{iow}$ are from Eq. 14 discussed in the text.

| Set of Compounds | $a$ | $b$ | log $K_{iow\ range}$ | $R^2$ | $N$[a] |
|---|---|---|---|---|---|
| Alkylated and Chlorinated benzenes, PCBs | 0.74 | 0.15 | 2.2 to 7.3 | 0.96 | 32 |
| Polycyclic Aromatic Hydrocarbons | 0.98 | -0.32 | 2.2 to 6.4 | 0.98 | 14 |

[a] Number of compounds used for establishing the relationship.

*Source:* Excerpted from table 9-2 of Schwarzenbach et al (2003)

*$K_{ow}$ controversy*

A note of caution about reported aqueous solubility and Kow values has been introduced by a recent very important study of the literature by Pontolillo and Eganhouse (2001). This has created quite a stir (Benner, 2002) since it impacts on environmental risk assessments, fate and transport models and sediment quality guidelines that use $K_{ow}$s to estimate factors such as sorption by particles and partitioning into organisms across membranes as discussed below. Pontolillo and Eganhouse (2001) reviewed about 700 publications from 1944 to 2001 and found that the reliability of the database for DDT and DDE solubility and $K_{ow}$ values is questionable. Egregious errors in reporting data and references and poor data quality and/or inadequate documentation of procedures were identified by the authors as two key problems. Original data make up only 6–26% of the reported values. The other data are "recycled" and often not properly referenced. They also noted significant errors of transcription that have been propagated through the later literature. This is unfortunate and sloppy science in my opinion, and Pontilillo and Eganhouse (2001) deserve credit for bringing this problem to the attention of the environmental sciences community and government agencies. It is a warning to all that a thorough check and understanding of the original literature, in addition to compilations and reviews, is the backbone of science and that adequate referencing is *sine qua non* for scientific publications.

**Uptake or Partitioning in Organisms.**

Hamelink, et al (1971) were among the first to propose a hypothesis that organisms in aquatic systems that obtained oxygen by passing water over gill surfaces would also exchange POPs and similar compounds from the water into their gill surfaces and from there would circulate them throughout their body tissues. One of the first uses of the $K_{ow}$ parameter of POPs and other organic contaminants with respect to environmental concerns was to estimate the extent to which compounds would be taken up from aqueous systems by aquatic organisms (Neeley et al, 1974). Bioconcentration factors (BCFs) are the extent to which an organisms bioaccumulates or concentrates the compound of interest in comparison to the concentration in the water until reaching equilibrium. This can be written as:

$$\mathrm{BCF} = \frac{C_i^{*\ \mathrm{organism}}}{\mathrm{C_{iw}}} \tag{15}$$

Where * indicates equilibrium considerations.

Following the approach of Schwarzenbach et al (2003), an organism's theoretical equilibrium bioaccumulation potential or *TBPi* can be estimated by:

$$C_{i\mathrm{bio}}{}^{*} = TBP_{(i)} \text{ (e.g. in mol} \cdot \text{kg}^{-1} \text{ dry organism)} = K_{i\mathrm{bio}} \cdot C_{i\mathrm{med}} \tag{16}$$

Where med is the medium of water (w) or sediment (*s*).

$K_{ibio}$ can be estimated from the $K_{iow}$ by the following equation:

$$\log K_{ibio} = a \cdot \log K_{iow} + b \tag{17}$$

As with the organic matter associated with particulate matter and sediments, there is a simplification in thinking that there is only one type of biochemical and associated partitioning in organism. In reality, organisms have various constituent biochemicals. It is possible to write equations that take into account the proportion of the organism made up of a given biochemical, e.g. lipid or protein, or consider parts of the organisms such as the liposomes of cells and include partition constants for each of these biochemicals or parts. These details are addressed by Schwarzenbach et al (2003). Given these caveats, use of $BCF_i$s or $TBP_i$s has proved to be useful in estimating the extent of bioaccumulation for given compounds. When there is a measured departure from the predicted equilibrium concentration, this has led to further understanding and research about processes governing such behavior.

Given the accumulation of POPs in sediments, it is important to know the biota –sediment concentration factors for animals living in or on the sediments. Assuming equilibrium, this is expressed as:

$$BSAF_i = \frac{C_{i\,\text{organisms}}}{C_{is}} \tag{18}$$

In some cases the relationship is expressed in terms of the concentration of $\text{POP}_i$ in the sediment normalized to organic carbon and the concentration in the organisms normalized to lipids.

$$BSAF_{i\,\text{lip-oc}} = \frac{C_{i\,\text{lip-organism}}}{C_{i\,\text{soc}}} \tag{19}$$

**The Role of Colloids and Dissolved Organic Matter**

The importance of colloids in the water column as a phase with which POPs would associate or sorb was reported by Means and Wijayaratne (1982) and Wijayaratne and Means (1984a,b). Colloidal interactions could occur where dissolved organic matter was in sufficiently high concentration that the higher molecular weight natural polymeric material would form colloids, or the organic matter might already be colloidal upon entering the aqueous system. Brownawell and Farrington (1985, 1986), Brownawell (1986), and Baker et al (1986) reported on the important role of colloidal organic matter in pore waters and for sediment-water partitioning. Brownawell (1986) and Brownawell and Farrington (1985,1986) noted that the apparent distribution coefficient between interstitial water and the solid phase of sediments for POPs and similar hydrophobic compounds could be described by an equation of the form:

$$K_{\text{d}} = \frac{C_{is}}{C_{i\text{w}} + C_{i\text{coc}}} \tag{20}$$

Where s denotes sediment, w denotes interstitial water, and coc denotes colloid.

The concentrations of the two sorbed phases can be estimated using a known or estimated $K_{ioc}$ by:

$$K_{id} = \frac{f_{ocs} \bullet K_{iocs} \bullet C_{iw}}{C_{iw} + f_{icoc} \bullet K_{icoc} \bullet C_{iw}} \quad (21)$$

In the case where $C_{iw}$ is small compared to $F_{icoc} \cdot K_{icoc} \cdot C_{iw,}$ Eq 19 simplifies to

$$K_{id} = \frac{f_{ocs} \bullet K_{iocs}}{f_{icoc} \bullet K_{icoc}} \quad (22)$$

Few actual values are available for $K_{icoc}$ and those that are available for natural coastal organic matter suggest that it is acceptable as a first approximation to assume that $K_{iocs} \approx K_{icoc}$. Thus, Eq. 22 further simplifies to:

$$K_{id} = \frac{f_{ocs}}{f_{icoc}} \quad (23)$$

Using this approach, Brownawell and Farrington (1985, 1986) and Brownawell (1986) were able to show good agreement between the model $K_{id}'$ and the actual measurements of chlorobiphenyl congeners in water and solid phases of sediments obtained by box cores of near shore coastal sediments of Buzzards Bay, MA USA. Specifically they assumed that colloidal organic carbon in the interstitial waters approximated dissolved organic carbon (DOC).

Dissolved organic matter, often expressed in units of dissolved organic carbon (DOC) also has an influence on the biological uptake or bioavailability of POPs. This is because the physical-chemical form of the POP in seawater or interstitial water is important when considering the uptake across membrane surfaces (i.e. gill surfaces) as a source of input to the organism in question. The uptake of PAH has been shown to be reduced when DOC- macromolecular material is present either in colloidal or dissolved form (e.g. Boehm, P. D. and J. G. Quinn, 1976; Leversee, et al, 1983; McCarthy, J. F. et al, 1985; Hamelink et al., 1994 and references therein). A simple example is presented in Table 15.11.

TABLE 15.11

Bioconcentration factors observed after 30hr exposure of *D. magna* to 3-methylcholantherene in the presence of different concentrations of dissolved humic material. (McCarthy et al, 1985). Reprinted with permission of Elsevier.

| DHM[a] concen. ($mg\ L^{-1}$) | Fraction of MC[b] bound to DHM | Fractional decrease in BCF due to presence of DHM. (±SD) | Observed BCF[c] |
|---|---|---|---|
| 0 | 0 | 13.209(±6.53) | --- |
| 0.15 | 0.05 | 12.171 (±1.94) | 0.07 |
| 1.5 | 0.32 | 8.121 (±1.61) | 0.38 |
| 15.0 | 0.82 | 2.311 (±1.27) | 0.82 |

[a] Dissolved Humic Material; b. MC is 3-methycholanthrene; c. BCF is bioconcentration factor.

**Photochemical Reaction Considerations.**
PAH are photoreactive in the upper part of the water column (Zepp and Schlotzhauewr, 1979; Payne and Philips, 1985; Ehrhardt et al 1992; NRC, 2003). Recently, Fashnacht and Blough (2002) reviewed the literature for photolysis of PAH in water and conducted experiments with twelve different PAH in aerated pure water, solutions of fulvic acid, and natural waters using polychromatic light (<290nm). In their review they noted that PAH may undergo either direct or sensitized photochemical reactions. They also noted that there have been suggestions from other studies that reactions might take place with chromophoric intermediates to enhance photochemical reactivity of PAH or that interactions and binding of PAH with DOM might either enhance or decrease photochemical reaction rates for PAH. They determine in their experiments that rate constants in pure water were essentially unchanged by additions of fulvic acid and did not differ from those in natural waters. Rate constants of decay of the PAH were first order over two half lives and varied from $3.2 \times 10^{-3}$ to $7.6 \ h^{-1}$. The presence of dissolved organic matter in these experiments seemed to have little influence.

Schwarzenbach et al (2003) present two excellent chapters on photochemistry and the principles to be understood. Pagni and Sigman (1999) are recommended by Schwarzenbach et al as providing information specific to PCBs and PAH, and Méallier (1999) is recommended for some pesticides.

### *Microbial Degradation Processes*

Only a brief summary is presented of this topic, which has an extensive scientific literature with respect to mechanisms. There was basic knowledge about the biodegradation of petroleum hydrocarbons by the late 1960s because of the importance of the subject to storage of petroleum and also because of the processes that might be a work in shallow oil reservoirs. (Davis, 1967). The subject was reviewed in 1975 as part of an assessment of inputs, fates, and effects of petroleum hydrocarbons in the marine environment (NRC, 1975), and twice since then by the NRC (1985 and 2003). Watkinson (1978) provides an interesting series of papers on the knowledge about biodegradation of hydrocarbons as of the mid 1970s. Cirniglia and Heitkamp, (1989) provided a very nice review of the subject. The recent NRC (2003) report also provides a summary of the state of knowledge and cites reviews by Leahy and Colwell (1990), Atlas and Bartha, (1992) and Heider et al (1999), among others.

In brief, microbial degradation under oxic conditions occurs with the n-alkanes degrading most rapidly, followed by the branched alkanes, especially the isoprenoid hydrocarbons such as pristane and phytane (Figure 15.2a). The oxidation usually involves steps of addition of oxygen through reactions that yield an alcohol, an aldehyde, and then carboxylation on the end carbon atom of the alkane chain. Then the process becomes one of the degradation of fatty acids, involving well known biochemical reactions. Since this involves removal of adjacent carbon atoms two at a time, there is a general sense that branched alkanes and especially isoprenoid alkanes such as pristane and phytane, are less readily degraded, because it is not easy to fit them into the beta oxidation sequence due to steric hindrance for the active site of key enzymes. Since both alkanes and branched alkanes and isoprenoid hydrocarbons are naturally biosynthesized and present in some biological systems in non-trivial concentrations, it is not surprising that these compounds are biodegraded.

Aromatic hydrocarbons are degraded at about the same rate or a little slower in general compared to the branched alkanes. The mechanism usually involves attachment of oxygen by way of an enzymatic dioxygenase reaction to adjacent unsubstituted carbon atoms to form a dioxygen intermediate. This is followed by addition of hydrogens to this intermediate with the resulting product being a *cis* diol. In turn, there is ring cleavage, and formation of various oxygenated products that will eventually lead to carbon dioxide. Some bacteria, fungi and algae also employ a Cytochrome P-450 monoxygenase pathway involving an epoxide formation and subsequent reactions, to yield phenols or diols that are then excreted or eliminated from the cells in some manner. Essentially this pathway is a detoxification mechanism (NRC, 2003). This has been demonstrated for various PAH in mesocosm and microcosm experiments with $^{14}C$ PAH (Hinga et al, 1980, 1986; Hinga and Pilson, 1987; McElroy et al, 1987, 1989 and references therein). These experiments also demonstrated that reaction products or intermediates in the pathway from the parent compounds and $CO_2$ accumulated in various amounts in the sediment and were still present at the end of the experiments—as much as 30 to 120 days later.

Recent research has demonstrated that aromatic hydrocarbons can be degraded under suflate reducing conditions (Coates et al, 1997; Hayes et al, 1999; Rothermich et al, 2002). Lower molecular weight PAHs degraded faster than the higher molecular weight four and five ring PAH (Rothermich et al, 2002). How fast and under what conditions this proceeds in the environment is dependent on environmental conditions. For example, samples taken 30 years after the West falmouth oil spill at a site that had been heavily oiled and monitored periodically over the years (Teal et al, 1992) still exhibit elevated levels of No. 2 fuel oil PAH in the marsh sediments (Reddy et al, 2002c). Presumably, the microbial community would have had time to acclimate to the presence of these PAH and degrade them. However, it is known that the presence of more easily degraded organic rich plant remains could cause competition for the available pool of electron receptors and otherwise decrease the anaerobic degradation rate of the PAH. It is interesting that Reddy et al (2002c) also reported the presence of relatively high concentrations of branched alkanes and cycloalkanes from No. 2 fuel Oil. It is important to note that this was only in one area of the marsh that had been the most heavily contaminated. Other areas had little or no No.2 fuel Oil hydrocarbons present 20 years after the spill (Teal et al, 1992).

Microbial degradation of PCBs under oxidative conditions is similar to the microbial degradation of PAH in that it involves the formation of the *cis* diol as described in papers and reviews by Brown et al, (1987) Abramowicz (1990), Bedard, (1990), Boyle et al (1992), Higson (1992), Mohn and Tiedje (1992), Haluska et al (1993), Sokol et al (1998) and Tiedje et al (1993). One reason for the environmental persistence of the PCBs, especially the higher chlorinated congeners, is that oxidative degradation by this mechanism is difficult. The chlorine substituents block the carbon atoms in the rings and as can be seen from examining structures in Figure 15.1b and c, there are few adjacent unsubstituted carbon atoms in the biphenyl rings. Nevertheless, there is slow oxidative degradation if conditions of nutrients, oxygen, bioavailability, and elevated concentrations of the PCBs in solution occur and other conditions are appropriate.

Anaerobic biotransformation of PCBs has been demonstrated. It is much different than aerobic biotransformation (Brown et al, 1987; Tiedje et al, 1993) and seems to proceed more effectively with higher chlorinated biphenyls. The PCBs are transformed by reductive dehalogenation under anaerobic conditions. This process completely changes the composition of the PCB mixture present in anaerobic sediments, but without changing the overall total concentrations of the PCBs. However, the biological effects of the overall mixture changes as a result of such biotransformations – some effects might be reduced and others might be enhanced (NRC, 2001). A likely scenario in environments with substantial PCB pollution in sediments would be a slow reductive dechlorination of the higher chlorinated chlorobiphenyls, producing less chlorinated biphenyls that could then be more readily degraded by aerobic degradation because more unsubstituted vicinal (adjacent) carbon atoms would be available.

There have been extensive studies of the microbial degradation of DDT. The product most likely to be found in environmental samples as a result of DDT being present in the environment and metabolized by animals is DDE (Figure 15.1), compound produced by dehydrochlorination. The product most likely to result from the microbial degradation of DDT is DDD (See Figure 15.1a). DDD can be further degraded by microorganisms to DDMU (structure not shown) by loss of a chlorine and loss of a hydrogen on the two ethane carbons to yield an ethane moiety. Although there are formulations of technical DDT that contain some DDE and DDD, usually, recent DDT use is indicated by a high ratio of DDT to DDE. As time progresses, much of the DDT is converted to DDE and DDD and then more slowly to other degradation or metabolic products. (MacGregor, 1974, 1976; Rhead, 1975).

Recently, there has been renewed interest in the further degradation of DDT, DDE and DDD under anaerobic conditions due to the burial of DDT, DDE and DDD in sediments near areas that had received substantial inputs of DDT in earlier years. For example, see the section describing the Palos Verdes Shelf below. Quensen et al (1998) noted the reductive dechlorination of DDE in cultures to yield DDMU. Thus, we can have DDMU produced by at least two different biodegradation pathways.

In the vast majority of instances of biodegradation in the coastal ocean, the important factors are presence of appropriate microbes; sufficient availability and mixing of the compound with nutrients and oxygen (for aerobic processes) or other electron acceptors for anaerobic conditions; bioavailability of the POP or PAH; and lack of competition with microbes degrading other, more easily reactive compounds e.g. natural organic compounds. McElroy et al (1989) point out that the fate of PAH—and by extension that of other organic contaminants such as POPs—is most likely significantly influenced by the interactivity of microbial degradation and photochemical reaction processes. Also both interact with metabolic processes of other marine organisms such as polychaetes, crustacea, and fish. This is a point that was noted earlier by van der Linden (1978) when considering the interaction of spilled oil photochemical reaction products and microbial degradation of the reaction products. At high concentrations, the photochemical reaction products seemed toxic to some microbes and at lower concentration were difficult to degrade.

*Biological Uptake and Elimination of POPs: Kinetic Considerations.*

Stegeman and Teal (1973) conducted a pioneering experiment in which they exposed oysters (*Crassostria virginica*) to petroleum hydrocarbons in an experimental system and after exposure, measured the release or elimination of the hydrocarbons by monitoring the tissue concentrations. They proposed that there were at least two compartments in the bivalve. In the first, petroleum hydrocarbons were taken up rapidly, in gills and circulating fluid that were in immediate contact with the hydrocarbons in the water.The second consisted of tissues in which there was a slower uptake as the circulatory fluid transported the hydrocarbons throughout the oyster, delivering the compounds to these tissues. Hydrocarbons in the rapidly exchanging gill tissues and circulatory fluid would be rapidly eliminated. Longer term, slower release occurs from the tissue in which the compounds had been "stored" or sequestered and needed to be released to the circulatory fluid and hence to the gills and seawater. This conceptual model has stood the test of time. Boehm and Quinn (1978) provided documentation of this longer term, slow release by examining the retention and release of petroleum hydrocarbons from the hard shell clam (*Mercenaria mercenaria)* chronically exposed to petroleum in the field and then moved to a cleaner system.

The consideration of uptake and elimination kinetics for a given POP in an organism or tissue of an organism can be expressed in an equation taken from Farrington and Westall (1986) of the form similar to that used by several others e.g. Schwarzenbach et al (2003).

The kinetics of biological uptake can be considered during the initial phase as a first order process:

$$dC_1/dt = -k_1C_1 + k_2C_2 \tag{24}$$

$$dC_2/dt = +k_1C_1 - k_2C_2 \tag{25}$$

$C_1$ is concentration in compartment 1 (e.g. food, water, sediment, etc.)
$C_2$ is concentration in compartment 2 ( body, tissue, etc.)

If we consider uptake from an infinite reservoir, i.e. the concentration of the chemical in the water does not change due to the small amount taken up by an organism relative the total amount in the water, and assuming $k_1C_1 \gg k_2C_2$ then:

$$dC_2/dt = k_1C_1 \tag{26}$$

$$C_2(t) = C_2(0) \exp k_1 t \tag{27}$$

Similarly, for elimination assume $k_2C_2 \gg k_1C_1$

$$C_2(t) = C_2(0) \exp -k_2 t \tag{28}$$

These equations are useful in describing the uptake or elimination in organisms for which there is little or no biological transformation of the specific POP to a metabolite such as occurs in bivalve mollusks for POPs and PAH (e.g. James, 1989; NRC, 1980). For multiple compartments of the Stegeman and Teal (1973) conceptual model type, we would simply write additive equations linking the compartments and their respective rate constants for uptake or elimination.

More detailed equations taking into account other processes such as may occur in fish can be written as noted by Schwarzenbach et al (2003) among others as:

$$(dC_{ifish}/dt = \underbrace{k_1(K_{ifishw}C_{iw-}C_{ifish})}_{\text{gill exchange}} + \underbrace{k_D(K_{ifishdiet}C_{idiet}}_{\text{uptake with food}} - \underbrace{k_E K_{ifishexcreta}C_{iecrerta})}_{\text{excretion}} - \underbrace{k_M C_{ifish}}_{\text{metabolism}} - \underbrace{k_G C_{ifish}}_{\text{growth}} \quad (29)$$

$k_1$, $k_E$, $k_M$, $k_G$ are the first order rate constants for the various processes indicated and the $K$ indicates the equilibrium constants for the partitioning of the chemical between the various combinations of fish, water, diet, and excreta.

The steady state special case (indicated by the superscript $\infty$), one often assumed in ocean sciences for several oceanic processes, can be written assuming that $dC_{ifish}/dt = 0$. Then we can solve Eq.29:

$$C^{\infty}_{ifish} = (k_1 K_{ifishw} C_{iw} + k_D K_{ifish\,diet} C_{idiet} - k_E K_{ifish\,excreta} C_{iecreta}) \,/\, (k_1 + k_M + k_G) \quad (30)$$

Dividing by the water concentration $C_{iw}$ we have the following:

$$\mathrm{BAF}^{\infty}_{\mathrm{ifish}} = \mathrm{C}^{\infty}_{\mathrm{ifish}} / \mathrm{C}^{\infty}_{\mathrm{iw}} \text{ or} \quad (31)$$

$$BAF^{\infty}_{ifish} = [k_1(K_{ifishw} + k_D K_{ifish\,diet} BAF_{idiet} - k_E(K_{ifish\,excreta}\, C_{iexcreta}/\, C_{iw})] \,/\, (k_1 + k_M + k_G)$$

Schwarzenbach et al (2003) go on to provide a time evolving solution to Eq. 29. This is beyond the scope of our considerations here. However, the above illustrates that if certain parameters are known or estimated, and concentrations of chemical *i* are known in some parts of the ecosystem or in an organism, then we can use these equations to predict what will happen when concentrations are increased or decreased in the water, or using similar equations, in the sediments.

A few simple cases suffice to illustrate this point. In 1978 and again in 1983 small spills of No.2 fuel oil occurred from oil barges transiting the Cape Cod Canal. The oil slick contaminated a subtitidal population of the common blue mussel (*Mytilus edulis*). The spills occurred on dates within a few weeks of each other, five years apart, in essence a "replicate experiment". Farrington et al (1978 and 1982) measured the fossil fuel hydrocarbons in the mussels on day 0 and then for up to 118 days after the event. The slicks rapidly disappeared because of the fast running tidal flows in the Cape Cod Canal. Therefore, we had a situation in which the concentration of the No. 2 fuel oil compounds in the sea water in which the contaminated mussels resided was effectively zero. Two example plots of concentrations of aromatic hydrocarbons in mussel tissue with time illustrate what happened to the No.2 fuel oil compounds (Figure 15.7a,b). There was a rapid elimination of the fuel oil chemicals. Using Eq. 28, a useful parameter, the *biological half-life,* the time it takes to eliminate one half of the concentration at time 0, can be defined:

$$t_{b1/2} = 0.693/k_2 \quad (32)$$

The $t_{b1/2}$ can be determined from the slope of the line of a semi log plot like those of Figure 15.7a and b, or calculated by curve fitting. There are different $t_{b1/2}$ for different No.2 fuel oil chemicals (Table 15.12 ) indicating the importance of chemical structure related to solubility and partitioning.

TABLE 15.12
Comparison of biological half lives ( $t_{b1/2}$) if No.2 fuel oil compounds in mussels (*Mytilus edulis)* from two small spills contaminating the same mussel populations (Farrington et al, 1986).

| **Compounds** | $t_{b1/2}$ *days* 1978 Day 1-21 | $t_{b1/2}$ *days* 1983 Day 3-29 |
|---|---|---|
| n-$C_{16}$ | 0.99[a] | 8.7[a] |
| Pristine | 7.7 | 6.3 |
| Phytane | 6.9 | 5.8 |
| n-$C_{23}$ | 4.6 | 5.8 |
| $\Sigma C_2$ - naphthalenes | 5.8 | 1.5 |
| $\Sigma C_3$ – naphthalenes | 7.7 | 5.8 |
| phenanthrene | 17 | 5.8 |
| $\Sigma C_1$ – phenanthrenes | 9.9 | 6.9 |
| $\Sigma C_2$ – phenanthrenes | 69[b] | 9.9 |

[a] there maybe some microbial degradation and/or interference of the naturally occurring hydrocarbons with these analyses.

[b] poor fit of data for these compounds to the equation.

TABLE 15.13
Biological half-lives of some individual PAHs and PCBs in *Mytilus edulis.* (adapted from Pruell et al, 1986).

| **Compound** | $t_{b1/2}$ *days* |
|---|---|
| Fluoranthene | 29.8 |
| Benz(a)anthracene | 17.8 |
| Benzo(e)pyrene | 14.4 |
| Benzo(a)pyrene | 15.4 |
| 2,3,4'-trichlorobiphenyl | 16.3 |
| 2,2',4,5,5'-pentachlorbiphenyl | 27.9 |
| 2,2'4,4'5,5'-hexachlorobiphenyl | 45.6 |
| 2,2',3,3',4,4'-hexachlorobiphenyl | 36.5 |

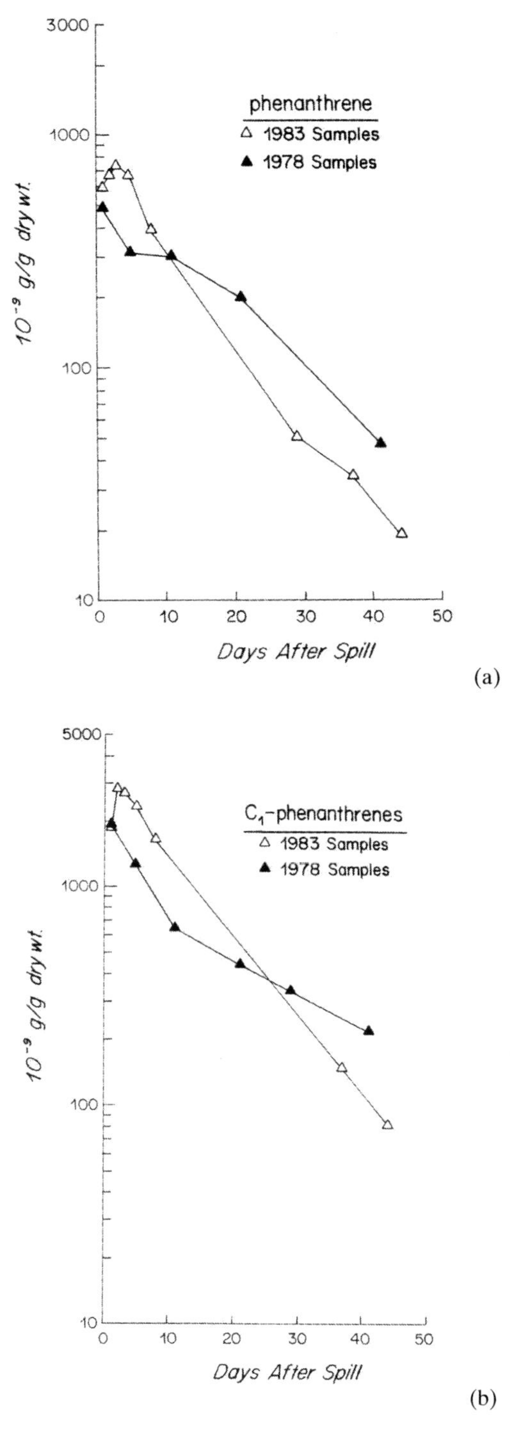


Figure 15.7 Elimination of PAHs by Mussels contaminated by a small No.2 fuel spill. (Farrington et al, 1982).

a Phenanthrene

b Methyl Phenanthrenes.

Pruell et al (1986) conducted a forty day experiment in which they exposed mussels (*Mytilus edulis*) to PCBs and PAH contaminated particulate matter associated with suspended sediments. They followed this by exposure of the mussels to clean seawater. The biological half lives were calculated using an equation of the form of Eq 32. (Table 15.13). The data of Tables 15.12 and 15.13 provide good estimates of the order of time for elimination of contaminants under conditions of short-term exposure of a day to weeks followed by removal of the contaminant source in the seawater. These conditions might be encountered in oil spill or short term dredging operations associated with contaminated sediments. As indicated above, placement in a clean environment after conditions of chronic long term exposure results in longer term slower release kinetics (Boehm and Quinn, 1977).

An elegant experiment by Rubenstein et al (1984) demonstrated that the dietary route of exposure was important as a source of PCB contamination for the demersal fish *Leistomos exanthurus* . In phase I, they exposed fish to the contaminant by placing the fish in water overlying contaminated sediment. They also exposed polychaetes to the contaminated mud. In phase II, they exposed control fish by feeding them contaminated worms and also feeding contaminated worms to fish previously exposed to the contaminated sediment through the water column. This experiment demonstrated unequivocally the importance of a food source as a route of transfer of contaminants, at least for benthic ecosystems.

TABLE 15.14

*Streblospio benedicti* Biota-Sediment Bioaccumulation Factors (BASFs) and BASF's relative to lipids and organic carbon ($BASF_{lip\text{-}oc}$)[a] in laboratory exposure and field collections (Ferguson and Chandler, 1998). Reprinted with permission of Elsevier.

| PAH | ←BASF (SD or range)[b]→ | | | ←$BASF_{lip\text{-}oc}$(SD or range)[b]→ | | |
|---|---|---|---|---|---|---|
| | Lab exposure | Diesel Creek | Marina Pipe | Lab exposure | Diesel Creek | Marina Pipe |
| Fluoranthene | 0.11<br>(0.04) | 0.33<br>(0.17-0.33) | 0.39<br>(0.27-0.59) | 0.25<br>(0.10) | 0.52<br>(0.26-0.85) | 1.36<br>(0.91-2.02) |
| Benz(a)anthracene | 0.43<br>(.13) | 0.13<br>(0.06-0.22) | 0.30<br>(0.22-0.50) | 0.98<br>(0.28) | 0.20<br>(0.09-0.34) | 1.01<br>(0.75-1.70) |
| Benzo(a)pyrene | 0.53<br>(0.17) | 0.14<br>(0.08-0.19) | 0.08<br>(0.02-0.16) | 1.20<br>(0.39) | 0.22<br>(0.13-0.30) | 0.29<br>(0.08-0.54) |

[a] The notation of BAF used by Ferguson and Chandler (1998) has been replaced with an equivalent meaning consistent with the notations for equations in this chapter.

[b] Means and standard deviations for laboratory exposures. Range of accumulation factors given for field sites.

McElroy et al (1989) in their review of the biogeochemistry of PAH in aquatic systems, noted the importance of considering trophic transfer and physiological conditions in the uptake, retention and release of PAH in marine organisms. Ferguson and Chandler (1998) measured PAH BASF (see Eqs. 18 and 19) for the polychaete worm *Streblospio benedicti* (Table 15.14). In the field measurements, BSAFs decreased with increasing log $K_{ow}$. The laboratory exposure experiments showed increasing BSAFs with increasing log $K_{ow}$. The disparities between the BSAFs from the field and laboratory in comparison to log $K_{ow}$ may be due to two primary factors noted by the authors: the need for longer equilibration time between the spiking and the experiment, and the fact that some PAH may not be as readily bioavailable for uptake, for reasons we will discuss next.

**PAH Associated with Particles and the Role of Soot or Black Carbon.** There are several sources of PAH in the environment, in general, and for the coastal ocean. These include: all the sources of petroleum discussed above, fossil fuel combustion, forest fires, grass fires, wood burning, road tar, wear of automobile and truck tires, creosote used for protecting wood, and a few natural diagenetic products found in low concentrations in sediments. Latimer and Zheng (2003) have just published an excellent review of the sources, transport, and fate of PAHs in the marine environment. For historical perspective, there are numerous other reviews, e.g. Blumer, (1976), Neff (1979) and NRC, (1983).

A key aspect of understanding the biogeochemical cycle of PAH is to understand that the various sources yield different mixtures of PAH. This was noted by Giger and Blumer (1974), Youngblood and Blumer (1975), and Hites and LaFlamme(1978) and references therein. PAH from pyrolytic sources such as combustion of fossil fuels, have a greater abundance of the unsubstituted aromatic hydrocarbons compared to the alkylated series of that parent compound. For example, phenanthrene relative to methyl phenanthrenes (Figure 15.8). The higher the temperature of combustion and the cleaner and more efficient the temperature, the more pronounced is the parent PAH compared to the alkyl substituted series. In comparison, the PAH in petroleum (formed at lower temperatures and over long periods of time) has more alkylated PAH compared to the parent compound as noted in the right panel in Figure 15.8. Of course the information in Figure 15.8 is idealized to some extent. PAH mixtures from the environment are often mixtures of pyrogenic and petroleum sources. However, in some cases the discrimination works remarkably well. Boehm et al (1985) noted in analyses of suspended matter from the New York Bight and also sediments from the dredge spoils site and the sewage sludge disposal site that the PAH in sludge were rich in petroleum PAH and the dredge spoils had more of a pyrogenic PAH source signature. Generally, most sediments in coastal areas have a pyrogenic signature, with the exception of some of the sediments in urban harbor areas where the signature indicates some contributions by petroleum sources. Historical records of cores in various bodies of water near or in coastal areas have provided an historical record of PAH inputs to sediments—presumably also indicating some aspect of PAH inputs to ecosystems (e.g. Latimer and Zheng, 2003; Hites et al, 1977, 1980; Wakeham and Farrington, 1980; Lima et al, 2003 and references therein). In all these cases the historical record for PAH shows a pyrogenic signature.

When Farrington et al (1982c, 1983) examined a large data set for PAH in mussels and oysters from the U. S. Coast (See "Mussel Watch" later in this chapter), they noted that mussels and oysters with elevated PAH concentrations near urban harbors had a general pattern of PAH composition more like a petroleum (or petrogenic) source. On the other hand, the literature on surface sediments from several areas in the general vicinity of the mussel or oyster samples reported a mainly pyrogenic compositional signature for the PAH. This led Farrington et al (1982,1983) to suggest that the pyrogenic PAHs in the sediments had entered the coastal environment tightly bound to soot or perhaps even entrapped within it, and were mostly not biologically available. Petroleum hydrocarbons could have been sorbed to marine organic particulate matter or associated with colloids, and more easily exchanged into the dissolved form to be taken up by the bivalves through feeding on particulate matter or via the gills, after being desorbed to dissolved form. This also would mean that the petrogenic PAH would be more readily available for microbial degradation. This hypothesis provided a satisfying answer to the vexing conundrum we, and perhaps others, faced at that time. With one group of colleagues we were recording an historical record of PAH in sediments that fit well with estimated emission time courses for fossil fuel consumption (Hites et al, 1977, 1980), suggesting that the PAH were not being markedly degraded. With other groups of colleagues we were recording the microbial degradation or environmental weathering of some of the same PAH compounds in sediments contaminated with spilled oil (e.g. Teal et al, 1978) and in mesocosm experiments (e.g. Hinga et el,1980; and Farrington et al, 1982b ). At the same time, Readman et al (1984) noted that a similar hypothesis was needed to explain the distributions of PAH they were seeing in the particulate matter and sediment data from an elegant study of the Tamar estuary in the U.K. They noted how PAH might be partitioned into several dissolved and particulate forms.

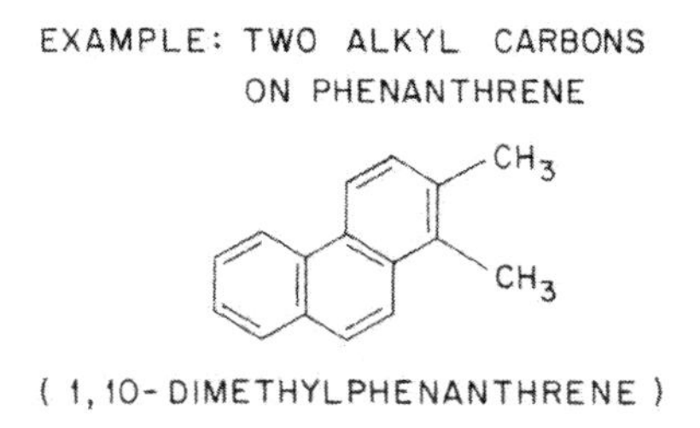


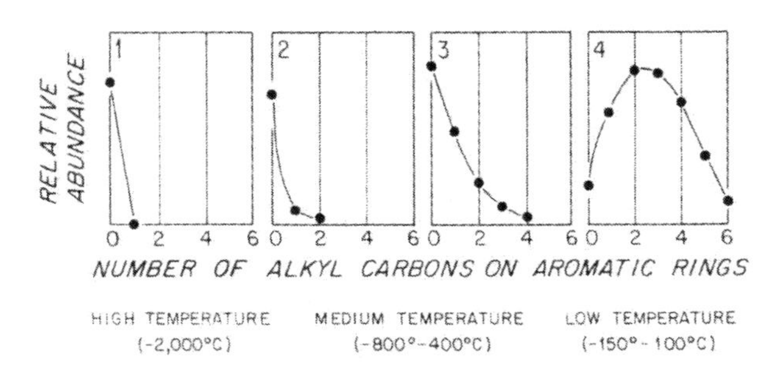


Figure 15.8 Examples of the distribution of alkylated PAH relative to the parent PAH for various temperatures of formation. (adapted from Blumer, 1976.)

Later, McGroddy and Farrington (1995) measured pore water PAH in box cores of sediments from Boston Harbor and showed that the observed distribution

$K'_d$ between pore waters and solid phase in the sediments was not well predicted by Eq. 14 while PCBs in the same samples were in general accord with Eq 14 modified as Eq. 21. Laboratory experiments measuring desorption of PAHs and PCBs from sections of the same box cores of sediment (McGroddy et al, 1996) confirmed these observations. The PAH were apparently not as available for equilibrium partitioning as were the PCBs. The data were again consistent with the hypothesis of tighter binding or association of the PAH with the soot in the coastal environment.

A breakthrough in understanding more completely the mechanism underlying the observations was proposed that involved measurement of the soot carbon in the sediments followed by using a theoretically estimated partitioning coefficient for the soot or black carbon for PAH. (e.g. Gustafsson and Gschwend, 1997; Gustafsson et al, 1997 and references therein). They noted that soot was a condensed, very high molecular weight aromatic and therefore the $\pi$ electrons of the PAH would promote good interaction with the surface of the soot. They calculated theoretical partition coefficients and modified Eq. 9 by adding a term for sorption by soot and the fraction of soot in the sediment:

$$K_d = F_{oc}K_{oc} + F_{sc}K_{sc} \tag{33}$$

Where the subscript *sc* designates soot carbon.

TABLE 15.15
Different PAH partitioning models applied to Boston Harbor sediments. (Adapted from Gustafsson et al 1997a)[a] Reprinted with permission from the American Chemical Society.

| | Log $K_d$ ($L_{porewater}/kg_{dry\ sed.}$) | |
|---|---|---|
| | phenanthrene | fluoranthene |
| **Fort Point Channel Box Core (7-9cm)** | | |
| Actual distributions in situ | 5.0 | 5.4 |
| Distributions predicted from $K_d = f_{oc}\ K_{oc}$ | 2.9 | 3.6 |
| Distributions predicted from Eq. 33 | 4.9 | 5.6 |
| Spectacle Island Box Core (0-2cm) | | |
| Actual distributions in situ | 5.9 | 5.2 |
| Distributions predicted from $K_d = f_{oc}\ K_{oc}$ | 2.8 | 3.4 |
| Distributions predicted from Eq. 33 | 4.5 | 5.2 |

[a] More details about this table are in Gustafsson et al (1997a)

Taking the data of McGroddy and Farrington (1985) for two sections of box cores from Boston Harbor, Gustaffson et al (1997a) measured soot carbon and used their estimated *K*sc values to arrive at the data presented in Table 15.15. Clearly, taking into account the soot sorption brings the observed and calculated log $K_d$ into agreement. Since those first publications there have been numerous studies in the literature further elucidating the role of black carbon or soot (e.g. Bucheli and Gustafsson, 2000).

**Metabolism by Marine Animals**

The literature related to the biochemistry, physiology, and molecular biology of how marine animals respond to and metabolize POPs and PAH is extensive and is beyond the scope of this review (Waid, 1986; Pritchard and Bend, 1991; Varanasi and Stein, 1991; Stegeman and Lech, 1991; NRC, 2001, 2003; WHO, 1993). Vertebrates have a great capacity for metabolizing aromatic hydrocarbons and also other POPs such as PCBs. The basic mechanism is by way of a cytochrome P450 mediated reaction with oxygen, forming an epoxide that is further metabolized to a diol. The diols can conjugate with other biochemicals to become more polar (water soluble) to facilitate elimination and excretion processes. The epoxides and diols may also be retained and passed up the food web (e.g., see review by McElroy et al, 1989). There are various forms of the cytochrome P450 and I refer the reader to the papers cited above as an initial guide to the extensive literature.

One important aspect of the metabolism by marine vertebrates is that there are relatively few elevated aromatic hydrocarbon concentrations reported for marine vertebrates. This is the result of elevated PAH exposure inducing the cytochrome P450 systems to metabolize the aromatic hydrocarbons. This is also true for PCBs, but here the molecular structure is an important factor, i.e. if there are vicinal carbons available for enzyme catalyzed epoxide formation. For this reason, there are interesting and important changes in PCB mixture composition when progressing from initial input to water to deposition to sediments, and simultaneous or subsequent transfer through the food web to higher predators (e.g. NRC, 2001; Farrington et al, 1986; Stegeman et al, 1986; Pruell et al, 2000 and references therein).

The experiment of Pruell et al (2000) is an excellent example of laboratory or microcosm experimentation that sorts out processes acting in the environment to influence the biogeochemical cycle of POPs or PAH. In this case, the focus is on potential changes in composition of POP mixtures as a result of transfer from contaminated sediment source to animal in multiple processes—sediment to polychaete, sediment to lobster, sediment to polychaete to lobster. Pruell et al exposed an infaunal polychaete, *Nereis virens,* to contaminated sediment collected from the Passaic River, New Jersey, USA for 70 days. The polychaetes were then fed to the American lobster, *Homarus americanus,* for up to 112 days. Lobsters were also exposed only to the contaminated sediments. The polychaetes accumulated PCBs , several chlorinated pesticides, and also polcychlorinated - *p* -dibenzodioxins, and polychlorinated dibenzofurans. We will focus here on the PCBs.

Both the polychaetes and the lobsters took up PCBs from the contaminated sediment. Also, lobsters accumulated PCBs from the polychaetes when they were fed only the contaminated polychaetes. Comparing BASFs for the CB congener mixtures in polychaetes, lobster muscle and lobster hepatopancreas (Figure 15.9) supports the following conclusions of Pruell and coworkers. The polychaete (*N. virens)* is preferentially metabolizing certain congeners of PCBs. Lobsters are also preferentially metabolizing certain congeners. The compositions of the PCB congeners in the lobster hepatopancreas and the muscle are markedly different from the polychaete composition and also different from each other. The reason for the latter is unclear. However, the differences between the congener mixture of the PCBs in the polychaete and the sediments is due to metabolism of primarily those PCBs that have adjacent vicinal carbons available for enzymatic addition of oxygen to the biphenyl ring—consistent with what we have discussed above about PCB metabolism. The lobsters seem to have an even more potent capacity to metabolize congeners selectively.

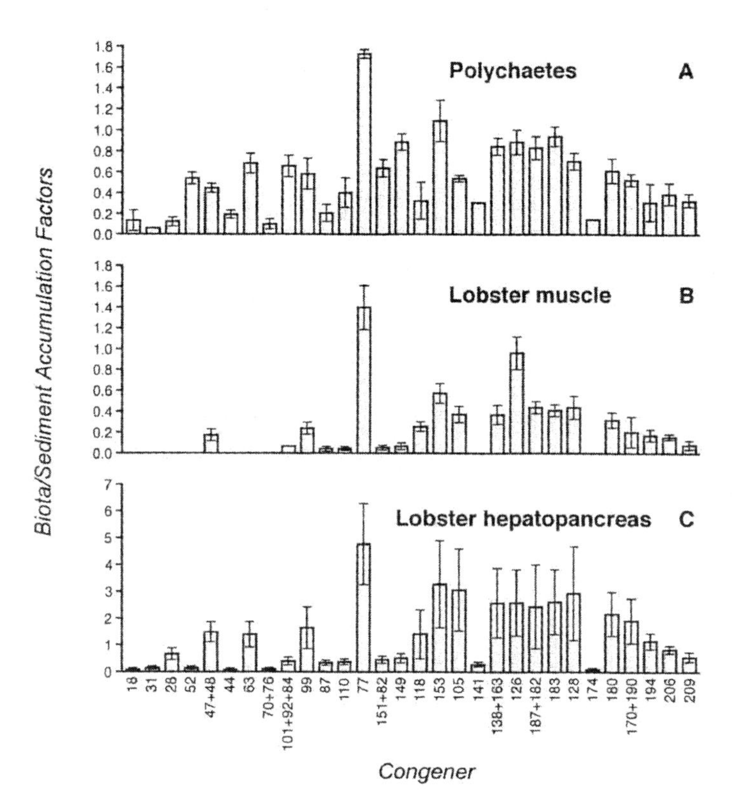


Figure 15.9 BASF for PCB congeners in polychaetes, lobster muscle, and lobster hepatopancreas after exposure to PCB contaminated sediments. (Fig 3. from Pruell et al, 2000). Reprinted with permission from Elsevier.

Congeners chlorinated in the *meta* and *para* positions (the carbon at the ands of the ring opposite where the two rings are joined and the carbon atom just adjacent to the end carbon atom—see Figure 15.2b and c—appear to be resistant to metabolism by both the polychaete and the lobster. This means that the more toxic coplanar congeners, numbers 77, 126, and 169, would be among the compounds enriched. Congener 169 is present in the sediments in only trace amounts. However, inspection of Figure 15.9 indicates that congener 77 and congener 126 are among those that are enriched. This has profound importance when considering the toxicity potential of the contaminants in the lobster, a food for human consumption.

**Summary**

Thus we have completed the course in assessing the important processes of the biogeochemistry of POPs and PAH in coastal ecosystems; from input, through the biogeochemical cycle, process by process, until we arrive at contaminated organisms one step away from consumption by people. The following sections of this chapter provide examples of what we know about some of the distributions of the

POPs and PAH in the environment and some examples of processes in the biogeochemical cycles active in coastal ecosystems from various parts of the world.

## 5. Examples from the coastal oceans worldwide

### *5.1 Chlorinated pesticides and PCBs.*

**Distribution in the atmosphere over the ocean and in surface sea water.** First it is useful to have a general picture of the situation in the global oceans and atmosphere prior to addressing coastal oceans in specifics. Harvey and Steinhauer (1974) interpreted their pioneering measurements of PCBs in surface sea water samples that spanned 36°S to 54°N as demonstrating co-distillation of the PCBs with water followed by latitudinal temperature dependent condensation. Because of controversies surrounding obtaining samples not contaminated by the sampling process and the very low concentrations in seawater and air, it was many years before further systematic measurements were made in the open ocean.

However, that has changed for surface ocean waters. Iawata et al (1993a) reported an extensive set of measurements of several of the POPs in oceanic air and surface sea water obtained on cruises between April of 1989 and August of 1990. Air sampling was conducted from the upper deck of a research vessel and seawater samples were obtained from either the underway ship's sea water system or by stainless steel bucket samples. In all cases, precautions were taken to avoid contamination from the ship. The POPs analyzed were α-HCH, γ-HCH, ΣHCHs, *trans* Chlordane, *cis* Chlordane, *trans* Nonachlor, ΣChlordanes, *p,p'*DDE, *o,p'* DDT, *p,p"*DDT, ΣDDT, and ΣPCBs (sum of 40 congeners, several as pairs not resolved by the GC methodology employed). Only a few key points from this large study are presented here.

ΣPCBs in an earlier study by Tanabe et al (1983), for open ocean sea water showed higher concentrations of Σ PCBs in the mid-latitudes compared to equatorial region areas. However, the 1989–1990 results showed a more uniform distribution of concentrations of ΣPCB for the mid-latitude and equatorial regions (Figure 15.10). Iwata et al (1993) commented that this change in concentrations "is likely to indicate the reduction of highly contaminated areas in developed countries and the expansion of PCB usage to the tropics during the recent decade." They also noted that comparing their data with the limited sets of data for atmospheric PCB concentrations over the ocean for the 1980s indicated little decrease in concentration.

The distribution of ΣDDT reported by Iwata et al (1993a) clearly indicates elevated concentrations for both air samples and surface water samples for the South and Southeastern Asia areas compared to all other regions sampled (Figure 15.11). They note that this is consistent with reported use of DDT for both anti-malarial purposes and to control crop pests in these regions. The concentrations of ΣHCHs in air (not shown) were similar to the distributions for ΣDDT near India in the Bay of Bengal and the Arabian Sea, and lower but still significant concentrations were observed in the northern North Pacific. Much higher concentrations of HCHs were found in the surface sea water in the higher latitude waters of the Chukchi Sea, Bering Sea, Gulf of Alaska and northern North Pacific (not shown). Iwata et al (1993a) noted that α-HCH to γ-HCH ratios were higher in the northern atmospheric samples and the values tend to decrease southward. They hypothesize that this may be mainly controlled in high latitude regions by the differential exchange between the atmosphere and the ocean for these two compounds as they are transported from the source regions. More details about the importance of the HCH isomer compositions and possible explanations for the observed distributions are presented in Iwata et al (1993b).

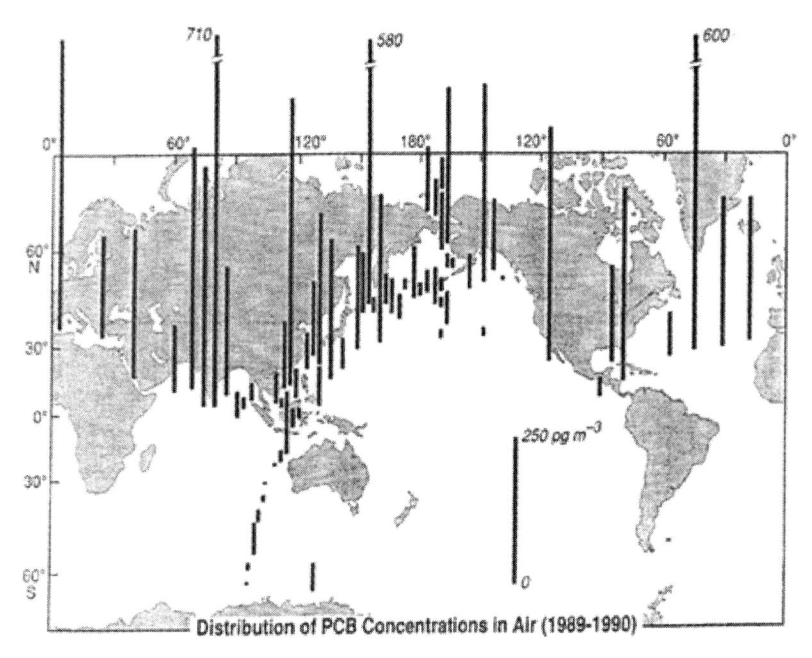


Figure 15.10a Distribution of PCBs in air, 1989-1990. (from Iwata et al, 1993b)

Iwata et al (1993a) calculated the apparent flux between the ocean and the atmosphere for various regions of the oceans they sampled using their data and equations of the same form as Eqs. 6 and 7. They made simplifying assumptions about conditions in the areas where they sampled with respect to thin films at the air-sea interface and the amount of the POP in surface seawater that was in the particulate phase. Keeping in mind that a negative flux is from the atmosphere to the ocean, they found that in most oceans and seas surveyed HCHs were transferred from the atmosphere to the water, particularly the Arabian Sea and Bay of Bengal (-2500 to -2600 x $10^{-9}$ g $m^{-2}$ $day^{-1}$), East China Sea ( -260 to -270 x $10^{-9}$ g $m^{-2}$ $day^{-1}$), and South China Sea (-350 to -360 x $10^{-9}$ g $m^{-2}$ $day^{-1}$) where the areas are proximal to tropical countries in which HCHs are still used. There were also considerable fluxes from atmosphere to ocean in northern high latitude waters such as the Chukchi Sea ( -56 to -190 x $10^{-9}$ g $m^{-2}$ $day^{-1}$), Bering Sea (-73 to -220 x $10^{-9}$ g $m^{-2}$ day 1), Gulf of Alaska (-43 to -190 x $10^{-9}$ g $m^{-2}$ day 1) and northern North Pacific (-150 to -240 x $10^{-9}$ g $m^{-2}$ $day^{-1}$). Throughout the data interpretation of their study, Iwata et al (1993) noted the importance of Henry's Law Constants in explaining major features of the global distribution of POPs in the atmosphere over the ocean and surface ocean waters, both for coastal and mid ocean regions away from the coast.

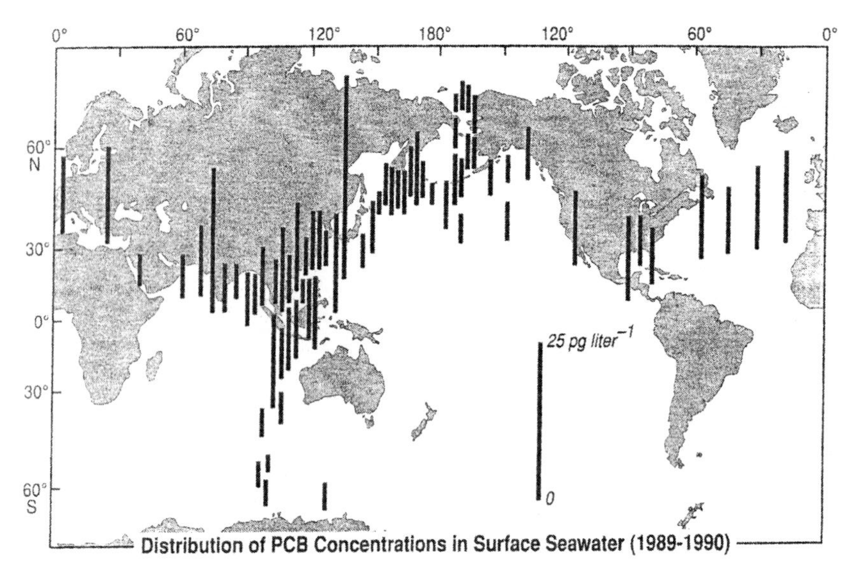


Figure 15.10b Distribution of PCBs in surface seawater, 1989-1990 (from Iwata et al, 1993b).

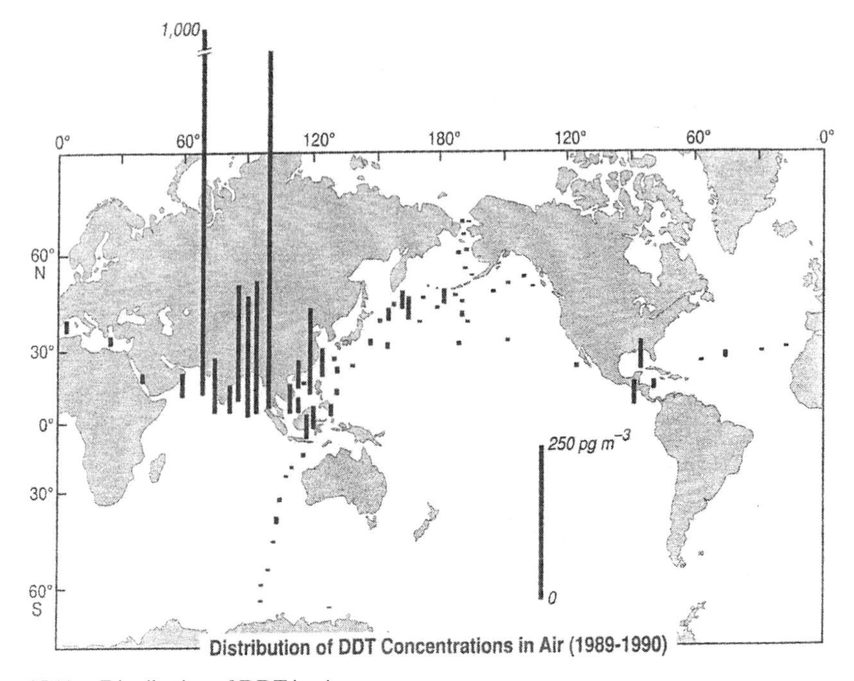


Figure 15.11a Distribution of DDT in air.

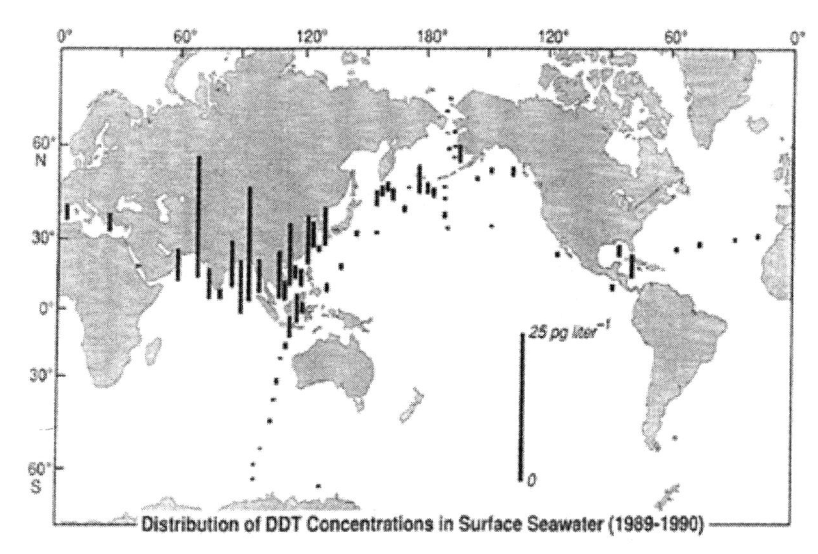


Figure 15.11b Distribution of DDT in surface seawater, 1989–1990 (from Iwata et al, 1993b). Reprinted with permission of the American Chemical Society

Recently, Sobek and Gustafsson (2004) in a study of PCBs in surface sea water along a transect 62°to 89°N, also reviewed the measurements in the atmosphere and atmosphere since the Harvey and Streinhauer (1974) results. The references cited in the Sobek and Gustaffson (2004) paper unequivocally demonstrate that latitudinal fractionation between the ocean and the atmosphere occurs and influences the global distribution of POPs. As noted by Sobek and Gustafsson (2004), some POPs with higher volatility than PCBs (e.g. hexachlorocyclohexanes (HCHs)) have higher concentrations in surface sea water toward the north in the northern hemisphere. Iwata et al (1993) and Lakaschus et al (2002) reported α – HCH concentrations in the Arctic that are the highest reported in the world oceans away from known sources of effluent inputs. Sobek and Gustafsson (2004) noted in their study that in contrast to α –HCH, PCB concentrations are decreasing with increasing latitude over the Arctic Ocean.

**PCBs in the North Sea and Baltic Sea**

Samples of North Sea water were obtained with a pump while underway on transects in the North Sea by Schulz-Bull et al (1991). Thirty-one CB congeners where analyzed for nine transects in February 1988 and six transects in August of 1988. Concentrations of individual chlorobiphenyl (CB) congeners in offshore waters were between < 0.5 and 2 $x10^{-12}$g $dm^{-3}$ (picograms per liter), and up to 40 x $10^{-12}$g $dm^{-3}$ in near coastal waters. Most particulate matter concentrations were below the limit of detection of 0.05 x $10^{-12}$g $dm^{-3}$ for a 100 $dm^{3}$ sample.

Schulz-Bull et al (1995) determined CB concentrations in solution and suspended matter in the Baltic Sea in November 1988, 1989 and March/April of 1991 using the same sampling track strategies as for the North Sea cruises and collecting samples of volumes that ranged between 65 and 1143 $dm^{3}$. Individual CB concentrations in solution were below or near $10^{-12}$g $dm^{-3}$ (picograms per liter) and ΣCBs varied between 2 and 237 x $10^{-12}$g $dm^{-3}$. The highest concentrations seemed to originate from local sources of input in the Belt Sea and the Gulf of Finland. There were regional differences in the composition of the CB mixtures and these compositional differences seemed to correlate with T-S diagram distinctions of water bodies. The authors suggested that phytoplankton blooms seemed to be responsible for the lowest ΣCB concentrations found in solution, of 2 to 14 x $10^{-12}$g $dm^{-3}$. In fact, the concentrations of CBs in suspended or particulate matter were very high in the Belt Sea-Kattegat area in spring 1991—up to 589 x $10^{-12}$g $dm^{-3}$ for individual CBs and up to 2859 x $10^{-12}$g $dm^{-3}$ for ΣCBs. The composition of the CBs in solution was different than the composition associated with the particulate phase. As might be expected from Eq. 11, there was a positive relationship between log $K'_d$ and log $K_{ow}$ for all transects except the spring transect. The authors describe the system as being in quasi-equilibrium during the autumn, but deviations take place from the quasi-equilibrium situation in the spring when the phytoplankton biomass was the highest. They note that phytoplankton activity may effectively remove CBs from the water column to the sediments and note that CBs stored in the sediments of the Baltic Sea are several orders of magnitude greater than CB amounts in the overlying water of the Baltic Sea.

Axelman et al (2000) and Bruhn and McLachlan (2002) measured CB concentrations in samples of surface waters and obtained concentration ranges for dissolved and particulate phases similar to those of Schulz-Bull et al (1995). Bruhn and McLachlan conducted eight cruises over a two-year period of time to measure water concentrations of PCBs at 5–10 m depth at several stations. They noted a seasonal variability to the data. They could not discern from their data whether or not the variability was due primarily to non-equilibrium partitioning during plankton blooms or variation in the quality of the organic matter and its influence on sorption.

Konat and Kowalewska (2001) measured PCBs in surface sediments from the Polish coastal areas of the Baltic Sea, near Gdansk and near the Pomeranian area and Szczecin Lagoon. The authors provide a summary of concentrations of PCBs in Baltic Sea surface sediments from their work and that of others and compare them to a few other areas worldwide. An important and intriguing aspect of their study was the collections of sediments for PCB analyses before and after the great flood of Poland in July/August of 1997. PCBs in surface sediments increased markedly after the flood, presumably because of washout and delivery of more highly contaminated sediments and soils from watershed areas and riverine sediments, and perhaps wash off from landfills and industrial sources due to the heavy rains and floods. The concentrations in the surface sediments decreased in the following years. Konat and Kowalewska suggest that this may be due to a combination of burial and dilution by freshly deposited sediment with lower PCB concentrations and perhaps some degradation of some of the CB congeners.

Sobek et al (2004) investigated whether or not the partitioning of PCBs to particulate matter dominated by planktonic primary productivity in the open Baltic Sea was kinetically limited or could be treated as an equilibrium process. Over a 25 month study they sampled at a well characterized time series station, BY31 (58°35'N, 18° 14'E), 40 km off the Swedish coast in 459 m water depth. Samples were obtained from the upper mixed layer at this location. During all 21 sampling occasions there was a linear relationship between log $K_{oc}$ and log $K_{ow}$, as would be expected from Eq. 11, despite a large variation in parameters such as particulate organic matter (POC) composition, primary production and phytoplankton species composition. The data were adjusted appropriately for the effects of salinity and temperature as per Eqs. 1 and 12 and related discussion above, and then described by an equation of the form of Eq 11.

$$\log K_{oc} = 0.88 \pm 0.07 \log K_{ow} + 0.90 \pm 0.47$$

The slope of the regression varied from 0.56 to 1.25. Sobek et al believed that these variations reflect the variability in the structural composition of the PIC pool. They concluded that the partitioning of PCBs was equilibrated and not kinetically controlled by rapid phytoplankton growth. This conclusion is supported by a study of the uptake of the polycyclic aromatic hydrocarbon phenanthrene by diatoms in culture (Fan and Reinfelder, 2003).

**Particulate Flux of PCBs and Accumulation of PCBs in Continental Shelf Sediments.** The question of the removal of PCBs from surface waters by particulate flux was addressed by Gustaffson et al (1997) in a study for the Northwestern Atlantic Ocean. They measured concentrations of CBs associated with particulate matter obtained from the center of the mixed layer using a submersible pump and obtaining samples of 165 liters for the station nearest to shore to 657 liters at the Sargasso Sea—Bermuda Atlantic Time Series (BATS) site. Concentrations of the measured CBs (Table 15.16) were in the FM (femtomolar) concentration range, equivalent to approximately 0.1 to 10 x $10^{-12}$g $dm^{-3}$ in the reporting format used by Schulz-Bull et al (1991, 1995) and in the range of CBs reported for particulate matter in the Baltic Sea (Schulz-Bull et al, 1995) and North Atlantic Ocean (Schulz et al, 1988). Gustaffson et al estimated fluxes by using $^{238}U - ^{234}Th$ disequilibria to calculate particle mass fluxes out of the mixed layer and then multiplying these by the concentration of particulate CBs measured. The more chlorinated CB congeners exhibited a more rapid decrease in concentration away from the coast than did the less chlorinated CBs (Table 15.16). Surface ocean export fluxes decreased with increasing distance offshore for all congeners measured. The flux of the more particle reactive congeners (higher chlorinated congeners) decreased more rapidly offshore than did the flux for the less chlorinated congeners.

Jönsson et al (2003) estimated the amount of PCBs accumulated in continental shelf sediments on a global basis. They compiled and interpreted data from 4214 distinct continental shelf sediment samples. Global inventories and burial fluxes are estimated for each of eight congeners common to the samples. The continental shelf areas are divided into 18 different areas. Calculations were performed for each area and samples were further classified according to distance from large population centers—<1km (Local), 1–10km (Regional), >10km (Remote). The Remote North Atlantic sub-area contains approximately half of the global shelf inventory for most of the CB congeners studied. Using CB52 as an example, Jönsson et al note that the remote North Atlantic, American Mediterranean, the Baltic Sea and the Mediterranean Sea have higher concentrations e.g. 0.1–6.5 x $10^{-9}$g $gdw^{-1}$ for CB 52. Lower concentrations, by about an order of magnitude, are typical for remote shelves in the Southern Hemisphere, the Arctic Ocean and the Arctic Mediterranean. The remote Asiatic Mediterranean and North Pacific shelves have intermediate concentrations of about 0.1 x $10^{-9}$g $gdw^{-1}$. Examples of the global shelf inventories and global burial flux for selected CBs are presented in Table 15.17.

TABLE 15.16

Particulate concentrations of individual chlorobiphenyl congeners in the surface ocean off the Northeast United States (from Gustafsson et al, 1997). Reprinted with permission of the American Chemical Society.

| | Concentrations in $10^{-15}$moles $l^{-1}$ or femtomolar (fM) | | | | | | |
|---|---|---|---|---|---|---|---|
| Station # → | 93-4 | 93-2 | 94-4 | 94-2 | 94-1 | 94-7 | 93-BATS |
| Station Location → | 43°40'N 70°14'W | 43°37'N 70°07'W | 43°40'N 70°14'W | 43°37'N 70°07'W | 42°38'N 69°36'W | 43°09'N 69°51'W | 31°50'N 64°10'W |
| Distance to urban center (km)→ | 2 | 15 | 2 | 15 | 110 | 125 | 1600 |
| Congener [a] ↓ | | | | | | | |
| 26 (3) | n/a[b] | n/a | 2.8 | 3.0 | 5.7 | 4.4 | 0.89 |
| 22 (3) | n/a | n/a | 1.5 | 3.7 | 4.9 | 1.7 | 0.40 |
| 52 (4) | 8.8 | 11 | 7.5 | 16 | 20 | 6.4 | 1.8 |
| 44 (4) | 6.2 | 6.7 | 4.8 | 9.6 | 12 | 4.2 | 1.1 |
| 66 (4) | 94 | 27 | 14 | 30 | 47 | 12 | 4.3 |
| 95 (5) | 37 | 4.7 | 9.5 | 20 | 29 | 8.0 | 1.7 |
| 99 (5) | 58 | 7.4 | 6.0 | 9.1 | 15 | 4.2 | 0.92 |
| 110 (5) | 81 | 8.0 | 16 | 24 | 39 | 13 | 3.5 |
| 146 (6) | 77 | 8.7 | 2.4 | 1.7 | 2.0 | 1.2 | 0.19 |
| 128 (6) | 24 | 1.9 | 2.6 | 1.1 | 1.5 | 0.97 | 0.17 |
| 187 (7) | 29 | 2.9 | 3.3 | 2.6 | 2.2 | 1.4 | 0.22 |
| 174 (7) | 12 | 1.4 | 1.6 | 1.3 | 1.4 | 0.78 | 0.10 |
| 177 (7) | 10 | 0.67 | 1.1 | 0.67 | 0.56 | 0.25 | 0.042 |
| 199 (8) | 6.4 | 0.78 | 2.1 | 1.5 | 0.61 | 0.39 | 0.049 |
| 194 (8) | 3.8 | 0.58 | 1.5 | 0.58 | 0.46 | 0.23 | 0.027 |

[a] Congener – numbers in parenthesis refer to number of chlorine substituents. See Schulz et al (1989) to link congener numbers with molecular structures.

[b] n.a. – not analyzed in these samples.

TABLE 15.17

Global inventories and burial fluxes in the continental shelf sediments. (From Jönsson et al, 2003). Reprinted with permission of the American Chemical Society.

| CB Congener | 28 | 52 | 153 | 180 |
|---|---|---|---|---|
| # of Chlorines | 3 | 4 | 6 | 7 |
| Global inventory (ton)[a] | 460 (290-910) | 700 (420-1500) | 1200 (720-2100) | 760 (380-2200) |
| Global burial flux (ton $yr^{-1}$)[b] | 9.6 (3.9-20) | 14 (5.1-31) | 24 (8.8-47) | 15 (4.7-46) |

[a] ( ) is lower and upper 95% confidence limits.

[b] ( ) is lower and upper 95% confidence limits propagated from the inventory, sedimentation rate, and sediment mixed layer depth.

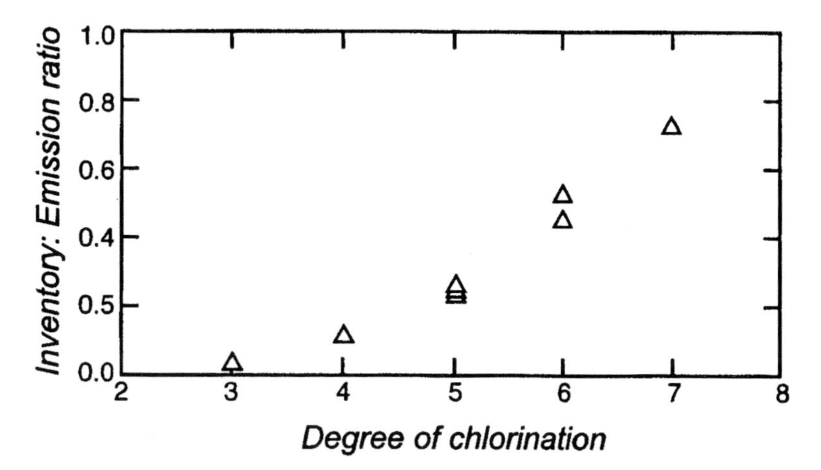


Figure 15.12 The fraction of recently estimated maximum cumulative historical emissions of PCBs (Breivik et al 2002) that are accounted in the inventories of the global shelf sediments as a function of the chlorination degree of the individual congeners. (From Jönnson et al, 2003). Reprinted with permission of the American Chemical Society.

The global shelf inventories represent about 1–6% of the global industrial production of these CBs according to Jönsson et al. They also summarized what is known of cumulative historical emissions of individual CB congeners. Even though these estimates of emissions have a large range of uncertainty, it is useful to compare them with the inventory of CBs in the continental shelf sediment. The comparison (Figure 15.12) reveals an interesting pattern for CBs with different levels of chlorination. The lower chlorinated CBs are favored for removal to some sink other than the continental shelf sediments. Jönsson et al suggest that perhaps the lower chlorinated biphenyls are reacting with hydroxyl radical, which is acting in the troposphere on the global recycling pool of these compounds. The continental shelf inventory accounts for about 80% of the total estimated emissions of the heptachlorinated CBs (Figure 15.12). An interesting finding of this study is that local hotspots such as US EPA Superfund sites do not contribute substantially to the global inventory of continental shelf sediments. The latter probably originate from historical diffuse discharges to rivers and the atmosphere (Jönsson et al, 2003). This does not mean that the superfund sites and similar locations worldwide are not locally important as sources for PCBs in contiguous ecosystems.

**Persistent Organochlorine Compounds in Black Sea Sediments.**
Samples of the upper 0–2 cm of sediments were collected in a cooperative effort at 45 coastal stations around the Black Sea in the Ukraine, the Russian federation, Turkey, and Romania between 1993 and 1995 (Fillman, et al, 2002) in collaboration with the Global Environmental Facility (GEF) Black Sea Environmental programme (BSEP). Samples were analyzed for Hexachlorobenzene (HCB), HCHs, DDT compounds, Heptachlor, Aldrin, Dieldrin, Endrin, and PCBs at the IAEA Marine Environmental Laboratory in Monaco. The following is the order of highest to lowest concentrations measured in terms of compound classes: DDTs > HCHs ≥ PCBs > HCB > cyclodienes. The authors note that Tanabe et al (1997) reported a similar ranking for POPs in organisms.

Concentrations of HCHs up to 40 x $10^{-9}$ g $gwd^{-1}$ along the Romanian coast were among the highest recorded in the world, similar to those reported for the North Coast of Vietnam and near some cities in India according to a compilation from the literature presented by these authors. The ratio of the α and γ isomers for the HCH along the Romanian coast was low indicating use of lindane (γ-HCH) as the primary source. The ratios elsewhere were consistent with an atmospheric delivery source and/or runoff of an atmospherically transported source. DDT and DDE concentrations were not unusually high compared to elsewhere in the world, but the ratio of DDE/DDT was low indicating fresh inputs from recent use of DDT. Concentrations of PCBs were low relative to other areas of the world.

**DDT in Zooplankton, Pelagic Fish, and Demersal Fish of the Arabian Sea.** There are several examples in the literature of measurements made of POPs in zooplankton and fish from various areas of the coastal ocean during different periods of time. I have chosen one from among these several credible studies to illustrate the environmental distribution of POPs and factors that may influence these distributions.

Sahilaja and Nair (1997) report on seasonal differences in organochlorine pesticide concentrations from zooplankton and fish from the Arabian Sea (Figure 15.13). Their study was designed in part to discern if there were differences associated with samples obtained pre monsoon and during the monsoon conditions. Total DDT (DDT+DDE+DDD) and Aldrin concentrations during the monsoon conditions were higher by factors of 4 and 5 respectively, for the zooplankton when compared with pre monsoon conditions. Total DDT concentrations for the fish were between 10 and 30 times higher during the monsoon than pre-monsoon and Aldrin concentrations were 3 to 40 times higher.

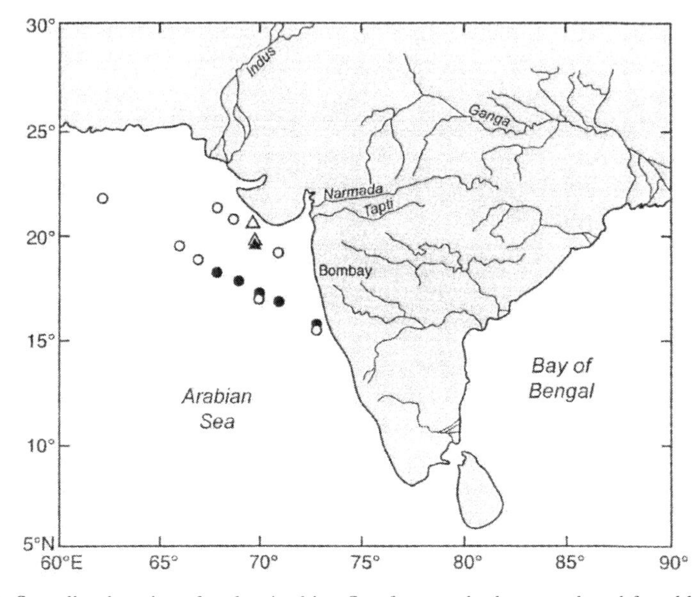


Figure 15.13 Sampling locations for the Arabian Sea for zooplankton analyzed for chlorinated pesticides. (open symbols March 1991; closed symbols, September 1991. (Shailaja and Nair, 1997). Reprinted with permission of Elsevier.

Significant amounts of primary DDT (i.e. not metabolized or degraded to DDE or DDD) were present during the monsoon season (Figure 15.14). The authors point out that the monsoon season coincides with primary agricultural activity on land and this may coincide with use of DDT. One other conclusion of note is that using lipid normalized concentrations, the pelagic varieties of fish had higher concentrations of pesticides than demersal fish during the pre-monsoon season, while the opposite was observed for the monsoon season.

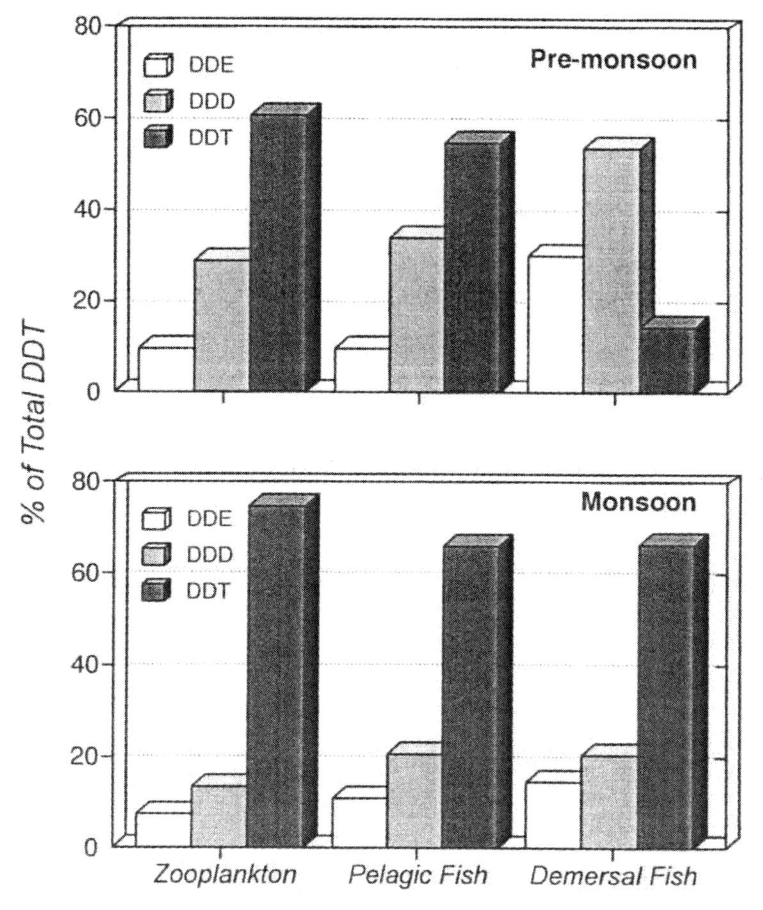


Figure 15.14 Relative proportions in percent of DDT, DDE, and DDD in zooplankton and fish from the Arabian Sea: Pre-monsoon and Monsoon conditions. (Shailaja and Nair, 1997). Reprinted with permission of Elsevier.

**Palos Verdes Shelf, Southern California—DDT Contaminated Sewage Effluent Discharge.**

The Palos Verdes shelf is a narrow continental shelf off the coast of California, offshore from Los Angeles. The Whites Point outfall has been discharging waste waters onto the outer continental shelf portion of the Palos Verdes Shelf for over sixty years. The world's largest producer of DDT was connected to the Los Angeles County sewer system that discharged through the White's Point outfall in the

1950s and 1960s, but ceased to be connected in 1960. The result was significant DDT and related products being discharged to the Palos Verdes shelf and accumulation of these compounds in Palos Verdes shelf sediments and in other components of the coastal and marine ecosystems (Lee and Wiberg, 2002).

Early documentation of the elevated concentrations of DDT and its metabolites in the Southern California coastal ecosystems came from a survey in 1965 of DDT residues in fish in California coastal and marine waters (Risebrough, 1969). DDT residues on the order of 1,000 ppm (μg/g) of lipids were present in Brown Pelicans, *Pelecanus occidentalis* (Risebrough, 1972; Anderson et al, 1975) and California Sea Lions, *Zalophus californianus* (Delong et al, 1973). Macgregor (1976) and Young et al (1976) and references therein, reported on the DDT residues in sediments of the area. These and other studies in the late 1960s and early 1970s of waste discharges to southern Califronia coastal waters led to the formation of the Southern California Coastal Water Research Project (SCCWRP). The SCCWRP annual reports and many peer reviewed scientific literature publications from these studies provided a long term record and increased understanding of the inputs, fates and effects of POPs and other pollutants in this area.

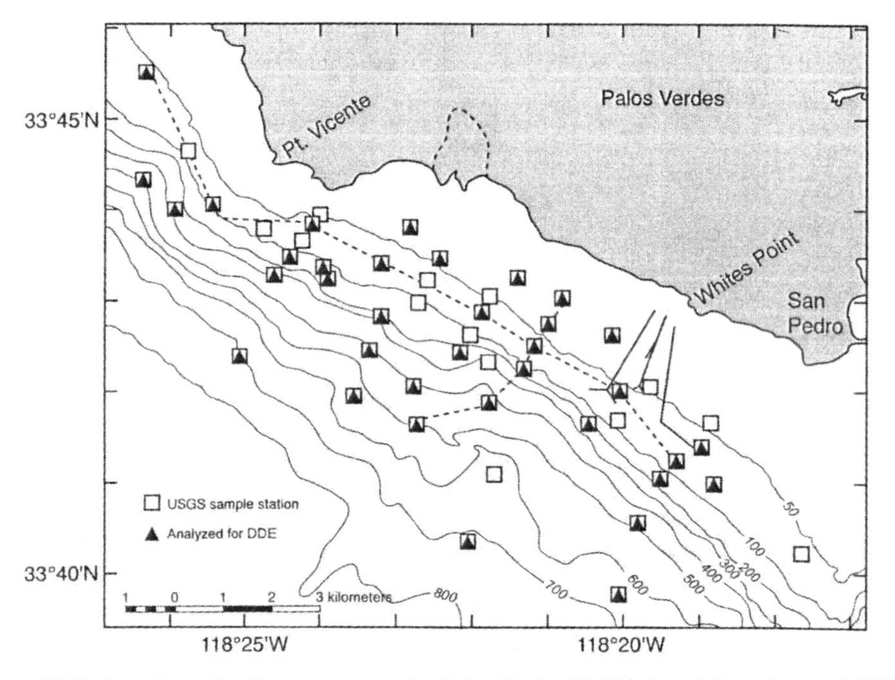


Figure 15.15 Locations of sediment core samples Palos Verdes Shelf (adapted from Lee at al, 2002). Reprinted with permission of Elsevier.

More recently, in the early 1990s, a large research effort in the Palos Verdes shelf and slope (margin) area was launched in support of a lawsuit filed by the U. S. Department of Justice against the DDT manufacturer (Lee and Wiberg, 2002). Numerous papers resulting from this research have been published in the peer reviewed scientific literature—too numerous to recount each one here. For example, 14 papers are published in *Continental Shelf Research* in one volume (Lee and

Wiberg, 2002). Arguably, this is one of the most complete studies to date of the input, fate and effects on POPs—in this case DDT (and degradation products/metabolites) and PCBs in coastal ocean ecosystems. The lawsuit and its related studies have generated some controversy (Benner,1998) as might be expected when the stakes are high in terms of costs of remediation and the science has some uncertainty.

The area in question is off Los Angeles County (Figure 15.15). Lee et al (2002) have provided detailed documentation of the distribution of DDT and its metabolites in the sediments. Their mapping showed that the deposit covers over $40km^2$, is up to 60cm thick and 9 million $m^3$ in volume (Figure 15.16) with "virtually all the deposit being contaminated with DDT and PCBs". Between 61 and 72 metric tons of p,p'DDE are found in the sediments as best can be determined (Lee and Wiberg, 2002).

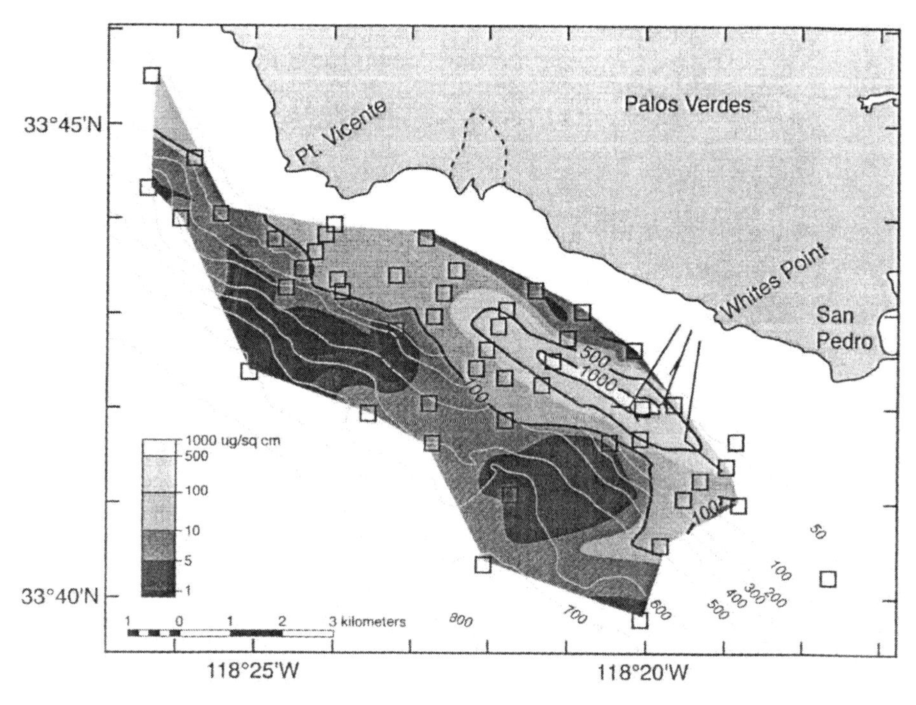


Figure 15.16 Contours of concentrations of DDT in sediments based on measurement of DDT in cores from station locations in Figure 15.14. (adapted from Lee et al, 2002). Reprinted with permission from Elsevier.

Eganhouse et el (2000) studied the diagenetic fate of organic contaminants on the Palos Verdes Shelf. The data from a core they analyzed is reproduced in Figure 15.16 which provides a down core profile of various DDT degradation products and also a comparison of the sum of these and DDE. Clearly DDE dominates here and elsewhere in the sediments of the area. The elevated concentrations between 25 and 42 cm correspond to the known times of maximum input from the White Points outfall discharges. Cores collected in triplicate in 1989 and 1993 from the area, and reproduced in Eganhouse et al (2000), show good agreement for depth profiles of DDE concentrations. The interpretations and conclusions of Eganhouse et al (2002) can be summarized as follows:

i. DDE is the dominant form of the DDT residues in the sediment and was probably formed from hydrolytic dehydrochlorination of the parent DDT discharged in the effluent during early stages of diagenesis.
ii. Approximately 9–23% of the DDE inventory in the sediments has been converted to DDMU since DDT discharge began in about 1953. This accounts for less than half the decline in *p,p'* DDE inventory observed at the site from 1981–1995.
iii. Remobilization by process such as sediment resuspension, contaminant desorption, and current advection accounts for most of the observed decreases in inventory of DDE.
iv. The *in situ* rates of conversion of DDE to DDMU are 100 to 1000 times slower than the laboratory microcosm rates reported by Quensen et al (1998).
v. Congener specific chlorobiphenyl concentration depth profiles show no evidence of reductive dechlorination in the same sediments for which this process has been noted to be active for DDE.

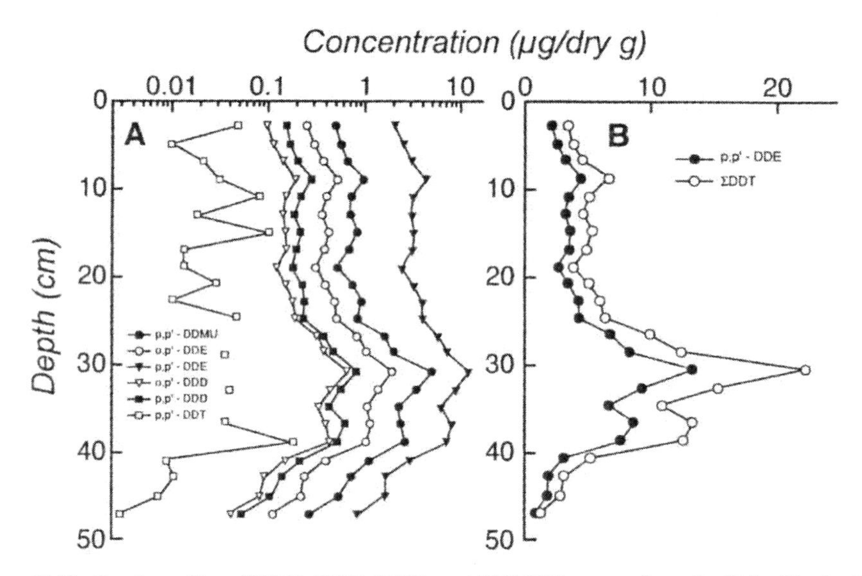


Figure 15.17 Depth profiles of DDT, DDE, DDD, and DDMU in a core from the Palos Verdes Shelf. (Fig. 3 in Eganhouse et al, 2000). Reprinted with permission of Elsevier.

In a related paper, Eganhouse and Pontolillo (2000) report that sedimentation and mass accumulation rates of sediments in the area of concern increased from 1955 to 1971 and then decreased from 1971 to 1981, in accord with historical information about emission of suspended solids from the outfall system. The sediment accumulation in the area was dominated by waste discharges. However, from the 1980s into the 1990s, mass accumulation rates again increased in the sediments in contrast to the decreasing discharges of suspended solids from the outfall system. Eganhouse and Pontolillo (2002) note several lines of evidence that support their contention that this latest increased mass accumulation rate is due to mobilization of debris from a land slide along the shoreline. The result is to bury the sediments heavily contaminated with DDT residues (and other organic contaminants) to greater sub-bottom depths, reducing contaminant availability for mobilization to the overlying water column.

The processes outlined in Figure 15.4 have been quantitatively or semi-quantitatively assessed and combined together in coupled models that follow DDT and DDE from the sediment into the processes of sediment resuspension and erosion, desorption, pellitization of sediment particles by animal excretion processes, biodiffusive flux from sediments to the overlying waters, advective transport, and transfer through the food web to higher trophic levels such as sea lions and birds (Wiberg and Harris, 2002; Wiberg et al, 2002; Sherwood et al, 2002; Drake et al, 2002; Connoly and Glaser, 2002a,b). The toxico kinetics and estimated effects on sea lions and birds were modeled by Connoly and Glaser (2000) and Glaser and Connoly (2000), respectively. Lee and Wiberg (2002) summarize the studies: "The papers show that the margin has partly recovered from extremely high levels of contamination present in the early 1970s but that relatively high levels of contamination remain and are impacting a number of animal species. Models show that natural recovery will proceed slowly."

### *5.2 Organic Chemical Contaminants as Tracers.*

There are organic chemical contaminants that can be used as molecular markers or tracers of human inputs of wastes to the oceans (Eganhouse, 1997). There are several examples of such chemicals (Figure 15.18). The linear alkyl benzenes (LABs) have been used extensively for such purposes and will serve to demonstrate by two case studies how they are used as tracers. LABs having 10 to 14 carbon normal alkyl chains are sulfonated in industrial production of alkylbenzene sulfonates which are widely used as anionic surfactants. Sulfonation is not complete and LABs are part of the LAS-type detergents that are disposed of by way of waste waters after use. LABs enter the environment sorbed to particles since they are highly hydrophobic. Thus, they become useful in tracking sewage effluent particulates and associated hydrophobic organic chemical contaminants, such as POPs and PAH.

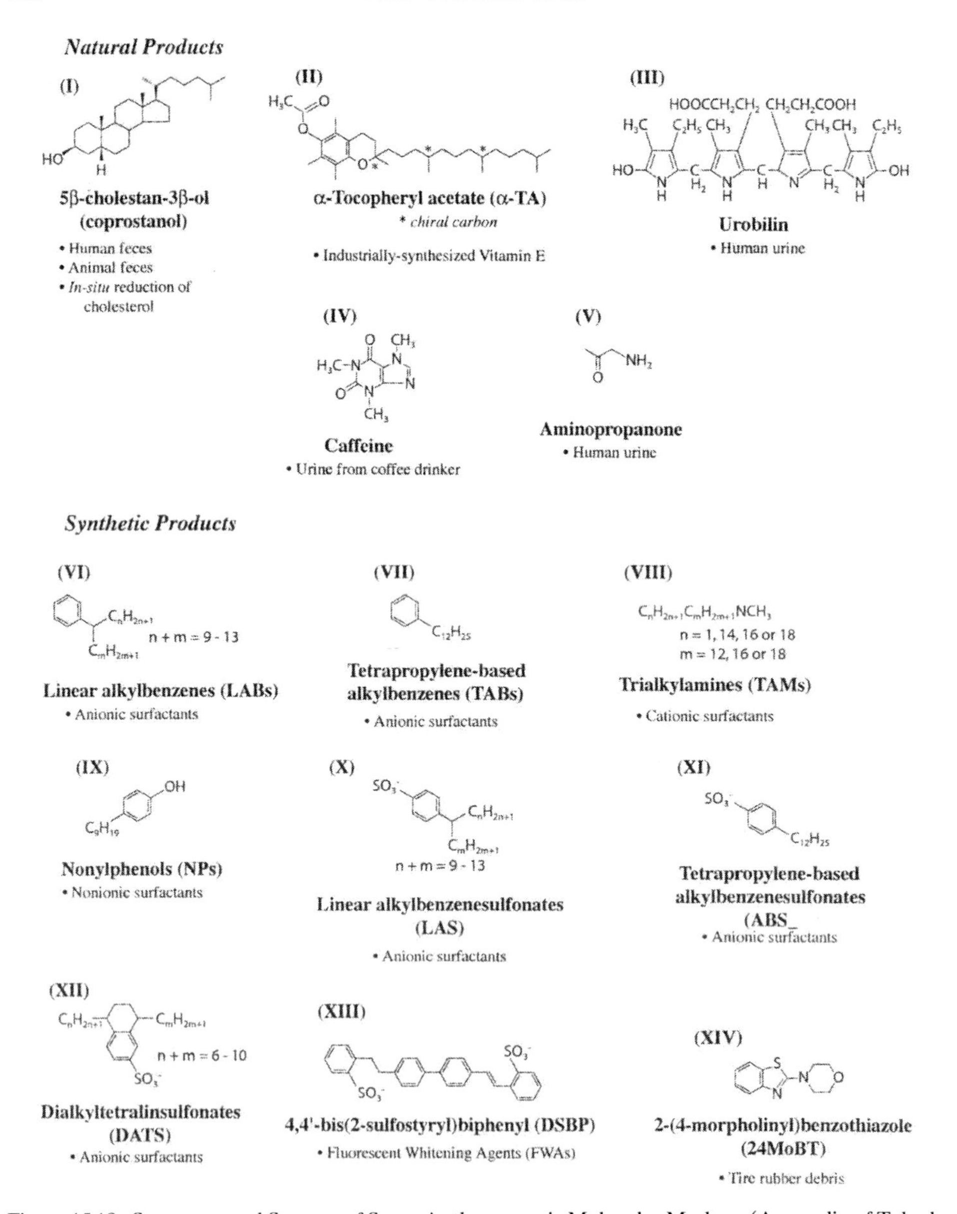


Figure 15.18 Structures and Sources of Some Anthropogenic Molecular Markers (Appendix of Takada et al, 1997).

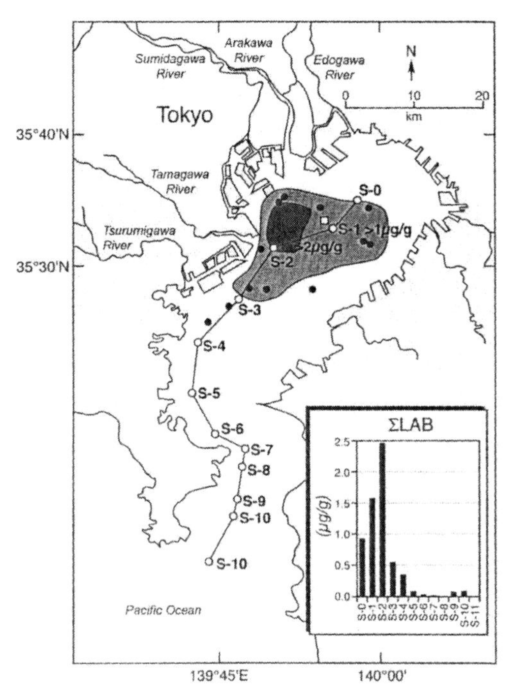


Figure 15.19 Horizontal distributions of LABs in surface sediments of Tokyo Bay. Contour lines based on data for 24 stations in the northern bay. Bar graph showing ΣLAB concentration along the transect from S-0 to S-11. (From Takada et al, 1997).

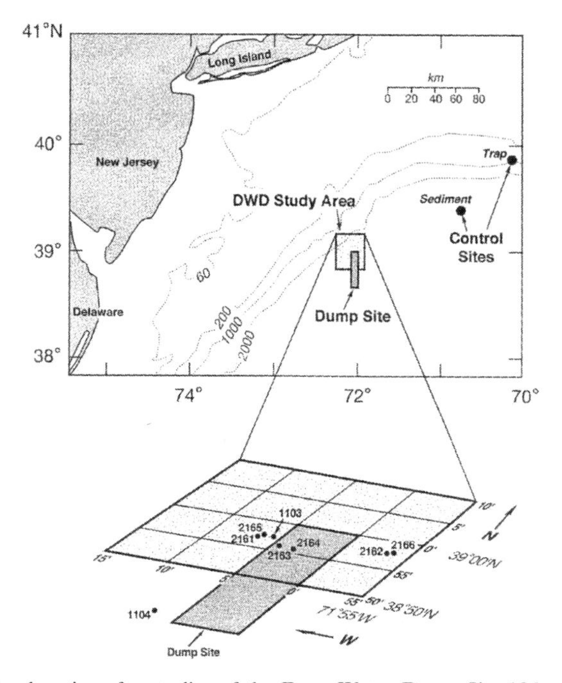


Figure 15.20 Sampling locations for studies of the Deep Water Dump Site 106 surface sediments and near bottom sediment trap particles. (Takada et al, 1994).

**Tokyo Bay and LABs.**

Takada et al (1997) have summarized some of the extensive data of Takada and co-workers regarding chemical contaminants in Tokyo Bay in Figure 15.19. The horizontal gradients of LAB concentrations in the sediments of Tokyo Bay are substantial. The highest concentrations are in areas 5–10 km off the mouths of major rivers flowing into the bay. Concentrations of LABs in bay surface sediments decrease along a transect to the south in the bay as indicated by the bar graphed data in Figure 15.19. The deposition of the LABs in a core taken near the S-1 station in Figure 15.19 indicates a strong correlation between the time course of the deposition of the LABs and the history of production of LABs in Japan (data not shown). This example using LABs is among many that are being published in the literature where molecular tracers such as LABs provide valuable data about sources of organic contaminant inputs and biogeochemical cycles (Eganhouse, 1997 among others).

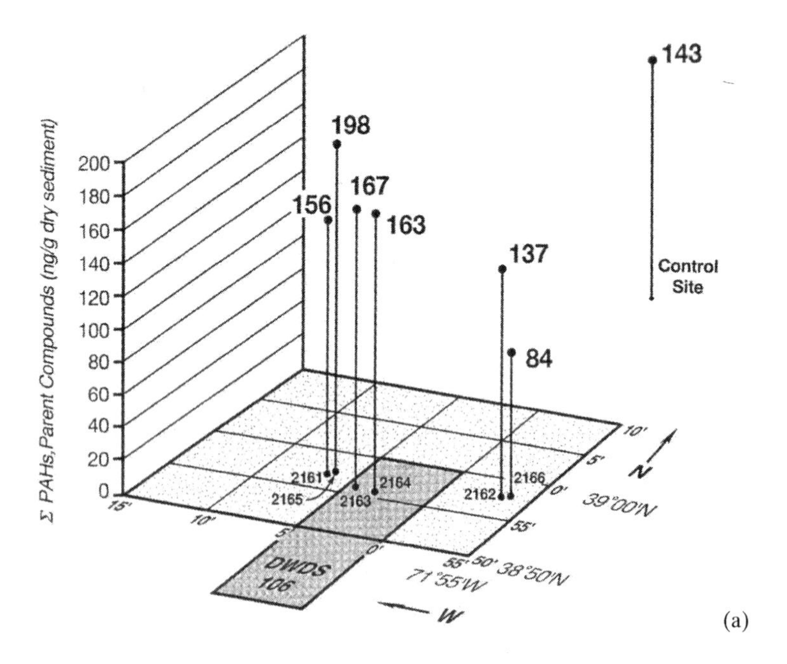


(a)

Figure 15.21 More detailed location chart from DWD study area in (Figure 15.22). The four digit numbers correspond to the *DSRV Alvin* Dive Numbers during which the cores were sampled. Stippled gray area is the sea surface water dumping area projected to the bottom. Tabular data from Takada et al (1994) were used to prepare these figures.

a) Concentrations of ΣPAHs (Parent Compounds only, alkylated PAH not included, DWDS

b) Coprostanol concentrations.

c) ΣLABs concentration.

(*figure continued next page*)

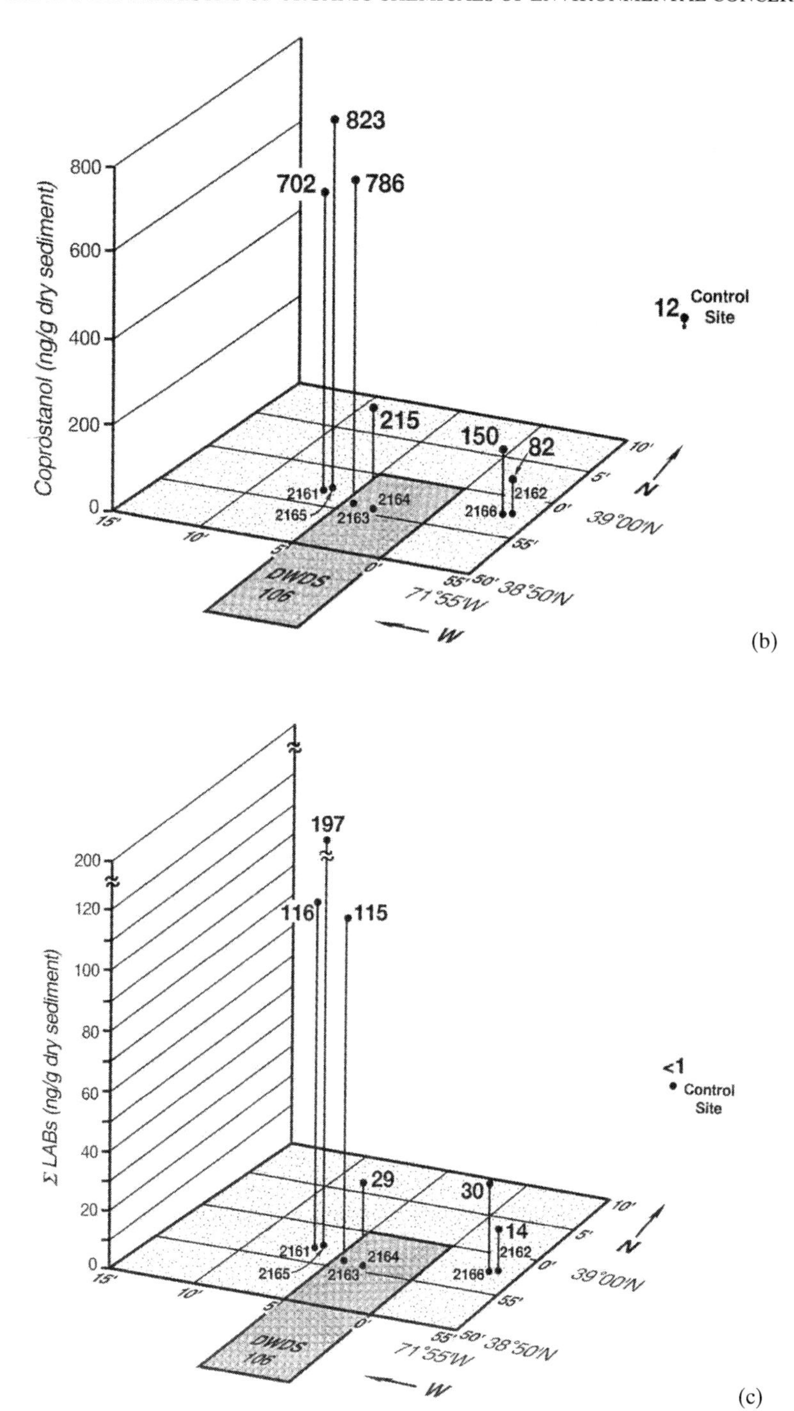


## Deep Water Dump Site 106 and LABs

The next example comes from the disposal of sewage sludge that occurred for several years at an area off the east coast of the United States. The disposal occurred at the surface of the ocean where the water depth was 2400–2900 meters on the continental slope. Figure 15.20. Despite the depth of the water column, the processes are probably similar to those occurring in shallower regions in the continental margin. DWDS (Deep Water Dump Site) 106 is 106 miles or 185 km off the coast of New Jersey. During 1988 to 1992 about 8 million tons of wet sludge was gradually discharged at the surface in batches by barges. The hope and plan was that the dispersion and microbial degradation in the ocean waters would prevent any build up of sludge contaminants at the sediment water interface. The study of Takada et al (1994, 1997) was part of a larger effort (see references therein) to discern if the plan and hope were realized. Samples of surface sediments, taken with careful navigation, were obtained in the DWDS 106 area by coring from the submersible *DSRV Alvin.* The numbers of the dives are indicated by the four digit number in Figure 15.21 a-c. Samples were analyzed for PAH, LABs and coprostanol. Later there were analyses for PCBs (Takada et al, 1997), but these will not be discussed here. Concentrations of PAH in surface sediments did not exhibit elevated concentrations above background PAH already present in surface sediments (Figure 15.21a). However, coprostanol did exhibit higher concentrations in the DWDS surface sediments in comparison to nearby control samples and other deep sea sediments reported in the literature (Figure 15.21b). LABs had definitely higher concentrations in comparison to non-detected at the control locations. A core taken within the DWDS from Dive No.2163 (location in Figure 15.20) showed elevated concentrations of coprostanol and markedly elevated concentrations of LABs in surface sediments, but the contaminant concentrations decreased to background or non-detectable concentrations by about 10 cm.

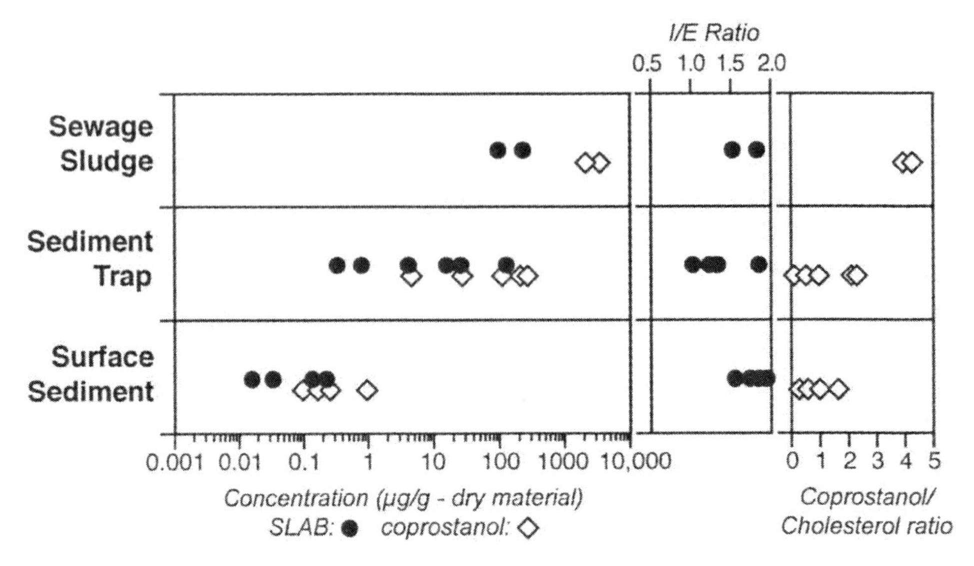


Figure 15.22 Concentrations of ΣLAB and Coprostanol, I/E Ratio of LABs (see text for an explanation) and Coprostanol/Cholesterol ratios for samples from DWD study of Takada et al, 1994. (From figure 7 of Takada et al, 1997).

These data demonstrated that organic contaminants in the sludge were reaching the bottom and that the sludge was being mixed into the surface sediments by bioturbation (Takada et al, 1994). Two sewage sludge samples and several sediment trap samples, from traps set in the area to collect settling particulate matter at 106m off the bottom, were analyzed. Comparison of coprostanol and LABs data from these samples with the surface sediment data, in terms of μg per g dry weight of material, shows that there is a definite decrease in concentration for these compounds as we progress from the dumped sludge concentrations to surface sediments (Figure 15.22). PAH data not shown indicates that the sludge contained a mix of pyrogenic and petrogenic PAH. As the particles settled through the water column and were deposited to the sediments, there appeared to be a loss of the petrogenic PAH, perhaps by solubilization and also some microbial degradation.

The coprostanol to cholesterol ratio decreases from the sludge to the particulate matter and is about the same in the particulate matter and the surface sediment (Figure 15. 22). Cholesterol is an abundant sterol in humans and in marine organisms. Most likely the ratio is decreased by increases in contributions of cholesterol from marine organisms to the particulate matter and surface sediment samples (Takada et al, 1994).

The I/E ratio is a measure of the relative proportions of the isomers of LABs that have the phenyl ring attached via an internal carbon atom in mid chain, rather than an external carbon atom. Observations and experimentation suggest that under some conditions in the marine environment, LABs with the phenyl ring attached to the external carbons are more rapidly degraded than their isomer counterparts with phenyl substitution on an internal carbon (Takada et al, 1997). The I/E data for the DWDS samples (Figure 15.22) show no changes or increases of I/E with depth, indicating a lack of biodegradation under these conditions. The concentrations and ratios of the compounds also provided insight about the biodegradation processes and sorption-desorption processes (or absence of appreciable biodegradation or desorption) for the LABS. Relatively rapid bioturbation to 10 cm was demonstrated. Other processes are discussed in Takada et al (1994 and 1997).

Overall, this study showed the power of molecular tracers to discern the possible inputs of organic contaminants from waste inputs in a deeper water situation.

### *5.3 Monitoring of the inshore edge of the coastal ocean using bivalve sentinel organisms. The 'Mussel Watch'.*

Professor Edward Goldberg of Scripps Institution of Oceanography articulated the need to have a longer term, wide ranging monitoring system for chemical contaminants in the coastal ocean to keep track of spatial and temporal trends of chemicals of environmental concern, such as the chlorinated pesticides, PCBs, and petroleum hydrocarbons (Goldberg, 1975, 1976). Shortly thereafter, the U. S. Mussel Watch Program was launched with bivalves collected from about 100 stations in various areas of the U. S. continental coastal area (Goldberg et al, 1978). Bivalves are good choices for sentinel organisms for a variety of reasons (Goldberg et al, 1978; NRC, 1978; among several others). The reasons include that the organisms occur in large populations that contain sustained sampling, they are sessile, they bioconcentrate many of the chemicals of concern and have limited capacity to metabolize them, they can withstand pollution stress better than many other species, they are cosmopolitan over large areas thereby reducing the need to compare concentrations between species, and they can provide an indication of contaminant bioavailability.

An overview of data for organic contaminants from the initial three-year sampling of mussels and oysters for the USA coast during 1976, 1977, and 1978 has been presented in Farrington et al (1983). The geographic distributions and ranges of concentrations of organic contaminants, with elevated concentrations near urban areas for PAH and for PCBs, has by now become a familiar picture for such analyses as indicated in Figure 15.23 for PCBs from the first three years of the NOAA Status and Trends Mussel and Oyster Watch Program, NOAA (1989). The prototype U. S. Mussel watch Program has been transformed into an operational monitoring program within NOAA. Similar programs exist in other countries, for example France (Claisse, 1989). A synopsis and interpretation of some of the USA data has been published by O'Connor (2002). The NOAA program data set can be accessed on the World Wide Web address in Table 15.18.

TABLE 15.18
Organic chemicals in oysters and mussels from the U. S. Coast 1986-1995: A comparison of trends in concentrations.

| | Number of Stations | | |
|---|---|---|---|
| Chemical | Increase | Decrease | No Trend |
| ΣChlordanes | 1 | 81 | 104 |
| ΣDDT | 1 | 38 | 147 |
| ΣDieldrin | 1 | 32 | 153 |
| ΣPCB | 1 | 37 | 148 |
| ΣPAH | 3 | 3 | 180 |

*Source:* National Oceanic and Atmospheric Administration (NOAA) 1998 (on-line) "Chemical Contaminants in Oysters and Mussels" by Tom O'Connor. NOAA's State of the Coast Report, Silver Spring, MD, NOAA, USA.
http://state-of-coast.noaa.gov/bulletins/html/ccom_05/ccom.html.

Elevated concentrations of DDT compounds (sum of DDT, DDE, and DDD, but mostly composed of DDE) are present in the Palos Verdes area of California (Figure 15.24). The use of mussels in local programs, California State programs and the USA national program to track the time course of the DDT contamination along the California coast is one of the success stories of the use of bivalve organisms as sentinels for coastal areas. Clearly the Palos Verdes area continues to be one of concern from the perspective of elevated DDE concentrations, caused mainly by the legacy of previous DDT discharges to the coastal ocean now residing in surface sediments, as noted above in this chapter.

PCB concentrations are high near urban areas and among the highest concentrations are in samples taken in Buzzards Bay (Figure 15.23). These samples are near, but not within, an area where there are elevated concentrations of PCBs in sediments at an EPA superfund site (e.g. see Brownawell and Farrington, 1985, 1986; Farrington et al, 1986). Concentrations of both PCBs and DDT are much lower in many of the sites sampled in remote coastal locations away from urban centers or runoff. Whatever DDE is present in the system is now residual from earlier manufacture and use. The same holds for most of the PCBs, since new production has been banned for years, but there are still large quantities in use or in landfill areas as well as those deposited to coastal sediments (NRC, 2001).

I have examined the data compilations on the NOAA web site and note that for the time interval 1986 to 1995 there were decreases in concentrations of several chlorinated pesticides and PCBs reported for many sampling locations (Table 15.18). However, for the PAH concentrations there was not a significant number of stations with a decrease in concentrations (Farrington, 1999). Why? Is this a message that increased use of fossil fuel compounds in the USA is continuing to contaminate our coastal ecosystems at high levels despite efforts to reduce PAH inputs? The recent historical record reported for estuarine and aquatic sediments by Lima et al (2002) and references therein argues for that scenario.

International efforts for a global "Mussel Watch" type program have been pursued. Phase I included South and Central America and the Caribbean areas (Sericano, et al, 1995). The ranges of concentrations of PCBs and PAH in samples collected from the IMW Phase I were similar to those found in the USA program. There were some sites with elevated chlorinated pesticides in the tropical coastal areas. Unfortunately, there was only one year of sampling and the program has not yet been repeated. Therefore trends are not being ascertained.

More recently, the effort shifted to Asia and South Asia. Monirith et al (2003) report data for chlorinated pesticides and PCBs in bivalves sampled from the coasts in the following Asian areas: Cambodia, China, Hong Kong, India, Indonesia, Japan, Korea, Malaysia, Philippines, Far East Russia, Singapore, and Vietnam. Data were reported for 1994, 1997, 1998, 1999, and 2001 to complete the collections in all the areas. A look at two figures reproduced from this set of data indicate the richness of the data (Figures. 15.25 and 15.26). DDT concentrations are quite elevated in several areas even if the concentrations are expressed on a lipid weight basis. Mussels from China, Hong Kong, Vietnam, and Far East Russia have higher proportions of *p,p'*DDT compared to *p,p'*DDE. This reflects the continued use of DDT as a pesticide in these areas. The PCB concentrations (Figure 15.27) were lower in Cambodia, Indonesia, Malaysia, Vietnam, India, and China than in Russia and Japan. The authors note that PCB use in Japan has been banned since 1972, but coastal waters continue to be contaminated by releases from old equipment still in use or in storage, and I suspect also by the legacy of previous releases present in coastal sediments. HCH (data not shown) concentrations are very high in coastal areas of India compared to other countries, reflecting the continuing use of HCH in India. The authors present a nice review of data for POPs in bivalves and sediments collected by other workers in local areas of the Asian-Pacific region. From the Asian-Pacific Mussel Watch and these other data, and comparison with bivalve concentrations in POPs data for other geographic locations, the authors conclude that concentrations of PCBs seem to be lower in Asian developing countries compared to developing countries, while chlorinated pesticides are much higher in the region than the global average.

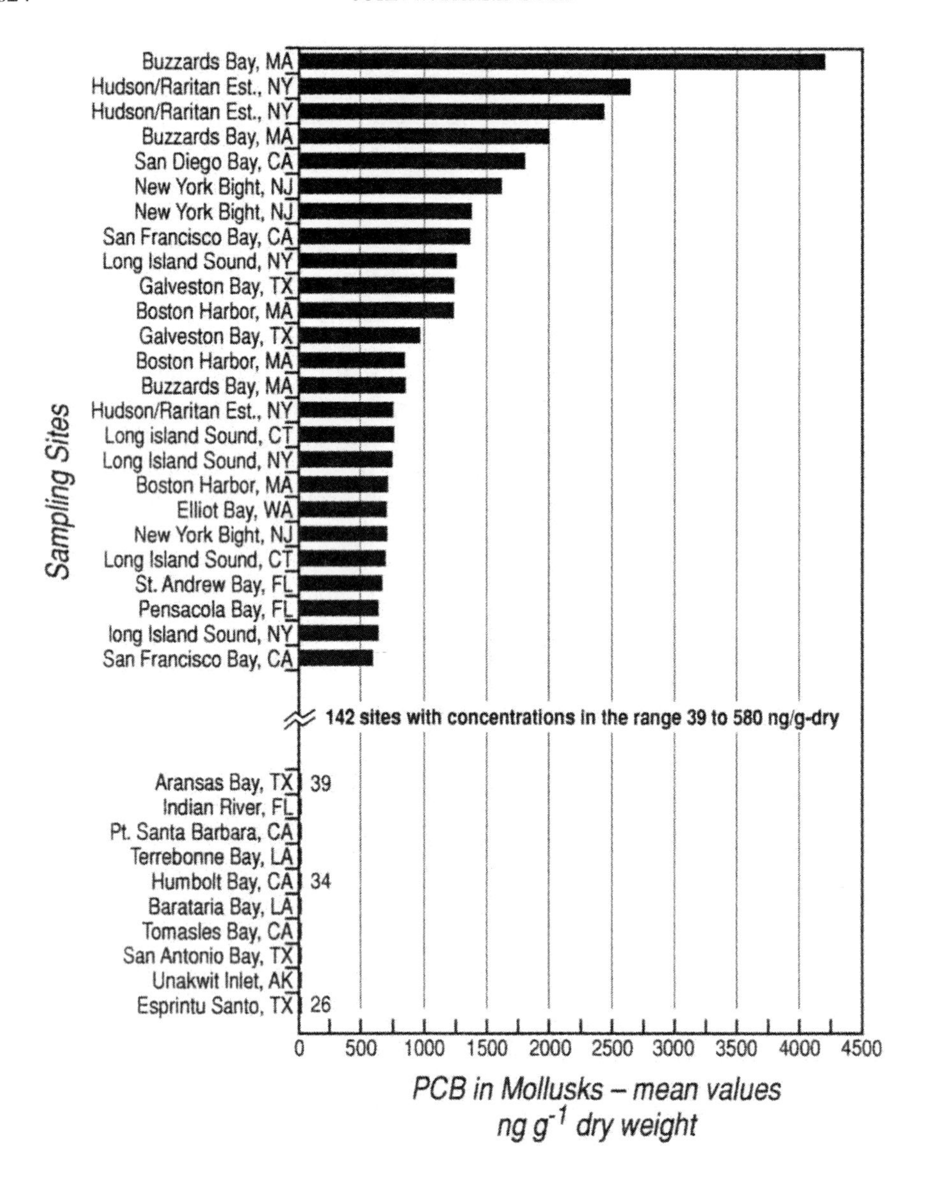


Figure 15.23 Example U. S. NOAA Status and Trends Program Mussel (and Oyster) Data for the U. S. Coast: PCB Concentrations. (See O'Connor, 2002, and the World Wide Web Site in Table for the source of the data.)

## 6. Discussion.

The details of the biogeochemical cycle of POPs and PAH in the coastal ocean are understood. Knowledge of the fundamental underlying physical and chemical processes are also reasonably well developed, as is the understanding of biodegradation and factors affecting bioavailability, mainly based on laboratory, microcosm, mesocosm, and near shore studies. Most of the field data come from near shore studies, because concentrations are highest in these areas and because the potential for human effects, through contamination of the food supply, is greatest. The ability to model coastal ecosystems to predict fields of concentrations and then to compare them to real world sets of data with enough resolving power in time and space, remains a challenge. We have a reasonable grasp of the equilibrium and steady state processes, but the biogeochemical cycles of POPs and PAH are not steady state as the field data demonstrate.

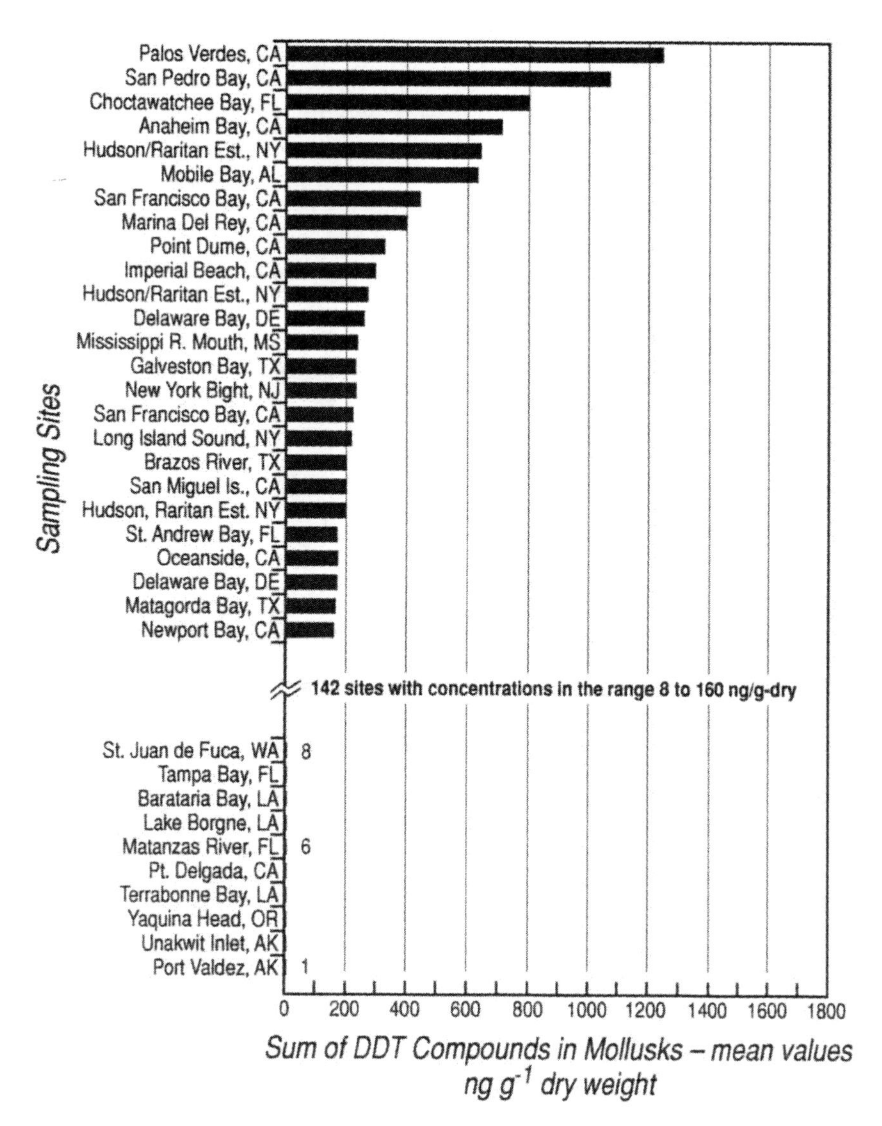


Figure 15.24 Example of U. S. NOAA Status and Trends Program Mussel (and Oyster) Date for the U. S. Coast: DDT Concentrations. (See legend of Figure 15.23 for source of data).

A disturbing trend apparent from limited data is that the use of POPs continues to contaminate coastal ecosystems of developing countries. In addition, the developed countries are dealing with the legacy of the previous use, which resulted in POP accumulations in coastal sediments that are continuing to contaminate the present coastal environment. The continued use of fossil fuels as a source of energy worldwide almost certainly means there will be continued PAH contamination of the coastal areas of developed countries and increased contamination of coastal areas of developing countries (e.g. Zakaria al, 2002).

The emerging concerns with the polybrominated diphenyl ethers mentioned at the beginning of the chapter (e.g. see Hites, 2004) remind us of the need to be ever vigilant with respect to emerging environmental concerns with chemicals that humans produce and use. All POPs were very effective for their intended application (except the chlorinated dibenzodioxins and the chlorinated dibenzofurans which were by-products), and a few continue to be used because they have no adequate, less costly, less harmful replacement. The lesson from the POPs guides us with respect to production and use of chemicals in the future. As stated in the beginning of this chapter, not all synthetic chemicals are of environmental concern. However, we now have enough knowledge to know which compounds have properties that promote persistence and transport within coastal ecosystems. Coupled with ecotoxicological information, we are in a good position to manage the situation to minimize the release of organic chemicals that could become the pollutants of the future.

The good news is that early warnings about compounds such as DDT and PCBs were heeded, albeit slowly in some areas of the developed world. Continued warnings are being heeded worldwide, albeit with continued use of these chemicals in several developing and emerging economy countries as is evidenced by the POPs accords noted at the beginning of the chapter. Eliminating or drastically curtailing use of POPs is vital, because once coastal ecosystems and the environment in general become contaminated, natural clean-up via biogeochemical cycles can take many centuries. This is illustrated by the slow decline in PCB and DDT concentrations in cod liver oil from the southern Baltic Figure 15.28 (Kannan et al, 1992). for the present, we have a legacy of past releases accumulated in high concentrations in sediments in several areas (NRC, 2001 and references therein). Where and when to proceed with remediation to cleanup these areas, or to allow "monitored natural attenuation" to take place, is a vexing problem in need of both further research and some thoughtful science-policy-management interactions (NRC, 2001).

I have not reviewed the literature about POPs and PAHs in Arctic coastal ecosystems since those are specialized ecosystems that I was not sufficiently prepared to review. However, it is clear from reading the literature, that a global process of import is underway with respect to several POPs and possibly PAH. Several POPs are traveling by a process that Wania and MacKay (1996) have termed the "grasshopping" from seawater to the atmosphere, back to seawater and through several of these cycles until arriving at high latitudes. Data review from the atmosphere and surface ocean reviewed earlier in this chapter are among those data documenting this process. In the Northern Hemisphere, the result of this process is contamination of the Arctic marine ecosystems with substantive concentrations of POPs in marine mammals and birds. Marine mammals are a source of food for several native peoples inhabiting the Artic region. There is serious, well founded concern, that the situation is a threat to the health of these native peoples (UNEP, 2004).

Thus, a further understanding of the details of biogeochemical processes of POPs and PAHs is of profound and immediate international importance.

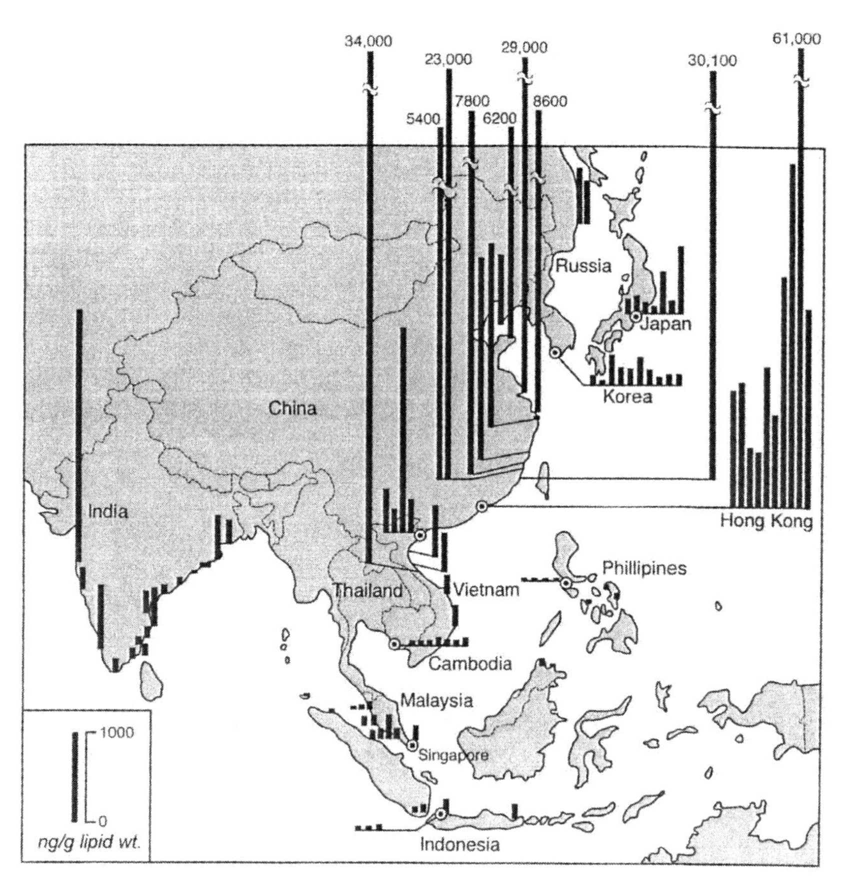


Figure 15.25 Distributions of concentrations of DDT in mussels colleted from coastal waters of some Asian countries during the Asian-Pacific Mussel Watch (Fig. 2. Monirith et al, 2003) Reprinted with permission of Elsevier.

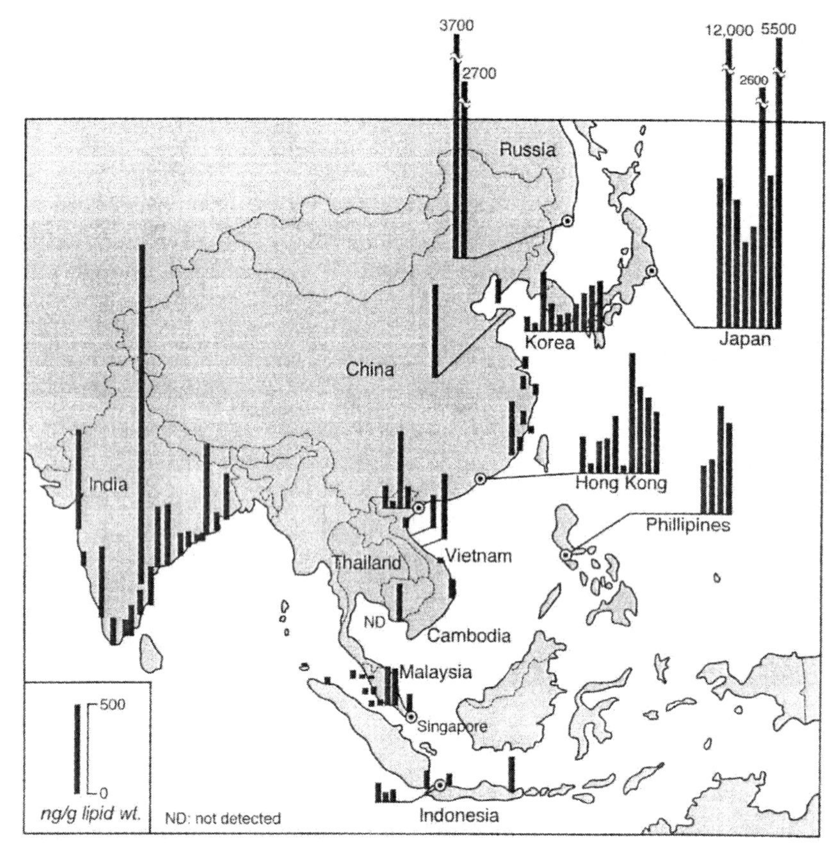


Figure 15.26 Distributions of concentrations of PCBs in mussels colleted from coastal waters of some Asian countries during the Asian-Pacific Mussel Watch (Fig. 6. Monirith et al, 2003) Reprinted with permission of Elsevier.

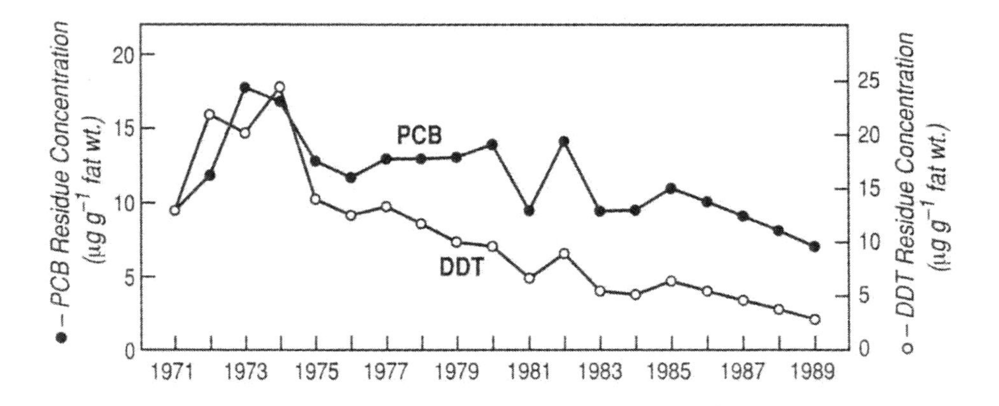


Figure 15.27 Time trends of DDT and PCB concentrations in cod liver oil from the southern Baltic, 1971-1989. (from Kannan et al, 1992). Reprinted with permission of Elsevier.

## Acknowledgements

I thank Professor James G. Quinn for launching my career in marine organic geochemistry and research in the biogeochemistry of organic chemicals of environmental concern in the oceans. Since that time students, postdocs, and colleagues too numerous to mention (they know who they are) have contributed substantively to my understanding of the topics covered in this chapter. for three years mid-career, I was at the University of Massachusetts Boston where I learned much about environmental quality issues and Boston Harbor. However, for the other thirty years of my post-Ph. D. career I have been fortunate to be at Woods Hole Oceanographic Institution where the Institution, colleagues, students, postdocs and coworkers have provided a rich and enjoyable intellectual environment. funding for my research over the years came from the Office of Naval Research, the National Science Foundation, NOAA, Sea Grant, EPA, Department of Energy, and private funding at WHOI. I thank WHOI Graphics Services and Janet Fields and Valerie Caron for assisting in putting this manuscript together. Last, but not least, my thanks to my family, this volume's editors, Ms. Gioia Sweetland, and other authors in this volume for their patience as I labored in the midst of other urgent duties to complete this manuscript. This is contribution number 11206 of Woods Hole Oceanographic Institution.

## Bibliography

Abramowicz, D. A., 1990. Aerobic and anaerobic biodegradation of PCBs: a review. *Crit. Rev.Biotechnol.* **10,** 241–251.

ACS, 1980. Guidelines for data aquisiton and data quality evaluation in environmental chemistry. American Chemical Society Committee on Environmental Improvement, Subcommittee on Environmental Analytical Chemistry. *Anal. Chem.***52,** 2242–2249.

Anderson, D. W., J. R. Jehl, Jr., R. W. Risebrough, L. A. Woods, Jr., L. R. Deweese, and W. G. Edgecomb, 1975. Brown pelicans improved reproduction off the southern California coast. *Science* **198,** 806–808.

Anonymous, 1966. Report of a new chemical hazard. *New Sci.* **32,** 612.

Atlas, R. M. and R. Bartha, 1992. Hydrocarbon biodegradation and oil spill bioremediation. *Advances in Microbial Ecology,* K. C. Marshal (ed.) Plenum Press, New York, pp. 287–238.

Atlas, E. and C. S. Giam, 1984. Sampling organic compounds for marine pollution studies. In, *Strategies and Advanced techniques for Marine Pollution Studies: Mediterranean Sea. C. S. Giam and H.J.-M.Dou (eds).* NATO Advanced Studies Series. Seris G. Ecological Sciences Vol. 9. Springer-Verlag, Berlin. pp. 209–230.

Axelman, J., D. Broman, and C. Näf, 200. Vertical fluxes and particulate/water dynamics of polychlorinated biphenyls (PCBs) in the open Baltic Sea. *Ambio* **29,** 210–216.

Baker, J. E., P. D. Capel, S. J. Esienreich, 1986. Influence of colloids on sediment-water partition coefficients of polychlorinated biphenyl congeners in natural waters. *Environ. Sci. Technol.* **20,** 1136–1143.

Banerjee, S., 1985. Caculation of water solubility of organic compounds with UNIFAC- derived parameters. *Environ. Sci. Technol.* **19,** 369–370.

Beddard, D. L. and M. L. Haberl, 1990. Influence of chlorine substitution on the degradation of polychlorinated biphenyls by 8 bacterial strains. *Microbiol. Ecol.* **20,** 87–102.

Benner, R., 1998. "Natural" remediation of DDT, PCBs debated. *Environ. Sci. Technol.* 360A-363A.

Benner, R., 2002. The Kow controversy. *Environ. Sci. Technol.* **36,** 411A-413A.

Biddleman, T. F. and C. E. Olney, 1974a. Chlorinated hydrocarbons in the Sargasso Sea atmosphere and surface water. *Science* **183,** 516–518.

Bidleman, T. F. and C. E Olney, 1974b. High volume collection of atmospheric polychlorinated biphenyls. *Bull. Environ. Cont. Toxicol.* **11,** 442–450.

Blumer, M., 1976. Polycyclic aromatic hydrocarbons in nature. *Scientific American* **234,** 35–45.

Blumer, M. and W. Youngblood, 1975. Polycyclic aromatic hydrocarbons in soils and Recent sediments. *Science* **188,** 53–55.

Breivik, K. A. Sweetman, J. M. Pacyna, and K. C. Jones, 2002a. Towards a global Emission inventory for selected PCB congeners – a mass balance approach. 1. Global production and comsumption. *Sci. Total. Eniron.* **290,** 181–198.

Breivik, K. A. Sweetman, J. M. Pacyna, and K. C. Jones, 2002a. Towards a global emission inventory for selected PCB congeners – a mass balance approach. 2. Emissions. *Sci. Total. Eniron.* **290,** 199–224.

Bodman, R.H., L. V. Slabaugh, and V. T. Bowen, 1961. A multi-purpose large volume sea-water sampler. *J. Mar. Res.* **19,** 141–148.

Boehm, P. D. and J. G. Quinn, 1976. The effect of dissolved organic matter in seawater on the uptake of mixed individual hydrocarbons and No.2 fuel oil by a marine filter feeding bivalve (*Mercenaria mercenaria*). *Estuar. Coast. Mar. Sci.* **4,** 93–105.

Boehn, P. D. and J. G. Quinn, 1977. The persistence of chronically accumulated hydrocarbons in the hard shell clam *Mercenaria mercenaria. Mar. Biol.* 44,227–233.

Boehm, P. D., S. Drew, T. Dorsey, J. Yarko, N. Mosesman, A. Jeffries, D. Pilson, and D. Fiest, 1985. Organic pollutants in New York Bight suspended particulates. Chapter 11 In. *Wastes in the Ocean Vol.6* (B. W. Ketchum, I. W. Duedall, J. M Capuzzo, P. K. Park, W. V. Burt, and D. R. Kester (eds.) John Wiley and Sons, New York. Pp252–279.

Boon, J. P., W.H. Lewis, M. R. Tjoen-A-Choy, C. R. Allchin, R. I. Law, J. De Boer, C. C. Ten Hallers-Tjabbes, and B. N. Zegers., 2002. Levels of polybrominated diphenyl ether (PDBE) Flame retardants in animals representing different trophic levels of the North Sea food web. *Environ. Sci. technol.* **36,** 4025–4032.

Boyle, A. W., C. J. Silvin, J. P. Hassett, J. P. Nakas, and S. W. Tanenbaum, 1992. Bacterial PCB biodegradation. *Biodegradation* **3,** 285–298.

Brown, J. F., Jr., R. E. Wagner, H. Feng, D. L. Bedard, M. J.Brennan, J. C. Carnahan, and R. J. May, 1987. Environmental dechlorination of PCBs. *Environ. Toxicol. Chem.* **6,** 579–594.

Brownawell, B. J., 1986. *The role of colloidal organic matter in the marine geochemistry of PCBs.* Ph, D, Dissertation. Massachusetts Institute of Technology/Woods Hole Oceanographic Institution Joint Program in Oceanography. Woods Hole, MA 318 pp.

Brownawell, B. J. and J. W. Farrington, 1985. Partitioning of PCBs in marine sediments. In: *Marine and Estuarine Chemistry.* A. C. Sigleo and A. Hattori (Eds.), Lewis Pub. Inc.,Chelsea, MI pp. 97–120.

Brownawell, B. J. and J. W. Farrington, 1986. Biogeochemistry of PCBs in interstitial waters of a coastal marine sediment. *Geochim. Cosmchim. Acta* **50,**157–169.

Bucheli, T. D. and O. Gustafsson, 2000. Quantification of the soot-water distribution coefficient of PAHs provides a mechanistic basis for enhanced sorption observations. *Environ. Sci. Technol.* **34,** 5144–5151

Buchholtz ten Brink, M.R., F. T. Manheim, E. L. Mecray, M. E. Hastings, and J. M. Currennce, along with J. W. Farrington, T. J. Fredette, S. H. Jones, M. L. Liebman, P. F. Larsen, W. Smith Leo, B. W. Tripp, G. T. Wallace, Jr., and L. G. Ward, 2002. Contaminated sediments database for the Gulf of Maine. U. S. Geological Survey Open –file report No. 02–403, Online at *http://pubs.usgs.gov/of/2002/of02–403/*

Butler, G. E. (ed), 1978. Principles of Ecotoxicology., SCOPE 13. John Wiley and Sons, New York, 350pp.

Carson, R., 1962. *Silent Spring* Houghton Mifflin, Boston, MA

Cerniglia, C. E. and M. A. Heitkamp,1989. Microbial degradation of PAH in the aquatic Environment. In. *Metabolism of Polynuclear Aromatic Hydrocarbons (PAHs) in the Aquatic Environment,* U. Varanasi, ed., CRC Press, Boca Raton, FL, pp 41–67 ..

Caisse, D. 1989. Chemical contaminants of the French coast. *Mar. Poll. Bull.* **20,** 523- 528.

Coates, J., J. Woodward, J. Allen, P. Philp, and D. Lovely, 1997. Anaerobic degradation of polycyclic aromatic hydrocarbons and alkanes in petroleum-contaminated marine harbor sediments. *Appl. Environ. Microbiol.* **63,** 3589–3593.

Connoly, J. P., and G. Glaser, 2002. *p,p'*DDE bioaccumulation in female sea lions of the California Channel Islands. *Contin. Shelf. Res.* **22,** 1059–1078.

Currie L. A., T. I. Eglinton, B. A. J. Benner, and A. Pearson, 1997. Radiocarbon "dating" of individual chemical compounds in atmospheric aerosol: first results comparing direct isotopic and multivariate statistical apportionment of specific polycyclic aromatic hydrocarbons. *Nucl. Instrum. Methods Phys. Res. B* **123,** 475–486

Davis, J. 1967. *Petroleum Microbiology.* Elsevier, New York.

De lappe, B. W., R. W. Risebrough, and W. Walker II, 1983. A large-volume sampling assembly for the determination of synthetic organic and petroleum compounds in the dissolved and particulate phases of seawater. *Can. J. Fish. Aquat. SWci.* **40,**322–336.

Delong, R. L., W. G. Gilmartin, and J. G. Simpson, 1973. Premature births in California sea lions: association with high organochlorine pollutant residue levels. *Science* **181,** 1168–1169.

Doney, S.C., W. J. Jenkins, and J. L. Bullister, 1997. A comparison of ocean tracer dating techniques on a meridional section in the eastern North Atlantic. *Deep Sea Res. Part I.* **44,** 603–626.

Drake, D. E., R. P. Eganhouse, and W. MacArthur, 2002. Physical and chemical effects of grain aggregates on the Palos Verdes margin, southern California. *Contin. Shelf. Res.* **22,** 967–986.

Duce, R. A., P. L. Parker, and C. S. Giam, (eds) 1974. Pollutant Transfer to the Marine Environment. , Deliberations and Recommendations of the NSF/IDOE Pollutant Transfer Workshop held in Port Aransas, Texas, January 11–12, 1974. National Science Foundation, Washington, DC . 55 pp.

Duce, R. A., J. D. Hem, M. G. Gross, and K. K. Turekian, 1976. Chapter 15. Transport Paths in Strategies for Marine Pollution Monitoring, E. D. Goldberg (ed). Wiley Interscience, John Wiley and Sons, New York. pp. 275–304.

Eganhouse, R. P. (Ed.), 1997. *Molecular markers in Environmental Chemistry.* ACS Symposium Series **671**. American Chemical Society, Washington, DC. 426 pp.

Eganhouse, R. P. and J. Pontolillo, 2000. depositional history of organic contaminants on the Palos Verdes Shelf, California. *Mar. Chem.* **70**, 317–338.

Eganhouse, R. P. , J. Pontolillo, and T. J. Leiker, 2000. Diagenetic fate of organic contaminants on the Palos Verdes Shelf, California. *Mar. Chem.* **70**, 289–315.

Eglinton T. I., L. I. Aluwihare, J. E. Bauer, E. R. M. Druffel, and A. P. McNichol, 1996. Gas chromatographic isolation of individual compounds from complex matrices for radiocarbon dating. *Anal. Chem.* **68**(5), 904–912.

Ehrhardt, M., K. Burns, and M. Bicego, 1992. Sunlight-induced compositional alterations in the seawater soluble fraction of a crude oil. *Mar. Chem.* **37,** 53–64.

Erickson, M. D., 1997. *Analytical Chemistry of PCBs, Second Edition.* CRC Press, Lewis Publishers, Boca Raton, FL, USA. 667 pp.

Fan, C-W., J. R. Reinfelder, 2003. Phenathrene kinetics in marine diatoms. *Environ. Sci. Technol.* **37,** 3405–3412.

Farrington, J.W. 1974. Some problems associated with the collection of marine samples and analysis of hydrocarbons. In: Proceedings of Conference/Workshop:Marine Environmental Implications of Offshore Drilling Eastern Gulf of Mexico U. S. Department of the Interior, Washington, D.C. USA. pp. 269–278.

Farrington, J.W., 1991. Biogeochemical processes governing exposure and uptake of organic pollutant compounds in aquatic organisms. *Environ. Health. Per.* **90,** 75–84.

Farrington, J. W. (1999). "Mussel Watch" and Chemical Contamination of the Coasts by Polycyclic Aromatic Hydrocarbons. Marine Pollution, Proceedings of a Symposium held in Monaco, 5–9 October 1998. International Atomic Energy Agency IAEA- TECDOC-1094. IAEA, Vienna, 1999. pp. 273–278.

Farrington, J. W. and P. A. Meyer, 1975. Hydrocarbons in the marine environment. Chapter 5. in *Environmental Chemistry* (G. Eglinton (ed). **Vol. 1,** 109–136. A Specialist Periodical Report, The Chemical Society, London.

Farrington, J. W. and B. W. Tripp, 1977. Hydrocarbons in western North Atlantic surface sediments. *Geochim. Cosmochim. Acta.* **41,** 1627–1641.

Farrington, J. W. and J. Westall, 1986. Organic chemical pollutants in the oceans and groundwater: A review of fundamental properties and biogeochemistry. In: *The role of the oceans as a waste disposal option.* G. Kullenberg (ed.) Proceedings NATO Advanced Research Workshop. D. Reidel Publishing Company, Boston, MA pp361- 425.

Farrington, J. W., J. M. Teal, G.C. Medeiros, K. A. Burns, E. A. Robinson Jr, J. G. Quinn, and T. L. Wade. 1976. Intercalibration of gas chromatographic analyses for hydrocarbons in tissues and extracts of marine organisms. *Anal. Chem.* **48,** 1711–1716.

Farrington, J. W., S. M. Henrichs, and R. Anderson, 1977. Fatty acids and Pb-210 geochronology of a sediment core from Buzzards Bay, Massachusetts. *Geochim. Cosmochim Acta,* **41,** 289–296.

Farrington, J.W., A.C. Davis, N.M. Frew, and K.S. Rabin (1982a). No. 2 fuel oil compounds in Mytilus edulis: Retention and release after an oil spill. *Marine Biology* **66,**15–26.

Farrington, J.W., B.W. Tripp, J.M. Teal, G. Mille, K. Tjessem, A.C. Davis, J.B. Livramento, N.A. Hayward and N.M. Frew, (1982b). Biogeochemistry of aromatic hydrocarbons in the benthos of microcosms. *Toxicology and Environmental Chemistry* **5,** 331–346.

Farrington, J.W., R.W. Risebrough, P.L. Parker, A.C. Davis, B. de Lappe, J.K. Winters, D. Boatwright, N.M. Frew, 1982c. Hydrocarbons, Polychlorinated Biphenyls, and DDE In Mussels and Oysters from the U.S. Coast, 1976–1978 - The Mussel Watch. Woods Hole Oceanographic Institution Technical Report 82–42. Woods Hole, MA

Farrington, J.W., E.D. Goldberg, R.W. Risebrough, J.H. Martin, and V.T. Bowen,1983. U.S. "Mussel Watch" 1976–1978. An overview of the trace metal, DDE, PCB, hydrocarbon and artificial radionuclide data. *Environ. Sci. and Technol.* **17,** 490–496.

Farrington, J. W. X. Jia, C. H. Clifford, B. W. Tripp, J. B. Livramento, A. C. Davis, N. M. Frew, and C. G. Johnson, 1986. No. 2 fuel oil compound retention and release by *Mytilus edulis.* Technical report No. 86–8 (CRC86–1), Woods Hole Oceanographic Institution, Woods Hole, MA

Farrington, J. W., A. C. Davis, B. J. Brownawell, B. W. Tripp, C. H. Clifford and J. B. Livramento, 1986. Some aspects of the biogeochemistry of polychlorinated biphenyls. In the Achusnet River estuary. In. *Marine Organic Geochemistry* ACS Symposium Series, No. 305. M.Sohn (ed), American Chemical Society, Washington, DC pp. 174–197

Farrington, J. W., A. C. Davis, N. M. Frew, and A Knap, 1988. ICES/IOC intercomparison exercise on the determination of petroleum hydrocarbons in biological tissues (mussel homogenate). *Mar. Poll. Bull.* **19,**372–380.

Fasnacht, M. P., and N. V. Blough, 2002. Aqueous photodegradation of polycyclic aromatic hydrocarbons . *Environ. Sci. Technol.* **36,** 4364–4369.

Ferguson, P. L. and G. T. Chandler, 1998. A laboratory and Field comparison of sediment Polycyclic aromatic hydrocarbon bioaccumulation by the cosmopolitan estuarine polychaete *Sterblosio benedicti* (Webster). *Mar. Environ. Res.* **454,** 387–401.

Fillmann, G., J. W. Readman, I Tolosa, J. Bartocci, J-P. Vileneuve, C. Cattini, and L. D. Mee, 2002. persistent organochlorine residues in sediments from the Black Sea. *Mar. Pollut. Bull.* **44,** 122–133.

Fowler, S. W., 1990. Critical review of selected heavy metal and chlorinated hydrocarbon concentrations in the marine environment. *Mar. Envir. Res.* **29**,1–64.

Frysinger, G. S., R. B. Gaines, C. M. Reddy, 2002. GC x GC – a new analytical tool for environmental forensics. *Environ. Forensics* **3,** 27–34.

Frysinger, G. S., Gaines, R. B., L. Xu, C. M. Reddy, 2003. resolving the unresolved complex mixture in petroleum-contaminated sediments. *Environ. Sci. Technol.* **37, 1653**–1662.

Galloway, W. B., J. L. Lake, D. K. Phelps, P. F. Rogerson, V. T. Bowen, J. W. Farrington, E. D. Goldberg, J. L. Laseter, G. C. Lawler, J. H. Martin, R. W. Risebrough, 1983. The Mussel watch intercomparison of trace level constituent determinations. *Environ. Toxicol. Chem.* **2,** 395–410.

GESAMP, 1993. Impacts of oil and related chemicals and wastes in the marine environment. Group of Experts on the Scientific Aspects of Marine Pollution, GESAMP Reports and Studies No.50. International Marine Organization, London, UK 180 pp ..

Gearing, P., J. Gearing, T. F. Lytle, and J. S. Lytle, 1976. Hydrocarbons is 60 Northeast Gulf of mexico Shelf sediments; a preliminary survey. *Geochim. Cosmohcim. Acta.* **40,** 1005–1017.

Gearing, J.N., P.J. Gearing, T.L. Wade, J.G. Quinn, H.B. McCarty, J.W. Farrington and R.F. Lee (1979). The rates of transport and fates of petroleum hydrocarbons in a controlled marine ecosystem and a note on analytical variability. In: Proceedings 1979 Oil Spill Conference, American Petroleum Institute, Washington, D.C., pp 555–564.

Gearing, P. J., J. N. Gearing, R.J. Pruell, T. L. Wade, and J. G. Quinn, 1980. Partitioning of No.2 fuel oil in controlled estuarine ecosystems. Sediments and suspended particulate matter. *Environ. Sci. and Technol.* **14,**1129–1136.

Giger, W. and M. Blumer, 1974. Polycyclic aromatic hydrocarbons in the environment: Isolation and characterization by chromatography, visible, ultraviolet and mass spectrometery. *Anal. Chem.* **46,** 1663–1671.

Glaser, G., and J. P. Connoly, 2002. A model of *p,p'* DDE and total PCB bioaccumulation in birds from the Southern California Bight. *Contin. Shelf, Res.***22,** 1079–110

Goldberg, E.D., 1972. (ed). *Marine Pollution Monitoring: Strategies for a National Program.* National Oceanic and Atmospheric Administration, U. S. Department of Commerce, Washington, D.C. USA

Goldberg, E.D., 1976. *The Health of the Oceans.* The UNESCO Press, Paris.

Goldberg, 1975. The Mussel Watch: a first step in global marine monitoring. *Mar. Poll. Bull.* **6,** 111.

Grice, G.D., G.R. Harvey, V. T. Bowen, and R. H. Backus, 1972. The collection and preservation of open ocean organisms for pollutant analysis. *Bull. Eviron. Cont. Toxicol.* **1,** 125–132.

Gruber, N. and J. Sarmineto, 2002. Large-sclae biogeochemical –physical interactions in elemental cycles. Chapter 9. in *The Sea,* **Volume 12,** 337–399 edited by A. R. Robinson, J. J. McCarthy, and B. J. Rothschild. John Wiley and Sons, Inc. New York.

Gustafsson, O., F. Haghseta, C. Chan, J. MacFarlane, P. M. Gschwend, 1997a. Quanitification of dilute sedimentary soot phase: implications for PAH speciation and bioavailability. *Environ. Sci. and Technol.* **31,** 203–209.

Gustafsson, O., and P. M Gschwend, 1997b. Soot as a strong sorption medium for polycyclic aromatic hydrocarbons in aquatic systems. Chapter 24. In. *Molecular markers in Environmental Chemistry.* R. P Eganhouse (ed). ACS Symposium Series 671. American chemical Society, Washington, DC. Pp. 365–381.

Hall, R. J., 1987. Impact of pesticides on bird populations. In G.J. Marco, R. M. Hollingworth, and W. Durham (eds) *Silent Spring revisited,* (Chapter 6), 85–111. American Chemical Society, Washington, DC

Hamelink, J. L. , R. C. Waybrant, R. C. Ball, 1971. A proposal: exchange equilibria control the degree chlorinated hydrocarbons are biologically magnified in lentic environments. *Trans. Am. Fish. Soc.***100,** 207–214.

Hamelink, J. L., P. F. Landrum, H. L. Bergman, W. H. Benson, 1994. Bioavailability. Physical, Chemical, and Biological Interactions. Proceedings of the Thirteenth Pellston Workshop. SEATC Special Publication Series. Lewis Publishers, Boca Raton, Florida. 239 pp.

Harvey, G.R. and W. G. Steinhauer, 1974. Atmospheric transport of polychlorinated biphenyls in the North Atlantic. *Atmos. Environ.* **8,** 387–388.

Hayes, L, K. Nevin, D. Lovely, 1999. Role of prior exposure on anaerobic degradation of naphthalene and phenanthrene in marine harbor sediments. *Org. Geochem.* **30,** 937- 945.

Heider, J, A, M. Spormann, H. R. Beller, and F. Widdel, 1999. Anaerobic bacterial metabolism of hydrocarbons. *FEMS Microbiol. Rev.***22,** 459–473.

Hinga, K.R., R.F. Lee, J.W. Farrington, M.E.Q. Pilson, K. Tjessem and A.C. Davis (1980). Biogeochemistry of benzanthracene in an enclosed ecosystem. *Environmental Science and Technology, 14:*1136–1143.

Hinga, K., Pilson, M, Almquist, G. and Lee, R. (1986). The degradation of 7,12-dimethylbenz[a]anthracene in an enclosed marine ecosystem. *Mar. Environ. Res ..* **18**, 79–91.

Hinga, K.R. and Pilson, M.E.Q. (1987) Persistence of benz[a]anthracene degradation products in an enclosed marine ecosystem. *Environ. Sci.Technol.* **21**, 648–653.

Hites, R. A., 2004. Polybrominated diphenyl ethers in the environment and in people: a meta-analysis of concentrations. *Environ. Sci. Technol.* **38,** 945–956.

Hites, R.A., R.E. LaFlamme and J.W. Farrington, 1977. Sedimentary polycyclic aromatic hydrocarbons: the historical record. *Science.* **198,** 829–831.

Hites, R.A., R.E. LaFlamme, J.G. Windsor, Jr., J.W. Farrington and W.G. Deuser, 1980. Polycyclic aromatic hydrocarbons in an anoxic sediment core from the Pettaquamscutt River, Rhode Island, U.S.A. *Geochim. Cosmochim. Acta.* **44,** 873–878.

Higson, F. K., 1992. Microbial degradation of biphenyl and its derivatives. *Adv. Appl. Microbiol.* **59,** 135–164.

Hoffman, E., G. Mills, J. S. Latimer, and J. G. Quinn, 1983. Annual inputs of petroleum hydrocarbons to the coastal environment via urban runoff. *Can. J. Fish. Aquat. Sci.* **40(Suppl.2),** 41–53.

Holliger, C., G. Wolfarth, and G. Diekert, 1998. reductive dechlorination in the energy metabolism of anaerobic bacteria. *FEMS Microbiol. Rev.***22,** 383–398.

Hopper, D. J., 1978. Microbial degradation of aromatic hydrocarbons. In *Developments in biodegradation of hydrocarbons* -1. Watkinson, R. J. (ed.), Applied Science Publishers, LTD. London. pp 85–134.

Hunt, J. M. , 1995. *Petroleum Geochemistry and Geology, Second Edition,* W. H. Freeman and Company, New York. 743p.

IDOE, 1972. *Baseline Studies of Heavy Metal, Halogenated Hydrocarbons, and Petroleum Hydrocarbon pollutants in the Marine Environment and Research Recommendations.*deliberations of the International decade of Ocean Exploration (IDOE) Baseline Conference May 24–26, 1972. New York, 1972. E. D. Goldberg, Convener. National Science Foundation, Washington, D. C. USA 54 p.

Islam, Md. S., and T. Tanaka, 2004. Impacts of pollution on coastal and marine ecosystems including coastal and marine fisheries and approach for management: a review and synthesis. *Mar. Pollut. Bull.* **48,** 624–649.

IOC, 1993. Chlorinated hydrocarbons in open ocean waters: sampling, extraction, clean-up and instrumental determination.*Tech. Rept. Series* **25,** UNESCO, Paris, 36pp.

Iwata, H., S. Tanabe, N. Sakai, and R. Tatusukawa, 1993a. Distribution of persistent organochlorines in the oceanic air and surface seawater and the role of the ocean in their global transport and fate. *Environ. Sci.Technol.* **27,** 1080–1098.

Iwata, H., S. Tanabe, and R. Tatsukawa, 1993b. A new view on the divergence of HCH isomer compositions in oceanic air. *Mar. Pollut. Bull.* **26,** 302–205.

James, M. O., 1989. Biotransformation and disposition of PAH in aquatic invertebrates. In: *Metabolism of Polynuclear Aromatic Hydrocarbons (PAHs) in the Aquatic Environment,* U. Varanasi, ed., CRC Press, Boca Raton, FL, pp 69–91.

Jensen, S. (1972). The PCB story. *Ambio* **1,** 123–131.

Johnson, R. G., 1974. Particulate matter at the sediment-water interface in coastal environments. *J. Mar. Res.* **32,** 313–330.

Jönnson, A., Ö. Gustafsson, J. Axelman, and H. Sundberg, 2003. Global accounting of PCBs in the Continental Shelf sediments. *Environ. Sci. Technol.* **37,** 245–255.

Keith, L. H. and W. A. Telliard, 1979. Environmental science and technology special report: priority pollutants. I. A perspective view. *Environ. Sci. Technol.* **13,** 416–423.

Ketchum, B. H., (ed) 1972. The Water's Edge. Critical problems of the Coastal Zone. Massachusetts Institute of Technology Press, Cambridge, MA 393 pp.

Koeman, J. H., De B. ten Noever, and R. H. De Vos, 1969. Chlorinated biphenyls in fish, mussels and birds from the river Rhine and the Netherlands coastal area. *Nature(London)* **221,** 1126–1128.

Kujawinski, E. B., J. W. Farrington, and J. W. Moffett, 2001. Marine protozoa produce organic matter with high affinity for PCBs during grazing. *Environ. Sci. Technol.,* **35,** 4060–4065.

LaCorte, S., M. Guillamon, E. Martinez, P. Viana, and D. Barcello, 2003. Occurrence and specific congener profile of 40 polybrominated diphenyl ethers in river and coastal sediments from Portugal. *Environ. Sci. Technol.* **37,** 892–898.

LaFlamme, R. E. and R. A. Hites, 1978. The global distribution of polycyclic aromatic hydrocarbons in recent sediments. *Geochim. Cosmochim. Acta* **42,** 289–303.

LaFleur, P. D. (ed.), 1976. Accuracy in trace analysis: sampling, sample handling, analysis. NBS Special Publication 422, U. S. National Bureau of Standards, Gaithersburg, MD 20899.

Lakaschuse, S., K. Weber, F. Wania, R. Bruhn, and O. Schrems, 2002. *Environ. Sci. Technol.* **36,** 138–145.

Latimer, J. S. and J. Zheng, 2003. The sources, transport, and fate of PAHs in the marine environment. Chapter 2 .. In. *PAHs: An Ecotoxi8cology perspective.* P.E. T. Douben (ed.), John Wiley and Sons, LTD. 9–34.

Law, R. and J. E. Portmann, 1980. Report on the first ICES intercomparison exercise on petroleum hydrocarbons. ICES, ACMP. International Council for the Exploration of the Seas, Copenhagen, Denmark.

Leahy, J. G. and Colwell, 1990. Microbial degradation of hydrocarbons in the environment. *Microbiol. Rev.* **54,** 305–315.

Lee, H. 1994. County Sanitation District of Los Angeles County (LACSD) multi-year measurements of chemical contaminants and bulk density of sediment on the Palos Verdes margin. Expert Report 2.F., Southern California Damage Assessment Witness Reports, October 4, 1994, 156pp.

Lee, H., and P. L. Wiberg, 2002. Character, fate, and biological effects of contaminated, effluent – affected sediment on the Palos Verdes margin, southern California: an overview. *Contin. Shelf Res.* **22,** 835–840.

Lee, H., C. R. Sherwood, D. E. Drake, B. D. Edwards, F. Wong, and M. Hamer, 2002. Spatial and temporal distribution of contaminated effluent-affected sediment on the Palos Verdes margin, southern California. *Cont. Shelf Res.* **22,** 859–880.

Lee M. L., G. P. Prado , J. B. Howard, and R. A. Hites,1977. Source identification of urban airborne polycyclic aromatic hydrocarbons by gas chromatographic mass spectrometry and high resolution mass spectrometry. *Biomed. Mass Spec.* **4(3),** 182–186.

Li, Y-F., 1999. Global technical hexachlorocyclohexane usage and its contamination consequences in the environment: from 1948 to 1997. *Sci. Total Environ.* **232,** 121–158.

Li, Y-F., M. T. Scholtz, and B. J. Van Heyst, 2003. Global gridded emission inventories of • – hexachlorocyclohexane. *Environ. Sci. Technol.* **37,** 3493–3498.

Lichtfouse E., H. Budzinski, P. Garrigues, and T. Eglinton, 1997. Ancient polycyclic aromatic hydrocarbons in modern soils: 13C, 14C and biomarker evidence. *Org. Geochem.* **26(5/6),** 353–359.

Lundgren, K., M. Tysklind, R. Ishaq, D. Broman, and B. Van Bavel, 2002. Polychlorinated naphthalene levels, distribution, and biomagnifcation in a benthic food chain in the Baltic Sea. *Environ. Sci. Technol.* **36,** 5005–5013.

Lyman, W. J., W. F. Reehl, D. H. Rosenblatt, 1982. *Handbook of chemical property estimation methods: environmental behavior of organic compounds.* McGraw-Hill Book Co., New York, NY. 960pp.

Macgregor, J. S., 1976. DDT and its metabolites in the sediments off southern California. *Fish. Bull.* **74,** 27–35.

Mackay, D., W. Y. Shiu, 1977. Aqueous solubility of polynuclear aromatic hydrocarbons. *J. Chem. Eng. Data.* **22,** 399–402.

May, W. E., 1980. The solubility behavior of polycyclic aromatic hydrocarbons in aqueous systems. In *Petroleum in the marine environment.*( L. Petrakis and F. T. Weiss (eds). American Chemical Society, Washington, DC pp. 143–192.

May, W. E., S. P. Wasik, D. H. Freeman, 1978. Determining the solubility behavior of some polycycli aromatic hydrocarbons in water. *Anal. Chem.* **50,** 997.

McCarthy, J. F., B. D. Jimenez, and T. Barbee, 1985. Effect of dissolved humic material on accumulation of polycyclic aromatic hydrocarbons: structure activity relationships. *Aquat. Toxicol.* **7,**15–24.

McElroy, A. E., B. W. Tripp, J. W. Farrington and J. M. Teal, 1987. Biogeochemistry of benz(a)anthracene at the sediment-water interface. *Chemosphere* **16,** 2429–2440.

McElroy, A. E., J. M. Teal, and J. W. Farrington, 1989. Bioavailability of polycyclic aromatic hydrocarbons in the aquatic environment. In: *Metabolism of Polynuclear Aromatic Hydrocarbons (PAHs) in the Aquatic Environment,* U. Varanasi, ed., CRC Press, Boca Raton, FL, pp 1–39.

Méallier, P., 1999. Phototransformation of pesticides in aqueous solution. In *The Handbook of Environmental Chemistry,* Vol. 2, Part L. P. Boule (ed.), Springer, Berlin. pp. 241–262.

Mohn, W. W., and J. M. Tiedje, 1992. Microbial reductive dehalogenation. *Microbiol. Res.* **56,** 482–507.

Monirith, I., D. Ueno, S. Takahshi, H. Nakata, A. Sudaryanto, A. Subramanian, S. Karuppiah, A. Ismail, M. Muchtar, J. Zehng, B. J. Richardson, M. Prudente, N.D. Hue, T. S Tana, A. V. Tkalin, S. Tanabe, 2003. Asia-Pacific mussel watch: monitoring contaminantion of persistent organochlorine compounds in coastal waters of Asian countries. *Mar. Pollut. Bull.* **46,** 281–300.

NAS, 1971. Radioacitvity in the Marine Environment. National Academy of Sciences, Wahington, D.C. 272 pp.

NAS, 1972. Marine Environmental Quality. Suggested Research Programs for Understanding Man's Effect on the Oceans. *Report of a special study held under the auspices of the Ocean Science Committee of the NAS-NRC Ocean Affairs Board, August 9–13, 1971.* National Academy of Sciences, Washington, D.C. 107 pp

Neeley, W. B., D. R. Branson, G. E. Bau, 1974. Partition coefficients to measure bioconcentration potential of organic chemicals in fish. *Environ. Sci. Technol.* **13,** 1113 -

Neff, J. 1979. Polycyclic aromatic hydrocarbons in the aquatic environment. Sources, fates and biological effects. Applied Science Publishers, LTD. London. 262 pp.

NRC, 1975. Petroleum in the Marine Environment. National Research Council, National Academy Press, Washington, DC.

NRC, 1980. The International Mussel Watch, Report of a Workshop. National Research Council, National Academies Press, Washington DC.

NRC, 1983. Polycyclic aromatic hydrocarbons: evaluation of sources and effects. National Academies Press, Washington, DC.

NRC, 1985. Oil in the Sea: Inputs, Fates and Effects. National Research Council, National Academy Press, Washington, DC.

NRC, 1989. Contaminated Marine Sediments – Assessment and Remediation. National Research Council, National Academies Press, Washington D.C. 493pp.

NRC, 1997. Contaminated Sediments in Ports and Waterways. National Research Council, National Academies Press, Washington D.C ..

NRC, 2001. A Risk-Management Strategy for PCB-Contaminated Sediments. National Research Council National Academies Press, Washington, D. C. 432 pp.

NRC, 2003. Oil in the Sea b III. Inputs, Fates, and Effects. National Research Council, National Academies Press, Washington, DC. 263 pp.

NTIS, 1976. Criteria Document for PCBs, Prepared for the EPA by the Massachusetts Audubon Society. Document Number PB 355 397. National Technical Information Service, Washington, D. C. 368 pp. plus appendices.

O'Connor, T. P. 2002. National distribution of chemical concentrations in mussels and oysters in the USA. *Mar. Environ. Res.* **53,** 117–143.

O'Malley V., T. Abrajano Jr, and J. Hellou, 1994. Determination of the $^{13}C/^{12}C$ ratios of individual PAH from environmental samples: can PAH sources be apportioned? *Org. Geochem.* **21**(6/7), 809–822

Okuda T., H. Kumata, H. Naraoka, R. Ishiwatari, and H. Takada, 2002. Vertical distributions and d13C isotopic compositions of PAHs in Chidorigafuchi Moat sediment, Japan. *Org. Geochem.* **33,** 843–848.

Pagni, R. M. and M. E. Sigman, 1999. The photochemistry of PAHs and PCBs in water and on solids. In *The Handbook of Environmental Chemistry,* Vol. 2, Part L. P. Boule(ed.), Springer, Berlin. pp. 139–179.

Payne, J. R., and Phillips, C. R., 1985. Photochemistry of petroleum in water. *Environ. Sci. Technol.* **19,** 569 -

Peltzer, E. T., J. B Alford, and R. B. Gagosian, 1984. Methodlogy for sampling and analysis of lipids in aerosols from the remote marine atmosphere. Technical Report No. WHOI-84–9, Woods Hole Oceanographic Institution, Woods Hole, MA 02543

Petrick, G., D. E. Schulz, J. C. Duinker, 1988. Clean-up of environmental samples by high performance liquid chromatography for analysis of organochlorine compounds by gas chromatography with electron capture detector. *J. Chromatogr.***435,** 241–248.

Petrick, G., D. E. Schulz-Bull, V. Martens, K. Scholz, and J. C. Duinker, 1996. An in-situ filtration/extraction systems for the recovery of trace organics in solution and on particles tested in deep ocean water. *Mar. Chem.* **54,** 97–105.

Pontolillo, J., nad R. P. Eganhouse, 2001. The search for Reliable Aqueous Solubility ($S_w$) and octanol-water partition coefficient ($K_{ow}$) data for hydrophobic organic compounds: DDT and DDE as a case study. *U. S. Geological Survey, Water Investigations Report 01–4201; USGS: Reston, VA.* 51 pp.

Pritchard, J. B., and J. R. Bend, 1991. Relative roles of metabolism and renal excretory mechanisms in xenobiotic elimination by fish. *Environ. Health. Perspect.* **90,** 85–92.

Pruell, R. J., J.L. Lake, W. R. Davis, J. G. Quinn, 1986. Uptake and depuration of organic contaminants by blue mussels (*Mytilus edulis*), exposed to environmentally contaminanted sediment. *Mar. Biol.* **91,** 497-

Pruell, R. J., B. K. Taplin, D.G. McGovern, R. McKinney, and S. B. Norton, 2000. Organic contaminant distributions in sediments, polychaetes (*Nereis virens)* and Amercian Lobster *(Homarus americanus)* from a laboratory food chain experiment. *Mar. Environ. Res.* **49,** 19–36.

Quensen, J.F., S. A. Mueller, M. K. Jain, and J. M. Tiedje, 1998. Reductive Dechlorination of DDE to DDMU in marine sediment microcosms. *Science* **280,** 772–724.

Reddy, C., A. Pearson A., L. Xu., A. McNichol, B. Benner Jr., S. Wise, G. Klouda, L. Currie L., and T. Eglinton, 2002a. Radiocarbon as a tool to apportion the sources of polycyclic aromatic hydrocarbons and black carbon in environmental samples. *Environ. Sci. Technol.* **36,** 1774–1782.

Reddy C., L. Xu, T. Eglinton, J. Boon, and D.Faulkner, 2002b. Radiocarbon content of synthetic and natural semi-volatile halogenated organic compounds. *Environ. Pollut.* **120,** 163–168.

Reddy, C., T. Eglinton, A. Hounsell, H. K.White, L. Xu, R. B. Gaines, and G. S. Frysinger, 2002c. The West Falmouth oil spill after thirty years: persistence of petroleum hydrocarbons in marsh sediments. *Environ. Sci. Technol.* **36,** 1774–1782.

Rhead, M. M., 1975. The fate of DDT and PCBs in the marine environment. Chapter 6 in *Environmental Chemistry* ( G. Eglinton (ed). **Vol. 1,**137–159 . A Specialist Periodical Report, The Chemical Society, London.

Risebrough, R. W. , 1969. Chlorinated hydrocarbons in marine ecosystems. In *Chemical Fallout,* (M. W. Miller and G.G. Berg, eds*)* Charles C. Thomas, Springfield, Ill. pp 5- 23.

Risebrough, R. W., 1972. Effects of environmental pollutants on animals other than man. In *Proceedings Sixth Berkeley Symposium Mathematical Statistics and Probability* (L. LeCam and E.L Scott, eds). University of California Press, Berkeley and LosAngeles, CA, pp. 443–463.

Risebrough, R. W., P. Reiche, D. B. Peakall, S.G. Herman, and M. N. Kirven, 1968. Polychlorinated biphenyls in global ecosystems. *Nature(London)* **229,** 1098–1102.

Rothermich, M., L. Hayes, D. Lovely, 2002. Anaerobic, sulfate dependent degradation of Polycyclic aromatic hydrocarbons in petroleum-contaminated harbor sediment. *Environ. Sci. Technol.* **36,** 4811–4817.

Rowland, S. T., and M. J. Molina, 1975. Chlorofluoromethane in the environment. *Rev. of Geophysics and Space Physics* **13,** 1–35.

Royal Commission on Environmental Pollution, 1981. Oil Pollution of the Sea. London, 307pp.

SCEP, 1970. Study of Critical problems of the Environment. *Man's impact on the global environment: assessment and recommendations for action.* The MIT Press, Cambridge, Mass. USA 319 p.

Schulz-Bull, D.E., G. Petrick, and J.C. Duinker, 1988. Chlorinated biphenyls in North Atlantic surface and deep water. *Mar. Pollut. Bull.* **19,** 526–531.

Schulz-Bull, D. E., G. Petrick, J. C. Duinker, 1989. Complete characterization of polcychlorinated biphenyl congeners in commercial Aroclor and Clophen mixtures by multi-dimensional gas chromatography-electron capture detection. *Environ. Sci. Tecnol.* **23,** 852–859.

Schulz-Bull, D.E., G. Petrick, and J. C. Duinker, 1991. Polychlorinated biphenyls in North Sea water. *Mar. Chem.* **36,**365–384.

Schulz D. E., G. Petrick, and J. C. Duinker, 1988. Chlorinated biphenyls in North Atlantic surface and deep water. *Mar. Pollut. Bull.* **19,** 526–531.

Schulz-Bull, D.E., G. Petrick, N. Kannan, and J. C. Duinker, 1995. Distribution of individual chlorobiphenyls (PCB) in solution and suspension in the Baltic Sea. *Mar. Chem.***48,** 245–270.

Schwarzenbach, R. P., P. M. Gschwend, D. M. Imboden, 2003.Environmental Organic Chemistry, Second Edition. Wiley Interscience, John Wiley and Sons, Inc. Hoboken, New Jersey. 1313 pp.

Seba, D. B. and E. F. Corcoran, 1969. Surface slicks as concentrators of pesticides in the marine environment. *Pestic. Monit. J.* **3,** 190.

Sericano, J. L., T. L. Wade, T. J. Jackson, J. M. Brooks, B. W. Tripp, J. W. Farrington, L. D. Mee, J. W. Readmand, J-P. Villeneuve and E. D. Goldberg. (1995) Trace organic contamination in the Americas: An overview of the US National Status and Trends and the International Mussel Watch Programmes. *Mar. Pollut. Bull.* **31,**214- 225.

Setschenow, J. 1889. Über die konstitution der salzlösungen auf grund ihres verhaltens zu kohlensäure. *Z.Phys. Chem. Vierer Band* **1,** 117–125.

Sherwood, C. R., D. E. Drake, P. L. Wiberg, R. A. Wheatcroft, 2002. Prediction of the Fate of p,p'DDE in sediment on the Palos Verdes shelf, California, USA. *Contin. Shelf. Res.* **22,** 1025–1058.

Shailaja, M. S. and M. Nair, 1997. Seasonal differences in organochlorine pesticide Concentrations of zooplankton and fish in the Arabian Sea. *Mar. Environ. Res.* **44,** 263–274.

Simonich, S. L. and R. A. Hites, 1995. Global distribution of persistent organochlorine compounds. *Science* **269,** 1851–1854.

Sobek, A. and O. Gustafsson, 2004. Latitudinal fractionation of polychlorinated biphenyls in surface seawater along a 62° N-89°N transect from the southern Norwegian Sea to the North Pole area. *Environ. Sci. Technol.* **38,** 2746–2747.

Sobek, A., O. Gustafsson, S. Hajdu, and U. Larsson, 2004. Particle-water partitioning of PCBs in the photic zone: a 25 month study in the open Baltic Sea. *Environ. Sci. Technol.* **38,** 1375–1382.

Sokol, R. C., C. M. Bethoney, and G. Y. Rhee, 1998. Effects of Aroclor-1248 concentration on the rate and extent of polychlorinated biphenyl dechlorination. *Environ. Toxicol. Chem.* **17,** 1922–1926.

Stegeman, J. J. and J. M. Teal, 1973. Accumulation, release and retention of petroleum hydrocarbons by the oyster, *Crassostria virginica. Mar. Biol.* **22,** 37–44.

Stegeman, J. J., P. J. Kloeper-Sams, and J. W. Farrington, 1986. Monoxygenase induction and chllorbiphenyls in the deep sea Fish *Coryphaenoides armatus. Science* **231,** 1287–1289.

Stegeman, J. J. and J. L. Lech, 1991. Cytochrome P-450 monoxygenase systems in aquatic species: carcinogen metabolism and biomarkers for carcinogen and pollutant exposure. *Environ. Health. Perspect.* **90,** 101–109.

Stuum, W., R. Schwartzenbach, and L. Sigg, 1983. From environmental analytical chemistry to toxicology – a plea for more concepts and less monitoring and testing. *Angew. Chem. Int. Ed. Enlg.* **22:** 380–389.

SRC, 2004. Syracuse Research Corporation, CHEMFACT Search Parameters, on the worldwide web at <http://esc.syrres.com/scripts/CHEMLISTegi.exe>.

Taylor, J. K., 1985. Principles of Quality Assurance of Chemical Measurements. NBSIR 85–3105. National Bureau of Standards, U. S. Department of Commerce, Gaithersburg, MD 20899. 71 pp.

Tanabe, S., T. Mori, R. Tatsukawa, N. Miyazaki, 1983. *Chemosphere* **12,**1269–1275.

Tanabe, S., B. Madhusree, A. A. Ozturk, R. Tatsukawa, N. Miyazaki, E. Osdamar, O. Aral, O. Samsun, B. Ozturk, 1997. Persistent organchlorine residues in harbor porpoise *(Phcoena phocoena)* from the Black Sea. *Mar. Pollut. Bull.* **34,** 712–780.

Takada, H. , F. Satoh, M. H. Bothner, B. W. tripp, C. G. Johnson, and J. W. Farrington, 1997). Anthrogenic molecular markers: Tools to identify the sources and transport pathways of pollutants. Chapter 12. in *Molecular Markers in Environmental Chemistry.* R. P Eganhouse (ed). ACS Symposium Series 671. American chemical Society, Washington, DC. Pp. 178–195.

Teal, J.M., K.A. Burns and J.W. Farrington, 1978. Analyses of aromatic hydrocarbons in intertidal sediments resulting from two spills of No. 2 fuel oil in Buzzards Bay, Massachusetts. *J. Fish Res. Bd. Canada,* **35,**510–520.

Teal, J. M., J. W. Farrington, K. A. Burns, J. J. Stegeman, B. W. Tripp, B. Wooden, and C. Phinney, 1992. the West Falmouth Oil Spill after 20 years: Fate of fuel oil compounds and effects on animals. *Mar. Pollut. Bull.***24,** 607–814.

Ter Schure, A.F.H., P. Larson, C. Agrell, and J. P. Boon, 2004. Atmospheric transport of polybrominated diphenyl ethers and polychlorinated biphenyls to the Baltic Sea. *Environ. Sci. Technol.* **38,** 1282–1295.

Tiedje, J. M., J. F. Quensen, III, J. Chee—Sanford, J. P. Schimel, and S. A. Boyd, 1993. Microbial reductive dechlorination of PCBs. *Biodegradation* **4,** 231–240.

Tissier, M. J. and J. L. Oudin, 1975. Influence de la pollution petroliere sur la reparation des hydrocarbures devases marines. In *Advances in Organic Geochemistry, 1973* (B. Tissot and F. Bienner, eds). pp. 1029–1041. Proceedings 6th International Meeting. Editions Technip. Paris, France.

UNEP/IOC/IAEA, 1992. Determination of petroleum hydrocarbons in sediments. *Reference Methods for Marine Pollution Studies,* **No.20,** UNEP, 1992. 75 pp. Report available from Marine Environmental Studies Laboratories, International Atomic Energy Agency, Marine Environmental Laboratory, B. P. No. 800-MC 98012, MONACO CEDCEX.

van der Linden, A. C., 1978. Degradation of oil in the marine environment. In D*evelopments in biodegradation of hydrocarbons* –1. Watkinson, R. J. (ed). Applied Science Publishers, LTD. London. 165–200.

Varanasi, U. and J. E. Stein, 1991. Dispositionof xenobiotic chemicals and metabolites in marine organisms. *Environ. Health. Perspect.* **90,** 93–100.

Vetter, W., U. Klobes, and B. Luckas, 2001. Distribution and levels of eight toxaphene congeners in different tissues of marine mammals, birds, and cod livers. *Chemosphere* **43,** 611–621.

Wakeham, S. G., A. C. Davis and J. L. Karas, 1983. Mesocosm experiments to determine the fate and persistence of volatile organic compounds in coastal seawater. *Environ. Sci. Technol.***17,** 611–617.

Waid, J. S. (ed). *PCBs and the environment.* Vols. 1–3. CRC Press, Boca Raton, FL.

Wania, F. and D. Mackay, 1996. Tracking the distribution of persistent organic pollutants. *Environ. Sci. Technol.* **30,** 390A -396A.

Watkinson, R. J. (ed.), 1978. *Developments in biodegradation of hydrocarbons* –1. Applied Science Publishers, LTD. London. pp232.

Wells, P. G., R. A. Duce, and M. E. Huber, 2002. Caring for the sea – accomplishments, activities and future of the United Nations GESAMP (the Joint Group of Experts on the Scientific Aspects of Marine Environmental Protection). *Ocean and Coastal Manag.***45,** 77–89.

Whitehous, B. G., 1984. The effects of temperature and salinity on the aqueous solubility of polynucelar aromatic hydrocarbons. *Mar. Chem.* **14,** 319–332.

WHO, 1976. Polychlorinated biphenyls. *Environmental Health Criteria 2* United Nations Environment Programme and the World Health Organization, World Health Organization, Geneva. 85 pp.

WHO, 1993. *Polychlorinated Biphenyls and Terphenyls, 2nd Ed. Environmental Health Criteria 140* S. Dobson and G. J. van Esch (eds.), World Health Organization, Geneva 682 pp.

WHO/IPCS, 1994. *Environmental Health Criteria 162, Polybrominated Diphenyl Ethers, 1st.ed.;* World Health Organization/ International Programme on Chemical Safety, Geneva, 1994; ISBN 92–4–157162–4.

Wiberg, P. and C. K. Harris, 2002. Desorption of p,p' DDE from sediment during Resuspension events on the Palos Verdes shelf, California: a modeling approach. *Contin. Shelf. Res.* **22,** 1005–1023.

Wiberg, P., D. E. Drake, C. K. Harris, M. A. Noble, 2002. Sediment transport on the Palos Verdes shelf over seasonal to decadal time scales. *Contin. Shelf. Res.* **22,** 987- 1004.

Wijayaratne, R. D. and J. C. Means, 1984a. Sorption of polycyclic aromatic hydrocarbons by natural estuarine colloids. *Mar. Environ. Res.* **11,** 77–89.

Wijayaratne, R. D. and J. C. Means, 1984b. Affinity of hydrophobic pollutants for natural estuarine colloids in aquatic environments. *Environ. Sci. Technol.* **18,** 121–123.

Wise, S. A, S. N. Chester, F. R. Guenther, H. S. Hertz, L. R. Hilpert, S. E. May, and R. M. Parris, 1980. interlaboratory comparison of determination of trace level hydrocarbons in mussels. *Anal. Chem.* **52,** 1828–1833.

Xie, W-H., W-Y. Shiu, and D. Mackay, 1997. A review of the effect of salts on the solubility of organic compounds in seawater. *Mar. Environ. Res.* **44,** 429–444.

Young, D. R., D. J. McDermott, and T. C. Heesen., 1976a. DDT in sediments and organisms around southern California outfalls. *J. Water Pollut. Control. Fed.* **48,** 1919- 1928.

Youngblood W. and M. Blumer, 1975. Polycyclic aromatic hydrocarbons in the environment: homologous series in soils and recent marine sediments. *Geochim. Cosmochim. Acta* **39,** 1303–1314.

Zakaria, M. P., H. Takada, S. Tsutsumi, K. Ohno, J. Yamada, E. Kouno, and H. Kumata, 2002. The distribution of polycyclic aromatic hydrocarbons (PAHs) in rivers and estuaries in Malaysia: A widespread input of petrogenic PAHs. *Environ. Sci. Technol.* **36,** 1907–1918.

Zegers, B. N., W. E. Lewis, K. Booij, R. H. Smittenberg, W. Boer, J. DeBoer, and J. P. Boon, 2003. Levels of Polybrominated diphenyl ether flame retardants in sediment cores from Western Europe. *Environ. Sci. Techno.* **37,** 3803–3807.

Zepp, R. G. and P. F. Schlotzhauer., 1979. Photoreactivity of selected aromatic Hydrocarbons in water. In: *Polynuclear Aromatic Hydrocarbons* (P. W. Jones and P. Leber (eds). Ann Arbor Science Publishers, Ann Arbor, MI.

# Part 3.
# MULTIPLE TIME SCALES OF VARIABILITIES

# Chapter 16. BIOLOGICAL CONSEQUENCES OF INTERANNUAL TO MULTIDECADAL VARIABILITY

FRANCISCO P. CHAVEZ

*Monterey Bay Aquarium Research Institute*

**Contents**



## 1. Introduction

Ecologists have often viewed the physical environment as a stable background against which biotic interactions drive population change and structure communities (May, 1973). However over the past two decades, strong El Niños, the ozone hole, and the looming specter of global warming forced the uncomfortable realization that the physical environment is changing, even on the relatively short timescales of ecological study, and that man's activities may affect climate in unforeseen ways. Climate and the physical environment have thus re-emerged as major themes in ecological science. In the oceans, it is clear that natural climate variability can have large impacts on ecosystem structure and biological productivity (Barber and Chavez, 1983; Chavez, 1987; McGowan, 1989). The correlation between climate variability and the productivity and structure of ocean ecosystems has been well established (Barber and Chavez, 1983; Chavez et al., 2003; Hare and Mantua, 2000; Hurrell et al., 2003). Climate-driven changes in ocean circulation, ocean mixing and/or dust deposition, can regulate the overall productivity of an ocean ecosystem by changing the supply of a limiting nutrient. Changes in primary productivity then cascade through every trophic level (Francis et al., 1998). Climate can also influence animal populations directly through effects on recruitment, competitive advantages or predation. Of particular interest are relationships between abiotic (bottom-up climate impacts on overall ecosystem productivity) and biotic (top-down climate impacts on competition and predation) effects (Ottersen et al., 2001).

*The Sea*, Volume 13, edited by Allan R. Robinson and Kenneth H. Brink
ISBN 0-674-01526-6 

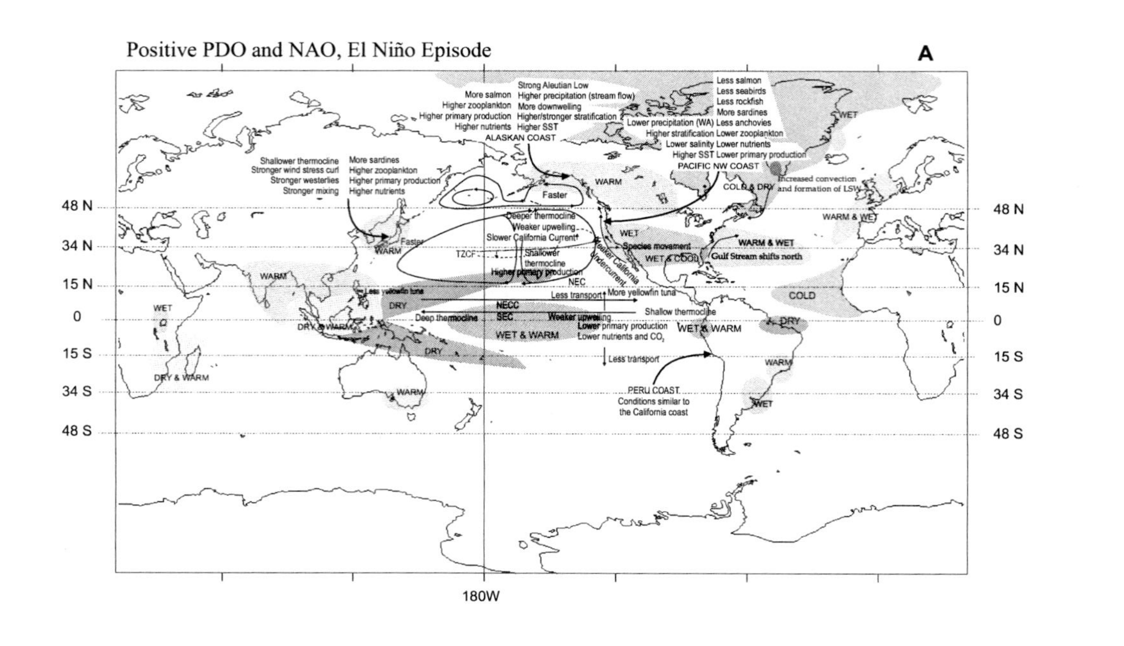

Positive PDO and NAO, El Niño Episode
A
48 N
34 N
15 N
0
15 S
34 S
48 S
180W
Strong Aleutian Low
More salmon
Higher zooplankton
Higher primary production
Higher nutrients
Higher precipitation (stream flow)
More downwelling
Higher/stronger stratification
Higher SST
ALASKAN COAST
Less salmon
Less seabirds
Less rockfish
More sardines
Less anchovies
Lower zooplankton
Lower nutrients
Lower primary production
Lower precipitation (WA)
Higher stratification
Lower salinity
Higher SST
PACIFIC NW COAST
Increased convection
and formation of LSW
Shallower thermocline
Stronger wind stress curl
Stronger westerlies
Stronger mixing
More sardines
Higher zooplankton
Higher primary production
Higher nutrients
Faster
Deeper thermocline
Weaker upwelling
Slower California Current
TZCF
Shallower
thermocline
Higher primary production
Weaker California
Undercurrent
NEC
Less yellowfin tuna
More yellowfin tuna
Less transport
NECC
SEC
Deep thermocline
Shallow thermocline
Weaker upwelling
Lower primary production
Lower nutrients and $CO_2$
Less transport
PERU COAST
Conditions similar to
the California coast
Species movement
WARM & WET
Gulf Stream shifts north
WET & COOL
COLD & DRY
WARM & WET
COLD
WET & WARM
WET & WARM
WET
WARM
DRY
DRY & WARM
DRY & WARM
WARM
WET
WET
WARM
DRY
WARM
WARM
WET

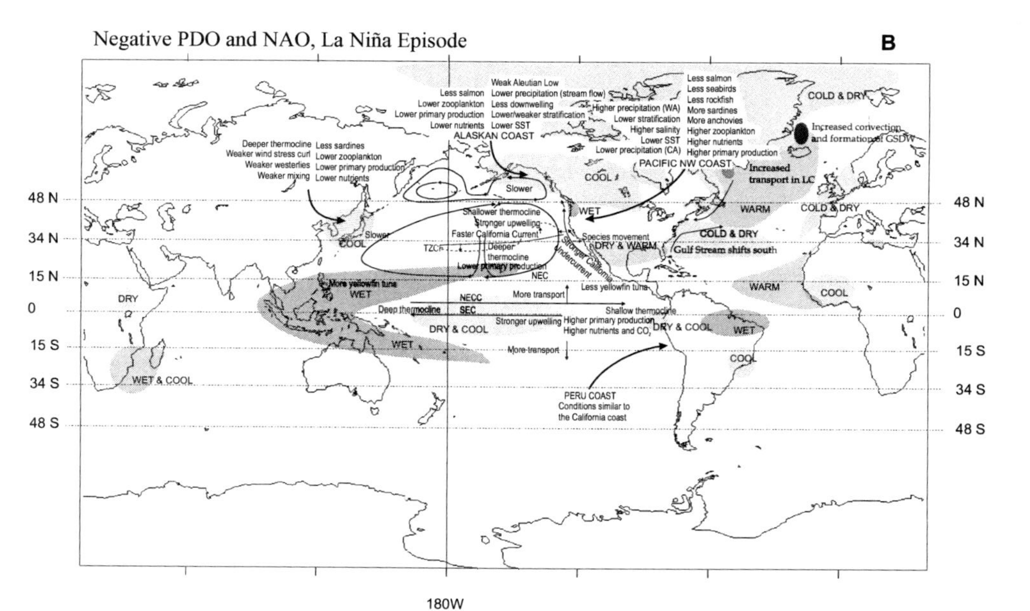


Figure 16.1 Summary of conditions during El Niño and the positive phases of the Pacific Decadal Oscillation (PDO) (or El Viejo) and the North Atlantic Oscillation (NAO) (A) compared to La Niña and the negative phase of the PDO (or La Vieja) and the NAO regime (B). During La Niña or La Vieja in the Pacific the eastern boundary and equatorial limbs of the subtropical gyre circulation are faster, leading to enhanced upwelling and primary productivity. Zooplankton populations increase as well. The anchovy flourishes in Peru and California. Off California and Oregon the salmon do well while they decline in Alaskan waters. Seabirds also increase in the upwelling systems. As a result of a faster gyre the thermocline and nutricline deepen in the subtropical gyre and productivity decreases. Production in the northeastern limb of the gyre boundary also increases. In the western and northwestern boundaries there is the suggestion of weaker circulation. The westerlies weaken in the northwest Pacific leading to lower Ekman pumping rates and mixed layers. Primary productivity and sardine abundance decline. In the warm waters of the northeastern tropical Pacific there are indications of lower yields of yellow fin tuna during the anchovy regime. A nearly opposite picture can be painted during the warm eastern Pacific El Niño and El Viejo phases. One notable difference between El Niño and El Viejo is production in the Gulf of Alaska which seems to be enhanced during El Viejo but reduced during El Niño. Similar large scale changes are reported in the Atlantic in association with the NAO. Not all effects can be shown or are understood but the reader should get the impression of the large scale nature of the variability. Compiled from Ropelewski and Halpert (1987), Chavez et al., (2003) and Hurrell et al., (2003).

Webster defines climate "as the average course or condition of the weather at a place usually over a period of years as exhibited by temperature, wind velocity, and precipitation." However, the average itself changes depending on the period used. For example, sea surface temperature (SST) for the coasts of Peru and California for the years 1993–1996 was warmer than during 1999–2003, and between during the 1997–98 El Niño, SST was extremely warm. These changes in temperature were accompanied by fluctuations in ocean productivity, with warmer years being less productive than cooler years. What causes climate to change from year to year and decade to decade? What are the consequences of this climate variability? Complete answers to these questions are still forthcoming although significant progress has been made over the past several decades, particularly in understanding the consequences of climate variability. For example, the warm SST observed during 1997–98 was a result of a strong El Niño that was well documented in the equatorial Pacific (McPhaden, 1999; Chavez et al., 1999) and along the west coast of the Americas (Chavez et al., 2002a; Tarazona and Castillo, 1999). El Niño is a prime example of the insight gained on the consequences of climatic variability. While El Niño had been recognized off Peru since the ancient civilizations of the Incas it was the large 1957–58 El Niño that brought international attention to the phenomenon (Sette and Isaacs, 1960). Following the 1982–83 El Niño it became clear that the oceanic perturbations in the tropical Pacific had global effects on climate (Rassmussen and Wallace, 1983). Oceanic effects were originally thought to be restricted to the tropical and eastern Pacific, but careful studies in the center of the Pacific Ocean close to Hawaii uncovered El Niño effects there as well (Karl et al., 1995). The past several decades has seen growing awareness of La Niña, the counterpart or opposite condition of El Niño.

Recently, focus has shifted to longer decade-scale changes that show remarkable basin-wide coherence and again, strong impacts on oceans ecosystems (Chavez et al., 2003; Hare et al., 2000, Hurrell et al., 2003). These multi-decadal changes help explain the differences in SST observed before and after the 1997–98 El Niño (Bond et al., 2003; Chavez et al., 2003; Hare and Mantua, 2000; Minobe, 2002; Peterson and Schwing, 2003; Schwing and Moore, 2000). The period prior to 1997–98 was associated with a warm quarter century, the period after may be a cool one.

This paper is a review of the biological consequences of interannual to multidecadal fluctuations in marine coastal communities. It is by no means comprehensive; we focus on three case studies: (1) El Niño/La Niña, (2) the Pacific Decadal Oscillation (PDO) or El Viejo/La Vieja and (3) the North Atlantic Oscillation (NAO). El Niño is treated in the greatest depth since it has been studied more extensively. Decadal-scale research has seen a recent acceleration of effort (Mantua and Hare, 2002) and we attempt to summarize that effort here. Oddly, Atlantic Ocean fluctuations are not as well understood, and the phenomenology and relation of the NAO to El Niño/La Niña and the PDO are examined briefly. The reader should consult the following books or special journal issues for greater detail on these three case studies (Arntz et al. 1985; Arntz and Fahrbach, 1996; Beamish, 1995; Chavez et al., 2002a; Diaz and Markgraf, 1992, 2000; Glynn, 1990, 2001; Hare et al., 2000; Hurrell et al, 2003; Kawasaki et al., 1991; McKinnell et al., 2001; Phi;ander, 1990; Robinson and del Pino, 1985; Sette and Isaacs, 1960; Stenseth et al., 2004; Tarazona and Castillo, 1999; Tarazona et al., 2001; Trillmich and Ono, 1991; Wooster and Fluharty, 1985). Studies of these natural climate variations are intrinsically important, since climate affects both man and the ecosystems he is dependent on. Additionally, such studies will provide a foundation for future work on global change. The magnitude of the natural fluctuations is of similar order as predicted to result, initially, from man-induced perturbations to global climate.

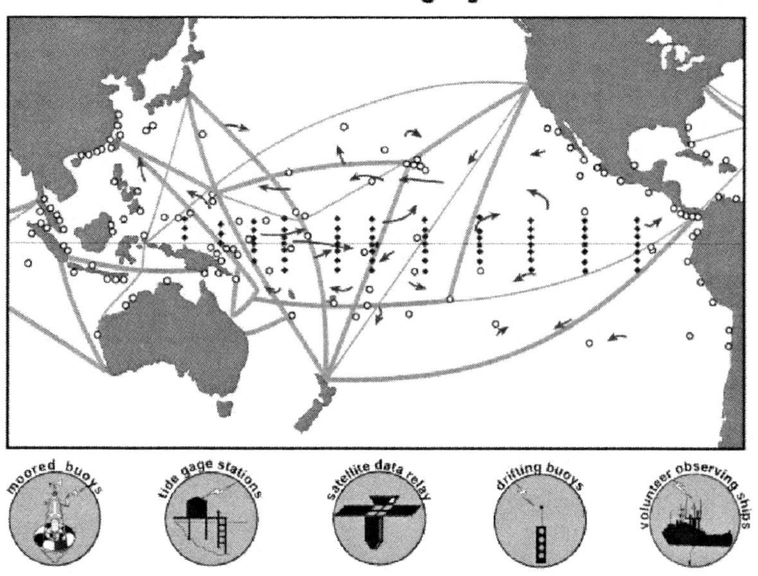


Figure 16.2 Array of observing systems in the tropical Pacific deployed to monitor the appearance of El Niño and La Niña. Redrawn from McPhaden et al., 1998.

## 2. Case Study: El Niño/La Niña

### *2.1 What is El Niño? Physical background.*

El Niño appears irregularly, once every three to seven years, and with varying intensity. El Niño was originally named by the fishermen of northern Peru in the late nineteenth century (Carranza, 1891) after a warm southward current that appeared every year around Christmas (The Christ child= El Niño). The fishermen recognized warm wet years during which their fisheries changed dramatically, floods caused damage and Peruvian deserts became grassland (Eguiguren, 1894). Images of these changes were left behind on the ceramics of pre-Inca civilizations. Years of unusually high rainfall in northern Peru were associated with what the locals thought was an intensification of the annual current. It was not until the 1960's that a Swedish scientist by the name of Bjerknes (Bjerknes, 1966, 1969), working at the Scripps Institution of Oceanography linked warming of the coastal ocean off Peru (and the equatorial Pacific) with larger scale climatic phenomena. It was then that a relatively small and inoffensive coastal current was associated with dramatic global weather disturbances (Chavez, 1986). Over the last 30 years the influence of El Niño, on oceanic and atmospheric conditions throughout the globe has been recognized widely. El Niño, of course, has been around for much longer than that, it can be seen in proxy data (such as lake sediments, coral growth rings, tree rings and ice cores) going back hundreds and even thousands of years (DeVries, 1987; Diaz and Markgraf, 2000; Quinn et al., 1987; Thompson et al., 1984). Today El Niño is associated with unusually warm water over large areas of the eastern Pacific Ocean (Cane, 1983; Philander, 1990; McPhaden, 1999). Because the ocean retains heat significantly longer than the atmosphere, the large scale changes in ocean temperature lead to persistent changes in global weather patterns via atmospheric teleconnections (Figure 16.1; Gill and Rasmusson, 1983; Rasmusson and Wallace, 1983). For example, hurricane activity in the North Atlantic and the southeastern United States is partly a function of tropical Pacific sea surface temperatures (Gray, 2002). Similar weather changes occur worldwide, making early warning of El Niño an important priority for many nations. The current basis for early prediction is an observing system consisting of satellites, moorings, drifters and ships (Figure 16.2).

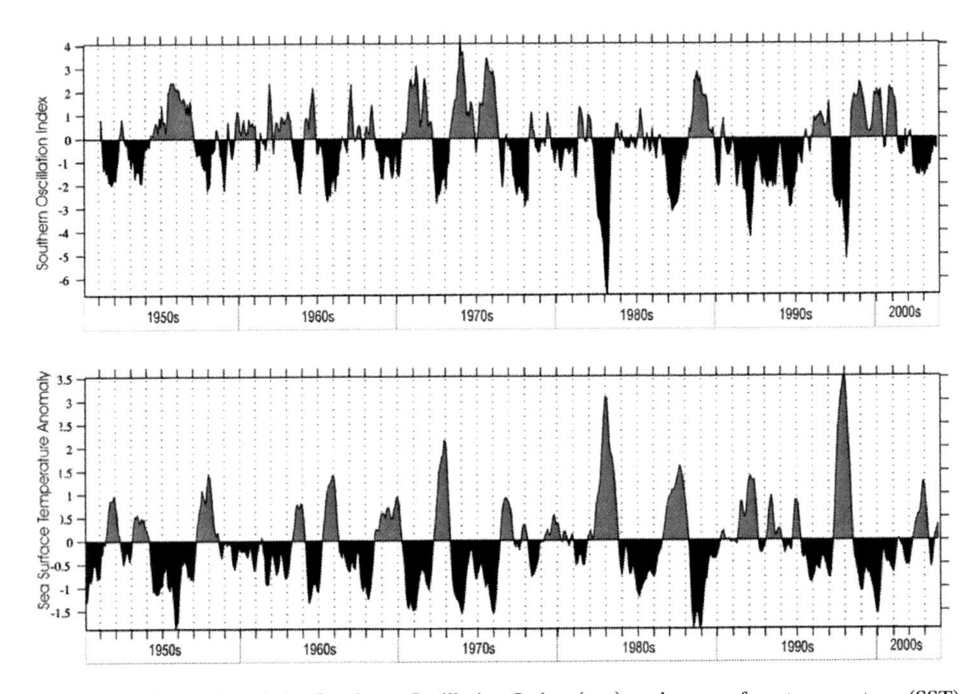


Figure 16.3 Time series of the Southern Oscillation Index (top) and sea surface temperature (SST) anomalies (bottom) for the equatorial Pacific. When there is a strong barometric pressure gradient (high in Tahiti and low in Darwin, Australia) the SOI is positive and SST is anomalously cold and La Niña conditions exist. When the opposite occurs we have El Niño. Strong El Niños occurred in 1982–83 and 1997–98.

El Niño is associated with an atmospheric pressure phenomenon called the Southern Oscillation—in fact El Niño is often called the El Niño/Southern Oscillation or ENSO (Philander, 1983). This nomenclature creates some confusion since ENSO most often refers to the full El Niño/La Niña cycle. The Southern Oscillation was discovered by a British scientist by name of Sir Gilbert Walker who in the 1920's observed that in years when barometric pressure was high over southern South America, it was low over Australia, and in other years the opposite occurred

(Walker, 1924). Walker named the phenomena the Southern Oscillation. Today a Southern Oscillation Index (SOI) is calculated as the difference between the barometric pressure at Tahiti and that at Darwin, Australia, and together with sea surface temperature in the tropical Pacific, the SOI is used by scientists to assess the occurrence and strength of El Niño (Figure 16.3). More recently, in the 1980's, Peruvian scientists coined the name La Niña for periods of anomalously cool temperature along the Peruvian coast. Although La Niña is recognized to have significant climatic impacts, it is less well understood than El Niño, and it is not yet clear if La Niña represents a symetrical, cool opposite to El Niño. Once underway, El Niño is relatively well understood and its phenomenology is at least generally predictable due the array of ocean observing systems that have been in place for the last 25 years (Figure 16.2). Although El Niño prediction is an active research area and progress is being made (Chen et al., 2004), the precise environmental factors that initiate El Niño and determine its periodicity and strength remain enigmatic, It has been suggested that the system is chaotic and therefore unpredictable (Vallis, 1986); on the other extreme El Niño has been linked to volcanic eruptions in the tropics (Adams et al., 2003). What is clear is that it is linked to a disruption in the *coupled* ocean-atmosphere system.

In order to better understand El Niño it is informative to first look at the so-called normal condition. As a consequence of the earth's rotation, an asymmetry in winds and sea surface temperatures (SST) develops in the tropical Pacific, such that the eastern equatorial Pacific is remarkably and to some extent abnormally cool. SST on the equator around the Galapagos Islands can be less than 20 C, despite continuous heating by the tropical sun. This cool water develops as a result of several inter-related processes. First, the easterly trade winds along the equator continuously move sun-warmed surface water to the western Pacific—so much so that the sea level off Indonesia is 30 centimeters higher than off Peru. This western flow produces the largest area of warm (~30 C) surface water in the world in the western Pacific and eastern Indian Oceans. This area is referred to as the western Pacific 'warm pool'. A second reason for the SST gradient—cool in the east and warm in the west—can be found in the vertical distribution of ocean temperature along the equator. A thin surface layer of warm surface water is typically separated from a thick reservoir of cool water below by another thin layer referred to as the thermocline (a region of rapid change in temperature). The thickness of the warm surface layer is not uniform. In the eastern Pacific the trade winds move surface water eastwards, and the warm layer thins to 50 m or less. But in the west, as heat and water accumulate, the warm layer thickens to ~200 meters. The final driver of the east-west asymmetry in temperature is associated with wind-driven upwelling. Upwelling refers to the vertical movement of water from about 60–100 meters depth to the surface, and along the Equator it is caused by the combined action of the easterly trades and the Coriolis effect. In the eastern Pacific where the warm layer is less than 60 meters deep, the upwelled water is very cold and forms what is known as the equatorial 'cold tongue'. However in the western Pacific where the warm pool is deep, even though upwelling continues SST is warm. The atmosphere responds directly to this east/west SST gradient. High pressure forms over the cool waters of the eastern Pacific, and low pressure forms over the warm western Pacific waters. This pressure difference intensifies the trades, which in turn reinforce the east/west SST gradient. This close coupling of atmosphere and ocean lies at the heart of equatorial climate dynamics (Philander, 1990). El Niño represents a temporary disruption to this system.

During El Niño, heat and water are redistributed to create more uniform equatorial surface temperatures, with warm water extending from Indonesia to Peru. This redistribution of heat results from a slackening or reversal of the easterly trades, particularly in the central and western Pacific (McPhaden, 1999). As the trades weaken or reverse, the east/west gradients of heat and sea surface height surge eastwards towards Peru. But given the chicken and egg nature of the ocean-atmosphere system, it is not clear what comes first—changes to wind or changes in the ocean? Once again the question becomes what initiates El Niño. One theory (Wyrtki, 1982) suggests that before El Niño can occur there must be a subtle but significant build-up of heat and water in the western Pacific. As the warm pool grows and expands eastward, often over several years, air over the warm pool warms and rises as atmospheric convective activity, eventually disrupting the easterly trades. The first notable signs of El Niño are strong westerly winds along the equator in the western Pacific between Indonesia and the date line. Not all westerly wind bursts result in El Niño but the ones that do are reinforced in the following manner. The wind bursts spurn Kelvin waves that propagate along the equator from west to east at speeds greater than 200 km per day, raising sea level and deepening the thermocline (Cane, 1983; Philander, 1990). When they reach South America, the waves are reflected poleward towards Alaska and Chile, now trapped against the coast by Corolis. El Niño is thus propagated through the ocean to mid and high latitudes off the Americas by waves that originate in the western Pacific (Enfield and Allen, 1980; Enfield et al., 1987; Shaffer et al., 1997).

Passage of the Kelvin waves affects a number of important feedbacks in the climate/ocean system of the tropical Pacific. In the eastern Pacific, a depressed thermocline produces warmer SSTs since warmer water now feeds upwelling in the equatorial cold tongue (this is also true in upwelling regions off California, Peru and Chile during El Niño). This warmer SST reduces the east/west atmospheric pressure gradient and trade winds, and further reinforcing the changes in thermal structure. In combination the weak or reversed trades and Kelvin waves with their associated thermal anomalies also lead to changes in the major currents. Waters from north and south of the equator converge in the equatorial upwelling region, and waters from the western Pacific warm pool migrate eastward. The extent of the eastward penetration of warm pool waters into the cold tongue region of the central and eastern tropical Pacific is a function of El Niño intensity.

A key component of the circulation of all upwelling regions is the presence of a subsurface undercurrent. These currents flow beneath the surface current, in the opposite direction, and are driven by pressure gradients established by along-current gradients in sea surface height. Along the equator during normal conditions, the equatorial undercurrent is 'pushed' rapidly eastward at or below the thermocline by pressure associated with the high sea level in the warm pool region (surface water is pushed in the opposite direction—west—by the trades). As with equatorial Kelvin waves, flow of the equatorial undercurrent jet is trapped along the equator by the Coriolis effect. During El Niño the equatorial undercurrent first accelerates with the passage of the Kelvin waves, and then slows as the pressure gradient between the eastern and western equatorial Pacific decreases. During strong El Niños, the undercurrent can disappear (McPhaden, 1999). Along the eastern boundaries of the Pacific, poleward-flowing undercurrents are typical of normal conditions, but El Niño's influences on these are not as well documented as those on the equatorial undercurrent. Some evidence supports changes that are similar—acceleration after the passage of Kelvin waves (Smith, 1983) followed by a weakening or disappearance (Chavez et al., 2002b; Huyer et al., 1987).

Recovery from El Niño begins with an unusual shoaling of the thermocline along the equator. Once the trades are re-established, SST drops dramatically since abnormally cool waters feed the wind-driven upwelling, signaling the end of El Niño. Cooler than average or La Niña years often follow El Niño as the system seems to "over-react".

The frequency and intensity of El Niños and La Niñas may be related to the background state of the climate (Fedorov and Philander, 2000). Longer term cycles like those associated with El Viejo (old man) and La Vieja (old woman) have an influence on El Niño. This particular cycle has a period of around 50 years. When the eastern Pacific is warmer than average, during El Viejo, El Niños may be more frequent and of greater intensity. Similarly, during La Vieja, La Niñas may be frequent or stronger. Longer time series, both historical and present, will enlighten us further on El Niño frequency and intensity (Enfield, 1989, Smith et al., 2001a) and the potential effects of anthropogenic perturbations (Herbert and Dixon, 2003).

The effects of El Niño are most easily described along the equator and the coast of Peru because these regions experience the most dramatic warming—the South American coast is intimately linked to equatorial dynamics and can experience SST warming of 10° C during El Niño (Barber and Chavez, 1983). Sea level and thermocline displacements associated with equatorial Kelvin waves have been well documented along the South American coast from Peru to Chile (Enfield and Allen, 1980; Shaffer et al. 1997, Ulloa et al., 2001) and are thought to be responsible for the observed warming. At higher latitudes in the north eastern Pacific the developmental sequences and drivers have not been as clear. The points and embayments in coastal topography north of the equator act to bleed energy from or trap the coastally trapped waves as they propagate northward. In particular the Gulf of California has been shown to seasonally trap propagating sea level anomalies during the 1997–98 El Niño (Strub and James, 2002). Given these complicated ocean dynamics some have argued that the El Niño warming in the northeast Pacific may be more closely related to atmospheric teleconnections via expansion of the Aleutian Low (Simpson, 1984a,b), which suppresses upwelling-favorable winds in the northeast Pacific. However, as onset of the northern El Niño precedes changes in the Aleutian Low (Chavez 1996, Huyer and Smith, 1985), remotely forced oceanic perturbations must still play a significant role. It is more than likely that both remote oceanic and local atmospheric processes lead to the observed physical and biological disturbances, particularly at temperate and high latitudes.

### 2.2. *Conceptual models of El Niño-driven biological effects*

The effects of El Niño on ocean ecosystems can be linked to the physical processes described above since the vertical distribution of nutrients parallels that of temperature. The thin warm upper layer is low in nutrients and the deeper cold layer beneath the thermocline has high levels. This pattern is a result of (1) the high absorption of light by water and (2) gravity. In a perfectly clear ocean the amount of light reaching 100 meters is not sufficient to drive appreciable levels of photosynthesis in an ocean with an average depth of 3800 meters. In this sunlit layer referred to as the euphotic zone (< 100 meters), new and larger particles are created by photosynthesis, phytoplankton growth and the consumption of phytoplankton by animals. These organic particles are eventually moved to the deep cold ocean by gravity, where they decay. In this sense the deep ocean is similar to compost —-a mixture of decaying organic material used as fertilizer—that an average household might create in their backyard. The organic material in compost decays to inorganic minerals, many of which are required for plant growth and thus are called nutrients. Processes that bring the cold, nutrient-rich deep water closer to the surface, such as upwelling, move nutrients into the euphotic zone and thus enhance photosynthesis and biological production. As a result of equatorial upwelling and creation of the cold plume, the eastern tropical Pacific and the eastern Pacific in general are regions of enhanced biological productivity (Barber and Chavez, 1983; 1986).

There are two important physical processes responsible for enhanced productivity in upwelling regions: (1) the thermocline, and more importantly the associated nutricline (nitrate), must be shallow, on the order of 40–80 m or less, so that (2) upwelling-favorable winds can draw nutrient-rich waters into the sunlit zone, thereby stimulating photosynthesis (Barber and Chavez, 1983). However during El Niño, the biological productivity of the South American coastal upwelling regions and the entire cold plume area of the eastern tropical Pacific (ETP) are dramatically reduced. Because local upwelling-favorable winds in the ETP are maintained and may even intensify (Enfield, 1981), the reduction of productivity during El Niño has been attributed to thermocline and nutricline deepening by Kelvin waves (Barber and Chavez, 1983). A second, not fully resolved, change in thermocline depth and upwelling is that associated with changes in wind stress curl (Halpern, 2002). The wind stress curl may reverse along the coast, going from upwelling (increasing offshore) to downwelling (decreasing offshore) favorable. Halpern argues this effect may be responsible for the long-term (6–18 months) deep thermocline along the eastern boundary. Whatever the mechanism, a deep coastal thermocline is established during El Niño and while coastal upwelling continues, the upwelled waters are low in nutrients and productivity declines sharply (Figures 16.4 and 16.5) (Barber and Chavez, 1983; Bograd and Lynn, 2001; Chavez et al., 2002b). The changes in upwelling are also reflected in the partial pressure of carbon dioxide, with strong reductions during El Niño (Friederich et al., 2002). Nutrient supply in the eastern Pacific can be reduced by an order of magnitude during strong events (Barber and Chavez, 1983; Chavez et al., 2002b), and decreases in primary production can be to one-fifth of normal have been reported for Peru and California (Barber and Chavez, 1983; Chavez et al., 2002b). The relationship between local wind, thermocline depth and nutrient supply that develops during El Niño suggests that global warming-related increases in winds alone (Bakun, 1990) may not lead to higher primary production in coastal upwelling systems—a shallow thermocline is critical for local winds to effectively pump nutrients into the euphotic zone.

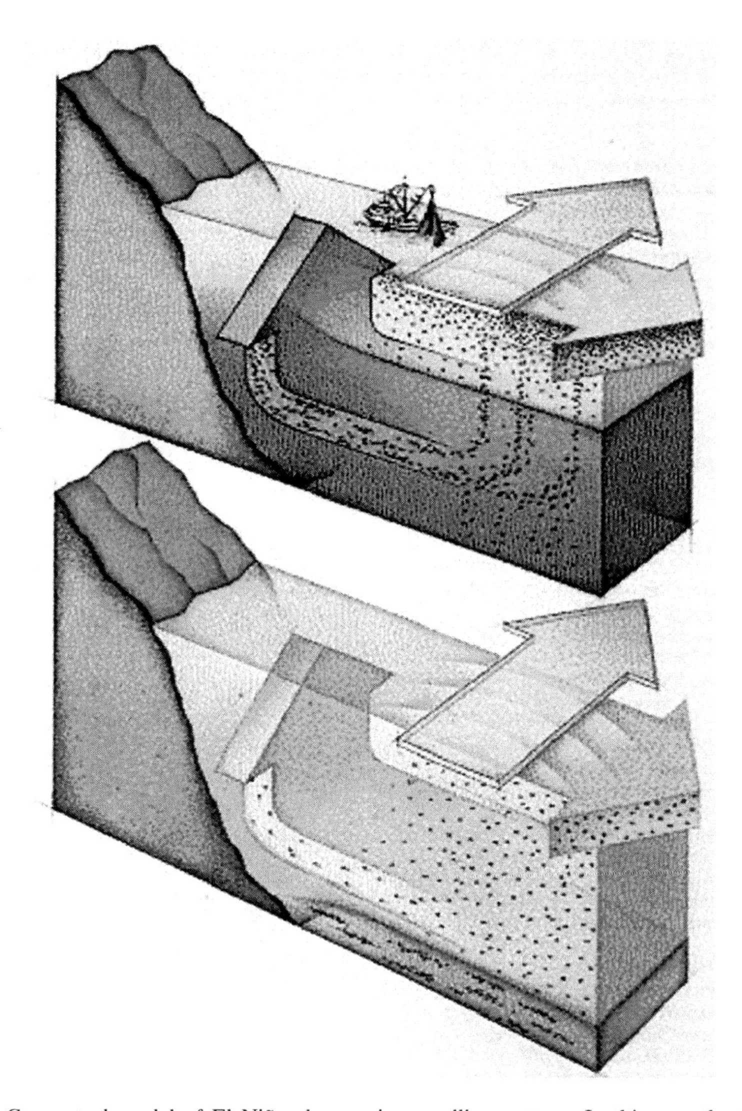

Figure 16.4 Conceptual model of El Niño changes in upwelling systems. In this case the California condition is represented with northwesterly winds driving coastal upwelling. During normal conditions (top) upwelled water rich in nutrients is recruited from below the thermocline leading to a very productive coastal ecosystem. In the El Niño condition the thermocline has deepened and the waters recruited are low in nutrients and productivity is dramatically reduced. Redrawn from Canby (1984).

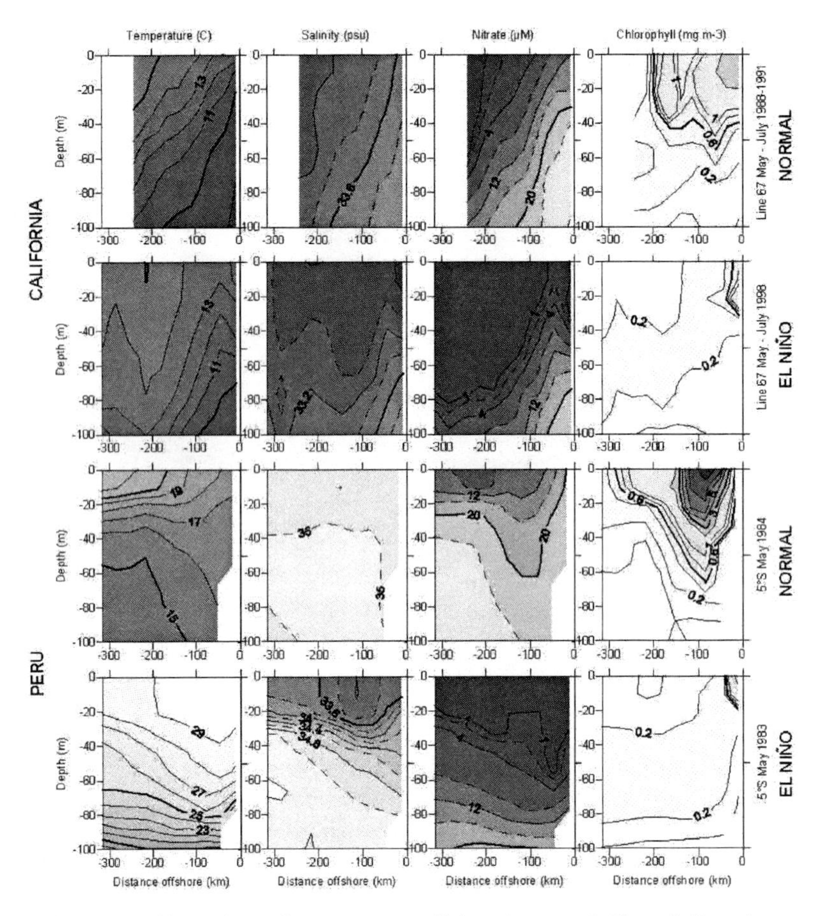


Figure 16.5 Oceanographic sections of temperature, salinity, nitrate and chlorophyll made perpendicular to the coasts of California at about 36°N (top) and Peru at about 5°S (bottom) during normal and El Niño conditions. In both cases the thermocline is shallow and breaks the surface near the coast during normal conditions. Surface waters are also colder, saltier and higher in nitrate and chlorophyll. During El Niño these conditions change dramatically to a deeper thermocline and warmer, fresher, low nitrate and low chlorophyll surface waters. Redrawn from Barber and Chavez (1983) and Chavez et al. (2002b).

## *2.3. Biological consequences of ENSO in the eastern Pacific—Primary producers*

The coastal oceans of the eastern Pacific are, on average, rich and productive. During active upwelling deep waters low in phytoplankton are brought to the surface at a rate that exceeds phytoplankton growth, and population numbers and biomass remain low at the site of upwelling. The phytoplankton increase offshore and downstream of the upwelling site and biological uptake consumes the surface nutrients. Nitrate depletion typically occurs several hundred kilometers from shore (Figure 16.6), delimiting the on/offshore dimension of the productive zone or habitat. Highest near-coast (0–25 km) concentrations of phytoplankton are found during weak upwelling or after strong upwelling-favorable winds relax. During El Niño, continued upwelling-favorable winds sustain some production in at the site of upwelling, but the size of the area of enhanced primary production is reduced dramatically. An overly simplistic conceptual model of pelagic ocean ecosystems is to categorize them into two general types (Chavez et al., 2002b). The first is a coastal ecosystem that can form dense phytoplankton concentrations or blooms and is dominated by small colonial centric diatoms. Biomass accumulates because at the beginning of the growth cycle, during upwelling, conditions are ideal and nutrients do not limit growth. Phytoplankton can outgrow their predators and bloom in patches. In this system diatoms are consumed directly by small pelagic fish like anchovies and sardines. In this very short food chain nutrients are transferred directly to fisheries resources and these are plentiful (Ryther, 1969). While the food chain is short the system is leaky in that a significant fraction of the primary production either (a) sinks to the bottom where it supports a rich benthic fauna and/or creates anoxic conditions (Margalef, 1978), or (b) is advected horizontally out of the upwelling system. The second ecosystem characterized as oceanic has at its base very small plants (less than 1 micron in diameter) referred to as picoplankton. The smallest and most abundant of these picoplankton are part of a group referred to as cyanobacteria. The most common (prochlorophytes) were not discovered until the 1980s yet represent the most abundant photosynthetic organisms in the ocean (Chisholm et al., 1988). Unlike diatoms, picoplankton thrive under low nutrient conditions. The consumers of picoplankton are also very small and have growth rates similar to the picoplankton, and therefore do not allow picoplankton biomass to accumulate as blooms. Together with heterotrophic bacteria the picoplankton and their consumers comprise the so-called microbial food web which is characterized as a complex and highly efficient but, due to low nutrient input, low production system (Azam et al., 1983; Pomeroy, 1974). Much fewer higher trophic level or fisheries resources are supported by picoplankton ecosystems because of (a) low nutrient input, (b) efficient internal recyling of the nutrients that are available, and (c) multiple trophic transfers are needed to transform picoplankton production to fisheries resources, so that only a small portion of microbial food web nutrients reach higher trophic levels. The changes associated with El Niño and these two ecosystems are depicted in Figure 16.6. A very narrow, diatom-dominated productive zone remains during El Niño but its areal extent is only a small fraction of what might be found during normal or La Niña conditions (Figure 16.5). The maintenance of the narrow coastal strip has further biological consequences. Small pelagic fish and krill concentrate in the cool and productive waters and become easy prey to fishermen and their natural predators.

There is evidence that phytoplankton that produce chemicals that are harmful to humans, such as dinoflagellates, can increase during El Niño. Large blooms of dinoflagellates create phenomena called red tides. Weak to moderate El Niño events are more likely to trigger red tides because of an optimum combination of continued but low nutrient supply and moderately increased stratification apparently favors toxin-producing phytoplankton. Such conditions also occur during late summer and fall of normal years. During strong El Niños,s nutrient supply falls below levels that can sustain the blooms and picoplankton ecosystems dominate. Pennate diatoms capable of producing domoic acid, another toxic chemical, have also been hypothesized to increase under similar conditons during El Niño (Fryxell et al., 1997). Warm water species of phytoplankton, in particular rare and solitary oceanic dinoflagellates, become notable during El Niño and have been suggested to be early indicators that El Niño is underway (Avaria and Munoz, 1987; Ochoa and Gomez, 1987). The appearance of warmer water species across all trophic levels in the normally cold-water eastern Pacific upwelling ecosystems is one of the dramatic effects of El Niño.

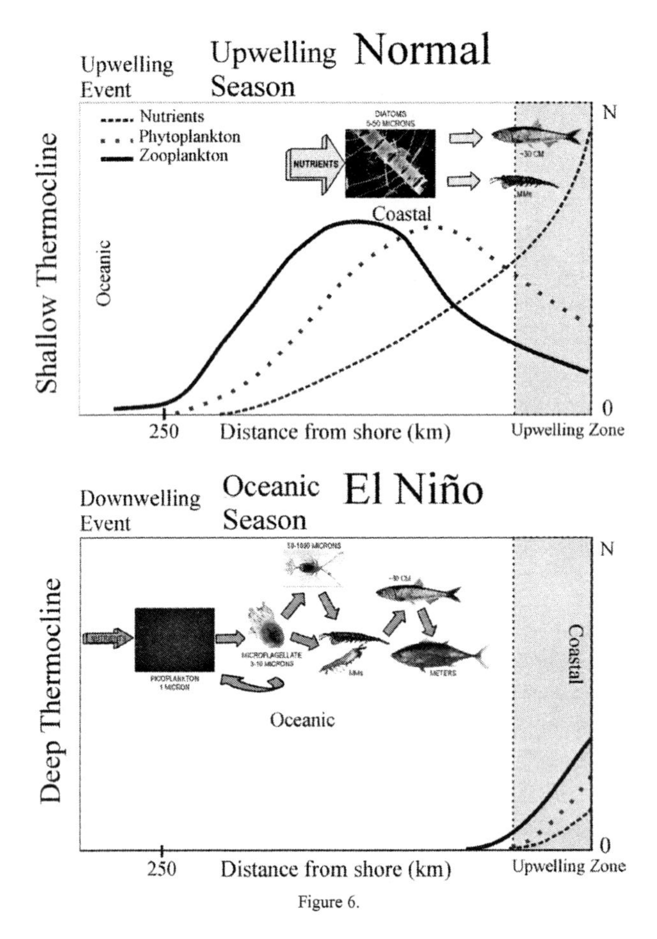


Figure 16.6 Conceptual model of offshore-onshore ecosystem changes associated with El Niño. The "coastal" ecosystem exhibits high biomass and primary productivity as a result of optimal light and nutrients (micro and macro) conditions. The coastal ecosystem is dominated by small colonial centric diatoms, large zooplankton (i.e. euphausids) and small pelagic fish that can graze directly on phytoplankton. During normal years this ecosystem occupies an area of ~200 km from shore with mesoscale filaments. There is a spatial and temporal separation of sources (nutrients, phytoplankton) and sinks (phytoplankton, zooplankton). This coastal ecosystem is thought to be leaky in that a significant fraction of the primary production may escape the upper mixed layer to midwater and the sediments. An "oceanic" nutrient-limited, low biomass and primary productivity ecosystem is found offshore of the coastal ecosystem. It is dominated by picophytoplankton whose grazers are protists with similar growth rates creating an efficient recycling system. A complex food web evolves with a smaller proportion of the primary production reaching the upper trophic levels. During warm years the productive coastal area is reduced dramatically and the oceanic ecosystem impinges on the shore. Alongshore advective redistributions further modify these communities by introducing tropical species during El Niño. Redrawn from Chavez et al. (2002b).

Macroalgal kelp species (*Macrosystis, Lessonia*) suffer from nutrient limitation during El Niño and their abundance decreases both in California (Tegner and Dayton, 1987) and Peru and Chile (Arnzt and Fahrbach, 1996). The strong wintertime storms associated with strong El Niños also contribute to the demise of these species. On the other hand green (*Ulva* sp.) and red (*Chondracanthus* sp.) algae can increase, presumably as a result of reductions in their typical predators (Arnzt and Fahrbach, 1996; Tarazona and Castillo, 1999).

### 2.4. *Biological consequences of ENSO in the eastern Pacific—Zooplankton to Seabirds*

The section above describes the bottom up (or nutrient supply) effects of El Niño on marine ecosystems. Lower levels of primary production lead to less zooplankton and eventually less food available to fish and other top consumers. Zooplankton abundance decreases in the eastern Pacific during El Niño have been reported off Peru; during 1982–83 levels decreased to one-fifth of the normal values (Carrasco and Santander, 1987). Similar decreases were reported for California during the 1997–98 El Niño (Hopcroft et al., 2002; Marinovic et al., 2002). Anomalous poleward and onshore currents are characteristic of El Niño and these have notable effects on zooplankton, with lower latitude and offshore species commonly appearing poleward and inshore of their normal species ranges (Carrasco and Santander, 1987; Marinovic et al. 2002, Peterson et al., 2002; Mackas and Galbraith, 2002). These advective changes make it difficult to quantitatively assess the decreases in secondary production during El Niño. Larger predatory zooplankton, such as chaetognaths, increase during the early stages of El Niño off Peru (Carrasco and Santander, 1987). Increases of large medusae can occur during some El Niños (Carrasco and Santander, 1987). These "new" dominant predators exert stronger top down effects on prey species that are already affected by bottom up effects.

El Niño-related changes in temperature, sea level, storms and currents also affect animal populations directly. This combination of physical disturbances has been shown to impact rocky intertidal populations off the Galapagos, Peru and Chile (Arntz and Fahrbach, 1996), where the typical communities (mussels, crabs etc.) practically disappear during the early stages and then slowly reappear. Warmer temperatures are also thought to be responsible for the dramatic coral bleaching that occurred in the eastern tropical north Pacific during the 1982–83 and 1997–98 El Niños (Glynn, 1984; Glynn, 2001). Although this may seem paradoxical given that corals are considered warm water species, they are killed or damaged by extreme warm waters. The warmer temperatures also allow other animals, like shrimp or the pelagic red crab (*Pleurocondes sp.*), to move into areas that they are typically excluded from due to low temperatures. Shrimp harvest increased dramatically off Peru during El Niño (Barber and Chavez, 1983; Velez et al., 1984), apparently because the shrimp or their larvae were advected into their new habitat by anomalous currents.

Fish species ranges also change during El Niño. For example warm water species, such as Mahi Mahi, tuna and marlin, become abundant off California and Peru during El Niño (Arntz and Tarazona, 1990; Barber and Chavez, 1983; Pearcy and Schoener, 1987; Pearcy, 2002; Velez et al., 1984). These voracious top predators consume local populations that are already impacted by lower food supplies and are concentrated in a reduced productive area. The small pelagic fish, like anchovy and sardine, that commonly dominate the waters off California and Peru, eventually move to deeper, cooler waters or migrate north towards Oregon (California) or south towards Chile (Peru). El Niño events seem to have greatest negative impact on anchovy populations (Valdivia, 1978). Sardine, jack and horse mackerel may be favored, and this effect seems to extend over the entire eastern Pacific (Arnzt and Fahrbach, 1996). Clear changes in the diet of these fish are observed during El Niño (Alamo and Bouchon, 1987).

El Niño effects on pinnipeds in the eastern Pacific have been well studied (Trillmich and Ono, 1991). Sea lion females typically leave their pups a week after birth to forage for several days before returning to nurse. During El Niño, the fish and squid that make up the majority of their diet move offshore and to deeper waters. In a study of sea lions off Peru Majluf (1991) showed that the duration of the foraging trips increased to up to 10 days and the dive depths increased considerably during El Niño. Warm ocean and land temperatures place further metabolic stress on the animals. The animals lose weight and during strong events there are large decreases in the adult populations. The young and old suffer the greatest losses. Effects are greatest closer to the equator and decrease poleward. A recent paper on northern elephant seal foraging during El Niño (Crocker et al., in press) showed changes similar to those reported for sea lions. Females increased the duration of their foraging trips and their mass gain decreased significantly during El Niño. This is of particular interest since these animals forage far into the North Pacific Ocean and at depths from 400 to 800 m, implying that El Niño decreases in production extend far from the eastern boundary and to great depths.

The dominant seabird off Peru is the cormorant *Phalacrox bougainvillii* whose common name is "guanay". These produce guano or seabird manure, an important resource, and their numbers decrease notably during El Niño (Jordan and Fuentes, 1966; Tovar and Garcia, 1982; Tovar and Cabrera, 1985). Some seabirds, like the boobie, *Sula variegata,* off Peru increase in numbers. Seabirds are mobile and follow their prey (small pelagic fish, krill) either north (California) or south (Peru). This increase in foraging area is similar to that reported for pinnipeds (above). Entire seabird year classes can be lost during strong El Niños as fecundity decreases and birds migrate away from rookeries. Seabirds apparently control their reproduction to match resources and cease or delay breeding during strong or weak events, respectively (Abraham and Sydeman, 2004; Sydeman et al., 2001). Dead seabirds often dot the beaches and some (primarily pelicans) enter human-populated areas in search of food. Increases in seabird parasite loads are also reported (Arntz and Fahrbach, 1996). In northern Peru large increases in rainfall also negatively affect bird populations. Interestingly, the increased rainfall washes guano into the ocean, providing local fertilization pockets for primary producers.

Fisheries and other resource exploitation dynamics are also affected during El Niño. The example of the Peruvian anchovy is probably the most notable. During normal conditions the cool productive area off Peru is large providing ample area for the fish to disperse away from their primary predator, the fisherman. During the early stages of El Niño, the small coastal pockets of cool water serve as refugia for the anchovy but concentrate them nearshore, where they are extremely vulnerable to fishing.

Poorly understood effects are those that occur in the deeper water communities. El Niño-related changes in mesopelagic communities (jelly-fish, myctophids from depths of 400–1000m) have been reported as have changes in benthic communities (Crocker et al., in press; Raskoff, 2001, Smith et al., 2001b). Are these a result of bottom-up, top-down or changes in currents and temperature? Most ecologists lean towards changes in food supply but the information we possess on these communities is incomplete.

Tidal and subtidal soft bottom communities off Peru change dramatically during El Niño. Many of the typical invertebrate species from the rocky intertidal and the sandy beaches like the clam, *Mesodesma donacium,* practically disappear. A few like the sand flea, *Emerita analoga,* increase. One of the most notable positive effects during the 1982–83 El Niño off Peru was the dramatic increase of the scallop, *Argopectens purpuratus.* Warmer temperatures appear to have accelerated the reproductive cycle and resulted in huge recruitment (Arntz and Fahrbach, 1996; Wolff, 1987). Deeper-water soft bottom communities off Peru increase dramatically in numbers and in number of species during El Niño as a result of increases in oxygen (Arntz and Fahrbach, 1996; Tarazona et al., 1988a,b). The oxygen effects are most notable in Peru and northern Chile where during normal conditions oxygen levels are among the lowest in the world ocean due to the decay of sinking phytoplankton from the highly productive surface water. The low oxygen places severe restrictions on animal growth. The benthic biota is restricted to organisms like the sulfur-oxidizing filamentous bacteria *Thioploca sp.* Bottom dwelling fish, such as hake, are normally restricted to northern Peru where oxygen levels are higher. Hake move southward and are favored during El Niño as oxygen levels increase. Warmer temperatures also allow colonization by immigrant tropical organisms (Tarazona et al., 1988a,b). These changes reflect both migrations and effects on local growth and survivorship.

At low oxygen levels nitrate becomes an electron donor for bacterial respiration and is converted to nitrite. As a result the eastern tropical Pacific is the largest contributor to global denitrification budgets (Codispoti et al., 2001). The anomalous oxygenation of the Peruvian coastal system during El Niño can have biological consequences due to supressed denitrification. Unusually large phytoplankton blooms may occur immediately following El Niño if nitrate levels are not reduced by denitrification and anomalously high levels are upwelled into the euphotic zone (Chavez, 1987).

Comparison of the biological effects of the 1982–83 and 1997–98 El Niños suggests that El Niño effects are not solely related to the intensity of temperature and nutrient anomalies. These two large events had similar intensity (i.e. SST anomalies of up to 10 C off Peru) but different biological consequences (Chavez et al., 2002b; Pearcy, 2002). The earlier event had more dramatic and long-lasting effects. These differences may be related to the background state of the Pacific. For example the effects may be related to the phase of multi-decadal variations described in the next case study.

### 2.5. *Biological consequences of ENSO in the western Pacific*

While El Niño is typically associated with declines in eastern Pacific biological productivity coastal areas far removed actually experience increases. Increases in primary production have been reported along the entire western Pacific, from Japan to New Zeland (see papers in this volume), presumably in response to a rising thermocline in the western Pacific that leads to increases in nutrient supply. Ship (Dandonneau, 1986) and satellite (Leonard and McClain, 1996; Murtugudde et al., 1999; Wilson and Adamec, 2001) studies of chlorophyll concentration show strong increases in the western tropical Pacific during El Niño. The changes in the western Pacific are more subtle than those in the eastern Pacific and the full gamut of El Niño effects has yet to be properly elucidated. A question of general interest is if the overall productivity of the Pacific actually decreases during El Niño or are decreases in the eastern Pacific balanced by increases in the western Pacific.

## 3. Case Study: Pacific Decadal Oscillation

### *3.1 What is the Pacific Decadal Oscillation*

In the 1990s a series of papers reported on Pacific climate, ocean and ecosystem variability with periods of 40–70 years (Ebbesmeyer et al., 1991; Francis and Hare, 1994; Mantua et al., 1997; Miller et al., 1994; Trenberth and Hurrell, 1994). A formal definition of the PDO was given in Mantua et al., (1997) and is based on sea surface temperatures in the North Pacific north of 20°N. More recently (Chao et al., 2000) show that this variability is also present in the South Pacific and is roughly symmetrical north and south of the equator. Almost ten years earlier a Japanese scientist (Kawasaki, 1983) had found synchronous variations in the landings of sardines off Japan, California, Peru, and Chile. Populations flourished for 20 to 30 years and then practically disappeared for similar periods. Periods of low sardine abundance were marked by dramatic increases in anchovy populations. The variability was later shown to extend to the Atlantic (Lluch-Belda et al., 1989) and to have occurred in pre-industrial times as evidenced by reconstructions from sediment cores (Baumgartner et al., 1992; Schwartzlose et al., 1999). The mechanism responsible for the global variations in sardines and anchovies becomes difficult to explain based on fishing and has been linked to the large-scale atmospheric and oceanic changes described in the 1990s (Chavez et al., 2003). Retrospective analysis suggests that from around 1925 to 1950 the Pacific was on average warm and dominated by sardines. A cool period from about 1950 to 1975 where anchovies dominated followed. Since 1975 the Pacific had been warm again until a recent cooling proposed for the late 1990s (Mantua and Hare, 2002; Chavez et al., 2003). Oceanographers have referred to the periods of rapid change between the warm and cool periods as regime shifts, a term coined by Isaacs (1975). As a result of the early response of ocean biota, it has been suggested that a regime or climate shift may even be best determined by monitoring marine organisms rather than climate (Hare and Mantua, 2000). The 25 year warm periods have been referred to as El Viejo (the old man or slang for father in Latin America, a play on El Niño) and the cool periods as La Vieja (the old woman). The pattern in SST anomalies associated with the multi-decadal changes is surprisingly similar to that observed during ENSO with an enhancement of high latitude anomalies relative to the tropics (Figure 16.7). Indeed the El Viejo period has been described as a prolonged mild El Niño (Mantua and Hare, 2002).

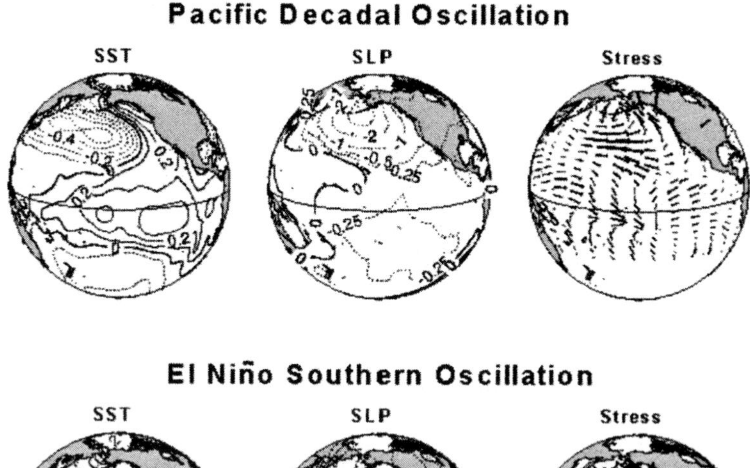

Figure 16.7 Comparison of SST, sea level pressure and wind stress anomalies for the Pacific Decadal Oscillation and El Niño. Note the similarity in the patterns and the relative amplification of the anomalies at higher latitudes during the PDO. From http://tao.atmos.washington.edu/pdo/.

A wide range of physical and biological time series in the Pacific and Atlantic Ocean basins show systematic variations on the multi-decadal time scale. Anomalies, representing deviations from the mean value, were negative from about 1950 to 1975 and positive from about 1975 to the mid to late 1990s (Figure 16.8). The underlying physical mechanism responsible for Pacific changes with a period of around 50 years remains the subject of significant debate (Barnett et al., 1999; Gu and Philander, 1997; Kleeman et al., 1999; McPhaden and Zhang, 2002; Schneider et al. 1999). It probably involves circulation in the so-called North (and South) Pacific Subtropical Cell. This cell is a shallow (<500 m) meridional overturning circulation in which water flows poleward out of the tropics in the surface layer and subducts under the subtropical gyre. It returns to the surface at high latitudes and once again subducts to be later (10–15 years) upwelled along the equator. The full cycle takes on the order of thirty years. Changes in the temperature of the water that downwells at high latitudes (Gu and Philander, 1997) or the speed of the Subtropical Cell (Kleeman et al., 1999) can produce temperature anomalies in model simulations that are similar to observations. The latter changes are consistent with the changes in atmospheric properties described by Barnett et al. (1999).

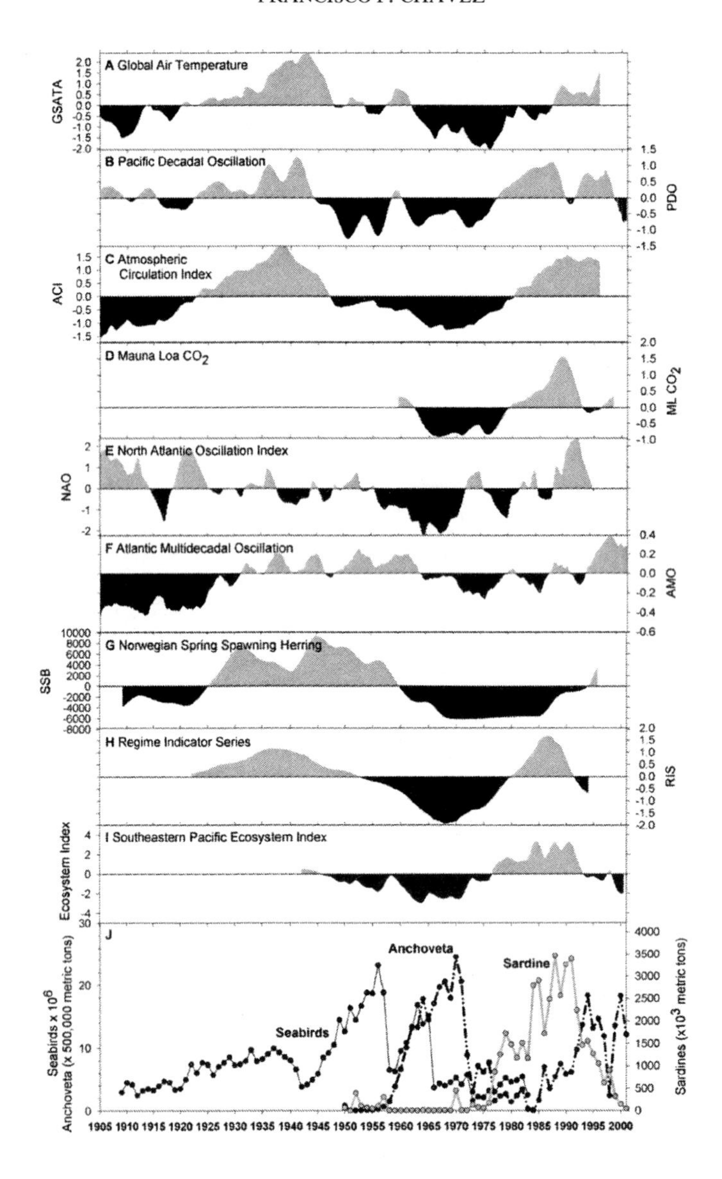


Figure 16.8 From top to bottom are time series of the anomalies of A) detrended global air temperature (Jones et al., 1999), B) the Pacific Decadal Oscillation (PDO), derived from principal component analysis of North Pacific SST (Mantua et al., 1997), C) the Atmospheric Circulation Index (ACI) describes the relative dominance of zonal or meridional transport in the Atlantic-Eurasian region (Klashyatorin, 2001), D) detrended atmospheric $CO_2$ measured at Mauna Loa (Keeling et al., 1989), E) the North Atlantic Oscillation (NAO), taken from the atmospheric sea level pressure difference between the Arctic and the subtropical Atlantic (Hurrell et al., 2003), F) The Atlantic Multidecadal Oscillation represents the 10 year running average of the detrended Atlantic sea surface temperature anomalies north of the equator (Enfield et al., 2001), G) anomalies in the Norwegian spring herring spawning biomass (Toresen and Ostvedt, 2000), H) the Regime Indicator Series (RIS) that integrates global sardine and anchovy fluctuations (Lluch-Belda et al., 1997), I) the Southeastern Tropical Pacific ecosystem (STPE) index based on , J) seabird abundance, anchoveta and sardine landings from Peru used to construct the STPE index. Series have been smoothed with a 3-year running mean. Modified from Chavez et al., (2003).

### *3.2 Biological consequences of El Viejo and La Vieja*

In a simplified conceptual view of the Pacific, the trade winds set up a basin-wide slope in sea level, thermal structure, and—importantly for biology—nutrient structure.

As described earlier for the ENSO consequences, the shallow thermocline in the eastern tropical Pacific leads to enhanced nutrient supply and productivity (Barber and Chavez, 1983). The multidecadal fluctuations have basin-wide effects on sea surface temperature (SST) and thermocline slope that are similar to El Niño and La Niña but on longer time scales (Figure 16.7). During the cool eastern boundary anchovy regime, the basin-scale sea level slope is accentuated (lower in the eastern Pacific, higher in the western Pacific)(Figure 16.9). Lower sea level is associated with a shallower thermocline and increased nutrient supply and productivity in the eastern Pacific; the inverse occurs in the western Pacific. In addition to thermocline and sea surface temperature, there are regime shift changes in the transport of boundary currents, equatorial currents, and of the major atmospheric pressure systems. Changes in the abundance of anchovies and sardines are only one of many biological perturbations associated with regime shifts, and these are reflected around the entire Pacific.

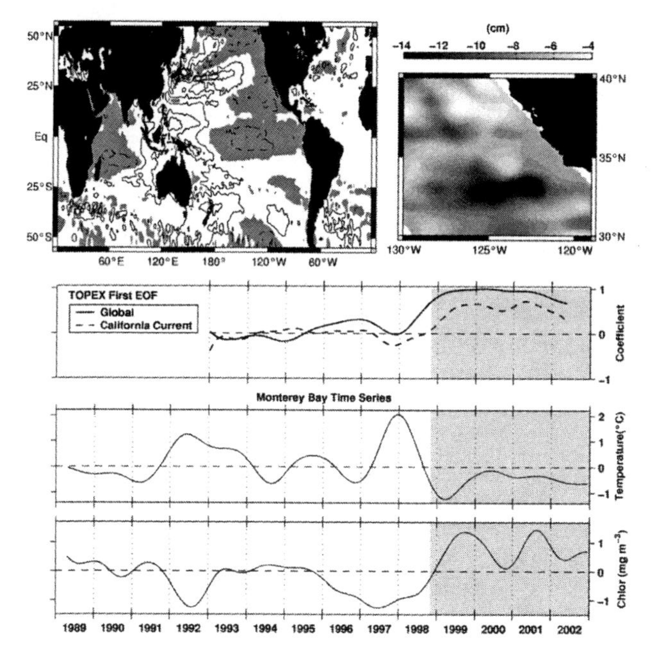


Figure 16.9 Top panel shows the first empirical orthogonal function (EOF) of global TOPEX sea surface height (SSH) to 60 degrees latitude, accounting for 31% of the variance in the time series. Input data were binned to 31 km along track and 30 days and were 18-month low-pass filtered. The coefficient of the global mode is shown in the second panel along with the coefficient and mapped first mode of SSH calculated over a region of the California Current The California Current time series was smoothed with a 90-day running mean rather than low-pass filtering, thus it retained the annual cycle. The first mode for this region accounts for 47% of the variance in the time series. The lower two panels show similarly filtered surface temperature and chlorophyll time series for the eastern margin of the California Current system in Monterey Bay. Notice the changes associated with El Niño (1992–93, 1997–98) and the shift after 1997–98. Modified from Chavez et al., (2003).

The northeast Pacific may be the most studied area in terms of regime shifts (Francis and Hare, 1994). Several recent reviews include those by Benson and Trites (2002), Hare and Mantua (2000) and Hollowed et al., (2001). An important physical change for this region is an intensification (sardine) or relaxation (anchovy) of the Aleutian Low (Miller et al., 1994; 2004). During the sardine regime from the late 1970s to the early 90s, zooplankton (Brodeur, 1993; Roemmich and McGowan, 1995) and salmon (Beamish et al., 1999) declined off Oregon and Washington but increased off Alaska (Hare and Mantua, 2000; Benson and Trites, 2002). The flip-flop between ecosystems in the Gulf of Alaska and the California Current is one of the conundrums associated with these longer term cycles. During El Niño the California Current and the Gulf of Alaska seem to be in phase in that productivity decreases in both locations (Whitney and Welch, 2002; Chavez et al., 2002b). The El Niño changes are associated with thermocline depth which decreases in both cases lowering nutrient supply and primary production. Spring mixed layer depth, on the other hand, decrease in the Gulf of Alaska while they increase in the California Current during the positive or warm phases of the PDO (Francis et al., 1998). These authors surmise that mixed layer decreases may favor primary production in the North Pacific by reducing light limitation. In the California Current the deeper mixed layers are associated with a deeper thermocline and lower primary production (Roemmich and McGowan, 1995). A possible explanation for the different responses of the California Current and the Gulf of Alaska to El Niño and El Viejo might be that: 1) during strong El Niños thermocline displacements dominate in both systems, reducing productivity at both locations; 2) weaker thermocline anomalies during El Viejo reduce productivity in the California Current but these effects are counteracted in the Gulf of Alaska by changes in mixed layer depth resulting in increases in productivity at this location. Similar changes are observed in the North Atlantic during the NAO (see following section). This suggests fundamental differences between low latitude stratified oceans where thermocline displacements and upwelling regulate and high latitude environments where mixing is the dominant process.

As with El Niño seabird populations decrease off California (Abraham and Sydeman, 2004; Hyrenbach and Veit, 2003; Veit et al., 1996) and Peru (Chavez et al., 2003) during El Viefo. The California Current weakens and moves shoreward during the warm phases and the subarctic gyre intensifies (Francis et al., 1998). Warmer temperature and lower salinity near the California coast support the weakening of the California Current (Roemmich and McGowan, 1995). A stronger and broader California Current, brought about during the cool La Vieja regime, is associated with a shallower coastal thermocline from California to British Columbia, leading to enhanced primary production (Figure 16.9). It should be noted that the multidecadal changes in the circulation of the California Current system are not fully resolved; Bograd and Lynn (2003) argue that regime-shift changes in the position rather than the intensity of the currents are the primary mechanism for some of the observed patterns. What is very clear is that the boundaries of warm (cold) water species move poleward (equatorward) during the warm (cool) regimes (Peterson and Schwing, 2003). In the Southeastern Pacific biological variability is similar to that observed in the Northeast Pacific albeit much less well documented.

In the northwest Pacific off Japan, the depth of the thermocline, nutricline and the winter mixed layer have shown changes on multi-decadal time scales (Miller et al., 1994; Minobe, 1997; Minobe, 2002). During the warm El Viejo regime, sea level dropped, the thermocline and nutricline shoaled, mixed layers deepened (Figure 16.1). Transport by the Kuroshio Current weakened (Deser et al., 1999; Joyce and Dunworth-Baker, 2003; Miller et al., 1998). Primary production increased and sardine populations expanded from coastal waters eastward across the North Pacific to beyond the International Date Line (Wada and Kashiwai, 1991). It remains unclear why sardines increased off Japan when local waters cool and become more productive, whereas they increased off California and Peru when those regions warm and become less productive (McFarlane et al., 2002).

In the warm pools of the western and northeastern tropical Pacific, physical variability has been harder to elucidate (Fiedler, 2002), partly because temperatures are warm and homogenous there. However, there is evidence of lower recruitment of yellow fin tuna during the cool La Vieja regime (Maunder and Watters, 2001). The northeastern tropical Pacific is surrounded by regions with strong multidecadal fluctuations (California Current, Peru Current, equatorial Pacific, subtropical gyre). Tuna in the warm waters of the western Pacific seem to be similarly affected. Populations of yellow fin tuna in the western Pacific may have increased during the cool regimes (Hampton and Fournier, 2001). Highly mobile organisms like the blue fin tuna migrate on basin scales, spending significant periods in areas altered by these large-scale climate and ocean changes. These organisms must respond in complex ways to regime shifts.

Episode to episode differences for El Viejo and La Vieja are just beginning to emerge (Bond et al., 2003). This should come as no surprise. After several iterations of El Niño it became clear that "no two El Niños (children) were alike" (see quote of Jordan in Chavez et al., 2002a). A telling example was provided by Rasmusson and Carpenter (1982) who described the development of a canonical El Niño in the tropical Pacific from data spanning from 1950 to 1975. Following this description (and the 1976 regime shift) no El Niño resembled the development of the canonical El Niño. It will be interesting if after the regime shift in the mid to late1990s, El Niño development once again follows that described by Rasmusson and Carpenter (1982); the first one, the moderate 2002–2003 event, has apparently not (McPhaden, 2004).

## 4. Case Study: North Atlantic Oscillation

The Atlantic Ocean is also subject to interannual to multidecadal variations. An Atlantic El Niño has been recognized although it occurs much more irregularly than in the Pacific (Gammelsrød et al., 1998; Philander, 1986). The Atlantic Multidecadal Oscillation (AMO) has been recently identified (Enfield et al. 2001) but it remains uncertain if this cycle is related or entirely different from the Pacific Decadal Oscillation (Figure 16.9). Changes in the abundance of anchovies and sardines occur off South Africa on similar time scales than the Pacific (Lluch-Belda et al., 1989; Schwartlose et al., 1999). Interestingly the changes are out of phase, i.e. sardine increases in the Pacific are associated with sardine decreases in the Atlantic. Longer and more comprehensive time series may show the fully global nature of multi-decadal variability of 40–70 year periods. For the Atlantic the most frequently cited climate driver is the North Atlantic Oscillation (NAO). The linkages between many of these climate indices remain unclear although some argue that ENSO, PDO, AMO and NAO are inter-related (Marshall et al. 2001).

The first account of NAO-like climate changes can be traced to missionaries more than two centuries ago. They reported year-to-year fluctuation in wintertime air temperatures in Scandinavia (van Loon and Rogers, 1978). Interestingly it was Walker (Walker and Bliss, 1932) who first reported the NAO in the scientific literature as 'expressing the tendency for pressure to be low near Iceland in winter when it is high near the Azores and south west Europe' and vice versa. The positive phase of the NAO is associated with below normal temperatures over Greenland and above normal temperatures in Scandinavia. In addition to temperature patterns of rainfall and sea level change as far as central Europe, southwards to subtropical West Africa and westwards to North America. Somewhat paradoxically it was Bjerknes (1964) that first noticed a relationship between the NAO and Atlantic SST. In addition to SST changes there are also notable changes in mixed layer depth and surface Ekman transport (Visbeck et al., 2003).

As with the PDO, the NAO seems to be associated with longer time scale gyre-scale changes in circulation as well as changes in deep water formation (Curry and McCartney, 2001; Greene et al., 2003; Visbeck et al., 2003). During the positive phase of the NAO the Gulf Stream shifts north and transport by the Labrador Current is reduced. The Labrador Sea has apparently cooled and freshened from 1966 to 1992; associated with these hydrographic changes local sea level (or steric height) is 8–10 cm lower. Changes in the Nordic Seas are similar and perhaps related. Warmer Atlantic waters have spread poleward to the Barents Sea and the Arctic, the Greenland Sea has become more stratified and the Nordic Seas have freshened in the upper 1000 m.

The biological response in the temperate and high latitude North Atlantic is complex probably because of the non-linear effects of wind mixing on primary production. This non-linear relationship has been used to explain the out-of-phase spatial changes (eastern Atlantic versus western Atlantic) in the relation between the NAO and phytoplankton/zooplankton abundance (Barton et al., 2003; Drinkwater et al., 2003; Greene et al., 2003). Positive correlations between phytoplankton and NAO have been found in the North Sea and negative correlations further west. These east-west differences are analogous to those occurring in the tropical Pacific during El Niño and El Viejo (i.e. lower production in the eastern Pacific, higher production in the western Pacific) but the mechanisms are entirely different. In the tropics the thermocline variations explain the changes while at high latitudes they seem to related to changes in mixed layers. Prior to 1996 a strong inverse relationship between the NAO and the abundance of the copepod *Calanus finmarchicus* in the North Sea was observed (Fromentin and Planque, 1996). The hypothesis was that stronger winter mixing during the positive phase delayed the spring phytoplankton bloom and adversely affected *Calanus finmarchicus.* Even though on average annual nutrient supply increased, light limitation by mixing reduced primary production during critical life-history stages. The abundance of *Calanus finmarchicus* seems to directly regulate the abundance of cod in the Barents Sea where the NAO and cod recruitment show positive correlations (Drinkwater et al., 2003). Given that *Calanus finmarchicus* is one of the principal forage species for right whales in the western North Atlantic a link between the NAO and whale abundance is postulated (Drinkwater et al., 2003). As with El Niño and El Viejo, NAO-driven advection of populations into and out of regions seems to play an important role in ecological dynamics (Heath et al., 1999). Changes in jellyfish (Lynam et al., 2003) and the benthos (Drinkwater et al., 2003) have also been linked to the NAO. The example of winter flounder along the U.S. east coast is worth noting. These have in decline since 1975 as the NAO moved from negative to positive values. Speculation is that perhaps warmer sea temperatures resulted in more phytoplankton consumption in the water column and therefore a lesser supply to benthic and/or bottom dwelling organisms.

Similar to anchovies and sardines in upwelling areas, small pelagics in the Nordic Seas oscillate from periods of high abundance to periods of almost complete absence (Alheit and Hagen , 1997, Toresen and Ostvedt, 2000, Figure 16.8). These fluctuations have been interpreted as ecosystem responses to prevailing wind speed and direction in relation to the NAO (Alheit and Hagen, 1997). Top predators such as the bluefin tuna in the eastern Atlantic have shown increases in recruitment during positive phases of the NAO but the northern albacore showed the inverse response.

As one can glean from the few examples given above the relationship between the NAO and ocean ecosystems is not as clear or well understood as in the Pacific. Ottersen et al., (2001) proposed that NAO effects be classified as either direct, indirect and integrated, primarily because of the inconclusive or different responses described above. What seems lacking for the NAO is a general theory for ocean-atmosphere coupling. Once this theory is established the biological consequences of NAO variability may become clearer.

## 5. Conclusions

This review has focused on interannual and multi-decadal variability in the Pacific and decadal variability in the Atlantic. Phenomena with strong linkages to coastal ocean ecosystems were chosen as case studies. Clearly variations between interannual and multi-decadal exist as do longer period fluctuations. The longer period fluctuations can at present only be determined from paleo-oceanographic records (Baumgartner et al., 1992; Diaz and Markgraf, 2000). Our understanding of the biological consequences of El Niño is relatively mature given that present day time series have captured a number of instances of these phenomena. The physical variability is well understood and conceptual models of how biology responds have been developed. There is growing interest in longer period fluctuations given the large impact they can have on living resources. Present day time series are limited to one or at the most two iterations of these phenomena so our views are likely to change significantly over time. Current conceptual models of decadal-scale variability for both physics and biology are more than likely over-simplistic. Given the growing interest in decadal-scale variability this review may become outdated rather quickly. A more complete understanding of climatic variability in general will depend on longer and more complete (in terms of parameters and space) time-series. When coastal ocean time series are put together with those from open oceans and terrestrial systems an integrated view of the earth system will be developed. It is then that we will likely uncover the interrelationships we think today might exist for example between El Niño, the PDO and the NAO. The next decades should be spent improving and expanding coastal observations and beginning the development of the integrated view of the earth system. This will require a new way of thinking about the environment and more than likely the development of a new generation of earth scientists.

The media typically refer to climatic phenomena negatively. For example in the journal *Nature* a prominent scientist wrote: "the 1997–98 El Niño severely disrupted global weather patterns and Pacific marine ecosystems, and by one estimate caused $33 billion in damage and cost 23,000 lives worldwide". While indeed these phenomena can lead to catastrophes there are "The two faces of El Niño" (Arntz, 1986). Examples of positive aspects are the dramatic increases in catch of shrimp and scallops off Peru and the reduction in hurricane activity in the Atlantic during El Niño. In many parts of the world, Peru and western United States for example, reservoirs used for drinking water and irrigation, are filled during rainy El Niño years. It is the extreme events, either a very warm or a very cold equatorial Pacific that can often lead to natural catastrophes. Given the recurring nature of El Niño and other natural climate variability it is clear that society needs to learn to live with them. Better understanding of the physical nature of climate variability and its ecological consequences will aid in this regard. Further, and perhaps more importantly, studies of these natural climate variations will be required for detection of human perturbations (Latif et al., 1997) and understanding of their consequences.

## Acknowledgements

This review was made possible by long-term support from the David and Lucile Packard Foundation. Support from NASA, NOAA and NSF is also recognized. Steven Bograd, Tim Pennington and an anonymous reviewer provided comments that greatly improved the article. David Johnston gave insights into the NAO and Reiko Michisaki created most of the figures. George Somero kindly provided unrestricted access to the library at Hopkins Station of Stanford University.

## Bibliography

Abraham, C.L. and W. J. Sydeman, 2004. Ocean climate, euphausiids and auklet nesting: inter-annual trends and variation in phenology, diet and growth of a planktivorous seabird, *Ptychoramphus aleuticus. Marine Ecology Progress Series,* **274,** 235–250.

Adams, J.B., M. E. Mann and C. M. Ammann, 2003. Proxy evidence for and El Niño-like response to volcanic forcing. *Nature,* **426,** 274–278.

Alamo, A. and M. Bouchon, 1987. Changes in the food and feeding of the sardine (*Sardinops sagax sagax*) during the years 1980–1984 off the Peruvian coast. *J. Geophys. Res.,* **92,** (C13), 14411–14415.

Alheit, J and E. Hagen, 1997. Long-term climate forcing of European herring and sardine populations. *Fisheries Oceanography,* **6,** 130–139.

Arntz, W.E., A. Landa and J. Tarazona, 1985. El Niño. Su impacto en la fauna marina. In *Bio. Inst. Mar.,* Peru-Callao, 222.

Arntz, W.E., 1986. The two faces of El Niño 1982–1984. *Meeresforsch.,* **31,** 1–46.

Arntz, W.E. and J. Tarazona, 1990. Effects of El Niño 1982–83 on benthos, fish and fisheries off the South American Pacific coast, In *Global ecological consequences of the 1982–83 El Niño-Southern Oscillation.,* P.W. Glynn, eds. Elsevier Oceanog. Series, 323–360.

Arntz, W. E. and E. Fahrbach, 1996. El Niño Experimento climatico de la naturaleza Causas fisicas y efectos biologicos. Trad. de C. Wosnitza-Mendo, J. Mendo. Fondo de Cultura Economica, Mexico, 312 pp.

Avaria, S. and P. Munoz, 1987. Effects of the 1982–1983 El Niño on the marine phytoplankton off northern Chile. *J. Geophys. Res.,* **92,** (C13), 14367–14382.

Azam F., T. Fenchel, J.G. Field, J.S. Gray, L.A. Meyer-Reil, and T.F. Thingstad, 1983. The ecological role of water-column microbes in the sea. *Mar Ecol Prog Ser* **10,** 257–263

Bakun, A., 1990. Global Climate Change and Intensification of Coastal Ocean Upwelling. *Science,* **247,** 198–202.

Barber, R.T. and F.P. Chavez, 1983. Biological consequences of El Niño. *Science,* **222,** 1203–1210.

Barber, R.T. and F.P. Chavez, 1986. Ocean variability in relation to living resources during the 1982–83 El Niño. *Nature,* **319,** 279–285.

Barnett, T. P., D. W. Pierce, M. Latif, D. Dommenget, and R. Saravanan, 1999. Interdecadal interactions between the tropics and midlatitudes in the Pacific basin. *Geophys. Res. Lett.,* **26,** 615–618.

Barton, A.D., C.H. Greene, B.C. Monger, A.J. Pershing, 2003. The Continuous Plankton Recorder survey and the North Atlantic Oscillation: Interannual- to Multidecadal-scale patterns of phytoplankton variability in the North Atlantic Ocean. *Progress in Oceanography.* **58,** 337–358.

Baumgartner, T. R., A. Soutar, and V. Ferreira-Bartrina, 1992. Reconstruction of the history of Pacific sardine and Northern Pacific anchovy populations over the past two millennia from sediments of the Santa Barbara basin. *CalCOFI Report,* **33,** 24–40.

Beamish, R. J. D., J. Noakes, G. A. McFarlane, L. Klyashtorin, V. V. Ivanov, and V. Kurashov, 1999. The regime concept and natural trends in the production of Pacific salmon. *Can.J.Fish.Aqua.Sci,* **56,** 516–526

Beamish, R. J. (ed.), 1995. Climate and Northern Fish Populations. *Can. Spec. Publ. Fish. Aquat. Sci.,* 121.

Benson, A.J. and Trites, A.W. 2002. Ecological effects of regime shifts in the Bering Sea and eastern North Pacific Ocean. *Fish and Fisheries.* **3,** 95–113.

Bjerknes, J., 1964. Atlantic air-sea interactions. *Advances in Geophysics,* **10,** 1–82.

Bjerknes, J., 1966. A possible response of the atmospheric Hadley circulation to equatorial anomalies of ocean temperature. *Tellus,* **18,** 820–829.

Bjerknes, J., 1969. Atmospheric teleconnections from the equatorial. *Pacific. Mon. Weather Rev.,* **97,** 163–172.

Bond, N. A., J. E. Overland, M. Spillane, and P. Stabeno, 2003. Recent shifts in the state of the North Pacific, *Geophys. Res. Lett.*, **30,** 23, 2183, doi:10.1029/2003GL018597.

Brodeur, R. D., 1993. Long-term variability in zooplankton biomass in the subarctic Pacific Ocean. *Fisheries Oceanography,* **1,** 32–38.

Bograd, S.J. and R.J. Lynn, 2001. Physical-biological coupling in the California Current during the 1997–99 El Niño-La Niña cycle. *Geophysical Research Letters,* **28,** (2), 275–278.

Bograd, S.J. and R.J. Lynn, 2003. Long-term variability in the Southern California Current System. *Deep-Sea Research Part II-Topical Studies in Oceanography,* **50,** (14–16), 2355–2370.

Canby, T.Y., 1984. El Niño's ill wind. *Natl. Geogr.,* **165,** 144–183.

Cane, M.A., 1983. Oceanographic events during El Niño. *Science,* **222,** 1189–1195.

Carranza, L., 1891. Contracorriente maritima observada en Payta y Pacasmayo. *Bol. Soc. Geogr. Lima,* **1,** 344–345.

Carrasco, S. and H. Santander, 1987. The El Niño event and its influence on the zooplankton off Peru. *J. Geophys. Res.,* **92,** (C13), 14405–14410.

Chao, Y., M. Ghil and J.C. McWilliams, 2000. Pacific interdecadal variability in this century's sea surface temperatures. *Geophysical Research Letters,* **27** (15), 2261–2264.

Chavez, F.P., 1986. The legitimate El Niño current. *Tropical Ocean-Atmosphere Newsletter,* **34,** 1.

Chavez, F.P., 1987. El Niño y la Oscilacion del Sur. *Investigacion y Ciencia,* **128,** 46–55.

Chavez, F.P., 1996. Forcing and biological impact of onset of the 1992 El Niño in central California. *Geophysical Research Letters,* **23,** 265–268.

Chavez, F.P., P.G. Strutton, G.E. Friederich, R.A. Feely, G.C. Feldman, D.G. Foley and M.J. McPhaden, 1999. Biological and chemical response of the equatorial Pacific Ocean to the 1997–98 El Niño. *Science,* **286,** 2126–2131.

Chavez, F.P, C.A. Collins, A. Huyer, and D. Mackas eds. 2002a. El Niño along the west coast of North America. *Progress in Oceanography,* **54,** 1–6.

Chavez, F.P., J.T. Pennington, C.G. Castro, J.P. Ryan, R.P. Michisaki, B. Schlining, P. Walz, K.R. Buck, A. McFayden and C.A. Collins, 2002b. Biological and chemical consequences of the 1997–98 El Niño in central California waters. *Progress in Oceanography,* **54,** 205–232.

Chavez, F.P., J.P. Ryan, S. Lluch-Cota and M. Ñiquen C., 2003. From anchovies to sardines and back-Multidecadal change in the Pacific Ocean. *Science,* **299,** 217–221.

Chen, D., M.A. Cane, A. Kaplan, S.E. Zebiak, D. Huang, 2004. Predictability of El Niño over the past 148 years, *Nature,* **428,** 734–736.

Chisholm S.W., R.J. Olson, E.R. Zettler, R. Goericke, J. Waterbury and N. Welschmeyer, 1988. A novel free-living prochlorophyte abundant in the oceanic euphotic zone. *Nature,* **334** (6180), 340–343.

Codispoti, L.A., J.A. Brandes, J.P. Christensen, A.H. Devol, S.W.A. Naqvi, H.W. Paerl and T. Yoshinari, 2001. The oceanic fixed nitrogen and nitrous oxide budgets: Moving targets as we enter the anthropocene? *Scientia Marina,* **65** (Suppl. 2), 85–105.

Crocker, D. E., D. P. Costa, B. J. Le Boeuf, P. M. Webb and D. S.Houser, in press. Impact of El Niño on the foraging behavior of female northern elephant seals, Marine Ecology Progress Series.

Curry, R.G. and M.S. McCartney, 2001. Ocean gyre circulation changes associated with the North Atlantic Oscillation. *J. Phys. Oceanog.,* **31,** 3374–3400.

Dandonneau, Y., 1986. Monitoring the sea surface chlorophyll concentrations in the tropical Pacific: Consequences of the 1982–83 El Niño, *Fish. Bull.,* **84,** 687–695.

DeVries, T.J., 1987. A review of geological evidence for ancient El Niño activity in Peru. *J. Geophys. Res.,* **92,** (C13), 14471–14479.

Deser, C. M., A. Alexander, and M. S. Timlin, 1999. Evidence for a wind-driven intensification of the Kuroshio Extension from the 1970s to the 1980s, *J. Climate,* **12,** 1697–1706.

Diaz, H. F., and V. Markgraf, 1992. El Niño: Historical and Paleoclimatic Aspects of the Southern Oscillation. Cambridge University Press, 476.

Diaz H .F., and V. Markgraf (Eds.), 2000. El Niño and the Southern Oscillation: Multiscale Variability and Global and Regional Impacts, Cambridge University Press, 496.

Drinkwater, K.F., A. Belgrano, A. Borja, A. Conversi, M. Edwards, C. H. Greene, G. Ottersen, A.J. Pershing, and H. Walker. 2003. The Response of Marine Ecosystems to Climate Variability Associated With the North Atlantic Oscillation, in *The North Atlantic Oscillation,* J. Hurrell, Y. Kushnir, G. Ottersen and M. Visbeck (eds.), AGU monograph, Washington, 211–234.

Ebbesmeyer, C.C., D.R. Cayan, D.R. McLain, F.H. Nichols, D.H. Peterson, and K.T. Redmond, 1991. 1976 step in the Pacific climate: Forty environmental changes between 1968–1975 and 1977–1984. In: *Proceedings of the Seventh Annual Pacific Climate (PACLIM) Workshop, April 1990, Pacific Grove, CA,* Betancourt, J.L. and V.L. Tharp (eds.), California Department of Water Resources, Interagency Ecological Studies Program, Sacramento, CA, USA, 115–126.

Eguiguren, V., 1894. Las lluvias de Piura. *Boln. Soc. Geogr. Lima,* **4,** 241–258.

Enfield, D. and J. S. Allen, 1980. On the structure and dynamics of monthly mean sea level anomalies along the Pacific Coast of North and South America. *J. Phys. Oceanogr.,* **10,** 557–578.

Enfield, D.B., 1981. Thermally driven wind variability in the planetary boundary layer above Lima, Peru. *Journal of Geophysical Research,* **86,** 2005–2016.

Enfield, D.B., M.P. Cornejo-Rodriguez, R.L. Smith, and P.M. Newberger, 1987. The equatorial source of propagating variability along the Peru coast during the 1982–1983 El Niño. *Journal of Geophysical Research,* **92**(C13):14,335–14,346.

Enfield, D.B., 1989. El Niño, Past and present. *Rev. Geophys.,* **27,** 159–187.

Enfield, D.B., A.M. Mestas-Nunez, and P.J. Trimble, 2001. The Atlantic Multidecadal Oscillation and its relationship to rainfall and river flows in the continental U.S. *Geophysical Research Letters,* **28,** 2077–2080.

Fedorov, A. V., and S. G. Philander, 2000. Is El Nino changing? *Science,* **288,** 1997–2002.

Fiedler, P.C., 2002. Environmental change in the eastern tropical Pacific Ocean: review of ENSO and decadal variability. *Marine Ecology-Progress Series,* **244,** 265–283.

Francis, R. C. and S.R. Hare, 1994. Decadal-scale regime shifts in the large marine ecosystems of the Northeast Pacific: a case for historical science. *Fish. Oceanogr.* **3,** 279–291.

Francis, R. C., S. R. Hare, A. B. Hollowed, and W. S. Wooster, 1998. Effects of interdecadal climate variability on the oceanic ecosystems of the NE Pacific. *Fish. Oceanogr.* **7,** 1–21.

Friederich, G., P. Walz, M. Burczynski and F.P. Chavez, 2002. Inorganic Carbon in the Central California Upwelling System During the 1997–1999 El Niño-La Niña Event. *Progress in Oceanography,* **54,** (1–4), 171–184.

Fromentin, J. and B. Planque, 1996. Calanus and the environment in the eastern North Atlantic. II. Influence of the North Atlantic Oscillation on *C. finmarchicus* and *C. helgolandicus. Marine Ecology Progress Series,* **134,** 111–118.

Fryxell, G.A., M.C. Villac, and L.P. Shapiro, 1997. Phycological Reviews 17. The occurrence of the toxic diatom genus *Pseudo-nitzschia (Bacillariophyceae)* on the West Coast of the U.S.A., 1920–1996: a review. *Phycologia* **36,** (6), 419- 437

Gammelsrød, T., C.H. Bartholomae, D.C. Boyer, V.L.L. Filipe and M. J. O'Toole, 1998. Intrusion of warm surface water along the Angolan-Namibian coast in February-March 1995: the 1995 Benguela Niño, Benguela Dynamics, *S. Afr. J. Mar. Sci.,* **19,** 51–56.

Gill, A.E. and E.M. Rasmusson, 1983. The 1982–83 climate anomaly in the equatorial Pacific. *Nature,* **306,** 229–234.

Glynn, P.W., 1984. Widespread coral mortality and the 1982–83 El Niño warming event. *Environmental Conservation* **11,** (2), 133–146.

Glynn, P .W. (Ed.), 1990. *Global Ecological Consequences of the 1982–83 El Niño-Southern Oscillation,* Elsevier, Amsterdam.

Glynn, P. W. (Ed.), 2001. A collection of studies on the effects of the 1997–98 El Niño Southern Oscillation event on corals and coral reefs in the eastern tropical Pacific. *Bull. Mar. Sci.,* **69,** 1–288.

Gray, W. M., 2002. Overview of twentieth century challenges and milestones in coping with hurricanes. Chapter 1, Coping with hurricanes, An historical analysis of 20th century progress. American Geophysical Union.

Greene, C.H., and A.J. Pershing, 2003. The Flip-Side of the North Atlantic Oscillation and modal shifts in Slope-Water circulation. *Limnology & Oceanography,* **48,** 319–322.

Greene, C.H., A.J. Pershing, A. Conversi, B. Planque, C. Hannah, D. Sameoto, E. Head, P.C. Smith, P.C. Reid, J. Jossi, D. Mountain, M.C. Benfield, P.H. Wiebe, E. Durbin, 2003. Trans-Atlantic responses of *Calanus finmarchicus* populations to basin-scale forcing associated with the North Atlantic Oscillation. *Progress in Oceanography.* **58,** 301–312.

Gu, D. and S.G.H. Philander, 1997. Interdecadal climate fluctuations that depend on exchanges between the tropics and extratropics, *Science,* **275,** 805–807.

Halpern, D. 2002. Offshore Ekman transport and Ekman pumping off Peru during the 1997–1998 El Niño. Geophys. Res. Let. 29, doi: 10.1029/2001GL014097.

Hampton, J. and D.A. Fournier, 2001. A spatially disaggregated, length-based, age-structured population model of yellowfin tuna (*thunnus albacares*) in the western and central Pacific Ocean. *Mar. Freshwater Res.,* **52,** 937–963

Hare, S. R., S. Minobe, and W.S. Wooster (Eds.), 2000. The nature and impacts of North Pacific climate regime shifts. *Progress in Oceanography,* **47,** 99–408.

Hare, S.R. and N.J. Mantua, 2000. Empirical evidence for North Pacific regime shifts in Pacific North America. *Progress in Oceanography,* **47:** 103–145.

Heath, M. R., J. O. Backhaus, K. Richardson, E. McKenzie, D. Slagstad, D. Beare, J. Dunn, J. G. Fraser, A. Gellego, D. Hainbucher, S. Hay, S. Jonasdottir, H. Madden, J. Mardaljevic, and A. Schacht, 1999b. Climate fluctuations and the spring invasion of the North Sea by *Calanus finmarchicus. Fisheries Oceanography,* **8,** Suppl. 1, 163–176.

Herbert, J.M. and R.W. Dixon, 2003. Is the ENSO phenomenon changing as a result of global warming? *Physical Geography,* **23,** (3), 196–211.

Hollowed, A.B., S.R. Hare and W.S. Wooster, 2001. Pacific Basin climate variability and patterns of Northeast Pacific marine fish production. *Progress in Oceanography,* **49,** (1–4), 257–282.

Hopcroft, R.R., S. Clark and F.P. Chavez, 2002. Copepod Communities in Monterey Bay During the 1997 to 1999 EL Niño and La Niña. *Progress in Oceanography,* **52,** (1–4), 251–264.

Hurrell, J.W., Y. Kushnir, G. Ottersen, and M. Visbeck (Eds.), 2003. The North Atlantic Oscillation Climate Significance and Environmental Impacts, Geophysical Monograph Series, 134.

Huyer, A. and R.L. Smith, 1985. The signature of El Niño off Oregon, 1982–1983. *Journal of Geophysical Research,* **90,** (C4), 7133–7142.

Huyer, A., R.L. Smith and T. Paluszkiewicz, 1987. Coastal upwelling off Peru during normal and El Niño times, 1981–1984. *J. Geophys. Res,* **92,** (C13), 14297–14308.

Hyrenbach, K. D. and R.R. Veit. 2003. Ocean warming and seabird communities of the southern California Current System (1987–98): response at multiple temporal scales *Deep-Sea Research II,* 50, 2537–2565.

Isaacs, J.D., 1975. Some ideas and frustrations over fishery science. *CalCOFI Report,* **28,** 34–43.

Jones, P.D., M. New, D.E. Parker, S. Martin and I.G. Rigor, 1999. Surface air temperature and its changes over the past 150 years. *Reviews of Geophysics,* **37,** (2), 173–199.

Jordan, R. and H. Fuentes, 1966. Las pobalciones de aves guaneras y su situacion actual. *Inf. Inst. Mar.,* **10,** 1–31.

Joyce, T.M. and J. Dunworth-Baker, 2003. Long-term hydrographic variability in the Northwest Pacific Ocean. *Geophysical Research Letters,* **30,** (2), 1043.

Karl, D. M., R. Letelier, D. Hebel, L. Tupas, J. Dore, J. Christian and C. Winn., 1995. Ecosystem changes in the North Pacificsubtropical gyre attributed to the 1991–92 El Niño. *Nature* **373,** 230–234.

Kawasaki, T., 1983. Why do some pelagic fishes have wide fluctuations in their numbers? Biological basis of fluctuation from theviewpoint of evolutionary ecology. *FAO Fish. Rep.* 2**91,** 1065–1080.

Kawasaki, T., S. Tanaka, Y. Toba, and A. Taniguchi (eds.), 1991. Long-term Variability of Pelagic Fish Populations and their Environment, Pergamon, Oxford, 402.

Keeling, C.D., R.B. Bacastow, A.l. Carter, S.C. Piper, T.P. Whorf, M. Heimann, W.G. Mook, and H. Roeloffzen, 1989. A three dimensional model of atmospheric CO2 transport based on observed winds: 1. Analysis of observational data. In: Aspects of Climate Variability in the Pacific and the Western Americas, Peterson D.H., (ed.), American Geophysical Union, Washington, DC, 165–236.

Klyashtorin, L.B., 2001. Climate change and long-term fluctuations of commercial catches: the possibility of forecasting. *FAO Fish.Tech.Pap.,* **410,** 86 p.

Kleeman, R., J. P. McCreary Jr., and B. A. Klinger, 1999. A mechanism for generating ENSO decadal variability, *Geophysical Research Letters,* **26,** 12, 1743–1746.

Latif, M., R. Kleeman and C. Eckert, 1997. Greenhouse warming, decadal variability, or El Niño? An attempt to understand the anomalous 1990s. *Journal of Climate,* **10,** (9), 2221–2239.

Leonard, C. L., and C. R. McClain, 1996. Assessment of interannual variation (1979–1986) in pigment concentrations in the tropical Pacific using the CZCS, *Int. J. Remote Sens.,* **17,** 721–732, 1996.

Lluch-Belda, D., R. Crawford, T. Kawasaki, A. MacCall, R. Parrish, R. Shwartzlose, P. Smith, 1989. Worldwide fluctuations of sardine and anchovy stock. The regime problem. *S. Afri. J. Mar. Sci.,* **8,** 195–205.

Lluch-Cota, D., S. Hernandez-Vazquez, S. Lluch-Cota, 1997. Empirical Investigation on the Relationship between climate and small pelagic Global Regimes and El Nino- Southern oscillation (ENSO), *Fisheries Circular,* **934,** 48p., Rome, Food and Agriculture Organization of the United Nations.

Lynam, C.P., S.J. Hay, and A.S. Brierley, 2004. Interannual variability in abundance of North Sea jellyfish and links to the North Atlantic Oscillation, *Limnol. Oceanogr.,* **49,** 3, 637–643.

Mackas, D.L. and M. Galbraith, 2002. Zooplankton community composition along the inner portion of Line P during the 1997–98 El Niño event. *Progress in Oceanography,* **52,** (1–4), 423–437.

Majluf, P., 1991. El Niño effects on pinnipeds in Peru, In *Pinnipeds and El Niño, Responses to environmental stress.,* F. Trillmich and K.A. Ono (eds.), Springer-Verlag, Berlin.

Mantua, N.J., S.R. Hare, Y. Zhang, J.M. Wallace and R.C. Francis, 1997. A Pacific interdecadal climate oscillation with impacts on salmon production. *Bulletin of the American Meteorological Society,* **78,** (6), 1069–1079.

Mantua, N.J. and S.R. Hare, 2002. The Pacific decadal oscillation. *Journal of Oceanography,* **58,** (1), 35–44.

Margalef, R., 1978. What is an Upwelling Ecosystem? In *Upwelling Ecosystems,* R. Boje and M. Tomczak (eds.), Springer-Verlag, Berlin Heidelberg New York, 12–14.

Marinovic, B., D.A. Croll, N. Gong, S. Benson and F. Chavez, 2002. Effects of the 1997–98 El Niño on Zooplankton within the Monterey Bay Coastal Upwelling System with Emphasis on the Euphausiid Community. *Progress in Oceanography,* **52,** (1–4), 265–277.

Marshall, J., H. Johnson, and J. Goodman, 2001. A study of the interaction of the North Atlantic Oscillation with the ocean circulation. *J. Climate,* **14,** 1399–1421.

Maunder, M.N. and G.M. Watters, 2001. Status of yellowfin tuna in the eastern Pacific Ocean. Inter-Amer. Trop. Tuna Comm., Stock Assessment Report, **1,** 5–86.

May, R.M., 1973. The stability and complexity of model ecosystems. Princeton University Press, 235.

McFarlane, G.A., P.E. Smith, T.R. Baumgartner and J.R. Hunter, 2002. Climate Variability and Pacific Sardine Populations and Fisheries. In Fisheries in a changing climate. N.A. McGinn (ed.), *American Fisheries Society Symposium,* **32,** 195.

McGowan, J.A., 1989. Pelagic ecology and Pacific climate. American Geophysical Union Monograph, **55,** 141–150.

McKinnell, S. M., R.D. Brodeur, K. Hanawa, A.B. Hollowed, J.J. Polovina, and C-I. Zhang, (eds.), 2001. Pacific climate variability and marine ecosystem impacts. *Prog. Oceanogr.* **49,** 1–6.

McPhaden, M.J., A.J. Busalacchi, R. Cheney, J.R. Donguy, K.S. Gage, D. Halpern, M. Ji, P. Julian, G. Meyers, G.T. Mitchum, P.P. Niiler, J. Picaut, R.W. Reynolds, N. Smith and K. Takeuchi, 1998. The tropical ocean global atmosphere observing system: A decade of progress. *Journal of Geophysical Research-Oceans,* **103,** (C7), 14169–14240.

McPhaden, M.J., 1999. Genesis and evolution of the 1997–98 El Niño. *Science,* **283,** (5404), 950–954.

McPhaden, M.J. and D.X. Zhang, 2002. Slowdown of the meridional overturning circulation in the upper Pacific Ocean. *Nature,* **415,** (6872), 603–608.

McPhaden, M.J., 2004. Evolution of the 2002/03 El Niño. *Bulletin of the American Meteorology Society,* doi: 10.1175/BAMS-85–5–677.

Miller, A. J., D. R. Cayan, T. P. Barnett, N. E. Graham and J. M. Oberhuber, 1994. The 1976–77 climate shift of the Pacific Ocean. *Oceanography,* **7,** 21–26.

Miller, A. J., D. R. Cayan, and W. B. White, 1998. A westward-intensified decadal change in the North Pacific thermocline and gyre-scale circulation, *J. Climate,* **11,** 3112–3127.

Miller, A.J., F. Chai, S. Chiba, J. R. Moisan and D. J. Neilson, 2004. Decadal-Scale Climate and Ecosystem Interactions in the North Pacific Ocean *Journal of Oceanography,* **60,** 163–188.

Minobe, S., 1997. A 50–70 year climatic oscillation over the North Pacific and North America. *Geophysical Research Letters,* **24,** 683–686.

Minobe, S., 2002. Interannual to interdecadal changes in the Bering Sea and concurrent 1998/99 changes over the North Pacific. *Progress in Oceanography,* **55,** (1–2), 45–64.

Murtugudde, R., S. Signorini, J. Christian, A. Busalacchi, C. R. McClain, and J. Picaut, 1999. Ocean color variability of the tropical Indo-Pacific basin observed by SeaWiFS during 1997–1998, *J. Geophys.Res.,* **104,** 18351–18366.

Ochoa, N. and O. Gomez, 1987. Dinoflagellates as indicators of water masses during El Niño, 1982–1983. *J. Geophys. Res.,* **92,** (C13), 14355–14367.

Ottersen, G., B. Planque, A. Belgrano, E. Post, P. C. Reid, and N. C. Stenseth. 2001. Ecological effects of the North Atlantic Oscillation. Oecologia **128,** 1–14.

Pearcy, W.G. and A. Schoener, 1987. Changes in the marine biota coincident with the 1982–1983 El Niño in the northeastern Subarctic. *J. Geophys. Res.,* **92,** (C13), 14417–14428.

Pearcy, W.G., 2002. Marine nekton off Oregon and the 1997–98 El Niño. *Progress in Oceanography,* **54,** 399–403.

Peterson, W.T., J. Keister and L. Feinberg, 2002. The effects of the 1997–98 El Niño event on hydrography and zooplankton off the central Oregon coast. *Progress in Oceanography,* **52,** (1–4), 381–398.

Peterson, W. T., and F. B. Schwing, 2003. A new climate regime in northeast Pacific ecosystems, *Geophys. Res. Lett.*, **30,** doi:10.1029/2003GL017528.

Philander, S.G.H., 1983. El Niño-Southern Oscillation phenomena. *Nature,* **302,** 295–301.

Philander, S.G.H., 1986. Unusual conditions in the tropical Atlantic Ocean in 1984. *Nature,* **322,** 236–238.

Philander, S.G.H., 1990. *El Niño, La Niña, and the Southern Oscillation.,* Academic Press, San Diego, 289.

Pomeroy, L. R., 1974. The ocean's food web, a changing paradigm. *Bioscience* **24,** 499–504.

Quinn, W.H., V.T. Neal and S.E. Antunez de Mayolo, 1987. El Niño occurrences over the past four and a half centuries. *J. Geophys. Res.,* **92,** (C13), 14449–14461.

Raskoff, K. A., 2001 The impact of El Nino events on populations of mesopelagic hydromedusae. *Hydrobiologia,* **451,** 121–129.

Rasmusson, E.M. and T.H. Carpenter, 1982. Variations in tropical sea surface temperature and surface wind fields associated with the Southern oscillation/El Nino. *Monthly Weather Review,* **110,** 354–384.

Rasmusson, E.M. and J.M. Wallace, 1983. Meteorological aspects of the El Niño/Southern Oscillation, *Science,* **222,** 1195–1202.

Ryther, J.H., 1969. Photosynthesis and fish production in the sea. *Science* **166,** 72–76.

Robinson, G. and E.M. del Pino, 1985. El Niño en las Islas Galapagos. El evento de 1982–83. *Fund. Charles Darwin,* Quito, 534.

Roemmich, D. and McGowan, J.A. 1995. Climatic warming and the decline of zooplankton in the California Current. *Science* **267,** 1324–1326.

Ropelewski, C.F. and M.S. Halpert, 1987. Global and regional scale precipitation patterns associated with the El Niño/Southern Oscillation. *Mon. Wea. Rev.,* **115,** 1606–1626.

Schneider, N., A.J. Miller, M.A. Alexander and C. Deser, 1999. Subduction of decadal North Pacific temperature anomalies: Observations and dynamics. *Journal of Physical Oceanography,* **29,** (5), 1056–1070.

Schwartzlose, R.A., J. Alheit, A. Bakun, T.R. Baumgartner, R. Cloete, R.J.M. Crawford, W.J. Fletcher, Y. Green-Ruiz, E. Hagen, T. Kawasaki, D. Lluch-Belda, S.E. Lluch-Cota, A.D. MacCall, Y. Matsuura, M.O. Nevarez-Martinez, R.H. Parrish, C. Roy, R. Serra, K.V. Shust, M.N. Ward and J.Z. Zuzunaga, 1999. Worldwide large-scale fluctuations of sardine and anchovy populations. *South African Journal of Marine Science-Suid-Afrikaanse Tydskrif Vir Seewetenskap,* **21,** 289–347.

Schwing, F.B. and C. Moore, 2000. A year without summer for California, or a harbinger of a climate shift. *EOS Transcript American Geophysical Union,* **81,** 27.

Shaffer, G., O. Pizarro, L. Djurfeldt, S. Salinas and J. Rutllant, 1997. Circulation and low-frequency variability near the Chilean coast: Remotely forced fluctuations during the 1991–92 El Niño, J. Phys. Oceanogr., **27,** 217–235.

Sette O. E. and J. D. Isaacs, (eds.), 1960. The Changing Pacific Ocean in 1957 and 1958. *CalCOFI Reports* **7,** 13–217.

Simpson, J.J., 1984a. El Niño-induced onshore transport in the California Current during 1982–1983. *Geophys. Res. Lett.,* **11,** 233–236.

Simpson, J.J., 1984b. A simple model of the 1982–83 Californian "El Niño". *Geophys. Res. Lett.,* **11,** 243–246.

Smith, R.L., 1983. Peru coastal current during El Niño: 1976 and 1982. *Science,* **221,** 1397–1399.

Smith, R.L., A. Huyer and J. Fleischbein, 2001a. The coastal ocean off Oregon, from 1961 to 2000: Is there evidence of climate change or only of Los Niños? *Progress in Oceanography,* **49,** (1–4 Special Issue), 63–93.

Smith, K.L., Jr., R.S. Kaufmann, R.J. Baldwin, A.F. Carlucci, 2001b. Pelagic-benthic coupling in the abyssal North Pacific: an 8-year time-series study of food supply and demand. *Limnol. Oceanogr.,* **46,** 543–556.

Stenseth, N., J. W. Hurrell, and A. Belgrano (eds.), 2004. Marine Ecosystems and Climate Variation - The North Atlantic. Oxford University Press, London, 272 p.

Strub, P.T. and C. James, 2002. The 1997–1998 Oceanic El Niño Signal Along the Southeast and northeast Pacific Boundaries - An Altimetric View. *Progress in Oceanography,* **54,** (1–4), 439–458.

Sydeman, W.J., M.M. Hester, J.A. Thayer, F. Gress, P. Martin and J. Buffa, 2001. Climate change, reproductive performance and diet composition of marine birds in the southern California Current system, 1969–1997. *Progress in Oceanography,* **49,** (1–4), 309–329.

Tarazona, J., H. Salzwedel and W.E. Arntz, 1988a. Oscillations of macrobenthos in shallow waters of the Peruvian central coast induced by El Niño 1982–83. *J. Mar. Res.,* **46,** 593–611.

Tarazona, J., H. Salzwedel and W.E. Arntz, 1988b. Positive effects of 'El Niño' on macrozoobenthos inhabiting hypoxic areas of the Peruvian upwelling system. *Oecologia,* **76,** 184–190.

Tarazona, J. and E. Castillo (eds.), 1999. El Nino 1997–98 y su Impacto sobre los Ecosistemas Marino y Terrestre. *Rev. peru. biol.* **Vol. Extraordinario,** 186p.

Tarazona, J., W. Arntz, E. Castillo de Maruenda, (eds.), 2001. El Niño en América Latina. Impactos biológicos y sociales, CONCYTEC, Lima, 1–423.

Tegner, M. J., and P. K. Dayton, 1987. El Niño effects on southern California kelp forest communities. *Advances in Ecological Research* **17,** 243–279.

Thompson, L.G., E. Mosely-Thompson and B. Morales Arnao, 1984. El Niño-Southern Oscillation events recorded in the stratigraphy of the tropical Quelccaya ice cap, Peru. *Science,* **226,** 50–53.

Toresen, R. and O.J. Ostvedt, 2000. Variation in the abundance of the Norwegian spring-spawning herring (*Clupea harengus,* Clupeidae) throughout the 20th century and the influence of climatic fluctuations. Fish and Fisheries, **1,** 231–256.

Tovar, H. and L. Garcia, 1982. Las poblaciones de aves guaneras durante 'El Niño' de 1957. *Bol. Lima,* **22,** 34–46.

Tovar, H. and D. Cabrera, 1985. Las aves guaneras y el fenomeno 'El Niño', In *El Niño. Su impacto en la fauna marina.,* W.E. Arntz, A. Landa and J. Tarazona, (eds.), *Bol. Inst. Mar.,* Peru-Callao, 181–186.

Trenberth. K.E. and J.W. Hurrell, 1994. Decadal Atmosphere-Ocean Variations in the Pacifc. *Climate Dynamics,* **9,** 303–319.

Trillmich, F., and K. A. Ono (eds.), 1991. Pinnipeds and El Niño, responses to environmental stress. Springer-Verlag, Berlin, Ecological studies **88,** 293.

Ulloa O, Escribano R, Hormazabal S, Quiñones R, González R, Ramos M., 2001. Evolution and biological effects of the 1997–98 El Niño in the upwelling ecosystem off northern Chile. *Geophys Res Lett* **28,** 1591–1594.

Valdivia, J., 1978. The anchoveta and 'El Niño'. *Rapp. P.-v. Reun. CIEM,* **173,** 196–202.

Vallis, G.K. 1986. El Niño: A chaotic dynamical system? *Science,* **232,** 243–245.

van Loon, H., and J.C. Rogers, 1978. The seesaw in winter temperatures between Greenland and northern Europe. Part I: General Description. *Mon.Wea.Rev.,* **106,** 296–310.

Veit, R.R.; J.A. McGowan, D.G. Ainley, et al. 1997 Apex marine predator declines ninety percent in association with changing oceanic climate. *Global Change Biology,* **3,** 23–28.

Velez, J.J., J. Zeballos and M. Mendez, 1984. Effects of the 1982–83 El Niño on fishes and crustaceans off Peru. *Trop. Ocean-Atmos. Newsl.,* **28,** 10ss.

Visbeck, M.H.; E. P. Chassignet, R.G., T.L. Delworth, R.R. Dickson and G. Krahmann. 2003. The ocean's response to North Atlantic Oscillation variability, in *The North Atlantic Oscillation,* edited by J. Hurrell, Y. Kushnir, G. Ottersen and M. Visbeck, AGU monograph, Washington, 113–145.

Wada, T. and M. Kashiwai, 1991. Changes in the growth and feeding ground of the Japanese sardine with fluctuation in stock abundance. In T. Kawasaki, S. Tanaka, Y. Toba, and A. Taniguchi (eds.), Long-term variability of pelagic fish populations and their environment, 181–190. Oxford, Pergamon Press.

Walker, G. T., 1924. Correlation in seasonal variations ofweather. IX. A further study of world weather. *Memoirs of the Indian Meteorological Department* **24** (Part 9) 275–332.

Walker, G.T. and E.W. Bliss, 1932. *World Weather V. Mem. Roy. Met. Soc.,* **4,** 53–84.

Whitney, F.A. and D.W. Welch, 2002. Impact of the 1997–8 El Nino and 1999 La Nina on nutrient supply in the Gulf of Alaska. *Progr.Oceanogr.,* **54,** 405–421.

Wilson, C. and D. Adamec, 2001. Correlations between Surface Chlorophyll and Sea-Surface Height in the Tropical Pacific during the 1997/1998 ENSO Event. *Journal of Geophysical Research,* **106,** 31,175–31,188.

Wolff, M., 1987. Population dynamics of the Peruvian scallop *Argopecten purpuratus* during the El Niño phenomenon of 1983. *Can. J. Fish Aquat. Sci.,* **44,** 1684–1691.

Wooster, W.S. and D.L. Fluharty (eds.), 1985. El Niño North. *Washington Sea Grant Program.* University of Washington, Seattle, 312.

Wyrtki, K., 1982. The Southern Oscillation, ocean-atmosphere interaction and El Niño. *Mar. Technol. Soc. J.,* **16,** 3–10.

# Chapter 17. EXTREME EVENTS TRANSPORTING SEDIMENT ACROSS CONTINENTAL MARGINS: THE RELATIVE INFLUENCE OF CLIMATE AND TECTONICS

JEFFREY D. PARSONS
CHARLES A. NITTROUER

*University of Washington*

**Contents**



## 1. Introduction

The first conceptual models of sedimentology attempted to demonstrate that the Earth has been a dynamic planet for millions of years. Uniformitarianism, the school of thought responsible for much of the early development of geology, suggests that the sedimentary record is made up of a slow, steady accumulation of sediment over time. Models based upon uniformitarian ideas have been successful in predicting the general distribution of material over millions of years. Continuity of sediment delivery is often implied, meaning that basic physical principles can be related to changing environmental variables. Sequence stratigraphy, the production of repeated patterns of sedimentary deposits associated with sea-level oscillations, provides an excellent example of a model where distribution of material can be predicted accurately based upon a few simple principles (e.g., Posamentier et al., 1988).

Despite the power of models based upon continuous, cyclic processes, intermittent, extreme events are ubiquitous in the sediment record. Extreme events can mean different things to different people. To geologists, this may mean unusual epochs in the geological record, such as periods of increased tectonic activity or

*The Sea*, Volume 13, edited by Allan R. Robinson and Kenneth H. Brink
ISBN 0-674-01526-6 ©2004 by the President and Fellows of Harvard College

dramatic swings in atmospheric composition and temperature (e.g., Snowball Earth: Hoffman et al., 1998). However, oceanographers, our intended audience, are more interested in shorter time scales and the processes that have recently affected the ocean. In this context, the most convenient definition of an extreme sedimentological event is a period of sustained, heightened sediment transport that produces a distinguishable, discrete deposit. From monsoonal floods to slide-generated turbidity currents, the marine sediment record of continental margins is usually composed of a series of these discrete deposits. Turbidite fans, the largest individual sedimentary features in the ocean, are good examples (Figure 17.1). Event beds in cores and outcrops define some of the most basic building blocks of process sedimentology and have been the focus of efforts for more than fifty years (e.g., Kuenen and Migliorini 1950; Heezen and Ewing, 1952; Bouma, 1962).

Figure 17.1 An approximately 30-m-thick (from the base of the outcrop to top of the cliff face) section of the Jackfork Formation in central Arkansas. The exposure shown is located near the DeGray Spillway. It has consistently been interpreted as a series of event deposits formed by sediment gravity flows (Shanmugam and Moiola, 1995; Bouma et al., 1997). Interpretations of the type of flow occurring during deposition have varied, but consensus in the geological community is that these were formed by turbulent suspensions (i.e., turbidity currents: Bouma et al., 1997).

Until recently, many of these dramatic events have not been observed directly, which has led some scientists to question the existence of certain types of behavior (e.g., turbidity currents: Shanmugam, 2003). However, turbidity currents also illustrate why understanding has come so slowly. Early observations, even in reasonably accessible environments, were often plagued by instrument loss (Inman et al., 1976). New work, which has relied on more easily detected indirect processes and comparison of these observations to associated behavior in laboratory and numerical models, has shown that these events occur sporadically and often in conditions hazardous to in situ measurement (Johnson et al., 2001; Puig et al., 2003; Khripounoff et al., 2003; Paull et al., 2003; Puig et al., 2004). Because of their seemingly unpredictable behavior, these transport episodes have been seen as stochastic events, dependent only upon chance. However, recent observations in both the field and the laboratory illustrate interrelationships between natural phenomena that were once thought to be unrelated (e.g., storms and turbidity currents: Puig et al., 2003). Understanding the relationships between these events is essential for accurate and reliable estimations of past environmental conditions.

In many instances, different phenomena in different environments are related to the same catastrophic event. The recent slide observed offshore of Papua New Guinea, which produced a deadly tsunami, provides an excellent example of the intertwining of seabed processes and coastal sedimentology (Tappin et al., 2001). Though the slide may have been fundamentally unpredictable, this type of event can possess a degree of periodicity and predictability. For example, the giant subduction earthquakes thought to produce both slides and turbidity currents on the Cascadian margin are often discussed in terms of an "average recurrence time"(Goldfinger et al., 2003).

Over time, these extreme events begin to dominate the sediment record. When the seabed is subject to successive periods of heightened shear stress, the dynamics of erosion into the bed and deposition from suspension dominate the preservation of strata. Entrainment into suspension ($E$) has been demonstrated to be a strong, nonlinear function of bed shear stress ($\tau_b$; $E \propto \tau_b^5$: Garcia and Parker, 1991). Deposition is also highly dependent on nonlinear, water-column variables (Winterwerp, 2001). Once a thick bed resulting from an extreme event is laid down, it requires an even larger event to remove it completely. On the other hand, small, common events are easily removed from the record. Deposits from these small, common events are replaced by subsequent larger sedimentation events. This process continues until the sediment record is strongly biased towards the largest, most unusual events.

The large spatial extent and preferential preservation of extreme events also causes their record to become biased in space. Most often, the energy transmitted to coastal areas (and ultimately the driving force behind sediment transport) during extreme events is distributed throughout the margin. This is true even above sea level, as in the case of tsunami deposits. Within the typical tidal prism, the continual processes of tides and wave breaking tend to obliterate any signal left by unusual events. At depth and above highest tide, sediment transport is rare and therefore the deposits can more easily remain unadulterated and subsequently buried. Therefore a discussion of extreme events, even when limited to coastal zones, naturally leads to discussion of deposits at the extremities of continental margins.

Considerable advances have been made since publication of the last catalog of extreme episodic, or convulsive, events (Clifton, 1988). In this chapter, we intend to provide a framework that will build on the base of knowledge developed within the last fifteen years from both observation and experiment. The framework should be helpful in guiding future work relating the different natural phenomena responsible for the formation of large sedimentary deposits. In order to address the most common issues encountered in sedimentology, we have omitted volcanic environments, particularly those near mid-ocean ridge crests, which are beyond the scope of this chapter.

## 2. Relationships between episodic events

To unravel the sediment record, it is useful to separate natural phenomena into broad categories. Settings dominated by only one type of geophysical phenomenon usually have been studied in order to simplify and isolate the understanding of that phenomenon (e.g., seismicity, Izu-Bonin trench: Hiscott et al., 1992; Beattie and Dade, 1996). However, it can be productive to identify regions where all types of behavior are important because it is in these areas where sedimentary signals are the largest. Therefore, to accurately assess the contribution of each type of process, the interactions between different sorts of events must be realized.

### *2.1. Climatic events*

We define a climatic event as any physical process that ultimately derives its energy from the Sun. Though this may seem like a restricted definition, it includes virtually any weather-related event. For instance, floods and storms are associated with the constantly changing input of solar energy and its transport through exchange of latent heat between water vapor and precipitation. As a result, the sediment record associated with catastrophic atmospheric events is capable of providing insight into past variability in climate at local, regional and global scales. In fact, if the sediment record is well resolved, it provides one of the only opportunities to understand terrestrial surface processes under conditions significantly different than the present.

### *2.2. Tectonic events*

Rather than being forced by climatic fluctuations, tectonic events are produced by the release of geophysically supplied potential energy. In other words, tectonic events occur as a result of energy provided by the Earth itself (i.e., ultimately from heat produced by radioisotope decay within its interior). The spatial and temporal distribution of the deposits associated with these events should reflect the variability of the tectonic motions themselves. Because these motions are not well understood, the sediment record associated with tectonic processes presents a potentially useful tool to describe these deadly events.

### *2.3. Relationships between processes*

In order to understand the relationships among different physical processes, it is important to separate the phenomenon that generates the event from the physical environment encountered during deposition. To do this, we have classified a variety of events into primary and secondary processes. Primary processes are those phenomena that derive their energy directly from climatic or tectonic forcing. Secondary processes are those that are generated from primary processes.

The relationships among processes rely upon a detailed physical understanding of the underlying mechanics regulating transport and deposition. Figure 17.2 illustrates the various interrelationships among the different events and processes to be discussed herein. Because understanding these interactions requires comprehensive data in space and time, our present knowledge of the transition between transport modes and interaction with complementary processes is still relatively poor. Laboratory modeling allows us to glimpse the interactions; however, these studies are limited by the size of the experimental facility. In the case of numerical modeling, rheology, the constitutive equations indicative of the relationship between stress and strain, often changes as one type of flow evolves into another. This change in the governing equations demands a reformulation of the numerical algorithm – a task beyond the scope of most numerical routines. Finally, transitional behavior (particularly in situations where nonlinear quantities like shear stress and sediment entrainment are important) can make the intermediate processes impossible to resolve, even when computational codes are well designed for process changes.

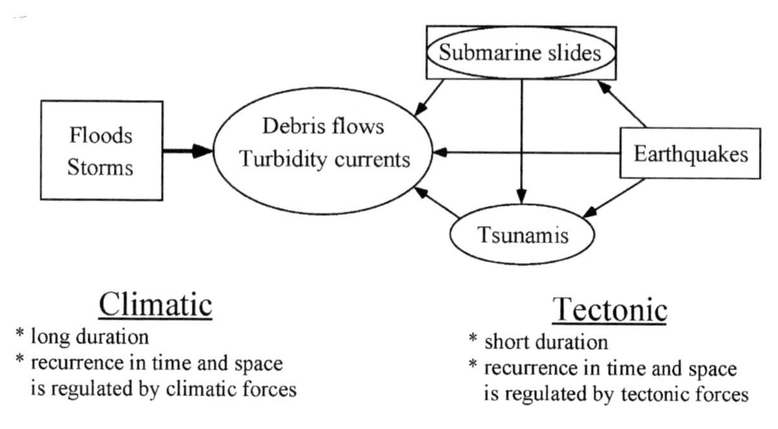


Figure 17.2 Diagram of the interrelationships of various types of episodic processes and their root cause. Primary processes are denoted by rectangles, while secondary processes are placed within ellipses. Slides can be either.

Despite these hurdles, work has begun to explore the relationships among different types of events – for example, the transformation of a debris flow into a turbidity current (Marr et al., 2001). The original studies describing turbidity currents and their deposits (Kuenen and Migliorini, 1950; Heezen and Ewing, 1952; Bouma, 1962) depended on the production of a turbidity current from a debris flow or slide. As a result, the need arose to understand this type of turbidity-current genesis. Recent laboratory experiments have shown that the transformation of debris flow to turbidity current is common, but rarely efficient (i.e., less than 10% of the debris-flow material becomes a turbidity current: Marr et al., 2001). That is, if these measurements are correct, and if debris flows were the only mechanism by which turbidity currents were formed, one would expect that debris-flow deposits should be much more extensive that turbidites (turbidity-current deposits). This lies in stark contrast with decades of observation into ancient deep-marine outcrops, which conclusively demonstrate the prevalence of turbidites in the sediment record (Normark et al., 1993; Kneller, 1995).

Computational modeling also offers a window into the stratigraphic relationships between different mass-flow processes. Novel computational strategies have been employed to model these processes in the production of margin-scale stratigraphy (*SedFlux*: Syvitski and Hutton, 2001). For instance, *SedFlux* is a group of programs able to incorporate a variety of interactions between different types of events, both climatic and tectonic (Syvitski and Hutton, 2001). These interactions, among the others outlined in Figure 17.2, are described in the rest of this chapter.

## 3. Evaluating episodic events

### *3.1. Observations*

Observational marine geology has advanced tremendously in recent years both in terms of measurement of water-column processes and of the seabed itself. Many of the advances can be attributed to improvement in the acquisition and processing of acoustic and seismic data. In the water column, processing backscattered acoustic energy has allowed us to assess not only water motions, but also sediment concentration to a degree not possible even ten years ago. Probably the most striking example of the ability to capture extreme events on the continental margin is the observation of large "fluid-mud" events associated with the conjunction of floods and storms on the northern Californian shelf (Figure 17.3: Traykovski et al., 2000; Ogston et al., 2000). These measurements provide information about one type of physical process that has progressively been recognized for inducing intermittent transport across continental margins (Wright et al., 1988; Cacchione et al., 1995).

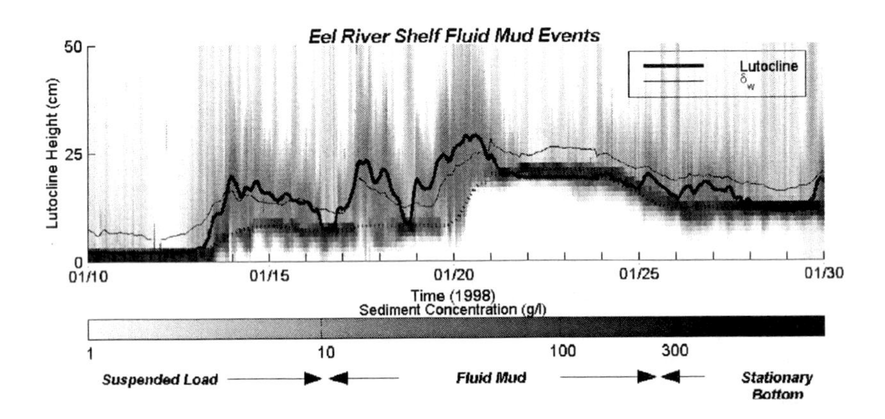


Figure 17.3 Time series of vertical sediment concentration made with an acoustic backscatter sensor (ABS) on the middle shelf (60-m water depth) of the Eel margin. The evolution of fluid mud can be observed as the result of combined flood and storm events (After Traykovski et al., 2000).

Quantification of the flux is a much more difficult problem. Though backscattered acoustic energy is related to sediment concentration, it is strongly dependent on the particle characteristics (e.g., size, mineralogy, floc state). In the case of an acoustic-backscatter sensor (ABS), the acoustic response of the water column is dependent on the nature of particles being transported and the vertical distribution of these characteristics. The result is that in situ samples are required to properly calibrate an ABS. Because it is usually impossible to acquire sediment samples at multiple depths/times during a single deployment, it is impossible to have firm constraints on the amount of material in suspension (and in transport), particularly in light of the strong concentration gradients often encountered in energetic environments (Traykovski et al., 2000; Lamb and Parsons, in press).

At larger scales and within the seabed, multi-channel seismic tomography has made it possible to identify extreme events in the recent sediment record (e.g., submarine slides: Figure 17.4). Using backscattered seismic (<<1 kHz) energy, variability of sound speed within the seabed can be recorded in time. This information is then deconvolved and displayed as a map of reflected amplitude. Seismic tomography has proved invaluable to the oil industry in their pursuit of "deep-water" oil reserves (Normark et al., 1993). New computational techniques have allowed geologists to "backstrip" seismic reflectors, enabling tomographic reconstruction of ancient seafloors (Nissen et al., 1999). These measurements have been useful in determining hazards, as well as identifying the characteristics of ancient environments. However, it is important to note that seismic imaging biases one towards episodic events. Seismic tomography only measures the derivative of seabed properties. That is, superimposed layers with a distinct contrast in acoustic impedance (e.g., the base of an erosive turbidity current) are preferentially recognized. This has a tendency to accentuate grain-size breaks and overlooks layers recording steady or gradually varying sedimentation.

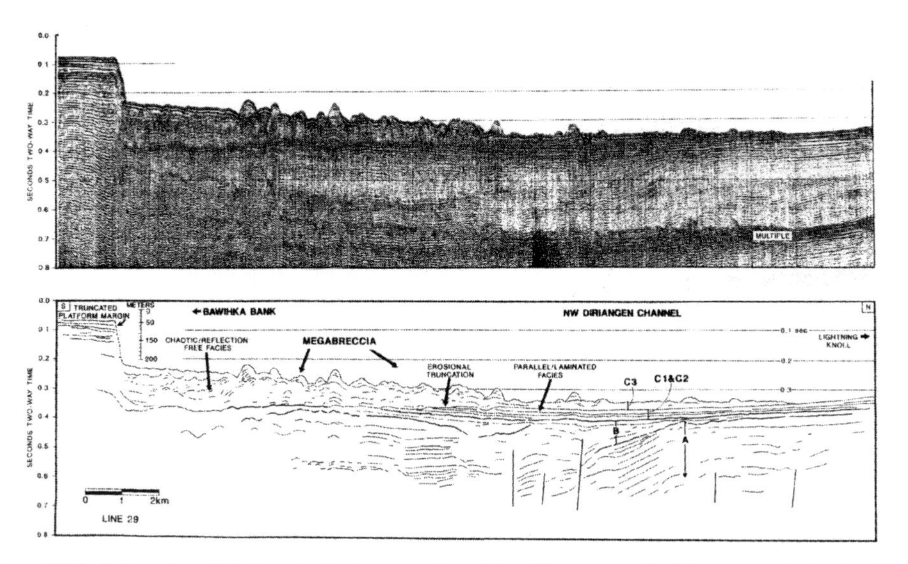


Figure 17.4 Submarine slide in seismic tomography of the Nicaraguan Caribbean margin (After Hine et al., 1992).

High-resolution bathymetric measurement (i.e., multibeam sonar), a close cousin of seismic tomography, has provided insight into the deposition and accumulation of modern sediments. Although sonar technology has existed for over fifty years, the ability to process the acoustic signal backscattered from the bed in multiple directions was only possible with the advent of portable high-speed computers. High-resolution bathymetry is now essential for the accurate placement of instrument systems and the retrieval of sediment cores in areas of complex bathymetry (e.g., submarine canyons, see Figure 17.5; Lomnicky et al., in review).

Of great importance for understanding other quantitative aspects of sedimentary processes are radioisotopes, which allow investigation of dynamic processes associated with the seabed (e.g., rates of burial, rates of reworking). Recently a number of naturally emplaced isotopes have been employed to investigate the seabed on time scales (days to decades) commensurate with the transport processes and measurements of these processes. The fundamental consideration for natural radioisotopes is their half-life; they can be used most effectively to evaluate a time scale equal to about 4-5 times their half-lives. Among the isotopes is $^{210}$Pb, which has a half-life of 22.3 years and can be used to investigate the characteristics of the seabed recorded during the past century (Nittrouer et al., 1979). Dissolved $^{210}$Pb is supplied to seawater from several sources, and is readily scavenged irreversibly to the surfaces of sediment particles. It then follows the paths of the particles through settling, lateral transport and burial. Commonly, a bomb-produced isotope, $^{137}$Cs (present since the early 1950s), is used as an independent tracer to check the interpretations from $^{210}$Pb profiles (Nittrouer et al., 1984).

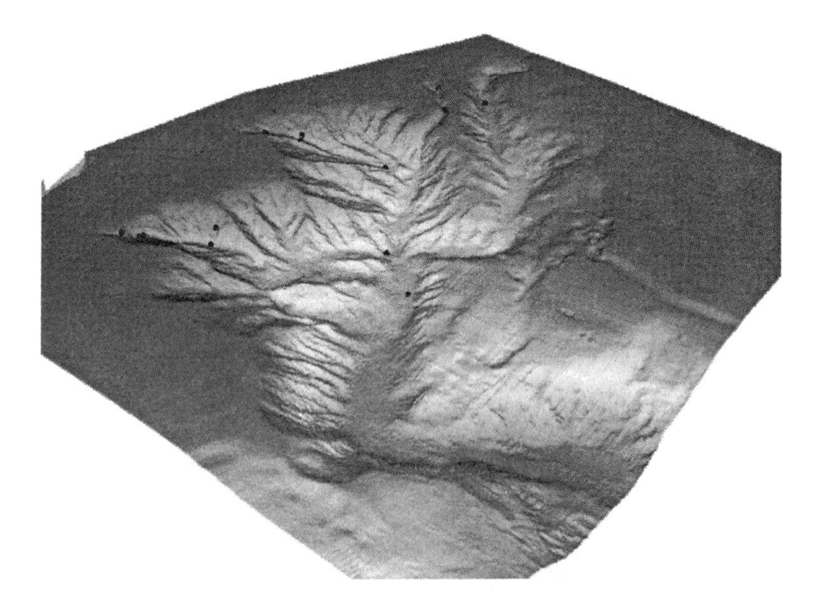

Figure 17.5 Multibeam bathymetry of the upper Eel Canyon with ROV-dive locations (black dots) where evidence of frequent gravity flows was observed (Lomnicky et al., in review).

Even shorter-lived isotopes have been valuable for investigating processes on time scales of days to months, typical of many extreme events. $^{7}$Be (half-life 53 days) is supplied to the earth's surface from cosmogenic sources, and is concentrated in the surface of soils. When heavy rainfall erodes the surface, sediments (with $^{7}$Be) are transported by rivers to the ocean and can be used to document flood deposits (Sommerfield et al., 1999). For rapid sedimentation of a less catastrophic nature (e.g., peak seasonal flows of large rivers), $^{234}$Th (half-life 23 days) is scavenged from seawater and has been used to document deposition rates in shallow coastal environments (McKee et al., 1983).

$^{14}C$ age dating is extremely valuable for documenting chronologies of extreme events that occur over time scales of $10^3$ years – e.g., for the Holocene record of turbidity currents. However, for measurement of historical extreme events, the resolution of $^{14}C$ dating is poor (usually $10^2$ years). This limitation comes from the long half-life of $^{14}C$ (5730 years: Godwin, 1962) and complications from atmospheric bomb testing.

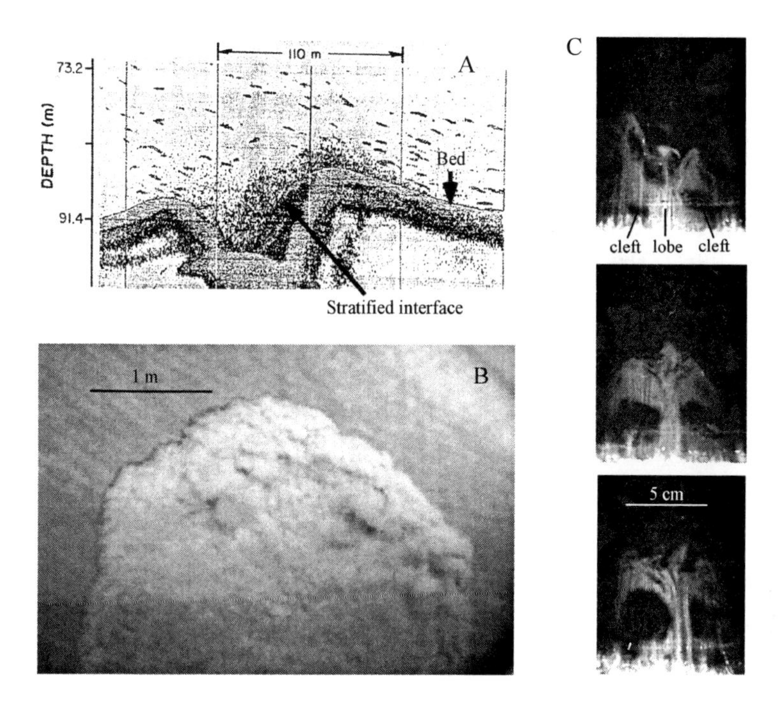


Figure 17.6 Illustration of the wide range of sampling techniques within modeled turbidity currents. A) Acoustic backscatter obtained from a mine-tailings turbidite channel in British Columbia (Hay, 1987). B) Laboratory turbidity current propagating through a sediment-free ambient fluid, viewed from above (Parsons et al., 2002). Note the lobe-cleft structure at the front of the flow. C) Laser-induced fluorescent imagery of the circulation of dense fluid (white) in a saline gravity-current front propagating perpendicular to the field of view through a quiescent ambient fluid (black). Photos were taken 0.1 s apart. Saline gravity currents are often used to model fine-grained turbidity currents.

### *3.2. Modeling*

Prediction has improved dramatically for flow characteristics during extreme episodic transport events and for the resulting sedimentary deposits. The steady increase in computational power has allowed for the solution of the basic governing equations associated with event-based sediment transport, particularly for small, well-constrained events. However, drawbacks remain. Most of these models focus upon a single flow type. Margin-scale models typically interchange single-flow-type subroutines depending on the physical environment being simulated (e.g., *SedFlux*: Syvitski and Hutton, 2001), but they cannot handle simultaneous processes. As a result, physics-based models are particularly good at capturing qualitative stratigraphic features dependent on gross morphologic controls, but reliable quantification requires knowledge of complicated particle-particle and particle-turbulence interactions. Numerical models incorporating a fundamental understanding of even the simplest of these interactions remain elusive (Brenner and Mucha, 2001).

Advances in computational power have also greatly aided the ability of laboratory experimentalists to understand the small-scale dynamics of sediment-laden flows. Combined with advances in instrumentation (both acoustic and electromagnetic), experiments have grown increasingly sophisticated. It is now possible to document the internal structure of most environmentally significant flows. Figure 17.6 illustrates flow visualization within laboratory gravity (turbidity) currents. In these flows, the lobe-cleft structure at the front of a turbidity current can be seen easily. The lobes and clefts have been shown to be important for regulating the speed and spatial distribution of sediment from episodic turbidity currents (i.e., dilute, particle-laden gravity currents: Parsons et al., 2002). After exploring a wide variety conditions, it is possible to formulate simple models that can be used to extrapolate the transport and distribution of sediment in extreme, natural events. The simple analytical models also establish a framework for the future numerical modeling of those processes.

### 3.3. *Ancient analogs*

Flysch deposits were once the primary source of knowledge about episodic turbidity currents in the ocean. Most of the early work on these extreme events was derived from the 1929 Grand Banks earthquake-slide and its deposit located on the Atlantic margin of Canada (Figure 17.7; Kuenen and Migliorini, 1950; Heezen and Ewing, 1952). The Grand Banks event, which began as a highly concentrated slope failure and subsequently evolved into a vigorous turbidity current, propagated considerably faster than any known oceanic current and continued for hundreds of kilometers. Later, Bouma (1962) associated a common sequence of bedding found in ancient marine rocks in the Pyrenees with this type of flow. This bedding (sometimes known as a Bouma sequence) has been identified in innumerable outcrops around the world (Normark et al., 1993). In fact, outcrops of turbidites (turbidity-current deposits) remain one of the best means to understand natural episodic events on continental margins, because of the difficulty in directly measuring these unpredictable, destructive flows (Inman et al., 1976). The sedimentary record provides examples of many climatic conditions and processes not found on Earth today, and this record can yield clues to how these environments and processes operated in the past.

The power of interpretation from outcrop needs to be carefully evaluated, however. The inverse problem associated with deducing flow and event characteristics is fundamentally non-unique, and therefore problematic, particularly considering the wide range of environments possible (and the lack of data constraining any physical parameter). Further, the finite size of most outcrops and basin shapes can yield a wide array of depositional sequences, even when flows producing them are similar (Kneller, 1995). However, as with laboratory and numerical modeling, advances in technology are providing innovative new ways of understanding the three-dimensional structure and relation of different facies (rock units) observed in outcrop (Corbeanu et al., 2001).

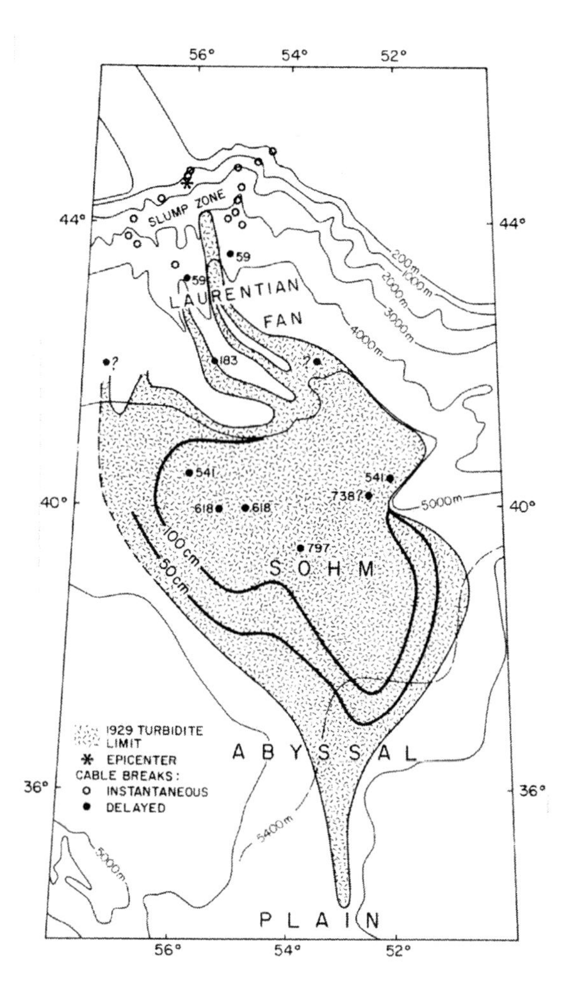


Figure 17.7 Extent of Grand Banks deposit and location of earthquake epicenter (After Piper et al., 1988).

## 4. Characterizing extreme episodic events

### *4.1. Climatic events*

One of the most powerful uses of the sediment record is as a "tape recorder" of past climate. A good example of climate events recorded in sedimentary deposits comes from the continental margin of California. In the north, strong floods associated with large mid-latitude cyclonic storms deposit extremely large amounts of sediment on the continental margin. Fresh, unconsolidated flood deposits are often reworked by subsequent storms, causing material to flow downslope in submarine canyons as wave-supported gravity currents (Puig et al., 2003). Farther south, canyons are related almost exclusively to strong, storm-wave-induced nearshore circulation of suspended sand (Inman et al., 1976). In the Monterey Bay of central California, where the canyon head is located near the modern-day coastline but where sediment supply is also substantial, turbidity currents can either be directly related to floods on land (Johnson et al., 2001) or associated with nearshore resuspension (Paull et al., 2003).

### 4.2. *Tectonic events*

Tectonic events can play an important role in the sediment record. Geophysical activity generally promotes production of sediment and the formation of a high-resolution sediment record. Because primary tectonic events and the secondary processes they produce have been poorly understood in the marine environment, scientists have attempted to use tectonically dominated deposits to understand both primary and secondary processes.

Hiscott et al. (1992), and later Rothman et al. (1994), used the sediment record obtained from the Izu-Bonin Fore-Arc to identify the scaling relationship between physical processes in turbidity currents. Because the Izu-Bonin Fore-Arc is a volcanic arc in the middle of the South Pacific Ocean, terrigenous (and therefore, climatic) influence over the deposits was presumed to be small. Rothman et al. (1994) found that the number of turbidite beds greater than a given thickness had a power-law relationship with that thickness. Rothman et al. (1994) derived a simple scale analysis to show that turbidity currents were scale-independent processes that produce features with a fixed fractal dimension. Beattie and Dade (1996) later used a stochastic analysis, which indicated that the turbidite thicknesses could have had a direct relationship to the sizes of the earthquakes that initiated them. Therefore, the sediment record may be a more effective tool for describing the original primary event (in the case of the Izu-Bonin data; the pattern of earthquakes on a volcanic arc), rather than the secondary process responsible for final deposition.

Several studies have investigated the interaction of seismically produced turbidites, or seismo-turbidites, but probably the strongest evidence has come from the Cascadia margin. Thirteen distinct turbidites in Holocene sediments have been identified on the Cascadia margin (Adams, 1990). Later work confirmed this result by identifying synchronous events from Vancouver Island to the Mendocino Triple Junction (Goldfinger et al., 2003). Each turbidite was hypothesized to be a result of plate-wide slip, or a "mega-seismic" event. Less dramatic and more localized earthquake events can cause distortions to bedding (i.e., growth faults: Novoa et al., 2000). Climatic events such as large storms could also produce synchronous features across a large area (Allan and Komar, 2002). On the Cascadia margin, however, the chronology of these events is also constrained by a high-resolution record of subsidence and associated tsunami deposits in nearshore areas (Atwater, 1987; Atwater, 1992; Atwater and Hemphill-Haley, 1997). The difficulty of inverting a single event bed into innumerable possibilities will continue to confound interpretations of climatic and tectonic features (Orange, 1999), but as the sediment record is uncovered, the potential exists for other types of independent corroboration.

### 4.3. *Duration and recurrence of events*

In addition to the environmental conditions the sediment record has recorded, the duration (usually related to thickness) of individual event beds can give some sense for the type of event that produced the deposit. Considering that there are physical limitations on the amount of material that can be deposited in a certain time (e.g., viscosity, vertical velocity), information relating to the duration of the event can be indicative of a particular source mechanism. In our framework, tectonic events are generally much shorter in duration than climatic events. The longest-lived secondary tectonic events, tsunamis, can travel halfway around the world in less than a day, and their greatest impact in any one location is usually limited in time as well (e.g., about two hours in the case of Chimbote Tsunami, Peru: Bourgeois et al., 1999).

Climatic events occur over a much longer period. The characteristic time scale of these events depends on the type of event, but most occur over a few days or more (e.g., El Nino can persist for months, even years). However, the input of energy in a climatic event over a given unit of time is generally smaller than in tectonic events. Because geophysical energy is generally stored for thousands of years or more, its catastrophic release in a short amount of time overwhelms the energy that can be stored due to climatic forces (usually in the form of latent heat). Notably, however, the locations where climatic signals are large commonly are the same locations as strong tectonic activity. This is because mountain building enhances latent heat release (i.e., precipitation) by uplifting warm moist air masses. Extreme precipitation events often produce intense sediment transport and large, thick deposits.

Recurrence patterns should also vary according to different driving forces. Climatic events may scale with Milankovitch periods or, at a shorter time-scale, with large-scale oscillations like El-Nino or the Pacific Decadal Oscillation (Allan and Komar, 2002). Deposits associated with this forcing should be distinguishable from impacts that are driven by tectonic processes. That is, if the sediment accumulation rate were known in a particular locale, it would be unlikely that the phenomenon was driven by a tectonic mechanism, if similar event beds were found to occur every five years.

Spatially, the differences between climatic and tectonic events are more complex. For instance, tectonic events presumably should be correlated with local plate motion. If earthquakes generate turbidites in a particular area, their frequency would be directly related to tectonic activity (Beattie and Dade, 1996). Likewise, if an area were dominated by hyperpycnal flows and associated turbidity currents (e.g., for convergent margins: Mulder and Syvitski, 1995), sediment deposits should be strongly correlated to local precipitation.

## 5. Physical processes, their linkages, and sedimentary signatures

### 5.1. *Floods*

Floods can deliver large quantities of sediment to the continental shelf. A common way to estimate the sediment delivery to the ocean is through the use of a rating curve. A rating curve is usually expressed by $Q_s = aQ^b$, where $Q$ is the discharge of water from a river mouth, $Q_s$ is the discharge of sediment and $a$ and $b$ are empirical constants. The relationship is nonlinear (i.e., $b$ is usually greater than unity), and sediment discharged in extreme events can be greatly increased during the peak of floods (Mulder and Syvitski, 1995). Because the sediment concentrations obtained during these releases can easily overwhelm the capacity of the water column to maintain the sediment suspension, floods can result in large, thick deposits.

Floods have been linked to deep-sea turbidity currents through the formation of hyperpycnal flows (river water with suspended sediment causing densities greater than seawater) from rivers during high discharge. Mulder and Syvitski (1995) argue that over geologic time, extreme floods eventually will occur that produce a river discharge capable of producing a hyperpycnal flow. The sediment concentration required to exceed oceanic salinity is approximately 40 g/l. Most rivers rarely produce concentrations this high, and the river systems with the largest water discharge (e.g., the Amazon) never do. However, during extreme floods, particularly in mountainous areas where the supply of sediment is potentially large, concentrations can exceed this limit. Traditional techniques (e.g., benthic tripods) have not yet been able to capture an event like this.

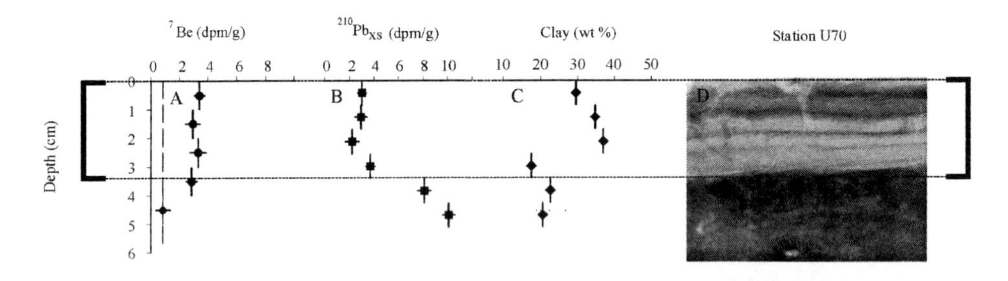


Figure 17.8 $^7$Be, $^{210}$Pb and percent-clay profiles plotted alongside an X-radiograph of a core taken on the middle shelf (70-m water depth) of the Eel margin (see Fig. 17.3). The presence of $^7$Be, low concentration of $^{210}$Pb, increased clay content, and physical sedimentary structures are indicative of this flood deposit (After Sommerfield et al., 1999).

More controversial is the idea that particle-turbulence interactions initiate convective instability of a statically buoyant, sediment-laden river plume, forming an intense turbidity current even for low river-mouth sediment concentrations (Parsons et al., 2001a). Convective instability has not been observed in the field, but in laboratory experiments it often is able to overwhelm the bottom boundary layer and create downslope-flowing turbidity currents. In oceanic environments, waves may act to enhance this process – both to maintain a concentrated suspension near the seabed and to enhance instability of the sediment-laden surface layer. Indirect support for this mechanism has come from canyons on steep margins, where turbidity currents appear to form due to sediment loading near the canyon head (Kineke et al., 2000; Johnson et al., 2001; Walsh and Nittrouer, 2003). However, other work has suggested that the direct connection between river mouths and modern canyon heads may be tenuous and dependent strongly on oceanic climate (Puig et al., 2004).

Environments where episodic floods occur (e.g., mountainous coastal settings) typically show correlation with storms in the ocean, and the signatures of flood deposits are a combination of flood and storm effects (Wheatcroft and Borgeld, 2000). However, the recent input (in high concentrations) of sediment from land can be documented by high activity of $^{7}$Be and low activity of $^{210}$Pb (Figure 17.8; Sommerfield et al., 1999). For episodic floods of rivers with large basins, the use of radioisotopes can be more complicated; because the initial sediment discharged to the ocean may be material stored previously in the river channel and not eroded from land surfaces (Palinkas et al., in press). However, in these environments, the flood is likely not to occur simultaneously with a storm, and the resulting deposit preserves a detailed record of variable pulses during the flood (Palinkas et al., in press; Wheatcroft et al., in press).

### 5.2. *Storms*

Storms, and associated large surface gravity waves and strong wind-driven currents, are probably the most obvious and common extreme event in the marine environment. Operation of instrumentation under these extreme conditions is difficult. However, measurement of storm-derived transport has been constrained by using benthic tripods with multiple sensor systems mounted on them. These measurements outside the wave boundary layer have been vital to constrain analytical models describing the mechanics of and transport associated with the wave boundary layer.

Field studies have been able to demonstrate that the predicted shear stresses imparted by surface waves are usually responsible for erosion of the seabed, and stresses from wind-driven currents control the direction of transport (Nittrouer and Wright, 1994). The direct impact of storm waves and currents is restricted to continental-shelf depths (e.g., wave base is typically $<50$ m). Net displacement of sediment is dominantly along shelf. Recent studies of transport climatology (i.e., multiple-year records) have demonstrated that the net direction of along-shelf transport can reverse from year to year (Ogston et al., 2004), perhaps in response to interannual climatic variations (e.g., El Nino events). Net transport also reveals a distinct seaward component, often the result of downwelling flows from surface water forced against the coastline. In other cases, density flows resulting from high concentrations of suspended sediment cause seaward movement down the shelf gradient (Traykovski et al., 2000; Wright et al., 2002). These may cause sediment to be transported to middle and outer portions of continental shelves (i.e., below wave base). Where the shelf is narrow (e.g., near the heads of submarine canyons), transport can continue to the continental slope (Puig et al., 2003). As a result, these unusual events are disproportionally important in the sediment record, particularly in deep-water environments where generation of sedimentary deposits is strongly regulated by sediment supply.

Considering that storms are a primary process and strongly indicative of climate, it is important to understand their manifestation in the sediment record. This record can provide a much longer time series of storms than any instrument system. For example, tropical-cyclone deposits have been studied extensively throughout the world (Snedden and Nummedal, 1991; Foster and Carter, 1997; Hayne and Chappel, 2001; Suckow et al., 2001). An important result of these major storms is that downwelling transports sediment (often sandy) and forms deposits well below wave base, where they are effectively preserved (commonly as sand beds in muddy middle- and outer-shelf deposits).

The most common physical structure in sandy environments related to storm waves is hummocky cross-stratification (HCS: Dott and Bourgeois, 1982; Myrow and Southard, 1996). HCS is characterized by upward curvature of laminations and low-angle curved lamina intersections (Figure 17.9; Walker, 1984). Though direct observations of hummocky bedding due to storms have been made in modern environments (e.g., Amos et al., 1996), ancient examples are often more common, extensive and thicker than modern analogs (Leckie and Krystinik, 1989; Myrow and Southard, 1996).

Figure 17.9 Ancient storm deposit (tempestite) from Coaledo Formation in southwestern Oregon. Darker material is finer-grained sediment, while lighter material is coarser sediment. The combination of depositional and erosional sedimentary structures is typical of storm deposits, when much sediment is in motion under conditions of high shear stresses. Photo appears courtesy of J. Crockett.

In muddy environments fluid-mud formation and transport appear to be the dominant mechanisms operating under storm conditions. Recent observations, with acoustic sensors and with optical sensors placed near the seabed, have documented the presence of high suspended-sediment concentrations (>10 g/l; Ogston et al., 2000; Traykovski et al., 2000) capable of moving down slope driven by forces resulting from their own density. The resuspension of sediment is the result of surface gravity waves, and the high concentrations are restricted to the wave boundary layer (<10 cm above seabed). The seaward transport does not appear to be erosional, and seaward flows become depositional as soon as wave resuspension is reduced (i.e., below wave base: Traykovski et al., 2000). If the gradient of the shelf is very gentle, there may be little or no flow (Traykovski and Geyer, in press).

### *5.3. Tsunamis*

Tsunamis are long wavelength waves generated from a large (>1 km), rapid (>10 m/s) movement of the Earth's crust. These movements are able to generate waves with extremely large wavelengths (>1 km). Because dissipation of wave energy is inversely proportional to the wavelength squared, tsunami waves are able to propagate thousands of kilometers with very little loss of energy. Their speed of propagation is also significantly greater than wind-generated waves. For example, wave speeds of 200 m/s are not uncommon. At this speed it takes approximately 24 hours to traverse the Pacific basin. Because the Pacific is surrounded by active margins, often with complicated coastlines capable of accentuating tsunami waves, most of the work on modern tsunamis (exceptions being, e.g., the Norwegian margin: Dawson, 1988; and the 1929 Grand Banks event: Hasegawa and Kanamori, 1987) comes from work on or around the continental margins bounding the Pacific.

One of the most common mechanisms suggested to initiate tsunamis has been earthquakes (Ben-Menahem and Rosenman, 1972). Much work on earthquake-generated tsunamis has been performed in the northeast Pacific (e.g., Atwater, 1987; Atwater 1992; Atwater et al., 1997), though research in other coastal settings has been done around the periphery of the Pacific Basin (e.g., Kamchatka: Pinegina et al., 2003; Chile: Bourgeois et al., 1999). These studies have focused primarily on large (M>8.0) earthquakes associated with subduction of oceanic crust. Coastal studies have been complemented by research farther out on the continental margin, where large seismic-subduction events have other, more indirect, manifestations in the seabed (e.g., Adams 1990; Goldfinger et al., 2003). All of these results suggest that the large sporadic subduction quakes produce large tsunamis, which can devastate areas both close to and extremely far from the source of the slip.

Initially geophysicists imagined that most tsunamis were caused by massive subaqueous slides (Milne, 1899; Okal, 2003b). These early predictions are most impressive because they were made without knowledge of seabed topography. Recent work has confirmed that slide-generated-tsunami deposits appear in the coastal sediment record (e.g., Gelfenbaum and Jaffe, 2003). In extreme cases, when a large slide originates on land, the volume of water displaced can be greatly enhanced. This is because an air pocket can form in the lee of the slide front, increasing the total displaced volume. In these cases, particularly when a slide occurs in a confined embayment, run-up can be enormous. These rare cases have been dubbed "mega-tsunamis". The archetypical example of a modern mega-tsunami is in Lituya Bay, Alaska. In July 1958, $3x10^7$ $m^3$ of material failed from the hillside adjacent to Lituya Bay, Alaska. The tsunami that resulted climbed as much as 200 m up the other side of the bay (Mader, 1999). Large slides associated with whole-sale collapse of island volcanoes also are suspected to have caused extreme run-up. The Hawaiian archipelago is bounded with seventeen large (>>1 $km^3$), and distinct slides (Moore et al., 1989). The Hulopoe Gravel provides a possible example of a mega-tsunamite generated from these extreme events. It is a coarse-gravel-to-boulder event deposit, which is 8 m thick at one location 200 m above sea level

(Moore and Moore, 1988). The extreme nature of the deposit and its internal complexity has caused some to reinterpret it as a more common sedimentological feature (fluvial: Felton et al., 2000; paleobeach: Keating and Halsey, 2003), but even these scientists admit that a catastrophic tsunami most likely resulted from slides surrounding the Hawaiian archipelago (Felton et al., 2000; Keating and McGuire, 2000).

There is clear evidence that tsunamis can be produced from slides on more conventional muddy continental margins (Tappin et al., 2001; Synolakis et al., 2002). The 1998 Sissano Tsunami was one of the best-studied, slide-induced tsunamis in history (Figure 17.10). Initial evidence suggested an earthquake source (Hurukawa et al., 2003). However, several researchers using a variety of techniques have conclusively demonstrated that the tsunami that inundated the north coast Papua New Guinea could not have resulted from an earthquake alone (Inamura and Hashi, 2003; Okal, 2003a). The Sissano tsunami was also one of the first events to have seismometer coverage sufficient to discriminate between the two types of events. Usually seismometers are not located close enough to the source region to be able to determine the nature of slip that induced the tsunami.

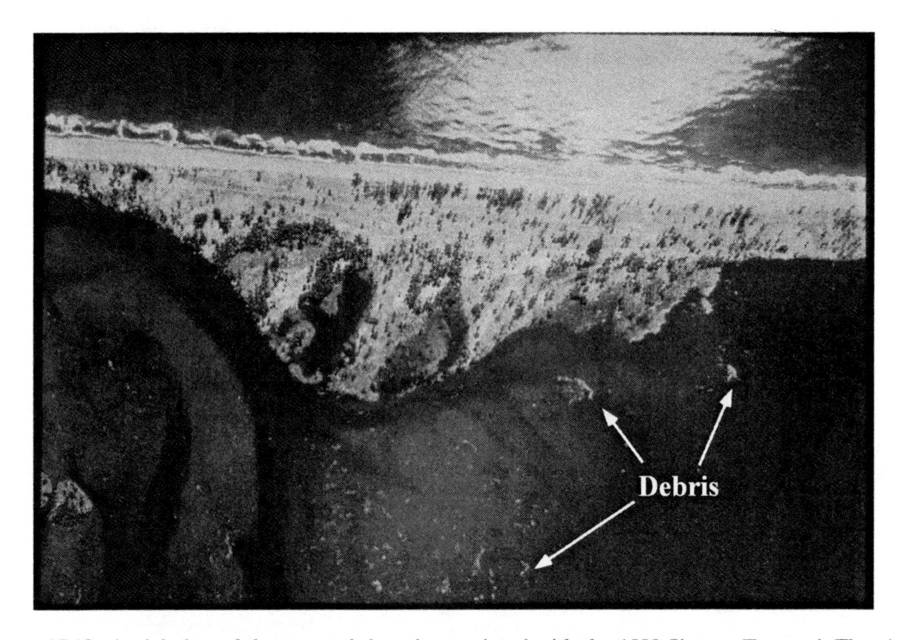


Figure 17.10 Aerial view of the tsunami deposit associated with the 1998 Sissano Tsunami. The view shown is of a spit that was completely overrun by the tsunami. The land appears light brown due to the thick sandy deposit. The dark region is the Sissano Lagoon, with debris scattered in it. The debris comes from the village of Otto and the trees surrounding it, which were completely destroyed during the event. Photo appears courtesy of H. Davies.

However, slides may not always produce tsunamis. The Atlantic margin of the US (a passive margin) is rife with large slide deposits and "en echelon" cracks (Driscoll et al., 2000). The cracks have a clear relationship to gas release (Hill et al., in press). However, the numerous slide deposits on the adjacent continental slope and rise indicate failures should be common. This conclusion conflicts with the subaerial geological record and historical data, which indicate that catastrophic tsunamis are relatively uncommon, or even nonexistent, on the east coast of the US. Though it is often assumed that failed material remains as a rigid plug as it is transported (Watts, 2000), internal deformation of the slide material strongly impacts its runout speed and thus its ability to produce a tsunami (Okal, 2003b). The tectonic and lithologic differences between the Atlantic and Pacific margin of the US may help explain the striking difference in tsunami occurrence.

Figure 17.11 A trench from a lagoonal environment in Kamchatka. The sediment record in this location is dominated by tsunamis (light brown, sand) and volcanic eruptions (white to gray fine-grained ash). Photo and interpretation appears courtesy of J. Bourgeois.

Work in coastal subaerial environments dominates the study of tsunami deposits. Most subaerial tsunami deposits are typified by upward fining and can be found best exposed in back-barrier lagoons (Figure 17.11). They are also generally sandy and have an appearance qualitatively similar to turbidites, but some deposits can be extremely coarse-grained, as in the 80 ton boulders found along the Ionian coast of Italy (Mastronuzzi and Sanso, 2000). In marine settings, tsunami deposits would have to be distinguished from turbidites (Bourgeois et al., 1988). In fact, experiments simulating energetic turbidity currents have used tsunami-like surges (Vrolijk and Southard, 1997). However, simple calculations of the energy imparted to the bed can discriminate between dramatic behavior associated with tsunamis (and the coarse material left on the bed after these events) and less energetic phenomena (Bourgeois et al., 1988).

### 5.4. *Subaqueous slides*

A range of events and processes can initiate catastrophic seabed failures along continental margins: earthquake motions, storm-wave agitation, or oversteepening from extensive deposition or erosion. These failures, or slides, can be distinguished from debris flows by their size, with slides generally being larger than $10^6$ $m^3$. Though the boundary between slides and debris flows is arbitrary, this length-scale is typically associated with a change in runout characteristics (Dade and Huppert, 1998) and their appearance in outcrop analogs (James and Stevens, 1986).

The dynamics of submarine slides are an active topic of research, and considerable effort has been focused on understanding the exotic interactions that occur in these flows. Debate centers on how similar the basal layer of a slide is to seismic faulting. Clearly a continuum of behavior is possible, with some slides being so large, deep and rigid that rapid deformation of the slide material only occurs in a relatively small zone near the base of the slide. Still others may be small enough that the entire slide volume is completely, but slowly, deformed; as in small muddy debris flows (Iverson, 1997).

In order to simplify the mechanics of these immense and complicated events, many theoretical investigations of slides have focused exclusively on dry flows. Summarized nicely by Campbell (1990), a host of particle-particle interactions related to the inertia of particle-particle collisions can segregate material and cause the frictional interactions to become muted. However, these effects will most likely be negligible in the ocean because the slopes encountered are rarely close to the angle of repose (the point at which granular effects become important: Iverson, 1997).

Other geophysical phenomena may play a role in the regulation of runout in the marine environment. Acoustic fluidization is a process whereby acoustic energy being generated at the time of failure (and during the course of motion) helps further mobilize material by reducing normal forces at the contacts between grains/clasts (Melosh, 1979). This work was originally performed in an effort to understand the long runout of debris in a vacuum (e.g., lunar impact craters), but the effect could easily explain the increase of dimensionless runout (distance traveled by the flow divided by the height of drop) with flow size observed in subaqueous slides (Hampton et al., 1996; Dade and Huppert 1998). Elastohydrodynamic lubrication also could cause an increase in runout with size, and potentially interact with hydroplaning (see next section; Brodsky and Kanamori, 2001). Despite the potential importance of these complexities, Iverson and Denlinger (2001) note that pore-pressure-friction interactions could possess scale effects that mimic the behavior seen in the data of Hampton et al. (1996). Future work should be able to identify the dominant mechanisms and provide analytical tools for prediction.

Regardless of their dynamics, slides are usually caused by local tectonism or crustal fluid flow (Orange, 1999). They can take place extremely quickly, sometimes producing secondary processes such as tsunamis (e.g., Dawson et al., 1988; Tappin et al., 2001; discussion in previous section). They also may generate areas where topography is oversteepened causing smaller subsequent debris flows. Because slides are usually a primary process, they are usually the source, rather than the result, of other events.

The role of slides in producing turbidity currents is less clear. Most of the discussion of the slide-generated turbidity currents has been associated with the 1929 Grand Banks event (Heezen and Ewing, 1952; Piper et al., 1999), though some work has examined the 1979 Nice (France) slide, which consumed a portion of the Nice International Airport (Mulder et al., 1997). Evidence from both of these events suggests that they quickly became turbulent gravity currents, which then eroded a significant portion of the open slope (Mulder et al., 1997). In the case of the Nice event, the turbidity current deposited material beginning with coarsest material near the toe of the slope. It is interesting that both of these events produced tsunamis, though the spatial extent of inundation in both cases was fairly limited.

The signature of slides in the seabed is dramatic, owing to their large, instantaneous energy release. They can be strongly erosive, particularly near the source, and form sharp basal discontinuities (Hine et al., 1992). Slides often transport large blocks of eroded material, leaving antecedent stratigraphy within the blocks intact. In ancient analogs, these features express themselves as large, "floating" clasts, sometimes ~100 m in size (Figure 17.12).

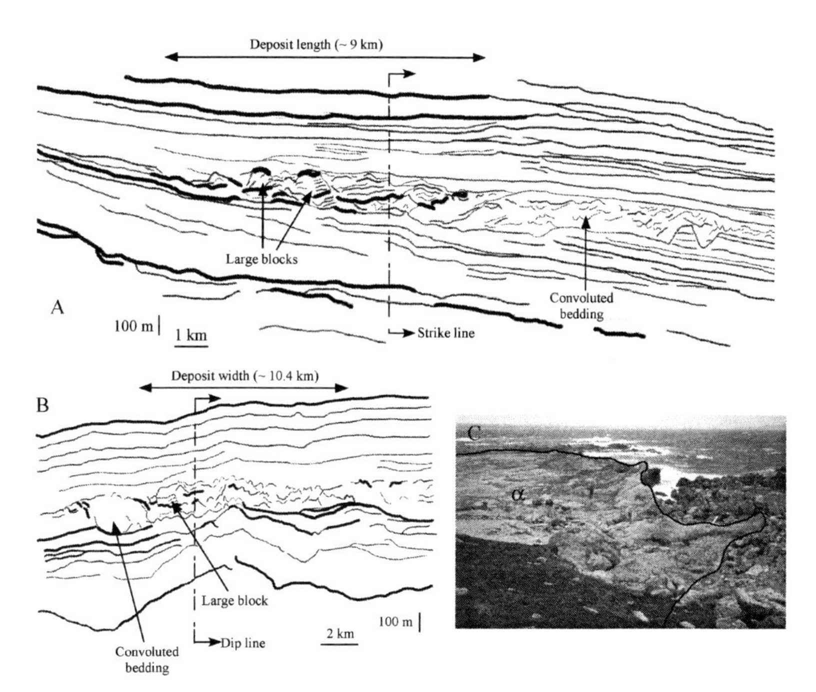


Figure 17.12 Depositional signatures of large subaqueous slides. A. Interpreted seismic dip (parallel to the direction of propagation) line from a large (>1 km$^3$) subaqueous slide on the Scotia margin. Convoluted bedding on the periphery of the slide deposit is a typical feature associated with large slide deposits. B. Interpreted seismic strike (perpendicular to the direction of propagation) line from the same slide deposit. Note the tilted bedding of a large block "floating" within the deposit. All original seismic data was obtained courtesy of L. G. Kessler, Marathon Oil. C. A clast named "α" in the Lower Head megabreccia in the Cow Head Formation of central Newfoundland. The Cow Head Formation is a carbonate apron formed during the Ordovician. The small dark speck is a geologist with his right arm held out.

### 5.5. *Subaqueous debris flows*

Debris flows are usually defined as highly concentrated slurries composed of a poorly sorted sediment, in which heightened shear stresses and/or pore-water pressures cause the slurry to fluidize. Debris flows usually encompass a volume smaller than $10^6$ $m^3$ (Dade and Huppert, 1996). The literature is widely varied with respect to the rheology of these flows. Rheology is the relationship between stress and strain within a material, but it also regulates the runout and the texture of the resulting deposit.

Considerable controversy has recently surfaced about whether most debris flows are dominated by grain-grain contacts, or whether they consist of a fluid-mechanical continuum (Iverson, 1997). Much of the controversy results from the grain-size difference between upland and continental-margin settings (Parsons et al., 2001b). Upland watersheds have been the focus of considerable research into debris flows, due to their potential hazard to human welfare and infrastructure (Costa, 1984). However, unlike upland environments, grain-size distributions observed on continental margins are usually confined to sand, silt and clay, with little or no gravel. Water contents are also substantially different. Upland watersheds rarely have water contents that are saturated; while on continental margins, the subsurface can be supersaturated from subaerial sources and crustal fluid circulation (Orange and Breen, 1992). As a result, description of subaqueous-debris-flow runout is usually characterized by a non-Newtonian-fluid continuum (Mohrig et al., 1998; Parsons et al., 2001b).

The maintenance of a fluid-mechanical continuum is also important for the transformation of subaqueous debris flows into turbidity currents. Recent research on the subject seems to indicate that all debris flows produce turbidity currents (Figure 17.13). However, the turbidity currents formed in this manner often consist of a negligible amount of the fine sediment present on the surface of the debris flow (<1%: J. Marr, personal communication). Mixing with ambient water during transport has been shown to be the most important control on the transformation processes. For instance, it has been suggested that mixing associated with a hydraulic jump in debris flows could be sufficient for the production of a dilute, turbulent turbidity current (Weirich, 1989; Piper et al., 1999).

In most settings, debris flows are secondary environmental events, initiated by an earthquake, failure or slide. The dynamics of debris flows are highly dependent on the rheological character of the sediment mass and water mixture, which is generally non-Newtonian. In marine environments, the rheology is a strong function of the silt-clay content and the degree of consolidation. Well consolidated (nearly lithified) and coarse-grained sediments will behave similar to coarse upland debris flows, while poorly consolidated, muddy material behaves as a Bingham fluid (i.e., possessing a yield strength, beyond which the material behaves as a Newtonian fluid).

The concentration of fine sediment also regulates the interactions with the bed. Muddy debris flows are usually not erosive (Mohrig et al., 1999), while coarser-grained flows can easily erode and remobilize antecedent deposits (Major, 1997). Large clasts or woody debris can reside on the surface of debris flows as a result of the high density of the flows and the dispersive pressure within the flow mass. Inverse grading (i.e., upward coarsening) is typically found in the resulting deposit, although homogenous and layered deposits are possible depending on the wave climate, the sediment supply from upslope, and other factors. Convoluted bedding is also common to debris flows, often resulting from the gentle, but continual shearing of the sediment column.

Marine debris flows have the unique characteristic that they can hydroplane. Hydroplaning occurs when the front of a debris flow begins to propagate upon a thin layer of water rather than a solid bed. Due to the reduction of shear associated with the thin layer, the flow accelerates downslope, often leaving the rest of the debris flow behind. Found only recently in a novel series of laboratory experiments (Mohrig et al., 1998), evidence of hydroplaning has subsequently been found in seismic slices of debris-flow deposits in three-dimensional, petroleum-industry seismic cubes (Figure 17.13; Nissen et al., 1999). The primary sedimentary signature of hydroplaning is the appearance of outrunner blocks, which are the remnants of the heads of debris flows that separated from their parent flows. Outrunner blocks have been previously observed in modern settings associated with large-scale slide features on collision margins (Prior et al., 1984).

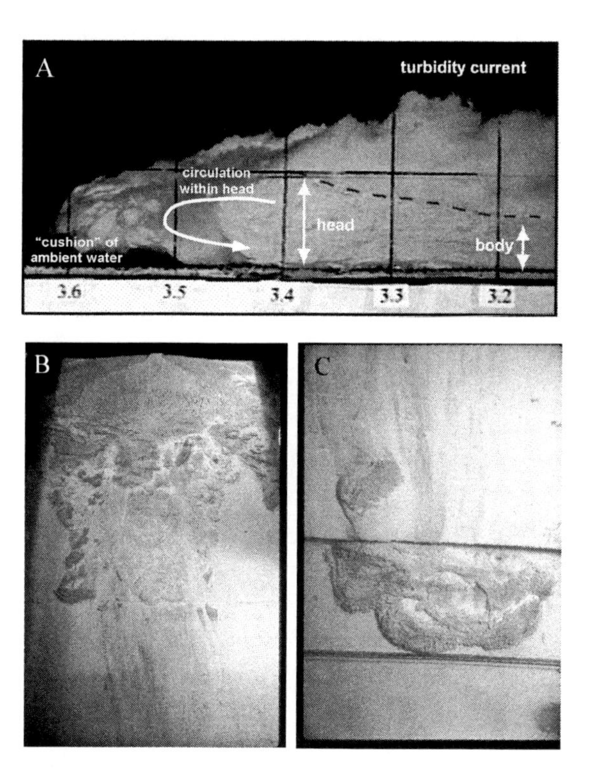


Figure 17.13 Depositional signatures of fine-grained, subaqueous debris flows. A. Diagram of a hydroplaning clay-rich, subaqueous debris flow. The numbers on the floor of the channel are in meters. B-C. Deposit of a three-dimensional, hydroplaning debris flow. The photograph was taken at the bottom/end of the tank where the head of the flow that separated from its parent flow finally came to rest. Photographs in A-C were obtained courtesy of D. Mohrig, J. Marr, and G. Parker.

### *5.6. Turbidity currents*

Turbidity currents are dilute, sediment-laden, gravity currents supported by turbulence within the flow itself. They are often contrasted with debris flows and appear to dominate the rock record for several reasons.

- They occur frequently (on geological time scales) in an area of continental margins (slope-rise) characterized by considerable sediment accretion.
- Turbidity currents have the ability to transport large amounts of material long distances with a relatively small amount of variability in deposit thickness.
- Turbidity currents are able to resuspend additional sediment, which can enable them to grow with time (i.e., referred to as ignition: Parker, 1982; Pantin, 2001).
- Turbidity currents can be a result of multiple processes: more concentrated flows (e.g., debris flows) mixing with ambient fluid during transport; direct supply of sediment from an upslope bottom boundary layer.

The underlying process responsible for their extreme efficiency in transporting sediment relies on the feedback of turbulence produced by gravity-induced motion. As the flow progresses, shear is produced along its periphery, resulting in turbulence. The turbulence entrains more material from the bed, thus supplying additional negative buoyancy, shear and turbulence. Theoretically speaking, the feedback can grow currents arbitrarily large (Parker, 1982). The constitutive grain-size distribution within a turbidity current is generally poorly sorted, with fine material supplying the density contrast required to maintain the suspension (Gladstone et al., 1998). The genesis of turbidity currents is interesting, particularly when they are formed from highly concentrated slides or debris flows. The complex mechanics associated with mixing of ambient fluid (presumably dominant in the transition of a debris flow into a turbidity current: Marr et al., 2001) are still poorly understood. One fact is for certain: the energies required to mix a large amount of sediment in dilute concentrations within the water column dictate that most turbidity currents are secondary events. Because of their very nature, secondary events are more difficult to relate directly to the forces that initiated them. However, overall accumulation of many turbidites can and should be useful in the determination of the intensity of sediment transport in a particular locale with time.

The destructive nature of turbidity currents makes progress towards understanding natural-scale turbidity currents extremely slow. Past studies have consistently lost equipment when they were placed to record turbidity-current activity in natural settings (Inman et al., 1976; Paull et al., 2003). Recent investigations have identified turbulent, gravity-driven transport in sandy environments during storms (Wright et al., 2002), though the offshore sediment flux measured by these researchers was small compared to the larger, more infrequent flows examined by Inman et al. (1976). Regardless, all of these studies have emphasized the complex interactions between turbulence production from waves and gravitational motion, and the entrainment and maintenance of suspension. From analysis of general trends in the ancient (and the hypothesis that turbidity currents are rare in the modern), the global occurrence of turbidity currents has generally been accepted to increase dramatically during low stands of sea level, when rivers supplied their loads to the tops of continental slopes (Muto and Steel, 2002). However, recent observations indicate that, on collision margins, turbidity currents may occur more often during high stands of sea level than previously thought (Kineke et al., 2000; Johnson et al., 2001; Puig et al., 2003; Walsh and Nittrouer, 2003).

The production of a turbulent suspension requires substantial energy from a primary process (storm, flood, slide, tsunami or earthquake). One of the first turbidity currents to be studied resulted from a large mass movement on the Grand Banks of the northwest Atlantic margin. After fifty years of study, debate continues about whether the flow was primarily generated by a slide, earthquake, or a transformation of a debris flow (Piper et al., 1999). Poor seismological records of the event itself will most likely keep the initiation of the Grand Banks turbidity current a mystery. Due to the intense mixing turbidity currents have with oceanic waters (Parsons et al., 2002); they are unlikely to transfer enough momentum to the water column as a whole to make a tsunami. However, a slide that creates a turbidity current may initially transfer enough momentum to the water column to generate a tsunami, in much the same way as simple rigid-body experiments (Watts, 2000). Therefore, tsunami and turbidity-current deposits might form simultaneously as secondary processes.

The connection of turbidity currents to climatic variability presents an exciting opportunity to use the sediment record to reconstruct past climate. Most of the connection between climatic events and turbidity currents has come from canyon-confined flows. Beginning with Inman et al. (1976), a number of scientists have presented evidence of turbidity-current activity in submarine canyons. In all cases, the flows were related to storm activity, either through sediment entrainment due to waves (Inman et al., 1976; Paull et al., 2003), failure due to wave loading (Puig et al., 2004) or due to heightened sediment supply by rivers (Johnson et al., 2001; Kineke et al., 2000). Climate-related turbidity currents have, in general, been slower than their slide-induced counterparts. Climatically driven turbidity currents generally propagate on the order of 1 m/s (Inman et al., 1976; Johnson et al., 2001; Khripounoff et al., 2003), while flows generated from tectonic forces may approach 30 m/s (Heezen and Ewing, 1952).

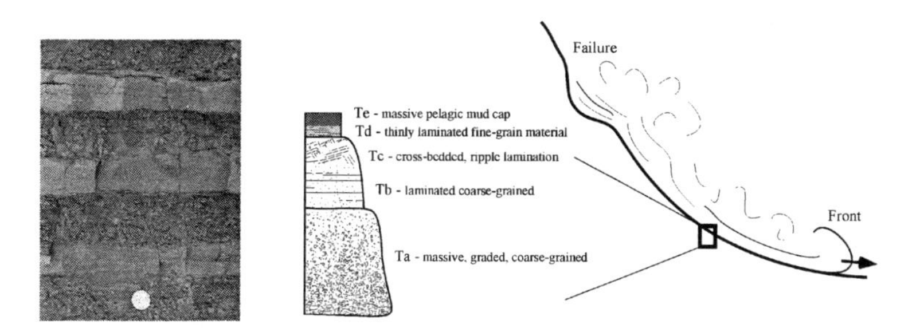


Figure 17.14 Schematic description of a Bouma sequence, with an accompanying field photograph from the Coaledo Formation of southwest Oregon. The field photograph illustrates three successive Bouma sequences (photo courtesy of J. Crockett). The fine-grained divisions within these sequences have been somewhat obscured by weathering. Good examples (like these) of Bouma sequences are often this size (5-10 cm).

The deposit of a turbidity current (i.e., a turbidite) is normally graded (i.e., upward fining), and the base can contain an erosional interface with tool marks and flutes, if the flow forming the deposit is particularly energetic (Myrow and Southard, 1996). First hypothesized by Bouma (1962), the sequence of deposits generated by a waning turbidity current now bears his name. The Bouma sequence, illustrated in Figure 17.14, possesses a massive sandy unit overlain by progressively finer material. These units are sequentially labeled with letters. However, depending on the temporal and spatial variability within the turbidite-forming turbidity current, any of these may be missing or altered (Kneller, 1995). In this case, the deposit reflects the flow variability in both streamwise and vertical dimensions (Kneller and McCaffrey, 2003).

Another unique signature of turbidite deposits is that they often form channel complexes (Ercilla et al., 1998). The cause of channelization is still a matter of debate. Imran et al. (1998) suggest that flow confinement causes interior portions of expanding flow to be erosional, while the periphery of the flow is depositional, owing to the competing and nonlinear effects of erosion and deposition. Imran et al. (1998) used numerical experiments to validate their hypothesis. In a series of laboratory experiments, Parsons et al. (2002) attribute a subtle channel form in the distal reaches of their fan deposit to the interaction between lobes and clefts, and antecedent topography. They proceed to attribute the mechanism described by Imran et al. (1998) to the transition between a confined jet and an unconfined plume, commonly observed at canyon mouths and referred to as a plunge pool (Lee et al., 2002). What is probably most interesting about turbidite channels is that channelforms thought to be a result of strictly unidirectional, continuous flow (as in subaerial rivers), can be formed exclusively from large, discrete events, as in the laboratory experiments of Parsons et al. (2002).

## 6. Conclusions

Extreme events represent a significant portion of the sediment record. With the interactions between different types of events clarified, it is possible to separate effects associated with tectonic activity from climatic variability. Untangling these two different types of events represents an important goal of sedimentology: predicting future climate based upon the known statistics of past events.

Because this exciting new age demands a different approach, it is useful to reassess the types of studies often mounted in the investigation of sedimentary processes. The most productive regions are where both climatic variability and tectonic activity are important. In the past, because each particular type of event was poorly understood and under sampled, it was important to investigate each process in isolation. However, with knowledge gained over the last several decades, importance has shifted to making measurements that begin to directly link seabed and water-column characteristics to primary processes. To calibrate models with these measurements, the process responsible for the sedimentary signal must be resolved sufficiently to differentiate it from other more localized effects. To enable a large sedimentary signal, heavy precipitation and a consistent, local depocenter is required. Regions of intense uplift near continental margins meet these stringent requirements. Ancient outcrops are also biased towards areas of tectonic activity, because it is only in these areas where marine rocks can be exposed on land. Thick sediment records usually provide greater resolution of formative processes. For all of these reasons, it is sensible to analyze areas where uplift is strong and located near a deep and predictable depocenter.

Much work remains to be done. The energy released in these extreme events cannot be duplicated in the laboratory or by using a numerical model. As a result, it is difficult to identify and quantify physical processes from first principles. The energy released also presents a problem for in situ measurements. Clever observations and new measurement strategies must be implemented to extract the data necessary to properly constrain the physical and numerical models. To do this, observationalists and modelers must work together to identify the crucial parameters to be measured and where to measure them.

## Acknowledgements

We gratefully acknowledge the financial support of ONR (N00014-03-1-0138, J.D.P.; N00014-99-1-0028, C.A.N.) and NSF (EAR-0309887, J.D.P.; OCE-0203351, C.A.N.) during the writing of this chapter. Jody Bourgeois supplied key photographs and engaging criticism of an early draft of the chapter. Bretwood "Hig" Higman also provided a helpful review of an early draft and informed our discussion of tsunami deposits. Paul Myrow assisted in the description of storm deposits and their relevance in ancient environments. Hugh Davies, Dave Mohrig, Jeff Marr, Gary Parker, John Crockett and Cindy Palinkas provided assistance with the photographs and data used in the figures. Ken Brink and Carl Friedrichs supplied helpful reviews that led to the final version of the chapter.

## Bibliography

Adams, J. 1990. Paleoseismicity of the Cascadia subduction zone: evidence from turbidites off the Oregon-Washington margin. Tectonics, vol. 9, p. 569-583.

Allan, J. C. and Komar, P. D. 2002. Extreme storms on the Pacific Northwest coast during the 1997-98 El Nino and 1998-99 La Nina. Journal of Coastal Research, vol. 18, p. 175-193.

Amos, C. L., Li, M. Z., and Chuong, K.-S. 1996. Storm generated hummocky stratification on the outer-Scotian shelf. Geo-Marine Letters, vol. 16, p. 85-94.

Atwater, B. F. 1987. Evidence for great Holocene earthquakes along the outer coast of Washington State. Science, vol. 236, p. 942-944.

Atwater, B. F. 1992. Geologic evidence for earthquakes during the past 2000 years along the Copalis River, southern coast Washington. Journal of Geophysical Research (Solid Earth), vol. 97, p. 1901-1919.

Atwater, B. F. and Hemphill-Haley, E. 1997. Recurrence intervals for great earthquakes of the past 3500 years at northeastern Willapa Bay, Washington. USGS Professional Paper No. 1576.

Beattie, P. D. and Dade, W. B. 1996. Is scaling in turbidite deposition consistent with forcing by earthquakes? Journal of Sedimentary Research, vol. 66, p. 909-915.

Ben-Menahem, A. and Rosenman, M. 1972. Amplitude patterns of tsunami waves from submarine earthquakes. Journal of Geophysical Research (Solid Earth), vol. 77, p. 3097-3128.

Bouma, A. H. 1962. *Sedimentology of Some Flysch Deposits: A Graphic Approach to Facies Interpretation*. Elsevier, Amsterdam, 159 p.

Bouma, A. H., DeVries, M. B., and Stone, C. G. 1997. Reinterpretation of depositional processes in a classic flysch sequence (Pennsylvanian Jackfork Group), Ouachita Mountains, Arkansas and Oklahoma: Discussion. AAPG Bulletin, vol. 81, p. 470-472.

Bourgeois, J., Hansen, T. A., Wiberg, P. L., and Kauffman, E. G. 1988. A tsunami deposit at the Cretaceous-Tertiary boundary in Texas. Science, vol. 241, p. 567-570.

Bourgeois, J., Petroff, C., Yeh, H., Titov, V., Synolakis, C. E., Benson, B., Kuroiwa, J., Lander, J., and Norabuena, E. 1999. Geologic setting, field survey and modeling of the Chimbote, northern Peru, Tsunami of 21 Febuary 1996. Pure and Applied Geophysics, vol. 154, p. 513-540.

Brenner, M. P. and Mucha, P. J. 2001. Fluid dynamics – That sinking feeling. Nature, vol. 409, p. 568-571.

Brodsky, E. E. and Kanamori, H. 2001. Elastohydrodynamic lubrication of faults. Journal of Geophysical Research (Solid Earth), vol. 106, p. 16357-16374.

Cacchione, D. A., Drake, D. E., Kayen, R. W., Sternberg, R. W., Kineke, G. C., and Tate, G. B. 1995. Measurements in the bottom boundary-layer on the Amazon subaqueous delta. Marine Geology, vol.125, 235-257.

Campbell, C. S. 1990. Rapid granular flows. Annual Review of Fluid Mechanics, vol. 22, p. 57-92.

Clifton, H. E. 1988. Sedimentologic relevance of convulsive geologic events. GSA Special Paper No. 29, p. 1-5.

Corbeanu, R. M., Soegaard, K., Szerbiak, R. B., Thurmond, J. B., McMechan, G. A., Wang, D. M., Snelgrove, S., Forster, C. B. and Menitove, A. 2001. Detailed internal architecture of a fluvial channel sandstone determined from outcrop, cores, and 3-D ground-penetrating radar: Example from the middle Cretaceous Ferron Sandstone, east-central Utah. AAPG Bulletin, vol. 85, p. 1583-1608.

Costa, J. E. 1984. Physical geomorphology of debris flows. In: *Developments and Applications of Geomorphology*, Springer, Berlin, p. 268-317.

Dade, W. B. and Huppert, H. E. 1998. Long-runout rockfalls. Geology, vol. 26, p. 803-806.

Dawson, A. G., Long, D., and Smith, D. E. 1988. The Storegga slides – Evidence from eastern Scotland for a possible tsunami. Marine Geology, vol. 82, p. 271-276.

Dott, R. H. and Bourgeois, J. 1982. Hummocky stratification – Significance of its variable bedding sequences. GSA Bulletin, vol. 93, p. 663-680.

Driscoll, N. W., Weissel, J. K., and Goff, J. A. 2000. Potential for large-scale submarine slope failure and tsunami generation along the US mid-Atlantic coast. Geology, vol. 28, p. 407-410.

Ercilla, G. et al. 1998. New high-resolution acoustic data from the "Braided System" of the Orinoco deep sea fan. Marine Geology, vol. 146, p. 243-250.

Felton, E. A., Crook, K. A. W., and Keating, B. H. 2000. The Hulopoe Gravel, Lanai, Hawaii: New sedimentological data and their bearing on the "giant wave" (mega-tsunami) emplacement hypothesis. Pure and Applied Geophysics, vol. 157, p. 1257-1284.

Foster, G. and Carter, L. 1997. Mud sedimentation on the continental shelf at an accretionary margin - Poverty Bay, New Zealand. New Zealand Journal of Geology and Geophysics, vol. 40, p. 157-173.

Garcia, M. and Parker, G. 1991. Entrainment of bed sediment into suspension. Journal of Hydraulic Engineering, vol. 117, p. 414-435.

Gelfenbaum, G. and Jaffe, B. 2003. Erosion and sedimentation from the 17 July 1998 Papua New Guinea tsunami. Pure and Applied Geophysics, vol. 160, p. 1969-1999.

Gladstone, C., Phillips, J. C., and Sparks, R. S. J. 1998. Experiments on bidisperse, constant-volume gravity currents: propagation and sediment deposition. Sedimentology, vol. 45, p. 833-843.

Godwin, H. 1962. Radiocarbon dating. Nature, vol. 195, p. 943-945.

Goldfinger, C., Nelson, C. H. and Johnson, J. E. 2003. Holocene earthquake records from the Cascadia subduction zone and northern San Andreas Fault based on precise dating of offshore turbidites. Annual Review of Earth and Planetary Sciences, vol. 31, p. 555-577.

Hampton, M., Lee, H.J., and Locat, J. 1996. Submarine landslides. Reviews of Geophysics, vol. 34, p. 33-59.

Hasegawa, H. S. and Kanamori, H. 1987. Source mechanism of the magnitude-7.2 Grand-Banks earthquake of November 1929 – Double couple or submarine landslide? Bulletin of the Seismological Society of America, vol. 77, p. 1984-2004.

Hay, A. E. 1987. Turbidity currents and submarine channel formation in Rupert Inlet British Columbia, 1. Surge observations. Journal of Geophsyical Research (Oceans), vol. 92, p. 2875-2881.

Hayne, M. and Chappell, J. 2001. Cyclone frequency during the last 5000 years at Curacoa Island, north Queensland, Australia. Palaeogeography Palaeoclimatology Palaeoecology, vol. 168, p. 207-219.

Heezen, B. and Ewing, M. 1952. Turbidity currents and submarine slumps, and the 1929 Grand Banks earthquake. American Journal of Science, vol. 250, p. 849-73.

Hoffman, P. F., Kaufman, A. J., Halverson, G. P., and Schrag, D. P. 1998. A Neoproterozoic snowball Earth. Science, vol. 281, p. 1342-46.

Hill, J. C., Driscoll, N. W., Weissel, J. K. and Goff, J. A. Large-scale elongated blowouts along the US Atlantic margin, Journal of Geophysical Research (Oceans) (in press)

Hine, A. C., Locker S. D., Tedesco, L. P., Mullins, H. T., Hallock, P., Belknap, D. F., Gonzales, J. L., Neumann, A. C., and Snyder, S. W. 1992. Megabreccia shedding from modern, low-relief carbonate platforms, Nicaraguan Rise. GSA Bulletin, vol. 104, p. 928-943.

Hiscott, R., Colella, A., Pezard, P., Lovell, M., and Malinverno, A. 1992. Sedimentology of deepwater volcanoclastics, Oligocene Izu-Bonin Forearc basin, based on formation microscanner images. Proceedings of the Ocean Drilling Program, Scientific Results, vol. 126, p. 75-96.

Hurukawa, N., Tsuji, Y., and Waluyo, B. 2003. The 1998 Papua New Guinea earthquake and its fault plane estimated from relocated aftershocks. Pure and Applied Geophysics, vol. 160, p. 1829-1841.

Imran, J., Parker, G., and Katapodes, N. 1998. A numerical model of channel inception on submarine fans: Journal of Geophysical Research (Oceans), vol. 103, p. 1219-1238.

Inman, D. L., Nordstrum, C. E., and Flick, R. E. 1976. Currents in submarine canyons: an air-sea-land interaction. Annual Review of Fluid Mechanics, vol. 8, p. 275-310.

Inamura, F. and Hashi, K. 2003. Re-examination of the source mechanism of the 1998 Papua New Guinea earthquake and tsunami. Pure and Applied Geophysics. vol. 160, p. 2071-2086.

Iverson, R. M. 1997. The physics of debris flows. Reviews of Geophysics, vol. 35, p. 245-296.

Iverson, R. M. and Denlinger, R. P. 2001. Flow of variably fluidized granular masses across three-dimensional terrain: 1. Coloumb mixture theory. Journal of Geophysical Research (Solid Earth), vol. 106, p. 537-552.

James, N. P. and Stevens, R. K. 1986. Stratigraphy and correlation of the Cambro-Ordivician Cow Head group, western Newfoundland. Bulletin of the Geological Society of Canada No. 366.

Johnson, K. S., Paull, C. K., Barry, J. P., and Chavez, F. P. 2001. A decadal record of underflows from a coastal river into the deep sea. Geology, vol. 29, p. 1019-1022.

Keating, B. H. and Helsley, C. E. 2002. The ancient shorelines of Lanai, Hawaii, revisited. Sedimentary Geology, vol. 150, p. 3-15.

Keating, B. H. and McGuire, W. J. 2000. Island edifice failures and associated tsunami hazards. Pure and Applied Geophysics, vol. 157, p. 899-955.

Khripounoff, A., Vangriesheim, A., Babonneau, N., Crassous, P. Dennielou, B., and Savoye, B. 2003. Direct observation of intense turbidity current activity in the Zaire submarine valley at 4000 m water depth. Marine Geology, vol. 194, p. 151-158.

Kineke, G. C., Woolfe, K. J., Kuehl, S. A., Milliman, J., Dellapenna, T. M., and Purdon, R. G. 2000. Sediment export from the Sepik River, Papua New Guinea: Evidence for a divergent dispersal system. Continental Shelf Research, vol. 20, p. 2239-2266.

Kneller, B. 1995. Beyond the turbidite paradigm: Physical models for deposition of turbidites and their implications for reservoir prediction. In: *Characterization of Deep Marine Clastic Systems*, Geological Society Special Publication No. 94, p. 31-49.

Kneller, B. C. and McCaffrey, W. D. 2003. The interpretation of vertical sequences in turbidite beds: The influence of longitudinal flow structure. Journal of Sedimentary Research, vol. 73, p. 706-713.

Kuenen, Ph. H. and Migliorini, C. I. 1950. Turbidity currents as a cause of graded bedding. Journal of Geology, vol. 58, p. 91-127.

Lamb, M. P. and Parsons, J. D. 2004. High-density suspensions formed from turbulent wave boundary layers. Journal Sedimentary Research (in press)

Leckie, D. A. and Krystinik, L. F. 1989. Is there evidence for geostrophic currents preserved in the sedimentary record of inner- to mid-shelf deposits? Journal of Sedimentary Petrology, vol. 59, p. 862-870.

Lee, S. E., Talling, P. J., Ernst, G. G. J., and Hogg, A. J. 2002. Occurrence and origin of submarine plunge pools at the base of the US continental slope. Marine Geology, vol. 185, p. 363-377.

Lomnicky, T. D., Nittrouer, C. A., and Mullenbach, B. L. Impact of local morphology on sedimentation in a submarine canyon: ROV studies in Eel Canyon. Journal of Sedimentary Research (in review)

Mader, C. L. 1999. Modeling the 1958 Lituya Bay mega-tsunami. Science of Tsunami Hazards, vol. 17, p. 57-67.

Major, J. J. 1997. Depositional processes in large-scale debris-flow experiments. Journal of Geology, 105, 345-366.

Marr, J. G., Harff, P. A., Shanmugam, G., and Parker, G. 2001. Experiments on subaqueous sandy gravity flows: the role of clay and water content in flow dynamics and depositional structures. Geological Society of America Bulletin, vol. 113, p. 1377-1386.

Mastronuzzi, G. and Sanso, P. 2000. Boulders transport by catastrophic waves along the Ionian coast of Apulia (southern Italy). Marine Geology, vol. 170, p. 93-103.

McKee, B. A., Nittrouer, C. A. and DeMaster, D. J. 1983. Concepts of sediment deposition and accumulation applied to the continental shelf near the mouth of the Yangtze River. Geology, vol. 11, p. 631-633.

Melosh, H. J. 1979. Acoustic fluidization: A new geologic process? Journal of Geophysical Research (Solid Earth), vol. 84, p. 7513-7520.

Milne, J. 1899. *Earthquakes and Other Earth Movements*. Appleton and Company, New York.

Mohrig, D., Whipple, K. X., Hondzo, M., Ellis, C., and Parker, G. 1998. Hydroplaning of subaqueous debris flows. Geological Society of America Bulletin, vol. 110, p. 387-394.

Mohrig, D., Elverhoi, A., and Parker, G. 1999. Experiments on the relative mobility of muddy subaqueous and subaerial debris flows, and their capacity to remobilize antecedent deposits. Marine Geology, vol. 154, p. 117-129.

Moore, J. G., Clague, D. A., Holcomb, R. T., Lipman, P. W., Normark, W. R., and Torresan, M. E. 1989. Prodigious submarine landslides on the Hawaiian Ridge, Journal of Geophysical Research (Solid Earth), vol. 94, p. 17465-17484.

Moore, J. G. and Moore, G. W. 1988. Large-scale bedforms in boulder gravel produced by giant waves in Hawaii. Geological Society of America Special Paper No. 29, p. 101-110.

Mulder, T., Savoye, B., and Syvitski, J. P. M. 1997. Numerical modelling of a mid-sized gravity flow: The 1979 Nice turbidity current (dynamics, processes, sediment budget and seafloor impact). Sedimentology, vol. 44, p. 305-326.

Mulder, T. and Syvitski, J. P. M. 1995. Turbidity currents generated at river mouths during exceptional discharges to the world oceans. Journal of Geology, vol. 103, p. 285-299.

Muto, T. and Steel, R. J. 2002. In defense of shelf-edge delta development during falling and lowstand of relative sea level. Journal of Geology, vol. 110, p. 421-436.

Myrow, P. M. and Southard, J. B. 1996. Tempestite deposition. Journal of Sedimentary Research, vol. 66, p. 875-887.

Nissen, S. E., Haskell, N. L, Steiner, C. T., and Coterill, K. L. 1999. Debris flow outrunner blocks, glide tracks, and pressure ridges identified on the Nigerian continental slope using 3-D seismic coherency. The Leading Edge, vol. 18, p. 595-599.

Nittrouer, C. A., DeMaster D. J., McKee, B. A., Cutshall, N. H., and Larsen, I. L. 1984 The effect of sediment mixing on Pb-210 accumulation rates for the Washington continental-shelf. Marine Geology, vol. 54, p. 201-221.

Nittrouer, C. A., Sternberg, R. W., Carpenter, R., and Bennett, J. T. 1979. Use of Pb-210 geochronology as a sedimentological tool – Application to the Washington continental-shelf. Marine Geology, vol. 31, 297-316.

Nittrouer, C. A. and Wright, L. D. 1994. Transport of particles across continental shelves. Reviews of Geophysics, vol. 32, p. 85-113.

Normark, W.R., Posamentier, H., and Mutti, E. 1993. Turbidite systems: State of the art and future directions. Reviews of Geophysics, vol. 31, p. 91-116.

Novoa, E., Suppe, J., and Shaw, J. H. 2000. Inclined-shear restoration of growth folds. AAPG Bulletin, vol. 84, p. 787-804.

Ogston, A. S., Cacchione, D. A., Sternberg, R. W., and Kineke, G. C. 2000. Observations of storm and river flood-driven sediment transport on the northern California continental shelf. Continental Shelf Research, vol. 20, p. 2141-2162.

Ogston, A. S., Guerra, J. V. and Sternberg, R. W. 2004. Interannual variability of nearbed sediment flux on the Eel River shelf, northern California. Continental Shelf Research, vol. 24, p. 117-136.

Okal, E. A. 2003a. T waves from the 1998 Papua New Guinea earthquake and its aftershocks: Timing the tsunamigenic slump. Pure and Applied Geophysics. vol. 160, p. 1843-1863.

Okal, E. A. 2003b. Normal mode energetics for far-field tsunamis generated by dislocations and landslides. Pure and Applied Geophysics, vol. 160, p. 2189-2221.

Orange, D. L. 1999. Tectonics, sedimentation, and erosion in northern California: submarine geomorphology and sediment preservation potential as a result of three competing processes. Marine Geology, vol. 154, p. 369-382.

Orange, D. L. and Breen, N. A. 1992. The effects of fluid escape on accretionary wedges: Part 2: Seepage force, slope failure, headless submarine canyons, and vents. Journal of Geophysical Research (Solid Earth), vol. 97, p. 9277-9295.

Palinkas, C.M., Nittrouer, C. A., Wheatcroft, R. A., and Langone, L. The use of Be-7 to identify event and seasonal sedimentation near the Po River delta, Adriatic Sea. Marine Geology (in press)

Pantin, H. 2001. Experimental evidence for autosuspension. International Association of Sedimentologists Special Publication No. 31, p. 189-205.

Parker, G. 1982. Conditions for the ignition of catastrophically erosive turbidity currents: Marine Geology, vol. 46, p. 307-327.

Parsons, J. D., Bush, J. W. M., and Syvitski, J. P. M. 2001a. Hyperpycnal plumes with small sediment concentrations. Sedimentology, vol. 48, p. 465-478.

Parsons, J. D., Scheweller, W. J., Stelting, C. W., Southard, J. B., Lyons, W. J., and Grotzinger, J. P. 2002. A preliminary experimental study of turbidite fan deposits. Journal of Sedimentary Research, vol. 72, p. 619-628.

Parsons, J. D., Whipple, K. X., and Simoni, A. 2001b. Laboratory experiments of the grain-flow, fluid-mud transition in well-graded debris flows. Journal of Geology, vol. 109, p. 427-447.

Paull, C. K., Ussler, W., Greene, H. G., Keaten, R., Mitts, P., and Barry, J. 2003. Caught in the act: the 20 December 2001 gravity flow event in Monterey Canyon. Geo-Marine Letters, vol. 22, p. 227-232.

Pinegina, T. K., Bourgeois, J., Bazanova, L. I., Melekestsev, I. V., and Braitseva, O. A. 2003. A millennial-scale record of holocene tsunamis on the Kronotskiy Bay coast, Kamchatka, Russia. Quaternary Research, vol. 59, p. 36-47.

Piper, D. J. W., Cochonat, P., and Morrison, M. L. 1999. The sequence of events around the epicentre of the 1929 Grand Banks earthquake: initiation of debris flows and turbidity current inferred from sidescan sonar. Sedimentology, vol. 46, p. 79-97.

Piper, D. J. W., Shor, A. N., and Hughes-Clarke, J. E. 1988. The 1929 "Grand Banks" earthquake, slump, and turbidity current. In: *Sedimentologic consequences of convulsive geologic events*, Geological Society of America Special Paper No. 29, p. 77-92.

Posamentier, H. W., Jervey, M. T. and Vail, P. R. 1988. Eustatic controls on clastic deposition – Conceptual framework. In: *Sea-Level Changes – An Integrated Approach*, Society of Sedimentary Geology (SEPM) Special Publication No. 42, p. 109-124.

Prior, D. B., Bornhold, B. D., and Johns, M. W. 1984. Depositional characteristics of a submarine debris flow. Journal of Geology, vol. 92, p. 707-727.

Puig, P., Ogston, A.S., Mullenbach, B.L., Nittrouer, C.A., Sternberg, R.W. 2003. Shelf-to-canyon sediment-transport processes on the Eel continental margin (northern California). Marine Geology, vol. 193, p. 129-149.

Puig, P., Ogston, A. S., Mullenbach, B. L., Nittrouer, C. A., Parsons, J. D. and Sternberg, R. W. 2004. Storm-induced sediment-gravity flows at the head of the Eel submarine canyon. Journal of Geophysical Research (Oceans), vol. 109, Article No. C03019.

Rothman, D. H., Grotzinger, J. P. and Flemings, P. 1994. Scaling in turbidite deposition: Journal of Sedimentary Research, vol. 64, p. 59-67.

Shanmugam, G. 2003. Preliminary experimental study of turbidite fan deposits – Discussion. Journal of Sedimentary Research, vol. 73, p. 838-839.

Shanmugan, G. and Moiola, R. J. 1995. Reinterpretation of depositional processes in a classic flysch sequence (Pennsylvanian Jackfork Group), Ouashita Mountains, Arkansas and Oklahoma. American Association of Petroleum Geologists Bulletin, vol. 79, p. 672-695.

Snedden, J. W. and Nummedal, D. 1991. Origin and geometry of storm-deposited sand beds in modern sediment of the Texas continental shelf. International Association of Sedimentologists Special Publication No. 14, p. 283-308.

Sommerfield, C. K., Nittrouer, C. A., and Alexander, C. R. 1999. Be-7 as a tracer of flood sedimentation on the northern California continental margin. Continental Shelf Research, vol. 19, p. 335-361.

Suckow, A., Morgenstern, U., and Kudrass, H. R. 2001. Absolute dating of recent sediments in the cyclone-influenced shelf area off Bangladesh: Comparison of gamma spectrometric (Cs-137, Pb-210, Ra-228), radiocarbon, and Si-32 ages. Radiocarbon, vol. 43, p. 917-927.

Synolakis, C. E., Bardet, J. P., Borrero, J. C., Davies, H. L., Okal, E. A., Silver, E. A., Sweet, S. and Tappin, D. R. 2002. The slump origin of the 1998 Papua New Guinea Tsunami. Proceedings of the Royal Society of London, vol. 458, p. 763-789.

Syvitski, J. P. M. and Hutton, E. H., 2001. 2D SEDFLUX 1.0C: An advanced process-response numerical model for the fill of marine sedimentary basins. Computers & Geosciences, vol. 27, p. 731-754.

Tappin, D. R., Watts, P., McMurtry, G. M., Lafoy, Y. and Matsumoto, T. 2001. The Sissano, Papua New Guinea tsunami of July 1998 – offshore evidence on the source mechanism. Marine Geology, vol. 175, p. 1-23.

Traykovski, P., Geyer, W. R., Irish, J. D., and Lynch, J. F. 2000. The role of wave-induced density-driven fluid mud flows for cross-shelf transport on the Eel River continental shelf. Continental Shelf Research, vol. 20, p. 2113-2140.

Traykovski, P. and Geyer, W. R. Observations and modeling of wave induced fluid-mud gravity flows on the Po prodelta. Continental Shelf Research (in press)

Vrolijk, P. J. and Southard, J. B. 1997. Experiments on rapid deposition from high-velocity flows. Geoscience Canada, vol. 24, p. 45-54.

Walker, R. G. 1984. Shelf and shallow marine sands. In: *Facies Models, Second Edition*. Toronto, Geoscience Canada, Reprint Series No. 1, p. 141-170.

Walsh, J. P. and Nittrouer, C. A. 2003. Contrasting styles of off-shelf sediment accumulation in New Guinea. Marine Geology, vol. 196, 105-125.

Watts, P. 2000. Tsunami features of solid block underwater landslides. Journal of Waterway, Port, Coastal and Ocean Engineering (ASCE), vol. 126, p. 144-152.

Weirich, F. H. 1989. The generation of turbidity currents by subaerial debris flows, California. Geological Society of America Bulletin, vol. 101, p. 278–291.

Wheatcroft, R. A. and Borgeld, J. C. 2000. Oceanic flood deposits on the northern California shelf: large-scale distribution and small-scale physical properties. Continental Shelf Research, vol. 20, p. 2163-2190.

Wheatcroft, R.A., Hunt, L. M., Stevenes, A., and Lewis, R. The initial distribution and properties of the October 2000 Po River flood deposit: Evidence from digital x-radiography. Continental Shelf Research (in press)

Winterwerp, J. C. 2001. Stratification effects by cohesive and noncohesive sediment. Journal of Geophysical Research (Oceans), vol. 106, p. 22559-22574.

Wright, L. D., Friedrichs, C. T., Scully, M. E. 2002. Pulsational gravity-driven sediment transport on two energetic shelves. Continental Shelf Research, vol. 22, p. 2443-2460.

Wright, L. D., Wiseman, W. J., Bornhold, B. D., Prior, D. B., Suhayda, J. N., Keller, G. H., Yang, Z. S., and Fan Y. B. 1988. Marine dispersal and deposition of Yellow-River silts by gravity-driven underflows. Nature, vol. 332, p. 629-632.

# Chapter 18. LONG TERM SEA LEVEL CHANGES AND THEIR IMPACTS

PHILIP L. WOODWORTH

*Pro7udman Oceanographic Laboratory*

JONATHAN M. GREGORY

*Hadley Centre for Climate Prediction and Research and Centre for Global Atmospheric Modelling, Department of Meteorology, University of Reading*

ROBERT J. NICHOLLS

*School of Civil Engineering and the Environment, Southampton University*

**Contents**



## 1. Introduction

The study of long term sea level change is of great scientific interest owing to its overlap with a wide range of important questions in geophysics, oceanography and climate change. It is also of great practical importance to coastal populations concerning long-term coastal evolution and the management of coastal hazards.

In 2001, the Intergovernmental Panel on Climate Change (IPCC) Third Assessment Report (TAR) reviewed recent publications in this field (Church et al., 2001). It concluded that global sea level increased at an average rate of 1–2 mm/year during the past 100 years (Figure 18.1b), with some evidence for a small acceleration in the rate of sea level rise between the $19^{th}$ and $20^{th}$ centuries. The magnitude of this recent change can be compared to the 120 m or so of sea level rise which occurred since the last glacial maximum (Figure 18.1a), and is comparable to the amplitude of fluctuations on timescales of a few hundred years or longer

*The Sea*, Volume 13, edited by Allan R. Robinson and Kenneth H. Brink
ISBN 0-674-01526-6 ©2004 by the President and Fellows of Harvard College

which could have taken place during the past 6000 years. The IPCC TAR also concluded that the rate of rise of sea level could increase significantly during the 21st century, with major impacts on the natural and socio-economic systems at the coast (McLean et al., 2001; Nurse et al., 2001).

In this paper, we review the status of global sea level monitoring activities. In addition, we revisit some of the discussion concerning explanations for past and possible future sea level change, with particular emphasis on research published since the TAR. Finally, we present an overview of likely impacts in coastal waters and at the coast of the increase in sea level (or water depth) which might accompany 21st century climate change. This includes a discussion of possible responses to adapt to these impacts.

## 2. Improvements Needed to the Global Data Sets and Monitoring Systems

Sea level changes have been measured at many locations in the northern hemisphere during the past two centuries (and at rather fewer locations in the southern hemisphere) by means of tide gauges. Most networks of tide gauges were installed for practical reasons such as flood warning, harbour operations or datum determination. However, many of them also do double-duty as providers of sea level information for scientific research. Their data are sent to national and international data banks, which perform quality control and make the information available to the scientific community. The same data banks also process information from special gauges that have been installed to serve the research aims of scientific programmes such as the World Ocean Circulation Experiment (WOCE) (Woodworth et al., 2002).

The organisation responsible for the collection, analysis and re-distribution of long term sea level information from tide gauges is the Permanent Service for Mean Sea Level (PSMSL), which is hosted by the Proudman Oceanographic Laboratory, UK under the auspices of the International Council for Science. The PSMSL data bank contains over 52,000 station-years of data from almost 2000 stations, and has been used extensively by researchers into regional and global sea level change during the last two centuries. Woodworth and Player (2003) provide a detailed description of the geographical coverage of PSMSL data, including in particular the availability of long records.

Although most of our knowledge of global sea level change during the 20th century comes from the PSMSL data set, there are two obvious problems associated with it: (1) the fact that a tide gauge measures sea level change relative to the land, which may also be moving, rather than 'real' sea level change; and (2) the fact that the historical data set is biased towards the northern hemisphere and necessarily contains information on coastal sea level change rather than that of the open ocean. Aspects of each of these problems are discussed below.

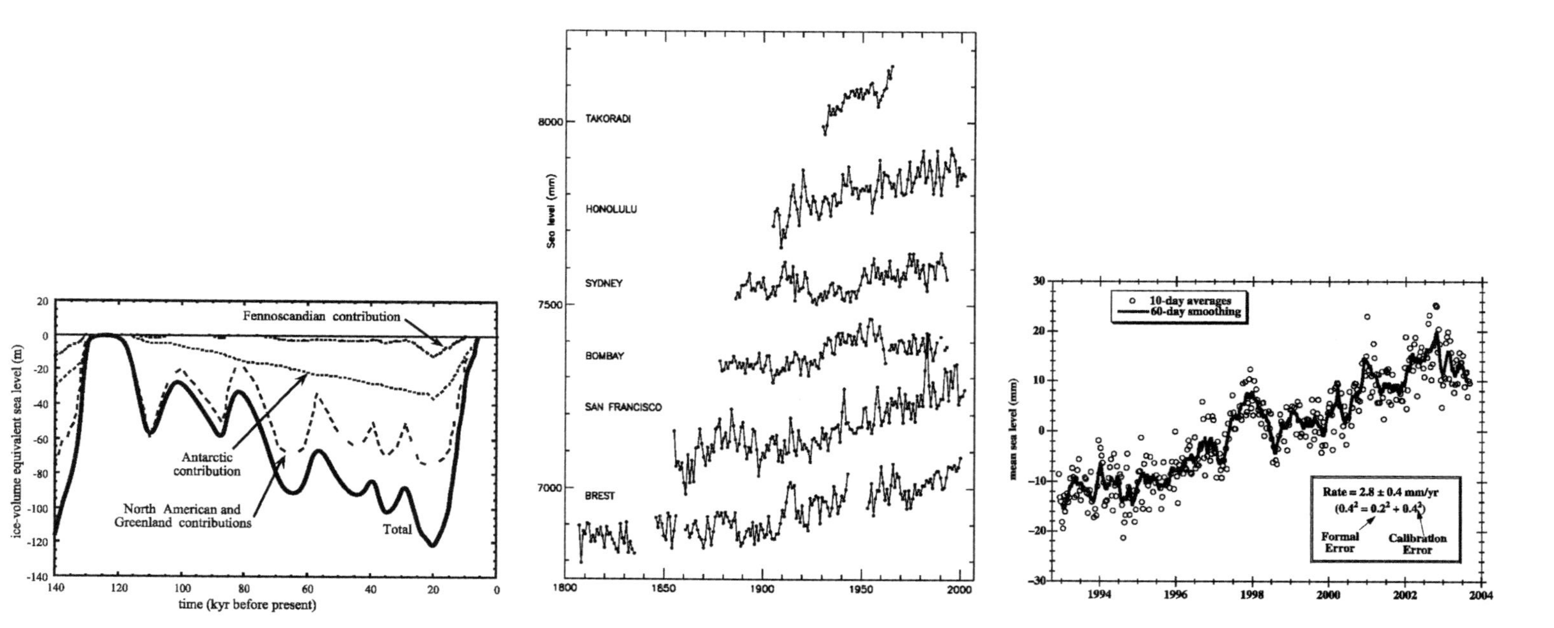


Figure 18.1 Sea level changes over different time scales: (a) global change and its components during the past 140,000 years (from Church et al., 2001); (b) long tide gauge records from the last century from each continent (Takoradi, Ghana, Africa; Honolulu, Pacific; Sydney, Australia; Bombay, India, Asia; San Francisco, N.America; and Brest, France, Europe. Each record has been offset vertically for presentation purposes. Observed trends for the twentieth century are 3.1, 1.5, 0.8, 0.9, 2.0 and 1.3 mm/year respectively. The effect of postglacial rebound as simulated by the Peltier ICE-3G model is less than or of the order 0.5 mm/year at each site); (c) global sea level change observed by the T/P and Jason-1 altimeters during the past decade without the ‘inverse barometer’ air pressure correction. Note the separation of estimated formal and calibration uncertainties on the trend (S. Nerem and E. Leuliette, private communication).

### *2.1 Vertical Land Movements*

All sea level data obtained from the Earth's surface, whether derived from tide gauges or from geological or archaeological information, take the form of measurements relative to the level of the nearby land upon a dynamic Earth (Mitrovica and Vermeersen, 2002). The tide gauge community is investing heavily in new geodetic techniques by means of which vertical land movements (VLMs) at the gauge benchmarks are monitored, thereby decoupling 'real' sea and land signals in the 'relative' tide gauge records. The main application of VLM data will probably be to the study of sea level data acquired in the future. However, if time series of VLM at certain sites prove to be monotonic (i.e. without earthquakes or other irregular features), then they should be as capable of application to study of sea level trends from past data.

The main geodetic technique is the Global Positioning System (GPS), with Absolute Gravity and Doppler Orbitography and Radiopositioning Integrated by Satellite (DORIS) as ancillary methods. However, this is still a developing field. Therefore, while the techniques have clearly shown their potential (e.g. Bingley et al., 2001; Cazenave et al., 1999; Williams et al., 2001), and while some regional experiments have obtained sea level trends corrected for VLMs by such means (notably the BIFROST experiment which determined a value of 2.1 +/- 0.3 mm/year for 20$^{th}$ century sea level change using Fennoscandian gauge records together with GPS data acquired near to the gauge sites, Milne et al., 2001), results on VLMs at gauges obtained this way have not so far been applied to a study of global sea level change.

Consequently, all authors of papers to date, including those summarised by the IPCC TAR, have applied VLM correction techniques which have been used for some years. One concerns the application of geodynamic numerical models of Glacial Isostatic Adjustment (GIA). GIA is the only geological process leading to VLM for which we have global models which can be applied to tide gauge data, although there are significant differences between the GIA models, for example due to differences in ice history or viscosity parameterisation (Mitrovica, 2003), and it is recognised that other geological processes (tectonics) can be of significantly greater importance locally. In most far-field locations, at some distance from the major areas of glaciation, the effect of applying a GIA correction to tide gauge data is to increase measured trends by several tenths mm/year (Peltier, 2001). However, there is a range of other geological processes which can lead to VLM in certain regions (e.g. active tectonics, Di Donato et al., 1999). The second concerns the use of geological (i.e. long term, typically since 6K BP) sea level data obtained from sites near to the gauges (e.g. Woodworth et al., 1999). By subtracting the long term rate of change of sea level, from whatever geological cause, from the tide gauge rate, one obtains estimates of sea level change due to present-day processes (e.g. climate change). The main difficulty with this method is that sufficiently reliable and copious geological data are not available worldwide.

Peltier (2001) and Church et al. (2001) contain detailed discussions of the relative merits of each of the VLM-correction techniques, and Woodworth (2003) discusses the limitations of present VLM modelling as applied to the long tide gauge records from the Mediterranean, which to a large extent reflects the global situation. It is clear that, until there have been significant improvements in VLM measurements through the use of GPS and other methods, and parallel developments in geodynamic modelling, a large part of the 1 mm/year range of uncertainty for global sea level rise quoted in the TAR will remain. Even though by 2010, one expects that many gauges will have acquired GPS land movement time series more than one decade long, not all gauges will have done so. In particular, it cannot be expected that the same density of GPS measurements will be undertaken in Africa, South America etc. as in Europe and USA. Consequently, GIA modelling will continue to be the only source of VLM information for many sites.

### *2.2 The GLOSS Programme*

One way in which the community is endeavouring to improve the geographical coverage of sea level information is through the Global Sea Level Observing System (GLOSS) programme (IOC, 1998). GLOSS is a Joint Technical Commission for Oceanography and Marine Meteorology (JCOMM) programme of the Intergovernmental Oceanographic Commission (IOC) and World Meteorological Organization (WMO), the aims of which are to improve the quality and quantity of data supplied to international oceanographic programmes (e.g. WOCE and CLIVAR) and for studies of long term sea level change. GLOSS was one of the first components of the Global Ocean Observing System (GOOS).

The GLOSS Implementation Plan calls for the development of a GLOSS Core Network (GCN) of 290 stations; a network of several tens of sites for on-going satellite altimeter calibration (GLOSS-ALT); a programme of investment in gauges with geodetic equipment (especially GPS) at sites with long records (the LTT, or long term trends, set); and the use of gauges at straits and other strategic locations for ocean circulation monitoring (the OC set).

In brief, the status of the GCN at the present time is near-identical to that for the past several years, with the network approximately two-thirds operational, if one uses data receipts by the PSMSL as a guide to operational status (Figure 18.2). (This apparent status will improve in the near future, when the question of the suitability of some stations in the GCN is revisited.) GCN status is somewhat better if one considers whether some kind of gauge exists at a site, no matter if that gauge is delivering data to the PSMSL or not. For example, in some parts of the world (e.g. Antarctica), it is sometimes impossible to record a time series of real sea level, but it is possible to maintain a simple under-ice sub-surface pressure sensor, the data from which can be very useful to oceanographic studies. From a recent survey, it was concluded that over 75% of GCN gauges now make hourly data (either real sea level or sub-surface pressure) available in 'delayed mode'.

However, these status summaries hide major problems in several regions, with expenditure in new tide gauge equipment in a number of countries, and the network improvements which result, balanced against the fact that many GLOSS stations in other countries are being terminated or require major upgrades. In addition, the investments made in gauges for the international programmes of the 1990s are unlikely to be repeated in future. This pessimism is contradicted to some extent by the stated requirements for investment in regional networks of coastal tide gauges by, for example, the GOOS Coastal Oceans Observations Panel. Therefore, GLOSS status may receive a boost in the long term from 'coastal', rather than 'oceanographic', applications. Whatever the scientific emphasis, investment in equipment and training will be a necessity in many countries. A detailed, recent review of GLOSS status can be found in IOC (2003).

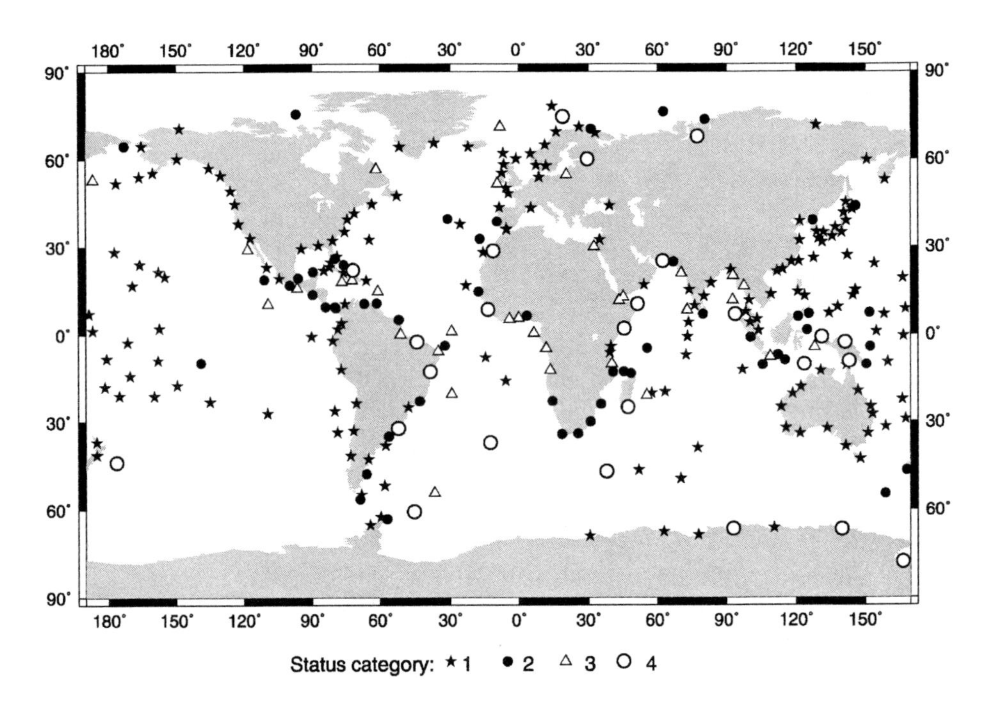


Figure 18.2 GLOSS Core Network status as of October 2003 in terms of delivery of mean sea level data to the PSMSL. Stations within Categories 1-4 can be considered: (1) fully up-to-date; (2) have relatively recent historical data only; (3) have older historical data only; and (4) have no data at all respectively. Over three-quarters of the total have a tide gauge of some kind, if only a simple pressure sensor, and make data available to international centres.

The World Ocean Circulation Experiment (WOCE) was a major international research activity of the World Climate Research Programme (WCRP). WOCE terminated at the end of 2002 and has been replaced to a large extent by the Climate Variability and Predictability Programme (CLIVAR). WOCE had its own tide gauge data collection activity, which was based on 'higher-frequency' (typically hourly) data collected in 'delayed mode' from over 160 ocean island and continental locations and, more recently, over 120 in 'fast' mode (i.e. data available within a month or so). The two centres responsible for this work were the British Oceanographic Data Centre (BODC) and the University of Hawaii Sea Level Center (UHSLC) respectively, with the PSMSL providing guidance to BODC. Woodworth et al. (2002) describe the scientific results derived from tide gauges during WOCE, including those on the interannual variability of sea level due to processes such as El Niño and the North Atlantic Oscillation.

In the future, it is intended that the WOCE delayed-mode functions will be taken care of through the GLOSS International Archiving Centres, and the

PSMSL will function as one of those Centres. Meanwhile, the WOCE 'fast' activity at UHSLC will metamorphose into a GLOSS 'fast' activity and become even faster, with data availability within a timescale of a week or so.

### 2.3. *Satellite Altimetry*

With the launch of TOPEX/Poseidon (T/P) in 1992 (and to a lesser extent the launch of ERS-1 in 1991), the community had access to a stream of near-real time and near-global sea level information. Reviews of the altimetric method and of science findings from altimetry are given by Fu and Cazenave (2001), which includes discussion of results long term sea level change measured by altimetry during the 1990s by Nerem and Mitchum (2001). A summary of other papers on this theme can be found in Church et al. (2001). In brief, sea level appears to have risen by almost 3 mm/year for the period 1993 to mid-2002 (Figure 18.1c, see also Nerem and Mitchum, 2001; Cabanes et al., 2001; Cazenave and Nerem, 2004), broadly in line with observations of ocean warming during the period (see discussion below). However, the altimetric record is as yet too short to be useful for the multi-decadal and century-timescale trends of interest in climate studies.[1]

Mitchum et al. (1999a,b) and Woodworth et al. (2002) updated the recommendations of earlier studies of altimeter requirements for oceanography (e.g. Koblinsky et al., 1992), in stressing the importance of continued high-quality T/P altimetric time series into the JASON-1 mission and beyond, JASON-1 being the T/P follow-on mission launched in December 2001. T/P and JASON-1 will together provide the accuracy necessary for basin- and global-scale sea level change studies, and will supply the reference altimeter data set with which other, less-precise altimeter data sets can be merged for studies across a range of spatial scales. The study groups also recommended the future provision of a family of second or third altimeter satellites (e.g. Geosat Follow On, Envisat) to complement the higher-flying JASON series, and potentially the development of alternative altimeter technology (e.g. swath altimetry). We understand that the space agencies have programmes in each of these areas.

### 2.4 *Continuing Need for Tide Gauges*

Now that altimetry is an established technique, a reasonable question is to ask why tide gauges are required in this 'Age of Altimetry'. This topic has been discussed in GLOSS documents (IOC, 1998, 2003) and by Mitchum et al. (1999a,b). The challenge is clearly through programmes such as GLOSS to construct a coherent global

[1] All the altimeter rates quoted in this paper, taken from publications by US and French researchers, should, in principle, be increased by approximately 0.3 mm/year, to account for the Glacial Isostatic Adjustment contribution to secular change in the geoid relative to the Earth centre of mass (Peltier, 2001), if comparisons to tide gauge-derived rates from the 20$^{th}$ century are to be made. However, given that GIA-corrected tide gauge data are used to correct for altimeter instrumental drift (Mitchum, 2000), it is unclear whether to make this relatively small correction. The standard error on the trend shown in Figure 18.1c (0.4 mm/year) includes both a formal error plus an estimate of the uncertainty in the calibration procedure. Standard errors on the trends computed by Cabanes and co-workers quoted in this paper are the smaller, formal errors only.

sea level monitoring system with both altimetry and tide gauge networks. The special reasons for continuing investment in tide gauge networks include:

- The general principle of continuity, and the relatively low cost of gauges compared to satellites. Should there be gaps in altimetric coverage in future, then the community will have to rely on the *in situ* network for monitoring the large scale variability of global sea level, in a similar way to which gauge data have been used to measure such variability before the altimeter era (e.g. Chambers et al., 2002; Church et al., 2004).
- The fact that long tide gauge records must be maintained for studies of secular trends and accelerations in mean sea level, tides and extreme sea levels (e.g. for input to IPCC assessments)
- The superiority of gauges over altimetry with regard to higher frequency sampling at important straits and other areas (e.g. Drake Passage, western boundary current margins) (Woodworth et al., 1996)
- The fact that most altimeter missions are not optimised to sample the high latitude oceans which are anyway ice covered for at least part of the year
- The need for gauges for calibration of altimeter data including the need to remove trends in measured altimetric data or in various orbital and environmental correction terms, any of which might affect a computation of the rate of long term sea level change (e.g. Mitchum, 2000; Dong et al., 2002)
- The need for large numbers of gauges for operational coastal applications (e.g. in the GOOS Coastal Module) of direct relevance to the increasing number of people who live near the coast.

However, there will be a continuing need to compare tide gauge and altimeter data sets at each location in order to study the reasons for any differences, probably within the context of coastal hydrodynamic models.

### *2.5 Space Gravity*

A second space technique which will be of great importance to sea level studies is space gravity (Woodworth and Gregory, 2003). The scientific case for the Gravity Field and Steady State Ocean Circulation Explorer (GOCE) mission was constructed to a great extent around the many benefits of better knowledge of the Earth's gravity field and geoid to the measurement and understanding of sea level change (Balmino et al., 1999). In particular, it was recognised that a better geoid would result in major gains in understanding the ocean circulation and solid Earth and glaciological processes which contribute to sea level change. Improved understanding of the processes then might lead to improvements in our ability to predict future changes. GOCE is planned for launch in 2005.

The case for a temporal gravity mission such as the Gravity Recovery and Climate Experiment (GRACE) was also constructed partly around the sea level change issue, by providing a better understanding of the global hydrological cycle, through the monitoring of changes in gravity associated with fluctuations in land water storage (both anthropogenic and natural) and ice caps, and of the ocean thermohaline circulation, through the monitoring of spatial variations in ocean water mass (or bottom pressure) (NRC, 1997; GRACE, 1998). GRACE, in combination with altimetry, should identify which sea level changes are due to steric (density) effects and those due to additional water in the ocean. GRACE was launched in March 2002.

Church et al. (2001) stressed the importance of altimetry and space gravity within a range of Earth Observation technologies required to reduce the uncertainties in sea level monitoring and understanding. Other techniques include space and airborne laser altimetry (e.g. the Ice, Cloud and land Elevation Satellite, ICESAT, launched in 2002) together with *in situ* GPS for monitoring ice sheets.

## 3. Explanation of Observed 20th Century Changes

### *3.1 Research on 20th Century Changes following the TAR*

The range of 1–2 mm/year (Church et al., 2001) for the rate of global sea level change for the past 100 years is rather wide, and it is important to try to reduce it, if one can, through, for example, better understanding of vertical land movements, as discussed above. With a more precise estimate of 20th century change, it will be possible to provide a better historical context within which present-day rates from altimetry can be discussed. In addition, with a more precise estimate of past change, it should be more straightforward to focus on which processes might have contributed to it. As a consequence, it might be easier to assess what 21st century change will be by providing better constraints on climate modelling.

TABLE 1

Estimated rates of sea level rise components from observations and models (mm/year) averaged over the period 1910 to 1990. (Note that the model uncertainties may be underestimates because of possible systematic errors in the models.) From Church et al. (2001).

| | Minimum (mm/year) | Central value (mm/year) | Maximum (mm/year) |
|---|---|---|---|
| Thermal Expansion | 0.3 | 0.5 | 0.7 |
| Glaciers and ice caps | 0.2 | 0.3 | 0.4 |
| Greenland – 20th century effects | 0.0 | 0.05 | 0.1 |
| Antarctica – 20th century effects | -0.2 | -0.1 | 0.0 |
| Ice sheets – Adjustment since LGM | 0.0 | 0.25 | 0.5 |
| Permafrost | 0.00 | 0.025 | 0.05 |
| Sediment deposition | 0.00 | 0.025 | 0.05 |
| Terrestrial storage (not directly from climate change) | -1.1 | -0.35 | 0.4 |
| **Total** | -0.8 | 0.7 | 2.2 |
| Estimated from observations | 1.0 | 1.5 | 2.0 |

Table 1 (copied from Table 11.10 of the TAR) lists possible processes together with minimum, mean and maximum estimates of their contribution to sea level change. For example, by considering an ensemble of hydrographic measurements for the past 100 years together with a set of results from Atmosphere-Ocean General Circulation Models (AOGCMs) forced by historical changes in the atmospheric concentrations of greenhouse gases and sulphate aerosols, the IPCC TAR concluded that thermal expansion could have contributed as little as 0.3 mm/year and as much as 0.7 mm/year to global sea level change. Other processes, including changes in glaciers and ice caps, in the Greenland and Antarctic ice sheets and permafrost, were considered to have contributed at a similar or lower rate. Anthropogenic terrestrial water storage, however, could be large, but its range of uncertainty is very wide (see below); changes in snow, soil moisture and groundwater were not considered but are probably small (cf Milly et al., 2003). The sum of the various processes included amounts to a hindcast of global sea level change for the past 100 years of +0.7 mm/year, within a range of –0.8 to +2.2 mm/year, which is consistent with the observed amount within the large uncertainties, if a little on the low side.

The TAR stressed that much work is needed to improve upon each component of the budget of Table 1, with a special requirement to rectify a lack of knowledge of terrestrial water storage changes on a global basis, for which measurements by GRACE and follow-on missions will be important. Another particular concern is that direct measurements of the mass balance and volume of the Greenland and Antarctic ice-sheets do not presently give strong constraints on their contributions, but in coming years this situation will be improved by means of ongoing satellite altimeter missions. The Table 1 estimates for the ice-sheets are based on indirect constraints from the geological record and numerical modelling, which indicate that the current ice-sheet contribution to sea level rise is small.

From the estimates of Table 1, the TAR inferred that 20$^{th}$ century climate change is very likely to have made a contribution to 20$^{th}$ century sea level rise, because the sum of terms not related to recent climate change (i.e. excluding thermal expansion, glaciers and ice caps, and ice sheet changes due to 20$^{th}$ century climate change) are insufficient to explain the observed rate. This falls short of a conclusive attribution of the observed rate to anthropogenic influence, although the model results suggest anthropogenic climate change is indeed a possible explanation. The geological record of sea level sets an upper bound on the size of natural variability, but the data are not good enough to exclude the possibility that the observed changes of the last two centuries are due to natural fluctuations. However, natural fluctuations in sea level need an attribution to possible physical causes, and climate model results suggest that fluctuations of this size in glacier volume (Reichert et al., 2002), thermal expansion or ice sheet mass balance (Lowe and Gregory, In preparation) are probably inconsistent with natural variability.

The IPCC TAR was published three years ago. Since then, further work has been done on the different processes responsible for past sea level change which we review here (for another recent review, see Cazenave and Nerem, 2004).

Mitrovica et al. (2001) and Tamisiea et al. (2001) discussed a ‘fingerprint’ method, wherein spatial variations in sea level trends are used to estimate global sea level change arising from either mountain glaciers, Greenland and ‘Antarctic plus other’. Their results suggested a considerably higher rate of Greenland melting than that suggested in the TAR, and might offer an explanation for the lower rates of sea level rise found in Europe than elsewhere (Woodworth et al., 1999). The authors assumed a spatially uniform signal for thermal expansion, although models suggest a pattern with strong spatial variability (see below). Therefore, it is possible that some of the observed spatial variations attributed to one or more of the three terms could be explained by ocean density and circulation changes, whether due to natural variability or to anthropogenic climate change. Plag and Jüttner (2001) adopted a similar approach to Tamisiea et al., finding the magnitude and sign of estimated Greenland and Antarctic contributions to be sensitive to the latitudinal selection of tide gauge trends employed in the analysis.

Cabanes et al. (2001) made use of the hydrographic data set of Levitus et al. (1998, 2000) to study two periods: 1993–98 (i.e. the recent period with T/P measurements) and 1955–95. For the more recent period, the observed T/P global rate was 3.2 +/- 0.2 mm/year, with this relatively large rate being a consequence of the El Niño around 1998 (see also Chambers et al., 2000 and see Figure.18.1.c for a more representative rate for the 1990s overall). This rate compares well to the 'thermosteric' rate (i.e. the steric rise due to temperature alone) of 3.1 +/- 0.4 mm/year, with any difference between these values not statistically significant, and with similar spatial patterns for the altimetric and thermosteric trends observed in each ocean basin, although Church (2001) questioned the reliability of such patterns in data sparse regions and thereby the reliability of the inferred global-average thermosteric trend. Cazenave et al. (2003) suggested that any possible small global-average excess of altimetric over thermosteric trend could be due to changes in the continental hydrological regime, discussed further below. (The global-average excess of 0.1 mm/year has to be increased to 0.4 mm/year if the Glacial Isostatic Adjustment correction is included.)

Cabanes et al. (2001) also estimated a rate of change in global sea level due to thermal expansion for 1955–95 of approximately 0.5 mm/year, with larger rates of thermosteric change near to a number of tide gauge sites with long records comparable to the rates observed by the gauges themselves. The implication was that the 'global' rate for the past 100 years summarized by the 1–2 mm/year range of the TAR, based on the available gauge data, may have been over-estimated, with a real global rate nearer to 0.5 mm/year, consistent with a thermosteric forcing alone. This conclusion was controversial and depended critically on the validity of using the sparse hydrographic data set in this way, particularly if only temperature and not salinity information is available (Douglas and Peltier, 2002). It was also unclear if the causes of apparent sea level change on century time scales, which are needed for IPCC-type studies, can be implied from shorter thermosteric and gauge records of only a few decades. Relatively low correlation observed between individual tide gauge and thermosteric trends in the Cabanes et al. study was improved by consideration of local salinity changes (Cabanes, 2002). Nevertheless, Miller and Douglas (2004) have demonstrated that the implication of over-estimation by Cabanes et al. was to a great extent an artefact of their use of the heavily spatially-smoothed Levitus hydrographic data set, and that the most reliable estimates of 20$^{th}$ century sea level change remain at the 1–2 mm/year level.

One point of interest of the Cabanes et al. study from a GLOSS perspective is that the authors noted that, if one considers thermosteric changes from the Levitus data set as a proxy for sea level change, then an extended global coastal network of tide gauges, such as that being developed by GLOSS, would sample global-average rates of change much as a truly-global altimetric system would; this emphasises once again the need for both sets of measurements in future. A further comment regarding the use of global hydrographic data is that simulations with climate models for the past few decades (Levitus et al. 2001; Lowe and Gregory, In preparation) show much less temporal and spatial variability in global ocean heat content than the data set of Levitus et al. (2000). It is unclear whether this is a deficiency of the models, or an artefact of the sparse sampling of the Levitus data. The latter explanation would imply large uncertainties in trends calculated from the hydrographic data.

Antonov et al. (2002) also made use of the Levitus data set, with particular attention to the role of salinity in sea level change. They concluded that during the second half of the 20$^{th}$ century, global-average salinity had decreased slightly, with two important implications. The first is that salinity changes will have compounded the temperature effects in producing a global-average steric trend approximately 10% larger than that computed by temperature alone (and as discussed by Cabanes et al., 2001). In most basins, temperature was confirmed as being the dominant forcing of steric change, with the major exception of the sub-polar North Atlantic, where temperature and salinity forcings largely cancel. The second implication is that, if the salinity decrease is not due to melting sea ice, which to a first approximation does not alter sea level, then a sea level rise of 1.3 +/- 0.5 mm/year due to fresh water input would be inferred, originating from changes in glaciers, polar ice sheets or continental water balance. The overall role of sea ice freshening has not been confirmed conclusively as sea ice volume changes are not well enough known from either observations or modelling. Reasonably quantitative statements can be made for the Arctic: Arctic sea ice is known to be thinning (Laxon et al., 2003) and a model study by Hilmer and Lemke (2000) indicates that reduction of Arctic sea ice could have been contributing the equivalent of only 0.3 mm/year of freshwater to the ocean over recent decades. Satellite observations show that Antarctic ice area has not been in decline (Folland et al., 2001).

Turning to glaciology, Meier and Wahr (2002)[2] concurred with Table 1 in concluding that for the period since 1961 small glacier wastage has been equivalent to 0.25–0.3 mm/year, but suggested that wastage since 1988 has accelerated to closer to 0.5 mm/year, if the major contributions from the Alaskan glaciers (which contribute almost half of the worldwide wastage) are properly accounted for (Arendt et al., 2002). The latter authors proposed that the Alaskan glaciers may have contributed a substantial fraction of the sea level rise suggested by altimetry during the 1990s; accelerations in the 1990s have also been observed from other, smaller ice fields (Rignot et al., 2003). This is inconsistent with the scenario suggested by Cabanes et al. (2001) and Cazenave et al. (2003), which accounts for the recent T/P trend in terms of thermosteric rise, plus perhaps a small additional water source such as a continental hydrological balance term, leaving little room to accommodate glacier wastage. Recent studies have confirmed the seemingly less important (order 0.15 mm/year) current roles of Greenland and Antarctica shown in Table 1 (Krabill et al., 2000; Rignot and Thomas, 2002; Douglas and Peltier, 2002). Attention is increasingly being focussed on the possible role of rapid dynamic changes in Antarctic ice discharge (Vaughan and Spouge, 2002; also discussed in the TAR), but at present there is no evidence that contributions from this mechanism are large (e.g. Shepherd et al., 2001).

[2] A minor issue concerns the use of the word 'eustatic' by different authors. Many authors, including the TAR, define eustatic sea level change to mean that which is caused by an alteration to the volume of the world ocean of whatever origin. Meier and Wahr (2002) and Munk (2002) define it in the sense of 'new water' to the ocean (e.g. from glacier wastage) which does not include steric sea level change.

Finally, we return to the large uncertainties of the terrestrial storage term of Table 1. Most of this component relates to anthropogenic influences on sea level change during the past 100 years due to processes such as dam construction and groundwater mining for agriculture, rather than modifications to the continental hydrological balance due to climate change. Milly et al. (2003) have studied the latter aspect and suggested a significant amount (approximately 0.12 mm/year) could have come from soil moisture, groundwater and snow cover changes since 1981. However, the anthropogenic factors remain with large uncertainties as discussed in the TAR.

Altogether these new results since the TAR was published infer similar or larger amounts of sea level change-equivalent from the various contributing terms than shown in Table 1. However, any sum of the contributors would remain on the low side of the range of observations, as shown by Table 1, unless the large amount of fresh water input inferred from the Antonov et al. (2002) findings were to be considered seriously. This in turn implies the need to identify a source for a fresh water input significantly larger than, for example, glacier melt; Munk (2003) discusses the detailed implications of the Antonov et al. findings. A major difficulty remains in accounting for the observed global sea level change, which arises from the non-simultaneity of measurements of the various contributing terms in addition to their individual uncertainties. Almost always one finds trends quoted for different periods when one knows that one is dealing with considerable interannual variability in all parameters; Douglas and Peltier (2002) stress this issue strongly. This emphasises the urgent need for further, and if possible more coordinated, research on each aspect of this topic.

In summary, there is what Walter Munk has called 'the enigma of 20$^{th}$ century sea level' (Munk, 2002), in that at face value (i.e. ignoring the large uncertainties in each quantity and the question of significant freshening) the rates of sea level change observed by tide gauges over the past 100 years have tended to be larger than can be accounted for at present by knowledge of the various contributing processes, and from complementary geophysical information including length of day and polar motion. Even in the recent period with T/P measurements, for which a considerably larger amount of data is available, important questions remain as to the causes of the observed changes. Whether these enigmas can be resolved by the availability of new types of data (e.g. by means of observation of the spatial dependence of the rate of geoid change by space gravity, Woodworth and Gregory, 2003) remains to be seen.

Another conclusion of the TAR was the need to learn more about changes in extreme sea levels, in addition to the relatively much-studied topic of mean sea level change. Woodworth and Blackman (2004) made a start in this direction by analysing a limited 'global' data set of 141 stations. They concluded that, if one considers the generality of the sea level information, then there is indeed evidence for an increase in extreme high water levels worldwide since 1975, and that the variations in extremes in this period are related to changes in regional climate. In many cases, the secular changes and the interannual variability in extremes are similar to those in mean sea level during the same period, and are, therefore, consistent with having the same magnitude and type of atmospheric and/or oceanic forcing. This is an important finding with regard to studies of the impacts of coastal sea level changes. If variations in extremes were to be a consequence of 'different physics' to those in the means, then uncertainties in the occurrence of extremes in future might be expected to be even larger than those in the mean levels, which the following section demonstrates are themselves considerable.

### *3.2 Sea Level Changes in the 1990s*

An important question relates to whether the observed T/P trend for the 1990s (of order 2.5–3 mm/year) can be considered significantly larger than the tide gauge rates reported for the past 100 years (1–2 mm/year) and, therefore, indicative of a recent major acceleration in the rate of change of global sea level. There is strong indirect evidence that sea level rose faster during the 1990s than during the second half of the 20th century overall. For example, we have referred above to the global-average thermosteric trends of Cabanes et al. (2001), based on Levitus data, which suggest a significant increase, while some of the glaciological evidence also supports an acceleration.

If one inspects the longer tide gauge records for evidence of recent acceleration, then one does indeed determine enhanced rates for the 1990s for several regions, although the geographical limitations of the PSMSL data set preclude a definite global conclusion (Holgate and Woodworth, 2004). It remains to be seen whether any 1990s accelerations persist into the 21st century.

TABLE 2

Estimated rates of sea level rise components (mm/year) for the 1990s. The climate-related terms (thermal expansion, glaciers and 20th century effects on ice sheets) are derived from the same models as used in Table 1. The terrestrial storage terms were estimated for the 1990s by Church et al. (2001) from the literature. The other terms are the same as in Table 1.

| | Minimum (mm/year) | Maximum (mm/year) |
|---|---|---|
| Thermal Expansion | 0.5 | 1.8 |
| Glaciers and ice caps | 0.3 | 0.5 |
| Greenland – 20th century effects | -0.1 | 0.1 |
| Antarctica – 20th century effects | -0.4 | -0.1 |
| Ice sheets – Adjustment since LGM | 0.0 | 0.5 |
| Permafrost | 0.00 | 0.05 |
| Sediment deposition | 0.00 | 0.05 |
| Terrestrial storage (not directly from climate change) | -1.9 | 1.0 |
| **Total** | **-1.6** | **3.9** |

If one adopts a primarily modelling approach, then as a contrast to Table 1, which refers to the period 1910–1990, one can employ the same set of models and published information included in the TAR for investigation of the more recent period of the 1990s. Our findings are shown in Table 2. The increased rates of change of the climate-related terms can be seen (thermal expansion, glaciers and 20th century effects on ice sheets). The central value for thermal expansion, for instance, is more than twice that for 1910–1990; Church et al. (2001) and Gregory et al. (2001) also comment on the acceleration in this term. It is, however, still substantially less than the estimates from hydrographic data. The spread between minimum and maximum rates are larger for the 1990s, because the rates themselves are larger. This is true also of the terrestrial storage terms, which are estimated by Church et al. (2001) to be larger for the 1990s than for the 20$^{th}$ century as a whole (their Table 11.8). Consequently the range of uncertainty on the total is broader than in Table 1. The middle of the range for the total of all terms is 1.2 mm/year, while excluding the terrestrial storage it is 1.6 mm/year. Although this is still somewhat below the T/P altimeter trends for the 1990s, the discrepancy might be accommodated within the plausible range for the contribution from terrestrial storage. Furthermore there will be a larger contribution (positive or negative) from variability in the figure for a single decade than for a century.

## 4. Future Changes in Sea Level

### *4.1 Global Average Change*

The climate change projected to take place during the 21$^{st}$ century is larger than that which occurred during the 20$^{th}$ century, so observed trends cannot be simply extrapolated into the future. Making projections of future sea level change depends on the ability to model the contributing physical processes. For the processes expected to be dominant during the 21$^{st}$ century, there is tolerable agreement between observational and model estimates for the last 100 years, as shown above, giving some confidence in the modelling. However, models which are adequate for present climate and recent changes are not guaranteed to give a reliable simulation for a markedly different future climate.

In the IPCC TAR, climate change projections for the 21$^{st}$ century were derived on the basis of results from the AOGCMs, which are three-dimensional global climate models representing the physics and dynamics of the climate system on scales typically of tens to hundreds of kilometres. There are many AOGCMs in use, differing in their formulations, and producing different results for future global and regional climate change, and for the sensitivity of glacier and ice sheet mass balance to climate change. Although many of them give a reasonable simulation of present-day climate and variability, they all have deficiencies, and it is not possible to use this as a basis for preferring any one of them. The range of AOGCM results thus represents a source of uncertainty in climate projections.

Another source of uncertainty arises from the need to adopt a scenario for future concentrations of greenhouse gases and aerosols. In the TAR, the concentrations were derived from a set of 35 scenarios for future emissions developed for the IPCC Special Report on Emissions Scenarios (SRES) (IPCC, 2000). These scenarios are based on four distinct 'storylines' which make a range of plausible but diverging assumptions about future demographic, societal, economic and technological change, but do not allow for any specific policy actions aimed at mitigating anthropogenic climate change. They span a wide range of possible futures. For instance, the carbon dioxide concentration in the atmosphere in the year 2100 ranges from 550 ppm by volume to 1000 ppm (the present-day concentration is about 370 ppm). Because AOGCMs are very expensive of computer time, it is only practicable to run them for a small number of scenarios. For the TAR, results from seven AOGCMs for global average temperature change and heat uptake by the ocean in response to a particular scenario were used to calibrate a simple climate model to reproduce the results of each of the AOGCMs. The simple climate model was run for all the SRES scenarios and AOGCMs (Cubasch et al., 2001).

Sea level change due to thermal expansion is generally the largest contribution. Projections of the glacier contribution were obtained using the simple model temperature projections and global glacier mass balance sensitivities to temperature change, derived from the AOGCM results, following previous work on the relationship of glacier mass balance sensitivity to local climatic regime. Substantial uncertainties exist in these estimates because only about forty of the worlds $10^5$ glaciers have long continuous mass balance measurements. For the Greenland and Antarctic ice sheets, an ice-sheet model was used to obtain mass balance sensitivities, which are also subject to some uncertainty. In the $21^{st}$ century, the contribution to sea level rise from the ice sheets is not expected to be large; it is likely that the Greenland ice sheet will lose mass on account of greater ablation, while the Antarctic ice sheet will gain mass as a result of increased precipitation (but negligible surface melting owing to the low temperature). It is now widely agreed that major loss of grounded ice from the West Antarctic ice sheet is very unlikely during the $21^{st}$ century, although on a longer timescale changes in ice dynamics resulting from enhanced basal melting could bring about significantly increased outflow of ice into the floating ice shelves and a retreat of the grounding line (Vaughan and Spouge, 2002).

The results for global average sea level rise by 2100 (relative to a baseline of 1990) span a wide range of 0.09 to 0.88 m (Figure. 11.12 of the TAR). The central value corresponds to a average rate of 4.8 mm/year, considerably more than the average of 1–2 mm/year during the $20^{th}$ century. The greater rate is consistent with the larger climate change, the projected global average temperature rise being 1.4–5.8°C over the same period. Moreover, projected sea level rise shows a continuous acceleration during the $21^{st}$ century. The uncertainty due to choice of scenario alone is relatively small. For the next few decades, in any particular model, the SRES scenarios all give rather similar results for sea level; by the end of the century, the scenario range approaches 50% of the central value. Much larger is the systematic uncertainty associated with modelling, arising from the choice of different AOGCMs and uncertainty in land ice mass balance. This points to a need for continued work in these areas aimed at understanding and resolving the differences among models and narrowing the uncertainties.

It may be noted that the bottom end of the range corresponds to 0.9 mm/year, which is below the lowest estimate of $20^{th}$ century sea level rise. Furthermore, a model predicting 0.9 mm/year for the $21^{st}$ century must predict even less rise for the $20^{th}$ century. However, model projections at the lower end of the range are nonetheless acceptable because their calculations take no account of anthropogenic changes in terrestrial water storage. The dominant terms are probably impoundment in reservoirs and extraction of groundwater; these terms are poorly known but possibly large (cf. Sahagian, 2000 and Gornitz, 2000; see discussion in Section 11.2.5 of the TAR). Their net average during the $20^{th}$ century is highly uncertain, even regarding its sign. It could have been a positive contribution of a few 0.1 mm/year, not included in the models, which could reconcile the lowest model estimates for the $20^{th}$ century with the observations (Table 1). If anthropogenic terms persist into the 21st century at their present rate, they would modify the projections by between –0.2 m and +0.1 m with, for instance, a reduction in the rate of reservoir construction tending to further increase sea level rise (cf. Sahagian, 2000). What may happen regarding these activities should properly be considered as an aspect of the choice of scenario, since it relates to societal change and choice which can be linked to specific emission scenarios via the SRES storylines (Arnell et al., 2003).

If greenhouse gas concentrations in the atmosphere were stabilised, sea level would nonetheless continue to rise for many 100s or even 1000s of years. Part of this 'sea level commitment' results from thermal expansion, which has a long time-scale characteristic of the weak diffusion and slow circulation processes which transport heat into the deep ocean, bringing it gradually into equilibrium with the new surface climate. After 500 years, thermal expansion might have reached only half of its eventual level, which models indicate could eventually reach 1–4 m for a carbon dioxide concentration of four times the pre-industrial level. Even this could be an underestimate if deep water production ceases at high latitudes, allowing warming to penetrate deeper into the ocean, as found, for instance, by Bi et al. (2001), whose model gave a final thermal expansion of about 4.5 m (D. Bi, personal communication) for a tripling of carbon dioxide.

Sinking in the north Atlantic is one of the important sources of cold deep water. On the basis mainly of results from simpler climate models, it has been suggested that anthropogenic climate change could cause a rapid shutdown of convection in the north Atlantic, the Atlantic overturning circulation and deep water production. Removal of the northward heat transport by the circulation would tend to cool the northern hemisphere. Although many AOGCMs do show a weakening of the overturning circulation during the 21st century (e.g. TAR Figure 9.21), the regional cooling from this effect is however outweighed by the greenhouse-gas warming. Since it would allow the deep Atlantic to warm up, a consequence of a collapse of the overturning would be a rise in sea level, of perhaps 0.5 m over several centuries (Section 11.5.4.1 of the TAR; Knutti and Stocker, 2000).

Time-scales of millennia are required for the slow adjustment of the ice sheets to climate change. A local annual-average warming of larger than 3°C in Greenland, if sustained, could lead to a virtual elimination of the ice sheet. This threshold is reached in practically all models and SRES scenarios by 2100. The timescale required for the melting depends on the warming; higher temperatures give faster rates. For instance, for a warming of 5.5°C, Greenland contributes about 3 m in 1000 years. The eventual global-average sea level rise from Greenland would be more than 6 m. It is moreover possible that the deglaciation of Greenland would be irreversible, even if the greenhouse gas concentration and climate were subsequently returned to pre-industrial conditions (Toniazzo et al., 2004).

### *4.2 Regional Change*

It is important to realise that the projected rise discussed in Section 4.1 is a global average, and that local sea level change may be more or less than this, which has important implications for impact estimates. Changes in sea level have not been, and will not be, the same at each location around the world.

The first reason for regional differences, which we have referred to above, concerns spatial variations in vertical land movements, with the only geological process for which we can provide estimates on a global basis being GIA. However, there is a range of other geological factors which can cause land levels to change. These include the major tectonic events due to earthquakes, consolidation in active sedimentary coastal systems such as deltas, and the consequences of industrial activities such as mining or ground water extraction. All of these factors will have much shorter spatial scales than GIA, necessitating local geological insight and programmes of GPS and other geodetic measurements mentioned above.

However, geology is not the only reason for the spatial dependence of sea level changes. The ocean is a fluid that is constantly readjusting to the changing forces applied to it, and sea level is determined by its density structure and circulation. Changes in density, circulation and sea level are obvious on seasonal and interannual (e.g. El Niño) timescales. Likewise, climate change will cause modifications in the surface fluxes of heat, freshwater (precipitation, evaporation and runoff from land) and momentum (wind stress), producing changes in interior density, circulation and sea level. Coastal impacts of sea level change relate to the level of the sea relative to the land; hence both land movement and climate change need to be taken into account.

The regional pattern of sea level response to climate change during the 21st century can be evaluated using AOGCMs. The variation is substantial, the spatial standard deviation being up to a third of the size of the global mean thermal expansion, so that some regions may experience very little sea level change, and others twice as much as the global average. Unfortunately, the patterns given by the various AOGCMs show few similarities. There are only three common features in the AOGCM results studied in the TAR. Sea level rise is greater than the global average in the Arctic Ocean, probably due to a freshening caused by increasing river inflow or precipitation. Sea level rise is less than the global average in the Southern Ocean, possibly because of the low thermal expansion coefficient of water at cold temperatures, changes in wind-stress, or transport of heat to lower latitudes. In the north Atlantic there is a dipole pattern comprising an enhanced rise to the north of the Gulf Stream extension and a reduced rise to the south. This could be related to a weakening of the circulation, which is a common AOGCM response to greenhouse warming. The lack of agreement among AOGCMs means that we can have little confidence in their regional projections for sea level change. Because of the practical importance of the issue, and the need for scenarios for impact assessment, this problem requires more attention.

The coarse spatial resolution which can practically be used in AOGCMs means that sea level changes in specific coastal areas cannot as yet be realistically simulated. Higher-resolution ocean models indicate that perturbations to circulation and sea level propagate rapidly via coastally trapped Kelvin waves (Bryan, 1996; Johnson and Marshall, 2002), with the signal penetrating much more slowly into the open ocean. For such reasons, sea level change near to the coast may differ from the open-ocean and global average values simulated by climate models. Another coastal effect which has so far been neglected in most projections of sea level change is the alteration to the gravity field and the elastic response of the solid Earth to the increased load of water in the ocean as glacial meltwater is added to it. At continental margins, this could offset as much as 30% of the global average rise. In the vicinity of land ice masses (or, in principle, reservoirs and aquifers) which undergo a large change in mass, there could be larger local effects, as discussed above in connection with the 'fingerprint' method of Tamisiea et al. (2001) and Plag and Jüttner (2001). Overall, this suggests that to be more useful to coastal planners, considerable improvements are required to the next generation of sea level change prediction models, particularly concerning regional processes.

## 5. Impacts of Future Sea Level and Related Climate Change

The TAR found that the rising sea levels observed in the 20$^{th}$ century is already having impacts, particularly in subsiding areas, and the projected rises for the 21$^{st}$ century discussed in Section 4 could have more serious impacts, particularly for small islands (McLean et al., 2001; Nurse et al., 2001). Subsidence is occurring in many densely populated coastal deltas and cities due to both natural and human factors and this will continue through the 21$^{st}$ century. As already discussed, additional change in sea level due to regional oceanic changes is highly uncertain, but a scenario of ±50% on global-mean rise is suggested (e.g. Hulme et al., 2002). Hence, taking the TAR range, in the worst case coastal sites in some regions could experience a rise of up to 1.3 m from 1990 to 2100, plus any effects due to land uplift/subsidence, even though global-mean rise will be about 0.9 m. In the best case scenario, sea levels might only change at a similar rate to the 20$^{th}$ century.

The TAR also recognised that sea level rise and related climate change will impact evolving natural and human systems, which co-exist in the coastal zone, and these systems will adjust/adapt to change (Figure 18.3). Therefore, in addition to the magnitude of sea level rise and climate change, impacts will depend on the sensitivity, exposure and adaptive capacity of each system. These are dynamic characteristics that will evolve through the 21$^{st}$ century due to factors such as demographic and economic changes and human management of the environment. Many impact studies have ignored integrated analysis and estimated worst-case (or potential) impacts on the existing world in the absence of any adjustment/adaptation.

The uncertainty about sea level rise, and many other relevant factors makes estimates of the impacts of sea level rise and climate change quite uncertain, and it is important to understand the assumptions and inputs used by any impact estimates. Similarly, the TAR recognised that economic costs of sea level rise are highly uncertain, depending on assumptions on a range of factors including impact potential and the feasibility of adaptation (see Table 4 in Neumann et al. (2001)). The TAR also recognised the importance of scale: while the impacts of sea level rise and climate change can be defined at broad scales, at more local scales a wide range of impacts are possible which are difficult to characterise in general terms. The impacts of sea level rise are now considered within the framework outlined in Figure 18.3. The natural system impacts are generally better understood than the socio-economic impacts, as noted in the TAR.

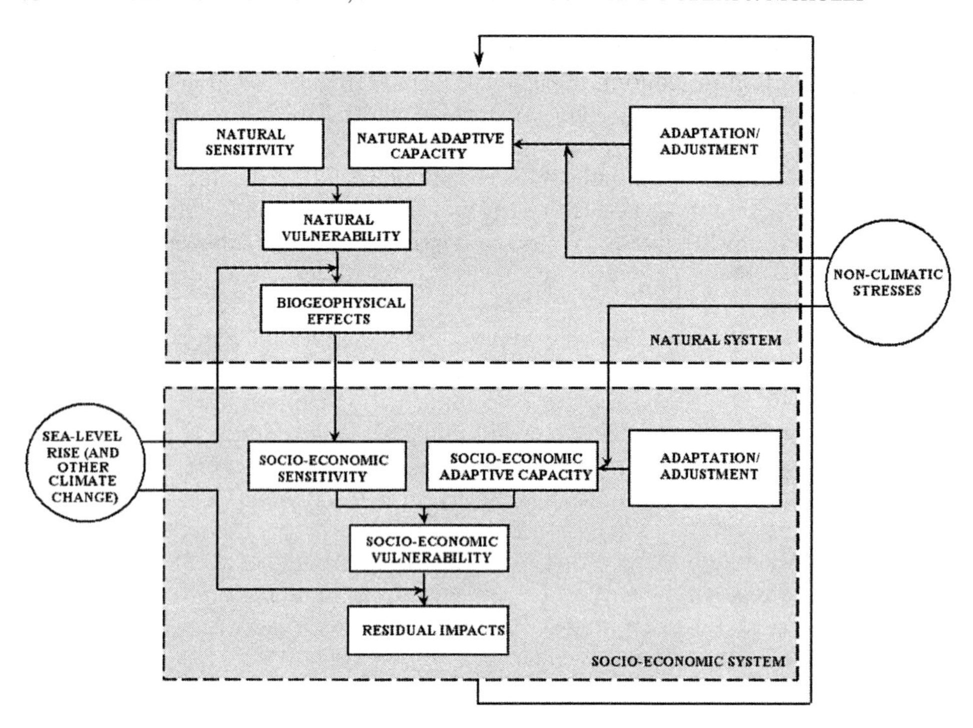


Figure 18.3 A conceptual framework for coastal impact and vulnerability assessment of sea level rise. This has been adapted from Klein and Nicholls (1999) to reflect the terminology used in the TAR.

### *5.1 Impacts in Coastal Waters*

While most impact studies focus on the inter-tidal and terrestrial areas of the coastal zone, if sea level (i.e. ocean depth) and climate (especially wind field) change in coastal areas, then there will also be associated changes in the ocean tides, storm surges and extreme sea levels, and waves (Sanchez-Arcilla et al., 2000). This will have important implications both for the coastal waters and the neighbouring coasts, including hazard management (e.g. flooding) and morphological evolution of the coastal zone, including coastal ecosystems. These processes have been studied most completely for the North West European continental shelf. For example, Figure 18.4 shows one simulation of possible future changes in extreme surge levels near London. Flather et al. (2001) found that a uniform 50 cm depth change around the UK would result in a more complicated spatial pattern of tidal change, increasing Mean High Waters by only 45 cm in some locations and as much as 55 cm in others. Such results are fairly robust, depending only upon well-understood tidal dynamics. However, possible future changes in the frequency and magnitude of storm surges and extreme still water levels are harder to assess as they depend primarily on predicted changes in regional meteorology, as well as depth changes. Consequently, it is interesting that changes computed in the spatial pattern of these quantities by Flather et al. (2001) and Lowe et al. (2001) with the use of different 'future weather' data sets provided by the ECHAM4 and HadCM2 climate models respectively are significantly different, even though the tide+surge components of the two modelling exercises are essentially the same, because of differences in the climatology of predicted regional wind fields.

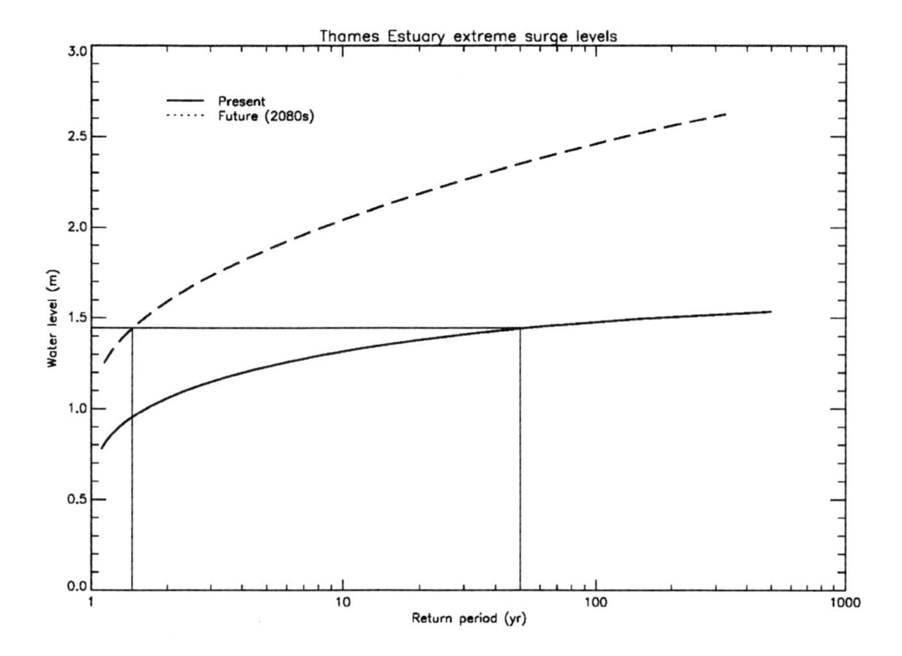


Figure 18.4 Extreme surge levels versus return period for the Thames Estuary for current (1961-1990) climate computed using the Hadley Centre AOGCM and the POL shelf tide-surge model (solid line). Extreme water levels would rise to the dashed line for a medium-high emission scenario for the 2080s using a central sea level rise estimate of 0.3 m. Storminess changes and vertical land movements are taken into account. The straight lines show the reduction in return period for an example extreme water level: the present day 50 year return period level becomes a 1.5 year level. (Computation by Jason Lowe, Hadley Centre, using methods described in Hulme et al., 2002).

Modelling of changes in wave conditions meanwhile has received comparatively little attention. Wave heights are known to have increased in recent decades in the North East Atlantic (but not significantly in the North Sea) as a consequence of the strengthening of the westerly wind field, which can be associated with the North Atlantic Oscillation (NAO) (Bacon and Carter, 1993). If the average winter NAO index increases during the 21st century as suggested by some studies (e.g. UK Climate Change Scenarios studies, Hulme et al., 2002), then North Atlantic wave heights might be expected to increase further. Waves will also break nearer the coast as a consequence of increased depth. However, reliable wave modelling requires the use of high resolution, local models within which the dynamics of tide-surge-wave interaction can be represented in detail.

To our knowledge, such modelling of changing wave conditions as a consequence of possible changes in regional meteorology has been investigated within only a small number of case studies (e.g. Wolf et al., 2002). Earlier work includes the study of the effect of a large negative NAO during the winter of 1995/96 as observed at Holderness on the east coast of England, compared with the large positive NAO during the preceding winter (1994/95) (Wolf, 1998). Further projects studied the transformation of extreme wave events to the coast using the SWAN wave model driven by statistics taken from *in situ* and satellite altimeter wave observations (Hargreaves et al., 2002).

A 3-D question in coastal seas concerns how much changing tidal currents might affect the pattern of high-productivity fronts which occur in summer due to stratification of shallower waters adjacent to deeper, mixed areas. The physics by means of which frontal adjustment takes place during the spring-neap cycle is complex (Sharples and Simpson, 1996). However, given the $\text{current}^3$ dependence of tidal mixing and the consequent order of magnitude ratio between spring and neap mixing rates, a depth change of order 1% due to climate change should have relatively little impact on currents and fronts. Possibly the only potential changes of that type one could envisage would be in very shallow seas close to resonance, or in regions of freshwater influence (ROFIs) and estuaries where there is a $\text{depth}^5$ dependence on the tendency to stratify against tidal mixing (Simpson et al., 1990). Overall, one would expect that frontal positions and their productivity should be little affected by any 21st century sea level change.

These changes in coastal waters could have important direct socio-economic impacts on activities such as shipping and navigation, fisheries and oil exploration but they have not been investigated in detail so far. Some of their links to coastal impacts are considered in Section 5.2.

### *5.2 Impacts on Coastal Areas*

The coastal zone is a major focus of human habitation and economic activity, as well as being important ecologically and in terms of global biogeochemical flux (McLean et al., 2001; Nicholls, 2002a). In 1990, it is estimated that 1.2 billion (or 23%) of the world's population lived in the near-coastal zone[3], at densities about three times higher than the global mean (Small and Nicholls, 2003). The population density also increases across the near-coastal zone, with the highest densities occurring below 20-m elevation. Coastal populations are widely reported to be growing more rapidly than the global mean, due to net coastal migration. This is accompanied by coastal urbanisation, which is producing a few large megacities together with many more numerous smaller cities and towns clustered close to the coastline. Therefore, the exposure of the human and natural coastal zone systems to sea level rise is significant and growing. At the same time, human-induced changes in coastal zones are widespread (see non-climatic stresses in Fig 18.3) and often profound, such as declining sediment and freshwater inputs due to increased catchment regulation, direct and indirect destruction of mangroves, and degraded corals due to a range of causes. Management of coastal zones is already perceived as a significant problem at the global scale. Therefore, sea level rise and climate change represent one of a number of stresses on the coastal zone as recognised both in the IPCC Second Assessment Report (SAR) (Bijlsma et al., 1996) and the TAR.

[3] The area both within 100 km horizontally and 100 m vertically of the coastline.

TABLE 3

The main natural system effects of relative sea level rise, including relevant interacting factors (adapted from Nicholls, 2002a). Some factors (e.g. sediment supply) appear twice as they may be influenced both by climate and non-climate factors.

| NATURAL SYSTEM EFFECT | | INTERACTING FACTORS CLIMATE-DRIVEN | NON-CLIMATE-DRIVEN |
|---|---|---|---|
| Inundation, flood and storm damage | Surge | Wave and storm climate, morphological changes, sediment supply | Sediment supply, morphological changes, land claim |
| | Backwater effect (river) | Run-off | Catchment management and land use |
| Wetland loss (and change) | | $CO_2$ fertilisation, sediment supply, air and sea temperature, salinity change[4] | Sediment supply, migration space, direct destruction and land claim |
| Coral reef degradation/loss[5] | | Atmospheric $CO_2$, sea temperature | Water quality and turbidity, overfishing, mining and direct destruction |
| Erosion | | Sediment supply, wave and storm climate | Sediment supply, beach mining |
| Saltwater Intrusion | Surface Waters | Run-off | Catchment management and land use |
| | Ground-water | Rainfall | Land use, aquifer use |
| Rising water tables/ impeded drainage | | Rainfall | Land use, aquifer use |

As discussed in Section 5.1, climate change and long-term sea level rise has a wide range of effects on coastal processes. In addition to raising the mean ocean level, rising sea level also modifies all the coastal processes that operate around sea level. Therefore, there are immediate effects of a rise in sea level, including submergence, increased flooding and storm damage of coastal land, as well as saltwater intrusion of surface waters. In addition, there are lagged effects including morphological and ecosystem changes, as well as saltwater intrusion into groundwater as the coast adjusts to the new environmental conditions. These morphological changes interact with the more immediate effects of sea level rise and will often exacerbate them. The most significant natural system effects of sea level rise are summarised in Table 3. As one might expect, low-lying coastal areas are most sensitive to sea level rise, including deltas, low-lying coastal plains, coral islands, beaches, barrier islands, coastal wetlands, and estuaries (e.g. Stewart et al., 1998).

Climate change will have other coastal implications such as higher sea surface temperatures, declining sea ice cover, and more speculatively, long-term changes to the tracks, intensity and/or frequency of storms (see Section 5.1).

[4] Sea saltwater intrusion

[5] Coral reefs can probably keep pace with sea level rise during the 21st century (McLean et al., 2001). Therefore the interacting factors are especially important.

The natural-system effects of sea level rise and climate change can have a range of direct and indirect socio-economic impacts (Figure 18.3), including the following identified by the TAR:

- Increased loss of property and coastal habitats
- Increased flood risk and potential loss of life
- Damage to coastal protection works and other infrastructure
- Loss of renewable and subsistence resources
- Loss of tourism, recreation, and transportation functions
- Loss of non-monetary cultural resources and values
- Impacts on agriculture and aquaculture through decline in soil and water quality

There has been a range of assessments of the potential impacts of sea level rise at local, national, regional and global scales as discussed in the SAR and TAR. These have included national-funded studies and organised programmes such as the US Country Studies Program, the Dutch Country Studies Programme, studies sponsored by the United Nations Environment Programme, and most recently the Assessment of Impacts of and Adaptation to Climate Change in Multiple Regions and Sectors (AIACC). These programmes were focussed on the impacts of climate change, of which coastal impacts are only one aspect. However, they have produced a range of useful assessments as described by the Synthesis and Upscaling of sea-level Rise Vulnerability Assessment Studies (SURVAS) Project and in a number of paper collections (e.g. Mimura, 1999).

To illustrate some of these issues, first selected impacts with a biophysical emphasis are considered. While the local/national studies provide detailed impacts for specific sites, they are difficult to synthesise due to differences in detailed methodology and/or underlying assumptions. Therefore, to illustrate the potential impacts of sea level rise, the following section draws on the more limited regional and global assessments. This scale of analysis is relevant to climate policy and related questions as to how much to reduce greenhouse gas emissions, versus adapt to the changes (e.g. Arnell et al., 2002).

**Deltas**

Deltas form near sea level where river-deposited sediment accumulates in the coastal zone, and hence provide a good example of threatened coastal lowlands. Subsidence leads to localised relative sea level rise without any global changes (e.g. Penland and Ramsey, 1990; Törnqvist et al., 2002). Deltaic areas have the potential to maintain elevation with rising sea level by riverborne sedimentation. However, delta management often excludes floods, while upstream dams reduce sedimentary inputs (Boesch et al., 1994; Sanchez-Arcilla et al., 1998). Hence, this is a good example of human management increasing vulnerability of coastal areas to sea level rise. Sustaining sediment supply and developing flood management approaches that allow sedimentation are two approaches that could assist delta survival under sea level rise based on a 'working with nature' philosophy.

**Wetland Loss (and Change)**

There is a wide range of coastal wetland, such as saltmarsh, brackish marsh, coastal fresh marsh, mangrove and unvegetated intertidal sediment, occurring in settings such as deltas, estuaries and lagoons. Their present location is controlled by recent sea level, so they are all sensitive to long-term sea level change. In response to sea level rise, coastal wetlands experience faster vertical accretion due to increased sediment input and/or organic matter preservation (Cahoon et al., 1995; 1999). If vertical elevation relative to sea level is maintained, the coastal wetland has the potential to grow upwards in place and survive *in situ*. If not, there might be scope for survival via inland migration. In areas without low-lying coastal areas, or in low-lying areas that are protected by humans to reduce coastal flooding, wetland migration cannot occur. This produces what is termed a 'coastal squeeze' between the fixed land/defences and rising sea levels (French, 1997). Therefore, adaptation to protect human use of the coastal zone may exacerbate wetland losses.

Erosion and other morphological changes also influence coastal wetlands, while salinity change may change the types of wetlands that are present (e.g. Boesch et al., 1994). Lastly, temperature rise could allow a poleward expansion of the distribution of mangroves at the expense of saltmarsh and other temperate coastal wetland types.

Sea level rise during the 20$^{th}$ century produced impacts in some coastal areas where the wetlands are well-protected from direct human pressure, such as the US Mid-Atlantic coast where relative sea level rise was about 3 to 4 mm/year (Leatherman et al., 1995; Hartig et al., 2001). Relative sea level rise of up to 10 mm/year in the Mississippi deltas was associated with substantial wetland losses: up to 100 km$^2$/year in the 1970s, falling to 65 km$^2$/year more recently (Boesch et al., 1994). These wetland losses have a range of long-term implications, particularly for the nationally important coastal fisheries in Louisiana. Coastal wetlands play an important role in the life cycles of many important fisheries species (McLean et al., 2001; Kennedy et al., 2002). Landings in Louisiana have been buoyant during the period of rapid wetland loss, but a rapid collapse appears inevitable if the wetlands continue to decline. Thus, the full impacts of relative sea level rise may not be apparent for many decades.

**Beach Erosion**

While still controversial (Komar, 1998), there is increasing field support of the Bruun Rule, which describes a direct link between sea level rise and shoreline recession on sandy beaches: shoreline recession is typically about 100 times the rise in sea level (Leatherman, 2001; Zhang et al., 2004). However, the Bruun Rule only describes one of a number of processes that shape sandy coasts over the long-term (decades to centuries) (Stive et al., 2002). A less-appreciated and additional erosional process linked to sea level rise is the indirect effect of sea level rise: as seas rise, estuaries and lagoons maintain equilibrium by raising their bed elevation in tandem, and hence act as a major sink of sand which is derived from the open coast (van Goor et al., 2001). This process potentially causes erosion of order ten times that predicted by the Bruun Rule in the vicinity of tidal inlets (e.g. along the North Holland coast due to the sink effect of the Wadden Sea).

Most of the world's sandy beaches had long-term erosional trends during the 20$^{th}$ century, and there is a continuing consensus that coastal erosion is a widespread phenomena (Bird, 1993). Whatever the cause of the erosion, any acceleration in sea level rise will exacerbate these erosional trends across the globe (McLean et al., 2001). The high and growing economic values along many beaches in terms of coastal property and tourism means that erosion could have significant impacts. As a response to this threat, Leatherman et al. (2003) have argued that systematic large-scale monitoring of coastal erosion is required to compliment the observations of sea level rise previously discussed in Section 2.

A challenge for future impact assessments of sea level rise is to link the simple concepts above into a model of the three-dimensional coastal system (Capobianco et al., 1999). This would allow more integrated analysis of changes to coastal morphology and ecosystems. It could also include a wider range of coastal types including soft cliffs, shingle beaches and muddy coasts. New models describing long-term coastal behaviour offer promise to improve prediction of future coastal evolution, integrating all the forces shaping the coast including sea level rise (Cowell et al., 2003).

**Selected Regional and Global Impacts of Sea level Rise**

Now some regional and global impacts of sea level rise are considered: flooding due to storm surge and losses of coastal wetlands The methods have been developed from the global vulnerability assessment of Hoozemans et al. (1992; 1993).

Coastal flooding due to storm surge is already a major hazard that will be exacerbated by sea level rise as shown in Fig 18.4. Even without more intense storms, a fairly small rise in sea level will have a significant impact on the frequency of coastal flooding (Smith and Ward, 1998; Gornitz et al., 2002). Scenarios of coastal population, GDP/capita as a surrogate for protection standards and relative sea level rise (including subsidence) allow future changes in flood risk to be explored (Nicholls et al., 1999; Nicholls, 2002b). These analyses explore the question "is sea level rise a significant problem if ignored?". Therefore, the defence standard is allowed to improve in some cases, but only assuming 1990 flood statistics, so that the additional impacts of sea level rise can be explored. Note that in such analyses storminess is considered constant, so any increase in storminess would increase the flood impacts described below.

Globally, it is estimated that about 200 million people lived in the coastal flood plain (below the 1 in 1,000 year surge-flood elevation) in 1990, or about 4% of the world's population (Hoozemans et al., 1993). Based on this data, it is estimated that on average 10 million people/year experienced coastal flooding in 1990 (Nicholls et al., 1999). Even without sea level rise, this number will change due to increasing coastal populations (i.e. exposure) and (generally) improving flood defences. Table 4 shows the relative impacts for the IS92a emissions scenario, imposed on an evolving world with consistent socio-economic characteristics. For a single emissions scenario, global-mean sea level rise scenarios range from 19 to 80 cm from 1990 to the 2080s (Warrick et al., 1996), reflecting the wide range of climate sensitivity to greenhouse forcing. The relative increase in the number of people flooded compared to the baseline without sea level rise is 30% to 110% under the low scenario and 12 to 27 times the baseline under the high scenario. The range depends on the protection scenario considered and under the high scenario, it is considered that >350 million people could experience flooding each year without adaptation (Nicholls, 2002b). Therefore, flood impacts could be severe during the 21st century, but there is a wide range of uncertainty even for a single emissions/socio-econcomic scenario, which hinders decisions on responses.

TABLE 4
Estimates of the number of people flooded for the full scenario range under the IS92a scenario in the 2080s. Constant protection is a pessimistic assumption that defence standards do not improve, while In Phase Evolving Protection is an optimistic assumption that defence standards improve at the same time as increasing GDP/capita, but with no allowance for relative sea level rise. A coastward migration of population is assumed, and all other climate factors are assumed to be constant. SLR – global sea level rise. See text for more discussion.

| Climate Scenario | SLR Relative to 1990 (cm) | Socio-Economic Scenario | **People Flooded (millions/year)** | | | | Source |
|---|---|---|---|---|---|---|---|
| | | | Constant Protection | | In Phase Evolving Protection | | |
| | | | No SLR | SLR | No SLR | SLR | |
| Low | 19 | IS92a | 36 | 77 | 13 | 17 | Nicholls |
| Mid | 45 | IS92a | 36 | 310 | 13 | 133 | (2002b) |
| High | 80 | IS92a | 36 | 426 | 13 | 353 | |

Stabilising the atmospheric concentration of greenhouse gases via mitigation policies is one precautionary response to these potential impacts. Arnell et al. (2002) have looked at the impacts of stabilising greenhouse gases at 750 ppm and 550 ppm, respectively on the same (IS92a) socio-economic scenarios (termed S750 and S550, respectively). In this case, only one sea level rise scenario per emissions scenario was available: that derived from the HadCM2 model for each emissions scenario (so the uncertainties concerning climate sensitivity are not captured). As shown in Table 5, the impacts are significantly reduced under stabilisation but they are not entirely avoided. Sea level rise also continues under stabilisation due to the sea level rise commitment discussed in Section 4.1. By the 2230s, sea levels are estimated to have risen by 91 and 73 cm compared to 1990 under the S750 and S550 scenarios, respectively. Collectively, this suggests that adaptation to sea level rise would also be required under stabilisation, and the most appropriate climate policy question is to decide on the appropriate mixture of mitigation and adaptation.

More recently, impacts have been developed for sea level rise under four of the SRES 'worlds' (A1FI, B1, A2 and B2) with the climate scenarios derived from the HadCM3 model (so again the uncertainties concerning climate sensitivity are not captured). The projected flood impacts are lower than in the unmitigated IS92a world considered above (Table 5). This primarily reflects lower projections of global sea level rise, higher GDP/capita scenarios, which are used to estimate flood defence standards, and in some cases lower exposed populations (Nicholls, 2004). While the A1FI world experiences the largest rise in sea level, the A2 world consistently has the greatest flood impacts due to sea level rise for all realistic scenario combinations. This reflects that the A2 world is most vulnerable to coastal flooding of the four SRES worlds considered, with the largest exposed population and the lowest GDP/capita. This clearly demonstrates that socio-economic development pathways will have a significant influence on vulnerability to sea level rise.

In all cases for flooding considered above, absolute impacts are generally largest in five regions: the southern Mediterranean, West Africa, East Africa, South Asia and South-East Asia. However, the small island regions of the Caribbean, Indian Ocean and Pacific Ocean tend to be most vulnerable in relative terms. Collectively, these results show that sea level rise could have a profound impact on the incidence of flooding, without any change in storminess. The implications for small island regions are particularly worrying.

TABLE 5
As Table 4, but showing people flooded as a function of stabilisation scenario (IS92a (unmitigated), S750 and S550) and SRES scenario (A1FI, A2, B1 and B2) based on scenarios developed from the HadCM2 and HadCM3 AOGCM.

| Climate Scenario (climate model) | SLR Relative to 1990 (cm) | Socio-Economic Scenario | People Flooded (millions/year) | | | | Source |
|---|---|---|---|---|---|---|---|
| | | | Constant Protection | | In Phase Evolving Protection | | |
| | | | No SLR | SLR | No SLR | SLR | |
| IS92a (HadCM2) | 37 | IS92a | 36 | 240 | 13 | 94 | Arnell et al (2002) |
| S750 (HadCM2) | 28 | IS92a | 36 | 132 | 13 | 35 | |
| S550 (HadCM2) | 25 | IS92a | 36 | 103 | 13 | 18 | |
| A1FI (HadCM3) | 34 | A1 | 25 | 127 | 0 | 10 | Nicholls (2003) |
| A2 (HadCM3) | 28 | A2 | 49 | 169 | 19 | 34 | |
| B1 (HadCM3) | 22 | B1 | 25 | 79 | 0 | 3 | |
| B2 (HadCM3) | 25 | B2 | 34 | 127 | 1 | 6 | |

Globally, coastal wetlands are declining quite rapidly (at 1%/year), largely due to indirect and direct human activities, with sea level rise only playing a minor role in the losses at this scale (Hoozemans et al., 1993). Arnell et al (2002) looked at potential impacts under unmitigated and stabilisation scenarios, taking the same IS92a scenarios as discussed under coastal flooding. Losses due to unmitigated sea level rise only are 12%[6] of the global stock by the 2080s with further losses expected in the 22$^{nd}$ century as the rate of rise is accelerating. Under S750 and S550, the losses are 8% and 6%, respectively. Sea level rise continues and accelerates slightly, so by the 2230s, wetland losses had increased to 12–13% and 8–9%, respectively. No further losses are expected due to sea level rise after the 2230s as the rate of sea level rise has peaked. When compared to existing trends of indirect and direct human destruction, sea level rise is not the primary concern for wetland conservation. Rather, it is a significant additional stress which worsens the prognosis for wetlands, although ironically, better conservation of wetlands will make sea level rise and climate change a more important issue for long-term management. Regional losses would be most severe on the Atlantic coast of North and Central America, the Caribbean, the Mediterranean, the Baltic and most small island regions (Nicholls et al., 1999).

[6] All wetland losses are mid estimates.

In conclusion, these studies illustrate that sea level rise could produce important impacts on people (via flooding) and coastal ecosystems during the 21st century. Reviews of national assessments also suggest that sea level rise might produce significant impacts for small islands, deltaic settings, all coastal wetlands, and developed sandy shorelines (McLean et al., 2001; Nurse et al., 2001). Further studies are required at national and smaller scales to identify specific impacts and response options (Section 5.3), and at regional and global scales for studies of mitigation, comparative vulnerability and targeting of adaptation efforts (Nicholls, 2002c). The Asia-Pacific Vulnerability Assessment (Mimura, 2000) and the Dynamic and Interactive Assessment of National, Regional and Global Vulnerability of Coastal Zones to Climate Change and Sea-Level Rise (DINAS-COAST) Project (Hinkel and Klein, 2003; Vafeidis et al., 2003) are examples of new impact models for this purpose.

**Post-2100 Impacts**

The 'sea level commitment' already discussed together with extreme or irreversible effects on sea level such as the melting of Greenland or the collapse of the West Antarctic Ice Sheet. This raises important long-term questions about impacts beyond the 21st century (Smith et al., 2001). Even with stabilisation, sea levels will continue to rise for hundreds of years. Therefore, many impacts of sea level rise may only be delayed, rather than avoided, particularly those that depend on the magnitude rather than the rate of rise. The larger rises in sea level that are possible after 2100 (Section 4.1) have not been investigated systematically, although some early often forgotten research considered the impacts of a large 5 to 6 m rise in sea level on the USA (Schneider and Chen, 1980). Such impacts are dramatic with large parts of the East and Gulf Coasts being submerged, particular in Texas, Louisiana, Florida and North Carolina. The post-2100 sea level rise scenarios that are being developed require more detailed exploration in terms of possible impacts. For example, when might the inhabitants of the Maldives and other atoll nations be forced to abandon their homes? In this context, the potential for adaptation, including international aid mechanisms needs to be considered. Equally, could any coastal society cope with worst-case abrupt rise scenarios of sea level?

### *5.3. Responding to Sea level Rise*

Based on the preceding discussion, a precautionary response to sea level rise and climate change in coastal areas would be to identify an appropriate mixture of mitigation and adaptation. Mitigation could greatly reduce the risks associated with sea level rise and climate change beyond the 21st century (Church et al., 2001). Adaptation acts to reduce the impacts of sea level rise and climate change, as well as other changes (as well as exploiting benefits) (Smit et al., 2001).

Adaptation was considered extensively in the IPCC TAR, both in a generic sense and in specific contexts such as coasts (e.g. McLean et al., 2001). Planned adaptation options to sea level rise are usually presented as one of three generic approaches (Klein et al., 2000; 2001):

- *(Planned) Retreat* – all natural system effects are allowed to occur and human impacts are minimised by pulling back from the coast;

- *Accommodation* – all natural system effects are allowed to occur and human impacts are minimised by adjusting human use of the coastal zone;
- *Protection* – natural system effects are controlled by soft or hard engineering, reducing human impacts in the zone that would be impacted without protection.

In practice, many responses may be hybrid and combine elements of more than one approach.

Given that many decisions at the coast have long-term implications, the coastal zone is an area where anticipatory adaptation needs to be carefully considered. Examples of anticipatory adaptation in coastal zones include upgraded flood defences and waste water discharges, higher levels for land claim and new bridges, and building setbacks to prevent development in hazardous areas (Bijlsma et al., 1996; McLean et al., 2001).

Importantly, studies of adaptation to coastal problems stress that it is a process, rather than just the implementation of technical options (Klein et al., 2000). Four stages in the adaptation process related to (1) *information and awareness building,* (2) *planning and design,* (3) *evaluation,* and (4) *monitoring and evaluation* were evident within multiple policy cycles, as well as various constraints on approaches to adaptation due to broader policy and development goals. With a few exceptions, climate change exacerbates existing pressures and problems, and it is fundamental to consider adaptation to climate change in the context of managing existing problems, rather than as a separate activity.

Broader measures in terms of enhancing adaptive capacity and creating an environment where adaptation can more easily occur are also vital (cf. Smit et al., 2001; Yohe and Tol, 2002). This should include more exploration of the links between impacts, adaptation and different development pathways (Nicholls, 2004). Even in Europe, it is striking how many countries take no account of sea level rise in coastal management, including historically observed trends (Tol et al., 2004). Capacity building should focus on developing coastal management as recognised in the IPCC TAR, most particularly in the more vulnerable areas such as the small islands (e.g. Barnett, 2001). In this regard international initiatives such as Caribbean Planning Adaptation to Climate Change (CPACC) and Pacific Islands Climate Change Assistance Programme (PICCAP) are especially welcome.

### 5.4. *Conclusions of Impacts*

Our understanding of the impacts of sea level rise has improved significantly over the last decade, but significant progress remains to be made. Key research questions for the future include:

- A better understanding of the consequences and impacts of sea level rise during the 20$^{th}$ century, including human responses to these changes.
- Better assessments of future impacts of sea level rise for the 21$^{st}$ century, including a focus on socio-economic impacts. This is required at a range of scales including global and regional assessment for climate policy and national and more local assessment to examine adaptation options.

- Better assessments of potential impacts due to extreme sea level rise or long-term rise due to the sea level commitment, as these impacts often seem to be ignored or downplayed in climate policy process.
- More assessment of the adaptation process, including comparative studies (e.g. Rupp and Nicholls, 2003).

Collectively, these studies would allow policy questions such as the cost of sea level rise, or the options for small islands to be considered in a much more comprehensive manner than has been possible to date.

## 6. Conclusions

Global tide gauge data sets, represented by the PSMSL and other data banks, have continued to expand in recent years, aided by regional and worldwide developments in the IOC GLOSS programme. This expansion has taken place alongside developments in new geodetic techniques for monitoring vertical land movements which will eventually provide estimates of 'absolute' sea level change at gauge sites. Together with these developments in *in situ* monitoring systems, there has been important progress in Earth Observation techniques. In particular, satellite radar altimetry has been shown to have the capability to provide truly-global sea level change estimates, rather than only at coastlines as for tide gauges, although gauges will continue to be required for continuity, local coastal use and for altimeter calibration. In addition, space gravity missions will provide precise measurements of the geoid and of temporal gravity variations with application to a wide range of sea level science. The challenge will be to combine the various *in situ* and space measurement techniques into one global monitoring system.

A synthesis of analyses of historical sea level data sets by the IPCC TAR showed that global sea level rose by 10–20 cm during the 20$^{th}$ century with a slow acceleration between the 19$^{th}$ and 20$^{th}$ centuries. Considerable study and debate remains as to the origin of the sea level change over this period, and it is clear that the budget estimate of Table 1 needs significant refinement. These tide gauge-derived estimates compare to a value of approximately 2.3 mm/year obtained from the T/P mission for the past decade, one major forcing of which appears to have been upper-ocean temperature change, although glacial and hydrological changes may also have contributed. The TAR studies also indicated that global sea level rise during the 21$^{st}$ century could be of order 50 cm, with a considerable amount of spatial variation about the average along coastlines due to vertical land movements and to regional adjustments of the ocean circulation.

While the study of past and potential future mean sea level change contains many uncertainties, the TAR identified the need for more research on an even more difficult topic: that of changes in extreme sea levels (and waves) as a result of climate change. It is extreme events which result in most coastal impacts, but the above discussion has demonstrated the difficulty in developing confident predictions of spatial and temporal evolution of the statistics of extremes and of 'risk'.

Impact analyses have mainly focussed on the impacts of sea level rise without considering regional effects or other climate change. Taking this simple view, the available analyses indicate that global-mean sea level rise scenarios could have important impacts on the world's coastal zones. For instance, the incidence of coastal flooding could increase substantially relative to the situation without sea level rise, although the impacts will be controlled by the magnitude of sea level rise, the socio-economic situation and the ability to adjust/adapt. In all cases, relatively high impacts in small island states are a particular concern. The available results suggest that despite the large uncertainties, a response to sea level rise would be prudent, rather than waiting for impacts to occur. Mitigation of greenhouse gas emissions will reduce but not stop the rise in sea level, so mitigation should be combined with adaptation. The role of adaptation received much greater attention in the IPCC TAR and there is a need to better understand the potential for and limitations of adaptation to sea level rise and climate change, especially in the more vulnerable areas such as small islands and subsiding deltas. Developing such a comprehensive understanding of impacts and adaptation to sea level rise will require a continuing partnership between the natural, engineering and social sciences.

The 'sea level commitment' referred to above also raises important longer-term questions for the 22nd century and beyond. Available model runs suggest that even with significant mitigation (i.e. deliberate reduction of greenhouse gas emissions), sea levels will continue to rise for hundreds of years. Therefore, many impacts of sea level rise may only be delayed, rather than avoided by mitigation. The possibility (even if an unlikely one) of an abrupt sea level rise, as discussed by the IPCC TAR, is also of great concern in terms of potential impacts, as rapid change is most likely to overwhelm the capacity of coastal nations to adapt and lead to major impacts. All of these factors have important implications for long term management of the coastal zone.

## Acknowledgements

We thank Anny Cazenave, Bruce Douglas, Vivien Gornitz, Gary Mitchum and David Pugh for useful comments on the paper. Jason Lowe (Hadley Centre) provided Figure 18.4. JMG's work was supported by the U.K. Department of the Environment, Food and Rural Affairs under contract PECD 7/12/37 and by the U.K. Government Meteorological Research Contract.

## Bibliography

Antonov, J.I., S. Levitus, and T.P. Boyer, 2002. Steric sea level variations during 1957–1994: importance of salinity. *J. Geophys. Res.,* **107(C12),** doi:10.1029/2001JC000964.

Arendt, A.A., K.A. Echelmeyer, W.D. Harrison, C.S. Lingle, and V.B. Valentine, 2002. Rapid wastage of Alaska glaciers and their contribution to rising sea level. *Science,* **297,** 382–386.

Arnell, N.W., M.G.R Cannell, M. Hulme, R.S. Kovats, J.F.B. Mitchell, R.J. Nicholls, M.L. Parry, M.T.J. Livermore, and A. White, 2002. The consequences of $CO_2$ stabilisation for the impacts of climate change. *Clim. Chang.,* **53,** 413–446.

Arnell, N.W., M.J.L. Livermore, S. Kovats, P. Levy, R. Nicholls, M.L. Parry, and S. Gaffin, 2003. Climate and socio-economic scenarios for climate change impacts assessments: characterising the SRES storylines. *Glob. Environ. Chang.* (in preparation).

Bacon, S. and D.J.T Carter, 1993. A connection between mean wave height and atmospheric pressure gradient in the North Atlantic. *Int. J. Climatol.,* **13,** 423–436.

Balmino, G., R. Rummel, P. Visser, and P. Woodworth, 1999. *Gravity Field and Steady-State Ocean Circulation Mission. Reports for assessment: the four candidate Earth Explorer Core Missions.* European Space Agency Report SP-1233(1), 217pp.

Barnett, J, 2001. Adapting to climate change in Pacific island countries: the problem of uncertainty. *World Development,* **29,** 977–993.

Bi, D., W.F. Budd, A.C. Hirst, and X. Wu, 2001. Collapse and reorganisation of Southern Ocean overturning under global warming in a coupled model. *Geophys. Res. Lett.,* **28,** 3927–3930.

Bijlsma, L., C.N. Ehler, R.J.T. Klein, S.M. Kulshrestha, R.F. McLean, N. Mimura, R.J. Nicholls, L.A. Nurse, H. Perez Nieto, E.Z. Stakhiv, R.K. Turner, and R.A. Warrick. 1996. Coastal zones and small islands. In *Impacts, adaptations and mitigation of climate change: scientific-technical analyses,* R.T. Watson, M.C. Zinyowera and R.H. Moss, eds. Cambridge University Press, Cambridge, pp. 289–324.

Bingley, R.M., A.H. Dodson, N.T. Penna, F.N. Teferle, and T.F. Baker, 2001. Monitoring the vertical land movement component of changes in mean sea level using GPS: results from tide gauges in the UK. *J. Geospatial Eng.,* **3,** 9–20.

Bird, E.C.F., 1993. *Submerging coasts: the effects of a rising sea level on coastal environments.* Wiley: Chichester,184pp.

Boesch, D.F., M.N. Josselyn, A.J. Mehta, J.T. Morris, W.K. Nuttle, C.A. Simenstad, and D.J.P. Swift. 1994. Scientific assessment of coastal wetland loss, restoration and management in Louisiana. *J. Coastal Res.,* **Special Issue No. 20,** 103pp.

Bryan, K., 1996. The steric component of sea level rise associated with enhanced greenhouse warming: a model study. *Clim. Dyn.,* **12,** 545–55.

Cabanes, C., A. Cazenave, and C. Le Provost, 2001. Sea level rise during past 40 years determined from satellite and in situ observations. *Science,* **294,** 840–842.

Cabanes, C. 2002. Les variations du niveau moyen global et régional de la mer: observation par altimetrie et marégraphie, analyse et interpretation physique. Ph.D. thesis, University of Toulouse, France.

Cahoon, D.R., D.J. Reed, and J.W.Day, Jr., 1995. Estimating shallow subsidence in microtidal saltmarshes of the southeastern United States: Kaye and Barghoorn revisited. *Mar. Geol.,* **128,** 1–9.

Cahoon, D.R., D.J. Reed, and J.W. Day, Jr., 1999. The influence of surface and shallow subsurface processes on wetland elevation: a synthesis. *Current Topics in Wetland Biogeochemistry,* **3,** 72–88.

Capobianco, M., H.J. De Vriend, R.J. Nicholls, and M.J.F. Stive, 1999. Coastal area impact and vulnerability assessment: a morphodynamic modeller's point of view. *J. Coastal Res.,* **15,** 701–716.

Cazenave, A., L. Soudarin, J-F. Cretaux, and C. Le Provost, 1999. Sea level changes from Topex-Poseidon altimetry and tide gauges, and vertical crustal motions from DORIS. *Geophys. Res. Lett.,* **26,** 2077–2080.

Cazenave, A., C. Cabanes, K.D. Minh, M-C. Gennero, and C. Le Provost, 2003. Present-day sea level change: observations and causes. *Space Science Reviews,* **108,** 131–144.

Cazenave, A., and R.S. Nerem, 2003. Present-day sea level change: observations and causes. *Reviews of Geophysics* **42,** RG 3001, doi: 10.1029/2003 RG000139.

Chambers, D.P., J. Chen, R.S. Nerem, and B.D. Tapley, 2000. Interannual mean sea level change and the Earth's water mass budget. *Geophys. Res. Lett.,* **27,** 3073–3076.

Chambers, D.P., C.A. Mehlhaff, T.J. Urban, D. Fujii, and R.S. Nerem, 2002. Low-frequency variations in global mean sea level: 1950–2000. *J. Geophys. Res.,* **107(C4),** doi 10.1029/2001JC001089.

Church, J.A., 2001. How fast are sea levels rising? *Nature,* **294,** 802–803.

Church, J.A., J.M. Gregory, P. Huybrechts, M. Kuhn, K. Lambeck, M.T. Nhuan, D. Qin, and P.L. Woodworth, 2001. Changes in sea level. In *Climate Change 2001: The Scientific Basis. Contribution of Working Group I to the Third Assessment Report of the Intergovernmental Panel on Climate Change,* J.T. Houghton, Y. Ding, D.J. Griggs, M. Noguer, P.J. van der Linden, X. Dai, K. Maskell and C.A. Johnson, eds. Cambridge University Press, Cambridge, 881pp.

Church, J.A., N.J. White, R. Coleman, K. Lambeck, and J.X. Mitrovica, 2004. Estimates of the regional distribution of sea-level rise over the 1950 to 2000 period. *J.Clim.* **17,** 2609–2625.

Cowell, P.J., M.J.F. Stive, A.W. Niedoroda, H.J. de Vriend, D.J.P. Swift, G.M. Kaminsky, and M. Capobianco, 2003. The coastal-tract (part 1): a conceptual approach to aggregated modelling of low-order coastal change. *J. Coastal Res.* **19,** 812–827.

Cubasch, U., G.A. Meehl, G.J. Boer, R.J. Stouffer, M. Dix, A. Noda, C.A., S.C.B. Raper, and K.S. Yap, 2001. Projections of future climate change. In *Climate change 2001: the scientific basis. Contribution of Working Group I to the Third Assessment Report of the Intergovernmental Panel on Climate Change,* J.T. Houghton, Y. Ding, D.J. Griggs, M. Noguer, P.J. van der Linden, X. Dai, K. Maskell and C.A. Johnson, eds. Cambridge University Press, Cambridge, 881pp.

Di Donato, G, A.M. Negredo, R. Sabadini, and L.L.A. Vermeersen, 1999. Multiple processes causing sea-level rise in the central Mediterranean. *Geophys. Res. Lett.,* **26,** 1769–1772.

Dong, X., P.L. Woodworth, P. Moore, and R. Bingley, 2002. Absolute calibration of the TOPEX/POSEIDON altimeters using UK tide gauges, GPS and precise, local geoid-differences. *Mar. Geod.,* **25,** 189–204.

Douglas, B.C. and W.R. Peltier, 2002. The puzzle of global sea-level rise. *Physics Today,* **55(March),** 35–40.

Flather, R.A, T.F. Baker, P.L. Woodworth, I.M. Vassie, and D.L. Blackman, 2001. *Integrated effects of climate change on coastal extreme sea levels.* POL Internal Document No.140, pp. 20.

Folland, C.K., T.R. Karl, J.R. Christy, R.A. Clarke, G.V. Gruza, J. Jouzel, M.E. Mann, J. Oerlemans, M.J. Salinger, and S.-W. Wang, 2001. Observed climate variability and change. In *Climate Change 2001: The Scientific Basis. Contribution of Working Group I to the Third Assessment Report of the Intergovernmental Panel on Climate Change,* J.T. Houghton, Y. Ding, D.J. Griggs, M. Noguer, P.J. van der Linden, X. Dai, K. Maskell and C.A. Johnson, eds. Cambridge University Press, Cambridge, 881pp.

French, P.W., 1997. *Coastal and estuarine management.* Routledge, London, 251pp.

Fu, L.-L. and A. Cazenave (eds.), 2001. *Satellite altimetry and earth sciences. A handbook of techniques and applications.* Academic Press, San Diego.

Gornitz, V., 2000. Impoundment, groundwater mining, and other hydrologic transformations: impacts on global sea level rise. In *Sea level rise: history and consequences,* B.C. Douglas, M.S. Kearney and S.P. Leatherman (eds.), Academic Press, pp. 97–119.

Gornitz, V., S. Couch, and E.K. Hartig, 2002. Impacts of sea level rise in the New York City metropolitan area. *Global and Planetary Change,* **32,** 61–88.

GRACE, 1998. *Gravity Recovery and Climate Experiment science and mission requirements document.* Revision A, JPLD-15928, NASA's Earth System Science Pathfinder Program.

Gregory, J.M., J.A. Church, G.J. Boer, K.W. Dixon, G.M. Flato, D.R. Jackett, J.A. Lowe, S.P. O'Farrell, E. Roeckner, G.L. Russell, R.J. Stouffer, and M. Winton, 2001. Comparison of results from several AOGCMs for global and regional sea-level change 1900–2100, *Climate Dynamics,* **18,** 241–253.

Hargreaves, J.C., D.J.T. Carter, P.D. Cotton, and J. Wolf, 2002. Using the SWAN wave model and satellite altimeter data to study the influence of climate change at the coast. *The Global Atmosphere and Ocean System,* **8,** 41–66.

Hartig, E.K., A Kolker, D. Fallon, and F. Mushacke, 2001. Wetlands. In *Climate change and a global city: the potential consequences of climate variability and change – Metro East Coast.* Report for the U.S. Global Change Research Program, National Assessment of the Potential Consequences of Climate Variability and Change for the United States. C. Rosensweig and W.D. Solecki, eds. Columbia Earth Institute, New York, pp. 67–86.

Hilmer, M. and P. Lemke, 2000. On the decrease of Arctic sea ice volume. *Geophys. Res. Lett.,* **27,** 3751–3754.

Hinkel, J., and R.J.T. Klein, 2003. DINAS-COAST: developing a method and a tool for dynamic and interative assessment. *LOICZ Newsletter,* **No. 27 (June 2003),** 1–4.

Holgate, S.J., and P.L. Woodworth, 2004. Evidence for enhanced coastal sea level rise during the 1990s. *Geophys. Rev. Lette.,* **31,** L07305, doi: 10.1029/2004 GL 019626.

Hoozemans, F.M.J., M. Marchand, H.A. Pennekamp, M.J.F. Stive, R. Misdorp, and L. Bijlsma 1992. The impacts of sea level rise on coastal areas: some global results. In *Global climate change and the rising challenge of the sea,* Proceedings of the IPCC Workshop held at Margarita Island, Venezuela, March 1992. J. O'Callahan, ed. NOAA, Silver Spring, Maryland, pp. 607–622.

Hoozemans, F.M.J., M. Marchand, and H.A. Pennekamp, 1993. *A global vulnerability analysis: vulnerability assessment for population, coastal wetlands and rice production on a global scale.* 2nd edition. Delft Hydraulics, The Netherlands.

Hulme, M., G. Jenkins, X. Lu, J.R. Turnpenny, T.D. Mitchell, R.G. Jones, J. Lowe, J.M. Murphy, D. Hassell, P. Boorman, R. McDonald, and S. Hill, 2002. *Climate change scenarios for the United Kingdom: The UKCIP02 Scientific Report.* Tyndall Centre for Climate Change Research, 120pp.

IOC, 1998. *Global Sea Level Observing System (GLOSS) implementation plan-1997.* Intergovernmental Oceanographic Commission, Technical Series, No. 50, 91pp. & Annexes.

IOC, 2003. *A Report on the Status of the GLOSS Programme and a Proposal for taking the Programme Forward.* Intergovernmental Oceanographic Commission report IOC/INF-1190, 41pp.

IPCC, 2000. *Special Report on Emissions Scenarios: A Special Report of Working Group III of the Intergovernmental Panel on Climate Change.* Cambridge University Press, Cambridge, 599 pp.

Johnson, H.L. and D.P. Marshall, 2002. A theory for surface Atlantic response to thermohaline variability. *J. Phys. Oceanogr.,* **32,** 1121–1132.

Kennedy, V.S., R.R. Twilley, J.A. Kleypas, J.H. Cowan Jr., and S.R. Hare, 2002. Coastal and marine ecosystems and global climate change: potential effects on U.S. resources. Pew Center on Climate Change, Arlington, Virginia, USA.

Klein, R.J.T. and Nicholls, R.J., 1999. Assessment of coastal vulnerability to sea-level rise. *Ambio,* **28,** 182–187.

Klein, R.J.T., J. Aston, E.N. Buckley, M. Capobianco, N. Mizutani, R.J. Nicholls, P.D. Nunn, and S. Ragoonaden, 2000. Coastal-adaptation technologies. In *IPCC special report on methodological and technological issues in technology transfer,* B. Metz, O.R. Davidson, J.W. Martens, S.N.M. van Rooijen and L.L. Van Wie McGrory, eds. Cambridge University Press, Cambridge, pp. 349–372.

Klein, R.J.T., R.J. Nicholls, S. Ragoonaden, M. Capobianco, J. Aston, and E.N. Buckley, 2001. Technological options for adaptation to climate change in coastal zones. *J. Coastal Res.,* **17,** 531–543.

Knutti, R., and T.F. Stocker, 2001. Influence of the thermohaline circulation on projected sea level rise. *J. Clim.,* **13,** 1997–2001.

Koblinsky, C.J., P. Gaspar, and G. Lagerloef, 1992. *The future of spaceborne altimetry: oceans and climate change. A long term strategy.* Joint Oceanographic Institutions Inc., 75pp.

Komar, P.D., 1998. *Beach and nearshore sedimentation.* Second Edition. Prentice Hall, Upper Saddle River, New Jersey, USA.

Krabill, W., W. Abdalati, E. Frederick, S. Manizade, C. Martin, J. Sonntag, R. Swift, R. Thomas, W. Wright, and J. Yungel, 2000. Greenland ice sheet: high-elevation balance and peripheral thinning. *Science,* **289,** 428–430.

Laxon, S., N. Peacock, and D. Smith, 2003. High interannual variability of sea ice thickness in the Arctic region. *Nature,* **425,** 947–949.

Leatherman, S.P. 2001. Social and economic costs of sea level rise. In *Sea-level rise: history and consequences,* B.C. Douglas, M.S. Kearney, and S.P. Leatherman, eds. Academic Press, San Diego, pp. 181–223.

Leatherman, S.P., R. Chalfont, E. Pendleton, S. Funderburk, and T. McCandles. 1995. *Vanishing Lands: Sea Level, Society and the Chesapeake Bay,* US Fish and Wildlife Service, Annapolis, Maryland.

Leatherman, S.P., B.C. Douglas, and J.L. LaBrecque, 2003. Sea level and coastal erosion requires large-scale monitoring. *EOS, Transactions of the American Geophysical Union,* **84,** 13 and 16 (14 January 2003).

Levitus, S., C. Stephens, J.I. Antonov, and T.P. Boyer, 1998. *World ocean database 1998.* U.S. Government Printing Office, Washington, D.C., 346pp.

Levitus, S., J.I. Antonov, T.P. Boyer, and C. Stephens, 2000. Warming of the world ocean. *Science,* **287,** 2225–2229.

Levitus, S., J.I. Antonov, J. Wang, T.L. Delworth, K.W. Dixon, and A.J. Broccoli, 2001. Anthropogenic warming of the Earth's climate system. *Science,* **292,** 267–270.

Lowe, J.A., J.M. Gregory, and R.A. Flather, 2001. Changes in the occurrence of storm surges around the United Kingdom under a future climate scenario using a dynamic storm surge model driven by the Hadley Centre climate models. *Clim. Dyn.,* **18,** 179–188.

Lowe, J.A., and J.M. Gregory, In preparation. Sea-level variability in the Hadley Centre coupled ocean-atmosphere model on interannual time scales: characteristics and comparison with observations.

McLean, R., A. Tsyban, V. Burkett, J.O. Codignotto, D.L. Forbes, N. Mimura, R.J. Beamish, and V. Ittekkot, 2001. Coastal Zone and Marine Ecosystems. In *Climate Change 2001: Impacts, Adaptation and Vulnerability,* J.J. McCarthy, O.F. Canziani, N.A. Leary, D.J. Dokken, and K.S. White, eds. Cambridge University Press, Cambridge, pp. 343–380.

Meier, M.F., and J.M. Wahr, 2002. Sea level is rising: do we know why? *Proceedings of the National Academy of Sciences,* **99,** 6524–6526.

Miller, L., and B.C. Douglas, 2004. Mass and volume contributions to 20th century global sea level rise. *Nature,* **428,** 406–409.

Milly, P.C.D., A. Cazenave, and M.C. Gennero, 2003. Contribution of climate-driven change in continental water storage to recent sea level rise. *Proc. Nat. Ac. Sci.* (submitted).

Milne, G.A., J.L. Davis, J.X. Mitrovica, H.-G. Scherneck, J. Johansson, M. Vermeer, and H. Koivula, 2001. Space-geodetic constraints on glacial isostatic adjustment in Fennoscandia. *Science,* **291,** 2381–2385.

Mimura, N. (ed.), 1999. National assessment results of climate change: impacts and responses. *Clim. Res.,* **12,** 77–230.

Mimura, N., 2000. Distribution of vulnerability and adaptation in the Asia and Pacific Region. In *Global Change and Asia-Pacific Coasts,* N. Mimura and H. Yokoki, eds. Proceedings of APN/SURVAS/LOICZ Joint Conference on Coastal Impacts of Climate Change and Adaptation in the Asia-Pacific Region, Kobe, Japan, November 14–16, 2000. Asian-Pacific Network for Global Change Research, Kobe and the Center for Water Environment Studies, Ibaraki University, pp. 21–25.

Mitchum, G.T., 2000. An improved calibration of satellite altimetric heights using tide gauge sea levels with adjustment for land motion. *Mar. Geod.,* **23,** 145–166.

Mitchum, G.T., R. Cheney, L.-L. Fu, C. Le Provost, Y. Menard, and P.L. Woodworth, 1999a. *The future of sea surface height observations.* Proceedings of the conference on The Ocean Observing System for Climate, St.Raphael, France, 18–22 October 1999.

Mitchum, G.T., R. Cheney, L.-L. Fu, C. Le Provost, Y. Menard, and P.L. Woodworth, 1999b. Sea surface height observations from altimeters and tide gauges. *CLIVAR Exchanges,* **4,** 11–16.

Mitrovica, J.X. and B.L.A. Vermeersen (eds.), 2002. *Ice sheets, sea level and the dynamic Earth. American Geophysical Union,* Washington, D.C.

Mitrovica, J.X., M.E. Tamisiea, J.L. Davis, and G.A. Milne, 2001. Recent mass balance of polar ice sheets inferred from patterns of global sea-level change. *Nature,* **409,** 1026–1029.

Mitrovica, J.X., 2003. Recent controversies in predicting post-glacial sea-level change. *Quaternary Science Reviews,* **22,** 127–133.

Munk, W., 2002. Twentieth century sea level: an enigma. *Proceedings of the National Academy of Sciences,* **99,** 6550–6555.

Munk, W., 2003. Ocean freshening, sea level rising. *Science,* **300,** 2041–2043.

Nerem, R.S. and G.T. Mitchum, 2001. Sea level change. In *Satellite altimetry and earth sciences. A handbook of techniques and applications,* L.-L. Fu and A. Cazenave, eds. Academic Press, San Diego, pp. 329–349.

Neumann, J.E., G. Yohe, R.J. Nicholls, and M. Manion, 2001. Sea level rise and its effects on coastal resources. In *Climate change: science, strategies and solutions,* E. Claussen, ed. Brill, Boston, pp. 43–62.

Nicholls, R.J., 2002a. Rising sea levels: potential impacts and responses. In *Global environmental change,* R. Hester and R.M. Harrison, eds. Issues in Environmental Science and Technology, Number 17, Royal Society of Chemistry, Cambridge, pp. 83–107.

Nicholls, R.J., 2002b. Analysis of global impacts of sea-level rise: a case study of flooding. *Physics and Chemistry of the Earth,* **27,** 1455–1466.

Nicholls, R.J., 2002c. Climate change and coastal zones. In *Global Climate Change and Sustainable Development,* House of Commons International Development Committee. Third Report of Session 2001–02, Volume II. Minutes of Evidence and Appendices (HC 518-II), Stationary Office, London, pp. EV46-EV48.

Nicholls, R.J., 2004. Coastal flooding and wetland loss in the 21st century: changes under the SRES climate and socio-economic scenarios. *Glob. Environ. Chang.* **14,** 69–86.

Nicholls, R.J., Hoozemans, F.M.J., and Marchand, M. 1999. Increasing flood risk and wetland losses due to global sea-level rise: Regional and global analyses. *Glob. Environ. Chang.,* **9,** S69-S87.

NRC, 1997. *Satellite gravity and geosphere.* National Academy Press, Washington, D.C.

Nurse, L., G. Sem, J.E. Hay, A.G. Suarez, P.P. Wong, L. Briguglio, and S. Ragoonaden, 2001. Small island states. In *Climate Change 2001: Impacts, Adaptation and Vulnerability,* J.J. McCarthy, O.F. Canziani, N.A. Leary, D.J. Dokken, and K.S. White, eds. Cambridge University Press, Cambridge, pp. 843–875.

Peltier, W.R., 2001. Global glacial isostatic adjustment. In *Sea level rise: history and consequences,* B.C. Douglas, M.S. Kearney, and S.P. Leatherman, eds. Academic Press, San Diego, pp. 65–95.

Penland, S. and Ramsey, K.E.,1990. Relative sea-level rise in Louisiana and the Gulf of Mexico: 1908–1988. *J. Coastal Res.,* **6,** 323–342.

Plag, H-P. and H-U. Jüttner, 2001. Inversion of global tide gauge data for present-day ice load changes. In *Proceedings of the Second International Symposium on Environmental Research in the Arctic and Fifth Ny-Alesund Scientific Seminar,* T. Yamanouchi, ed. Memoirs of the National Institute of Polar Research, Special Issue No. 54, 301–317.

Reichert, B.K., L. Bengtsson, and J. Oerlemans, 2002. Recent glacier retreat exceeds internal variability. *J. Clim.,* **15,** 3069–3081.

Rignot, E. and R.H. Thomas, 2002. Mass balance of polar ice sheets. *Science,* **297,** 1502–1506.

Rignot, E., A. Rivera, and G. Casassa, 2003. Contribution of the Patagonian icefields of South America to sea level rise. *Science,* **302,** 434–437.

Rupp, S. and R.J. Nicholls, 2003. The application of managed retreat: A comparison of England and Germany. In *Dealing with flood risk,* M. Marchand, K.V. Heynert, H. van der Most, H. and W.E. Penning, eds. Proceedings of an interdisciplinary seminar on the regional implications of modern flood management, Delft Hydraulics Select Series 1/2003. Delft University Press, Delft, 42–51.

Sahagian, D., 2000. Global physical effects of anthropogenic hydrological alterations: sea level and water redistribution. *Global Planet. Change,* **25,** 39–48.

Sanchez-Arcilla, A., J. Jimenez, and H.I. Valdemoro, 1998. The Ebro delta: morphodynamics and vulnerability. *J. Coastal Res.,* **14,** 754–772.

Sanchez-Arcilla, A., P. Hoekstra, J. Jimenez, E. Kaas, and A. Maldonado, 2000. Climate change implications for coastal processes. In *Sea level change and coastal processes. Implications for Europe,* D. Smith, S.B. Raper, S. Zerbini, and A. Sanchez-Arcilla, eds. EUR 19337, European Commission, Brussels, pp. 173–213.

Schneider, S.H. and Chen, R.S., 1980. Carbon dioxide flooding: physical factors and climatic impact. *Annual Review of Energy,* **5,** 107–140.

Sharples, J. and J.H. Simpson, 1996 The influence of the springs-neaps cycle on the position of shelf sea fronts. In *Buoyancy effects on coastal and estuarine dynamics,* D.G. Aubrey and C.T. Friedrichs eds. American Geophysical Union, Washington D.C., pp. 71–82.

Shepherd, A., D.J. Wingham, J.A.D. Mansley, and H.F.J. Corr, 2001. Inland thinning of Pine Island Glacier, West Antarctica. *Science,* **291,** 862–864.

Simpson, J.H., J. Brown, J. Mathews, and G. Allen, 1990. Tidal straining, density currents, and stirring in the control of estuarine stratification. *Estuaries,* **13,** 125–132.

Small, C., and R.J. Nicholls, 2003. A global analysis of human settlement in coastal zones. *J. Coastal Res.,* **19,** 584–599.

Smit, B., O. Pilifosova, I. Burton, B. Challenger, S. Huq, R.J.T. Klein, and G. Yohe, 2001 Adaptation to climate change in the context of sustainable development and equity. In *Climate change 2001: impacts, adaptation and vulnerability,* J.J. McCarthy, O.F. Canziani, N.A. Leary, D.J. Dokken, and K.S. White, eds. Cambridge University Press, Cambridge, pp. 877–912.

Smith, J.B., H-J. Schellnhuber, M.M.Q. Mirza, S. Fanknauser, R. Leemans, L. Erda, L. Ogallo, B. Pittock, R. Richels, C. Rosenzweig, U. Safriel, R.S.J. Tol, J. Weyant, and G. Yohe, 2001. Vulnerability to climate change and reasons for concern: a synthesis. In *Climate change 2001: impacts, adaptation and vulnerability,* J.J. McCarthy, O.F. Canziani, N.A. Leary, D.J. Dokken, and K.S. White, eds. Cambridge University Press, Cambridge, pp. 913–967.

Smith, K. and R. Ward, 1998. *Floods: physical processes and human impacts,* Wiley, Chichester, 382pp.

Stewart, R.W., B.D. Bornhold, H. Dragert, and R.E. Thomson, 1998. Sea level change. *The Sea,* **10,** 191–211.

Stive, M.J.F. S.J.C. Aarninkoff, L. Hamm, H. Hanson, M. Larson, K. Wijnberg, R.J. Nicholls, and M. Capobianco, 2002. Variability of shore and shoreline evolution. *Coastal Engineering,* **47,** 211–235.

Tamisiea, M.E., J.X. Mitrovica, G.A. Milne, and J.L. Davis, 2001. Global geoid and sea level changes due to present-day ice mass fluctuations. *J. Geophys. Res.,* **106,** 30849–30863.

Tol, R.S.J., Klein, R.J.T., and Nicholls, R.J., 2004. Towards successful adaptation to sea level rise along Europe's coasts. *J. Coastal Res.* (in press).

Toniazzo, T., J.M. Gregory, and P. Huybrechts, 2004. Climatic impact of a Greenland deglaciation and its possible irreversibility. *J. Clim.,* **17,** 21–33.

Törnqvist, T.E., J.L. González, L.A. Newson, K. van der Borg, and F.M. de Jong, 2002. Reconstructing "background" rates of sea-level rise as a tool for forecasting coastal wetland loss, Mississippi delta. *EOS, Transactions of the American Geophysical Union,* **83** , 525 and 530–531 (12 November 2002).

Vafeidis, A., R.J. Nicholls, and L. McFadden, 2003. Developing a database for global vulnerability analysis of coastal zones: The DINAS-COAST project and the DIVA tool. In *Proceedings of EARSEL 2003* (European Association of Remote Sensing Laboratories), Ghent, Belgium, June 2003.

Vaughan, D., and J. Spouge, 2002. Risk estimation of collapse of the West Antarctic ice sheet. *Clim. Chang.,* **52,** 65–91.

Van Goor, M.A., M.J.F. Stive, Z.B., Wang, and T.J. Zitman, 2001. Influence of relative sea level rise on coastal inlets and tidal basins. *Proceedings of Coastal Dynamics 2001,* ASCE, New York, pp. 242–251.

Warrick, R.A., J. Oerlemans, P.L. Woodworth, M.F. Meier and C. Le Provost, 1996. Changes in sea level. In *Climate Change 1995: The Science of Climate Change,* J.T. Houghton, L.G. Meira Filho and B.A. Callander, eds. Cambridge University Press, Cambridge, pp. 359–405.

Williams, S.D.P., T.F. Baker, and G. Jeffries, 2001. Absolute gravity measurements at UK tide gauges. *Geophys. Res. Lett.,* **28,** 2317–2320.

Wolf, J., 1998. Waves at Holderness: results from in-situ measurements. In *Proceedings of Oceanology '98, Brighton, UK, March 1998,* pp. 387–398

Wolf, J., S.L. Wakelin, and R.A. Flather, 2002. Effects of climate change on wave height at the coast. In *Proceedings of the Twelfth International Offshore and Polar Engineering Conference, Kitakyushu, Japan, May 26–31, 2002,* **3,** 135–142.

Woodworth, P.L., J.M. Vassie, C.W. Hughes, and M.P. Meredith, 1996. A test of the ability of TOPEX/POSEIDON to monitor flows through the Drake Passage. *J. Geophys. Res.,* **101,** 11935–11947.

Woodworth, P.L., M.N. Tsimplis, R.A. Flather, and I. Shennan, 1999. A review of the trends observed in British Isles mean sea level data measured by tide gauges. *Geophys. J. Int.,* **136,** 651–670.

Woodworth, P.L., C. Le Provost, L.J. Rickards, G.T. Mitchum, and M. Merrifield, 2002. A review of sea-level research from tide gauges during the World Ocean Circulation Experiment. *Oceanography and Marine Biology: An Annual Review,* **40,** 1–35.

Woodworth, P.L., 2003. Some comments on the long sea level records from the northern Mediterranean. *J. Coastal Res.,* **19,** 212–217.

Woodworth, P.L. and R. Player, 2003. The Permanent Service for Mean Sea Level: an update to the 21st century. *J. Coastal Res.,* **19,** 287–295.

Woodworth, P.L. and J.M. Gregory, 2003. Benefits of GRACE and GOCE to sea level studies. *Space Science Reviews,* **108,** 307–317.

Woodworth. P.L., and D.L. Blackman, 2004. Evidence for systematic changes in extreme high waters since the mid-1970s. *J. Clim.,* **17,** 1190–1197.

Yohe, G., and Tol, R.S.J., 2002. Indicators for social and economic coping capacity – moving toward a working definition of adaptive capacity. *Glob. Environ. Chang.,* **12,** 25–40.

Zhang, K., B.C. Douglas, and S.P. Leatherman, 2004. Global warming and coastal erosion. *Clim. Chang.,* **64,** 41–58.

# Part 4.
# SCIENTIFIC ISSUES FOR APPLICATIONS

# Chapter 19. OVERVIEW OF SCIENCE REQUIREMENTS

THOMAS C. MALONE

*Horn Point Laboratory, University of Maryland Center for Environmental Science*

TONY KNAP

*Bermuda Biological Stations for Research, Inc.*

MICHAEL FOGARTY

*National Marine Fisheries Service, Woods Hole*

**Contents**



## 1. Introduction

"The overarching principles of ecosystem-based management of fisheries are an extension of the conventional principles for sustainable fisheries....They aim to ensure that, despite variability, uncertainty, and likely natural changes in the ecosystem, the capacity of aquatic ecosystems to produce fish food, revenues, employment, and, more generally, other essential services...is maintained indefinitely for the benefit of present and future generations....This implies conservation of ecosystem structures, processes and interactions through sustainable use."
"Basic Principles of Ecosystem Management", FAO Fisheries Atlas

### *1.1 Phenomena of Interest in the Coastal Ocean*

The combined effects of climate change and human alterations of the environment are especially pronounced in the coastal zone where inputs of energy and matter from land, sea and air converge. On a global scale, ecosystem goods and services

*The Sea*, Volume 13, edited by Allan R. Robinson and Kenneth H. Brink
ISBN 0-674-01526-6 

(Table 19.1) are concentrated in the coastal zone (Costanza et al., 1997; Daily et al., 2000), a circumstance that has made this region a center of increasing human activity for millennia. Conflicts in the use of the oceans began to emerge early in the 19$^{th}$ century when the global population numbered about 1 billion (Soares, 1998; Pauly et al., 2002). Today, over 5 billion people inhabit the earth with nearly 40% concentrated in a narrow band within 100 km of the coastline at altitudes less than 100 m above mean low tide (Small and Nicholls, 2003). This translates into a mean population density of 112 people km$^{-2}$ in the coastal zone, nearly three times the global mean (Nicholls and Small, 2002). These numbers are predicted to more than double by 2050 with the greatest increases occurring in the coastal zone (United Nations, 1996). As the number and density of people living in the coastal zone increases, (1) the risks of natural disasters and (2) conflicting demands on coastal systems (to support shipping, living marine resources, tourism, recreation, and living space and to receive, process, and dilute the effluents of human society) will continue to grow.

TABLE 19.1
Ecosystem goods and services provided by coastal marine and estuarine ecosystems in rank order of estimated value (adapted from Costanza *et al.*, 1997).

| Rank | Ecosystem Service | Ecosystem Functions | Examples |
|---|---|---|---|
| 1 | Nutrient cycling | Nutrient storage & processing | Nutrient cycles |
| 2 | Waste treatment | Breakdown & removal of excess nutrients & contaminants | Pollution control, detoxification |
| 3 | Disturbance regulation | Buffer impacts of extreme weather & climate change | Storm protection, flood control, drought recovery |
| 4 | Recreation | None | Boating, sport fishing, swimming |
| 5 | Food production | Fraction of plant production extractable as food | Harvest of fish & macroalgae |
| 6 | Refugia | Sustain habitats & biodiversity | Nurseries, refuge from predators |
| 7 | Cultural | None | Aesthetic, artistic, spiritual, research |
| 8 | Biological control | Trophic dynamics & biodiversity | Keystone predators, pest control |
| 9 | Raw materials | Portion of plant production extractable as raw materials | Fuel, lumber |
| 10 | Gas regulation | Chemical composition of the atmosphere | $CO_2$, $O_3$, $SO_x$, $CH_4$ |

Coastal ecosystems are experiencing unprecedented changes as indicated by the frequency or magnitude of a wide diversity of phenomena that affect the safety and efficiency of marine operations, the susceptibility of coastal populations to natural hazards, public health risks, the health of coastal marine and estuarine ecosystems, and the sustainability of living marine resources (Table 19.2). The phenomena of interest represent a broad spectrum of temporal variability (hours - decades) from changes in sea state and harmful algal blooms to habitat loss and global sea level rise. In addition, although many phenomena of interest are local in spatial scale, they occur in coastal ecosystems throughout the globe, a pattern that suggests they are, more often than not, local expressions of larger scale forcings of natural origin, anthropogenic origin, or both. Apparent increases in the occurrence of many of these phenomena indicate profound changes in the capacity of coastal ecosystems to support goods and services. They are making the coastal zone more susceptible to natural hazards, more costly to live in, and of less value to national economies (GESAMP, 2001). Thus, it is likely that, in the absence of a system for improved detection and prediction of the phenomena of interest and their environmental and socio-economic consequences, conflicts between marine commerce, recreation, development, environmental protection, and the management of living resources will become increasingly contentious and politically charged. The social and economic costs of uninformed decisions will increase accordingly.

TABLE 19.2

Natural and anthropogenic drivers of change (forcings) and their expression in terms of phenomena of interest in coastal marine and estuarine ecosystems that affect the safety and efficiency of marine operations, the susceptibility of human populations to natural hazards and global climate change, public health risks and ecosystem health, and the sustainability of living marine resources. Natural drivers of change occur in the absence of human intervention but may be altered or enhanced by human activities. With the exception of introductions of human pathogens and chemical contamination, anthropogenic drivers fall into the latter category.

| | **FORCINGS OF INTEREST** |
|---|---|
| "Natural" | •Global warming, sea level rise |
| | •Natural hazards (extreme weather, seismic events) |
| | •Currents, waves, tides & storm surges |
| | •River & groundwater discharges, sediment inputs |
| Anthropogenic | •Alteration of hydrological & nutrient cycles |
| | •Inputs of chemical contaminants & human pathogens |
| | •Harvesting natural resources (living & nonliving) |
| | •Physical alterations of the environment |
| | •Introductions of non-native species |
| | **PHENOMENA OF INTEREST** |
| Climate & weather | •Variations in sea surface temperature; surface fluxes of momentum, heat & fresh water; sources & sinks of carbon; sea ice |
| Marine operations | •Variations in water level, bathymetry, surface winds, currents & waves; sea ice; susceptibility to natural hazards |
| Natural hazards | •Storm surge & coastal flooding; coastal erosion & loss of buffer habitats; sea level; public safety & property loss |
| Public health | •Risk of exposure to human pathogens, chemical contaminants, and biotoxins (contact with water, aerosols, seafood consumption) |
| Healthy Ecosystems | •Habitat modification, loss of biodiversity, cultural eutrophication, harmful algal events, invasive species, diseases in & mass mortalities of marine organisms |
| Living marine resources | •Fluctuations in spawning stock size, recruitment & natural mortality; changes in spatial extent & condition of essential habitat; food availability & hydrographic conditions |

These concerns have led to international conventions and agreements intended to control and mitigate the effects of human activities on the environment and living marine resources. Prominent among these are (1 ) the 1982 UN Convention on the Law of the Sea (UNCLOS, including the 1995 U.N. Agreement on Straddling Fish Stocks and Highly Migratory Fish Stocks), (2) Regional Seas Conventions, (3) the Jakarta Mandate, (4) the Global Plan of Action on Land-Based Sources of Pollution, and (5) two conventions and a program of action signed at the 1992 UN Conference on Environment and Development Rio de Janeiro, the Framework Convention on Climate Change, the Convention on Biodiversity, and the Program of Action for Sustainable Development or *Agenda 21* (Keckes, 1997). Achieving the goals of these agreements depends on repeated, routine and timely provision of accurate and quantitative assessments of the state of coastal ecosystems and the goods and services they support on national to global scales.

Major anthropogenic drivers of change include (1) extractions of living marine resources (Pauly et al., 1998; Jackson et al., 2001); (2) land-use practices that alter inputs of water, sediments, nutrients, human pathogens, and chemical contaminants from coastal drainage basins (see Ogston et al., Chapter 4; Mackenzie et al., Chapter 7 and Farrington, Chapter15; Limburg and Schmidt, 1990; Smith et al., 1997; Vitousek et al., 1997; Jickells, 1998; Howarth et al., 1991, 2000; Shuval, 2001; GESAMP. 2001; NRC, 2000a); (3) physical alteration of habitats (see Kaiser et al., Chapter 25); (4) the globalization of marine commerce (Hinrichsen, 1998; Carlton, 1996); and (5) the release of greenhouse gases (Trabalka, J.R., 1985; Boesch et al., 2000; Najjar et al., 2000).

Changes in the state of marine systems occur through natural processes as well. Thus, many of the changes occurring in coastal ecosystems are related to extreme weather (Parsons and Nittrouer, Chapter 17) and to large scale, natural processes such as the El Niño-Southern Oscillation (ENSO) (see Chavez, Chapter 16; Barber and Chavez, 1986; Dayton and Tegner, 1984; Wilkinson et al., 1999; Kudela and Chavez, 2000; Arcos et al., 2001); the Pacific Decadal Oscillation (PDO) (e.g., Belgrano et al., 1999; Francis and Hare, 1994), and the North Atlantic Oscillation (NAO) (e.g., Pearce and Frid, 1999; Reid et al., 2001; Beaugrand et al., 2002).

Anthropogenic and natural drivers of change and their expressions in terms of the phenomena of interest (Table 19.2) are not independent of each other. Coastal ecosystems are subject to multiple drivers of change and any given driver may have multiple effects that are exacerbated by other drivers of change and their effects. For example, phenomena such as harmful algal blooms (Zingone and Wyatt, Chapter 22) and coastal eutrophication (Rabalais, Chapter 21) occur naturally, but anthropogenic nutrient loading associated with human alterations of the nitrogen cycle (Vitousek et al., 1997; Howarth et al., 2000; NRC, 2000a) increase the probability of blooms and the duration and spatial extent of bottom water hypoxia; fluctuations in fish catch in a given region or ecosystem can be related to the effects basin scale oscillations on recruitment and natural mortality (e.g., Francis and Hare, 1994; Reid et al., 2001; Beaugrand and Reid, 2002), but the effects of fishing mortality can be exacerbated by the effects of natural variability resulting in the collapse of a fishery (Pauly et al., 2002); harvesting benthic fish stocks not only causes declines in stocks, it modifies benthic habitats (Kaiser et al., Chapter 23) and can exacerbate the effects of anthropogenic nutrient inputs on water quality (Cloern, 1982; Cloern 2001; Jackson et al., 2001); and loss or modification of essential fish habitat can reduce the carrying capacity of ecosystems for fish populations and increase the susceptibility of the coastline to erosion and of human populations to coastal flooding (Mitsch et al., 1994; Short et al., 2001). Examples such as these make a compelling case for a more holistic approach to managing human uses and mitigating their affects on the coastal ocean.

### *1.2 Physics and Ecosystem Dynamics*

Coastal ecosystems are constrained by irregular coastlines and a shallow, highly variable bathymetry. Within these complex systems, interactions among intertidal, benthic and pelagic communities enhance nutrient cycles, primary productivity and the capacity of coastal ecosystems to support goods and services relative to oceanic systems. Physical and biological processes resonate over shorter time scales (higher frequencies) and over a broader spectrum of variability compared to both terrestrial and deep, open ocean systems (Steele, 1985; Powell, 1989). Consequently, populations and processes in coastal ecosystems are more variable on smaller space and shorter time scales than is typical for either oceanic or terrestrial ecosystems.

With the important exception of biologically structured habitats such as coral reefs, sea grass beds and tidal wetlands, the structure and function of marine ecosystems are, to a great extent, defined by physical processes (cf. Denman and Powell, 1984; Levin, 1992; Franks, 1997). Changes in biological, chemical and geological properties and processes are related through a hierarchy of physical-ecological interactions that can be represented by robust models of ecosystems dynamics (Nihoul and Djenidi, 1998; Hofmann and Lascara, 1998; Rothschild and Fogarty, 1998; Robinson, 1999; Liu et al., 2000; Cloern, 2001; Mantua et al., 2002). The time-space relationships of turbulent mixing, generation times of marine organisms, life histories, home ranges, and trophic dynamics suggest a close coupling between physical and biological processes over a broad range of time (hours–decades) and space (1–1000 km) scales (Figure 19.1). Biological and physical processes exhibit characteristic scales of variability that are related in a multidimensional continuum of time, space and ecological complexity (Costanza et al., 1993; Gardner et al., 2001 and references therein) so that large spatial scales tend to be associated with long time scales, the abundance of large organisms with long generation times, and greater biodiversity while small spatial scales tend to be associated with short time scales, small organisms with short generation times, and less biodiversity (e.g., Sheldon et al., 1972; Steele, 1985; Powell, 1989). On the scale of the ocean basins and their circulations, the distribution and abundance of species are related to water mass distributions, large scale current regimes, mesoscale eddies, and frontal systems. At smaller scales, the abundance and distribution of organisms are related to interactions between turbulent mixing and biological attributes such as motility, mechanisms of nutrient uptake and feeding, and patterns of reproduction and development. Thus, the scale-dependent linkage of ecological processes to the physical environment is fundamental to understanding and predicting spatial and temporal variability and pattern (e.g., variance spectra and fragmentation, time-space substitution) and size-dependent trophic interactions from small organisms with short generation times (e.g., viruses, bacteria, phytoplankton and zooplankton) to large organisms with longer generation times (e.g., macrobenthic, fish populations and marine mammals).

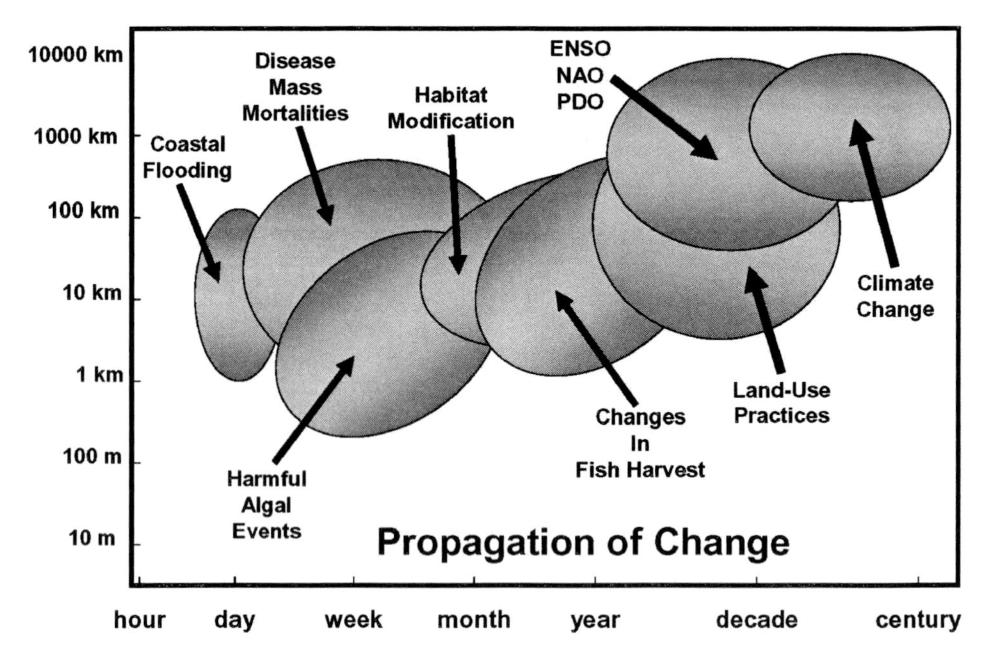


Figure 19.1 Temporal and spatial scales for selected forcings and phenomena of interest in coastal marine and estuarine ecosystems. The forcings and phenomena exhibit characteristic scales of variability that form a broad continuum from hours to decades and meters to thousands of kilometers. Most of the phenomena of interest are local expressions of larger scale forcings of natural origin, anthropogenic origin, or both. Thus, a broad spectrum of variability from global to local scales must be observed to quantify the state of marine systems and to predict changes in state.

### *1.3 Managing Human Use in the Context of Natural Variability*

All coastal ecosystems are subject to multiple forcings from both natural and anthropogenic sources, and the effect of one is often exacerbated by others. Changes in the frequency of or secular trends in the magnitude of the phenomena of interest reflect both the dynamics of coastal ecosystems and the nature of the external forces that impinge on them directly or indirectly via the propagation of variability among scales (global ſ regional ſ local). However, current efforts to manage human uses and mitigate their impacts typically focus on specific human activities (e.g., fishing mortality, nutrient pollution, physical alteration of habitats), specific habitats and places (e.g., mangroves, marshes, coral reefs, estuaries) or individual species (e.g., single species fisheries management, identification of endangered species) without due consideration of the propagation of variability and change across multiple scales in time, space and ecological complexity (Botsford et al., 1997; Gardner et al., 2001). As environmental research reveals the complexities of ecosystem dynamics, it is becoming increasingly clear that managing human uses and mitigating their effects with the goals of sustaining and restoring healthy ecosystems and the goods and services they support can best be achieved through ecosystem-based strategies that consider ecosystems and their state changes in a regional context where a *region* is defined as the next larger scale that must be observed to understand and predict the local scale of interest (Nixon, 1996; NRC, 1999a, 2000b; Sherman and Duda, 1999; UNEP, 2001).

The goal of formulating ecosystem-based approaches to managing human activities and mitigating their effects begs the question of how to define ecosystems and specify boundary conditions in marine environments that are not constrained by geographically fixed boundaries. Large marine ecosystems (LMEs) provide a good first approximation (Sherman and Duda, 1999). LMEs encompass large areas of the coastal ocean (> 200,000 km$^2$) and are characterized by distinct hydrographic regimes, submarine topography, productivity and trophic structures (Sherman and Alexander, 1986). Although porous, the boundaries of LMEs are based on the concept that critical processes controlling the structure and function of marine ecosystems are best addressed in a regional context (Ricklefs, 1987; Nixon, 1996; NRC, 2000b). They represent geographic regions defined by ecological criteria rather than political boundaries, i.e., they are natural ecological units for ecosystem assessments and ecosystem-based, adaptive management (Sherman, 1998).

In addition to establishing initial boundary conditions, the development of an ecosystem-based approach involves a shift from highly focused, short-term sectoral approaches as now practiced to a more holistic approach that spans multiple scales in time, space and ecological complexity (Costanza et al., 1993; Lubchenko, 1994; Sherman, 1998). Implementing ecosystem-based approaches depends on the ability to engage in adaptive management in which decisions (e.g., control of point and diffuse inputs of nutrients, specifying catch quotas for fisheries) are influenced by knowledge of the current state of marine ecosystems and natural environmental variability (Botsford et al., 1997; NRC, 1999a; Sætre et al., 2001). This requires the capacity to routinely and rapidly assess environmental conditions, detect changes, and provide timely predictions of likely future states. *We do not have this capacity today.*

## 2. Assessing the State of Coastal Marine Ecosystems

This reality is clearly illustrated by recent attempts to assess current ecosystem states. Adaptive management of human activities to protect, sustain, and restore healthy ecosystems and the goods and services they provide depend on the capability to repeatedly assess and anticipate changes in the status of coastal ecosystems and living resources on local to global scales (Nixon, 1996; Murawski, 2000; Clark et al., 2001). Efforts to provide such assessments increased dramatically in the 1990s (Parris, 2000). The recently released Pilot Analysis of Global Ecosystems (PAGE) (www.wri.org/wr2000), a precursor to the Millennium Ecosystem Assessment) and the State of the Nation's Ecosystems (O'Malley and Wing 2000; www.heinzctr.org/ecosystems) are used here to illustrate our current capacity to provide timely access to data and information required to quantify indices of ecosystem state at rates that are tuned to the time scales on which management decisions *should* be made.

In 1995, the U.S. Office of Science and Technology Policy initiated an effort to develop a set of indicators to quantitatively assess the status of coastal ecosystems for the United States as a whole. Representatives from industry, non-governmental organizations, the academic research community, and government agencies were brought together in a working group and tasked to specify a maximum of 16 scientifically credible and objective indicators. The assessment was crafted to be comprehensive and relevant to U.S. citizens and policy makers. Assessments were developed for six general types of ecosystems (forests, grasslands and shrublands, freshwater, farmlands, urban and suburban, and coastal marine) based on four categories of indicators for each ecosystem type: system dimensions (extent), chemical and physical condition, biological components, and human uses. In addition to the goal of providing the first comprehensive assessment of ecosystems on a national scale, the exercise was conducted to identify gaps in knowledge and deficiencies in current monitoring programs and data management systems. Results of the combined efforts of two working groups over 5 years were released in 2002 and are summarized in Table 19.3.

The World Resources Institute released the results of a preliminary global assessment of coastal ecosystems in April, 2001 (Burke et al., 2001). The assessment involved collaboration with intergovernmental organizations, agencies, research institutions, and individual experts from more than 25 countries. PAGE had three major objectives: (1) to provide a preliminary overview of the condition of coastal ecosystems on global and continental scales, (2) to identify serious data and information gaps that limit the ability to assess the condition of coastal ecosystems, and (3) to support the launch of the Millennium Ecosystem Assessment. The analysis considered the status of coastal, agricultural, forest, freshwater and grassland ecosystems. Recognizing the diversity of coastal marine and estuarine ecosystems, six broad categories were used to assess their status: extent and change, shoreline stabilization, water quality, biodiversity, commercial fish landings, and tourism and recreation. Results are summarized in Table 19.4.

TABLE 19.3

Summary of the results of the Heinz Center project to assess the condition of coastal marine and estuarine ecosystem on a national scale (no - data were unavailable to calculate the indicator; partial - some data available but insufficient to quantify the indicator; yes - indicator quantified).

| Category | Indicator | Data |
|---|---|---|
| System Dimension | Area of tidal wetlands, seagrasses, coral reefs & shellfish beds | Partial |
| | Proportion of shoreline that has been armored or consists of sandy beaches, rocky shores, cliffs, mudflats, salt marshes, & mangroves | Partial |
| Chemical & Physical Condition | Proportion of bottom area of estuaries and the EEZ within 40 km of the coast line that is anoxic (0 ppm), hypoxic (< 2 ppm), low (2–4 ppm), and adequate (> 4 ppm) | No |
| | Deviation of seasonal mean maximum surface temperature from the long-term mean for surface waters within 40 km of the coast line | Yes |
| | Proportion of shoreline managed and natural & proportion of natural shoreline that is eroding, accreting, and stable | No |
| Biological Components | Proportion of sediments that exceed federal sediment quality guidelines for concentration of pesticides, PCBs, PAHs, & metals | Partial |

| | | |
|---|---|---|
| | Proportion of species that are identified as vulnerable to extinction, imperiled, or critically imperiled | No |
| | Occurrence of non-native species measured in terms of both spatial extent (proportion of estuarine surface area affected) and number of species | No |
| | Proportion of bottom area for which the condition of benthic communities is classified as "undegraded", "moderately degraded", and "degraded" | Partial |
| | Annual frequency of "unusual" strandings and mass mortalities of large marine organisms (fish, mammals, turtles, birds) | Partial |
| | Seasonal mean maximum chlorophyll-a concentration in surface waters of the EEZ; proportion of estuarine surface area with concentrations < 5 ug/liter, 5–20 ug/liter, and > 20 ug/liter | Partial |
| | Annual incidence of harmful algal events classified as high, medium and low intensity in terms frequency and spatial and temporal extent | No |
| Human Uses | Annual fish harvest (weight) by region and major groups (shellfish, bottom fish, anadromous fish, etc.) from the U.S. EEZ | Yes |
| | Proportion of commercially important fish stocks that are increasing, decreasing or stable | Partial |
| | Concentrations of PCBs, DDT and mercury in the edible tissue of fish and shellfish | No |
| | Proportion of swimmable coastal waters that have high, moderate and low concentrations of enteric bacteria or are not monitored | No |

A most noteworthy result of both assessments is that, despite the length of time devoted to each effort (3–5 years) and the relative simplicity of the indicators, only 2 of 16 for U.S. waters (sea surface temperature, annual fish landings) and 4 of 22 globally (natural and altered land cover, number of marine protected areas, commercial fish landings, percent change from peak catch) could be estimated with some degree of confidence.

Although there are obvious differences in the two assessments, e.g., the U.S. analysis reported results when sufficient data were available to make computations that were considered to be scientifically and statistically sound while while the global analysis often did not specify quantitative indicators or relied on regional examples, the message is clear:

> If assessments of the status of coastal ecosystems and resources are to be quantitative and comprehensive, and if they are to be repeated in a timely fashion for decision makers and the public at regular intervals, major improvements are needed in the kinds, quality and quantity of data collected and in the efficiency with which data are disseminated, managed, and analyzed.

Data required to compute quantitative indicators on national and global scales were generally unavailable, inadequate, or nonexistent; and for those indicators that could be calculated, the time required to complete the analysis was far too long to be a useful tool for adaptive management. These exercises reveal fundamental problems in the conduct of research and monitoring programs that make comprehensive, quantitative, timely and routine assessments of coastal ecosystem status on regional, national or global scales virtually impossible.

TABLE 19.4
Results of the PAGE analysis for coastal ecosystems.
See Table 19.3 for explanation of results.

| Category | Indicator | Results |
|---|---|---|
| Extent and Change | Regional extent of sea ice; mangroves & non-mangrove wetlands; estuaries, barrier islands; coral reefs; mountainous & hilly narrow shelves; plains & hilly wide shelf; & unclassified | Partial |
| | Natural & altered land cover within 100 km of the coastline | Yes |
| | Disturbance of benthic communities | Partial |
| Shoreline Stabilization | Shoreline position | Partial |
| | Rates of erosion & accretion | Partial |
| | Susceptibility to natural hazards | Partial |
| Water Quality | Indicators of eutrophication (high nutrient concentrations, high phytoplankton biomass, oxygen depletion, loss of seagrasses, reduced light penetration) | Partial |
| | Harmful algal events | Partial |
| | Closures of shellfish beds and beaches | Partial |
| | Concentrations of persistent organic pollutants and heavy metals in marine organisms | Partial |
| | Oil spills (frequency and volume) | Partial |
| Biodiversity | Species richness | Partial |
| | Number of threatened species | Partial |
| | Rate of habitat loss or degradation | Partial |
| | Rate of non-native species invasins | Partial |
| | Number of marine protected areas | Yes |
| Marine Fisheries | Commercial fish landings (capture fisheries and aquaculture) | Yes |
| | Percent change in catch from peak year | Yes |
| | Changes in trophic structure of fish landings | Partial |
| Tourism & Recreation | Value of tourism in dollars and jobs | Partial |
| | Value relative to the GDP | Partial |
| | Tourist arrivals by region | Partial |

The most important problems that require immediate attention are the following:

1. inefficient and ineffective data management and communications procedures that do not provide rapid access to diverse data from many different sources (e.g., data from remote and *in situ* sensing, data collected by different government agencies and programs);
2. lack of common standards and protocols for measurements and data exchange and management;
3. under-sampling (insufficient resolution in time and space to make a statistically meaningful calculation), especially in the southern hemisphere;
4. lack of spatially and temporally synoptic observations of key physical, chemical and biological variables;
5. lack of operational models for assimilating and analyzing data with acceptable speed and skill.

A new approach is needed that enables adaptive management through routine, continuous and rapid provision of data and information over the broad spectrum of time-space scales required to link ecosystem (local) scale changes to regional and global scale forcings of both natural and anthropogenic origin.

### 3. Linking Science and Management

Effective environmental management and sustainable use of natural resources depends (1) on efficient coupling between advances in the environmental sciences and their application for the public good (Figure 19.2) and (2) on our understanding of the interdependency of ecological and socio-economic systems. Today, there are unacceptable gaps between these processes on both counts (Malone et al., 1193; Bowen and Riley, 2003). Although the challenges are many, successful establishment of the Global Ocean Observing System will fill the current void between science and management through the routine and repeated provision of scientifically credible, quantitative assessments of the status of coastal ecosystems on local, regional and global scales.

The observing system for the World Weather Watch is a case in point. The development of an operational system for global monitoring of the atmosphere and numerical weather prediction, the World Weather Watch (WWW), began over 300 years ago with the initiation of regular meteorological observations (Daley, 1991). The first national weather service with a permanent observing network was in established in France in the mid-1800s. By the early 1900s, a real-time global observing system was in place that consisted of a sparse network of unevenly spaced land-based sites for monitoring meteorological variables. An upper atmosphere network was established in the mid-1900s and by the late 1960s satellite-borne radiometers were providing global coverage. Today, the WWW consists of three closely linked subsystems: (1) a global monitoring network of sensors and platforms (surface and radiosonde networks and aircraft- and satellite-based sensors to monitor wind velocity, atmospheric pressure, and air temperature and moisture content from the earth's surface to the outer limits of the troposphere); (2) a global telecommunications subsystem (GTS) for data telemetry; and (3) a global network of data centers that collect, process, archive and disseminate data and information in near real-time. The observing system for the WWW not only provides the data and information required to nowcast and forecast weather patterns, the data streams are critical to advances in meteorology and improved prediction of global climate change. In this model, the relationship between meteorological research and weather forecasting is institutionalized. The WWW observing system supplies and manages the data required for Numerical Weather Predictions, and meteorologists both benefit from the data streams required for weather prediction and enable improved forecasting skill through advances in sensor technologies and understanding of the causes of atmospheric variability on global to local scales (Figure 19.2). This arrangement not only sustains the integrity of meteorological research (hypothesis-driven, research projects that are finite in duration), it strengthens it.

The WWW observing system is a useful model of an operational, "end-to-end" system. However, unlike the WWW which has a singular purpose, the GOOS is a multi-purpose system, the development of which depends on and benefits a broad spectrum of scientific disciplines. Development of an observing system that effectively links scientific advances in many disciplines to the information needs of multiple user groups will require a sustained effort by many groups that do not have history of collaboration to achieve common ends. Thus, many of the challenges are cultural, not technical.

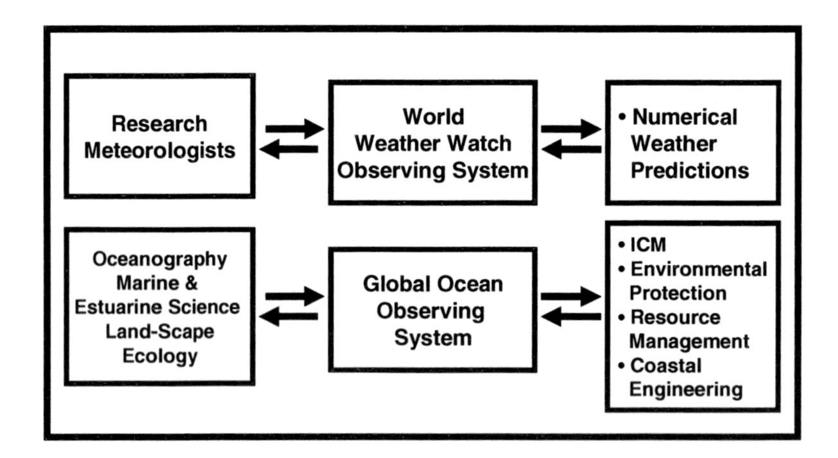


Figure 19.2 The observing system for the World Weather Watch is operational (routine, continuous provision of meteorological data and data-products of known quality) with guaranteed data streams and products (nowcasts forecasts of the weather and weather patterns on local to global scales). The operations system benefits from meteorological research, but numerical weather prediction are not dependent on the meteorological research community directly. The operational system also contributes to advances in the science of meteorology, but it's primary purpose (and motivation for government funding) is to predict the weather for the public good, e.g., to improve social and economic conditions. A similar but more complex system, the Global Ocean Observing System, is needed for coastal and ocean environments (ICM – Integrated Coastal Management).

## 4. The Coastal Module of the Global Ocean Observing System

We are on the cusp of a revolution that will make possible routine, timely and quantitative assessments of ecosystem state for the purposes of ecological forecasting and adaptive management. The revolution is occurring on two related fronts:

1. advances in observing and modeling capabilities (e.g., Haidvogel and Beckman, 1998; Franks, 1997; Cerco, 2000; Chapelle et al., 2000; Dudley et al., 2000; Moll and Radach, 2001; Hofmann and Friedrichs, 2002) and
2. the emergence of operational oceanography in the form of the Global Ocean Observing System (GOOS)(IOC, 1998; NRC 1999b; Nowlin et al., 1996, 2000).

Key drivers of rapid increases in observing and modeling capabilities are advances in data communications and computing power (Haidvogel et al., 2000; Walstad and McGillicuddy, 2000); remote and *in situ* sensing (Dickey, 2002; Dickey et al., 1998; Glenn et al., 2000a); the capacity to measure key physical, biological and chemical variables synoptically in time and space (e.g., Glenn et al., 2000a, b); and methods for linking observations to models (e.g., Lawson et al.,

1996; Robinson, 1999; Robinson and Lermusiaux, 2002). These advances are, for the first time, making it possible to visualize changes in four dimensions in near-real time and to quantify the effects of anthropogenic and climatic forcings.

The time is right to establish an integrated ocean observing system that capitalizes on current and emerging technologies and knowledge to develop and implement ecosystem-based, adaptive management strategies. Successful implementation of the observing system will increase the value to society of research and monitoring in marine and estuarine ecosystems by providing the data and information required to meet the conditions of existing international treaties and conventions and by providing the means to routinely assess and anticipate changes in the status of marine systems on national, regional and global scales.

### *4.1 Purpose and Scope*

The purpose of the GOOS is to establish a sustained and integrated ocean observing system that makes more effective use of existing resources, new knowledge, and advances in technology to continuously provide data and information in forms and at rates required to more effectively achieve six related societal goals:

1. improve the safety and efficiency of marine operations;
2. control and mitigate the effects of natural hazards;
3. improve the capacity to detect and predict the effects of global climate change on coastal ecosystems;
4. reduce public health risks;
5. protect and restore healthy ecosystems; and
6. restore and sustain living marine resources.

Achieving these goals depends on developing the capacity to assess the status of marine systems and to detect and predict changes in them rapidly and routinely. Although each goal has unique requirements for data and information, they have many data needs in common. Likewise, the requirements for data communications management are similar across all six goals. Thus, an integrated approach to the development of a multi-use, multi-disciplinary observing system is feasible, sensible and cost-effective.

GOOS is a movement to integrate, enhance and supplement existing research and monitoring activities for rapid data acquisition, dissemination, and analysis in response to the needs of governments, industries, science, education, and the public for information on marine and estuarine environments (IOC, 1998). The System is envisioned as a sustained and integrated global network that routinely and systematically acquires and disseminates data and data products on past, present and future states of the marine environment, ecosystems and the goods and services they provide. Under the oversight of the GOOS sponsors[1], the observing system is being organized in two related and convergent modules: (1) the global ocean module being developed by the Ocean Observations Panel for Climate and (2) the

[1] The Intergovernmental Oceanographic Commission (IOC), World Meteorological Organization (WMO), United Nations Environment Programme (UNEP), Food and Agriculture Organization (FAO); and the International Council for Science (ICSU).

coastal module being developed by the Coastal Ocean Observations Panel. The former is primarily concerned with changes in and the effects of the ocean-climate system on physical processes of the upper ocean and on the global carbon budget. The latter is primarily concerned with the effects of climate and human activities on coastal ecosystems and socio-economic systems of coastal nations including marine operations. The design of the coastal module and the science requirements for ecosystem-based approaches to managing human uses of the coastal ocean are the focus of the remainder of this paper.

### *4.2 A. Global System for the Coastal Ocean*

The design of coastal GOOS must take into account the changing mix of ecosystem types that constitute the coastal environment in different regions of the world and the time-space scales that characterize the phenomena of interest within them. In this context, design and implementation must also consider (1) the need to address a broad diversity of phenomena encompassed by the 6 goals (Table 19.2); (2) although the six goals of GOOS have unique requirements for data, data management, and analysis, they have many requirements in common; (3) the phenomena of interest tend to be local expressions of larger scale forcings; (4) ecosystem theory posits that the phenomena of interest are related through a hierarchy of interactions; and (5) the kinds of ecosystems and resources that constitute the coastal ocean and priorities for detection and prediction differ among regions. In addition, the design of the coastal module must take into consideration the following (UNEP, 2001):

- Most international agreements and conventions that target marine pollution and living marine resources are regional in scope.
- National and regional bodies provide the most effective venue for identifying user groups and specifying their data and information requirements.

Thus, the design plan for the coastal module (http://ioc.unesco.org/goos/) takes into account regional differences and leaves the design and implementation of regional observing systems to stakeholders in their respective regions. At the same time, to the extent that the six goals of coastal GOOS have data requirements in common, a global coastal network (GCN) of observations provides economies of scale that minimizes redundancy and allows regional observing system to be more cost-effective. To these ends, the coastal module includes both global and regional components that link global, regional and local scales of variability through a hierarchy of observations, data management and models (Figure 19.3). *Such a linked hierarchy can best be established through the formation of a global organization of national and regional observing systems that provides the means* by which national GOOS programmes and GOOS Regional Alliances (GRAs) can ensure the development of a global coastal network that benefits the national and regional programmes and play significant roles in the formulation of common standards and protocols, transfer of technology and knowledge, and setting of priorities for capacity building.

The importance of physical processes in structuring the pelagic environment and of scale dependent linkages of ecological processes to the physical environment suggests there is a relatively small set of variables that, if measured with sufficient resolution for extended periods over sufficiently large areas, will serve many needs from forecasting changes in sea state (e.g., surface wave fields) and the effects of tropical storms and harmful algal events on short time scales (hours to days) to predicting the environmental consequences of human activities and climate change on longer time scales (years to decades). These are the "common" variables that are required by most regional systems and are to be measured and processed as part of the global coastal network (Table 19.5). Depending on national and regional priorities, GOOS Regional Alliances (GRAs) may increase the resolution at which common variables are measured, supplement common variables with additional variables, and provide data and information products that are tailored to the requirements of stakeholders in the respective regions. Thus, GRAs both contribute to and benefit from the global network.

TABLE 19.5

Common variables recommended by the Coastal Ocean Observations Panel to be measured as part of the global coastal network (GCN). This provisional list of common variables is a first step in the process of determining what variables to measure as part of the global coastal observing system. The list will change as the GCN comes into being. The procedure for selecting the common variables is described in more detail in the "The Integrated, Strategic Design Plan for the Coastal Ocean Observations Module", the software for which may be downloaded from www.phys.ocean.dal.ca/~lukeman/COOP/.

| | |
|---|---|
| Physical | Sea level, Bathymetry & Shoreline position |
| | Temperature & Salinity |
| | Currents & Surface Waves |
| | Sediment grain size |
| | Attenuation of solar radiation |
| Chemical | Sediment organic content |
| | Dissolved inorganic nitrogen, phosphorus, & silicon |
| | Dissolved oxygen |
| Biological | Benthic biomass |
| | Phytoplankton biomass |
| | Fecal indicators |

Additional variables must be measured to quantify external forcings of coastal ecosystems. These include large scale ocean processes and inputs from atmospheric and land-based sources to be measured as part of the overall Integrated Global Observing Strategy (Table 19.6).

It must be emphasized that the global network will not, by itself, provide all of the data and information required to detect and predict changes in or the occurrence of many of the phenomena of interest (Table 19.2). There are categories of variables that are important globally, but the variables measured change from region to region. These include species-specific stock assessments for fisheries management; biologically structured habitats (coral reefs, sea grass beds, tidal marshes and mangrove forests); species of harmful algae, marine mammals, turtles and birds; and chemical contaminants. Decisions on what variables to measure (e.g., species of fish to be targeted for management purposes, species of chemical contaminants), the time and space scales of measurements, and the mix of observing techniques to be used are best made by stakeholders in the regions affected. Thus, the establishment of regional observing systems will be critical to detecting and predicting most of the phenomena of interest in the public health, ecosystem health and living marine resources categories.

TABLE 19.6

Variables needed from other observing systems that comprise the Integrated Global Observing Strategy. These include variables recommended for earth observation by the Integrated Global Observing Strategy (IGOS), World Meteorological Organizations (WMO), Global Climate Observing System (GCOS), Ocean Observations Panel for Climate (OOPC), Committee on Earth Observation Satellites (CEOS), and the International Ocean Colour Coordinating Committee (IOCCG).

| Meteorological | Atmosphere-Ocean | Sea Surface | Land-Based Inputs |
|---|---|---|---|
| Air temperature | $pCO_2$ | Sea surface temperature | Water |
| Vector winds | | Sea surface salinity | Nutrients |
| Humidity | | Surface waves | Sediments |
| Wet and dry precipitation | | Surface currents | Contaminants |
| | | Surface chlorophyll | |
| Incident solar radiation | | Attenuation of solar radiation | |

In addition to economies of scale and improved cost-effectiveness, the global network will establish, maintain, and improve the observational, data management and modelling infrastructure that benefits national and regional observing systems in several important ways:

- provide a network of reference and sentinel stations and sites to establish long-term time-series observations, provide advanced warnings of events and trends, and enable adaptive monitoring for improved detection and prediction;
- establish internationally accepted standards and protocols for measurements, data dissemination, management, and models;
- optimize data and information exchange;
- link the large scale network of observations for the ocean-climate module to the local scales of interest in coastal ecosystems and provide information on open boundary conditions and atmospheric forcings;
- provide the means for comparative ecosystem analysis required to understand and predict variability on local scales of interest; and
- facilitate capacity building.

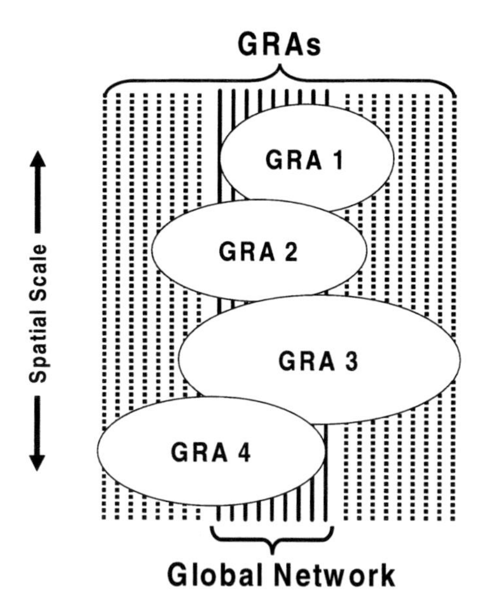


Figure 19.3 Schematic relationship between the global coastal network of coastal observations and GOOS Regional Alliances (GRAs) established to oversee the design, implementation, operation and development of regional observing systems. Solid vertical lines represent the common variables to be measured and processed as part of the global coastal network. Dashed lines represent regional enhancements by GRAs (discs), such as sea ice in polar regions, or coral characteristics in the tropics. Geographic boundaries of regional systems are not be fixed in space; they may overlap; and they are determined by the time and space scales of the phenomena of interest that are high priorities in each region.

### *4.3 Elements of an End-To-End Observing System*

Both detection and prediction depend on the development of an integrated and sustained observing system that effectively links measurements to data management and analysis for more timely access to data and delivery of environmental information (Nowlin *et al.,* 1996, 2001; Malone, 2003; Malone and Cole, 2000). The system must be integrated to effectively link the interdependent processes of monitoring and modeling and to provide multi-disciplinary (physical, geological, chemical and biological) data and information to many user groups. Linking user needs to measurements to form an end-to-end, user-driven system requires a managed, two-way flow of data and information among three essential subsystems (Fig. 19.4):

- The observing subsystem (networks of platforms, sensors, sampling devices, and measurement techniques) to measure the required variables on the required time and space scales;
- The communications network (data dissemination and access) and data management subsystem (telemetry, protocols and standards for quality assurance and control, data dissemination and exchange, archival, user access); and
- The data assimilation, analysis and modeling subsystem.

The *observing subsystem* for the global coastal network measures the common variables and transmits data to the communications network and data management subsystem. The infrastructure consists of the mix of platforms, samplers, and sensors required to measure the common variables with sufficient spatial and temporal resolution to capture important scales of variability in four dimensions. This will require the synthesis of data from remote sensing and *in situ* measurements involving various combinations of six categories of monitoring elements: (1) a network of coastal laboratories; (2) the global network of coastal tide gauges (GLOSS); (3) fixed platforms, moorings, drifters and underwater vehicles; (4) research and survey vessels, ships of opportunity (SOOP) and voluntary observing ships (VOS); (5) remote sensing from satellites and aircraft; and (6) remote sensing from land-based platforms. Many of these observing technologies are already deployed to some extent, and a global approach is needed to obtain the necessary integration.

*Data communications and management* link measurements to applications. The six monitoring elements must be linked from the beginning by a common data management structure. The objective is to develop a system for both real-time and delayed mode data that allows users to exploit multiple data sets and data products from disparate sources in a timely fashion. The development of this component of the system should be the highest priority for implementation.

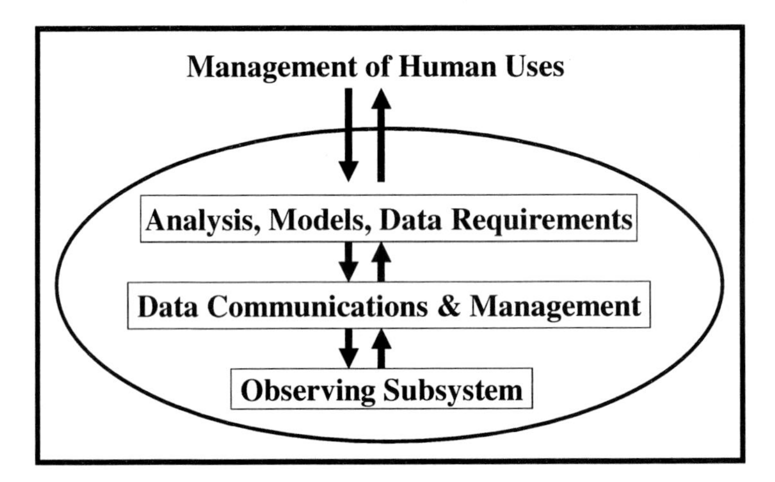


Figure 19.4 The observing system consists of three subsystems (inside the oval), the development of which is driven by user-requirements, technical capabilities, and the sustainable investments in infrastructure (capitalization) and operations (including the required technical expertise). Depending on capabilities and needs, user groups may access data from any one or all of the subsystems directly.

*Data assimilation and modeling* are critical components of the observing system in that they enable and drive the development of an integrated system. Real-time data from remote and *in situ* sensors will be particularly valuable in (1) producing more accurate estimates of the distributions of state variables, (2) developing, testing and validating models, and (3) initializing and updating models for improved forecasts of coastal environmental conditions and, ultimately, changes in the phenomena of interest in the categories of ecosystem and public health and living marine resources. A variety of modeling approaches (conceptual models, statistical relationships and constructs, dynamical models based on first principles, and coupled models of biotic and abiotic processes of marine ecosystems) will be required.

The main practical application of models is in the estimation of quantities that are not observed directly. Models can be used to interpolate between, and extrapolate, observations that are sparsely distributed in space or time. The more advanced interpolation-extrapolation methods use data assimilation techniques to blend, in an optimal fashion, observations and dynamical models. This can lead to high resolution gridded reconstructions of conditions that prevailed during earlier periods. Hindcasting has many applications. One is the reconstruction of past population sizes of exploited fish species using information on observed catches and some basic biology. Another application that is especially relevant to GOOS is the development of 'climatologies' for specified periods. The determination of the mean, second and higher moments of variability are of fundamental importance to environmental science and are required products for engineering design (Koblinsky and Smith, 2001). Climatologies are also required for effective management of coastal resources and as a background against which to interpret recent changes in marine ecosystems. Hindcasts from physical models can be used to synthesize scattered measurements of variables such as water temperature, salinity, sea level, nutrients and radio-nuclides, and produce consistent estimates of the three dimensional state of coastal systems. Such information can be used to interpret, for example, changes in the abundance of commercially important fish stocks, the frequency and magnitude of harmful algal blooms, trends in the loss of coral reefs, and the spatial and temporal extent of bottom water oxygen depletion.

Models can also be used to estimate the present and future states of the coastal ocean and its living resources. Nowcasting provides an estimate of the present state of coastal marine ecosystems and living resources, using all information available up to the present time. For example, nowcasts are used to assess the current status of exploited fish stocks, to ensure safe navigation in shallow water and can be used to guide adaptive sampling of the marine ecosystems based on real-time estimates of current conditions. Forecasting the future state of the coastal marine ecosystem and living resources is arguably the most important application of models. For example, models have been used for many years to forecast storm surges and wave spectra with a lead-time of hours to days. Models have also been used for decades in fisheries science to forecast sustainable catch limits with lead times of years. Models can also be used to generate plausible climate change 'scenarios', taking into account both natural variability and anthropogenic forcing.

### *4.4 One System, Six Goals*

Although this treatment has focused on managing human uses of coastal ecosystems and sustaining ecosystem goods and services, GOOS is intended to provide the data and information required for a much broader spectrum of products and services (section 4.1). Given that operational meteorology and oceanography are far more advanced for predicting extreme weather and the physical conditions of the upper ocean than for predicting changes in public health risks, ecosystem health, and the sustainability of living marine resources, GOOS will be implemented in phases (Figure 19.5). The phased implementation of the coastal module will be guided by many considerations including the following:

1. The data requirements for improved coastal marine services are, for the most part, common to all of the goals addressed by the coastal module. Safe and efficient coastal marine operations and the mitigation of natural hazards require accurate nowcasts and timely forecasts of storms and coastal flooding; of coastal current-, wave-, and ice-fields; and of water level, temperature and visibility. The set of variables that must be measured and assimilated in near real time include barometric pressure, surface wind vectors, air and water temperature, sea level, stream flows, surface currents and waves, and ice extent.
2. In addition to these variables, minimizing public health risks and protecting and restoring coastal ecosystems require timely data on environmental variables needed to detect and predict changes habitats and in biological, chemical and geological properties and processes, e.g., distributions of habitat types, concentrations of nutrients, suspended sediments, contaminants, biotoxins and pathogens; attenuation of solar radiation; biomass, abundance and species composition of plants and animals; and habitat type and extent. Mitigating the effects of natural hazards and reducing public health risks also requires a predictive understanding of the effects of habitat loss and modification (coral reefs, barrier islands, tidal wetlands, sea grass beds, etc.) on the susceptibility of coastal ecosystems and human populations to them.
3. In addition to data on the state of marine ecosystems (e.g., hydrography, currents, distribution and condition of critical habitats), the demands of sustaining living marine resources and managing harvests (of wild and farmed stocks) in an ecosystem context require timely information on population (stock) abundance, distribution, age- (size) structure, fecundity, recruitment rates, migratory patterns, and mortality rates (including catch statistics).

Figure 19.5 provides a conceptual framework for the phased implementation of the coastal module recognizing that (1) all of the major goals can and must be addressed from the beginning and (2) current operational capabilities from sensors to models dictate initial emphasis on marine services and natural hazards. Thus, the system will be designed to evolve and incorporate biological and chemical variables as new technologies, knowledge and operational models are developed.

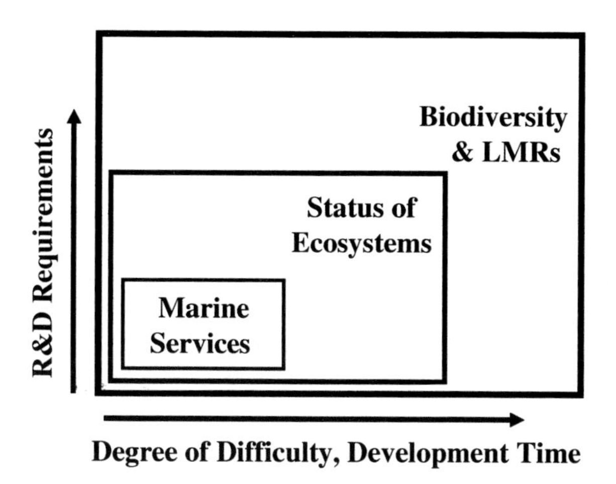


Figure 19.5 Time-dependent development of a fully integrated observing system. Predictions of most of the phenomena of interest require the same physical and meteorological data acquired for effective marine services and forecasting natural hazards. Those elements of an operational observing system required for improved marine services and forecasts of natural hazards are most developed (including the required operational models) while those required for ecosystem-based management of living resources are least developed. Given current capabilities and the importance of physical processes to the phenomena of interest in the ecosystem health and living marine resources categories, initial development of the global system will target the measurement and processing of physical variables *and* those biological and chemical variables that can be measured routinely and for which products can be clearly defined. To the extent that they are operational, non-physical variables should be incorporated into the observing system now. As technologies and procedures for rapid measurement and analysis of additional biological and chemical properties are developed, they will be incorporated into the system.

## 5. Research to Operational Oceanography

An operational ocean observing system, such as the World Weather Watch, is a new concept for oceanographers and marine ecologists, and there is an ongoing debate over what constitutes an operational system and what does not (Nowlin et al., 2001). The controversy is exacerbated by the multi-purpose goals of the GOOS (Table 19.2) and the broad spectrum of scientific disciplines upon which its development depends (Figure 19.2). Closing the gap between advances in the environmental sciences and application of new technologies and knowledge to achieve societal goals (Table 19.2) depends on the establishment of mechanisms for efficiently migrating new knowledge and technologies from research to an operational mode. An iterative process is needed by which advances in technology and knowledge are identified, selected, incorporated, and evaluated over time. The selection process by which candidate technologies, data management techniques and models become incorporated into an operational system can be conceptualized as in four stages as follows (Nowlin et al., 2001; Nowlin and Malone, 2003):

*Research Projects:* Observational (platforms, sensors, measurement protocols, data telemetry), data management and communications, and analytical (e.g., models and algorithms) techniques are developed by research groups.

*Pilot Projects:* Acceptance of techniques by research and operational communities is gained through repeated testing and pilot projects designed to demonstrate their utility and sustainability in a routine, operational mode. Techniques that show promise as potential elements of the operational system or sustained observations for research are tested repeatedly over a range of conditions. This exposes weaknesses, provides opportunities to address those weaknesses, and permits a better understanding of how they may be applied. Research groups, with involvement of operational groups, are primarily responsible for this stage.

*Pre-Operational Projects:* Research and operational communities collaborate to ensure that incorporation of new techniques from pilot projects into the operational system are likely to lead to a value added product (is more cost-effective, improves or expands existing capabilities) and will not compromise the integrity and continuity of existing data streams and product delivery of the operational system. Operational groups, with the involvement of researchers, are primarily responsible for this stage.

*The Operational System:* Routine and sustained provision of data and data products in forms and at rates specified by user groups are performed by operational groups with researchers functioning as advisors and users. This stage is improved through the incorporation of elements that are successful in a pre-operational mode. The appropriate government agency, ministry or GOOS Regional Alliance is responsible for the coordinated incorporation of such elements into the operational system, i.e., successful pre-operational projects, or elements thereof, are transferred to an operational agency, office, center or GRA for incorporation into the operational system.

Although presented as a linear sequence, in practice all four stages will be in play simultaneously with feedback among all stages. Research and development projects (stages 1–3) may focus on elements of the system (a particular sensing technology, development of sampling protocols, model development, data management and communications protocols, etc.) or on the development of an integrated system (e.g., end-to-end, regional observing systems). Successful pilot projects, or elements thereof, may be incorporated into long-term time series observations for scientific purposes, may become pre-operational, or both. Examples of Observing System elements that represent operational, pre-operational, and pilot project stages include (Nowlin et al., 2001):

- The ENSO observing system has progressed through research, pilot project and pre-operational stages to become operational. In the 1980s, the Tropical Ocean-Global Atmosphere (TOGA) research project began to develop detection capabilities, scientific understanding and models required to predict ENSO events. This developed into a multi-national pilot project during the 1980s and early 1990s. With the successful development of predictive skill based on routine observations, the pilot project became pre-operational in 1994 and in 1999 became an *operational* component of GOOS.
- The TOPEX/Poseidon satellite altimeter mission for precise measurements of sea surface height is an example of an observing system technology that has been proven in the research community and is now *pre-operational.*
- The Global Ocean Data Assimilation Experiment (GODAE) is a one-time pilot project to demonstrate the feasibility and practicality of real-

time global ocean data assimilation and numerical modelling for short-range open ocean forecasts, boundary conditions to extend predictability of coastal regimes, initialize climate forecast models; and research. A related sampling technology that is in the pilot project stage is the Argo project which is deploying several hundred autonomous profiling floats to measure and telemeter temperature and salinity data for the upper ocean (0 - 2000 m).

From sensing capabilities to models, operational capabilities are most developed for safe and efficient marine operations, forecasting extreme weather and its impacts on coastal populations, and predicting long-term climate change (Figure 19.5). Thus, the initial GOOS is primarily concerned with improving forecasts of marine weather, natural hazards, and surface currents and waves and with predicting global climate change with greater skill. Developing those aspects of the observing system concerned with ecosystem health and living marine resources will require synergy between research and the evolution of operational oceanography with an emphasis on in situ and remote sensing of biological and chemical properties, the formulation of climatologies for chemical and biological properties, and the development of data assimilation techniques and operational models that link physical and ecological processes for routine nowcasts and forecasts of phenomena of interest relevant to reducing public health risks and sustaining and restoring healthy ecosystems and the natural resources they support in an ecosystem context.

## ACKNOWLEDGEMENTS

This contribution is based on and has been enriched by discussions with Keith Thompson, John Cullen, Bob Bowen, Julie Hall, Worth Nowlin, Jr., and the entire Coastal Ocean Observing Panel including Dagoberto Arcos, Bodo von Bodungen, Alfonso Botello, Lauro Calliari, Mike Depledge, Eric Dewailly, Juliusz Gajewski, Johannes Guddal, Hiroshi Kawamura, Coleen Moloney, Nadia Pinardi, Hillel Shuval, Vladimir Smirnov, and Mohideen Wafar. The Panel's work has been supported by the Intergovernmental Oceanographic Commission and its member States.

## BIBLIOGRAPHY

Arcos, D.F., L.A. Cubillos and S.P. Núñez. 2001. The Jack mackerel fishery and El Niño 1997–98 effects off Chile. *Progress in Oceanography,* 49: 597–617.

Barber, R.T. and F.P. Chavez. 1986. Ocean variability in relation to living resources during the 1982–83 El Niño. *Nature,* 319: 279–285.

Beaugrand, G., P.C. Reid, F. Ibañez, J.A. Lindley, and M. Edwards. 2002. Reorganization of North Atlantic marine copepod biodiversity and climate. *Science,* 296: 1692–1694.

Belgrano, A., O. Lindahl, and B. Hernroth. 1999. North Atlantic Oscillation primary productivity and toxic phytoplankton in the Gullmar Fjord, Sweden (1985–1996). *Proc. Royal Soc. Lond. B,* 266: 425–430.

Boesch, D.F., J.C. Field, and D. Scavia (eds). 2000. The Potential Consequences of Climate Variability and Change on Coastal Areas and Marine Resources: Report of the Coastal Areas and Marine Resources Sector Team, U.S. Global Change Research Program, NOAA Coastal Ocean Program, Decision Analysis Series No. 21, Silver Spring, MD, 163 pp.

Botsford, L.W., J. C. Castilla, and C.H. Peterson. 1997. The management of fisheries and marine ecosystems. *Science,* 277: 509–515.

Bowen, R.E. and C. Riley. 2003. Socio-economic indicators and integrated coastal management. Ocean and Coastal Management (www.sciencedirect.com)

Burke, L., Y. Kura, K. Kassem, C. Revenga, M. Spalding, and D. McAllister. 2001. Pilot Analysis of Global Ecosystems: Coastal Ecosystems, World Resources Institute, Washington, D.C., 93 pp. (http://www.wri.org/wr2000)

Carlton, J.T. 1996. Marine bioinvasions: the alteration of marine ecosystems by nonindigenous species. *Oceanography,* 9: 36–45.

Cerco, C. 2000. Chesapeake Bay eutrophication model, pp. 363–404, In: J. E. Hobbie (ed.) Estuarine science: A synthetic approach to research and practice. Island Press, New York.

Chapelle, A., A. Menesguen, J. M. DeslousPaoli, P. Souchu, N. Mazouni, A. Vaquer, and B. Millet, 2000. Modelling nitrogen, primary production and oxygen in a Mediterranean lagoon. Impact of oysters farming and inputs from the watershed. *Ecological Modelling,* 127: 161–181.

Clark, J.S., S.R. Carpenter, M. Barber, S. Collins, A. Dobson, J.A. Foley, D.M. Lodge, M Pascual, R. Pielke, W. Pizer, C. Pringle, W.V. Reid, K.A. Rose, O. Sala, W.H. Schlesinger, D.H. Wall, and D. Wear. 2001. Ecological forecasting: an emerging imperative. *Science,* 293: 657–660.

Cloern, J.E. 1982. Does the benthos control phytoplankton biomass in South San Francisco Bay? *Mar. Ecol. Prog. Ser.,* 9: 191–202.

Cloern, J.E. 2001. Our evolving conceptual model of the coastal eutrophication problem. *Mar. Ecol. Prog. Ser.,* 210: 223–253.

Costanza, R., W.M. Kemp and W.R. Boynton. 1993. Predictability, scale, and biodiversity in coastal and estuarine ecosystems: implications for management. *Ambio,* 22: 88–96.

Costanza, R., R. d'Arge, R. de Groots, S. Farber, M. Grasso, B. Hannon, K. Limburg, S. Naeem, R.V. O'Neill, J. Paruelo, R.G. Raskin, P. Sutton and M. van den Belt. 1997. The value of the world's ecosystem services and natural capital. *Nature,* 387:253–260.

Daily, G.C., T. Soderqvist, S. Aniyar, K. Arrow, P. Dasgupta, P. Ehrlich, C. Folke, A.M. Jansson, B-O. Jansson, N. Kautsky, S.A. Levin, J. Lubchenko, K-G. Maler, D. Simpson, D. Starrett, D. Tilman, and B. Walker. 2000. The value of nature and the nature of value. *Science,* 289: 395–396.

Daley, R. 1991. Atmospheric Data Analysis, Cambridge Atmospheric and Space Science series. Cambridge University Press, 457 pp.

Dayton, P.K. and M.J. Tegner. 1984. Catastrophic storms, El Niño, and patch stability in a southern California kelp community. *Science,* 224: 283–285.

Denman, K.L. and T.M. Powell. 1984. Effects of physical processes on planktonic ecosystems in the coastal ocean. *Oceanography and Marine Biology: Annual Review,* 22: 125–165.

Dickey, T. 2002. Emerging instrumentation and new technologies in the early 21st century, Oceans 2020: Science for Future Needs, J.G. Field, G. Hempl, and C.P. Summerhayes (eds), Island Press, Washington, DC, 213–256.

Dickey, T., Frye, D., Jannasch, H., Boyle, E., Manov, D., Sigurdson, D., McNeil, J., Stramska, M., Michaels, A., Nelson, N., Siegel, D., Chang, G., Wu, J. and Knap, A. 1998. Initial results from the Bermuda Testbed Mooring Program. *Deep-Sea Res.,* 771–794.

Dudley, R. W., V. G. Panchang, and C. R. Newell, 2000. Application of a comprehensive modeling strategy for the management of net-pen aquaculture waste transport. *Aquaculture,* 187: 319–349.

Francis, R.C. and S.R. Hare. 1994. Decadal-scale regime shifts in the large marine ecosystem of the north-east Pacific: a case for historical science. *Fish. Oceanogr.* 3: 279–291.

Franks, P.J.S. 1997. Coupled physical-biological models for the study of harmful algal blooms. *Ocean Res.,* 19: 153–160.

Gardner, R.H., W.M. Kemp, V.S. Kennedy and J.E. Petersen (eds). 2001. Scaling Relations in Experimental Ecology, Columbia University Press, New York, 373 pp.

GESAMP. 2001 Scoping Activities: Ballast Water Management; Management and Treatment of Ballast Water to Reduce Risks of Alien Species Transfer (R.G.V. Boelens). GESAMP XXXVI/11, 20 June 2001. 10 pp.

Glenn, S.M., Boicourt, W., Parker, B. and Dickey, T.D. 2000a. Operational observation networks for ports, a large estuary and an open shelf. *Oceanography,* 13: 12–23.

Glenn, S.M., T.D. Dickey, B. Parker, and W. Boicourt. 2000b. Long-term, real-time coastal ocean observation networks. *Oceanography,* 13: 24–34.

Haidvogel, D.B. and A. Beckman, 1998. Numerical models of the coastal ocean. In: *The Sea.* The global coastal ocean. Processes and methods. K.H.Brink and A.R.Robinson (eds), Volume 10, J.Wiley & Sons, p. 457–482.

Hinrichsen, D. 1998. *Coastal Waters of the World: Trends, Threats and Strategies.* Island Press, Washington, D.C., 275 pp.

Hofmann,E.E. and C.M.Lascara, 1998. Overview of interdisciplinary modeling for marine ecosystems. In: *The Sea,* Vol. 10, The global coastal ocean. Processes and methods. K.H.Brink and A.R.Robinson (eds), J.Wiley & Sons, p. 507–540.

Hofmann,E.E. and M. A. M. Friedrichs, 2002. Predictive Modelling for Marine Ecosystems. In: *The Sea.* Biological-Physical Interactions in the sea. A. R. Robinson, J. J. McCarthy, and B. J. Rothschild (eds), Volume 12, J.Wiley & Sons.

Howarth, R.D., J.R. Fruci and D. Sherman. 1991. Inputs of sediment and carbon to an estuarine ecosystem: influence of land use. *Ecol. Appl.,* 1: 27–39.

Howarth, R., D. Anderson, J. Cloern, C. Elfring, C. Hopkinson, B. Lapointe, T. Malone, N. Marcus, K. McGlathery, A. Sharpley and D. Walker. 2000. Nutrient pollution of coastal rivers, bays and seas. *Issues in Ecology, Ecological Society of America,* 7: 1–15.

IOC. 1998. The GOOS 1998, GOOS Publication No. 42, IOC, Paris, 144 pp.

Jackson, J.B.C., M.X. Kirby, W.H. Berger, K.A. Bjorndal, L.W. Botsford, B.J. Bourque, R.H. Bradbury, R. Cooke, J. Erlandson, J.A. Estes, T.P. Hughes, S. Kidwell, C.B. Lange, H.S. Lenihen, J.M. Pandolfi, C.H. Peterson, R.S. Steneck, M.J. Tegner, and R.R. Warner. 2001. Historical overfishing and the recent collapse of coastal ecosystems. *Science,* 293: 629–643.

Jickells, T.D. 1998. Nutrient biogeochemistry of the coastal zone. *Science,* 281: 217–222.

Keckes, S. 1997. Global Maritime Programs and Organizations: An Overview. Maritime Institute of Malaysia, 211 pp.

Koblinsky, C. and N.R. Smith. 2001. Observing the Oceans in the 21st Century. GODAE Project Office, Bureau of Meteorology, Melbourne, Australia, 604 pp.

Kudela, R.M. and Chavez, F.P. 2000. Modeling the impact of the 1992 El Niño on new production in Monterey Bay, California. *Deep-Sea Res.* II, 47: 1055–1076.

Lawson, L., E. Hofmann and Y. Spitz. 1996. Time series sampling and data assimilation in a simple marine ecosystem model. *Deep-Sea Research* 43: 625–621.

Levin, S.A. 1992. The problem of pattern and scale in ecology. *Ecology,* 73: 1943–1967.

Limburg, K.E. and R.E. Schmidt. 1990. Patterns of fish spawning in Hudson River tributaries: response to an urban gradient. *Ecology,* 71: 1238–1245.

Liu, K.K., K. Iseki, and S.Y. Chao. 2000. Continental margin carbon fluxes, p. 187–239. In R.B. Hanson, H.W. Ducklow and J.G. Field. The Changing Ocean Carbon Cycle, Cambridge University Press.

Lubchenko, J. 1994. The scientific basis of ecosystem management: framing the context, language, and goals. In Ecosystem Management – Status and Potential. 103rd Congress, 2nd Session, Committee Print. U.S. Government Printing Office, Superintendent of Documents. ISBN 0–16–046423–4, p. 33–39.

Malone, T.C. 2003. The coastal component of the U.S. integrated ocean observing system. *Environmental Monitoring and Assessment,* 81: 51–62.

Malone, T.C. and Cole, M .. 2000. Toward a global scale coastal ocean observing system. *Oceanography,* 13 (1): 7–11.

Malone, T.C., W. Boynton, T. Horton, and C. Stevenson. 1993. Nutrient loadings to surface waters: Chesapeake Bay case study. In Keeping Pace with Science and Engineering: Case Studies in Environmental Regulation. M.F. Uman and C. O'Melia (eds.), National Academy Press, p. 8–38.

Mantua, N., D. Haidvogel, Y. Kushnir, and N. Bond. 2002. Making the climate connections: bridging scales of space and time in the U.S. GLOBEC program. *Oceanography,* 15: 75–86.

Mitsch, W.J., R.H. Mitsch, and R.E. Turner. 1994. Wetlands of the old and new worlds: ecology and management. In Gobal Wetlands: Old and New, W.J. Mitsch (ed), Elsevier Science B.V., p. 3–56.

Moll, A. and G. Radach. 2001. Synthesis and new conception of North Sea Research. Working Group 6: Review of three-dimensional ecological modeling related to the North Sea shelf system. Hamburg, Zentrum fur Meeres- und Klimaforschung der Universitat Hamburg, 225 pp.

Murawski, S.A. 2000. Definitions of overfishing from an ecosystem perspective. *ICES J. Mar. Sci.,* 57: 649–658.

Najjar, R.G., H.A. Walker, P.J. Anderson, E.J. Barron, R.J. Bord, J.R. Gibson, V.S. Kennedy, C.G. Knight, J.P. Megonigal, R.E. O'Connor, C.D. Polsky, N.P. Psuty, B.A. Richards, L.G. Sorenson, E.M. Steele, and R.S. Swanson. 2000. The potential impacts of climate change on the mid-Atlantic coastal region. *Climate Res.,* 14: 219–233.

National Research Council. 1999a. Sustaining Marine Fisheries. National Academy Press, Washington, D.C., 164 pp.

National Research Council. 1999b. Global Ocean Science, Toward an Integrated Approach. National Academy Press, Washington, D.C., 165 pp.

National Research Council. 2000a. Clean Coastal Waters: Understanding and Reducing the Effects of Nutrient Pollution. National Academy Press, Washington, D.C. 405 pp.

National Research Council. 2000b. Bridging Boundaries Through Regional Marine Research. National Academy Press, Washington, D.C., 115 pp.

Nicholls, R.J. and C. Small. 2002. Improved estimates of coastal population and exposure to hazards. *Eos Transactions,* 83: 301–305.

Nihoul, J.C.J. and S. Djenidi.1998. Coupled physical, chemical and biological models. In: *The Sea,* Vol. 10, The global coastal ocean. Processes and methods. K.H.Brink and A.R.Robinson (eds), J.Wiley & Sons, p. 483–506.

Nixon, S.W. 1996. Regional coastal research — What is it? Why do it? What role should NAML play? *Biol. Bull.,* 190: 252–259.

Nowlin, W.D., Smith, N., Needler, G., Taylor, P.K., Weller, R., Schmitt, R., Merlivat, L., Vezina, A, Alexiou, A., McPhaden, M. and Wakatsuchi, W. 1996. An ocean observing system for climate. *Bull. Am. Meteor. Soc.,* 77: 2243–2273.

Nowlin, W.D., M. Briscoe, N. Smith, M.J. McPhaden, D. Roemmich, P. Chapman, and J.F. Grassle. 2001. Evolution of a sustained ocean observing system. *Bull. Amer. Met. Soc.,* 82: 1369–1376.

Nowlin, W.D. and T.C. Malone. 2003. Research and GOOS. J.Mar.Tech.Soc., 37: 42–46.

O'Malley, R. and K. Wing. 2000. Forging a new tool for ecosystem reporting. *Environment,* 42: 20–32.

Pauly, D., V. Christensen, J. Dalsgaard, R. Froese and F. Torres, Jr. 1998. Fishing down marine food webs. *Science,* 279: 860–863.

Pauly, D., V. Christensen, S. Guénette, T.J. Pitcher, U.R. Sumaila, C.J. Walters, R. Watson and D. Zeller. 2002. Toward sustainability in world fisheries. *Nature,* 418: 689–695.

Parris, T.M. 2000. Tracking down state of the environment reports. *Environment,* 42: 3–4.

Pearce, K.F. and C.L J. Frid. 1999. Coincident changes in four components of the North Sea ecosystem. *J. Mar. Biol. Ass. U.K.,* 79: 183–185.

Powell, T. 1989. Physical and biological scales of variability in lakes, estuaries, and the coastal ocean. In Perspectives in theoretical ecology, J. Roughgarden, R.M. May, and S.A. Levin (eds.). Princeton University Press, Princeton, N.J. p. 157–180.

Reid, P.C., M. F. Borges, and E. Svendsen. 2001. A regime shift in the North Sea circa 1988 linked to changes in the North Sea horse mackerel fishery. *Fisheries Res.,* 50: 163–171.

Ricklefs, R.E.1987. Community diversity: relative roles of local and regional processes. *Science,* 235: 167–171.

Robinson, A.R. 1999. Forecasting and simulating coastal ocean processes and variabilities with the Harvard ocean prediction system. In Coastal Ocean Prediction, AGU Coastal and Estuarine Studies, 56: 77–99

Robinson, A.R. and Glenn, S.M. 1999. Adaptive sampling for ocean forecasting. *Naval Res. Rev.,* 51: 26–38.

Robinson, A. R. and P. F. Lermusiaux, 2002. Data assimilation for modeling and predicting coupled physical-biological interactions in the sea. In: *The Sea,* Vol. 12, Biological-Physical Interactions in the Sea. A. R. Robinson, J. J. McCarthy, and B. J. Rothschild (eds), J.Wiley & Sons,

Rothschild, B.J. and M.J. Fogarty. 1998. Recruitment and the population dynamics process. In: *The Sea,* Vol. 12, Biological-Physical Interactions in the Sea. A. R. Robinson, J. J. McCarthy, and B. J. Rothschild (eds), J.Wiley & Sons, 293–325.

Sætre, R. H.R. Skjodal, N.C. Flemming and F. van Beek. 2001. Toward a North Sea ecosystem component of GOOS for assessment and management, Report from a strategic workshop in Bergen 5–7 September 2001, Fisken Og Havet, No. 11.

Sheldon, R.W., A. Prakash and W.H. Sutcliffe. 1972. The size distribution of particles in the ocean. *Limnol. Oceanogr.,* 17: 327–340.

Sherman, K. 1998. Large marine ecosystems as science and management units. In Workshop on the Ecosystem Approach to the Management and Protection of the North Sea, Oslo, Norway (www.odin.dep.no/md/html/conf/workshop/1998/report.html).

Sherman, K. and L.M. Alexander (eds.) 1986. Variability and management of large marine ecosystems. AAAS Selected Symposium, 99. Boulder, CO, USA.

Sherman, K. and A.M. Duda. 1999. An ecosystem approach to global assessment and management of coastal waters. *Mar. Ecol. Prog. Ser.,* 190: 271–287.

Short, F.T., R.G. Coles and C. Pergent-Martini. 2001. Global seagrass distribution. In F.T. Short and R.G. Coles (eds). Global Seagrass Methods. Elsevier Science B.V.

Shuval, H., 2001. "Thalassogenic Disease"- Human disease caused by wastewater pollution of the marine environment with special reference to the Mediterranean, Rapp. Comm. Int. Mer Medit., 36: 221–223 ( 36th CIESM Congress Proceedings- Monte-Carlo September, 2001)

Small, C. and R.J. Nicholls. 2003. A global analysis of human settlement in coastal zones. *J. Coastal Res.* (in press)

Smith, R.A., G.E. Schwarz and R.B. Alexander. 1997. Regional interpretation of water-quality monitoring data. *Water Resources Res.,* 33: 2781–2798.

Soares, M. (ed.) 1998. The Ocean Our Future, Report of the Independent World Commission on the Oceans, Cambridge University Press, 247 pp.

Steele, J.H. 1985. A comparison of terrestrial and marine ecological systems. *Nature,* 313: 355–358.

Trabalka, J.R. (ed.) 1985. Atmospheric Carbon Dioxide and the Global Carbon Cycle, United States Department of Energy, Department of Commerce, Springfield, VA, 315 pp.

United Nations. 1996. Country population statistics and projections 1950–2050. Report, Food and Agricultural Organization of the United Nations, Rome, Italy.

UNEP. 2001. Ecosystem-based management of fisheries: opportunities and challenges for coordination between marine Regional Fishery Bodies and Regional Seas Conventions. UNEP Regional Seas Reports and Studies, No. 175, 52 pp.

Vitousek, P.M., J.D. Aber, R.W. Howarth, G.E. Likens and others. 1997. Human alterations of the global nitrogen cycle: causes and consequences. *Ecol. Appl.,* 7: 737–750.

Wilkinson, C., Lindén, O., Cesar, H., Hodgson, G., Rubens, J. and Strong, A.E. 1999. Ecological and socioeconomic impacts of the 1998 coral mortality in the Indian Ocean: an ENSO impact and a warning of future change? *Ambio,* 28: 188–196.

# Chapter 20. FUNCTIONAL DIVERSITY AND STABILITY OF COASTAL ECOSYSTEMS

JOHN H. STEELE

*Woods Hole Oceanographic Institution*

JEREMY S. COLLIE

*Graduate School of Oceanography, University of Rhode Island*

**Contents**



## 1. Introduction

A dominant theme in marine research has been the close coupling between physical and biological processes, especially the relationship between the distribution and productivity of planktonic species (from phytoplankton to fish) to currents and mixing rates. These links have been described for a wide range of scales, from patch dynamics (Levin et al., 1993), through eddies (Robinson, 1983) to the ecological geography of the oceans (Longhurst, 1998). This approach has proved very successful in elucidating temporal and spatial patterns for basic biogeochemical processes, and for dominant marine species. Much remains to be done, but the data and the models that codify this approach, provide a basis for a renewed attack on the ecological interactions underlying the biogeochemical processes and the dynamics of marine communities. These interactions, typified by concepts such as diversity and stability, provide the theme for this review.

The primacy of physical-biological coupling as an explanatory tool for marine systems can be contrasted with the emphasis on ecological interactions in terrestrial environments (Steele, 1985). A longstanding controversy in terrestrial ecology concerns the relations between diversity, stability and productivity. Tilman (2001)

*The Sea*, Volume 13, edited by Allan R. Robinson and Kenneth H. Brink
ISBN 0-674-01526-6 

proposed that stability of individual plant populations may decrease with increasing diversity but overall community stability increases. Such inferences are very relevant to the trade-offs in the management of terrestrial systems, especially the allocation of "countryside" (Daily, 2001) to farming or forests.

In the sea, the same issues and trade-offs are becoming the focus of our management, not only for coastal and continental shelf environments but also for the open sea. There are, however, major differences in our attitudes and particularly in our knowledge. For the countryside "humanity always has been and always will be part of nature" (Daily, 2001). For the seas, we are much more ambivalent about our role. Many see us as intruders whose impacts should be reduced to the absolute minimum. On land we advocate eating from the bottom of the trophic pyramid, but in the sea "fishing down the food web" (Pauly et al., 1998) is considered a recipe for decreasing diversity that "may have a major effect on ecosystem stability." The history of marine fisheries is one of serial exploitation. In nearly all mature fisheries, a "huge reduction" in fishing capacity is required (Pauly et al., 2002). But, given this requirement, what are the criteria for an optimal balance? For marine environments, especially pelagic ecosystems, we do not have a body of work on the relations between diversity, stability and productivity that can assist us in regulating the uses of marine resources (Worm and Duffy, 2003). Further, is it apparent that terrestrial concepts are not transferable – and certainly not the attitudes.

Unlike terrestrial systems where the focus is predominantly on plant competition, marine concerns are with higher trophic levels and prey-predator interactions. Specifically the issues involve top-down consequences from fishing and bottom-up effects predominantly from eutrophication or climatic change. The question of how and where in the food web these effects interact is central to arguments about ecosystem impacts on biodiversity and stability. Physical-biological coupling at all trophic levels implies that marine systems are much more responsive to physical variability, especially at decadal scales; but, therefore, they must also be much more adaptable to such changes. In consequence this review focuses on the need to consider diversity and stability as dynamic rather than static concepts. In turn these dynamic considerations have implications for our approach to management of resources within these ecosystems.

In summary, this review explores the following issues:

*Physical-biological coupling.* We describe the close links between spatial patterns and time scales of physical processes and marine organisms. These interactions explain major features of the distribution and abundance of microbial populations, of zooplankton species and of larval fish.

*Environmental perspective.* However they exclude most of the trophic interactions, especially those determining longer-term stability of ecosystems. In particular, we consider the distinctive role of the microbial loop (e.g. Fasham et al., 1990) in coastal environments because it is export from this loop that feeds the higher trophic levels, and some of the effects of climatic change will operate through this filter.

*Dynamics of diversity.* We discuss the links between diversity, productivity and stability in continental shelf ecosystems. We need concepts to address the restructuring of communities resulting from combinations of environmental changes and overfishing.

*Modeling diversity and dynamics.* The balance between species and dynamic complexity in models is central to interpretation of such issues as "fishing down the food web", or the relative roles of top-down and bottom-up forcing. We analyse two approaches to these issues: the use of linear networks to describe the equilibrium states of communities, and the use of non-linear modeling to represent switching between alternative states of these communities. We report on recent applications of bifurcation models to fisheries situations (Spencer and Collie, 1996).

*Resolving complexity.* We propose that a sequence of periods with relatively stable structure, separated by shorter intervals with rapid transitions, or "regime shifts", is a useful approximate description of the observed phenomena.

*Human Relevance.* Such a description can illuminate the gap and suggest the bridges between the complexity of ecosystem dynamics and operational requirements for management tools.

## 2. Physical-biological Coupling

The direct control of marine populations by physical processes has been the basic assumption underlying much oceanographic research, and is captured in the meaning of the term "plankton". Even for non-planktonic organisms such as fish or benthos the larval planktonic phase is usually considered the critical period for survival and the prime determinant of abundance. The primacy of physical control is seen at all trophic levels, in all dimensions and at all scales.

### *Background*

The major distinctions between pelagic marine and terrestrial systems can be expressed in terms of space and time scales (Fig. 20.1). This forms a basis for the general conclusion that marine components are closely coupled to physical processes. Conversely the terrestrial systems show very much longer time scales for ecological versus physical processes, suggesting that, in some sense, the terrestrial communities have adapted to eliminate the short and medium-term effects of environmental change. This adaptation is apparent in the physiology and reproduction of vertebrates. Fish are cold blooded and nearly all marine species broadcast their eggs. Terrestrial vertebrates invest a lot of their available energy in temperature control and parental care. A further observation from Fig. 20.1 is that the primary producers on land are very long-lived, much longer than the vertebrates, including humans. The pelagic marine system has a very good concordance between time scale and trophic level, with all the scales being shorter than most terrestrial plants.

All these aspects help to explain the difference between the natural systems as well as our response to them at our human time scales. The separate emphases on ecological interactions on land and physical forcing at sea are natural responses to a world with relatively slowly varying physical environments and relatively small-scale exploitation of the sea. Now, however, both of these conditions are changing. On land we talk of doubling $CO_2$ in less than a century. A glance at the global time scales for soil and forests in Fig. 20.1 illuminates the reasons for concern at ecosystem disruption by too rapid environmental change (Davis, 1981).

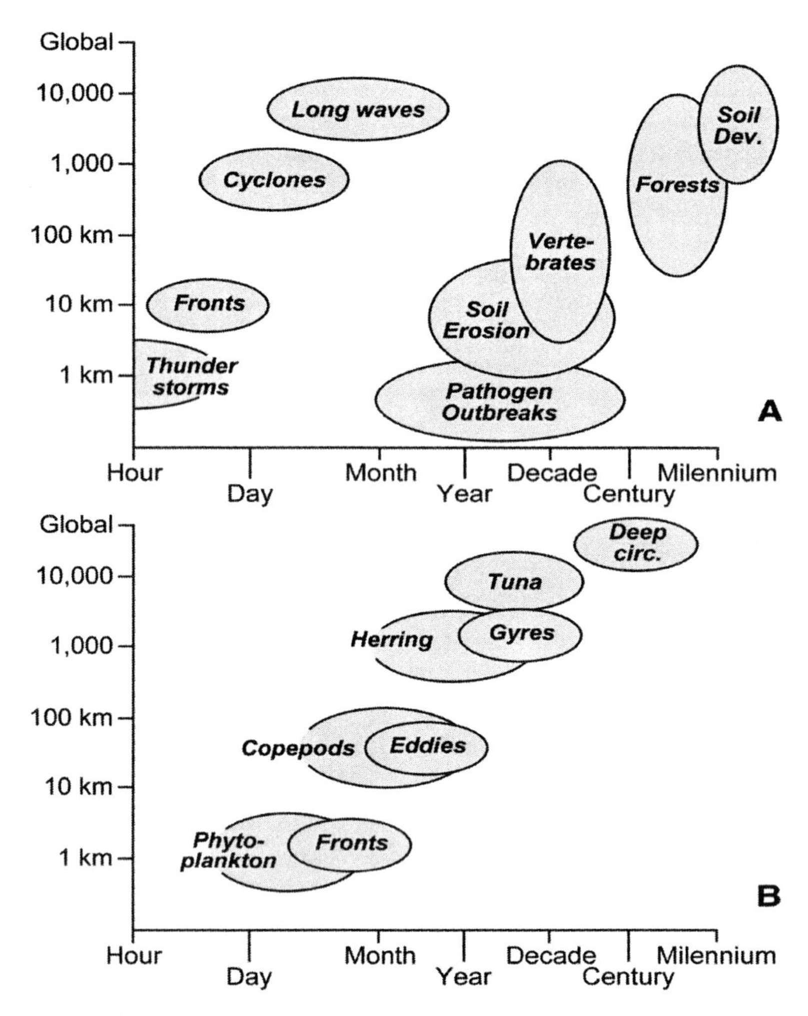


Figure 20.1 Space and time scales: (a) atmospheric processes and terrestrial groups, demonstrating their marked separation in time scale, and (b) similar scales for ocean physics and for major groups in pelagic ecosystems (adapted from Steele, 1991).

In the sea there is a different concern. We would expect the systems to adapt relatively easily to climate change, as indicated by the paleological record (Thomas, 2001). But would this adaptation be far-reaching; with changes at all trophic levels? And how do lower trophic levels respond to pervasive and gross overfishing? These interactions raise the vexing question of attributing causality. Such questions, in turn, require for their answers a much better insight into the dynamics of food webs, and for management, some ability to specify the implications of trade-offs between diversity and productivity in a rapidly changing environment.

*Lower trophic levels*

Early plankton models (Riley et al, 1949) used vertical mixing rates across the thermocline as the factor determining the rate of primary production. Cushing (1989) linked the magnitude of vertical stratification to the extent and complexity of the microbial loop (Fig. 20.2). These studies provide the basis for a specific relation between stability, productivity and diversity illustrated by Fig. 20.2. Physical stability, in the sense of persistent vertical stratification, will result in low "new" production limited by slow upward mixing of nutrients. However the proportionately large recycling of nutrients within the microbial loop will generate relatively higher total (= new + recycled) production and require diversified communities (Fig. 20.2). These interrelations are opposite in direction and different in underlying process from the proposed terrestrial approach (Tilman, 2001). The marine concept is encapsulated in the $f$-ratio = (new/total) production which varies from $0.8 > f > 0.1$. This large range determines how much of the total primary production is available to higher trophic levels in the food web.

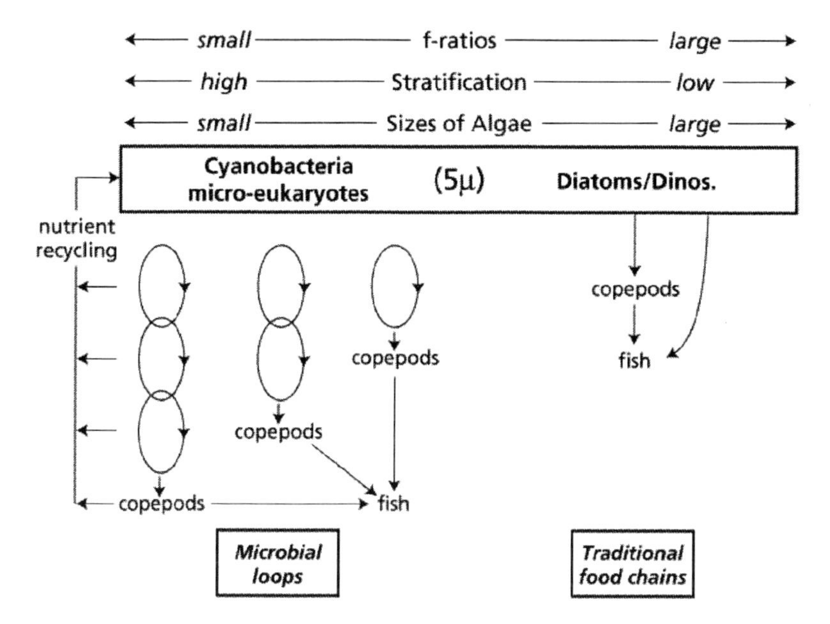


Figure 20.2 Changes in stratification cause change in community structure exemplified by the f-ratio (adapted from Cushing, 1992).

The field work that forms the basis for these generalizations has been confined to open ocean systems, with the lowest $f$-ratios found in mid-ocean gyres (Laws et al., 2000). A major question is how far these concepts will transfer to coastal systems especially ones where the water column is well mixed top-to-bottom. Also, how will the different patterns seen in microbial communities translate to higher trophic levels?

### *Higher Trophic Levels*

The highest fishery yields are obtained in upwelling systems from simple linear food chains in which much of the production is channeled through one or two forage species (Ryther, 1969; Cury et al., 2000). High production is thus associated with low trophic level and low species diversity. These productive food chains also exhibit high variability (Spencer and Collie, 1997b), driven from the bottom up by changes in oceanographic conditions such as El Nino. Thus there is also a negative relationship between production and stability, defined as persistence. More stable and diverse fish communities are generally found at lower latitudes. Here, habitat associations enhance species diversity, and communities are structured more by species interactions. The most productive fish communities have low species diversity yet high functional diversity. Production is dominated by one or two species with highly variable abundance. Species dominance changes with a shift in the oceanographic regime, yet high levels of production are maintained by the alternative species. Examples include the alternation between sardine and anchovy fisheries in upwelling systems and fish and benthic invertebrates in sub-boreal ecosystems (Myers and Worm, 2003).

There have been many attempts to correlate directly fish populations or, more usually, fishery yields, with physical factors such as temperature, salinity, currents, or wind strength and direction. One recent example (Solow, 2002) relates Norwegian cod recruitment to the NAO (North Atlantic Oscillation). Data from the last two decades (Fig. 20.3a) show a strong relation but Solow's retrospective analysis (Fig. 20.3b) revealed a lack of any relation for the preceding decades. A correlation between plankton populations and NAO was noted over two decades (Reid et al., 1998) but this, also, broke down in later observations. There are other examples of fishery-environment correlations changing with time (Drinkwater and Myers, 1987). Taken together, such results suggest that changes at decadal scales *within* the ecosystem can turn on and off these physical-biological couplings with fish stocks (Skud, 1982). Two likely reasons for these shifting correlations are that the underlying relationships are nonlinear and involve additional environmental covariates.

The most comprehensively documented interactions between climate and marine communities, principally salmon stocks, have been in the North Pacific. Various statistical analyses by Hare and Mantua (2000), suggest that there were "regime shifts" in 1977 and 1989 in concatenations of both physical and biological time series, although Rudnick and Davis(2003) criticize some of the methods used. Hare and Mantua conclude, "the large marine ecosystems of the North Pacific and Bering Sea appear to filter climate strongly and respond non-linearly to climate forcing". The Pacific populations of sardine are subject to very large fluctuations at decadal or longer time scales. These oscillations have been linked to climatic factors and to other species (Klyashtorin, 1998). However, McFarlane et al. (2002) point out that these "changes in sardine abundance, distribution and behavior provide an example of the need to consider ecosystem changes as distinct reorganizations rather than simply cycles or oscillations".

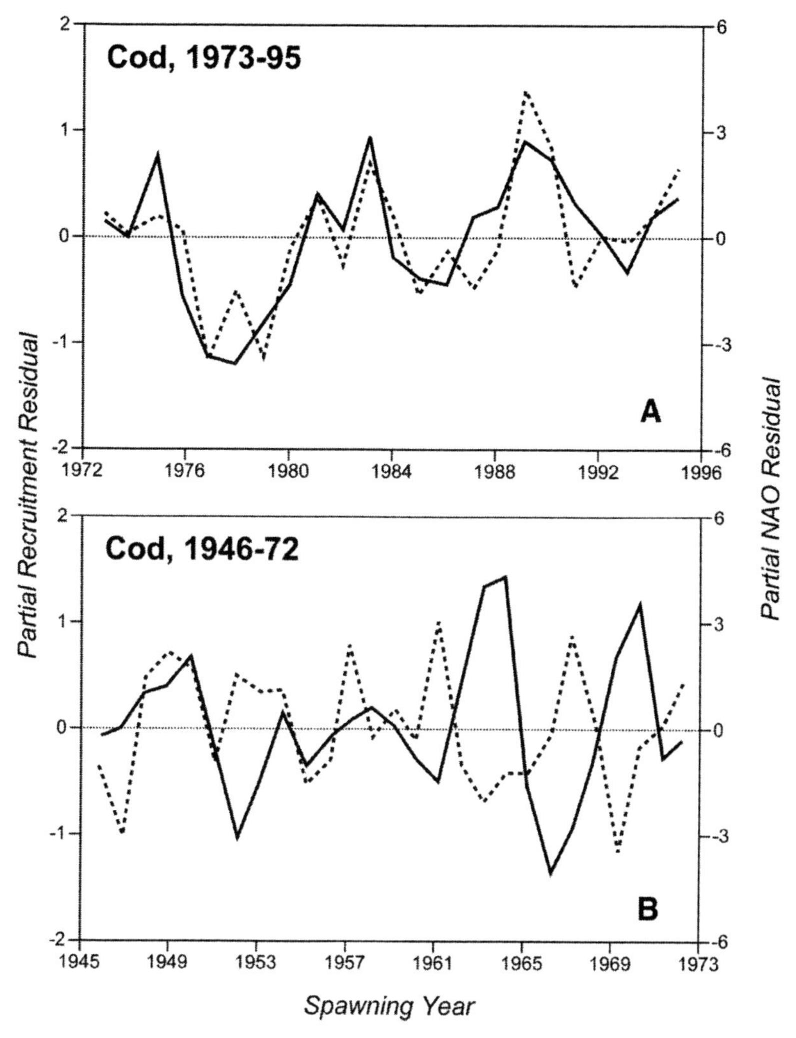


Figure 20.3 Time series of residuals of a recruitment index for Northeast Arctic cod (solid) and the NAO index three years earlier (dashed) for the spawning years (a) 1973-1995 and (b) 1946-1972 (from Solow, 2002).

The "cyclical" approach depends on the, often implicit, assumption that the population variability in space and time is, effectively, a consequence of variability in the physical processes. This may be a reasonable postulate for time scales less than one year. However without some form of density dependence, a system subject to stochastic forcing could walk randomly to extinction. This is the technical reason for an ecosystem perspective, in addition to an interest in the well-being of the non-commercial communities.

### *Physical-biological Models*

The last two decades have seen major advances in models coupling physical and biological processes. These developments can be expressed in cartoon form (Fig. 20.4). The traditional separation of physical and biological oceanography allowed each discipline to focus on the coupling between processes at different scales (Wunsch, 1981; Haury et al., 1978). Recent advances with coupled models have required an orthogonal approach, concentrating on the interactions between physical and biological processes in separable space-time regions (Fig. 20.4). This led to separate modeling activities and distinct national and international efforts (Steele, 1998). The underlying assumptions are:

1. physical processes are the first-order determinant of ecological dynamics in the sea;
2. these causal relations operate directly on separable populations or trophic groups. As a consequence;
3. the dominant physical factors are at the same time (and therefore space) scales as the ecological processes.

One specific approach has been to use detailed 3-D simulations of the complex physical regimes in areas such as Georges Bank (Werner et al., 2001) or the North Sea (Bartsch et al., 1989) to track the advection and dispersal of fish larvae relative to their copepod food supply. In this approach an individual-based model of feeding and growth is embedded in a finite-element circulation model. Such studies can show how larvae may, or may not, be transported to their nursery grounds (Corten, 1986). They can also demonstrate mismatches in space and time between the larvae and their preferred food (Lynch et al., 2001). These models usually treat the organisms as passive particles with possible additions of vertical migration and/or fixed mortality rates. The spatial scale of the circulation model (5–20km) is typically four orders of magnitude larger than the length scales important to feeding of larval fish. Small changes in the biological rate parameters can have large consequences on cohort survival. Some investigators have therefore simplified the biological model by making larval growth rates direct functions of physical variables such as temperature and day length (Heath and Gallegos, 1998).

To the extent there is scaling up or down, this is predominantly in terms of the physical interactions. For example, an 0-D model of the microbial loop (Fasham et al., 1990) can be embedded in a basin scale physical simulation (Sarmiento et al., 1993). However the comparison with remotely sensed observations depends critically on the simulation of vertical mixing rather than the horizontal components of the general circulation. Conversely, simulations of larval fish advection in the North Sea (Bartsch et al., 1989) or Georges Bank/Gulf of Maine (Werner et al., 2001) require the introduction of diel vertical migration to accommodate the effects of vertical structure in the current systems.

All these studies demonstrate the tremendous advantages in constraining the bounds on a particular question (Medawar's (1967) art of the soluble). But they also illustrate how such success generates new questions, usually at these boundaries. These questions concern the assumption that one can neglect the inputs to a particular "box" in Fig. 20.4b from the higher and lower trophic groups in Fig. 20.4a and, by implication, from their longer and shorter time scales. While physical processes are likely to be the first-order determinants of ecological dynamics, this coupling is likely to be mediated by food web processes.

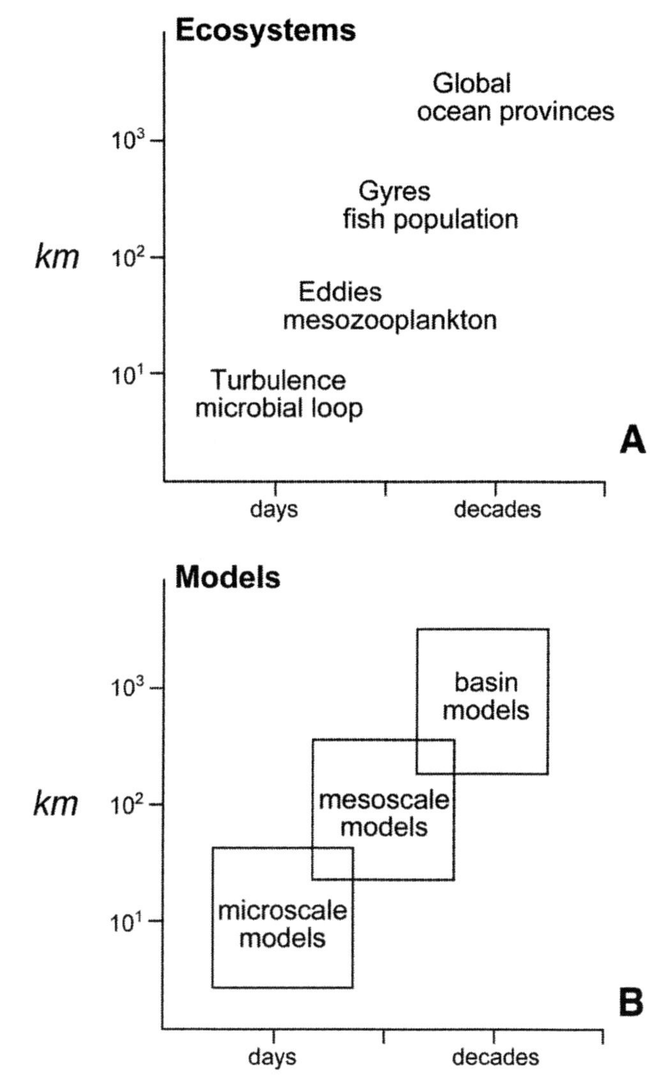


Figure 20.4 Physical-biological programs and models: (a) processes (see Fig.1 and text), (b) scales of coupled models.

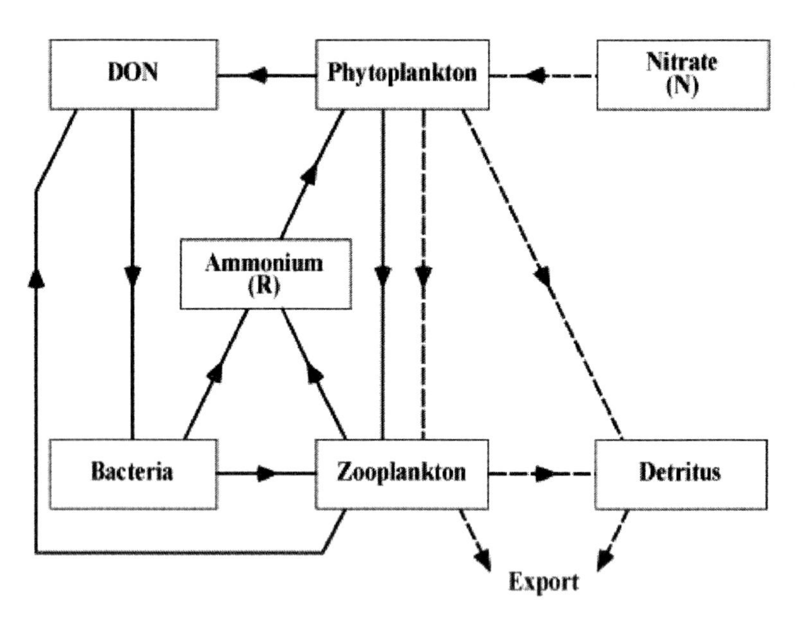


Figure 20.5 Flow diagram for cycling between bacteria, phytoplankton, and zooplankton (after Fasham et al., 1990). The dashed lines are the traditional transfers to other trophic levels. Solid lines depict recycling through microbial components.

## 3. The Ecosystem Perspective

The desire for "holistic" management of marine ecosystems stems primarily from our concerns that extreme over fishing on the shelf not only can drive fish stocks to commercial extinction but can also alter the rest of the food web by direct impacts in the sea bed (Collie et al., 2000) or through top-down changes in community structure (Jackson et al., 2001). Bottom-up issues include the increasing eutrophication of coastal waters where the immediate effects seen in "dead zones" (Rabalais and Turner, 2001) may be indicators of larger-scale consequences for food webs, such as the Baltic Sea. (Elmgren, 1989). In the open ocean the main concern has been with the role of the biological pump in transporting carbon to deeper layers. This has focussed attention on the microbial food web (Fig. 20.5) in the euphotic zone. But, as interest extends to longer time scales, there is a need to know more about the "export" processes (Fig. 20.5) to deeper water and higher trophic levels.

In the open ocean the *f*-ratio (new/total) typically increases with total production and decreases with temperature (Eppley and Peterson, 1979; Laws et al., 2000). The changes can be highly non-linear (Fig. 20.6). In fisheries, we are interested in that fraction of "new" production that is exported to higher trophic levels (Fig. 20.5). Potentially, the non-linearity (Fig. 20.6) could result in a large multiplicative factor when deducing the effects of changes in primary production on fisheries yield (Iverson, 1990 ). This factor could be further amplified when we consider the combination of increased temperature and greater stratification—both apparently decreasing the *f*-ratio. As we pointed out earlier, these trends in physics and productivity will induce changes in community structure at several trophic levels, and could be expected to affect the species composition as well as the total fish yield.

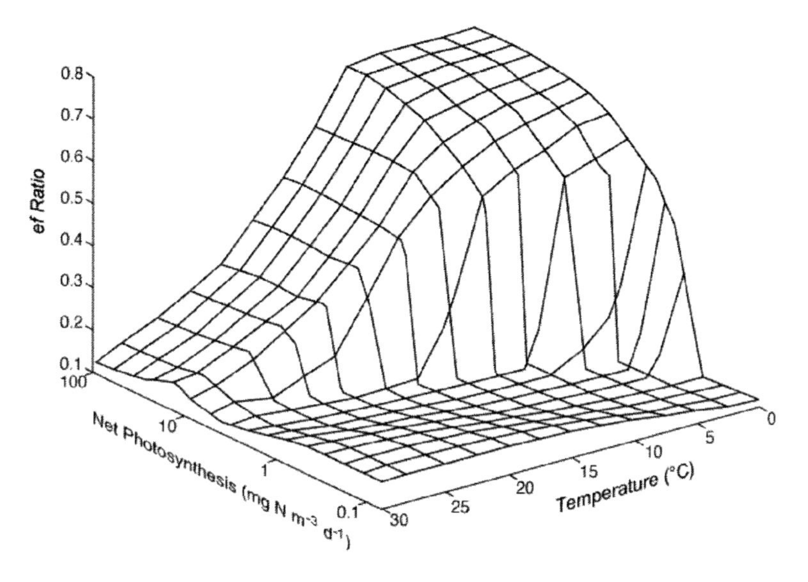


Figure 20.6 Results of a simulation model of the microbial loop giving the export-ratio (export/total production) as a function of photosynthesis and temperature (adapted from Laws et al., 2000).

### *New versus total production on the shelf*

Nearly all our insight into the functions of the microbial food web has been based on studies in the open ocean. These input/output relations epitomized by the *f*-ratio, can be as critical for processes on the continental shelf. An unsupported expectation from Fig. 20.6 that, at high production rates on the shelf, most of the production will go to higher trophic levels, could lead to excessive estimates of fish yields in the past, or the future (Christensen and Pauly, 1998). How well does the assumption that the *f*-ratio increases with primary production work when one compares different physical regimes on the shelf? There is increasing evidence that higher production levels in shallower waters on the shelf may be the result of more cycling rather than input of new nutrients (Richardson and Pederson, 1998; Franks and Chen, 2001). Further, on the shelf, the physical processes for new nutrients involve lateral intrusion as well as vertical exchange.

The critical factors are the relative time scales of nitrification versus vertical mixing, with the benthic food web as an important trophic component. The time scales of the ecological processes (Fig. 20.7a) go from days to years. Nitrification of organic matter to $NO_3$ is usually considered a relatively slow process. "Marine nitrifying bacteria are ubiquitous in the world ocean... However they are never very abundant and, at least for those species in culture, grow very slowly. Certain heterotrophic bacteria can also oxidize $NH_4$ to $NO_2$ and $NO_3$ (but) very little is known about heterotrophic nitrification in the sea" (Karl and Michaels, 2001). These authors quote experimental work by von Brand and Rakestraw more than

50 years ago that shows a two-month time lag from $NH_4$ to $NO_3$ (Fig. 20.7b). Observations of phosphate regeneration below the euphotic zone in the North Sea indicated the same time scale (Steele, 1956).

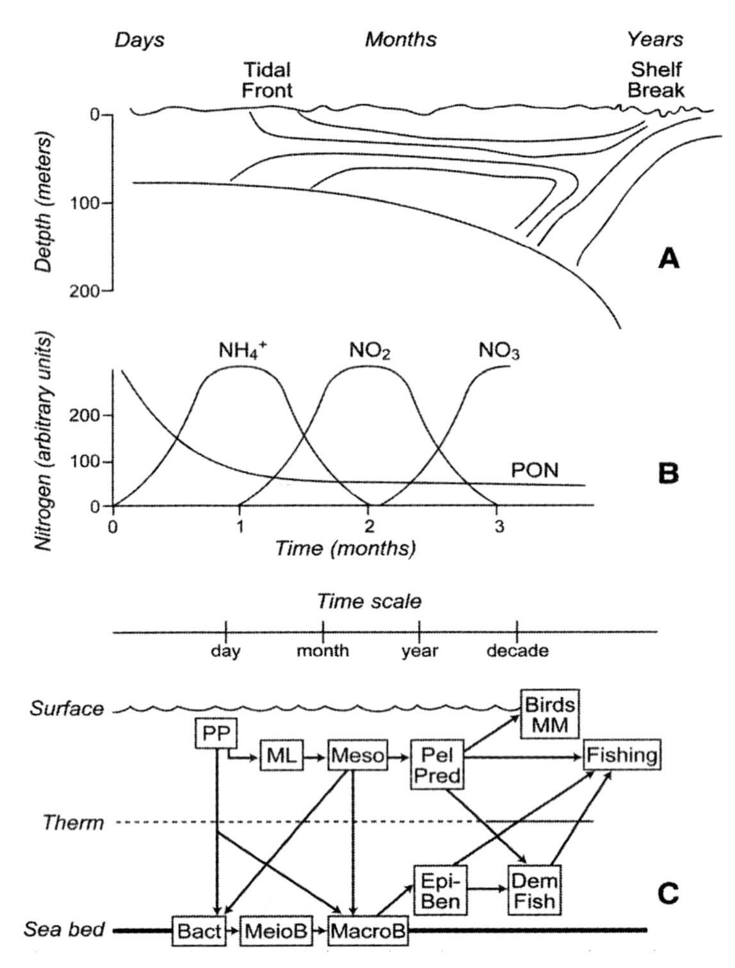


Figure 20.7 Coastal processes: (a) time scales – days to years - of top-to-bottom physical mixing are related to frontal systems determined primarily by depth and tidal currents, (b) regeneration of organic matter appears to have time scales of weeks to months, (c) food web expressed in terms of time scales.

The time scales for physical exchange of water and nutrients between the upper euphotic layers and deeper water also vary from years to days as one goes from deep water onto the shelf and then into shallow tidally mixed areas. These time scales have two near discontinuities—at the shelf-break front and the tidally mixed front (Fig. 20.7c). These changes can be seen going from the Norwegian Sea to the Dogger Bank, from the northwest Atlantic to Georges Bank and from the Bering Sea to the Alaskan shelf. Comparing the different processes in Fig. 20.7 reveals the inadequacy of the simple "oceanic" dichotomy between new and recycled production, measured by the uptake ratio $NO_3/(NO_3+NH_4)$. In particular, the benthic components will play a major role involving both recycling and production near the tidal front. This front can have significant horizontal displacement at lunar time scales (Pingree, 1978) that correspond to the presumed nitrification rates.

The benthic communities on the shelf have two relatively distinct roles in relation to water column stratification. In well stratified water columns (right hand side of Fig. 20.7a) the benthos is effectively isolated from primary production in the euphotic zone and dependent on detrital material sinking out. Nutrients regenerated from this community are resupplied for primary production mainly during convective winter overturn. Moreover many benthic invertebrates have an annual pelagic phase in their life cycle. Thus benthic communities in a stratified water column are coupled to primary production on an annual time scale.

At the left hand side of Fig. 20.7a, where mixing is dominant, benthic suspension feeders occupy a niche similar to filter feeding plankton with the advantages of a sedentary life style. Such "tethered zooplankton" in coastal ecosystems can consume a substantial fraction of the primary production. This "short circuiting" of the traditional oceanic food chain is most apparent in estuaries. For example, it has been estimated that former oyster reefs could filter, in three days, an equivalent to the entire volume of Chesapeake Bay (Newell, 1988). While less dramatic, suspension-feeding bivalves can also consume a substantial amount of production on continental shelves. For example, Hermsen et al. (2003) estimated scallop production on northern Georges Bank of up to 280 kcal $m^{-2}$ $yr^{-1}$. Assuming a trophic efficiency of 10%, dense aggregations of scallops and other bivalves could filter a substantial amount of the water column productivity according to the energy budget of Sissenwine et al. (1984).

The transition from suspension to detrital feeding corresponds to a decrease in natural physical disturbance and to a change in habitat from sand through mud interspersed with gravel and boulders. This gradient in physical structure corresponds to a trend to more fragile fauna inhabiting more complex habitats (Collie et al., 2000). Unfortunately these latter communities are increasingly subject to human disturbance by trawls and dredges, resulting in simplification of the community structure. Generally, large epifaunal species are more affected by bottom fishing (e.g. Prena et al., 1999) than smaller macrofaunal (e.g Kenchington et al., 2001) or meiofaunal taxa (Schratzerger et al., 2001). In areas impacted by bottom fishing, reductions in macrofaunal (Jennings et al., 2001) and megafaunal production (Hermsen et al., 2003) have been measured. It is unclear where in the food web the excess energy or carbon is going, though it is likely to be channeled to smaller organisms. A decrease in epifauna can clearly affect the predatory fauna, especially demersal fish. It is less likely that such changes will alter the overall nutrient cycling at annual time scales, since the most sensitive habitats usually occur under seasonally stratified water columns that separate the euphotic and benthic webs.

### *Cascade Theories*

We have shown how bottom-up processes observed in the open ocean become more complicated on the continental shelf. Conversely, are top-down processes evident in the sea? So called "trophic cascades" have been described in lake studies (Carpenter et al., 1985). The basic concept is that in lakes with three or four trophic levels—phytoplankton, zooplankton, planktivorous fish, piscivorous fish—natural or experimental changes at the top trophic level can cascade to the bottom, usually by alternating high and low abundances at decreasing trophic levels. These manipulations are most effective in small lakes with relatively simple food webs. There is considerable disagreement about evidence for this effect in terrestrial systems (Pace et al., 1999; Polis et al., 2000) even though the original idea came from land (Hairston et al., 1960). Our interest here is in the evidence for cascades in the sea and especially in switches in high/low patterns. The most frequently cited illustration (Estes and Duggins, 1995) involves sea otters, sea urchins, kelp beds and more recently, killer whales (Estes et al., 1998). More generally rocky shores are viewed as examples of top-down control (Menge, 2000). Coral reefs show phase shifts in dominance from corals to macro-algae but the relative importance of fishing, eutrophication and hurricanes is arguable (Nystrom et al., 2000; Knowlton, 1992; McManus et al.,2000). These examples lie between smaller lakes and the open ocean, geographically and ecologically. When one turns to pelagic food webs, the evidence for cascades is equivocal.

There are relatively few cases of possible cascades across three levels of pelagic systems. Some are experimental manipulations and others are merely interpretations of gross pattern in field communities without actual switching (Verity and Smetacek, 1996). Daskalov (2002) describes inverse switches in abundance at three trophic levels in the Black Sea, but the relative consequences of fishing versus eutrophication and invading gelatinous predators are difficult to disentangle. The general shortage of evidence is borne out by statistical studies. Micheli (1999) found, from a meta-analysis of mesocosms and natural systems, that "nutrients generally enhance phytoplankton, and carnivores depress herbivore biomass. However resource and consumer effects attenuate through marine pelagic food webs". The lack of general cascade sequences in the ocean has led to speculation that there is a bottleneck in the system at the planktonic herbivore/carnivore link; the "wasp-waist" hypothesis (Cury et al., 2000) seen specifically in the low diversity of small pelagic fish in upwelling systems.

The main tests should be found on the continental shelves through the inadvertent large-scale manipulations arising from the massive changes in fish populations, yet they provide little positive support for cascades. Although there are major changes in tropical fisheries (Pauly et al, 1998) the most studied systems are at high latitudes. The huge switch in North Sea stocks at the time of the "gadoid outburst" (Fig. 20.8) shows a five-year reduction in total copepods but no obvious indications of a cascade. Similarly, the dramatic changes in fish stocks on Georges Bank and the Gulf of Maine (Fig. 20.9) are not matched by switches in plankton abundance (Sherman et al., 1998). These two "experiments" suggest that the consequences of removal of major predatory groups (pelagic or demersal) do not propagate far down the food web. Rather there appears to be replacement by adjacent trophic groups. This is seen especially for commercially harvested groups, including cephalopods (Piatkowski et al., 2001) but may extend to non-commercial predators such as jellyfish (Jackson et al., 2001). In one case, the disappearance of cod around Newfoundland, the replacement species appear to be, in part, shellfish such as crab and shrimp (Rice, 2002). All of these examples imply great flexibility in the food web.

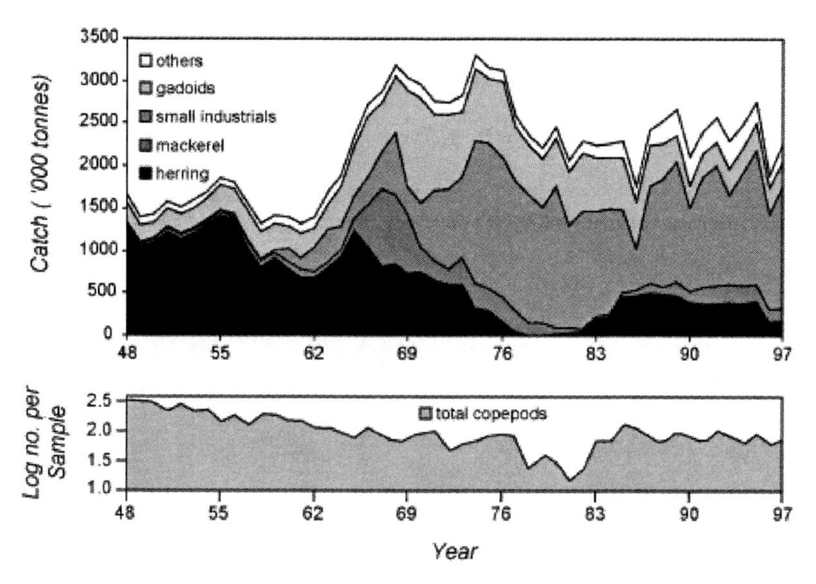


Figure 20.8 Time series in the North Sea: (top) landings of selected fish species, (bottom) copepod abundance (adapted from Edwards et al, 2001).

## 4. The Dynamics of Species Diversity

*Diversity in marine ecosystems*

The dominant perceptions of marine biodiversity fall into two categories: those shallow near-shore waters where the primary producers are at the human scale—coral reefs and kelp forests; and in the open sea, the megafauna at higher trophic levels—marine mammals, sea birds and large fish such as tuna. Coral reefs are often compared to rain forests in terms of the biotic structures that support primary production (Connell, 1978). So measures for the conservation of reef diversity, such as reserves, may have much in common with forest systems. Kelp "forests" also appear similar, although they have much shorter life spans and different reproductive life cycles.

Denizens of the coastal seas, on the other hand, have quite different life styles. The longest lived biota are at the top rather than the bottom of the trophic ladder and even then the dominant fish populations have life spans of a few decades at most (Fig. 20.1, deep-sea fish are not considered here). There is an even greater contrast between the primary producers, with the phytoplankton having life spans at least a thousand times less than trees. We are only now appreciating the tremendous diversity in the microbial loop. Most food web diagrams ignore this complexity. Even those that focus on this aspect use broad taxonomic groupings, such as diatoms, coccolithophores, flagellate, ciliates, or size categories—nanoplankton, picoplankton, microplankton.

Given these facts about marine systems: that there is close coupling between physical and biological variability, and that the time scales are much shorter than terrestrial ones, we can expect marine systems to be much more sensitive to alterations in their environment but also to be much more adaptable. Thus, there needs to be a different perspective on diversity in the sea and its relations to natural and anthropogenic forcing.

On land, conservation of diversity is usually seen as, essentially, preservation of species number or "richness". This is linked to the need to preserve habitats, especially those plant communities which are the long lived basis for the ecosystem structures. There is accumulating evidence that the diversity of these communities is related, positively, to their productivity and stability (Tilman, 2001). Thus biodiversity can be seen as having net benefits to the human societies that live within these systems by contributing to "ecosystem services" (Costanza et al., 1997).

In coastal seas, species richness is undoubtedly a significant property of any ecosystem, but it appears much more difficult to quantify this aspect and to relate it to other attributes of the system. According to Ray and Grassle (1991), "Whole theories [of marine diversity] have been predicated on latitudinal differences, depth differences, patterns of upwelling, physical energy in the system, the degree of pelagic-benthic coupling, proximity and magnitude of riverine input, water mass distribution, and interactions with adjacent land masses. None of these theories have been adequately tested because largescale sampling programs have not been continued over long enough time periods." We shall not attempt to reiterate these theories. New programs such as Ocean Biogeographical Information Systems (OBIS) are proposing to improve the database for major groups such as fish, cephalopods, corals reefs etc. (Grassle, 2000).

### *Functional diversity*

How we define diversity depends on how we see its relation to other ecological attributes such as stability and resilience. Holling (1973) defined resilience as the capacity of a system to buffer disturbance. This role for diversity in maintaining robustness was termed "functional diversity" (Folke et al., 1996). Tilman (2001) in a review of functional diversity defines it as "those components of biodiversity that determine how an ecosystem functions". This definition is primarily static and concerned with persistence. Sudden switches are considered to be "catastrophic" (Scheffer et al., 2001) and imply a damaging loss of resilience.

In marine systems these switches can be regarded as the natural response to environmental change at longer time scales (Steele, 1985). The switches demonstrate the *overall* resilience of the system through its ability to absorb large changes in its environment, including fishing. This distinction imposes the need for a different approach to diversity via complexity as a dynamic attribute. Further, it raises questions about the relevance of stability in its usual meaning of persistence. Given the very different and shorter time scales of adaptability in the sea, we need a different and dynamic category of diversity in the sea. To denote this dynamic aspect of marine ecosystems, Steele (1991) defined the term "functional diversity" as the variety of different responses that any marine ecosystem can make to environmental change. Similarly, resilience is used in terrestrial ecology to denote resistance to switching between alternative community structures; whereas we consider marine systems to be "robust" when they *can* adapt in this way. These alternative uses of the term "functional diversity" reflect the contrasting perceptions of the two systems.

In the last decade these different responses by marine systems have been codified by the term "regime shifts" (Brodeur and Ware, 1992; Lluch-Belda et al., 1989; Steele, 1998). The underlying concept is that, at decadal time scales, marine ecosystems display patterns that could be typified as punctuated equilibria. In this context, functional diversity would correspond to robust systems responding to environmental change by switching to alternative community structures. "Environmental change" would include climatic, fishery and contaminant forcing; i.e. top-down as well as bottom-up. This approach would replace the relatively static concept of diversity (as used on land) with the inherent dynamics of change in marine ecosystems. A loss of functional diversity would result in a loss of the ability to adapt to environmental change by switching between different regimes.

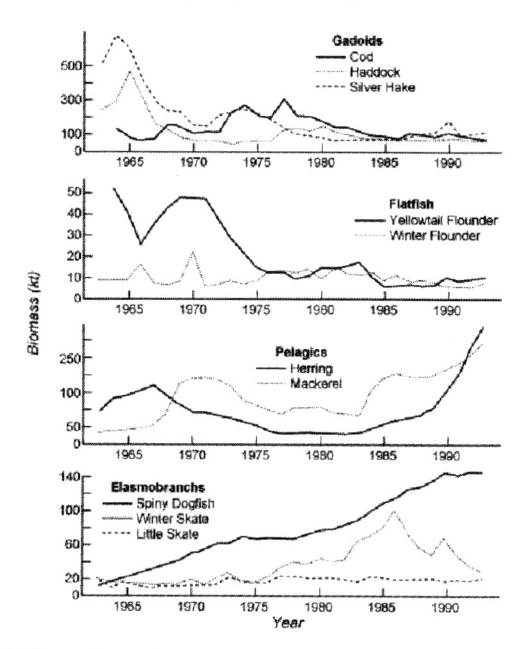


Figure 20.9 Biomass of fish species on Georges Bank (from Collie & DeLong, 1999).

The best known "natural" example is given by the alternating dominance, at decadal scales, of sardine and anchovy populations at several locations in the world's oceans (Chavez et al., 2003). However the extreme overfishing of gadoid species, especially cod, in the North Atlantic provides another perspective. Fig. 20.9 from Georges Bank shows the replacement of gadoids first by elasmobranches and then by herring and mackerel. In the Newfoundland fishery (Worm and Myers, 2003) the decline in cod was followed by increase in flatfish and then invertebrates such as shrimp and crab. In the North Sea, Fig. 20.8, the major replacement species were small "industrial" species such as sandeels. These examples of functional diversity also illustrate the potential consequences of successive depletion of the replacement communities or stocks by continued overfishing.

This definition of functional diversity involves a number of assumptions. For example, the concept of alternative communities implies the existence of multiple equilibria expressed in terms of highly non-linear model systems. Further, these model systems involve the coupling of several different trophic levels as well as forcing by "external" factors such as climatic change or fishing pressure. Such a conceptual basis is necessary if data on species (or genetic) diversity are to be related to other measures such as fish production—just as competition theory is used to relate plant diversity to productivity on land (Lehman and Tilman, 2000).

## 5. Modeling: Diversity or Dynamics?

The previous sections described how the structure of food webs and their responses to perturbation are intimately related. But the nature of these relations is not obvious and, certainly, not simple. Past experience has demonstrated that, technically, it is easy to construct large non-linear computer simulations of ecosystem dynamics but these have not enhanced our insight into how such natural systems work. So when we turn to theory, the usual approaches require some separation into a consideration of diversity as a relatively static description of the complex structure of food webs, or of stability in terms of the dynamics of comparatively simple systems. Sometimes this is posed as the choice between realism and understanding. We shall review recent work from both aspects then consider how they may be combined.

### *Food web structures and fluxes*

There is a large literature on the purely structural aspects of food webs (Pimm, 1982 Cohen et al., 1990; Polis and Winemiller, 2001). Much of this literature discusses observations and theories concerning the actual versus potential number of links in food webs. The number of links between species, or trophic groups, S, could be $0(S^2)$ but is usually closer to 0(S). This complements the classic paper by May (1971) who showed that, in simple Lotka-Volterra type models, complexity does not enhance stability, defined in terms of the response to small perturbations. In the present context the main, qualitative, message is that *functionally* food webs will be relatively simple, even though very large species-by-species networks can be constructed (Hardy, 1924; Link, 2002). In the sea, because of the dominance of the physical-biological coupling, the most marked correspondence is between the amplitude and frequency of physical disturbance and dominance of a few species, particularly at lower trophic levels (Fig. 20.2). This correspondence is illustrated by the contrast between productive upwelling areas and oligotrophic mid-ocean gyres, or between benthic communities on the continental shelf and in the deep ocean. Thus the functional groupings tend to be defined by physical factors—temperature and currents—or by related nutrient availability, phosphate, nitrate, iron, silicate, as well as by predator/prey relations.

The interest here is in the flux of "food" through these webs portrayed as links between functional groups, where food can be defined in units of biomass, energy, carbon or essential nutrients such as nitrogen. The choice of unit varies with the purpose of the study; biomass and energy being used for upper trophic levels; carbon and nitrogen for microbial webs. The earliest energy budget by Lindeman (1942) was for a fresh-water ecosystem. An early marine one (Fig. 20.10) illustrates the principles and the problems. The input, primary production, determines the flows up the food web when combined with (a) the transfer (or ecological) efficiency for each box and (b) the proportions of each output that go to other boxes (see Box A).

Such food web models can accommodate many taxa and can readily be solved with the methods of linear algebra, but there are two obvious problems. Firstly, the microbial loop is given only perfunctory notice as the contribution of detritus to benthos compared with Fig. 20.5. More critical is the fact that the values for the consumption fractions must be specified a priori. These are either poorly known, or completely unknown, especially for the boxes in the middle of the web. This is critical if the response of the system to external stress is to redistribute these allocations. There are similar problems for a top-down calculation such as ECOPATH (Christensen and Pauly, 1993) starting from fish production and estimating the "output" as detritus and plankton production. These terms are notoriously difficult to constrain. Increasing the complexity of the web does not eliminate these problems.

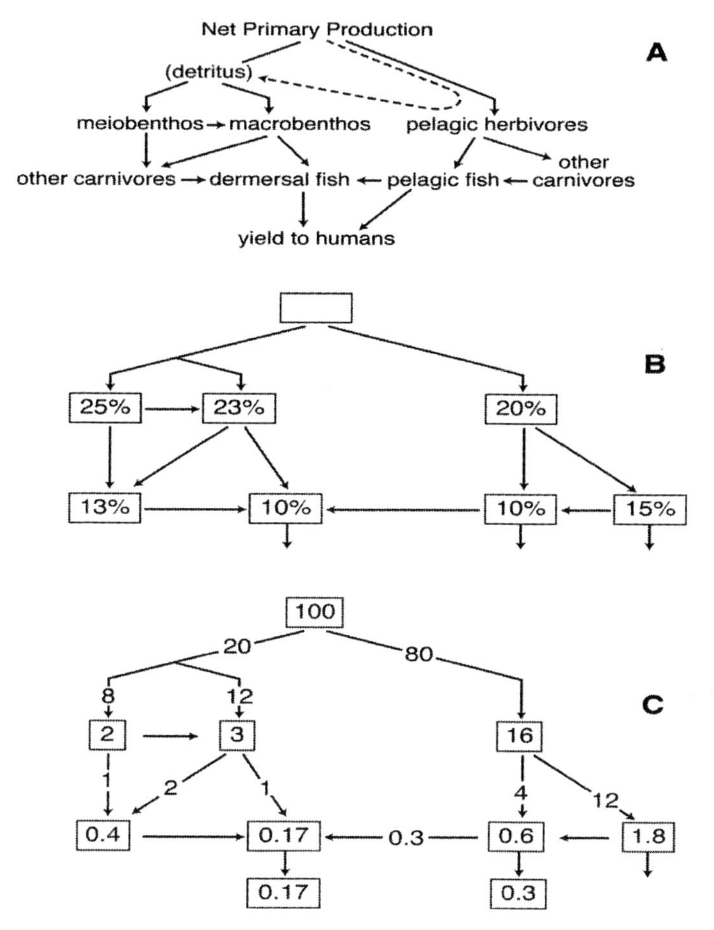


Figure 20.10 Carbon fluxes through a food web for the North Sea, assuming new production of 100g $m^{-2}$ $yr^{-1}$. (a) a simplified food web, (b) transfer, or ecological efficiencies expressing output as percent of input, (c) each box indicates production available to competitors and arrows determine the allocation (adapted from Steele,1965).

The inverse modeling approach (Eq. A2) resolves this problem by treating the food-web links as state variables, constrained within biologically realistic intervals. Any data on internal fluxes, as well as input and output rates, can be used as constraints. Also the efficiency is usually constrained by upper and lower bounds; these can be bounds on the ratios *R*/*E*, and so on. However these constraints are insufficient to determine the system and some optimization (usually a minimization) on all the variables must be introduced (Eq. A4). This is a critical step since such a criterion acts, essentially, as a general ecological principle for this system.

A further problem with these budgets for a full food web from phytoplankton to fish is the assumption of mass balance when we know that there is significant variability at daily, seasonal, annual and decadal time scales. For this reason separate budgets for lower trophic levels have been developed for different seasons (Jackson and Eldridge, 1992).

### *BOX A: Energy budget models*

Production of each food-web component, *i*, can be calculated from a set of linear equations,

$$e_i \cdot C_i = \Sigma\, a_{ij} \cdot C_j + g_i\,, \tag{A1}$$

where $e_i$ = efficiency, the "$a_{ij}$"s are the fractions of production going to consumption, $C_j$; and the "g"s are the external input/output rates. Such food web models can accommodate many taxa and can readily be solved with the methods of linear algebra.

An alternative formulation, developed by Vezina and Platt (1988), recognizes our inadequate quantitative knowledge of the links in the food web by making them into state variables, $G_{ij} = a_{ij} . C_j$. Also, efficiency is specified indirectly by introducing respiration, *R*, and particulate excretion, *E*. Then consumption is written as,

$$C_i = \Sigma\, G_{ij} + R_i + E_i\,, \tag{A2}$$

and efficiency as,

$$e_i = \Sigma\, G_{ij}/C_i\,. \tag{A3}$$

The flexibility provided by these additional variables has to be balanced by additional constraints since there is no increase in the number of equations. In network models of marine systems (Vezina and Platt, 1988; Jackson and Eldridge, 1992) the criterion has been to minimize the sum of squares of all the variables:

$$\text{Min}\{\Sigma\,(\Sigma\, G_{ij}^2) + R_i^2 + E_i^2\} \tag{A4}$$

This criterion was derived from inverse modeling in physical oceanography (Wunsch, 1996) and it is not clear how much basis it has in an ecological context (Odum and Pinkerton, 1955).

### *Dynamic ecosystem and multispecies models*

Food-web models can describe trophic interactions during periods of relative stability. Obviously, marine ecosystems are not at equilibrium and can exhibit dramatic shifts in relative abundance of species or trophic levels. It is necessary to introduce time dynamics into food web models by replacing the steady-state equations with differential or difference equations. Dynamic models of ecosystems are known to have a wide range of possible responses to perturbation, from rapid damping, through limit cycles to chaos (Hassell et al., 1976). One possible choice is to assume that communities can switch from one quasi-stable state to another in response to external forcing (Holling, 1973; May, 1977).

The first key question is how many species or taxa to include. Inclusion of many species gives more biological realism but at the cost of more interaction terms to estimate. Aggregating species into taxonomic or functional groups allows the modeler to focus on key trophic interactions with fewer parameters to estimate. However, there is a risk of aggregating species with dissimilar dynamics and of overlooking important feedback mechanisms. There do not seem to be clear guidelines for choosing the optimal model complexity, but an obvious prerequisite is to have a good understanding of the feeding interrelationships. Hereafter we use "species" to refer to either a single species or group of species.

Compensatory density-dependence is necessary to stabilize the ecosystem model (Murray, 1989) and to ensure that all species will persist at equilibrium. Otherwise the simulated ecosystem will self-simplify with one or more species going extinct. One way to ensure species persistence is to partition each species into one pool that is available to predators and another that is unavailable (Matsuda et al., 1991; Walters et al., 1997). Then, even with high predator abundance, a prey population can persist in the unavailable pool. In this approach, implemented in the popular ECOSIM program (Walters et al., 1997), stabilization is thus provided by the prey dynamics. ECOSIM has been used to investigate the dynamics of many coastal pelagic ecosystems. However, the ECOSIM approach does not intrinsically incorporate the complex nonlinear dynamics that are consistent with regime shifts that are observed in marine ecosystems. For example, an ECOSIM model of the eastern Bering Sea did not reproduce the shift from a marine mammal dominated ecosystem to a pollock dominated system (Trites et al., 1999). An extrinsic forcing function would be required to simulate this type of community shift.

An alternative approach is to include prey self limitation and nonlinear predator functional responses (Holling, 1965, Hassell et al., 1976). The Holling type-II functional response (hyperbolic curve) is inversely density dependent. At high prey densities, the predators become saturated and the per capita prey mortality is reduced. However, at low prey densities per capita predation increases and the prey may go extinct. With a Holling type-III functional response (sigmoid curve), per capita mortality is reduced at low prey densities, allowing the prey population to persist. Thus the type-III functional response is equivalent to having a refuge at low prey densities. In this case, stabilization of the ecosystem model depends on the predator dynamics.

Models with nonlinear functional responses can exhibit complex dynamic behavior. Abrupt shifts in prey and predator abundance can result from gradual changes in prey productivity or predator abundance (Steele and Henderson, 1981) as a consequence of environmental changes propagating from the bottom up or the top down. In fish communities, such shifts can also be triggered by fishing pressure. Spencer and Collie (1996) showed that environmental changes (patterned on SST data) and fishing could act in concert to cause flips in predator-prey systems.

Shifts in species abundances can be illustrated with a simple pair of equations that simulate the top of the food web in Fig. 20.10 for a pelagic, $X$, and a demersal, $Y$, fish stock (see Box B). Both stocks have food sources derived from production at lower trophic levels through mesozooplankton and detritus feeders, respectively; and linear fishing mortality. These formulations correspond to the simplest single-species fishery models (Rothschild, 1986). But we assume that the demersal fish (e.g. cod) feed opportunistically on the pelagics (e.g. herring or sand-lance) with an s-shaped, Holling type-III, relation (Holling, 1965). The equilibrium solutions are the intersections of the isoclines in Fig. 20.11 (a,b). The solid lines illustrate those regions of parameter space where the system bifurcates to give alternative solutions for $X$ *and* $Y$. If the parameter, $r_1$, is varied across the bifurcation, Fig. 20.11a, to represent environmental changes, the equilibrium values of $X$ and $Y$ switch from relatively high/high to low/low. If, however, the fishing rate on pelagics, $c$, is varied, then the switch in $X$ and $Y$ is from high/low to low/high (Fig. 20.11b). Thus, alternative causes of variability can induce opposite correlations. Thus top down as well as bottom up control can induce switching—or regime shifts—but with different effects on fish stocks.

*BOX B A simple food web model*

Assume that $X$ (pelagic) and $Y$ (demersal) populations have independent hyperbolic food supplies and linear fishing mortality. Assume $Y$ preys opportunistically on $X$ with an S-shaped functional response (Holling, 1965). Then

$$\frac{dX}{dt} = \frac{r_1 X}{a+X} - \frac{fYX^2}{e^2+X^2} - cX$$

$$\frac{dY}{dt} = \frac{r_2 Y}{b+Y} + \frac{kfYX^2}{e^2+X^2} - dY$$

The stochastic values for any parameter, $p$, are given by an autoregressive relation that generates red noise,

$$p_t = p_{t-1}(1-\lambda) + \lambda p_0 + \eta\varepsilon_t$$

Where $\varepsilon_t$ is a random variable with mean = 0 and variance = 1 and unit time is one year.

Values used here are $a$=2, $b$=10, c=0.2, $d$=0.2, $e$=0.5, $f$=0.6, $k$=0.1, $r_1$=5, $r_2$=2.

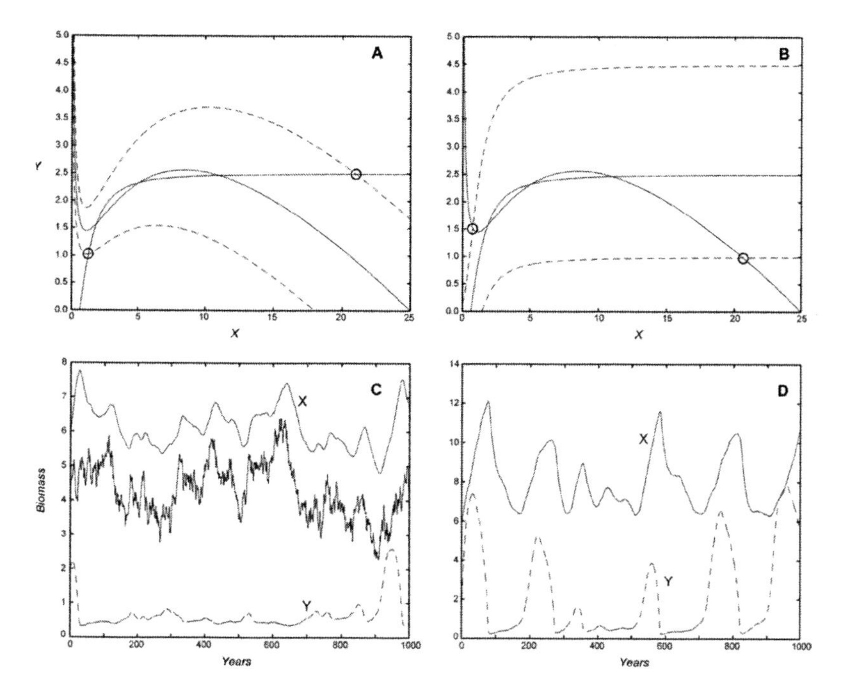


Figure 20.11 A model of pelagic-demersal $(X,Y)$ interactions.
Top: Isoclines for the model in Box B with (a) variable "environmental" input, $r_1$, (b) variable demersal fishing effort, $d$.
Bottom: Time series with $r_1$ and $r_2$ varying stochastically; (c) high fishing mortality and the stochastic forcing (d) fishing mortality across the bifurcation range.

To illustrate these features, simulations with this model Figs. 20.11 (c,d) were run with both "zooplankton", *r1,* and "benthos", *r2,* varying stochastically with the same red noise forcing. With high fishing mortality on pelagic stocks, *X,* (Fig. 20.11c) these stocks are relatively low and the demersals, *Y,* follow the stochastic input. If, however, the fishing effort is halved, so that the system is in the bifurcation domain, then *X and Y* interact significantly, and there is no obvious relation to the stochastic food supply (Fig. 20.11d). If other parameters are varied, similarly complex relations occur. This simple model shows how (a) the effects of bottom-up and top-down changes may be difficult to distinguish and (b) changes in parameters, such as fishing mortality, can move the system from dependence on environmental inputs to a dominant role for internal, ecological, interactions. This broad concordance between observations, such as those in Fig. 20.3, and simple models indicates how such non-linear food web interactions contribute to the observed patterns.

The challenge is to replicate the temporal patterns observed in real ecosystems and the mechanisms that lead to regime shifts. The Georges Bank fish community provides a well-studied example of large-amplitude shifts in species abundances (Fig. 20.9). Primarily as a result of over-fishing, the gadoids and flatfish were replaced by pelagics and elasmobranches in the 1980s and 1990s. Collie and DeLong (1999) fit a series of multi-species models to the aggregate biomass of the four species groups on Georges Bank. With linear or type-II functional responses, the system would self simplify such that one or more species would go extinct at equilibrium. The model with type-III functional responses fit statistically better than the other two models and also resulted in the persistence of all four species groups when simulated to an equilibrium level. This is one example in which a dynamic model with nonlinear functional responses can explain a large-amplitude shift in species abundances.

## 6. Resolving Complexity

A major theme of this article is the great natural variability, on annual to decadal time scales, of marine communities and ecosystems, in contrast with terrestrial systems. Recent descriptions of this variability have been in terms of physical/biological coupling based on forcing at the longer, climatic, time scales. Past analyses of the relations between climate and fisheries (Cushing, 1982) have drawn on direct observations, such as the Russell Cycle in the English Channel as well as on historical patterns such as the cod and herring fluctuations off Norway since the 16$^{th}$ century (Oeistad, 1994), using general statements about climatic change.

However, the lack of consistent long-term correlations between physical and population time series implies more than straightforward physical causes leading to biological effects. Further it seems intrinsically unlikely that the only interaction would be between one physical process and one population or trophic group. As an example, the time series depicted in Fig. 20.8 has been intensively analyzed with the conclusion, "There does appear to be some correlation between the gadoid stock and the North Atlantic Oscillation" (ICES 1999). But also, "It is likely that the two biotic factors (gadoid increase; herring decrease) are related in some way but a causal link between them does not seem to be justifiable."

The patterns in Fig. 20.8 would suggest that the gadoid outburst represents the dis-equilibrium between two different regimes. Jones (1984) calculated that the same basic productivity could support the "before" and "after" fish stocks but with very different patterns of energy flux involving a switch from large to small (industrial) pelagic fish. Similar rapid non-linear transitions may be used to explain the large changes in the N.W. Atlantic fisheries (Collie and DeLong, 1999; Skud, 1982).

All these examples, and the corresponding numerical simulations, suggest that these ecosystems can be considered to have two general modes;

1. *variability* within a particular community structure, and
2. *switching* between communities.

This distinction between two modes would permit a corresponding analytical separation into quasi-steady-state networks subject to small, linear, perturbations, and non-linear dynamic simulations with relatively few, large "blocks" of species. There are two major obstacles to such a separation:

*I. Many data sets that do not appear to display this dichotomy.*

Caddy and Gulland (1983) placed historical patterns of fish stocks into four groups: I. Steady state (e.g. North Sea turbot, II. Cyclic (e.g. Californian Dungeness crabs), III. Irregular (e.g. juvenile Norwegian herring), IV. Intermittent (e.g. Californian sardine/anchovy). All these patterns can be simulated with a simple model (Spencer and Collie, 1997) with different red noise forcing across the bifurcation range of parameter values. Steele and Henderson (1984) showed how stress from increased fishing pressure could lead to more frequent switching, which is, at the least, in line with comparisons of historical (Oiestad, 1994) and recent (Collie and DeLong, 1999) data. Thus statistical analyses of the patterns in time series of individual species are not necessarily the best evidence for determining underlying responses when the mechanisms involve non-linear relations between species or trophic groups.

The assumption that the dynamics can be decomposed into two separable responses is not new. May (1977) reviewed models that produced multiple stable states and Ludwig et al. (1978) applied these principles to explain outbursts of spruce budworms. This methodology has been applied to plankton and fish (Steele and Henderson 1981, 1984). For example, a multiple-equilibrium model can explain the high (1931–1970) and low (1971–1991) production regimes of Georges Bank haddock (Fig. 20.12). These applications, however, assume that the switching involves a small number of units or "boxes". What constrains the system between switches? The earlier analysis of food webs left open the question of optimization criteria for the "regimes." Are there alternatives to a least squares fit (Box B)?

One approach is to use the concept of resilience for this purpose, defined as the rate of return to equilibrium after perturbation (Cropp and Gabric, 2002). Laws et al. (2001) proposed that planktonic communities adapt their structure to maximize resilience. This criterion enabled parameter values to be defined for their ecosystem. The implication is that, in general, such communities are near equilibrium. A similar idea was suggested by studies of the dynamics of insect populations (Hassell et al, 1976). Possibly, such concepts could provide optimal criteria for the structure of food webs.

A different problem arises in relating simple switching models to complex food webs. If such non-linear mechanisms occurred at several positions in the web, then there could be a plethora of stable states producing a regime indistinguishable from random noise. However as May (1977) sums up, "the one kindly light in this encircling gloom is that many complex communities are, arguably, made up of loosely coupled lower-order systems to which (these) simple models are pertinent." The ecological problem is to define these groupings (Garrison and Link, 2000).

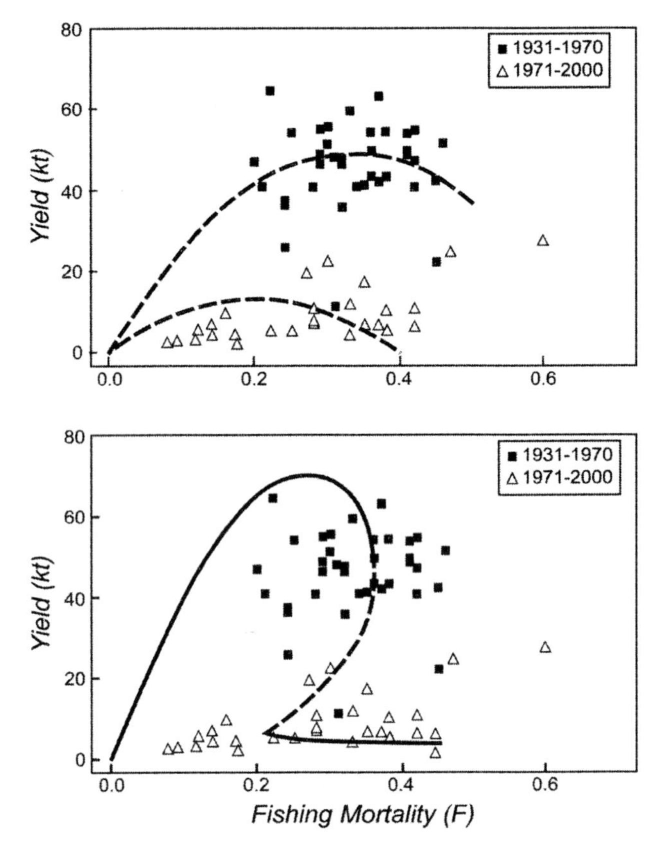


Figure 20.12 Yield vs. fishing mortality of haddock. (top) Schaefer curves with different parameters. (bottom) model with switching between different equilibria. (from Spencer & Collie, 1997b)

*II. The switching between major fish stocks, especially pelagic and demersal, would appear to require very large changes in their food organisms.*

One common feature of the observations is the great variability in abundance and composition of many marine communities at inter-annual and decadal time scales. These changes do not appear to be ecologically deleterious, but they imply great flexibility within the food web, for example in the alternation of different pelagic species—anchovy/sardine or herring/mackerel (Cushing 1982) or of fish and crustacean (Worm and Myers, 2003). This flexibility is a particular issue where fish communities switch between pelagic and demersal dominance (see Figs. 20.9,20.11).

The differing food requirements for such changes could be met in three ways:

1. *By changes in the output from primary production through the microbial food web to detritus and mesozooplankton.* Could systems somehow utilize the large pool of detritus observed in present day euphotic zones? For the Eastern Tropical Pacific, Christensen and Pauly (1998) consider that detritus "mainly leaks out of the system" and "with a better utilization of detri-

tus, a tenfold increase of tuna and marine mammal biomass can easily be accommodated within the system."

2. *By greatly reducing mid-level components of the food web.* In many systems, predatory "jellies" such as ctenophores dominate (Jackson et al., 2001). Eliminating them could increase the food for fish by a factor of 2–3 (Steele, 1965). This implies that the food webs supporting large fish stocks in historically pristine environments were *less* diverse than present webs.
3. *By re-allocating significant pelagic resources to demersal fish* (or vice-versa). The critical phase for recruitment to fish populations is usually in the juvenile stages when many demersal species are pelagic, and when their energy requirements are significant (Jones, 1984; Sissenwine et al., 1984).

These three solutions are not mutually exclusive. Historical reconstructions with very large biomass of fish such as cod (Kurlansky 1999) present the most difficult cases and might involve all three (Steele and Schumacher, 2000). Jackson et al. (2001) utilize the first two to depict changes in estuaries such as Chesapeake Bay, subject to heavy fishing and eutrophication. The overall problem is the lack of data on present and past concentrations and fluxes of detritus, jellies and pre-recruit fish. It is too easy to propose pristine scenarios or simulations (Steele and Schumacher, 2000). The lack of evidence for linear cascades (Verity and Smetacek, 1996) can be due to the paucity of data, but would suggest that the energy needs occasioned by large shifts in top predators are absorbed by "non-linear" adjustments at adjacent trophic levels. Similarly the success of the relatively independent bottom-up biogeochemical models of the microbial food web implies there is little feed back from higher trophic levels.

These gaps in our knowledge of food web structures highlight the tension between the temptation to use manageable but abstract concepts such as large linear food webs alternating with simpler non-linear dynamics, against the need to seek greater insights into the specific behavior of specific components of these webs (Verity and Smetacek, 1996). In an ideal world these distinct pathways would illuminate each other.

Lastly, there is the implicit assumption that, nearly everywhere in coastal waters, the ecosystems can adapt to the stresses that we impose in ways that are ecologically acceptable if not economically advantageous. Of course there are cases where whole ecosystems do break down, such as highly polluted estuaries. Yet examples for open coastal waters are relatively rare and isolated. The eutrophication of the shallow waters around the mouths of the Rhine and Elbe rivers in the southern North Sea is an outstanding example (NSTF, 1993). The "dead zone" off the estuary of the Mississippi River is another (Rabalais & Turner 2001). A present concern is with the benthic component where overfishing and technology have resulted in greater and more widespread disturbance of the sea bed (NRC, 2002). There is increasing evidence of alteration to habitats and therefore to the fauna. Whether the changes degrade the overall system is a matter for discussion.

## 7. Human Relevance

As discussed in the Introduction, we are highly ambivalent about our human role in the oceans and especially the coastal seas. Given the press of population toward the coast we cannot expect to have a pristine marine environment. Our increasing populations can never completely avoid adding nutrients and we shall continue to take fish from the sea. Additional stressors include pollution, species introductions and climate change. Yet we seem unwilling or unable to accept complete responsibility. As Jared Diamond (1997) has pointed out, it was the development of cultivated land that provided the basis for much larger and more complex social systems than those of hunter-gatherers. The substitution of monocultures of annual plants for diverse perennial communities greatly increases food supply but exposes the system to attendant risks from overuse and short-term environmental variability. This is the underlying trade-off in the management of terrestrial ecosystems and, after centuries of experience, it remains complicated, and controversial.

### *Options for ecosystem management*

There is a general desire for "holistic" management of marine ecosystems (NMFS,1999; NRC, 1999), but as others have pointed out, we do not manage ecosystems, we manage people. The greatest obstacles to coherent management, either as hunter-gatherers, or as farmers and city dwellers, arise from the need for major social changes corresponding to the far-reaching advances in marine technology. Resolutions to these issues are outside the scope of this science review.

In the sea, do we still have a choice between hands-on management of the marine environment, or should we strive for minimal impacts? The benefits and problems with the former can be seen in mariculture, particularly for finfish (Naylor et al. 2000). The farming of salmon has converted a rare and expensive delicacy into a staple food, comparable with chicken. The comparison is in more than price and availability. Salmon need their own seafood, which is taken by "industrial fishing" for small fish that are nearer the base of the food web and so more abundant. As Pauly et al. (1998) have pointed out, "fishing down the food web" can yield much larger harvests, but will also disrupt the natural system, as agriculture does on land. There are other advantages such as reliable stocking of the fish ponds and disadvantages from problems with disease and genetic modification. Potentially, other species, such as tuna with high commercial value, or haddock, with highly irregular yearly recruitment, could be farmed.

The alternative is to maintain a hunter-gatherer approach to harvesting the seas. Unfortunately we cannot do this completely laissez-faire. Even in medieval times, hunting for deer in forests was licensed to a privileged few; others were prohibited. We are reluctant to do this for the sea, and there is world-wide evidence of the consequences. Effectively all mature fisheries have capacity greatly in excess of needs. Manifestly, open access does not work and we cannot devise solutions to an apparently simple problem. The response on land was enclosure of common grazing, but this seems unacceptable for the oceans.

### *Examples of management trade-offs*

At present our management options try to combine extreme aspects of open-access and hands-on management alternative approaches. As an example, consider the ecosystem effects of bottom fishing, an issue that has garnered international attention. The first-order effects of bottom fishing on benthic communities are quite well understood (NRC 2002). Habitats can be mapped and ranked according to their vulnerability to fishing disturbance. The effects of bottom fishing on fish production and the indirect effect on the ecosystem are less well understood, but in principle are scientifically tractable. Ultimately we need to decide what we want coastal ecosystems to look like. Is the goal to maximize fish production or biodiversity? In some cases the goals are compatible and reinforcing, for example when juvenile fish survival depends on complex benthic habitat. On the other hand, some goals may be incompatible. A highly productive scallop fishery may otherwise have low biodiversity. An area of productive fish pens may degrade the habitat for resident species.

One way to resolve these trade-offs is to zone the continental shelf. There is great emphasis on Marine Protected Areas (MPAs), analogous to Nature Reserves on land. Yet these proposals largely ignore the difficult social and economic issues in managing the remainder of coastal and shelf seas (NRC 2001). The logical corollary to a program defining who can and cannot use the MPAs is to determine who will and will not have access to the surrounding seas. This would extend beyond MPAs and catching fish to energy (wind farms), eutrophication (nutrient discharge) and recreation (viral/bacterial release). This zoning, in turn, will demand a different attitude to our use of the seas. Such schemes have to be combined with other aspects of management of coastal waters. The Japanese have attempted integrated plans for the seaward extension of cities, particularly airports, energy development, release of nutrient wastes as fertilizer, and farming of seaweeds as well as fish. This view—appropriate to a small, densely populated country—is quite different from the American relation with the sea. Historically, the Dutch carried this view to its logical conclusion by converting estuaries and marshes to prime agricultural land.

## *Conclusions*

It is becoming clear that we cannot manage entire ecosystems; we can only regulate first-order impacts without being able to predict second-order consequences. A main theme of this article is that functional diversity is a dynamic and not a static property of marine ecosystems. If, as seems likely, the dynamics of marine ecosystems are highly nonlinear, there are several consequences for making predictions and decisions. One is that coastal ecosystems can shift among alternative quasi-steady states, some of which may be more desirable than others. Obviously, strictly linear models are insufficient for making predictions and the key nonlinear interactions need to be identified. Even so, there are limits to the predictability of marine ecosystems (Steele, 1998). As a consequence, human beings must learn to be more adaptable to large-amplitude shifts that can be expected, though not accurately predicted. If some regime shifts are human induced, management measures may need to be more deliberate than gradual to reverse the shift.

The marked changes in haddock yields on Georges Bank, Fig. 20.12, exemplify these issues. It is possible to explain the marked decrease in yield around 1970 as due to some shift in the physical or nutrient environment causing generally lower productivity, Fig. 20.12a. On the other hand, fishing pressure may have initiated a switch from a high to a low stable state for this fish stock, Fig. 20.12b. The upturn in other stocks, Fig. 20.9, would support the latter explanation, and is the basis for the model (Spencer and Collie, 1997). To choose between these explanations requires information about the rest of the food web and about the physical processes. Our ability to distinguish between these alternative—or complementary—causes is also critical for management.

An obvious conclusion is that a hands-on approach requires a much more comprehensive knowledge of ecosystem dynamics, particularly when our management involves large-scale changes to the system. Again, comparison with terrestrial problems is useful both for the similarities and the differences. The trend on land has been to move from a focus on selected, usually endangered, species to management of their habitats. Often the forests or savannahs are themselves of direct environmental or economic interest. There are certain habitats in shallow inshore water of direct appeal—coral reefs and kelp forests—that we protect. In the open sea the concerns have been mainly about charismatic fauna at higher trophic levels—whales, turtles, seabirds; as well as commercial species—swordfish, tuna, cod or haddock. Their "habitat" comprises the whole food web supporting them, as well as the physical environment. Thus, whatever management options are put in place, an appreciation of the consequences of ecosystem changes for stability and diversity will be needed.

## Acknowledgements

Both authors acknowledge support from the GLOBEC Georges Bank Program through NSF/OCE grant 0217399 to Dian Gittord and NA17RJ-1223 to JHS. J.H.S. acknowledges travel support from the U.S. JGOFS program through NSF/OCE grant #0049009 to Michael Landry and Robert Armstrong.

## Bibliography

Bartsch, J., K. Brander, M. Heath, P. Munk, K. Richardson, and E. Svendsen. 1989. Modelling the advection of herring larvae in the North Sea. *Nature* **340:**632–636.

Brodeur, R. D. and D. M. Ware. 1992. Long term variability in zooplankton biomass in the subarctic Pacific Ocean. *Fish. Oceanogr.* **1:**32–39.

Caddy, J. F. and J. A.Gulland. 1983. Historical patterns of fish stocks. *Marine Policy* **83:** 267–278.

Carpenter, S. R., J. K. Kitchell, and J. R. Hodgson. 1985. Cascading trophic interactions and lake productivity. *Bioscience* **35:** 634–639.

Chavez, F. P., J. Ryan, S. E. Lluch-Cota, and M. Niquen C. 2003. From anchovies to sardines and back: multidecadal change in the Pacific Ocean. *Science* **299:**217–221.

Christensen, V. and D. Pauly, (eds.) 1993. Trophic models of aquatic ecosystems. *ICLARM Conf. Proc.* **26** (390pp).

Christensen, V. and D. Pauly. 1998. Changes in models of aquatic ecosystems approaching carrying capacity. *Ecol. Appl.* **8S:**104–109

Cohen, J. E., F. Briand and C. M. Newman. 1990. *Community Food Webs.* Springer-Verlag, New York.

Collie, J.S. & DeLong, A.K. 1999. Multispecies interactions in the Georges Bank fish community. In: Alaska Sea Grant College Program, Ecosystem Approaches for Fisheries Management, pp. 187–210. Alaska Sea Grant Publication AK-SG-99-01.

Collie, J. S., S. J. Hall, M. J. Kaiser, and I. R. Poiner. 2000. A quantitative analysis of fishing impacts on shelf-sea benthos. *J. Anim. Ecol.* **69:**785–798.

Connell, J. H. 1978. Diversity in tropical rain forests and coral reefs. *Science* **199:** 1302–1310.

Corten, A. 1986. On the causes of recruitment failure of herring in the North Sea in the years 1972–1975. *J. Cons. Int. Explor. Mer* **41,** 281–294.

Costanza R, et al. 1997. The value of the world's ecosystem services and natural capital. *Nature* **387:** 253–260.

Cropp, R. and A. Gabric. 2002. Ecosystem adaptation: do ecosystems maximize resilience? *Ecology* **83:**2019–2026.

Cury, P., A. Bakun, J. M. Crawford, A. Jarre, R. A. Quiñones, L. J. Shannon, and H. M. Verheye. 2000. Small pelagics in upwelling systems: patterns of interaction and structural changes in "wasp-waist" ecosystems. *ICES J. Mar. Sci.* **57:**603–618.

Cushing, D.H. 1982. *Climate and Fisheries.* London, Academic Press, 373 pp.

Cushing, D.H. 1989. A difference in structure between eco-systems in strongly stratified waters and in those that are only weakly stratified. *J. Plankton Res.* **11:**1–13.

Daily, G.C. 2001. Ecological forecasts. *Nature* **411:** 245.

Daskalov, G.M. 2002. Overfishing drives a trophic cascade in the Black Sea. *Mar. Ecol. Prog. Ser.* 225: 53–63.

Davis, M.B. 1981. Quaternary history and the stability of forest communities. In: *Forest Succession, Concepts and Application* (West, Shugart & Botkin, eds) New York: Springer Verlag.

Diamond, J. 1997. *Guns, germs and steel.* Random House, New York. 480 pp.

Drinkwater, K. F. and R. Myers. 1987. Testing predictions of marine fish and shellfish landings from environmental variables. *Can. J. Fish. Aquat. Sci* **44:**1569–1573.

Edwards, M., P. Reid, and B. Planque. 2001. Long-term and regional variability of phytoplankton biomass in the Northeast Atlantic (1960–1995*). ICES J. Mar. Sci.* **58:**39–49.

Elmgren, R. 1989. Man's impact on the ecosystem of the Baltic Sea: energy flows today and at the turn of the century. *Ambio* **18:**326–332.

Eppley, R.W. and B. J. Peterson. 1979. Particulate organic matter flux and planktonic new production in the deep ocean. *Nature* **282:**677–680.

Estes J. A. and D. O. Duggins. 1995. Sea otters and kelp forest in Alaska: Generality and variation in a community ecological paradigm. *Ecol. Monogr.* **65:**75–100.

Estes, J. A., M. T. Tinker, T. M. Williams, and D. F. Doak. 1998. Killer whale predation on sea otters linking oceanic and nearshore ecosystems. *Science* **282:**473–476.

Fasham, M. J. R, H. W. Ducklow, and S. M. McKelvie. 1990. A nitrogen based model of plankton dynamics in the oceanic mixed layer. *J. Mar. Res.* **48:**591–639.

Folke, C., C. S. Holling, and C. Perrings. 1996. Biological diversity, ecosystems, and the human scale. *Ecol. Appl.* **6:** 1018–1024.

Franks, P. J. S. and C. Chen. 2001. A 3-D prognostic numerical model study of the Georges Bank ecosystem. Part II: biological-physical model. *Deep-Sea Research II* **48:**457–482.

Garrison, L. P. and J. S. Link. 2000. Fishing effects on spatial distribution and trophic guild structure in the Georges Bank fish community. *ICES J. Mar. Sci.* **57:**723–730.

Grassle, J. F. 2000. The ocean biogeographic system (OBIS): an on-line, worldwide atlas for accessing, modeling and mapping marine biological data in a multidimensional geographic context. *Oceanography* **13:**5–7.

Hairston, N. G., F. E. Smith, and L. B. Slobodkin. 1960. Community structure, population control and competition. *Amer. Natur.* **94:** 421–425.

Hardy, A. E., 1924. The herring in relation to its animate environment. *Fish. Investig.* London. Ser. 2. **7:**1–53.

Hare, S. R. and N. J. Mantua. 2000. Empirical evidence for North Pacific regime shifts in 1977 and 1989. *Prog. Oceanogr.* **47:** 103–145.

Hassell, M.P., J. H. Lawton, and R. M. May. 1976. Patterns of dynamical behaviour in single-species populations. *J. Anim. Ecol.* **45,** 471–486.

Haury, L. R., J. A. McGowan, and P. H. Wiebe. 1978. Patterns and Processes in the Time Space Scales of Plankton Distributions. In J. H. Steele (ed.), *Spatial Patterns in Plankton Communities,* Plenum, New York.

Heath, M.R. and A. Gallegos. 1998. Bio-physical modelling of the early life stages of haddock, *Melanograpmmus aeglefinus,* in the North Sea. *Fish. Oceanogr.* **7:**110–125.

Hermsen, J., J. S. Collie, and P. C. Valentine. 2003. Mobile fishing gear reduces benthic megafaunal production on Georges Bank. *Mar. Ecol. Prog. Ser.* **260:**97–108.

Holling, C. S. 1965. The functional response of predators to prey density and its role in mimicry and population regulation. *Memoirs of the Entomological Society of Canada* **45:**1–60.

Holling C. S. 1973. Resilience and stability of ecological systems. *Annu. Rev. Ecol. Syst.* **4:**1–23.

International Council for the Exploration of the Sea (ICES). 1999. Report on gadoid stocks in the North Sea. *ICES CM* 1999/C:**15.**

Iverson, R.L. 1990. Control of marine fish production. *Limnology and Oceanography,* **35**(7):1593–1604.

Jackson, G. A. and P. M. Eldridge. 1992. Food web analysis of a planktonic system off Southern California. *Prog. Oceanog.* **30:**223–258.

Jackson, J. B. C. (and 18 other authors). 2001. Historical overfishing and the recent collapse of coastal ecosystems. *Science* **293:**629–638.

Jones, R. 1984. Some observations on energy transfer through the North Sea and Georges Bank food webs. *Rapport et Proces-verbeaux, Reunion Conseil Int. Explor. Mer* **183:** 204–217.

Jennings, S., T. A. Dinmore, D. E. Duplisea, K. J. Warr, and J. E. Lancaster. 2001. Trawling disturbance can modify benthic production processes. *J. Anim. Ecol.* **70:**459–475.

Karl, D.M., and A. F. Michaels. 2001. Nitrogen Cycle. *Encyclopedia of Ocean Sciences.* (ed. J. H. Steele, S. A. Thorpe, and K. K. Turekian) 1876–1884.

Kenchington, E. L. R., J. Prena, K.D. Gilkinson, D. C. Gordon, K. MacIsaac, C. Bourbonnais, P. J. Schwinghamer, T.W. Rowell, D. L. McKeown, and W. P. Vass. 2001. Effects of experimental otter trawling on the macrofauna of a sandy bottom ecosystem on the Grand Banks of Newfoundland. *Can. J. Fish. Aquat. Sci.* **58:**1043–1057.

Klyashtorin, L. B. 1998. Long-term climate change and main commercial fish production in the Atlantic and Pacific. *Fisheries Research* **37:** 115–125.

Knowlton, N. 1992. Thresholds and multiple stable states in coral reef community dynamics. *Amer. Zool.,* **32:**674–682.

Kurlansky, M. 1997. *Cod: A biography of the fish that changed the world.* Walker and Co., New York.

Laws, E.A., P. G. Falkowski, W. O. Smith, H. Ducklow, and J. J. McCarthy. 2000. Temperature effects on export production in the open ocean. *Global Biogeochem. Cycles* **14:**1231–1246.

Lehman, C.L. and D. Tilman. 2000. Biodiversity, Stability, and Productivity in Competitive Communities. *Amer. Natur.* **156:** 534–552.

Levin, S. A., T. M. Powell, and J. H. Steele, (eds). 1993. *Patch Dynamics.* Springer Verlag Berlin. 307pp.

Lindeman, R.L. 1942. The trophic-dynamic aspect of ecology. *Ecology* **23:**399–418.

Link, J. 2002, Does food web theory work for marine ecosystems? *Mar. Ecol. Prog. Ser.* **230:**1–9.

Lluch-Belda, D., R. J. M. Crawford, T. Kawasaki, A. D. McCall, R. H. Parrish, R. A. Schwartzlose, and P. E. Smith. 1989. Worldwide fluctuations of sardine and anchovy stocks: the regime problem. *South African Journal of Marine Science* **8:**195–205.

Longhurst, A. 1998. *Ecological geography of the sea,* Academic Press, 398 pp.

Ludwig, D., D. Jones, and C. S. Holling. 1978. Qualitative analysis of insect outbreak systems: the spruce budworm and the forest. *J. Anim. Ecol.* **47:**315–32.

Lynch, D.R., C. V. Lewis, and F. E. Werner. 2001. Can Georges Bank larval cod survive on a calanoid diet? *Deep-Sea Research II* **48:**609–630.

Matsuda, H., T. Wada, Y. Takeuchi, Y. Matsumya. 1991. Alternative models for species replacement of pelagic fishes. *Res. Popul. Ecol.* **33:**41–56.

May, R.M. 1977. Thresholds and breakpoints in ecosystems with a multiplicity of stable states. *Nature* **269:**471–477.

McFarlane, G. A., P. E. Smith, T. R. Baumgartner,and J. R. Hunter. 2002. Climate variability and Pacific sardine populations. *Amer. Fish. Soc. Symp. 32:* 195–214.

McManus, J. W., A. B. M. Lambert, K. N. Kesner-Reyes, S. G. Vergara, and M. C. Ablan. 2000. Coral reef fishing and coral-algal phase shifts: implications for global reef status. *ICES J. Mar. Sci.* **57:**572–578.

Medawar, P. B. 1967. The art of the soluble. Methuen, London.

Menge, B. A. 2000. Top-down and bottom-up community regulations in marine rocky intertidal habitats. *J. Exp. Ma. Biol. & Ecol.* **250:**257–289.

Micheli, F. 1999. Eutrophication, fisheries and consumer-resource dynamics in marine pelagic ecosystems. *Science* **285:**325–326.

Murray, J.D. 1989. *Mathematical biology*. Springer-Verlag. Berlin.

Myers, R.A. and B. Worm. 2003. Rapid worldwide depletion of predatory fish communities. *Nature,* **423:**280–283

National Marine Fisheries Service (NMFS). 1999. Ecosystem-based fishery management. *A report to Congress by the Ecosystems Principles Advisory Panel. U.S. Department of Commerce*, Silver Spring, MD.

NRC (National Research Council). 1999. *Sustaining marine fisheries.* National Academy Press, Washington, D.C.

National Research Council (NRC). 2001. *Marine protected areas.* National Academy Press, Washington, D.C. 272 p

National Research Council (NRC). 2002. *Effects of trawling and dredging on seafloor habitat.* National Academy Press, Washington, D.C. 126 p.

Naylor, R. L., R. J. Goldberg, J. H. Primavera, N. Kautsky, M. C. M. Beveridge, J. Clay, C. Folke, J. Lubchenko, H. Mooney, and M. Troell. 2000. Effect of aquaculture on world fish supplies. *Nature* **505:**1017–1024.

Newell, R. I. E. 1988. Ecological changes in the Chesapeake Bay: are they the result of overharvesting the American oyster, *Crassostrea virginica?* P. 536–546 in, Understanding the estuary; advances in Chesapeake Bay research. M.P. Lynch and E.C. Krome. *Chesapeake Research Consortium,* Publication **129** CBP/TRS 24/88. Gloucester Point, Virginia.

NSTF, 1993. North Sea Task Force: Quality status report. Olsen & Olsen, Denmark, 132pp.

Nystrom, M., C. Folke and F. Mobert. 2000. Coral reef disturbance and resilience in a human-dominated environment. *TREE* **15,** no. 10.

Odum, H. T. and R.C Pinkerton. 1955. Times speed regulator: optimum efficiency for maximum power output in physical and biological systems. *Amer. Sci.* **43:**331–343.

Oiestad, V. 1994. Historic changes in cod stocks and cod fisheries: Northeast Arctic cod In Cod and Climate Change Symposium Proceedings, *ICES Mar. Sci. Symp. 1***98:**17–30.

Pace, M. L., J. J. Cole, S. R. Carpenter, and J. F. Kitchell. 1999. Trophic cascades revealed in diverse ecosystems. *TREE* **14:**483–488.

Pauly, D., V. Christensen, J. Dalsgaard, R. Froese, and F. Torres Jr. 1998. Fishing down marine food webs. *Science* **279:**860–863.

Piatkowski, U., G. J. Pierce, M. Morais da Cunha. 2001. Impact of cephalopods in the food chain and their interaction with the environment and fisheries: an overview. *Fisheries Research* **52:**5–10.

Pimm, S. L. 1982. *Food webs.* Chapman & Hall, London.

Pingree, R. D. 1978. Mixing and stabilization of phytoplankton distributions on northwest European continental shelf, in J. H. Steele (ed) in *Spatial Patterns in Plankton Communities.* Plenum. New York.

Polis, G.A., A. L. W. Sears, G. R. Huzel, D. R. Strong, and J. Maron. 2000. When is a trophic cascade a trophic cascade? *TREE* **15:** 473–475.

Polis, G. A. and K. Winemiller. 2001. *Food webs: integration of patterns & dynamics.* Kluwer Academic. Boston

Prena, J., P. J. Schwinghamer, T. W. Rowell, D. C. Gordon, K. D. Gilkinson, W.P. Vass, and D.L. McKeown. Experimental otter trawling on a sandy bottom ecosystem of the Grand Banks of Newfoundland: analysis of the trawl bycatch and effects on epifauna. *Mar. Ecol. Prog. Ser.* **181:**107–124.

Rabalais, N. N. and R. E. Turner. 2001. Coastal hypoxia: Consequences for living resources and ecosystems. *Coastal and Estuarine Studies,* **58.** American Geophysical Union, Washington, DC. vii + 463 p.

Ray, G. C. and J. F. Grassle. 1991. Marine Biological Diversity. *Bioscience* **41:**453–457

Reid, P. C., B. Planque, and M. Edwards. 1998. Is observed variability in the long-term results of the CPR survey a response to climate change? *Fish. Oceanogr.* **7:**282–288.

Rice, J. 2002. Changes to the page marine ecosystem of the Newfoundland-Labrador shelf. In, *Large Marine Ecosystems of the North Atlantic* (eds. K. Sherman and H.J. Skjoldal) Elsevier Science.

Richardson, K. and B. F. Pederson. 1998. Estimation of new production in the North Sea. *ICES J. Mar. Sci.* **55:** 574–580.

Riley, G. A., H. Stommel, and D.F. Bumpus. 1949. Quantitative ecology of the plankton of the western North Atlantic. *Bull. Bingham Oceanogr. Col.* **12:**1–169.

Robinson, A. R. 1983. *Eddies in Marine Science,* Springer-Verlag, Germany.

Rothschild, B. J. 1986. *Dynamics of marine fish populations.* Cambridge, Massachusetts: Harvard University Press.

Rudnick, D.L. and R.E. Davis. 2003. Red noise and regime shifts. Deep Sea Res. **50:**691–699 ..

Ryther, J. H. 1969. Photosynthesis and fish production in the sea. *Science* **166,** 72–76.

Sarmiento, J. L., R. D. Slater, M. J. R. Fasham, H. W. Ducklow, J. R. Toggweiler, and G. T. Evans. 1993. A seasonal three-dimensional ecosystem model of nitrogen cycling in the North Atlantic euphotic zone. *Global Biogeochem. Cy.* **7:** 417–450.

Scheffer, M, J. Carpenter, J. A. Foley, C. Folke, and B. Walker. 2001. Catastrophic shifts in ecosystems. *Nature* **413:** 591–596.

Schratzberger, M., A. Dinmore, and S. Jennings. 2001. Impacts of trawling on the diversity, biomass and structure of meiofauna assemblages. *Marine Biology* **140:**83–93.

Sherman, K., A. Solow, J. Jossi, and J. Kane. 1998. Biodiversity and abundance of the zooplankton of the Northeast Shelf ecosystem. *ICES J. of Mar. Sci,* **55:** 730–738.

Sissenwine, M. P., E. B. Cohen, and M. D. Grosslein. 1984. Structure of the Georges Bank Ecosystem: *Rapp. P.-v.* Reun. *Cons. int. Explor. Mer* **183:** 243–254.

Skud, B.E. 1982. Dominance in fishes: the relation between environment and abundance. *Science* **216:** 144–149.

Solow, A.R. 2002. Fisheries recruitment and the North Atlantic Oscillation. *Fisheries Research* **54:** 295–297.

Spencer, P. and J. S. Collie. 1996. A simple predator-prey model of exploited marine fish populations incorporating alternative prey. *ICES J. Mar. Sci.* **53:**615–628.

Spencer, P.D. and J. S. Collie. 1997a. Patterns of population variability in marine fish stocks. *Fish. Oceanogr.* **6:**188–204.

Spencer, P. D. and J. S. Collie. 1997b. Effect of nonlinear predation rates on rebuilding the Georges Bank haddock (*Melanogrammus aeflefinus*) stock. *Can. J. Fish. Aquat. Sci.* **54:**2920–2929.

Steele, J. H. 1956. Plant production on the Fladen Ground. *J. Mar. Biol. Ass. UK* **35:**1–33.

Steele, J.H. 1965. Some problems in the study of marine resources. *In International Commission for the Northwest Atlantic Fisheries,* pp. 563–576. ICNAF Special Pub.**6.**

Steele, J.H. 1974. *The structure of marine ecosystems.* Harvard University Press. Cambridge, Mass. 128 p.

Steele, J.H. 1985. A comparison of terrestrial and marine ecological systems. *Nature* **313:** 355–358.

Steele, J.H. 1998. Regime shifts in marine ecosystems. *Ecol. Appl.* **8** (Suppl. 1):S33-S36.

Steele, J.H. 1998. From carbon flux to regime shift. *Fish. Oceanogr.* **7:**176–181.

Steele, J.H. 1991. Marine functional diversity. *Bioscience* **41:** 470–474.

Steele, J.H. and E. W. Henderson. 1981. A simple plankton model. *Amer. Natur.* **117:** 676–691.

Steele, J.H. & Henderson, E.W. 1984. Modeling Long-Term Fluctuations in Fish Stocks. *Science* **224:** 985–987.

Steele, J.H. & Schumacher, M. 2000. Ecosystem structure before fishing. *Fisheries Research* **44:**201–205.

Thomas, E. 2001. Paleoceanography *Encyclopedia of Ocean Sciences.* (ed. J. H. Steele, S. A. Thorpe, and K. K. Turekian) 2077–2082.

Tilman, D. 2001. Functional Diversity. *Encyclopedia of Biodiversity* **3:** 109–120.

Trites, A. W., P. A. Linvingston, M. C. Vasconcellos, S. Mackingson, A. M. Springer, and D. Pauly. 1999. Ecosystem considerations and the limitations of ecosystem models in fisheres management: Insights from the Bering Sea. In, *Ecosystem Approaches for Fisheries Management.* University of Alaska Sea Grant, AK-SG-99-01: 609–619.

Verity, P. G. and V. Smetacek. 1996. Organism life cycles, predation and the structure of marine ecosystems. *Mar. Ecol. Prog. Ser.* **130:**277–293.

Vezina, A. R. and T. Platt. 1988. Food web dynamics in the ocean I. Best estimates of flow networks using inverse methods. *Mar. Ecol. Prog. Ser.* **42:**269–287.

Walters, C., V. Christensen, and D. Pauly. 1997. Structuring dynamic models of exploited ecosystems from trophic mass-balance assessments. *Reviews in Fish Biology and Fisheries* **7:**139–172.

Werner, F. E., B. R. Mackenzie, R. I. Perry, R. G. Lough, C. E. Naimie, B. O. Blanton, and J. A. Quilan. 2001. Larval trophodynamics, turbulence, and drift on Georges Bank: a sensitivity analysis of cod and haddock. *Scientia Marina* **65** (Suppl. 1):99–115.

Worm, B. and J. E. Duffy. 2003. Biodiversity, productivity and stability in real food webs. *Trends in Ecology and Evolution* **18:** 628–632.

Worm, B. and R. A. Myers. 2003. Meta-analysis of cod-shrimp interactions reveals top-down control in oceanic food webs. *Ecology* **84:**162–173.

Wunsch, C. 1981. Low frequency variability of the sea, in *Evolution of Physical Oceanography* (ed. B. A. Warren. And C. Wunsch). MIT Press. 362–375.

Wunsch, C. 1996. *The ocean circulation inverse problem.* Cambridge University Press, New York.

# Chapter 21. EUTROPHICATION

NANCY N. RABALAIS

*Louisiana Universities Marine Consortium*

**Contents**



## 1. Introduction

There is little doubt that human population growth and its associated activities have altered the landscape, hydrologic cycles, and the flux of nutrients essential to plant growth at accelerating rates over the last several centuries (Vitousek et al., 1997; Galloway and Cowling, 2002; Galloway et al., 2003). In an effort to support human population and to address the need for economic growth, humans have increased significantly the flux of nitrogen and phosphorus to aquatic and terrestrial ecosystems through alterations of global cycles of those nutrients. Excess nutrients are finding their way to the coastal ocean in increasing amounts especially during the last half of the $20^{th}$ century. There are thresholds of nutrient loading above which the nutrient inputs no longer stimulate entirely positive responses from the ecosystem such as increased fisheries production. Instead, land-based sources of nutrients are causing problems, for example, poor water quality, noxious algal blooms, oxygen depletion and in some cases, loss of fisheries production. Over the last four decades it has become increasingly apparent that the effects of excess nutrients that lead to eutrophication are not minor and localized, but have large-scale implications and are spreading rapidly.

### *1.1 Definitions*

Eutrophication is the increase in the rate of carbon production and carbon accumulation in an aquatic ecosystem (modified from Nixon, 1995). The definition was

*The Sea*, Volume 13, edited by Allan R. Robinson and Kenneth H. Brink
ISBN 0-674-01526-6 

developed initially as a description for the natural aging process of freshwater systems, and has been more recently applied to estuarine and coastal systems. The source of the increased organic carbon may come from within the system (autochthonous) or from outside the system (allochthonous). This distinction is relevant when management strategies are developed to reverse eutrophication and to identify the sources and mechanisms of carbon accumulation. For example, a coastal system could become eutrophic from an increased delivery of organic carbon from terrestrial sources or from nutrient-enhanced primary production resulting from increased nutrient loads. Reducing organic loading from riverine sources would require different management strategies than those required to reduce nutrient loads.

The definition of eutrophication given above recognizes that eutrophication is not a trophic state, but a process involving changes leading to higher ecosystem production. The causes of eutrophication should not be confused with the process itself. The causes may include changes in physical characteristics of the system such as changes in hydrology, changes in biological interactions such as reduced grazing, or an increase in the input of organic and inorganic nutrients. For example, upwelling systems cycle through phases of increased nutrient availability, high primary and secondary productivity, and often oxygen depletion in the lower water column. The trophic status of upwelling systems would be considered 'eutrophic'—an organic carbon supply of 300–500 g C $m^{-2}$ $y^{-1}$, as defined by Nixon (1995). Upwelling systems, however, are not undergoing eutrophication any more than mid-ocean oxygen minimum zones, which follow a similar process of organic matter accumulation and subsequent organic decomposition resulting in oxygen depletion that affects mid-water plankton and benthos where the zones impinge on continental shelves and slopes and sea mounts (Levin et al., 1991, 2000; Levin and Gage, 1998). There is sufficient evidence that coastal ecosystems are experiencing eutrophication, i.e., changes in the rate of primary production over long periods, with subsequent effects on multiple trophic levels. While the causes may be multiple and interactive, eutrophication in the coastal ocean and in the $20^{th}$ and $21^{st}$ centuries is more often caused by increased loads of nutrients that would otherwise limit the growth of phytoplankton.

A variety of responses, such as noxious algal blooms, fish kills, oxygen depletion, or seagrass losses, should also not be confused with the process of eutrophication. The responses are multiple and may often result in 'increases' or 'decreases' of components of coastal ecosystems, to which humans often ascribe beneficial or detrimental values. There is little doubt that there have been ecosystem-level changes in coastal systems as a result of eutrophication.

### *1.2 Focus*

Coastal systems extend from the barrier island shoreface to the edge of the continental shelf, but coastal eutrophication is most likely to occur within the 100-m isobath. Often, but not always, shallower waters are not susceptible to eutrophication because turbid waters caused by resuspension of sediments or delivery of sediments from coastal rivers inhibits the penetration of sunlight needed for photosynthesis. Some shallow, but clearer, waters that support macroalgal or seagrass growth are susceptible to eutrophication when increased nutrient loading occurs where there is a physical system that supports long water residence times. Compared to shallow waters, deeper more ‘open’ waters are usually not as susceptible to eutrophication because nutrients are diluted, less primary production occurs, less organic matter is exported, and deeper depths allow for more remineralization of fluxed carbon (Suess, 1980). Coastal areas are susceptible to the process of eutrophication where there is a physiography that enhances water residence times so that nutrients from land-based runoff or atmospheric deposition are less likely to dilute as quickly as an open coast. Because of these features, many of the semi-enclosed seas will be discussed here. This chapter focuses on the coastal ocean, but information from estuaries is included where appropriate.

This chapter falls within a broader category of “Scientific Issues for Applications” and therefore focuses on the idea that eutrophication is a process that can be altered by human intervention. Thus, this chapter will focus on anthropogenic eutrophication—eutrophication caused by human activities, such as increased nutrient loading or over-enrichment—and causes that would be conducive to management actions to reduce undesirable effects while sustaining living resources that society deems valuable. Reversing unwanted effects requires a reasonable understanding of the multiple causes leading to the identifiable effects observed in an altered ecosystem. Identifiable causes, in theory, can be manipulated to minimize detrimental effects. Most of the effects are ‘local’ in the sense that they are smaller in scale than the regions defined for this book. Yet these local problems occur worldwide. Comparing systems in the search of commonalities should lead to a more in-depth mechanistic and quantitative understanding of the cause-and-effect relationships that might be favorably adjusted.

A proportionally greater part of this chapter deals with the negative aspects of eutrophication that are induced by human activities, versus the positive aspects. Coastal ecosystems that have been substantially changed as a result of eutrophication exhibit a series of identifiable symptoms, such as reduced water clarity, excessive, noxious, and, sometimes, harmful algal blooms, loss of critical macroalgal or seagrass habitat, development or intensification of oxygen depletion in the form of hypoxia (lowered dissolved oxygen levels) or anoxia (the absence of dissolved oxygen), and, in some cases, loss of fishery resources. The more subtle responses of coastal ecosystems to eutrophication include shifts in phytoplankton and zooplankton communities, shifts in the food webs that they support, loss of biodiversity, changes in trophic interactions, and changes in ecosystem functions and biogeochemical processes. Nutrient mitigation activities, however, may produce unintended results that should be considered in any nutrient management strategy.

## 2. Process of Eutrophication

### *2.1 Changes in Nutrient Inputs to Coastal Waters*

Reactive nitrogen ($N_r$) has increased substantially over the last 150 years through the artificial fixation of nitrogen into fertilizers, the emission of nitrous oxides from the consumption of fossil fuels, and the volatilization of reduced forms from fertilizers and animal wastes. The creation of reactive nitrogen increased by a factor of 20 since 1860 to the present anthropogenic production of $N_r$ of ~150–165 Tg N $y^{-1}$ (Galloway and Cowling, 2002; Galloway et al., 2003). The forms of reactive nitrogen that affect aquatic ecosystems include inorganic dissolved forms (nitrate and ammonium), a variety of dissolved organic compounds (amino acids, urea, and composite dissolved organic nitrogen), and particulate nitrogen. Phytoplankton and higher plants utilize different forms of nitrogen preferentially, and the relative proportion, as well as the load, of nitrogen forms may differentially influence phytoplankton growth, size structure, and community composition (e.g., Antia et al., 1991; Peierls and Paerl, 1997).

Phosphorus additions to the landscape enter via phosphorus-containing fertilizers manufactured from mined phosphorus, animal manures, and waste products from animals supplemented with phosphorus-enriched feed, and enter rivers and streams via wastewater effluents and soil erosion. As phosphorus inputs to the land exceed phosphorus outputs in farm products, phosphorus is accumulating in the soil with important implications for increased runoff from the landscape to surface waters (Bennett et al., 2001). The increased flux of phosphorus eroded from the landscape or carried in wastewater effluents to the world's rivers has increased the global flux of phosphorus to the oceans almost 3-fold above historic levels of ~ 8 million Tg P $y^{-1}$ to current loadings of ~22 Tg $y^{-1}$ (Howarth et al., 1995; Bennett et al., 2001). The P accumulation in landscapes of developed countries is declining somewhat, but increasing in developing countries (Bennett et al., 2001).

Whereas the riverine concentrations of nitrogen and phosphorus have increased in recent times, the concentration or loads of dissolved silicon have remained the same or decreased (Figure 21.1). The result is that the relative proportions of these nutrients more closely approximate the Redfield ratio of Si:N:P of 16:16:l (Justić et al., 1995a,b; Turner et al., 2003a,b). The percentage of world rivers approaching the Si:N ratio of 1:1 will increase as nitrate flux increases with further economic development (Turner et al., 2003a,b).

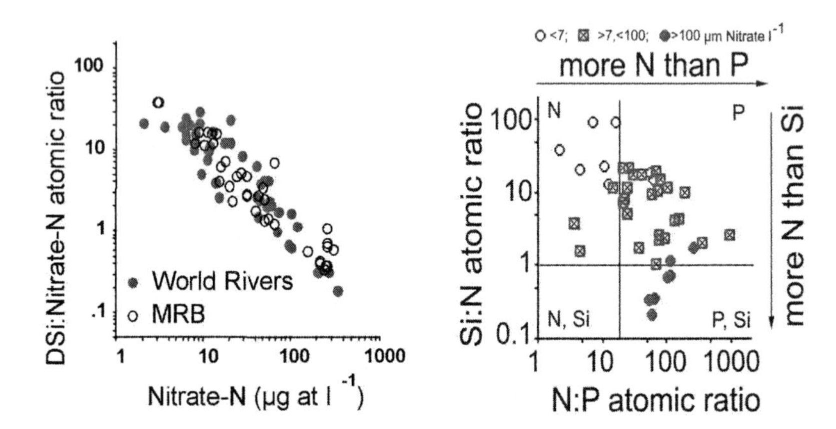


Figure 21.1 Left panel: dissolved nitrate-N concentration for the world's rivers and sub-basins of the Mississippi River (MRB) versus dissolved silicate to nitrate ratio (from Turner et al., 2003a; used with the permission of the author and Kluwer Academic Publishers). Right panel: molar ratios of dissolved Si and nitrate (Si:N) and dissolved nitrate and phosphate (N:P) in large rivers (from Turner et al., 2003b; reproduced with permission of Elsevier). Si:N and N:P atomic ratios of 1:1 and 16:1 are indicated by horizontal and vertical lines. Nutrient limitations (assuming sufficient concentration) within each quadrant are indicated by letters. World rivers are shifting from N to Si and P limitation as nitrate-N increases; the Mississippi was historically in the upper left, and is now in the lower right quadrant.

### 2.2 *Sources of Nutrients*

Eutrophication is becoming a major, global environmental problem in estuarine and coastal waters linked to watersheds with population growth, a focusing of the populace in coastal regions, and agriculture expansion in major river basins. Humans have altered the global cycles of nitrogen and phosphorus over large regions and increased the mobility and availability of these nutrients to marine ecosystems (Peierls et al., 1991; Howarth et al., 1995, 1996; Vitousek et al, 1997). As the sources of increased nutrient inputs are identified in watersheds, it becomes clearer what nutrient reduction mechanisms are necessary to return to less human-influenced levels. Human-influenced inputs derive from the increase in human populations and their activities, particularly from the application of nitrogen and phosphorus fertilizers, nitrogen fixation by leguminous crops, and atmospheric deposition of oxidized nitrogen from fossil-fuel combustion.

Knowledge of the sources contributing to coastal nutrient loads helps pinpoint appropriate nutrient management strategies for specific watersheds. The sources of human-influenced nutrients are commonly described as being from point sources or non-point sources. Point sources include wastewater treatment or industrial discharges and are usually considered less important as nutrient contributors than non-point sources on broad regional scales (National Research Council, 2000). In some areas, however, point sources can be a large proportion of the total load. For example, wastewater from New York City contributes 67 percent of the nitrogen inputs to Long Island Sound on an annual basis (National Research Council, 2000). Point sources may be important in estuaries where population centers are located, such as Narragansett Bay, Rhode Island (Nixon and Pilson, 1983), or where specific industrial activities contribute large loads, such as historic phosphate mining in the watershed of the Albemarle-Pamlico estuarine system in North Carolina. In other regions wastewater point sources contribute less of the total load to a receiving coastal system. About 25 percent of the nitrogen and phosphorus inputs to Chesapeake Bay come from wastewater and other point sources (Boynton et al., 1995), and less than 10 percent from similar sources make up the total annual nitrogen load of the Mississippi River (Goolsby et al., 1999).

Non-point sources are of a more diffuse origin than point sources and may come off the landscape in stream flow, from the atmosphere in both dry and wet deposition, and from coastal aquifers. Fertilizer application, both nitrogen and phosphorus, remains a major contributor to non-point nutrient pollution (Howarth et al., 1995). The total load from non-point sources has stabilized in some watersheds, i.e., the Mississippi basin (Goolsby et al., 1999), but continues to increase on a worldwide basis (Seitzinger et al., 2002). There is often a direct relationship in large river basins between fertilizer applications and riverine nitrogen and phosphorus fluxes and concentrations (Turner and Rabalais, 1991; Caraco, 1995; Smil, 2001). Coupled with artificially drained land, such as occurs over much of the mid-west of the United States, the landscape serves as an efficient conduit for subterranean movement of dissolved inorganic nitrogen applied in the form of fertilizers (Randall and Mulla, 2001). Other agricultural activities—planting of leguminous crops, application of manure, and waste from confined animal operations—also contribute large loads of both nitrogen and phosphorus to the environment.

Rivers play a crucial role in the delivery of nutrients to the ocean. These rivers terminate in estuaries or the nearshore coastal ocean, where the effects of nutrient over-enrichment are most pervasive. In the sub-basins to the North Atlantic Ocean and specifically in the Baltic catchments and the watershed of the Mississippi River, the inputs of anthropogenic nitrogen via rivers far exceed those from other sources—atmospheric, coastal point sources, and nitrogen fixation (Howarth et al., 1996; Goolsby et al., 1999; Grimvall and Stålnacke, 2001). Phosphorus loads, likewise, come mostly from rivers (Bennett et al., 2001; Grimvall and Stålnacke, 2001).

On local to global scales, one of the most rapidly increasing sources of nutrients to both fresh waters and the coastal zone is the atmosphere. In watersheds leading to coastal waters, up to 40 percent of the nitrogen inputs are from atmospheric origin, originating from industrial, agricultural, and urban sources (Duce, 1986). In other watersheds, such as the Mississippi River basin, the proportion is much lower, less than 10 percent (Goolsby et al., 1999). Nitrogen enters the atmosphere in the form of nitrous oxides generated in the burning of fossil fuels, either for energy production or transportation. Besides direct leaching or runoff from farm fields, fertilizer nitrogen can also enter the environment through the atmosphere. Globally, 40 percent of the inorganic nitrogen fertilizer that is applied to fields is volatilized as ammonia, either directly from fertilizer or from animal wastes after crops grown with the fertilizer have been fed to chickens, pigs or cattle. Along the eastern U.S. coast and eastern Gulf of Mexico, atmospheric nitrogen sources currently account for 10 to over 40 percent of 'new' nitrogen loading to estuaries (Paerl, 1995, 1997; Paerl et al., 2001). For the northern Gulf of Mexico affected by the effluent of the Mississippi River, the contribution is only 1 percent (Goolsby et al., 1999).

Groundwater and offshore nutrient supplies may supplement watershed and atmospheric loads but are probably the least well quantified components. For the northeastern Gulf of Mexico, the total nitrogen input from groundwater to offshore Apalachee Bay slightly exceeded the combined flux of total nitrogen from the Ochlockonee and Sopchoppy Rivers (Fu and Winchester, 1994). Groundwater inputs to the South Atlantic Bight were estimated (based on summer conditions) to be about 40 percent of the river-water flux to the coastal waters (Moore, 1996). Similar inputs to the Louisiana coastal zone were predicted to be minimal (Rabalais et al., 2002b). The relative contribution of offshore sources of nutrients from upwelled waters differs by continental shelf area. Onwelling of nitrate from deeper waters may be important in shelf edge (100 m depth range) cycling of carbon and nitrogen (Walsh, 1988, 1991).

Clearly there are multiple pathways of increased inputs of nutrients to aquatic systems. The management of these nutrients must be multi-faceted, cross numerous boundaries, and reflect the relative sources within a watershed (Howarth et al., 2002). In coastal areas, such as most of the northeastern United States that receive inputs from watersheds with little agricultural activity, atmospheric deposition of nitrogen from fossil fuel combustion can account for up to 90 percent or more of the nitrogen contributed by non-point sources. Other estuaries, such as Chesapeake Bay, receive significant contributions of both agricultural sources and fossil fuel sources. Agricultural, not atmospheric, sources dominate the export of nitrogen in the Mississippi River basin.

### 2.3 *Nutrient-Enhanced Productivity*

The focus of nutrient limitation in this chapter is macronutrients, although micronutrients such as iron may be important in many oceanic systems. Most coastal systems susceptible to eutrophication are influenced by rivers, which carry sufficient iron so that this micronutient is not considered limiting to phytoplankton growth (Johnson et al., 1999). Macronutrients that limit the growth of phytoplankton, macroalgae, and vascular plants in marine systems are nitrogen and phosphorus, and silicon additionally limits the growth of diatoms (Howarth, 1988; Vitousek and Howarth, 1991; Conley et al., 1993). If a nutrient is limiting, the addition of that nutrient will enhance phytoplankton growth and production. Phosphorus is usually considered the primary limiting nutrient in most lakes and reservoirs (Hecky and Kilham, 1988), while nitrogen is considered the primary limiting nutrient in marine waters (Ryther and Dunstan, 1971; Howarth, 1988). Most would agree that this dichotomy of a single nutrient limitation in either system is an oversimplification (as is ignoring other external factors, such as physics and trophic interactions, see Section 3). Howarth (1988) summarized that many estuarine and coastal marine ecosystems are probably limited by nitrogen, but phosphorus may limit production in some systems (e.g., Turner et al., 1990) and evidently during certain seasons, and may also secondarily limit production in combination with nitrogen. Over-enrichment with nitrogen and phosphorus alters the ratios of nitrogen, phosphorus, and silicon to each other, such that silicon limitation may occur on a more frequent basis (Dortch and Whitledge, 1992; Conley et al., 1993; Justić et al., 1995a,b; Turner et al., 1998). Many estuarine and coastal systems display variability in phosphorus, nitrogen or silicon limitation, or combinations of these, by season and by location along a fresh-to-marine gradient (Conley, 2000; Elmgren, 2001; Mee, 2001; Yin, 2001; Rabalais et al., 2002a). Understanding these distinctions in a coastal system are important in efforts to target, by time and type, nutrient loads to the system.

High biological productivity in the immediate and extended plume of the Mississippi River is mediated by high nutrient inputs and regeneration, favorable light conditions, and suitable temperature, salinity and mixing rates (Lohrenz et al., 1990, 1994). Primary production in shelf waters near the delta and to some distance from it was significantly correlated with nitrate and nitrite concentrations and fluxes over the period 1988 to 1994 (Lohrenz et al., 1997). There was also a high degree of coherence between Mississippi River nitrate fluxes and net production at a 20-m water depth station in the core of the hypoxic zone 90 km down-plume from the Mississippi River influence for 1985–1992 (Justić et al., 1997). Increases in diatom production in surface waters and diatom-based carbon accumulation in sediments, overall carbon accumulation in sediments, and worsening conditions of oxygen parallel decadal changes in the flux of nitrate from the Mississippi River to the coastal ocean (see Section 4).

Algal blooms in the Yellow Sea increased in frequency over the past several decades as atmospheric deposition of nutrients to the coastal sea increased in addition to direct nutrient runoff (Zhang, 1994). It is estimated that a typical rain event over the Yellow Sea may supply sufficient nitrogen, phosphorus, and silicon to account for over half and up to 100 percent of the primary production of a harmful algal bloom event (Zhang, 1994).

Boynton and Kemp (2000) demonstrated strong relationships between river flow into the Chesapeake Bay (strongly correlated with nutrient flux) and phytoplankton production and biomass and the subsequent fate of that production in spring deposition of chlorophyll *a*. They further demonstrated a strong relationship between the sedimented chlorophyll *a* and the seasonal decline of deep-water dissolved oxygen. While these relationships do not negate the importance of water-column stratification and physical processes, they do clearly link quantitatively nutrient flux with eventual degradation of water quality in the form of chronic, seasonal oxygen problems.

### *2.4 Sedimentary Processes*

Excess organic material, in the form of dead and senescent algae, zooplankton fecal pellets, and marine aggregates, sinks to the lower water column and the seabed where the carbon is remineralized by aerobic and anaerobic processes or buried. In coastal areas susceptible to eutrophication, bottom currents are often minimal, water column mixing is reduced, and development of hypoxia and anoxia is common below the pycnocline. As aerobic bacteria decomposition proceeds, the dissolved oxygen concentration overlying the sediments approaches anoxia. In this transition, numerous biological and geochemical processes in the sediments shift to produce a negative feedback into declining oxygen levels or a stimulation for increased production in the upper water column.

With continued carbon production, concentrations of organic carbon and nitrogen and microbial biomass may increase, and the microbial decomposition potential of substrates and community oxygen consumption rise, but not in simple linear relationships (Meyer-Reil and Köster, 2000). An outline of changes occurring is that the redox potential discontinuity layer migrates upward to the sediment-water interface, sulfate respiration replaces oxygen respiration, hydrogen sulfide, which is toxic to metazoans, is generated from the sediments, and oxygen penetrates less deeply into the sediments as the bioturbation potential of the macroinfauna decreases due to their demise from sulfide toxicity or oxygen deficiency. The sediments become less cohesive, more susceptible to resuspension that increases turbidity of the overlying water, which in turn reduces the potential for growth of the photosynthetic microphytobenthic community and generation of oxygen into the lower water column. This series of events is not continuous, but changes with strength of stratification, rate of carbon accumulation, and resuspension from mixing events. There are coastal areas, however, e.g., northern Gulf of Mexico, where stratification and oxygen depletion below the pycnocline are maintained for periods of several weeks to several months (Rabalais and Turner, 2001).

The nitrification/denitrification cycle of estuarine and continental shelf sediments, which returns $N_2$ to the atmosphere, is an ameliorating mechanism to remove excess $N_r$, but is disrupted by the limited availability of oxygen in sediments. Denitrification proceeds much of the year but is dependent on the nitrate supplied by nitrification, a process that is dependent on the presence of oxygen, which may be absent for extended periods and over broad areas (Kemp et al., 1990; Childs et al., 2002). With the shift in redox potential, there is an increase in the flux of inorganic nutrients, ammonium and phosphate into the overlying water. These inorganic nutrients become available to fuel further phytoplankton production in the overlying water. The degree to which these nutrients diffuse upward through the water column and across a strong pycnocline is not well known, but this process is not apparent in nutrient profiles in a 20 m stratified, hypoxic water column in the Gulf of Mexico.

### *2.5 Coastal Systems Susceptible to Eutrophication*

Not all coastal systems with elevated nutrient loads will undergo eutrophication. Many rivers deliver large quantities of fresh water laden with nutrients and organic carbon, but zones of nutrient-enhanced productivity and or bottom-water hypoxia/anoxia do not develop. Increased phytoplankton biomass, carbon accumulation, and oxygen depletion are more likely to occur in coastal systems characterized by longer water residence times and stratified water columns. The amount of suspended sediment delivered to a coastal system may also influence whether enhanced carbon production will result from high nutrient inputs.

The volume of freshwater discharge, exclusive of the nutrient load, can influence residence time, stratification, turbidity, and nutrient dilution. For example, higher chlorophyll biomass and more widespread bottom-water hypoxia occur when the Mississippi River is in high flow, in part because of the resulting intensified stratification on the continental shelf (Rabalais et al., 1998). In contrast, higher discharges from the Hudson River cause lower residence time, increased turbidity, less stratification, lower primary production and less eutrophication in the estuary (Howarth et al., 2000) but increased stratification, chlorophyll biomass, and bottom-water hypoxia in the New York bight (Swanson and Sindermann, 1979; Whitledge, 1985).

The coastal systems adjacent to many large rivers display the same gradient of lower productivity in the nearfield nutrient-rich, turbid, light-limited plume, higher production and chlorophyll biomass at intermediate salinities where water clarity improves and nutrients remain sufficient, and more oligotrophic conditions in the far field with much improved water clarity but depleted nutrients (Rhoads et al., 1985; Turner et al., 1990; Lohrenz et al., 1990; Rabalais et al., 2002a). Swift currents that move materials away from the river delta and that do not permit the development of stratification do not allow for the accumulation of biomass or depletion of oxygen, for example in the Amazon and Orinoco plumes. Similar processes off the Changjiang (Yantze River) and high turbidity in the plume of the Huanghe (Yellow River) were once thought to be reasons why hypoxia did not develop in those coastal systems. Incipient symptoms of anthropogenic eutrophication are becoming evident at the terminus of both these systems as nutrient loads increase (Li et al., unpubl. data on Changjiang hypoxia; Zhang, 1994; Liu et al., 2003). The likelihood that more and more coastal systems will become eutrophic, especially in developing countries, is worrisome.

## 3. Responses to Increased Nutrients

It is difficult to separate the rate of carbon production from the effects of increased nutrient inputs from other processes such as grazing, complex trophic interactions, physical accumulation, and dispersal. Marine and coastal pelagic systems, like lake ecosystems, appear to be influenced by both higher trophic levels (top-down) and prey quality and density (bottom-up) (McQueen and Post 1988; McQueen et al. 1989). Micheli's (1999) meta-analysis of marine food webs supports the top-down/bottom-up theory of food web controls in which (1) nutrient enrichment does not reach higher trophic levels, and (2) top-down effects can significantly reduce zooplankton, but not increase phytoplankton growth. Both processes are confirmed by recent studies (Duffy, 2002; Kimmerer, 2002; Bundy at al., 2003; Worm and Myers, 2003). These straight-line food webs do not take into account, however, an intraguild predator that utilizes basal prey and an intermediate predator, i.e., omnivory. The pelagic food web structure of the coastal ocean could well be represented by the intraguild predator model, because the most abundant planktivorous fishes (menhaden, herring, anchovy, bumper) can feed on both zooplankton and phytoplankton (Govoni et al. 1983; Ditty 1986; Deegan 1990; Castillano-Rivera et al., 1996). Multiple trophic interactions and the coupling of physics and biology are acknowledged but are beyond the scope of this chapter. Instead a simplified model of system response to nutrients, mediated by trophic interactions, is presented in Figure 21.2.

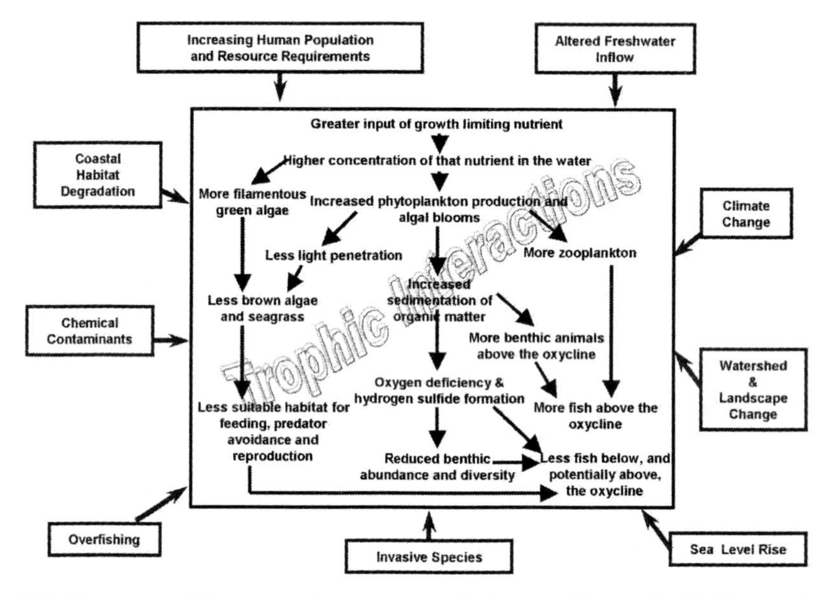


Figure 21.2 Responses within a coastal ecosystem to the increased input of a limiting nutrient amid multiple stressors (not inclusive) that may interact with or dominate over increases in nutrient loads. [original figure, permission not needed]

The initial response of an aquatic system to a higher loading of limiting nutrients is an increase in phytoplankton growth rate and biomass accumulation. A change in the relative proportion of one nutrient over others may also favor a shift in the composition of the phytoplankton community responding to the increased nutrient loading. These primary responses propel a cascade of ecological interactions, including shifts in system metabolism and changes in trophic interactions that result in a series of direct and indirect responses (Figure 21.2; Cloern, 2001).

These responses are set within a larger framework of multiple, human-induced stressors that also affect water quality and living resources (Paine et al., 1998; Breitburg et al., 1999; Daskalov, 2003; Mills et al., 2003). Stressors, particularly at the community or ecosystem level, such as fishing patterns, interactions with chemical contaminants, and invasive species, will modify the rate of carbon production and integration into higher trophic level production. Understanding the responses of coastal systems to anthropogenic eutrophication will require understanding the linkages with multiple stressors and how they interact with responses to increased nutrients.

### *3.1 Initial Response to Increased Nutrients*

When a nutrient, multiple nutrients or micronutrients that are limiting to the growth of phytoplankton are supplied to an aquatic ecosystem, the initial response is an increase in primary production, usually with some relational increase in phytoplankton or plant biomass, including sea grasses, filamentous algae, and macroalgae. Positive relationships exist between nutrient-loading rates and microalgal production and biomass (Boynton et al., 1982; Nixon et al., 1996; Lohrenz et al., 1999; Cloern, 2001). These relationships are often nonlinear because of large differences among coastal waters in the rates or pathways through which external nutrients are converted into algal biomass (Cloern, 2001).

Many examples from individual coastal ecosystems follow the same pattern shown in Figure 21.3, which is an increase in phytoplankton biomass (typically measured as chlorophyll *a*), as nutrient loads increase (e.g., Himmerfjärden Bay, Sweden, Elmgren and Larsson, 2001; Irish Sea, Allen et al., 1998; Dutch Wadden Sea, Cadée, 1992; northern Adriatic Sea, Šolić et al., 1997; Chesapeake Bay, Harding and Perry, 1997; northern Gulf of Mexico, Rabalais et al., 2002a).

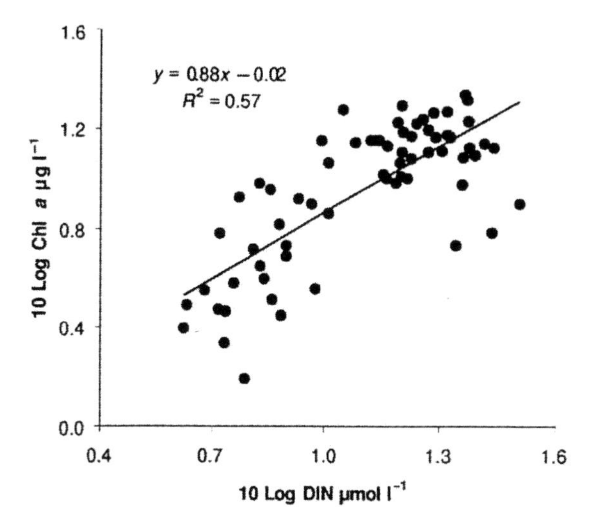


Figure 21.3 Maximal spring chlorophyll *a* concentration in Himmerfjärden Bay, Sweden as a function of maximum inorganic nitrogen concentration (DIN) in January/February before the start of the spring bloom, five stations in 1977-1994. (From Elmgren and Larsson, 2001; used with permission of the authors.)

### *3.2 Shifts in Plankton Communities*

Phytoplankton are not only affected by the quantity of nutrient loading, but also by the relative supply of nutrients. Redfield et al. (1963) postulated that nutrient ratios indicated the possibility for stoichiometric and physiological limits to phytoplankton growth. Changes in the relative proportions of the macronutrients may exacerbate eutrophication, favor noxious algal blooms, and aggravate conditions of oxygen depletion (Conley et al., 1993; Justić et al., 1995a,b; Officer and Ryther, 1980; Turner et al., 1998). Global patterns of the ratios of dissolved nitrogen, phosphorus, and silica in large rivers indicate that primarily nitrate flux controls the ratios of these nutrients as delivered to the coastal ocean (Turner et al., 2003a).

As the N:P ratio exceeds 16:1, phosphorus limitation of phytoplankton growth is implied, and in a similar fashion a Si:N ratio below 1:1 implies silica limitation (Figure 21.1) (Turner et al., 2003b). When the Si:N atomic ratio is near 1:1, aquatic food webs leading from diatoms to fish may be compromised, and the frequency or size of harmful or noxious algal blooms may increase (Officer and Ryther, 1980; Turner et al., 1998). Officer and Ryther (1980) hypothesized that if the minimal Si:N proportion of 1:1 for diatoms was not met, then a phytoplankton community of non-diatoms may be competitively enabled. The effect of Si limitation on species composition is important because diatoms dominate the biomass, particulate flux to the bottom in fecal pellets, and percentage of direct sinking phytoplankton (Dortch et al., 2001) and play a pivotal role in trophodynamics and carbon flux (Turner et al., 1998). If the decrease of the Si:N ratio in world rivers continues, the consequences for food webs, trophic structure, and communities could be substantial.

Results from field and laboratory studies suggest that the lack of silica or phosphorus relative to nitrogen can control phytoplankton community composition (e.g., Dortch and Whitledge, 1992; Egge and Aksnes, 1992; Rabalais et al., 1996; Dugdale and Wilkerson, 1998; Dortch et al., 2001). A series of eutrophication-related ecological changes in the Baltic Sea and Kattegat were caused primarily by increased anthropogenic nutrient inputs (both nitrogen and phosphorus) mainly after World War II (reviewed by Elmgren and Larsson, 2001). One ecosystem response was an increase of toxic or noxious algal blooms. Cyanobacterial blooms in the open sea, particularly of the toxic, nitrogen-fixing genus *Nodularia* became a problem, and a number of fish kills by the prymnesiophyte *Prymnesium pravum* were reported from the Baltic coastal zone. The former species is stimulated by low inorganic N:P ratios; the latter species is thought to be favored by high N:P ratios. Closer attention in the future should be paid to the importance of balanced nutrient composition as well as nutrient supply dynamics for the development of eutrophication versus efficient trophic transfer and fish production in nutrient-enriched systems (Kristiansen and Hoell, 2002).

In the case of the Louisiana shelf near the Mississippi River delta, this turns out to be a correct prediction where the Si:N ratio now fluctuates just above and below 1:1 (Table 23.1). Below 1:1 there was a drastic decline in zooplankton abundance. The percent copepod abundance of the zooplankton declined from about 80 to 20 percent. The fecal pellet production of copepods and the relative proportion of carbon fluxed from the upper water column to the lower water column via fecal pellets also declined. Copepod fecal pellets, containing many partially decomposed diatoms, sink much more rapidly than individual phytoplankton cells, and there was relatively less decomposition en route to the bottom. Thus, more of the organic material sinking to the bottom layer is respired there if copepods dominate the zooplankton community, than if the Si:N ratio was <1:1, which occurred when copepods were relatively scarce. These predictions are less precise than when eutrophication occurs in the presence of higher diatom production (when Si:N > 1:1). In this scenario there will be more fecal pellet production, more carbon sedimentation to the bottom layer, higher respiration rates in the lower water column, and development of hypoxia in a stratified water column.

TABLE 21.1
Summary of observations and probable consequences with Si:N atomic ratios above and below 1:1 for the northern Gulf of Mexico (from Turner, 2001).

| Si:N | < 1:1 | > 1:1 |
|---|---|---|
| *Observed:* | | |
| % copepod of meso-zooplankton | 20 % | >80 % |
| % carbon in sediment trap that are fecal pellets | 10 % | >80 % |
| % fecal pellets of primary production | 10 % | >70 % |
| Respiration per chlorophyll *a* in bottom waters ($\mu$g oxygen $l^{-1}$ $h^{-1}$)/($\mu$g chl *a* $l^{-1}$) | <1 | >8 |
| Respiration losses in bottom waters | lower | higher |
| *Implications:* | | |
| Potential for flagellated algal blooms, HABs | higher | lower |
| Bottom water hypoxia zone | smaller | continuing |

The size distribution of organisms will certainly change if the copepod density diminishes for a period longer than their generation time (when the Si:N ratio falls below 1:1). There is a good relationship between the size of copepods and their prey size (Hansen et al., 1994). If the food size and quality changes, then the distribution of biomass in different size categories will also change (e.g., Petipa, 1973; Sprules et al., 1990). Subsequent predictions (Turner, 2001) are that a reformed phytoplankton community of smaller cells will be grazed by an also new community of smaller prey of different escape velocities, growth rates, aggregation potential, and palatability. Smaller organisms have a higher production per biomass, but also a higher respiration rate. The implication of these size relationships is that the length of the food chain will diminish as carbon produced at one level is lost in each predator-prey interaction and that fish production will decline. This sequence of events has been clearly documented on the northwestern shelf of the Black Sea as nitrogen and phosphorus loads increased and silica decreased (Zaitsev, 1992; Lancelot et al., 2002).

### *3.3 Secondary Production*

There is a variety of evidence that nutrients stimulate secondary production and sometimes fishery yields in marine ecosystems (Nixon, 1988; Caddy, 1993; Colijn et al., 2002). Nixon and Buckley (2002) summarized several examples of coastal systems where anthropogenic additions of nutrients stimulated secondary production. Multiple-year fertilization experiments were carried out over fifty years ago in Scottish sea lochs that showed dramatic increases in the abundance of benthic infauna and greatly enhanced growth of fish as a result of inorganic nitrogen and phosphorus additions. Historical comparisons for the responses of the Baltic Sea to nutrient enrichment and eutrophication have shown that the weight of benthic animals per unit area above the halocline in the Baltic is presently up to ten or twenty times greater than it was in the early 1920s and that the total fish biomass in the system may have increased eight fold between the early part of the 1900s and the 1970s. There is convincing evidence that the growth rates of plaice, sole, and other species have increased in the highly enriched central and southern North Sea since the 1960s or 1970s.

In an historical analysis of the Nile River and nutrient flux and marine fisheries in the adjacent Mediterranean Sea, Nixon (2003) described how the construction of the Aswan High Dam reduced flow from the Nile by 90 percent, which was followed by a diminished diatom bloom and a collapse in the sardine and prawn fisheries. Following 15 unproductive years, the fishery began a dramatic recovery during the 1980s, which is consistent with increased fertilizer use, expanded agricultural drainage, increased human population and expanded urban water supplies and sewage collection systems. In the case of the Nile, anthropogenic replacement of nutrients withheld by dams replenished a coastal ecosystem. Colijn et al. (2002) summarized the various responses of the North Sea ecosystem to increased nutrient concentrations and loadings. They found that the phytoplankton concentration and production rate increased, phytoplankton species composition changed, and there was an increase in biomass of macrozoobenthos. A concomitant increased of higher trophic levels, such as fish and shrimp, was difficult to link directly to the eutrophication process. Before 1960 and eutrophication in the Seto Inland Sea, red sea bream were abundant (Nagai, 2003). The biomass of anchovy increased during the period of eutrophication from 1960–1990. During excessive eutrophication post-1990, there were high N:P ratios, shifts in phytoplankton communities, and increased abundance of jellyfish. In the case of the Seto Inland Sea, nutrient additions or deletions did not result in linear responses, but rather shifts in nutrient ratios along with reductions in phosphorus and continued high nitrogen loads resulted in complex trophic interactions that did not result in higher fish production (Nagai, 2003; Yamamoto, 2003).

A meta-analysis of 47 marine food webs, including open systems, was conducted by Micheli (1999). Some of these studies experimentally added nutrients and others manipulated the food web by adding or removing prey or predators. The analysis of mesocosm studies revealed that the addition of zooplankton resulted in lower densities of their mesoplankton prey, but not phytoplankton. The addition of nutrients and zooplankton resulted in slightly higher phytoplankton and much lower mesoplankton biomass. The effect of adding nutrients to ecosystems modified to have either two or three trophic levels was to increase the phytoplankton biomass, but not the primary grazers of the phytoplankton. Micheli (1999) also analyzed various time series data for open marine systems (summarized in Turner, 2001). The availability of nitrogen and the primary production rate were strongly correlated to the accumulation of phytoplankton, but not of higher trophic levels. Micheli's analyses demonstrated a weak coupling between phytoplankton, mesoplankton and zooplankton for closed and manipulated systems. If the increased primary production that accompanies nutrient enrichment does not result in increased macro-consumer biomass of higher trophic levels then a higher proportion of the total carbon flow must be shunted to smaller consumers/decomposers or it is buried. An example of this result may be found offshore of the Mississippi River, where carbon burial rates increased this century as eutrophication occurred (Eadie et al., 1994); some of the excess carbon may also have entered the microbial food web throughout the water column.

### *3.4 Detrimental Effects*

Water quality degradation occurs with detrimental effects on components of the ecosystem and on ecosystem functioning, when loads of nutrients to the coastal ecosystem exceed the capacity for assimilation.

**Water Clarity**

As phytoplankton biomass increases as a result of nutrient enrichment, water clarity decreases with a decrease in the penetration of light through the water column. These relationships are evident in many coastal areas where long-term data sets derived primarily from Secchi disk depth readings show a decline over time or with increased nitrogen or phosphorus loading. Well-documented examples exist for the northern Adriatic Sea, the Baltic Sea, and the northern Gulf of Mexico (Justić, 1988; Elmgren and Larsson, 2001; Rabalais et al. 2002a) (Figure 21.4).

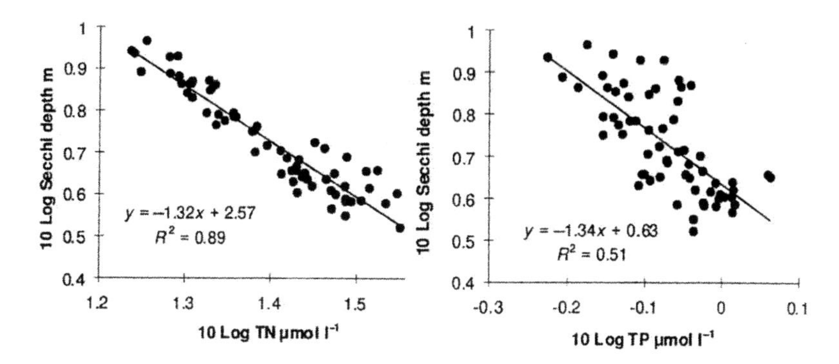


Figure 21.4 Mean Secchi disk depth from Himmerfjärden Bay, Sweden for the period 1977-1994 compared with total nitrogen and total phosphorus concentrations (from Elmgren and Larsson, 2001; used with permission of the authors).

Increased turbidity also decreases the light availability for photosynthetic, oxygen-producing microphytobenthos. Compared to the southeastern Louisiana shelf, benthic production of oxygen is greater on the southwestern shelf where light penetration is greater (Bierman et al., 1994). The penetration of light to the seabed on the Louisiana shelf and benthic production of oxygen is thought to be the mechanism by which the system remains hypoxic, albeit extremely low, and not anoxic (Dortch et al., 1994). Excess nutrients leading to decreased water clarity would be a negative feedback to a system where the oxygen budget is already unbalanced (Figure 21.2).

**Macrophyte Responses**

Where macrophytes are a dominant component of the marine ecosystem, there is a fairly predictable series of shifts in species components as eutrophication increases (reviewed by Schramm, 1999). In un-eutrophied marine or brackish shallow coastal waters, the dominant producers are usually long-lived seagrasses on soft bottoms and seaweeds on hard substrata. The growth of bloom-forming planktonic microalgae and fast-growing, short-lived epiphytic macroalgae begin to dominate over slow-growing, long-lived macroalgae in the initial stages of eutrophication. Perennial macrophytes gradually decline along with changes in structure (species composition, coverage, or depth distribution limits) and function (production and reproduction). With increased nutrient loads leading to a eutrophic state, free-floating macroalgae, in particular 'green tide' forming taxa such as *Ulva* and *Enteromorpha* alternate with dense phytoplankton blooms in dominance and replace the perennial and slow-growing benthic macrophytes until their extinction. Under hypereutrophic conditions, phytoplankton constitute the dominant primary producers, and benthic macrophytes disappear completely (Johansson and Lewis, 1992; Deegan et al., 2002; Hauxwell et al., 2003). The changes in trophic status and shifts in algal communities are not gradual but occur in a stepwise fashion and with sudden shifts.

Schramm (1999) provided numerous examples of this sequence throughout European waters. Prolonged and persistent brown tides (*Aureoumbra lagunensis* in the Laguna Madre, Texas and *Aureococcus anophagefferens* in estuaries from Narragansett Bay, Rhode Island, to Barnegat Bay, New Jersey) detrimentally affected seagrass beds and suspension-feeding bivalves, including bay scallops (Bricelj and Lonsdale, 1997; Stockwell et al., 1996). Clear water and rocky shores with dense growths of the brown seaweed bladderwrack (*Fucus vesiculosus*) that provided spawning and nursery grounds for many fish (Jansson and Dahlberg, 1999) characterized the Baltic Sea in the 1940s. Today, filamentous green and brown algae shade the bladderwrack and may even totally replace it. The cause of these changes is increased plankton blooms that reduced light penetration by 3 m compared to the first half of the century (Sandén and Håkansson, 1996). The lower growth limit moved up by about 3 m since the 1940s, and the bladderwrack does not now grow as densely as before (Kautsky et al., 1986). Bladderwrack habitat functions of refuge from predators, source of prey, and location as spawning and nursery grounds are compromised and no longer support many species previously found there. A similar loss of massive beds of red macroalgae occurred on the northwestern shelf of the Black Sea with concomitant loss of fisheries stocks that were supported by that habitat (Mee, 1992).

**Harmful Algal Blooms**

Excessive phytoplankton growth in response to nutrient increases or shifts in nutrient ratios or both may result in a bloom of a single species that has some negative ecosystem impacts. These events are typically called harmful algal blooms, or HABs, and include red tides, brown tides, and toxic and noxious blooms. Toxic forms may directly affect a variety of life forms, such as macroalgae, invertebrates, and vertebrates, including humans, and indirectly cause impacts through the consumption of toxins accumulated in fish and shellfish. Less obvious impacts of toxic forms are reduced grazing that leads to increased flux of organic matter leading to hypoxia.

There has been some debate as to whether the frequency of harmful algal blooms increased last century, and several researchers suggest a clear global expansion (Smayda, 1990; Hallegraeff, 1993) while others are more equivocal (Anderson et al., 2002; Sellner et al., 2003) (see also Chapter 22). Temporal trajectories of increased occurrence of HABs with increased nutrient loads are repeatable worldwide, but other factors, such as increased awareness and reporting, changes in freshwater inflow and circulation patterns, and worldwide transport via ship ballast water, may also be important.

Harmful algal blooms cannot always be linked to nutrient over-enrichment, but several lines of evidence are worth mentioning. Compelling evidence points to a linkage between nutrient loading and the often-cited increased frequency of harmful algal blooms in the Seto Inland Sea (Cherfas, 1990; Yamamoto, 2003). Red-tide outbreak frequency increased by an order of magnitude (from 40 to more than 300 annually) between 1965 and 1975 as nutrient loading increased. Bloom frequency was reduced by half in subsequent years following a 50 percent reduction in phosphorus loading that started in 1972 and has remained 30 percent less than the numbers during their peak of abundance. Among these high biomass algal blooms were some toxic blooms, a small fraction, but with a similar increasing and decreasing trend with nutrient enrichment and then decrease. The types of harmful algal blooms, however, have shifted as nitrogen levels remained stable, phosphorus was reduced, and the N:P ratio moved well above 16:1 (Yamamoto, 2003). Other lines of evidence link several HAB species to anthropogenic eutrophication (Burkholder, 1998).

The neurotoxin-producing forms of the diatom *Pseudo-nitzschia* occur in the northern Gulf of Mexico (Parsons et al., 1999) often in bloom proportions. The seasonal abundance correlates with high dissolved inorganic nitrogen flux from the Mississippi River (Dortch et al., 1997). Limited historical data, based on community composition from the 1950s, 1970s, and 1990s, suggest that *Pseudo-nitzschia* abundance has increased in the northern Gulf of Mexico since the 1950s (Rabalais et al., 1996). The shift in abundance from heavily silicified diatoms in the 1950s to less silicified species, including *Pseudo-nitzschia* at present, coincides with a reduction of the silicate:nitrate ratio from 4:1 to 1:1. The increasing abundance of *Pseudo-nitzschia* since the 1950s is clearly seen in dated sediment cores from the Mississippi River bight (Figure 21.5) (Parsons et al., 2002) and is coincident with human-related increases in riverine nitrogen flux and decreases in the Si:N ratio. This study provides evidence for the linkage of coastal eutrophication to some toxic forms that kill or debilitate higher organisms.

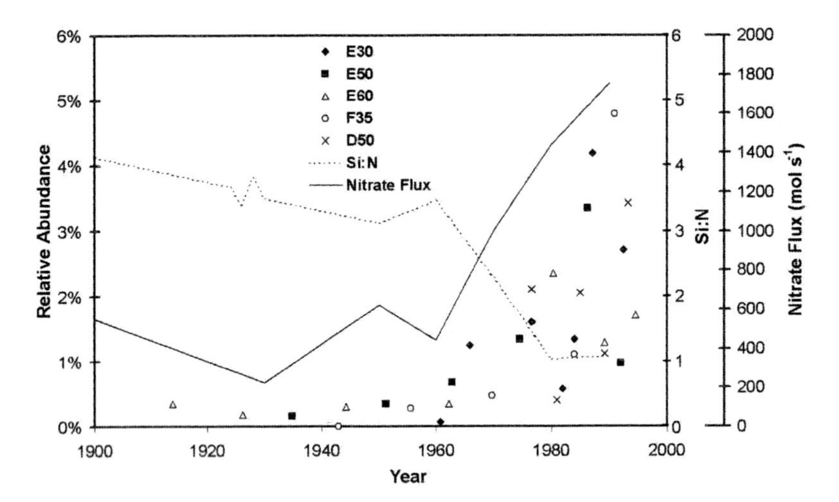


Figure 21.5 Relative (% versus total diatoms) abundance of total *Pseudo-nitzschia* from five sediment cores by estimated year ($^{210}$Pb-derived core dates) compared with nitrate flux (mol $s^{-1}$) from the Mississippi River and the silicate to nitrate (Si:N) ratio from the lower Mississippi River (replotted from data in Parsons et al., 2002).

## Hypoxia and Anoxia

Aerobic bacteria consume oxygen during decomposition of the excess carbon that sinks from the upper water column to the seabed. There will be a net loss of oxygen in the lower water column, if the consumption rate is faster than the diffusion of oxygen from surface waters to bottom waters. Hypoxia is more likely when stratification of the water column occurs and will persist as long as oxygen consumption rates exceed those of resupply. Oxygen depletion occurs more frequently in coastal areas with longer water residence times, with higher nutrient loads, and with stratified water columns (National Research Council, 2000).

While hypoxic environments have existed through geologic time and are common features of the deep ocean or adjacent to areas of upwelling, their occurrence in estuarine and coastal areas is increasing, and the trend is consistent with the increase in human activities that result in nutrient over-enrichment. Diaz and Rosenberg (1995) noted that no other environmental variable of such ecological importance to estuarine and coastal marine ecosystems around the world has changed so drastically, and in such a short period of time, as dissolved oxygen. They found that the severity of hypoxia (either duration, intensity, or size) increased where hypoxia occurred historically or that hypoxia existed now when it did not occur previously. The severity of hypoxia has increased in the northern Gulf of Mexico according to paleoindicators (see Section 4.2; Rabalais et al., 1996, 2002b), and the size and frequency of occurrence have increased as the flux of nitrate increased according to models that relate nitrate flux, offshore nitrogen concentrations, and size of the hypoxic area (Justić et al., 2003a, Scavia et al., 2002; Turner et al., 2005).

Some of the largest hypoxic zones are in the coastal areas of the Baltic Sea, the northern Gulf of Mexico, and the northwestern shelf of the Black Sea (reaching

84,000 $km^2$, 22,000 $km^2$, and 40,000 $km^2$ (until recently, see Section 4.1), respectively (Rosenberg, 1985; Mee 2001; Rabalais et al., 2002b) (see Figure 21.6 for representative Gulf of Mexico bottom-water hypoxia distribution). Hypoxia existed on the northwestern Black Sea shelf historically, but anoxic events became more frequent and widespread in the 1970s and 1980s (Tolmazin, 1985; Zaitsev, 1992; Mee, 2001), reaching over areas of the seafloor up to 40,000 $km^2$ in depths of 8 to 40 m. There is also evidence that the suboxic zone of the open Black Sea enlarged towards the surface by about 10 m since 1970. Similar declines in bottom-water dissolved oxygen have occurred elsewhere as a result of increasing nutrient loads and anthropogenic eutrophication, e.g., the northern Adriatic Sea (Justić et al., 1987), the Kattegat and Skaggerak (Rosenberg, 1985; Andersson and Rydberg, 1988), Chesapeake Bay (Officer et al., 1984; Cooper and Brush, 1991), the German Bight and the North Sea (Dethlefsen and von Westernhagen, 1983), Long Island Sound (O'Shea and Brosnan, 2000), and New York Bight (Swanson and Sindermann, 1979; Whitledge, 1985).

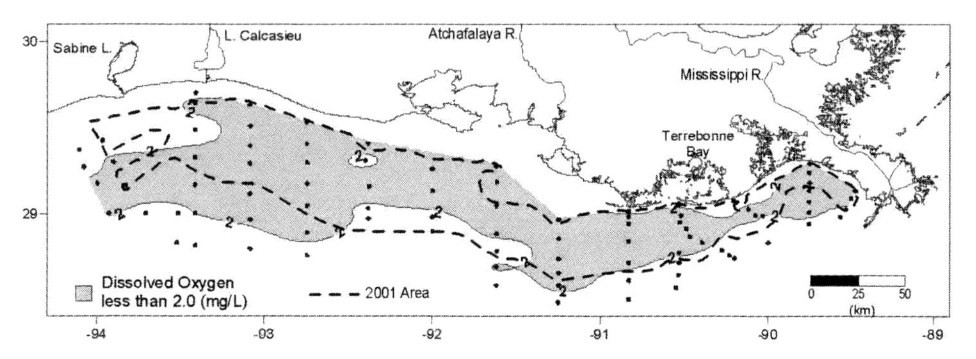


Figure 21.6 Similar size and expanse of bottom water hypoxia in mid-July 2002 (shaded area) and in mid-July 2001 (outlined with dashed line). (Data source: N. Rabalais et al., unpubl. data)

The obvious effects of hypoxia/anoxia include the displacement of pelagic organisms and selective loss of demersal and benthic organisms (Rabalais and Turner, 2001). These impacts may be aperiodic if recovery occurs, may occur on a seasonal basis with differing rates of recovery, or may be permanent so that there is a long-term shift in ecosystem structure and function. As the oxygen concentration falls from saturated or optimal levels towards depletion, a variety of behavioral and physiological impairments affect the animals that reside in the water column or in the sediments or attached to hard substrates (Burnett and Stickle, 2001; Gray et al., 2002). Recent research indicates that aquatic hypoxia is an endocrine disruptor with adverse effects on reproductive performance of fishes (Wu et al., 2003), and reduced reproductive capacity may therefore be a widespread environmental consequence of hypoxia. Mobile animals, such as shrimp, fish, and some crabs, flee waters where the oxygen concentration falls below 3 to 2 mg $l^{-1}$. Movements of animals onshore can result in "jubilees" where stunned fish and shrimp are easily captured, or in massive fish kills (Rabalais et al., 2001a). As dissolved oxygen concentrations continue to fall, less mobile organisms become stressed and move up out of the sediments, attempt to leave the seabed, and often die (Rabalais et al., 2001a). As oxygen levels fall from 0.5 mg $l^{-1}$ towards 0 mg $l^{-1}$, there is a fairly linear decrease in benthic infaunal diversity, abundance, and biomass (Rabalais et al., 2001b).

Entire taxa may be lost in the severely stressed seasonal hypoxic/anoxic zones. Larger, longer-lived burrowing infauna are replaced by short-lived, smaller surface deposit-feeding polychaetes, and certain typical marine invertebrates are absent from the fauna, for example, pericaridean crustaceans, bivalves, gastropods, and ophiuroids (Rabalais et al., 2001b). Long-term trends for the Skagerrak coast of western Sweden in semi-enclosed fjordic areas experiencing increased oxygen stress (Rosenberg, 1990) showed declines in the total abundance and biomass of macroinfauna, abundance and biomass of mollusks, and abundance of suspension feeders and carnivores. These changes in benthic communities result in an impoverished diet for bottom-feeding fish and crustaceans and contributes. Karlson et al. (2002) estimated a 3 million metric ton reduction in benthic macrofaunal biomass in waters of Scandinavia and the Baltic during the worst years of hypoxia occurrence. This loss, however, may have been partly compensated for by the biomass increase that occurred in well-flushed organically enriched coastal areas not subject to hypoxia.

**Decreased Secondary Production**

Documenting the loss of fisheries related to the secondary effects of eutrophication, such as the loss of seabed vegetation and extensive bottom-water oxygen depletion is complicated by poor fisheries data, inadequate economic indicators, increase in overharvesting that occurred at the time that habitat degradation progressed, natural variability of fish populations, shifts in harvestable populations, and climate variability (Caddy, 2000; Jackson et al., 2001; Boesch et al., 2001; Rabalais and Turner, 2001). Caddy (1993), however, illustrated how an increase in nutrient input results in a continuum of fisheries yield up to a maximum, after which there is a decline in various compartments of the fishery yield as seasonal hypoxia and permanent anoxia become features of semi-enclosed seas.

Eutrophication often leads to the loss of habitat (rooted vegetation or macroalgae) or low dissolved oxygen, both of which may lead to loss of fisheries production. In the deepest bottoms of the Baltic proper, animals have long been scarce or absent because of low oxygen availability. This area was 20,000 $km^2$ until the 1940s (Jansson and Dahlberg, 1999). Since then, about a third of the Baltic bottom area has intermittent oxygen depletion (Elmgren, 1989). Lowered oxygen concentrations and increased sedimentation have changed the benthic fauna in the deeper parts of the Baltic, resulting in an impoverished diet for bottom fish. Above the halocline in areas not influenced by local pollution, benthic biomass has increased due mostly to an increase in mulluscs (Cederwall and Elmgren, 1990). On the other hand, many reports document instances where local pollution resulting in severely depressed oxygen levels has greatly impoverished or even annihilated the soft-bottom macrofauna. In many areas the species composition and dominance changed as a result of eutrophication (Cederwall and Elmgren, 1990).

Eutrophication of surface waters accompanied by oxygen deficient bottom waters can lead to a shift in dominance of fish stocks from demersal to pelagic species. In the Baltic Sea and Kattegatt where eutrophication-related ecological changes occurred mainly after World War II (reviewed by Elmgren and Larsson,

2001), changes in fish stocks have been both positive, due to increased food supply (e.g., pike perch in Baltic archipelagos) and negative (e.g., oxygen deficiency reducing Baltic cod recruitment and eventual harvest). Similar shifts are hinted at with limited data on the Mississippi River-influenced shelf with the increase in selected pelagic species in bycatch from shrimp trawls and a decrease in certain demersal species (Chesney and Baltz, 2001). In the case of commercial fisheries in the Black Sea, it is difficult to discern the impact of eutrophication through the loss of macroalgal habitat and oxygen deficiency or a shift in the system from dominance by benthic production to pelagic production as eutrophication advanced, amid the possibility of overfishing and increased predation of zooplankton by an invasive species. After the mid-1970s, benthic fish populations collapsed (e.g., turbot), and pelagic fish populations started to increase (small pelagic fish, such as anchovy and sprat). The commercial fisheries diversity declined from 25 fished species to about five in twenty years (1960s to 1980s), while anchovy stocks and fisheries increased rapidly (Mee, 2001). The point on the continuum of increasing nutrients versus fishery yields remains vague as to where benefits are subsumed by environmental problems that lead to decreased landings or reduced quality of production and biomass.

**Coral reefs**
Coral reefs represent a coastal habitat of critical concern as more and more reef tracts decline throughout the tropical oceans. There is quite a bit of controversy concerning the cause of these declines. Many human activities affect the 'health' of coral reefs worldwide (LaPointe, 1997; Knowlton, 2001; Szmant, 2002; Wolanski et al., 2003). Increased urbanization, deforestation and expanded agricultural activities contribute sediment, organic carbon, human wastes and nutrient loads to coastal waters. Sediments can directly smother corals. Untreated or partially treated sewage is discharged or permeates into waters surrounding many coral reefs. Destruction through construction projects, removal of specimens and overfishing have direct or indirect effects on reefs. Increasing water temperature is the primary factor in coral bleaching and subsequent diseases, and follows trends in global warming. Coral diseases are a serious cause of coral decline and may be aggravated by excess nutrients. The interaction of these many human activities and their resultant nutrient, sediment, and pollutant loads make it difficult to understand how increasing nutrient loads alone impact coral reefs. The effects of nutrient enrichment on coral reefs are likely to be more evident in bays and confined water bodies rather than well-flushed oceanic reefs. Nutrients are probably less important than other factors in causing reef declines in most locations, but they may be aggravating the negative effects of other factors in ways that are difficult to assess.

## 4. History of Anthropogenic Eutrophication

Changes in nutrients as well as the symptoms of anthropogenic eutrophication follow similar time courses over several decades on a global scale among developed countries. This same sequence of events is becoming more evident in developing countries.

Data tracking increased nutrient loading over long periods are more numerous than data on the effects or symptoms of eutrophication over similar time scales. Also, data on the symptoms are seldom collected in routine water quality sampling. Symptoms may be subtle in the initial stages of anthropogenic eutrophication, possibly mimicking natural variability of the system. It is not until there are observations on continuing or worsening symptoms, multiple symptoms or all of these features, along with recorded changes in water quality conditions, that a conclusion can be drawn that anthropogenic eutrophication is occurring

As scientists began documenting the sequence of symptoms progressing towards eutrophication and correlating them with changes in water quality, particularly nutrient loads, the patterns and similar trajectories became evident globally in developed countries of Europe, North America, and Asia. This pandemic anthropogenic eutrophication was well established before scientists started to recognize the patterns of coastal eutrophication and implications for those ecosystems (Rosenberg, 1985; Nixon, 1995; Cloern, 2001). Consequently, at the beginning of the 21st century Boesch (2002) could clearly demonstrate the "explosive and synchronous intensification of eutrophication" in susceptible coastal ecosystems over the narrow period of 1960 to 1980. These changes are consistent with population growth, industrialization, urbanization, increased combustion of fossil fuel, and increased use of mined and manufactured fertilizers in the post World War II era. (Figure 21.7).

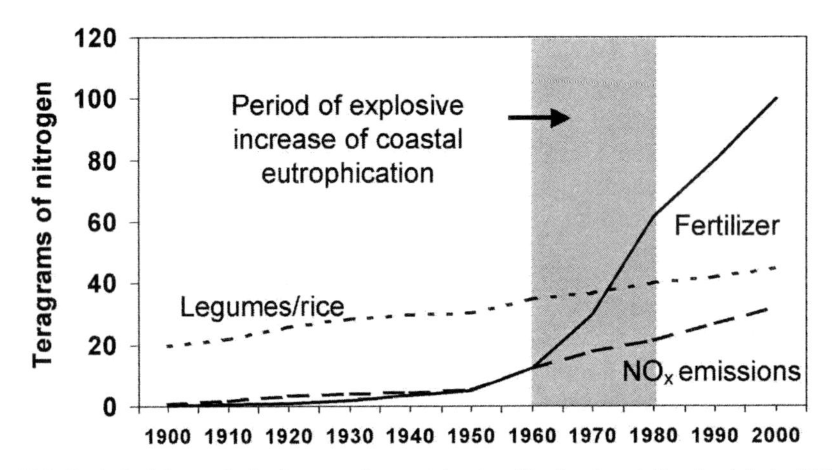


Figure 21.7 Period of the explosive increase in coastal eutrophication in relation to global additions of anthropogenically fixed nitrogen (Boesch, 2002. Fig. 2; reprinted with permission of the Estuarine Research Federation).

Human activity that altered the delivery of nutrients to coastal waters did not begin just within the last 50 years, but has been a long-term feature of any landscape modified by humans (Turner and Rabalais, 2003; Turner et al., 2003a,b). The acceleration within the last 50 years, however, has been a feature of anthropogenic eutrophication in many developed countries. It is likely that coastal eutrophication will continue to expand globally given the trajectories of future nutrient loads both in developed and developing countries (Seitzinger et al., 2002). Populations will continue to expand with higher consumptive requirements for food and fuel. The current trend is for fertilizer use to escalate as the industrialization of agriculture intensifies in the developed world and spreads even more rapidly in developing countries. Coastal ecosystems in many parts of the world, where conditions are favorable to stratification and retention of water, are at high risk for developing eutrophication and the associated direct and indirect responses, bringing to bear Rosenberg's (1985) prediction for coastal eutrophication to become the future 'coastal nuisance' and supporting the National Research Council's (2000) statement that "Given the growing magnitude of the problem and the significance of the resources at risk, nutrient over-enrichment represents the greatest pollution threat faced by the coastal marine environment." In hindsight the sequence of ecosystem changes seems obvious, but the summation of symptoms needed to reach pandemic levels before they were recognized as a global human-driven process that deserved human attention.

### *4.1 Measured Parameters*

Some parameters can be measured, while other changes need to be inferred. Those that can be measured are those that can be incorporated into the monitoring component of an adaptive management plan. Data collected initially for other purposes, but over a long period, have became the basis of analyses verifying eutrophication in several coastal systems.

When Fr. Pietro Angelo Secchi, an astrophysicist and scientific advisor to the Pope, was requested by Commander Cialdi, head of the Papal Navy, to develop a method to measure water transparency in the Mediterranean Sea in 1865, he devised a round white disk lowered through the water until no longer visible to the human observer. [Present-day Secchi disks are 20-cm in diameter with black and white quarters.] Secchi disk depth provides a rough estimate of light penetration in the water column and subsequently light limitation for photosynthesis. The application of this methodology in the northern Adriatic Sea in 1911 through the present with few interruptions of data collection provided a measure of water transparency that could be translated into surface water productivity. These data coupled with surface and bottom water dissolved oxygen content determined by Winkler titrations and nutrient loads outline the sequence of eutrophication in the northern Adriatic Sea (Figure 21.8) (Justić et al., 1987; Justić, 1988, 1991). Similar Secchi disk data sets but not for as great a temporal resolution are available for areas of the Baltic Sea (Figure 21.4) and the northern Gulf of Mexico (Rabalais et al., 2002a).

Bien et al. (1958) first documented the dilution and non-conservative uptake of silicate in the Mississippi River plume by sampling from the river mouth seaward in 1953 and 1955. A notable characteristic of the mixing diagram is that the concentration of silicate often falls below the conservative mixing line, thus indicating biological uptake. Similar measurements supplemented those of Bien et al. (1958) for a total of 31 data sets (collected between 1953 and 1990) (Turner and Rabalais, 1994a). Turner and Rabalais (1994a) found that, although the concentration of silicate in the 20 psu mixing zone declined in the last several decades, the net uptake (at 30 psu, normalized for river end-member) above dilution was higher after 1979 than before. This suggests that net silicate uptake in the dilution gradient from river to sea in the plume of the Mississippi River has increased and indicates enhanced diatom production. Similar lines of evidence for changes in phytoplankton biomass have been documented for Chesapeake Bay where long-term increases in Chesapeake Bay water column chlorophyll biomass were documented since the 1950s (Harding and Perry, 1997) as nutrient loads to the system increased.

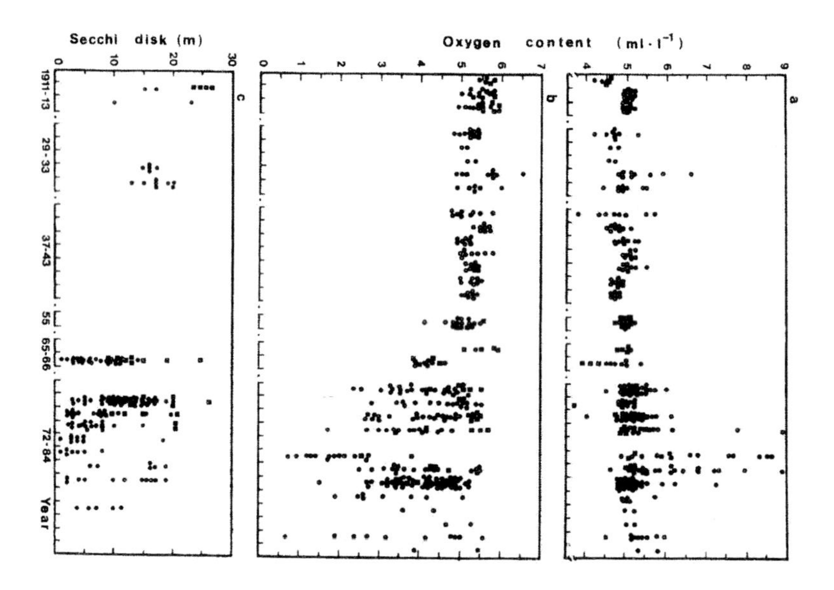


Figure 21.8 Oxygen content during August-September and Secchi disk depth during April-May in the northern Adriatic Sea from 1911 to 1984: a, surface layer; b, 2 m above the bottom. Periods without data are omitted. Statistical trends in all three sets of data are significant at $P = 0.05$ (from Justić, 1991). Used with permission of the author.

The near century-long oxygen content data available for the northern Adriatic Sea are an exception among coastal systems. Other records of oxygen data range more on the order of three to four decades at the most, but where available show a definite trend of increasing surface-water oxygen and decreasing bottom-water oxygen with increases in nutrient loading (Figs. 21.8 and 21.9).

Bottom-water hypoxia on the northwestern shelf of the Black Sea adjacent to the Danube River was first documented in 1973 (Figure 21.10; Zaitsev, 1992). By 1990 the areal extent of hypoxia on the bottom of hypoxia was 40,000 km$^2$ (Figs. 21.10 and 21.11). There is substantial evidence that eutrophication in the Black Sea is the result of large increases in the discharge of nitrogen and phosphorus to the Black Sea from the 1960s and 1970s (Mee, 2001). The typical scenario followed with increased nutrients triggering dense phytoplankton blooms, a decrease in seawater transparency, and an increase in the load of organic detritus reaching the seafloor (Tolmazin, 1985; Mee, 1992, 2001; Zaitsev, 1992). The high organic loading led to the expansion of oxygen-deficient waters over the northwestern shelf in depths of 8 to 40 m and over areas of the seafloor up to 40,000 km$^2$. As a result of the economic collapse of the former Soviet Union and declines in subsidies for fertilizers, the decade of the 1990s witnessed a substantially decreased input of nutrients to the Black Sea (Figure 21.12; Mee, 2001; Lancelot et al., 2002). For the first time in several decades oxygen deficiency was absent from the northwestern shelf of the Black Sea in 1996, and receded to an area less than 1,000 km$^2$ in 1999 (Figure 21.11).

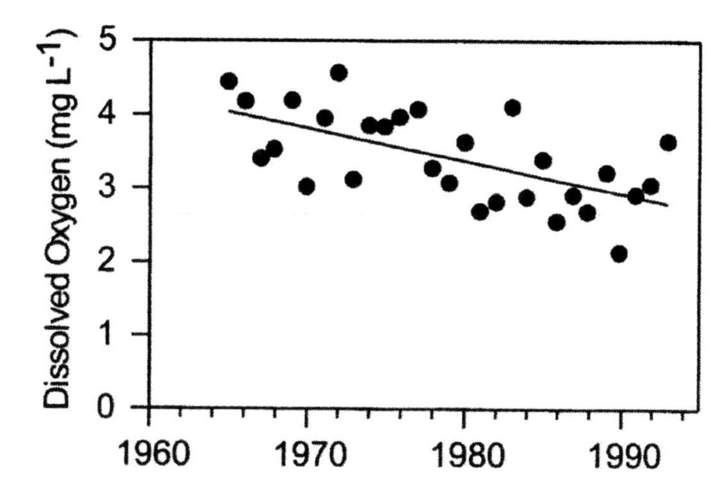


Figure 21.9 Trends of decreasing dissolved oxygen concentrations in bottom waters of the southern Kattegat (redrawn from Christensen, 1998; in Cloern, 2001; used with permission of Inter-Research).

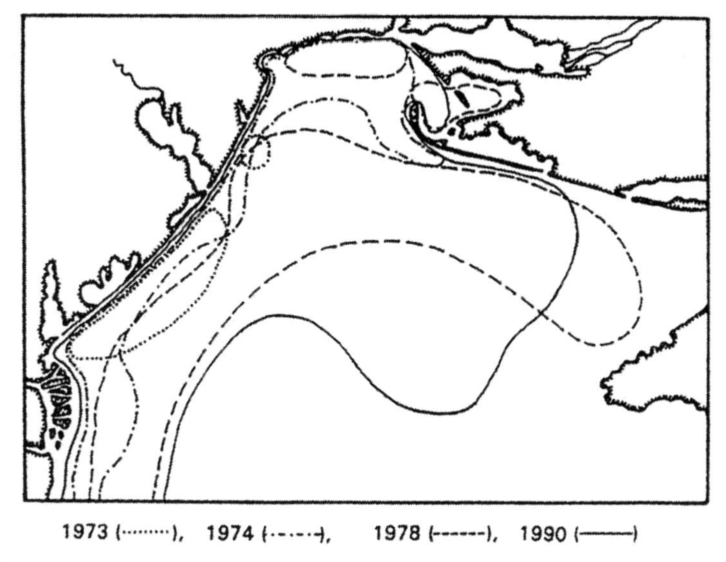


Figure 21.10 Development of hypoxia on the northwestern shelf of the Black Sea (from Zaitsev, 1992; permission fee paid to Blackwell Publishing).

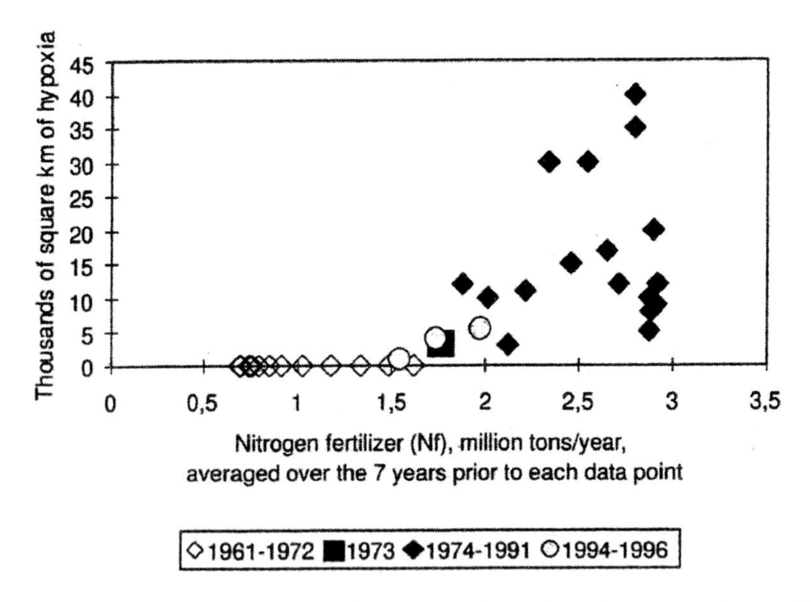


Figure 21.11 Change in the areal extent of bottom-water hypoxia on the northwestern shelf of the Black Sea in relationship to groupings of years with differing use of nitrogen fertilizer in the Danube River watershed (from Mee, 2001; used with permission of the author).

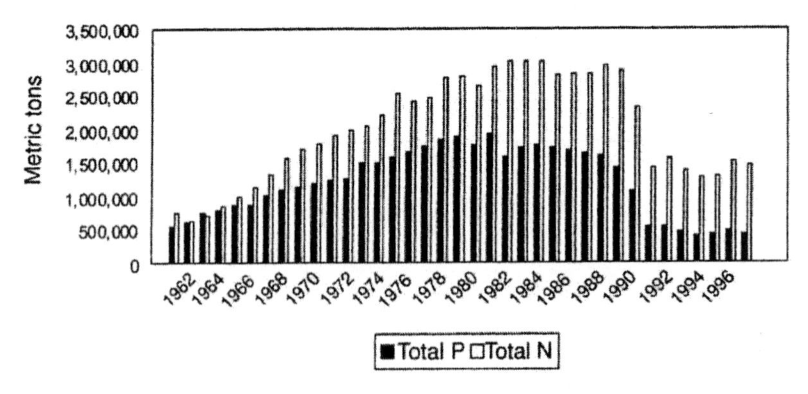


Figure 21.12 Nitrogen and phosphorus fertilizer trends for use in the watersheds draining to the northwestern shelf of the Black Sea (from Mee, 2001; used with permission of the author). The nitrogen and phosphorus loads correspond to fertilizer use.

### 4.2 *Paleoindicators*

Relevant data for analysis of historic conditions of either productivity in coastal waters or oxygen conditions in bottom-waters are seldom long-term or complete and cannot be assessed directly. An improvement is to examine the sediment record for paleoindicators of long-term transitions related to eutrophication. Biological, mineral or chemical indicators of plant communities, level of productivity or conditions of hypoxia preserved in sediments, where sediments accumulate, provide clues to prior hydrographic and biological conditions. Looking down-core through marine sediments is equivalent to looking back in time. The records reflect conditions extant in bottom waters at the time the sediment was deposited and thus provide clues to temporal changes in biogeochemical conditions.

Multiple lines of evidence from sediment cores indicate an overall increase in phytoplankton productivity and continental shelf oxygen stress (in intensity or duration) in the northern Gulf of Mexico adjacent to the plume of the Mississippi River especially in the last half of the 20th century. The changes in these indicators are consistent with the increases in riverine nutrient loading during that same period and landscape use changes, and are more apparent in areas of chronic hypoxia. This evidence comes as an increased accumulation of diatom remains (biologically bound silica, BSi) and marine-origin carbon accumulation in the sediments (Figure 21.13). The sediment data for increased diatom production in recent decades confirm the conclusions of Turner and Rabalais (1994a) of an increase in silicate uptake in the Mississippi River plume.

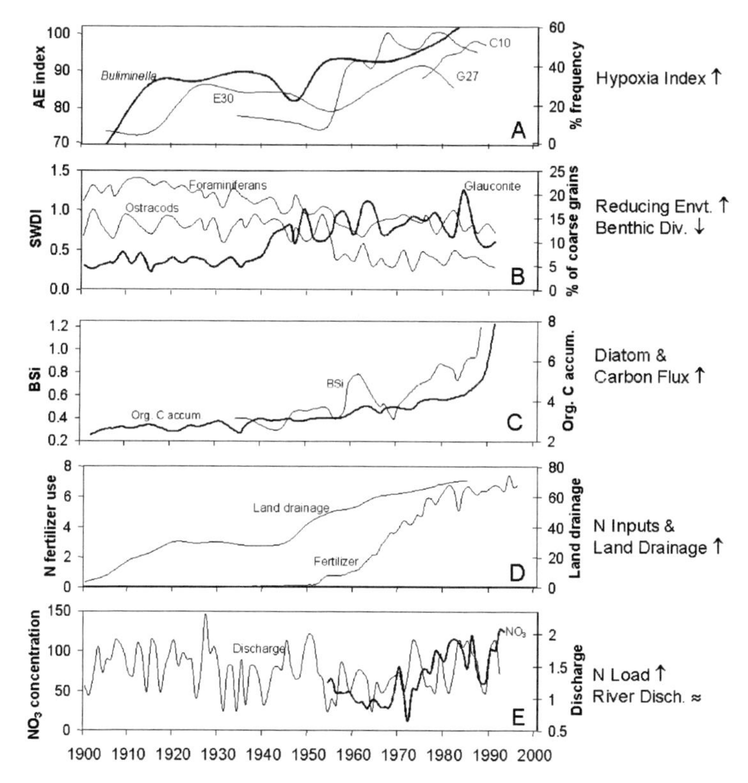


Figure 21.13 Indices of historical change under the Mississippi river plume (modified from Rabalais et al., 2002b; permission not needed as per policy of AIBS for their authors in *BioScience*). A: A–E index (foraminiferal hypoxia index); % frequency of *Buliminella* (hypoxia tolerant foraminiferan). B: SWDI (Shannon-Wiener Diversity Index) for foraminiferans and ostracods; % glauconite of coarse grains. C: BSi (biologically bound silica, frequency); organic carbon accumulation (mg C $m^{-2}y^{-1}$). D: Nitrogen fertilizer use in the Mississippi River basin ($10^6$ mt $y^{-1}$); land drainage (millions of acres). E: Nitrate concentration (µM) in the lower Mississippi River; lower Mississippi River discharge ($10^4 m^3 s^{-1}$).

Figure 21.13 also shows time courses for several surrogates for oxygen conditions: glauconite abundance and benthic foraminiferans and ostracods. The mineral glauconite forms under reducing conditions in sediments regardless of the oxygen in overlying waters, but its relative increase over time is an indication of decreasing water-column oxygen concentrations over time. The average glauconite abundance in the coarse fraction of sediments was ~5.8% from 1900 to a transition period between 1940 and 1950, when it increased to ~13.4% (Figure 21.13; Nelsen et al., 1994), suggesting that hypoxia *may* have existed at some level before the 1940–1950 time period, but that it worsened since then.

Benthic foraminiferans and ostracods are useful indicators of reduced oxygen levels because oxygen stress decreases their overall diversity and causes shifts in community composition. Foraminiferal and ostracod diversity decreased since the 1940s and early 1950s, respectively (Figure 21.13; Nelsen et al., 1994, unpublished data). While present-day foraminiferal diversity is generally low in the Mississippi River bight, comparisons among assemblages from areas of different oxygen depletion indicate that the dominance of *Ammonia parkinsoniana* over *Elphidium* spp. (A-E index) was much more pronounced in oxygen-depleted compared to well-oxygenated waters and increased after the 1950s (Figure 21.13) (Sen Gupta et al., 1996). The A-E index has also proven to be a strong, consistent oxygen-stress signal in other coastal areas (e.g., Chesapeake Bay, Karlsen et al., 2000; Long Island Sound, Thomas et al., 2000). *Buliminella morgani,* a hypoxia-tolerant species, known only from the Gulf of Mexico, dominates the present-day population (> 50 percent) within areas of chronic seasonal hypoxia, and has increased markedly in recent decades (Figure 21.13). *Quinqueloculina* sp. (a hypoxia-intolerant foraminiferan) was a conspicuous member of the fauna from 1700 to 1900 but not since (not illustrated, Rabalais et al., 1996), indicating that oxygen stress was not a problem prior to 1900.

The remains of bacteria and phytoplankton in the form of chemical pigment signatures can also be used to document changes in productivity and oxygen depletion. In sediment cores from the Mississippi River bight, there was a general increase in chlorophyll *a,* pheopigments (chlorophyll degradation products), zeaxanthin (indicative of cyanobacteria), fucoxanthin (indicative of diatoms), and most carotenoids (various phytoplankton taxa) over time, with the change gradual from 1955 to 1970, followed by a fairly steady increase to 1997, the time the cores were collected (Rabalais et al., 2004). The increasing pigments and greater concentrations in areas where hypoxia is more likely to occur indicate an increase in eutrophication or a worsening of hypoxia or both. Chen et al. (2001) used pigments from two anoxygenic phototrophic brown-pigmented green sulfur bacteria to verify that hypoxia was not detected on the southeastern Louisiana shelf prior to the 1900s and that hypoxia worsened between 1960 and the time of core collection in 1998–1999, similar to the foraminiferan results reported by Rabalais et al. (1996) and Sen Gupta et al. (1996).

Several lines of investigation with paleoindicators documented anthropogenic changes within Chesapeake Bay. Since European settlement there has been a gradual eutrophication of Chesapeake Bay with an acceleration of these changes in the last 50 years. The flux of biogenic silica (BSi) to sediments as an index of diatom productivity was relatively low before European settlement of the watershed (Colman and Bratton, 2003). Human activities in the succeeding 300–400 years, led to a four- to five-fold increase in biogenic silica flux, which still appears to be increasing. Shifts from primarily forest vegetation to agricultural crops, as indicated by ragweed pollen, occurred at different periods in various subestuaries and tributaries of the Chesapeake, and the most dramatic changes occurred after 25 to 30 percent deforestation (Brush, 2001). Increased eutrophication attributable to an intensification of human activities and worsening hypoxia in the last half century is recognizable in the sedimentary record of chlorophyll, diatom diversity and species composition, total organic carbon, total organic nitrogen and phosphorus, sulfur, pyrite, stable isotopes, and lipid biomarker signatures (Brush, 1984; Cooper and Brush, 1991; Cooper, 1995; Karlsen et al., 2000; Zimmerman and Canuel, 2002).

The vertical distribution of benthic foraminiferans in a sediment core from the northern Adriatic Sea under the plume of the Po River indicates a gradual increase in eutrophication at the end of the 19th century with an acceleration post 1930, the beginning of seasonal hypoxia in 1960, and more intense or prolonged hypoxia beginning in 1980 to present (Table 23.2) (Barmawidjaja et al., 1995). These changes related to increased nutrient loads followed human-induced changes in the main outflow canals of the Po River (canals, levees, and redistribution of discharge to different distributaries) and a reduction in the marine vegetative cover in the delta. These data are consistent with those outlined earlier—doubling or better of dissolved inorganic phosphorus and nitrogen during the 1970s-1980s (Harding et al., 1999), the decrease in Secchi disk depth (Justić, 1988), and changes in surface water productivity and bottom water oxygen depletion (Justić et al., 1987)

TABLE 21.2

Summary of eutrophication history inferred from changes in benthic foraminiferans from a core in the northern Adriatic Sea adjacent to the Po River delta (modified from Barmawidjaja et al., 1995).

| Event | Source of Change |
|---|---|
| 1830 AD: base of core | |
| 1849 AD: increase sediment load | diversion of Po outflow |
| 1870 AD: final disappearance of vegetation due to mud load and natural eutrophication | diversion of Po outflow to present position at Po della Pila |
| 1880–1920 AD: gradual increase of eutrophication | human-induced increase of Po nutrient load |
| 1930 AD: acceleration of eutrophication | modern farming methods, increase in urbanization and industrialization |
| 1960 AD: increase in seasonal hypoxia | intensification of eutrophication due to human-induced increase of Po nutrient load |
| 1981–1987 AD: maximum anoxia | intensification of eutrophication due to human-induced increase of Po nutrient load |

A similar sequence of longer-term human settlement and modification of the landscape as seen in Chesapeake Bay and the northern Adriatic also occurred in the Mississippi River basin but lagged into the middle of the 19th century, when European settlers began to move across the Ohio, Mississippi, and Missouri valleys (Figure 21.14) (Turner and Rabalais, 2003). Biogenic silica, as an indicator of offshore marine productivity, increased at the time of land clearance in the basin. The nitrogen released from the watershed in the middle of the 19th and early 20th centuries was likely organic forms remineralized as row cropping and deforestation expanded and as lands were drained and the effectiveness of denitrification decreased. This period was followed by a dramatic decrease in biogenic silica before a major increase in the middle of the 20th century that was accelerated by additional inputs of nutrients post World War II. The more recent influence on nutrient loading, from intense and widespread farming and especially from fertilizer use, has had a more significant effect on water quality than the conversion of native vegetation to cropland and grazing pastures, or of land drainage.

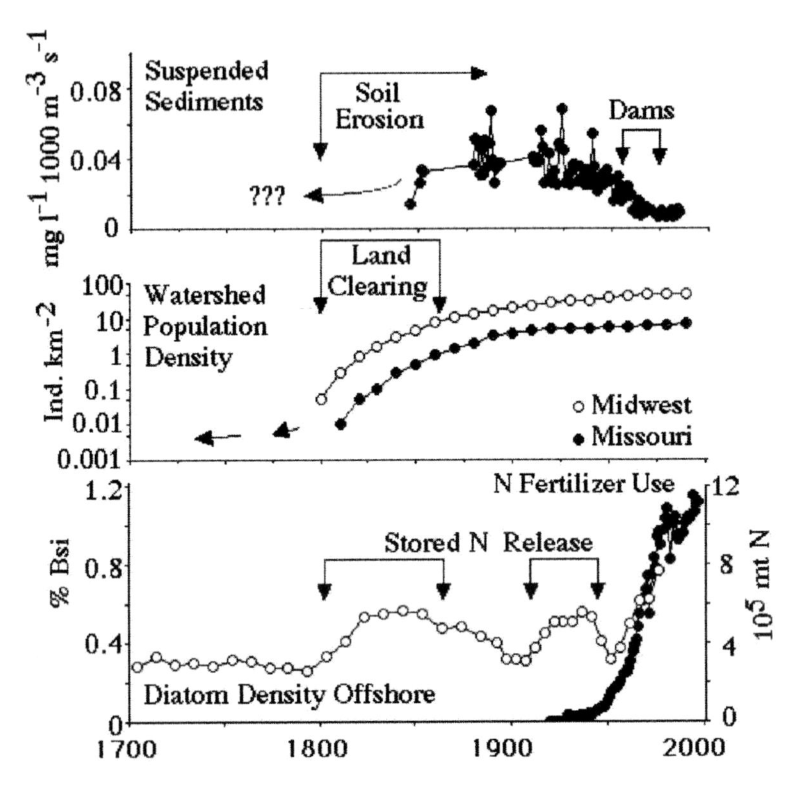


Figure 21.14 Summary interpretation of the relationships among population growth, land conversion to agriculture, and fertilizer use in the Mississippi River basin and coastal diatom production (from Turner and Rabalais, 2003). [permission not needed as per policy of AIBS for their authors in *BioScience*]

## 5. Management Challenges

The effects of two centuries of land use in the Mississippi River watershed (Figure 21.14) are seen in the water quality of the basin and the offshore coastal waters (Rabalais et al., 2002b; Turner and Rabalais, 2003), as is the case for many other systems. The water quality reflects both conversion of native forests and rangeland to cropland and grazing pastures, drainage of wetlands, artificial drainage of agricultural lands, and 50 years of artificial fertilizer applications. Management of both inland and offshore ecosystems is thereby linked. The landscapes of most watersheds in developed countries are mostly occupied by humans and influenced by their activities; those of developing countries are being pressured by similar human endeavors. As human population increases (and it will) that growth alone will cause new water quality changes in both developed and developing countries.

The landscapes and the ecosystems that depend upon them are held within social structures that can be modified for the betterment of the ecosystems and the humans that depend upon them, or for the worse. The social infrastructure supporting the human-made landscape is non-trivial and important to many. It involves food, fiber and fuel supplies, national political agendas, international trade, and global climate change. It will take political and societal will for scientists, citizens, land and water managers, agriculturalists, industrialists and environmentalists to work toward a mutually-satisfying equilibrium of interests that is fair, sustainable, and socially-responsive (Turner and Rabalais, 2003).

### *5.1 System Response and Inertia*

Any nutrient management strategy applied to the landscape or the chemicals applied to the landscape to reduce excess nutrients will likely result in changes in the reverse of those seen over the last decades and centuries, but the change will probably be gradual because of the strong buffering capacity of the soil ecosystem. It will likely take the many decades it took for the present system to develop to succeed in water quality rehabilitation. The strong buffering capacity of the soils is becoming evident (Stålnacke et al., 1999). There is an inertia in terrestrial systems and rivers and streams with regard to losses from land to sea following nutrient reductions actually achieved or planned (Grimvall et al., 2000). Nitrate leaching from a grain field in Sweden continued almost unabated 13 years after fertilizers were no longer added (Löfgren et al., 1999). Data on river water quality following the collapse (circa 1990) of agriculture in the former Soviet republics of Estonia, Latvia, and Lithuania demonstrated that although fertilizer application fell to the level of the 1950s, the concentration of inorganic phosphorus and nitrogen was the same in 1994 as in 1987 (Löfgren et al., 1999).

Public and private funds have been expended within the Chesapeake Bay watershed to reduce the controllable sources of nitrogen and phosphorus entering the bay by 40 percent by the year 2000 (Boesch, 2001). Efforts targeted both point sources, such as treated sewage discharges, and nonpoint sources, especially those from agriculture, or to trap the nutrients in the watershed by wetland and riparian zone restoration. Assessing whether the reduction targets were reached was difficult, but it appears that the goal was nearly met for phosphorus, but nitrogen loadings, although reduced, did not achieve the goal. Similarly, Grimvall et al. (2000) reported that there was a remarkable lack of response in eastern European river nutrient loads in response to the dramatic decrease in the use of commercial fertilizers that started in the late 1980s. In western Europe, while studies of decreased phosphorus emissions have shown that riverine phosphorus loads can be rapidly reduced from high to moderate levels, a further reduction, if achieved at all, may take decades.

The accumulated loads of organic matter and the internal load of inorganic and organic nutrients in the sediments underlying eutrophic waters perpetuate conditions of eutrophication as they continue to be processed by normal geochemical processes in the sediments (Rabalais, 2002). Within estuaries and coastal systems, a decrease in the external nutrient loads may not produce an immediate shift in the eutrophic state of the system, in part because of the continued remineralization of labile carbon and releases of regenerated nutrients. Boynton and Kemp (2000) suggested a 'nutrient memory' over time scales of a year rather than seasonal periods as suggested by Chesapeake Bay water residence times. Justić et al. (1997) suggested that at least a year of continued carbon respiration following high deposition of carbon in a flood year contributed to oxygen demand on the Louisiana continental shelf in the subsequent summer. Besides an inherent lag in system responses, climatic variability may mask any attempts to reduce nutrients within a watershed. For the Kattegat, a statistically significant detectable improvement in oxygen conditions, given the variability present in the system will take 18 years (Richardson, 1996). These predictions are relevant to management strategies to mitigate nutrient loads to estuaries and coastal waters, and the perceived projection for 'recovery.'

As a result of the economic collapse of the former Soviet Union and declines in subsidies for fertilizers, the decade of the 1990s witnessed a substantially decreased input of nutrients to the Black Sea, with resulting signs of recovery in some aspects of the pelagic and benthic ecosystems (Mee, 2001; Lancelot et al., 2002). There is a recovery in zoobenthos species diversity, phytoplankton biomass has declined by about 30 percent of the 1990 maxima, there is some recovery of the diatoms, the phytoplankton are more diverse, the incidence of intense blooms has declined, and there is a limited recovery of some zooplankton stocks and diversity in limited geographic areas. There has been no recovery of benthic macroalgae. Oxygen deficiency decreased (Figure 21.11). Most fish stocks in the northwestern Black Sea are still depleted (Mee, 2001). There should be little doubt of the strong relationships among human activities, Black Sea eutrophication, and demise of pelagic and benthic coastal ecosystems (Lancelot et al., 2002), as well as similar linkages in the partial recovery of those systems following reduced nutrients. While the mediator of the nutrient reductions from the watershed of the Black Sea was economic hardship and decline, i.e., it was not a preferable means of reducing nutrient loads, the resilience of the coastal ecosystem within periods of a few years to a decade is heartening because it is a clear example of biological resiliency when nutrient loading is reduced. Researchers within the narrow, coastal inlets of the Bodden are less optimistic about system recovery where nutrients to that sector of the Baltic were reduced greatly during the last decade of the $20^{th}$ century, but the expected improvement of water quality has not been demonstrated (Meyer-Reil and Köster, 2000).

### 5.2 *Targeting Nutrients for Management*

It is usually a great oversimplification to assume that a single nutrient limits marine systems. Howarth (1988) noted that, given our understanding of nutrient limitation and knowledge of increasing inputs of nutrients to coastal systems, to control eutrophication in these systems requires nitrogen controls, but that phosphorus controls also make sense. There is a need to distinguish between the processes of biomass limitation and growth rate limitation. If the concern with increased nutrient loads is eutrophication, then biomass accumulation is the important issue, and accumulation of biomass in most coastal marine systems is nitrogen limited, as shown for Chesapeake Bay (Malone et al., 1996; Harding and Perry, 1997), over the broad region of the northern Gulf of Mexico influenced by the Mississippi River and subject to hypoxia (Rabalais et al., 2002a) and in the Baltic proper (reviewed by Elmgren, 2001). Unfortunately, our knowledge of the relative importance of nitrogen and phosphorus to phytoplankton growth is focused on temperate systems, and tropical systems, while more frequently limited by phosphorus (Downing et al., 1999), are less well studied and their management less prescribed.

From the late 1960s to the mid-1980s the nutrient inputs to the Baltic increased by at least a factor of four for nitrogen and eight for phosphorus (Elmgren, 2001). Initially the focus there was on phosphorus as the key element in the eutrophication process, but this was soon questioned. Eventually both algal growth-potential experiments and analysis of the water chemistry implicated nitrogen as the main limiting nutrient in the open Baltic proper, as well as in open Danish waters. The spring phytoplankton bloom in the Baltic proper is clearly nitrogen limited, but summer production may sometimes be phosphorus limited. Silicate concentrations in the Baltic have declined in the last few decades, but not yet to levels expected to be limiting to diatom production (Elmgren, 2001). Recent paleoecology studies indicate that nitrogen-fixing cyanobacterial blooms occurred in the Baltic for over the last 7000 years and that summer nitrogen limitation is a 'natural' condition (Bianchi et al., 2000). The seasonal variation of nutrient limitation and time-course of nutrient changes in the Baltic raise practical issues related to the management of nutrients.

Fourteen nations have agreed to restore water quality in the Baltic by reducing loads of both nitrogen and phosphorus, and changing the goal requires consent from all signatories. Reducing the input of phosphorus alone would not restore the Baltic. Instead a phosphorus-limited Baltic would have primary production more concentrated to the spring bloom than at present, more sedimentation of oxygen-consuming organic matter, and less fish production for the same total production. In addition to the complexities of nutrient limitation, there is evidence that the nutrient loads from the land have been rather constant since 1970 but that nutrient concentrations in the Baltic gradually increased up to the late 1980s. This disconnect in loads and concentrations is likely related to landscape alterations that allow for a quicker transport of nitrogen from field to sea and lower denitrification losses en route (e.g., lowering of lakes, hydrologic modifications including dams that trap silicate, draining of wetlands, and straightening of waterways. Thus, nutrient management involves both repair to the landscape and reductions in inputs (Mitsch et al., 2001; Paludan et al., 2002).

### *5.3 Management Scenarios*

Reducing excess nutrient loading to estuarine and marine waters requires individual, societal, and political will. Proposed solutions may be controversial and may appear to extract societal and economic costs. Yet, multiple, cost-effective methods of reducing nutrient use and delivery can be integrated into a management plan that results in improved habitat and water quality, both within the watershed and the receiving waters (National Research Council, 2000; Greening and Elfring, 2002).

Successful plans with successful implementation and often with successful results span geopolitical boundaries, for example the Chesapeake Bay Agreement, the Comprehensive Conservation and Management Plans developed under the National Estuary Program for many U.S. estuaries, a Long Island Sound agreement for New York and Connecticut, the Hypoxia Action Plan for the Mississippi River basin, and international cooperation among the nations fringing the Baltic Sea as part of the Helsinki Commission. These efforts are usually more successful in reducing point sources of nitrogen and phosphorus than with the multiple non-point sources of high solubility and growing atmospheric inputs of nitrogen. Nutrient management has been successful for coral reefs in Kaneohe Bay (Smith, 1981) and for the improved water clarity and recovery of seagrass beds in Tampa and Sarasota Bays (Johansson and Lewis, 1992; Sarasota Bay National Estuary Program, 1995). Successful programs set goals for restoration, determine nutrient reductions needed to meet goals, monitor for change, and incorporate adaptive management that considers research, monitoring and modeling feedbacks (Greening and Elfring, 2002). While not considered a social and economic success, the reduction of nutrients into the northwestern Black Sea demonstrated the capacity of a large coastal ecosystem to recover from decades of excess nutrients.

The growing decline of coastal water quality and expansion of symptoms of eutrophication, and also the proven successes of reducing nutrients, are reasons enough for us to continue and to expand efforts to reduce nutrient over-enrichment. Countermeasures may take decades for effective restoration to be realized, and it may take decades for management intervention to be realized in the landscape and in the coastal ecosystem. Elmgren (2001) noted that it took two decades from discovery of Baltic proper eutrophication to political agreement on effluent reductions, which have yet to become effective. He recommended that environmental management decisions can hardly wait for scientific certainty, and further recommended adaptive management strategies that treated environmental management decisions as experiments to be monitored, learned from and modified as needed.

## 6. Future Expectations

The continued and accelerated export of nitrogen and phosphorus to the world's coastal ocean is the trajectory to be expected unless societal intervention in the form of controls or changes in culture are pursued. Seitzinger et al. (2002) modeled future projections of dissolved inorganic nitrate (DIN) export from world rivers in a Business-as-Usual scenario and predicted that DIN export rates increased from approximately 21 Tg N $y^{-1}$ in 1990 to 47 Tg N $y^{-1}$ by 2050. Increased DIN inputs to coastal systems in most world regions were predicted by 2050. The largest increases were predicted for southern and eastern Asia, associated with predicted large increases in population, increased fertilizer use to grow food to meet the dietary demands of that population, and increased industrialization.

Trends for future nitrogen export are not necessarily entrained on accelerated increases. Results of an analysis of alternative scenarios for North America and Europe in 2050 indicate that reductions in the human consumption of animal protein could reduce fertilizer use and consequently result in substantial decreases in DIN export rates by rivers in those regions (Seitzinger et al., 2001). McIsaac et al. (2001, 2002) predicted that a 12 to 14 percent decrease in the application of fertilizer in the Mississippi River basin could reduce the net anthropogenic nitrogen inputs by 30 percent without a reduction in crop production. The results of model scenarios indicate that this level of reduction in flux to the northern Gulf of Mexico, under 'normal' climate conditions, would be sufficient to reduce the frequency of hypoxia by 37 percent (Justić et al., 2003b) and approach the size of the hypoxic zone to the agreed-upon goal of 5,000 $km^2$ (Scavia et al., 2003). In another scenario for 2050 (Seitzinger et al., 2002), future air pollution controls in Europe that would reduce atmospheric deposition of nitrogen oxides in watersheds were predicted to decrease DIN export by rivers, particularly from Baltic and North Atlantic watersheds.

With or without nutrient management scenarios, the projected increasing nutrient export trajectory will occur within a likely scenario of global climate change. The implications for coastal eutrophication and subsequent ecosystem changes such as worsening conditions of oxygen depletion are significant. A modeling study that examined the impacts of global warming on the annual discharge of the 33 largest rivers of the world (Miller and Russell, 1992) suggested that the average annual discharge of the Mississippi River would increase 20 percent if the concentration of atmospheric $CO_2$ doubled. If so, hypoxia would intensify and expand on the Louisiana continental shelf (Justić et al., 1996, 1997, 2003a,b). Other studies have shown that the runoff estimates for many U.S. rivers, including the Mississippi, differed greatly between the Canadian model and the Hadley model (Figure 21.15) (Wolock and McCabe, 1999). In the former the Mississippi River discharge would decrease by 30 percent and would increase by 40 percent in the latter by the year 2099. Whichever the case, the increase or decrease in flow, flux of nutrients, and changes in water temperature are likely to have important but as yet not clearly identifiable influences on the process of eutrophication as water residence times, physical conditions, and biological processes change.

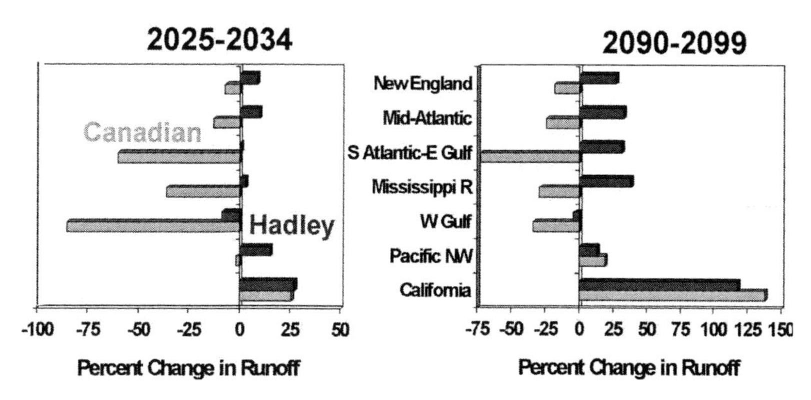


Figure 21.15 Projected changes in average annual runoff for basins draining to coastal regions of the U.S. according to the Canadian and Hadley climate models (data from Wolock and McCabe, 1999). [original figure by D. Justić from published data, permission not needed]

The consequences of eutrophication are disquieting when one reviews the trajectory of nutrient inputs and coastal ecosystem responses that occurred during the last half of the 20th century and when one then considers projections for population growth and resource demands projected for the 21st century. Still, we understand much more about the process, the causes, and the responses of the system to excess nutrients presently than in 1985 when Rosenberg (1985) predicted that eutrophication was to become the "future coastal nuisance." Eutrophication has surpassed a categorization of 'nuisance,' and nutrient over-enrichment has been labeled "the greatest pollution threat faced by the coastal marine environment" (National Research Council, 2000). At the same time scientists, resource managers, and the public are becoming increasingly aware of the environmental, social, and political issues surrounding eutrophication and have developed multi-faceted, multi-level, integrated, and sound management plans to reduce nutrient loads to coastal waters and alleviate the symptoms of eutrophication. If we can collectively reduce nutrients to the level of success that pesticides such as DDT or sources of atmospheric deposition that led to acid rain were reduced, then there remains the potential to reverse current coastal water quality degradation while sustaining renewable resources and the well-being of society.

## Acknowledgments

Funding for the preparation of this manuscript was provided by the National Oceanic Atmospheric Administration, Coastal Ocean Program, project nos. NA06OP0528 and NA03NOS4780037. B. Cole and A. Sapp helped with graphics. I thank R. E. Turner and two anonymous reviewers for comments on the manuscript.

## Bibliography

Allen, J. R., D. J. Slinn, T. M. Shammon, R. G. Hartnoll and S. J. Hawkins. 1998. Evidence for eutrophication of the Irish Sea over four decades. *Limnol. Oceanogr.,* **43,** 1970–1974.

Anderson, D. M., P. M. Glibert and J. M. Burkholder. 2002. Harmful algal blooms and eutrophication: nutrient sources, composition, and consequences. *Estuaries,* **25,** 704–726.

Andersson, L. and L. Rydberg. 1988. Trends in nutrient and oxygen conditions within the Kattegat: effects on local nutrient supply. *Estuar. Coastal Shelf Sci.,* **26,** 559–579.

Antia, N., P. Harrison and L. Oliveira. 1991. The role of dissolved organic nitrogen in phytoplankton nutrition, cell biology, and ecology. *Phycologia,* **30,** 1–89.

Barmawidjaja, D. M., G. J. van der Zwaan, F. J. Jorissen and S. Puskaric. 1995. 150 years of eutrophication in the northern Adriatic Sea: evidence from a benthic foraminiferal record. *Mar. Geol.,* **122,** 367–384.

Bennett, E. M., S. R. Carpenter and N. F. Caraco. 2001. Human impact on erodable phosphorus and eutrophication: a global perspective. *BioScience,* **51,** 227–234.

Bianchi, T. S., E. Engelhaupt, P. Westman, T. Andrén, C. Rolf and R. Elmgren. 2000. Cyanobacterial blooms in the Baltic Sea: natural or human-induced? *Limnol. Oceanogr.,* **45,** 716–726.

Bien, G. S., D. E. Contois and W. H. Thomas. 1958. The removal of soluble silica from fresh water entering the sea. *Geochim. Cosmochim. Acta,* **14,** 35–54.

Bierman, Jr., V. J., S. C. Hinz, D. Zhu, W. J. Wiseman, Jr., N. N. Rabalais and R. E. Turner 1994. A preliminary mass balance model of primary productivity and dissolved oxygen in the Mississippi River Plume/Inner Gulf shelf region. *Estuaries,* **17,** 886–899.

Boesch, D.F. 2001. Science and integrated drainage basin coastal management. Chesapeake Bay and Mississippi Delta. Pp. 37–50 in *Science and Integrated Coastal Management,* von Bodungen, B., and R. K. Turner (eds.), Dahlem University Press, Berlin.

Boesch, D. F. 2002. Challenges and opportunities for science in reducing nutrient over-enrichment of coastal ecosystems. *Estuaries,* **25,** 886–900.

Boesch, D. F., E. Burreson, W. Dennison, E. Houde, M. Kemp, V. Kennedy, R. Newell, K. Paynter, R. Orth and R. Ulanowicz. 2001. Factors in the decline of coastal ecosystems. *Science,* **293,** 1589–1590.

Boynton, W. R. and W. M. Kemp. 2000. Influence of river flow and nutrient loads on selected ecosystem processes. A synthesis of Chesapeake Bay data. Pp. 269–298 in *Estuarine Science: A Synthesis Approach to Research and Practice,* Hobbie, J. E. (ed.), Island Press, Washington, D.C.

Boynton, W. R., W. M. Kemp and C. W. Keefe. 1982. A comparative analysis of nutrients and other factors influencing estuarine phytoplankton production. Pp. 69–90 in Kennedy, V.S. (ed.), *Estuarine Comparisons,* Academic Press, New York.

Boynton, W. R., J. H. Garber, R. Summers and W. M. Kemp. 1995. Inputs, transformations, and transport to nitrogen and phosphorus in Chesapeake Bay and selected tributaries. *Estuaries,* **18,** 285–314.

Breitburg, D., S. Seitzinger and J. Sanders (eds.). 1999. The effects of multiple stressors on freshwater and marine ecosystems. *Limnol. Oceanogr.,* **44,** 739–972.

Bricelj, V. M. and D. J. Lonsdale. 1997. *Aureococcus anophagefferens:* Causes and ecological consequences of brown tides in US mid-Atlantic coastal waters. *Limnol. Oceanogr.,* **42,** 1023–1038.

Brush, G. S. 1984. Stratigraphic evidence of eutrophication in an estuary. *Water Resources Res.,* **20,** 531–541.

Brush, G. S. 2001. Natural and anthropogenic changes in Chesapeake Bay during the last 1000 years. *Hum. Ecol. Risk Assess.,* **7,** 1283–1296.

Bundy M. H., D. L. Breitburg and K. G. Sellner. 2003. The responses of Patuxent River upper trophic levels to nutrient and trace element induced changes in the lower food web. *Estuaries,* **26,** 365–384.

Burkholder, J. M. 1998. Implications of harmful microalgae and heterotrophic dinoflagellates in management of sustainable marine fisheries. *Ecol. Appln.,* **8(1) Suppl.,** S37-S62.

Burnett, L. E. and W. B. Stickle. 2001. Physiological responses to hypoxia. Pp. 101–114 in Rabalais, N. N. and R. E. Turner (eds.), *Coastal Hypoxia: Consequences for Living Resources and Ecosystems.* Coastal and Estuarine Studies **58,** American Geophysical Union, Washington, D.C.

Caddy, J. F. 1993. Toward a comparative evaluation of human impacts on fishery ecosystems of enclosed and semi-enclosed seas. *Rev. Fish. Sci.,* **1,** 57–95.

Caddy, J. F. 2000. Marine catchment basin effects versus impacts of fisheries on semi-enclosed seas. *ICES J. Mar. Sci.,* **57,** 628–640.

Cadée, G. C. 1992. Phytoplankton variability in the Marsdiep, the Netherlands. *ICES Mar. Sci. Symp.,* **195,** 213–222.

Caraco, N. F. 1995. Influence of human populations on P transfers to aquatic systems: A regional scale study using large rivers. Pp. 235–247 in Tiessen, H. (ed.), *Phosphorus in the Global Environment.* SCOPE **54.** John Wiley & Sons Ltd., New York.

Castillano-Rivera, M., A. Kobelkowsky and V. Zamayoa. 1996. Food resource partitioning and trophic morphology of *Brevoortia gunteri* and *B. patronus. J. Fish. Biol.,* **49,** 1102–1111.

Cederwall, H. and R. Elmgren. 1990. Biological effects of eutrophication in the Baltic Sea, particularly the coastal zone. *Ambio,* **19,** 109–112.

Chen, N., T. S. Bianchi, B. A. McKee and J. M. Bland. 2001. Historical trends of hypoxia on the Louisiana shelf: application of pigments as biomarkers. *Organic Geochem.,* **32,** 543–561.

Cherfas, J. 1990. The fringe of the ocean—under siege from land. *Science,* **248,**163–165.

Chesney, E. J. and D. M. Baltz. 2001. The effects of hypoxia on the northern Gulf of Mexico coastal ecosystem: A fisheries perspective. Pp. 321–354 in Rabalais, N. N. and R. E. Turner, (eds.), *Coastal Hypoxia: Consequences for Living Resources and Ecosystems.* Coastal and Estuarine Studies **58,** American Geophysical Union, Washington, D.C.

Childs, C. R., N. N. Rabalais, R. E. Turner and L. M. Proctor. 2002. Sediment denitrification in the Gulf of Mexico zone of hypoxia. *Mar. Ecol. Prog. Ser.,* **240,** 285–290, & 2003, Erratum **247,** 310.

Christensen, P. B. (ed.). 1998. The Danish marine environment: has action improved its state? Marine Research Programme HAV90, Havforskning fra Miljøstyrelsen Nr. 62. Danish Environmental Protection Agency, Copenhagen.

Cloern, J. E. 2001. Review. Our evolving conceptual model of the coastal eutrophication problem. *Mar. Ecol. Prog. Ser.,* **210,** 223–253.

Colijn, F., K.-J. Hesse, N. Ladwig and U. Tillmann. 2002. A large-scale uncontrolled fertilisation experiment carried out in the North Sea. *Hydrobiol.,* **484,** 133–148.

Colman, S. M. and J. F. Bratton. 2003. Anthropogenically induced changes in sediment and biogenic silica fluxes in Chesapeake Bay. *Geology,* **31,** 71–74.

Conley, D. J. 2000. Biogeochemical nutrient cycles and nutrient management strategies. *Hydrobiologia,* **410,** 87–96.

Conley, D. J., C. L. Schelske and E. F. Stoermer. 1993. Modification of the biogeochemical cycle of silica with eutrophication. *Mar. Ecol. Prog. Ser.,* **101,** 179–192.

Cooper, S. R. 1995. Chesapeake Bay watershed historical land use: impact on water quality and diatom communities. *Ecol. Appln.,* **5,** 703–723.

Cooper, S. R. and G. S. Brush. 1991. Long term history of Chesapeake Bay anoxia. *Science,* **254,** 992–996.

Daskalov, G. M. 2003. Long-term changes in fish abundance and environmental indices in the Black Sea. *Mar. Ecol. Prog. Ser.,* **255,** 259–270.

Deegan L. A., B. J. Peterson and R. Portier. 1990. Stable isotopes and cellulose activity as evidence for detritus as a food source for juvenile menhaden. *Estuaries,* **13,** 14–19.

Deegan, L. A., A. Wright, S. G. Ayvazian, J. T. Finn, H. Golden, R. R. Merson and J. Harrison. 2002. Nitrogen loading alters seagrass ecosystem structure and support of higher trophic levels. *Aquatic Conserv.: Mar. Freshw. Ecosyst.,* **12,** 193–212.

Dethlefsen, V. and H. von Westernhagen. 1983. Oxygen deficiency and effects on bottom fauna in the eastern German Bight 1982. *Meeresforch.,* **60,** 767–775.

Diaz, R. J. and R. Rosenberg. 1995. Marine benthic hypoxia: A review of its ecological effects and the behavioural responses of benthic macrofauna. *Ocean. Mar. Biol. Ann. Rev.,* **33,** 245–303.

Ditty J. G. 1986. Ichthyoplankton in neritic waters of the northern Gulf of Mexico off Louisiana: composition, relative abundance, and seasonality. *Fish. Bull.,* **84,** 935–944.

Dortch, Q. and T. E. Whitledge. 1992. Does nitrogen or silicon limit phytoplankton production in the Mississippi River plume and nearby regions? *Cont. Shelf Res.,* **12,** 1293–1309.

Dortch, Q., N. N. Rabalais, R. E. Turner and G. T. Rowe. 1994. Respiration rates and hypoxia on the Louisiana shelf. *Estuaries,* **17,** 862–872.

Dortch, Q., R. Robichaux, S. Pool, D. Milsted, G. Mire, N. N. Rabalais, T. M. Soniat, G. A. Fryxell, R. E. Turner, R. E. and M. L. Parsons. 1997. Abundance and vertical flux of *Pseudo-nitzschia* in the northern Gulf of Mexico. *Mar. Ecol. Prog. Ser.,* **146,** 249–264.

Dortch, Q., N. N. Rabalais, R. E. Turner and N. A. Qureshi. 2001. Impacts of changing Si/N ratios and phytoplankton species composition. Pp. 37–48 in Rabalais, N. N. and R. E. Turner (eds.), *Coastal Hypoxia: Consequences for Living Resources and Ecosystems.* Coastal and Estuarine Studies **58,** American Geophysical Union, Washington, D.C.

Downing, J. A., M. McClain, R. Twilley, J. M. Melack, J. Elser, N. N. Rabalais, W. M. Lewis, Jr., R. E. Turner, J. Corredor, D. Soto, A. Yanez-Arancibia and R. W. Howarth. 1999. The impact of accelerating land-use change on the N-cycle of tropical aquatic ecosystems: Current conditions and projected changes. *Biogeochem.,* **46,** 109–148.

Duce, R. A. 1986. The impact of atmospheric nitrogen, phosphorus, and iron species on marine biological productivity. Pp. 497–529 in P. Baut-Menard (ed.), *The Role of Air-Sea Exchange in Geochemical Cycling.* Reidel.

Duffy, J .E. 2002. Biodiversity and ecosystem function: the consumer connection. *Oikos* **99,** 201–219.

Dugdale, R. C. and F. P. Wilkerson. 1998. Silicate regulation of new production in the equatorial Pacific upwelling, *Nature,* **391,** 270–273.

Eadie B. J., B. A. McKee, M. B. Lansing, J. A. Robbins, S. Metz and J. H. Trefry. 1994. Records of nutrient-enhanced coastal productivity in sediments from the Louisiana continental shelf. *Estuaries,* **17,** 754–765.

Egge, J. K. and D. L. Aksnes. 1992. Silicate as regulating nutrient in phytoplankton competition. *Mar. Ecol. Prog. Ser.,* **83,** 281–289.

Elmgren, R. 1989. Man's impact on the ecosystem of the Baltic Sea: energy flows today and at the turn of the century. *Ambio,* **18,** 326–332.

Elmgren, R. 2001. Understanding human impact on the Baltic ecosystem: changing view in recent decades. *Ambio,* **30,** 222–231.

Elmgren, R. and U. Larsson. 2001. Eutrophication in the Baltic Sea area. Integrated coastal management issues. Pp. 15–35 in von Bodungen, B., and Turner, R.K. (eds.), *Science and Integrated Coastal Management,* Dahlem University Press, Berlin.

Fu, J.-M. and J. W. Winchester. 1994. Sources of nitrogen in three watersheds of northern Florida, USA: Mainly atmospheric deposition. *Geochim. Cosmochim. Acta,* **58,** 1581–1590.

Galloway, J. N. and E. B. Cowling. 2002. Reactive nitrogen and the world: two hundred years of change. *Ambio,* **31,** 64–71.

Galloway, J. N., J. D. Aber, J. W. Erisman, S. P. Seitzinger, R. W. Howarth, E. B. Cowling and B. J. Cosby. 2003. The nitrogen cascade. *BioScience,* **53,** 341–356.

Goolsby D. A., W. A. Battaglin, G. B. Lawrence, R. S. Artz, B. T. Aulenbach, R. P. Hooper, D. R. Keeney and G. J. Stensland. 1999. Flux and Sources of Nutrients in the Mississippi-Atchafalaya River Basin, Topic 3 Report for the Integrated Assessment of Hypoxia in the Gulf of Mexico. NOAA Coastal Ocean Program Decision Analysis Series No. 17. Silver Spring, Maryland: NOAA Coastal Ocean Program.

Govoni, J. J., D. E. Hoss and A. J. Chester. 1983. Comparative feeding of three species of larval fishes in the northern Gulf of Mexico: *Brevoortia patronus, Leiostomus xanthurus,* and *Micropogonias undulatus. Mar. Ecol. Prog. Ser.,* **13,** 189–199.

Gray, J. S., R. S.-S. Wu and Y. Y. Or. 2002. Review. Effects of hypoxia and organic enrichment on the coastal marine environment. *Mar. Ecol. Prog. Ser.,* **238,** 249–279.

Greening, H. and C. Elfring. 2002. Local, state, regional, and federal roles in coastal nutrient management. *Estuaries,* **25(4b),** 838–847.

Grimvall, A. and Stålnacke, P. 2001. Riverine inputs of nutrients to the Baltic Sea. Pp. 113–131 in Wulff, F. V., L. A. Rahm and P. Larsson (eds.), *A Systems Analysis of the Baltic Sea,* Ecological Studies Analysis and Synthesis **148,** Springer, Berlin.

Grimvall, A., P. Stålnacke and A. Tonderski. 2000. Time scales of nutrient losses from land to sea—a European perspective. *Ecol. Engineering,* **14,** 363–371.

Hallegraeff, G. M. 1993. A review of harmful algal blooms and their apparent global increase. *Phycologia,* **32,** 79–99.

Hansen, B., P.K. Bjørnsen and P. J. Hansen. 1994. The size ratio between planktonic predators and their prey. *Limnol. Oceanogr.,* **39,** 395–403.

Harding, Jr., L. W. and E. S. Perry. 1997. Long-term increase of phytoplankton biomass in Chesapeake Bay, 1950–94. *Mar. Ecol. Prog. Ser.,* **157,** 39–52.

Harding, Jr., L. W., D. Degobbis and R. Precali. 1999. Production and fate of phytoplankton: Annual cycles and interannual variability. Pp. 131–172 in Malone, T. C., A. Malej, L. W. Harding, Jr., N. Smodlaka and R. E. Turner (eds.) *Ecosystems at the Land-Sea Margin. Drainage Basin to Coastal Sea.* Coastal and Estuarine Studies **55,** American Geophysical Union, Washington, D.C.

Hauxwell, J., J. Cebrián and I. Valiela. 2003. Eelgrass *Zostera marina* loss in temperate estuaries: relationship to land-derived nitrogen loads and effect of light limitation imposed by algae. *Mar. Ecol. Prog. Ser.,* **247,** 59–73,

Hecky, R. E. and P. Kilham. 1988. Nutrient limitation of phytoplankton in freshwater and marine environments: a review of recent evidence on the effects of enrichment. *Limnol. Oceanogr.,* **33,** 796–822.

Howarth, R. W. 1988. Nutrient limitation of net primary production in marine ecosystems. *Ann. Rev. Ecol. System.,* **19,** 898–910.

Howarth, R. W., H. S. Jensen, R. Marino and H. Postma, 1995. Transport to and processing of P in near-shore and oceanic waters. Pp. 323–356 in Tiessen, H. (ed.), *Phosphorus in the Global Environment.* SCOPE 54, John Wiley & Sons Ltd., Chichester.

Howarth, R. W., G. Billen, D. Swaney, A Townsend, N. Jaworski, K. Lajtha, J. A. Downing, R. E. Elmgren, N. Caraco, T. Jordan, F. Berendse, J. Freney, V. Kudeyarov, P. Murdoch and Z.-L. Zhu, 1996, Regional nitrogen budgets and riverine N & P fluxes for the drainages to the North Atlantic Ocean: Natural and human influences. *Biogeochem.,* **35,** 75–139.

Howarth, R. W., A. Sharpley and D. Walker. 2002. Sources of nutrient pollution to coastal waters in the United States: implications for achieving coastal water quality goals. *Estuaries* **25,** 656–676.

Howarth, R. W., D. P. Swaney, T. J. Butler and R. Marino. 2000. Climate control on eutrophication of the Hudson River estuary. *Ecosystems* **3,** 210–215.

Jackson, J. B. C., M. X. Kirby, W. H. Berger, K. A. Bjorndal, L. W. Botsford, B. J. Bourque, R. H. Bradbury, R. Cooke, J. Erlandson, J. A. Estes, T. P. Hughes, S. Kidwell, C. B. Lange, S. L. Hunter, J. M. Pandolfi, C. H. Peterson, R. S. Steneck, M. J. Tegner and R. R. Warner. 2001. Historical overfishing and the recent collapse of coastal ecosystems. *Science,* **293,** 629–638.

Jansson, B.-O. and K. Dahlberg. 1999. The environmental status of the Baltic Sea in the 1940s, today, and in the future. *Ambio,* **28,** 312–319.

Johansson, J. O. R. and R. R. Lewis III. 1992. Recent improvements of water quality and biological indicators in Hillsborough Bay, a highly impacted subdivision of Tampa Bay, Florida, USA. Pp. 1199–1215 in Vollenweider, R. A., R. Marchetti and R. Viviani (eds.), *Marine Coastal Eutrophication. The Response of Marine Transitional Systems to Human Impact: Problems and Perspectives for Restoration,* Proceedings, International Conference, Bologna, Italy, 21–24 March 1990, Elsevier, Amsterdam.

Johnson, K. S., F. P. Chavez and G. E. Friederich. 1999. Continental-shelf sediment as a primary source of iron for coastal phytoplankton. *Nature,* **398,** 697–700.

Justić, D. 1988. Trend in the transparency of the northern Adriatic Sea 1911–1982. *Mar. Pollut. Bull.,* **19,** 32–35.

Justić, D. 1991. Hypoxic conditions in the northern Adriatic Sea: historical development and ecological significance. Pp. 95–105 in Tyson, R. V. and T. H. Pearson (eds.), *Modern and Ancient Continental Shelf Anoxia,* Geological Society Special Publication **58,** The Geological Society, London.

Justić, D., T. Legović and L. Rottini-Sandrini. 1987. Trends in oxygen content 1911–1984 and occurrence of benthic mortality in the northern Adriatic Sea. *Estuar. Coastal Shelf Sci.,* **25,** 435–445.

Justić, D., N. N. Rabalais and R. E. Turner. 1995a. Stoichiometric nutrient balance and origin of coastal eutrophication. *Mar. Pollut. Bull.,* **30,** 41–46.

Justić, D., N. N. Rabalais, R. E. Turner and Q. Dortch. 1995b. Changes in nutrient structure of river-dominated coastal waters: Stoichiometric nutrient balance and its consequences. *Estuar. Coast. Shelf Sci.,* **40,** 339–356.

Justić, D., N. N. Rabalais and R. E. Turner. 1996. Effects of climate change on hypoxia in coastal waters: a doubled $CO_2$ scenario for the northern Gulf of Mexico. *Limnol. Oceanogr.*, **41,** 992–1003.

Justić, D., N. N. Rabalais and R. E. Turner. 1997. Impacts of climate change on net productivity of coastal waters: Implications for carbon budget and hypoxia. *Climate Res.*, **8,** 225–237.

Justić, D., N. N. Rabalais and R. E. Turner. 2003a. Simulated responses of the Gulf of Mexico hypoxia to variations in climate and anthropogenic nutrient loading. *J. Mar. Syst.*, **42,** 115–126.

Justić, D., R. E. Turner and N. N. Rabalais. 2003b. Climatic influences on riverine nitrate flux: Implications for coastal marine eutrophication and hypoxia. *Estuaries,* **26,** 1–11.

Karlsen, A. W., T. M. Cronin, S. E. Ishman, D. A. Willard, R. Kerhin, C.W. Holmes and M. Marot. 2000. Historical trends in Chesapeake Bay dissolved oxygen based on benthic foraminifera from sediment cores. *Estuaries,* **23,** 488–508.

Karlson, K., R. Rosenberg and E. Bonsdorff. 2002. Temporal and spatial large-scale effects of eutrophication and oxygen deficiency on benthic fauna in Scandinavian and Baltic waters – a review. *Oceanogr. Mar. Biol. Ann. Rev.*, **40,** 427–489.

Kautsky, N., H. Kautsky, U. Kautsky and M. Waern. 1986. Decreased depth penetration of *Fucus vesiculosus* (L.) since the 1940's indicates eutrophication of the Baltic Sea. *Mar. Ecol. Prog. Ser.*, **28,** 1–8.

Kemp, W. M., P. Sampou, J. Caffrey, M. Mayer, K. Henriksen and W. Boynton. 1990. Ammonium recycling versus denitrification in Chesapeake Bay sediments. *Limnol. Oceanogr.*, **35,** 1545–1563.

Kimmerer, W. J. 2002. Effects of freshwater flow on abundance of estuarine organisms: physical effects or trophic linkages. *Mar. Ecol. Prog. Ser.*, **243,** 39–55.

Knowlton, N. 2001. The future of coral reefs. *Proc. Natl. Acad. Sci., USA,* **98,** 5419–5425.

Kristiansen, S. and E. E. Hoell. 2002. The importance of silicon for marine production. *Hydrobiologia,* **484,** 21–31.

Lancelot, C., J.-M. Martin, N. Panin and Y. Zaitsev. 2002. The north-western Black Sea: a pilot site to understand the complex interaction between human activities and the coastal environment. *Estuar. Coast. Shelf Sci.*, **54,** 279–283.

LaPointe, B. E. 1997. Nutrient thresholds for eutrophication and macroalgal blooms on coral reefs in Jamaica and southeast Florida. *Limnol. Oceanogr.*, **42,** 1119–1131.

Levin, L. A., C. L. Huggett, and K. F. Wishner. 1991. Control of deep-sea benthic community structure by oxygen and organic-matter gradients in the eastern Pacific Ocean. *J. Mar. Res.*, **49,** 763–800.

Levin, L. A. and J. D. Gage. 1998. Relationships between oxygen, organic matter and the diversity of bathyal macrofauna. *Deep-Sea Res. II,* **45,** 129–163.

Levin, L. A., J. D. Gage, C. Martin and P. A. Lamont. 2000. Macrobenthic community structure within and beneath the oxygen minimum zone, northwest Arabian Sea. *Deep-Sea Res. II,* **47,** 189–226.

Liu, S. M., J. Zhang, H. T. Chen, Y. Wu, H. Xiong and Z. F. Zhang. 2003. Nutrients in the Changjiang and its tributaries. *Biogeochem.*, **62,** 1–18.

Löfgren, S., A. Gustafson, S. Steineck and P. Stählnacke. 1999. Agricultural development and nutrient flows in the Baltic states and Sweden after 1988. *Ambio,* **28,** 320–327.

Lohrenz, S. E., M. J. Dagg and T. E. Whitledge. 1990. Enhanced primary production at the plume/oceanic interface of the Mississippi River. *Cont. Shelf Res.*, **10,** 639–664.

Lohrenz, S. E., G. L. Fahnenstiel and D. G. Redalje. 1994. Spatial and temporal variations in photosynthesis parameters in relation to environmental conditions in coastal waters of the northern Gulf of Mexico. *Estuaries,* **17,** 779–795.

Lohrenz, S. E., G. L. Fahnenstiel, D. G. Redalje, G. A. Lang, X. Chen and M. J. Dagg. 1997. Variations in primary production of northern Gulf of Mexico continental shelf waters linked to nutrient inputs from the Mississippi River. *Mar. Ecol. Prog. Ser.*, **155,** 435–454.

Lohrenz, S. E., G. L. Fahnenstiel, D. G. Redalje, G. A. Lang, M. J. Dagg, T. E. Whitledge and Q. Dortch. 1999. The interplay of nutrients, irradiance and mixing as factors regulating primary production in coastal waters impacted by the Mississippi River plume. *Cont. Shelf Res.,* **19,** 1113–1141.

McIsaac, G. G., M. B. David, G. Z. Gertner and D. A. Goolsby. 2001. Nitrate flux in the Mississippi River. *Nature,* **414,** 166–167.

McIsaac, G. G., M. B. David, G. Z. Gertner and D. A. Goolsby. 2002. Relating net N input in the Mississippi River basin to nitrate flux in the lower Mississippi River: A comparison of approaches. *J. Environ. Qual.,* **31,** 1610–1622.

McQueen, D. J., M. R. S. Johannes and J. R. Post. 1989. Bottom-up and top-down impacts on freshwater pelagic community structure. *Ecol. Monogr.,* **59,** 289–309.

McQueen, D. J. and J. R. Post. 1988. Cascading trophic interactions - uncoupling at the zooplankton-phytoplankton link. *Hydrobiol.,* 159, 277–296.

Mee, L. D. 1992. The Black Sea in crisis: A need for concerted international action. *Ambio,* **21,** 278–286.

Mee, L. D. 2001. Eutrophication in the Black Sea and a basin-wide approach to its control. Pp. 71–91 in von Bodungen, B. and R. K. Turner (eds.), *Science and Integrated Coastal Management,* Dahlem University Press, Berlin.

Meyer-Reil, L.-A. and M. Köster. 2000. Eutrophication of marine waters: effects on benthic microbial communities. *Mar. Pollut. Bull.,* **41,** 255–263.

Micheli, F. 1999. Eutrophication, fisheries, and consumer-resource dynamics in marine pelagic ecosystems. *Science,* **285,** 1396–1399.

Miller, J. R. and G. L. Russell. 1992. The impact of global warming on river runoff. *J. Geophys. Res.,* **97,** 2757–2764.

Mills, E. L., J. M. Casselman, R. Dermott, J. D. Fitzsimons, G. Gal, K. T. Holek, J. A. Hoyle, O. E. Johannsson, B. F. Lantry, J. C. Makarewicz, E. S. Millard, I. F. Munawar, M. Munawar, R. O'Gorman, R. W. Ownes, L. G. Rudstam, T. Schaner and T. J. Stewart. 2003. Lake Ontario: food web dynamics in a changing ecosystem (1970–2000). *Can. J. Fish. Aquat. Sci.,* 50, 471–490.

Mitsch, W. J, J. W. Day, Jr., J. W. Gilliam, P. M. Groffman, D. L. Hey, G. W. Randall and N. Wang. 2001. Reducing nitrogen loading to the Gulf of Mexico from the Mississippi River basin: Strategies to counter a persistent ecological problem. *BioScience,* **15,** 373–388.

Moore, W. S. 1996. Large groundwater inputs to coastal waters revealed by 226 Ra enrichments. *Nature,* **380,** 612–614.

Nagai, T. 2003. Recovery of fish stocks in the Seto Inland Sea. *Mar. Pollut. Bull.,* **47,** 126–131.

National Research Council. 2000. *Clean Coastal Waters – Understanding and Reducing the Effects of Nutrient Pollution.* National Academy Press, Washington, D.C.

Nelsen. T. A., P. Blackwelder, T. Hood, B. McKee, N. Romer, C. Alvarez-Zarikian and S. Metz. 1994. Time-based correlation of biogenic, lithogenic and authigenic sediment components with anthropogenic inputs in the Gulf of Mexico NECOP study area. *Estuaries,* **17,** 873–885.

Nixon, S. W. 1995. Coastal marine eutrophication: A definition, social causes, and future concerns. *Ophelia,* **41,** 199–219.

Nixon, S. W. 1988. Physical energy inputs and the comparative ecology of lake and marine ecosystems. *Limnol. Oceanogr.,* 3**3,** 1005–1025.

Nixon, S.W. 2003. Replacing the Nile: are anthropogenic nutrients providing the fertility once brought to the Mediterranean by a great river? *Ambio,* **32,** 30–39.

Nixon, S. W. and B. A. Buckley. 2002. "A strikingly rich zone"—nutrient enrichment and secondary production in coastal marine ecosystems. *Estuaries,* **25,** 782–796.

Nixon, S. W. and M. E. Q. Pilson. 1983. Nitrogen in estuarine and coastal marine ecosystems. Pp. 565–648 in Carpenter, E. J. and D. G. Capone (eds.), *Nitrogen in the Marine Environment.* Academic Press, New York.

Nixon, S.W., J. W. Ammerman, L. P. Atkinson, V. M. Berounsky, G. Billen, G., W. C. Boicourt, W.R. Boynton, T. M. Church, D. M. DiToro, R. Elmgren, J. H. Garber, A. E. Giblin, R. A. Jahnke, N. J. P Ownes, M. E. Q. Pilson and S. P. Seitzinger. 1996. The fate of nitrogen and phosphorus at the land-sea margin of the North Atlantic Ocean. *Biogeochem.,* **35,** 141–180.

Officer, C. B. and J. H. Ryther, 1980, The possible importance of silicon in marine eutrophication. *Mar. Ecol. Prog. Ser.,* **3,** 83–91.

Officer, C. B., R. B. Biggs, J. L. Taft, L. E. Cronin, M. Tyler and W. R. Boynton. 1984. Chesapeake Bay anoxia: origin, development and significance. *Science,* **223,** 22–27.

O'Shea, M. L. and T. M. Brosnan. 2000. Trends in indicators of eutrophication in western Long Island Sound and the Hudson-Raritan estuary. *Estuaries,* **23,** 877–901.

Paerl, H. W. 1995. Coastal eutrophication in relation to atmospheric nitrogen deposition: current perspectives. *Ophelia,* **41,** 237–259.

Paerl, H. W. 1997. Coastal eutrophication and harmful algal blooms: Importance of atmospheric deposition and groundwater as "new" nitrogen and other nutrient sources. *Limnol. Oceanogr.,* **42,** 1154–1165.

Paerl, H. W., R. L. Dennis and D. R. Whitall. 2001. Atmospheric deposition of nitrogen: implications for nutrient over-enrichment of coastal waters. *Estuaries,* **25,** 677–693.

Paine, R. T., M. J. Tegner and E. A. Johnson. 1998. Compounded perturbations yield ecological surprises. *Ecosystems,* **1,** 535–545.

Paluden, C. F. E. Alexeyev, H. Drews, S. Fleischer, A. Fugisang, T. Kindt, P. Kowalski, M. Moos, A. Radlowki, G. Stromfors, V. Estberg and K. Wolter. 2002. Wetland management to reduce Baltic Sea eutrophication. *Water Sci. Technol.,* **45,** 87–94.

Parsons, M.L., C. A. Scholin, P. E. Miller, G. J. Doucette, C. L. Powell, G. A. Fryxell, Q. Dortch and T. M. Soniat. 1999. *Pseudo-nitzschia* species (Bacillariophyceae) in Louisiana coastal waters: Molecular probe field trials, genetic variability, and domoic acid analyses. *J. Phycol.,* **35,** 1368–1378.

Parsons, M., Q. Dortch and R. E. Turner. 2002. Sedimentological evidence of an increase in *Pseudo-nitzschia* (Bacillariophyceae) abundance in response to coastal eutrophication. *Limnol. Oceanogr.,* **47,** 551–558.

Peierls, B. and H. Paerl. 1997. Bioavailability of atmospheric organic nitrogen deposition to coastal phytoplankton. *Limnol. Oceanogr.,* **42,** 1819–1823.

Peierls, B. L., N. Caraco, M. Pace and J. Cole. 1991. Human influence on river nitrogen. *Nature,* **350,** 386–387.

Petipa, T.S., E.V. Pavlova and G.N. Mironov. 1973. The food web structure, utilization and transport of energy by trophic levels in the planktonic communities, Pp. 142–167 in Steele, J. H. (ed.), *Marine Food Chains,* Oliver and Boyd, Edinburgh, reprinted by Otto Koeltz Antiquariat, Koenigstein-Ts./B.R.D.

Rabalais, N. N. and R. E. Turner (eds.). 2001. *Coastal Hypoxia: Consequences for Living Resources and Ecosystems.* Coastal and Estuarine Studies **58,** American Geophysical Union, Washington, D.C., 454 pp.

Rabalais N. N., R. E. Turner, D. Justić, Q. Dortch, W. J. Wiseman, Jr. and B. K. Sen Gupta. 1996. Nutrient changes in the Mississippi River and system responses on the adjacent continental shelf. *Estuaries,* **19,** 386–407.

Rabalais, N. N., R. E. Turner, W. J. Wiseman, Jr. and Q. Dortch. 1998. Consequences of the 1993 Mississippi River Flood in the Gulf of Mexico. *Regulated Rivers: Research & Management,* **14,** 161–177.

Rabalais, N. N., D. E. Harper, Jr. and R. E. Turner. 2001a. Responses of nekton and demersal and benthic fauna to decreasing oxygen concentrations. Pp. 115–128 in Rabalais, N. N. and R. E. Turner (eds.), *Coastal Hypoxia: Consequences for Living Resources and Ecosystems.* Coastal and Estuarine Studies **58,** American Geophysical Union, Washington, D.C.

Rabalais, N. N., L. E. Smith, D. E. Harper, Jr. and D. Justić. 2001b. Effects of seasonal hypoxia on continental shelf benthos. Pp. 211–240 in Rabalais, N. N. and R. E. Turner (eds.), *Coastal Hypoxia: Consequences for Living Resources and Ecosystems.* Coastal and Estuarine Studies **58,** American Geophysical Union, Washington, D.C.

Rabalais, N. N., R. E. Turner, Q. Dortch, D. Justić, V. J. Bierman, Jr. and W. J. Wiseman, Jr. 2002a. Review. Nutrient-enhanced productivity in the northern Gulf of Mexico: past, present and future. *Hydrobiol.,* **475/476,** 39–63.

Rabalais, N. N., R. E. Turner and D. Scavia. 2002b. Beyond science into policy: Gulf of Mexico hypoxia and the Mississippi River. *BioScience,* **52,** 129–142.

Rabalais, N. N., N. Atilla, C. Normandeau and R. E. Turner. 2004. Ecosystem history of Mississippi River-influenced continental shelf revealed through preserved phytoplankton pigments. *Mar. Pollut. Bull.,* **49;** 537–547.

Randall, G. W. and D. J. Mulla. 2001. Nitrate-N in surface waters as influenced by climatic conditions and agricultural practices. *J. Environ. Qual.,* **30,** 337–344.

Redfield, A. C., B. H. Ketchum and F. A. Richards. 1963. The influence of organisms on the composition of seawater. Pp. 26–77 in Hill, M. N. (ed.), *The Sea, Vol. 2.* Interscience Publishers, John Wiley, New York.

Rhoads, D. C., D. F. Boesch, Z. Tang, F. Xu, L. Huang and K. J. Nilsen. 1985. Macrobenthos and sedimentary facies on the Changjiang delta platform and adjacent continental shelf, East China Sea. *Cont. Shelf Res.,* **4,** 189–213.

Richardson, K. 1996. Conclusions, research and eutrophication control. Pp. 243–267 in Jørgensen, B. B. and K. Richardson (eds.), *Eutrophication in Coastal Marine Ecosystems.* Coastal and Estuarine Studies **52,** American Geophysical Union, Washington, D.C.

Rosenberg, R. 1985. Eutrophication – the future marine coastal nuisance? *Mar. Pollut. Bull.,* **16,** 227–231.

Rosenberg, R. 1990. Negative oxygen trends in Swedish coastal bottom waters. *Mar. Pollut. Bull.,* **21,** 335–339.

Ryther, J. H. and W. M. Dunstan. 1971. Nitrogen, phosphorus, and eutrophication in the coastal marine environment. *Science,* **171,** 1008–1013.

Sandén, P. and B. Håkansson. 1996. Long-term trends in Secchi depth in the Baltic Sea. *Limnol. Oceanogr.,* **41,** 346–351.

Sarasota Bay National Estuary Program 1995. *Sarasota Bay: The Voyage to Paradise Reclaimed.* Southwest Florida Water Management District, Brooksville, Florida.

Scavia, D., N. N. Rabalais, R. E. Turner, D. Justić, and W. J. Wiseman, Jr. 2003. Predicting the response of Gulf of Mexico hypoxia to variations in Mississippi River nitrogen load. *Limnol. Oceanogr.,* **48,** 951–956.

Schramm, W. 1999. Factors influencing seaweed responses to eutrophication: some results from EU-project EUMAC. *J. Appl. Phycol.,* **11,** 69–78.

Seitzinger, S. P., C. Kroeze, A. F. Bouwman, N. Caraco, F. Dentener and R. V. Styles. 2002. Global patterns of dissolved inorganic and particulate nitrogen inputs to coastal systems: recent conditions and future projections. *Estuaries,* **25,** 640–655.

Sellner, K. G., G. J. Doucette and G. J. Kirkpatrick. 2002. Harmful algal blooms: causes, impacts and detection. *J. Ind. Microbiol. Biotechnol.,* **30,** 383–406.

Sen Gupta, B. K., R. E. Turner and N. N. Rabalais. 1996. Seasonal oxygen depletion in continental-shelf waters of Louisiana: Historical record of benthic foraminifers. *Geology,* **24,** 227–230.

Smayda, T. J. 1990. Novel and nuisance phytoplankton blooms in the sea: evidence for global epidemic. Pp. 29–40 *in* Granéli, E., B. Sundstrom, R. Edler and D. M. Anderson (eds.), *Toxic Marine Phytoplankton.* Elsevier Science, New York.

Smil, V. 2001. *Enriching the Earth: Fritz Haber, Carl Bosch, and the Transformation of World Food.* The MIT Press, Cambridge.

Smith, S. V. 1981. Responses of Kaneohe Bay, Hawaii, to relaxation of sewage stress. Pp. 391–410 in Neilson, B. J. and L. E. Cronin (eds.), *Estuaries and Nutrients.* Humana Press, Inc., Clifton, New Jersey.

Šolić, M., N. Krustolović, I. Marasović, A. Baranović, T. Pucher-Petković and T. Vučetić. 1997. Analysis of time series of planktonic communities in the Adriatic Sea: distinguishing between natural and man-induced changes. *Oceanol. Acta,* **20,** 131–143.

Sprules, W. G., S. B. Brandt, D. J. Steward, M. Munawar, E. H. Jin and J. Love. 1990. Biomass size spectrum of the Lake Michigan pelagic food web. *Can. J. Fish. Aquatic Sci.,* **48,** 105–115.

Stålnacke, P., N. Vagstad, T. Tamminen, P. Wassmann, V. Jansons and E. Loigu. 1999. Nutrient runoff and transfer from land and rivers to the Gulf of Riga. *Hydrobiol.,* **410,** 103–110.

Stockwell, D.A., T. E. Whitledge, E. J. Buskey, H. DeYoe, K. C. Dunton, G. J. Hold and S. A. Holt. 1996. Texas coastal lagoons and a persistent brown tide. Pp. 81–84 in McElroy, A. (ed.), Brown Tide Summit, Ronkonkoma, NY (USA, 20–21 Oct 1995), New York Sea Grant Program Publ. No. NYSGI-W-95–001.

Suess, E. 1980. Particulate organic carbon flux in the oceans: surface productivity and oxygen utilization. *Nature,* **288,** 260–263.

Swanson, R. L. and C. J. Sindermann (eds.). 1979. Oxygen Depletion and Associated Benthic Mortalities in New York Bight, 1976. NOAA Professional Paper 11, U.S. Department of Commerce, National Oceanic and Atmospheric Administration, U.S. Government Printing Office, Washington, D.C., 345 pp.

Szmant, A. M. 2002. Nutrient enrichment on coral reefs: is it a major cause of coral reef decline? *Estuaries,* **25,** 743–766.

Thomas, E, T. Gapotchenko, J. C. Varekamp, E. L. Mecray, and M. R. B. ter Brink. 2000. Benthic Foraminifera and environmental changes in Long Island Sound. *J. Coast. Res.,* **16,** 641–645.

Tolmazin, R. 1985. Changing coastal oceanography of the Black Sea. I. Northwestern shelf. *Progr. Oceanogr.,* **15,** 2127–276.

Turner, R. E. 2001. Some effects of eutrophication on pelagic and demersal marine food webs. Pages 371–398 in Rabalais, N. N. and R. E. Turner (eds.), *Coastal Hypoxia: Consequences for Living Resources and Ecosystems.* Coastal and Estuarine Studies **58,** American Geophysical Union, Washington, D.C.

Turner, R. E. and N. N. Rabalais. 1991. Changes in Mississippi River water quality this century. Implications for coastal food webs. *BioScience,* **41,** 140–148.

Turner, R. E. and N. N. Rabalais. 1994a. Changes in the Mississippi River nutrient supply and offshore silicate-based phytoplankton community responses. Pp. 147–150 in Dyer, K. R. and R. J. Orth (eds.), *Changes in Fluxes in Estuaries: Implications from Science to Management.* Proceedings of ECSA22/ERF Symposium, International Symposium Series, Olsen & Olsen, Fredensborg, Denmark.

Turner, R. E. and N. N. Rabalais. 1994b. Coastal eutrophication near the Mississippi river delta. *Nature,* **368,** 619–621.

Turner, R. E. and N. N. Rabalais. 2003. Linking landscape and water quality in the Mississippi River basin for 200 years. *BioScience,* **53,** 563–572.

Turner, R. E., N. N. Rabalais and Z.-N. Zhang. 1990. Phytoplankton biomass, production and growth limitations on the Huanghe (Yellow River) continental shelf. *Cont. Shelf Res.,* **10,** 545–571.

Turner, R. E., N. Qureshi, N. N. Rabalais, Q. Dortch, D. Justić, R. F. Shaw and J. Cope. 1998. Fluctuating silicate:nitrate ratios and coastal plankton food webs. *Proc. Natl. Acad. Sci., USA,* **95,** 13048–13051.

Turner, R. E., N. N. Rabalais, D. Justić and Q. Dortch. 2003a. Global patterns of dissolved N, P and Si in large rivers. *Biogeochem.,* **64,** 287–317.

Turner, R. E., N. N. Rabalais, D. Justić and Q. Dortch. 2003b. Future aquatic nutrient limitations. *Mar. Pollut. Bull.,* **46,** 1028–1030.

Turner, R. E., N. N. Rabalais, E. M. Swenson, M. Kasprzak and T. Romaire. 2005. Summer hypoxia in the northern Gulf of Mexico and its prediction from 1978 to 1995. *Mar. Envtl. Res.,* **59,** 65–77.

Vitousek, P. M. and R. W. Howarth. 1991. Nitrogen limitation on land and in the sea: how can it occur? *Biogeochem.,* **13,** 87–115.

Vitousek, P.M., J. D. Abler, R. W. Howarth, G. E. Likens, P. A. Matson, D. W. Schindler, W. H. Schlesinger and D. G. Tilman. 1997. Human alterations of the global nitrogen cycle: Sources and consequences. *Ecol. Appln.,* **7,** 737–750.

Walsh, J. J. 1988. *On the Nature of Continental Shelves.* Academic Press, Inc., New York, 508 pp.

Walsh, J. J. 1991. Importance of continental margins in the marine biogeochemical cycling of carbon and nitrogen. *Nature,* **350,** 53–55.

Whitledge, T. E. 1985. Nationwide Review of Oxygen Depletion and Eutrophication in Estuarine and Coastal Waters. Executive Summary. Report to National Ocean Service, National Oceanic and Atmospheric Administration. Brookhaven National Laboratory, Upton, New York, 28 pp.

Wolanski, E., R. Richmond, L. McCook and H. Sweatman. 2003. Mud, marine snow and coral reefs. *Amer. Sci.,* **91,** 44–51.

Wolock, D. M. and G. J. McCabe. 1999. Estimates of runoff using water-balance and atmospheric general circulation models. *J. Amer. Water Res. Assoc.,* **35,** 1341–1350.

Worm, B. and R. A. Meyers. 2003. Meta-analysis of cod-shrimp interactions reveals top-down control in oceanic food webs. *Ecology,* **84,**162–173.

Wu, R. S. S., B. S. Zhou, D. J. Randall, N. Y. S. Woo and P. K. S. Lam. 2003. Aquatic hypoxia is an endocrine disruptor and impairs fish reproduction. *Environ. Sci. Technol.,* **37,** 1137–1141.

Yamamoto, T. 2003. The Seto Inland Sea—eutrophic or oligotrophic? *Mar. Pollut. Bull.,* **47,** 37–42.

Yin, K., P.-Y. Qian, M. C. S. Wu, J. C. Chen, L. Huang, X. Song and W. Jian. 2001. Shift from P to N limitation of phytoplankton growth across the Pearl River estuarine plume during summer. *Mar. Ecol. Prog. Ser.,* **221,** 17–28.

Zaitsev, Y. P. 1992. Recent changes in the trophic structure of the Black Sea. *Fish. Oceanogr.,* **1,** 180–189.

Zhang, J. 1994. Atmospheric wet depositions of nutrient elements: Correlations with harmful biological blooms in the Northwest Pacific coastal zones. *Ambio,* **23,** 464–468.

Zimmerman, A. R. and E. A. Canuel. 2002. Sediment geochemical records of eutrophicatiosohaline Chesapeake Bay. *Limnol. Oceanogr.,* **47,** 1084–1093.

# Chapter 22. HARMFUL ALGAL BLOOMS: KEYS TO THE UNDERSTANDING OF PHYTOPLANKTON ECOLOGY

ADRIANA ZINGONE

*Stazione Zoologica 'Anton Dohrn', Villa Comunale, 80121 Napoli, Italy*

TIM WYATT

*Instituto de Investigaciones Marinas, Eduardo Cabello 6, 36208 Vigo, Spain*

**Contents**



## 1. Introduction

Human health problems and economic disturbance are at times brought about by the presence and/or the massive proliferation of microalgae in coastal waters. Harmful Algal Blooms (HABs) are hence commonly considered as a unified category of events based on these negative effects. In fact, from a scientific point of view, the only feature shared by these events is that they are generally caused by microalgae. A wide diversity of such organisms can cause negative effects, therefore any appraisal of HABs has to deal with an array of different organisms and with their individual distributions, physiological properties and ecological characteristics. Since blooms occur in all kinds of coastal waters, physical, chemical and ecological processes involved in their dynamics and in the manifestation of harmful effects are also distinct from case to case.

Intoxications now known to be of algal origin, deriving from the ingestion of seafood, have been reported at least since the sixteenth century (ciguatera in the greater Antilles). Some members of the Captain Vancouver's expedition died in what is now British Columbia from Paralytic Shellfish Poisoning (PSP). Yet until forty to fifty years ago (LoCicero, 1975) harmful phenomena due to algae were mostly ascribed to a couple of dozen dinoflagellate species belonging to 3–4 genera

*The Sea*, Volume 13, edited by Allan R. Robinson and Kenneth H. Brink
ISBN 0-674-01526-6 

and a few cyanobacteria. At that time, problems posed by high biomass blooms and ichthyotoxins were already known, while only ciguatera, PSP toxins and brevetoxins were recognised as health threats of algal origin. The term *red tides* was originally used for sea-water discolorations caused by microorganisms. Its meaning was subsequently extended to cover all harmful phenomena and any significant concentration increases of non-harmful algae. Nowadays, the term *harmful algal blooms* is generally used to cover all detrimental events, while the list of harmful species has nearly quadrupled, and now includes diatoms, pelagophytes, raphidophytes and prymnesiophytes; new and serious threats to human health have been identified (Amnesic Shellfish Poisoning, ASP, Azaspiracid Poisoning, AZP, Diarrhetic Shellfish Poisoning, DSP, Neurotoxic Shellfish Poisoning, NSP, etc.), as well as problems posed to the expanding aquaculture and tourism industries (Table I). About 90 eukaryotic microalgal species are now included in a confirmed list of toxic phytoplankton (Moestrup, 2003), but the number rises to 170 if marine and brackish cyanobacteria and a number of unconfirmed toxicity events are considered (Landsberg, 2002). To these lists, several other non-toxic species are to be added (Zingone and Enevoldsen, 2000) that are harmful by producing high biomass or mucilages or causing mechanical damage to farm fish. Clearly, the total number of harmful algae is conservative, as it constantly grows by the addition of newly described species and the discovery of toxicity or other harmful effects in already known species. Further growth of this list can be anticipated as interactions between man and coastal environments intensify.

Along with the increase of the harmful microalgal species list, the need is growing to exploit and protect effectively coastal marine resources. Mitigation operations have so far mainly consisted of intense monitoring activities of causative organisms and/or toxins, activities which have spread worldwide, in order to ensure safe seafood consumption and trade. The next challenge is represented by the development of alert systems based on automated observations coupled with predictive models that can expand the lead-time to harmful events, so as to allow more cost effective mitigation operations. Progress in modelling is however seriously hampered by the lack of knowledge on the basic mechanisms underlying the development of specific algal blooms.

Though the number of harmful species is limited, their diversity in terms of taxonomy, morphology, physiology and ecology is high. An example of morphological diversity is presented in Figure 22.1, which shows some representatives of four of the six algal classes that include harmful species. Investigations on the occurrence, proliferation and mechanisms of nuisance of harmful algae offer a variegated sample-book of species-specific adaptations and bloom dynamics that comprises most of the processes relevant to phytoplankton in a wide range of coastal ecosystems. For their potential contribution to deeper insights into algal bloom dynamics, HABs have been defined as the *Rosetta Stone* of phytoplankton ecology (Smayda, 1997), a comparison that points to the possibility of deciphering the poorly understood language of microalgal ecology through the study of biological processes underlying harmful algal blooms.

TABLE 1.

Harmful effects of microalgae in coastal and brackish waters (modified from Zingone and Enevoldsen, 2000). Four categories of resources are identified which can undergo negative effects caused by marine HABs. A single harmful event may have consequences for more than one category of resources, as in the case of brown tides of *Aureococcus anophagefferens.*

| EFFECTS ON DIFFERENT RESOURCES | EXAMPLES OF CAUSATIVE ORGANISMS | |
|---|---|---|
| *HUMAN HEALTH* | | |
| Paralytic shellfish poisoning (PSP) | Dinoflagellates | *Alexandrium spp., Pyrodinium bahamense, Gymnodinium catenatum* |
| | Cyanobacteria | *Anabaena circinalis, Aphanizomenon flos-aquae* |
| Diarrhetic shellfish poisoning (DSP) | Dinoflagellates | *Dinophysis* spp., *Prorocentrum* spp. |
| Neurotoxic shellfish poisoning (NSP) | Dinoflagellates | *Karenia brevis* |
| Amnesic shellfish poisoning (ASP) | Diatoms | *Pseudo-nitzschia* spp., *Nitzschia navis-varingica* |
| Azaspiracid poisoning (AZP) | Dinoflagellates | *Protoperidinium crassipes* |
| Ciguatera fish poisoning (CFP) | Dinoflagellates | *Gambierdiscus toxicus* |
| Respiratory problems and skin irritation | Dinoflagellates | *Karenia brevis, Pfiesteria* spp. |
| | Cyanobacteria | *Nodularia spumigena, Lyngbya majuscula* |
| Hepatotoxicity | Cyanobacteria | *Microcystis aeruginosa, Nodularia spumigena* |
| *NATURAL AND CULTURED MARINE RESOURCES* | | |
| Haemolytic, hepatotoxic, osmoregulatory effects and other unspecified toxicity | Dinoflagellates | *Gymnodinium spp., Cochlodinium polykrikoides, Pfiesteria spp. (?), Gonyaulax spp., Karenia spp.* |
| | Raphidophytes | *Heterosigma akashiwo, Fibrocapsa japonica* |
| | Prymnesiophytes | *Chrysochromulina spp., Prymnesium spp.* |
| | Pelagophytes | *Aureococcus anophagefferens* |
| | Cyanobacteria | *Microcystis aeruginosa* |
| Mechanical damage | Diatoms | *Chaetoceros spp.* |
| Gill clogging and necrosis | Prymnesiophytes | *Phaeocystis spp.* |
| | Diatoms | *Thalassiosira spp., Cerataulina pelagica* |
| *TOURISM AND RECREATIONAL ACTIVITIES* | | |
| Production of foams, mucilages, discoloration, repellent odours | Dinoflagellates | *Noctiluca scintillans, Prorocentrum spp.* |
| | Prymnesiophytes | *Phaeocystis spp.* |
| | Diatoms | *Cylindrotheca closterium, Coscinodiscus wailesii* |
| | Pelagophytes | *Aureococcus anophagefferens* |
| | Cyanobacteria | *Nodularia spumigena, Microcystis aeruginosa Lyngbya spp., Aphanizomenon flos-aquae* |
| *MARINE ECOSYSTEM* | | |
| Hypoxia, anoxia | Dinoflagellates | *Noctiluca scintillans, Heterocapsa triquetra* |
| | Diatoms | *Skeletonema costatum, Cerataulina pelagica* |
| | Prymnesiophytes | *Phaeocystis spp.* |
| Negative effects on feeding behaviour, reduction of water clarity | Pelagophytes | *Aureococcus anophagefferens, Aureoumbra lagunensis* |
| Toxicity to wild marine fauna | Dinoflagellates | *Karenia brevis, Alexandrium spp.* |
| | Diatoms | *Pseudo-nitzschia australis* |

Harmful events have three basic prerequisites: first, a harmful species must be present in a given area, second, it has to reach a critical concentration and, third, it has to manifest its harmfulness. Hence, the study of harmful algal blooms includes three main steps: the assessment of the geographic ranges of harmful species, the discovery of mechanisms underlying temporal fluctuations of harmful species abundances, also including *exceptional events,* and understanding of the mechanisms of impact (Figure 22.2).

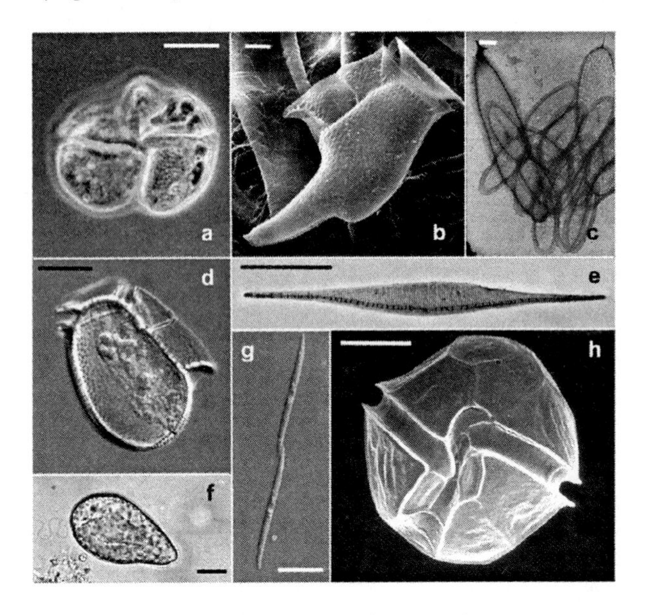


Figure 22.1 Some of the most common harmful species. a) *Karenia brevis,* an unarmoured dinoflagellate causing NSP, Light Microscopy (LM); (b) *Dinophysis caudata,* a thecate dinoflagellate producing okadaic acid, a toxin causing DSP, Scanning Electron Microscopy (SEM); (c) Organic scales covering the cell body of the prymnesiophyte *Chrysochromulina polylepis,* responsible for fish mortality, Transmission Electron Microscopy (TEM); (d) the DSP-producer *Dinophysis fortii,* LM; (e) the colonial diatom *Pseudo-nitzschia galaxiae,* which produces the toxin domoic acid (DA), responsible of ASP, TEM; (f) the ichthyotoxic raphidophyte *Chattonella subsalsa,* LM; (g) *Pseudo-nitzschia multistriata,* another DA-producing diatom, LM; (h) *Alexandrium tamarense,* a thecate dinoflagellate which can cause PSP, SEM. Scale bars 10 μm, except for (c), where it is 0.2 μm.

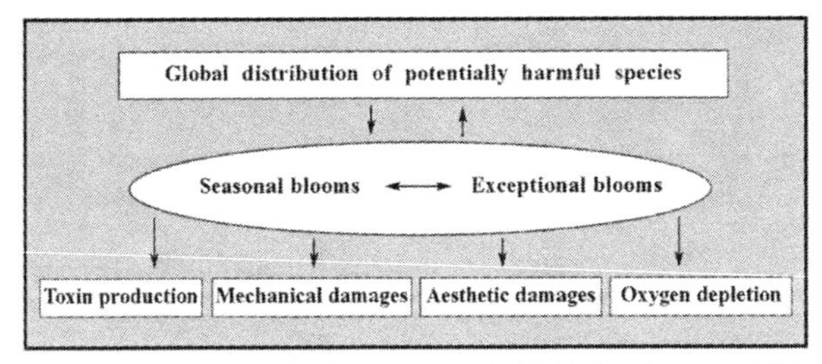


Figure 22.2 Schematic representation of the three steps of HAB development. The global distribution, bloom dynamics and mechanisms of impact of harmful microalgae are the three basic components of HABs discussed in this chapter. Marine microalgae can exert an array of negative impacts on human activities and on the marine ecosystem (Table I). For this to occur, three conditions must coincide, first that a harmful species is present in an area, second that it reaches a critical concentration, and third that a target organism or resource is affected either directly or through a vector.

## 2. Biogeography

### *2.1. The Apparent Cosmopolitanism of Marine Microalgae*

Although not all coastal sites undergo the negative impacts of microalgae, it is more and more evident that all coastal regions host a number of potentially toxic or harmful species. Clearly, information on the distributions of harmful species and of their changes over time is useful for risk assessment and planning of the exploitation of marine resources.

The biogeography of phytoplankton species is in general in a very primitive state. Apart from a number of papers dealing with open water species distribution (e.g. Semina, 1997), or with individual taxa or selected areas (e.g. Dodge, 1996; Okolodkov, 1999), few attempts have been made in recent years to reconstruct geographic ranges of marine microalgae and to interpret different patterns and dispersal routes. This probably depends on the limited amount of information available on a global scale, since wide areas are totally or partially unexplored for phytoplankton distribution. Even at places where phytoplankton is currently studied, lists of species are far from complete. Indeed, the seasonality of most species, coupled with their vertical and horizontal patchiness, strongly impair sampling effectiveness, especially for species that do not reach high concentrations.

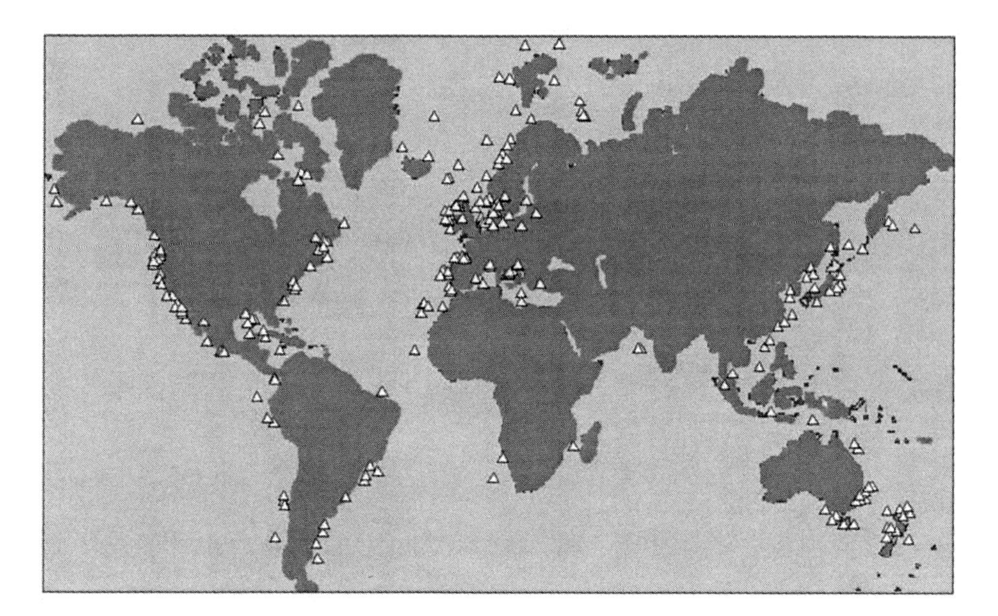

Figure 22.3 Distribution map of the nine toxic species of the genus *Pseudo-nitzschia*, showing the cosmopolitanism of the genus. Data from Hasle (2002).

Some harmful species are considered cosmopolitan, whereas others have more restricted ranges. A few toxic microalgae, including those causing ciguatera, the PSP-producing dinoflagellate *Pyrodinium bahamense* and some *Alexandrium* species, have a range which is restricted to tropical/subtropical areas. Very few toxic species are found in Polar Seas. Nine *Pseudo-nitzschia* species are known to produce the toxin domoic acid, which causes the human syndrome ASP. Only two of these, namely, *P. seriata* and *P. delicatissima* reach 80° latitude N, and only *P. seriata* is restricted to cold waters of the Northern Hemisphere. Similarly, only a few *Alexandrium* species have been recorded beyond 50° of latitude, though the associated PSP represents a problem in some of these areas. This situation could reflect scarcity of investigations and, mainly, lower human population density and lack of economic resources that could be affected in higher latitudes.

With exceptions of the above-mentioned cases of truly tropical and cold water species, the majority of the other harmful species belong to the wide category of temperate species, which do not show any particular temperature constraint to their distribution. As a matter of fact, most species of the diatom genus *Pseudo-nitzschia* are found in all latitudes (Figure 22.3), yet their occurrence is not uniform within their geographic range (Hasle, 2002) and it is scarcely predictable based e.g. on geographical or ecological factors. The resulting composition of local microflora may differ from site to site. The same applies to *Alexandrium minutum* (Figure 22.4), a species producing PSP toxins, which has a wide geographic range but is apparently absent in some intensively monitored areas (Hansen et al., 2003).

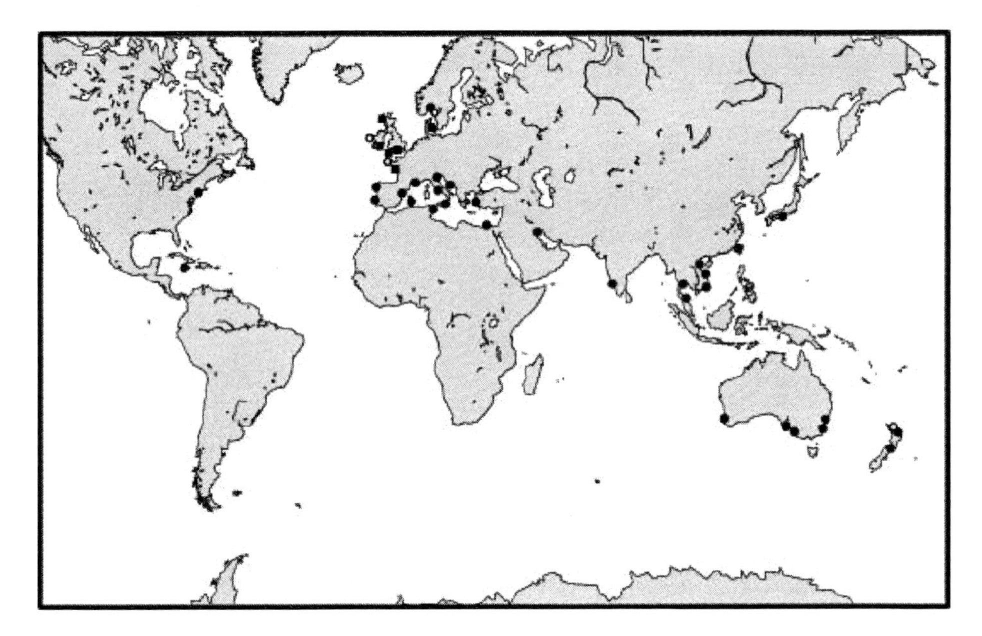

Figure 22.4 Distribution map of *Alexandrium minutum*. The absence of the species from some areas, such as off parts of Africa and South America, may be due to lack of monitoring. The scarcity of records along north American coasts could be real, considering that extensive monitoring carried on for several decades at different coastal sites (from Hansen et al., 2003), but could also reflect the difficulty of identifying *Alexandrium* species in the light microscope.

As for salinity, high values are a limiting factor for the distribution of several diazotrophic cyanobacteria as well as for species of *Pfiesteria*, so far only found in estuarine waters (Burkholder et al., 2001).

Based on the global distribution of freshwater ciliated protozoa, it has been suggested that all free living eukaryotic microbes are potentially cosmopolitan, since they generally reach high concentrations and no physical barrier can limit their dispersal so as to restrict their geographical range (Finlay et al., 1996; Finlay, 2002). The relative scarcity of species richness in eukaryote microbes as compared e.g. to terrestrial plants would apparently support this view. However, the idea of a limited number of widely distributed unicellular organisms is based mainly on the morphological concept of species, which is regarded as rather weak for microalgae (Mann, 2001). In addition, the low taxonomic resolution of most phytoplankton studies places a severe constraint on our ability to interpret biogeographical patterns in microalgae. Thus species which are difficult to identify are believed to be far more cosmopolitan than those that are not, and phytoplankton species obviously are among the most difficult to identify (Wyatt and Carlton, 2002). A clear example is represented by the case of *Karenia brevis* (syn. *Gymnodinium breve*), which in the past was reported from many different sites. In recent years, a better knowledge of this taxon has been acquired (Daugbjerg et al., 2000) and several other *Karenia* species have been described (Chang et al., 2001; Yang et al., 2001, Haywood et al., 2004). Based on this new information, it is now clear that *K. brevis* is indeed only found at a few localities.

For a number of harmful species, a closer look into intraspecific diversity has demonstrated a high degree of genetic and physiological variability. Within the PSP producing dinoflagellates of the *Alexandrium tamarense* species complex, phylogenetic relatedness has been found to be higher among strains of different morphospecies from the same geographic site than among strains of a single morphospecies over its geographic range (Anderson et al., 1994; John et al., 2003). Other species of *Alexandrium* such as *A. minutum* (Figure 22.4) are in contrast more homogeneous phylogenetically over their ranges (Hansen et al., 2003). A lack of correspondence between morphospecies and molecular species has been highlighted for other harmful (e.g. *Pseudo-nitzschia,* in Lundholm et al., 2003) as well as non-harmful (e.g. *Scrippsiella* spp., in Montresor et al., 2003) phytoplankton taxa. On the other hand, elucidation of the polymorphic life cycle of such harmful species as *Dinophysis* (Reguera and González-Gil, 2001) and *Prymnesium* (Larsen and Edvardsen, 1998) has revealed that morphotypes thus far classified as distinct species can sometimes be life stages of a single species. All these studies show that we are still far from a reliable assessment of taxonomic units that are phylogenetically significant and safely usable in ecological and physiological investigations. Indeed, a high number of cryptic species with distinct geographic ranges could be hidden within each morphospecies, casting doubt on the view of a few taxonomic entities distributed worldwide. In addition to molecular tools, mating compatibility experiments on species over their ranges represent a further natural way to assess the boundaries between taxonomic units and the relatedness of spatially distant populations of a single morphospecies (Blackburn et al., 2001). In conclusion, a more reliable biogeography of harmful and non harmful protists is needed that should be soundly supported by exhaustive sampling in coastal areas and strongly rooted in genetic, life cycle and mating compatibility information.

### 2.2. *Changes in Spatial Distribution of Harmful Species*

On short (years) and intermediate (decades) time scales, the range of a species is the integrated result of local colonisations and extinctions; we can usually recognize core areas, where the species is a regular annual component of the plankton, and expatriate areas which are colonized from the core areas with varying frequency. On longer time scales (centuries and more), distributions of harmful algal species are subject to range modifications. Though very few long-term datasets are available for this kind of analysis, it is clear that the geographic range of harmful and non-harmful species can expand or contract depending on several possible natural or human-induced factors. Climatic changes can affect species distributions either directly, through temperature variations, storms and catastrophic events, or, indirectly, through their periodic or long-term effects on oceanographic conditions, e.g. changes in stratification period or in coastal circulation patterns.

Species can be transported to new areas by currents, but other routes of inoculation are possible, such as the rinsing of spores from the feet of migratory birds, aerial arrival of viable stages, or introductions due to human commerce. Species translocations via ship's ballast waters, fouling organisms or livestock shipments have often been advocated to explain changing species distribution, especially in cases when range expansions of species or strains were improbable on the basis of hydrodynamic features (McMinn et al., 1997; Matsuyama, 1999). For the reasons stated above, to establish whether a microalga has been transported is more difficult than for larger, easily identifiable organisms that can be detected without problems. Indeed, it can rarely be excluded that a suspected phytoplanktonic invader was already present as part of the hidden, unsampled microflora, nor are the ecological consequences of a microalgal invasion as evident as for benthic or nektonic species. These difficulties are probably at the origin of the extremely low number of proven invaders among phytoplankters (Wyatt and Carlton, 2002), as compared e.g. to macrophytes (Boudouresque and Verlaque, 2002). On the other hand, if the theory of a few widespread species (Finlay, 2002) is rejected in favour of an underestimated microalgal diversity, human mediated transport assumes a very important role in species translocation from one coastal site to another, and its actual role is certainly largely underestimated.

Some categories of species are more easily transported. These include those species which can resist unfavourable conditions, e.g. by producing resting stages, among which several harmful species are included. A very long dark survival is possible also for species which do not form cysts, such as diatoms (Peters, 1996) and the pelagophyte *Aureococcus anophagefferens* (Popels and Hutchins, 2002). Species reaching high concentrations are more likely to be effectively transported since they can provide larger inocula. Finally, human-mediated transport is enhanced especially between sheltered areas, such as harbours and aquaculture sites, where many dinoflagellates thrive. All these reasons make many harmful species good candidates for transport, and the low number of invasions so far demonstrated is most probably a very conservative estimate of the real number.

The application of molecular techniques in the comparison of local populations, coupled with sound knowledge of large scale hydrography, can help shed light on species transport. The study of phylogenetic relationships among isolates of different geographic origin has in fact been used in recent years in the attempt to track range modifications and dispersal routes in harmful species such as the *Alexandrium tamarense* species complex (Scholin et al., 1995; Medlin et al., 1998) and *Gymnodinium catenatum* (Bolch et al., 1999). This approach has also revealed a recent range expansion of the ichthyotoxic flagellate *Fibrocapsa japonica* based on the genetic homogeneity of strains from several different areas (Kooistra et al., 2001). In addition, the possible allochthonous origin and a high chance of human transportation has been proposed for Mediterranean strains of *Alexandrium catenella,* first found in 1998 in the Thau lagoon (France), based on their genetic similarity with the Japanese ribotype of the same species (Lilly et al., 2002). In fact, the species had been detected in the Balearic basin in 1983 (Margalef and Estrada, 1987) and in Barcelona harbour since 1996 (Vila et al., 2001), therefore its presence in Thau lagoon could rather be seen as a range expansion from other Mediterranean sites. Even so, the high genetic homogeneity of strains of *A. catenella* from several Mediterranean localities, all almost identical to a Japanese strain (A. Penna, pers. comm.), would still support the hypothesis of a relatively recent introduction.

It should be considered that, in addition to the translocation of a microalgal species into an area where it was absent, allochthonous strains can be introduced into areas where a species is already present, which would allow for genome exchange among resident and invader strains leading to increased genetic diversity within the species and possibly to fitness enhancement. The relevance of these cryptic invasions again highlights the importance of investigations at the genetic level aimed at tracking dispersal pathways and genetic make-up of local populations.

For an invasion to occur, a series of factors must be considered in addition to the transport vector itself (Hallegraeff, 1998). Microalgae in quiescent or resting phases can resist a number of weeks in the adverse conditions offered by ballast waters, yet the season, ecological conditions and numbers of cells at the time of release may be such that the species fails to settle and develop in the receiving area. In addition, as Darwin wrote, *we should never forget that to range widely implies not only the power of crossing barriers, but the more important power of being victorious in distant lands in the struggling for life with foreign associates.*

Local extinctions have generally received less attention than range expansions. Transport by ballast waters can account for the introduction but not for the disappearance of a species at a given site, unless the species is substituted by a competing invader. In a 20-year investigation in the Gulf of Naples (Mediterranean Sea), *Pseudo-nitzschia subpacifica* was commonly found in spring from 1984 to 1987, but has not been recorded since then (unpublished data). Ecological conditions often change in a gradual way so as to become unfavourable for a species. The progressive diminution of a species abundance may reach a critical threshold below which the population is too small to be able to survive, leading to local extinction. Some of the processes which might be involved in local extinction and bloom potential of *Alexandrium* are explored in a model by Eilertsen and Wyatt (2000) which draws attention to the potential importance of cyst bed dynamics and inoculation *strategies.* If the balance between cyst germination and cyst recruitment is negative, the cyst bed will become depleted, leading eventually to a situation where a local source of inoculation is absent; if the balance is positive, scarce concentrations may be augmented and eventually reach threshold values which can seed blooms.

### 2.3. Geographical Distribution of HABs

The presence of harmful species at given sites is a necessary but not sufficient condition for the development of harmful algal blooms, so that the geographic distributions of HABs do not necessarily strictly reflect those of the causative species (Figure 22.5). A potentially harmful species does not always exert a negative impact because it often includes both toxic and non-toxic strains whose distribution is not homogeneous over the range of the species. Other causes for the lack of negative impact of potentially harmful species are that they do not always reach critical concentrations, or the vector and target organisms may be absent, or potential seafood resources are neither exploited nor monitored. Yet subtle environmental shifts, human-made or natural modifications and change of exploitation patterns of coastal areas may suddenly evoke the expression of harmfulness. One of the most striking examples is the case of *Chrysochromulina polylepis,* a species which blooms from time to time in the Skagerrak, and which in 1988 produced mortalities of farmed fish and wild marine fauna. The reasons for the remarkable toxicity in that specific year are still largely unknown (Gjosaeter et al., 2000).

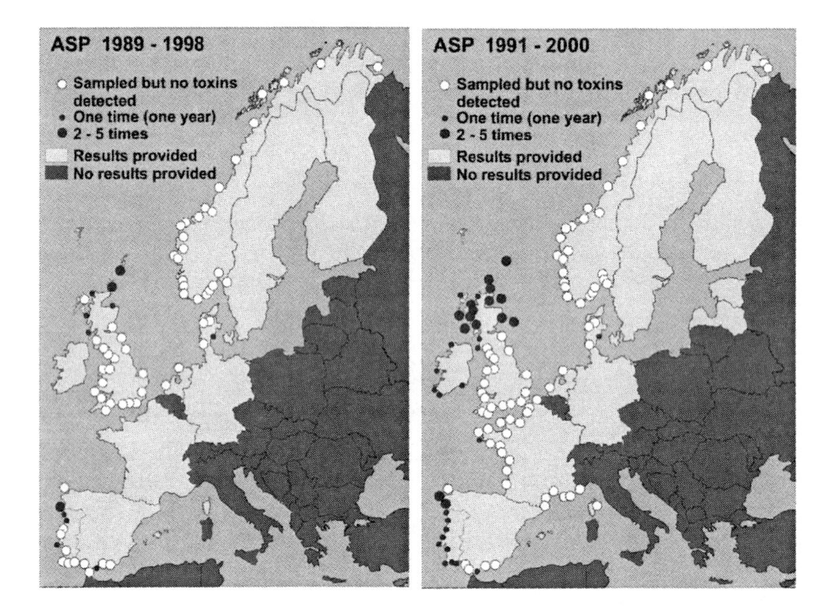


Figure 22.5 Distribution of ASP along European coasts, redrawn from ICES-IOC decadal maps of harmful events along North Atlantic and north Eastern Pacific coasts (ICES-IOC, 2003). The increase in records in the decade 1991–2001 reflects both the increase in monitoring sites (e.g. Wales) and new records in previously sampled areas (Portugal). While potentially toxic *Pseudo-nitzschia* species are widely distributed along the Atlantic coasts (Fig. 3.024), ASP is only recorded in certain areas.

Given the variety of potentially harmful organisms, and of mechanisms leading to the attainment of critical concentrations, it is not surprising that HABs are reported from all kinds of coastal ecosystems, from eutrophic and polluted to pristine, from physically isolated water bodies to those dominated by mesoscale processes. Almost all the coastal regions of the world's seas suffer from some kinds of problem related to HABs, although a considerable diversity exists in the geographic distribution of single types of HABs. For example, the ICES-IOC decadal maps of harmful events along North Atlantic and north Eastern Pacific coasts (ICES-IOC, 2003) show that PSP problems are very widespread, though some areas are unaffected. DSP is apparently a problem only along European coasts and in Atlantic Canada. Within the ICES region, cyanobacterial blooms are typical of the Baltic Sea, whereas ciguatera is confined to selected sites in the Gulf of Mexico. Fish kills are reported for sites scattered along the coasts included in these maps, while azaspiracid toxins are so far only known from north eastern Atlantic waters.

The heterogeneous distribution of HAB types may reflect the use of inadequate analytical techniques, as in the case of AZP, which is not distinguished from other lipophylic toxins by the mouse bioassay, or the lack of specific routine monitoring, as is the case for the absence of DSP reports from the United States. A particular use of coastal areas may also reveal specific kinds of HABs, discolouration representing a problem only in coastal areas used for tourism and recreational activities. In other cases, the irregular distribution of HAB types is based on ecological and geographical features, i.e. species causing ciguatera are tropical/subtropical, while cyanobacteria require low salinity and stable water column conditions of e.g. the Baltic Sea in summer in order to bloom. NSP is caused by *Karenia brevis,* a dinoflagellate with a narrow geographic range, and is also restricted in its spatial occurrence. However, no ecological explanation is found so far for the localization of some phenomena such as ASP, AZP, DSP and PSP in some areas more than in others, in spite of the wide distribution of the causative organisms.

## 3. Bloom Dynamics of Harmful Algae

### 3.1. *Definitions*

Plankton blooms can be defined numerically and dynamically in several different ways depending on whether we wish to emphasize a) their timing, b) their abundance, or c) their spatial patterns. In principle, an elementary definition of an algal bloom can be based on some specified change in the abundance, so, for example, we can say that a bloom is taking place when the cell concentration at a locality exceeds some threshold concentration at that locality by an arbitrary amount. This is the intuitive meaning in phrases such as *the spring bloom,* and is adequate for many purposes (Smayda, 1997). But it might sometimes be useful to specify what the threshold level is and by how much it should be exceeded (one standard deviation?) to qualify as a bloom. Then blooms may be judged *exceptional* on two scores. If such events are annual, but the numbers are much higher than expected (two standard deviations?), or if the frequency of blooms is very high compared with expected population changes, then we can judge them exceptional. Thus the two measures we need to define blooms numerically (ignoring spatial extent) are abundance and frequency. Both these measures have been difficult to estimate in the past, but can now be obtained from the time series of species counts accumulating as a result of algal monitoring programmes in many regions at risk from toxic algae. Even with these elementary requirements, devising an acceptable definition is already more demanding, and will need to refer to the bloom species by name, since each has its own unique patterns of abundance. Although these considerations are elementary, they are hardly ever made explicit.

A misleading view regarding HABs is that they are always irregular, unpredictable and conspicuous events involving the accumulation of highly concentrated populations. In fact, many of the highly toxic species often constitute a regularly occurring component of normal phytoplankton populations and can exert their impact at low cell concentrations (100–1000 cells·$l^{-1}$). These species would probably remain undiscovered in the absence of vectoring organisms used as food resources. A clear example of lack of coupling between HABs and high biomass is provided by the case of the Gulf of Maine (Figure 22.6), where in three different cruises in summer 1998 phytoplankton biomass showed high values alongshore while the toxic *Alexandrium* displayed maximum cell concentrations offshore (Townsend et al., 2001). In these cases, the underlying scientific questions are very general, and concern the factors shaping spatial and temporal patterns of phytoplankton distribution.

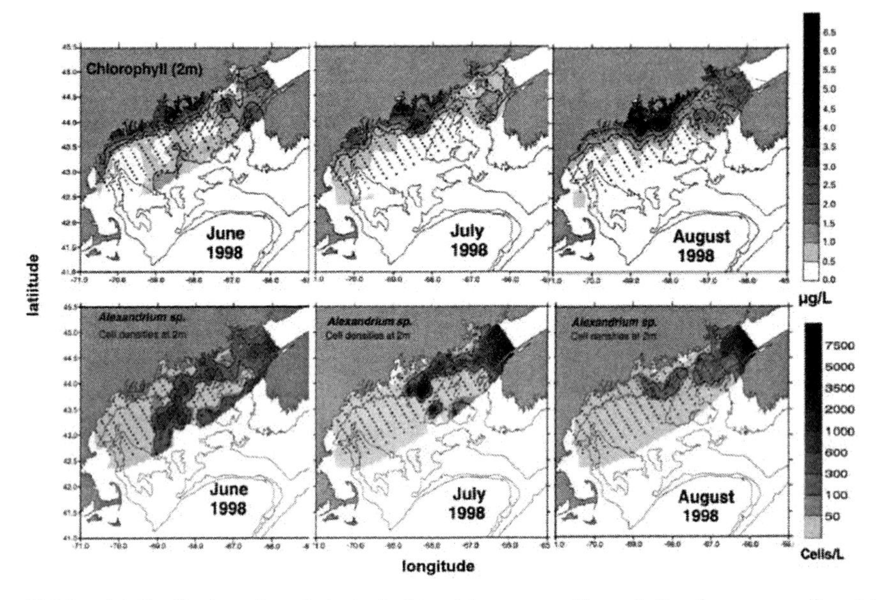


Figure 22.6 Spatial distribution of total phytoplankton biomass, as chlorophyll *a,* (upper panel) and the toxic dinoflagellate *Alexandrium* sp. (lower panels) along the coasts of Maine (USA) (from Townsend et al., 2001). Maximum phytoplankton biomass values are found in inshore waters, whereas *Alexandrium* sp. shows the highest concentrations offshore.

All microphytoplankton species vary in their abundance over the seasons, and in many cases they are undetectable for the major part of the time. For harmful species, this implies that they are not equally dangerous throughout the year, rather they have generally one/several predictable/unpredictable periods of the year when they may exert their harmful effects. This applies to both species reaching high biomass and species which require only few thousand cells per litre or less to be harmful. In both cases, we will use the word *bloom,* assuming that a biomass increase against a baseline is always needed. Cases in which toxicity arises with no biomass increase are also possible due to the changing proportion of toxic and non-toxic strains within a population, or to the stimulation of toxin production or accumulation by changing environmental conditions (see section 5).

There are no reasons to suppose that the population dynamics of harmful algal species are in any way distinctive, which means that their study can illuminate the ecology of other species and, also, that we can use examples from non harmful species to explain harmful bloom dynamics. The only exception to this statement could be represented by the possible role of toxins as allelopathic and grazing deterrent substances, yet, as discussed below, non-toxic algae may also produce allelopathic substances which can confer comparable advantages. So, the processes of interest in this context are those which either allow populations to increase in numbers and form blooms, or which inhibit them from doing so, or which return them to non-bloom conditions. When algal numbers are plotted on time, we can ideally, and to some extent too by analogy with batch cultures, recognize inoculation, pre-bloom, bloom, and post-bloom phases (Fogg, 1965; Steidinger, 1983). The real difference from batch cultures is that the sequential series resulting from the balance between rates of increase and decrease in nature can be interrupted or reversed at any time, so that blooms do not inevitably follow inoculation. The exponential phase can be extended (or reinitiated) by addition of a limiting resource such as a nutrient, or by dilution if self-shading is limiting. High dilution rates can be expected to return populations to earlier phases, or even eliminate them (Skellam, 1951; Kierstead and Slobodkin, 1953; Martin, 2000). We can assume that the fluctuating abundance of a species in the sea is due in part to these well understood processes.

### 3.2. *The Phases of Phytoplankton Blooms*

If the species of interest should appear at some moment in the water where it was previously absent or undetected, then we can recognize that inoculation has taken place; this will in most cases have been a result either of advection of cells from elsewhere, or of hatching of resting stages from the sediments or from deeper in the water column. We must also recognize that some cells may be there in the water all the time, but that their concentrations are too low to be detected by routine sampling (operationally absent). This would be the case for any species if it were known not to form resting stages.

The inoculum may not lead immediately to population growth - in batch cultures there is frequently a lag phase - and this is sometimes ascribed to the need to bring about some subtle change in the water, *conditioning,* as a prerequisite for normal cell division to occur (Fogg, 1965), or to an Allee effect (Allee, 1932; Berryman, 2003). Allee effects have in common that there is an absence of *cooperation* at low population densities which constrains the cell division rate below its intrinsic value. It may be that something akin to *quorum sensing* (Miller and Bassler, 2001) is involved. This expression refers to the process whereby bacteria sense the density of their conspecific neighbours, leading to a density dependent gene regulation process which coordinates activities such as colony formation and coordination by means of cell-to-cell signals. At some critical cell density the intracellular concentration of signal molecules reaches a threshold, and the signal is diffused into the medium. When the concentration reaches a critical threshold, it is able to react with a transcriptional activator protein, which leads to its autocatalytic production. The complex formed by the signal molecule and the protein binds to a region of the DNA and switches on the genes which regulate the process to be coordinated. Neighbouring cells can then sense their density from the external concentration of the signal molecule.

It has not been established that analogous social effects occur in phytoplankton, but such could be one explanation for the observation that good growth in cultures often requires some kind of conditioning, and that filtrates from older cultures are sometimes effective. It seems likely that some such mechanism lies behind processes leading to colony formation where cells are embedded in extracellular matrices as in some *Phaeocystis* species, and in the accumulations of large slimy colonies which occur e.g. in blooms of *Coscinodiscus wailesii,* a large centric diatom which forms dense blooms in the North Sea and English Channel, and *Tasman Bay slime* in New Zealand (Chang, 1984). Communication *between* species may also be analogous to quorum sensing in situations where *cooperation* leads to relatively stable microbial communities such as aggregates or marine snow (Leppard, 1999), or to the phenomenon known as *mare sporco* characteristic of the northern Adriatic (Degobbis et al., 1995). Extracellular polysaccharide matrices can reduce the hazards to such communication caused by shear (Jenkinson and Wyatt, 1992).

But in nature, a failure of algal numbers to increase with time might alternatively indicate that the population is in a dynamic steady state due to a combination of positive and negative regulatory processes. The apparent similarity to the lag phase recognized in cultures would then be deceptive.

The end of the lag phase is marked by the beginning of the exponential growth phase. We can proceed at first by ignoring the spatial component of blooms. The simplest way to represent changes in a population is to say that the rate of change in numbers is equal to the product of a rate, $r$, and the stock, $P$. We write

$$dP/dt = rP \tag{1}$$

and can define $r$ in different ways, either as the *intrinsic* capacity for increase (often called, following Fisher, 1930, the Malthusian parameter, and then written $r_{max}$), or as the *natural* or realized growth rate which is the result of both cell divisions and losses. In either case, $r$ is a mean value for the population, and all the individuals represented by $P$ are considered identical. These are routine constraints in phytoplankton models but can be relaxed if more complex questions are to be asked. As we know, for any positive value of $r$, this equation leads to exponential growth, and can therefore only be an appropriate representation of the growth curve for limited periods. If $r$ is treated as the natural growth rate, then it can be divided into different components, cell division and concentration by hydrodynamic mechanisms on the one hand, and losses due to grazing, dispersion, etc., on the other, a procedure first used for modelling phytoplankton dynamics by Fleming (1939).

One way to tackle the problem of unrestricted growth in a model population was resolved by Verhulst (1838). He began by writing

$$dP/dt = rP - mP^2 \tag{2}$$

where $m$ is a loss rate. The quadratic loss term ensures that its impact is small at low population densities and increasingly effective as numbers rise; the power of two is arbitrary, since any power >1 will have a similar effect. Verhulst factorised by $rP$ and let $r/m = K$ to obtain

$$dP/dt = rP(1 - P/K) \quad (3)$$

which is the familiar *logistic* equation in which $r$ has become density-dependent. Thus $K$ represents the equilibrium size of the population ($P_{max}$), but is nowadays usually referred to as the *carrying capacity,* and equated with some limiting resource (e.g. light, nutrients, space) so that attention is directed to the potential regulation of the population by environmental factors rather than by the population itself. A broader way to interpret $K$ is as a competitive interaction or interference between the individuals comprising the population.

Historically, far more attention has been devoted to environmental control of growth, but this is now changing as the view gains ground that resource levels are regulated by biological activity, and that environmental attributes are *emergent* properties of population dynamics (e.g., Watson and Lovelock, 1983; Wilkinson, 1999). It is also true that most studies of algal dynamics have concentrated on the asexual growth phase, but life history imperatives are now receiving more focussed attention (Garcés et al., 2002). Here again we have density-dependent factors and Allee effects, whereby low densities can affect the probability of finding mates; this cannot be the case in asexually reproducing phytoplankton, but must matter in phases of blooms when gametogenesis takes place. If the onset of gamete formation is triggered by cell density, then not only is intercellular signalling implied, but also a means of autoregulation of numbers brought about by a life-history requirement. It is not conventional in the context of phytoplankton to label such mechanisms *social,* though effectively they are.

Numbers increase, but this does not mean that they are necessarily unconstrained by losses to grazers, etc. If the apparent lag phase mentioned above is due to a balance between cell division and grazing, then the onset of the exponential phase might be due to either an increase in the former or a decrease in the latter, or indeed to changes in both; the change in the dynamics is then most simply represented by introducing a grazing term into the elementary formulation of population growth (equation 1) given earlier. We write

$$dP/dt = rP - \alpha PZ \quad (4)$$

here $Z$ represents the grazer population and $\alpha P$ is the rate of capture. If we further add an equation for the grazers

$$dZ/dt = \beta PZ - \gamma Z \quad (5)$$

where $\beta$ is the conversion efficiency of algae to zooplankton and $\gamma$ is the grazer death rate, we arrive to the well known Lotka-Volterra predator-prey model (Lotka, 1925; Volterra, 1926). It is worth noting that in this formulation of the algal dynamics, algae are limited in numbers only by the grazing rate, the encounter rate is proportional to the abundance of both (principle of mass action), and prey are eaten at a fixed proportion of the encounter rate. There is no control of the algae by resource limitation, which while counter to the intuitions of many algal ecologists is in accord with much observational and experimental evidence.

Many models of population dynamics combine the logistic and Lotka-Volterra approaches, so that populations can in certain circumstances reach their carrying capacity. These circumstances are that other regulatory processes have failed to constrain algal numbers.

As the exponential phase ends, a stationary phase may occur, in which cell numbers oscillate around some maximum concentration, usually presumed to be limited by resources, although the only reasonable conclusion in the absence of additional information is that production and consumption rates are in balance. Or, more commonly, numbers may decline to pre-bloom levels. The decline phase in nature is taken to indicate that loss rates due to grazing, spore formation or encystment, cell adhesion, sinking, parasitism, autolysis, viral lysis or dispersal exceed the division rate or other augmentative processes such as swarming and physical concentration.

If all cells in a bloom more or less simultaneously undergo gametogenesis, mate, and produce fast sinking cysts, the decline of the vegetative population can be very rapid indeed, even in the absence of other losses. In *Alexandrium,* the zygote first forms a motile planozygote (Anderson, 1998), and the decline in planktonic abundance may then be more gradual.

High population abundance attracts parasites, and harmful algae are no exception; viruses, bacteria, fungi, and flagellates have all been implicated in the decline of phytoplankton populations, and some are of considerable interest as control agents for HABs (e.g., Anderson, 1997b).

Most studies of marine viruses so far have focused on bacteriophages, but there is evidence that some viruses lyse microalgae too (Van Etten et al., 1991; Zingone, 1995). High percentages of cells which contain viruses have been found in declining blooms of *Heterosigma akashiwo* (Nagasaki et al., 1994) and *Aureococcus anophagefferens* (Sieburth et al., 1988). Viral infection is sometime thought to be characteristic of the stationary phase of algal growth, but in one species of *Phaeocystis* in the North Sea it is the solitary cells which precede the colonial phase which are attacked (Brussaard et al., 1995). The list of nuisance and toxic species that can be attacked by viruses has slowly grown over the last few years, and so far includes also the prymnesiophytes *Phaeocystis pouchetii* (Jacobsen et al., 1996) and *P. globosa* (Bratbak et al., 1998), the filamentous cyanobacterium *Lyngbya majuscula* (Hewson et al., 2001), the dinoflagellates *Alexandrium catenella* (Onji et al., 2001), *Heterocapsa circularisquama* (Tarutani et al., 2001) and possibly *Karenia brevis* (Paul et al., 2002). Since viruses multiply at a rate several orders of magnitude higher than that of the host, a single infection may result in a chain reaction that leads to the massive lysis of the populations attacked and hence to the sudden termination of a bloom. This simple consideration has opened the way to investigations on the possibility of using viruses to control HABs, which started with experiments on freshwater cyanobacteria. The role of viruses in the collapse of algal blooms should however be reconsidered taking into account the high intraspecific variability of both microalgal sensitivity to viral infections and viral infectivity (Nagasaki and Yamaguchi, 1998; Zingone et al., 1999), which implies that a highly dynamic interaction exists between virus and host populations within an algal species. This leaves little hope for the use of viruses in mitigation. On the other hand, the extreme strain-specificity of viral infection could play an important role in the genetic make-up of the algal populations by positively selecting resistant strains of the host species. Clearly, lysis of mass blooms also causes the release of large amounts of DOC (and potential anoxia), and has important consequences for the flow of materials in the microbial loop, and for the course of succession. These processes obviously work in the same way for harmful and non harmful species, yet some peculiarity could exist e.g. the substantial release of toxins into the environment or the contribution to large amounts of transparent exopolymer particles (TEP) following massive lytic processes of high biomass blooms.

Several bacteria (*Alteromonas, Cytophaga, Flavobacterium, Pseudoalteromonas,* etc.) are known which are capable of lysing harmful algae, including several dinoflagellate and raphidophyte species (e.g., Kim et al., 1998; Lovejoy et al., 1998; Imai et al., 2001) and the diatom *Coscinodiscus wailesii* (Nagai and Imai, 1998) either following direct contact, or by secretion of complex extracellular compounds of poorly known nature. In some cases, the effects of the algicidal activity may be enhanced by bacteria swarming onto the host cell (Lovejoy et al., 1998). As in the case of viral infections, bacterial infections are generally characterized by high species-specificity, enhanced bacterial growth rate and rapid algicidal activity, which make bacteria potentially very effective in HAB regulation. Algicidal bacteria increase their concentration in the latest phases of the bloom (Imai et al., 1998; Kim et al. 1998; Imai et al., 2001), but their effective contribution to algal mortality in natural systems is still to be assessed. As in the case for viruses, the possibility of using bacterial infections to terminate specific HABs is probably hindered by variations or loss of algicidal ability, resistant host strains and complex intra-bacterial community interactions (Mayali and Doucette, 2002; Skerratt et al., 2002).

Infection rates and host mortality rates induced by infection can be high, and the dynamics complex, also in parasitic interactions, as for example demonstrated by experiments with the parasitic dinoflagellate *Amoebophrya* (Coats and Park, 2002); susceptibility to infection varies between the three hosts tested (species of *Akashiwo, Gymnodinium,* and *Karlodinium*), and variations exist in the virulence of dinospores (transmission phase of life history) from different hosts. Infections can of course alter the course of host population trajectories at all stages through their impact on photosynthesis, vertical migration, and other processes (Park et al., 2002a; Park et al., 2002b). As compared to *Amoebophrya,* the lethal infection caused by the flagellate *Parvilucifera infectans* to some harmful *Alexandrium* and *Dinophysis* species is less specific but possibly more efficient, due to the high infection rates and the release of several hundreds of new parasites 48 hours or less after the infection of one host cell (Delgado, 1999, Norén et al., 1999). However, while we lack detailed information of the biology of the infecting organisms, the use of any kind of infections in mitigation practices hides the risk of undesirable side effects such as damage to other organisms, including commercial species.

### *3.3. Do Toxins Deter Grazers?*

Nowhere in the preceding do we find any hint that the population dynamics of harmful algae are in any way different in principle from innocuous species. But there are persistent views that in some species the toxins themselves provide an advantage *vis a vis* grazing, for example that zooplankton grazing is either modified or inhibited by the presence of toxins (Turner and Tester, 1997), or that grazers resort to alternative food sources in response to the presence of toxins (Huntley et al., 1987). Either mechanism can be expected to enhance the net growth rate of the toxin producer and thus lead to more rapid bloom formation by lowering grazing rates. It is also well known that some of these toxins can poison and kill vertebrates (VanDolah, 2000; Landsberg, 2002), so that higher trophic levels might also have an influence on algal population dynamics if their impact on grazers is reduced. It has also been shown that some toxic algae *e.g. Chrysochromulina polylepis* and some *Alexandrium* species have allelopathic propensities (Myklestad et al., 1995; Arzul et al., 1999; Schmidt and Hansen, 2001; Tillmann and John, 2002), so they may gain a competitive advantage over other species.

Neither grazing depression nor allelopathy are necessarily mediated by the identified known toxins, and both properties may be due to metabolites which have so far remained unidentified, as in the case of the brown-tide former pelagophytes *Aureococcus anophagefferens* and *Aureoumbra lagunensis* (Buskey and Stockwell, 1993; Bricelj and Lonsdale, 1997; Bricelj et al., 2001). Also, many of the experimental results have been established with algal species and grazers which do not necessarily co-occur in nature, and should therefore be treated as suggestive rather than indicative. It is established that some toxic dinoflagellates (e.g., *Alexandrium*) can be grazed by zooplankton and passed up the food chain to poison fish (White, 1981) without causing any apparent damage to the vectors. There is evidence that some grazers (*Acartia tonsa, Eurytemora herdmani* feeding on *Alexandrium*) are not affected by the toxins (Teegarden and Cembella, 1996), a result confirmed for other copepods (*Calanus finmarchicus, Acartia hudsonica, Metridia lucens, Pseudocalanus* spp.) by Hassett (2003). Other studies indicate that copepods (*Calanus pacificus, Paracalanus parvus*) may respond to toxins by rejecting cells (Huntley et al., 1986; Huntley et al., 1987) and by increase in heart rate (Sykes and Huntley, 1987).

The responses of copepods to the domoic acid (DA) produced by some species of *Pseudo-nitzschia* depend on whether the toxin is presented to the grazers in particulate or dissolved form, and of course on the dose. DA seems to have little effect on copepod feeding when provided as cells. Nanomolar concentrations in dissolved form seem to be harmless to *Acartia* and *Temora* (Lincoln et al., 2001), while micromolar levels of dissolved DA kill *Tigriopus.* The feeding behaviour of *Euphausia pacifica* is intermittent on a diet of toxic *Pseudo-nitzschia multiseries,* continuous on non-toxic *P. pungens,* but the euphausiids suffer no apparent harm from the toxin (Bargu et al., 2003).

The toxic prymnesiophytes *Chrysochromulina polylepis* and *Prymnesium parvum/patelliferum* have been shown to reduce grazing rates, the former of *Favella ehrenbergi* (Carlsson et al., 1990), and at high concentrations of a heliozoan, *Heterophrys marina* (Tobiesen, 1991) and a dinoflagellate *Oxyrrhis marina* (John et al., 2002), the latter of *Eurytemora affinis* (Koski et al., 1999).

Evidence is accumulating that diatom diets, e.g. *Thalassiosira rotula* and *Skeletonema costatum,* can cause lowered fertility and reproductive failure in cultured copepods (Poulet et al., 1994; Miralto et al., 1999); the putative agents for these effects are aldehydes which have been isolated from *Pseudo-nitzschia delicatissima, Thalassiosira rotula* and *Skeletonema costatum.* On the other hand, an extensive comparison of copepod hatching success and diatom concentrations from a variety of coastal and upwelling regions around the world failed to extend this model to all natural environments (Irigoien et al., 2002). But the experiments remind us that the toxins well known to students of harmful algae are not necessarily the only metabolites which need to be taken into account in the mediation of trophic relationships. In addition, even if such effects could be recognized in natural situations, the ecological significance of metabolites which deter grazing would remain ambiguous given the difference in time scales between bloom dynamics and copepod generation times, though this argument would not apply to grazers whose generation times are similar to those of phytoplankton. The functions of the known toxins would then have to be sought elsewhere. In one case, it has been argued that the gonyautoxins serve as pheromones (Wyatt and Jenkinson, 1997).

### *3.4. The Timing of HABs*

Is it possible to predict the timing and potential magnitude of phytoplankton blooms? Prediction of the spring bloom has been an important component of biological oceanography almost since its inception more than a century ago (Wyatt and Jenkinson, 1993). But attempts to predict the occurrence of named phytoplankton species are rather rare. The general problems are well understood, and probably most major processes which lead to the growth and decline of phytoplankton populations have been identified. But we are in more uncertain territory when we start to ask questions about these processes *vis a vis* particular species, harmful or otherwise. On the other hand the timing and duration of single blooms are key elements in determining harmfulness, as well as for HAB prediction and mitigation. In this context, exogenous and endogenous mechanisms driving the occurrence and seasonal succession of HA are of extreme relevance.

Many harmful species are normal components of the seasonal succession at some sites. *Pseudo-nitzschia* species show recurrent seasonal peaks e.g. at places along the West coast of USA (Fryxell et al., 1997) and in the Gulf of Naples (Zingone et al., 2003). Based on an 11-year investigation, *A. tamarense* blooms in the St. Lawrence region, have been shown to be associated with one seasonal phase of the typical phytoplankton succession in the area (Blasco et al., 2003). These species can hence be considered as part of the *main sequence* of events (Figure 22.7) described in the classic Margalef's Mandala (Margalef, 1978). In the Mandala, and in its further elaborations (Margalef, 1997a; Reynolds and Smayda, 1998; Cullen et al., 2002), distinct phytoplankton life-forms are selected as a response to a specific set of environmental parameters, which contribute to build up a habitat template. Since these templates are seasonally recurrent, specific life forms are recurrent too. Dinoflagellate *red tides* and harmful blooms constitute a parallel sequence that occurs in areas where nutrient supply is decoupled from mixing and turbulence (Margalef et al., 1979). It has subsequently been argued that dinoflagellates are quite a diverse group in terms of ecological requirements, therefore distinct groups of species occupy different positions along a gradient of environmental parameters including nutrients, turbulence and euphotic zone depth (Smayda and Reynolds, 2001).

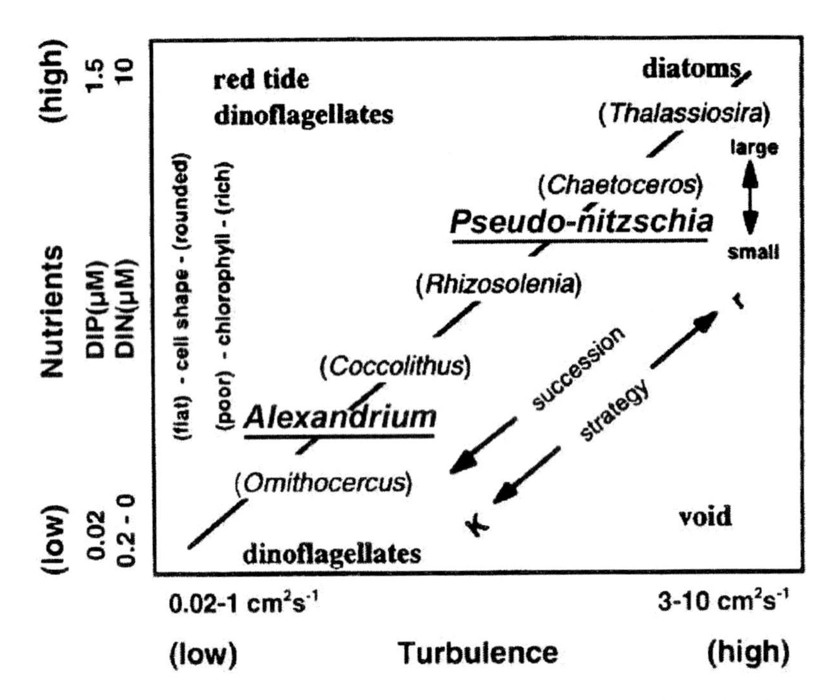


Figure 22.7 The Margalef Mandala (Margalef 1997a) modified to include toxic species along the main sequence of different phytoplankton life-forms which succeed one to another following changes of turbulence and nutrient concentrations. While dinoflagellate red tides can be associated with special conditions of low turbulence and high nutrient availability (upper left area of the Mandala), many other harmful species are part of the normal seasonal succession of phytoplankton species.

In this kind of paradigm, species succession is hence driven by exogenous forces which select phylogenetic groups or functional groups or life forms that may include harmful species. The prediction of the occurrence of one of the functional groups is hence possible based on the knowledge and forecasting of basic oceanographic conditions. It should be noted however that, in this view, while the occurrence of a functional group, that can be equated to the phytoplankton community, is reliably predictable, single species belonging to this group are not, due to a highly stochastic component of the biological response at the specific level (Smayda and Reynolds, 2003).

On the other hand, for a number of phytoplankton species there is evidence of seasonal rhythms of occurrence that can be relatively independent from such environmental conditions as turbulence and nutrients. Some species of the diatom genus *Chaetoceros* inoculate the water column from benthic resting stages when day length reaches 12 hours, i.e., at the spring equinox, and others when it reaches 14 h (Eilertsen et al., 1995). Some dinoflagellates exhibit the same behaviour, e.g., *Alexandrium* (Anderson and Keafer, 1987) and *Scrippsiella* (Costas and Varela,

1989). In Narragansett Bay (USA), the flagellate *Heterosigma akashiwo* appears in mid-May with a notable punctuality (Figure 22.8), despite interannual variability of oceanographic parameters such as temperature (Li and Smayda, 2000). Following the inoculum, the exact timing of maximum population abundance depends on the net growth rate achieved after the inoculation of the bloom, but in general we expect the annual timing of such an event to have low variance, of the order of a few weeks for species in this category. In the case of *Heterosigma akashiwo,* a peak is reached after six weeks from inoculation (Li and Smayda, 2000). Literature suggests that both species that produce heteromorphic resting stages and species that apparently do not can display a remarkable punctuality in their annual occurrence (Zingone, 2002). A wide range of mechanisms could favour this punctuality. These include mandatory dormancy for resting cysts, responses to changing photoperiod and circa-annual clocks of the growth rate. In contrast to the case of exogenous control, endogenous control of species occurrence would increase the probability of successful predictions of single species over those of the whole community.

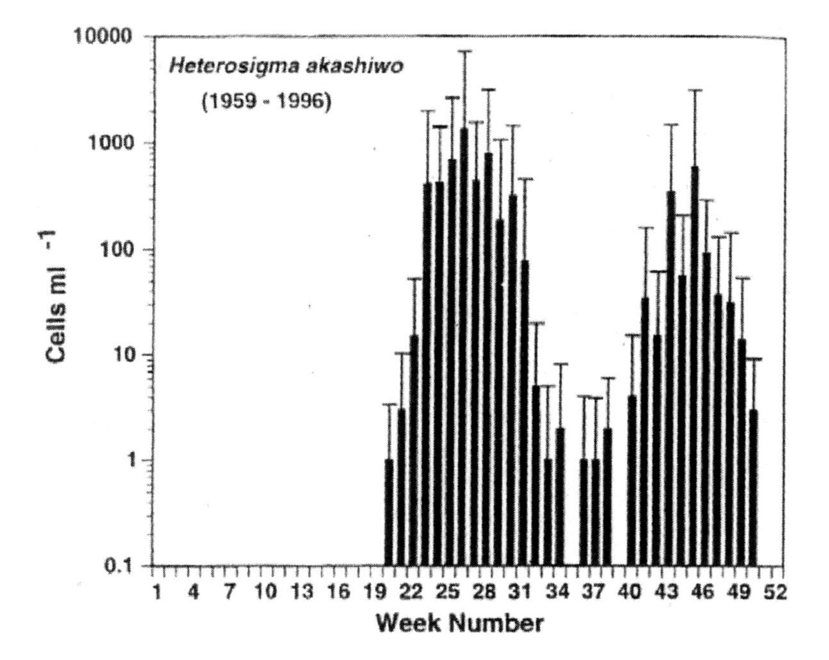


Figure 22.8 Recurrent seasonal occurrence of *Heterosigma akashiwo* based on a long-term (40 yr) study in Narragansett Bay (Li and Smayda, 2000). The first annual bloom occurs every year, whereas the autumn bloom is recorded 30% of the years. Both summer and autumn blooms are regular in their timing, with an 88% probability that the first annual maximum develops during weeks 23 to 28.

Several harmful and non harmful species can bloom more than once a year, apparently under different environmental conditions. This is again the case for *Heterosigma akashiwo* (Li and Smayda, 2000) which, in addition to the bloom around June, shows another increase in autumn (Figure 22.8). Although this secondary pulse is not recorded in 30% of the years sampled, the timing of its occurrence is as regular as the one of the summer bloom. A particularly interesting case of multiple blooms is that of the domoic acid-producing diatom *Pseudo-nitzschia galaxiae,* which in the Gulf of Naples (Mediterranean Sea) shows three growth periods over the year, in March, May and August (Cerino et al., 2005). The cell size of the populations during each bloom phase is distinct, the three blooms being formed by small, medium- and large-sized cells, respectively (Figure 22.9). This temporal pattern highlights a high physiological and ecological plasticity in this species and a complicate relationship between its life cycle and the seasonality of the blooms. In fact, diatom size gradually decreases during the asexual division phase of their life cycles, and abruptly increases to the maximum cell size only in correspondence with sexual reproduction, which generally occurs within a low cell-size range and less frequently than once a year. The regular occurrence of different size classes could hence be the result of yearly synchronised sexuality. On the other hand, the different blooms can hardly be attributed to a single population, which otherwise should be every time smaller than the preceding one and not the opposite.

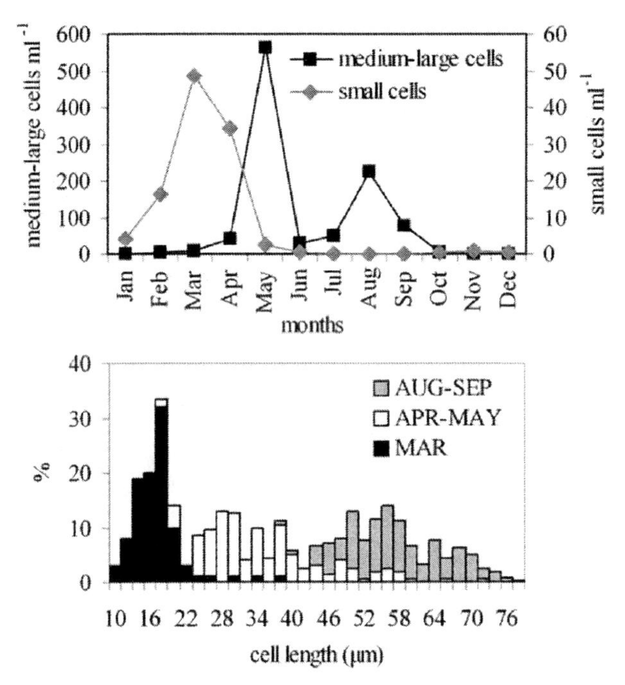


Figure 22.9 Seasonal pattern and size frequency distribution of the toxic diatom *Pseudo-nitzschia galaxiae* in the Gulf of Naples. (a) Monthly averages (1984–1991 and 1995–2000); (b) cell lengths (transapical axis) during late winter, spring and summer blooms of the species (from Cerino et al., 2005). The three bloom periods are a recurrent feature of the species over a 14 yr time series. The different size ranges indicate different stages of the life cycle, the largest cells being the youngest, i.e. those found immediately after sexual reproduction. In fact, diatom size progressively decreases during the asexual reproduction phase until a sexual event, during which maximum cell size is restored. Differences in size, probably implying differences in physiological characteristics, allow *P. galaxiae* to colonize distinct ecological niches over the year.

It is a relatively simple matter, at least in principle, to define exceptional blooms of punctual species in terms of say the standard deviation of the annual timing of their peak abundance. In contrast, there are species (e.g., the diatoms *Skeletonema costatum, Corethron criophilum, Nitzschia grunowii,* and the haptophyte *Phaeocystis pouchetii*) which can apparently bloom at any time of the year (there is however a possibility that these taxa include many distinct cryptic species), and any estimates of the variance of timing in this group will be very high. For the brown tides of *Aureococcus,* and other unpredictable flagellate blooms, Smayda and Villareal (1989) introduced the *open niche* concept. After the spring summer-bloom, a period of potential flagellate blooms would occur in Narragansett Bay. This niche can be occupied by different species that show a high interannual variability. Thus for the microplankton of any given locality we anticipate a fairly regular successional sequence of some species against a background of more loosely timed weed-like species. We can assume that these different strategies reflect different possibilities offered by the environment, and that any change in the latter will alter the relative frequency with which the different species form blooms each year.

We also need to consider, on the time scales of blooms, the periods during which population numbers exceed the threshold, and how long the exponential growth and decline phases last. What is the appropriate measure of time in this context? We tend to automatically assume that calendar time will do, but if we scale by some measure of biological time such as generation time, we will see that algal blooms are comparable to fishing periods and other phenomena. Starting with an inoculum of 100 cells $l^{-1}$, with a division rate of 1 $d^{-1}$ and no diffusion, the population can reach nearly $10^6$ $l^{-1}$ in 13 d (= 13 generations), which is equivalent in numbers of generations to the sort of build up periods of *booms* in some pelagic fish stocks. If gametogenesis intervenes at a concentration of $10^5$ cells $l^{-1}$ (Wyatt and Jenkinson, 1997), then only 10 d are required; if $r>1$, even less. The blooms are ephemeral from our viewpoint, but congruent with life history strategies.

### 3.5. *Physical-Biological Interactions in HAB Dynamics*

HABs encompass all space and time scales relevant to phytoplankton biology, ecology and distribution, from the level of molecules and short-lived subcellular processes to global biogeographical ranges and long term fluctuations of natural populations. At all these space and time scales, interactions with the surrounding physical, chemical and biological environment take place and constitute sources of variability, resulting in a high degree of unpredictability of the occurrence, extension and noxious effects of these phenomena. In order to clarify algal distribution, bloom dynamics and toxicity mechanisms, a number of relevant chemical and physical processes should be taken into consideration, the relative importance of which changes in relation to the species and the biological process involved, as well as to the ecosystem considered.

Physical-biological interactions that are relevant to harmful bloom development occur over a wide range of scales, spanning from turbulence to upwelling, from thin-layered structures to large scale currents (Donaghay and Osborn, 1997). In many cases it appears that species-specific properties and behaviour can modulate these interactions, whereby the respective role of the endogenous/biological and exogenous/physical components is difficult to disentangle.

The relevance of small-scale physical processes to phytoplankton ecology is clearly demonstrated by the responses evoked by fluid motion at several distinct levels. Diatoms have been shown to rapidly perceive changes in fluid motion among other chemical and physical stimuli. The response, registered as a change in calcium concentration - activation in signal transduction – is detected only to the first of two consecutive stimuli, demonstrating the capability of a quick adaptation to a new physical regime (Falciatore et al., 2000).

Turbulence can induce significant morphological changes in some dinoflagellates (Zirbel et al., 2000) which confirms that complex cellular shapes represent an adaptive response to the physical environment (Margalef, 1997b). At the physiological level, the growth rate of many species is affected by turbulence (Thomas and Gibson, 1990) through a decline in cell division and at times by mortality. Decline in growth rates has been shown to depend on both the growth phase and the flow conditions, with decline in cell division prevailing at low shear rates and cell death, possibly caused by an apoptosis-like program, which dominates at high shear rates (Juhl et al., 2000; Juhl and Latz, 2002). At energy dissipation rates comparable to those of ocean turbulence, the shear is probably more effective on populations in their later stages of growth. Tolerance to shear stress varies from one species to another (Estrada and Berdalet, 1998), and some harmful species (*Gymnodinium catenatum* and *Alexandrium fundyense)* even increase their growth rate with relatively high turbulence treatments (Sullivan and Swift, 2003), which could explain how some dinoflagellate bloom can develop in early-spring high-mixing conditions or in frontal or upwelling zones (Smayda, 2000; 2002).

Turbulence patterns in the sea influence several aspects of phytoplankton dynamics, not only through their role in the distribution of temperature and other scalars, but also via influences on the nutrient environment, and the ability of grazers to capture cells via encounter rates and perceptive abilities (Rothschild and Osborn, 1988; and see Yamazaki et al., 2002, for a review). In addition to local hydrodynamic forces, contact rates between planktonic organisms depend on several different factors which include their relative densities and their velocities due to both their swimming abilities and behavioural mechanisms which allow them to increase or decrease these rates. The problem was modelled by Gerritsen and Strickler (1977) in a predator/prey context; they assumed their predators to be at the centres of spheres each with encounter radius $R$, swimming in random directions at constant velocity $v$ through a diffuse cloud of evenly distributed prey particles ($N$ per unit volume), moving in random directions with velocity $u$; each predator thus *sweeps* a certain volume of water per unit time and can interact with the prey particles within it.

Analysis showed that the model is very sensitive to the value of $R$, which was assumed to vary with the strength of the mechanical disturbance produced by the prey, the distance from the sense organs of the predator, the sensitivity of the receptors, and the noise level created by turbulence and the predator's movements. The predator encounter rate, $Z$, is the product of the volume swept ($\pi R^2 N$) and the velocity component $A$, which is equal to $(u^2 + 3v^2)/3v$ when $v \geq u$ and $(v^2 + 3u^2)/3u$ when $u \geq v$ (equations 6 and 7 in Gerritsen and Strickler, 1977). In this scheme, $Z$ is an increasing function of $R$, $u$, $v$, and $N$, so that more food can be obtained by the predator if it increases its perceptual volume (larger $R$) or its swimming velocity ($v$); the prey in turn can reduce the probability of being encountered and hence captured by reducing its velocity ($u$) or by reducing the predator's $R$, for example by minimizing the clues which allow it to be perceived.

It has been postulated several times that the direct effect of local turbulence on contact rates and trophic links in the planktonic environment is very important. This was made explicit in a revised version of Gerritsen and Strickler's model by Rothschild and Osborn (1988). They replaced $u$ and $v$ in the earlier formulation by terms which incorporate the root-mean-square turbulent velocity, $w$, at the appropriate length scale; $u$ by $\sqrt{(u^2 + w^2)}$ and $v$ by $\sqrt{(v^2 + w^2)}$. Any increase in the turbulent kinetic energy dissipation rate, $\varepsilon$, then *drives the turbulence effect to smaller scales and increases its effect on particle contact rates.* In this revised model, the impact of turbulence on contact rates is greatest for the lowest values of $u$ and $v$. Two important points are i) that experimental studies of feeding rates in which turbulence is weak or absent will underestimate the real rates possible in the natural environment with wind or tidally induced turbulence present, and ii) that in such circumstances, the *additional* food made available by turbulence can be obtained at no extra cost to the predator, so that the energetic budgets made by optimal foraging theory will be affected.

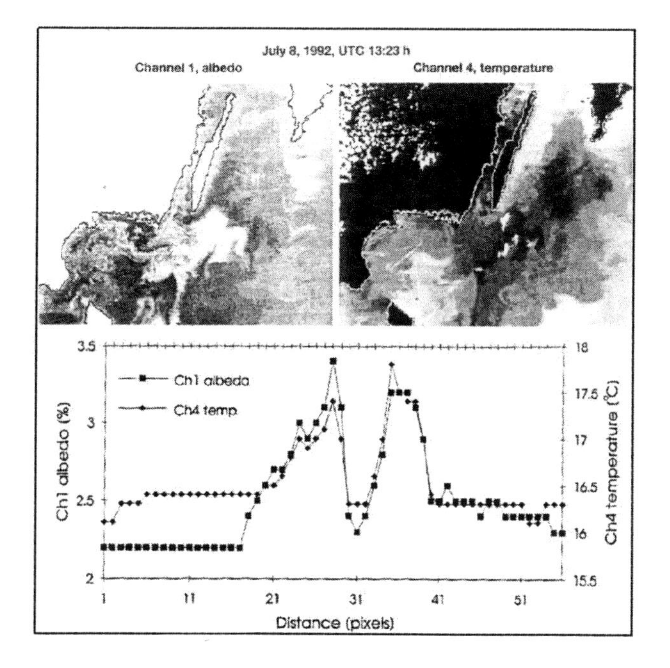


Figure 22.10 A cyanobacterial bloom in the Baltic Sea detected by AVHRR (from Kahru *et al.* 1993). The patch of cyanobacteria detected as albedo by channel 1 corresponds to an increase in temperature detected by channel 4. The diagram below, obtained with the data stretched along a line, shows the high correlation between the two signals. The cyanobacterial bloom causes an increase in seawater temperature which is dissipated at night by cooling.

The analysis was taken a step further by MacKenzie et al. (1994); here the predator must pursue the selected prey, focus on it, and adopt the correct posture to make an attack; this sequence requires some minimum time, *t*, which will depend on how far the prey is moved by turbulence during it. The model shows, as expected, that the encounter rate increases as turbulence increases, but that the probability of capture decreases in a sigmoid fashion along the same axis; the product of these two curves, the capture rate relative to non-turbulent conditions, is a dome-shaped curve with a peak at intermediate values of the turbulent velocity ($\omega$) at the scale appropriate to the pursuit process. At high values of $\omega$ the model predicts that the capture rate will be lower than in the absence of turbulence. Thus the increased encounter rates generated by turbulence are offset by the reduction in capture efficiency. As in the Gerritsen and Strickler's model, the predator gains advantage from turbulence by increasing the value of *R*, but even more by increasing *t*, its speed of pursuit. While the focus of these models has mainly concerned food capture rates, the obverse is algal population dynamics, so that turbulence is just as central to the latter as it is to zooplankton feeding.

Enhancement of encounter rates can also be effective in favouring mating, which may be mediated through the release of specific substances acting as chemical signals. Toxins themselves have been suspected to act as pheromones favouring encounter rates (Wyatt and Jenkinson, 1997). Gamete encounter in phytoplankton is one of the processes where the relative importance of physical versus biological control should be investigated.

Another approach to the problem of plankton contact rates was taken by Jenkinson and Wyatt (1992), who borrowed the idea of Deborah numbers from the field of rheology. They looked at the problem of encounter rates in steady simple shear flow, initially at scales smaller than the Kolmogorov length, and showed that there is a dome-shaped relationship between shear and encounter rates. Jenkinson (1995) reviews the two approaches.

At the scale of whole populations, physical discontinuities along the vertical axis of the water column, associated with large-scale hydrographic processes (e.g. seasonal river outflow), can entrain actively growing populations of harmful species within thin (< 5 m) subsurface or near-bottom layers. Thin phytoplankton layers are common but have not often been sampled effectively, can persist over days and expand over kilometres (Dekshenieks et al., 2001). Discontinuities in vertical distribution are known for several phytoplankton species, including among others the harmful diatoms of the genus *Pseudo-nitzschia* in a fjord of the San Juan Island, Washington, USA (Rines et al., 2002), the ichthyotoxic dinoflagellate *Karenia mikimotoi* along the Eastern North Atlantic coasts (Richardson and Kullenberg, 1987; Gentien, 1998), and *Dinophysis* in the Baltic Sea (Carpenter et al., 1995). From the standpoint of harmfulness, thin layers could be more easily exploited by filter-feeders and render them toxic much more quickly, or hit the target pen-fish more or less effectively. Large aggregations of harmful microalgae have ecological implications, since they could either attract or be avoided by potential grazers. In the case of *Akashiwo sanguinea* (syn. *Gymnodinium sanguineum* and *G. splendens*), a clear avoidance reaction was documented for several different zooplankton species off Californian coasts (Fiedler, 1982). Reduced grazing could therefore enhance local phytoplankton concentrations, contributing to the persistence of thin layers. As for other mechanisms of formation, a passive entrainment of a surface microlayer downwards has been demonstrated for thin layers of *Pseudo-nitzschia* (Rines et al., 2002), while autoaggregative behaviour possibly mediated by chemical substances may play a role in the vertical distribution of *Karenia mikimotoi* (Gentien, 1998).

*Karenia mikimotoi* is among the dinoflagellates that can excrete large amounts of exopolymers (Jenkinson, 1993), which could enhance aggregation by modifying physical and rheological properties of seawater. Indeed, physical-biological interactions are not only to be considered as the passive, one-way mechanisms for phytoplankton to endure physical variability. Mucous secretion, which is widespread in several bloom-former dinoflagellates, has been suggested as an engineering strategy through which the organisms can actively modify seawater viscosity (Jenkinson and Wyatt, 1995) and hence the turbulence field (Smayda, 2002). Another example of physical change of the environment induced by phytoplankton is provided by the increase in sea surface temperature during massive blooms of non harmful (Sathyendranath et al., 1991) as well as harmful species (Kahru et al., 1993) (Figure 22.10).

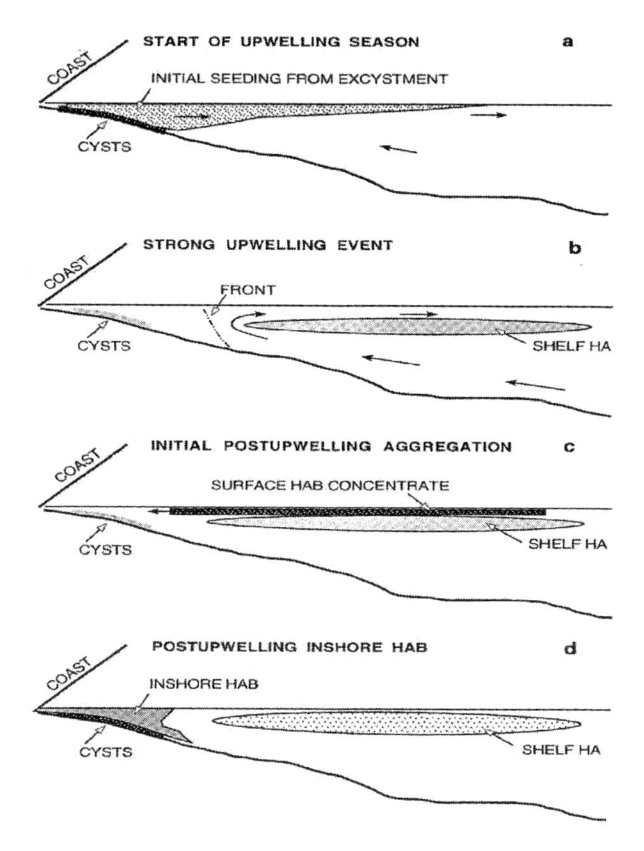


Figure 22.11 Conceptual model of the interplay of the life cycle of a cyst-forming harmful microalga with hydrodynamics in an upwelling area (from Donaghay and Osborn, 1997). (a) At the start of the upwelling season, cyst germination seeds the bloom; (b) following a strong upwelling event, the bloom is displaced offshore and entrained at subsurface depth; (c) after the relaxation of the upwelling, the bloom concentrates in surface waters; (d) the bloom is moved inshore by across-shelf currents. An analogous conceptual model, which couples the physical environment with the migratory behaviour and physiological adaptation of dinoflagellates, has been developed for South African upwelling areas (Pitcher et al. 1998).

Physical conditions in upwelling areas are spread over a range of spatial and temporal scales, so that an array of different processes implying physical-biological interactions illustrated in this section could occur in these areas. Thirty-three flagellate species have been reported to cause toxic blooms or massive accumulations in upwelling areas, in disagreement with the common view that flagellates should only thrive in calm, stratified conditions (Smayda, 2000). A rather common feature in upwelling conditions is that the most notable harmful effects coincide with the relaxation phase of upwelling, whereby populations developing offshore of the upwelling front during the active phase are advected to coastal sites by across-shelf and alongshore currents (Pitcher et al., 1998). Donaghay and Osborn (1997) provide a conceptual model of bloom development for species that produce resting stages in upwelling areas (Figure 22.11). Cyst germination would seed the bloom at the start of the upwelling season. Following a strong upwelling event, the bloom would be displaced offshore and entrained at subsurface depth. The relaxation of the upwelling would allow the bloom to concentrate in surface waters and hence across-shelf currents would transport it inshore. The increased swimming velocities attained by *Gymnodinium catenatum* and other chain-forming dinoflagellates would provide a selective advantage for these species to cope with the high vertical velocities in upwelling areas (Fraga et al., 1988; Smayda, 2002).

## 4. Trends and Fluctuations

### *4.1. Are Harmful Algal Blooms Increasing?*

At the global scale, it is evident that more harmful phenomena are recorded these days than in the past (Smayda, 1990; Hallegraeff, 1993). This increase in records is shown in distinct ways, e.g. in the range expansions of some harmful algae (see above section 2), or in the increase of coastal areas affected by HABs (Figure 22.5), or in a higher bloom frequency or magnitude, or still in increased harmfulness due to higher toxicity levels or more severe anoxic conditions. For each of these cases, a wide range of explanations can be hypothesised, which generally fall in two distinct categories: those implying the occurrence of environmental changes driving an effective increase, and those ascribing the observed trend to changes in the perception of the problems, such as the increased awareness of HABs or the more intensive and extended use of coastal resources.

As discussed in the previous section, natural causes and transport can both be responsible for expanding the geographic range of harmful and non-harmful species. *Alexandrium* provides several examples of changes in range or abundance. As judged by the very rare occurrence of PSP cases in the British medical literature, *Alexandrium* seems to have been relatively uncommon prior to 1968 in British coastal waters, or at least in those where mussel consumption is traditional; there was a major PSP outbreak in 1968 in northeast England, and mussel monitoring which began there that year has revealed that the risk of poisoning has been high most years since (Wyatt and Saborido-Rey, 1993). A similar shift is recognized in the western Gulf of Maine, where *Alexandrium* was only rarely detected prior to 1970 (Mulligan, 1975), but has presented a major risk since then. There has also been a large northward extension of toxic *Alexandrium* in Chile between the 1970s and 1990s (Guzmán et al., 2002). In recent years, field studies and modelling in the Gulf of Maine have provided much more detailed insights into *Alexandrium* ecology in this region than hitherto (Anderson, 1997a; Anderson, 1998; McGillicuddy et al., 2003).

Very few time series of adequate length are available to confirm trends of HABs (Wyatt, 1995), and in some cases the trends themselves tend to turn into fluctuations as the time series lengthen. In a time series from 1942 to 1984, PSP toxicity showed rather synchronous patterns along British Columbia coasts (Western Canada), despite differences at individual locations. A suggestion of an overall periodicity of 7–8 years indicated the influence of some large scale control mechanism, in which ENSO events may be implicated (Gaines and Taylor, 1985). A great interannual variability of PSP outbreaks caused by *Alexandrium tamarense* is also found in other areas, e.g. along the north east coast of UK (Wyatt and Saborido-Rey, 1993; Joint et al., 1997). Long-term fluctuations of *Phaeocystis globosa* are evident in a 30-year time series in the Marsdiep (Dutch Wadden Sea), where blooms of the species are reported since the end of the 19th century (Cadée and Hegeman, 2002). Trends of this kind are much more difficult to recognize in other harmful species because neither the taxonomy nor the toxins have been as well known for sufficient periods of time as they have with *Alexandrium. Karenia mikimotoi,* formerly identified as *Gyrodinium aureolum,* was first found forming large brownish patches in Norwegian waters in 1966 (Braarud and Heimdal, 1970), and soon thereafter in the Western Approaches along the Ouessant front (Brittany, France); while not impossible, it is unlikely that this organism could have escaped detection earlier unless it was extremely rare.

There are some preliminary indications that the frequency of HABs may respond to basin scale climatic fluctuations such as El Niño-Southern Oscilllations (ENSO) or the North Atlantic Oscillation (NAO). A link between ENSO and blooms of *Pyrodinium bahamense* in the west central Pacific was first suggested by Maclean (1989). In the Atlantic, many studies have demonstrated a relation between the NAO and plankton dynamics; so far there has only been one specific proposal linking the abundance of a harmful algal species to this index (Belgrano et al., 1999) but there seems little doubt that more attention will increase this number, and we regard this as a fruitful avenue for future research. We can expect further information on climate/HAB links to emerge as science and monitoring continue to progress. But the proximate causes of these connections are still not understood in detail, though they are beginning to appear within our grasp.

A very useful tool to track long term fluctuations in plankton abundance is provided by information stored in the sediments in the form or either chemical signatures or siliceous or calcareous cell walls and fossilizable resting stages of microalgae. Fossil pigments have revealed that cyanobacterial blooms are a peculiar feature of the Baltic Sea, being as old as its brackish water phase, i.e. ca 7 000 years (Bianchi et al., 2000), although sedimentation rates intensified in the '60s (Poutanen and Nikkilä, 2001).

The sedimentary records of dinoflagellate resting stages can be considered on different time scales: from the past few years to hundreds of years, where a comparison with measured environmental parameters is possible, to assess e.g. the human impact; from thousands to a few millions of years, to detect the effects of climate change; back to hundreds of million of years, to trace evolutionary changes (Wall, 1965; Reid and Harland, 1977; Dale, 2001). A well studied example is that of the reticulated cysts found in sediment cores from the Kattegat (Dale et al.,

1993; Thorsen and Dale, 1998), originally ascribed to the toxic *Gymnodinium catenatum,* but more likely belonging to the non-toxic species, *G. nolleri* (Ellegaard and Oshima, 1998). No matter what the species was, the principle is not in doubt that very large changes in the distribution and abundance of this cyst type have occurred in these waters during the Holocene which are reflected in cyst sedimentation rates, and this must be true for phytoplankton species currently regarded as harmful as well as others. In the Oslofjord, the increase in total cyst number and in the relative proportion of *Lingulodinium polyedrum* in cored sediments are interpreted as signals of the cultural eutrophication occurring in the area from the mid-1800's to the present (Dale et al., 1999).

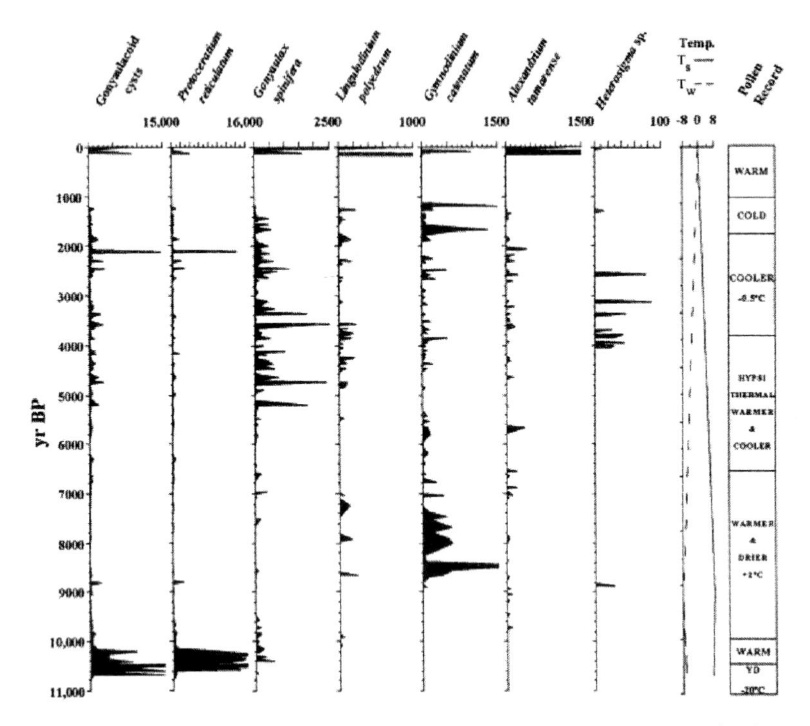


Figure 22.12 Dinoflagellate and raphidophyte (*Heterosigma* sp.) cyst fluxes (cysts $cm^{-2}$ $yr^{-1}$) over the last 11 000 years at Saanich Inlet (eastern North Pacific Ocean). In this basin, high sedimentations rates coupled with anoxic bottom waters allow the deposition of varved sediments. Cyclical patterns of cyst flux abundances are evident for the total of gonyaulacoid cysts and for single species, yet the large modern deposits differ from earlier blooms in the simultaneous occurrence of several HAB species (from Mudie et al., 2002).

A cyclical pattern of dinoflagellate blooms, recorded as the abundance of *Alexandrium* cysts, was revealed in historical series of varved sediments from Saanich Inlet (Western Canada) (Mudie et al., 2002). In this area, resting cysts of harmful species made their first appearance around 9500 yr BP, and were abundant at intervals during the periods from 7000 to 5000 yr BP and 4000 to 1000 yr BP (Figure 22.12). Mudie et al. (2002) noted that, on decadal time-scale, species cycles start very abruptly – in ten years or less – and that most persist at intermittent intervals for about 100–1000 yr before returning to background population sizes. For eastern Canada, Mudie et al. (2002) examined cores from two basins (La Have and Emerald) on the Scotian shelf. The La Have core reaches back to 10 500 yr BP, and its oldest section, before 10 000 yr BP contains major peaks of *A. excavatum*-type cysts and minor amounts of *A. tamarense*-type; these types were much less common in the interval 10 000 to 7000 yr BP, and there were even fewer between 7000 and 2500 yr BP. The Emerald basin core reaches back to 14 000 yr BP and shows a major *A. tamarense*-type peak from 13 000 to 11 000 yr BP. The abrupt nature of these changes to which Mudie et al. (2002) draw attention is important since similarly rapid response times are now emerging from palaeoclimatic studies of ice cores. The suddenness of changes in the abundance of *Alexandrium* in the twentieth century already mentioned have their counterparts in the more distant past before human activity had significantly impacted coastal seas.

It is generally accepted that, independently from an actual increase in HABs, the perception of these events has notably increased in the last decades (Hallegraeff, 1993). The need to protect human health and aquaculture operations has stimulated monitoring operations all around the world's seas. A more widespread use of specific identification techniques (electron microscopy, molecular techniques) has been made, and the knowledge about the taxonomy and morphology of harmful species has notably increased. An increase in PSP episodes reported from tropical regions might be ascribed to cultural changes amongst seafood consumers rather than to ecological changes in the causative species; links between red tides and poisoning were formerly well recognized e.g. in Papua New Guinea, and shellfish consumption was taboo in these circumstances (Riroriro and Sims, 1989); such taboos are usually lost as people migrate to urban areas, and thus fail to provide protection. Also, green mussels (*Perna viridis*) are not a traditional food in the tropics; in the Philippines, mass cultivation did not begin until the 1970s, so that poisoning due to this vector would have been rare prior to then.

The apparent expansion of coastal areas affected by HABs has provoked comparison with epidemics (Smayda, 1990). Mechanisms of algae and pathogen diffusion are distinct in that planktonic algae, in contrast to most pathogens, do not need a host organism to multiply and to be transported. However, assuming that for these organisms the transporting and the multiplying factors are decoupled, and that nutrient availability at a coastal site acts as the multiplier host, epidemiological models can be applied to account for the diffusion of microalgae (Wyatt, 1989). The epidemiological analogy holds true in that the affected organism (man) has notably increased in abundance and activity, which has enhanced greatly the encounter probability. Indeed, the steadily increasing use of coastal ocean resources detects and amplifies any signals of detrimental effect, and it is not therefore necessary to invoke an increase of the pathogen (microalgae) to account for an increasing trend of HAB impacts, which can be easily extrapolated to the future.

In conclusion, while positive trends of HAB records are known at locations around the world, the occurrence and intensity of these events may in many cases reflect natural fluctuations of the causative organisms, and not necessarily more sinister trends caused by human impacts. The intensive monitoring programs that are now conducted in several coastal areas around the world should in the long run (if the political will to continue them is sustained) provide a more objective answer to the hitherto unresolved question of whether the HAB problem is on the increase. At the same time, the parallel analysis of time-series of plankton and environmental data and a more extended use of sedimentary records may provide new and interesting insights into the mechanisms driving trends and fluctuations in the marine pelagic.

### *4.2. HABs and Eutrophication*

Positive trends in the frequency, magnitude or extension of HABs have been attributed to the eutrophication of coastal systems (Smayda, 1990). As discussed in the previous section, whether a global trend in HABs has taken place is still an open question, although it is clear that nutrient inputs have notably increased in recent decades in different regions due to human activities (Vitousek et al., 1997). At a local scale, there are several reports of positive trends in HABs that parallel trends in eutrophication (see Anderson et al., 2002 for a review). Among the most cited examples are the cases of Tolo Harbour in Hong Kong (Lam and Ho, 1989), the China Sea (Zhang, 1994), the Seto inland Sea (Japan) (Okaichi, 1989) and the Baltic Sea (Poutanen and Nikkilä, 2001). However, even in these cases, a relationship between HABs and increased nutrient inputs of anthropogenic origin has never been convincingly demonstrated. New evidence and alternative interpretations provided in recent years do not unambiguously support the conclusions of earlier investigations. In Tolo Harbour, chlorophyll *a* did not increase over the 1982–1992 period despite continuously increasing nutrient concentrations, and there was no evidence of a replacement of diatoms by dinoflagellates (Yung et al., 1997). Major *red tides,* defined as blooms causing sea water discolouration, were recorded in areas with low nutrient levels, suggesting that active and passive aggregation are responsible for these blooms. In turn, accumulation was mainly related to hydrographic processes and the influence of monsoon (Yin, 2003). In the Gulf of Mexico, accumulation rates of frustules of the potentially toxic diatom *Pseudo-nitzschia* spp. have increased in the sediments over the same period when nutrient inputs have increased (Parsons and Dortch, 2002), which may indicate that the balance between cell division rate and grazing rate has altered; we do not know whether either or both.

There is no generally accepted reason to suppose that increased nutrient inputs should favour harmful algae more than other species. Not even the cases of biomass accumulation, which obviously require a certain amount of nutrients to develop, can always be related to increased nutrient inputs of anthropogenic origin. This conclusion might be surprising, however it is obvious that coastal areas are complex systems, and nutrient additions clearly have non-linear and unpredictable consequences (Cloern, 2001). Comparable nutrient inputs may have totally distinct impacts. In Chesapeake Bay, the decomposition of high biomass spring blooms may lead to hypoxic or anoxic conditions (e.g., Malone et al., 1988; Glibert et al., 1995). In San Francisco Bay, DIN concentrations comparable to those of Chesapeake Bay and DIP concentrations ten times higher correspond to chlorophyll *a* values and primary production 20 times lower than in Chesapeake Bay (Cloern, 2001). Moderate increases in nutrient inputs may not lead to a noticeable change in stock because the grazers respond by eating it. But this response is overwhelmed by very high nutrient loading (hypertrophication), whereby phytoplankton biomass accumulates, and can lead to anoxic events and mortalities.

The Devonian decimation of the tropical marine benthos (and simultaneous widespread bottom water anoxia and increased carbon burial rates) has been attributed to intensified nutrient and sediment fluxes following a) the evolution of deep roots, which accelerated soil formation, and b) the evolution of seeds, which allowed colonization of dry upland environments (Algeo and Scheckler, 1998). This hypothesis implies that massive algal blooms caused the aerobic/anaerobic boundary to move intermittently upwards from the sediments into the water column, and that this process was driven by nutrient supply rates. It is frequently argued that anthropogenic eutrophication of coastal waters carries the same dangers, and the northern Adriatic, northwestern Black Sea, and other regions which receive large inputs of nutrients are cited as examples. But the data available do not allow us unequivocally to impute the observed biological changes which have been recorded in these regions to increased nutrient inputs, and sometimes (as in the case of silica inputs to the Black Sea) indicate that elementary enrichment hypotheses (of the form *more nutrients must lead directly to more phytoplankton stock*) should be more rigorously analysed (Wyatt, 1998; and see below).

Eutrophication due to farming, waste disposal, and other industries is also accompanied by changes in runoff patterns. Burchard (1998) suggests that increased drainage and runoff, as opposed to increased precipitation, may have accompanied the disturbance of savanna brought about as wild African ruminants were replaced by domestic stock, and that this may have been the primary cause of high lake levels in the last 130 ka. Increased runoff and the consequent increased stabilization of coastal waters provides a potential mechanism to increase production which is additional to that which might be ascribed to increased nutrients.

Simple models (see section 3) can be used to explore the potential impact of increased nutrient supplies on phytoplankton abundance ($P$). In such models, $dP/dt$ is usually controlled by the logistic expression in which the so-called *carrying capacity* ($K$) is equated with a limiting nutrient, and a functional response which represents the losses due to grazing by zooplankton ($Z$). Steady state solutions to these sets of equations can be found in which $dP/dt$ is regulated by grazing, and in which threshold changes in the forcing by the algal growth rate, $r$, caused by an increase in $K$, can allow $P$ to escape from grazer control and reach a new level of biomass (a bloom) regulated by $K$. If grazing *catches up,* algal biomass is returned to its former equilibrium (Truscott, 1995), if not, decay and anoxia may result.

An early stimulus to the potential effect of nutrient enrichment on predator-prey models was provided by Rosenzweig (1971). He explored a variety of model formulations, and found that in all cases, high $K$ led to limit cycles of high amplitude and period, and thus to a strong increase in the probability of extinction. He called this the *paradox of enrichment.* Empirical studies have not borne out these predictions; for example in freshwater lakes, it has been shown that small amplitude cycles do not increase with increasing nutrient levels (McCauley and Murdoch, 1990), and that there is only a small increase in equilibrium phytoplankton abundance, about 1.8, for a tenfold increase in phosphorus (Watson et al., 1992). In the latter study, for which data from more than 100 lakes were compiled, there was however a very significant change in the phytoplankton composition, with a sixfold increase in the stock of algae larger than 50 μm diameter. In the sea, primary production is generally proportional to stock (e.g., Vernet, 1991), and algal stock levels are more convincingly related to grazing by copepods and microzooplankton than to nutrient levels (Eilertsen et al., 1989; Kristiansen et al., 1994; Lewitus et al., 1998).

Blooms can also be generated in such models without altering $K$, by any mechanism (e.g., mixing, upwelling) which decouples $P$ and $Z$. Rovinski et al. (1997) provide a very interesting example in which decoupling is generated by diurnal vertical migration (DVM) of the zooplankton in shear flow. This leads to *travelling waves* of $P$ and $Z$ biomass along the axis of the shear field, and to enhancement of their productivity and mean stock levels relative to the no shear situation. This enhancement does not depend on a change in nutrient status, and the model response is reduced or even reversed when nutrient limitation occurs. This study illustrates the potential importance of introducing space into one dimensional models. In short, *the basic idea that primary productivity is itself independent of trophic structure demands verification, because there is ample theoretical ground for doubt* (Lundberg and Fryxell, 1995).

It is well known that in some ecosystems consumers enhance the supply of their resources (e.g., Hylleberg, 1975; Goldman et al., 1979; Williams and Carpenter, 1988). This positive feedback between trophic levels is now usually ascribed to nutrient regeneration via the microbial loop. Thus the older view that a species changes the nutrient composition of the medium and thus leads (via Michaelis-Menten kinetics) to its own replacement (*succession*) by species with different $r_{max}$ values and half saturation coefficients needs to be revised.

The travelling waves which emerge from Rovinski et al.'s model are an example of an *emergent structure,* and are equivalent to the expanding rings generated by the model of Dubois (1975) which combined predator-prey dynamics with diffusion. As both $P$ and $Z$ *benefit from this mechanism of patch formation, the condition for its occurrence, namely the periodic, differential migration, may be considered a strategy that benefits both the predator and prey species* (Rovinski et al., 1997). This suggests that selection for DVM might operate at the system level rather than or in addition to operating at the level of individuals. It also raises the question as to why some autotrophic organisms have evolved the capacity of vertical migration, and we might ask for example, whether this allows them to place themselves in the troughs between the waves of higher phytoplankton biomass generated by this model, and if so, whether this favours their access to nutrients or light.

Emergent structures generated by simple models were first explored in the context of morphogenesis. It is now widely recognized that they add a new dimension to pattern generation in ecology as well as morphology, and that a *structuralist* view in which interactions between the micro- and macro-dynamics are in both directions can complement the *reductionist* view in which the dynamics flow from micro- to macro- but not in the reverse direction, since such a flow is sometimes considered to contravene the tenets of darwinian selection (e.g., Bascompte and Solé, 1995). A well known model which provides an escape from this dilemma, and which combines both positive and negative feedback mechanisms, is the Daisyworld model (Watson and Lovelock, 1983; Saunders, 1994) in which two populations adjust their relative and absolute abundance to a common limiting factor. The important features of this model in the present context are that both species are significantly more abundant at the same resource level than they would be if the other were absent, that neither lives at its optimal resource level, and that the latter is restrained within habitable bounds over a wide range of forcing. Within this range, changes in the relative abundance of the two species occur, but must be very severe before one or other species becomes extinct. In this way, free nutrient concentrations become emergent properties of the system as opposed to forcing agents.

In addition to these general considerations, a number of case-specific and organism-specific processes are known for HABs in which the nutrient issue does not appear to be an over-riding issue. In general, the most toxic species, like some *Alexandrium* and *Dinophysis*, exert their impact when present at low concentrations, hence do not depend on nutrient concentrations to reach the threshold at which they become harmful. In many other cases, such as the blooms of *Karenia brevis* in the Gulf of Mexico (Tester and Steidinger, 1997), *K. mikimotoi* along the western European coasts (Gentien, 1998; McMahon et al., 1998; O'Boyle et al., 2001), *Alexandrium* spp. in the Gulf of Maine (Anderson, 1998; Townsend et al., 2001; McGillycuddy et al., 2003) and in the St. Lawrence region (Blasco et al., 2003), blooms develop in offshore waters, which are not affected by eutrophication.

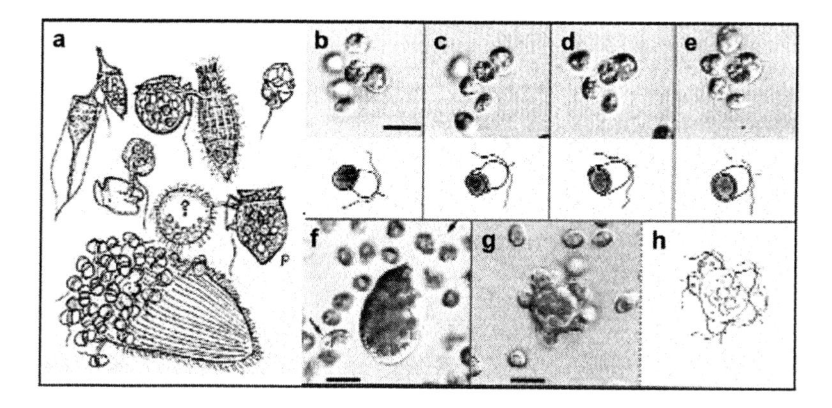


Figure 22.13 Examples of heterotrophic and mixotrophic microalgae feeding on other microalgae or ciliates. (a) Feeding mechanisms in dinoflagellates at times imply the use of specialised structures such as the peduncle (from Jacobson, 1999 and authors cited therein); in some cases the prey, i.e. the ciliate in the lower part of the figure, may be attacked by numerous dinoflagellates; (b)-(e) a sequence of events in prey ingestion by the ichthyotoxic species *Prymnesium patelliferum;* (f)-(h) *Prymnesium patelliferum* attacks the dinoflagellate *Oxyrrhis marina.* Membranes of single cells appear to fuse, forming a common food vacuole around the prey (from Tillmann, 1998).

A number of harmful species are heterotrophic. Examples are the dinoflagellates *Noctiluca,* causing massive blooms in many temperate, subtropical and tropical areas (Elbrächter and Qi, 1998), *Protoperidinium,* recently associated with the production of the toxin azaspiracid (James et al., 2003), the estuarine, ambush-predator *Pfiesteria* (Burkholder et al., 2001) and some *Dinophysis* species. Facultative or obligate mixotrophy (Figure 22.13) (Granéli and Carlsson, 1998; Tillmann,

1998; Willis et al., 1999; Koike et al., 2000), rather common in flagellates, are specialized nutritional strategies that can allow a species to thrive even in limiting nutrient conditions. A variety of different food uptake mechanisms have been studied (Jacobson, 1999). The heterotrophic dinoflagellate *Dinophysis rotundata* is normally ingested by the ciliate *Tiarina fusus,* which is about twice its own size, but can in turn eat the ciliate by emptying it with a feeding tube (Hansen, 1991). Some ichthyotoxic species like *Prymnesium patelliferum* also can ingest organisms which are much larger than themselves, and exhibit *social* behaviour, forming a kind of consortium around the prey (Tillmann, 1998).

The capability of making vertical migrations may allow motile phytoplankton species to cope with the vertical segregation of light and nutrients in stratified waters (Hasle, 1950; Macintyre et al., 1997; Cullen and MacIntyre, 1998). In this case, migration speed is increased by the development of a colonial habit, e.g. in the dinoflagellate *Gymnodinium catenatum* (Fraga et al., 1988). Indeed, processes controlling colony formation in phytoplankton have not been given much attention, and their adaptive benefits even less. It has been suggested for diatoms that coiled chains can influence diffusion rates and thus exert some control over nutrient availability. In the case of *mare sporco* occurring in the Adriatic Sea, which is a multispecies phenomenon involving phytoplankton, bacteria and other organisms, Azam et al. (1999) describe how events within the mucilaginous masses can sustain primary production in the absence of free nutrients. At the ecosystem level, complex trophic relationships, including microbial interactions and excretion by grazers, may indeed enhance nutrient availability and growth rates under apparently oligotrophic conditions.

To complicate the picture, nutrient enrichment is most often accompanied by other effects of human activities which lead to profound modifications of coastal waters and of the their adjacent watersheds. It has been postulated that overexploitation of predators and filter-feeders can result in excess biomass accumulation and in a series of symptoms of eutrophication such as enhanced blooms, hypoxia and anoxia (Jackson et al., 2001). The building of dams, canalisation or deviation of rivers, and other forms of freshwater exploitation can dramatically change the amount and chemical composition of freshwater conveyed to the shelf, and its seasonal distribution (Friedl and Wüest, 2002). Mechanisms of action of such changes are often complicated: the building of the Three Gorges Dam on the Yangtze River (China) may lead to reduced nutrient supplies to coastal waters since the resulting lower freshwater ouflow will reduce cross-shelf exchange and hence the nutrient supply originating from the onshore advection of the subsurface Kuroshio current (Chen, 2000). Docks and harbours modify the coastline over large areas directly and through changes in sediment accumulation and removal rates, and interfere with coastal currents while creating sheltered areas which could favour the occurrence of massive blooms, a mechanism proposed to explain the increase of HABs in Catalan harbours (Vila et al., 2001). While nutrients attract a lot of attention, the effects of a wide range of organic and inorganic compounds produced by agriculture and factories are often ignored. In all these cases, increases of harmful species, either toxic or producing anoxia, will only be the most visible aspects of profound changes. At the same time, more subtle effects such as the loss of phytoplankton diversity or shifts in the specific composition of microalgal communities could sensibly influence the fate of primary production and modify the shape of the trophic web in coastal areas, with potential consequences for fisheries and exploitation of other natural resources.

## 5. Harmful Effects and their Amplification in the Ecosystem.

Blooms of marine microalgae can lead to a wide spectrum of negative effects. These range from toxicity to humans and marine animals to anoxia, reduction of water clarity and poorly understood negative interactions with both marine flora and fauna. The diversity of impacts reflects the biodiversity of causative organisms (Table 1) and mechanisms of impact (Zingone and Enevoldsen, 2000) and the variability of the coastal environment.

In this section we focus on the role of physical, chemical and biological interactions which modulate the impact of harmful algae. Indeed, not only must harmful species reach critical concentrations, but also the target resources must be affected for a bloom to cause harm. In this respect, the interactions between the biology and external environmental factors not only affect the presence and abundance of harmful species, but also the operational boundary between harmless and harmful. Environmental conditions can directly modulate the toxicity of a species, while currents and meteorological events can accumulate/disperse a bloom, or simply keep it far from vectors or target organisms, or disrupt anoxic conditions before the local fauna is dead. Therefore, in addition to the occurrence, duration and magnitude of harmful blooms, our prognoses should consider whether *potentially* harmful blooms will result in harmful events. As detailed in the following section, the detrimental effects of specific algal blooms are in many cases under the control of small and large-scale hydrographic processes, which highlights the relevance of integrated monitoring systems in coastal waters (GOOS) to the predictability of HABs (see Malone et al., this volume). On the other hand, such biological interactions as grazing, bacterial activity or viral lysis can also affect harmful events, by terminating blooms before they impinge on the target resource.

### *5.1. Toxin Production and Propagation*

The most notorious harmful microalgae are those which produce substances that menace human health. The biochemical and toxicological characteristics of algal toxins are very heterogeneous, as are their target organisms and impact mechanisms. More than 100 different toxins are produced by marine microalgae (Daranas et al., 2001), which based on their mode of action can be grouped into nine categories: ASP, AZP, DSP, PSP, NSP and ciguatera toxins, spirolides, hepatotoxins, and carcinogenic toxins. With the exception of spirolides (Cembella et al., 2000), whose potential effects on human health are still unknown, all these toxins are associated with distinct human syndromes (Codd et al., 1999; VanDolah, 2000; Daranas et al. 2001; Lewis, 2001). Another group of substances, the ichthyotoxins, are only detrimental to marine invertebrate or fish (Landsberg, 2002). In addition to the effects on human health, ASP, NSP, and PSP toxins may also directly affect marine vertebrates, such as fish, whales, sea lions, and birds (Landsberg and Steidinger, 1998; Scholin et al., 2000; Cembella et al., 2002; Landsberg, 2002; Shumway et al., 2003) or invertebrates (Shumway, 1990; and see section on grazing).

Toxins can be concentrated in different organisms as they pass through the food web, so that their impact is amplified for sensitive consumers; this process is well illustrated by PSP toxins, concentrated by zooplankton and shellfish and passed on to vertebrates (whales, sea birds, man). A particular case of detrimental impact through the food web is the unpleasant taste of Atlantic cod feeding on zooplankton that had grazed on the DMS-producer *Phaeocystis globosa* (blackberry feed, Levasseur et al., 1994).

In addition to toxins, many biologically active compounds are normally produced by microalgae that can have a wide array of effects on the growth and physiological activities of another phytoplankton species (Legrand et al., 2003) or of marine organisms in general (Cembella, 2003); the aldehydes produced by some diatoms (see above) are a clear example, and their structure and possible effects are now described (Miralto et al., 1999). Substances other than PSP toxins are involved in the negative impact exerted by *Alexandrium* on caged salmon or on heterotrophic dinoflagellates (Cembella et al., 2002; Tillmann and John, 2002). A so far unidentified substance produced by the brown-tide pelagophyte *Aureococcus anophagefferens* induces reproductive failure and reduced feeding activity in invertebrates, (Bricelj and Lonsdale, 1997).

In some cases, mechanisms by which algae are harmful are so far elusive. The recently (1988) discovered dinoflagellates *Pfiesteria piscicida* and *P. shumwayae* are associated with mass mortalities of menhaden, *Brevoortia tyrannus,* in estuaries of the western Atlantic, leading to the hypothesis that an ichthyotoxic mechanism is involved. Affected fish often exhibit skin ulcers which have also been attributed to *Pfiesteria,* although the pathogenic oomycete, *Aphanomyces invadans* is a more likely agent (Kiryu et al., 2002). Skin irritation and a neurological syndrome involving memory loss in humans have also been attributed to exposure to water containing *Pfiesteria,* and to aerosols contaminated by it (Burkholder et al., 2001; Burkholder and Glasgow, 2002). No incontrovertible evidence for such toxins has so far emerged (Berry et al., 2002), the mechanism by which *Pfiesteria* may kill fish remains obscure, and there are other candidate killers in the same estuaries including some poorly characterized "*Pfiesteria*-like" organisms, other fungi and protozoans, and an undescribed species of *Chattonella* which produces the known fish-killing brevetoxin (Bourdelais et al., 2002). Noga et al. (1996) and Vogelbein et al. (2002) describe how swarming dinospores of *Pfiesteria* erode the epidermis of fish; presumably heavy infestations are necessary to cause death. Given the taxonomic diversity of potential fish pathogens in these estuaries, and the possible complexity of some of their life cycles – it is claimed that *Pfiesteria* itself may have as many as twenty four distinct life stages, though this is disputed (Burkholder and Glasgow, 2002; Litaker et al., 2002a; Litaker et al., 2002b) – it may be some time before the real significance of *Pfiesteria* is fully resolved.

Toxicity exhibits a wide range of phenotypic variability along with growth phases and with all physical and chemical factors directly or indirectly involved in cell growth and physiological status. Toxin content per cell results from the balance between toxin production rates (toxin-μ) and cell division rates (cell-μ), therefore, parameters known to affect division rates will also influence toxin content per cell. Toxin content of dinoflagellates can be influenced by temperature, light, salinity as well as by nitrogen and phosphorus limitation, and high ammonium concentrations, yet only for nutrient stress is the effect clearly independent of growth (Anderson et al., 1990; Flynn et al., 1994; Flynn et al., 1996a; Flynn et al., 1996b; Hamasaki et al., 2001). Domoic acid (DA) production in diatoms increases with decreasing growth rate and is enhanced by phosphate or silicate depletion (Pan et al., 1998). In addition, production of this toxin increases under trace metal stress. DA potentially has an important physiological role in binding such trace metals as iron and copper, thus increasing the availability of the former and reducing the toxicity of the latter (Rue and Bruland, 2001; Maldonado et al., 2002). All these results on toxin production variability have been obtained in laboratory experiments, and their ecological significance has yet to be fully elucidated. But such findings, together with genotypic variability of toxin production, may in the long run help us to understand more fully the variations in toxicity often observed for blooms of the same species. As a matter of fact, toxin concentration measured during blooms is generally much higher that expected based on toxin content per cell in laboratory studies (Trainer et al., 2002).

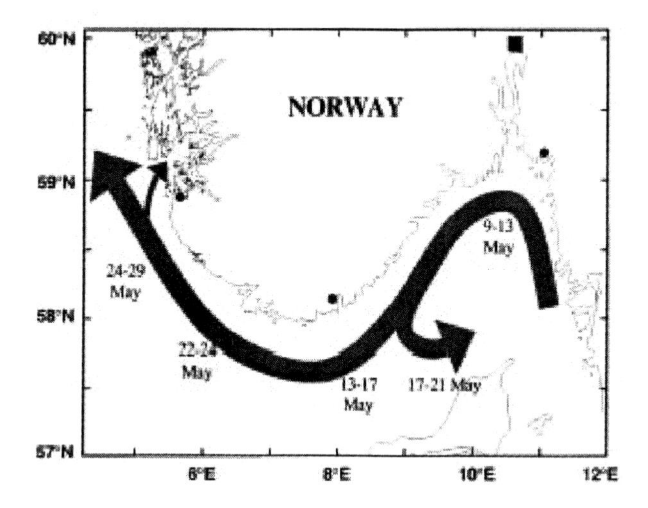


Figure 22.14 Progression of the *Chrysochromulina polylepis* bloom of 1978 along the Swedish and Norwegian coast (from Gjosaeter et al., 2000, modified after Dahl et al., 1989). The propagation of the bloom followed the northward flow of the Baltic current and of its continuation as the Norwegian coastal current. Calculations based on the current velocity indicate that the bloom originated in the Kattegat. The bloom impacted the entire littoral and sublittoral, affecting a large number of different species.

Detrimental effects of toxins are controlled by environmental parameters, especially hydrodynamic conditions. As explained in previous sections, many harmful blooms are transported by buoyant plumes and coastal currents (Figure 22.14), or are aggregated in subsurface microlayers, or are advected to coastal areas following the relaxation of upwelling. Along the western coast of Ireland, variability in wind forcing of the coastal currents is at the origin of the advection of harmful populations to aquaculture sites (O'Boyle et al., 2001). Numerous other cases are known where the harmful organisms are advected close to the target organism areas by buoyant plumes and alongshore currents (Franks and Anderson, 1992). In all these cases the harmfulness is directly related to small and large scale oceanographic processes and indirectly to meteorological conditions that force the local hydrodynamic regimes (e.g. Hallegraeff et al., 1995; Trainer et al., 2002). These HABs are thus more suitable to predictions based on models that can be fed with oceanographic and meteorological data.

In the case of benthic harmful microphytes, the physical and biological mechanisms involved in toxin distribution and propagation are very distinct from those operating in the plankton. Toxic benthic species mainly belong to two genera, *Prorocentrum,* which can produce DSP toxins, and *Gambierdiscus,* which is responsible for ciguatera. *Gambierdiscus* spp. colonise macrophytes in all tropical and subtropical areas. Small fish which feed primarily on macrophytes of coral reefs, and incidentally on their epiphytic microphytes, are preyed upon by larger fish which act as vectors of the toxins to human consumers. The role of environmental factors on the spatial and temporal distribution of *Gambierdiscus* spp. and other associated species is poorly known. It is suggested that the temporal dynamics and the spatial variability within one site are mainly dictated by water movements (Tindall and Morton, 1998). Toxicity of *G. toxicus* shows high intraspecific variability, which can explain why the severity of ciguatera events does not correlate with the abundance of the species (Chinain et al., 1999).

### 5.2. *Mucilage and Foams*

Exopolymeric substances are produced by a variety of marine microalgae, to which they presumably confer selective advantages in terms of functional size, buoyancy, protection from viral lysis, control of nutrient gradients, and others (Decho, 1990). In addition to their adaptive functions, these substances may accumulate locally producing a number of negative effects, such as fish and shellfish gill clogging and mortality (Takano, 1965; Potts and Edwards, 1987), net clogging and unpleasant water quality.

Many phytoplankton species produce mucous substances which directly affect marine fauna. These include both several diatom species and a number of dinoflagellates and other flagellates. The *Karenia mikimotoi* (syn. *Gyrodinium aureolum*) blooms off southwest Ireland can form enormous masses of mucus (Jenkinson and Connors, 1980).

The production and accumulation of mucilage are not always related to a single causative species. In the Adriatic Sea, mucilage events have been reported at least since 1770 under the name of *mare sporco* (dirty sea). Their recurrence, which is sporadic and unpredictable, has been ascribed to a number of biological mechanisms, including excess polysaccharide exudation by benthic and planktonic diatoms under nutrient limitation (e.g. Malej, 1995), grazing inhibition (Malej and Harris, 1993), viral lysis, and complex bacteria-organic matter interactions (Azam et al., 1999). In particular, bacteria may contribute to mucilage formation by sustaining high primary production rates through P regeneration and producing long-lived polysaccharides (Azam et al., 1999). In addition to these biological causes, peculiar circulation patterns and weather conditions may ultimately determine the irregular appearance of this mucilage (Degobbis et al., 1995; Schuster and Herndl, 1995). While the formation of marine snow and transparent exopolymer particles (TEP) is well known in all world seas (Passow and Alldredge, 1994; Passow, 2002), in no other place does it occurs so massively and frequently as in the Adriatic Sea. This typical localization of the phenomenon indicates that the geographic and hydrographic peculiarities of the area rather than its biological properties should be better investigated to be able to understand and predict the mechanisms underlying mucilage accumulation in Adriatic coastal waters.

Visible change of seawater quality is produced by foams generated from extracellular polymeric materials. The debris of *Phaeocystis* colonies is a common source of such foams, In contrast to Adriatic mucilage, in this case the causative organism is well known, and recurrent and predictable events occur in certain areas such as the North Sea coasts of the German Bight (Lancelot et al., 1998). Also in this case, it is a combination of biological properties of the species, ecological conditions associated with colony disruption at the end of blooms, and physical conditions which determine the extent, duration and impact of the phenomenon. Both Adriatic mucilage and *Phaeocystis* foams produce considerable impacts on tourism and fisheries, but negative effects on the rest of the ecosystem have not been reported. This points to the fact that a number of natural phenomena are considered negative only in relation to human use of the sea.

In some cases the production of mucilage and sticky substances offers a possibility to mitigation through the use of clays and powders which cause a bloom to sink (Kim, 1998; Yu et al., 2000; Sengco et al., 2001). However, only a few studies have taken into account effects on the rest of the ecosystem (Lewis et al., 2003; Shumway et al., 2003; Sengco and Anderson, 2004), including any possible positive and negative feedback which can eventually render the mitigation useless or even more detrimental. As in the case of large-scale ocean fertilisation, these kinds of invasive human interventions can lead to final results that are at present unpredictable, due to the lack of a thorough knowledge of the ecosystem equilibria that might be disrupted (Chisholm et al., 2001).

### 5.3. *Other Physical Effects*

Damage to gills of fish and invertebrates can also occur as a direct effect of phytoplankton cell accumulation, without mucilage production, due to the effects of hard structures such as silica spines and setae. Mortality in pen salmon is sometimes due to gill mucus production and bleeding, with consequent blood hypoxia or secondary bacterial infections. As with toxins, these phenomena are highly species-specific, so far being ascribed only to the diatoms *Chaetoceros convolutus* and *C. concavicornis*. Despite the difference in the mechanism of action, an interesting parallel with toxicity exists in that very low microalgal concentrations cause significant damage. Cell numbers as low as a few cells per ml can induce massive mortality, whereas at lower concentrations the farmed fish become more susceptible to other kinds of diseases (Yang and Albright, 1992; Yang and Albright, 1994). Information on the physiology and ecology of these *Chaetoceros* species is scanty (Harrison et al., 1993), but their presence and attainment of critical concentrations are not related to the chemical environment found in thin layers (Rines et al., 2000). The extent of the damage caused is presumably related to the positions of these thin layers and the depths at which fish are reared.

### *5.4. Discolouration and Reduction of Clarity*

Tourism and recreational activities in coastal areas are at times affected by changes in the colour of the sea. In fact, discolouration by algae ranges from milky white (*Emiliania huxleyi*) to tomato-red (*Noctiluca scintillans*), from brown (*Aureococcus anophagefferens*) to golden-yellow (*Alexandrium* and *Scrippsiella*), greenish-brown or olive (diatoms) and emerald-green (prasinophytes). A very dark colour is assumed by sea water as a secondary effect of anoxia and production of hydrogen sulphide caused by blooms (see section 5.5.), e.g. in South African waters (Matthews and Pitcher, 1996).

In addition to the loss of aesthetic values, seawater discolourations may have a negative effect on the benthic vegetation by filtering out the light. A progressive change in the summer compensation depth in the northwestern Black Sea between the 1950s and 1980s, caused by a 40-fold increase in phytoplankton stocks, was probably responsible in large part for the disappearance there of *Phyllophora* meadows (Zaitsev, 1993). Brown tides formed by *Aureococcus anophagefferens* are typically persistent over period of weeks, and can therefore limit the growth of the underlying benthic vegetation (Bricelj and Lonsdale, 1997). In some cases the discolouration is caused by a toxic species, which can considerably enhance the negative impact.

Visible changes in seawater colour due to phytoplankton indicate high biomass levels, derived either from *in situ* growth or from physical accumulation. In the case of South African waters, the entrapment of a bloom dominated by non toxic species which lead to anoxia was ascribed to an abnormal wind pattern indicating passive accumulation (Matthews and Pitcher, 1996). Active aggregation in surface waters is also possible as a phototactic response, which at least in the case of *Karenia brevis* is allowed by a peculiar resistance to UV (Evens et al., 2001). In stratified waters, *Alexandrium tamarense* may actively aggregate in very thin and dense layers at the surface during the day (Macintyre et al., 1997).

### *5.5. Anoxia, Hypoxia and $H_2S$ Formation*

Hypoxic (oxygen concentration $< 2$ ml $l^{-1}$) and anoxic ($< 0.5$ ml $l^{-1}$) conditions may lead to more or less temporary damage to the local fauna and flora which is unable to escape from the affected area. Anoxic phenomena are often associated with the production of $H_2S$. The mortality of benthic fauna and flora is higher in this case, due to toxicity of the resulting sulphides. Partial or total negative imbalance in the oxygen budget occurs in all cases when consumption processes prevail over the biological production plus the external supply. The first term of the equation tends to increase when *in situ* produced biomass and exogenous organic matter are decomposed. This excess decomposition implies that a significant part of the organic matter produced or supplied to the system is not exported to the secondary compartment of the trophic web. The external supply term generally tends to decrease when water masses are trapped in deep layers in absence or reduction of physical exchange.

Anoxic conditions are associated with high primary production levels, and hence with high nutrient loading and eutrophication (see Rabalais, this volume). Yet some sites are more prone than others to hypoxic/anoxic conditions, and flushing rates are clearly important. This is the case of deep fjords or shallow waters with reduced water exchange. Recurrent anoxic conditions are reported in upwelling-dominated area, as a response to high production in surface areas and corresponding oxidative processes at depth (Margalef, 1975). Along the south-western African coasts, the quiescent phases of upwelling are generally accompanied by dinoflagellate blooms that are transported onshore along with the frontal system. These phenomena can be particularly intense with abnormal wind patterns (Matthews and Pitcher, 1996). (Some areas here are permanently anoxic, the so-called azoic zones). Huge mass mortalities of fish and invertebrates are frequently observed on this coast (Pitcher and Cockroft, 1998); an early event, dated to about 1828, records a mass mortality of two colonies of southern fur seals (Wyatt, 1980). A related phenomenon is also reported from Peru, where it is named *the Callao painter,* since $H_2S$ blackens the paint work on ships exposed to anoxic/sulphurous waters in this locality.

In other cases, anoxic conditions are aperiodic events. In the New York Bight, massive blooms of *Ceratium tripos* in 1976 lead to 1000s of $km^2$ of dead benthos (Swanson and Sindermann, 1979). The bloom occurred in a year of unusually high runoff from the river Hudson, yet the hypoxic conditions might well have occurred in any case, as a result of the natural physical forcing and the biological response, even without carbon loading from the coast (Falkowski et al., 1980). Thus here, as in other HAB-related events, several different processes are strongly linked in generating the observed phenomena.

## 6. Concluding Remarks

We have provided a far from exhaustive account of events primed by harmful algal blooms which indicates that these are not marginal biological phenomena. They involve a wide variety of organisms, physiological and ecological processes, and physical-biological interactions. The important role of environmental variables over a wide range of scales is clearly shown at each of the three fundamental levels of HAB development, i.e. the global distribution of harmful microalgae, the development of blooms and the manifestation of detrimental effects. The physiological, ecological, chemical and physical processes involved at each of the three levels of HAB development can vary greatly depending on the species and on the area involved, and on the mechanism of interaction with man's activities.

We conclude in general that no distinctions can be made between the processes which govern the ecology of harmful and non-harmful or beneficial species of microalgae. This conclusion is in a sense trivial, since the distinctive character of the harmful category is based on how particular species interact with man's activities. But if we pursue the matter a little further, we see that the last point is the key. *All* algal species are potentially harmful depending precisely on those interactions, and as we have seen earlier, the number of harmful species has increased greatly in recent decades. This is due partly to the increasing attention which these phenomena now receive, but it must be mainly due to the increasing intensity and diversity of interactions of man with the coastal environment in his insatiable pursuit of food and wealth.

For HABs as in general for phytoplankton, the respective roles of the physics versus the biology, endogenous versus exogenous control, genetic versus phenotypic responses are poorly understood. Do resting cysts take advantage of resuspension by bottom currents to germinate? Or, rather, does the seasonally changing light environment act as a source of information that triggers excystment? Or, is a purely endogenous clock that dictates the timing for a species to appear on the scene at the right time? Are turbulence or chemical communication or both at the base of mating encounters between gametes? Is aggregation in thin layers passive or active? These are examples of questions that can be formulated for the majority of processes which underlie the development of harmful and non harmful blooms.

Understanding the distribution and bloom dynamics of harmful species is identified as the first necessary step towards the prediction and mitigation of HABs. To this end, IOC and SCOR have launched an international scientific programme, The Global Ecology and Oceanography of Harmful Algal Blooms (GEOHAB, 2001). The *scientific goal* of GEOHAB is to improve prediction of HABs by determining the ecological and oceanographic mechanisms underlying their population dynamics, integrating biological, chemical and physical studies supported by enhanced observation and modelling systems.

The substantial lack of difference in the ecology of harmful and non harmful species has another important implication. Much scientific effort in the last fifty years has focused on the assessment of biogeochemical cycles and global carbon budgets where phytoplankton has been equated with bulk parameters like chlorophyll concentration, fluorescence, or remotely sensed ocean colour. Much less progress has been made on species-specific contributions to biogeochemical cycles, on phytoplankton species succession, and on biological and ecological mechanisms which control species abundance. Biogeochemical cycles seem to vary significantly in relation to the species-specific properties of the dominant organisms, therefore community shifts are likely to affect important marine ecosystem functions, such as carbon fluxes and food-web structure (Tallberg and Heiskanen, 1998; Boyd and Newton, 1999; Tortell et al., 2002; Rost et al., 2003), DMS production and climate regulation (Charlson et al., 1987; Malin, 1997; Malin and Kirst, 1997; Scarrat et al., 2002), nitrogen fixation (Carpenter and Romans, 1991; Karl et al., 1997) and fisheries (Legendre, 1990; Smetacek et al., 2002; Karl et al., 2001). A reasonably high risk exists that any attempt to formulate reliable budgets and to improve predictive capabilities on e.g. climate change effects in marine ecosystems, fisheries or artificial fertilisation of high nutrient-low chlorophyll (HNLC) areas may be ineffective, unless we develop an adequate knowledge of the organisms playing the game. In this respect, interdisciplinary research in the field of harmful algae and their bloom dynamics can be valuable for a wider range of objectives and allow HABs to unlock the complexity of phytoplankton ecology as the Rosetta Stone did for Egyptian hieroglyphics. Studying species-specific properties and their interactions with the environment will be invaluable in providing keys to the interpretation of e.g. species succession, community shifts, response to physical forcing and other processes relevant to biological oceanography. An interdisciplinary and across-boundary scientific approach is desirable, to include knowledge spanning from the watershed to the offshore boundary and integrating all the components of the system analysed at matching scales of variability. The evidence of multiscale physical-chemical-biological interactions in plankton dynamics requires that the scales of the relevant biological processes dictate the scales of physical processes to be investigated. This implies a considerable amount of creative energy in designing research programmes, in steering technological progress for analytical instrumentation, and in developing new models which incorporate the complexity of marine organisms.

## References


Algeo, J. T. and S. E. Scheckler, 1998. Terrestrial-marine teleconnections in the Devonian: links between the evolution of land plants, weathering processes, and marine anoxic events. *Phyl. Trans. R. Soc. Lond. B,* **353,** 113–130.

Allee, W. C., 1932. *Animal Aggregations: A Study in General Sociology.* University of Chicago Press, Chicago.

Anderson, D. M., 1997a. Bloom dynamics of toxic *Alexandrium* species in the northwestern U.S. *Limnol. Oceanogr.,* **42,** 1009–1022.

Anderson, D. M., 1997b. Turning back the harmful red tide. *Nature,* **388,** 513–514.

Anderson, D. M., 1998. Physiology and bloom dynamics of toxic *Alexandrium* species, with emphasis on life cycle transitions. In *Physiological Ecology of Harmful Algal Blooms,* D. M. Anderson, A. D. Cembella and G. M. Hallegraeff, eds. Springer-Verlag, Berlin, pp. 29–48.

Anderson, D. M., P. M. Glibert and J. M. Burkholder, 2002. Harmful algal blooms and eutrophication: nutrient sources, compositions, and consequences. *Estuaries,* **25,** 704–726.

Anderson, D. M. and B. A. Keafer, 1987. An endogenous annual clock in the toxic marine dinoflagellate *Gonyaulax tamarensis. Nature,* **325,** 616–617.

Anderson, D. M., D. M. Kulis, G. J. Doucette, J. C. Gallagher and E. Balech, 1994. Biogeography of toxic dinoflagellates in the genus *Alexandrium* from the northeastern United States and Canada. *Mar. Biol.,* **120,** 467–478.

Anderson, D. M., D. M. Kulis, J. J. Sullivan, S. Hall and C. Lee, 1990. Dynamics and physiology of saxitoxin production by the dinoflagellate *Alexandrium* spp. *Mar. Biol.,* **104,** 511–524.

Arzul, G., M. Seguel, L. Guzman and E. Erard-Le Denn, 1999. Comparison of allelopathic properties in three toxic *Alexandrium* species. *J. Exp. Mar. Biol. Ecol.,* **232,** 285–295.

Azam, F., S. Fonda-Umani and E. Funari, 1999. Significance of bacteria in the mucilage phenomenon in the northern Adriatic Sea. *Ann. Ist. Super. Sanità,* **35,** 411–419.

Bargu, S., B. Marinovic, S. Mansergh and M. W. Silver, 2003. Feeding responses of krill to the toxin-producing diatom *Pseudo-nitzschia. J. Exp. Mar. Biol. Ecol.,* **284,** 87–104.

Bascompte, J. and R. V. Solé, 1995. Rethinking complexity: modelling spatiotemporal dynamics in ecology. *Trends Ecol. Evol.,* **10,** 361–366.

Belgrano, A., O. Lindahl and B. Hernroth, 1999. North Atlantic Oscillation primary productivity and toxic phytoplankton in the Gullmar Fjord, Sweden (1985–1996). *Proc. R. Soc. Lond. B,* **266,** 425–430.

Berry, J. P., K. S. Reece, K. S. Rein, D. G. Baden, L. W. Haas, W. L. Ribeiro, J. D. Shields, R. V. Snyder, W. K. Vogelbein and R. E. Gawley, 2002. Are *Pfiesteria* species toxicogenic? Evidence against production of ichthyotoxins by *Pfiesteria shumwayae. Proc. Natl. Acad. Sci. USA,* **99,** 10970–10975.

Berryman, A. A., 2003. On principles, laws and theory in population ecology. *Oikos,* **103,** 695–701.

Bianchi, T. S., E. Engelhaupt, P. Westman, T. Andrén, C. Rolff and R. Elmgren, 2000. Cyanobacterial blooms in the Baltic Sea: natural or human-induced? *Limnol. Oceanogr.,* **45,** 716–726.

Blackburn, S. I., C. J. S. Bolch, K. A. Haskard and G. M. Hallegraeff, 2001. Reproductive compatibility among four global populations of the toxic dinoflagellate *Gymnodinium catenatum* (Dinophyceae). *Phycologia,* **40,** 78–87.

Blasco, D., M. Levasseur, E. Bonneau, R. Gelinas and T. T. Packard, 2003. Patterns of paralytic shellfish toxicity in the St. Lawrence region in relationship with the abundance and distribution of *Alexandrium tamarense. Sci. Mar.,* **67,** 261–278.

Bolch, C. J. S., S. I. Blackburn, G. M. Hallegraeff and R. E. Vaillancourt, 1999. Genetic variation among strains of the toxic dinoflagellate *Gymnodinium catenatum* (Dinophyceae). *J. Phycol.,* **35,** 356–367.

Boudouresque, C. F. and M. Verlaque, 2002. Biological pollution in the Mediterranean Sea: invasive versus introduced macrophytes. *Mar. Pollut. Bull.,* **44,** 32–38.

Bourdelais, A. J., C. R. Tomas, J. Naar, J. Kubanek and D. G. Baden, 2002. New fish-killing alga in coastal Delaware produces neurotoxins. *Env. Health Persp.,* **110,** 465–470.

Boyd, P. W. and P. P. Newton, 1999. Does planktonic community structure determine downward particulate organic carbon flux in different ocean provinces? *Deep-Sea Res. I,* **46,** 63–91.

Braarud, T. and B. R. Heimdal, 1970. Brown water on the Norwegian coast in autumn 1966. *Nytt Mag. Bot.,* **17,** 91–97.

Bratbak, G., A. Jacobsen, M. Heldal, K. Nagasaki and F. Thingstad, 1998. Virus production in *Phaeocystis pouchetii* and its relation to host cell growth and nutrition. *Aquat. Microb. Ecol.,* **16,** 1–9.

Bricelj, V. M. and D. J. Lonsdale, 1997. *Aureococcus anophagefferens:* Causes and ecological consequences of brown tides in U.S. mid-Atlantic coastal waters. *Limnol. Oceanogr.,* **42,** 1023–1038.

Bricelj, V. M., S. P. MacQuarrie and R. A. Schaffner, 2001. Differential effects of *Aureococcus anophagefferens* isolates ('brown tide') in unialgal and mixed suspensions on bivalve feeding. *Mar. Biol.,* **139,** 605–615.

Brussaard, C. P. D., R. Riegman, A. A. M. Noordeloos, G. C. Cadée, H. Witte, A. J. Kop, G. Nieuwland, F. C. V. Duyl and R. P. M. Bak, 1995. Effects of grazing, sedimentation and phytoplankton cell lysis on the structure of a coastal pelagic food web. *Mar. Ecol. Prog. Ser.,* **123,** 259–271.

Burchard, I., 1998. Anthropogenic impact on the climate since man began to hunt. *Palaeogeogr. Palaeoclimatol. Palaeoecol.,* **139,** 1–14.

Burkholder, J. M. and H. B. Glasgow, 2002. The life cycle and toxicity of *Pfiesteria piscicida* revisited. *J. Phycol.,* **38,** 1261–1267.

Burkholder, J. M., H. B. Glasgow and N. Deamer-Melia, 2001. Overview and present status of the toxic *Pfiesteria* complex (Dinophyceae). *Phycologia,* **40,** 186–214.

Buskey, E. J. and D. A. Stockwell, 1993. Effects of a persistent "brown tide" on zooplankton populations in the Laguna Madre of southern Texas. In *Toxic Phytoplankton Blooms in the Sea,* T. J. Smayda and Y. Shimizu, eds. Elsevier, Amsterdam, pp. 659–666.

Cadée, G. C. and J. Hegeman, 2002. Phytoplankton in the Marsdiep at the end of the 20th century; 30 years monitoring biomass, primary production, and *Phaeocystis* blooms. *J. Sea Res.,* **48,** 97–110.

Carlsson, P., E. Granéli and P. Olsson, 1990. Grazer elimination through poisoning: one of the mechanisms behind *Chrysochromulina polylepis* blooms? In *Toxic Marine Phytoplankton,* E. Granéli, B. Sundstrom, L. Edler and D. M. Anderson, eds. Elsevier, New York, pp. 116–122.

Carpenter, E. J., S. Janson, R. Boje, F. Pollhene and J. Chang, 1995. The dinoflagellate *Dinophysis norvegica:* biological and ecological observations in the Baltic Sea. *Eur. J. Phycol.,* **30,** 1–9.

Carpenter, E. J. and K. Romans, 1991. Major role of the cyanobacterium *Trichodesmium* in nutrient cycling in the North Atlantic Ocean. *Science,* **254,** 1356–1358.

Cembella, A. D., 2003. Chemical ecology of eukaryotic microalgae in marine ecosystems. *Phycologia,* **42,** 420–447.

Cembella, A. D., N. I. Lewis and M. A. Quilliam, 2000. The marine dinoflagellate *Alexandrium ostenfeldii* (Dinophyceae) as the causative organism of spirolide shellfish toxins. *Phycologia,* **39,** 67–74.

Cembella, A. D., M. A. Quilliam, N. I. Lewis, A. G. Bauder, C. Dell'Aversano, K. Thomas, J. Jellet and R. R. Cusack, 2002. The toxigenic marine dinoflagellate *Alexandrium tamarense* as the probable cause of mortality of caged salmon in Nova Scotia. *Harmful Algae,* **1,** 313–325.

Cerino, F., L. Orsini, D. Sarno, C. Dell'Aversano, L. Tartaglione and A. Zingone, 2005. The alternation of different morphotypes in the seasonal cycle of the toxic diatom *Pseudo-nitzschia galaxiae. Harmful Algae,* **in press,**

Chang, F. H., 1984. The ultrastructure of *Phaeocystis pouchetii* (Prymnesiophyceae) vegetative colonies with special reference to the production of new mucilaginous envelope. *N. Z. J. Mar. Freshw. Res.,* **18,** 303–308.

Chang, F. H., S. M. Chiswell and M. Uddstrom, 2001. Occurrence and distribution of *Karenia brevisulcata* (Dinophyceae) during the 1998 summer toxic outbreak on the central east coast of New Zealand. *Phycologia,* **40,** 215–222.

Charlson, R. J., J. E. Lovelock, M. O. Andreae and S. G. Warren, 1987. Oceanic phytoplankton, atmospheric sulphur, cloud albedo and climate. *Nature,* **326,** 655–661.

Chen, C.-T. A., 2000. The Three Gorges Dam: reducing the upwelling and thus productivity in the East China Sea. *Geophys. Res. Lett.,* **27,** 381–384.

Chinain, M., M. Germain, X. Deparis, S. Pauillac and A.-M. Legrand, 1999. Seasonal abundance and toxicity of the dinoflagellate *Gambierdiscus* spp. (Dinophyceae), the causative agent of ciguatera in Tahiti, French Polynesia. *Mar. Biol.,* **135,** 259–267.

Chisholm, S. W., P. G. Falkowski and J. J. Cullen, 2001. Dis-crediting ocean fertilization. *Science,* **294,** 309–310.

Cloern, J. E., 2001. Our evolving conceptual model of the coastal eutrophication problem. *Mar. Ecol. Prog. Ser.,* **210,** 223–253.

Coats, D. W. and M. G. Park, 2002. Parasitism of photosynthetic dinoflagellates by three strains of *Amoebophrya* (Dinophyta): parasite survival, infectivity, generation time, and host specificity. *J. Phycol.,* **38,** 520–528.

Codd, G. A., S. G. Bell, K. Kaya, C. J. Ward, K. A. Beattle and J. S. Metcalf, 1999. Cyanobacterial toxins, exposure routes and human health. *Eur. J. Phycol.,* **34,** 405–415.

Costas, E. and M. Varela, 1989. A circannual rythm in cyst formation and growth rates in the dinoflagellate *Scrippsiella trochoidea* Stein. *Chronobiologia,* **16,** 265–270.

Cullen, J. J., P. J. Franks, D. M. Karl and A. Longhurst, 2002. Physical influences on marine ecosystem dynamics. In *The Sea: Biological Physical Interactions in the Sea,* A. R. Robinson, J. J. McCarthy and B. J. Rothschild, eds. John Wiley & Sons, New York, pp. 297–336.

Cullen, J. J. and J. G. MacIntyre, 1998. Behaviour, physiology and the niche of depth regulating phytoplankton. In *Physiological Ecology of Harmful Algal Blooms,* D. M. Anderson, A. D. Cembella and G. M. Hallegraeff, eds. Springer-Verlag, Berlin, pp. 559–579.

Dahl, E., O. Lindahl, E. Paasche and J. Throndsen, 1989. The *Chrysochromulina polylepis* bloom in Scandinavian waters. In *Novel Phytoplankton Blooms,* E. M. Cosper, V. M. Bricelj and E. J. Carpenter, eds. Springer-Verlag, Berlin, pp. 384–405.

Dale, B., 2001. The sedimentary record of dinoflagellate cysts: looking back into the future of phytoplankton blooms. *Sci. Mar.,* **65,** 257–272.

Dale, B., A. Madsen, K. Nordberg and T. A. Thorsen, 1993. Evidence for prehistoric and historic 'blooms' of the toxic dinoflagellate *Gymnodinium catenatum* in the Kattegat-Skagerrak region of Scandinavia. In *Toxic Phytoplankton Blooms in the Sea,* T. J. Smayda and Y. Shimizu, eds. Elsevier, Amsterdam, pp. 47–52.

Dale, B., T. A. Thorsen and A. Fjellsa, 1999. Dinoflagellate cysts as indicators of cultural eutrophication in the Oslofjord, Norway. *Estuar. Coast. Shelf Sci.,* **48,** 371–382.

Daranas, A. H., M. Norte and J. J. Fernández, 2001. Toxic marine microalgae. *Toxicon,* **39,** 1101–1132.

Daugbjerg, N., G. Hansen, J. Larsen and Ø. Moestrup, 2000. Phylogeny of some of the major genera of dinoflagellates based on ultrastruture and partial LSU rDNA sequence data, including the erection of three new genera of unarmoured dinoflagellates. *Phycologia,* **39,** 302–317.

Decho, A. W., 1990. Microbial exopolymer secretions in ocean environments: their role(s) in food webs and marine processes. *Oceanogr. Mar. Biol. Annu. Rev.,* **28,** 75–153.

Degobbis, D., S. F. Umani, P. Franco, A. Malej, R. Precali and N. Smodlaka, 1995. Changes in the northern Adriatic ecosystem and the hypertrophic appearance of gelatinous aggregates. *Sci. Tot. Environ.,* **165,** 43–58.

Dekshenieks, M. M., P. L. Donaghay, J. M. Sullivan, J. E. B. Rines, T. R. Osborn and M. S. Twardowski, 2001. Temporal and spatial occurrence of thin phytoplankton layers in relation to physical processes. *Mar. Ecol. Prog. Ser.,* **223,** 61–71.

Delgado, M., 1999. A new 'diablillo parasite' in the toxic dinoflagellate *Alexandrium catenella* as a possibility to control harmful algal blooms. *Harmful Algae News,* **19,** 1–3.

Dodge, J. D., 1996. Biogeography of the dinoflagellate *Ceratium* in the Indian Ocean. *Nova Hedwigia Beih.,* **112,** 423–436.

Donaghay, P. L. and T. R. Osborn, 1997. Toward a theory of biological physical control of harmful algal bloom dynamics and impacts. *Limnol. Oceanogr.,* **42,** 1283–1296.

Dubois, D., 1975. A model of patchiness for prey-predator plankton populations. *Ecol. Modelling,* **1,** 67–80.

Eilertsen, H. C., K. S. Tande and J. P. Taasen, 1989. Vertical distributions of primary production and grazing by *Calanus glacialis* Jashnov and *C. hyperboreus* Krøyer in Arctic waters (Barents Sea). *Polar Biol.,* **9,** 253–260.

Eilertsen, H. C. and T. Wyatt, 2000. Phytoplankton models and life history strategies. *S. Afr. J. mar. Sci.,* **22,** 323–338.

Eilertsen, H. C. H. R., S. Sandberg and H. Tollefsen, 1995. Photoperiodic control of diatom spore growth: a theory to explain the onset of phytoplankton blooms. *Mar. Ecol. Prog. Ser.,* **116,** 303–307.

Elbrächter, M. and Y.-Z. Qi, 1998. Aspects of *Noctiluca* (Dinophyceae) population dynamics. In *Physiological Ecology of Harmful Algal Blooms,* D. M. Anderson, A. D. Cembella and G. M. Hallegraeff, eds. Springer-Verlag, Berlin, pp. 315–335.

Ellegaard, M. and Y. Oshima, 1998. *Gymnodinium nolleri* Ellegaard et Moestrup sp. ined. (Dinophyceae) from Danish waters, a new species producing *Gymnodinium catenatum*-like cysts: molecular and toxicological comparisons with Australian and Spanish strains of *Gymnodinium catenatum. Phycologia,* **37,** 369–378.

Estrada, M. and E. Berdalet, 1998. Effects of turbulence on phytoplankton. In *Physiological Ecology of Harmful Algal Blooms,* D. M. Anderson, A. D. Cembella and G. M. Hallegraeff, eds. Springer-Verlag, Berlin, pp. 601–618.

Evens, T. J., G. J. Kirkpatrick, D. E. Millie, D. J. Chapman and O. M. E. Schofield, 2001. Photophysiological responses of the toxic red-tide dinoflagellate *Gymnodinium breve* (Dinophyceae) under natural sunlight. *J. Plankton Res.,* **23,** 1177–1193.

Falciatore, A., M. R. d'Alcalà, P. Croot and C. Bowler, 2000. Perception of environmental signals by a marine diatom. *Science,* **288,** 2363–2366.

Falkowski, P. G., T. S. Hopkins and J. J. Walsh, 1980. An analysis of factors affecting oxygen depletion in the New York Bight. *J. Mar. Res.,* **38,** 479–506.

Fiedler, P. C., 1982. Zooplankton avoidance and reduced grazing responses to *Gyrodinium splendens* (Dinophyceae). *Limnol. Oceanogr.,* **27,** 961–965.

Finlay, B. J., 2002. Global dispersal of free-living microbial eukaryote species. *Science,* **296,** 1061–1063.

Finlay, B. J., J. O. Corliss, G. Esteban and T. Fenchel, 1996. Biodiversity at the microbial level: the number of free-living ciliates in the biosphere. *Q. Rev. Biol.,* **71,** 221–237.

Fisher, R. A., 1930. *The Genetical Theory of Natural Selection.* Clarendon Press, Oxford.

Fleming, R. H., 1939. The control of diatom populations by grazing. *J. Cons. Int. Explor. Mer.,* **14,** 210–227.

Flynn, K., J. M. Franco, P. Fernández, B. Reguera, M. Zapata, G. Wood and K. Flynn, 1994. Changes in toxin content, biomass and pigments of the dinoflagellate *Alexandrium minutum* during nitrogen refeeding and growth into nitrogen or phosphorus stress. *Mar. Ecol. Prog. Ser.,* **111,** 99–109.

Flynn, K., K. J. Jones and K. J. Flynn, 1996a. Comparisons among species of *Alexandrium* (Dinophyceae) grown in nitrogen- or phosphorus-limiting batch culture. *Mar. Biol.,* **126,** 9–18.

Flynn, K. J., K. Flynn, E. H. John, B. Reguera, M. I. Reyero, and J. M. Franco, 1996b. Changes in toxins, intracellular and dissolved free aminoacids of the toxic dinoflagellate *Gymnodinium catenatum* in response to changes in inorganic nutrients and salinity. *J. Plankton Res.,* **18,** 2093–2111.

Fogg, G. E., 1965. *Algal Cultures And Phytoplankton Ecology.* University of Wisconsin Press.

Fraga, S., D. M. Anderson, I. Bravo, B. Reguera, K. A. Steidinger and C. M. Yentsch, 1988. Influence of upwelling relaxation on dinoflagellates and shellfish toxicity in Ría de Vigo, Spain. *Estuar. Coast. Shelf Sci.,* **27,** 349–361.

Franks, P. J. S. and D. M. Anderson, 1992. Alongshore transport of a toxic phytoplankton bloom in a buoyancy current: *Alexandrium tamarense* in the Gulf of Maine. *Mar. Biol.,* **112,** 153–164.

Friedl, G. and A. Wüest, 2002. Disrupting biogeochemical cycles - Consequences of damming. *Aquat. Sci.,* **64,** 55–65.

Fryxell, G. A., M. C. Villac and L. P. Shapiro, 1997. The occurrence of the toxic diatom genus *Pseudo-nitzschia* (Bacillariophyceae) on the West Coast of the USA, 1920–1996: a review. *Phycologia,* **36,** 419–437.

Gaines, G. and F. J. R. Taylor, 1985. An exploratory analysis of PSP patterns in British Columbia 1942–1984. In *Toxic Dinoflagellates,* D. M. Anderson, A. W. White and D. G. Baden, eds. Elsevier, New York, pp. 439–444.

Garcés, E., A. Zingone, M. Montresor, B. Reguera and B. Dale, eds., 2002. *LIFEHAB: Life History of Microalgal Species Causing Harmful Blooms.* European Commission, Brussels, http://www.icm.csic.es/bio/projects/lifehab/.pdf.

Gentien, P., 1998. Bloom dynamics and ecophysiology of the *Gymnodinium mikimotoi* species complex. In *Physiological Ecology of Harmful Algal Blooms,* D. M. Anderson, A. D. Cembella and G. M. Hallegraeff, eds. Springer-Verlag, Berlin, pp. 155–173.

GEOHAB, 2001. *Global Ecology and Oceanography of Harmful Algal Blooms.* SCOR and IOC, Baltimore and Paris.

Gerritsen, J. and J. R. Strickler, 1977. Encounter probabilities and community structure in zooplankton: a mathematical model. *J. Fish. Res. Board Can.,* **34,** 73–82.

Gjosaeter, J., K. Lekve, N. C. Stenseth, H. P. Leinaas, H. Christie, E. Dahl, D. S. Danielssen, B. Edvardsen, F. Olsgard, E. Oug and E. Paasche, 2000. A long-term perspective on the *Chrysochromulina* bloom on the Norwegian Skagerrak coast 1988: a catastrophe or an innocent incident? *Mar. Ecol. Prog. Ser.,* **207,** 201–218.

Glibert, P. M., D. J. Conley, T. R. Fisher, L. W. Harding and T. C. Malone, 1995. Dynamics of the 1990 winter/spring bloom in Chesapeake Bay. *Mar. Ecol. Prog. Ser.,* **122,** 27–43.

Goldman, J. C., J. J. McCarthy and D. G. Peavey, 1979. Growth rate influence on the chemical composition of phytoplankton in oceanic waters. *Nature,* **279,** 210–215.

Granéli, E. and P. Carlsson, 1998. The ecological significance of phagotrophy in photosynthetic flagellates. In *Physiological Ecology of Harmful Algal Blooms,* D. M. Anderson, A. D. Cembella and G. M. Hallegraeff, eds. Springer-Verlag, Berlin, pp. 539–557.

Guzmán, L., H. Pacheco, G. Pizarro and C. Alarcón, 2002. *Alexandrium catanella* y veneno paralizante de los mariscos en Chile. In *Floraciones Algales Nocivas en el Cono Sur Americano,* E. A. Sar, M. E. Ferrario and B. Reguera, eds. Instituto Español de Oceanografía, Vigo, Spain, pp. 237–256.

Hallegraeff, G. M., 1993. A review of harmful algal blooms and their apparent global increase. *Phycologia,* **32,** 79–99.

Hallegraeff, G. M., 1998. Transport of toxic dinoflagellates via ships' ballast water: bioeconomic risk assessment and efficacy of possible ballast water management strategies. *Mar. Ecol. Prog. Ser.,* **168,** 297–309.

Hallegraeff, G. M., M. A. McCausland and R. K. Brown, 1995. Early warning of toxic dinoflagellate blooms of *Gymnodinium catenatum* in southern Tasmanian waters. *J. Plankton Res.,* **17,** 1163–1176.

Hamasaki, K., M. Horie, S. Tokimitsu, T. Toda and S. Taguchi, 2001. Variability in toxicity of the dinoflagellate *Alexandrium tamarense* isolated from Hiroshima Bay, Western Japan, as a reflection of changing environmental conditions. *J. Plankton Res.,* **23,** 271–278.

Hansen, G., N. Daugbjerg and J. M. Franco, 2003. Morphology, toxin composition and LSU rDNA phylogeny of *Alexandrium minutum* (Dinophyceae) from Denmark, with some morphological observations on other European strains. *Harmful Algae,* **2,** 317–335.

Hansen, P. J., 1991. *Dinophysis* - a planktonic dinoflagellate genus which can act both as a prey and a predator of a ciliate. *Mar. Ecol. Prog. Ser.,* **69,** 201–204.

Harrison, P. J., P. A. Thompson, M. Guo and F. J. R. Taylor, 1993. Effects of light. temperature and salinity on the growth rate of harmful marine diatoms, *Chaetoceros convolutus* and *C. concavicornis* that kill netpen salmon. *J. Appl. Phycol.,* **5,** 259–265.

Hasle, G. R., 1950. Phototactic vertical migration in marine dinoflagellates. *Oikos,* **2,** 162–175.

Hasle, G. R., 2002. Are most of the domoic acid-producing species of the diatom genus *Pseudo-nitzschia* cosmopolites? *Harmful Algae,* **1,** 137–146.

Haywood, A. J., K. A. Steidinger, E. W. Truby, P. R. Bergquist, P. L. Bergquist, J. Adamson and L. Mackenzie, 2004. Comparative morphology and molecular phylogenetic analysis of three new species of the genus *Karenia* (Dinophyceae) from New Zealand. *J. Phycol.,* **40,** 165–179.

Hassett, R. P., 2003. Effect of toxins of the 'red-tide' dinoflagellate *Alexandrium* spp. on the oxygen consumption of marine copepods. *J. Plankton Res.,* **25,** 185–192.

Hewson, I., J. M. O'Neal and W. C. Dennison, 2001. Virus-like particles associated with *Lyngbya majuscula* (Cyanophyta; Oscillatoriacea) bloom decline in Moreton Bay, Australia. *Aquat. Microb. Ecol.,* **25,** 207–213.

Huntley, M., P. Sykes, S. Rohan and V. Marin, 1986. Chemically-mediated rejection of dinoflagellate prey by the copepods *Calanus pacificus* and *Paracalanus parvus:* mechanism, occurence and significance. *Mar. Ecol. Prog. Ser.,* **28,** 105–120.

Huntley, M., K. Tande and H. C. Eilertsen, 1987. On the trophic fate of *Phaeocystis pouchetii* (Hariot). II. Grazing rates of *Calanus hyperboreus* (Kroyer) on diatoms and different size categories of *Phaeocystis pouchetii. J. Exp. Mar. Biol. Ecol.,* **110,** 197–212.

Hylleberg, J., 1975. Selective feeding by *Abarenicola pacifica* with notes on *Abarenicola vagabunda* and a concept of gardening in lugworms. *Ophelia,* **14,** 113–137.

ICES-IOC, 2003. Maps of Harmful Events Related to Phytoplankton Blooms in Western Europe and North America. IFREMER, France, *http://www.ifremer.fr/envlit/documentation/dossiers/ciem/aindex.htm.*

Imai, I., M. C. Kim, K. Nagasaki, S. Itakura and Y. Ishida, 1998. Relationships between dynamics of red tide-causing raphidophycean flagellate and algicidal micro-organisms in the coastal sea of Japan. *Phycol. Res.,* **46,** 139–146.

Imai, I., T. Sunahara, T. Nishikawa, Y. Hori, R. Kondo and S. Hiroishi, 2001. Fluctuations of the red tide flagellates *Chattonella* spp. (Raphidophyceae) abd the algicidal bacterium *Cytophaga* sp. in the Seto Inland Sea, Japan. *Mar. Biol.,* **138,** 1043–1049.

Irigoien, X., R. P. Harris, H. Verheye, P. Joly, J. Runge, M. Starr, D. Pond, R. Campbell, R. Shreeve, P. Ward, A. Smith, H. Dam, W. Peterson, R. Davidson et al., 2002. Copepod hatching success in marine ecosystems with high diatom concentrations. *Nature,* **419,** 387–389.

Jackson, J. B. C., M. X. Kirby, W. H. Berger, K. A. Bjorndal, L. W. Botsford, B. J. Bourque, R. H. Bradbury, R. Cooke, J. Erlandson, J. A. Estes, T. P. Hughes, S. Kidwell, C. B. Lange, H. S. Lenihan, J. M. Pandolfi, C. H. Peterson, R. S. Steneck, M. J. Tegner and R. R. Warner, 2001. Historical overfishing and the recent collapse of coastal ecosystems. *Science,* **293,** 629–638.

Jacobsen, A., G. Bratbak and M. Heldal, 1996. Isolation and characterization of a virus infecting *Phaeocystis pouchetii* (Prymnesiophyceae). *J. Phycol.,* **32,** 923–927.

Jacobson, D. M., 1999. A brief history of dinoflagellate feeding research. *J. Eukaryot. Microbiol.,* **46,** 376–381.

James, K. J., C. Moroneya, C. Rodena, M. Satakeb, T. Yasumoto, M. Lehanea and A. Fureya, 2003. Ubiquitous 'benign' alga emerges as the cause of shellfish contamination responsible for the human toxic syndrome, azaspiracid poisoning. *Toxicon,* **41,** 145–151.

Jenkinson, I. R., 1993. Viscosity and elasticity of *Gyrodinium* cf. *aureolum* and *Noctiluca scintillans* exudates in relation to mortality of fish and damping of turbulence. In *Toxic Phytoplankton Blooms in the Sea,* T. J. Smayda and Y. Shimizu, eds. Elsevier, Amsterdam, pp. 757–762.

Jenkinson, I. R., 1995. A review of two recent predation-rate models: the dome-shaped relationship between feeding rate and shear rate appears universal. *ICES J. Mar. Sci.,* **52,** 605–610.

Jenkinson, I. R. and P. P. Connors, 1980. The occurrence of the red tide organism, *Gyrodinium aureolum* Hulbert (Dinophyceae), around the south and west of Ireland in August and September, 1979. *J. Sherkin Island,* **1,** 127–146.

Jenkinson, I. R. and T. Wyatt, 1992. Selection and control of Deborah numbers in plankton ecology. *J. Plankton Res.,* **14,** 1697–1721.

Jenkinson, I. R. and T. Wyatt, 1995. Does bloom phytoplankton manage the physical environment? In *Harmful Marine Algal Blooms,* P. Lassus, G. Arzul, E. Erard, P. Gentien and C. Marcaillou, eds. Technique et Documentation - Lavoisier, Intercept ltd., Paris, pp. 603–613.

John, U., R. A. Fensome and L. K. Medlin, 2003. The application of a molecular clock based on molecular sequences and the fossil record to explain biogeographic distributions within the *Alexandrium tamarense* 'species complex' (Dinophyceae). *J. Mol. Evol.,* **20,** 1015–1027.

John, U., U. Tillmann and L. K. Medlin, 2002. A comparative approach to study inhibition of grazing and lipid composition of a toxic and non-toxic clone of *Chrysochromulina polylepis* (Prymnesiophyceae). *Harmful Algae,* **1,** 45–57.

Joint, J., J. Lewis, J. Allen, R. Proctor, G. Moore, W. Higman and M. Donald, 1997. Interannual variability of PSP outbreaks on the north east UK coasts. *J. Plankton Res.,* **19,** 937–956.

Juhl, A. R. and M. I. Latz, 2002. Mechanisms of fluid shear-induced inhibition of population growth in a red-tide dinoflagellate. *J. Phycol.,* **38,** 683–694.

Juhl, A. R., V. Velasquez and M. I. Latz, 2000. Effect of growth conditions on flow-induced inhibition of population growth of a red-tide dinoflagellate. *Limnol. Oceanogr.,* **45,** 905–915.

Kahru, M., J. M. Leppanen and O. Rud, 1993. Cyanobacterial blooms cause heating of the sea surface. *Mar. Ecol. Prog. Ser.,* **101,** 1–7.

Karl, D., R. Letelier, L. Tupas, J. Dore, J. Christian and D. Hebel, 1997. The role of nitrogen fixation in the biogeochemical cycling in the subtropical North Pacific Ocean. *Nature,* **388,** 533–538.

Karl, D. M., R. R. Bidigare and R. M. Letelier, 2001. Long-term changes in plankton community structure and productivity in the North Pacific Subtropical Gyre: the domain shift hypothesis. *Deep-Sea Res. II,* **48,** 1449–1470.

Kierstead, H. and B. Slobodkin, 1953. The size of water masses containing plankton blooms. *J. Mar. Res.,* **12,** 141–147.

Kim, H. G., 1998. *Cochlodinium polykrikoides* blooms in Korean coastal waters and their mitigation. In *Harmful Algae,* B. Reguera, J. Blanco, M. L. Fernández and T. Wyatt, eds. Xunta de Galicia and Intergovernmental Oceanographic Commission of UNESCO, Santiago de Compostela, Spain, pp. 227–228.

Kim, M. C., I. Yoshinaga, I. Imai, K. Nagasaki, S. Itakura and Y. Ishida, 1998. A close relationship between algicidal bacteria and termination of *Heterosigma akashiwo* (Raphidophyceae) blooms in Hiroshima bay, Japan. *Mar. Ecol. Prog. Ser.,* **170,** 25–32.

Kiryu, Y., J. Shields, W. Vogelbein and Z. D, 2002. Induction of skin ulcers in Atlantic menhaden by injection and water-borne exposure to the zoospores of *Aphanomyces invadans. J. Aquat. Anim. Health,* **14,** 11–24.

Koike, K., K. Koike, M. Takagi, T. Ogata and T. Ishimaru, 2000. Evidence of phagotrophy in *Dinophysis fortii* (Dinophysiales, Dinophyceae), a dinoflagellate that causes diarrhetic shellfish poisoning (DSP). *Phycol. Res.,* **48,** 121–124.

Kooistra, W. H. C. F., M. K. de Boer, E. G. Vrieling, L. B. Connell and W. W. C. Gieskes, 2001. Variation along ITS markers across strains of *Fibrocapsa japonica* (Raphidophyceae) suggests hybridisation events and recent range expansion. *J. Sea Res.,* **46,** 213–222.

Koski, M., J. Engström and M. Viitasalo, 1999. Reproduction and survival of the calanoid copepod *Eurytemora affinis* fed with toxic and non-toxic cyanobacteria. *Mar. Ecol. Prog. Ser.,* **186,** 187–197.

Kristiansen, S., T. Farbrot and P. A. Wheeler, 1994. Nitrogen cycling in the Barents Sea. Seasonal dynamics of new and regenerated production in the marginal ice zone. *Limnol. Oceanogr.,* **39,** 1630–1642.

Lam, C. W. Y. and K. C. Ho, 1989. Red tides in Tolo harbour, Hong Kong. In *Red Tides: Biology, Environmental Science and Toxicology,* T. Okaichi, D. M. Anderson and T. Nemoto, eds. Elsevier Science, New York, pp. 49–52.

Lancelot, C., M. D. Keller, V. Rousseau, W. O. Smith Jr. and S. Mathot, 1998. Autoecology of the marine haptophyte *Phaeocystis* sp. In *Physiological Ecology of Harmful Algal Blooms,* D. M. Anderson, A. Cembella and G. M. Hallegraeff, eds. Springer-Verlag, Berlin, pp. 209–224.

Landsberg, J. H., 2002. The effects of harmful algal blooms on aquatic organisms. *Rev. Fish. Sci.,* **10,** 113–390.

Landsberg, J. H. and K. A. Steidinger, 1998. A historical review of *Gymnodinium breve* red tides implicated in mass mortalities of the manatee (*Thrichechus manatus latirostris)* in Florida, USA. In *Harmful Algae,* B. Reguera, J. Blanco, M. L. Fernández and T. Wyatt, eds. Xunta de Galicia and Intergovernmental Oceanographic Commission of UNESCO, Santiago de Compostela, Spain, pp. 97–103.

Larsen, A. and B. Edvardsen, 1998. Relative ploidy levels in *Prymnesium parvum* and *P. patelliferum* (Haptophyta) analyzed by flow cytometry. *Phycologia,* **37,** 412–424.

Legendre, L., 1990. The significance of microalgal bloom for fisheries and for the export of particulate organic carbon in oceans. *J. Plankton Res.,* **12,** 681–699.

Legrand, C., K. Rengefors, G. O. Fistarol and E. Granéli, 2003. Allelopathy in phytoplankton - biochemical, ecological and evolutionary aspects. *Phycologia,* **42,** 406–419.

Leppard, G. G., 1999. Structure/function/activity relationships in marine snow. Current understanding and suggested research thrusts. *Ann. Ist. Super. Sanità,* **35,** 389–395.

Levasseur, M., M. D. Keller, E. Bonneau, D. D'Amours and W. K. Bellows, 1994. Oceanographic basis of a DMS-related Atlantic cod (*Gadus morhua*) fishery problem: blackberry feed. *Can. J. Fish. Aquat. Sci.,* **51,** 881–889.

Lewis, M. A., D. D. Dantin, C. C. Walker, J. C. Kurtz and R. M. Greene, 2003. Toxicity of clay flocculation of the toxic dinoflagellate, *Karenia brevis,* to estuarine invertebrates and fish. *Harmful Algae,* **2,** 235–246.

Lewis, R. J., 2001. The changing face of ciguatera. *Toxicon,* **39,** 97–106.

Lewitus, A. J., E. T. Koepler and J. T. Morris, 1998. Seasonal variation in the regulation of phytoplankton by nitrogen and grazing in a salt-marsh estuary. *Limnol. Oceanogr.,* **43,** 636–646.

Li, Y. and T. J. Smayda, 2000. *Heterosigma akashiwo* (Raphidophyceae): on prediction of the week of bloom initiation and maximum during the initial pulse of its bimodal bloom cycle in Narragansett Bay. *Plankton Biol. Ecol.,* **47,** 80–84.

Lilly, E. L., D. M. Kulis, P. Gentien and D. M. Anderson, 2002. Paralytic shellfish poisoning toxins in France linked to a human-introduced strain of *Alexandrium catenella* from the western Pacific: evidence from DNA and toxin analysis. *J. Plankton Res.,* **24,** 443–452.

Lincoln, J. A., J. T. Turner, S. S. Bates, C. Léger and D. A. Gauthier, 2001. Feeding, egg production, and egg hatching of copepods *Acartia tonsa* and *Temora longicornis* on diets of the toxic diatom *Pseudo-nitzschia pungens. Hydrobiologia,* **453/454,** 107–120.

Litaker, R. W., M. W. Vandersea, S. R. Kibler, V. J. Madden, E. J. Noga and P. A. Tester, 2002a. Life cycle of the heterotrophic dinoflagellate *Pfiesteria piscicida* (Dinophyceae). *J. Phycol.,* **38,** 442–463.

Litaker, R. W., M. W. Vandersea, S. R. Kibler, E. J. Noga and P. A. Tester, 2002b. Reply to comment on the life cycle and toxicity of *Pfiesteria piscicida* revisited. *J. Phycol.,* **38,** 1268–1272.

LoCicero, V. R. ed., 1975. *Proceedings of the First International Conference on Toxic Dinoflagellate Blooms.* The Massachusetts Science and Technology Foundation, Wakefield, Massachusetts.

Lotka, A. J., 1925. *Elements of Physical Biology.* William and Wilkins, Baltimore.

Lovejoy, C., J. P. Bowman and G. M. Hallegraeff, 1998. Algicidal effects of a novel marine *Pseudoalteromonas* isolate (Class *Proteobacteria,* Gamma Subdivision) on harmful algal bloom species of the genera *Chattonella, Gymnodinium,* and *Heterosigma. Appl. Environ. Microbiol.,* **64,** 2806–2813.

Lundberg, P. and J. M. Fryxell, 1995. Expected population density versus productivity in ratio-dependent and prey-dependent models. *Am. Nat.,* **146,** 153–161.

Lundholm, N., Ø. Moestrup, G. R. Hasle and K. Hoef-Emden, 2003. A study of the *Pseudo-nitzschia pseudodelicatissima/cuspidata* complex (Bacillariophyceae): what is *P. pseudodelicatissima? J. Phycol.,* **39,** 797–813.

Macintyre, J. G., J. J. Cullen and A. D. Cembella, 1997. Vertical migration, nutrition and toxicity in the dinoflagellate *Alexandrium tamarense. Mar. Ecol. Prog. Ser.,* **148,** 201–216.

MacKenzie, B. R., T. J. Miller, S. Cyr and W. C. Leggett, 1994. Evidence for a dome-shaped relationship between turbulence and larval fish ingestion rates. *Limnol. Oceanogr.,* **39,** 1790–1799.

Maclean, J. L., 1989. Indo-Pacific red tides, 1985–1988. *Mar. Pollut. Bull.,* **20,** 304–310.

Maldonado, M. T., M. P. Highes and E. L. Rue, 2002. The effect of Fe and Cu on growth and domoic acid production by *Pseudo-nitzschia multiseries* and *Pseudo-nitzschia australis. Limnol. Oceanogr.,* **47,** 515–526.

Malej, A., 1995. Gelatinous aggregates in the northern Adriatic Sea. *Bull. Inst. Oceanogr. (Monaco),* **15,** 149–157.

Malej, A. and R. P. Harris, 1993. Inhibition of copepod grazing by diatom exudates: a factor in the development of mucus aggregates? *Mar. Ecol. Prog. Ser.,* **96,** 33–42.

Malin, G., 1997. Sulphur, climate and the microbial maze. *Nature,* **387,** 857–859.

Malin, G. and G. O. Kirst, 1997. Algal production of dimethil sulfide and its atmospheric role. *J. Phycol.,* **33,** 889–896.

Malone, T. C., L. H. Crocker, S. E. Pike and B. W. Wendler, 1988. Influences of river flow on the dynamics of phytoplankton production in a partially stratified estuary. *Mar. Ecol. Prog. Ser.,* **48,** 235–249.

Malone, T. C., T. Knap and M. Fogarty, 2004. Overview of science requirements. In *The Sea: The Global Coastal Ocean: Multiscale Interdisciplinary Processes,* A. R. Robinson and K. Brink, eds. Harvard University Press, Cambridge.

Mann, D. G., 2001. Book review. *Phycologia,* **40,** 387–389.

Margalef, R., 1975. External factors and ecosystem stability. *Schweiz. Z. Hydrol.,* **37,** 102–107.

Margalef, R., 1978. Life-forms of phytoplankton as survival alternatives in an unstable environment. *Oceanol. Acta,* **1,** 493–509.

Margalef, R., 1997a. *Our Biosphere.* Ecology Institute, Oldendorf/Lue.

Margalef, R., 1997b. Turbulence and marine life. *Sci. Mar.,* **61,** 109–123.

Margalef, R. and M. Estrada, 1987. Synoptic distribution of summer microplankton (Algae and Protozoa) across the principal front in the western Mediterranean. *Invest. Pesq.,* **51,** 121–140.

Margalef, R., M. Estrada and D. Blasco, 1979. Functional morphology of organisms involved in red tides, as adapted to decaying turbulence. In *Toxic Dinoflagellate Blooms,* D. L. Taylor and H. H. Seliger, eds. Elsevier, Amsterdam, pp. 89–94.

Martin, A. P., 2000. On filament width in oceanic plankton. *J. Plankton Res.,* **22,** 597–602.

Matsuyama, Y., 1999. Harmful effect of the dinoflagellate *Heterocapsa circularisquama* on shellfish aquaculture in Japan. *JARQ,* **33,** 283–293.

Matthews, S. G. and G. C. Pitcher, 1996. Worst recorded marine mortality on the South African coast. In *Harmful and Toxic Algal Blooms,* T. Yasumoto, Y. Oshima and Y. Fukuyo, eds. UNESCO, Sendai Kyodo Printer, Sendai, pp. 89–92.

Mayali, X. and G. J. Doucette, 2002. Microbial community interactions and population dynamics of an algicidal bacterium active against *Karenia brevis* (Dinophyceae). *Harmful Algae,* **1,** 277–293.

McCauley, E. and W. W. Murdoch, 1990. Predator-prey dynamics in environments rich and poor in nutrients. *Nature,* **343,** 455–457.

McGillicuddy, J. D. J., R. P. Signell, C. A. Stock, B. A. Keafer, M. D. Keller, R. D. Hetland and D. M. Anderson, 2003. A mechanism for offshore initiation of harmful algal blooms in the coastal Gulf of Maine. *J. Plankton Res.,* **25,** 1131–1138.

McMahon, T., R. Raine and J. Silke, 1998. Oceanographic control of harmful phytoplankton blooms around southwestern Ireland. In *Harmful Algae,* B. Reguera, J. Blanco, M. L. Fernández and T. Wyatt, eds. Xunta de Galicia and Intergovernmental Oceanographic Commission of UNESCO, Santiago de Compostela, Spain, pp. 128–130.

McMinn, A., G. H. Hallegraeff, P. Thomson, A. V. Jenkinson and H. Heijnis, 1997. Cyst and radionuclide evidence for the recent introduction of the toxic dinoflagellate *Gymnodinium catenatum* into Tasmanian waters. *Mar. Ecol. Prog. Ser.,* **161,** 165–172.

Medlin, L. K., M. Lange, U. Wellbrock, G. Donner, M. Elbrächter, C. Hummert and B. Luckas, 1998. Sequence comparisons link toxic European isolates of *Alexandrium tamarense* from the Orkney Islands to toxic American stocks. *Eur. J. Protistol.,* **34,** 329–335.

Miller, M. B. and L. B. Bassler, 2001. Quorum-sensing in bacteria. *Annu. Rev. Microbiol,* **55,** 165–199.

Miralto, A., G. Barone, G. Romano, S. A. Poulet, A. Ianora, G. L. Russo, I. Buttino, G. Mazzarella, M. Laabir, M. Cabrini and M. G. Giacobbe, 1999. The insidious effect of diatoms on copepod reproduction. *Nature,* **402,** 173–176.

Moestrup, Ø., ed., 2003: *IOC Taxonomic Reference List of Toxic Algae.* Intergovernmental Oceanographic Commission of UNESCO. http://www.bi.ku.dk/ioc/default.asp.

Montresor, M., S. Sgrosso, G. Procaccini and W. H. C. F. Kooistra, 2003. Intraspecific diversity in *Scrippsiella trochoidea* (Dinophyceae): evidence for cryptic species. *Phycologia,* **42,** 56–70.

Mudie, P. J., A. Rochon and E. Levac, 2002. Palynological records of red tide-producing species in Canada: past trends and implications for the future. *Palaeogeogr. Palaeoclimatol. Palaeoecol.,* **180,** 159–186.

Mulligan, H. F., 1975. Oceanographic factors associated with New England red-tide blooms. In *Proceedings of the First International Conference on Toxic Dinoflagellate Blooms,* V. R. LoCicero, ed. The Massachusetts Science and Technology Foundation, Wakefield, Massachusetts, pp. 23–40.

Myklestad, M. M., B. Ramlo and S. Hestmann, 1995. Demonstration of strong interaction between the flagellate *Chrysochromulina polylepis* (Prymnesiophyceae) and a marine diatom. In *Harmful Marine Algal Blooms,* P. Lassus, G. Arzul, P. Gentien and C. Marcaillou, eds. Technique et Documentation - Lavoisier, Intercept ltd., Paris, pp. 633–638.

Nagai, S. and I. Imai., 1998. Killing of a giant diatom *Coscinodiscus wailesii* Gran by a marine bacterium *Alteromonas* sp. isolated from the Seto Inland Sea of Japan. In *Harmful Algae,* B. Reguera, J. Blanco, M. L. Fernández and T. Wyatt, eds. Xunta de Galicia and Intergovernmental Oceanographic Commission of UNESCO, Santiago de Compostela, Spain, pp. 402–405.

Nagasaki, K., M. Ando, S. Itakura, S. Imai and Y. Ishida, 1994. Viral mortality in the final stage of *Heterosigma akashiwo* (Raphidophyceae) red tide. *J. Plankton Res.,* **16,** 1595–1599.

Nagasaki, K. and M. Yamaguchi, 1998. Intra-species host specificity of HaV (*Heterosigma akashiwo* virus) clones. *Aquat. Microb. Ecol.,* **14,** 109–112.

Noga, E. J., L. Khoo, J. B. Stevens, Z. Fan and J. M. Burkholder, 1996. Novel toxic dinoflagellate causes epidemic disease in estuarine fish. *Mar. Pollut. Bull.,* **32,** 219–224.

Norén, F., Ø. Moestrup and A.-S. Rehnstam-Holm, 1999. *Parvilucifera infectans* Norén et Moestrup gen. nov. et sp. nov. (Perkinsozoa phylum nov.): a parasitic flagellate capable of killing toxic microalgae. *Eur. J. Protistol.,* **35,** 233–254.

O'Boyle, S., G. Nolan and R. Raine, 2001. Harmful phytoplankton events caused by variability in the Irish coastal current along the west of Ireland. In *Harmful Algal Blooms 2000,* G. Hallegraeff, S. I. Blackburn, C. J. S. Bolch and R. J. Lewis, eds. IOC UNESCO, Paris, pp. 145–148.

Okaichi, T., 1989. Red tide problems in the Seto Inland sea, Japan. In *Red Tides: Biology, Environmental Science and Toxicology,* T. Okaichi, D. M. Anderson and T. Nemoto, eds. Elsevier Science, New York, pp. 137–142.

Okolodkov, Y. B., 1999. Species range types of recent marine dinoflagellates recorded from the Arctic. *Grana,* **38,** 162–169.

Onji, M., T. Sawabe and Y. Ezura, 2001. Purification of a virus-like paticle causing growth suppression of *Alexandrium catenella. Bull. Fac. Fish. Hokkaido Univ.,* **52,** 135–138.

Pan, Y., S. S. Bates and A. D. Cembella, 1998. Environmental stress and domoic acid production by *Pseudo-nitzschia:* a physiological perspective. *Nat. Toxins,* **6,** 127–135.

Park, M. G., S. K. Cooney, J. S. Kim and D. W. Coats, 2002a. Effects of parasitism on diel vertical migration, phototaxis/geotaxis, and swimming speed of the bloom-forming dinoflagellate *Akashiwo sanguinea. Aquat. Microb. Ecol.,* **29,** 11–18.

Park, M. G., S. K. Cooney, W. H. Yih and D. W. Coats, 2002b. Effects of two strains of the parasitic dinoflagellate *Amoebophrya* on growth, photosynthesis, light absorption, and quantum yield of bloom-forming dinoflagellates. *Mar. Ecol. Prog. Ser.,* **227,** 281–292.

Parsons, M. L. and Q. Dortch, 2002. Sedimentological evidence of an increase in *Pseudo-nitzschia* (Bacillariophyceae) abundance in response to coastal eutrophication. *Limnol. Oceanogr.,* **47,** 551–558.

Passow, U., 2002. Production of transparent exopolymer particles (TEP) by phyto- and bacterioplankton. *Mar. Ecol. Prog. Ser.,* **236,** 1–12.

Passow, U. and A. L. Alldredge, 1994. Distribution, size and bacterial colonization of transparent exopolymer particles (TEP) in the ocean. *Mar. Ecol. Prog. Ser.,* **113,** 185–198.

Paul, J. H., L. Houchin, D. Griffin, T. Slifko, M. Guo, B. Richardson and K. Steidinger, 2002. A filterable lytic agent obtained from a red tide bloom that caused lysis of *Karenia brevis* (*Gymnodium breve*) cultures. *Aquat. Microb. Ecol.,* **27,** 21–27.

Peters, E., 1996. Prolonged darkness and diatom mortality. II. Marine temperate species. *J. Exp. Mar. Biol. Ecol.,* **207,** 43–58.

Pitcher, G. C., A. J. Boyd, D. A. Horstman and B. A. Mitchell-Innes, 1998. Subsurface dinoflagellate populations, frontal blooms and the formation of red tide in the southern Benguela upwelling system. *Mar. Ecol. Prog. Ser.,* **172,** 253–264.

Pitcher, G. C. and A. C. Cockroft, 1998. Low oxygen, rock lobster strandings and PSP. *Harmful Algal News,* **17,** 1–3.

Popels, L. C. and D. A. Hutchins, 2002. Factors affecting dark survival of the brown tide alga *Aureococcus anophagefferens* (Pelagophyceae). *J. Phycol.,* **38,** 738–744.

Potts, G. W. and J. M. Edwards, 1987. The impact of a *Gyrodinium aureolum* bloom on inshore young fish populations. *J. Mar. Biol. Assoc. U.K.,* **67,** 293–297.

Poulet, S. A., A. Ianora, A. Miralto and L. Meijer, 1994. Do diatoms arrest embrionic development in copepods? *Mar. Ecol. Prog. Ser.,* **111,** 79–86.

Poutanen, E.-L. and K. Nikkilä, 2001. Carotenoid pigments as tracers of cyanobacterial blooms in recent and post-glacial sediments of the Baltic Sea. *Ambio,* **30,** 179–183.

Rabalais, N. N., in press,. Eutrophication. In *The Sea: The Global Coastal Ocean: Multiscale Interdisciplinary Processes,* A. R. Robinson and K. Brink, eds. Cambridge.

Reguera, B. and S. González-Gil, 2001. Small cell and intermediate cell formation in species of *Dinophysis* (Dinophyceae, Dinophysiales). *J. Phycol.,* **37,** 318–333.

Reid, P. C. and R. Harland, 1977. Studies of Quaternary dinoflagellate cysts from the North Atlantic. In *Contributions on Stratigraphic Palinology 1. Cenozoic Palynology,* C. Elsik, ed. Amer. Assoc. Stratigraphic Palynology, pp. 147–169.

Reynolds, C. S. and T. J. Smayda, 1998. Principles of species selection and community assembly in the phytoplankton: further explorations of the Mandala. In *Harmful Algae,* B. Reguera, J. Blanco, M. L. Fernández and T. Wyatt, eds. Xunta de Galicia and Intergovernmental Oceanographic Commission of UNESCO, Santiago De Compostela, Spain, pp. 8–10.

Richardson, K. and G. Kullenberg, 1987. Physical and biological interactions leading to phytoplankton blooms: a review of *Gyrodinium aureolum* blooms in Scandinavian waters. *Rapp. P.-v. Réun. Cons. int. Explor. Mer,* **187,** 19–26.

Rines, J. E. B., P. L. Donaghay, M. M. Dekshenieks, J. M. Sullivan and M. S. Twardowski, 2002. Thin layers and camouflage: hidden *Pseudo-nitzschia* spp. (Bacillariophyceae) populations in a fjord in the San Juan Islands, Washington, USA. *Mar. Ecol. Prog. Ser.,* **225,** 123–137.

Rines, J. E. B., J. M. Sullivan, P. L. Donaghay and M. M. Dekshenieks, 2000. Where are the harmful algae? Thin layers of phytoplankton, and their implications for understanding the dynamics of harmful algal blooms. Abstracts of the 9th Conference on Harmful Algal Blooms, Hobart, Tasmania.

Riroriro, K. and L. D. Sims, 1989. Management of red tides in Papua New Guinea. In *Biology, Epidemiology and Management of Pyrodinium Red Tides,* G. M. Hallegraeff and J. M. Maclean, eds. ICLARM, Manila, pp. 149–151.

Rosenzweig, M. L., 1971. Paradox of enrichment: destabilization of exploitation ecosystems in ecological time. *Science,* **171,** 385–387.

Rost, B., U. Riebesell and S. Burkhardt, 2003. Carbon acquisition of bloom-forming marine phytoplankton. *Limnol. Oceanogr.,* **48,** 55–67.

Rothschild, B. J. and T. R. Osborn, 1988. Small-scale turbulence and planktonic contact rates. *J. Plankton Res.,* **10,** 465–474.

Rovinski, A. B., H. Adiwidjaja, V. Z. Yakhnin and M. Menzinger, 1997. Patchiness and enhancement of productivity in plankton ecosystems due to the differential advection of predator and prey. *Oikos,* **78,** 101–106.

Rue, E. and K. Bruland, 2001. Domoic acid binds iron and copper: a possible role for the toxin produced by the marine diatoms *Pseudo-nitzschia. Mar. Chem.,* **76,** 127–134.

Sathyendranath, S., A. D. Gouveia, S. R. Shetye, P. Ravindran and T. Platt, 1991. Biological control of surface temperature in the Arabian Sea. *Nature,* **349,** 54–56.

Saunders, P. T., 1994. Evolution without natural selection: implications of the daisyworld parable. *J. Theor. Biol.,* **166,** 365–373.

Scarrat, M. G., M. Levasseur, S. Michaud, G. Cantin, M. Gosselin and S. J. de Mora, 2002. Influence of phytoplankton taxonomic profile on the distribution of dimethylsulfide and dimethylsulfoniopropionate in the northwest Atlantic. *Mar. Ecol. Prog. Ser.,* **244,** 49–61.

Schmidt, L. E. and P. H. Hansen, 2001. Allelopathy in the prymnesiophyte *Chrysochromulina polylepis:* effect of cell concentration, growth phase and pH. *Mar. Ecol. Prog. Ser.,* **216,** 67–81.

Scholin, C. A., F. Gulland, G. J. Doucette, S. Benson, M. Busman, F. P. Chavez, J. Cordaro, R. DeLong, A. D. Vogetaere, J. Harvey, M. Haulena, K. Lefebvre, T. Lipscomb, S. Loscutoff, L. J. Lowenstine, R. Marin III, P. E. Miller, W. A. McLellan, P. D. R. Moeller, C. L. Powell, T. Rowles, P. Silvagni, M. Silver, T. Spraker, V. Trainer and F. M. Van Dolah, 2000. Mortality of sea lions along the central California coast linked to a toxic diatom bloom. *Nature,* **403,** 80–84.

Scholin, C. A., G. M. Hallegraeff and D. M. Anderson, 1995. Molecular evolution of the *Alexandrium tamarense* species complex (Dinophyceae): dispersal in the North American and West Pacific region. *Phycologia,* **34,** 472–485.

Schuster, S. and G. J. Herndl, 1995. Formation and significance of trasparent exopolymeric particles in the northern Adriatic Sea. *Mar. Ecol. Prog. Ser.,* **124,** 227–236.

Semina, H. J., 1997. An outline of the geographical distribution of oceanic phytoplankton. *Adv. Mar. Biol.,* **32,** 527–563.

Sengco, M. R. and D. M. Anderson, 2004. Controlling harmful algal blooms through clay flocculation. *J. Eukaryot. Microbiol.,* **51,** 169–172.

Sengco, M. R., A. Li, K. Tugend, D. Kulis and D. M. Anderson, 2001. Removal of red- and brown-tide cells using clay flocculation. I. Laboratory culture experiments with *Gymnodinium breve* and *Aureococcus anophagefferens. Mar. Ecol. Prog. Ser.,* **210,** 41–53.

Shumway, S. E., 1990. A rewiew of the effects of algal blooms on shellfish and aquaculture. *J. World Aquac. Soc.,* **21,** 65–104.

Shumway, S. E., S. M. Allen and P. Dee Boersma, 2003. Marine birds and harmful algal blooms: sporadic victims or under-reported events? *Harmful Algae,* **2,** 1–17.

Shumway, S. E., D. M. Frank, L. M Ewart and J. Ward, 2003. Effect of yellow loess on clearance rate in seven species of benthic, filter-feeding invertebrates. *Aquaculture Res.* **34,** 1391–1402.

Sieburth, J. M., P. W. Johnson and P. E. Hargraves, 1988. Ultrastructure and ecology of *Aureococcus anophagefferens* gen. et sp. nov. (Chrysophyceae): the dominant picoplankter during a bloom in Narragansett Bay, Rhode Island, summer 1985. *J. Phycol.,* **24,** 416–425.

Skellam, J. G., 1951. Random dispersal in theoretical populations. *Biometrika,* **38,** 196–218.

Skerratt, J. H., J. P. Bowman, G. Hallegraeff, S. James and P. D. Nichols, 2002. Algicidal bacteria associated with blooms of a toxic dinoflagellate in a temperate Australian estuary. *Mar. Ecol. Prog. Ser.,* **244,** 1–15.

Smayda, T. J., 1990. Novel and nuisance phytoplankton blooms in the sea: evidence for a global epidemic. In *Toxic Marine Phytoplankton,* E. Granéli, B. Sundstrom, L. Edler and D. M. Anderson, eds. Elsevier, New York, pp. 29–40.

Smayda, T. J., 1997. What is a bloom? A commentary. *Limnol. Oceanogr.,* **42,** 1132–1136.

Smayda, T. J., 2000. Ecological features of harmful algal blooms in coastal upwelling systems. *S. Afr. J. mar. Sci.,* **22,** 219–253.

Smayda, T. J., 2002. Turbulence, watermass stratification and harmful algal blooms: an alternative view and frontal zones as "pelagic seeds banks". *Harmful Algae,* **1,** 95–112.

Smayda, T. J. and C. S. Reynolds, 2001. Community assembly in marine phytoplankton: application of recent models to harmful dinoflagellate blooms. *J. Plankton Res.,* **23,** 447–461.

Smayda, T. J. and C. S. Reynolds, 2003. Strategies of marine dinoflagellate survival and some rules of assembly. *J. Sea Res.,* **49,** 95–106.

Smayda, T. J. and T. A. Villareal, 1989. The 1985 'brown-tide' and the open phytoplankton niche in Narragansett Bay during summer. In *Novel Phytoplankton Blooms.,* E. M. Cosper, V. M. Bricelj and E. J. Carpenter, eds. Springer-Verlag, Berlin, pp. 159–187.

Smetacek, V., M. Montresor and P. Verity, 2002. Marine productivity: footprints in the past and steps into the future. In *Phytoplankton Productivity. Carbon Assimilation in Marine and Freshwater Ecosystems,* P. J. le B. Williams, D. N. Thomas and C. S. Reynolds, eds. Blackwell Science, Ltd., Oxford, pp. 350–369.

Steidinger, K., 1983. A re-evaluation of toxic dinoflagellate biology and ecology. In *Prog. Phycol. Res.,* F. E. Round and D. J. Chapman, eds. Elsevier Science Publishers B.V., pp. 147–188.

Sullivan, J. M. and E. Swift, 2003. Effects of small-scale turbulence on net growth rate and size of ten species of marine dinoflagellates. *J. Phycol.,* **39,** 83–94.

Swanson, R. L. and C. J. Sindermann, 1979. Oxygen depletion and associated mortalities in the New York Bight. *NOAA Prof. Pap.,* **11,** 1–345.

Sykes, P. A. and M. E. Huntley, 1987. Acute physiological reactions of *Calanus pacificus* to selected dinoflagellates: direct observations. *Mar. Biol.,* **94,** 19–24.

Takano, H., 1965. New and rare diatoms from Japanese marine waters - I. *Bull. Tokai Reg. Fish. Res. Lab.,* **42,** 1–13.

Tallberg, P. and A.-S. Heiskanen, 1998. Species-specific phytoplankton sedimentation in relation to primary production along an inshore-offshore gradient in the Baltic Sea. *J. Plankton Res.,* **20,** 2053–2070.

Tarutani, K., K. Nagasaki, S. Itakura and M. Yamaguchi, 2001. Isolation of a virus infecting the novel shellfish-killing dinoflagellate *Heterocapsa circularisquama. Aquat. Microb. Ecol.,* **23,** 103–111.

Teegarden, G. J. and A. D. Cembella, 1996. Grazing of toxic dinoflagellates, *Alexandrium* spp. by adult copepods of coastal Maine: implications for the fate of paralytic shellfish toxins in marine food webs. *J. Exp. Mar. Biol. Ecol.,* **196,** 145–176.

Tester, P. A. and K. A. Steidinger, 1997. *Gymnodinium breve* red tide blooms: initiation, transport, and consequences of surface circulation. *Limnol. Oceanogr.,* **42,** 1039–1051.

Thomas, W. H. and C. H. Gibson, 1990. Effects of small-scale turbulence on microalgae. *J. Appl. Phycol.,* **2,** 71–77.

Thorsen, T. A. and B. Dale, 1998. Climatically influenced distribution of *Gymnodinium catenatum* during the past 2000 years in coastal sediments of southern Norway. *Palaeogeogr. Palaeoclimatol. Palaeoecol.,* **143,** 159–177.

Tillmann, U., 1998. Phagotrophy by a plastidic haptophyte, *Prymnesium patelliferum. Aquat. Microb. Ecol.,* **14,** 155–160.

Tillmann, U. and U. John, 2002. Toxic effects of *Alexandrium* spp. on heterotrophic dinoflagellates: an allelochemical defence mechanism independent of PSP-toxin content. *Mar. Ecol. Prog. Ser.,* **230,** 47–58.

Tindall, D. R. and S. L. Morton, 1998. Community dynamics and physiology of epiphytic/benthic dinoflagellates associated with ciguatera. In *Physiological Ecology of Harmful Algal Blooms,* D. M. Anderson, A. D. Cembella and G. M. Hallegraeff, eds. Springer-Verlag, Berlin, pp. 293–313.

Tobiesen, A., 1991. Growth rates of *Heterophrys marina* (Heliozoa) on *Chrysochromulina polylepis* (Prymnesiophyceae). *Ophelia,* **33,** 205–212.

Tortell, P. D., G. R. DiTullio, D. M. Sigman and F. M. M. Morel, 2002. $CO_2$ effects on taxonomic composition and nutrient utilization in an Equatorial Pacific phytoplankton assemblage. *Mar. Ecol. Prog. Ser.,* **236,** 37–43.

Townsend, D. W., N. R. Pettigrew and A. C. Thomas, 2001. Offshore blooms of the red tide dinoflagellate, *Alexandrium* sp., in the Gulf of Maine. *Cont. Shelf. Res.,* **21,** 347–369.

Trainer, V. L., B. M. Hickey and R. A. Horner, 2002. Biological and physical dynamics of domoic acid production off the Washington coast. *Limnol. Oceanogr.,* **47,** 1438–1446.

Truscott, J. E., 1995. Environmental forcing of simple plankton models. *J. Plankton Res.,* **17,** 2207–2232.

Turner, J. T. and P. A. Tester, 1997. Toxic marine phytoplankton, zooplankton grazers, and pelagic food webs. *Limnol. Oceanogr.,* **42,** 1203–1214.

Van Etten, J. L., L. C. Lane and R. H. Meints, 1991. Viruses and viruslike particles of eukaryotic algae. *Microbiol. Rev.,* **55,** 586–620.

VanDolah, F., 2000. Marine algal toxins: origins, health effects, and their increased occurrence. *Env. Health Persp.,* **108,** 133–142.

Verhulst, P. F., 1838. Recherches mathématiques sur la loi d'accroissement de la population. *Nouveaux Mémoires de l'Académie Royale des Sciences et des Belles-Lettres de Bruxelles,* **18,** 1–38.

Vernet, M., 1991. Phytoplankton dynamics in the Barents Sea estimated from chlorophyll budget models. *Polar Res.,* **10,** 129–145.

Vila, M., J. Camp, E. Garcés, M. Masó and M. Delgado, 2001. High resolution spatio-temporal detection of potentially harmful dinoflagellates in confined waters of the NW Mediterranean. *J. Plankton Res.,* **23,** 497–514.

Vila, M., E. Garces, M. Maso and J. Camp, 2001. Is the distribution of the toxic dinoflagellate *Alexandrium catenella* expanding along the NW Mediterranean coast? *Mar. Ecol. Prog. Ser.*, **222**, 73–83.

Vitousek, P. M., J. D. Aber, R. W. Howarth, G. E. Likens, P. A. Matson, D. W. Schindler, W. H. Schlesinger and D. G. Tilman, 1997. Human alteration of the global nitrogen cycle: Sources and consequences. *Ecol. Appl.*, **7**, 737–750.

Vogelbein, W. K., V. J. Lovko, J. D. Shields, K. S. Reece, P. L. Mason, L. W. Haas and C. C. Walker, 2002. *Pfiesteria shumwayae* kills fish by micropredation not exotoxin secretion. *Nature*, **418**, 967–970.

Volterra, V., 1926. Fluctuations in the abundance of a species considered mathematically. *Nature*, **118**, 558–560.

Wall, D., 1965. Modern hystrichosperes and dinoflagellate cysts from the Woods Hole region. *Grana Palynologica*, **6**, 297–314.

Watson, A. J. and J. E. Lovelock, 1983. Biological homeostasis of the global environment: the parable of Daisyworld. *Tellus*, **35B**, 284–289.

Watson, S., E. McCauley and J. A. Downing, 1992. Sigmoid relationships between phosphorus, algal biomass, and algal community structure. *Can. J. Fish. Aquat. Sci.*, **49**, 2605–2610.

White, A. W., 1981. Marine zooplankton can accumulate and retain dinoflagellate toxins and cause fish kills. *Limnol. Oceanogr.*, **26**, 103–109.

Wilkinson, D. M., 1999. Is Gaia really conventional ecology? *Oikos*, **84**, 533–536.

Williams, S. L. and R. C. Carpenter, 1988. Effects of unidirectional and oscillatory water flow on nitrogen fixation (acetylene reduction) in coral reef algal turfs, Kaneohe Bay, Hawaii. *J. Exp. Mar. Biol. Ecol.*, **226**, 293–316.

Willis, B. M., K. C. Hayes, J. M. Burkholder, H. B. Glasgow jr., P. M. Gilbert and M. K. Burke, 1999. Mixotrophy and nitrogen uptake by *Pfiesteria piscicida* (Dinophyceae). *J. Phycol.*, **35**, 1430–1437.

Wyatt, T., 1980. Morrell's seals. *J. Cons. Int. Expl. Mer*, **39**, 1–6.

Wyatt, T., 1989. Modelling the spread of red tides - A comment. In *Toxic Marine Phytoplankton*, E. Granéli, B. Sundstrom, L. Edler and D. M. Anderson, eds. Elsevier, New York, pp. 253–254.

Wyatt, T., 1995. Global spreading, time series, models and monitoring. In *Harmful Marine Algal Blooms*, P. Lassus, G. Arzul, E. Erard, P. Gentien and C. Marcaillou, eds. Technique et Documentation - Lavoisier, Intercept ltd., Paris, pp. 755–764.

Wyatt, T., 1998. Harmful algae, marine blooms, and simple population models. *Nature & Resources*, **34**, 40–51.

Wyatt, T. and J. T. Carlton, 2002. Phytoplankton introductions in European coastal waters: why are so few invasions reported? In *Alien Marine Organisms Introduced by Ships in the Mediterranean and Black Seas.*, F. Briand, ed. CIESM Workshop Monographs n°20, Monaco, pp. 41–46.

Wyatt, T. and I. R. Jenkinson, 1993. The North Atlantic turbine: views of production processes from a mainly North Atlantic perspective. *Fish. Oceanogr.*, **2**, 231–243.

Wyatt, T. and I. R. Jenkinson, 1997. Notes on *Alexandrium* population dynamics. *J. Plankton Res.*, **19**, 551–575.

Wyatt, T. and F. Saborido-Rey, 1993. Biogeography and time-series analysis of British PSP records, 1968 to 1990. In *Toxic Phytoplankton Blooms in the Sea*, T. J. Smayda and Y. Shimizu, eds. Elsevier, Amsterdam, pp. 73–78.

Yamazaki, H., D. L. Mackas and K. L. Denman, 2002. Coupling small-scale physical processes with biology. In *The Sea: Biological Physical Interactions in the Sea*, A. R. Robinson, J. J. McCarthy and B. J. Rothschild, eds. John Wiley & Sons, New York, pp. 51–112.

Yang, C. Z. and L. J. Albright, 1992. Effects of the harmful diatom *Chaetoceros concavicornis* on respiration of rainbow trout *Oncorhynchus mykiss*. *Dis. Aquat. Org.*, **14**, 105–114.

Yang, C. Z. and L. J. Albright, 1994. The harmful phytoplankter *Chaetoceros concavicornis* causes high mortalities and leucopenia in chinook salmon (*Oncorhynchus tshawytscha*) and coho salmon (*O. kisutch*). *Can. J. Fish. Aquat. Sci.*, **51**, 2493–2500.

Yang, Z. B., I. J. Hodgkiss and G. Hansen, 2001. *Karenia longicanalis* sp. nov. (Dinophyceae): a new bloom-forming species isolated from Hong Kong, May 1998. *Bot. Mar.,* **44,** 67–74.

Yin, K., 2003. Influence of monsoons and oceanographic processes on red tides in Hong Kong waters. *Mar. Ecol. Prog. Ser.,* **262,** 27–41.

Yu, Z., X. Sun and J. Zou, 2000. Progress of harmful algal bloom (HAB) mitigation with clays in China. In *Harmful Algal Blooms,* G. M. Hallegraeff, S. I. Blackburn, C. J. Bolch and R. J. Lewis, eds. IOC UNESCO, Paris, pp. 484–487.

Yung, Y.-K., C. K. Wong, M. J. Broom, J. A. Ogden, S. C. M. Chan and Y. Leung, 1997. Long-term changes in hydrography, nutrients and phytoplankton in Tolo Harbour, Hong Kong. *Hydrobiologia,* **352,** 107–115.

Zaitsev, Y. P., 1993. Impact of eutrophication on the Black Sea fauna. *General Fisheries Council for the Mediterranean (FAO), Studies and Reviews,* **64,** 63–86.

Zhang, J., 1994. Atmospheric wet depositions of nutrient elements: Correlations with harmful biological blooms in the Northwest Pacific coastal zone. *Ambio,* **23,** 464–468.

Zingone, A., 1995. The role of viruses in the dynamics of phytoplankton blooms. *G. Bot. It.,* **129,** 415–423.

Zingone, A., 2002. The role of alternate life stages in the timing and succession of phytoplankton blooms. In *LIFEHAB: Life History of Microalgal Species Causing Harmful Blooms,* E. Garcés, A. Zingone, M. Montresor, B. Reguera and B. Dale, eds. European Commission, Brussels, http://www.icm.csic.es/bio/projects/lifehab.pdf, pp. 92–94.

Zingone, A. and H. O. Enevoldsen, 2000. The diversity of harmful algal blooms: a challenge for science and management. *Ocean Coast. Manag.,* **43,** 725–748.

Zingone, A., P. Licandro and D. Sarno, 2003. Revising paradigms and myths of phytoplankton ecology using biological time series. In *Mediterranean Biological Time Series,* F. Briand, ed. CIESM Workshop Monographs n° 22, Monaco, pp. 109–114.

Zingone, A., D. Sarno and G. Forlani, 1999. Seasonal dynamics of *Micromonas pusilla* (Prasinophyceae) and its viruses in the Gulf of Naples (Mediterranean Sea). *J. Plankton Res.,* **21,** 2143–2159.

Zirbel, M. J., F. Veron and M. I. Latz, 2000. The reversible effect of flow on the morphology of *Ceratocorys horrida* (Peridiniales, Dinophyta). *J. Phycol.,* **36,** 46–58.

# Chapter 23. HABITAT MODIFICATION

MICHEL J. KAISER

*School of Ocean Sciences, University of Wales-Bangor*

STEPHEN J. HALL

*World Fish Center*

DAVID N. THOMAS

*School of Ocean Sciences, University of Wales-Bangor*

**Contents**



## 1. Introduction

The ecosystem function of marine habitats and the services that they provide to human societies have a critical role in the maintenance of the global natural and economic environment (Costanza et al., 1997, 1998). A habitat is the environment in which organisms grow, reproduce and die, although many organisms utilise different habitats at various stages of their life-history. Organisms that are associated with particular habitats may have become so well adapted to specific habitat conditions that they are considered endemic to that habitat (e.g. true estuarine biota or hydrothermal vent fauna). Other, less specifically adapted species may use a range of habitats that share common features, e.g. sedimentary habitats or those that provide structure on the seabed. Some taxa may be associated with particular habitats at juvenile stages of their life-history and then become more generalised in their habitat use with age or size. Habitats formed by biota (e.g. oyster reefs, mussel beds, coral reefs, mangroves, seagrasses) have a key role in the sequestration of organic carbon, minerals and nutrients from the water column. They also

*The Sea*, Volume 13, edited by Allan R. Robinson and Kenneth H. Brink
ISBN 0-674-01526-6 

act as a sink for pollutants and can provide a source of primary and secondary production for other habitats. Habitats also provide services to humans, for example, reefs, sandbanks, gravel bars and mangrove forests dissipate wave action and provide protection from coastal erosion in many areas of the world.

Sedimentary habitats are probably the most extensive of all habitats found in coastal shelf seas. They provide a substratum in and on which biota live. Emergent epifauna increase topographic complexity and function as feeding stations for other organisms such as crustacea and fishes and redirect organic matter from the water column to the seabed via their feeding activities. Infauna greatly increase the complexity of the sub-surface sediment habitat. The bioturbating fauna found in mud habitats create subterranean structures that greatly increase the sub-surface area of sediment across which gaseous exchange and biogeochemical cycling can occur. Those habitats formed by organisms are among the most structurally complex. Reefs of oysters and mussel beds provide intricate microhabitats for a large number of organisms. Of all the biogenic habitats studied, coral reefs are undoubtedly the best recorded and are estimated to support *ca.* 423 000 species of plants and animals. More importantly, large aggregations of filter-feeding organisms can remove and process a large proportion of the primary production and suspended sediment in the overlying water column. At the coastal margins, mangrove ecosystems have high measured levels of productivity and although much of this is recycled within a complex food web, a substantial component is also exported to adjacent sublittoral areas as leaf litter, detritus and dissolved organic matter. The mangrove root system acts as a sediment trap which enhances bacterial activity prompting the role of mangrove mud as an important carbon sink.

Given the global ecological and economic importance of shelf sea habitats, it is essential to understand both the natural and anthropogenic processes that ultimately lead to habitat modification. It is important to quantify the relative magnitude and frequency of natural versus anthropogenic disturbance processes if we are to be able to assess their relative influence with respect to habitat modification. Physical perturbations, chemical changes in the environment, and biological changes within the associated communities can all result in modification of the habitat at a range of temporal and spatial scales.

## 2. Defining habitat and modification

Before we consider more fully the specific modification processes that occur within habitats it is necessary to define habitat and to consider the natural versus human sources of modification for the purposes of this chapter. Within shelf sea systems, a habitat is a three dimensional entity that is influenced to varying degrees by interactions between the substratum (seabed), the water column and atmospheric processes, many of which are covered in detail elsewhere in this volume. The physical extent of an organim's habitat will determine the immediacy of the impact of processes that occur at different scales. Thus on a daily basis, the surrounding few metres of habitat may have the greatest impact on an attached organism such as a sponge, whereas much larger scale processes may impact on a highly mobile taxon such as pelagic tuna fish that can swim 10s km within a day. The land mass adjacent or remote from a habitat and the geological processes that occur beneath the substratum within the underlying strata will have impacts at much larger temporal and spatial scales (Figure 23.1). Changes in any of the latter four components of the habitat can bring about an ecological disturbance of the system (see Hall 1994). Thus the scale at which we consider the habitat for a particular taxon will vary according to the extent of time under consideration.

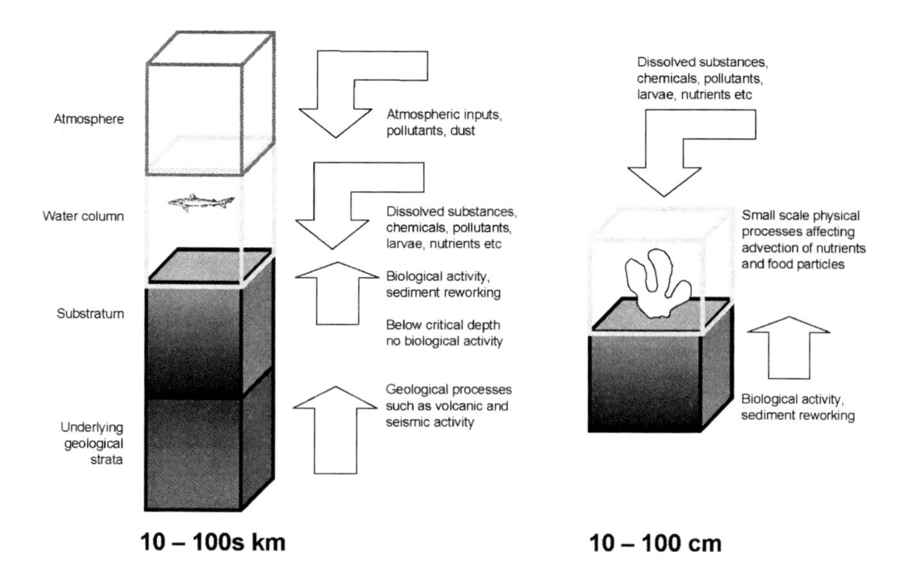


Figure 23.1 The different components that constitute or influence specific habitats in shelf seas. Some influences are very localised (e.g. seismic activity) while others are spread across a much wider area (e.g. oceanic influences).

Scale is one of the most important considerations when defining the boundaries that delimit a habitat. At the largest scale, habitat can be defined by geographical disposition: i.e. polar, temperate, tropical. Within these areas there will be further regional subdivisions of habitat. At a more local scale, habitats begin to be defined by their physical constitution (rock, gravel, sand) and their associated flora and fauna. At this point, habitats can be categorised into distinct biotopes that have sufficient unique biological and physical characteristics to set them apart from other biotopes. Within each biotope, there are a plethora of micro-habitats, each with their own unique attributes. Many biotopes are interlinked forming functional ecological units (eg coral reefs, seagrass beds, mangroves) and ecosystems (eg. shallow tropical marine environments). Each biotope is strongly influenced by its immediate aquatic environment, but will also be under the influence of other proximate and distant habitats (e.g. that are a source of larvae) and regional and global environmental factors.

There are two sources of habitat modification that need to be defined for the purposes of this chapter; natural sources of habitat modification, and those that originate from human activities. Large-scale natural disturbances, such as seasonal storms and strong tidal currents, form a background against which other smaller disturbances occur, such as those induced by predator feeding activities (Von

Blaricom, 1982; Oliver and Slattery, 1985; Hall et al., 1994). Hall *et al.* (1994) suggested that frequent small-scale predator disturbances may have a considerable additive effect on benthic communities, creating a long-term mosaic of patches in various states of climax or recolonization (Grassle and Saunders, 1973; Connell, 1978). They concluded, however, that while it was possible to detect short-term effects of predator disturbance, large-scale effects could not be inferred. This implies that the effects of small-scale disturbance events, even when frequent, are over-whelmed by a background of large-scale disturbances. Alternatively, the rapid recolonisation that occurs after small-scale disturbances means that large-scale effects never become apparent (Figure 23.2). Thus while small-scale modifications of habitat may be relevant over short periods of time, at longer time-scales they are far less important than the consideration of larger-scale sources of natural disturbance.

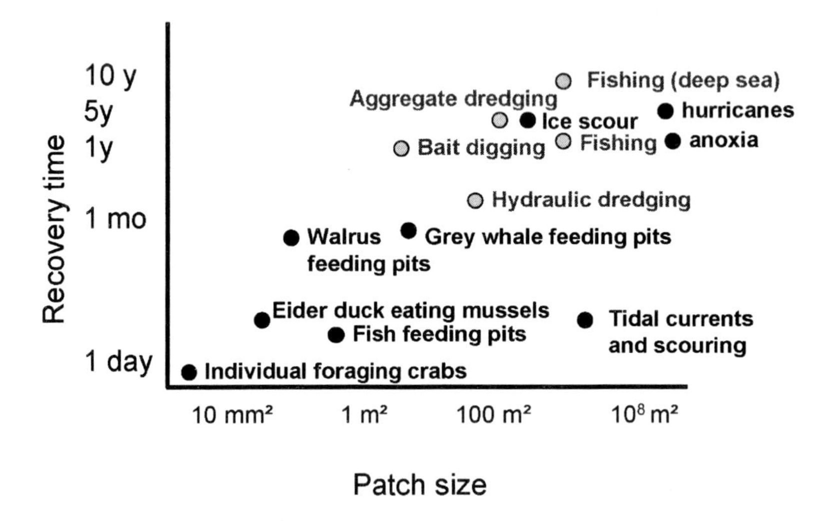


Figure 23.2 The relative recovery rate of different scales of disturbance that occur in the marine environment. The figure shows various forms of fishing activity compared with naturally occurring disturbances such as predation effects and physical sources of disturbance (after Hall 1994).

The imposition of human activities upon the marine environment can lead to disturbances of habitats that occur in addition to the natural disturbances that occur. Therefore natural sources of habitat modification form the backdrop against which the significance of human sources of disturbance should be assessed. These human sources of disturbance are only likely to become significant if their scale, frequency or intensity exceeds that of the natural sources of habitat modification that prevail in a particular environment. For example, the additive effects of an entire fishing fleet may reach such a threshold. However, while human activities affect all of the world's continental shelf areas, they do not occur uniformly but are usually highly aggregated. Hence while the additive effects of human activities may create severe localised habitat modification they may affect only a limited proportion of a particular habitat resource. The extent of natural disturbance regimes that modify marine habitat is reflected in the stability and characteristics of the habitat. Hence it is appropriate at this juncture to consider key factors that influence habitat stability. While we have focussed on influences on seabed habitat stability some of the same considerations apply to the water column (e.g. the influences of land run-off).

### *2.1 Habitat stability*

The physical characteristics of the habitat will, in part, determine its stability. Most shallow subtidal habitats are exposed to the influence of wave action and seabed currents to a greater or lesser extent. Considering the nature of the seabed from the extreme of exposed bedrock to fine muds, we can assign a level of stability to each type of substratum. Clearly bedrock will be the most stable substratum except in areas of volcanic activity. In general, but not always, one finds decreasing stability from boulders, through cobbles, to gravel, to sands, which are usually the least stable environment. The greatest sediment stability is achieved with a mixture of very fine (clay), mud and sand particles as a result of a combination of chemical, physical and biological processes (Dernie et al., 2003). Each substratum is exposed to a range of natural disturbing forces such as ice scour, wave action and tidal currents. Hence, although the stability of bedrock is the same in a wave exposed and sheltered coastline, the physical forces that impact upon that habitat will vary greatly and hence the biota that can live there. The nature of the sediment reflects the hydrodynamic forces that control deposition and erosion. For example, sand-wave systems are created by near-bed currents (Figure 23.3), and may be subjected to periodic sediment resuspension by wave action. This contrasts with deep (> 100 m) basins and fjordic systems where tidal currents and wave action have far less influence. Here, sedimentation of fine particles can occur and typically leads to the formation of muddy sediments.

The cohesive nature of both mud and sands is increased by microbial action that produces extracellular polymeric substances (EPS). These substances act like glue that binds the sediment together and reduces its erodability. Erosion of sediments most commonly occurs as a result of wave action or through currents that flow over the seabed. Glacial scour in polar regions can remove the sediment right down to the bedrock. As water flows over a surface it forms layers (laminae) that flow at different speeds; this is termed laminar flow. As water flows over the seabed it forms a boundary layer that flows slower than the layers of water above due to the friction between the water particles and the sediment surface. Boundary layer flow in an experimental flume tank with smooth surfaces would be described as laminar. The introduction of surface roughness (e.g. emergent plants or animals) will induce turbulent flow that breaks down the boundary layer. This has important consequences for both the sediment and organisms and plants that live in that habitat. Increasing turbulence and shear stress accelerates the advection of particles (phytoplankton and sediment) to the seabed and hence influences the rate of supply of food to filter-feeding and microbial fauna (Widdows et al., 1998).

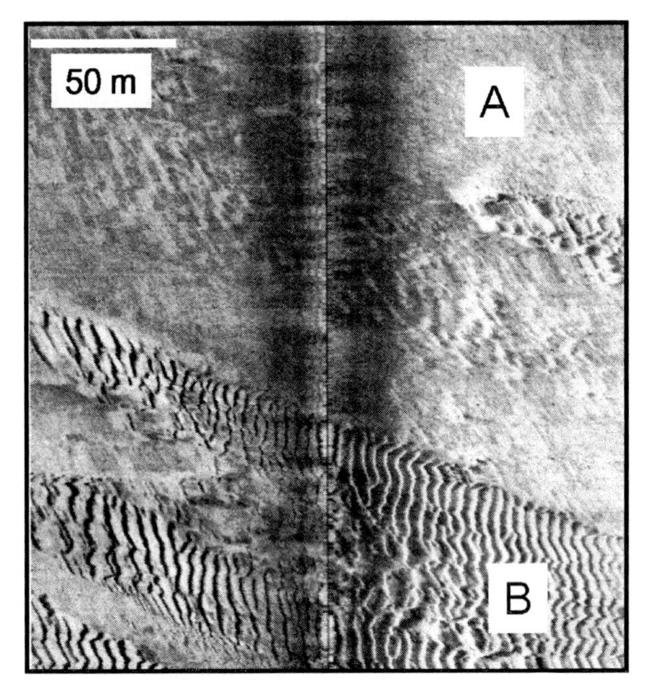


Figure 23.3 Seabed currents are important physical agents of habitat modification that structure bed forms such as mega-ripples. This in turn leads to a high diversity of habitat types within small areas of the seabed. The figure demonstrates the striking difference between a stable gravel lag deposit with a high diversity of fauna (A) and a mobile sand habitat with a relatively impoverished fauna (B) within an area of a few hundred m of seabed.

At the smallest temporal and spatial scales (μm, ms) chemical reactions between sediment particles, flocculation of organic matter, viral and bacterial processes occur (Table 1). The development of bacterial communities and microalgal mats generate extracellular polymeric substances that have an import role in modifying the structural stability of the habitat (Paterson and Black, 1999). At the next scale up, minute by minute disturbances occur as the result of the feeding activities of meiofauna and macrofauna and as a result of their movements upon or through the sediment creating burrows or furrows in their wake (Schafer, 1972). At slightly larger scales up to one metre, larger-scale bioturbating processes become important (e.g. faecal mound formation, burrow chamber excavation) and the feeding activities of megafauna such as fishes and crabs become important agents of disturbance that occur every tidal cycle (Raffaelli et al., 1987; Warwick et al., 1980; Thrush et al., 1995; Austen et al., 2000). Beyond this scale the influence of natural processes such as storms and ice-scour operate.

In addition to natural disturbances, any consideration of habitat modification processes needs to incorporate an appreciation of the role of human activities and how these impact upon habitat stability. Among these must be included, watershed modifications that alter drainage, sediment and contaminant inputs to the sea, sediment extraction and fishing activities. For the purposes of this chapter the examples given below are not intended to be exhaustive but are indicative of the range of factors that effect habitat modification, the causes of which are given more detailed consideration elsewhere in this volume.

TABLE 1

The spatial and temporal scale at which processes that lead to the disturbance of the habitat occur. The list is illustrative and not designed to be comprehensive. Further this table illustrates the frequency with which such processes occur and does not indicate the time taken for the disturbed system to recover to a condition similar to that of an undisturbed area.

| Spatial scale | Temporal scale (frequency) | Processes |
|---|---|---|
| µm | milliseconds - seconds | chemical reactions, virus and bacterial driven processes |
| mm - cm | seconds - minutes | meiofaunal processes, macrofaunal sediment reworking, predation, herbivory, faecal production |
| 0.1 m – 1 m | minutes - days | bioturbation, diatom mat formation, megafaunal disturbances (e.g. feeding pit excavation), precipitation of minerals (e.g. calcium carbonate mounds) |
| 1 m – 100 m | days - months | Recolonisation and redistribution of small macrobiota, biomass and population fluctuations, sediment resuspension and settlement, bedload transport, tidal scour and currents |
| 100 – 10000 m | months – years | Hurricane and storm events, Iceberg scouring, submarine landslide events, seismic activity, recolonisation of large macrobiota |
| > 10000 m | years – decades | volcanic and seismic activity, anoxic events, submarine landslides, global warming, El Niño events, coral bleaching, global warming, recolonisation of large slow-growing macrobiota |

## 2.2 *Influence of biota on habitats*

Once organisms and plants colonise a substratum they will begin to modify its structure, either through physical disturbance of the habitat as they move through or grow into it, or they may modify its surface topography by protruding from or growing upon the substratum. Thus a seabed habitat is not just defined by the composition of its geological components, it is an amalgamation of these plus the characteristics conferred upon it by its associated fauna and flora.

The growth of plants within the sediment generally has the effect of further increasing its stability. The rhizomes of seagrasses help to bind sediment together as they form an intricate network through the sediment. The blades of the seagrass also reduce current flow and enhance sedimentation and the collection of organic material among the seagrass bed. Maerl is a form of calcareous algae that grows in twiglets and branches. These interlock to form a living sediment matrix with large interstitial spaces that permit deep penetration of oxygenated water into the maerl substratum. As a result of its structural complexity, maerl communities often have highly diverse community assemblages associated with them. Maerl is slow growing (1 $mm.y^{-1}$) hence these communities are particularly vulnerable to damage by anthropogenic activities such as fishing (see Hall-Spencer and Moore 2000). The holdfasts of subtidal kelp species can contribute to the erosion of the substratum, if for example, the kelp is torn from the substratum by storm action. If the kelp is attached to friable rock this can be sheared off with the holdfast while small boulders are more likely to be moved by wave action due to the drag associated with any attached kelp which acts like a drogue.

Animals can have both a stabilizing and destabilizing role within sediments. Palaeoecologists have studied the manner in which live animals perturb sediment structures. A seminal publication by Schaefer (1972) describes in detail the different modes of sediment disruption that occur as a result of animal activities. These vary from the surface 'bull-dozers' such as spatangoid sea urchins and whelks, to the feeding pits of starfish and rays, to the burrow labyrinths and chambers created by burrowing crustacea. All of these animal related sediment disturbances are known as bioturbation (Schafer 1972).

Bioturbators perform a key role in seabed systems as they rework surface sediments and enhance the passage of oxygenated water deeper into the sediment than it would otherwise penetrate by passive diffusion between sediment particles (Figure 23.4). The burrows themselves increase the surface area available for oxygen and ionic exchange that encourages enhanced microbial activity in the burrow walls. Some animals such as callianasid shrimps and polychaetes (e.g. *Arenicola* spp.) take advantage of this microbial growth and utilize it as a food source. Animals living within the sediment convey surface sediments to deeper layers within the sediment and then rework deep sediments and pass them back up the surface. This process recycles minerals and nutrients by transferring them to the surface of the seabed as sediment mounds and faecal pellets.

Figure 23.4 The physical disturbance created by individual fauna such as burrow dwelling crabs may be small in scale (cms), but the additive effects of the disturbance created by the entire population within an estuary can lead to the complete reworking of the surface sediment in just one tide (copyright MJ Kaiser).

## 3 Direct agents of change

Changes in sea level, ocean temperature and water circulation cause changes in habitats and their associated fauna over large areas usually over the long-term (> 20 years). On shorter time scales natural abiotic phenomena such as cyclones and hurricanes have regional impacts on a wide range of marine habitats (Hall 1994). At the same scale, biotic factors such as the naturally occurring outbreaks of ecosystem engineering organisms (organisms that modify the habitat through their activities such as grazing or predation) such as starfishes and sea urchins can lead to the alteration of habitat (e.g. Done 1982; Coleman and Williams 2002). Changes in ocean temperature may lead to the proliferation of non-indigenous species that can alter the existing habitat structure or associated assemblage (e.g. slipper limpets and some invasive marine algae) (Kaiser et al. 2001). Human activities undertaken on the continental shelf have the greatest immediate impacts on marine habitats. Such activities include total shoreline modification through sea defence construction and other coastal installations, hydrocarbon exploitation, marine mining for products such as diamonds, aggregate extraction, fishing with towed bottom-fishing gears, dynamite fishing on coral reefs, the use of cyanide in reef fisheries, tourism, dumping of waste materials, and anchor damage (see Jennings and Kaiser 1998).

### *3.1 Climatic and regional oceanographic influences*

Hydrographic processes exert a strong influence on the structure of continental shelf seabed ecosystems through their direct influence on seabed sediment composition and distribution and through their influence on the supply of organic matter to the seabed. Oceanographic features such as current patterns, river run-off, frontal systems and convective mixing combine to influence the nature of the seabed and the biomass and production of the associated communities. For example, Cruetzberg (1985) undertook a long-term study of an area of the North Sea in which a frontal system developed between mixed water and summer stratified water. The seabed habitat recorded in this area altered from sand to muddy sediment concomitant with the transition from the mixed to the stratified area. In addition, a 15 km band of sediment north of the boundary area was also found to be organically enriched and associated with a high biomass benthic assemblage. Cruetzberg (1985) interpreted the wider-scale influence of the frontal system to occur as the result of tidal advection and deposition of organic matter through the prevailing tidal currents.

El Niño and other large-scale hydrographic phenomena are well known to exert a strong influence on both fish and benthic biology leading to regime shifts in certain ecosystems (e.g. Cury et al. 2003). The effects of El Niño events can even be detected in the responses of meiofaunal assemblages. For example, Neira et al. (2001) found that meiofauna were more abundant at mid-shelf locations during an El Niño event off Chile. This was partly attributable to the deeper penetration of oxygen within the sediment matrix which no doubt expanded the available habitat for meiofauna and may have altered microbial production.

Large-scale oceanographic circulation patterns drive long-term hydrographic features such as the Gulf Stream which have formed the basis for the evolution of migration pathways in addition to other associations for fish and marine mammals. Nevertheless, short-term dynamics in frontal systems associated with the Gulf Stream can affect the distribution of predatory species of fish that are generally found to the northern side of the Gulf Stream (Magnuson et al. 1981). Other large-scale hydrographic processes such as salinity anomalies and wind-forcing factors have been linked to variable recruitment success in species such as cod *Gadus morhua* in the North Sea. Certainly, it is clear that long-term trends in fish populations can be tracked very closely to large-scale climate fluctuations (e.g. Bakun 1990; O'Brien et al. 2000; Cury et al. 2003). While the consequences of these fluctuations are relatively clear in the populations of species that are harvested, the effects at lower trophic levels may go undetected, but can manifest themselves in terms of habitat change if the organisms affected are habitat engineers (e.g. outbreaks of coral eating starfishes) (for review see Coleman and Williams 2002).

Seasonal climatic changes affect circulation patterns of the ocean and are related to the occurrence of storms, hurricanes, typhoons and monsoons that are responsible for the resuspension of bottom sediments and their transport from estuaries to the outer edge of the continental shelf. While these disturbance events can affect large areas of the seabed, their cyclic nature means that many of the biota affected have evolved life strategies or behaviours that enable them to reduce the negative effects of these events. For example, Posey et al. (1996) reported few ecological changes to the benthic fauna in the Gulf of Mexico that were exposed to hurricane disturbance, as only the top 6 cm of the sediment were disturbed. The fauna living in the top 6 cm of the sediment are in general motile and capabable of rapid burrowing, while those found deeper within the sediment were not affected by the physical activities of the hurricane.

### *3.2 Riverine input*

The composition and quantity of riverine discharge into the shelf seas will affect habitat composition and biology in the immediate area of discharge. Globally, this is an important process given that 70% of the sediments input to the sea are from riverine sources (Milliman 1991). However, the coastal zones of arid regions have little riverine discharge (e.g. the coast of Namibia), although these areas are often adjacent to up-welling systems that input nutrients from deeper oceanic waters. Since an understanding of the quantity and rate of riverine input is important for determining the consequences of dam construction, fluvial management regimes and the effects of alteration of patterns in precipitation linked to global climate change, it is surprising to note the paucity of studies that address this topic (Hall 2002). In general, however, the amount of sediment discharged by a river system will depend upon the size of the associated drainage basin; smaller drainage basins have a smaller surface area in which to store sediments which, for an equivalent flow rate, results in a 7-fold increase in sediment discharge for every 10-fold decrease in basin area (Milliman 1991). Thus smaller rivers discharge a proportionately greater load of sediment per unit of water discharged than much larger systems.

In regions where river water and sediment discharge are significant (e.g. the Amazon and Papuan shelves) they are known to play an important role in structuring the topography and sedimentary conditions of the immediate shelf. In addition, the input of organic matter fuels benthic food webs and microbial decomposition processes (e.g. Alongi and Robertson 1985). Rhoads et al. (1985) proposed a generalised model of the ecology of these systems. Close to the river mouth the water column is dominated by river effluent; here, distinct layering of sediment occurs which reflects discrete periods of deposition and erosion. Consequently the fauna is highly impoverished in this environment and tends to be dominated by bacteria. Further offshore, the sediment burden restricts primary production, but sedimentation rates are moderate and the increasing depth of water induces greater habitat stability and hence the environment became more suitable for benthic macrofauna. Further offshore still, high primary productivity occurs with high rates of nutrient release to the water column from benthic activity. Phytodetritus supplements the food source to the benthos and encourages bioturbating fauna which tend to be surface deposit feeders. Microbial activity is enhanced further increasing nutrient cycling. Beyond this region in the open sea, primary production in low despite low turbidity, sediment accumulation is minimal and benthic biomass is low.

Rhoads et al.'s (1985) model appears to be supported by a more recent empirical study which demonstrated that benthic community structure altered in phase with strong seasonal and spatial patterns and that this trend covaried with shelf-wide physical processes (Aller and Stupakoff 1996). The body size, abundance and depth distribution of benthic fauna were found to be at their lowest right across the shelf at times that coincided with periods of peak riverine discharge and maximum trade-wind stress. During the periods of low discharge and minimal wind stress, macrofauna abundance was highest and bacterial biomass increased by a factor of two.

Radiographs of sediment structure on the continental shelves of the Amazon, East China Sea and Indonesia reveal a large number of primary structures (e.g. erosional or burial discontinuities in subsurface sediments) and disrupted biogenic traces. This indicates that the periodic elimination of benthic fauna by physical disturbance events is a natural feature of river-dominated tropical and sub-tropical benthic assemblages (Alongi and Robertson 1995). Examination of invertebrate death assemblages in subsurface deposits off the Amazon indicate periodic surface exposure and burial at depths of up to 40 m which indicates the large-scale nature of such disturbances (Aller 1995).

### *3.3 The influence of land-use*

Deforestation and other agricultural practices have lead to substantial soil erosion in certain parts of the world. This erosion is estimated to be 3.7 times greater now than 2500 years ago, prior to the period when humans first started to clear forests for agriculture. For example, the Yellow River discharges ten times more sediment at present than it did prior to the widespread cultivation of land in northern China. It seems likely that sediment input to shelf seas will continue to increase with the sustained demand for greater areas of agricultural land to support subsistence farming in developing countries (FAO 2000).

High nutrient loads in river discharge are associated with phytoplankton blooms that generate mainly diatomaceous phytodetritus. This material fuels rapid rates of community respiration and hence leads to the depletion of oxygen at the sediment-water interface. Such events are usually periodic and tend to coincide with periods of calm weather when inshore waters undergo stratification. The precise timing and extent of such events will vary inter-annually according to the exact climatic conditions that prevail and upon the quality and quantity of river discharge (Alongi 1998). The Baltic Sea, the Adriatic and the Gulf of Mexico are well documented examples of locations where such events occur on a regular basis (Diaz and Rosenberg 1985; Justic 1991; Turner and Rabalais 1994). Systems that are particularly susceptible to eutrophication would include those with restricted tidal inundation, poor water exchange or those susceptible to stratification. Typically such systems include fjordic areas, enclosed bays and areas with relatively deep water.

While anoxic events would appear to be natural events associated with areas of river discharge, the extent and severity of these events would appear to be related to human influences. Nitrogen loading had doubled in the Mississippi between the 1900s and 1980s and this has prompted some to suggest that the increasing scale of anoxic events in the Gulf of Mexico are linked to eutrophication (Turner and Rabalais 1994). This supposition is supported by studies of stable isotope signatures and organic tracers in sediment cores which indicate that nutrient levels started to increase in the 1950s and finally levelled off in the 1980s. This coincides with a three fold increase in the use of chemical fertilizers in the latter half of the $20^{th}$ Century (Eadie et al. 1994). The sources of nitrogen that flow into the Mississippi would appear to be diffuse and can originate as much as 1000 km up stream.

Inputs of nitrogen are not only derived from land run-off into river systems. There is increasing recognition of the importance of atmospheric sources of nitrogen. Fossil fuel consumption appears to be the primary source of atmospheric inputs of nitrogen which have increased by 50 to 200% over the last 50 years (Paerl 1995).

In addition to concern regarding the effects of elevated nutrient inputs into coastal waters, there are many other contaminants that persist in the marine environment, particularly in sediments. Many of these contaminants, such as radionuclides and organic substances, are derived from industrial and heat generation sources. Polychlorinated Biphenols (PCBs) are particularly persistent and accumulate through the food chain with negative effects expressed in predators at the top of food chains. For example, PCB contamination of prey has been linked to hatching failure for a number of avian predators and mercury occurs in higher concentration in the feathers of seabirds sampled in the later half of the $20^{th}$ Century compared with those sampled from museum collections (Thompson et al. 1998; Arcos et al. 2002). While the effects of such contaminants on apex predators have been documented in a number of cases, the effects on organisms at lower trophic levels is less well studied. Although invertebrates such as amphipods are used in lethal toxicity tests, the incidence of contaminant mortality associated with bioengineering organisms that affect habitat structure remains unknown. The effects of contaminants may be subtle in that they do not necessarily cause direct mortality but may have negative effects on recruitment processes and larval viability (e.g. McCarthy et al. 2003).

### *3.4 Reduction of water discharge*

Clearly the inputs derived from riverine discharge have an important influence on shelf sea ecosystem processes. However, the damming of rivers for the purposes of diverting water to agricultural and industrial usage has severely lowered the discharge of some river systems with negative consequences for local ecosystems. For example, in the Adriatic Sea, output of the River Po and adjacent river systems has been lowered by 12% in the recent years. The reduction in nutrient inputs has been associated with a decrease in primary production in the local shelf water mass (Alongi 1998). The structure of many river delta regions are maintained by the supply of suspended river born sediment. However, human interference with the supply of suspended sediment can lead to alteration in the structure of deltaic regions. For example, the construction of the barrage on the River Nile in 1868 led to the reversal of the accretion of the Nile delta, a situation later worsened by the construction of the Aswan dam. This led to coastal erosion rates of between 5 – 240 m per year. The associated reduction in nutrients transported out to sea was linked with a decline in the catches of sardines. In later years, fishery production increased as the reductions in nutrients supplied through the Nile discharge were replaced by increasing coastal urbanisation (Caddy 2000).

### *3.5 Petroleum exploration and production*

The impacts of oil production on shelf sea ecosystems has been well reviewed (Kingston 1992; Olsgard and Gray 1995). The ecological effects of the contaminants of drilling muds typically lead to a reduction in species diversity, the effects of which ameliorate with increasing distance from the drilling platform (e.g. Gray et al. 1990; Warwick and Clarke 1995). In the North Sea, the chronic use of drilling muds over a period of 6–9 years was associated with community effects at a distance of up to 6 km from the drilling platform, although in cases when water-based muds were used, the ecological effects were reduced (Olsgard and Gray 1995). In other areas the effects may be much more localised as in the case of gas platforms in the Gulf of Mexico where the ecological effects of the drilling muds was confined to a distance of 100 m from the drilling platform (Montagna and Harper 1996).

### *3.6 Aquaculture*

Inland and marine aquaculture is a rapidly expanding industry which grew by 5% per year between 1950 and 1969 and by about 8% per year during the 1970s and 1980s. This growth rate has increased further to 10% per year since 1990 (FAO 2000b). This has resulted in an overall production increase of 20 million tonnes over the last decade. Most of the mariculture component is coastal, and as such affects only a limited area of the continental shelf *per se.* Nevertheless at the shoreward margin, local impacts on seabed communities caused by inputs from shore-based aquaculture facilities (e.g. ponds) or marine cages have been clearly documented for both finfish cultivation (e.g. Findlay et al. 1995) and shellfish production (Kaiser 2001). Most of the important habitat modification processes related to aquaculture relate to pollution from excessive organic inputs and lead to classic responses in the benthic fauna (Pearson and Rosenberg 1978) and in extreme cases can result in anoxic sediments in the immediate area of cultivation and a benthic community dominated by bacteria such as *Beggiatoa.* More specific details of some of the mechanisms of habitat modification associated with aquaculture practices are given below.

### 3.6.1 Bivalve aquaculture

There is an increasing awareness of the environmental effects that may result from the various stages of bivalve cultivation processes. Most notably, adverse effects have been associated with mussel and oyster farms on the continent of Europe (Tenore et al., 1985, Castel et al., 1989). Environmental problems occurred at sites where hydrographical conditions were unsuitable for high density cultivation of bivalves and the carrying capacity of the local marine environment was exceeded (Castel et al., 1989). One of the main attractions of bivalve species for aquaculture, is that they are on-grown to market size in the natural environment without the requirement for supplemental or artificial feeds. Nevertheless, on-growing usually requires the introduction of structures into the marine environment on or from which the bivalves are either supported or suspended. The introduction of such structures has an immediate effect on local hydrography and provides a new substratum upon which other epibiota can settle and grow. In addition, the introduction of high densities of cultivated organisms increases local oxygen demand and elevates the input of organic matter into the immediate environment. At high stocking densities of bivalves the larval settlement of other benthic species may be reduced as their larvae are filtered and digested or become entrained within pseudofaeces (Baldwin et al., 1995).

In suspended cultivation systems, bivalves are either attached directly to ropes or placed in net bags attached to ropes suspended in the water column from floats or rafts. The bivalves provide a surface on which many epibiotic organisms attach and grow. Thus a large biomass of biota develops in suspended culture systems and this in turn has a major effect on phytoplanktonic, benthic and hydrographic processes in close proximity to the cultivation site. As with other bivalves, mussels provide a complex surface area on which epifaunal communities consisting of over 100 species can develop (Tenore and Gonzalez, 1976). As organisms die and fall off the ropes, they can provide a significant food resource for species found in the vicinity of the culture site. Small portunid crabs, *Pisidia longicornis,* were found to be abundant among fallen mussels beneath rafts in the Spanish rias. These, in turn, were fed upon opportunisitically by several fish species that normally consume polychaete worms (Lopez-Jamar et al., 1984). *P. longicornis* were so abundant at cultivation sites that their larvae dominate (90% of the biomass) the zooplankton community that would be normally characterised by copepods (Alvarez-Ossorio, 1977). Mussels excrete high levels of ammonia (Tenore and Gonzalez, 1976) which promotes high levels of productivity in algae attached to mussel lines. It is calculated that this is equivalent to algal production in local intertidal systems (Lapointe et al., 1981). So great is the productivity associated with mussel lines in the Spanish rias, that Tenore et al., (1982) speculated that inshore fisheries were potentially enhanced by the bedload transport of organic rich sediment into coastal areas.

Cultivation sites that are well flushed by tidal currents, as in the Spanish Rias, do not encourage the accumulation of pseudofaeces beneath mussel rafts that can lead to the development of anoxic conditions (Rodhouse and Roden, 1987). The relatively beneficial effects that occur in the Spanish rias contrast sharply with the effects observed by Dahlbäck and Gunnarsson (1981) in Sweden. They demonstrated organic sedimentation rates of 2.4–3.1 g organic C $m^{-2}$ $d^{-1}$ beneath mussel longlines that was twice as much as found in adjacent uncultivated areas. This excessive organic enrichment was lead to the development of anoxic sediment conditions. Mats of bacteria, *Beggiatoa* spp., then developed beneath the the longlines at their study site. In this situation, the benthic infauna had low diversity and biomass which is a well documented response to highly polluted sites (Pearson and Rosenberg, 1978). Similarly, the productivity of densely stocked Japanese oyster grounds was detrimentally affected by the generation of large quantities of pseudofaeces and high filtration rates (Ito and Imai, 1955; Kusuki, 1977). Pseudofaeces production beneath oyster cultivation rafts that it was at least equivalent to natural sources of sedimentation (Mariojouls and Kusuki, 1987). Many of the problems associated with suspended mussel culture also apply to suspended fish cage culture and there is presently a desire to move towards automated cultivation systems that can be sited well offshore where the environmental effects would be greatly reduced due to the larger carrying capacity of open waters.

### 3.6.2 Shrimp farming

Despite the fact that shrimp farming has become a multi-billion dollar business worldwide, the sustainability of this sector of the aquaculture industry is currently affected by widespread crop failures and associated negative environmental impacts that result from the farming practices. Water exchange between shrimp farms and their surrounding environment is a standard practice to avoid excessive build up of waste products, avoid excessive eutrophication within ponds and to maintain healthy planktonic blooms. Extensive cultivation systems require a daily water exchange rate of up to 5% whereas intensive systems require an exchange rate of up to 30% (Clifford, 1985). Calculations of the nutrient budgets for such systems indicated that more than 76% of the nitrogen and more than 87% of the phosphorous input is retained within the pond water and sediments (Robertson and Philips, 1995). These nutrients, are discharged into the surrounding ecosystem whenever the pond water is exchanged, harvesting occurs or when the ponds are dredged out. However, the receiving body of water will have a limited capacity to assimilate concentrated and persistent pulses of highly nutrient enriched waters and sediments. At some point, the critical load of nutrients will be attained beyond which the quality of the receiving water will begin to deteriorate with negative consequences for the local flora and fauna. Mangroves are an obvious sink for such nutrient outputs and Robertson and Philips (1995) calculated that approximately 3 ha of mangrove was required to assimilate the nutrient load generated by a 1 ha semi-intensive shrimp farm. This rose to 22 ha of mangrove for 1 ha of intensive shrimp pond. While the outputs from isolated shrimp farms do not tend to these critical levels, a successful farming operation often attracts the development of other farms in close proximity. As the number of farms in a restricted area increases so the water quality begins to deteriorate as the concentration of nutrient discharges increases. This inevitably leads to elevated stress levels in the shrimp and increased susceptibility to infection by pathogens. Proactive management can help to avoid these situations occurring. For example in Canada the capacity of proposed salmon farm sites to assimilate nutrients is examined prior to consent for planning, while within enclosed bodies of water, salmon farms are required to be spaced at least 3 km apart to minimise environmental impacts (Black and Truscott, 1994).

### 3.6.3 Shrimp farming and mangrove destruction

Mangroves are an important habitat, acting as nursery areas for estuarine fishes and invertebrates, and provide a feeding and breeding ground for birds and mammals. They provide protection against storm activity and prevent soil erosion and provide an important source of income for poor coastal communities (Bailey, 1988). In the initial stages of the shrimp industry, mangroves were considered to be areas of low commercial value that were ideal for development for shrimp farming. Furthermore, mangroves are the natural habitat for many of the cultivated shrimp species hence they seemed to be the ideal location for shrimp farms (Fegan, 1996). However, mangroves are actually very poor sites for shrimp farms as their acid sulphate soils have acidity of pH 3–4 when dried out. On a global scale, shrimp farming has been responsible for less than 5% of mangrove destruction to date. However, on a localised scale the impact of shrimp farming may be far more severe, with almost complete clearance in some areas (Phillips et al., 1993). Aquaculture pond construction projects have destroyed 20% of the mangrove forests in some areas of Ecuador (Snedaker et al., 1986) while in Indonesia the majority of the 300 000 ha of mangrove forest cleared to date is currently used for shrimp cultivation.

Shrimp farming has a number of other impacts on the environment such as the intrusion of saltwater into neighbouring agricultural areas and may even pollute groundwater supplies. Many of the chemical treatments used to control outbreaks of disease are occasionally used in excess and pollute surrounding waters or alternatively they may build up within sediments to high concentrations (e.g. copper compounds). The collection of post-larvae and reproductive adults from the wild for the aquaculture industry supports artisanal fisheries in some areas. Although most shrimp production is now supported by hatcheries, one of the major cultivated species *Penaeus monodon,* still requires the capture of adult broodstock from the wild. In some areas the collection of wild juveniles may also have effects on other biota as shrimp are important predators of juvenile fish and benthos in coastal margins.

## *3.7 Direct effects of fishing on the seabed*

It is now well established that bottom-fishing activities that involve the use of mobile gear have a physical impact upon the seabed and the biota that lives there. However, the ecological significance of these fishing disturbances for the immediate and wider ecosystem are coloured by differing views on their importance (e.g. Dayton *et al.,* 1995; Kaiser 1998; Watling and Norse 1998; Norse and Watling 1999; Auster and Langton 1999; Kaiser and De Groot 2000; Kaiser and Jennings 2001; Thrush *et al.* 2002). The ubiquity of trawling is revealed by a recent analysis where catches were mapped to country continental shelves (Hall et al. unpublished data); trawling and dredging activity was reported from 19,984,200 ($km^2$) of national continental shelves, which represents about 75% of the global continental shelf. Although there is significant variation in catch density (catch by a country divided by the area of continental shelf), it is clear that demersal trawling and dredging is a ubiquitous global activity on continental shelves. However, although mobile fishing gears are the major concern on a global scale, other physically destructive practices such as dynamite fishing or the "muro-ami" fishery of the Philippines

(Carpenter, 1977), which employs stones, chains or poles to break up coral and drive fish out into nets, are also locally important.

The majority of seabed (demersal) fishing activity is undertaken in shallow seas on the continental shelf at depths of less than 200 m. However, as traditional stocks of fish dwindle, fishers have moved their attention to previously unexploited species. As a result, bottom trawling occurs around sea mounts and on the continental shelf slope at depths greater than 1000 m. Benthic communities experience continual disturbance at various scales in time and space (for more detail see Hall 1994 and Hall et al. 1994). In general, shallow continental-shelf sea environments experience more frequent disturbances than those deeper sea environments that are not exposed to wave action and strong currents (Figure 23.5). Large-scale natural disturbances, such as seasonal storms and regular (daily) scouring by tidal currents, form a background against which other smaller disturbances occur, such as those induced by predator feeding activities (Figs. 2 and 4). The additive effects of many small-scale disturbances may be obliterated by larger, but less frequent, disturbance events (e.g. Hall et al. 1993). Thus for fishing disturbance of the seabed to have an ecologically significant impact it must exceed the background levels and frequency of natural disturbance. It is important to consider the relative scale at which fishing disturbance occurs. Given a similar habitat, very intensive but highly localised fishing disturbance may have fewer ecological implications than less intense, but wide-spread fishing disturbance.

Figure 23.5 Examples of four habitats that differ in their stability and their exposure to natural forms of disturbance which influences their vulnerability to human activities. Surf beaches (A) are exposed to high levels of chronic natural disturbance by wave action and have an unconsolidated coarse sediment structure and low species abundance and diversity. Gravel lag (B) sediments can provide a stable substratum but can occur in areas of high tidal scour resulting in an impoverished fauna as in this example. Granite bedrock at a depth of 140 m provides a stable substratum protected from wave action, upon which slow growing sessile biota can become established such as the sponges in this illustration. Reef building fauna develop into a stable substratum and enhance local productivity through the production of faeces as seen in this rich horse mussel habitat. Photograph A copyright MJ Kaiser, photograph B,C,D copyright E.I.S. Rees with permission.

Shallow-water communities on exposed coastlines are likely to be the most resilient to physical disturbance from bottom fishing (Figure 23.5). For example, Posey et al. (1996) recently demonstrated that even large-scale disturbances, such as hurricanes, have relatively short-term effects on shallow water communities adapted to frequent physical disturbance. Nevertheless, there are situations in which the associated fauna can increase the inherent stability of supposedly unstable habitats. For example, dense aggregations of spionid worms can increase the stability of intertidal sediments through their ability to bind sediment particles together (Thrush et al. 1996). There is no doubt that as habitat stability increases the relative effects of fishing will also increase as will the longevity and severity of the its ecological effects (Collie et al. 2000).

### 3.7.1 Alteration of surface topography

Perhaps the most obvious change that trawling can cause is in surface topography and most current studies support the general conclusion is that trawling increases surface roughness, owing to the furrowing caused by trawl doors (e.g. Schwinghamer et al. 1996). An issue of scale is important here, however, since at a slightly smaller scale trawls generally lower surface topography by smoothing ripples, and mounds and other structures created either by fauna or the physical environment. This combination of changes should, perhaps, be viewed as the replacement of a landscape with widespread, small scale, low relief topographic features (ripples and mounds) with a rather smoother landscape, interspersed with higher relief, but less frequent features caused by the ploughing of trawl doors.

Paradoxically, it may be the loss of smaller scale features that should be of most concern. Currie and Parry (1996), for example, reported clear visible changes following scallop dredging which persisted for about a year. For the most part, these changes were associated with the destruction of mounds and depressions caused by the burrowing activities of Callianasid shrimps. In turn, this flattening led to the removal of unattached weed and seagrass which tends to accumulate around such features. The spatial heterogeneity that mounds and patches of organic matter provide is often considered to be an important factor controlling the diversity and species composition of benthic infaunal communities (e.g. Hall, 1994), so their destruction would be expected to have effects. In addition, we know from field observations and experimental studies that juveniles of demersal fish on continental shelves might benefit from a high abundance of relatively small physical features (sponges, empty shells, small rocks etc) (Gotceitas and Brown 1993, Juanes and Walters 1993; Auster et al. 1997). Over time trawling can be expected to gradually lower the physical relief of the habitat with potentially deleterious consequences for some fish species.

### 3.7.2 Effects of sediment resuspension

The direct physical contact of fishing gear with the substratum can lead to the resuspension of sediments and the fragmentation of rock and biogenic substrata. To date few studies have directly examined the potential ecological consequences of sediment resuspension attributed to trawling activities. The resuspension, transport and subsequent deposition of sediment may affect the settlement and feeding of the biota in other areas. Sediment resuspended as a result of bottom fishing will have a variety of effects including: the release of nutrients held in the sediment

(Duplisea et al. 2002), exposure of anoxic layers, release of contaminants increasing biological oxygen demand (Reimann and Hoffman 1991), smothering of feeding and respiratory organs. The quantity of sediment resuspended by trawling depends on sediment grain size and the degree of sediment compaction which is higher on mud and fine sand than on coarse sand.

Resuspended sediments must subsequently resettle, either in situ or after transport by water currents. Only a few estimates of the magnitude of these processes have been made (e.g. Churchill, 1989; Pilskaln et al. 1998). Churchill (1989) for example, estimated that coarse sand was typically penetrated to a depth of 1 cm by otter boards, which re-suspended approximately 39 kg sediment. $sec^{-1}$, whereas the figures for fine sand and muddy sand were 2 cm (78 $kg.sec^{-1}$) and 4 cm (112 kg. $sec^{-1}$), respectively. After monitoring salinity and suspended sediment load over a three month period at a 125 m deep site in the Middle Atlantic Bight, Churchill (1989) concluded that most of the suspended sediment load was advected from inshore. Storms in shallower water accounted for most of the suspended sediment pulses, except for the most dramatic events during the fishing season, which coincided with intense fishing activity.

Transmissiometers that measure background light levels in water, frequently recorded the highest levels of turbidity during periods of trawling activity off the north east coast of the United States (Churchill 1989). In deeper water where storm-related bottom stresses have less influence, otter trawling activity contributed significantly to the resuspension of fine material. Churchill (1989) calculated the sediment budget for certain areas of the mid-Atlantic Bight and concluded that trawling was the main factor that accounted for the offshore transport of sediment at depths of between 100 and 140 m. However, Churchill (1989) calculated that the transport of sediment that resulted from fishing activities would not produce significant large-scale erosion over a period of a few years. However, Churchill (1989) made no inferences regarding the potential biological impact of this sediment transport.

Planques et al. (2001) have undertaken the first observations of the sedimentological consequences of trawling on continental shelf sediments. They used moored instruments and transmissiometers to quantify the effect of an experimentally fished otter trawl on the fine-mud sediment in water 20–40 m deep off the coast of Barcelona, Spain. They found that the disruption of the surface layers of the sediment led to elevated levels of tidally resuspended sediment for up to 5 days after the trawl disturbance event. The furrows made by the otter boards remained evident for at least one year after the initial disturbance which corroborates other similar examinations of trawl marks in muddy sediments (e.g. Tuck et al. 1998). The furrows made by trawling in these sediments have the potential to significantly increase topographic complexity. The ecological significance of this change to seabed surface is unknown.

Given that bottom trawling can lead to large-scale resuspension and transport of sediment it is reasonable to ask whether fishing changes the particle size distribution or the internal structure of sediments. One might imagine, for example, that finer sediments would be washed out and transported further by water currents, leading to a gradual coarsening of median sediment grain size. Little data are available to address this question, but efforts to use acoustic methods for evaluating changes to sediment structure have recently been developed (Schwinghamer et al. 1996). The results of an experimental trawling study on hard-packed sand described in Schwinghammer et al. (1996) indicated that trawling decreased, the fractal geometry (structural complexity) of the internal sediment structure at mm scales—a characteristic that was not revealed by analysis of bulk sediment properties. The authors suggested that such changes might be due to collapse of burrow structures and sediment voids caused by fauna and that the structural change might affect exchange processes with overlying water column. In follow up studies, however, acoustic estimates of internal sediment structure gave inconsistent results that were difficult to interpret with respect to experimental trawling (Schwinghammer et al. 1998). Similarly, changes in bulk sediment properties showed large spatial and temporal variability. On balance it would appear that the kinds of changes in sediment structure due to trawling may be rather subtle compared to changes due to other factors.

### 3.7.3 Effects on biogeochemical processes

Bottom trawling is a key source of physical disturbance in shallow shelf seas, but little is known of the effects of trawling disturbance on functional processes, despite the expectation that sediment community function, carbon mineralisation and biogeochemcial fluxes will be strongly affected by trawling disturbance. This is because trawling reduces the abundance of bioturbating macrofauna that play a key role in biogeochemical processes and because the physical mixing by trawling unlike the mixing by macrofauna does not contribute directly to community metabolism. Duplisea et al., (2001) used an existing simulation model of a generalised soft sediment system to examine the effects of trawling disturbance on carbon mineralisation and chemical concentrations. They contrasted the effects of a natural scenario, where bioturbation increases as a function of macrobenthos biomass, with those of a trawling disturbance scenario where physical disturbance results from trawling rather than the action of bioturbating macrofauna (which are killed by the action of the trawl gear). Simulation results suggest that the effects of low levels of trawling disturbance will be similar to those of natural bioturbators but that high levels of trawling disturbance cause the system to become unstable due to large carbon fluxes between oxic and anoxic carbon compartments. The presence of macrobenthos in the natural disturbance scenario stabilises sediment chemical storage and fluxes, because the macrobenthos are important participants in the total community metabolism. In soft sediment systems, where physical disturbance due to waves and tides is low, they suggested that intensive trawling disturbance may destabilise benthic system chemical fluxes, and that this instability had the potential to propagate more widely through the marine ecosystem.

### 3.7.4 Effects of fishing in different habitats

Several authors have suggested that the relative ecological importance of fishing disturbance will be related to the magnitude and frequency of background of natural disturbances that occur in a particular marine habitat (Kaiser 1998; Auster and Langton 1999). Certainly, it makes intuitive sense that organisms that inhabit unconsolidated sediments should be adapted to periodic sediment resuspension and smothering. Similarly, it seems plausible that organisms living in seagrass beds rarely experience repeated intense physical disturbances or elevated water turbidity as created by bottom fishing gears. Indeed, such intuition has been the cornerstone of hypotheses about impacts and recovery dynamics for benthos (eg Hall, 1994; Jennings and Kaiser, 1998). However, Collie et al. (2000) found that their initial impact results with respect to habitat were somewhat inconsistent among analyses. While the initial responses to fishing disturbance of taxa in sand habitats were usually less negative than in other habitats, a clear ranking for expected impacts did not emerge. Such inconsistencies may reflect interactions between the factors arising from the unbalanced nature of the data, with many combinations of gear and habitat absent. For example, the relatively low initial impact on mud habitats may be explained by the fact that most studies were done with otter trawls. If data were also available for the effect of dredgers on mud substrata a more negative response for this habitat may have been observed. Nevertheless, it should be borne in mind that initial effects of disturbance may be hard to detect in mud communities that often have low abundances of biota which tend to be burrowed deep (10–200 cm) within the sediment. Presumably, the deeply burrowed fauna would be relatively well protected from the physical effects of disturbance although the passage of the gear will cause their burrows to collapse. Whether these inconsistencies can be explained in this way can only await further study. It is also important to note that it is important not to classify habitats by the particular nature of the sediment. For example, intertidal sandflats inhabited by high densities of tubiculous worms such as spionids will be more stable (and hence more adversely affected by fishing) than sandflats with relatively little infauna (Thrush et al. 1996).

### 3.7.5 Habitat recovery

The short-term effects of bottom-fishing disturbance on habitats and their biota are of interest but of less ecological importance than the issue of the potential for recovery or restoration. The short-term outcome of disturbance experiments is open to misinterpretation. Unfortunately, relatively few studies of trawl disturbance have included a temporal component of sufficient duration to address longer-term changes that occur as a result of bottom fishing disturbance. This is almost certainly a result of the conflict between financial resources, project duration, statistical and analytical considerations. Nevertheless, Collie et al. (2000) were able to incorporate studies that included a recovery component into their analysis. This permitted them to speculate about the level at which physical disturbance becomes unsustainable in a particular habitat. For example, their study suggested that sandy sediment communities are able to recover within 100 days which implies that they could perhaps withstand 2–3 incidents of physical disturbance per year without changing markedly in character. If these recovery rate estimates for sandy habitats are realistic, this would suggest that areas of the seabed that are trawled more frequently than three times per year are held in a permanently altered state by the physical disturbance associated with fishing activities. As we discuss later, such levels of fishing disturbance exist in areas such as the North Sea and this outcome has important implications for predicting the outcome of management systems that may cause changes in the spatial pattern of seabed disturbance. This expectation is supported by a recent study that links the size and species composition of North Sea benthic communities to patterns of chronic beam trawling disturbance (Jennings et al. 2000b). There was minimal evidence for trawling effects on size composition or benthic production in a series of sites trawled up to 2.3 times year$^{-1}$. However, at another series of sites trawled up to 6.5

times year$^{-1}$, the most heavily trawled sites were characterised by a fauna of low biomass and low production, that consisted of very small individuals. Larger bivalves and burrowing sea-urchins, that can dominate the biomass in infrequently trawled areas, were effectively absent (Jennings et al. 2001b).

### 3.7.6 Effects of fishing on coral reefs and other biogenic habitats

The benthic communities which human beings identify most strongly with are those that are characterised by a rich epifauna, that provide abundant biogenic structure. Coral reefs are the exemplars, but sponge gardens, calcareous algae, or maerl beds and various hard substratum communities are all valued targets for conservation. There is little doubt that put in the path of a trawl, or subjected to dynamite or "muro-ami" fishing, these communities are at risk. A good example of the magnitude of effect is provided by Hall-Spencer and Moore (2000) who showed that scallop dredges have profound effects on calcareous algae (maerl) beds, with up to 70% of thalli in dredge tracks killed through burial. Similarly, Poiner et al. (1998) report the results of a trawl depletion experiment in the inter-reef areas of the Great Barrier Reef, which showed that each trawl removed and caught between 5 and 20% of the available biomass of sessile fauna, with 70–90% removed after 13 trawls. Note that the above estimate does not include fauna that were detached from the seabed, but not caught. However, video analysis of the effects of the trawl ground rope undertaken by Sainsbury (1988) for the Australian northwest shelf, indicate that about 89% of encounters lead to dislodgment of sponges and almost certainly subsequent death.

With respect to other forms of fishing that affect habitat structure, dynamite and cyanide fishing on coral reefs are probably the most obvious, but the removal of fish themselves may also affect the nature of the available habitat. Species that act as ecosystem engineers are being increasingly recognized as playing an important role in the marine systems. In the Gulf of Mexico, for example, where unconsolidated sediment overlies hard rock substratum, fish such as the red grouper have been shown to create burrows, or dig pits, down to the rock (Coleman and Williams 2002). This seabed excavation in turn allows a rich epifaunal community to colonise. In such circumstances, depletion of the fish resource will lead to concomitant effects on the biodiversity of the benthos. The extent of such phenomena is currently unknown.

It could certainly be argued that by effecting changes to biogenic structure that fishing is most likely to influence the benthic communities of marine systems. Although the data are relatively sparse and well executed studies of effects of mobile bottom-fishing gears on many biogenic habitats are difficult to find (e.g. Collie et al. 2000), it seems self evident that destroying the physical integrity of reefs or other biogenic structures will have profound consequences, both for fish populations and the other taxa. Indeed, on coral reefs, some of the most complex of biogenic habitats, there are significant positive relationships between fish biomass and topographic complexity (Luckhurst and Luckhurst 1978; Roberts and Ormond 1987). What is true for coral reefs is almost certainly true for other biogenic habitats. The issue here is not only that marked and undesirable effects ensue when trawling, dynamite fishing, or other physically disruptive practices take place, but the extent of the fishing activity and the distribution of sensitive habitats. The lack of high resolution (± 100 m) maps of benthic habitats and biota is probably the biggest current impediment to effective protection of vulnerable habitat from fishing activities. Only by combining such data with micro-scale data on the distribution and frequency of trawling disturbance for major fishing grounds can we accurately assess the extent of impact of fishing on benthic habitats. Such data will also provide a sound basis for developing mitigation strategies.

Since trawling disturbance reduces habitat complexity, this may reduce the total production of the associated community. However, it has also been argued that frequent trawling disturbance may lead to the proliferation of smaller benthic species with faster life histories that can withstand the mortality imposed by trawling and are favoured as food by commercially fished species. Since smaller species are more productive, limiting trawling disturbance may 'farm the sea', with knock-on benefits for consumers, including some fish populations.

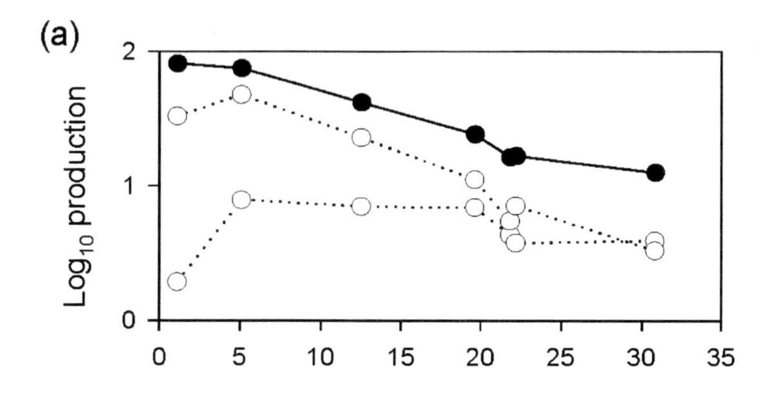


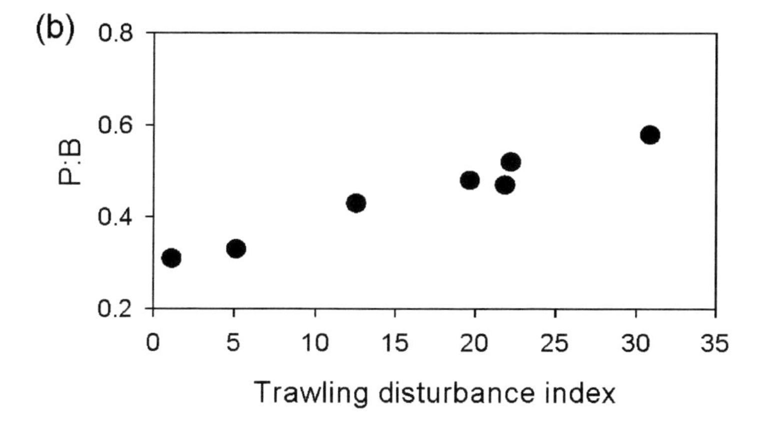


Figure 23.6 (a) The relationship between trawling disturbance and production for infauna in the Silver Pit region of the North Sea. The continuous line shows the relationship for the whole community, while the upper broken line shows the relationship for bivalves/spatangoids and the lower broken line shows the relationship for polychaetes (not significant). (b) The relationship between trawling disturbance and production to biomass ratios for the same infaunal community (After Jennings et al., 2001b).

### 3.7.7 Effects of fishing on benthic production

There have been few empirical tests of the impact of trawling disturbance on production, but one study of trawling impacts on a soft-sediment community suggests that the effects may be negative. In this study, benthic production was assessed across quantified gradients of trawling disturbance on real fishing grounds in the North Sea (Jennings et al. 2001b). Biomass fell with increased levels of trawling disturbance (Figure 23.6). Production was estimated from size-spectra, using an allometric relationship between body mass and the production to biomass (P:B). In heavily fished areas, the abundance of larger individuals was depleted more than smaller ones, as reflected by the positive relationship between the slope of the normalised size spectra and trawling disturbance. Relative infaunal production (production per unit biomass) rose with increased trawling disturbance, and this was largely attributable to the dominance of smaller animals in the disturbed communities. The significant increase in relative production (P:B) did not compensate for the loss of total production that resulted from the depletion of large individuals (Figure 23.6). Multivariate analyses have also been used to show that bottom trawling disturbance had a dominant effect on the size structure and production of the fauna of this soft-sediment benthic community and that the role of other environmental variables such as sediment particle size and depth was relatively weak (Duplisea et al. unpublished). Given that trawling disturbance led to reductions in production in this relatively simple and unstructured habitat, it is expected that the effects may be even more profound when trawling leads to reductions in the complexity of highly structured biogenic habitats and their associated fauna.

## 4. Changes in trophic interactions

The gradual decline in stocks of target species is the most obvious response to human harvesting activity. However harvesting activities affect far more than just the target species (Pace *et al.*, 1999). Target species fulfil their own functional role within the ecosystem of which they are a component e.g. they may be predators, prey, scavengers, competitors, form habitat structures and maintain patchiness and habitat heterogeneity. The consequences of harvesting a particular species will depend on their role and dominance within that ecosystem. Hence harvesting a top predator might relieve predation on its prey species while harvesting prey species will increase predation pressure on alternative prey types. Yodzis (1998) highlights the pitfalls in these over-simplistic predictions of the outcome of predator removal in marine ecosystems when he describes the arguments usually put forward for marine mammal culling in the Benguela ecosystem: *"We harvest the target species; marine mammals eat the target species; if marine mammal populations are reduced by some amount X, then we can increase our harvest by the amount those X marine mammals would have eaten, without any net decrease in target species stock."* What falls out of Yodzis' (1998) study is the inescapable fact that despite a well quantified ecosytem the outcome of culling top predators is highly variable for which there were two possible explanations. Either random variation in the outcome of predictive models reflected ignorance of the system due to limitations in the measurements of parameters taken, or it could reflect random influences that are experienced in the system that cannot be eliminated (e.g. climatic variation).

With respect to benthic systems, Jennings et al. (2001a) found that chronic trawling disturbance led to dramatic reductions in the biomass of infauna and epifauna, but that these reductions were not reflected in changes to the mean trophic level of the community, or the relationships between the trophic levels of different size classes of epifauna. Despite order of magnitude decreases in biomass of infauna, and a shift from a community dominated by bivalves and spatangoids to one dominated by polychaetes, the mean trophic level of these communities differed by less than one trophic level between sites and differences were not linked to levels of fishing disturbance. The trophic structure of the benthic invertebrate community in this part of the North Sea may have been quite robust, thus ensuring the efficient processing of production within those animals that have sufficiently high intrinsic rates of population increase to withstand the levels of mortality imposed by trawling. The lack of changes in the trophic level of the benthos could imply that species less vulnerable to disturbance are taking the trophic roles of larger more vulnerable species. It would be worthwhile to undertake an explicit study of whether smaller species with faster life histories begin to fill the trophic functions vacated by larger species with slower life histories because the latter cannot withstand the high mortality rates imposed by repeated trawling. Jennings et al (2001) emphasised that their results were only applicable to the free-living fauna of mobile substrates and that it was unlikely that they could be extrapolated to deeper areas with lower natural disturbance where many habitat forming species are found (Auster et al. 1996). Here, loss of habitat would have important consequences for many species, and stability in trophic structure is unlikely to be observed as biomass falls.

On the west coast of North America, the effective conservation of sea otter *Enhydra lutris* may have helped to conserve kelp beds that provide important habitats for many fish and invertebrates. Historically, the hunting of sea-otters so depleted their populations that they no longer exerted any predatory control over urchin populations and the urchins began to destroy kelp beds. Ultimately, the kelp beds became urchin barren grounds, where the rock was colonised by small epilithic algal species and urchins. In the late twentieth century, otter populations have grown following hunting bans and improved conservation, so they have exerted in turn more predation pressure on the urchins. This has allowed new kelp beds to grow on areas of barren ground. Clearly, these effects are confounded by many other factors such as oceanographic effects on urchin recruitment and interactions between urchin and abalone fisheries. Nevertheless, in general terms, the effective conservation of otters has helped to conserve the kelp beds that provide important habitats for many species of fish and invertebrates (Estes and Duggins, 1995; Estes and VanBlaricom, 1985).

### *4.1 Effects of removal of key taxa*

The overexploitation or effective conservation of fishes on tropical reefs can affect community structure and a range of ecosystem processes. Many reefs are intensively fished since they provide the main sources of protein and income for coastal people with few other opportunities for fishing, farming or hunting. Reef fishers target species from all trophic groups, and on many fished reefs, the abundance of herbivorous and invertebrate feeding fishes has been reduced by an order of magnitude or more (Russ, 1991).

The main groups of algal consumers on reefs are herbivorous fishes and sea urchins. The abundance of sea urchins is regulated by recruitment success, food supply and natural mortality due to predation and disease. The main urchin predators are fishes such as emperors (family Lethrinidae) and triggerfishes (Balistidae) that are also targeted by fishers (McClanahan, 1995b). On some reefs in the Caribbean and East Africa, fish predation appears to play a key role in controlling the abundance of urchin populations and they have proliferated following the overfishing of their predators (McClanahan and Muthiga, 1988; McClanahan, 1992, 1995a).

Once urchins have become abundant, they graze the majority of algal production. Urchins can tolerate low algal biomass because they have low consumption and respiration rates. This allows them to outcompete herbivorous fishes that have higher consumption and respiration rates and reach maximum biomass levels an order of magnitude higher (McClanahan, 1992). Since the herbivorous fishes are poor competitors, they may not recover to former levels of abundance when fishing is stopped (McClanahan, 1995a).

As they graze, urchins erode the reef matrix and prevent the settlement and growth of coral recruits. Unless recruitment failure or urchin disease leads to a collapse of urchin populations, other intervention is needed to promote recovery of the reef ecosystem. (McClanahan *et al.*, 1996) attempted such intervention on a small scale by deliberate removal of urchins. When they removed urchins from unfished experimental plots on Kenyan reefs, there were significant increases in algal cover and fish abundance within one year. However, on fished reefs, herbivorous fishes were less abundant, and the algae rapidly overgrew corals as they proliferated. The ecosystem shifts that McClanahan *et al.* (1996) induced in fished areas by urchin removal were remarkably similar to those observed in the heavily fished Caribbean when there was mass urchin mortality following disease. Here, the loss of urchins led to heavy growths of algae that soon dominated the reef community (Carpenter, 1985; Lessios, 1988).

The effects of predator removal on urchin populations contrast with the effects of piscivore removal on reef fish populations. While many studies have shown that the abundance of piscivorous reef fishes is dramatically reduced by fishing, there is little evidence for a corresponding increase in the abundance of their prey. We will give some examples of this effect, and consider why the response of prey fish communities is so weak.

Several studies have documented significant decreases in the abundance of piscivorous target species following fishing and yet there was no evidence for a corresponding increase in the abundance of their prey (Bohnsack, 1982; Jennings and Polunin, 1997; Russ, 1985). The reasons for this are likely to be linked to the structure of reef fish communities, where phylogenetic groupings contain many species, with a wide range of life history traits, behavioural differences and feeding strategies (Hiatt and Strasburg, 1960; Hixon, 1991; Parrish *et al.*, 1985, 1986). Moreover, most fish species undergo marked ontogenetic changes in diet, and act as the prey and predators of other species in the course of their life history. As a result, while the collective impacts of predators are large, the impacts of individual predator species on the dynamics of their prey are minor. This effect was termed diffuse predation by Hixon (1991).

It is worth noting that on much smaller scales ($m^2$ rather than $km^2$) there is some evidence for the role of predation as a structuring force, particularly when habitat or refuge space is directly limited. Thus Caley (1993), Hixon and Beets (1993) and Carr and Hixon (1995) have conducted elegant studies which demonstrated that experimental reductions in piscivore abundance lead to detectable decreases in the abundance and diversity of their prey. However, even at these scales, it is widely accepted that recruitment variation has a more significant impact on population structure (Doherty, 1991; Doherty and Fowler, 1994; Sale, 1980).

### 4.2 *Why systems respond differently to predator removal*

In marine ecosystems, the loss of fish predators can have limited consequences for prey species. While there are well known exceptions, such as those we have cited, scientists have looked very hard for interactions in terrestrial and aquatic ecosystems and failed to find them. For example, Milchunas *et al.* (1998) synthesised the literature on experimental manipulations and comparative studies of the effects of livestock grazing on shortgrass steppes and its relationship to ecosystem function. While they found that some groups, particularly birds, were strongly affected by even light grazing pressure. Nevertheless, at the community level, trophic structure composition did not vary greatly across grazing treatments. Even when there were relatively large effects of grazing on consumer groups this did not alter ecosystem processes such as primary production or soil nutrient pool and cycling rates. Since an effect is perceived as a positive result, while no effect may go unreported, the literature is strongly biased towards significant interactions.

In many of the more species rich marine ecosystems the biomass spectrum is extended and there is more variance in size within the main phylogenetic groupings. This is particularly apparent on tropical reefs. Moreover, the phylogenetic groupings tend to contain more species, with a wider range of life history traits, behavioural differences and feeding strategies (Jennings and Kaiser, 1998). As a result, predation is diffuse (Hixon, 1991) and the overall effect of all piscivores on their prey can be substantial although the impact of any individual species, or small group of species, is minor. In lake ecosystems, conversely, a few keystone species dominate the biomass within a trophic group and the dynamics of the species are often closely linked.

## 5. Climate Change and shelf communities

One of the major concerns over coming decades is the prospect of global climate change with its potentially significant consequences for marine systems and activities that depend upon them (e.g. Maul 1993; Mooney et al. 1993; Peterson and Estes 2001). Kennedy (1990) identified five environmental factors that might be modified as the climate warms into the 21st Century: sea level rise, water-column warming, precipitation, wind and water column circulation. To this must be added the potential for an enhanced frequency and intensity of storms (Muller-Karger 1993). The list of factors that may alter in response to global warming continues to expand.

For two of the above changes there is general agreement about the direction of change in different regions; sea level rise is expected to increase at an accelerating pace everywhere (mainly through thermal expansion of water and the melting of ice sheets) and warming is expected to be greater towards the poles than the tropics (IPCC 2000). The direction of change in different regions for precipitation, wind and water circulation and storm frequency and intensity is less certain.

### *5.1 Temperature effects*

Current assessments suggest that global warming is likely to have a marked effect on biological processes and biodiversity in the oceans. In general it would appear that we will see, in addition to an overall rise in temperature, a reduction in the equator to poleward temperature gradient (IPCC 2000). The frequency of El Nino-like conditions in the tropical Pacific is likely to increase, with the eastern tropical Pacific warming more than the western (IPCC 2000).

To examine the possible importance of temperature effects, many have examined the distribution patterns of organisms for clues (see references in Kennedy 1990; Ware 1995; Lehody et al. 1997; Peterson and Estes 2001). In conjunction with hydrographic conditions, temperature plays a significant role in determining the biogeographic distribution of species and we can undoubtedly expect to see distribution changes as the climate changes (Hayden and Dolan 1976). For example, Kennedy (1990) drawing on earlier work by Bousfield and Thomas (1975), hypothesises that a northward extension of warmer conditions may lead to merging of the currently disjunct distribution of warm temperate species such as the commercially important hard-clam *Mercenaria mercenaria* and the eastern oyster *Crassostrea virginica.* At present the northern limit of the main distribution of these species is Cape Cod on the Eastern seaboard of the United States, but there is a second concentration found in the warmer waters of the Gulf of St Lawrence.

Observations on a reef off North Carolina also provide evidence of likely warming effects for fish communities (Parker and Dixon 1998). Two surveys separated by 15 years, revealed a marked shift in community composition, coincident with a marked warming trend. Mean monthly bottom water temperatures in winter were 1–6 degrees higher in the later survey, when two new families and 29 new species of tropical fishes were recorded in the area. An increase in fish-cleaning symbiosis was also especially noticeable (Parker and Dixon 1998). While it seems plausible to expect similar changes to those already observed with any future increase in temperature, Kennedy (1990) highlights the cautionary notes of Bakun (1990) who stresses that, if the causal mechanisms for warming are different, other associated influences are also likely to differ.

Coral bleaching is an increasing global phenomenon with important consequences for local economies (e.g. Hughes 1994; Done et al. 1996; Smith et al. 1996; Fagoonee et al. 1999; Harvell et al. 1999). Coral bleaching occurs when either, the symbiotic zooxanthellae are expelled, algal pigmentatioin is lost, or both scenarios occur, giving the coral tissues a 'bleached' appearance. These events have been linked with exceptionally high seawater temperatures, elevated doses of ultraviolet radiation, infection by pathogens and the alteration of local salinity regimes.

The paleontological record can also provide clues as to what might happen. For example, rapid warming associated with the inflow of the warm Tsushima Current during the transition from Pleistocene glacial to interglacial stages has been shown to be associated with major changes in the benthic mollusc fauna of the Sea of Japan continental shelf (Kitamura et al. 2000). The fossil record suggests that there were two stages of faunal change. The first occurred when warm-water species migrated into the Sea of Japan and co-existed with cold-water species, which coincided with a northward shift in species ranges. The second stage involved the further range expansion of warm-water molluscs shortly after the local extinction of cold-water species. During this phase, it is speculated that benthic mollusc communities with very low diversity and density existed temporarily and locally at inner shelf depths (<100 m). Such a community has no modern analogue, but may have resulted from a marine climate with a higher seasonality than occurs today (Kitamura et al. 2000). These findings suggest that another period of rapid warming might have severe impacts on offshore benthic communities and lead to community patterns that are not represented by modern faunal distributions. Importantly, range expansions may not be as rapid as range contractions because the former requires the conjunction of numerous factors for successful introductions (e.g. transport, absence of predators and pathogens). We can expect, therefore, to see new mixes of species arising, perhaps leading to shifts in the patterns of biological control (i.e. competition and predation) in these communities and the occurrence of alternative ecologically stable states.

Temperature effects on fisheries are of particular concern since water temperature has a dominant effect, not only on fish distributions (Lehody et al. 1997), but also on spawning and survival of larvae and fish growth (Heath 1992). It seems likely that important changes in fish distributions could occur, but current assessments suggest that overall productivity will be unaffected (Wood and McDonald 1997; IPCC 2000). To make specific predictions at regional scales, however, decadal scale shifts in hydrographic conditions must also be taken into account; improvements in General Circulation Models will be required to achieve this (IPCC 2000).

In general, the effects of temperature change are likely to be more profound at higher latitudes, especially when the effects on ice sheets are taken into account. Sea ice covers approximately 11% of the ocean, depending on season and it is predicted that both the extent, thickness and duration will all decline (IPCC 2000). We have already seen a 10 to 15% decline in ice cover the northern hemisphere since the 1950s and with some areas that were closed to navigation in the past are now permanently open (IPCC 2000). This change in access may lead to changes in patterns of ocean resource use and to the geographical location of production zones (Peterson and Estes 2001). It also seems likely that further declines in ice cover or later freezing may lead to changes in nearshore ice dynamics, seabed scouring and sediment transport. In particular, the storms occurring on shoreline sediments that are no longer bound together by ice is likely to lead to much greater sediment inputs into the nearshore (IPCC 2000).

Another area of particular concern with respect to warming in semi-enclosed water bodies is the potential for increased incidences of anoxia owing to the lower oxygen carrying capacity of water at higher temperatures and the increased oxygen demand of the biota. Coupled with the increased likelihood of such events due to eutrophication, this may become increasingly important for coastal fisheries. In the coastal waters of the USA, for example, a reduction in the availability of oxygen rich cooler water may have consequences for striped bass(Coutant 1990) and for the commercially important blue crab (*Callinectes sapidus*)(Stickle et al. 1989).

### *5.2 Changes in Rainfall*

Given the dominating influence of river and land run-off on coastal systems (see above), effects of climate change on rainfall, particularly monsoon rains in the tropics, are also likely to be important. The inter-model consistency in rainfall predictions for the globe (IPCC 2000) suggest little change in SE Asia and South America, where riverine influences are especially important, but potential small increases (5 – 20%) for the Indian sub-continent, East Asia, Canada and Northern Europe potentially large increases (>20%) for Northern Asia and the Sahara. Even if rainfall estimates are correct, however, determining influences on continental shelves is fraught with difficulty, owing to the confounding effects of land clearing and water abstraction. It seems likely, however, that recruitment in Asian fishes and penaeid shrimp of economical importance (Garcia and Le Reste 1981; Pauly and Navaluna 1983) may be markedly affected. Changes in rainfall will also have an effect on buoyancy driven water currents, although the likely consequences of this have yet to be assessed.

### *5.3 Changes in water circulation*

Of the effects of climate change that may already be taking place, perhaps the most notable is the recorded increased upwelling intensity which results from intensified alongshore wind stress on the ocean surface (Bakun 1990). Since the intensity of upwelling is critical for many of the most productive marine regions in the world, the potential for climate change to affect fisheries seems, therefore, to be profound, although speculation to date has focused primarily on pelagic resources. Apart from the possibility of upwelling intensification owing to increase in wind stress, which is likely to affect pelagic fisheries most (Bakun 1990), other effects resulting from wind and circulation changes are also possible. For example, there are many cases in the literature of the possible importance of wind-induced currents for larval transport (e.g. Heath 1992). Without clearer predictions of how circulation patterns might change, it will be difficult to make the next step in predictions might anticipate consequences for fisheries.

### *5.4 Changes in storm frequency*

A change in climate is likely to affect wind and current regimes in a region, and in tropical regions this may be manifest as an increase in the heat content of the ocean with a consequent increase in wind activity and the frequency of storms (Muller-Karger 1993). In this respect Gray (1993) notes a relationship between increased sea-surface temperature and both the frequency and intensity of hurricanes, implying that tropical coastal seas may expect greater hurricane activity in future. Unfortunately, however, current models used to predict climate change are incapable of simulating or predicting the generation of hurricanes so quantifying the magnitude of the likely increase is not possible (Muller-Karger 1993).

If storm intensity does increase we can certainly expect impacts on benthic communties. One study that examines this issue is that of Rothlisberg et al. (1990) who considered the effects on Penaeid prawn species, one of Australia's most valuable fishery resources. The conclusion was that increased banana prawn populations are to be expected if global warming leads to an increase in sea level, rainfall and cyclone activity (Rothlisberg et al. 1990). This would occur due to the inundation of tidal flats. In contrast, a decrease in tiger prawn populations would be expected because seagrass beds would be destroyed by cyclones. Destruction of 20% of the seagrass beds in the Gulf of Carpentaria (some 183km$^2$) in 1986 led to a 30% reduction in tiger prawn catch for the area, because the beds form the nursery grounds for the species (Rothlisberg et al. 1990). Increases in the frequency of such effects of habitat destruction are to be expected both here and elsewhere in the tropics. Given the immense value of shrimp fisheries to the economies of some countries, such effects could be profound.

### *5.5 Changes in Polar regions*

It is arguable that the most profound effects of global climate change will be first manifested in the high latitude Arctic and Antarctic. These oceans are characterized by the ephemeral cover by sea ice, which varies in thickness from just a few mm's to over 10m. Sea ice is a feature, not just of polar-regions but also of the Baltic, Caspian and Okhotsk Seas, covering at its maximum extent 13% of the worlds surface area. It is therefore one of the major biomes on the planet on a par, in terms of aerial extent, to that of deserts and tundra (Lizotte, 2001; Comiso, 2003). Seasonal sea ice formation, consolidation and subsequent melt drives ocean circulation, air-ocean exchange, weather patterns and of course is a fundamental aspect in both pelagic and benthic components of these oceans and coastal regions.

The physical and mechanical properties of sea ice have been a focus of intense research activity for the shipping industry and engineers for many decades. Despite it being recognised as a unique habitat for a plethora of microorganisms over 150 years ago, it is only in the past 25 years that the biology and chemistry of sea ice have become the focus of systematic investigations (Brierley and Thomas, 2002; Thomas and Dieckmann, 2002). Many plankton organisms depend on food reserves contained within sea ice, notably the krill in the southern Ocean and amphipods and polar cod in the Arctic. These in turn are key food sources for higher predators. The complex range of factors that affect the productivity and availability of ice-associated productivity could therefore cascade up the food webs to seabirds, cetaceans, ice seals, and in the Arctic polar bears (Ainley et al. 2003).

Phytoplankton primary production in high latitudes is constrained by the short windows of adequate light conditions. Retreating sea ice in the marginal ice zone is key to stimulating ice edge blooms due to the melt water enhancing stratification and stabilization of the surface water layers (reviewed by Leventer, 2003). These phytoplankton blooms are predominantly seeded from sea ice assemblages released from the melting ice. When ice melts there are resulting pulses of organic matter export to the underlying benthos, both from the sea ice itself and the phytoplankton blooms, often mediated by intense zooplankton grazing and packaging of phytoplankton into rapid sinking feacal pellets (Leventer, 2003).

In these waters of relatively low overall annual primary production, these highly seasonal export events are key to the bentho-pelagic coupling and the maintenance of viable benthic assemblages (Arntz and Gili, 2001). Any changes to the seasonal sea ice cover due to climate changes will have profound impacts, not only to the organisms living within and upon the ice, but also to those underlying the ice.

Over two decades of satellite data show that there has been a reduction in the average thickness of sea ice and changes in the overall characteristics of the Arctic ice cover since the 1980's that correspond with warming of surface temperatures. The most dramatic changes are the reduction in perennial summer sea ice cover, which is predicted to disappear within the next 100 years. The trends in the Antarctic are not so clear-cut, and there is evidence of a slight decadal increase in ice cover. However, there have been significant regional shifts in the distribution of sea ice, with decreases in the perennial ice cover of the Amundsen and Bellingshausen Seas being matched by similar magnitude increases in the Ross Sea (Comiso, 2003).

Such changes in sea ice extent are confused by other large-scale cyclical phenomena. One of these is the Antarctic Circumpolar Wave (ACW) propagating eastward around the periphery of the Southern Ocean sea ice zone (White and Peterson, 1996). This results in increases followed by decreases in ice extent with a periodicity of 3 to 4 years. Given the substantial inter-annual variation in such systems it makes predicting the consequences of global climate change a difficult task to accomplish.

Recruitment in some organisms, such as the Antarctic krill, has been directly linked to the extent of sea ice cover (reviewed by Brierley and Thomas, 2002). Reduced incidence of extensive sea-ice cover may have been the cause of the significantly lower krill population sizes observed during the late 1980s and early 1990s. However, recent line transect acoustic surveys under ice by the *Autosub-2* autonomous underwater vehicle (Brierley *et al.,* 2002) have shown that it is the ice edge, rather than ice extent *per se,* that is important for krill. Even a 25% reduction in sea-ice area would equate to only a 9% reduction in sea-ice edge extent.

Changes in the thickness and concentration of sea ice, as well as variability in the seasonal and spatial extent of ice cover, have consequences for mammals and seabirds, possibly producing species-specific alterations in demography, range, and population size (Ainley et al., 2003). The loss of ice suitable for resting, molting, and breeding will all combine to have direct and immediate effect on all those species that need to haul out or that use sea ice for migration. Barbraud and Weimeskirch (2001), note that paradoxically varying sea-ice extent has some opposing impacts on emperor penguins. Increased ice extent improves adult survival by increasing food availability, but is detrimental to hatching success because it increases the distance that adults have to walk between the breeding colony and feeding grounds. Similarly, many species of whales and seals forage in polar waters and any change in seasonal plankton productivity will potentially influence seasonal migrations of these animals. Whales and seals in particular are restricted in sea ice covered areas by access to areas of open water for breathing. Any sea ice changes that increase the extent of open water within the sea ice zone will clearly increase the accessibility to these mammals (reviewed by Ainley et al. 2003).

It is possible that reduced sea ice cover and warmer and thinner ice floes may result in increased primary production by sea ice photoautotrophs within the ice and greater inoculum of 'seeding organisms' into the water on ice melt. Naturally increased food sources in the ice will enhance zooplankton stocks in under ice waters. It is also possible that extended periods of primary production may result in the summer open waters, especially in the Southern Ocean where inorganic nutrients are available in excess. In the Arctic increased river runoff and increased nutrient loading into the coastal waters where ice forms will have significant effects on the biology of these waters, which coupled with the decrease in summer sea ice must result in large-scale habitat change throughout the Arctic Ocean.

A particular feature of sea ice influenced coastal habitats is the effect of ice scour. Clearing of benthic habitats by seasonal drifting ice can reach down to depths of 10 m or more in extreme cases, and ice disturbance is possibly the major structuring element of polar nearshore biological communities. Both intertidal and sublittoral communities can be highly disturbed, at times showing characteristics of a highly ephemeral flora and fauna (Gutt, 2001).

An extreme example of ice scour disturbance resulting in severe habitat transformation is that caused by icebergs and in shallow coastal regions in both the Arctic and Antarctic the grounding of icebergs causes considerable disturbance to benthic communities (Gutt 2001; 2000; Gerdes et al. 2003). Sessile organisms are eradicated and pioneer species begin to grow in high abundances on the devastated substratum. In some areas major iceberg scour events have been estimated to take place over periods of every 50 to 200 years and because of the very slow growth of many species, particularly in Antarctica, areas disturbed in this manner are likely to be characterised by a continuous natural fluctuation between destruction and recovery. Communities can be held at early successional stages, or even completely removed by scouring, and these effects occur from the intertidal to depths up to 500 m in Antarctica. The wide scales of disturbance intensity are thought to contribute to the overall high levels of Antarctic benthic biological diversity (Gutt 2001). The significance of iceberg disturbance is likely to increase with accelerated melting of ice shelves, although it must be stressed that the ice shelve calving that often receives much attention by the media, often cannot be related to global climate warming. Iceberg scour effects are less prevalent in the Arctic due to there being far fewer icebergs there compared to the Antarctic.

## 6. Management considerations

An implicit assumption of the concept of habitat management is the thought that human intervention has the capability to influence ecosystem processes with a predictable outcome. However, given the large-scale background changes that are apparent in marine systems any attempt at management needs to define clear objectives that are appropriate given a naturally changing environment. Factors that have negative effects on marine systems such as fishing activity, agricultural run-off, dredging, tourism, can all be relaxed or even prevented entirely (given the political will). However, such a change of habitat use may not lead to the desired 'recovery' of the community or the ecosystem if habitat change has achieved an alternative stable state. In such cases, major biological events are necessary to achieve an appropriate phase shift that would lead to recovery. Marine protected areas (MPAs) and fishery no take zones (NTZs) can achieve many of the habitat management objectives above. However, there is great debate over the exact size, number and configuration of MPAs or NTZs required to achieve ecologically meaningful protection. Furthermore, the successful implementation of such measures requires an entire water-shed approach as detrimental effects of some riverine discharges could negate any positive effects accrued by a MPA or NTZ.

In certain circumstances it is possible to reverse the deleterious effects of human activities. For example, mangrove trees are amenable to reforestation, particularly those species which produce propagules which can easily be planted (e.g. *Rhizophora* spp). However, the restoration of the mangrove ecosystem to full biodiversity and productivity is likely to require establishment of multispecies forests stands and a long period of secondary succession. So long as the coastal protection function of mangroves is maintained or restored though management of intact shoreline forests, aquaculture and other land use can be integrated with reforestation.

Perhaps the most difficult area for management is the scenario when the effects of management on one species, has subtle consequences for other organisms at a different trophic level (Redford et al. 2001). From a management perspective, it is necessary to understand when cascade effects are likely to occur. Kaiser and Jennings (2001) identified several rules of thumb that should be considered in a management context. Very few trophic cascades involve more than three interactions between different species, i.e. they are simple systems regardless of the fact that they may be embedded within highly complex systems of high biodiversity. These mini-systems within larger more complex systems are perhaps to be expected in high diversity ecosystems in which there is a greater tendency towards specialist feeding interactions and are perhaps less likely to occur in temperate areas. The interactions within the cascade are between organisms that are assigned to major trophic levels within a system e.g. predator - herbivore - primary producer, very few involve intermediate trophic levels (Table 2). This contrasts sharply with the food-web of the Benguela system in which there are at least seven interactions between the bottom and the top of the system (Yodzis 1998). The key predators or herbivores in reported trophic cascades are most usually the dominant organisms at their trophic level. Hence in the case of significant decrease in the population of predator or herbivore, there are few others to take their place in the short-term. This contrasts sharply with many of the more "open" marine ecosystems where there may be three or more predators that exert similar levels of predation on one or more species (e.g. sharks, cetaceans and marine mammals all eat pilchard in the Benguela system). Cannibalism is rare within species involved in systems prone to trophic cascades in sharp contrast to many marine ecosystems in which cannibalism is common. Even in the absence of major predators, cannibalism can limit species density.

Thus the prognosis for our ability to manage shelf seas from an ecosystem perspective is perhaps somewhat bleak given our inability to predict many of the key physical and chemical processes that are likely to alter ecosystems in the future and the uncertainty surrounding the effects of species manipulations on other parts of the ecosystem. A more realistic approach may be to move away from species based approaches to management and to focus on maintenance of functional processes that are perhaps more stable and relevant measures of system integrity.

TABLE 2

Examples of studies in which trophic cascades have been identified. In all cases the linkages between different trophic levels were short. These examples are illustrative and are not meant to be an exhaustive list (adapted from Pace *et al.*, 1999)

| Ecosystem | Cascade | Evidence | Effects | References |
|---|---|---|---|---|
| **Marine** | | | | |
| Oceanic | Salmon-zooplankton-phytoplankton | Ten-year time series | Increased phytoplankton when salmon abundant | Shiomoto *et al.*, 1997 |
| Coastal | Whales-otter-urchins-kelp | Long-term data and behavioural studies | Predation by whales on otters leads to urchin population expansion and increased grazing of kelp | Estes *et al.*, 1998 |
| Intertidal | Birds-urchins-macroalgae | Manipulation of bird predation on urchins | Algal cover greatly increased when urchins reduced | Wooton, 1995 |
| **Freshwater** | | | | |
| Streams | Fish-invertebrates-periphyton | 1° and 2° production affected by predation of invertebrate populations | Annual 1° production affected by 6 fold difference | Huryn, 1998 |
| Shallow lake | Fish-zooplankton-phytoplankton | Observations of lakes under clear and turbid conditions | Reductions in fish abundance led to shift in zooplankton size-structure with consequent effects on phytoplankton | Jeppesen *et al.*, 1998 |

## Bibliography

Ainley, D.G., Tynan, C.T., and Stirling, I. (2003). Sea ice: A critical habitat for marine mammals. In Sea Ice - An Introduction to its Physics, Chemistry, Biology and Geology (eds D.N. Thomas and G.S. Dieckmann), pp. 240–266. Blackwell Publishing, Oxford.

Aller, J.Y. (1995) Molluscan death assemblages on the Amazon shelf: implication for physical and biological controls on benthic populations. Paleogeography, Paleoclimatology and Paleoecology, 118, 181–212.

Aller, J.Y. and Stupakoff, I. (1996) The distribution and seasonal characteristics of benthic communities on the Amazon shelf as indicators of physical processes. Continental Shelf Research, 16.

Alongi, D.M. (1990) The ecology of tropical soft-bottom benthic ecosytems. Oceanography and Marine Biology: An Annual Review, 28, 381–496.

Alongi, D.M. (1998) Coastal ecosystem processes CRC Press, Cambridge MA.

Alongi, D.M. and Robertson, A.I. (1995) Factors regulating benthic food chains in tropical river deltas and adjacent shelf areas. Geology and Marine Letters, 15, 145–152.

Alvarez-Ossorio, M. (1977) Un estudia de la Ria de Muros en Noviembre de 1975. Bolletin Instituto Espanol Oceanografia, 2, 1–223.

Arcos, J.M., Ruiz, X., Bearhop, S., and Furness, R.W. (2002) Mercury levels in seabirds and their fish prey at the Ebro Delta (NW Mediterranean): the role of trawler discards as a source of contamination. Marine Ecology Progress Series, 232, 281–290.

Arntz, W.E. and Gili, J.M. (2001) A case for tolerance in marine ecology: let us not put out the baby with the bathwater. Scientia Marina, 65, 283–299.

Austen, M.V.C., Widdicombe, S., and Villano-Pitacco, N. (1998) Effects of biological disturbance on diversity and structure of meiobenthic nematode communities. Marine Ecology Progress Series, 174, 233–246.

Auster, P., Malatesta, R., and Donaldson, C. (1997) Distributional responses to small-scale habitat variability by early juvenile silver hake, *Merluccius bilinearis.* Environmental Biology of Fishes, 50, 195–200.

Auster, P.J. and Langton, R.W. (1999). The effects of fishing on fish habitat. In Fish habitat: essential fish habitat and restoration (ed L. Benaka), Vol. Symposium 22, pp. 150–187. American Fisheries Society, Bethesda, Maryland.

Auster, P.J., Malatesta, R.J., Langton, R.W., Watling, L., Valentine, P.C., Donaldson, C.L., Langton, E.W., Shepard, A.N., and Babb, I.G. (1996) The impacts of mobile fishing gear on seafloor habitats in the Gulf of Maine (Northwest Atlantic): implications for conservation of fish populations. Reviews in Fisheries Science, 4, 185–202.

Bailey, C. (1988) The social consequences of tropical shrimp mariculture development. Ocean and Coastal Management, 11, 31–44.

Bakun, A. (1990) Global climate change and intensification of coastal upwelling. Science, 247, 198–201.

Barbraud, C. and Weimeskirch, H. (2001) Emperor penguins and climate change. Nature, 411, 183–186.

Bax, N.J. (1991) A comparison of the biomass flow to fish, fisheries and mammals in six marine ecosystems. ICES Marine Science Symposia, 193, 217–224.

Black, E.A. and Truscott, J. (1994) Strategies for regulation of aquaculture site selection in coastal areas. Journal of Applied Icthyology, 10, 294–306.

Bohnsack, J.A. (1982). Effects of piscivorous predator removal on coral reef fish community structure. In Gutshop '81: Fish Food Habits and Studies (eds G.M. Caillet and C.A. Simenstad), pp. 258–267. University of Washington, Seattle.

Bousfield, E.L. and Thomas, M.L.H. (1975) Postglacial changes in distribution of littoral marine invertebrates in the Canadian Atlantic region. Proceedings N.S. Institute Science, 27, 47–60.

Brierley, A.S., Fernandes, P.G., Brandon, M.A., Armstrong, N.W., McPhail, S.D., Stevenson, P., Pebody, M., Perrett, J., Squires, M., Bone, D.G., and Griffiths, G. (2002) Antarctic krill under sea ice: elevated abundance in a narrow band just south of ice edge. Science, 295, 1890–1892.

Brierley, A.S. and Thomas, D.N. (2002) The ecology of Southern Ocean pack ice. Advances in Marine Biology, 43, 171–278.

Caddy, J.F. (2000) Marine catchement basin effects versus impacts of fisheries on semi-enclosed seas. ICES Journal of Marine Science, 57, 628–640.

Caley, M.J. (1993) Predation, recruitment and the dynamics of communities of coral-reef fishes. Marine Biology, 117, 33–43.

Carpenter, K.E. and Alcala, A.C. (1977) Philippine coral reef fisheries resources, Part 2, Muro-ami and kayakas reef fisheries, benefit or bane? Philippine Journal of Fisheries, 15, 217–235.

Carpenter, R.C. (1985) Sea urchin mass-mortality: effects on reef algal abundance, species composition and metabolism and other coral reef herbivores. Proceedings of the Fifth International Coral Reef Symposium, 4, 53–60.

Carr, M.H. and Hixon, M.A. (1995) Predation effects on early post-settlement survivorship of coral-reef fishes. Marine Ecology Progress Series, 124, 31–42.

Castel, J., Labourg, P.-J., Escaravage, V., Auby, I., and Garcia, M. (1989) Influence of seagrass beds and oyster parks on the abundance and biomass patterns of meio- and macrobenthos in tidal flats. Estuarine, Coastal and Shelf Science, 28, 71–85.

Churchill, J. (1989) The effect of commercial trawling on sediment resuspension and transport over the Middle Atlantic Bight continental shelf. Continental Shelf Research, 9, 841–864.

Clifford, H.C. (1985). Semi-intensive shrimp farming. In Texas Shrimp Farming Manual (eds G.W. Chamberlain, M.G. Haby and R.J. Miget), Vol. 4, pp. 15–42. Texas Agricultural Extension Sercice, College Station, Texas.

Coleman, F.C. and Williams, S.L. (2002) Overexploiting marine ecosystem engineers: potential consequences for biodiversity. Trends in Ecology and Evolution, 17, 40–44.

Collie, J.S., Hall, S.J., Kaiser, M.J., and Poiner, I.R. (2000) A quantitative analysis of fishing impacts on shelf-sea benthos. Journal of Animal Ecology, 69, 785–799.

Comiso, J.C. (2003). Large-scale characteristics and variability of the global sea ice cover. In Sea ice - An Introduction to its Physics, Chemistry, Biology and Geology (eds D.N. Thomas and G.S. Dieckmann), pp. 112–142. Blackwell Publishing, Oxford.

Connell, J.H. (1978) Diversity in tropical rain forests and coral reefs. Science, 199, 1302–1310.

Costanza, R., D'Arge, R., De Groot, R., Farber, S., Grasso, M., Hannon, B., Limburg, K., Naeem, S., O'Neill, R.V., Paruelo, J., Raskin, R.G., Sutton, P., and Van den Belt, M. (1997) The value of the world's ecosystem services and natural capital. Nature, 387, 253–260.

Costanza, R. and Greer, J. (1998). The Chesapeake Bay and its watershed: A model for sutainable ecosystem management. In Ecosystem Health (eds D. Rapport, R. Costanza, P.R. Epstein, C. Gaudet and R. Levins), pp. 261–312. Blackwell Science, Oxford.

Coutant, C.C. (1990) Temperature-oxygen habitat for freshwater and coastal striped bass in a changing climate. Transactions of the American Fisheries Society, 119, 240–253.

Creutzberg, F. (1985). A persistent Chlorophyll *a* maximum coinciding with an enriched benthic zone. In Proceedings of the 19th European marine Biolgoy Symposium (ed P.E. Gibbs), pp. 97–108. Cambridge University Press, Cambridge.

Currie, D.R. and Parry, G.D. (1996) Effects of scallop dredging on a soft sediment community: a large-scale experimental study. Marine Ecology Progress Series, 134, 131–150.

Cury, P., Shannon, L., and Shin, Y.-J. (2003). The functioning of marine ecosystems: a fisheries perspective. In Responsible fisheries in the marine ecosystem (eds M. Sinclair and G. Valdimarsson), pp. 103–124. CABI Publishing, Cambridge, MA.

Dalhback, B. and Gunnarson, L. (1981) Sedimentation and sulfate reduction under a mussel culture. Marine Biology, 63, 269–275.

Dayton, P.K., Thrush, S.F., Agardy, M.T., and Hofman, R.J. (1995) Environmental effects of marine fishing. Aquatic Conservation, 5, 205–232.

Dernie, K.M., Kaiser, M.J., and Warwick, R.M. (2003) Recovery rates of benthic communities following physical disturbance. Journal of Animal Ecology, 72, 1043–1056.

Diaz, R.J. and Rosenberg, R. (1995) Marine benthic hypoxia: a review of its ecological effects and the behavioural responses of benthic macrofauna. Oceanography and Marine Biology: An Annual Review, 33, 245–303.

Doherty, P. (1991). Spatial and temporal patterns in recruitment. In The ecology of fishes on coral reefs (ed P. Sale), pp. 261–293. Academic Press, San Diego.

Doherty, P. and Fowler, T. (1994) An empirical test of recruitment limitation in a coral reef fish. Science, 263, 935–939.

Done, T. (1992) Phase-shifts in coral-reef communities and their ecological significance. Hydrobiologia, 247, 121–132.

Done, T.J., Ogden, J.C., Wiebe, W.J., and Rosen, R.R. (1996). Biodiversity and Ecosystem Function of Coral Reefs. In Functional roles of biodiversity: A global perspective (eds H.A. Mooney, J.H. Cushman, E. Medina, O.E. Sala and E.D. Schulze). John Wiley and Sons Ltd, New York.

Duplisea, D.E., Jennings, S., Malcolm, S.J., Parker, R., and Sivyer, D. (2001) Modelling the potential impacts of bottom trawl fisheries on soft sediment biochemistry in the North Sea. Geochemical Transactions, in press.

Duplisea, D.E., Jennings, S., Warr, K.J., and Dinmore, T.A. (2003) A size-based model for predicting the impacts of bottom trawling on benthic community structure. Canadian Journal of Fisheries and Aquatic Science.

Eadie, B.J., McKee, B.A., Lansing, M.B., Robbins, J.A., Metz, S., and Trefry, J.H. (1994) Records of nutrient-enhanced coastal ocean productivity in sediments from the Louisiana continental shelf. Estuaries, 17, 754–765.

Estes, J.A. and Duggins, D.O. (1995) Sea otters and kelp forests in Alaska: generality and variation in a community ecology paradigm. Ecological Monographs, 65, 75–100.

Estes, J.A. and Van Blaricom, G.R. (1985). Sea otters and shellfisheries. In Marine mammals and fisheries (eds J.R. Beddington, R.J.H. Beverton and D.M. Lavigne), pp. 187–235. George Allen and Unwin, London.

Fagoonee, I., Wilson, H.B., Hassell, M.P., and Turner, J.R. (1999) The dynamics of zooxanthellae populations: a long-term study in the field. Science, 283, 843–845.

FAO (2000a) Global forest resource assessment 2000 FAO Forestry Paper. Report 140, Food and Agriculture Organisation of the United Nations, Rome, Italy.

FAO (2000b) The State of World Fisheries and Aquaculture Food and Agriculture Organisation of the United Nations, Rome, Italy.

Fegan, D.F. (1996) Sustainable shrimp farming in Asia: vision of pipedream? Asian Aquaculture, 2, 22–24.

Findlay, R.H., Watling, L., and Mayer, L.M. (1995) Environmental impact of salmon net-pen culture on marine benthic communities in Maine: a case study. Estuaries, 18, 145–179.

Garcia, S. and Le Reste, L. (1981) Life cycle, dynamics, exploitation and management of coastal penaeid shrimp stocks. Food and Agriculture Organisation of the United Nations, Rome.

Gerdes, D., Hilbig, B., and Montiel, A. (2003) Impact of iceberg scouring on macrobenthic communities in the high-Antarctic Weddell Sea. Polar Biology, 26, 295–301.

Gotceitas, V. and Brown, J. (1993) Substrate selection by juvenile Atlantic cod (*Gadus morhua*): effects of predation risk. Oecologia, 93, 31–37.

Grassle, J.F. and Saunders, H.L. (1973) Life histories and the role of disturbance. Deep Sea Research, 20, 643–659.

Gray, C.R. (1993). Regional meteorology and hurricanes. In Climate Change in the Intra-Americas Seas (ed G.A. Maul), pp. 162–192. Edward Arnold, London.

Gray, J.S., Clarke, K.R., Warwick, R.M., and Hobbs, G. (1990) Detection of initial effects of pollution on marine benthos: an examination from the Ekofisk and Eldfisk oilfields, North Sea. Marine Ecology Progress Series, 66, 285–299.

Gutt, J. (2000) Some 'driving forces' structuring communities of the sublittoral Antarctic macrobenthos. Antarctic Science, 12, 297–313.

Gutt, J. (2001) On the direct impact of ice on marine benthic communities, a review. Polar Biology, 24, 553–564.

Hall, S. and Raffaelli, D. (1993) Food webs: theory and reality. Advances in Ecological Research, 24, 187–239.

Hall, S.J. (1994) Physical disturbance and marine benthic communities: life in unconsolidated sediments. Oceanography and Marine Biology Annual Review, 32, 179–239.

Hall, S.J. (2002) The continental shelf benthic ecosystem: current status, agents for change and future prospects. Environmental Conservation, 29, 350–374.

Hall, S.J., Raffaelli, D., and Thrush, S.F. (1994). Patchiness and disturbance in shallow water benthic assemblages. In Aquatic Ecology (eds P.S. Giller, A.G. Hildrew and D. Raffaelli), pp. 333–376. Blackwell Science, Oxford.

Hall, S.J., Robertson, M.R., Basford, D.J., and Fryer, R. (1993) Pit-digging by the crab *Cancer pagurus:* a test for long-term, large-scale effects on infaunal community structure. Journal of Animal Ecology, 62, 59–66.

Hall-Spencer, J.M. and Moore, P.G. (2000). Impact of scallop dredging on maerl grounds. In Effects of fishing on non-target species and habitats: biological, conservation and socio-economic issues (eds M.J. Kaiser and S.J. De Groot), pp. 105–118. Blackwell Science, Oxford.

Harvell, C.D., Kim, K., Burkholder, J.M., Colwell, R.R., Epstein, P.R., Grimes, D.J., Hofmann, E.E., Lipp, E.K., Osterhaus, A.D.M.E., Overstreet, R.M., Porter, J.W., Smith, G.W., and Vasta, G.R. (1999) Emerging marine diseases-climate links and anthropogenic factors. Science, 285, 1505–1510.

Hayden, B.P. and Dolan, R. (1976) Coastal marine fauna and marine climates of the Americas. Journal of Biogeography, 3, 71–81.

Heath, M.R. (1992). Field Investigations of the early Life Stages of Marine Fish. In Advances in Marine Biology (eds J.H.S. Blaxter and A.J. Southward), pp. 2–133. Academic Press, London.

Hiatt, R.W. and Strasburg, D.W. (1960) Ecological relationships of the fish fauna on coral reefs of the Marshall Islands. Ecological Monographs, 30, 65–127.

Hixon, M. (1991). Predation as a process structuring coral reef fish communities. In The ecology of fishes on coral reefs (ed P. Sale), pp. 475–508. Academic Press, San Diego.

Hixon, M.A. and Beets, J.P. (1993) Predation, prey refuges and the structure of coral reef fish assembleges. Ecological Monographs, 63, 77–101.

Hughes, T. (1994) Catastrophes, phase shifts and large-scale degradation of a Caribbean coral reef. Science, 265, 1547–1551.

IPCC (2000) Climate Change 2001: The Scientific Basis. Contribution of Working Group I to the Third Assessment Report of the Intergovernmental Panel on Climate Change Cambridge University Press, New York, USA.

Ito, S. and Imai, T. (1955) Ecology of oyster bed I.: On the decline of productivity due to repeated culture. Tokoku Journal of Agricultural Research, 5, 251–268.

Jennings, S., Dinmore, T.A., Duplisea, D.E., Warr, K.J., and Lancaster, J.E. (2001a) Trawling disturbance can modify benthic production processes. Journal of Animal Ecology, 70, 459–475.

Jennings, S. and Kaiser, M. (1998) The effects of fishing on marine ecosystems. Advances in Marine Biology, 34, 201–352.

Jennings, S., Pinnegar, J.K., Polunin, N.V.C., and Boon, T.W. (2001b) Weak cross-species relationships between body size and trophic level belie powerful size-based trophic structuring in fish communities. Journal of Animal Ecology, 70, 934–944.

Jennings, S., Pinnegar, J.K., Polunin, N.V.C., and Warr, K.J. (2001c) Impacts of trawling disturbance on the trophic structure of benthic invertebrate communities. Marine Ecology Progress Series, 213, 127–142.

Jennings, S., Pinnegar, J.K., Polunin, N.V.C., and Warr, K.J. (2002) Linking size-based and trophic analyses of benthic community structure. Marine Ecology Progress Series, 226, 77–85.

Jennings, S. and Polunin, N. (1997) Impacts of predator depletion by fishing on the biomass and diversity of non-target reef fish communities. Coral Reefs, 16, 71–82.

Justic, D. (1991). Hypoxic conditions in the northern Adriatic Sea: historical development and ecological significance. In Modern and Ancient Continetal Shelf Anoxia (eds R.V. Tyson and T.H. Pearson), Vol. 58, pp. 95–102. Geological Society Special Publication, London, UK.

Kaiser, M.J. (1998) Significance of bottom-fishing disturbance. Conservation Biology, 12, 1230–1235.

Kaiser, M.J. (2001). Ecological effects of shellfish cultivation. In Environmental impacts of aquaculture (ed K. Black), pp. 51–75. Sheffield University Press, Sheffield.

Kaiser, M.J. and De Groot, S.J. (2000) Effects of fishing on non-target species and habitats: biological, conservation and socio-economic issues Blackwell Science, Oxford.

Kaiser, M.J. and Jennings, S. (2001). An ecosystem perspective on conserving targeted and non-targeted species. In Conservation of Exploited Species (eds J.D. Reynolds, G.M. Mace, K.H. Redford and J.G. Robinson), pp. 345–369. Cambridge University Press, Cambridge.

Kingston, P.F. (1992) Impact of offshore oil production installations on the benthos of the North Sea. ICES Journal of Marine Science, 49, 45–53.

Kitamura, A., Omote, H., and Oda, M. (2000) Molluscan response to Early Pleistocene rapid warming in the Sea of Japan. Geology and Marine Letters, 28, 723–726.

Kusuki, Y. (1977) Fundamental studies on the deterioration of oyster grazing grounds II: organic content of faecal materials. Bulletine of the Japanese Society of Science and Fisheries, 43, 167–171.

Lapointe, B., Niell, F., and Fuentes, J. (1981) Community structure, succession and production of seaweeds associated with mussel-rafts in the Ria de Arosa, N.W. Spain. Marine Ecology Progress Series, 5, 243–253.

Lehody, P., Bertignac, M., Hampton, J., Lewis, A., and Picaut, J. (1997) El Nino Southern Oscillation and tuna in the western Pacific. Nature, 389, 715–717.

Lessios, H.A. (1988) Mass mortality of *Diadema antillarum* in the Caribbean: what have we learned? , Annual Reviews Ecology and Systematics, 19.

Leventer, A. (2003). Particulate flux from sea ice in polar waters. In Sea ice - An Introduction to its Physics, Chemistry, Biology and Geology (eds D.N. Thomas and G.S. Dieckmann), pp. 303–332. Blackwell Publishing, Oxford.

Lizotte, M.P. (2001) The contributions of sea ice algae to Antarctic marine primary production. American Zoologist, 41, 57–73.

Lopez-Jamar, E., Iglesias, J., and Otero, J. (1984) Contribution of infauna and mussel-raft epifauna to demersal fish diets. Marine Ecology Progress Series, 15, 13–28.

Luckhurst, B.E. and Luckhurst, K. (1978) Analysis of the influence of substrate variables on coral reef fish communities. Marine Biology, 49, 317–323.

Mariojouls, C. and Kusuki, Y. (1987) Appréciation des quantités de biodépôts émis par les huîtres en élevage suspendu dans la baie d'Hiroshima . Haliotis, 16, 221–231.

Maul, G.A., ed. (1993) Climate Change in the Intra-Americas Seas, pp 211 pp. Edward Arnold, London.

McCarthy, I.D., Fuiman, L.A., and Alvarez, M.C. (2003) Aroclor 1254 affects growth and survival skills of Atlantic croaker Micropogonias undulates larvae. Marine Ecology Progress Series.

McClanahan, T. (1995a) A coral-reef ecosystem-fisheries model- impacts of fishing intensity and catch selection on reef structure and processes. Ecological Modelling, 80, 1–19.

McClanahan, T. (1995b) Fish predators and scavengers of the sea urchin *Echinometra mathaei* in Kenyan coral-reef marine parks. Environmental Biology of Fishes, 43, 187–193.

McClanahan, T., Kakamura, A., Muthiga, N., Yebio, M., and Obura, D. (1996) Effects of sea-urchin reductions on algae, coral and fish populations. Conservation Biology, 10, 136–154.

McClanahan, T.R. (1992) Resource utilization, competition and predation: a model and example from coral reef grazers. Ecological Modelling, 61, 195–215.

McClanahan, T.R. and Muthiga, N.A. (1988) Changes in Kenyan coral reef community structure and function due to exploitation. Hydrobiologia, 166, 269–276.

Menasveta, P. (2002) Improved shrimp growout systems for disease prevention and environmental sustainability in Southeast Asia. Reviews in Fisheries Science, 10, 391–402.

Michunas, D.G., Lauenroth, W.K., and Burke, I.C. (1999) Livestock grazing: animals and plant biodiversity of shortgrass steppe and the relationship to ecosystem function. Oikos, 83, 65–74.

Milliman, J.D. (1991). Flux and fate of fluvial sediment and water in coastal seas. In Ocean Margin Processes in Global Change (eds F.C. Mantoura, J.-C. Martin and R. Wollast), pp. 69–89. John Wiley and Sons, Chichester, UK.

Montagna, P. and Harper, D.E.J. (1996) Benthic infaunal long-term response to offshore production platforms in the Gulf of Mexico. Canadian Journal of Fisheries and Aquatic Sciences, 53, 2567–2588.

Mooney, H.A., Fuentes, E.R., and Kronberg, B.I., eds. (1993) Earth System Responses to Global Change: Contrasts between North and South America. Academic Press, London.

Muller-Karger, F.E. (1993). River discharge variability including satellite observed plume-dispersal patterns. In Climate Change in the Intra-Americas Seas (ed G.A. Maul), pp. 162–192. Edward Arnold, London.

Neira, C., Sellanes, J., Soto, A., Gutierrez, D., and Gallardo, V.A. (2001) Meiofauna and sedimentary organic matter off Central Chile: response to changes caused by the 1997–1998 El Nino. Oceanologica Acta, 24, 313–328.

O'Brien, C.M., Fox, C.J., Planque, B., and Casey, J. (2000) Climate variability and North Sea cod. Nature, 404, 142.

Oliver, R.S. and Slattery, P.N. (1985) Destruction and opportunity on the sea floor: effects of gray whale feeding. Ecology, 66, 1965–1975.

Olsgard, F. and Gray, J.F. (1995) A comprehensive analysis of the effects of offshore oil and gas exploration and production on the benthic communities of the Norwegian continental shelf. Marine Ecology Progress Series, 122, 277–306.

Pace, M.L., Cole, J.J., Carpenter, S.R., and Kitchell, J.F. (1999) Trophic cascades revealed in diverse ecosystems. Trends in Ecology and Evolution, 14, 483–488.

Paerl, H.W. (1995) Coastal eutrophication in relation to atmospheric nitrogen deposition: current perspectives. Ophelia, 41, 237–259.

Paez-Osuna, F. (2001) The environmental impact of shrimp culture; causes, effects and mitigating circumstances. Environmental management, 28, 131–140.

Parker, R.O. and Dixon, R.L. (1998) Changes in a North Carolina Reef Fish Community after 15 Years of Intense Fishing - Global Warming Implications. Transactions of the American Fisheries Society, 127, 908–920.

Parrish, J., Norris, J., Callahan, M., Magarifugi, E., and Schroeder, R. (1986). Piscivory in a coral reef community. In Gutshop '81: Fish Food Habits and Studies (eds G. Caillet and C. Simenstad), pp. 73–78. University of Washington, Seattle.

Parrish, J.D., Callahan, M.W., and Norris, J.E. (1985) Fish trophic relationships that structure reef communities. Proceedings of the Fifth International Coral Reef Symposium, 4, 73–78.

Paterson, D.M. and Black, K.S. (1999) Water flow, sediment dynamics and benthic ecology. Advances in Ecological Research, 29, 155–193.

Pauly, D. and Navaluna, N.A. (1983). Monsoon-induced seasonality in the recruitment of Phillipine fishes. In Proceedings of the Expert Consultation to Examine Changes in Abundance and Species Composition of Neritic Fish Resources, San Jose, Costa Rica (eds G.D. Sharp and J. Csirke), Vol. Report 291, pp. 823–835. FAO Fisheries, Rome, Italy.

Pearson, R. (1981) Recovery and recolonisation of coral reefs. Marine Ecology Progress Series, 4, 105–122.

Pearson, T. and Rosenberg, R. (1978) Macrobenthic succession in relation to organic enrichment and pollution of the marine environment. Oceanography and Marine Biology: An Annual Review, 16, 229–311.

Peterson, C.H. and Estes, J.A. (2001). Conservation and Management of Marine Communities. In Marine Community Ecology (eds M.D. Bertness, M.E. Hay and S.D. Gaines), pp. 469–507. Sinauer Associates, Sunderland, Mass.

Phillips, M.J., Lin, C.K., and Beveridge, M.C.M. (1993) Shrimp culture and the environment: lessons from the world's most rapidly expanding warm water aquaculture sector. In Environment and Aquaculture in Developing Countries (ed Anonymous), Vol. 31. ICLARM Conference Proceedings, Manila, Philippines.

Pilskaln, C.H., Churchill, J.H., and Mayer, L.M. (1998) Resuspension of sediment by bottom trawling in the Gulf of Maine and potential geochemical consequences. Conservation Biology, 12, 1223–1229.

Planques, A., Guilen, J., and Puig, P. (2001) Impact of bottom trawling on water turbidity and muddy sediment of an unfished continental shelf. Limnology and Oceanography, 46, 1100–1110.

Poiner, I.R., Glaister, J., Pitcher, C.R., Burridge, C.Y., Wassenberg, T.J., Gribble, N., Hill, B.J., Blaber, S.J.M., Milton, D.M., Brewer, G.D., and Ellis, N. (1998) The environmental effects of prawn trawling in the far northern section of the Great Barrier Reef: 1991–1996. Final Report to GBRMPA and FRDC. CSIRO Division of Marine Research Queensland Department of Primary Industries Report CSIRO, Canberra.

Pongthanapanich, T. (1996) Economic study suggests management guidelines for mangroves to derive optimal ecomonic and social benefits. Asian Aquaculture, 1, 16–17.

Posey, M., Lindberg, W., Alphin, T., and Vose, F. (1996) Influence of storm disturbance on an offshore benthic community. Bulletin of Marine Science, 59, 523–529.

Raffaelli, D.G. and Milne, H. (1987) An experimental investigation of the effects of shorebird and flatfish predation on estuarine invertebrates. Estuarine, Coastal and Shelf Science, 24, 1–13.

Rhoads, D.C., Boesch, D.F., Tang, Z.-C., Xu, F.-S., Huang, L.-Q., and Nilsen, K.J. (1985) Macrobenthos and sedimentary facies on the Changjiang delta platform and adjacent continental shelf, East China Sea. Continental Shelf Research, 4, 189-.

Riemann, B. and Hoffmann, E. (1991) Ecological consequences of dredging and bottom trawling in the Limfjord, Denmark. Marine Ecology Progress Series, 69, 171–178.

Roberts, C.M. and Ormond, R.F.G. (1987) Habitat complexity and coral reef fish diversity and abundance on Red Sea fringing reefs. Marine Ecology Progress Series, 41, 1–8.

Robertson, A.I. and Philips, M.J. (1995) Mangroves as filters of shrimp pond effluent: predictions and biogeochemical research needs. Hydrobiologia, 295, 311–321.

Rodhouse, P. and Roden, C. (1987) Carbon budget for a coastal inlet in relation to intensive cultivation of suspension-feeding bivalve molluscs. Marine Ecology Progress Series, 36, 225–236.

Rothlisberg, P., Staples, D., Poiner, I.R., and Wolanski, E. (1990). Possible effects of climatic and sea level changes in prawn populations in the Gulf of Carpentaria. In Greenhouse planning for climate change (ed G. Pearman), pp. 216–227. Brill, E.J., Leiden, The Netherlands.

Russ, G.R. (1991). Coral reef Fisheries: effects and yields. In The ecology of fishes on coral reefs (ed P.F. Sale), pp. 601–635. Academic Press, San Diego.

Sainsbury, K.J. (1987). assessment and management of the demersal fishery on the continental shelf of northwestern Australia. In Tropical Snappers and Groupers - Biology and Fisheries Management (eds J.J. Polovina and S. Ralston), pp. 465–503. Westview Press, Boulder, Colorado.

Sainsbury, K.J. (1988). The ecological basis of multispecies fisheries and management of a demersal fishery in tropical Australia. In Fish Population Dynamics (ed J.A. Gulland), pp. 349–382. John Wiley, London.

Sale, P.F. (1980) The ecology of fishes on coral reefs. Oceanography and Marine Biology Annual Review, 18, 367–421.

Schafer, W. (1972) Ecology and palaeoecology of marine environments University of Chicago Press, Chicago.

Schwinghamer, P., Guigné, J., and Siu, W. (1996) Quantifying the impact of trawling on benthic habitat structure using high resolution acoustics and chaos theory. Canadian Journal of Fisheries and Aquatic Science, 53, 288–296.

Sherman, K., Jones, C., Sullivan, L., Smith, W., Berrien, P., and Ejsymont, L. (1981) Congruent shifts in sandeel abundance in western and eastern North Atlantic ecosystems. Nature, 291, 486–489.

Smith, G.T., Ives, L.D., Nagelkerken, I.A., and Ritchie, K.B. (1996) Caribbean sea fan mortalities. Nature, 383, 487.

Stickle, W.B., Kapper, M.A., Liu, L.L., Gnaiger, E., and Wang, S.Y. (1989) Metabolic adaptations of several species of crustaceans and molluscs to hypoxia: tolerance and microcalorimetric studies. Biological Bulletin, 177, 303–312.

Tenore, K., Boyer, L., Cal, R., Corral, J., Garcia, F., Gonzalez, M., Gonzalez, E., Hanson, R., Iglesias, J., and Krom, M. (1982) Coastal upwelling in the Rias Bajas, NW Spain: Constrasting the benthic regimes of the Rias de Arosa and Muros. Journal of Marine Research, 40, 701–772.

Thomas, D.N. and Dieckmann, G.S. (2002) Antarctic sea ice - a habitat for extremophiles. Science, 295, 641–644.

Thompson, D.R., Bearhop, S., Speakman, J.R., and Furness, R.W. (1998) Feathers as a means of monitoring mercury in seabirds: insights from stable isotope analysis. Environmental Pollution, 101, 193–200.

Thrush, S.F. and Dayton, P.K. (2002) Disturbance to marine benthic habitats by trawling and dredging - implications for marine biodiversity. Annual Review of Ecology and Systematics, 33, 449–473.

Thrush, S.F., Hewitt, J.E., Cummings, V.J., and Dayton, P.K. (1995) The impact of habitat disturbance by scallop dredging on marine benthic communities: what can be predicted from the results of experiments? Marine Ecology Progress Series, 129, 141–150.

Thrush, S.F., Whitlatch, R.B., Pridmore, R.D., Hewitt, J.E., Cummings, V.J., and Wilkinson, M.R. (1996) Scale-dependent recolonization: the role of sediment stability in a dynamic sandflat habitat. Ecology, 77, 2472–2487.

Tuck, I., Hall, S., Roberston, M., Armstrong, E., and Basford, D. (1998) Effects of physical trawling disturbance in a previously unfished sheltered Scottish sea loch. Marine Ecology Progress Series, 162, 227–242.

Turner, R.E. and Rabalais, N.N. (1994) Coastal eutrophication near the Mississippi River delta. Nature, 368, 619.

V.S., K. (1990) Anticipated effects of climate change on estuarine and coastal fisheries. Canadian Journal of Fisheries and Aquatic Sciences, 15, 16–24.

Von Blaricom, G.R. (1982) Experimental analysis of structural regulation in a marine sand community exposed to oceanic swell. Ecological Monographs, 52, 283–305.

Walters, C.J. and Juanes, F. (1993) Recruitment limitation as a consequence of natural selection for use of restricted feeeding habitats and predation risk taking by juvenile fishes. Canadian Journal of Fisheries and Aquatic Science, 50, 2058–2070.

Ware, D.M. (1995) A century and a half of change in the climate of the North East Pacific. Fisheries Oceanography, 4, 267–277.

Warwick, R.M. and Clarke, K.R. (1995) New 'biodiversity' measures reveal a decrease in taxonomic distinctness with increasing stress. Marine Ecology Progress Series, 129, 301–305.

Warwick, R.M. and Uncles, R.J. (1980) Distribution of benthic macrofauna associations in the Bristol Channel in relation to tidal stress. Marine Ecology Progress Series, 3, 97–103.

Watling, L. and Norse, E.A. (1998) Disturbance of the seabed by mobile fishing gear: a comparison to forest clearcutting. Conservation Biology, 12, 1180–1197.

White, W.B. and Peterson, R.G. (1996) An Antarctic circumpolar wave in surface pressure, wind, temperature and sea-ice extent. Nature, 380, 699–702.

Widdows, J., Brinsley, M.D., Salkeld, P.N., and Elliott, M. (1998) Use of annular flumes to determine the influence of current velocity and biota on material flux at the sediment-water interface. Estuaries, 21, 552–559.

Wood, C.M. and McDonald, D.G., eds. (1997) Global Warming: Implications for Freshwater and Marine Fish, pp 425. Cambridge University Press, Cambridge, UK.

Woodhouse, C. (1996) Farms avoid new U.S. curb on shrimp imports. Fish Farming International, 23, 24.

Yodzis, P. (1998) Local trophodynamics and the interaction of marine mammals and fisheries in the Benguela ecosystem. Journal of Animal Ecology, 67, 635–58.

# Chapter 24. REGIME SHIFTS

ANDREW BAKUN

*Rosenstiel School of Marine and Atmospheric Science*
*University of Miami, Miami, Florida, USA*

**Contents**



## 1. Introduction

The issue of regime shifts in ocean ecosystems has become a very hot topic in marine ecology and fisheries science. As well as being of major scientific interest in itself, it is a subject of great potential importance to fisheries management, to protection of marine biodiversity, and to resolving worrisome trends from within the background "noise level" of normal natural marine ecosystem variability.

It is an issue that has arisen rather recently. Throughout most of the twentieth century, marine ecosystems had been generally regarded as quasi-stable entities. But by the early 1990s, the conventional wisdom began to be shaken by the march of events and by the steady accumulation of longer and longer data series. In particular, unprecedented shifts in population abundances, productivities and distributions in the Pacific Ocean had evidently been launched in the early to mid-1970s. However, it was not until a decade or more later that the extent of these changes began to be appreciated. Venrick et al. (1987) reported a doubling of chlorophyll in the subtropical North Pacific starting in the mid-1970s. Brodeur and Ware

*The Sea*, Volume 13, edited by Allan R. Robinson and Kenneth H. Brink
ISBN 0-674-01526-6 

(1992, 1995) found a doubling of zooplankton, pelagic fish and squid biomasses in the subarctic North Pacific in the 1980s compared to a reference period several decades earlier. Roemmich and McGowan (1995) reported a 70% decrease in biomass of large zooplankton in the California Current over roughly the same period. The collapse in the early 1970s of the Peruvian anchoveta, by far the largest exploited fish population that had ever existed on earth, coincided with a 60% to 70% decline in biomass of large zooplankton in the habitat (Loeb and Rojas, 1988; Carrasco and Lozano, 1989; Alheit and Bernal, 1993). Sardine populations in far corners of the Pacific Ocean began simultaneously growing their populations (Kawasaki, 1983; Schwartzlose et al., 1999), while synchronous salmon production trends were noticed all around the subarctic Pacific basin (Beamish and Boulion, 1993). And on the other side of the world in the Atlantic Ocean off tropical west Africa, the locally important "Ghana herring" (*Sardinella aurita*) fishery collapsed, followed by a remarkable outbreak of enormous numbers of triggerfish that spread progressively along the African coast (see next section).

But what really seems to have awakened many marine scientists and fishery managers was a rash of declines in important managed fish populations after the mid-1980s, notably many of the valuable cod stocks of the North Atlantic, but also some of the most important exploited fish stocks of the North Pacific: Alaskan pollock, Pacific cod, Japanese sardine, Pacific hake, sablefish, chinook salmon, coho salmon, and many others.

Then in 1992, an international conference on "Climate and Northern Fish Populations" (Beamish 1995) seems to have marked a "watershed" shift in opinion regarding the inherent stability of marine ecosystems. Suddenly, the term "regime shift", with strong implications of its being a response to large-scale climatic variation (see Sections 3 to 5) had, for better or for worse, become established as a routine part of the conversational vernacular of marine ecologists and fishery scientists, at least in the region of the North Pacific. In the decade that has followed, the notion has spread around the world and has served both as inspiration and excuse (it has become not uncommon to hear, when data series have not fit a hypothesis to an anticipated degree, or when an ostensibly well managed fishery resource may have collapsed, that "a regime shift must have occurred"). But the evidence that radical changes are a natural element of the dynamics of ocean ecosystems has indeed become compelling (see following sections). Clearly, the notion of regime shifts in marine ecosystems is one that is here to stay.

But, inevitably, the new appreciation of natural variability in marine populations has in some ways "muddied the waters" in the debate to assign responsibility for collapses in fishery resource stocks. Indeed it does appear that climatic and ecosystem effects that may be rather independent of fishing itself have been significantly involved in many declines of fishery resource populations. But overfishing, over-capitalization of the worlds fishing fleets and associated industries, and adverse environmental impacts of fishing constitute continuing serious threats to fishery resource and marine ecosystem sustainability (Garcia and Newton, 1997; Christy, 1997, Garcia and de Leiva Moreno, 2003; Sinclair and Valdimarson, 2003). Since one could say that scientific advance, in bringing to light the major role of natural variability, might thereby have unearthed a stumbling block to the clear societal consensus that would be needed to motivate effective remedial actions with respect to the aspects that human beings do control, it would seem to be incumbent on the marine science community to seek particularly rapid progress in untangling the various issues involved.

### 1.1 *Definition of the term "regime shift"*

The term "regimes", as applied in this context, seems to have been first coined by John Isaacs (e.g., Issacs, 1975) in discussing the increase in anchovy abundance off California following the famous collapse of the California sardine population. Isaacs' perception in this instance appears to have reflected the then prevalent concept of marine ecosystems as being rather stable entities, leading to the idea that when drastic change was observed, it might reflect shifts between different potential quasi-stable configurations. Isaacs speculated that intensifying fishing of anchovies might promote recovery of the collapsed California sardine population. (Implementation of this idea was later tried, and ultimately failed, in a different region of the world; see Section 11.)

Later the term "regime", and the idea that regimes could change, was resolutely pushed forward in a series of three papers on global sardine and anchovy population variations produced over the time span of a decade by largely the same group of authors (Lluch-Belda et al. 1989, 1992, Schwartzlose et al. 1999). However, the attention of this "Regime Group", as they called themselves (the largest component of the group were from the California Current region and so naturally adopted Isaac's verbiage), was centered on the evident global-scale synchrony in large-amplitude, multidecadal populations of these species (see Section 4), with the result that the conceptual focus shifted away from interspecies competition toward the synchronizing effects of climatic variability. In this process, the connotation of the term "regime" as implying a rather durable, self-maintaining state was gradually transformed to an implication simply of markedly different levels of stock productivities and/or abundances characterizing different periods of time. This conceptual modification was undoubtedly given impetus by the growing awareness of the remarkable period of strong, persistent trends in climatic characteristics and in marine population responses in the period following the early to mid-1970s.

Currently, some colleagues in the Pacific region would prefer that the term "regime shift" be strictly reserved for ecosystem changes that may be demonstrably, or at least assumedly, linked to climatic changes (e.g., Alheit and Niquin, 2004; Beamish et al., 2004; Wooster and Zhang, 2004). However, other colleagues have been employing the term without an implied limitation to obviously climate-associated changes (e.g., Steele, 1998; Collie et al., 2004; Currie and Shannon, 2004; Laws et al., 2004; Vézina at al., 2004). Moreover, it is the opinion of this author that such a limitation tends to constrain the scientific discussion, and furthermore carries the danger of promoting a mindset that all such radical changes must be primarily forced by some manner of climatic change (a dangerous mindset given the pressures for overexploitation of "common property" fishery resources). It also seems important that, if the term is to have any real operational utility, it must connote changes that are not transitory and ephemeral but are "sticky", in the sense that they endure recognizably over substantial (i.e., multi-annual to multi-decadal) periods (i.e., "locked-in, persistent episodes" as expressed by Beamish et al., 2000). Also its usage has up to this time been with respect to systems of quite large (regional or basin-scale) geographical extent.

Since it is a term already burdened with quite a load of historical baggage and already may mean somewhat different things to different people, it seems important to avoid broadening and diffusing its meaning even more. However, since the term has already become part of the vernacular in various regional settings, and since some fishery managers and public agencies seem now to actually be beginning to take the idea seriously into account, it seems better to avoid the confusion and potential disenchantment that might result from attempts to re-specify its meaning more precisely or narrowly than its current usage. Accordingly, in this chapter the term "*regime shift*", as applied to marine ecosystems, will be used in the quite general sense of connoting *a persistent* (i.e., "sticky"*) radical shift in typical levels of abundance or productivity of multiple important components of the marine biological community structure, occurring at multiple trophic levels and on a geographical scale that is at least regional in extent* (i.e., affecting in a substantial manner the species hierarchies and trophic flow configurations of one or more regional 'large marine ecosystems', or 'LMEs', as defined by Sherman, 1994).

In that defined sense of the term, much of the available information on regime shifts in ocean ecosystems comes from the fisheries. Consequently, a large part of the discussion in this chapter is based on fish populations, although this is not meant to in any way imply that unexploited ecosystem components may not sometimes be of equal or greater importance. Where relevant information on other species groups, including primary producers and zooplankton, may be available for the chosen example situations, it is included. The intention is to present a series of examples that can serve to illustrate the evident general aspects of regime shift phenomena in marine ecosystems and to resolve common patterns pointing to key mechanisms and potentially useful unifying concepts.

## 2. Geographical redistributions

Major alterations in the locations and extents of habitat intensively occupied by key populations have been a quite typical aspect of most of the episodes that fit the above "regime shift" description. Since it seems probable that important linkages exist between the distributional shifts and the abundance/productivity shifts, as well as a belief of this author that these may be valuable keys to developing useful management responses to certain types of regime shift phenomena, illuminating and briefly pondering these linkages will be one major objective of this chapter. Accordingly, as a basis for the discussion to follow, it may be useful at the outset to cite a few illustrative examples of such plausibly-linked changes in abundance and geographical distribution.

### Pacific sardines

Figure 24.1 illustrates the changes in distributions that were observed in association with the population expansions and contractions of the three historically largest populations of Pacific sardines. These changes seem to have involved much more than mere congruent responses to expansion and contraction of areas of suitable conditions (e.g., temperature, etc.) for the species involved (Crawford et al., 1991; McFarlane and Beamish, 2001).

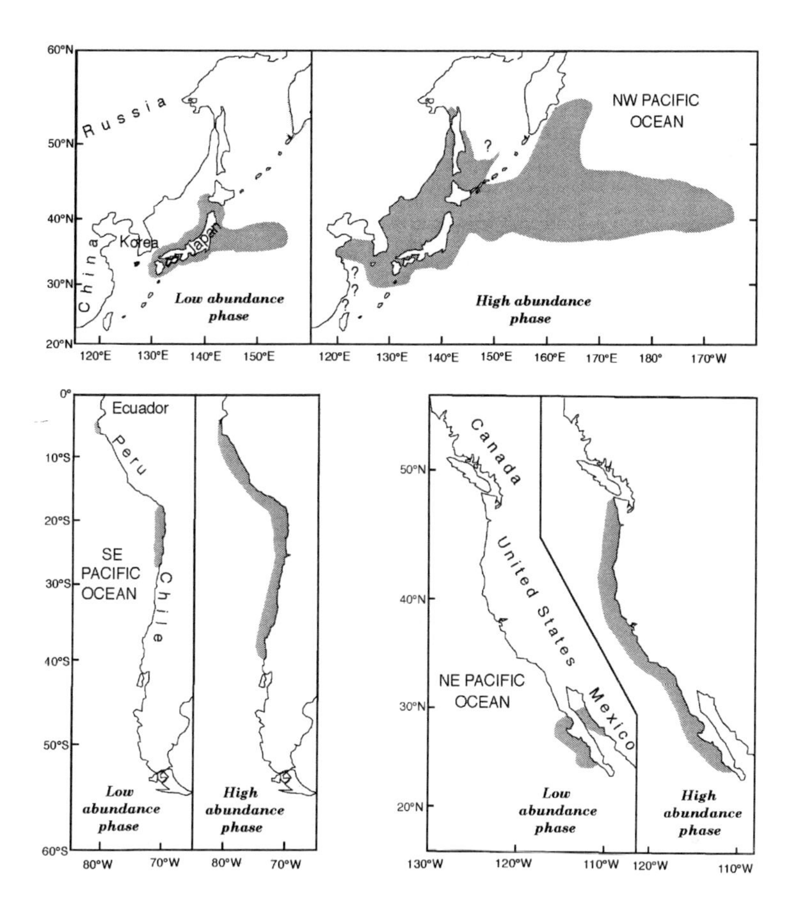


Figure 24.1 Changes in distribution corresponding to low and high abundance phases of the three largest populations of Pacific sardines (re-drawn from Lluch-Belda et al., 1989 and from Kuroda, 1991).

The contraction of habitat involved in the California sardine collapse of the 1940s seems to have proceeded gradually as the population collapsed southward, occurring earliest off Canada (the Vancouver Island fishery), several years later off Central California (the San Francisco and Monterey fisheries), and then finally in the reproductive habitat itself within the southern California Bight (the San Pedro fishery), leaving only a smaller separate, nearly non-migratory, nearly unexploited remnant southern subpopulation off the southern portion of the Baja California peninsula (Radovich, 1982).

But in the more recent 1990s period of renewed population growth there are indications that migrational habits and reproductive habitat selection have been durably altered relative to the period prior to the initial decline. In the earlier (1930–40s) period, the population operating to the north of central Baja California is believed to have functioned essentially as an interconnected unit. For example, although there were dissenting views (e.g., Radovich, 1982), the scientific consensus (e.g., Murphy, 1966) held that the bulk of the catches off Canada were of sardines that had migrated there as adults from the primary reproductive habitat off southern California and northern Mexico, nearly two thousand kilometers to the south. However, in the recent expansion, there are indications of formation of more local, less migratory stocklets at several locations within the distributional envelope of the earlier high abundance period. For example, local reproduction was established in Canadian waters in the late 1990s (McFarlane and Beamish, 2001), spawning occurred off Oregon in 1994 (Bentley et al., 1996), etc.

Off Japan, the situation apparently involved more than just expansion and contraction, but actual abandonment of an earlier reproductive habitat. Wada and Oozeki (1999) indicate that during the 1950s-60s period of low abundance the Japanese sardine utilized two rather separate regional spawning grounds which were directly imbedded in local nursery grounds inshore of the main flow of the Kuroshio (Figure 24.2). In the period of expansion in the 1970s and early 1980s, the more northerly (downstream) of these spawning grounds disappeared entirely while the other expanded greatly in extent to extend into the main Kuroshio flow, and the nursery area was displaced many hundreds of kilometers downstream to the transition area of the Kuroshio and Oyashio Extensions (Figure 24.2). Recently, even though the sardine population has returned to a very low abundance level, there remains only that single upstream spawning center, which is compressed to much smaller size in very nearshore waters of the far southwestern part of the Pacific coast of Japan (T. Wada, pers. comm.; K. Kuroda, pers. comm .. ).

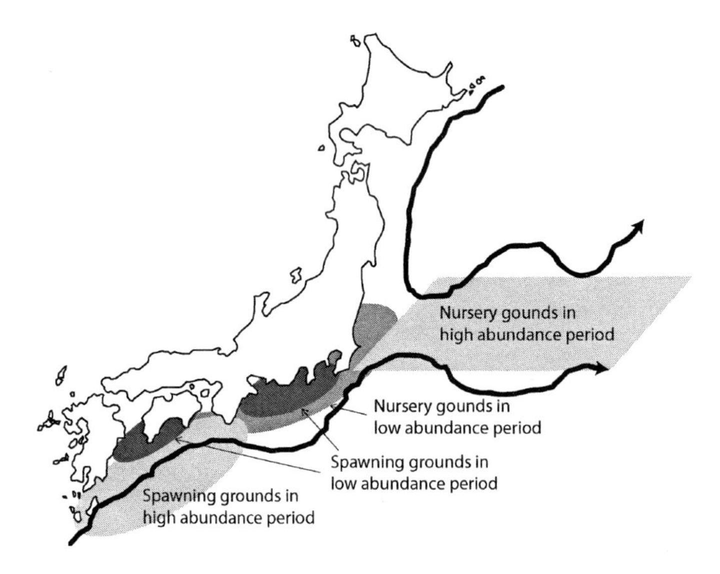


Figure 24.2 Shift of spawning and nursery grounds of the Japanese sardine from the low abundance period before the 1970s shift to the high abundance period of the early 1980s (redrawn from Wada and Oozeki, 1999).

## Triggerfish outbreak off west Africa

The triggerfish *Balistes capriscus* (sometimes designated *B. carolinensis*) is a deep-bodied fish of a body form normally associated with coral reef systems. Before the

1970s, the species was rather rare in the eastern Atlantic. But in winter of 1971–72, sudden increases off Ghana were noted. (Figure 24.3) The increases quickly spread eastward to Togo and Benin and westward to Cote d'Ivoire. In 1981, 500,000 tons of *Balistes* were estimated to exist off Ghana and Côte d'Ivoire between the 10 m and 200 m isobaths. This was 83% of the total estimated pelagic biomass of that area. Several years after the initial outbreak, a new center of outbreak appeared far to the northwest along the African coast off the country of Guinea. In 1978–79 the biomass in the zone stretching from Sierra Leone to Guinea-Bissau was 450,000 tons, increasing to 700,000 tons in 1980, to a million tons in 1981, and to 1.3 million tons in 1982. But by 1986, the population had fallen to less than 220,000 tons in that zone. This corresponded to a similar decrease in the other zone inside the Gulf of Guinea, where by 1986 it had essentially disappeared from Togo-Benin and was not more than 230,000 tons and 29% of total pelagic biomass off Côte d'Ivoire and Ghana. Particular decreases in both zones appear to have corresponded to the year 1986 (Caverivière, 1991), which is the year when fish that may had been spawned during the intense "Atlantic El Niño" episode of 1984 (Shannon et al. 1986, Hisard, 1988), and had survived the experience, should have been augmenting the population.

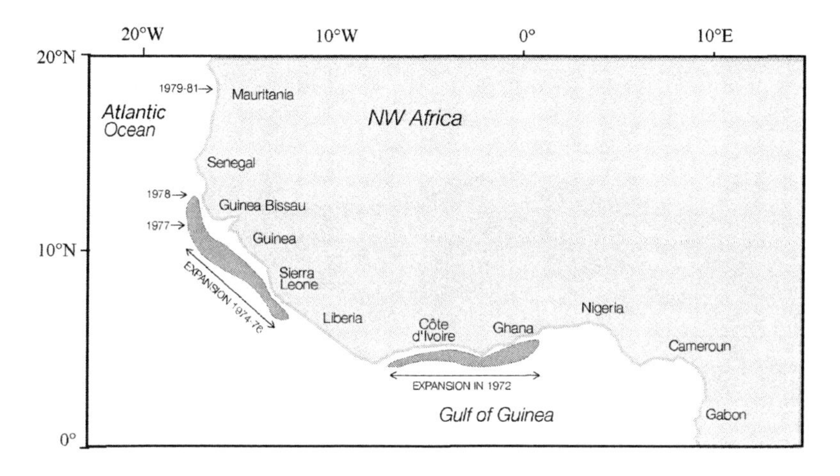


Figure 24.3 The expansions of *Balistes carolinensis* along the west coast of Africa (redrawn from Carivière, 1991).

**Sardines and anchovies in the Sea of Cortez**

In the 1970s, an important Mexican fishery on a growing population of sardines developed inside the Gulf of California (i.e., on the opposite side of the Baja California peninsula from the remaining remnant of the California Current sardine population that was discussed in the initial paragraphs of this section; see Figure 24.4). Landings grew to a peak in 1988 of nearly a quarter of a million tons, whereupon the fishery collapsed abruptly to less than one-third of that amount the following year. This was already several years after the population is thought to have reached a peak in actual biomass (Schwartzlose et al., 1999). Suddenly, anchovies (*Engraulis mordax*) appeared in abundance, a fish that in the knowledge of the current inhabitants had never before existed there. The first time an anchovy was ever reliably recorded in the Gulf was in 1986 (Hammann and Cisneros-Mata, 1989), just as the earlier increasing trend in sardine population abundance had evidently reversed to an abrupt decreasing trend. The anchovy population established itself and quickly grew to dominate the Gulf fishery (Lluch-Belda et al., 1992), while sardines nearly disappeared completely from the landings by the 1991/92 fishing season. But the story did not finish there. In the face of the heavy exploitation pressure focused directly upon it, the anchovy population thereupon abruptly declined, with catches falling to zero in 1996/97. Meanwhile, the sardine population rapidly increased (returned?), supporting a catch of over two hundred thousand tons in 1996/97, the same year that anchovies ultimately entirely disappeared from the catches. Moreover, Holmgren-Urba and Baumgartner (1993), in comparing their paleo-sedimentary scale deposit time series for the Gulf to that produced by Soutar and Isaacs (1974) for a California Current site in the Santa Barbara Channel, found indications of a pattern migratory shifts between the two systems occurring at multi-decadal periods.

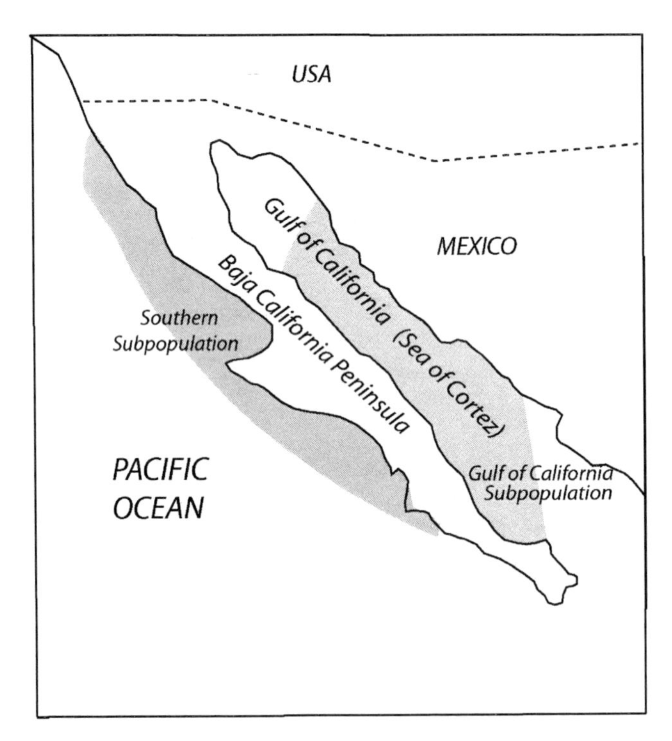


Figure 24.4 Map of the Gulf of California (Sea of Cortez) region. Ranges of the Southern and Gulf of California subpopulations of California sardine are indicated by shading.

Earlier, it had been thought that the Baja California Peninsula, extending far south into quite tropical waters, constituted an effective barrier to exchange of temperate sardine and anchovy populations between waters of the Gulf and those of the California Current on the other (Pacific Ocean) side of the peninsula. But recent analysis of catches of these species by the tuna fleet, which uses them as bait, have shown that both species regularly frequent waters south of the southern extremity of the peninsula (Rodríguez-Sánchez, et al., 2002), actually occurring as far south as the Islas Revillagigedo. Thus the Peninsula is far from being a total barrier. This fact is reflected in the findings of Holmgren-Urba and Baumgartner (1993) of indications of migratory shifts between the two systems at multi-decadal periods in paleo-sedimentary scale deposit time series (Soutar and Isaacs, 1974).

**Bohuslän Herring**

Records of the herring fishery in the Skagerrak, off the Bohuslän coast of Sweden (to the north of Gothernburg) date back to the 10th century (Alheit and Hagen, 2002). It is known that, even in the times of the Vikings, large numbers of herring migrated in certain periods to overwinter off the Bohuslän coast. These periods lasted for 20–50 years and the fisheries on them were large and very important to the economy of the region (270,000 MT landing in one year during a period in the 18th century). However the periods of abundance were separated by even longer intervals (50–70 years) of absence during which relative prosperity turned to abject poverty, and even starvation, for the inhabitants. There have been nine periods of abundance identified, the most recent having been in the late 19th and very early 20th centuries.

Evidently, the Bohuslän herring were part of a North Sea stock. Why or how they periodically altered their migration and overwintering habits in this manner is currently not well understood. As possible clues, Alheit and Hagen (1997) have pointed out a rough correspondence of Bohuslän herring periods with local maxima in a "winter severity" index time series (Lamb, 1972) related to the North Atlantic Oscillation (NAO), although the correspondence is not one that would indicate a simple temperature response (for example, that cold temperatures consistently initiate the periods and warm temperatures terminate them, etc.). They also observe that the Bohuslän periods have been roughly out of phase with the periods of abundance off Norway of the Atlanto-Scandanavian herring.

**Northwest African sardines**

Over the period in which there have been major fisheries operating, sardines (*Sardina pilchardus*) off northwestern Africa have appeared to transform their habitat geography on two notable scales. Over a multi-decadal time scale, a gradual 2000-km southward expansion along the West African coast seems to have occurred (Demarque,1998, Benet at al., 1998, Kifani, 1998). In the 1920s, the southward limit of the species is reported to been been northern Morocco. However, by the 1950s sardines had become abundant as far south as Mauretania (Belvéze, 1984), and by the mid-1970s the species was fairly common even so far south as Senegal (Conand, 1975; Fréon, 1988). Later, the range again contracted northwards for a time, sardines disappearing nearly entirely from all waters south of Morocco in 1982–84. Subsequently, the range again expanded far southward, such that sardines were once again being caught in Senegal during the 1990s (Benet at al., 1998). Certain aspects of these movements have corresponded to interannual temperature trends, notably a gradual decreasing trend of SST in the region from the 1950s to the mid-1970s, coinciding with an increasing trend in upwelling-favorable winds (Demarque,1998, Benet at al., 1998, Kifani, 1998) that accompanied the initial southward expansion. However, the latest reversal back toward the south appears to have occurred in the opposite situation of increasing temperature trend and decreasing upwelling-favorable wind trend.

Within this larger regional scale, there appear to exist three rather autonomous, locally reproducing stocks (Belveze and Erzini, 1983). The central stock, located off central Morocco, was the mainstay of the Moroccan fishery from the 1950s until the 1980s. This fishery developed at the northern end of the seasonal migration from a primary reproductive zone to the south (situated in the large coastal bight in the vicinity of the town of Tantan) to adult feeding grounds off the zone of strong coastal upwelling (Parrish et al, 1983) that exists near the major coastal capes of Cap Sim and Cap Ghir. Major fishing ports and extensive canning industries were established in this zone at ports such as Safi, Essaouira and Agadir. However, since the late 1980s, the sardines have ceased migrating to these northern fishing ports, and by the mid-1990s, what was left of the active Moroccan fishery on this stock was operating to the south, directly within the reproductive grounds near Tantan. This left the northern fishing ports in a state of economic depression and the Moroccan Government considering an extremely expensive, socially-dislocative move of a segment of industry and associated population and infrastructure to a nearly inhabited and infrastructure-bereft area of the Sahara Desert coast in order to target the less impacted southern stock (Do Chi and Kiefer, 1996). In this case, the same decreasing temperature trend (and associated increasing trend in wind-induced upwelling) cited to explain the earlier part of the larger time and space scale southward expansion discussed in the previous paragraph has been cited in the opposite sense. In this case, the apparent decrease in upwelling intensity in recent decades (showing up as the decreasing trend in alongshore wind intensity and an associated increasing trend in SST cited in the previous paragraph) is proposed as an explanation for the cessation of the earlier established northward feeding migration of the central sardine stock (Kifani, 1998; Benet, 1998).

## 3. Anchovy–sardine alternations

Many marine ecosystems of the world share one striking aspect in the configuration of their biological community structures. They typically contain a very large number of species at the lower (e.g., planktonic) trophic levels. They also contain a large number of species (e.g., predatory fishes, large coelenterates, seabirds, marine mammals, etc.) which, as adults at least, feed at the apex or near-apex levels. However, in many of the more richly productive ecosystems of the world, there is often a crucial intermediate trophic level, occupied by small, plankton-feeding pelagic fishes, that is typically dominated by only one, or at most several, species.

For example, the fish biomass of temperate coastal upwelling systems tends to be dominated by one species of sardines and one species of anchovy. Similarly, temperate ecosystems influenced by western ocean boundary currents may be most of the time dominated by a single sardine species (e.g., *Sardinops sagax* in the Kuroshio region near Japan), an anchovy species (e.g., *Engraulis anchoita* in the Falkand/Malvinas Current of eastern South America), or a menhaden species (e.g., *Brevoortia tyrannus* in the Gulf Stream region off the eastern U.S.) Tropical analogs of these temperate ocean ecosystems tend to be dominated by rather analogous tropical species: sardinellas, anchovellas, thread herrings, etc. Other examples are the herring, sandeels or capelin of boreal shelf ecosystems. The krill of the Southern Ocean may represent an invertebrate analog. Because of this typical community configuration, featuring many species at the bottom, many at the top, but constricted to a very few dominant species at a mid-level (remiscent of the body form of a wasp, in which flows of information and material between complex, multifunctional, relatively large thoraxial and abdominal body segments must pass through a very narrow tubular waist segment), these ecosystems have been referred to as "wasp-waist" ecosystems (Rice 1995, Bakun 1996, Cury et al. 2000, Cury et al. Chapter 15, this vol.). Modeling studies have shown that variability in the trophic dynamics of these ecosystems tend to be largely dominated by variations in these mid-trophic level *wasp-waist* populations (Rice 1995). Typically these populations of small pelagic planktivores vary radically in size, and these variations may have major effects on the trophic levels above, which depend on the *wasp-waist* populations as their major food source, and also on the trophic levels below which face heavy consumption by the massive *wasp-waist* populations.

In the period of modern recorded fisheries, two *wasp-waist* species groups, anchovies (genus *Engraulis*) and sardines (genera *Sardinops* or *Sardina*) have tended, to a remarkable degree, to fluctuate in abundance in opposite phase (Schwartzlose et al., 1999). This out-of-phase pattern of abundance not only occurs in the rich California and Peru-Humboldt upwelling regions of the eastern Pacific, but also in the dynamically very different type of ecosystem of the northwestern Pacific near Japan, also in the Atlantic in the Benguela Current upwelling region off southern Africa, and even within such an enclosed, rather oligotrophic peripheral sea as the Mediterranean. (Puzzlingly, this clear pattern of out-of-phase alternation is largely absent in the paleosedimentary fish scale time series of Soutar and Isaacs, 1974. This may be because the sedimentary series represent fixed points in space, and as has been seen above, changes in abundance tend to interact with changes in distribution which cannot be effectively tracked with samples taken at a single fixed point in space. )

In the case of more tropical *wasp-waist* analogs, there are indications that a similar pattern of alternation may hold. For example, off Angola, *Sardinella madurensis* became the dominant species in the mid-1980s, in contrast to the situation in earlier years when *Sardinella aurita* dominated the catches. However, an opposite switch occurred off Congo (along the same coast but much nearer the equator), where *S. aurita* displaced *S. madurensis* (Cochrane and Tandstad, 2000, Binet et al., 2001). *S. aurita* is the more migratory of the two and prefers upwelling areas, whereas *S. madurensis* is more sedentary, inhabiting coastal, often brackish, waters (Binet et al., 2001). Thus *S. aurita* seems more an analog to the sardine and *S. madurensis* to the anchovy.

The reason for the pattern of sardine-anchovy alternation has been a puzzle. Anchovies and sardines do not seem to directly compete. The particle size ranges that each is able to filter are quite different. They tend to occupy differing spatial distributions. They have very different degrees of migrating capacity and tendency, sardines being strong migrators and anchovies much less so[1]. Finally, although they predate upon each other's eggs and larvae, each species appears to cannibalize its own eggs and larvae as readily as predating upon the other's, evidently removing interspecies predation as a potential mechanism by which one species might preferentially depress the other.

A considerable amount of otherwise paradoxical information appears to point to a mechanism wherein sardines and anchovies may be adapted to utilize different non-contemporaneous types of "loopholes[2]" in the pervasive fields of predators that exert heavy mortalities on their eggs and larvae (Bakun and Broad, 2003). Sardines may take advantage of perturbed conditions, utilizing their migratory capabilities and their ability to glean some degree of sustenance from the very small food particle size-spectra that may characterize less productive regions through which they may have to migrate, to access "loopholes" in the fields of larval predators that may be opened as a result of the perturbed conditions. On the other hand, anchovies may respond in a superior manner during non-perturbed conditions for several other reasons. As a result of their "stay at home" tendencies, anchovies do not expend much energy in migrating. Moreover, they do not require size-related hydrodynamic advantages to the degree that the more migratory sardines do and consequently have less need to grow as large. Consequently, they can have shorter life cycles and become highly fecund earlier in life. In addition, because of the much coarser filters in their gillrakers, anchovies can effectively deploy a much larger "filter basket" and so can be particularly efficient in gathering the large nutritious food particles which may be abundant in "good times". Consequently, anchovies can take advantage of these relative efficiencies to allocate relatively more trophic energy to producing sufficiently copious quantities of reproductive product to allow an adequate number survivors to pass through the increased predation levels that may exist in the "good times" periods between environmental perturbations (such as El Niño episodes, etc.). Accordingly, anchovies might increase in abundance while facing a level of predation on its larvae that might be more than sardines are able to withstand. Thus one may think of the anchovy's chosen loophole as being the "good times" loophole that exists between environmental perturbations, whereas the sardine is adapted to seeking out and utilizing loopholes in predation on its vulnerable early life stages that may be opened during the particularly difficult times of the perturbation itself. (Bakun and Broad cite a stock market analogy, portraying anchovies as "bulls" who do well in "good times" and sardines as contrarian "bears," who during "bad times" are able to turn adversity into opportunity.)

[1] An exception occurs in the very special conditions of the southernmost zone of the extremely energetic Benguela upwelling system, where both anchovies and sardines migrate around the nearby southern end of the African continent to spawn, probably to avoid heavy offshore losses of reproductive product in the extremely strong offshore surface transport field of that region, which may in fact pose relatively greater problems to the weak-swimming anchovies.

[2] Meaning: a way of escaping a difficulty, especially an omission or ambiguity, as in the wording of a contract or law, that provides a means of evasion; also, a small hole or slit in a wall, especially one through which small arms may be fired (The American Heritage Dictionary).

## 4. The 1970s regime shift in Pacific Ocean ecosystems

As outlined in the Introduction, the event that has undoubtedly most transformed conventional thinking about variability in marine populations is the remarkably widespread rash of steep decadal-scale population increases, as well as some decreases, that occurred, particularly in the Pacific but with some suggestive "echos" also in the Atlantic, from the early to mid-1970s to the mid-1980s.

Early on, Kawasaki (1983) had drawn attention to the fact that if one compares the annual sardine catches of the major sardine fisheries of the Pacific Rim (e.g., Figure 24.5), one sees a pattern of synchronous rise and fall that is remarkable, given that the populations are so far apart that any significant biological linkage would seem to be impossible (they are so far apart, in fact, that at that time there was general belief that they must each represent separate, distinct species, although more recently it has been suggested by Parrish et al., 1989, that they may in actuality all represent the same single species). When Kawasaki first published the result, his data extended only as far as 1983 (indicated in Figure 24.5 by a vertical dotted line). But several years later, the synchronous pattern was reconfirmed as the three populations underwent yet another additional simultaneous reversal in population trend in the mid-1980s (Bakun, 1996).

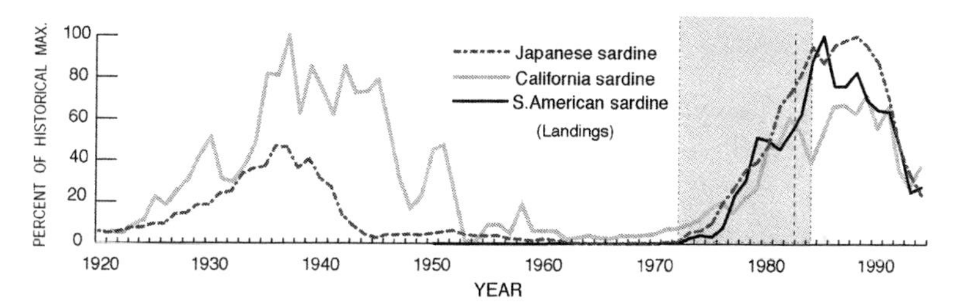


Figure 24.5 Sardine landings from fisheries on the three largest regional sardine stocks of the Pacific (as percent of the maximum value in each series). Shaded rectangle indicates the early 1970s to late-1980s period of steep decadal increase (modified from Bakun and Broad, 2003).

The improbability of direct biological linkages over such enormous distance scales would seem to implicate climatic teleconnections as the most likely linkage acting to synchronize these variations (Kawasaki and Omori, 1988; Lluch-Belda et al., 1989, 1992; Bakun 1996, 1998; Schwartzlose et al., 1999; McFarlane et al., 2000; Klyashtorin, 2001; Chavez et al., 2002). Consequently, there has been a continuing preoccupation with trying to identify climatic (or other) data series varying in the same two-peaked cyclical pattern (period of about 60 years, extrema in the 1920's and in the 1980s, etc.) even up to the present time. However, there are always problems involved in attempting to interpret fishery landings as proxies for population abundance. It seems clear that the portions of these catch series after 1970 do fairly reliably represent the real multi-annual trends in population variability. The collapse of the Peruvian anchoveta in the early 1970s, and the sardine declines of the late 1980s were surely due to lack of fish to catch. Likewise the sudden rise of the Japanese sardine fishery in the 1970s clearly reflected explosive population growth. There is also evidence (Schwartzlose et al., 1999) that in the 1970s sardine stocks were increasing in abundance in the Gulf of California (See Section 2 above) as well as off western South America. However, there seems less reason for confidence that the earlier portions of the series represent acceptable proxies for abundance. The rise in sardine catches before the Second World War must in large part have reflected the growth of fishing effort off both California and Japan in response to the development of fishing and canning technology and of associated markets for tinned sardines; there seems to be no hard evidence that the catch data actually connote substantial growth in population abundance during this period. And while the declines of the two fisheries following the earlier peaks undoubtedly reflected actual population declines under heavy exploitation, the two declines are in no way coincident, being offset from one another by about a decade.

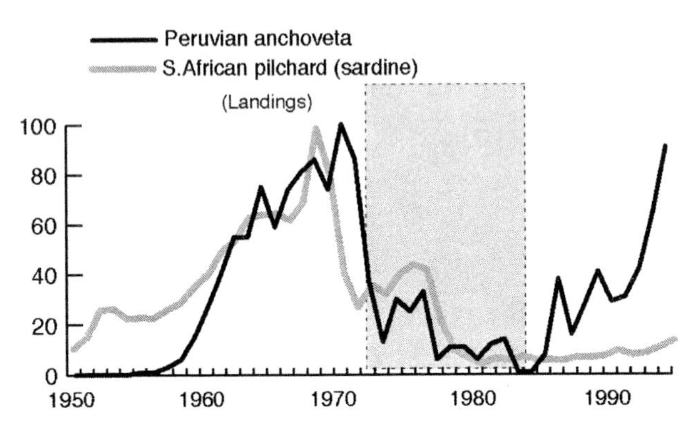


Figure 24.6 Landings by the Peruvian anchoveta fishery and by the southern Benguelan (South African) pilchard (i.e.,sardine) fishery (modified from Bakun and Broad, 2003). Shaded rectangle indicates the early 1970s to late-1980s period similarly denoted in Fig. 24.5.

So, in terms of population dynamics, it appears that the credible parts of the small pelagic series shown in Figure 24.5 are mainly those after about 1970. The simultaneous sardine population increases in widely distant corners of the Pacific in the 1970's and early 1980's, and the suppression of the Peruvian anchoveta population during the same period (Figure 24.6), thus appear to be real and correctly depicted in the landings data series. Moreover, this is a feature that seems to be reflected in the abundance trends of a great number of other populations of marine organisms in the Pacific (Beamish, 1993). A few examples are plotted in Figure 24.7; for many more examples from the northeastern Pacific and Bering Sea, one may consult the extensive list presented by Benson and Trites (2002) which is compiled from number of recent earlier papers, notably those by Hare and Mantua (2000), Beamish et al. (2000), King at al. (2000), McFarlane et al. (2000), McFarlane et al. (2001). Species therein represented include (in addition to those shown in Figure 24.7) sablefish, herring, Chinook salmon, coho salmon, halibut, hake, mackerel, thornyhead, Pacific Ocean perch, Dover sole, English sole, rock sole, rockfish of several species, various species and populations of marine mammals, etc.; other recent collections of examples can be found in Wooster and Zhang (2004) and in Beamish et al. (2004). Furthermore, in the tropical areas of the Pacific, stocks of yellowfin and skipjack tunas evidently increased during this same early-1970s to late 1980s period, while albacore tuna declined (Lehodey et al., 2003). Lobsters, sea birds, seals, and coral reef fishes in the northwestern Hawaiian Islands all seem to experienced increased production during the period (Polovina et al., 1994), whereas in the years after the mid-1980s, lobster landings from that area dropped by two-thirds (Anon. 1993). Patterns of abundance of both phytoplankton and zooplankton in various regions of the Pacific also seem to have responded to the pattern (Sugimoto and Tadokoro, 1997, 1998; Nakata and Hidaka, 2003).

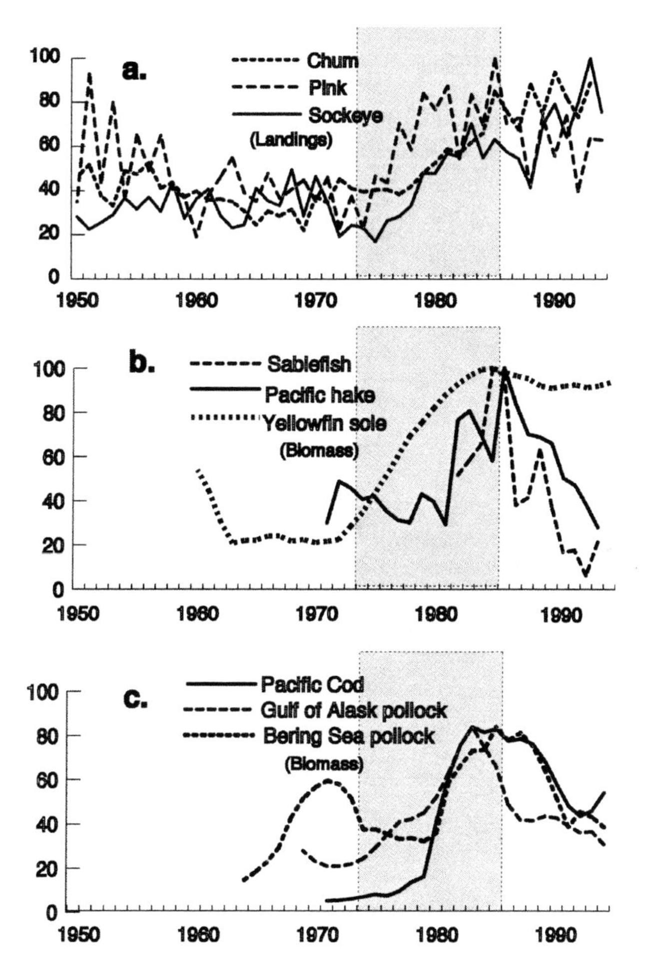


Figure 24.7 (a) Landings of the three salmon species accounting for the bulk of the salmon production in the North Pacific (as percent of the maximum value in each series). Shaded rectangle indicates the early 1970s to late-1980s period of steep decadal-scale increase. (b & c) Fishery independent biomass estimates of important stocks of groundfish in the Bering Sea and Gulf of Alaska (modified from Bakun and Broad, 2003).

## 5. Climate-related trends and switches

The period from the early 1970s to the mid-1980s was indeed characterized by particularly strong decadal-scale trends in several of the most-prominent Pacific basin-scale climatic index series, and even in certain global-scale climatic index series such as globally-averaged air temperatures, etc. (Figure 24.8). The direction of these trends (e.g., long-term decline in the Southern Oscillation Index, intensification of the Aleutian Low, etc.), plus the fact that the period contained three strong El Niño episodes without being countered by any substantial La Niñas (Trenberth, 1990), lend a decidedly "heightened El Niño" character to that period.

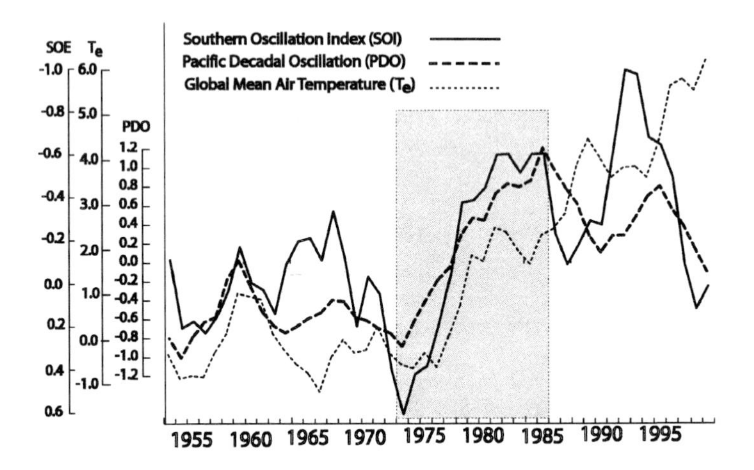


Figure 24.8 Low-passed time series of several large-scale climatic indices (3-yr running means of annually-averaged values). Shaded rectangle indicates the early 1970s to late-1980s period of steep decadal-scale trends (modified from Bakun and Broad, 2003).

It happened that a number of climate scientists had been invited by the then newly formed North Pacific Marine Science Organization (PICES) to participate in the international conference on "Climate and Northern Fish Populations" (Beamish 1995) that was cited in Section 1. The evidence for a drastic "regime shift" in several marine ecosystems of the North Pacific, that seemed only explicable as having been climate-driven, evidently also impressed these climatologists some of whom had already been aware of shifts in climatic characters related to the steep climatic trends of the 1970s (Nitta and Yamada, 1989; Trenberth, 1990). Soon a corresponding regime shift in the North Pacific regional climate, occurring about 1976, was identified by several authors. This has resulted in a current situation of the use of the term *regime shift* in two separate senses: (1) in indicating drastic large time- and space-scale shifts in the abundances of major components of marine biological communities, and (2) in describing apparent transitions between periods of significantly differing average climatic characteristics. At this point, in order to avoid confusion it seems necessary to qualify one's use of the term "regime shift" as either "*ecosystem regime shift*", the main subject of this chapter, or "*climatic regime shift*" to connote the sense that many colleagues are now employing the term.

The climatic shift feature occurring in the 1970s (most often identified with the year 1976 which marked the most rapid changes in many northeastern Pacific time series) has been recognized in a large number of scientific papers (e.g., Ebbesmeyer et al., 1991; Trenberth and Hurrell, 1994; Miller et al., 1994; Polovina et al., 1994; Parrish et al., 2000; Hare et al., 2000, etc.). A more recent climatic regime shift occurring in the North Pacific region in about 1998–99 has also been suggested (e.g., Minobe, 2002; Bograd et al. 2002; Nakata and Hidaka, 2003; Beamish et al., 2004). Other possible climatic regime shifts may have occurred around 1945–47, 1970–71, and 1988–89 (Minobe, 1997; Masuda and Akitomo, 2002; Peterson and Schwing, 2002; Alheit and Niquen, 2004). It is currently not well understood how these shifts might be produced, although they clearly must be a product of the global scale ocean—atmosphere feedback processes that operate in this frequency band of the "red noise" climate variability spectrum.

### *5.1 Abrupt switches or more gradual development?*

In normal English language usage, the term "regime shift", imparts the notion of a rather sudden, abrupt switch. Moreover, climatic shifts are often illustrated in the literature as differences in mean values of prior and subsequent segments of various time series, transmitting the impression of an instantaneous switch between different rather stationary mean states. This seems to have led to a widespread notion that regime shifts generally materialize during a single year or slightly more, and to the practice of trying pinpoint a specific year or years as the precise point at which a regime shift occurred. This impression has probably been strengthened by the "ADS" methodology (Ebbesmeyer at al., 1991; Hare and Mantua, 2000; Mantua, 2004) that has been very influential in popularizing the regime shift concept, but that may overemphasize (or even manufacture) step-like features in a set of multiple time series that exhibit the mixtures of periods and phase relationships characterizing normal "red noise" climatic variability (Rudnick and Davis, 2003).

The 1976 El Niño episode, while relatively weak in the equatorial zone, had a particularly large extension to the high-latitude regions of the North Pacific. Thus, when superimposed on the strong decadal-scale climatic trends of the 1970s, which were trending from less "El-Niño-like" to more "El-Niño-like" character as the period progressed (Figure 24.8), the most rapid changes in various northern Pacific climatic time series tended to occur during that particular year. Accordingly, the point of greatest change seems to have become the focus of attention for many scientists studying the event in high-latitude regions of the Pacific. (And since both the low-passed SOE and PDO anomaly series (Figure 24.8), as well as both first and second principal components presented in the very influential work of Hare and Mantua (2000), cross their "zero" lines near that point, a degree of preoccupation with "zero-crossings" seems to have arisen.) On the other hand, it seems that scientists working in the subtropical upwelling regions, where the effects of the 1976 El Niño were less pronounced, have tended to focus more on the "turning points" (Schwartzlose et al., 1999) when major long-term trends seem to be initially established (i.e., a focus on inflexion points rather than crossings of a long-term mean line). For example, Alheit and Niquen (2004) show evidence that the key ecosystem transition off Peru associated with the events of the 1970s actually occurred at the very beginning of that decade, or even a year or two earlier than that. Thus it would seem that scientists wishing to productively incorporate the idea of regime shifts in data analysis and interpretation at this point might reasonably decide not to be too slavishly bound to anyone's published pronouncements as to specific dates of "system transition" or to rigid preconceptions about a proposed regime shift's nature or effect. This point is underscored by Mantua's (2004) demonstration that different statistical tests, including those proposed by Ebbesmeyer et al. (1991), Fath et al. (2003), and Solow and Beet (in press), when applied to similar data may fail to yield similar candidates for regime shift transition dates.

A more general issue is whether it may be more productive to view a regime shift as (1) a sudden system reorganization that occurs largely within an individual year, and then may dominate a system for a decade or more until another abrupt shift to a different organizational state occurs, or (2) one that continues to evolve over a number years with the same essential processes and mechanisms perhaps acting in a rather continuous manner both to initiate and to terminate a period of particular character. One would think that this should be somewhat a matter of scale. In small confined systems such as lakes (Scheffer et al., 2001; Scheffer and van Nes, 2004), viewpoint #1 may be entirely appropriate. But with respect to climatic regime shifts and large scale open ocean marine ecosystem regime shifts, the much larger scales, and more open nature, of the system involved may render viewpoint #2 more relevant. This set of issues will undoubtedly become clearer as additional careful scientific studies accumulate.

### 5.2 *Amplification occurring within the marine biotic system*

A third proposed possibility for what we are calling a *climatic regime shift* is that it may not involve any essential shift in system organization or function at all but be merely an expected occasional feature of normal "red-noise" climatic variability (Rudnick and Davis, 2003). But if so, it is in distinct contrast to what we are calling a marine *ecosystem regime shift,* which clearly must involve much more than simple reinforcement patterns among different scales of "quasi-normal" variability. Within the ocean, biological variations are often of much greater amplitude than corresponding variations in climatic indices (Hare and Mantua, 2000). Evidently, mechanisms operate in marine ecosystems that are able to amplify rather subtle climatic effects occurring on these multi-annual to inter-decadal time scales to produce relatively much more dramatic changes in marine population abundance and community composition.

In view of the highly leveraged reproductive strategies that are employed by a wide variety of fishes and other marine organisms (whereby a single adult may spawn tens of thousands, or hundreds of thousands, of potential offspring), one can easily imagine how such amplification might occur. For example, simply because of the enormous implied early life mortality, quite small percentage changes in composite early life mortality rates will produce large differences in survivorship. (For example, a net pre-recruit mortality rate near 0.99998 is about what would required to stabilize a population of a species producing 100,000 eggs over the life of a female. In such a case, reduction in from of that rate to 0.99990, which is a reduction of less than 1/100 of 1%, would result in a 500% increase in numbers of individuals surviving to enter the adult population.) Such magnified population fluctuations, perhaps engendered by some "loophole" (see Section 3) having been opened in the fields of predation on early stages, could then cascade in similar manner through the trophic community by altering the early mortality rates of other similarly highly leveraged species. The result would be a major reorganization of the essential community trophic structure (i.e., an essential change in underlying "regime") producing the amplified composite response, acting across multiple trophic levels, that we recognize as a marine ecosystem regime shift.

## 6. Climate—ecosystem linkages

All of these various issues not withstanding, the Pacific episode occurring during the 1970s and early 1980s, as indicated by the shaded rectangles in Figs. 24.6, 24.7 and 24.8 and embodied in the various references cited in Section 4 of this chapter, provides quite a quintessential example that clearly meets the definition marine ecosystem regime shift adopted in Section 1.1., and also is clearly climate-related. As such, it provides a very useful case in point for consideration of mechanisms available for transmission of regime-scale climate variability to ocean ecosystems, to produce the altered survival rates, etc., that may lead to such a essential shift in ecosystem structure and function. A particular interpretive advantage is the distinctive "El Niño"-like character of this period (see previous section), as represented by the decadal-scale decline of the SOE, the corresponding intensification of the mean strength of the Aleutian Low pressure cell, etc. El Niño is not only the most important type of short-term climatic variation affecting large areas of the world, but is also the type for which the underlying mechanisms are best described and understood. Whether (1) the nature of this mid-1970s to mid-1980s period in the Pacific area may realistically reflect actual decadal-scale analogs to the same basic mechanisms that determine an annual-scale El Niño, or whether (2) its distinctive character may more appropriately be viewed merely as the composite mean of an unusual frequency of major, minor, and "quasi" annual El Niño episodes coupled with an unusual paucity of offsetting La Niño episodes of equal strengths, is as yet unclear. But all living organisms must live, die, and reproduce entirely in "real time", without regard for any large-scale averages of conditions that the individuals do not themselves sense or experience. Accordingly the effects on actual individual organisms must be similar in either case. This provides an opportunity to understand the linkages of physical influences to biological consequences on the basis of prior understanding acquired though experiencing and studying the much more numerous realizations that have been available in the accumulating record of El Niño episodes.

The discussion in this section (generally following analogous discussions in Bakun, 1996 and Bakun, 1999) will move from high to low latitudes, starting from the far north of the subarctic Pacific, and then moving south to the tropics via a transitional discussion of effects on the Pacific salmon habitats which stretch along the eastern Pacific boundary from the Bering Sea through much of the subtropical California Current region. While it may be natural to think of temperature as being the a predominant factor in climatic influences, it may become evident that direct effects of temperature (sensible heat transfers) might to be of considerably less importance than transfers of momentum, mechanical energy, and (fresh) water. (In order to avoid excessive length in the discussion in this section, it will be necessary to wield a fairly "broad brush", resulting in somewhat cursory treatment of many very interesting aspects.)

### 6.1. *The Subarctic region*

The Aleutian Low pressure cell that overlies the subarctic North Pacific is one of the most intense quasi-permanent atmospheric systems on earth. It represents a center of violent atmospheric cyclogenesis during the winter half of the year. During that period of the year, the intense cyclonic weather systems that underlie the Aleutian Low feature exert a very powerful pattern of cyclonic wind stress curl on the ocean surface that drives a radially-directed Ekman transport field. This produces very strong divergence in the ocean surface flow field, driving upwelling of deeper waters toward the sea surface and resulting abrupt upward ridging of the density structure (Favorite et al., 1976) in the interior of the gyre. This dome-like "ridge" structure supplies the baroclinic foundation in which the energetic ocean currents of the region are imbedded. The Ekman transport responding to the furious winds drives the surface waters against the coastal boundary of the eastern and northern Gulf of Alaska (Figure 24.9), driving the density structure precipitously downward and intensifying the geostrophic currents that flow from east to west along the northern continental margins of the Pacific basin.

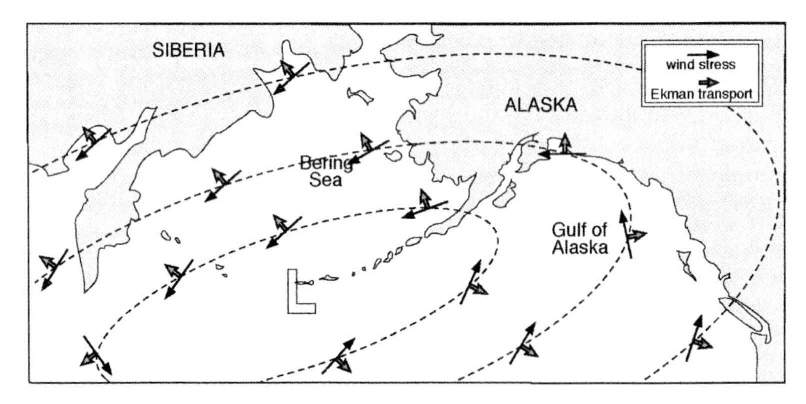


Figure 24.9 Diagram illustrating the effect of the Aleutian low pressure cell in driving coastward Ekman transport toward the periphery of the Gulf of Alaska and other regions of the subarctic Pacific Ocean (modified from Bakun, 1996). Dashed lines indicate pressure isobars surrounding the low pressure center. (Note that whereas the higher-level geostrophic winds closely parallel isobars of pressure, the winds near the sea surface are deflected by friction to the left of the trend of the isobars.)

Thus the ocean flow system in the Gulf of Alaska Large Marine Ecosystem is dominated by strong cyclonic (anti-clockwise) circulation around its periphery. The northward Alaskan Current flow on the eastern side of the Gulf is broad, diffuse, and rather slow, as is typical of eastern ocean boundary currents (Wooster and Reid, 1963). But as the flow veers around the northern boundary of the Gulf and turns southwestward, it narrows and intensifies to become the swift western boundary flow known as the Alaskan Stream (Favorite, 1967; Thomson, 1972). The Alaskan Stream flow then runs rapidly westward along the Aleutian Island chain, with branches flowing into the Bering Sea through the deeper passes between islands. There are also intermittent exchanges with the Subarctic Current which flows eastward just to the south. As a result of the large inflows of surface waters into the Bering Sea through Aleutian Islands passes, the surface layer of the Bering Sea is composed primarily of waters from the Alaskan Stream and Subarctic Current systems (Favorite et al. 1976), i.e., from within the North Pacific Ocean proper lying to the south of the island chain.

The other major driving force for the local system is the copious coastal runoff that occurs in the region (Royer, 1979). The entire continental coastal region in many ways resembles a massive, continuous runoff plume spread along the coast. Geostrophy tends to confine the less dense, runoff-influenced surface waters against the coast, which adds to the upward sea surface slope from the ocean toward the coast produced by the wind-driven coastward Ekman transport and so further enhances the alongshore circulation. Much of the runoff in the region comes from summer melting of precipitation stored during the cold seasons as snow and ice. This continues to reinforce the flow system during the summer season when the wind forcing nearly vanishes in comparison to its winter intensity level. The general intensification of the Aleutian Low pressure cell that occurred during the mid-70s to mid-80s period would thus act to intensify the entire system, in that the increased cyclogenesis (storm production) that this implies would act to intensify both the wind forcing and the precipitation. One effect would be to increase the advection of warm waters into the northern part of the region, generally raising ocean temperatures all along the coast.

In terms of the "*ocean triad*" elements that may regulate habitat favorability (Bakun, 1996), enhanced *enrichment* of the system would occur during this period as a result of strengthened wind stress curl-driven upwelling in the interior of the gyre, perhaps augmented by correspondingly increased winter wind mixing. The intensified Alaskan Stream flow would tend to increase the production of cyclonic frontal vortices, and associated vortex-driven upwelling, along its offshore edge. The tendency for eastward displacement of the anomalous intensified Low pressure cell into the Gulf of Alaska proper during El Niño conditions would cause the Ekman transport pattern to be even more intensely directed toward the coast in the eastern and northern parts of the Gulf, and in fact extend the area of coastward Ekman transport much further westward toward the Aleutian Island chain. This would tend to intensify the transport of enriched surface waters toward the coastal habitat.

*Concentration* mechanisms would also be enhanced by such an intensified system. Increased runoff would favor stronger and more widespread frontal formations between runoff-influenced and oceanic surface waters. Intensified Alaskan Stream flow would generate increased frictional dissipation which would act to break down geostrophy in the frontal zones, producing density-driven convergence both at the outer and inner edges of the current jet. Superimposed on this, the intensification and westward extension of the shoreward wind-driven Ekman transport pattern would act to increase the piling of oceanic surface waters against the coast, enhancing coastal downwelling and associated formation of convergent frontal structures.

The general pattern of coastward Ekman transport would carry larvae toward, and favor their *retention* within, the coastal habitat. The intensified geostrophic flow system would rapidly spread larvae and juveniles along the coast and westward along the chain of Aleutian Islands, where those near the ocean surface would tend to be carried onto the shelf or into the proximity of island passes by the wind-driven surface Ekman transport. Once on the shelf, where the flows are somewhat decoupled from the flows offshore of the continental shelf break, they might find favorable juvenile habitat, or they may be injected through island passes and carried by the eastward coastal flow on the north side of the Aleutian Chain onto the broad shelf habitat of the eastern Bering Sea.

### *6.2 North Pacific Salmon*

Trends in Pacific salmon catches from 1925 to 1988 have been similar to the smoothed trends in the strength of the "Aleutian Low" pressure system (Beamish and Bouillon, 1993). This correlation indicates that salmon populations in the subarctic regions of the Pacific Ocean have generally done well during eras when the tropical Pacific has tended toward intensified El Niño-like conditions. Conversely, salmon populations feeding in the subtropical coastal upwelling zone off the western USA have appeared to do poorly during El Niño episodes (Pearcy, 1992) and many declined severely during the prolonged El Niño-like period following the mid-1970s (Hare and Francis, 1985)

Along the latitudinal gradient from Alaska south to California, the northernmost stocks of each salmon species would naturally tend to approach the cool limit of their range while the southernmost stocks would be toward the warm limit. One would expect the optimum to be somewhat intermediate. There is no question that El Niño tends to warm the coastal ocean and adjacent coastlands of the northeastern Pacific. In such a case, El Niño conditions would tend to push this "temperature optimum" northward. And salmon may well be particularly sensitive in this respect. Salmon spend their earliest life stages in shallow freshwater habitats that are much more subject to drastic year-to-year temperature anomalies than are habitats within the ocean. For example, severely cold winter conditions in Alaska may even freeze salmon spawning streams completely down to the stream bed, thereby totally depleting the temperature buffering capacity of the liquid phase of water and allowing the temperature to drop to lethal limits for the buried eggs. Extremely warm conditions in California may lead to early loss of snowpack and reduced flow rates later in the season. Both reduced flow rates and higher temperatures would, among other things, tend to lower the dissolved oxygen levels available to egg and larval stages.

Beyond direct temperature effects, the decadal intensification of dynamical aspects of the subarctic system described in the previous section could have distinctly beneficial effects for growth and survival of salmon smolts, as well as of later life stages. Moreover, salmon enter the ocean after an extended period of early life in freshwater and have consequently grown to substantial size and associated swimming power by the time they enter the marine system. Thus they may be relatively well able to cope with the enhanced levels of turbulence and advection that typify an intensified dynamic system. On the other hand, in the California Current LME where salmon stocks did not fare well during the period, El Niño conditions tend to feature a thickened layer of warm, nutrient-depleted surface water such that upwelling is much less effective in enriching the trophic system. Accordingly, El Niño conditions are characterized by drastic reductions in primary production and zooplankton abundance (Bernal, 1981; Miller et al., 1985). One might not be surprised that salmon smolts, being zooplankton feeders, may not survive well during a decadal period dominated by El Niño and quasi-El Niño-type conditions.

### *6.3 The equatorial zone*

The long-term decline in the Southern Oscillation Index (Figure 24.8) during the mid-1970s to mid-1980s period signifies a decadal-scale relaxation of the trade wind circulation in the near-equatorial zone of the Pacific. Since it is the trade winds that drive equatorial upwelling, one expects that the equatorial upwelling should slow down in response, resulting in lowered nutrient input and associated poor conditions at the low end of the food chain. Over time, relatively less baroclinic upwarping and downwarping of the subsurface density structure should result, leading to generally less intense shear zones between the oppositely-directed high speed current filaments of the equatorial current system. This in turn implies less intense divergence (*enrichment*) and convergence (*concentration*) in the frontal eddy fields, i.e., less robust "*ocean triad*" conditions for the early stages of tunas and other large highly migratory fishes able to solve the remaining "third element" of the triad (*retention*) through sheer swimming power of the adult fish (Bakun, 1996).

All this should have resulted in poor food densities for tuna larvae and early juveniles. Nonetheless, this was a period when the Pacific yellowfin and Pacific skipjack tuna populations, which spawn in the near-equatorial band (Cole, 1980; Foresberg, 1980), exhibited impressive upward population trends (Lehodey et al., 2003). On the other hand, Pacific albacore tuna which spawn somewhat to the north and south of the equatorial zone (Foreman, 1980), experienced poor reproductive success (Lehodey et al., 2003) and a consequent population decline (Anon., 1992; FAO, 1992). Note that this off-equatorial zone may become anomalously divergent during the annual-scale decline of the SOI which signals an El Niño episode (Philander and Lau, 1988), in contrast to the near-equatorial band where divergence and resulting upwelling decrease during El Niños. To the extent that the same tendency may hold during a decadal-scale SOI decline, this might have increased the productivity of the zone where albacore spawn.

So why should yellowfin and skipjack have done well reproductively when their reproductive habitat was presumably less productive at the low end of the trophic pyramid, and the albacore have done poorly when, in the same period, their reproductive habitat was more productive? Bakun and Broad (2003) suggested that the apparent paradox may be due too the intervention of a mobile mid-level "wasp-waist" population.

Frigate mackerels (*Auxis* sp., etc.) are small-sized, fast-swimming, voraciously feeding tuna-like fishes, which although not extensively fished, are apparently enormously abundant in the tropical oceans (Richards, 1984; Fonteneau and Marcille, 1993; Colette and Aadland, 1996). They are likely to be major predators on young tunas. If the poorly productive conditions during this period caused the local frigate mackerel populations, and whatever other species may exert particular predation mortality on tuna early life stages, to either collapse in abundance or to shift their distributions away from the near-equatorial zone, this could perhaps have opened a sufficient survival "loophole" to have enhanced reproductive success of yellowfin and skipjack through this period. Conversely, the increased productivity of the off-equatorial zone where albacore spawn might have supported growth of the local frigate mackerel populations or alternatively have produced an anomalous shift in the area distribution of those populations away from the anomalously less productive tropical band and toward the anomalously more productive zones to the north and the south. This might have effectively shifted predation pressure from the equatorial spawning zone of yellowfin and skipjack, to the slightly higher-latitude zone where albacore tend to concentrate their spawning. In such a case, the benefits to the yellowfin and the skipjack may have been at the cost of the albacore. (If indeed this was the case, it may have troubling implications in that a situation of adults preying on the predators of their offspring implies a potentially dangerous feedback instability in the system, such as seems to have operated in the examples of adverse regime shifts discussed later in Sections 8 and 11 of this chapter.)

Beyond such direct effects on marine ecosystems of the equatorial zone, ocean anomalies in the equatorial region (e.g., generated by the long-term decline of the SOE shown in Figure 24.8 and the long-term decline in the momentum transfer from the trade winds to the ocean that such a decline represents, etc.) may have particularly important remote influences on other parts of the ocean. This is due to (1) climatic teleconnections that act through the atmosphere via the large scale redistributions of zones of rising and sinking air masses, and of the horizontal momentum transported in their connecting air flows, that are associated with changes in patterns of heat exchange through the sea surface in the equatorial zone of the Pacific (Bjerknes, 1966; Horel and Wallace, 1981), and (2) to the propagation of geophysical waves (Kelvin waves, etc.) eastward along the equatorial wave guide and then northward and southward along the eastern ocean continental boundary (Yoshida, 1967; Gill and Clarke, 1974; Wyrtki, 1975; Allen, 1975).

## 7. Pacific-Atlantic teleconnections

Although considerably less clear and dramatic than within the Pacific Basin itself, there have been, as mentioned in the previous section, certain "suggestive echos" of similar population increases or declines in the Atlantic Ocean that roughly correspond to this same early 1970s to late 1980s period. Lluch-Belda et al. (1989, 1992) opened this issue by pointing out that the sardine landings in the South Atlantic off the west coast of South Africa, have varied in quite directly opposite phase to that of the three large regional sardine fisheries of the Pacific (Fig 24.5). Thus abundance of the South African sardine has covaried with that of the Peruvian anchoveta in a substantial degree of multi-decadal synchrony (Figure 24.6). In the Canary Current region, which is the eastern ocean upwelling system in the North Atlantic that is analogous to the Benguela system in the South Atlantic, a period of anomalously low catches of sardines from 1977–1986 (Roy and Reason 2001) suggests a similar opposite phase variability with respect to the Pacific sardine populations.

Moreover, one can find many other patterns of abundance in the Atlantic that seem to mimic the general patterns that are so prominent in the Pacific. For example, the period of the early 1970s to the mid-1980s was the period of the triggerfish (*Balistes*) outbreak off the tropical Atlantic coast of Africa, treated above in Section 2, and an associated temporary suppression and subsequent rebound of the regional population of *Sardinella aurita* in that same region, discussed below in Section 7.1. It was also a period of increasing survival of Greenland halibut (Serebryakov, 1992), a period of dramatic growth in the lobster landings in eastern Canada (Pezzack, 1992), and a period of large increase in Newfoundland spawning northern cod stock (Paz and Larrañeta, 1992). Roughly contemporaneous to a proposed later Pacific regime shift occurring about 1988–89, was a somewhat controversial (Taylor, 2002; Solow and Beet, in press) regime shift in the North Sea ecosystem, involving major changes in a suite of phytoplankton, zooplankton and planktivorus fish abundances, that may have been driven by variation in inflow of oceanic waters into the North Sea (Reid et al., 2001, 2003), but also may have reflected changes in the ecosystem structure over a much larger area of the northeastern Atlantic Ocean (Beaugrand, 2003, 2004).

Of course, one must be cautious in attempting to infer actual inter-ocean mechanistic linkages from such temporal correspondences (Fréon et al., 2003). Moreover, the fishing pressure in the North Atlantic has been so heavy as to tend to obscure or distort natural multi-annual abundance swings. In particular, extremely rapid growth in fishing pressure during the 1960s and early 1970s on both sides of the North Atlantic (Sherman, 1989) tends to interfere with recognition of real trends and inflection points in the early to mid-1970s when the most widespread transitions occurred in Pacific ecosystems.

But climatic linkages between the two oceans certainly do exist. The earth's ocean-atmosphere system is an interconnected dynamic entity. Any truly major dynamic shifts in one region must affect other regions in one way or another. Because the Pacific Ocean is so large and its heat capacity so enormous, the massive back and forth swings of the Pacific ENSO system constitute the dominant component of climate variability occurring on inter-annual time scales over essentially the entire earth. The global-scale nature of the shift that occurred during the 1970s might be epitomized if it were involved, as proposed by Beamish et al. (1999, 2000, 2004), in observed changes in the rotation rate of the solid earth itself.[3] Research has shown, for example, that sea surface temperature (SST) anomalies associated with ENSO events in the Pacific affect both wind speed and SST in the Atlantic as well. This linkage occurs in two regions (Carton et al. 1996, Enfield and Meyer, 1997; Sutton et al. 2000), the principal one being the zone of NE trade winds west of 40°W along a 10°-20°N latitude band and extending into the Caribbean, and another one being associated with the Intertropical Convergence Zone (ITCZ). In

[3] This would signify that the associated change in the angular momentum of the earth's ocean-atmosphere system as a whole was sufficient, in order that the angular momentum of the total system of "planet earth" be conserved, to have forced a measurable opposing change in the angular momentum of the enormously more massive solid body of the earth (including its enclosed "liquid core"). Such an implied balance would of course be altered, or perhaps even dynamically perturbed (Beamish et al., 2004), to the extent that negatively or positively reinforcing variations in the angular momentum of the interior liquid core of the earth might somehow have been involved.

both areas, the Atlantic response has been found to lag the Pacific by about 5 months. SST is also somewhat correlated with ENSO in the western South Atlantic along the 20–25°S latitude band. Evidently related to this mode of inter-ocean teleconnection, Roy and Reason (2001) linked Pacific ENSO variability to upwelling intensity off NW Africa (10°-26°N) at a 4–5 month lag and associated the low catches of sardines from 1977–1986 with positive SST (low upwelling) anomalies occurring in the years of hatching and growth of the fish in question.

### *7.1 Contemporaneous gyrations in the "wasp waist" of the Guinea Current LME*

There is a suggestive pattern of correspondence between cold anomalies (strong upwelling) in the Guinea Current upwelling zone and Pacific El Niños (Bakun, 1995), that fits mechanistically well with the ENSO-associated relaxation of Atlantic trade wind circulation (described in the previous paragraph) which, because of the primarily east-west trend of the coast and of its near proximity to the equator, acts to oppose the local coastal upwelling that appears to result at least partly from Kelvin wave propagation of effects originating either further eastward along the African coast or actually propagated from the western tropical Atlantic in the equatorial wave guide (e.g., Picaut, 1985). Periods of intense local upwelling have been associated both with strongly increased local *Sardinella aurita* availability and with low rainfall periods in the adjacent coastlands (Bakun, 1978). Figure 24.10 is suggestive of action of these linkages also on the longer "regime" time scale characterized by the early 1970s to mid 1980s period of heightened El Niño character in the Pacific that has been the focus of much of the discussion in this chapter.

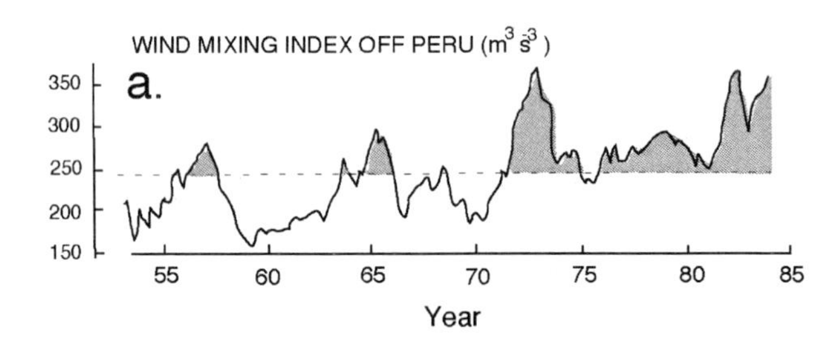


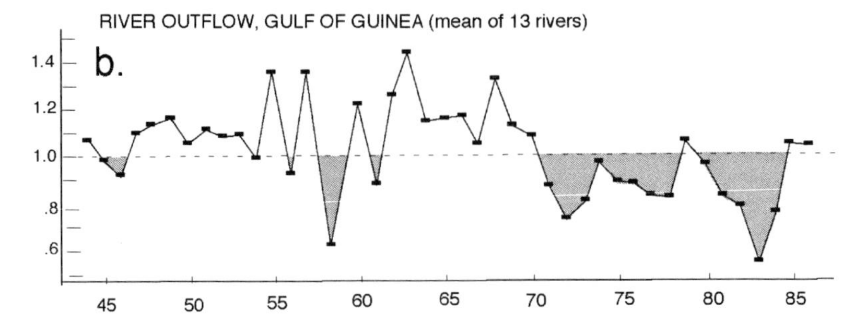


Figure 24.10 (a) Wind-induced turbulent mixing index off Peru (increases in wind-generated turbulence being a characteristic of El Niño influence on the ocean off Peru) juxtaposed with (b) variations in river outflow (mean of 13 rivers) into the Gulf of Guinea (after Mahé, 1991). (Figure redrawn from Bakun, 1995).

*Sardinella aurita* was the major species harvested before the 1970s, with the bulk of the early production being in Ghanian waters. Until 1972 the relationship between upwelling indices, plankton production, and *S. aurita* landings had been rather linearly positively correlated (Mensah, 1991). However, during 1972, a year of particularly strong El Niño in the tropical zone, the fish appear to have been exceptionally available to the the local artisanal fishery (Bakun, 1978; Binet, et al., 1991) leading to catches some two and one-half times higher than in any previous year (Koranteng, 1991). Following this unusually high catch, the population collapsed and this formerly dominant species essentially vanished from the catches. This coincided with initiation of the triggerfish (*Balistes carolinensis*) outbreak (Section 2) which started in this same area.

It seems possible that the juvenile triggerfish, being pelagic plankton feeders of a species capable of rapid population response, somehow took advantage of the collapse of the *Sardinella aurita* population to temporarily replace it at the "wasp waist" of this ecosystem. But in the late 1970s, a period characterized by anomalously strong upwelling (as suggested by the low coastal runoff during this period, see Figure 24.10, above), *Sardinella aurita* began to rebuild its population. In the 1980s, production grew to a point where annual catches comparable to the catch in 1972, which was followed by population collapse, became routine and apparently sustainable. Associated with this apparent increase in stock productivity were changes both in fishing areas and in fishing seasons. In the earlier period, the bulk of the catch came from Ghana during the upwelling season; later, the catch was spread also over the coast of Côte d'Ivoire to a much greater degree and was much more continuous though the entire year (Binet, et al., 1991). This then offers yet another instance where abundance changes evidently left enduring impacts on subsequent geographic distribution and migrational behavior.

## 8. The Baltic Sea — a simplified system

In viewing a complex situation such as the global-scale shifting circumstances of the 1970s, one may experience a sort of "informational overload" where it may be difficult to isolate the key questions to be addressed, or even to decide how to arrange the observations and experiences in a way that can reveal the sorts of informative patterns that might lead to fruitful insights. Thus it may be useful to build a frame of reference by examining some examples that are set in more bounded, more simplified systems, where the crucial processes and mechanisms may be more accessible.

The Baltic Sea is one of the largest brackish water bodies in the world. It is connected to the Atlantic Ocean by two very narrow and very shallow (< 20 m) passages. It is a quite shallow sea, having a mean depth of 57m, but contains a series of deeper basins with maximum depths in the range of 105 meters to 459 meters (Kullenberg, 1986). Its large drainage area situated in a region of quite high precipitation results in a low density surface water layer which generally overflows to the Atlantic over the shallow entrance sills; a strong halocline overlies higher density, more saline waters near the sea bottom. The brackish conditions result in a very simplified fish fauna such that only three species, cod, herring and sprat, constitute some 95% of the fishery landings, which however are substantial and valuable. Cod is the major predator in the system. Herring and sprat are plankton feeders.

The stagnant waters in the deeper basins are renewed only at quite long intervals, between which the oxygen concentration in the deeper zone becomes progressively lowered. Atlantic water can only enter the Baltic when meterological conditions (strong westerly winds) cause high sea levels outside to coincide with low sea levels inside the entrance straits so as to overcome the pressure head produced by the excess input of freshwater. Since the shallow sill areas are quite wide, considerably mixing with the fresher Baltic Sea surface water takes place before the inflowing water can reach the interior basins. This normally lowers the density below that required to displace the higher salinity waters occupying the deeper areas. Thus it is only when the inflows are particularly large and extended that the stagnant deeper waters are renewed. A major renewal event took place in 1975/76. This was followed by a 16-year period of stagnation of the central Baltic deep water, which ended with another major inflow of highly saline water associated with strong westerly gales that occurred in January, 1993 (Matthäus and Lass, 1995).

Cod eggs have a density such that in most of the ocean areas in which cod occur, their eggs are neutrally buoyant in the upper 20 to 50 meters. In the Baltic however, the lowered salinity causes the eggs to reach neutral buoyancy much deeper, where they may be exposed to the lowered oxygen conditions in the stagnant near-bottom waters. Cod eggs will not develop, or hatch successfully, at very low oxygen levels (Mackenzie et al., 2000). Throughout the 1980s, as oxygen availability in the deep layers declined, cod have been subject to declining reproductive success (Köster and Möllmann, 2000). This, combined with overfishing, has caused the cod stock in the Baltic to wane to historically low levels in the early 1990s. Accordingly, abundances of herring and sprat, which are preyed upon by cod, increased over the same period. However, although hydrographic conditions conducive for survival of early life stages of cod returned during the 1990s as a result of the bottom water renewal that occurred in 1993, recruitment of cod remained "locked in" to far below average levels (Köster and Möllmann, 2000). Thus, this appears to have constituted the type of "sticky" self-perpetuating regime shift that, once established, may tend to resist reverting smoothly back to previous circumstances.

In this case, a feedback mechanism accounting for the "stickiness" of the shift appears to have been the growth in abundance of the sprat and herring populations resulting from the decline in abundance of their major predator, the cod. These plankton-feeding clupeid species have been identified as major predators on cod eggs and larvae in the Baltic (Köster and Möllmann, 2000), particularly since the cod eggs and larvae occur in dense aggregations within or just below the halocline, which is precisely the zone where the clupeids tend to concentrate feeding activities during their main daytime feeding period (Köster and Schnak, 1994). Thus the "loophole" that had opened for clupeid population growth as a result of the overfishing of the the cod during a climate-related period of reproductive difficulty caused by extended lack of bottom water renewal, may have durably closed the "loophole" that could otherwise have allowed the cod population, after the renewal eventually had taken place, to rebound and reestablish the earlier trophic configuration of the Baltic Sea ecosystem.

### 9. The Black Sea—incursions at the "wasp waist"

The Black Sea is a deep basin connected to the Mediterranean Sea by only a narrow, shallow channel. Its catchment basin is considerably larger even than that of the Baltic Sea, being more than twenty times greater than the entire surface area of the Black Sea itself (Prodanov et. al. 1997). Consequently the Black Sea receives a very large excess of freshwater input compared to evaporation. This causes particularly strong stratification of the waters and effective isolation of the deeper layers from the lower-density, diluted surface waters that exit the Black Sea and flow into the Mediterranean. As a result, the Black Sea is permanently anoxic at depths deeper than about 120 to 150 m.

In spite of its subsurface anoxia, the Black Sea ecosystem has historically been a highly productive fishery system. Several decades ago it contained a balance of ecological groups, including migrant upper trophic level predators, (e.g., bonito, blue fish and mackerel, each of which supported economically important fisheries), a "wasp waist" of small pelagic fishes, and a productive supporting base of zooplankton and phytoplankton. But in the last several decades, the Black Sea has been wracked by a succession of transitions caused by a sequence of disruptive human interventions into the workings of the natural system.

The basin contains some of Europe's largest river systems and drains many heavily populated and highly developed regions. Excessive nutrient inputs from agricultural, industrial, and domestic activities into the Black Sea have resulted in severe eutrophication of the ecosystem (Mee, 1992; Caddy, 2000). This may initially have helped the growth of fishery landings, in tonnage if not in value, by increasing planktonic food densities and helping to shift the system into dominance by fast growing, planktivorous, small pelagic fishes (Caddy, 1993). Annual landings of anchovy of several hundred thousand tons become the norm up to the early 1980s (FAO, 1992). In the late 1980s landings increased to levels approaching nine hundred thousand tons per year. At their maximum in 1988, the catches of anchovy represented more than 60% of the total fishery catches taken from the Black Sea. As a result of this particularly heavy exploitation, in the following year the anchovy spawning biomass had declined by more than 85% (Shiganova, 1998).

These intense removals at the "wasp waist" by the fishery, probably in conjunction with the increasingly severe eutrophication, have apparently made room in the system for less desirable components (Caddy and Griffiths, 1990; Daskalov, 2002). This seems to have been initiated in the 1970s, when there were large blooms of *Noctiluca miliaris,* a zooplankton species with very low food value for higher trophic levels. However, since the mid-1980s there has been a more drastic change to a situation in which the ecosystem has been burdened with enormous basinwide biomasses of "jelly predators".

First it was the jellyfish (*Aurelia aurita*). This medusa built up biomasses estimated to be as much as 450 million tons (a really amazing number, since the area of the Black Sea is only about one six-hundredth of the total area of the worlds oceans, while 450 million tons represents nearly six times the total marine fish catch for the world as a whole). Later, the medusa was itself displaced by another "jelly predator", the ctenophore *Mnemiopsis leidyi,* which became even more abundant. This ctenophore is an exotic species that had previously been absent from the eastern side of the Atlantic. It is believed to have been introduced into the Black Sea ecosystem in discharges of ballast water from ships coming from the American side of the ocean. It was first noted in the Black Sea in 1982 (Pereladov, 1988). But in 1989, the year following the drastic reduction in anchovy biomass by the fishery, zooplankton biomass increased, assumedly in response to the resulting reduction in consumption by anchovies (Shiganova, 1998). It was at this point, that *Mnemiopsis* abundance explosively increased, reaching biomasses that have been estimated as high as a billion tons. These events led to a nearly total collapse of fisheries in the Black Sea. Ctenophores, besides competing for zooplanktonic food, are proficient predators of fish larvae (Monteleone and Duguay, 1988). By 1991, landings of anchovy, the last major fishery in the Black Sea, were only about 15% of those a few years earlier (FAO, 1992).

However in 1992, the *Mnemiopsis* plague abated, and zooplankton abundance and species diversity once again began to increase. Fish abundance also increased. By 1994, landings of Black Sea anchovy again reached levels approaching those of the mid-1980s. But their selected spawning habitat had evidently changed, such that the greatest abundance of anchovy eggs and larvae were found in the southern Black Sea, in contrast to the situation that had predominated earlier. In consequence, Turkey became the only significant exploiter of the Black Sea stock of anchovy. Other valuable fishery species (e.g., bonito, mackerel, chub mackerel, horse mackerel, as well as several demersal species) also rebounded, but likewise mainly in the south (Shiganova, 1998), where impacts of eutrophication, etc., are less pronounced.

Most recently, the situation is again impacted by a new alien ctenophore, probably also arriving via ballast water discharges from ships coming from the American side of the Atlantic Ocean (Dumont and Shiganova, 2003). This ctenophore, *Beroe ovata,* is an obligate predator on other ctenophores such as *Mnemiopsis.* It appears to have reduced *Mnemiopsis* to low abundance levels. Dumont and Shiganova foresee a possibility, since *Beroe* is a "hyperspecialist" with no alternative prey to *Mnemiopsis,* of alternating linked cycles of explosion and collapse of the two species. If so, the "wasp-waist" of the Black Sea ecosystem might, in the future, be mired in a situation of cyclic radical transformations and associated alternations between total fishery collapse and short periods of more or less productive fisheries on certain particularly rapidly responding resource species such as anchovy that might be able to successfully rebuild a degree of abundance in the periods in which *Beroe* would temporarily dominate.

## 10. Pandemic phase shift in coral reef systems

Recent years have seen an increasing number of coral reef areas shifting from a dominant coverage by the species of hard corals that build the reef structures to coverage by fleshy macroalgae. This is happening in essentially all regions of the world's tropical oceans, but at least in the Caribbean region the shift has become so pervasive (Nystrom et al., 2000) as to clearly qualify as a regime shift according to the criteria proposed in Section 1. It appears that the proximate cause is the stripping of living coral from the reef surface, either (1) by natural disturbances such as intense storms, (2) by biological processes such as direct predation on coral polyps (e.g., by crown-of-thorns starfish, parrotfish, etc.) or erosion due to activities of mobile organisms (e.g., scouring of the coral surface by sea urchins, etc.), or

(3) human activities (dynamite fishing and other fishing practices that inflict direct damage on the surface coral cover, ship groundings, activities of divers, etc.). In addition, episodes of coral bleaching related to climatic warming have been increasingly frequent in recent years (Hoegh-Guldberg, 1999); such bleaching may lead to death of coral polyps (Glynn, 1993). Once free space is opened on the reef species, it may be rapidly colonized by fast-growing algae. These algae do not appear to be superior competitors to adult corals (McCook, 2001), but once established act to inhibit the settling of coral recruits to recolonize the space. Pollution is another obvious suspect. But the coral-to-algae phase shift has occurred in many areas where pollution seems not to be an issue. Rather, pollution-related increases in plant nutrients may help the macro-algae to grow more quickly, but do not seem to lead to the removal of coral cover so as to adequately explain the observed changes. Moreover, healthy populations of herbivorous fish and invertebrates may be well capable of reacting to increase consumption to balance nutrient-related increases in algal growth rates (McCook, 1999; Nyström et al., 2000).

Like other examples of regime shifts treated in this chapter, overfishing might have been a key ingredient in the coral reef phase shifts (Hughes, 1994; McManus et al., 2000; Nyström et al., 2000; McManus and Polsenberg, 2004). Large predatory fishes, which tend to be the most marketable (i.e., groupers, snappers, etc.), tend to be targeted initially by local fishers. Later as the predator fishes become scarce, fishers may shift attention to herbivorous fishes (parrotfish, surgeonfish, etc.). This then may present an opportunity for increases in abundance of herbivorous invertebrates. In the Caribbean, for example, the explosive growth of large populations of long-spined sea urchins (*Diadema antellarum*) may itself have resulted in significant damage to living coral as the overabundant urchins scoured the reef surface for algae. Much destruction of surface coral cover is known also to have been inflicted by storms in that period; for example, very extensive damage was done to Jamaican reefs by Hurricane Allen, a category 5 hurricane that struck in 1980 (Hughes, 1994). At that point, the accumulation of very high concentrations of urchins may have set the stage for the outbreak of a lethal infectious disease that in 1983–84 spread rapidly throughout the Caribbean virtually eliminating *Diadema antellarum* from its entire range (Knowlton, 2001). This mass mortality, ultimately mediated by the earlier removal by the fisheries of (1) the predatory fish that preyed upon the urchins and (2) the herbivorous fishes that competed with them, appears to have essentially eliminated grazing pressure on the seaweeds, allowing them to overgrow the extensive damaged areas and fend off their recolonization by settling corals.

## 11. The Northern Benguela: overfishing, ecosystem collapse, toxic eruptions

The northern Benguela Current marine ecosystem, located off the west coast of southern Africa, features the strongest sustained coastal upwelling of any of the "classical" eastern ocean coastal upwelling zones of the world's oceans (Parrish et al., 1983; Bakun, 1996). Because of this massive upwelling, the area is characterized by very high rates of primary organic productivity (Carr, 2001).

In the mid-1960s, the sardine biomass in the northern Benguela is estimated to have been about 10 million tonnes; catches were at annual levels of about 1.5 million tonnes (Boyer, 1996). Then, under very heavy fishery exploitation, the sardine resource abruptly collapsed. In the following four decades, up to the present time, both biomass and catches have consistently been "stuck" at levels not exceeding a tenth of those that were typical earlier, and recently have fallen even more. This is in spite of the establishment of modern fishery resource management procedures and their careful application in recent years. In the process, the "wasp waist" of that ecosystem has shifted from exploitable pelagic fish species to domination by a combination of "jelly predators" (medusas, etc.) and low-valued pelagic gobies (Boyer and Hampton, 2001).

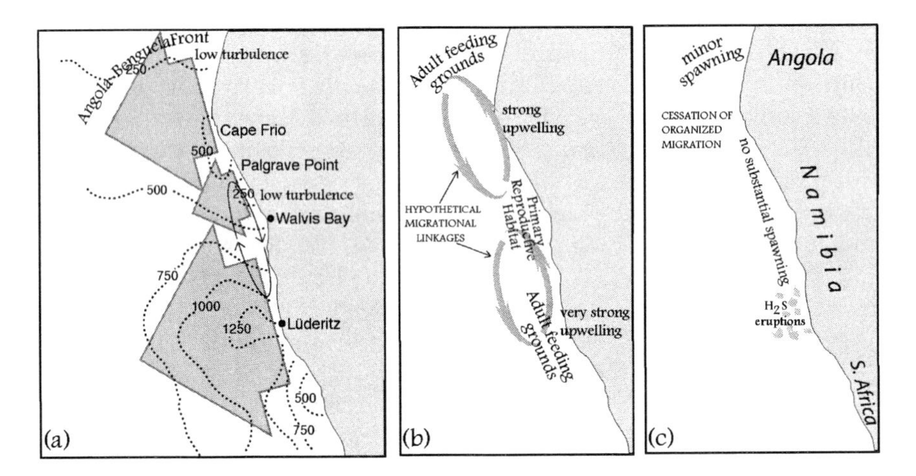


Figure 24.11 (**a**) Diagram of characteristic summer features of the environment off Namibia. Broad shaded arrows indicate surface Ekman transport (arrow symbols scaled so that linear dimensions are roughly proportional to transport magnitude). The existence of a more or less enclosed gyral circulation near the coast south of Palgrave point is suggested by the oval-shaped arrow pattern in that location. Contours indicate the seasonal mean distribution of wind-mixing index ($m^3s^{-3}$). (**b**) Hypothesized original adaptive sardine migration pattern that may have linked optimal adult feeding grounds to optimal reproductive habitat, thereby maximizing stock productivity. (**c**) Recently observed situation.

Based on the evident distributional shifts that have accompanied observed major shifts in productivity of other "wasp waist" fish populations, this author has hypothesized the following speculative scenario (Bakun, 2001; Bakun, in press). In the earlier higher-biomass period, much of the stock biomass and the bulk of the reproductive output is believed to have been situated in the broad, somewhat sheltered coastal indentation between Palgrave Point and Walvis Bay that lies between the two intense upwelling cores centered near Cape Frio and near Lüderitz (Figure 24.11(a)). This area appears to constitute the best available local approximation to the "ocean triad" type of reproductive habitat typically preferred by coastal pelagic fishes inhabiting eastern ocean boundary upwelling regions (Bakun, 1996, 1998).

It is a characteristic pattern in coastal ocean upwelling systems that adult sardines utilize their strong swimming abilities to migrate out of the relatively sheltered primary reproductive habitats to gain access to particularly favorable adult feeding grounds near the exposed upwelling centers (Bakun, 1996). If the same pattern occurred in the northern Benguela as has been the case elsewhere, it seems that adults during the high biomass years would have migrated for feeding from the main reproductive grounds in the coastal bight off Walvis Bay, southward to the convergent edges of the intense upwelling plumes formed off Lüderitz and northward to the area of the Angola-Benguela Front where the imbedded frontal structures would have acted to concentrate the production occurring in recently upwelled waters being transported offshore (Figure 24.11(b)).

The intense fishery that became established near Walvis Bay was concentrated directly within the primary reproductive habitat itself. The highly productive interlinked system was thereby faced with an extensive mechanized fishery, well positioned to immediately sweep up the compact, migrating schools of large, reproductively-ready adults entering the primary reproductive area. According to the "school-mix feedback" theory (Bakun, 2001; Bakun, in press), the arriving schools would have tended to contain a large representation of individuals possessing specific *affinities,* either genetically inherent or perhaps imprinted in early life (Cury, 1994), to this primary reproductive zone; the larger the representation of individuals with affinities to the zone contained in a particular school, the more likely it would be for the school to have migrated there and become vulnerable to the fishery. The fishery would therefore act to steadily drain the affinities to that zone from the population. Progressively, fewer and fewer schools would contain large representation of individuals with special affinity to the zone, and fewer and fewer schools would be impelled to migrate to that zone. Eventually, fish that were the product of spawning that happened to take place in the Angola-Benguela Frontal zone, which happens to be far enough toward the tropics that the turbulence input level falls within the 250 $m^3s^{-3}$ value identified by Bakun (1996) as characterizing favorable reproductive habitat (Figure 24.11(a)) would increase in relative abundance and progressively dominate the school dynamics to greater and greater degree. More and more reproduction would have taken place either in this secondary reproductive zone to the north or else rather randomly throughout the habitat (Figure 24.11(c)). In such a case, production of individuals possessing an affinity to the naturally more favorable reproductive zone would tend eventually to fall to levels too low to substantially affect migrational behavior of the schools in which they had become entrained. This might have "broken" the adapted migration circuit for this population, and thereby robbed the population of its synergistic use of the available segments of habitat. Indeed, in the most recent years, systematic adult migration appears to have largely ceased; in at least part of the recent period, it appears that nearly all of the stock biomass and essentially all of its reproductive output has been situated in the Angola-Benguela frontal zone (Figure 24.11(c)). This conforms to the observed reality that the stock has remained bogged down in a low biomass and low productivity state for an unexpectedly long time after the implementation of modern management practices.

One might further hypothesize an additional suite of associated feedback mechanisms that might be involved in keeping the system "stuck" in its less favorable altered state (Bakun and Weeks, 2004). Adult sardines, owing to the particularly fine-meshed filtering apparatus in their gillrakers (Van der Lingen, 1994), are able to directly consume phytoplankton. They also possess the strong swimming capability needed to avoid being swept away by the strongly divergent surface flow conditions that drive the upwelling process. Thus, sardines are able to access and consume the associated production in the vicinity of the upwelling cells, before it can diffuse outward to zones of less intensely divergent flow conditions where it may be available to concentrations of more weakly swimming herbivorous zooplankton, or alternatively may sink unutilized to the floor of the continent shelf/slope. Removal of this competition for available phytoplankton production by the formerly large concentrations of adult sardines, as well as their direct predation on the zooplankton themselves, may have encouraged increases in zooplankton abundance. This in turn could have stimulated the observed growth of populations of zooplanktivorous pelagic gobies and medusas (in much the same as may have occurred in the Black Sea situation, discussed above), which thereupon could have inflicted increased rates of predation mortality on pelagic early life stages of sardines and other fish species, tending to amplify and perpetuate the shift initiated by the overfishing.

Note that the overall effect of this proposed series of alterations of the trophic system is a shift in relative abundance from herbivores to carnivores (i.e., increases at trophic levels 1 and 3 linked to suppression of trophic level 2). This is interesting in view of the recent identification of massive eruptions of toxic hydrogen sulfide gas occurring in the area stretching from the Lüderitz upwelling center to somewhat north of Walvis Bay (Figure 24.11(a)), caused by thick deposits of decomposing diatoms that accumulate unoxidized on the sea floor (Weeks et al., 2002; Weeks et al., 2004). Individual eruption episodes have been observed to cover as much as 20,000 km$^2$ of ocean surface, and one or more such episodes were evident in the region more often than not during the initial one-year period of observation. The effects on the ecosystem of such extensive toxic gas eruptions in immediately eliminating living organisms in the affected area, and at the same time having the more durable effect of having stripped dissolved oxygen from the water column, can only be guessed. And while it is clear that hydrogen sulfide production in this region is not a new phenomenon, the various observed changes in the system that seem linked to the overfishing of several decades ago could have done nothing other than to exacerbate the situation.

## 12 What makes a shift "sticky"?

A set of case histories has been presented. It includes some particularly straightforward examples that have occurred in bounded, enclosed situations, e.g., the Baltic Sea and the Black Sea, where the system is simplified, the signals of change are strong, and consequently the feedback processes accounting for the "stickiness" are fairly accessible. It also includes some indication of the intriguingly counterintuitive aspects of the grandiose climate-driven marine ecosystem regime shifts that seem to occur over very broad scales in the world's oceans. Certain common themes emerge. For example, each of the examples has involved heavy fishing, combined with some triggering impulse, either climatic or produced by additional human interventions, leading to durable (i.e., "sticky") regime shifts. Although it appears that the proximate triggering mechanism may sometimes be *overfishing* itself (e.g., the coral reef and Benguela examples, Sections 10 and 11), it seems that it is the external triggering mechanisms that may often act to transform *heavy fishing* to *overfishing*. Consequences range from declines or collapses of resource populations, to durable long-term degradations of stock productivities, to general collapses in important ecosystem services (Bakun and Weeks, 2004).

It has been broadly noted that after a major fishery decline, fish stocks tend often not to rebound as expected following reduction or termination of fishing pressure, but rather tend to remain in a durably depressed state (Steele and Schumacher, 2000; Hutchens, 2000). It seems clear that somehow fisheries are doing something to the fish stocks or to their ecosystem to impair the basic productivity of the resource production system in a manner that may be more durable than explainable by the mere attendant reductions in spawning biomasses.

The possibility that a level of fishing that may have been sustainable in one regime may constitute overfishing in another has motivated many fishery scientists to try to track various climatic, environmental and biological time series to try to recognize when a regime shift might be occurring (much as a technical stock market analyst might try to judge the future behavior of a given stock on the basis of current market behavior). For example, Hare and Mantua (2000) present a time series of the first principal component of one hundred and one separate time series of various types. Unfortunately, at this point one probably can only confidently recognize a regime shift in open ocean systems well after the fact, when fish stocks have already declined, when species are already replaced, or when an ecosystem already exhibits signs of serious trouble. To recognize an important shift in time to take effective remedial actions, a better understanding of the mechanisms involved is needed (even stock market analysts have the advantage of a basic understanding of the mechanisms involved in their field). In particular, we really need to have a good idea of what it is that serves to make a shift "sticky". It would seem that there are at least three possible classes of mechanisms that could be involved.

### A. Multiple stable states

It may be that a system shifts between two (or at most a few) discrete, rather stable potential states (Scheffer et al., 2001; Collie et al., 2004; Knowlton, 2004). This is the preconception that underlies the term "regime shift" itself. It seems to have been the idea that led John Isaacs to apply the term to shifts in dominance of sardines and anchovies off California (see Section 1, above). A quite clear example of such a mechanism is that proposed in the Baltic Sea example (Section 8, above). Among the examples discussed in this chapter, the Black Sea, coral reef, and northern Benguela examples (Sections 8, 9 and 10) also imply involvement of mechanisms of this general type. An attractive aspect is that mechanisms of this class, when understood, tend to provide a rather solid framework for conceiving active remedial management actions.

### B. Direct control by low frequency climatic variation

Another common preconception is that ecosystem variations that may be responses to climate fluctuations should be directly driven by, be in phase with, and have essentially the same periods as associated climatic variations. There is, in fact, a peak in variance in the spectral band near the 50-year period in climatic time series covering the last century (Minobe, 2000) that roughly corresponds to the duration of the most recent sardine—anchovy cycle in the Pacific (Lluch-Cota et al., 1997; Chavez at al., 2002). If direct climate driving should be the source of the "stickiness" of the regimes (i.e., if regimes are directly held in place by the long durations of particular low-frequency climate components controlling them), then there would seem little for management to do but wait for the climate to "work its will"; the management responses would have to be essentially passive (i.e., to attempt to perceive climate shifts in order to apply more precautionary management methodologies during periods of transition).

Arguing against this class of mechanism is the fact that the multidecadal components of climatic variability tend, at any specific location, to be quite subtle compared to more energetic higher frequency variations. Of course, living organisms do not contain spectral analyzers in their response systems that could enable them to isolate subtle low frequency variations within a more energetic "noise level" of higher frequency variability. And, certainly, no individual ever experiences global-scale or basin-scale averages. An organism only senses the actual ambient conditions to which it is exposed (i.e., the summed total of the spectral components into which the actual series of ambient conditions can be decomposed). Moreover, (as elaborated in Section 4) similar species in widely separated locations have appeared to respond synchronously to climatic signals, even though the local ecosystem circumstances and observed responses in terms of specific environmental properties and processes have been very different. Although many, including this author, have tried, it has been very hard to build convincing scenarios involving continuing direct local control by large-scale, low-frequency climatic variations.

**C. Transient response to a climatic impulse**

A third possibility for explaining the apparent "stickiness" might be that once a regime-type process is set in motion (e.g., by a climate impulse, or perhaps in a negative sense by a pulse of serious overfishing, etc.), there might be a substantial natural response time for the perturbation to run its course within the interacting components of the ecosystem.

Consider the following analogy to a spring plankton bloom in a temperate system. Winter conditions have left the surface layers primed with nutrients and rather purged of herbivorous zooplankton. The triggering impulse is the establishment of sufficient stability in the water column for phytoplankton to grow and multiply (Sverdrup, 1953), whereupon the phytoplankton bloom commences. This provides sufficient concentrations of food for zooplankton to begin to feed efficiently and thus to reproduce. The zooplankton have longer generation times than the phytoplankton and so initially lag and allow the bloom to continue to grow. But as their abundance grows, the zooplankton progressively consume larger and larger amounts of the potential reproductive capacity of their food resource, and eventually graze the bloom down, even if the plant nutrients in the water have not been exhausted. Now, substitute small pelagic fish for the phytoplankton in our analogy, and their higher trophic level predators (predatory fishes, seabirds, etc.) for the zooplankton. Assume, for the sake of the example, that the phytoplankton may have had a characteristic generation time of about one day, the zooplankton may have had a generation time of about five days, and the bloom may have lasted for about 50 days. If the small pelagic fish in our new scenario have a generation time of one year, and their predators an average generation time of about five years, then by direct analogy, a bloom of small pelagic fishes should last about fifty years (which matches the recent cyclic pattern discussed above).

Thus a period of rapid climate trend might simply open a "loophole" (see Section 3, above), that allows the prey to temporally take advantage of an associated relaxation in predation pressure to temporally outstrip its "biological controls". But eventually, the predators react, in one way or other (population growth, adaptive relocation of feeding habitat, etc.), to re-establish their suppressive effect such that both trophic levels are again forced to progressively revert toward the less productive initial state.

Item 'C', here, would be similar to item 'B', above, in that management might be able to do little in terms of controlling the occurrence of an initiating climatic impulse. But it might also be more like item 'A' in that once the mechanistic process is understood, active management actions might be taken to promote maximum beneficial effect of the released productivity to the various affected trophic levels by phasing the exploitation of the various levels in such a way as to support maximum net production of most valuable components, optimum economic yield to fisheries, maximum societal benefit, etc.

## 13. Generalizations

So it appears that we may potentially to be speaking of one or more of perhaps several somewhat different things when we cite a *marine ecosystem regime shift* (i.e., an event meeting the definition offered in Section 1.1). But it is a term that has evolved progressively through different contexts, and so it may be understandable that this is where we currently find ourselves. Even so, it continues to be an evocative expression, and is undoubtedly a useful categorization, encapsulating a very important suite of allied ideas. Probably, most important is that a clear distinction be maintained as to whether one is speaking of a *climatic regime shift* (which may or may not be an actual persistent shift in operation or effect of essential mechanisms and processes; see Section 5) or a *marine ecosystem regime shift,* and in doing so to adhere to a consistent definition, such as the one proposed here, even if necessarily rather broad.

Three major types of external influences have emerged in this discussion as having evidently acted in producing regime shifts in marine ecosystems. These are (1) climatic variability, (2) fishery exploitation, and (3) anthropogenic habitat modification (including introduction of exotic species). Type 1, climatic variability, seems generally beyond the control of preventive or remedial management (except of course in the case of anthropogenically-induced climate change, serious resolution of which currently seems to be blocked by economic considerations and lack of political will).

Type 2, habitat modification, has appeared particularly prominently in the Black Sea (Section 9) and coral reefs (Section 10) examples. While this type of external influence does not seem yet to have been a major influence open-ocean ecosystems, it has had obvious effects in many coastal or insular situations, particularly within enclosed seas. The issues involved are in most cases fairly obvious. But, again, mitigation actions or adequate measures for avoidance are often a question of economic considerations that are often major and of political will that is sometimes not major enough.

Type 3, fishery exploitation, is the external driving force for marine ecosystem change that seems most amenable to effective management by enlightened human actions. Certainly, in nearly every one of the examples treated in the earlier sections of this chapter, fishery exploitation has appeared as being at least an important contributing factor. In the examples where the most clearly undesirable shifts have occurred, serious overfishing has been a consistent ingredient. But, in some cases (e.g., the Baltic Sea example, Section 8), state-of-the-art fishery management procedures have not prevented an extremely costly shift from occurring. In that particular case, it seems to have been the interaction between a climatic influence (i.e., the irregularity of renewal of bottom waters) and continued fishing at rates not appropriately adjusted to account for that influence, that appears to have triggered the undesirable (and once established, persistent) shift in that system.

On the other hand, in the Black Sea (Section 9) and coral reef (Section 10) examples, it has been interaction of fishing with anthropogenic habitat alterations and/or exotic species introductions that seems to have led, or at least contributed, to those undesirable "sticky" shifts. Finally, in the northern Benguela example (Section 11), it was hypothesized that fishing may have altered the geographical utilization of the regional ecosystem by the key "wasp waist" fish species. Such changes in geographical occupation of the ecosystem, particularly by the lowest trophic-level mobile (i.e., "wasp waist") species, have been evident in many of the examples presented here, as have been the resulting transformations in species dominance when these shifts have resulted in new patterns and levels of predation on early life stages of formerly dominant ecosystem components.

One thing that has perhaps been lacking in the past has been an adequate appreciation of the potential for such adverse self-enhancing feedbacks within marine ecosystems. The hypothetical feedback mechanisms that have been described here seem not to be highly complex, and it seems that we may be at the point where we have learned enough to begin to infer at least potential action of particular classes of feedback mechanisms within specific ecosystem configurations, and to begin to manage with particular attention toward avoiding setting in motion the most potentially damaging of these. We certainly have the advantage of some striking clues to guide the quest, notably certain remarkable regularities, including (a) specific species seeming to alternate dominance at the key "wasp-waist" position in whatever type of system they occur, even though the systems in which they occur are very different and function via very different ecosystem dynamics (Section 3), (b) involvement of "wasp-waist" populations as key ingredients in the process in most of the shifts we have examined, as well as others we have not examined here, (c) synchronous responses of similar species in widely separated locations to rather subtle basin-scale or global-scale climatic signals, even though the local ecosystem circumstances and observed responses in terms of specific environmental properties and processes have been very different (Section 4); (d) a common pattern in which changes in selected habitat geography, particularly by the key "wasp waist" species, tend to accompany important regime shifts.

Moreover, the changes in selected habitat that have been outlined, which do not seem simply relatable to existence of any smooth, readily-sensed gradients in such properties as temperature, food availability, etc., seem to imply an ability of mobile fish populations to develop adaptive responses, which persist across generations, much more rapidly than could ever be possible through standard genetic selection processes (Bakun, 2001). If true, this would seem to constitute a potential "philosopher's stone" for marine ecosystem/resource management in that once the mechanisms behind such rapidly evolving responses were identified, they could perhaps be utilized to repair damage done by overfishing, or even to "nudge" the system toward an inferred state of enhanced fishery productivity or of enhanced delivery of other desirable ecosystem goods and services (Bakun, in press).

With a degree of confidence in a specific mechanism having been established, a variety of response actions might be conceived. For example, one could envision routine monitoring of reproductive habitat selection, using more automated versions of current "egg pump" techniques (Checkley et al. 1997), etc. When reproductive habitat selection were observed to begin to shift or "wander", one could try on the basis of the acquired understanding of the assumed mechanism to infer whether it appeared to represent a positive or negative development. Then one could further consider how adjusting the pattern of exploitation might serve to "nudge" the stock—habitat system toward what might be believed to be its most productive configuration. Actions of this type could herald an auspicious transition from an era of reactive management that has relied largely on a conceptual framework of terrestrial analogies and has yielded disappointing results, to a new era of keenly proactive marine ecosystem management based on acquired scientific understanding of processes and mechanisms that may be unique to marine systems.

Finally, it may be appropriate to acknowledge that the predominant focus on exploited fishery resources that has characterized the discussion in this chapter has, admittedly, been somewhat arbitrary. Clearly, regime shifts in marine ecosystems must likewise involve unexploited species including planktonic species. The choice of presented material was dictated by the fact that the fished systems offer the most information currently available with which to piece together a coherent, revealing, accessible "story", as well as having been the major impetus for the recent emergence of the "marine ecosystem regime shifts" issue in marine science. Moreover, it can also be acknowledged that rather than a complete, comprehensive review of all published material on marine ecosystem regime shifts, which would have been impossible to do in any coherent, readable manner, this chapter is primarily a presentation of the experience, viewpoint and current stage of understanding possessed by its author.

## References


Alheit, J. and P. Bernal, 1993. Effects of physical and biological changes on the biomass yield of the Humboldt Current ecosystem. In *Large Marine Ecosystems - Stress, Mitigation, and Sustainability*. K. Sherman, L. M. Alexander, and B. Gold, eds., American Association for the Advancement of Science, pp. 53–68.

Alheit, J. and E. Hagen, 1997. Long-term climate forcing of European hering and sardine populations. *Fish. Oceanogr.* **6,** 130–139.

Alheit, J. and E. Hagen, 2002. Climate Variability and Historical NW European Fisheries. In *Climate Development and History of the North Atlantic Realm.* G. Wefer, W. H. Berger, K.-E. Behre and E. Jansen, eds., Springer-Verlag, Berlin, pp. 435–445.

Alheit, J. and M. Niquen, 2004. Regime shifts in the Humboldt Current ecosystem. *Prog. Oceanogr.* **60,** 201–222.

Allen, J.S. 1975. Coastal trapped waves in a stratified ocean. *J. Phys. Oceanogr.* **5,** 300–325.

Anonymous, 1992. *Oceanic interdecadal climate variability. IOC Tech. Ser.,* **40,** Intergovernmental Oceanographic Commission, UNESCO, Paris, 40 pp.

Anonymous, 1993. Why lobster boom is over. *Fishing News International,* March, 1993, p. 4.

Bakun, A. 1978. Guinea Current upwelling. *Nature* **271,** 147–150.

Bakun, A. 1995. Global climate variations and potential impacts on the Gulf of Guinea sardinella fishery. In *Dynamics and Use of Sardinella Resources from Upwelling off Ghana and Ivory Coast.* F.X. Bard and K.A. Korentang, eds., ORSTOM Editions. Paris, pp. 60 -84.

Bakun, A. 1996. *Patterns in the Ocean: Ocean Processes and Marine Population Dynamics.* University of California Sea Grant, San Diego, California, USA, in cooperation with Centro de Investigaciones Biológicas de Noroeste, La Paz, Baja California Sur, Mexico. 323 pp.

Bakun, A. 1998. Ocean Triads and Radical Interdecadal Stock Variability: Bane and Boon for Fishery Management Science. In *Reinventing Fisheries Management,* T.J. Pitcher, P.J.B. Hart and D. Pauly, eds., Kluwer Academic Publishers, Dordrecht, Netherlands, pp. 331–358.

Bakun, A. 1999. A dynamical scenario for simultaneous "regime-scale" marine population shifts in widely separated large marine ecosystems of the Pacific. In *Large Marine Ecosystems of the Pacific Rim: Assessment, Sustainability and Management,* K. Sherman and Q. Tang, eds., Blackwell Science Inc., Malden, Mass, pp. 2–26.

Bakun, A. 2001. 'School-mix feedback': a different way to think about low frequency variability in large mobile fish populations. *Progress in Oceanography* **49,** 485–511.

Bakun, A. in press. Seeking a broader suite of management tools: the potential importance of rapidly-evolving adaptive response mechanisms (such as "school-mix feedback". *Bull. Mar. Sci.* **76(2).**

Bakun, A. and K. Broad. 2003. Environmental loopholes and fish population dynamics: Comparative pattern recognition with focus on El Niño effects in the Pacific. *Fish. Oceanogr.* **12,** 458–473.

Bakun, A. and S.J. Weeks (2004) Greenhouse gas buildup, sardines, submarine eruptions, and the possibility of abrupt degradation of intense marine upwelling ecosystems. *Ecol. Lett.* **7,** 1015–1023.

Beamish, R.J., ed. 1995. *Climate change and northern fish populations. Can. Spec. Pub. Fish. Aquat. Sci.,* ***121.***

Beamish, R.J. 1993. Climate and exceptional fish production off the west coast of North America. *Can. J. Fish. Aquat. Sci.* **50,** 2270–2291.

Beamish, R.J. and D.R. Bouillon. 1993. Pacific salmon production trends in relation to climate. *Can. J. Fish. Aquat. Sci.* 50, 1102–1116.

Beamish, R.J., D.J. Noakes, G.A. McFarlane, L. Klyashtorin, V.V. Ivanov, and V. Kurashov, 1999. The regime concept and natural trends in the production of Pacific salmon. *Can. J. Fish. Aquat. Sci.* 65(3), 516–526.

Beamish, R.J., G.A. McFarlane and J.R. King, 2000. Fisheries climatology: understanding decadal scale processes that naturally regulate British Columbia fish populations. In *Fisheries oceanography: an Integrative approach to Fisheries Ecology and Management.* Harrison, P.J. and T.R. Parsons, eds., Blackwell Science, Oxford, pp. 94–145.

Beamish, R.J., A. J. Benson, R. M. Sweeting and C. M. Neville, 2004. Regimes and the history of the major fisheries off Canada's west coast. *Prog. Oceanogr.* **60,** 355–385.

Beaugrand, G. 2003. Long-term changes in copepod abundance and diversity in the north-east Atlantic in relation to fluctuations in the hydroclimatic environment. *Fish. Oceanogr.* **12,** 270–283.

Beaugrand, G. 2004. The North Sea regime shift: Evidence, causes, mechanisms and consequences. *Prog. Oceanogr.* **60,** 245–262.

Belvèze, H. 1984. *Biologie et dynamique des populations de sardine peuplant les Côtes atlantiques marocaines et propositions pour un aménagement des pêcheries.* Doctoral Thesis, Université de Bretagne Occidental, Brest, France, 532 pp.

Belvèze,H. and K. Erzini, 1983. Influence of hydroclimatic factors on the availability of sardine (*Sardina pilcahrdus* Walb.) in the Moroccan fisheris in the Atlantic. In *Proceedings of the Expert Consultation to Examine Changes in Abundance and Species Composition of Neritic Fish Resources.* Sharp, G.D. and J. Csirke, eds., *FAO Fish. Rep.* **291**(2), pp. 285–328.

Bentley, P.J., R.L. Emmett, N.C.H. Lo and H.G. Moser, 1996. Egg production of the Pacific sardine (*Sardinops caerulea*) off Oregon in 1994. *CalCOFI Rep. 37*, 193–200.

Benson, A.J., and A.W. Trites, 2002. Ecological effects of regime shifts in the Bering Sea and eastern North Pacific Ocean. *Fish and Fisheries* **3,** 95–113.

Bernal, P. A. 1981. A review of the low-frequency response of the pelagic system in the California Current. *CalCOFI Rep. 12,* 49–62.

Binet, D., E. Marchal and O. Pezennec, 1991. *Sardinella aurita* de Côte-d'Ivoire et du Ghana: fluctuations halieutiques et changements climatiques. In *Pêcheries Ouest-Africaines Variabilité, Instabilité et Changement,* P. Cury and C. Roy, eds., Editions de l'ORSTOM, Paris, pp. 320–342.

Binet, D., B. Samb, M.T. Sidi, J.J. Levenez and J. Servain, 1998. Sardine and other pelagic fisheries changes associated with multi-year trade wind increases in the southern Canary current. In *Global versus local changes in upwelling systems,* Durand, M.H., P. Cury, R. Mendelssohn, C. Roy, A. Bakun and D. Pauly, eds., ORSTOM Editions, Paris, pp. 211–233.

Binet, D., B. Gobert and L. Maloueki, 2001. El Niño-like warm events in the eastern Atlantic (6°N, 20°S) and fish availability from Congo to Angola (1964–1999). *Aquat. Living Resour.* **14,** 99–113.

Bjerknes, J. 1966. A possible response of the atmospheric Hadley circulation to equatorial anomalies of ocean temperature. *Tellus* **18,** 820–829.

Bograd, S., R.J. Lynn and J.A. McGowan, 2002. Interdecadal physical-biological coupling in the southern California Current system. *Abstracts Volume, 11th Annual Meeting, North Pacific Marine Science Organization (PICES), Qingdao, China, October 2002.* p. 97.

Boyer, D. (1996). Stock dynamics and ecology of pilchard in the northern Benguela. In *The Benguela Current and comparable eastern boundary upwelling ecosystems.* M.J. O'Toole, ed. Deutsche Gesellschaft für Techhnische Zusammenarbeit (GTZ) GmbH. Eschborn, Germany, pp. 79–82.

Boyer D.C. and I. Hampton, 2001. An overview of the living marine resources of Namibia. *S. Afr. J. mar. sci.,* **23,** 5–35.

Brodeur, R.D., and D.M. Ware, 1992. Long-term variability in zooplankton biomass in the subarctic Pacific Ocean. *Fish. Oceanogr.* 1, 32–38.

Brodeur, R.D., and D.M. Ware, 1995. Interdecadal variability in distribution and catch rates of epipelagic necton in the North Pacific Ocean. In. *Climate Change and Northern Fish Populations.* Beamish, R.J., ed., *Can. Spec. Publ. Fish. Aquat. Sci.* 121. pp. 329–356.

Caddy, J.F. 1993. Towards a comparative evaluation of human impacts on fisheries and ecosystems of semi-enclosed seas. *Rev. Fisheries. Sci.* **1,** 57–95.

Caddy, J.F. 2000. Marine catchment basin effects versus impacts of fisheries on semi-enclosed seas. *ICES J. mar. Sci.* **57,** 628–640.

Caddy, J.F. and R. Griffiths, 1990. Recent trends in fisheries and environment in the General Fisheries Council for the Mediterranean (GFCM) area. *Studies and Reviews, General Fisheries Council for the Mediterranean,* **63,** FAO, Rome. 71 pp.

Carr M.E. 2001. Estimation of potential productivity in eastern boundary currents using remote sensing. *Deep-Sea Res.* **49,** 59–80.

Carrasco, S. and O. Lozano, 1989. Seasonal and long-term variations of zooplankton volumes in the Peruvian sea, 1964–1987. In *The Peruvian Upwelling Ecosystem: Dynamics and Interactions.* D. Pauly, P. Muck, J. Mendo and 1. Tsukayama, eds., International Center for Living Aquatic Resources Management (ICLARM), Manila. pp. 82–85.

Carton, J. A., X. Cao, B. S. Giese, and A. M. da Silva, 1996. Decadal and interannual SST variability in the tropical Atlantic Ocean. *J. Phys. Oceanogr.,* **26,** 1165–1175.

Caverivière, A. 1991. L'explosion démographique du baliste (*Balistes carolinensis*) en Afrique de l'Ouest et son évolution en relation avec les tendaces climatique. *In* P. Cury and C. Roy (eds.) *Pêcheries Ouest-Africaines Variabilité, Instabilité et Changement.* Editions de l'ORSTOM, Paris. pp. 354–367 ..

Chavez, F.P., J. Ryan, S. Lluch-Cota and M. Ñiquen C. 2002. From anchovies to sardines and back : multidecadal change in the Pacific Ocean. *Science* **299,** 217–221.

Checkley, D.M. Jr., Ortner, P.B., Settle, L.R,. & Cummings, S.R. 1997. A continuous underway fish egg sampler. *Fish. Oceanogr.* **6,** 58–73.

Christy, F., 1997. Economic waste in fisheries: Impediments to change and conditions for improvement. In *Global Trends: Fisheries Management.* E.L. Pikitch, D.D. Huppert and M.P. Sissenwine, eds. *Amer. Fish. Soc. Symp. 20.* American Fisheries Society, Bathesda, Maryland. pp. 28–39.

Cochrane, K.L. and M. Tandstad, eds. 2000. Report of the Workshop on the Small Pelagic Resources of Angola, Congo and Gabon, Luanda, Angola, Nov. 1997. *FAO Fish. Rep. 6***18,** 149 pp.

Cole, J.S. 1980. Synopsis of biological data on the yellowfin tuna, *Thunnus albacares* (Bonneaterre, 1788) in the Pacific Ocean. *IATTC Spec. Rep. 2,* 71–150.

Collette, B.B. and C.R. Aadland, 1996. Revision of the frigate tunas (Scombridae, *Auxis*), with descriptions of two new subspecies from the eastern Pacific. *Fish. Bull. US,* **94,** 423–441.

Collie, J.S., K. Richardson and J. H. Steele, 2004. Regime shifts: Can ecological theory illuminate the mechanisms? *Prog. Oceanogr.* **60,** 281–302.

Conand, F. 1975. Distribution aet abondance des larves de clupeids au large des côtes du Sénégal et de la Mauritanie en septembre, octobre et novembre 1972. *ICES, C.M/1975/J:***4,** 9 pp.

Crawford, RJM., L.G. Underhill, L.V. Shannon, D. Lluch-Belda, W.R. Siegfried and C.A. Villacastin-Herrero, 1991. An empirical investigation of trans-oceanic linkages between areas of high abundance in sardine. In *Long term variability of pelagic fish populations and their environments,* T. Kawasaka, , S. Tanaka, Y. Toba and A. Tanaguchi, eds., Pergammon Press, Tokyo, pp. 319–322.

Cury, P. 1994. Obstinate nature: an ecology of individuals: thoughts on reproductive behavior and biodiversity. *Can. J. Fish. Aquat. Sci.* **51,** 1664–1673.

Cury P., A. Bakun, R.J.M. Crawford, A Jarre-Teichmann, R. A. Quiñones, L.J. Shannon, H. M. Verhey, 2000. Small pelagics in upwelling systems: patterns of interaction and structural changes in "wasp-waist" ecosystems. *ICES J. Mar. Sci.,* **211:** 603–618.

Cury, P. and L. Shannon, 2004. Regime shifts in upwelling ecosystems: observed changes and possible mechanisms in the northern and southern Benguela. *Prog. Oceanogr.* **60,** 223–243.

Daskalov, G. M. 2002. Overfishing drives a trophic cascade in the Black Sea. *Mar. Ecol. Prog. Ser.* **225,** 53–63.

Demarcq, H., 1998. Spatial and temporal dynamics of the upwelling off Senegal and Mauritania: local change and trend. In *Global versus local changes in upwelling systems,* M.H. Durand, P. Cury, R. Mendelssohn, C. Roy, A. Bakun and D. Pauly, eds., ORSTOM Editions, Paris, pp. 149–165.

deYoung, B., R. Harris, J. Alheit, G. Beaugrand, N. Mantua and L. Shannon, 2004. Detecting regime shifts in the ocean: Data considerations. *Prog. Oceanogr.* **60,** 143–164.

Do Chi, T., and D.A. Kiefer, eds. 1996. *Report of the Workshop on the Coastal pelagic Resources of the Upwelling System off Northwest Africa: research and predictions. FI:TCP/MOR/4556(A) Field Document 1.* FAO, Rome.

Dumont, H. and T. Shiganova, 2003. The invasion of the Black, Mediterranean and Caspian Seas by the ctenophore, Mnemiopsis leidyi: a NATO Worksop held in Baku (Azerbaizan) on 24–26 June 2002. *GLOBEC Int. Newsletter,* **9**(1) (April), 18–20.

Ebbesmeyer, C. C., D. R. Cayan, D. R. McLain, F. H. Nichols, D. H. Peterson, and K. T. Redmond, 1991. 1976 step in the Pacific climate: forty environmental changes between 1968–1975 and 1977–1984. In *Proceedings of the seventh annual Pacific climate (PACLIM) Workshop, April 1990.* J. L. Betancourt and V. L. Tharp, eds. pp. 115–125. *California Dept. of Water Resources Ecol. Stud. Prog. Tech. Rep. 2***6.**

Enfield, D.B. and D.A. Mayer, 1997. Tropical Atlantic sea surface temperature variability and its relation to El Niño - Southern Oscillation. *J. Geophys. Res.,* **112,** 929–945.

FAO, 1992. Review of the State of World Fishery Resources. *FAO Fish. Circ.* **711,** Rev. 8. 114 pp.

Fath, B.D., H. Cabezas and C.W. Pawlowski, 2003. Regime changes in ecological systems: An information theory approach. *J. Theo. Biol.,* **222,** 517–530.

Favorite, F. 1967. The Alaskan Stream. *Int. N. Pac. Fish. Comm. Bull.* **21,** 1–20.

Favorite , F., A.J. Dodimead and K. Nasu, 1976. *Oceanography of the Subarctic Pacific region, 1960–71. Int. N. Pac. Fish. Comm. Bull.* **33.** 187 pp.

Fonteneau, A. and J. Marcille, 1993. *Resources, fishing and biology of the tropical tunas of the Eastern Central Atlantic. FAO Fish. Tech. Pap. 2***92.** 354 p.

Foreman, T.J., 1980. Synopsis of biological data on the albacore tuna, Thunnus alalunga (Bonneaterre, 1788) in the Pacific Ocean. IATTC Spec. Rep. 2, 17–70.

Forsberg, E.D. 1980. Synopsis of biological data on the skipjack tuna, *Katsuwanus pelamis* (Linaeus, 1758) in the Pacific Ocean. *IATTC Spec. Rep. 2,* 295–360.

Fréon, P., 1988. *Réponses et adaotation des stocks de clupéidésd'Afrique de l'puest à la variability du milieu et de l'exploitation. Analyse et réflexion à partir de l'exeple du Sénégal.* ORSTOM, Coll. Etudes et Thèses, Paris, 287 pp.

Fréon, P., C. Mullon and B. Voisin, 2003. Investigating remote synchronous patterns in fisheries. *Fish. Oceanogr.* **12,** 443–457.

Garcia, S. and C. Newton, 1997. Current situation, trends and prospects in world capture fisheries. In *Global Trends: Fisheries Management.* E.L. Pikitch, D.D. Huppert and M.P. Sissenwine, eds. *Amer. Fish. Soc. Symp. 20.* American Fisheries Society, Bathesda, Maryland. pp. 3–27.

Garcia, S. and I. de Leiva Moreno, 2003. Global overview of marine fisheries. In *Responsible Fisheries in the Marine Ecosystem.* M. Sinclair and G. Valdimarson, eds. FAO, Rome, Italy and CABI Publishing, Wallingford, Oxon, UK and Cambridge, Mass., USA, pp. 1–24.

Gill, A.E., and A.J. Clarke, 1974. Wind-induced upwelling, coastal currents, and sea level changes. *Deep-Sea Res.,* **21,** 325–345.

Glynn, P.W. 1993. Coral-reef bleaching – ecological perspectives. *Coral Reefs* **12,** 1–17.

Hammann, M. G. and M.F. Cisneros-Mata, 1989. Range extension and commercial capture of the northern anchovy, *Engraulis mordax* Girard, in the Gulf of California, Mexico. *Calif. Fish. Game.* **75**(1), 49–53.

Hare, S.R. and R.C. Francis, 1995. In: *Climate change and northern fish populations,* Beamish, R.J., ed. *Can. Spec. Pub. Fish. Aquat. Sci.,* ***121.*** pp 357–369.

Hare, S.R. and N.J. Mantua, 2000. Empirical evidence for North Pacific regime shifts in 1977 and 1989. *Prog. Oceanogr.* **47,** 103–145.

Hare, S.R., S. Minobe, W.S. Wooster and S. McKinnell, 2000. An introduction to the PICES symposium on the nature and impacts of North Pacific climate regime shifts. *Prog. Oceanogr.* **47,** 99–102.

Hisard, P. 1988. El Niño response of the tropical Atlantic Ocean during the 1984 year. In *Int. Symp. on Long Term Changes in Marine Fish Populations.* T. Wyatt and M.G. Larrañeta, eds. Vigo, Spain. pp. 273–290.

Hoegh-Guldberg, O. 1999. Climate change, coral bleaching and the future of the world's coral reefs. *Mar. Freshw. Res.* **8,** 839–866.

Holmgren-Urba, D. and T.R. Baumgartner, 1993. A 250-year history of pelagic fish abundances from the anaerobic sediments of the central Gulf of California. *CalCOFI Rep. 3***4,** 60–68.

Horel, J.D. and J.M. Wallace, 1981. Planetary scale atmospheric phenomena associated with the Southern Oscillation. *Mon. Weather Rev.,* **119,** 813–829.

Hughes, T.P. 1994. catastrophies, phase shifts, and large-scale degredation of a Carabbean coral reef. *Science* 265, 1547–1551.

Hutchings, J.A. 2000, Collapse and recovery of marine fishes. *Nature,* **406,** 882–885.

Isaacs, J.D., 1975. Some ideas and frustrations about fishery science. *CalCOFI Rep.,* **18,** 34–43.

Kawasaki, T. 1983. Why do some pelagic fishes have wide fluctuations in their numbers? - biological basis of fluctuation from the viewpoint of evolutionary ecology. In *Reports of the Expert Consultation to Examine Changes in Abundance and Species Composition of Neritic Fish Resources.* G.D. Sharp and J. Csirke, eds. *FAO Fish. Rep.* **291**(3). pp. 1165–1180.

Kawasaki, T. and M. Omori, 1988. Fluctuations in the three major sardine stocks in the Pacific and the global trend in mean temperature. In *Int. Symp. on Long Term Changes in Marine Fish Populations.* T. Wyatt and M.G. Larrañeta, eds., Vigo, Spain. pp. 273–290

Kifani, S. 1998. Climate dependent fluctuations of the Moroccan sardine and their impact on fisheries. In *Global versus local changes in upwelling systems,* M.H. Durand, P. Cury, R. Mendelssohn, C. Roy, A. Bakun and D. Pauly, eds., ORSTOM Editions, Paris, pp. 235–248.

King, J.R., G.A. McFarlane and R.J. Beamish, 2000. Decadal-scale patterns in the relative year class success of sablefish (*Anoplopoma fibria*). *Fish. Oceanogr.* **9,** 62–70.

Klyashtorin, L.B. 2001. Climate change and long-term fluctuations of commercial catches: the possibility of forecasting. *FAO Fish.Tech.Pap.,* **411.** 86 p.

Knowlton, N. 2001. Sea urchin recovery from mass mortality: New hope for Carabbean reefs? *Proc. Nat. Acad. Sci.,* **98,** 4822–4824.

Knowlton, N. 2004. Multiple "stable" states and the conservation of marine ecosystems. *Prog. Oceanogr.* **60,** 387–396.

Koranteng, K.A. 1991. Some aspects of the Sardinella fishery in Ghana. In *Pêcheries Ouest-Africaines Variabilité, Instabilité et Changement.* P. Cury and C. Roy, eds., Editions de l'ORSTOM, Paris. pp. 269–277.

Kullenberg, G. 1986. Long-term changes in the Baltic ecosystem. In *Variability and Management of large Marine Ecosystems.* K. Sherman and L.M. Alexander, eds., AAAS Selected Symposium 99. American Association for the Advancement of Science, Wash., D.C. pp 19–31.

Köster, F.W., and C. Möllmann, 2000. Trophodynamic control by clupeid predators on recruitment success in Baltic cod? *ICES J. Mar. Sci.,* **57:** 311–323.

Köster, F.W. and D. Schnack, 1994. The role of predation on early life stages of cod in the Baltic. *Dana* **11,** 179–201.

Kuroda, K. 1991. Studies on the recruitment process focusing on the early life history of the Japanese sardine, *Sardinops melanosticus* (Schlegel). *Bull. Natl. Res. Inst. Fish. Sci.* **3,** 25–278.

Lamb, H.H. 1972. *Climate: Past, Present and Future. I. Fundamentals and Climate Now.* Methuen, London. 613 pp.

Laws, E. 2004. Export flux and stability as regulators of community composition in pelagic marine biological communities: Implications for regime shifts. *Prog. Oceanogr.* **60,** 343–354.

Lehodey, P., F. Chai, and J. Hampton, 2003. Modelling climate-related variability of tuna populations from a coupled ocean-biogeochemical-populations dynamics model. *Fish. Oceanogr.* **12,** 483–494.

Lluch-Belda, D., R.J.M. Crawford, T. Kawasaki, A.D. MacCall, R.H. Parrish, R.A. Schwartzlose and P.E. Smith (1989) World-wide fluctuations of sardine and anchovy stocks: the regime problem. *S. Afr. J. mar. Sci.* 8, 195–205.

Lluch-Belda, D., R.A. Schwartzlose, R. Serra, R.H. Parrish, T. Kawasaki, D. Hedgecock, and R.J.M. Crawford, 1992a. Sardine and anchovy regime fluctuations of abundance in four regions of the world oceans: a workshop report. *Fish. Oceanogr.* 1, 339–347.

Lluch-Cota, D.B., S. Hernández-Vázquez and S.E. Lluch-Cota, 1997. Emprical investigation on the relationship between climate and small pelagic global regimes and El Niño—Southern Oscillation (ENSO). *FAO Fish. Circ.* **934,** 48 pp.

Loeb, V.J. and O. Rojas, 1988. Interannual variation of ichthyoplankton composition and abundance relations off northern Chile, 1964–83. *Fish. Bull., U.S.* **86,** 1–14.

MacKenzie, B. R., Hinrichsen, H.-H., Plikshs, M., Wieland, K., and Zezera, A. S. 2000: Quantifying environmental heterogeneity: estimating the size of habitat for successful cod egg development in the Baltic Sea. *Mar. Ecol. Prog. Ser.* **193,** 143–156.

Mahé, G. 1991. La variabilité des apports fluviaux au golfe de Guinea utilisée comme indice climatique. In *Pêcheries Ouest-Africaines Variabilité, Instabilité et Changement.* P. Cury and C. Roy, eds., Editions de l'ORSTOM, Paris. p p. 343–353.

Mantua, N. 2004. Methods for detecting regime shifts in large marine ecosystems: a review with approaches applied to North Pacific data. *Prog. Oceanogr.* **60,** 165–182.

Masuda, S. and K. Akitomo, 2002. A model of regime transitions in the North Pacific. *Abstracts Volume, 11th Annual Meeting, North Pacific Marine Science Organization (PICES), Qingdao, China, October 2002.* pp. 82–33.

Matthäus, W., and H.U. Lass, 1995. The recent salt inflow into the Baltic Sea. *J. Phys. Oceanog.* **25,** 280–286.

McCook, L.J. 1999. Macroalgae, nutrients and phase shifts on coral reefs: scientific issues and management consequences for the Great Barrier Reef. *Coral Reefs* **18,** 357–367.

McCook, L.J. 2001. Competition between corals and algal turfs along a gradient of terrestrial influence in the nearshore central Great barrier Reef. *Coral Reefs* **19,** 400–417.

McFarlane, G.A. and R.J. Beamish, 2001. The reoccurrence of sardine off British Columbia characterizes the dynamic nature of regimes. Prog. Oceanogr. **47,** 147–169.

McFarlane, G.A., J.R. King and R.J. Beamish, 2000. Have there been recent changes in climate? Ask the fish. Prog. Oceanogr. 47, 147–169.

McFarlane, G.A., R.J. Beamish and J. Schweigert, 2001. Common factors have opposite impacts on pacific herring in adjacent ecosystems. In *Herring: Expectations for a New Millennium,* F. Funk, J. Blackburn, D. Hay, A,J. Paul, R. Stephenson, R. Toreson and D. Withell, eds., Alaska Sea Grant College Program AK-SC-01-04, pp. 51–67.

McManus, J.W. and J. F. Polsenberg, 2004. Coral–algal phase shifts on coral reefs: Ecological and environmental aspects. *Prog. Oceanogr.* **60,** 263–279.

McManus, J.W., L.A.B. Meñez, K.N. Kesner-Reyes, S.G. Vergara and S.M.C. Ablan 2000. Coral reef fishing and coral-algal phase shifts: implications for global reef status. *ICES J. Mar. Sci.* **57,** 572–578.

Mee, L.D., 1992. The Black Sea in crisis: a need for concerted international action. *Ambio,* **21,** 278–286.

Mensah, M.A. 1991. The influence of climatic changes on the coastal oceanography of Ghana. p. 67–79. In *Pêcheries Ouest-Africaines Variabilité, Instabilité et Changement.* P. Cury and C. Roy, eds., Editions de l'ORSTOM, Paris. 524 pp.

Miller, A.J., Cayan, D.R., Barnett, T.P., Graham, N.E. and Oberhuber, J.M., 1994. The 1976–77 climate shift of the Pacific Ocean. *Oceanography* **7,** 21–26.

Miller, C.B., H.P. Batchelder, R.D. Brodeur, and W.G. Pearcy. 1985. Response of zooplankton and icthioplankton off Oregon to the El Niño event of 1985. In *El Niño North,* W.S. Wooster and D.L. Fluharty, eds., University of Washington Sea Grant Program, Seattle, pp. 185–187.

Minobe, S. (1997) A 50–70 year climate oscillation over the North Pacific and North America. *Geophys. Res. Lett.* **24,** 683–686.

Minobe, S. 2000. Spatio-temporal structure of the pentadecadal variability over the North Pacific. *Prog. Oceanogr.* **47,** 381–408.

Minobe, S. 2002, Interannual to interdecadal changes of water temperature, sea-level displacement, and sea-ice distribution in the Bering Sea and associated atmospheric circulation changes. *Prog. Oceanogr.* **55:** 45–64.

Monteleone, D.M. and L.E. Duguay, 1988. Laboratory studies of predation by the ctenophore *Mnemiopsis leidyi* on the early stages in the life history of the bay anchovy *Anchoa mitchilli. J. Plankton Res.* **11,** 359–372.

Murphy, G.I. 1966. Population biology of the Pacific sardine (*Sardinops caerulea*). Proc. *Calif. Adad. Sci.* 34, 1–84.

Nakata, K. and K. Hidaka, 2003. Decadal scale variability in the Kuroshio marine ecosystem in winter. *Fish. oceanogr.* **12,** 234–244.

Nitta, T. and S. Yamada, 1989. Recent warming of tropical sea surface temperature and its relationship to the Northern Hemisphere circulation. *J. Meteorol. Soc. Japan* **67,** 375–383

Nyström, M. C. Folk and F. Moberg, 2000. Coral reef disturbance and resilience in a human-dominated environment. *Trends Ecolog. Evol.* **15,** 413–417.

Parrish, R.H., A. Bakun, D.M. Husby and C.S. Nelson, 1983. Comparative climatology of selected environmental processes in relation to eastern boundary current pelagic fish reproduction. In *Proceedings of the Expert Consultation to Examine Changes in Abundance and Species Composition of Neritic Fish Resources.* G.D. Sharp and J. Csirke, eds., *FAO Fish. Rep.* **291**(3), pp. 731–778.

Parrish, R.H., R. Serra, and W.S. Grant, 1989. The monotypic sardines, *Sardina* and *Sardinops:* their taxonomy, distribution, stock structure, and zoogeography. *Can. J. Fish. Aquat. Sci.* **46,** 2019–2036.

Parrish, R.H., F.B. Schwing, and R. Mendelssohn, 2000. Mid-latitude wind stress: the energy source for climatic shifts in the North Pacific Ocean. *Fish. Oceanogr.* **9,** 224–238.

Paz, J., and M.G. Larrañeta. 1992, Cod year-class variations and abundance of other commercial fish in NAFO Divisions 3NO. *J. Northw. Atl. Fish. Sci.* **14,** 129–134.

Pearcy, W.G., 1992. Ocean Ecology of North Pacific Salmonids. Washington Sea Grant Program. Univ. Washington press, Seattle and London, 179pp.

Peterson, W.T. and F.B. Schwing, 2002. Recent changes in climate and carrying capacity in the northern California Current shelf waters suggest a regime shift was initiated in July 1998. *Abstracts Volume, 11th Annual Meeting, North Pacific Marine Science Organization (PICES), Qingdao, China, October 2002.* p. 83.

Pereladov, M.V. 1988. Some observations for biota of Sudak Bay of the Black Sea. In *The Third All-Russian Conference on Marine Biology.* Naukova Dumka, Kiev, pp. 237–238. (In Russian.)

Pezzack, D.S. 1992. A review of lobster (*Homarus americanus*) landing trends in the Northwest Atlantic, 1969–88. *J. Northw. Atl. Fish. Sci.* **14,** 115–128.

Philander, S.G.H. and N.-C. Lau, 1988. Predictability of El Niño. In *Physically-Based Modelling and Simulation of Climate and Climatic Change — Part 2,* Schlesinger, M.E. , ed., Kluwer Academic Press, Dordretch, pp. 967–982.

Picaut, J. 1985. Major dynamics affecting the Eastern Tropical Atlantic and Pacific Oceans. *CalCOFI Rep.* **26,** 41–50.

Polovina, J.J., G.T. Mitchum, N.E. Graham, M.P. Craig, E.E. DeMartini and E.N. Flint, 1994. Physical and biological consequences of a climate event in the central North Pacific. *Fish. Oceanogr.* **3,** 15–21.

Prodanov, K., K. Mikhailov, G. Daskalov, K. Maxim, A. Chashchin, A. Arkhipov, V. Shlyakhov and E. Ozdamar, 1997. Environmental impact on fish resources in the Black Sea. In *Sensitivity of North Sea, Baltic Sea and Black Sea to Anthropogenic and Climatic Changes,* E. Ozsoy and A. Mikaelyan, eds., Kluwer Academic Press, Dordrecht, pp. 163–181.

Radovich, J. 1982. The collapse of the California sardine fishery; what have we learned? *CalCOFI Rep.* **23,** 56–58.

Reid, P.C., M. F. Borgesb and E. Svendsen, 2001. A regime shift in the North Sea circa 1988 linked to changes in the North Sea horse mackerel fishery. *Fish. Res.* **50,** 163–171.

Reid, P.C., M. Edwards, G. Beaugrand, M. Skogen and D. Stevens. Periodic changes in the zooplankton of the North sea during the twentieth century linked to oceanic inflow. *Fish. Oceanogr.* **12,** 260–269.

Rice, J. 1995. Food web theory, marine food webs, and what climate change may do to northern fish populations. In *Climate Change and Northern Fish Populations,* Beamish, R.J., ed., *Can. Spec. Publ. Fish. Aquat. Sci.* **121.** pp. 561–568.

Richards, W.J. 1984. *Kinds and abundances of fish larvae in the Caribbean Sea and adjacent areas.* U.S. Dep. Commer. *NOAA Tech. Rep.* **NMFS-SSRF-776.** 54 pp.

Rodríguez-Sánchez, R., D. Lluch-Belda, H. Villalobos, and S. Ortega-García, 2002. Dynamic geography of small pelagic fish populations in the California Current System on the regime time scale (1931–1997) *Can. J. Fish. Aquat. Sci.* **59,** 1980–1988.

Roemich, D. and J. McGowan, 1995. Climatic warming and the decline of zooplankton in the California Current. *Science* **267,** 1324–1326.

Roy, C. and C. Reason, 2001. ENSO-related modulation of coastal upwelling in the eastern Atlantic. *Prog. Oceanogr.* **49,** 245–255.

Royer, T.C. 1979. On the effect of precipitation and runoff on coastal circulation in the Gulf of Alaska. *J. Phys. Oceanogr.* **9,** 555–563.

Rudnick, D.L. and R.E. Davis, 2003. Red noise and regime shifts. *Deep-Sea Res. I,* **50,** 691–699.

Scheffer, M. and E. H. van Nes, 2004. Mechanisms for marine regime shifts: Can we use lakes as microcosms for oceans? *Prog. Oceanogr.* **60,** 303–319.

Scheffer, M., S. Carpenter, J.A. Foley, C. Folke and B. Walker, 2001. Catastrophic shifts in ecosystems. *Nature* **413,** 591–596.

Schwartzlose, R.A., J. Alheit, A. Bakun, T. Baumgartner, R. Cloete, R.J.M. Crawford, W.J. Fletcher, Y. Green-Ruiz, E. Hagen, T. Kawasaki, D. Lluch-Belda, S.E. Lluch-Cota, A.D. MacCall, Y. Matsuura, M.O. Nevares-Martinez, R.H. Parrish, C. Roy, R. Serra, K.V. Shust, N.M. Ward, N.M. and J.Z. Zuzunaga, 1999. Worldwide large-scale fluctuations of sardine and anchovy populations. *S. Afr. J. mar. Sc.* **21,** 289–347.

Serebryakov, V.P., A.K. Chumakov and I.I. Tevs, 1992. Spawning stocl, population fecundity and year-class strength of Greenland halibut (*Reinhardtius hippoglossoides*) in the Northwest Atlantic, 1947–86. *J. Northw. Atl. Fish. Sci.* **14,** 117–114.

Sinclair, M. and G. Valdimarson, 2003. *Responsible Fisheries in the Marine Ecosystem.* FAO, Rome, Italy and CABI Publishing, Wallingford, Oxon, UK and Cambridge, Mass., USA, 426 pp.

Shannon, L. V., A. J. Boyd, G. B. Brundrit, and J. Taunton-Clark, 1986. On the existence of an El Niño phenomenon in the Benguela system. *J. Mar. Res.* **44,** 495–520.

Sherman, K, 1994. Sustainability, biomass yields, and health of coastal ecosystems: An ecological perspective. *Marine Ecol. Progr. Ser.* **112,** 277–301.

Sherman, K, 1989. Biomass flips in large marine ecosystems. In *Biomass yields and geography of large marine ecosystems,* Sherman, K. and L.M. Alexander, eds. Westview Press, Boulder, Colorado, pp. 327–333.

Sherman, K., J. Kane, S. Murawski, W. Overholtz and A. Solow, 2002. The U.S. Northeast Shelf large marine Ecosystem: zooplankton trends in fish biomass recovery. In *Large marine Ecosystems of the North Atlantic: Changing States and Sustainability,* Sherman, K. and H.R. Skjoldal, eds., Elsevier, Amsterdam, pp. 195–215.

Shiganova, T. A. 1998. Invasion of the Black Sea by the ctenophore *Mnemiopsis leidyi* and recent changes of pelagic community structure. *Fishery Oceanography* **7,** 305–311.

Steele. J.H. 1998. Regime shifts in marine ecosystems. *Ecolog. Appl.* **8**(1) Supplement, S33-S36.

Steele, J.H. and M. Schumacher, 2000. Ecosystem structure before fishing. *Fisheries Res.* **44,** 201–205.

Solow, A.R. and A.R. Beet. in press. A test for regime shift. Submitted to *Fish. Oceanogr.*

Soutar, A.,and J.D. Isaacs, 1974. Abundance of pelagic fish during the 19$^{th}$ and 20$^{th}$ centuries as recorded in anaerobic sediment off the Californias. *Fish. Bull., US* **72,** 257–273.

Sugimoto T. and K. Tadokoro, 1997. Interannual-interdecadal variations in zooplankton biomass, chlorophyll concentration and physical environment in the subArctic Pacific and Bering Sea. *Fish. Oceanogr.* **6,** 74–93.

Sugimoto, T. and K. Tadokoro, 1998. Interdecadal variations of plankton biomass and physical environment in the North Pacific. *Fish. Oceanogr.* **7,** 289–299.

Sutton, R.T., S.P. Jewson, and D.P. Rowell, 2000. The Elements of Climate Variability in the Tropical Atlantic Region. *J. Climate,* **13,** 3261–3284.

Sverdrup, H.U. 1953. On conditions for vernal blooming of phytoplankton. *J. Cons. int. Explor. Mer* **18,** 287–295.

Taylor, A.H. 2002. North Atlantic climatic signals and the plankton of the European continental shelf. In: *large Marine Ecosystems of the North Atlantic: Changing States and Sustainability,* Sherman, K and H.R. Skjoldal, eds. Elsevier Science, Amsterdam pp. 3–26.

Thomson, R.E. 1972. On the Alaskan Stream. *J. Phys. Oceanogr.* **2,** 363–371.

Trenberth, K.E. 1990. Recent observed interdecadal climate changes in the Northern Hemisphere. *Bull. Amer. Meteorol. Soc.* **71,** 988–993.

Trenberth, K.E. and J.W. Hurrell, 1994. Decadal atmospheric-ocean variations in the Pacific. *Clim. Dyn.* **9,** 303 319.

Van der Lingen C.D. 1994. Effect of particle size and concentration on the feeding behavior of adult pilchard, *Sardinops sagax. Mar. Ecol. Prog. Ser.* **119,** 1–13.

Venrick, E.L., J.A. McGowan, D.R. Cayan and T.L. Hayward, 1987. Climate and chlorophyl a: long-term trends in the Central North pacific ocean. *Science* **238,** 70–72.

Vézina, A.F., F. Berreville and S. Loza. 2004. Inverse reconstructions of ecosystem flows in investigating regime shifts: impact of the choice of objective function. *Prog. Oceanogr.* **60,** 355–385.

Wada, T. and Y. Oozeki, 1999. A populations dynamics model for the Japanese sardine – why the sardine shows such large population fluctuations. *Bull. Tohoku Natl. Fish. Res. Inst.* **62,** 171–180.

Weeks, S.J., B. Currie and A. Bakun, 2002. Massive emissions of toxic gas in the Atlantic. *Nature* **415,** 493–494.

Weeks S.J., B. Currie, A. Bakun and K.R. Peard, 2004. Hydrogen sulphide eruptions in the Atlantic Ocean off southern Africa: Implications of a new view based on SeaWiFS satellite imagery. *Deep-Sea Res. 1,* **51,** 153–172.

Wooster, W. S., and J. L. Reid, 1963. Eastern boundary currents. In *The Sea, vol. 2,* Hill, M.N., ed., Interscience Pub., New York, pp. 253–280.

Wooster, W.J. and C. I. Zhang, 2004. Regime shifts in the North Pacific: early indications of the 1976–1977 event. *Prog. Oceanogr.* **60,** 183–200.

Wyrtki, K. 1975. El Niño—the dynamic response of the equatorial Pacific. *J. Phys. Oceanogr.,* **5,** 572–584.

Yoshida, K. 1967. Circulation in the eastern tropical oceans with special reference to upwelling and undercurrents. *Japan. J. Geophys.,* **4,** 1–75.

# INDEX